Handbook of Hazardous Materials

Handbook of Hazardous Materials

Edited by
Morton Corn
Division of Environmental Health Engineering
The Johns Hopkins University
School of Hygiene and Public Health
Baltimore, Maryland

Academic Press, Inc.
A Division of Harcourt Brace & Company
San Diego New York Boston London Sydney Tokyo Toronto

This book is printed on acid-free paper. ♾

Academic Press, Inc.
1250 Sixth Avenue, San Diego, California 92101-4311

United Kingdom Edition published by
Academic Press Limited
24–28 Oval Road, London NW1 7DX

Library of Congress Cataloging-in-Publication Data

Handbook of hazardous materials / edited by Morton Corn.
p. cm.
Includes index.
ISBN 0-12-189410-X
1. Toxicology. 2. Environmental health I. Corn, Morton W.
[DNLM; 1. Hazardous Substances--toxicity. 2. Environmental Exposure--adverse effects. WA 465 S464 1993]
RA1211.S38 1993
615.9'02--ed20
DNLM/DLC
for Library of Congress 93-2111
CIP

PRINTED IN THE UNITED STATES OF AMERICA
93 94 95 96 97 98 EB 9 8 7 6 5 4 3 2 1

Contents

Contributors

Numbers in parentheses indicate the pages on which the authors' contributions begin.

Marlene Absher (661)
Department of Medicine
University of Vermont
Burlington, Vermont 05405

Bengt Akesson (701)
Department of Occupational and Environmental Medicine
University Hospital
S-221 85 Lund
Sweden

Rolf Altenburger (15)
Institut für Zellbiologie, Biochemie, und Biotechnologie
Universität Bremen
2800 Bremen 33
Germany

Jean-Claude Amiard (733)
Faculté de Pharmacie de Nantes
Service d'Ecotoxicologie
Centre National de la Recherche Scientifique
44035 Nantes Cedex
France

Claude Amiard-Triquet (733)
Faculté de Pharmacie de Nantes
Service d'Ecotoxicologie
Centre National de la Recherche Scientifique
44035 Nantes Cedex
France

Melvin E. Andersen (577)
Chemical Industry Institute of Toxicology
Research Triangle Park
North Carolina 27709

H. Babich (463)
Stern College
Yeshiva University
New York, New York 10033

Asim K. Bej (309)
Department of Biology
University of Alabama at Birmingham
Birmingham, Alabama 35294

Wolfgang Boedeker (15)
Institut für Zellbiologie, Biochemie, und Biotechnologie
Universität Bremen
2800 Bremen 33
Germany

Robert S. Boethling (55)
Office of Pollution Prevention and Toxics
U.S. Environmental Protection Agency
Washington, D.C. 20460

György M. Böhm (265)
School of Medicine
University of Sao Paulo
Sao Paulo 01246
Brazil

David L. Brown (233)
Department of Biology
University of Ottawa
Ottawa, Ontario
Canada K1H 8M5

Monique Cadrin (233)
Department of Pathology and Laboratory Medicine
University of Ottawa
Ottawa, Ontario
Canada K1H 8M5

John Cairns, Jr. (627)
Department of Biology
and University Center for Environmental and Hazardous Materials Studies
Virginia Polytechnic Institute and State University
Blacksburg, Virginia 24061

Johnny L. Carson (527)
Department of Pediatrics
Department of Cell Biology and Anatomy
The University of North Carolina at Chapel Hill
Chapel Hill, North Carolina 27599

Vincent Castranova (213)
National Institute for Occupational Safety and Health
Morgantown, West Virginia 26505

Clarence R. Collier (521)
School of Medicine
University of Southern California
Los Angeles, California 90033

Deborah A. Cory-Slechta (411)
Department of Environmental Medicine
University of Rochester
School of Medicine and Dentistry
Rochester, New York 14627

John E. Craighead (399)
Department of Pathology
University of Vermont
Colchester, Vermont 05446

Brenda Cuccerini (505)
School of Medicine
George Washington University
Washington, D.C. 20037

Devra Lee Davis (463)
Office of the Assistant Secretary for Health
U.S. Department of Health and Human Services
Washington, D.C. 20201

José L. Domingo (705)
Laboratory of Toxicology
University of Barcelona
Barcelona 43201
Spain

Alan M. Ducatman (81)
Institute of Occupational and Environmental Health
West Virginia University School of Medicine
Morgantown, West Virginia 26506

Daniel Dziedzic (99)
Biomedical Science Department
General Motors Research Laboratories
Warren, Michigan 48090

Bernd Elsenhans (431)
Walther Straub-Institut für Pharmakologie und Toxikologie
Ludwig-Maximilians-Universität Munchen
D-8000 Munich 2
Germany

A. J. Englande, Jr. (333)
Department of Environmental Health Sciences
School of Public Health and Tropical Medicine
Tulane University
New Orleans, Louisiana 70112

Anna M. Fan (563)
Pesticide and Environmental Toxicology Section
Office of Environmental Health Hazard Assessment
California Environmental Protection Agency
Berkeley, California 94704

Michael Faust (15)
Institut für Zellbiologie, Biochemie, und Biotechnologie
Universität Bremen
2800 Bremen 33
Germany

Wolfgang Forth (431)
Walther Straub-Institut für Pharmakologie und Toxikologie
Ludwig-Maximilians-Universität Munchen
D-8000 Munich 2
Germany

Robert A. Frakes (293)
Maine Department of Human Services
Augusta, Maine 04333

Silvana Galassi (491)
Water Research Institute
Brugherio, Italy
Department of Biology
University of Milan
Milan, Italy

John F. Gamble (183)
Occupational Health and Epidemiology Division
Exxon Biomedical Sciences, Inc.
East Millstone, New Jersey 08875

Christopher J. Gordon (693)
Neurotoxicology Division
Health Effects Research Laboratory
U.S. Environmental Protection Agency
Research Triangle Park, North Carolina 27711

William F. Grant (591)
Department of Plant Science
MacDonald Campus of McGill University
Montreal, Quebec
Canada H9X 3V9

G. D. Griffin (681)
Health and Safety Research Division
Oak Ridge National Laboratory
Oak Ridge, Tennessee 37831

L. Horst Grimme (15)
Institut für Zellbiologie, Biochemie, und Biotechnologie
Universität Bremen
2800 Bremen 33
Germany

Kenneth B. Gross (99)
Biomedical Science Department
General Motors Research Laboratory
Warren, Michigan 48090

P. J. (Bert) Hakkinen (145)
Paper Technology Division
The Procter & Gamble Company
Cincinnati, Ohio 45224

Darryl Hawker (45)
Faculty of Environmental Sciences
Griffith University
Nathan, Queensland 4111
Australia

Edwin E. Herricks (69)
Department of Civil Engineering
University of Illinois at Urbana-Champaign
Urbana, Illinois 61801

Lebelle R. Hicks (293)
Maine Department of Agriculture
Augusta, Maine 04333

G. Richard Hogan (649)
College of Science and Technology
St. Cloud State University
St. Cloud, Minnesota 56301

Katherine Hunting (505)
School of Medicine
George Washington University
Washington, D.C. 20037

Gary E. Isom (161)
Department of Pharmacology and Toxicology
Purdue University
West Lafayette, Indiana 47907

Troyce D. Jones (321)
Oak Ridge National Laboratory
Managed by Martin Marietta Energy Systems, Inc.
Oak Ridge, Tennessee 37831

William Jones (213)
National Institute for Occupational Safety and Health
Morgantown, West Virginia 26505

Armanda Jori (723)
Istituto di Richerche Farmacologiche
20157 Milan
Italy

Sadanobu Kagamimori (443)
Department of Community Medicine
Toyama Medical and Pharmaceutical University
Toyama City 930 01
Japan

Han K. Kang (201)
Veterans Health Administration
U.S. Department of Veterans Affairs
Washington, D.C. 20026

Terutaka Katoh (443)
Department of Community Medicine
Toyama Medical and Pharmaceutical University
Toyama City 930 01
Japan

Kannan Krishnan (577)
Département de Médecine du Travail et Hygiène du Milieu
Université de Montréal
Montreal, Quebéc
Canada H3C 3J7

Michael D. Lebowitz (285)
Respiratory Sciences Center
University of Arizona College of Medicine
Tucson, Arizona 85724

Bruce E. Lehnert (475)
Cell Growth, Damage and Repair Group
Life Sciences Division
Los Alamos National Laboratory
Los Alamos, New Mexico 87545

Carola Lidén (387)
Department of Occupational Dermatology
Karolinska Hospital
S-104 01 Stockholm
Sweden

William Lijinsky (241)
NCI—Frederick Cancer Research Development Center
Frederick, Maryland 21701

Frederick W. Lipfert (1)
Environmental Consultant
Northport, New York 11768

Dominique Lison (153)
Industrial Toxicology and Occupational Medicine Unit
Catholic University of Louvain
1200 Brussels
Belgium

Ingvar Lundberg (387)
Department of Occupational Health
Karolinska Hospital
S-104 01 Stockholm
Sweden

Jane Y. C. Ma (213)
National Institute for Occupational Safety and Health
Morgantown, West Virginia 26505

Joseph K. H. Ma (213)
National Institute for Occupational Safety and Health
Morgantown, West Virginia 26505

Meena H. Mahbubani (309)
School of Dentistry
University of Alabama at Birmingham
Birmingham, Alabama 35294

Eduardo Massad (265)
School of Medicine
University of Sao Paulo
Sao Paulo 01246
Brazil

Paul V. McCormick (627)
University Center for Environmental and Hazardous Materials Studies
Virginia Polytechnic Institute and State University
Blacksburg, Virginia 24061

John G. Mohler (521)
School of Medicine
University of Southern California
Los Angeles, California 90033

Yuchi Naruse (443)
Department of Community Medicine
Toyama Medical and Pharmaceutical University
Toyama City 930 01
Japan

A. T. Natarajan (453)
Departmemt of Radiation Genetics and Chemical Mutagenesis
State University of Leiden
2 333 Leiden
The Netherlands

Günter Obe (453)
Department of Genetics
University GH Essen
D-445117 Essen 1
Germany

M. Hema Prasad (639)
Institute of Genetics
Hospital for Genetic Diseases
Osmania University
Hyderabad 500 016
India

Alfredo Provini (491)
Water Research Institute
Bruherio, Italy
Department of Biology
University of Milan
Milan, Italy

James J. Quackenboss (285)
Environmental Monitoring Systems Laboratory—Las Vegas
U.S. Environmental Protection Agency
Las Vegas, Nevada 89193

P. P. Reddy (639)
Institute of Genetics
Hospital for Genetic Diseases
Osmania University
Hyderabad 500 016
India

Kenneth R. Reuhl (233)
Department of Pharmacology and Toxicology
Rutgers University
Piscataway, New Jersey 08855

Noel R. Rose (377)
Department of Immunology and Infectious Diseases
School of Hygiene and Public Health
The Johns Hopkins University
Baltimore, Maryland 21205

Sheldon H. Roth (367)
Faculty of Medicine
University of Calgary
Calgary, Alberta
Canada T2N 4N1

Paulo H. N. Saldiva (265)
School of Medicine
University of Sao Paulo
Sao Paulo 01246
Brazil

David J. Schaeffer (69)
Department of Veterinary Biosciences
University of Illinois at Urbana-Champaign
Urbana, Illinois 61801

Irene Scheunert (223)
GSI—Institut für Bodenökölogie
D-8042 Neuherberg
Germany

Klaus Schümann (431)
Walther Straub-Institut für Pharmakologie und Toxikologie
Ludwig-Maximilians-Universität Munchen
D-8000 Munich 2
Germany

Raghubir P. Sharma (173)
Center for Environmental Toxicology
Utah State University
Logan, Utah 84322

Glen Shaw (611)
National Research Centre for Environmental Toxicology
Brisbane, Australia

Lance L. Simpson (91)
Division of Environmental Medicine and Toxicology
Jefferson Medical College
Philadelphia, Pennsylvania 19107

Frank A. Smith (277)
Department of Environmental Medicine
University of Rochester School of Medicine and Dentistry
Rochester, New York 14642

Elizabeth T. Snow (127)
Norton Nelson Institute of Environmental Medicine
New York University Medical Center
Tuxedo, New York 10987

Donald L. Sparks (671)
Department of Plant and Soil Sciences
University of Delaware
Newark, Delaware 19717

Katherine S. Squibb (127)
Program in Toxicology
University of Maryland School of Medicine
Baltimore, Maryland 21201

Anna Steinberger (113)
Department of Obstetrics, Gynecology, and Reproductive Sciences
University of Texas Medical School at Houston
Houston, Texas 77030

William T. Stott (545)
Toxicology Research Laboratory
Dow Chemical Company
Midland, Maryland 48674

David L. Swift (255)
Division of Environmental Health Engineering
The Johns Hopkins University
Baltimore, Maryland 21205

Emanuela Testai (119)
Biochemical Toxicology Unit
Istituto Superiore di Sanità
I 00161 Rome
Italy

Lennart Torstensson (351)
Department of Microbiology
Swedish University of Agricultural Sciences
S-750 07 Uppsala
Sweden

Luciano Vittozzi (119)
Biochemical Toxicology Unit
Istituto Superiore di Sanità
I 00161 Rome
Italy

Lance A. Wallace (713)
Office of Research and Development
U.S. Environmental Protection Agency
Warrenton, Virginia 22186

A. P. Watson (681)
Health and Safety Research Division
Oak Ridge National Laboratory
Oak Ridge, Tennessee 37831

James S. Webber (29)
Wadsworth Center for Laboratories and Research
New York State Department of Health
Albany, New York 12201

Laura S. Welch (505)
School of Medicine
George Washington University
Washington, D.C. 20037

Candace S. Wheeler (99)
Biomedical Science Department
General Motors Research Laboratory
Warren, Michigan 48090

Hanspeter Witschi (539)
ITEH
University of California, Davis
Davis, California 95616

Ronald E. Wyzga (1)
Electric Power Research Institute
Palo Alto, California 94303

Preface

The *Handbook of Hazardous Materials* provides brief, effective summaries of relevant information and the prevailing "state of knowledge" on a range of relevant topics in the field of hazardous materials. Written by experts, the articles are keyed to the knowledgeable scientific/technical reader who needs a summary or precis in a field of endeavor other than his or her own. In many cases, the non-scientific reader will derive the same benefit from reading the article.

This book is needed because there has been an explosion of enabling federal statutes and consequent regulatory agency promulgations since the mid-1970s, all seeking to assess and, if need be, regulate the risk(s) of potentially hazardous materials in our society. For example, the Clean Air Act, the Clean Water Act, the Toxic Substances Control Act, the Occupational Safety and Health Act, and the Resources Conservation Recovery Act are a few of the statutes that have sensitized the public to hazardous materials issues and have directed the efforts of large numbers of scientists and engineers. The interests of involved readers will vary from assessment of problems to potential health effects and remediation. This volume does not pretend to be comprehensive; rather, it is eclectic. If, as anticipated, it meets the needs of many readers, the effort can be continued to include additional topics in subsequent volumes.

This book differs from others in that it can also serve as interesting reading for those without "a need to know." Articles are eminently readable and will serve as primers for the interested reader. The ubiquitous nature of artificial, potentially hazardous materials in the environment, coupled with increased concern for the quality of the environment, demands that the general public learn more about these materials, their occurrence in the environment, and appropriate approaches to minimize risks to human health and the environment.

In general, topics addressed are familiar and of concern in the sense of recent past or current emphasis by the government, the scientific community, the media, or all three. This book is designed to be useful to those interested in practical consequences.

Finally, concern for potentially toxic substances is inherent in all fields of science and technology. We must always concern ourselves with minimizing the risks introduced by new and exciting scientific discoveries and technologies, if they are to be successfully integrated into society. Safety is acceptable risk, and the U.S. public has increasingly demanded more, not less, environmental and occupational safety (i.e., less risk). This book will assist those intent on understanding and/or contributing to the achievement of that goal.

Morton Corn

Ambient Acidic Aerosols

Frederick W. Lipfert
Environmental Consultant

Ronald E. Wyzga
Electric Power Research Institute

Glossary

Acid gas Water-soluble gas that yields protons (H^+) in solution.

Aerosol Gaseous suspension of solid or liquid particles.

Bronchoconstriction Resistance to breathing caused by narrowing or obstruction of the airways.

Clearance Ability of the cilia to expel foreign particles from the airways.

FEV_1 Amount of air expelled from the lungs in the first second of a forced expiratory maneuver.

Strong acid Acid that dissociates completely in water.

Weak acid Acid that dissociates only partially, depending on other substances present.

ACIDIC AEROSOLS *are gaseous suspensions of solid or liquid particles that yield electrically charged hydrogen atoms (H^+) when dissolved in water. The gas (ambient air) in which the particles are suspended may or may not include acid gases such as hydrochloric (HCl) or nitric (HNO_3) acids. Sulfuric acid (H_2SO_4) and its various ammonium salts [NH_4HSO_4, $(NH_4)_2SO_4$, $(NH_4)_3H(SO_4)_2$] are the most commonly found acidic particles, although $(NH_4)_2SO_4$ is a very weak acid. Some organic acids can exist as particles; these are also weak acids. Because of its low vapor pressure, H_2SO_4 normally exists in the atmosphere in the particulate phase (as fine droplets). Since most of the extant information on the biologic effects of acid aerosols has been obtained from experiments involving sulfate compounds, this article emphasizes H_2SO_4 and its ammoniated salts. There are no national ambient air quality standards for acidic aerosols or for suspended sulfates.*

I. Sources of Atmospheric Acids

A. Direct Emissions

Acid gases and particles can be emitted from certain manufacturing processes including the manufacture of H_2SO_4 and explosives as well as other chemical processes. The physical form of H_2SO_4 emitted from such processes tends to be as *acid mist*, characterized by median particle diameters of several micrometers (μm) and larger. In addition, the combustion of sulfur-bearing fuels normally results in the direct emission of a small fraction (<5%) of fuel sulfur in the form of H_2SO_4 or as metal sulfate salts. These direct stack emissions are referred to as ''primary'' sulfates. Another minor source of ambient acids is from oxidation of the sulfur dioxide (SO_2) in gasoline (which contains about 0.05% sulfur) as the exhaust gases pass through the catalytic converter. When chloride-containing coal is burned, the Cl^- is normally emitted as HCl (gas). Hydrofluoric acid (HF) from industrial sources was one of the agents suspected of contributing to the 1930 air pollution disaster in Meuse Valley, Belgium.

Handbook of Hazardous Materials

B. Secondary Particles

Since most of the sulfur released to the atmosphere from anthropogenic activities is in the form of SO_2, the subsequent oxidation of SO_2 to H_2SO_4 is the most important source of acidic aerosols in regions where sulfur-bearing fuels are burned. These particles are hygroscopic and of submicrometer size; as the aerosol travels and ages in the atmosphere, median diameters tend to increase to the order of 0.7–1 μm. Chemical transformations are discussed in Section II.

C. Natural Sources of Ambient Acids

Sulfur compounds are emitted from volcanoes, soil, vegetation, and wetlands, and include SO_2, hydrogen sulfide (H_2S), carbonyl sulfide (COS), and other compounds including sulfates (SO_4^{2-}). Their route of removal from the atmosphere is through slow oxidation to water-soluble sulfate particles (some of which may be acidic), which are then scavenged by precipitation. Natural emissions are an important part of the global sulfur cycle because of the large extent of the emitting areas. However, they constitute only a minor part of the total atmospheric sulfur in most urban or industrial regions where the density of sulfur emissions may be several orders of magnitude higher than found in natural areas. (For a ground level source or group of sources, ambient air quality is directly proportional to emissions density.) Although marine aerosols are essentially neutral (NaCl), they can react with strong acids (H_2SO_4 or HNO_3) to release HCl.

II. Transport and Transformations

It is commonly assumed that all of the sulfur in fuel is oxidized to SO_2 during fuel combustion in modern installations and that, during the combustion process, varying amounts of atmospheric N_2 are oxidized to nitric oxide (NO), which is the primary oxide of nitrogen species emitted. As the combustion gas plume travels downwind in the atmosphere, it is subject to various chemical transformations and removal processes which can have important bearings on the nature of any adverse environmental effects. The pollutants that are removed from the air can accumulate in surface deposits, which may have

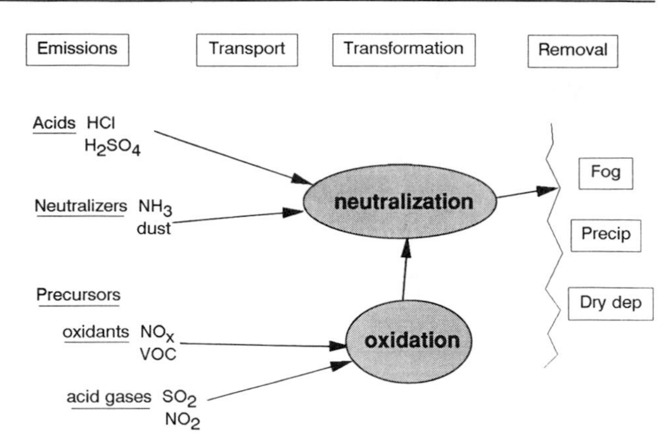

Figure 1 Schematic illustrating the main atmospheric processes involved for ambient acidic aerosols. [From Lipfert (1991); courtesy of Springer-Verlag.]

varying effects upon their receptors. This chain of processes is depicted schematically in Figure 1.

A. Transformations

1. Sulfur Compounds

Sulfur dioxide (SO_2) can be oxidized to form H_2SO_4 either in the gas phase through photochemical reactions, or in the aqueous phase in cloud water. When clouds evaporate, sulfate particles are left behind. The photochemical reactions involving ozone (O_3) are fairly slow (1–3%/hr) but operate during all daylight hours. Aqueous reactions are faster, but require the presence of clouds; hydrogen peroxide is thought to be the most important aqueous phase oxidant. The relative importance of these two sulfate formation pathways will vary with season and climatic factors.

Secondary sulfate particles are usually associated with ammonium; sulfate particles involving calcium or other crustal elements are also found, but tend to be larger in particle size (>1 μm). The acidity of sulfate particles depends on their source and the extent of contact with ambient ammonia. The high correlation between SO_4^{2-} and NH_4^+ in both aerosol and precipitation may be due in part to the inverse dependencies of their water solubilities on solution pH. Sulfur dioxide dissolves only slightly in water, according to physical (Henry's Law) solubilities; its solubility is enhanced considerably by dissolution to form bisulfite ion, but this solubility is pH dependent and quite limited for pH <5. As a cloud droplet or surface film acidifies due to the dissolution of SO_2, continued SO_2 dissolution depends on the oxidation of the SO_2 in solution. However, ammonia (NH_3)

can buffer the solution and NH_3 solubility increases as the pH drops. Thus, in the presence of moisture, the two gases have an "affinity" for each other; sulfate aerosols can be precipitated to the atmosphere as these solutions evaporate.

The low vapor pressure of H_2SO_4 and ammonium sulfates is an important property, which ensures that they remain as particles under normal atmospheric conditions. In contrast, nitric and hydrochloric acids exist as vapors, and ammonium nitrate tends to be unstable.

2. Nitrogen Compounds and Oxidants

Nitric oxide (NO) is nearly insoluble in water and is unreactive in solution; therefore, little is deposited on surfaces or oxidized by reactions in cloud water. Nitric oxide is oxidized to NO_2 by reaction with ozone (O_3) or other oxidants on a time scale of hours. NO_2 is subject to dry deposition and can be further oxidized to form nitric acid by gas phase reaction with the OH^- radical. Nitric acid is quite volatile, highly soluble in cloud and rain water, and deposits on virtually all surfaces.

Ozone formation is dependent on both nitrogen oxides and volatile organic compounds (VOCs), which are hydrocarbons emitted from motor vehicles, solvent usage, and natural sources. Ozone and other oxidants participate in complex photochemical cycles, with peak concentrations occurring on summer afternoons.

3. Neutralization of Acids

As discussed further in Section V, the toxicity or irritant potential of an acid aerosol is thought to be due chiefly to its hydrogen ion content (H^+). The importance to toxicity of the remainder of the molecule is an open question. In clinical experiments on animals and humans, H_2SO_4 has been shown to be the most potent sulfate compound; ammonium bisulfate (NH_4HSO_4) is somewhat less potent (even for the same H^+ dose), and ammonium sulfate $(NH_4)_2SO_4$ is virtually innocuous. The sulfates found in ambient air are not likely to exist as pure compounds *per se* but, at sufficiently high relative humidity, tend to exist as internal mixtures of H^+, SO_4^{2-}, and NH_4^+. These mixtures may reflect the composition of the aerosol when it was first formed, or as modified by contact with other atmospheric compounds during transport. Urban aerosols have been shown to be less acidic than their rural counterparts, for example. Neutralization is also increased near sources of ammonia, which include human and animal wastes, coal combustion, and manufacturing plants. The 1970 *British Report on Air Pollution and Health* by the Royal College of Physicians recommended placing open bottles of dilute ammonia in the rooms of patients with chronic bronchitis or chronic heart disease to neutralize acid mists.

B. Atmospheric Removal Mechanisms

The solubility of SO_2 in water is central to its removal from the atmosphere; this removal is responsible for the maintenance of a stable background concentration (rather than a continuous atmospheric build-up in response to continuous emissions). Removal mechanisms are termed "wet" if associated with hydrometeors; "dry," if otherwise. SO_2 can readily be removed by dry deposition to vegetation and other moist surfaces, but as discussed earlier, the uptake of SO_2 by surface moisture depends on its buffering capacity. Sea water is highly buffered and may be the "perfect" absorber of SO_2. Uptake of SO_2 into cloud water or surface moisture layers may be greatly enhanced by aqueous-phase oxidation, especially by hydrogen peroxide (H_2O_2). In surface moisture, attack of the surface by dissolved SO_2 (H_2SO_3) or H_2SO_4 can provide buffering.

Removal of airborne sulfate particles is dominated by wet removal processes, mostly by dissolution into cloud water concurrent with cloud formation, followed by deposition in precipitation. Wet removal processes are further controlled by precipitation types and rates. Dry deposition processes on surfaces are affected by atmospheric transport rates that mix fresh pollutant into the surface boundary layers and by the physical properties of particles. Dry deposition of sulfate particles is limited by the low diffusion coefficient of small aerosol particles. Removal of NO_2 and ozone is controlled more by chemical reactions than by direct deposition due to their limited solubility in water. For SO_2, dry removal processes are generally considered slightly more important than wet, and dry deposition rates of (gaseous) SO_2 far exceed those of its sulfate transformation products. This is a result of the characteristic SO_4^{2-} particle sizes (typically 0.1–1 μm) being near the minimum for sedimentation and other physical deposition processes. One of the results of this difference in deposition velocities is that sulfur particles will remain airborne and thus travel further from their original sources than SO_2.

Fog offers still another opportunity to form or

modify atmospheric processes (fog is defined as a cloud in contact with the ground). Since the common sulfate particles are quite hygroscopic, they are readily scavenged by the relatively large (~5–50 μm) fog water droplets, as is nitric acid vapor. In addition, if gaseous SO_2 is absorbed into the droplets, it may be oxidized to form H_2SO_4 by any of several chemical reactions involving either oxidants or catalysts within the droplet. These processes constitute one of the natural "sinks" for SO_2 since the larger fog particles tend to settle gravitationally and deposit at rates much faster than the "dry" aerosol. After fog water droplets evaporate and the fog clears, sulfate aerosol particles may be left behind; this also happens with clouds. These particles tend to be larger (~0.7 μm) than the aerosol particles formed by condensation of gas phase precursors.

C. Atmospheric Transport Processes

If an atmospheric constituent is not removed, it will be transported by the winds, which vary greatly in time and space. The vertical dimension is particularly important, not only because transport speeds increase with height but because the probability of encountering clouds increases with height and the likelihood of being trapped by a ground-based inversion is decreased. Aircraft measurements show that concentrations of water-soluble species decrease much faster with altitude than do insoluble species such as ozone. Electric utilities tend to use tall stacks, up to about 300 m in North America and 200 m in Europe. Many European cities use sulfur-bearing fuels for space heating; these ground-level releases are more likely to be trapped by atmospheric inversions than are emissions from tall stacks.

III. Measurement Methods

A. Air Sampling and Analysis

Air pollution is traditionally measured in either mass or volumetric units per volume of air; such concentrations are determined using specific "reference" methods of sample preparation and chemical analysis, as defined by regulatory agencies, for example. Although a reference method has not been defined for acid aerosols, methods of monitoring have evolved over the years into a complex technology. The methodology now in most common use involves the following steps:

- Selecting the desired range of particle sizes
- Capturing a representative sample
- Protecting the collected sample from inadvertent neutralization or other chemical reactions or physical losses during transport to the laboratory or while awaiting analysis
- Preparing an aqueous solution of the collected particles
- Determining the chemical composition of this solution

The difficulties encountered in each of these steps depend on the objectives of the monitoring. For example, characterizing the long-term average properties of the atmosphere is much easier than developing integrated exposure data for free-living populations with *representative* samples. The latter task requires information on microenvironments for indoors (home, work, transit) as well as outdoors, by time of day and season. In addition, data are required for a range of particle sizes, since different sizes tend to deposit in different regions of the respiratory system and thus may result in different types of health effects.

In most current research programs, acid aerosol monitoring is based on 12–24 hr filter samples of particles smaller than about 2–3 μm. The samples are protected against inadvertent neutralization by ambient ammonia in two ways. First, the sampling apparatus strips the ammonia from the gas stream to prevent a subsequent gas flow from neutralizing previously collected particles on the filter. After removal from the sampler, the filters are further protected against neutralization and loss of acidity by use of a citric acid atmosphere during transit and while in the laboratory. The NH_3 removal process has the disadvantage of altering the phase equilibrium of the gas stream and might not be required for shorter sampling times. However, using shorter sampling times and a smaller particle size cut-off (such as one in which most of the acid particles are smaller than 1 μm) would reduce the mass of collected material, which can create analytical difficulties. In addition, shorter sampling times would result in a higher laboratory analysis load.

Extraction of the collected particles from the filter into solution is now a well-developed process with no particular problems. Determination of the concentrations of the major ions in solution is also straightforward, with ion chromatography the preferred technology. However, direct determination of the acidity of the solution has several options. The simplest method uses a pH meter with a glass electrode and measures the free acidity of the solution. This will equal the strong acid content if no buffering agents are present in the sample. Determination of the total acidity requires titration to a specified end point (i.e., total titratable acidity). For example, titration to pH 3 yields strong acidity; to pH 7, strong and weak acidity; to pH 10 will include very weak acids such as ammonium sulfate.

This technology is now in routine use by several research groups but yields no direct information on the chemical species present. To the extent that the effects of, say, H_2SO_4, differ from those of NH_4HSO_4 at the same pH, for example, this is an important unfulfilled need. Methods of selectively extracting H_2SO_4 from filters are not in common use because of uncertainties regarding their efficiency. Methods of determining H_2SO_4 without using filters were first developed in the 1970s and have been refined since then. The basic method uses the flame-photometric SO_2 analyzer, modified to accept small particles. Thermal volatilization is used to separate H_2SO_4 from the ammonium salts, which unfortunately cannot be further speciated through this technique. If the ionic composition is known and the sulfuric acid content has been determined independently, the split between ammonium bisulfate and ammonium sulfate may be inferred.

Another technology for speciation of ammonium salts is based on differentiating their infrared spectral signatures. This method offers fast response and nondestructive analysis, but thus far its application has been limited to the laboratory where it was developed.

An approximate measure of aerosol acidity levels may be obtained from ion charge balances if reliable measures of sulfate, nitrate, and ammonium ion concentrations are available. This method has often been used to check the validity of pH determinations in precipitation samples. While sulfate ion concentrations are usually reliable, there can be problems with both ammonium and nitrate. Ordinary filter sampling will likely capture some gaseous nitric acid on the filter, resulting in an overstatement of particulate nitrate. However, in the eastern United States, particulate nitrate levels are sufficiently low so that this error does not significantly affect the ion balance, especially at high levels of acidity. Ammonium concentrations may be subjected to either positive or negative artifacts. Ambient ammonia may add to the ammonium present in the air sample if allowed to contact acidic particles, as discussed earlier. If ammonium sulfate contacts basic particles (which are usually in the larger size fractions), ammonia gas may be liberated, resulting in a negative artifact. This is more likely to occur in non-size-selective sampling such as high-volume sampling, and is a likely explanation for the inordinately low NH_4^+ levels typically found in some of the older ambient databases.

B. Measures of Dose or Effect

Multiplying concentration by the rate of breathing (typically about 20 cubic meters per day or 14 liters per minute, for quiet breathing) gives the amount of pollutant inspired (i.e., the nominal pollutant dose). This is straightforward for a well-defined stable substance such as carbon monoxide. However, for chemically or physically reacting substances, it is more difficult to predict the effective dose actually received at a specific location or region of the airways.

For example, strong acids are characterized chemically by their pH or by the concentration of hydrogen ions (H^+). For fog, concentrations are based on the liquid content; for other aerosols, on the volume of air sampled. The two may be related by the liquid water content (LWC) of the aerosol, which ranges from about 0.01 to 1 g H_2O per cubic meter of air for fogs and about four orders of magnitude less for clear air aerosols, depending on the relative humidity. For this reason, the same mass concentrations of acid fog particles and acid aerosol particles represent greatly different ionic strengths and pH values. For example, a fog with a liquid water content of 1 g/m^3 and pH of 3.7[1] corresponds to an air concentration of 10 μg/m^3 as sulfuric acid. That same concentration in a clear air aerosol could correspond to pH <1. These considerations refer to the particles before they enter the respiratory tract, where dilution will take place through hygroscopic growth.

Both chemical and physical changes take place during breathing. The respiratory tract is characterized by high humidity (~98%) and varying amounts of endogenous ammonia (NH_3), which can neutralize acids. More ammonia is found in the mouth, probably because of the bacteria found there. Thus, the amount of aerosol neutralization depends on

[1] At a pH of 3.7, the concentration of $H^+ = 10^{(6-pH)} = 10^{2.3}$ = 200 μeq/liter of water. Since there is 1 g of water per cubic meter of air, it requires 1000 m^3 of air to supply 1 liter of fog water. Thus the H^+ concentration on an air basis is 200 neq/m^3. Expressed in mass units of H_2So_4, this corresponds to 9.8 μg/m^3.

whether the nose or the mouth is the primary route of entry into the respiratory tract. In addition to being neutralized by ammonia, hygroscopic particles can absorb moisture; this increases their average diameter and reduces the pH (but not the mass of H^+). The particle size is very important in determining the specific region of the respiratory tract in which the particles deposit. Water-soluble gases (such as SO_2 or HNO_3) are scrubbed out by the moist surfaces of the upper respiratory tract, where they may react with some of the NH_3. Submicrometer particles may penetrate deeper into the lung and some of them will be exhaled. Larger particles (up to 5 μm) are more likely to deposit higher in the airways, although some of them may also deposit in the pulmonary region. Research is in progress to model these chemical and physical processes to develop a better understanding of the relationships between various measures of ambient concentrations and the actual dosage delivered to various parts of the respiratory system. It is likely that the physical properties of inhaled particles are important in this regard, a phenomenon which may help explain some of the differences in health effects observed between particles of the same H^+ concentration but differing composition.

IV. Current Ambient Levels

A. Methods of Reporting Ambient Data

There are several metrics in common use for reporting ambient acidity data. Mass ($\mu g/m^3$) or volumetric (ppb) concentrations are used for specific compounds. Acidity may be reported as

- pH (usually only for fog or precipitation)
- Mass units expressed as $\mu g/m^3$ of H_2SO_4, as if H_2SO_4 were the only acidic species and no ammonium were present in the mixture (not to be confused with direct measurements of H_2SO_4 *per se*)
- Concentration of chemical (H^+) equivalents (equivalent weight = molecular weight/valence)

The last metric is usually expressed as nanoequivalents per cubic meter of air sampled, or neq/m^3. H^+ concentrations can be converted from neq/m^3 to "equivalent" mass units expressed as H_2SO_4 in $\mu g/m^3$ by dividing by 20.4. Chemical equivalents provide a useful means of comparing the acid potency of different compounds.

Measurements of aerosol H_2SO_4, *per se,* have been reported in the literature, but if the aerosol is internally mixed, these data must represent a portion of the mixture rather than free sulfuric acid particles.

B. Ambient Measurements

No routine monitoring databases have been developed for aerosol acidity, but a number of research campaigns have accumulated useful data. As with most other air pollutants, concentrations of acid aerosols vary with time of day and season and are influenced by sampling and averaging times. Current ambient levels in the eastern United States range from about 1–2 $\mu g/m^3$ (expressed as $\mu g/m^3$ of equivalent H_2SO_4) on an annual average basis to about 10–40 $\mu g/m^3$ on a 24-hr basis. Frequency distributions vary by location and typically consist of periods of low levels or even negative values (i.e., at alkaline pH) punctuated by "episodic" periods of a few days' duration (Figure 2). As a result, the frequency distribution of H^+ doses over a year or an entire season tends to be dominated by these few episodic values, more so than ozone, SO_2, or sulfates, for example. The episodic nature of H^+ exposure is shown in Figure 3; in a location near large SO_2 sources in the summer (Figure 3a), about 50% of

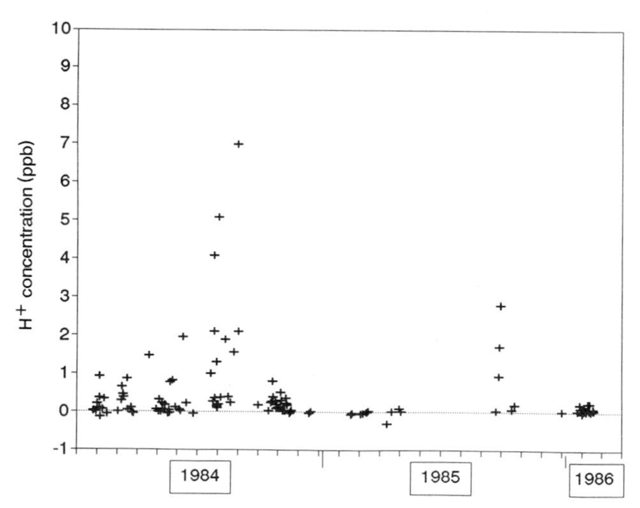

Figure 2 Time history of aerosol acidity at Whiteface Mountain, New York, based on intermittent sampling. [Modified from Lipfert (1991); courtesy of Springer-Verlag. Data from T. J. Kelly, Trace gas and aerosol measurements at Whiteface Mountain, NY. Brookhaven National Laboratory Reports BNL 37110 (Sept. 1985) and BNL 39464 (Jan. 1987).]

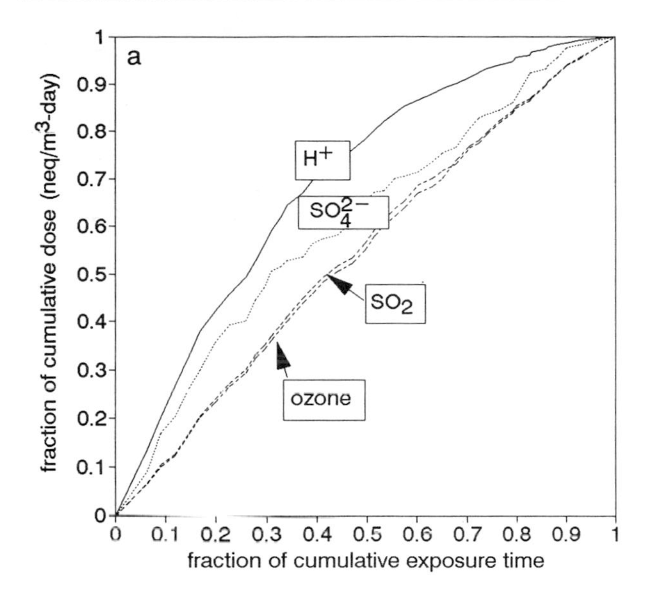

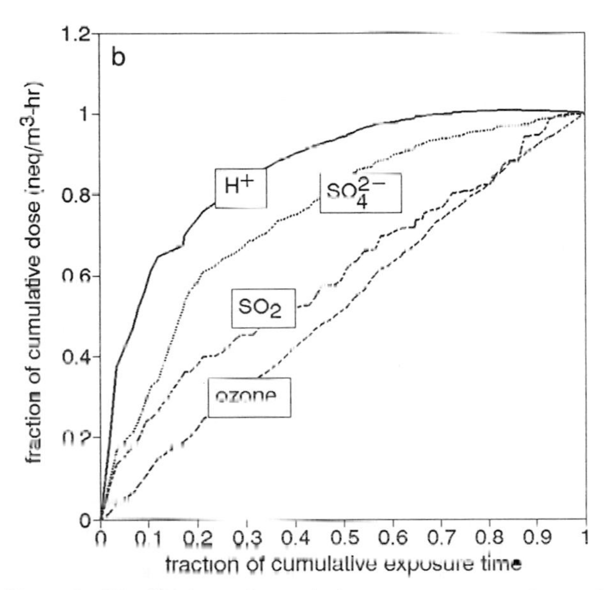

Figure 3 Distributions of cumulative exposures to various pollutants as a function of exposure time. Samples have been ranked according to the H^+ dose. (a) Allegheny Mountain, PA, August 1983. [Data from W. R. Pierson *et al.* (1989). Atmospheric acidity measurements on Allegheny Mountain and the origins of ambient acidity in the northeastern United States. *Atm. Env.* **23**, 431–459.] (b) Whiteface Mountain, NY, 1984. [Data from T. J. Kelly, Trace gas and aerosol measurements at Whiteface Mountain, NY. Brookhaven National Laboratory Reports BNL 37110 (Sept. 1985) and BNL 39464 (Jan. 1987).]

the total H^+ exposure occurs in only 20% of the total time, whereas the exposure to ozone is more nearly uniform throughout the time period. At a more remote location and considering the entire year (Figure 3b), H^+ exposure is even more episodic: 50% of the exposure occurs in less than 10% of the time. Note that at this location, H^+ values are alkaline about 30% of the time. The highest short-term acid peaks tend to occur during summer afternoons, apparently because the higher levels of oxidants outweigh the higher levels of ammonia typically present then.

Since routine monitoring of aerosol acidity has not yet begun, data on aerosol acidity are sparse and have been gathered mainly from various research campaigns. Figures 4 and 5 present eastern U.S. annual average and maximum 24-hr levels of aerosol acidity, expressed as $\mu g/m^3$ of H_2SO_4. Given the limited data available, the patterns are surprisingly consistent. For both measures, acid concentrations are higher near and east of the Appalachian Mountains. Twenty-four-hour values reach a maximum in southwestern Pennsylvania, whereas the few annual data suggest a peak in the southern Appalachians. To the west, $[H^+]$ decreases dramatically; since sulfate levels drop less rapidly, this is apparently due to increased neutralization. Acid levels on the west coast are also quite modest (except perhaps for local areas in the Los Angeles basin). The range in peak acidity values appears to be due more to variability in neutralization than in sulfate concentrations. Data on the values of average molar ratios suggest that compositions close to free sulfuric acid are rarely present in ambient air, and that the average aerosol

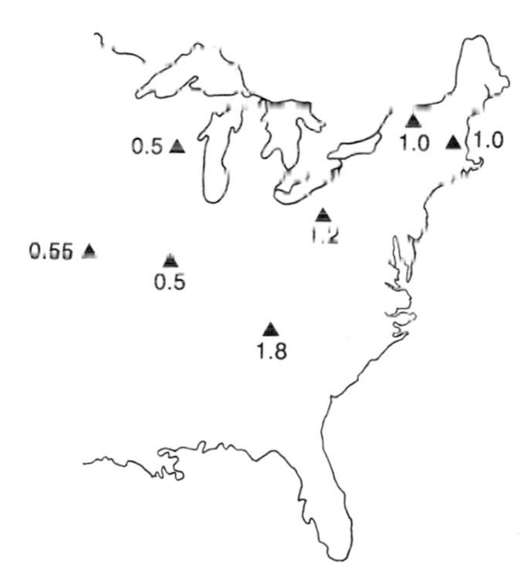

Figure 4 Annual average measured aerosol acidity levels at various locations in the Eastern United States (H^+ in $\mu g/m^3$ as H_2SO_4). [From Lipfert (1991); courtesy of Springer-Verlag. Data from J. D. Spengler *et al.* (1989). Exposures to acidic aerosols. *Env. Health Perspect.* **79**, 43–52; and from T. J. Kelly, Trace gas and aerosol measurements at Whiteface Mountain, NY. Brookhaven National Laboratory Reports BNL 37110 (Sept. 1985) and BNL 39464 (Jan. 1987).]

Figure 5 Maximum 12–24 hr average aerosol acidity levels (H^+ in $\mu g/m^3$ as H_2SO_4). [Data from J. D. Spengler *et al.* (1990). Acid air and health. *Envir. Sci. Tech.* **24**, 946–956.]

composition corresponds to a mixture slightly less acidic than ammonium bisulfate. Concentrations of acid gases can be higher than those of acid particles on a molar equivalent basis. Most of the data shown in Figures 4 and 5 were obtained from rural and suburban sites, where the aerosol tends to be less neutralized than in urban locations.

California fog composition data are compared in Figure 6, on the basis of air concentrations (fog water concentrations multiplied by liquid water content). This places the values in the same range as for clear air aerosols in the eastern United States, except for the South Coast Air Basin (SCAB) data. These data are based on the averages of several sampling campaigns. Fog water at locations in the eastern United States and in Europe tends to contain less nitrate than in southern California, but a much higher proportion of nitrate than typically found in aerosol at the same locations because of the absorption of nitric acid vapor.

C. Coincidence with Ozone

Figure 7 plots aerosol acidity versus ozone at various diverse sites, with a linear regression line shown for each site. Most of these data fall into a common envelope defined by $H^+ \leq 50\ \mu g/m^3$ (as H_2SO_4) at an ozone level of 200 $\mu g/m^3$ (0.1 ppm) (24-hr averages for both parameters), suggesting that the ozone

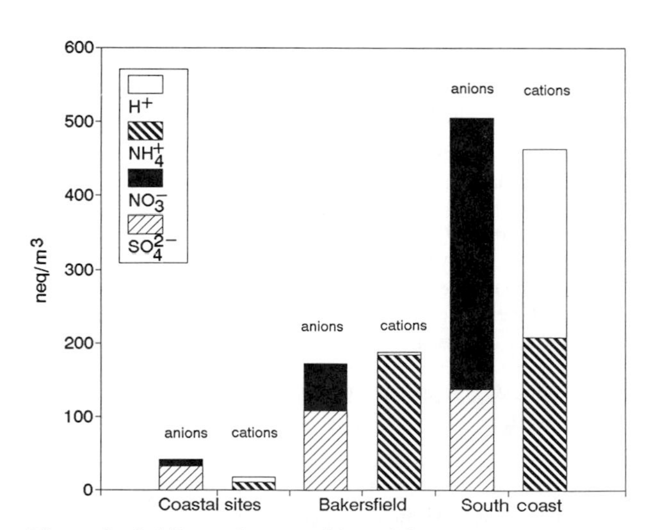

Figure 6 Acidity and composition of fog in California. [From Lipfert (1991); courtesy of Springer-Verlag. Data from various field measurements (see F. W. Lipfert, An assessment of acid fog. Paper IU-8.09, presented at the 9th World Clean Air Conference, Montreal, Canada, September 1992).]

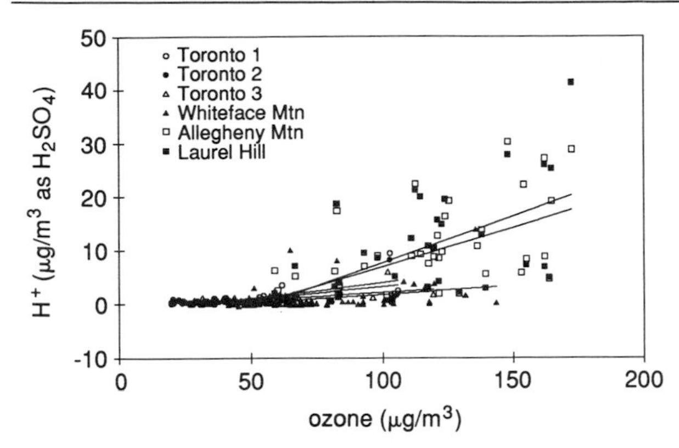

Figure 7 Aerosol acidity versus ozone, for various locations and time periods. [From Lipfert (1991); courtesy of Springer-Verlag. Data from J. M. Waldman *et al.* (1990). Spatial and temporal patterns in summertime sulfate acidity and neutralization within a metropolitan area. *Atm. Env.* **24B,** 115–126; T. J. Kelly, Trace gas and aerosol measurements at Whiteface Mountain, NY. Brookhaven National Laboratory Reports BNL 37110 (Sept. 1985) and BNL 39464 (Jan. 1987); and W. R. Pierson *et al.* (1989). Atmospheric acidity measurements on Allegheny Mountain and the origins of ambient acidity in the northeastern United States. *Atm. Env.* **23,** 431–459.]

level at a given site has an influence on the maximum aerosol acidity. Note that ozone is reduced in many urban areas because of titration by nitric oxide. The aerosol acidity levels at Allegheny Mountain and Laurel Hill, Pennsylvania, stand out as significantly higher than the other sites (which can also be seen in Figure 5).

D. Indoor Acidity Levels

Indoor levels of aerosol acidity have been found to be much lower than outdoor levels, except in the presence of acid aerosol sources such as unvented kerosene heaters. Indoor sources of ammonia, including cleansers and pets, are thought to be responsible for the reduced levels of indoor acidity.

V. Suspected Health Effects

A. Grounds for Health Concerns

The best evidence that air pollution can seriously affect health comes from the historical air pollution disasters of the Meuse Valley (1930), Donora, Pennsylvania (1948), and London (1952), in which the increases in rates of mortality and morbidity were incontrovertible. Air pollution measurements were sparse during these episodes but airborne acids were implicated indirectly in all of them. Direct emissions sources of H_2SO_4 were present in the Meuse Valley and Donora (with acid gases also likely), and it was later hypothesized that the foggy conditions in London, together with soot and metal catalysts, could have oxidized a substantial portion of SO_2 to H_2SO_4.

Concerns about airborne acids have been exemplified in recent years by the debate on the effects of acidic precipitation. Recently, this issue has broadened to include the direct health effects of breathing acidic pollutants, especially fine particles. In 1989, the Environmental Protection Agency (EPA) began the process leading to the listing of acid aerosols as a "criteria" air pollutant, with specific reference to acidic particulates. The first steps in that process include establishing a reference measurement method and assessing ambient concentration levels of acid aerosols.

The literature on respiratory effects of H_2SO_4 is extensive and includes animal toxicology, chamber experiments on human volunteers, and studies of occupational exposures. Studies of mixed sulfate aerosols typical of actual ambient exposures are scarce, however.

B. Experimental Evidence

1. Animal Toxicological Studies

Animal studies have found several endpoints responsive to acid aerosol exposure, ranging from changes in lung clearance rates to changes in lung morphology and rates of cell turnover. Guinea pigs appear to be the most sensitive species, with pulmonary function changes for exposures as low as 1 hr at 100 $\mu g/m^3$. Larger particles can be more lethal, however; an 8 hr exposure to 2.7-μm particles at 27 mg/m^3 killed half of the test animals, while 60 mg/m^3 was required to achieve this response with 0.8-μm particles. Studies of bronchial clearance rates in various animals have shown increased clearance at low doses (similar to or higher than present ambient levels) and decreased clearance at higher doses; evidently, acid aerosols stimulate clearance defenses at low doses but inhibit them at the higher exposure levels.

2. Human Clinical Studies

Experimental exposures of human volunteers have emphasized transient lung function and clearance effects in response to sulfuric acid at concentrations greater than current ambient levels. Early studies on normal subjects found responses only after exposure to very high levels (often in excess of 1000 $\mu g/m^3$) of H_2SO_4. More recent work has reported transitory responses by asthmatic subjects at lower levels (68–100 $\mu g/m^3$), lasting only a few minutes. These experiments have also shown that the pulmonary response to H_2SO_4 is enhanced if the subjects take steps to reduce their oral ammonia levels immediately before the test, which supports the hypothesis that the active agent is the H^+ content of the aerosol. Exposure times for the reported laboratory studies vary from single doses of less than an hour to repeated doses of a few hours per day. The magnitude of the inspired dose also depends on the breathing rate, which can be elevated by exercise.

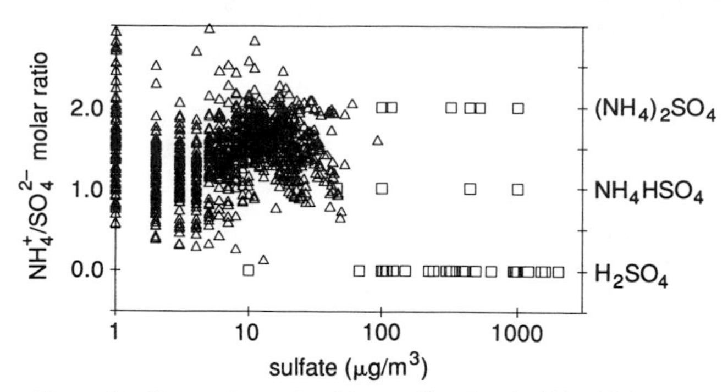

Figure 8 Comparison of ambient sulfate levels (△) with human clinical exposures (□). [Ambient data from F. W. Lipfert (1988). Exposure to acidic sulfates in the atmosphere. EPRI EA-6150, Electric Power Research Institute, Palo Alto, CA.]

C. Unresolved Issues for Experimental Studies

1. Effects of Specific Sulfate Mixtures

Monitoring data show that the average composition of the mixed sulfate aerosol typically found in the eastern United States is approximately that of ammonium bisulfate; sulfuric acid is not the major acid constituent in the ambient environment. It has generally been assumed that H^+ is the agent of principal concern and that biological responses to other acidic species may be estimated by using the H^+ contribution as a measure of relative potency, based on responses to H_2SO_4. Some experimental evidence is consistent with this hypothesis; however, there have also been contradictions: NH_4HSO_4 exposures to rabbits had significantly less effect on alveolar clearance than did equivalent concentrations of H^+ delivered as H_2SO_4. These findings were confirmed by comparisons of phagocytic activity of alveolar macrophages in rabbits. Thus the question of predicting human response to identical H^+ loads delivered by means of differing aerosol compositions remains an important subject for research. The extant body of experimental research on acid aerosol health effects deals almost entirely with sulfuric acid exposures.

2. Effects at Realistic Doses

The disparity between experimental exposures and current ambient levels is illustrated in Figure 8, which plots aerosol composition (ratio of NH_4^+ to SO_4^{2-}, or the degree of neutralization, where a value of 2.0 represents complete neutralization and 0.0 represents sulfuric acid) against sulfate concentration level. The plot shows little overlap between the exposure levels used for experimental human clinical studies (right-hand portion of the figure, which features H_2SO_4 at high concentrations) and the ambient exposure levels (left-hand portion of the figure, with highly neutralized sulfate at much lower concentrations).

3. Separate Effects of Concentration and Exposure Time on Effective Dose

Results from animal studies have been used to support the hypothesis that the total integrated exposure or dose (concentration × time [$C \times T$]) may be the most appropriate acid aerosol dose measure. For effects on lung clearance, there are limited supporting data; however, these data are also consistent with an effect due to variations in concentration alone, since concentration and integrated exposure were highly correlated in these experiments. More recent results that examined responses of rabbits exposed to the same total dose ($C \times T$) of H_2SO_4, but with varying concentrations (C) and exposure times (T), suggest that there may be a threshold concentration below which acute responses are not observed.

Figure 9 contrasts recent U.S. ambient data with concentrations and exposure times for clinical exposures of asthmatic subjects. Note that acidity concentrations do not continue to rise with shorter averaging times (shown by the circles and the best-fit curvilinear regression line). The clinical exposures (triangles) are characterized by high concentrations and short times; the two regimes do not overlap. The solid triangles in the figure represent conditions under which a statistically significant (reversible) lung function change was observed; the lowest of these is somewhat problematic since subsequent experiments have failed to replicate this outcome and the

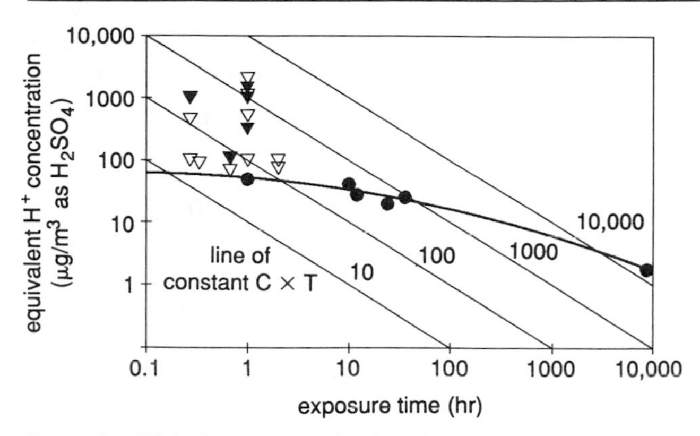

Figure 9 Clinical exposures of asthmatics to sulfuric acid versus maximum observed levels of aerosol acidity. ▼, significant FEV_1 response; ▽, nonsignificant response; ●, maximum observed ambient. [Data from J. Koenig, An assessment of pulmonary function changes and oral ammonia levels after exposure of adolescent asthmatic subjects to sulfuric or nitric acid. Paper 89-92.4, presented at the 82nd Annual Meeting of the Air and Waste Management Association, Anaheim, CA, June 1989; Linn *et al.* (1989). Effect of droplet size on respiratory responses to inhaled sulfuric acid in normal and asthmatic volunteers. *Am. Rev. Respir. Dis.* **140,** 161–166; and Linn *et al.* (1990). Respiratory response of young asthmatics to sulfuric acid aerosol (abstract). *Am. Rev. Respir. Dis.* **141,** A74.]

responses typically lasted only a few minutes. The open triangles represent clinical exposures to asthmatics that did *not* produce significant responses. By following the diagonal lines of constant $C \times T$ in the figure, it appears that exposure times of 12 hr or more at the maximum observed ambient concentration levels (about 30–40 $\mu g/m^3$) would be needed to inspire a total acidity dose similar to those reliably inducing lung function changes.

The validity of the $C \times T$ hypothesis needs to be investigated more systematically, especially at lower concentrations and longer exposure times, conditions for which thresholds due to neutralization by endogenous ammonia may be important. Note that elevated breathing rates induced by exercise are unlikely to be sustained long enough to substantially influence doses greater than a few hours.

4. Ammonia Neutralization

Most clinical experiments have only considered concentration levels substantially higher than ambient. It is not clear how to extrapolate responses observed under these conditions to more realistic (lower) ambient conditions. Such an extrapolation may be particularly confounded by endogenous (breath) ammonia, since experiments have determined that artificial depletion of oral NH_3 (by dental brushing or ingesting acids) can enhance the lung function responses obtained at a given sulfuric acid dose level:

- The responses of exercising asthmatics increased about 150% when they gargled with lemon concentrate before exposure to 350 $\mu g/m^3$ H_2SO_4.
- Lung function responses of adolescent exercising asthmatics to about 70 $\mu g/m^3$ H_2SO_4 were statistically significant only when the subjects drank lemonade before exposure.

The ability to predict responses to current ambient acid levels requires further understanding of the role of endogenous ammonia and the extent to which it may offer some protection from breathing acids. Also, exercise will complicate the neutralization process, by changing the breathing mode, breathing rate, and the residence times of particles.

5. Effects of Other Acidic Species

Although the emphasis has been on acid sulfate particles, it is possible that gaseous acids (HNO_3, organic acids) might be important. Significant lung function decrements in adolescent asthmatics exposed to 0.05 ppm HNO_3 have been reported, and a significant synergistic biological response was associated with joint exposures to ozone and NO_2, which in combination will yield HNO_3 and HNO_2. Finally, acid gases may react with endogenous ammonia that would otherwise be available to neutralize acid particles.

Mixtures may be important since acidic ambient aerosols do not occur in isolation and peak levels may coincide with those of other pollutants. Animal studies have suggested that synergism can occur between high doses of acid aerosols and other pollutants, such as ozone.

6. Effects of Particle Size

The effects of particle size are not well defined. Those epidemiological studies that can reasonably be interpreted as showing acid aerosol health effects all involve larger particles (fog and mist). Large particles have also been shown to exacerbate symptomatic responses at high concentrations, but asthmatics seem to respond more to submicron aerosols.

7. Clarification of "Adverse" Health Effects

The medical significance of the different symptomatic and biological responses observed in health studies is not always clear and may differ according to the subject's health status. Examples include small transitory decrements in lung function and small transient increases and decreases in lung clearance rates, which are difficult to interpret. Research to achieve a better understanding of biological mechanisms of action and their relationships to clearly adverse responses in humans would be one way to clarify significance. Long-term studies of animals may be useful in this regard.

D. Epidemiology

Definitive (positive) associations of ambient acidic aerosols with adverse human health responses by the general population have yet to be demonstrated in any epidemiological studies completed. In the past, this outcome was partially due to the lack of appropriate exposure measurements. However, a more recent time-series study of daily mortality in several U.S. cities found consistent linear dose–response relationships for PM_{10}, but not for H^+ concentrations. Several previous observational studies have linked sulfates with excess mortality and with hospital admissions, but these studies could not rule out other pollutants as causal factors. In addition, the relationship between SO_4^{2-} and H^+ can be quite variable, so that it is problematic to assume that H^+ was the causal factor on the basis of the sulfate relationships in these studies.

E. Occupational Studies

It has been estimated that over 800,000 workers in the United States have some potential occupational exposure to H_2SO_4. These exposures tend to involve concentrations much higher than ambient (the occupational health standard for H_2SO_4 is 1 mg/m^3), and median particle sizes are larger than typically found in community air. However, the range in particle sizes in industrial exposures is large (geometric standard deviations from 2 to 7), so that small particles are likely to be found in these exposures as well. Occupational studies necessarily deal with a selected subset of the population from which susceptible individuals are likely to have been removed through self-selection.

Lead-acid battery workers exposed to H_2SO_4 concentrations from 80 to 350 μg/m^3 were found to have no acute lung function changes associated with daily exposures. The prevalences of various acute respiratory symptoms related to work were less than 7% for both high and low exposure groups; cough, eye irritation, wheeze, chest pains, and nausea were more frequent in the high exposure group; nasal congestion and headache were more frequent in the low exposure group. The upper range of H_2SO_4 exposures was about 1 mg/m^3 and particulate lead concentrations ranged up to about 600 μg/m^3. Median acid particle sizes were around 5 μm, but at two of the five plants studied, about 20% of the acid particles were of submicrometer size. A similar study in a British battery factory also failed to find significant decrements in lung function over the course of an 8-hr shift that could be attributed to acid exposure. These data sets extend the range of human exposure ($C \times T$) data to about 8–11 mg/m^3 hr.

Chronic effects were examined in a companion study, which found no changes in symptoms and only slightly reduced (around 5%) lung function parameters in the high exposure (210 μg/m^3) compared to a low exposure group (100 μg/m^3). There may have been particle size differences between these two groups as well. Smoking and age differences were controlled in these studies; the lung function decrements due to acid exposure were slightly less than those due to smoking. The most obvious chronic health effects associated with acid exposure were tooth etching and erosion. It is particularly interesting that no upper respiratory effects were noted in these studies, which involved particles large enough to affect these regions.

A mortality study of steel workers exposed to pickling acids showed a significant excess in lung cancer deaths, beyond what could reasonably be expected due to higher than average rates of smoking. Excesses were shown for exposure to both sulfuric acid mist and to "other" acids (mainly hydrochloric acid vapors). Exposures to other air pollutants were not taken into account, which could include high concentrations of metals. Deaths due to other respiratory diseases and to circulatory causes were within expected limits. Although some of the sulfuric acid particles involved in these exposures could be expected to penetrate to the pulmonary region, where clearance could be affected, this is less likely for the acid vapors. When the mortality comparison was based on the experience of all steel workers in the county rather than on the general U.S. population, only the "other acid" group showed a significant excess in lung cancer.

VI. Other Environmental Effects

A. Atmospheric Visibility

Atmospheric visibility, the ability to see distant objects clearly, depends on the transmission of light and the ability of the observer to distinguish an object from its background. The contrast with background depends on scattering and absorption of light by the atmosphere. Both particles and gases in the atmosphere can scatter and absorb light. The scattering and absorption of light by aerosols depends on their size distribution, refractive index, and shape, as well as the relative humidity. For visible light, particle diameters in the range 0.1–1.0 μm are effective light scatterers. Thus, sulfate aerosols have been associated with visibility reductions, in both the eastern and western United States. This property is essentially independent of particle chemistry (except for hygroscopic properties and size changes in response to relative humidity).

B. Damage to Materials

In theory, deposition of acidic aerosols can add to the rate of atmospheric corrosion of certain susceptible materials, including carbonate stone, galvanized steel, and paints containing acid-soluble compounds. However, rates of deposition of submicrometer aerosols tend to be much lower than those of acid gases (SO_2 and HNO_3), both because the particle (dry) deposition velocities are about an order of magnitude lower and because air concentrations of acid aerosols are usually lower than SO_2. Thus, in practice, the contributions of acidic aerosols to atmospheric corrosion can usually be neglected.

C. Ecological Acidification

Since acidic aerosols tend to be distributed over large regions, including natural areas, and their main sink is deposition to the surface, they can cause acidification of poorly buffered natural areas. Because of its reactions in soil, NH_4^+ is also acidifying, so that the chemical composition of the aerosol does not play a major role in such acidification. However, scavenging of aerosol particles is one of the pathways for acidifying precipitation, and wet deposition at sufficiently low pH can cause direct damage to some species of vegetation. The possibility of acidification of drinking water supplies in remote areas also exists, which can cause indirect adverse health effects by leaching toxic substances out of the water delivery systems if the pH is not controlled.

VII. Control Techniques

A. Additional Neutralization

Figure 1 suggests several possibilities for controlling atmospheric acidity. The most direct method would be to increase ambient levels of neutralizing substances. For example, it has been suspected that one of the reasons for higher precipitation pH levels in the 1950s was higher ambient levels of alkaline dust from unpaved roads. While it is feasible to increase indoor ammonia levels by various means, our environmental ethics generally do not permit adding to the (outdoor) ambient pollutant burden as a means of controlling its chemistry.

B. Reduced Oxidation

The fraction of SO_2 oxidized to sulfate depends on the operative mechanism and the supply of reagents. For example, in winter, the process may be limited by the supply of oxidants, because of reduced photochemical activity. The summer of 1988 was hotter and dryer than normal, for example, and both ozone and sulfate levels increased slightly. Control of oxidants by reductions in emissions of precursors (NO_x and VOCs) thus may provide some additional benefits by also reducing sulfate levels.

C. Reduced Direct and Precursor Emissions

Controlling direct emissions of acids from manufacturing processes requires only the use of appropriate collection devices and should be encouraged. Reductions in primary sulfate from fuel oil combustion (for a given sulfur content) can be achieved by reducing the amount of excess combustion air; the limit to this process is the creation of excess smoke and CO emissions.

Reducing the atmospheric sulfur burden by cutting back on SO_2 emissions should also have a nearly direct benefit on ambient sulfate levels. This is the stated intention of the acid deposition provisions of the 1990 Clean Air Act Amendments. However, the effects of oxidant limitations may change the spatial distribution of sulfates and acidic aerosols. For example, SO_2 emissions in the United States increased from 1950 to the mid-1970s, after which they decreased slightly. Ambient SO_2 levels have tended to follow suit, except that reductions in cities using sulfur-bearing fuels for space heating have generally been larger. Ambient sulfate levels have changed very little over this time period, however, especially if one considers the likelihood that the earlier levels were biased high because of the types of collection filters used. Thus, the implication is that in more recent years, a higher fraction of the emitted SO_2 has been oxidized to sulfate. If sulfate levels are controlled by the available oxidant supply, one would expect little change in sulfate levels near the SO_2 sources, but larger changes further away.

VIII. Summary

Acidic aerosols are found in industrial atmospheres and in both rural and urban areas in the eastern United States. Average concentrations in community air are about two orders of magnitude lower than in industrial exposures. Much remains to be done to characterize the extent of population exposures and their potential health significance. It appears that the major gaps in ambient exposure characterization include detailed particle size resolution and speciation for periods less than 24 hours, sampling in major eastern population centers, and indoor sampling.

The major health effects question is: What are the dose–response relationships at conditions similar to current actual population exposures? The fact that only minor chronic respiratory effects have been found for industrial exposures suggests that effects in healthy, nonsensitive individuals exposed to community air are likely to be minor. This conclusion is supported by the results of lung function tests of normal subjects at high acidity concentrations. For asthmatics and other sensitive groups, further experiments are needed at longer exposure times involving realistic ambient mixtures and appropriate concentration levels before adequate judgments can be made about their health risks.

Related Articles: MONITORING INDICATORS OF PLANTS AND HUMAN HEALTH FOR AIR POLLUTION; NITRIC OXIDE AND NITROGEN DIOXIDE, TOXICOLOGY; OXIDES OF NITROGEN; PLANTS AS DETECTORS OF ATMOSPHERIC MUTAGENS.

Bibliography

CASAC (1988). Subcommittee on Acid Aerosols. Report on Acid Aerosol Research Needs, EPA–SAB/CASAC-89-002. U.S. Environmental Protection Agency. Washington, D.C. Oct. 19, 1988.

Acid Aerosols, special issue of *Environmental Health Perspectives*. **79,** 3–205 (1989).

Graham, J. A. *et al.* (1990). Direct Health Effects of Air Pollutants Associated with Acidic Precursor Emissions. Report 22, Vol. III. National Acid Precipitation Assessment Program. Washington, D.C.

Lee, S. D., Schneider, T., Grant, L. D., and Verkerk, P. K., eds. (1986). "Aerosols: Research, Risk Assessment, and Control Strategies." Lewis Publishers, Chelsea, Michigan.

Lipfert, F. W. (1991). Acid aerosols. *In* "Acid Deposition, Origins, Impacts, and Abatement Strategies" (J. W. S. Longhurst, ed.), Springer-Verlag, Berlin.

Lipfert, F. W., Morris, S. C., Wyzga, R. E. (1989). Acid aerosols: The next criteria pollutant. *Env. Sci. Tech.* **23,** 1316–1322.

Morris, S. C, and Lipfert, F. W. (1989). Health Effects of Acid Aerosols and Their Precursors. Brookhaven National Laboratory Report to the U.S. Department of Energy. Brookhaven National Laboratory, Upton, New York.

National Acid Precipitation Assessment Program (1990). State of Science and Technology Reports, Washington, D.C.

National Research Council of Canada (1982). Effects of Aerosols on Atmospheric Processes. NRCC No. 18473. Ottawa, Canada.

Tanner, R. L. (1989). The measurement of strong acid in atmospheric samples. *In* "Methods of Air Sampling and Analysis" (J. P. Lodge, ed.), 3rd ed., Lewis Publishers, Chelsea, Michigan.

U.S. Environmental Protection Agency (1988). Acid Aerosols Issue Paper. EPA-600-8-88-005a. Washington, D.C.

Aquatic Toxicology, Analysis of Combination Effects

Rolf Altenburger, Wolfgang Boedeker, Michael Faust, and L. Horst Grimme
University of Bremen, Germany

Glossary

Additivity Combined effect of agents expected from the effects of the individual substances. Expected effects are calculated using biometrical models based on either the concept of *effect summation, effect multiplication* or *concentration addition*.

Concentration addition A concept for combination effects based on the assumption that any constituent of a mixture can be replaced totally or in part by the equieffective amount of another while the effect of the mixture remains constant.

Effect multiplication (independence) A concept for combination effects based on the assumption that the effect of a mixture equals the product of the effects of the components applied singly.

Effect summation A concept for combination effects based on the simple idea that the effect of a mixture equals the sum of the effects provoked by the individual constituents.

Synergism/antagonism Terms denoting effects of mixtures which are greater or less than expected on the basis of one of the concepts mentioned above.

***THE ANALYSIS** of combination effects in aquatic toxicology is focused on the efforts to predict the effects of chemical mixtures from the effects of single substances. With this scope, terms of widespread use like* synergism, potentiation, antagonism, *and* additivity *are explained on the basis of underlying concepts. The necessity to provide a reasonable reference for judgments on combined effects is shown in order to avoid confusion of terminology. The requirements for experimental studies concerning the amount, type, and quality of data needed to allow for meaningful statements are considered. Biometrical models for the assessment of combined effects are discussed and recommendations concerning their use are given. Finally, an overview is supplied on combination effect studies of aquatic pollutants.*

I. Introduction and Scope

Simultaneous or sequential exposure of organisms to two or more chemicals may alter biological responses in qualitative and quantitative ways relative to that for single compounds. The analysis of combination effects resulting from an exposure to mixtures of chemicals has therefore been the subject of numerous studies in pharmacology, toxicology, and biometrics. These long-lasting efforts led to various procedures for hazard management (see Table I). The main focus of those regulatory activities has been human toxicology. In contrast, although surface waters are subject to pollution with numerous chemicals, water quality standards are commonly based on risk extrapolations from studies with single substances.

Table I Examples of Regulations and Recommendations on Chemical Mixtures

Purpose	Country/organization	Method/recommendation
Food and water residue levels	Canada, European Community	Total pesticide residues in drinking water <100 μg/liter Total pesticide residues in drinking water <0.5 μg/liter
	Germany, Sweden, United States	Sum of grouped pesticide residues in food must not exceed certain maximum level
	Australia, Finland, Taiwan, United States	The sum of ratios of the measured concentrations to the permissible level of each pesticide present in food should not exceed 1 in certain groups; the same is true for water
Threshold limits in the work environment	American Conference of Governmental Industrial Hygienists (ACGIH), International Commission on Radiation Protection (ICRP), International Labour Office (ILO), Ministry of Work, Germany, USSR, World Health Organization (WHO)	For mixtures of several (airborne) contaminants, for which summation of effects has been established, the exposure limit concentration is to be calculated by the formula: $c_1/M_1 + c_2/M_2 + \cdots + c_n/M_n \leq 1$, where c_i are the measured concentrations and M_i are the respective exposure limits
Hazard classification	European Community	Mixtures of chemicals are subject to various recommendations for classification and labeling
	World Health Organization (WHO)	Hazard classification of mixtures of pesticides should be done by applying the formula $c_1/T_1 + c_2/T_2 + \cdots + c_n/T_n = 100/T_m$ with c_i as above and T_i denoting the EC50 values of the constituents and T_m of the mixture M, respectively
Water pollution control	European Inland Fisheries Advisory Commission (EIFAC)	For describing the joint effects of mixtures of toxicants on aquatic organisms, the following formula is held to be appropriate: $c_1/T_1 + c_2/T_2 + \cdots + c_n/T_n \leq 1$, with c_i and T_i as above

Experimental studies dealing with the aquatic toxicity of mixtures may be differentiated according to their scope:

1. The toxicity of a specific complex mixture, whose components are often not precisely known, is assessed in much the same way as for single toxicants. Examples might be a specific industrial waste, a sewage plant effluent, or a typical composition of pollutants found in a specific surface water. As far as the results are compared with the effects of individual chemicals, the aim is to identify constituents which dominate the overall toxicity of the mixture.
2. The modification of the toxicity of a single compound by the presence of other substances or a change in environmental conditions is functionally analyzed. Typical examples are alterations of a biological response by water quality characteristics like pH, hardness, salinity, temperature, and nutrients. The aim is to determine the variation of toxicity under realistic environmental conditions and to identify major toxicity-modifying factors.
3. The effects of definite combinations of substances are set in relation to the effects of the individual compounds. An expected effect, formulated or calculated on the basis of single toxicity data, serves as a criterion for the assessment of the observed mixture toxicity. The aim is to identify types of combination effects, to improve the pharmacological understanding of such effects, and to develop tools for reliable prediction of the toxicity of mixtures of pollutants.

With the focus on the effort to predict the effect of combined substances on the basis of the effects of the single substances, the main task is to decide whether the magnitude of the observed effect of a mixture of agents is smaller, equal to, or even greater than expected from the concentration effect relationships of the individual agents. The assessment cannot be done without making use of some kind of mathematical model developed to calculate expected responses from toxicity data of single substances. The validity of results obtained from the use of any mathematical model depends on reasonable assumptions about expected effects. Therefore, knowledge of the inherent features of the different models is necessary in order to make a rational choice. The purpose of this article is to give a comprehensive and structured survey of mathematical modeling of combination effects with special emphasis on applications in aquatic toxicology.

II. Terminology and Concepts

There are different terms used to label possible types of combination effects, but these terms are used inconsistently. Despite the efforts of several authors, a standardization in terminology is not in sight. Table II gives a survey of frequently found terms used to classify combination effects.

First of all, it should become apparent that any statement on the type of combination effect observed requires an answer to the question: What is the expected response? It is now generally held that any combination of substances is expected to be more active than any of the substances alone. Denotation of this situation, however, is difficult. Nevertheless, *additivity* is the most frequently used term of reference. On the one hand, the usage of the term *additivity* might be confusing because it is sometimes taken to mean that the effects of such combinations may be obtained by adding the effects of their constituents. On the other hand, combined effects termed *independence, indifference,* or *noninteraction* will be easily misinterpreted as (eco-) toxicologically irrelevant, since sometimes they are used with the meaning that the effect of a combination is equal to the effect of the most active component alone. Much confusion also exists concerning the understanding of antagonism and synergism. Furthermore, there is no proper differentiation between synergism and potentiation. Sometimes potentiation is used synonymously with synergism, while commonly the former is used for a characterization of an effect stronger than synergistic.

Table II Classification Terms for Combination Effects

Expression indicates	Combined effect is
Augmentation, enhancement, potentiation, sensitization, superadditivity, supra-additivism, synergism, synergy	Greater than expected
Additivity, additivism, independence, indifference, noninteraction, summation, zero-interaction	Expected
Antagonism, antergism, depotentiation, desensitization, infra-additiveness, negative synergy, noninteraction, subadditivity, zero-interaction, no addition	Less than expected

Second, the question of how to calculate expected additive effects arises. To meet this challenge, various mathematical methods and models have been invented more or less implicitly based on three different concepts and understandings of additivity.

The concept of *effect summation* has greatly influenced and confused the discussion about the assessment of combination effects. However, consensus seems to be achieved that additive effects do not simply equal the sum of the single effects. Effect summation seems to lack a pharmacological apprehension of additivity; therefore, there is no clear reference for the determination of combination effects.

The concept of *effect multiplication,* which is sometimes called *response addition* or *independence,* is based on the assumption that the toxicants of a mixture act on different biological subsystems within the same organism (e.g., have different sites and modes of actions). However, it should be noted that the concept of effect multiplication originally was brought forward by biometricians and that, despite its convincing mathematical features, the pharmacological inference for a nonreceptor level is far from being clear.

Finally, the concept of *concentration addition* should be mentioned. This concept gained widespread acceptance because of its plausible pharmacological understanding. Concentration addition in the most simple case is said to occur when one substance acts like a dilution of another, so any effect can be obtained by replacing one substance totally or in part by the equi-effective amount of another. There is widespread belief that mixtures obeying this concept show similar action (e.g., by a common molecular target site). There is still no agreement whether concentration addition is a suitable concept even in cases of pharmacologically dissimilar behavior of substances.

Generally, we recommend labeling of different types of combination effects only when making clear reference to a defined and pharmacologically plausible concept. In order to avoid further confusion and misunderstanding, enhanced or weakened effects should be named super- or subadditive, respectively, whenever additivity is the reference.

III. Experimental Design

The experimental requirements concerning the amount, type, and quality of data needed for meaningful statements on combination effects depend, first of all, on the purpose of the intended study—analysis of the mode of combined action or screening on toxicological effects. Furthermore, adequate choice of effect parameters, response levels, and combination ratios depends on the decision as to whether simultaneous or sequential exposures of binary or multiple mixtures are to be studied.

Second, the assessment model to be used for data analysis evokes prerequisites for the experimental design. Thus, if for instance regression techniques are involved, the choice of appropriate concentration levels and combination ratios will considerably improve the precision for parameter estimations of fitted response functions. Furthermore, the optimal spacing between concentration levels is dependent on the transformation of scales used for analysis. More generally speaking, the different assessment models comprise specific requirements which have to be met.

Third, studies on combination effects in principle underlie resource constraints which ask for optimization of data generation. Concerning selection of concentration levels, range-finding tests prior to excessive testing are suitable to obtain first knowledge of the concentration response relationship. Statistical techniques should generally be applied for an optimal allocation of a minimum number of test subjects as well as for the discrimination between observed and expected responses.

Combination effects have often been studied using an experimental design wherein the substances S_1 and S_2 are tested alone at concentrations c_1 and c_2 and subsequently in the combination $c_1 + c_2$ ("2×2 design"). Employing this approach one is left to compare the measured combined effects with those of the single substances on the assumption that any such combination behaves like a new compound. Obviously, the disadvantage of this approach is that valid conclusions can only be drawn for very specific situations, with the result that every combination of substances in every combination ratio and every concentration level becomes a test case for itself.

Experimental approaches in which only one substance is tested at various concentration levels while the others are kept constant are an extension of the above outlined ("$2 \times n$ design"). The interpretation of results from studies conducted with this design is often based on the discussion of changes in concentration response curves. Criteria for this, however, are contradictory.

Determining the concentration response curves of one substance at various concentration levels of others leads to a design which allows for data analysis with biometrical models ("$n \times n$ design"). Unfortunately, efficient planning of concentration levels for the substance with varying concentration is difficult unless the concentration response curves of the individual substances are considered.

Incorporation of prior knowledge of the effects of the single substances is achieved in a design where the combination ratio is chosen on the basis of fractions of effective concentrations (e.g., toxic units) ("diagonal design"). Here, in contrast to all above-mentioned designs, the combination ratio is kept constant and concentration response curves of fixed mixtures are determined.

For combination effect studies with more than one active substance, a minimum requirement is the knowledge of the concentration response curves of the single substances. To allow a statistical estimation of parameters, at least five different concentrations centered on the EC_{50} should be tested. Experimental designs which do not determine the concentration required by the individual agents to achieve the effect of the mixture are inadequate to detect any combination effect other than antagonism. Therefore, they are usually unsuitable for combination effect studies in aquatic toxicology. Finally, a warning: Whatever design may be preferred for a combination of substances, there should be no more than one factor varying at a time (i.e., either the ratio varies by using different concentration levels of one substance only or the concentration level of the mixture changes while the ratio is kept constant).

IV. Models for Assessment

Various mathematical methods and models for the assessment of the effects of combinations of agents have been developed, some dating back to the 19th century. The different approaches were not developed subsequently, but they reflect diverse points of view of physicians, mathematicians, and pharma-

Table III Categories of Biometrical Models on Combination Effects

Category	Model	Assessment term[a]
Graphical methods and algebraic equivalents	Isobolographics	Shape of isoboles (see Fig. 1)
	Toxic unit summation	$S = c_1/EC_1 + c_2/EC_2$
	Additivity index	$AI = (1/S) - 1$, for $S \leq 1$
		$AI = -S + 1$, for $S \geq 1$
	Mixture toxicity index	$MTI = 1 - \log S/(\log S - \log(\max(c_i/EC_i)))$
Generalized receptor models	Median effect model	$CI = c_1/EC_1 + c_2/EC_2$
		$CI = c_1/EC_1 + c_2/EC_2 + c_1c_2/EC_1EC_2$
Input/output models	Effect summation	$E(c_{1,2}) = E(c_1) + E(c_2)$
	Effect multiplication	$E(c_{1,2}) = E(c_1)E(c_2)$
Quantal response models	Simple similar action	$P_{1,2} = F(\theta_1 + \theta_2 \log(c_1 + \theta_3 c_2))$
	Joint independent action	$P_{1,2} = 1 - \int_{c1}^{\infty} \int_{c2}^{\infty} f(c_1, c_2, \phi)\, dc_1\, dc_2$

[a] For details and abbreviations see the text; nomenclature as proposed by original authors.

cologists. In order to facilitate a structured survey, we distinguish between four different categories (Table III): graphical methods, generalized receptor models, input/output models, and quantal response models.

A. Graphical Methods and Algebraic Equivalents

The history of the scientific assessment of combination effects started with illustrative graphical treatments of experimental data, leading to the still valuable method of isobolograms originally developed for drug assessment in pharmacology.

In biometrical terms, graphical methods are nonparametrical since they do not employ any mathematical or statistical determination of the underlying concentration response relationships. The intention is to detect deviations of experimental data from additivity, which is graphically represented by specific reference lines.

Plotting quantitative effects of two substances A_1 and A_2 combined in various proportions against their individual concentrations leads to the presentation of the concentration response relationship as a concentration response surface in three-dimensional space. A reduction to a two-dimensional plot is then achieved by fixing a defined response level (e.g., EC_{50}). All concentrations of the combined substances giving that response are depicted in the plane which is spanned by the axes representing the concentrations of the individual substances. The curve connecting these points is called an *isobole,* a curve of equivalent response. Figure 1 shows the three typical shapes of isoboles which are used to distinguish between additivity, synergism, and antagonism, referenced to the con-

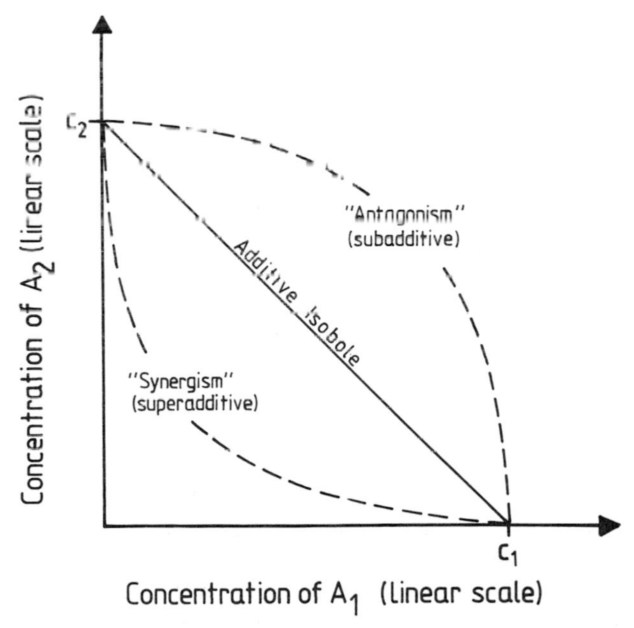

Figure 1 Isobologram after S. Loewe showing different types of combination effects: Additive isoboles are lines connecting equi-effective concentrations c_1 and c_2 of the agents A_1 and A_2 applied singly. Subadditive effects are indicated by elliptical isoboles located above the additivity line and superadditive effects by hyperbolic isoboles located below the additivity line.

cept of concentration addition. Isobolographics is still a valuable tool for combination effect studies, especially because the results are readily accessible to interpretations.

However, there has been immense comment on the statistical validity of isobolographics. The problem derives from the necessity to decide whether eventual departure from the additivity line is systematic or due to chance. Even by applying tests of statistical significance the results attained by isobolographics are restricted to combination ratios and response levels which have actually been investigated experimentally. Another disadvantage of the graphical method of isobolograms is its restriction to binary mixtures.

This last restriction is easily overcome by the algebraic equivalents of isobolographics which are given by the so-called toxic unit summation. This method is widely accepted as a basic rule in national and international recommendations on risk assessments of mixtures of chemicals (see Table I). The toxic unit denotes the ratio of the individual concentration c_i of a chemical A_i in a mixture to its effect concentration of the same studied response level. Concentration addition is then said to occur when the sum of the toxic units S equals 1, and antagonism or synergism occurs when S is larger or smaller than 1, respectively. In order to transform S to a mathematically more consistent range, different indices have been developed and are commonly used in the literature (see Section VI).

B. Generalized Receptor Models

Generalized receptor models originate from studies of drug receptors and enzyme kinetics. For the assessment of combination effects they turn out to be procedures requiring some knowledge about kinetics of receptor/effector interactions. Only the model derived form the so-called median effect principle attained some practical importance. The basis of this model is derived from the mass action law in studies of enzyme inhibitors. Extending this approach to the analysis of combined effects, it is differentiated between "mutually exclusive" and "mutually nonexclusive" kinds of effects, depending on parallelism of the concentration response curves. The combined effects of two substances can then be assessed from the so-called combination index CI, with CI $=$ 1 for additivity and CI <1 or >1 for synergism or antagonism, respectively. For mutual exclusiveness again this model is in agreement with the concept of concentration addition.

The model may be adapted to combination effect studies with more than two substances. Unfortunately, this model does not provide any tests for statistical significance of the results.

C. Input/Output Models

Nonparametrical, nongraphical approaches have been developed and employed by authors who suspected that sufficient knowledge of the concentration response relationships and of the pharmacokinetical and pharmacodynamical behavior of agents is often lacking. Therefore, these models take the biological systems as black boxes. Without implicating any biological premises they just account for the concentrations administered as input and the observed effects as output. With input/output models the assessment of combination effects is based either on summation or on multiplication of effects.

Assessments according to the model of effect summation are based on the naive approach of comparing the observed fractional effects $E(c_{1,2})$ of some concentration $c_{1,2}$ of combined substances with the sum of the effects $E(c_1)$ and $E(c_2)$ provoked by each single agent. This model is not restricted to the investigation of combinations of binary mixtures but rather can easily be generalized to any number of substances. So, $E(c_{1,2,n}) = \Sigma_{i=0}^{n} E(c_i)$ indicates additivity, and $E(c_{1,2,n}) < \Sigma_{i=0}^{n} E(c_i)$ or $> \Sigma_{i=0}^{n} E(c_i)$ means antagonism or synergism, respectively.

The model of effect multiplication, which sometimes is referred to as the multiplicative survival model, is based on the assumption that in the case of additivity the combined effects of substances equal the product of the effects obtained by the substances applied singly. So $E(c_{1,2}) = E(c_1)E(c_2)$ for two substances and

$$E(c_{1,2,\ldots,n}) = \prod_{i=1}^{n} E(c_i)$$

for n substances, with abbreviations as above.

Synergism or antagonism will then be stated if the observed combined effect is greater or smaller, respectively, than the product of the single effects.

The model of effect multiplication has been invented in phytopharmacology, but the multiplicative rule was originally founded in quantal response models. However, in contrast to quantal response models no parametrization of concentration response curves is obtained by the model of effect multiplication. According to its probabilistic origin, wherein statistical independence is assumed for the effects of the combined substances, effect multiplication does not cover the assessment of all kinds of combinations of substances.

D. Quantal Response Models

Quantal response models were derived from probability theory. The biological justification originally brought forward assumed that a response of an organism to a concentration of a substance will occur if the concentration exceeds a certain threshold of tolerance. The responses considered are thought of in terms of "all or none" (e.g., dead or alive) and are called quantal responses. Consequently, in a randomized population these thresholds are regarded as statistically normally distributed, and the typically S-shaped log concentration response curves are taken as distribution functions of the thresholds. Therefore, quantal response models are restricted to monotonous concentration response curves. The aim of these models is to achieve a concise mathematical description of the concentration response curves of combined substances by statistical estimation of the determining parameters. Combination effects are assessed in terms of underlying mode of actions. In general, quantal response models for the assessment of the effects of combined agents have been intensively discussed in the biometrical literature for about 50 years, thereby giving a large contribution to the understanding of the mathematical implications of combined effects. These models are not limited to investigations of binary mixtures but can be extended to the study of multiple mixtures of chemicals and allow for extrapolation of the effects expected from any concentration of combined substances.

In mathematical notation the general concentration response relation in quantal response models is given by

$$P(c) = F(c \mid \theta_1, \ldots, \theta_s),$$

where P means the probability of the effect expected to occur from the concentration c of a substance A; $\theta_1, \ldots, \theta_s$ are unknown parameters given by the specific model. As a suitable function F, the normal distribution function is usually chosen. Within the family of quantal response models simple similar action and independent action are the most important approaches.

Simple similar action is said to occur in combinations of substances having the same site of action as well as a similar mode of action. This implies that every effect of one substance can be achieved by a proportional amount of the other, leading to the following formula of the combined effect $P_{1,2}$:

$$P_{1,2}(c_{1,2}) = F(\theta_1 + \theta_2 \log (c_1 + \theta_3 c_2)).$$

For the model of simple similar action the (probit-) transformed functions are linear and parallel. Thereby lines of the combined substances are located between the lines of each single substance. The aim of this model originally was (i) to allow statements concerning the site and mode of action of agents, (ii) to parametrize a concentration response curve and enable the extrapolation of responses, and (iii) to give a measure for the "relative potency" of combinations. Therefore, although simple similar action is conceptually closely related to concentration additivity, verdicts are not.

In the case of independent action, substances are assumed to have different sites and modes of actions. This leads to a representation of the concentration response relationship of the combined substances by the formula cited in Table III. Herein, in addition to the above-mentioned abbreviations, $f(c_1,c_2,\phi)$ is the bivariate density function of the normal distribution. Depending on the correlation coefficient ϕ, at least three special cases of independent action can be derived from the model, including the most interesting case of simple independent action ($\phi = 0$), by the formula

$$P_{1,2} = P_1 + P_2 - P_1 P_2.$$

Simple independent action is canonically best derived from statistical assumptions of independent behavior of agents and has been discussed long since as the counterpart model to isobolographics, founding the concept of effect multiplication. However, the aim of this quantal response model again is to parametrize the concentration response relationship and thereby to enlighten the underlying pharmacological behavior of mixtures rather than just statements on strengthened or weakened combination effects.

E. Interrelations of Models

Despite their diverse origin, the different mathematical models often lead to similar assessment terms, as can be seen from Table III. So isobolographics, toxic unit summation, the median effect model (for mutual exclusiveness), and even the model of simple

Table IV Biometrical Models on Combination Effects in Relation to Sites and Modes of Action of Substances

Kind of action of substances	Models to choose
Common site and mode of action	Isobolographics, toxic unit summation, simple similar action, mixture toxicity index, median effect model
Different site and mode of action	Joint independent action, effect multiplication, median effect model, mixture toxicity index

Table V Choice of Biometrical Models on Combination Effects in Relation to Concentration Response Curves

Model to choose	Shape of curves			
	Linear	Exponential	Sigmoid	Misc.
Isobolographics	X	X	X	X
Toxic unit summation	X	X	X	X
Simple similar action	—	—	X	—
Independent action	—	—	X	—
Median effect model	—	—	X	—
Effect summation	X	—	—	—
Effect multiplication	—	X	—	—

similar action (shown by simple algebraic transformation) are equivalent and implicitly based on the concept of concentration addition. Futhermore, the concept of effect multiplication is inherent in the homonymous input/output model as well as in the equivalent quantal response model of simple independent action. However, as pointed out earlier, the mathematical treatment of concentration response relationships as well as the original intention of the inventors are different.

At last, there seem to be different ways of grouping the approaches to the assessment of effects of combined substances. First, a differentiation could be made by the pharmacological understanding of the underlying concentration response relationship, distinguishing models for mixtures of substances with common sites and modes of actions from mixtures with dissimilar acting substances (see Table IV). Second, a differentiation could be based on the shape of the concentration response curves, provided that assessments are based on the concept of concentration addition. In this case particular models are valid for particular curves only (see Table V). Under the premise of concentration additivity, the models of effect summation, effect multiplication, and the median effect are special cases of isobolographics and its algebraic equivalents.

V. Choosing a Model

In general, the assessment of the effects of combined substances can be devoted to four different pharmacological situations. These are, with respect to a definite effect:

1. Only the combination is active, not any of the substances applied singly.
2. Applied singly, only one of the substances is active, but in combination the response is altered.
3. All substances are active and show similar concentration response curves (are within the same family of mathematical functions).
4. All substances are active and show dissimilar concentration response curves.

The choice of the right model for the assessment of combined effects depends on which situation is found. In cases 1 and 2 there is no need for a mathematical model at all. The assessment is achieved sufficiently by demonstrating that the corresponding substances are inert and for case 2 respectively that the alterations are statistically significant. In situations 3 and 4 the biometrical models can be chosen based on the differentiations outlined above. However, in most experimental studies there will be little or no knowledge about the specific sites and modes of actions of toxicants, so that it may not seem suitable to base the choice of a model on these unknown pharmacological features. Instead, we recommend the choice of a model with reference to the concept of concentration addition by inspection of the concentration response curves according to Table V. However, the experimenter should be aware that an application of the mentioned models has to follow an explicit decision on the shape of the concentration response curves. This can be ex-

Table VI Studies on Combination Effects of Aquatic Pollutants

Toxicants	Mixtures[a]	Species	Effect[b]	Models/criteria[c]	Assessment	Reference
Metals and other inorganics		**Fish**				
Cu, Zn, Pb, As	Multiple (4)	*Thymallus arcticus, Oncorhynchus kisutch, Oncorhynchus mykiss*	LC_{50} (96 hr)	Toxic unit summation, additivity index	Less than additive	Buhl, K. J., and Hamilton, S. J. (1990). *Ecotoxicol. Environ. Saf.* **20,** 325
Hg, Cd, Ag, As, Pb, Cu, Zn, Mn, Cr, B, Mo, Ni, V, selenite, selenate	Multiple (13–15)	*Oncorhynchus tshawytscha*	LC_{50} (96 hr)	Toxic unit summation, mixture toxicity index, additivity index	Additive to young; less than additive to older fish	Hamilton, S. J. and Buhl, K. J. (1990). *Ecotoxicol. Environ. Saf.* **20,** 307
Cd, Ag	Binary	*Pseudopleuronectes americanus*	Viability of embryos	Multiple regression analysis	Interaction	Voyer, R. A. and Heltshe, J. F. (1984). *Water Res.* **18,** 441
Cr, Ni	Binary	*Poecilia reticulata*	LC_{50} (96 hr); LT_{50}	Toxic unit summation, additivity index	Additive	Khangerot, B. S. and Ray, P. K. (1990). *Bull. Environ. Contam. Toxicol.* **44,** 832
		Algae and cyanobacteria				
Cr, SN, Ca, Mg, Ni, Co, Zn	Binary	*Anabaena doliolum*	Inhibition of growth (15 days); nutrient uptake (8 hr), nitrate reductase and glutamine synthetase activity	Not stated	Antagonistic effect of Ca, Mg, Mn; synergistic effect of Ni, Co, Zn on Cr- or Sn-toxicity	Rai, L. C. and Dubey, S. K. (1989). *Ecotoxicol. Environ. Saf.* **17,** 94
Cd, Zn	Binary	*Chlorella pyrenoidosa*	Changes in maximum specific growth rate	Not stated	"Amelioration"	Bennett W. N. and Brooks, A. S. (1989). *Environ. Toxicol. Chem.* **8,** 877
Ni, Cr, Pb	Multiple (3)	*Nostoc muscorum*	Inhibition of growth (15 days), nutrient uptake (76 hr), photosynthesis, nitrogenase activity	Not stated	Mostly antagonistic; synergistic for some exposure times, parameters, and combinations	Rai, L. C., and Raizada, M. (1989). *Ecotoxicol. Environ. Saf.* **17,** 75
		Other species and multispecies tests				
Hg, Cu	Binary	*Gammarus duebeni* (amphipod)	LC_{50} (95 hr); urine production rate; Hg accumulation	Not stated	Less than additive	Moulder, S. M. (1980). *Mar. Biol.* **59,** 193
Pb, Hg, Zn	Multiple (3)	*Uronema marinum* (ciliate protozoa)	Inhibition of cell division	Multiple regression analysis	No interaction	Voyer, R. A. and Heltshe, J. F. (1984). *Water Res.* **18,** 441
Chlorine, ammonia	Binary	Periphyton communities on artificial substrates	Inhibition of species richness of protozoans (7 days), algal biomass, community metabolism	Toxic unit summation, mixture toxicity index, multiple regression analysis	Less than additive or no significant interaction, depending on response parameter	Cairns, J. Jr. *et al.* (1990). *Aquat. Toxicol.* **16,** 87
Organics		**Fish**				
11 Chlorophenols, 10 chlorobenzenes, 11 chlorotoluenes, 19 aliphatic chlorohydrocarbons, 4 alcohols, 9 glycol derivatives, 5 other related compounds	Multiple (10, 11, or 50)	*Poecilia reticulata*	LC_{50} (7 or 14 days)	Toxic unit summation	Additive for chlorophenols and the 50 component mixture; less than additive for chlorobenzenes	Könemann, H. (1980). *Ecotoxicol. Environ. Saf.* **4,** 415

(*continues*)

Table VI (*Continued*)

Toxicants	Mixtures[a]	Species	Effect[b]	Models/criteria[c]	Assessment	Reference
17 Chloro- and alkylanilines	Multiple (6–17)	*Poecilia reticulata*	LC_{50} (14 days)	Mixture toxicity index	Additive	Hermens, J. *et al.* (1984). *Ecotoxicol. Environ. Saf.* **8**, 388
11 Nonreactive, nonionized organics; 11 chloroanilines; 11 chlorophenols; 9 reactive organic halides	Binary, multiple (9–33)	*Poecilia reticulata*	LC_{50} (14 days)	Mixture toxicity index	Partially additive to additive	Hermens, J. *et al.* (1985). *Ecotoxicol. Environ. Saf.* **9**, 321
27 Alcohols, ketones, ethers, alkyl halides, benzenes, nitriles, tertiary aromatic amines	Binary, multiple (3–14)	*Pinephales promelas*	LC_{50} (96 hr)	Isobolographics, toxic unit summation, additivity index, mixture toxicity index	Near additive	Broderius, S. and Kahl, M. (1985). *Aquat. Toxicol.* **6**, 307
		Daphnids				
25 Alcohols and chlorohydrocarbons	Multiple (10–25)	*Daphnia magna*	Inhibition of growth and reproduction	Mixture toxicity index	Additive	DeWolf, W. *et al.* (1988). *Aquat. Toxicol.* **12**, 39
50 Nonreactive organic compounds	Multiple (5–50)	*Daphnia magna*	IC_{50} immobilization (48 hr), LC_{50} (16 days), EC_{50} reproduction (16 days)	Mixture toxicity index	Additive for 48-hr IC_{50} and 16-day LC_{50}; partially additive for 16-day EC_{50} reproduction	Hermens, J. *et al.* (1984). *Aquat. Toxicol.* **5**, 143
		Algae and cyanobacteria				
Toluene, *m*-xylene, benzene	Binary, multiple (3)	*Selenastrum capricornutum*	EC_{50} growth inhibition (8 days)	Effect multiplication	Close to additive, additive, or synergistic, depending on concentration level	Herman, D. C. *et al.* (1990). *Aquat. Toxicol.* **18**, 87
		Other Species				
21 Nonreactive organic chemicals	Multiple (21)	*Photobacterium phosphoreum* (microtox text)	Inhibition of bioluminescence (15 min)	Mixture toxicity index	Partially additive	Hermens, J. *et al.* (1985). *Ecotoxicol. Environ. Saf.* **9**, 17
18 Alkanes, alkenes, alkyl-substituted benzenes, naphthalenes, phenols, pyridines	Binary, multiple (3–7)	Mixed marine bacterial culture	EC_{50} growth inhibition (16 hr)	Mixture toxicity index, additivity index	Approximately additive for homologous compounds; synergistic for nonhomologous compounds	Warne M. St. J. *et al.* (1989). *Ecotoxicol. Environ. Saf.* **18**, 121
Pesticides		**Fish**				
Dinoseb, monosodium metharsonate (MSMA)	Binary	*Ictalurus punctatus, Pinephales promelas*	LC_{50} (96 hr)	Additivity index	Greater than additive	Skelley, J. R. (1989). *Environ. Toxicol. Chem.* **8**, 623
Carbaryl, phenthoate	Binary	*Channa punctatus*	Inhibition of acetylcholinesterase; accumulation of acetylcholine	Not stated	Synergistic	Rao, K. R. S. S. and Rao, K. V. R. (1989). *Ecotoxicol. Environ. Saf.* **17**, 12
		Algae and cyanobacteria				
Permethrin, 3 permethrin degradation products	Binary, multiple (3–4)	*Anabaena inaequalis*	EC_{50} photosynthesis inhibition (3 hr); nitrogenase activity (5 hr); and growth inhibition (12–14 days)	Effect multiplication	Synergistic, antagonistic, or additive interaction, depending on parameter and mixture components	Stratton, G. W. and Corke, C. T. (1982). *Environ. Pollut.* **A29**, 71

Permethrin, atrazine	Binary	*Anabaena inequalis*	Inhibition of growth, photosynthesis, and nitrogenase activity	Effect multiplication	Additivity, antagonism, or synergism, depending on parameter, toxicant ratio, and concentration	Stratton, G. W. (1983). *Bull. Environ. Contam. Toxicol.* **31,** 297
Atrazine, metribuzin, amitrole, glufosinate	Binary	*Chlorella fusca*	Inhibition of growth (14 hr), reproduction (24 hr), and photosynthesis (15 min)	Isobolographics	Additivity for atrazine and metribuzin, subadditivity for amitrole and glufosinate	Altenburger, R. *et al.* (1990). *Ecotoxicol. Environ. Saf.* **20,** 98
Atrazine, copper	Binary	*Selenastrum capricornutum*	IC_{50} growth (3–7 days)	Multiple regression analysis	No interaction	Roberts, S. *et al.* (1990). *Water Res.* **24,** 485
		Other species				
Copper, 7 dithiocarbamates	Binary	*Colpidium campylum* (ciliate protozoan)	Growth inhibition	Multiple regression analysis	Additive or synergistic, depending on concentration and type of dithiocarbamate	Bonnemain, H. and Dive, D. (1990). *Ecotoxicol. Environ. Saf.* **19,** 320
Atrazine, copper	Binary	*Photobacterium phosphoreum* (Microtox test), *Colpidium campylum* (protozoan)	IC_{50} bioluminescence (30 min) IC_{50} growth (24 hr)	Multiple regression analysis	Synergistic effect on bacteria; no interaction in protozoans	Roberts, S. *et al.* (1990). *Water Res.* **24,** 485
Miscellaneous		**Fish**				
24 Organic chemicals with anticipated different modes of action (pesticides and others)	Multiple (8–24)	*Poecilia reticulata*	LC_{50} (14 days)	Mixture toxicity index	Partially additive or additive	Hermens, J. and Leeuwang, P. (1982). *Ecotoxicol. Environ. Saf.* **6,** 302
61 Different industrial organic chemicals, 2 pesticides, 2 inorganics	Multiple (3–50)	*Poecilia reticulata*	LC_{50} (14 days)	Mixture toxicity index	Additive for simple similarly acting chemicals; partially additive for dissimilar acting agents	Könemann, H. (1981). *Toxicology* **19,** 229
		Daphnids				
9 Organics with presumably different modes of action (pesticides and others)	Multiple (9)	*Daphnia magna*	NOEC growth (16 days)	Toxic unit summation	Less than additive	Deneer, J. W. *et al.* (1988). *Ecotoxicol. Environ. Saf.* **15,** 72

[a] Tested mixtures are classified by number of components. For multiple mixtures number of agents is given in parentheses. For the number of different mixtures tested, mixture ratios and concentration levels see the original literature.

[b] EC_{50}, median effect concentration; IC_{50}, median inhibitory concentration; LC_{50}, median lethal concentration; LT_{50}, median period of survival; NOEC, no observed effect concentration.

[c] See Table III.

tremely difficult, especially with poor data. Making use of isobolographics or the algebraic equivalents instead of the apparently convenient input/output models might bypass this problem. For situation 4 a recommendation is difficult. There is still dispute as to whether the concept of concentration addition has to be applied. So in the case of binary mixtures, isobolographics again might be the adequate choice. In general, for all situations cited above the assessments of combination effects should be backed by statistical methods.

VI. Survey of Reported Effects

In 1980 the European Inland Fisheries Advisory Commissions (EIFAC) Working Party On Water Quality Criteria for European Freshwater Fish reviewed the literature "on combined effects on freshwater fish and other aquatic life of mixtures of toxicants in water." This examination of 188 publications was updated on the basis of 19 further reports in 1986 (see the Bibliography). The special topic of the joint action of pesticide mixtures on fish was surveyed by A. S. Murty, covering literature up to 1982 (see the Bibliography). Table VI may serve as a guide to literature on combination effects of aquatic pollutants published during the last decade. It is intended to provide a cross section on the types of studies performed; toxicants, species, and test systems investigated, and concepts and models used. Further literature is accessible from the references given.

The majority of studies dealt with mixtures of heavy metals. Systematic work has been done with regard to the combined effects of industrial organic chemicals with an anticipated unspecific mode of action. Analysis of combination effects of pesticides and other specific acting agents are rare. Most information is available on the toxicity of mixtures to fish and daphnids. Relatively few studies investigated combination effects on other aquatic species or multispecies microcosm test systems.

Most of the studies in Table VI refer to definite concepts, methods, or models when assessing the toxicity of mixtures. The mixture toxicity index (MTI) has gained particularly rapid acceptance and wide application, since it provides a simple means for the analysis of combination effects with multiple chemical mixtures. Some of the studies apply different indices, methods, or models to the same data set. However, systematic analysis of the interrelationships between the results obtained with different types of models as well as of the interdependencies of experimental design and the validity of assessments is still lacking.

Reported combination effects are often rather contradictory, especially with heavy metals. For some toxicants any type of interaction can be detected, depending on the specific conditions, methods, and criteria used. Most congruent results have been obtained for the acute mixture toxicity of nonionized, nonreactive organics with a similar narcotic type of action. Concentration addition has been shown to be a reasonable predictor for the effects of this type of mixture. Attempts to clarify the relationships between types and degrees of combination effects and several determining factors (chemical nature, mode of action, concentration and mixture relations of toxicants, exposure time, environmental conditions, and type and level of complexity of biosystems and toxicity parameters) have in particular been brought forward.

Despite these efforts and advances made during the last decade, there is still an ongoing controversary on two crucial points:

- How valid are water quality criteria derived from single toxicant bioassays for conditions of multiple exposure?
- Does every chemical in any concentration contribute to the overall toxicity of a mixture or do threshold values exist?

Thus, a lot of further research is needed. However, the basis for systematic approaches has been laid and promising first steps have been taken.

Related Articles: ECOTOXICOLOGICAL TESTING; ORGANIC MICROPOLLUTANTS IN LAKE SEDIMENTS; PROTISTS AS INDICATORS OF WATER QUALITY IN MARINE ENVIRONMENTS; SELENIUM.

Acknowledgment

Financial support by the Federal Environmental Agency, Berlin, is gratefully acknowledged.

Bibliography

Berenbaum, M. C. (1985). The expected effect of a combination of agents: The general solution. *J. Theor. Biol.* **114,** 413–431.
Bödeker, W., Altenburger, R., Faust, M., and Grimme, L. H. (1990). Methods for the assessment of mixtures of pesticides:

Mathematical analysis of combination effects in phytopharmacology and ecotoxicology. *Nachrichtenbl. Deut. Pflanzenschutzd. (Braunschweig)* **42,** 70–78.

European Inland Fisheries Advisory Commission (EIFAC) (1987). "Water Quality Criteria for European Freshwater Fish. Revised Report on Combined Effects on Freshwater Fish and Other Aquatic Life of Mixtures of Toxicants in Water." EIFAC Technical Paper No. 37, Rev. 1.

Murty, A. S. (1986). "Toxicity of Pesticides to Fish," Vol. 2. CRC Press, Boca Raton, Florida.

National Research Council (NRC) (1980). "Principles of Toxicological Interactions Associated with Multiple Chemical Exposures." National Academy Press, Washington, D.C.

Vouk, V. B., Butler, G. C., Upton, A. C., Parke, D. V., and Asher S. C., eds. (1987). "Methods for Assessing the Effects of Mixtures of Chemicals." Wiley, Chichester.

Asbestos-Contaminated Drinking Water

James S. Webber
New York State Department of Health

Glossary

Asbestosis Nonmalignant, dose-related, lung disease (pneumoconiosis) marked by interstitial fibrosis and reduced pulmonary efficiency.

Aspect ratio Fiber length divided by fiber width.

Mesothelioma Rare malignant tumor of the body linings (pleura, peritoneum, or pericardium) which grows diffusely as a thick sheet covering the viscera.

Micrometer (μm) One thousandth of a millimeter.

pH Negative logarithm (base 10) of the H^+ concentration in water. Water with pH <7 is considered acidic and water with pH >7 is considered alkaline.

Standardized mortality ratio (SMR) Ratio of mortality in a study group divided by the expected mortality from a control group. The ratio is usually expressed as percentile, with values less than 100 indicating lowered mortality among the study group and values greater than 100 indicating higher mortality among the study group.

Transmigration Ability of particles to pass from the digestive tract through the intestinal wall and into the circulatory system.

ASBESTOS *is a collective term that describes naturally occurring hydrated silicate minerals that break into fibers. Asbestos is naturally present in some water bodies but its accelerated use in the twentieth century has increased its concentration in many other water systems. Although ingested asbestos does not appear to be as hazardous as airborne asbestos, some ingested asbestos undoubtedly reaches the bloodstream and a few studies have revealed a possible link between this migration and tumors at areas remote from the pulmonary or digestive systems. Some epidemiologic studies of populations drinking asbestos-contaminated water have detected increased gastrointestinal cancer. A prudent course of action is to minimize asbestos-contamination in drinking water through proven, cost-effective methods.*

I. Mineralogy and History

A. Semantics, Mineralogy, and Properties

The ancient Greeks called these flexible, fire-resistant minerals ασβεστος, meaning inextinguishable or unquenchable. Asbestos characteristically breaks into very thin fibers, often as narrow as 0.025 μm in diameter, which have high resistance to heat and have enormous tensile strength. Definitions for asbestos have changed over the centuries as new varieties have been identified and as classification systems have evolved. Controversy over precise definitions continue today because of differences in regulatory and commercial perspectives.

Six mineral types are currently recognized as asbestos (Table I). The sole serpentine variety, chrysotile, is the most common type of asbestos, accounting for more than 90% of worldwide production. Chrysotile is the most flexible of the asbestos types and is the most vulnerable to leaching by acids. Its structure is unique among asbestos varieties—a magnesium hydroxide layer overlies a

Table I Characteristics of Asbestos Minerals

Mineral type	Variety	Chemical formula
Serpentine	Chrysotile	$Mg_3(Si_2O_5)(OH)_4$
Amphibole	Amosite (grunerite)	$(Fe,Mg)_7(Si_8O_{22})(OH)_2$
	Crocidolite (riebeckite)	$Na_2Fe_5(Si_8O_{22})(OH)_2$
	Anthophyllite	$(Mg,Fe)_7(Si_8O_{22})(OH)_2$
	Tremolite	$Ca_2Mg_5(Si_8O_{22})(OH)_2$
	Actinolite	$Ca_2(Mg,Fe)_5(Si_8O_{22})(OH)_2$

silicate layer, forming an elongated scroll. The five amphibole asbestos types have more typical structure, with double-chain silicates sandwiching a cation (Fe, Mg, Na, or Ca) layer. Crocidolite is an iron- and sodium-rich riebeckite asbestos that has higher tensile strength than any other asbestos type. Amosite, an iron- and magnesium-rich grunerite fiber, is highly resistant to leaching. Anthophyllite is a magnesium-rich amphibole that has the highest fusion point. Tremolite and actinolite are in a calcium–iron replacement series and are of little commercial significance.

B. History of Asbestos and Humans

Although asbestos was woven into fireproof cloth by ancient Greeks and used as lamp wicks in medieval times, its utilization was fairly limited until the beginning of the twentieth century. At about that time, people began to recognize the diverse technological applications of the unique properties of asbestos, namely its thermal and electrical resistance, tremendous tensile strength, chemical resistance, flexibility, and low cost. In response to escalating demands, asbestos production increased more than 100-fold during the first eight decades of the century. Thousands of uses have been described for asbestos, ranging from sprayed-on fireproofing and insulation to brake linings, floor tiles, shingles, asbestos–cement pipes, and missile covers.

An increase in disease was inevitable as the exploitation of asbestos accelerated during the twentieth century. In some instances, however, these elevations were not noted immediately because of the latency period, usually measured in decades, between the time of exposure and the manifestation of symptoms. Asbestosis was first documented around 1930 in asbestos workers and shortly thereafter, increased incidences of lung cancer were also noted. Following World War II, an alarming increase in mesothelioma, a previously rare disease, was noted not only in asbestos workers but also in persons marginally exposed to asbestos. Because these diseases were associated with inhalation of airborne asbestos fibers, increasingly stringent regulations have been enacted during the past few decades to reduce exposure of workers and the public to airborne fibers.

Does asbestos-contaminated drinking water pose a health hazard similar to airborne asbestos? Most studies have indicated that, while there may be some health risks associated with ingestion of asbestos, the health hazards are substantially less than those posed by inhalation. This is discussed in more detail in Section III.

II. Occurrence

A. Large-Scale Surveys

The majority of water-supply systems in North America contain less than one million asbestos fibers per liter (MFL). A survey of several hundred systems in the United States revealed that more than 80% had less than one MFL while only 10% had more than ten MFL (Table II). Similarly, in Canada 5% of the population receive more than one MFL while less than 1% receive more than 100 MFL. Comparable surveys in Europe have revealed similar patterns.

Table II Distribution of Reported Asbestos Concentrations in Drinking Water from 406 Cities in 47 States, Puerto Rico, and the District of Columbia[a]

Highest asbestos concentration (10^6 fibers/L)	Number of cities	Percentage
Below detectable limits	117	28.8
<1	216	53.2
1–10	33	8.1
>10	40	9.9
Total	406	100

[a] From Millette, J. R., Clark, P. J., Stober, J., and Rosenthal, M. (1983). Asbestos in water supplies of the United States. *Environmental Health Perspectives,* **53,** 45. Reproduced by permission.

B. Episodes of Gross Contamination

In spite of the generally low asbestos concentrations in public water-supply systems, there have been several instances where concentrations have been drastically elevated. While some of the highest concentrations were from non-point sources, point-source contamination was also significant on a localized level.

1. Natural Erosion

Some of the highest waterborne concentrations of asbestos have been measured in areas of serpentinized bedrock where erosional processes often entrain chrysotile into drinking water. In eastern North America, such outcroppings have produced concentrations of more than 1000 MFL in Quebec and Newfoundland. Likewise, erosion of serpentinized deposits in California have generated concentrations in excess of 10,000 MFL. Further north in Washington State, chrysotile concentrations have exceeded 100 MFL due to erosion. Chrysotile fibers produced by natural erosion are generally small, usually less than 1 μm in length. Elevated amphibole asbestos concentrations are rarely encountered in natural waters of North America.

2. Deteriorated AC Pipe

Asbestos fibers are susceptible to sloughing off the inner surfaces of deteriorated AC pipes during transmission of water. Corrosive water often accelerates this process by dissolving the cement matrix of the AC pipe. An Aggressiveness Index (A.I.) has been proposed by the American Water Works Association to predict the potential for water to corrode AC pipe. This is defined as

$$\text{A.I.} = \text{pH} + \log_{10}[\text{alkalinity (mg/L as } CaCO_3) + \text{hardness (mg/L as } CaCO_3)].$$

Indices less than 10 are considered highly aggressive. In a South Carolina drinking water system, corrosive water's disintegration of AC pipe was blamed for chrysotile concentrations in excess of 500 MFL. In Woodstock, New York, concentrations of chrysotile and crocidolite in excess of 1000 MFL were attributed to long-term degradation of AC pipe by acidic water. Factors such as seasonal use, drastic changes in water pressure, and improper tapping and cutting operations can also contribute to elevated concentrations.

3. Anthropogenic Disturbances

One of the most widely reported contamination episodes was discovered in Duluth, Minnesota during the 1970s. An iron-mining operation had been disposing of taconite tailing wastes into western Lake Superior for two decades; amphibole fibers from these tailings eventually turned up in the unfiltered drinking water of Duluth. During times of turbulence, concentrations would reach 600 MFL. This dumping was eventually stopped and Duluth instituted filtration to remove residual fibers from the drinking water. In another instance, concentrations of up to 74 MFL were blamed on an old asbestos waste pile in Kentucky.

Table III Some Size Characteristics of Asbestos Fibers Found in Various Water Supplies[a]

Source	Type of fiber	Number of fibers measured	Average length (μm)	Average width (μm)	Average aspect ratio	Maximum length found (μm)
Reservoir with natural erosion (WA)	Chrysotile	289	0.8	0.034	25 : 1	3
Reservoir with natural erosion (CA)	Chrysotile	644	1.3	0.04	39 : 1	10
Cistern with asbestos tile roof (VI)	Chrysotile	342	2.3	0.04	62 : 1	25
Distribution sites from five asbestos cement pipe systems (SC, PA, FL)	Chrysotile	1440	4.3	0.044	121 : 1	80
Lake Superior (MN)	Amphibole	468	1.5	0.18	11 : 1	14

[a] From Millette, J. R., Clark, P. J., Pansing, M. F., and Twyman, J. D. (1980). Concentration and size of asbestos in water supplies. *Environmental Health Perspectives* **34,** 22, Reproduced by permission.

C. Characteristics of Waterborne Asbestos

As previously discussed, chrysotile is the dominant asbestos type identified in water samples. This reflects its widespread use in industry and its abundance in nature. Chrysotile, with its unique positive surface charge, is vulnerable to attack by acids and may degrade in low-pH water systems. Amphiboles, on the other hand, are more impervious to dissolution but are rarely encountered in drinking water samples. Amphiboles are often minor cocontaminants with chrysotile when deteriorated AC pipes are the asbestos source. Asbestos-like mineral fibers such as halloysite, rutile, and palygorskite have been occasionally detected in water systems but little is known about potential biological effects of ingestion of these fiber types.

Asbestos fibers resulting from AC pipe or anthropogenic activity usually have larger dimensions than fibers from natural erosion sources (Table III). Individual fiber length, width, and especially mass (length $\times$ width^2 $\times$ density) are often distinctly larger from human-derived sources.

III. Health Hazards

Although evidence that inhalation of asbestos can cause serious health problems is overwhelming, health risks posed by ingestion of asbestos are apparently much lower.

A. Ingestion

1. Animal Toxicity Studies

Most investigations of asbestos ingestion by laboratory animals have failed to produce evidence of toxicity or consistent, reproducible, organ-specific carcinogenicity. The largest series of studies was performed by the National Toxicology Program (NTP), under the auspices of the National Institute for Environmental Health Sciences, in which large numbers of diverse rodent populations were fed a variety of asbestos types. With the exception of one study that revealed an increase in large-intestine tumors, all other NTP studies revealed no significant health differences between experimental and control populations. While most other studies have similarly failed to demonstrate pathogenicity of ingested asbestos (Table IV), there have been indications of induced abnormalities such as impeded permeability of the intestines. One drawback in relating these feed studies to ingestion of waterborne asbestos is that most animal studies were performed with much longer fibers than are normally found in asbestos-contaminated drinking water. In addition, asbestos fibers embedded in food pellets used in the animal studies may behave differently from the loose fibers characteristic in drinking water.

2. Transmigration Studies

Asbestos fibers typically found in contaminated drinking water are small enough to pass through the wall of the small intestine and into the blood and lymph fluid. Most studies that have investigated this possibility have found that fibers do indeed reach the circulatory system, probably at a ratio of 10^{-7} to 10^{-3} of the fiber concentrations ingested. Elevated concentrations of amphibole fibers were measured in the urine of Duluth residents who had been drinking contaminated Lake Superior water. Several studies on asbestos workers have also detected increased asbestos levels in urine. Elevated asbestos concentrations have been detected in various tissues and organs of laboratory animals that have ingested asbestos. Many of these human and animal studies have found a preferential transmigration of small fibers (usually shorter than 2 μm), the size range most common to drinking water. Some case studies of cancers of digestive and urinary organs in asbestos workers have revealed asbestos fibers in these organs. However, a larger body of evidence with controls will have to be accumulated before definitive linkages can be established.

3. Epidemiologic Studies

a. Ecological Studies

More than a dozen epidemiologic studies investigated possible correlations between asbestos-contaminated drinking water and cancer mortality. Certain investigations have revealed statistically significant correlations with various organs and tissues, but most have not revealed a definitive association (Tables V and VI). Some of the most significant studies have focused on the San Francisco Bay area, where long-term contamination of water by chrysotile has resulted from natural erosion of serpentinized bedrock. A large population and population-based cancer registries in the Bay Area have increased statistical sensitivity and enhanced meaningful analyses. Several investigations there have revealed positive correlations for combined

Table IV Summary of Asbestos Ingestion Studies[a]

Study (author [date])	Species	Test material	Dose	Exposure time	Study duration	Number of animals (initial/ examined)	Malignant tumors		
							Number	Location	Type[b]
Bonser [1967]	Rat	Crocidolite	0.15% in diet *ad libitum*	To 78 wk	To 78 wk	40/12	0		
		Control	0	0	To 86 wk	65/25	1	Liver	S
Gross [1974]	Rat	Chrysotile	5% in diet *ad libitum*	21 mo	21 mo	10/10	0		
		Control	0	0	21 mo	5/5	0		
		Chrysotile	10 mg/wk	16 wk	To 1.5 yr	31/31	2	Breast	C
		Crocidolite	5 mg/wk	16 wk	To 1.5 yr	33/33	0		
		Crocidolite	10 mg/wk	16 wk	To 1.5 yr	34/34	1	Node	L
		Control	0	0	To 1.5 yr	24/24	5	3 Breast Thigh Node	C S L
		Crocidolite (2 sources)	10 mg/wk	18 wk	To 1.5 yr	63/63	0		
		Control	0	0	To 1.5 yr	24/24	0		
Gibel [1976]	Rat	Chrysotile	20 mg/day	Life	441[c]	50/42	12	Lung 4 Kidney 3 Node 4 Liver	C C L C
		Talc	20 mg/day	Life	649[c]	50/45	3	Liver	C
		Control	0	0	702[c]	50/49	2	Liver	C
Cunningham [1977]	Rat	Chrysotile	1% in diet *ad libitum*	To 24 mo	To 24 mo	10/7	6	Brain Pituitary Node 2 Kidney Peritoneum	S C L C S
		Control	0	0	6 24 mo	[illegible]0/8	1	Peritoneum	S
		Chrysotile	1% in diet *ad libitum*	To 24 mo	To 30 mo	[illegible]0/36	11	2 Thyroid Thyroid Liver Chemodectoma jugular body	C S C

(*continues*)

Table IV *(Continued)*

Study (author [date])	Species	Test material	Dose	Exposure time	Study duration	Number of animals (initial/examined)	Malignant tumors: Number	Location	Type[b]
		Control	0	0	To 30 mo	40/32	11	Colon Ileum Adrenal 2 Node Bone Thyroid Liver 2 Adrenal Kidney Node 5 Fat	C S C L S C C C C L S
Wagner [1977]	Rat	Chrysotile	100 mg/day	101 days/5 mo	619[c]	32/32	3	Node Stomach Uterus	L S S
		Talc	100 mg/day	101 days/5 mo	614[c]	32/32	3	2 Uterus Stomach	S S
		Control	0	0	641[c]	16/16	0		
Smith [1980]	Hamster	Amosite	0.5 mg/L *ad libitum*	To 23 mo	To 23 mo	60/60	1	Lung	C
		Amosite	5 mg/L *ad libitum*	To 23 mo	To 23 mo	60/60	3	2 Stomach Peritoneal mesothelioma	C
		Amosite	50 mg/L *ad libitum*	To 23 mo	To 23 mo	60/60	0		
		Taconite tailings	0.5 mg/L *ad libitum*	To 23 mo	To 23 mo	60/60	1	Uterus	S
		Taconite tailings	5 mg/L *ad libitum*	To 23 mo	To 23 mo	60/60	0		
		Taconite tailings	50 mg/L *ad libitum*	To 23 mo	To 23 mo	60/60	0		
		Control	0	0	To 23 mo	120/120	1	Node	L
Bonham [1980]	Rat	Chrysotile	10% in diet *ad libitum*	To 32 mo	To 32 mo	240/189	4	3 Colon Abdominal mesothelioma	C
		Cellulose fiber	10% in diet *ad libitum*	To 32 mo	To 32 mo	242/197	2	Colon	C
		Control	0	0	To 32 mo	121/115	3	Colon	C

Ward [1980]	Rat	Azoxymethane[d]	7.4 mg/kg wk	10 wk	34 wk	21/21	12	5 Ileum 7 Colon	C C
		Azoxymethane plus amosite	7.4 mg/kg wk 10 mg 3/wk	10 wk	34 wk	21/18	10	3 Ileum 7 Colon	C C
		Azoxymethane plus chrysotile	7.4 mg/kg wk 10 mg 3/wk	10 wk	34 wk	21/21	10	4 Ileum 6 Colon	C C
Ward [1980]	Rat	Amosite	10 mg 3/wk	10 wk	34 wk	21/21	0		
		Chrysotile	10 mg 3/wk	10 wk	34 wk	21/21	0		
		Saline	1.0 mL 3/wk (gavage)	10 wk	34 wk	21/21	0		
		Untreated	—	0	34 wk	21/21	0		
		Azoxymethane	7.4 mg/kg wk	10 wk	To 95 wk	50/48	39	12 Ileum 27 Colon	C C
		Azoxymethane plus amosite	7.4 mg/kg wk 10 mg 3/wk	10 wk	To 95 wk	50/48	44	15 Ileum 29 Colon	C C
		Saline plus amosite	1/wk (SC) 10 mg 3/wk	10 wk	To 95 wk	50/49	17	Ileum 16 Colon	C C
Hiding [1981]	Rat	Filtered Duluth tapwater	1 MFL *ad libitum*	690[c]	690[c]	28/27	3	Lung Ovary Forestomach	C C C
		Unfiltered Duluth tapwater	100 MFL *ad libitum*	960[c]	960[c]	30/28	4	Salivary gland Skin Uterus Mediastinum	C C S L
		Lake Superior water sediment	5,000 MFL *ad libitum*	840[c]	840[c]	22/22	3	Lung Skin Uterus	C C S
		Taconite tailings	100,000 MFL *ad libitum*	870[c]	870[c]	30/30	3	Neck Chest wall Mediastinum	S S L

(*continues*)

Table IV *(Continued)*

Study (author [date])	Species	Test material	Dose	Exposure time	Study duration	Number of animals (initial/examined)	Malignant tumors		
							Number	Location	Type[b]
		Chrysotile/amosite	20 mg/day	870[c]	870[c]	30/30	6	Breast 2 Fibrous histiocytoma Skin Mediastinum Pleural mesothelioma	C C L
		Amosite	300 mg/day	750[c]	750[c]	20/20	1	Leukemia	
		Diatomaceous earth	20 mg/day	840[c]	840[c]	30/30	5	Salivary gland 2 Uterus Skin Peritoneal mesothelioma	C S C
Bolton [1982]	Rat	Amosite	250 mg/wk	25 mo.	Life	24/24	1	Stomach	S
		Crocidolite	250 mg/wk	25 mo.	Life	22/22	1	Adrenal	C
		Chrysotile	250 mg/wk	25 mo.	Life	22/22	5	Fat Pleural histiocytoma 2 Adrenal Plasma cell tumor	S S C
		Margarine control	0	0	Life	24/24	4	2 Adrenal Bladder Peritoneum	C C S
		Control	0	0	Life	23/23	2	Fat Lymphoma	 S

[a] From Condie, L. W. (1983). Review of published studies of orally administered asbestos. *Environmental Health Perspectives* **53,** 4–7. Reproduced by permission.
[b] Type C, carcinoma; S, sarcoma, L, lymphoma.
[c] Mean survival time in days.
[d] Azoxymethane given subcutaneously; saline administered by oral gavage or subcutaneously.

Table V Summary of Studies of Gastrointestinal Cancer Risk in Relation to Ingested Asbestos by Cancer Site[a,b]

Gastrointestinal cancer site (ICD 7th revision codes)	Study (site/author)												
	Duluth			Connecticut		Quebec		Bay Area, CA			Utah	Puget Sound, WA	
	Mason	Levy	Sigurdson	Harrington	Meigs	Wigle	Toft	Kanarek	Conforti	Tarter	Sadler	Severson	Polissar
All sites combined (150–159)	(++)	(−−)	(00)	ns	ns	(00)	(+0)	(++)	(++)	(++)	ns	(00)	ns
Esophagus (150)	(+−)	(00)	(00)	ns	ns	(00)	(00)	(0+)	(++)	ns	ns	ns	(00)
Stomach (151)	(++)	(+0)	(00)	(00)	(00)	(+0)	(+0)	(++)	(++)	ns	(00)	(00)	(00)
Small intestine (152)	ns	(00)	(00)	ns	ns	ns	ns	(00)	(00)	ns	(00)	ns	(++)
Colon (153)	(00)	(−−)	(00)	(00)	(00)	(00)	(00)	(00)	(+0)	ns	(0−)	(−−)	(00)
Rectum (154)	(++)	(00)	(00)	(00)	(00)	(00)	(00)	(00)	(00)	ns	(00)	ns	(00)
Biliary passage/liver (155–156A)	(00)	(00)	(00)	ns	ns	ns	ns	(00)	(00)	ns	ns	ns	(00)
Gallbladder (155.1)	ns	(00)	(00)	ns	ns	ns	ns	(0+)	(00)	ns	(0+)	ns	(00)
Pancreas (157)	(0+)	(++)	(0+)	ns	(+0)	(0+)	(00)	(0+)	(++)	ns	(00)	ns	(00)
Peritoneum (158)	ns	(00)	(00)	ns	ns	ns	ns	(++)	(0+)	ns	(00)	ns	(00)

[a] From Marsh, G. M. (1983). Critical review of epidemiologic studies related to ingested asbestos. *Environmental Health Perspectives* **53,** 50. Reproduced by permission.

[b] (Male, female): association with ingested asbestos; +, positive; 0, none; −, negative; ns, not studied.

Table VI Summary of Studies of Nongastrointestinal Cancer Risk in Relation to Ingested Asbestos by Cancer Site[a,b]

Nongastrointestinal cancer site (ICD 7th revision codes)	Study (site/author)												
	Duluth			Connecticut		Quebec		Bay Area, CA			Utah	Puget Sound, WA	
	Mason	Levy	Sigurdson	Harrington	Meigs	Wigle	Toft	Kanarek	Conforti	Tarter	Sadler	Severson	Polissar
Buccal cavity and pharynx (140–148)	ns	ns	ns	ns	ns	(00)	(00)	ns	ns	ns	ns	ns	(00)
Bronchus, trachea, lungs (162, 163)	(+0)	ns	(00)	ns	(00)	(+0)	(+0)	(+0)	(00)	ns	ns	ns	(00)
Pleura (162.2)	ns	ns	ns	ns	ns	ns	ns	(0+)	(0+)	ns	ns	ns	ns
Prostate (177) (males only)	ns	ns	ns	ns	ns	0	0	0	+	ns	ns	ns	+
Kidneys (180)	ns	ns	ns	ns	(00)	(00)	(00)	(0+)	(00)	ns	(+0)	(00)	(00)
Bladder (181)	ns	ns	ns	ns	(00)	(00)	(00)	(00)	(00)	ns	ns	ns	(00)
Brain/CNS (193)	(00)	ns	ns	ns	ns	(00)	(00)	(00)	(00)	ns	ns	ns	(+−)
Thyroid (194)	ns	ns	ns	ns	ns	ns	ns	(00)	(00)	ns	ns	ns	(++)
Leukemia, aleukemia (204)	(00)	ns	ns	ns	ns	(00)	(00)	(00)	(00)	ns	(+0)	ns	(+−)

[a] From Marsh, G. M. (1983). Critical review of epidemiologic studies related to ingested asbestos. *Environmental Health Perspectives* **53,** 50.
[b] (Male, female): association with ingested asbestos; +, positive; 0, none; −, negative; ns, not studied.

gastrointestinal sites, esophagus, stomach, peritoneum, and pancreas. Epidemiologic studies of the Duluth population, exposed to fibrous amphiboles discharged as taconite tailings, have yet to reveal significant elevations in cancer rates.

b. Occupational Studies

Epidemiologists have also evaluated gastrointestinal cancer incidence among asbestos workers predicated on the assumption that a significant proportion of inhaled fibers is trapped in the upper respiratory system, cleared to the throat by the mucociliary escalator, and swallowed. Many of these studies have revealed significantly elevated gastrointestinal cancers in a variety of asbestos-related occupations. A recent meta-analysis that pooled data from previous studies (Table VII) found that the incidence of gastrointestinal cancer mortality was significantly correlated with level of exposure.

4. Risk Assessments

On the basis of ecological epidemiologic studies, a working group of the Department of Health and Human Services calculated that an exposure to 100 MFL produced an average lifetime risk of 3.3×10^{-3} excess deaths. The Safe Drinking Water Committee of the National Academy of Sciences used gastrointestinal cancer data from occupation ally exposed workers to estimate relative risks from drinking asbestos-contaminated water and calculated that drinking 0.1 to 0.2 MFL for a 70-year lifespan would cause one excess gastrointestinal cancer death in a population of 100,000. This exposure estimate was similar to an earlier estimate of 0.5 MFL for the Ambient Water Quality Criteria for Asbestos.

B. Inhalation

There are mechanisms by which waterborne asbestos can become airborne and inhalable, a proven pathogenic pathway. Asbestos-contaminated drinking water can evaporate, leaving asbestos exposed and available for airborne entrainment. In Woodstock, New York, where drinking water was grossly contaminated by deteriorated AC pipes, airborne asbestos levels in affected houses were significantly elevated compared with control houses. However, concentrations did not exceed the range expected in ambient environments. Ultrasonic humidifiers may also generate airborne asbestos from contaminated water. As the tiny water droplets released by these humidifiers evaporate, tremendously elevated levels of airborne minerals are generated from ordinary tapwater. Fibers in generated droplets may also become airborne when asbestos-contaminated water is used.

C. Regulations

In 1991 the U.S. Environmental Protection Agency (EPA) promulgated drinking-water standards that included a Maximum Contaminant Level (MCL) for asbestos of 7 MFL. This MCL was based solely on fibers longer than 10 μm because of a study of laboratory animals that showed increased intestinal tumors when exposed to long-fiber asbestos. Public water supply systems that are considered at risk (e.g., AC pipe carrying corrosive water) are required to monitor waterborne concentrations on a nine-year cycle. The EPA in 1989 issued a final rule that will effectively phase out the manufacture of AC pipe by 1996. This rule was not related to health concerns associated with drinking water but rather was part of an overall ban on asbestos use.

IV. Analysis

Traditional analytical chemistry techniques are not capable of detecting asbestos in water because of the ubiquity of its principal elements (Na, Mg, Si, Ca, Fe) in all waters. X-ray diffraction is also poorly suited because of its lack of sensitivity and its inability to distinguish fibrous minerals from their nonfibrous counterparts. Polarized-light microscopy lacks the ability to detect the submicrometer fibers characteristically found in water. Scanning electron microscopy also has resolution limitations and is unable to measure crystalline dimensions, essential for positive asbestos identification. Transmission electron microscopy (TEM) then, is the only method for reliable identification and quantitation of asbestos in water. TEM is well suited for detecting the very narrow (0.025 μm) fibers typically found in water, while TEM's electron-diffraction capability, coupled with energy-dispersive x-ray spectroscopy, allows definitive identification of all asbestos species. Water samples are prepared by filtration through 0.1-μm-pore polycarbonate filters on which waterborne particles (including asbestos) are trapped. Filter sections are carbon-coated, placed on TEM grids, and the polycarbonate dissolved to leave particles suspended in a carbon film for TEM

Table VII Summary of Asbestos Workers and Gastrointestinal Cancer Incidence Used in Meta-analysis[a]

Cohort No.	Study[b] (author [year])	Number location	Observed/expected counts and SMR for various sites: Subcohort	Esophagus	Stomach	Colon–rectum	GI tract (ICD 150–154)	All GI, incl. meso.	All GI, excl. meso.	Lung	All cancer	Nonmalignant respiratory disease	All causes	Cancer proportional mortality	Latency
Group 1: Cement workers															
1.1	Lacquet [1980]	2,650 Belgium						17/11.8 (144)		21//22.3 (94)	48/55.4 (87)		201/213 (94)	.24	0
1.2	Thomas [1982]	1,592 Wales					14/14.1 (99)			28/33.0 (85)	58/61.0 (95)		261/243.2 (107)	.22	15
1.3	Finkelstein [1983, 1984]	428 Ontario	Production				8/2.8 (285)			21/4.1 (512)	44/11.6 (379)	9/2.8 (321)	86/47.5 (181)	.51	20
		107 Ontario	Maintenance				1/0.8 (118)			5/1.2 (413)	12/3.6 (333)	4/0.9 (444)	22/14.7 (150)	.54	20
1.4	Ohlson & Hogstedt [1985]	1,176 Sweden			1/5.9 (17)	11/5.9 (186)				11/9.0 (123)	44/50.0 (88)	13/8.5 (153)	220/214 (103)	.20	0
1.5	Gardner [1986]; Gardner and Powell [1986]	1,510 England					23/23 (100)			35/38 (92)	95/97 (98)		384/408 (94)	.25	0
1.6	Hughes & Weill [1980]; Hughes [1987]	2,565 Louisiana	Plant 1	5/5.0 (100)	8/7.5 (107)	10/8.3 (120)				48/41.2 (117)	127/113.9 (112)		477/522.2 (91)	.27	20
		4,366 Louisiana	Plant 2	7/7.9 (89)	15/12.0 (125)	11/15.0 (73)				107/74.3 (144)	226/199.1 (114)		874/922.7 (95)	.26	20
1.7	Albin [1983]	1,373 Sweden					19/10.8 (176)			16/6.6 (244)	56/37.8 (148)		172/157.4 (109)	.33	10
Group 2: Insulators															
2.1	Selikoff [1979]	17,800 U.S./Canada		17/6.5 (264)	18/12.7 (142)	54/34.0 (159)				397/93.7 (424)	845/277.1 (305)	177/53.8 (329)	1,946/1,376.0 (141)	.43	20
Group 3: Miners and millers															
3.1	Meurman [1974, 1979]	1,092 Finland						2/2.1 (95)		8/2.4 (333)	na		na	na	10
3.2	McDonald [1980]	11,379 Quebec					79/101 (78)			230/184 (125)	759/688.2 (110)	213/198.5 (107)	3,291/3,019.3 (109)	.23	20
3.3	Amandus & Wheeler [1987]	575 Montana						6/8.1 (74)		20/9.0 (223)	38/28.4 (134)	20/8.2 (244)	161/146.1 (110)	.24	0
Group 4: Textile workers															
4.1	Enterline & Kendrick [1967]	1,843 U.S.						11/7.5 (146)		14/6.1 (229)	35/23.6 (148)	21/5.3 (396)	186/153.7 (121)	.19	0
4.2	McDonald [1982]. Also Mancuso and El-Attar [1967]	4,137 Pennsylvania	Low-dose					47/44.9 (105)		42/47.5 (88)	159/147 (108)	39/31.9 (122)	791/773.1 (102)	.20	20
			High-dose					7/2.95 (237)		11/2.6 (416)	30/8.8 (341)	31/2.4 (1292)	104/48.3 (215)	.29	20
4.3	McDonald [1983]. Also Dement [1983]	2,534 S. Carolina	Low-dose					18/13.9 (129)		36/24.4 (148)	85/61.1 (139)	13/14.9 (87)	431/369 (117)	.20	20
			High-dose					8/2.7 (301)		23/5.7 (404)	35/11.3 (310)	26/4 (650)	139/77.9 (178)	.25	20

4.4	Peto [1977, 1985]	145 England	0/0.5 (0)	4/2.3 (17[illegible])	3/2.1 (145)			20/5.6 (361)	40/15.5 (258)	24/9.9 (242)	123/67.9 (181)	.32	0
		283 England	0/0.3 (0)	2/1.1 (18[illegible])	4/2.0 (198)			4/1.9 (211)	19/16.3 (116)		49/49.4 (99)	.39	0
		3,211 England	11/6.6 (167)	29/29.0 (10[illegible])	20/26.7 (75)			132/100.5 (131)	260/245.1 (106)	176/129.7 (136)	1,113/972.9 (114)	.23	0
Group 5: Shipyard workers													
5.1	Puntoni [1977, 1979]	4,274 Italy	2/5.0 (40)	38/31.0 (12[illegible])	32/22.2 (144)			123/33.9 (362)	304/171.1 (178)	148/69.2 (214)	1,073/777.5 (138)	.28	0
5.2	Rossiter & Coles [1980]	6,292 England				63/[illegible].3 (76)		84/119.7 (70)	265/282.1 (94)		1,038/1,081.2 (96)	.26	0
Group 6: Other industrial workers													
6.1	Enterline & Kendrick [1967]	12,402 U.S.				36/40.4 (89)		46/35.3 (130)	117/133.0 (88)	45/30.1 (150)	840/943.8 (89)	.14	0
6.2	Enterline & Kendrick [1967]	7,510 U.S.				36/30.3 (119)		35/25.9 (124)	98/96.1 (102)	33/21.9 (151)	567/637.1 (89)	.17	0
6.3	Weiss [1977]	264 U.S.				4/3.8 (105)		4/4.3 (93)	13/17.4 (75)		66/108.8 (61)	.20	0
6.4	Henderson & Enterline [1979]	1,348 U.S.				55/39.9 (138)		63/23.3 (270)	173/108.8 (159)	68/39.3 (173)	781/648.7 (120)	.22	0
6.5	Jones [1980]; Wignall & Fox [1982]	951 U.K.					10/20.3 (49)	12/6.3 (190)	74/66.0 (112)		na	na	0
6.6	Acheson [1982]	757 Leyland, UK		5/4.4 (114)				13/6.2 (210)	66/54 (123)		219/185 (118)	.30	0
		570 Blackburn, UK		4/3.1 (129)				6/4.8 (125)	44/41 (109)		177/128 (138)	.25	0
6.7	Berry & Newhouse [1983]; Newhouse [1982]	13,460 England				132/134.6 (98)		149/150.8 (99)	419/433.1 (97)		1,638/1,689.8 (97)	.26	10
6.8	Ohlson [1984]	3,297 Sweden		30/51.1 (59)	15/20.9 (72)			27/25.7 (105)	144/166 (87)		586/727 (81)	.25	20
6.9	McDonald [1984]	3,641 Connecticut				59/51.6 (114)		73/49.1 (149)	202/159.7 (127)	37/29.0 (128)	803/740 (109)	.25	20
6.10	Acheson [1984]	5,969 England	2/2.0 (100)	7/7.5 (9[illegible])	10/7.6 (132)			57/29.1 (196)	109/75.5 (144)	38/29.1 (130)	333/298.8 (111)	.33	0

(continues)

Table VII *(Continued)*

Cohort No.	Study[b] (author [year])	Number location	Subcohort	Esophagus	Stomach	Colon–rectum	GI tract (ICD 150–154)	All GI, incl. meso.	All GI, excl. meso.	Lung	All cancer	Nonmalignant respiratory disease	All causes	Cancer proportional mortality	Latency
			Observed/expected counts and SMR for various sites												
6.11	Newhouse & Berry [1979]; Newhouse [1985a]; Newhouse [1973]	3,000 England	Low-dose male						25/23.3 (107)	48/29.7 (162)	108/72.5 (149)		319/302.7 (105)	.34	10
			High-dose male						42/24.7 (170)	110/33.5 (328)	229/78.7 (291)		499/310 (161)	.46	10
		700 England	Low-dose female						4/2.4 (167)	2/0.8 (250)	12/7.3 (164)		40/29.1 (137)	.30	10
			High-dose female						19/10.8 (176)	35/4.2 (833)	117/36.8 (318)		234/126.3 (185)	.50	10
		1,400 England							11/8.1 (136)	38/10.7 (355)	72/26.5 (272)		157/103.4 (152)	.46	10
6.12	Seidman [1986]	933 New Jersey		1/2.1 (49)	9/5.8 (156)	22/11.9 (185)				102/20.5 (498)	197/74.2 (265)	50/19.4 (532)	593/355.9 (167)	.33	5
6.13	Hodgson & Jones [1986]	31,150 England	Pre-1969	6/8.8 (68)	24/24.9 (96)	14/27.4 (51)				152/112.1 (136)	na		834/931.6 (89)	na	10
			Post-1969	0/0.9 (0)	0/0.2 (0)	0/0.3 (0)				1/1.0 (96)	na		9/9.5 (95)	na	10
6.14	Woitowitz [1986]	3,070 Germany	Subcohort I			5/6.34 (79)				22/15.3 (144)	57/50.0 (114)		185/194.7 (95)	.31	9
		665 Germany	Subcohort II			3/1.4 (215				9/2.6 (347)	22/10.3 (213)	6/2.6 (231)	71/40.7 (175)	.31	9

[a] From Frumkin, H., and Berlin, J. (1988). Asbestos exposure and gastrointestinal malignancy review and meta-analysis. *American Journal of Industrial Medicine*, **14,** 79–95.
[b] For complete reference, see original source.

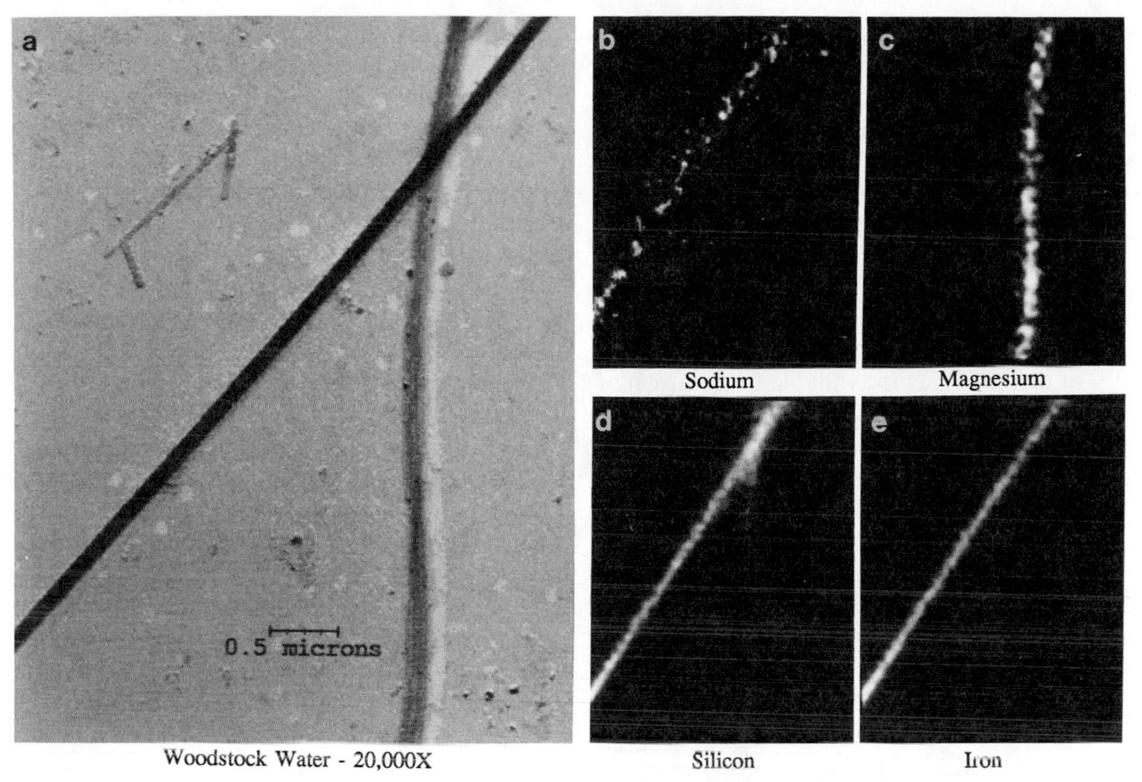

Figure 1 (a) Digitized scanning transmission electron micrograph of asbestos fibers collected in Woodstock, New York, drinking water in 1985. (b–e) X-ray maps depicting relative element concentrations. The crocidolite fiber is distinguished by sodium, silicon, and iron while the chrysotile fiber is distinguished by magnesium and silicon.

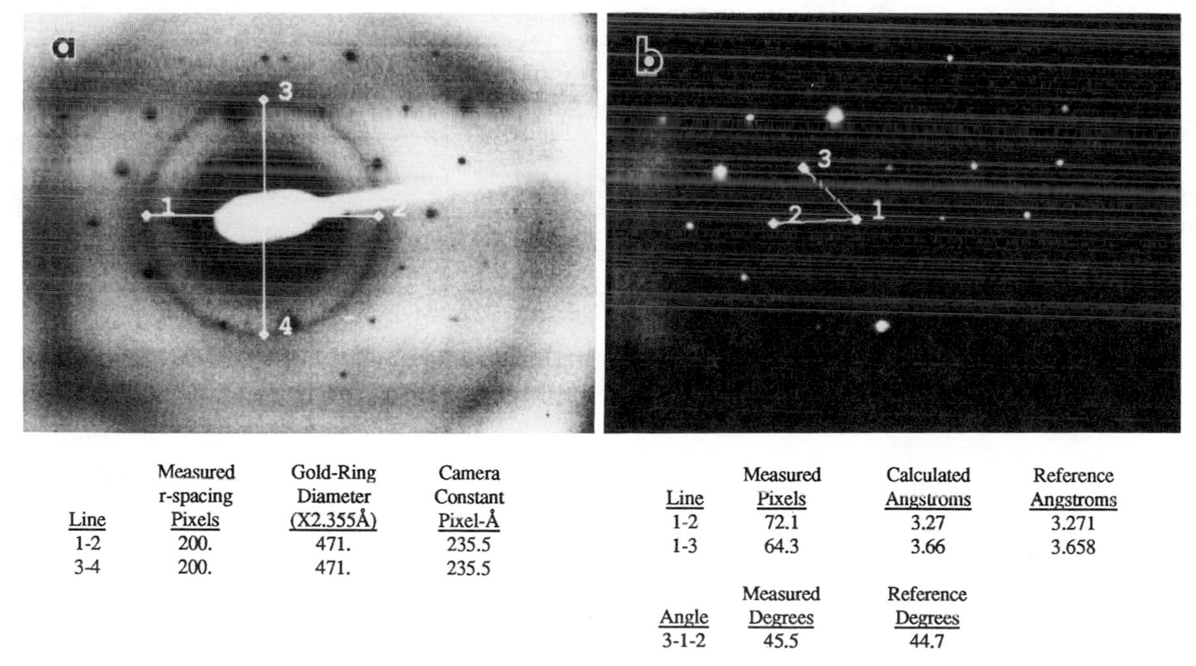

Line	Measured r-spacing Pixels	Gold-Ring Diameter (X2.355Å)	Camera Constant Pixel-Å
1-2	200.	471.	235.5
3-4	200.	471.	235.5

Line	Measured Pixels	Calculated Angstroms	Reference Angstroms
1-2	72.1	3.27	3.271
1-3	64.3	3.66	3.658

Angle	Measured Degrees	Reference Degrees
3-1-2	45.5	44.7

Figure 2 Digitized [212] zone–axis electron diffraction pattern from a crocidolite fiber in the Woodstock sample. (a) Unprocessed CCD-camera pattern with 2.355-Å gold (internal standard) diffraction ring. (b) Processed pattern with measured *r*-spacings, calculated *d*-spacings, and reference *d*-spacings.

analysis. Figures 1 and 2 illustrate TEM-generated characteristics used in identifying chrysotile and crocidolite asbestos in a drinking-water sample. The current EPA methods, published in 1983, requires analysis of all fibers longer than 0.5 μm and thus is not fully compatible with the EPA's 1991 MCL, which is based on long (>10 μm) fibers.

V. Control

A variety of technologies is available for removal of asbestos from drinking water. Specific applications will depend on the source of the contamination.

A. Natural Erosion

When source water is contaminated by asbestos from bedrock, conventional treatments utilizing one or several of the sedimentation, flocculation, coagulation, filtration, or equivalent direct filtration methods usually remove 95 to 99% of waterborne asbestos if properly optimized. Dredging of asbestos-laden sediments in rivers or reservoirs may also reduce asbestos concentrations in certain situations.

B. Deteriorated AC Pipes

Deteriorated AC pipes often contribute asbestos contamination on a localized level. In several episodes, complaints about fibers clogging faucet strainers were the first hints of AC-deterioration problems. Deterioration and resultant contaminant problems may not always be detectable using TEM analysis and the EPA's MCL because of the sporadic nature of the problem and the fact that more than 95% of waterborne asbestos fibers are usually shorter than 10 μm. For example, tapwater in Woodstock, New York, which had a visible fiber suspension and contained more than 1000 MFL (based on all fiber sizes), was below the 10-μm-length-based MCL of 7 MFL. TEM analysis for all fiber sizes and a visual inspection of AC pipes should be part of routine monitoring.

In instances where deterioration is caused by corrosive water, chemical treatments may be applied to reduce the corrosivity. Lime treatment applied in a consistent manner is often effective in reducing acidity, a common promoter of AC-pipe deterioration. Chemicals may also be added to form a protective lining inside the pipe; zinc salts have proven effective in some cases. Pipes may also be physically cleaned and lined with asphalt or rubberized compounds to minimize deterioration. Future contamination problems can be minimized by following recommended practices when installing or modifying AC pipe. Finally, AC pipes may be bypassed or replaced by non-AC pipes.

Bibliography

Frumkin, H., and Berlin, J. (1988). Asbestos exposure and gastrointestinal malignancy: Review and meta-analysis. *American Journal of Industrial Medicine,* **14,** 79–95.

Schreier, H. (1989). "Studies in Environmental Science 37: Asbestos in the Natural Environment." Elsevier, Amsterdam.

Summary workshop on ingested asbestos, (1983). *Environmental Health Perspectives,* **53,** 1–204.

Toft, P., Meek, M. E., Wigle, D. T., and Meranger, J. C. (1984). Asbestos in drinking water. *CRC Critical Reviews in Environmental Control,* **14**(2), 151–197.

Webber, J. S., and Covey, J. R. (1991). Asbestos in water. *CRC Critical Reviews in Environmental Control,* **21**(4), 331–371.

Bioconcentration of Persistent Chemicals

Darryl Hawker
Griffith University, Australia

Glossary

Bioaccumulation Uptake of a contaminant from water and food by an aquatic organism.

Bioconcentration factor Ratio of the concentration of a contaminant in an organism to that in the ambient medium (usually water) at equilibrium.

Fugacity Thermodynamic function related to chemical potential that has units of pressure and may be regarded as the escaping tendency of a chemical from a given phase.

Hydrophobic Having little affinity for water.

Lipophilic Having an affinity for lipids.

BIOCONCENTRATION *is generally defined as the uptake and accumulation of a chemical by biota from the surrounding medium only. Note that this definition encompasses bioconcentration of gaseous phase compounds by terrestrial biota. The uptake of chemicals by biota is of significant concern in the hazard assessment and management of chemicals. This uptake may occur via different pathways (e.g., from ingested food or from ambient media such as water or air). A wide variety of terms such as bioamplification, biomagnification, bioaccumulation, and bioconcentration are used in a sometimes inconsistent manner in the literature to describe these pathways. Although there is increasing interest in this area, most work has concerned bioconcentration of aqueous phase chemicals by aquatic organisms. For this reason, this article focuses on bioconcentration from water. Uptake from water and food by aquatic organisms is termed bioaccumulation.*

I. Introduction

Persistent chemicals are those that are relatively resistant to physical degradation processes (e.g., hydrolysis, phototransformation) and more importantly in the context of bioconcentration, relatively resistant to biodegradation. Examples of persistent chemicals include highly chlorinated aliphatic and aromatic hydrocarbons. Some characteristics of chemicals that may be regarded as persistent and capable of significant bioconcentration are summarised in Table I.

The bioconcentration factor (K_B) is defined as the ratio of the concentration of chemical in the organism to that in the surrounding water at equilibrium. It is a measure of the tendency of a chemical to accumulate or concentrate in an organism.

The bioconcentration process for hydrophobic compounds is often regarded as a partitioning or distribution between the surrounding water and the organism's lipid tissues. This is because most of the contaminant body burden for hydrophobic compounds is found in lipid tissues. As a result, K_B may be expressed on a whole weight or lipid weight basis, reflecting the basis on which the contaminant concentration in the organism is measured. If the units for biotic and aqueous concentrations are the same, then K_B is dimensionless. Often however, biotic concentrations are expressed in mg kg^{-1} (or equivalent units) and aqueous concentrations in mg l^{-1} (or equivalent units). This means that in such circum-

Table I Characteristics of Organic Compounds That Can Exhibit Substantial Bioconcentration

Chemical structure	High proportion of C—Cl bonds Absence of polar functional groups
Molar volume	Less than approximately 230 cm^3 mol^{-1}
$\log K_{OW}$	2 to 6
Aqueous solubility	$10^{-3.40}$ mol m^{-3} to $10^{1.50}$ mol m^{-3}
Minimum molecular internal cross-section	<1.05 nm

Source: Modified from Connell, D. (1988). Bioaccumulation behaviour of persistent organic chemicals with aquatic organisms. *Rev. Environ. Contam. Toxicol.* **101,** 117–154.

stances, K_B strictly has units of l kg^{-1}. To convert values of K_B with these units to dimensionless values, it is necessary to multiply by the density of the biota involved. A density of 1.0 g cm^{-3} or 1 kg l^{-1} is often assumed, but the units (if any) associated with K_B, as well as the basis on which the biotic concentration is expressed should be considered when comparing data from different sources.

While the bioconcentration factor K_B represents an equilibrium position, the rate at which this is reached is often an important factor. For persistent, nondegradable compounds, bioconcentration has been successfully modeled by a simple two compartment (water/biota) model, such as that depicted in Figure 1, assuming first order uptake (k_1) and clearance (k_2) rate constants. Thus, the rate of change of biotic concentration (C_B) with time is given by

$$\frac{dC_B}{dt} = k_1 C_W - k_2 C_B, \tag{1}$$

where C_W is the aqueous concentration of the chemical. At equilibrium, $dC_B/dt = 0$, and therefore

$$\frac{k_1}{k_2} = \frac{C_B}{C_W} = K_B. \tag{2}$$

A model for bioconcentration based on fugacity with the driving force being fugacity differences of the bioconcentrating chemical between aqueous and biotic phases has also been developed. This model has been shown to be equivalent to the kinetic or rate constant approach outlined above.

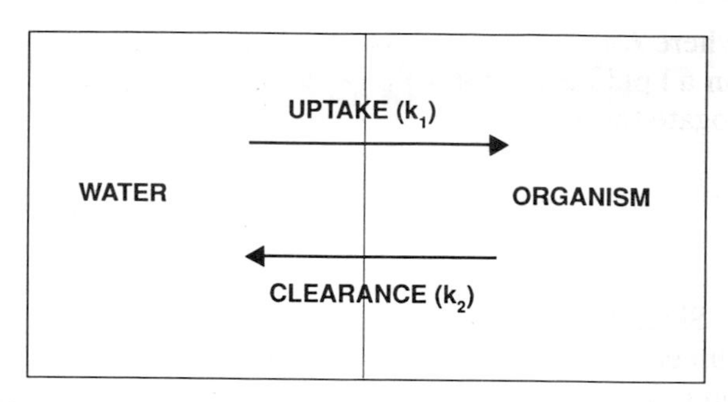

Figure 1 A simple two-compartment model for bioconcentration.

II. Factors Influencing Bioconcentration

A. Hydrophobicity

The most widely utilized measure of the hydrophobicity of a chemical is the octan-1-ol/water partition coefficient (K_{OW}). Relationships that describe K_B simply in terms of K_{OW} are based on three primary assumptions: (1) that bioconcentration may be regarded as a passive distribution process between water and biotic lipid that is analogous to partitioning between water and octan-1-ol; (2) lipids are the ultimate repository of bioconcentrated chemicals and other tissues can effectively be ignored; and (3) octan-1-ol is a good model for biotic lipid. Based on these assumptions, and employing the concept of fugacity, theoretical relationships between K_B and K_{OW} may be derived. Fugacity is a thermodynamic function related to chemical potential. The fugacity (f) of a chemical in a given phase is defined as the ratio of the chemical's concentration in that phase to the fugacity capacity constant (Z) of the chemical in that phase:

$$f = C/Z. \tag{3}$$

At equilibrium, the fugacity of the chemical in the lipid phase (f_L) of biota is equal to that in water (f_W):

$$f_W = \frac{C_W}{Z_W} = f_L = \frac{C_L}{Z_L}, \tag{4}$$

where C_W and C_L are the chemical's concentration in water and lipid respectively. Rearranging Eq. 4,

$$\frac{C_L}{C_W} = \frac{Z_L}{Z_W} = K_{BL}, \tag{5}$$

where K_{BL} is the bioconcentration factor expressed on a lipid basis. Assuming octan-1-ol is a good surrogate for lipid:

$$\frac{C_L}{C_W} = \frac{C_O}{C_W} = K_{OW} = K_{BL}. \tag{6}$$

Since the concentration in nonlipid tissues is assumed to be negligible, the contaminant concentration in the whole organism (C_B) is given by $C_B = lC_L$, where l is the decimal fraction of lipid in the organism. Thus,

$$K_B = lK_{OW}, \tag{7}$$

or expressed logarithmically,

$$\log K_B = 1.00 \log K_{OW} + \log l. \tag{8}$$

There have been a number of relationships between log K_B and log K_{OW} presented in the literature. It has been found that despite similar compounds being bioconcentrated, different relationships are found with fish, molluscs, and crustaceans. Most work has been carried out with fish, and a compilation of relationships between log K_B and log K_{OW} based on original and modified data sets is presented in Table II.

Inspection of the slopes and intercepts of these expressions reveals substantial variability. While a number of slopes are close to unity as predicted from Eq. 8, many are significantly less than one. Based on the theoretical derivation above, a negative intercept should result, but the data contained in Table II show that this is not always the case. Overall, it seems that correlations in accordance with Eq. 8 are usually obtained with data from related groups of chemicals, such as chlorinated hydrocarbons, and then only for compounds with log K_{OW} of 2 to approximately 6. Extreme caution should be exercised when using these relationships to predict K_B for chemicals or organisms of a different type from those employed in deriving the relationship.

The use of octan-1-ol as a model or surrogate for biotic lipid has been criticized. A recent study found that octan-1-ol was a suitable surrogate phase for biological lipid only for compounds with molar volumes between 100 and 230 cm^3 mol^{-1}. For larger compounds, a distinctive difference in their solubilities in lipid phases (such as dimyristoyl phosphatidyl choline) and in octan-1-ol has been observed. Additionally, thermodynamic analysis of partitioning of hydrophobic compounds between fish and water, and between octan-1-ol and water revealed that the two processes were governed by different factors. Favorable entropy is apparently the driving force in the case of fish/water partitioning, whereas octan-1-ol/water partitioning results largely from a favorable enthalpy change. Thus, any correspondence between K_B and K_{OW} may be coincidental and without a firm theoretical basis.

Expressions such as Eq. 7 are derived assuming that lipid phases are the ultimate and only significant repository of bioconcentrated chemicals. For relatively hydrophilic compounds capable of achieving

Table II A Compilation of Published Relationships between log K_B and log K_{OW} for Fish of the Form log $K_B = a \log K_{OW} + b$

a	*b*	Number of data (*n*)	Correlation coefficient (*r*)
0.54	0.12	8	0.95
0.63	0.73	11	0.79
1.16	0.75	9	0.98
0.85	−0.70	55	0.95
0.77	−0.97	36	0.76
0.94	−1.50	26	0.87
0.98	−0.06	6	0.99
0.83	−1.71	8	0.98
0.46	0.63	25	0.63
0.76	−0.23	84	0.82
1.02	−1.82	9	0.98
1.00	−1.32	44	0.95
0.79	−0.40	122	0.93
1.02	−0.63	11	0.99
0.60	0.19	31	0.75
0.89	0.61	18	0.95
0.61	0.69	11	0.84
0.94	−1.10	49	0.89
0.75	−0.32	32	0.87
0.78	−0.35	22	0.95
1.00	−0.87	11	0.99
0.96	−0.56	16	0.98
0.89	0.61	18	0.95
0.98	−1.30	20	0.90

Source: Compiled from Schuurmann, G., and Klein, W. (1988). Advances in bioconcentration prediction. *Chemosphere* **17**, 1551–1574; and Connell, D. (1988). Bioaccumulation behaviour of persistent organic chemicals with aquatic organic organisms. *Rev. Environ. Contam. Toxicol.* **101**, 117–154. Reproduced with permission of the publisher.

reasonable concentrations in aqueous biotic phases, overall biotic concentrations will not be given by lC_L. For such compounds (which have log K_{OW} values less than approximately 2), K_B will not be simply proportional to K_{OW}. From the previous discussion, this is also the case for large compounds with molar volumes greater than 230 cm^3 mol^{-1}, which corresponds roughly to compounds with log $K_{OW} > 6$. It has been suggested that K_B would be better described in terms of triolein/water partition coefficients instead of K_{OW}. At present though, there is insufficient relevant data to test this proposal.

The occurrence of different relationships between log K_B and log K_{OW} for various aquatic organisms and different chemical groups together with their limited utility in terms of log K_{OW} range suggests that hydrophobicity is only one factor involved in bioconcentration. It has been argued that hydrophobicity models in general ignore factors such as steric hindrance by the gill as well as the importance of blood flow in determining uptake and distribution of chemicals.

B. Aqueous Solubility

There have been numerous published relationships between measures of hydrophobicity such as K_{OW} and aqueous solubility. A sound theoretically based expression is only obtained when the aqueous solubility of compounds that are solid at the temperature of interest are expressed as the subcooled liquid aqueous solubility. On such a basis a relationship for chlorobenzenes and PCBs, for example, has been found to be

$$\log K_{OW} = -0.81 \log S_W + 3.23, \tag{9}$$

where S_W is the liquid or subcooled liquid aqueous solubility (mol m^{-3}). For those compounds for which K_{OW} is found to be an adequate descriptor of K_B, aqueous solubility should also be useful. Relationships of an inverse linear form between log K_B and log S_W are commonly observed.

C. Lipid Solubility

For a group of compounds with a sufficiently wide range of log K_{OW}, relationships with log K_B are often curvilinear as shown in Figure 2. This has prompted suggestions that factors other than chemical hydrophobicity may be important in bioconcentration.

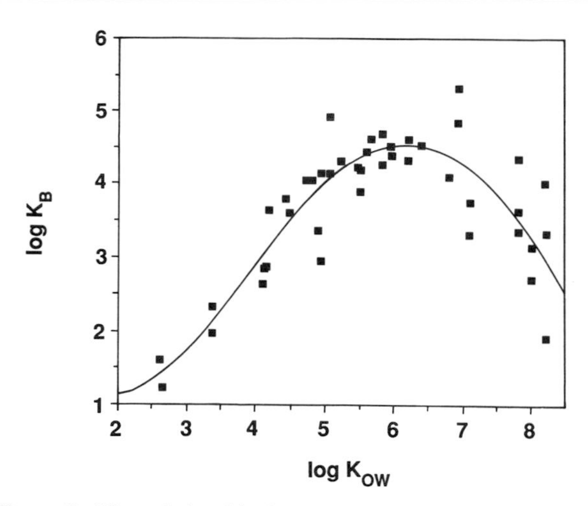

Figure 2 The relationship between log K_B and log K_{OW} for chlorinated hydrocarbons and related compounds possessing a wide range of log K_{OW} values together with the line of best fit. [From Hawker, D. W. (1990). Description of fish bioconcentration factors in terms of solvatochromic parameters. *Chemosphere* **20,** 467–477.]

Large hydrophobic chemicals that would be expected to bioconcentrate to a considerable degree based solely on their hydrophobicity have usually been assumed to be lipophilic. Recent studies have shown that this may not be the case. Because of the relatively structured nature of bulk lipid phases, cavity formation and ultimately solution in them is energetically expensive, particularly for larger solutes. While solubility in lipid is roughly constant for solutes of 100 to 200 cm^3 mol^{-1} molar volume, solubility actually decreases at a rate that increases with molar volume, for larger solutes. This in contrast to solvents such as octan-1-ol and hexane for which solubility decreases more slowly and linearly with molecular size. Relatively large hydrophobic solutes such as highly substituted PCBs and chlorodibenzodioxins seem to be poorly soluble in both water and lipid. In all cases, however, lipid solubility is still greater than aqueous solubility. For large compounds of sufficient concentration in ambient water, bioconcentration by aquatic organisms may be actually limited by achievement of maximal solubility in biotic lipid.

D. Molecular Size

As discussed earlier, the size of the bioconcentrating molecule may influence bioconcentration by reaching the saturation point of lipid solubility. Molecular size may also influence bioconcentration in other ways. It has been observed for instance that most of a series of polychlorinated napthalenes bioconcentrated to an extent consistent with their hydrophobicity. The most highly substituted substrates (the two heptachloronaphthalenes and octachloronaphthalene) did not bioconcentrate to any measurable degree, however. It has been suggested that this was due to significant or perhaps complete retardation of membrane permeation (for compounds with a minimum internal cross-section of 0.95 nm or greater). A subsequent investigation with a wider range of compounds found that the minimum internal cross-section above which bioconcentration would not occur was 1.05 nm.

According to the hydrophobicity model of bioconcentration, K_B represents a thermodynamic equilibrium between a compound in the biotic lipid phase and the surrounding water. Kinetic factors such as a reduction in the uptake rate should not alter the final equilibrium position. The simple model in Figure 1 predicts that uptake and clearance rate constants should be controlled by aqueous or lipid phase diffusion, depending on the compound. The theoretical relationship between k_1, k_2 and K_{OW} is depicted in Figure 3. Studies have shown that for hydrophilic ($K_{OW} \leq 10$) compounds, uptake and clearance are controlled by diffusion and flow processes in lipid phases of fish. For more hydrophobic compounds ($K_{OW} \geq 10^3$), however, uptake and clearance are governed by diffusion and flow processes in aqueous phases of fish. Reduced bioconcentration because of molecular size would therefore probably involve reduced transport in an aqueous phase rather than increased resistance in a lipid membrane phase.

A slightly more sophisticated model for bioconcentration involving chemical removal from an aquatic organism by egestion to feces characterized by a rate constant k_E (as well as clearance) results in the following expression for K_B:

$$K_B = \frac{k_1}{k_2 + k_E} = \frac{Z_f}{Z_W} \frac{D_f}{D_f + D_e}, \qquad (10)$$

where D_f is a fish/water transport parameter (mol Pa^{-1} hr^{-1}) and D_E is the transport parameter associated with chemical elimination into the feces. Transport parameters relate differences in fugacity to interphase flux. Equation 10 shows that the bioconcentration factor is a measure of fish/water partitioning only when $D_E << D_f$, or $k_E << k_2$. If these conditions are not met, and resistance [$(1/D)$ in the fecal egestion pathway] is comparable to clearance (possibly due to molecular size) then K_B

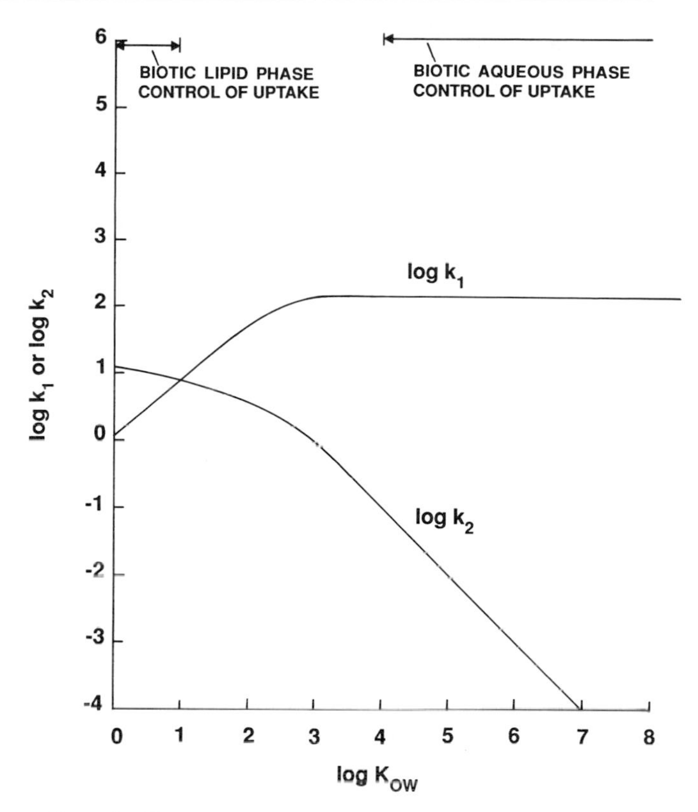

Figure 3 Theoretical relationship between the logarithms of uptake and clearance rate constants with log K_{OW}; also showing the controlling biotic phase for uptake with compounds of varying log K_{OW}. [Modified from Gobas, F. A. P. C., and Mackay, D. (1987). Dynamics of hydrophobic organic chemical bioconcentration in fish. *Environ. Toxicol. Chem.* **6**, 495–504.]

will be less than k_1/k_2 and Z_f/Z_W. Some results indicate that for chemicals with log K_{OW} less than approximately 6.5, clearance to water via the gills is the primary depuration mechanism. For chemicals with higher K_{OW} values, clearance becomes such a slow process and fecal elimination becomes the predominant process. Even when this factor is taken into account, however, measured K_B values of superhydrophobic compounds are lower than predicted.

To account for the influence of molecular size, particularly for chemicals with log $K_{OW} > 5.5$ and molar volume >230 cm^3 mol^{-1}, lipid/water partition coefficients have been described as a function of both molar volume and log K_{OW}. Some authors have nominated thresholds above which negligible bioconcentration will occur purely in terms of molecular weight. While this may be successful for a limited group of compounds, it has no theoretical justification and would likely fail for a wider range of compounds.

E. Exposure Time

Bioconcentration does not occur instantaneously. The time required for a steady-state concentration to be achieved in an aquatic organism varies with the compound being bioconcentrated. In some cases, it can be a relatively lengthy time period. Integration of Eq. 1 assuming a constant aqueous concentration affords

$$C_B = C_W \frac{k_1}{k_2} (1 - e^{-k_2 t}) \tag{11}$$

or

$$K_B = \frac{k_1}{k_2} (1 - e^{-k_2 t}), \tag{12}$$

where C_B is the concentration in the organism on a whole weight basis.

According to this expression, it theoretically requires an infinite time period to attain equilibrium and K_B. The time period required to reach 0.99 K_B is however finite. It may be regarded as the effective time period for attainment of equilibrium (t_{eq}) and is given by

$$\log t_{eq} = 0.663 - \log k_2. \tag{13}$$

There have been a significant number of studies concerning bioconcentration kinetics. In these and related investigations, a relationship between log k_2 and log K_{OW} of the form shown in Figure 3 is usually found. Because k_2 decreases with K_{OW}, t_{eq} increases with the hydrophobicity of the bioconcentrating chemical. For large, extremely hydrophobic compounds, the time period required to achieve effective equilibrium may be quite lengthy. This has been attributed to increasing resistance to chemical transport in aqueous phases of the fish as mentioned previously. If K_B values are measured directly (i.e., from a ratio of biotic and ambient aqueous chemical concentrations), exposure time may be insufficient to attain equilibrium, particularly for superhydrophobic ($K_{OW} > 10^6$) compounds. Under such circumstances, the observed bioconcentration factor will be less than the true equilibrium based value.

It may be of interest to predict nonequilibrium bioconcentration factors ($K_{B.obs}$). If the true equilibrium K_B is related to K_{OW} by an expression of the general form

$$K_B = bK_{OW}^a \tag{14}$$

and k_2 is related to K_{OW} by an expression of the general form

$$\frac{1}{k_2} = xK_{OW} + y, \tag{15}$$

then from Eq. 11:

$$\begin{aligned} K_{B.obs} &= K_B[1 - e^{-t/(xK_{OW}+y)}] \\ &= bK_{OW}^a[1 - e^{-t/(xK_{OW}+y)}]. \end{aligned} \tag{16}$$

There exist threshold K_{OW} values such that all compounds possessing a lower K_{OW} value achieve equilibrium within a given exposure time. This threshold K_{OW} value is given by

$$K_{OW} = (\frac{t}{4.605} - y)\frac{1}{x}, \tag{17}$$

where t is the exposure time. For compounds with K_{OW} values higher than this threshold, $K_{B.obs} < K_B$. Values of $K_{B.obs}$ can be calculated from Eq. 11 or Eq. 16, where $(1 - e^{-k_2 t})$ is the fraction that $K_{B.obs}$ is of K_B. It has been observed that laboratory derived K_B values are useful predictors of field K_B data except for hydrophobic chemicals with small clearance rate constants and relatively lengthy equilibration times. For these compounds, contaminated food may be the main contributor to the observed biotic concentration rather than bioconcentration.

F. Organism Lipid Content

According to the hydrophobicity model of bioconcentration, when expressed on a whole weight basis, K_B is given by the product of K_{OW} and the decimal mass fraction of the organism that is lipid. This organismal lipid content is an important determinant of bioconcentration. A published relationship describing log K_B in terms of log K_{OW} for fish is

$$\log K_B = 1.00 \log K_{OW} - 1.32. \tag{18}$$

Based on the hydrophobicity model, this implies

that an average lipid content for the fish investigated is $10^{-1.32}$ or 4.8%. Reviews have concluded that while lipid content of aquatic organisms is one factor influencing bioconcentration, it may not be as important as previously thought. Different organisms have different lipid contents and even a single species of organism has a variable lipid content depending on sex, age, diet, and seasonal variations. It is not surprising then that specific relationships should exist between log K_B and log K_{OW} for different organisms. The precision of these relationships may in fact be improved by the consideration of factors such as those noted earlier.

Fish, for example, may vary in lipid content by a factor of five from species to species and by more than a factor of two within a single test species due to seasonal variation. Furthermore, within an aquatic organism, lipid material is by no means homogenous in composition. Significant differences exist between lipids of various organs in a given organism.

Improved precision in relationships between log K_B and log K_{OW} is generally observed by normalizing with respect to lipid content and expressing K_B on a lipid basis. This reduces variability in K_B due to differing lipid content of the test organisms. Overall, it would appear that while organism lipid content is an important determinant of bioconcentration, there are other influencing factors. This is particularly true for superhydrophobic compounds.

G. Organism Type

The bioconcentration of many hydrophobic chemicals has often been shown to be species dependent to some degree. As mentioned previously, different relationships between K_B and K_{OW} are observed with different aquatic organisms, such as fish, molluscs, and crustaceans. This may be due to differing lipid content and composition. Physiological differences between organisms may influence uptake and clearance rates, internal distribution, storage, and metabolism of a chemical. Despite this, one study has found little difference between marine and freshwater fish as far as bioconcentration is concerned. It has been suggested that chemical transport resistances are dependent on fish size. It seems that processes such as diffusion and rates of ventilation and circulation increase as fish size increases, but at a rate proportionally less than the volume increase. This is reflected in distinct kinetic data in relation to K_{OW} for different organisms (e.g., guppies, goldfish, and rainbow trout).

The range of organisms for which relationships between K_B and physicochemical parameters of chemicals (such as K_{OW}) have been developed is quite diverse. Apart from fish, molluscs, and crustaceans, organisms utilized in bioconcentration investigations include algae (*Selenastrum capricornutum*) and oligochaete worms. There have also been a number of reviews of bioconcentration by microorganisms. It is interesting to note that it has been found that bioconcentration occurs to roughly the same extent whether the microorganisms are living or dead. A possible rationale for this behavior is that bioconcentration occurs via a simple passive diffusion process.

H. Organism Growth

One factor often overlooked with bioconcentration is growth of the bioconcentration organism. Simple models for bioconcentration such as that illustrated in Figure 1 do not take this into account. For relatively hydrophobic compounds, which exchange slowly with water and have extended equilibrium times, tissue concentrations may appear to level off with time as a result of growth dilution rather than equilibrium attainment. If organism growth fits a model such as

$$M = M_o(1 + At) \tag{19}$$

where M_o is initial mass, A a constant (time^{-1}), and t time, then observed concentrations can be corrected for growth by multiplying by $1 + At$. Since growth rates tend to decline with the size of the organism, growth dilution effects are most important with smaller organisms, or organisms of lower trophic levels.

I. Bioavailability

It is generally accepted that only dissolved chemicals can traverse the gill structure and undergo bioconcentration. The presence of dissolved or particulate organic material may alter, significantly in some instances, the amount of truly dissolved chemical available for bioconcentration. The unbound bioavailable fraction is often much less than the total or nominal chemical concentration in the water. A simple mass balance analysis enables the true aqueous concentration (C_W) to be determined from the total (sorbed and dissolved) concentration (C_T), the sorbent/water partition coefficient (K_{SW}) and the volume fraction (V) of sorbent in the system by

$$C_W = C_T (1 + K_{SW}V)^{-1}. \quad (20)$$

While only dissolved compounds can be bioconcentrated, sorbed materials can still be bioaccumulated by other pathways such as oral ingestion.

The capacity of hydrophobic chemicals to bioconcentrate in the presence of dissolved or particulate organic material is determined by a number of factors. In the presence of initially uncontaminated dissolved organic matter (humic acid), reductions in the concentration of dissolved contaminant and equivalent reductions in uptake efficiency have been observed. For contaminated low-organic carbon particulate matter in contact with dissolved contaminant, the effect has been shown to vary with the hydrophobicity of the contaminant. For the more hydrophilic compounds (e.g., 1,2,3,4-tetrachlorobenzene; log K_{OW} = 4.55) investigated, no alteration was found in uptake rate. This may be due to relatively small amounts of sorbed compared with dissolved compound. For compounds with log K_{OW} values from 5 to 6, however, an increased uptake rate compared with the system containing no particulates results. In this situation, low aqueous solubility together with relatively high rates of desorption enable the particles to act as an extra source of chemical for bioconcentration. With more hydrophobic compounds (log $K_{OW} > 6$), rates of desorption are generally too slow to influence aqueous concentration and hence bioconcentration, at least for exposure periods of less than 20 days. As a caveat, it should be remembered that the particulates employed in this particular study had a low organic carbon content which may not be representative of natural particulates.

Factors contributing to the apparent relatively low K_B values for superhydrophobic chemicals (Fig. 2), then, may include nonnegligible clearance via egestion to the feces, or decreased lipid solubility, as well as contaminant sorption to organic matter. This latter factor results in overestimation of the dissolved aqueous concentration, and hence underestimation of K_B. The available solute fraction for bioconcentration (calculated as the ratio of observed to expected uptake rate constants) has been found to range between approximately 0.5 and 1.0 for compounds with a molar volume of less than 300 cm^3 mol^{-1}. For large compounds, the logarithm of the available solute fraction has been determined to fall linearly with molar volume.

J. Water Quality Characteristics

Apart from the presence of dissolved and suspended solids that may affect the bioavailability of chemicals as outlined above, other water quality characteristics that influence bioconcentration include temperature and pH. The bioconcentration factor is often viewed as a lipid/water partition coefficient. Thermodynamically, all biphasic partition coefficients have a temperature dependence. The values for log K_{OW} of a compound, for example, change by about 0.01 log units per degree near room temperature. Uptake and clearance rates will both tend to increase with temperature, and the effect on K_B will depend on the relative effect of each process.

For chemicals capable of ionization, K_{OW} and hence K_B change with the pH of the aqueous phase. Ionization implies the occurrence of polar functional groups within the molecule. Most nonpolymeric persistent chemicals do not possess polar functional groups, which are sites for potential physical and biological degradation processes. In general, persistent compounds are not likely to be greatly directly influenced by variation in the pH of the ambient water. Indirect effects such as alteration in suspended or dissolved solids levels are, however, possible.

III. Measurement of Bioconcentration Factors

The bioconcentration factor for a particular chemical may be defined as the ratio of the concentration in the organism, to that in the surrounding water at equilibrium. Conceptually, the simplest method of determination is to expose an organism to an aqueous solution of the compound of interest until equilibrium characterized by steady biotic and aqueous concentrations) is achieved. Under these conditions, continuous sources of loss of the compound (e.g., to the atmosphere) should be identified and eliminated. The duration of such a test is dependent upon the equilibrium time, which may be quite lengthy. It is also desirable to employ at least two different exposure concentrations of the test compound to identify any concentration-dependent effects. These concentrations should be sufficiently low as to avoid toxic effects on accumulation. This method of K_B determination is known as the "steady-state" or "plateau" method.

A simple kinetic model for bioconcentration is shown in Figure 1. From the relevant mathematical formulae in Eqs. 1 and 2, it can be seen that K_B can also be derived from the ratio of the first-order uptake and clearance rate constants. Consequently, a second method for measuring K_B is from measurement of these rate constants. This method is commonly known as the kinetic method. Experimentally, the test organisms are exposed to an aqueous solution of the compound for a relatively brief time period. During this time, the clearance rate is negligible compared with the uptake rate and as an approximation $dC_B/dt = k_1C_W$. This enables evaluation of the uptake rate constant (k_1). When the organisms have accumulated or concentrated the test compound to a sufficient degree, they are transferred to clean, uncontaminated water and allowed to depurate. Initially, since C_W is small, the kinetic situation is described by $dC_B/dt = -k_2C_B$. The magnitude of k_2 can then be simply determined.

A critical comparison of these two bioconcentration test methods has found that the bioconcentration potential of a wide variety of chemicals was adequately determined by either method. The kinetic method does offer some experimental advantages in that it requires shorter exposure times, and obviates difficulties arising from factors such as extended equilibrium times and growth dilution. More recent studies found that comparable results were only obtained for relatively hydrophilic compounds with short equilibrium times, such as 2,5-dichlorobiphenyl, lindane, and γ-chlordane. For more hydrophobic chemicals, the magnitude of the kinetic-based K_B values was consistently greater than for those values derived from the steady-state procedure.

With both experimental methods, the aquatic organisms may be exposed to a single volume of contaminated water (a static technique), or a continuous controlled addition system designed to ensure a turnover of the contaminated water in the aquarium (the so-called "flow-through" technique). Of these exposure techniques, the flow-through method is generally preferred. The static system can cause a buildup of potentially toxic metabolites and waste products, and cannot maintain a constant aqueous concentration of the contaminant. With the flow-through system however, these problems may be overcome.

IV. Prediction and Estimation of Bioconcentration Factors

Laboratory based bioconcentration tests are relatively expensive to undertake. Given this and the proliferation of new chemicals, it is important that methods capable of satisfactory prediction and estimation of bioconcentration factors be developed. As outlined earlier, values of log K_B may be predicted from linear relationships with parameters such as log K_{OW}, and the logarithm of the maximum aqueous solubility, log S_W. In general, those relationships based on more limited and less diverse groups of compounds are more precise. A critical investigation found there was no significant advantage in using K_{OW} instead of S_W. Studies using various molecular descriptors and parameters such as molecular weight (MW), solvent accessible molecular surface area (SASA), solvent accessible molecular volume (SAV), and molar refraction (MR) have shown that, for certain chlorinated and polyaromatic hydrocarbons, SASA and MR are better predictors of log K_B than log K_{OW}. Expressions describing these relationships are

$$\log K_B = 1.39 \times 10^{-2}\,\text{SASA} - 1.89$$
$$\log K_B = 5.54 \times 10^{-2}\,\text{MR} + 0.19$$
$$\log K_B = 6.94 \times 10^{-3}\,\text{SAV} - 0.77.$$

Other molecular descriptors including molecular topological indices such as connectivity indices have also been shown to correlate with bioconcentration factors.

In some environmental situations, field based K_B values are difficult to obtain because of factors such as the variable levels of contaminants or toxic levels of some contaminants, as well as short lifespans and mobility of resident organisms. To address some of these problems, passive sampling devices such as solvent-filled membrane bags to simulate bioconcentration have recently been employed as a surrogate for aquatic organisms. Use of dialysis membranes containing solvents such as hexane have resulted in accumulation of hydrophobic contaminants in a manner qualitatively similar to aquatic organisms. Osmotic losses to the surrounding aqueous phase can, however, be large over time.

Polymers such as low density polyethylene enclosing synthetic lipids such as triolein or extracted fish lipid may hold greater promise as an abiotic method to simulate bioconcentration. Preliminary investigations have shown contained lipid/water partition coefficients to be quantitatively similar to values of K_{OW}. Current problems include retention of the test substance in the membrane matrix and growth of microorganisms on the exterior surface of the membranes.

Related Articles: Ecotoxicological Testing; Ethanol Fuel Toxicity; Industrial Solvents; Organic Micropollutants in Lake Sediments; Organic Solvents, Health Effects; Polychlorinated Biphenyls (PCBs), Effects on Humans and the Environment; Zinc.

Bibliography

Barron, M. G. (1990). Bioconcentration. *Environ. Sci. Tech.* **24,** 1612–1618.

Clark, K. E., Gobas, F. A. P. C., and Mackay, D. (1990). Model of organic chemical uptake and clearance by fish from food and water. *Environ. Sci. Tech.* **24,** 1203–1213.

Connell, D. W. (1988). Bioaccumulation behaviour of persistent organic chemicals with aquatic organisms. *Rev. Environ. Contamin. Toxicol.* **101,** 117–154.

Connell, D. W., and Hawker, D. W. (1988). Use of polynomial expressions to describe the bioconcentration of hydrophobic chemicals by fish. *Ecotoxicol. Environ. Safety* **16,** 242–257.

Connolly, J. P., and Pedersen, C. J. (1988). A thermodynamic-based evaluation of organic chemical accumulation in aquatic organisms. *Environ. Sci. Tech.* **22,** 99–103.

Gobas, F. A. P. C., Muir, D. C. G., and Mackay, D. (1988). Dynamics of dietary bioaccumulation and Faecal elimination of hydrophobic organic chemicals in fish. *Chemosphere* **17,** 943–962.

Gobas, F. A. P. C., Shiu, W. Y., and Mackay, D. (1987). Factors determining partitioning of hydrophobic organic chemicals in aquatic organisms. *In* "QSAR in Environmental Toxicology," Vol. II (K. L. E., Kaiser ed.), pp. 107–123. D. Reidel, Dordrecht.

Hawker, D. W. (1990). Description of fish bioconcentration factors in terms of solvatochromic parameters. *Chemosphere,* **20,** 467–477.

Schüürmann, G., and Klein, W. (1988). Advances in bioconcentration prediction, *Chemosphere* **17,** 1551–1574.

Biodegradation of Xenobiotic Chemicals

Robert S. Boethling
U.S. Environmental Protection Agency

Glossary

Acclimation Process in which exposure of a microbial population to a chemical results in more rapid transformation of the chemical than initially observed.

Cometabolism Transformation of a nongrowth substrate in the obligate presence of a growth substrate or another transformable compound.

Mineralization Conversion of an organic compound to inorganic compounds (CO_2, H_2O, and the oxides or mineral salts of any other elements in the compound).

Persistence Ability of a chemical substance to remain in a particular environment in an unchanged form.

Primary biodegradation Any biologically induced structural transformation of the parent compound that changes its molecular intregrity.

Ultimate biodegradation Any biologically mediated conversion of an organic compound to inorganic compounds and other products associated with normal metabolic processes (this term is similar to mineralization, but acknowledges that some proportion of the parent compound will always be used for synthesis of new cell material).

Xenobiotic chemical Man-made (anthropogenic) chemical that has a chemical structure to which microorganisms have not been exposed in the course of evolution.

BIODEGRADATION, *the transformation of chemical compounds by living organisms, is one of the most important processes that causes the breakdown of organic compounds. It is a major loss mechanism in aquatic and terrestrial environments, and provides the very foundation of the modern wastewater treatment plant. The eventual mineralization of organic compounds—their conversion to inorganic substances such as* CO_2 *and water—can be attributed almost entirely to biodegradation. In contrast, photochemical, hydrolytic, and other chemical processes in nature seldom if ever proceed to this extent, and products of unknown toxicity and/or persistence may be generated. Persistent intermediates can result from partial biodegradation of a compound, but this is the exception rather than the rule in the environment, where microbial populations work together to transform organic matter for growth and energy.*

I. Microbial Basis of Biodegradation

Biodegradation is not a process confined to the microbial world. Nevertheless, microorganisms (primarily bacteria and fungi) are by far the most important agents of biodegradation in nature, in terms of the mass of material transformed as well as the extent to which it is degraded. An abundance of evidence exists to show that microorganisms are responsible for the degradation of many organic chemicals that cannot be altered significantly by higher organisms. For the most part, animals simply excrete chemicals that do not fit into their normal metabolic pathways, usually after conjugation with biogenic compounds, and plants tend to convert chemicals into water-insoluble forms that can be easily stored. In contrast, microbial populations are characterized by rapid growth rates in the presence of food, high metabolic activity, catabolic versatility, and species diversity. In sheer numbers they also overwhelm, as the earth's microbial biomass has been estimated to exceed that of all animals combined.

A. Biochemistry of Biodegradation

Any of several responses may be observed when a microbial population is exposed to a xenobiotic chemical: (i) it may be essentially unreactive; that is, not toxic to the microorganisms present nor degraded by them. The potential causes of molecular recalcitrance are numerous, and will be discussed later in this article; (ii) it may have an adverse effect on the microbial population; that is, it may act as an antimicrobial; (iii) it may be degraded immediately, due to the presence of constitutive (always present) or rapidly inducible enzymes; (iv) it may be degraded by a process called cooxidation, or, more generally, cometabolism; or (v) it may be degraded only after a period of acclimation that may vary from days to weeks or more, depending upon the nature of the chemical and the environmental circumstances. Acclimation is thought to involve one or more of the following: (1) induction of enzymes necessary for degradation; (2) growth of an initially low population of the degrading microorganisms; or (3) acquisition of new catabolic capabilities by gene transfer or mutation. This list is not exhaustive, nor are all alternatives necessarily mutually exclusive.

Organic compounds and microorganisms are distributed throughout aquatic and terrestrial environments. The basic catabolic processes employed by microbial populations vary with the environmental circumstances and represent evolutionary adaptations to prevailing conditions. Nevertheless, as Gibson has stated,

> Although the reaction sequences used by microorganisms to degrade organic compounds are quite diverse, they are all directed toward[s] a common goal; that is, the production of carbon and energy for growth.

Energy-yielding metabolism is generally characterized by oxidation of organic or inorganic substrates, in which the electrons derived from the reactions are used to regenerate reduced pyridine nucleotides (NADH and NADPH). The reduced pyridine nucleotides provide needed reducing power for biosynthetic reactions, and in the case of microorganisms that carry out aerobic or anaerobic respiration, they are also coupled to phosphorylation of ADP to generate ATP, via cellular electron transport chains. In aerobic respiration the ultimate electron acceptor is oxygen (O_2), whereas in anaerobic respiration it may be nitrate or sulfate, and the end products are water, nitrite or nitrogen, and sulfide, respectively. In fermentation organic compounds serves both as substrates (source of electrons) and the ultimate electron acceptors, and reduced pyridine nucleotides and ATP are generated in "substrate-level" reactions. Thus, in aerobic environments, where O_2 is abundant, aerobic respiration is the predominant catabolic process, but in anaerobic environments any of several processes may occur. Local conditions determine which predominates.

B. Biodegradation Pathways

Our knowledge of how microorganisms bring about the degradation of organic substrates derives largely from studies of pure cultures (single strain or species of microorganism) able to grow at the expense of the selected compound. Through such studies, mechanisms of biodegradation have been defined by the characterization of intermediates and the enzymes that catalyze successive steps in the degradation pathway. Naturally occurring as well as man-made organic compounds vary greatly in structural complexity, but their utilization by microorganisms always involves the same basic strategy. That strategy is stepwise degradation to yield one or more intermediate products capable of entering the central pathways of metabolism. Thus, the study of biodegradation pathways is largely the study of the initial degradative steps.

To serve as a source of carbon and energy, an organic compound must first enter the microbial cell by passing through the cell wall and cytoplasmic membrane. A compound may enter the cell by passive diffusion or with the assistance of specific transport systems, the latter probably being the more common situation in aquatic and terrestrial environments, which are frequently characterized by very low levels of organic substrates and other nutrients. In some cases, such as with large polymeric substrates like proteins and polysaccharides, biodegradation is initiated by extracellular enzymes, the action of which yields fragments that can be transported into the cell.

Once inside the cell, the reactions that a compound may undergo are determined by its chemical structure. Almost all of the reactions involved in biodegradation can be classified as oxidative, reductive, hydrolytic, or conjugative. Examples of the first three types of reactions are shown in Table I. As of 1975, at least 26 oxidative, 7 reductive, and 14 hydrolytic transformations of pesticides had been described, and the list is no doubt much longer now.

In addition to serving as the terminal electron acceptor in aerobic environments, oxygen plays a vital role in the initial degradative steps for many organic compounds. Compounds such as linear and cyclic alkanes and benzene are essentially inert until oxygen is introduced into the structure. Under aerobic conditions, such reactions are carried out by enzymes known as oxygenases, which use O_2 as a cosubstrate. The ability to catalyze oxidations using

Table I Examples of Biodegradation Reactions

Type of reaction	Examples of chemicals subject to reaction
β-oxidation[a]: 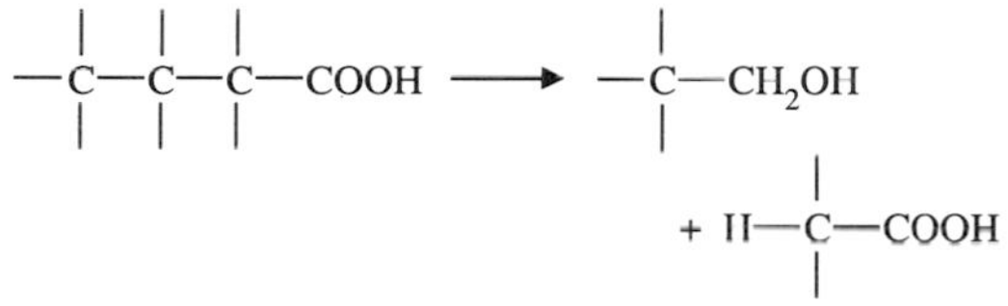 $-C-C-C-COOH \longrightarrow -C-CH_2OH + H-C-COOH$	fatty acids
Epoxidation: 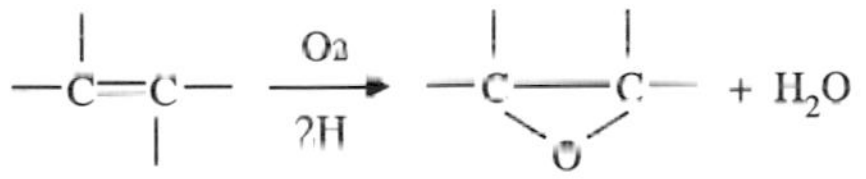 $-C{=}C- \xrightarrow[2H]{O_2} -C\overset{O}{—}C- + H_2O$	aldrin, heptachlor, styrene
Nitro reduction[a]: $R-NO_2 \longrightarrow R-NH_2$	parathion, other nitro compounds
Reductive dehalogenation: 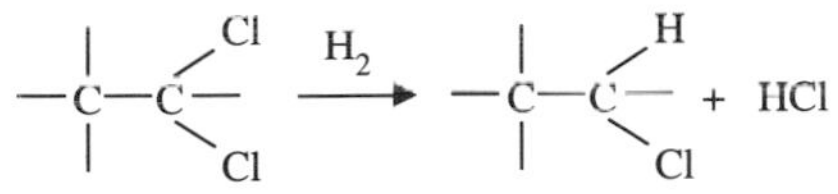 $-C-CCl_2- \xrightarrow{H_2} -C-CHCl- + HCl$	DDT, BHC
Hydrolytic dehalogenation: 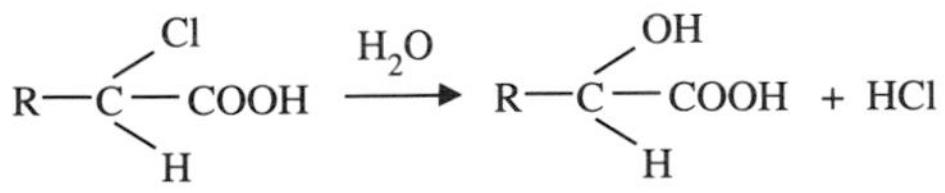 $R-CHCl-COOH \xrightarrow{H_2O} R-CH(OH)-COOH + HCl$	chlorobenzoates, dalapon
Hydrolysis (ester shown): 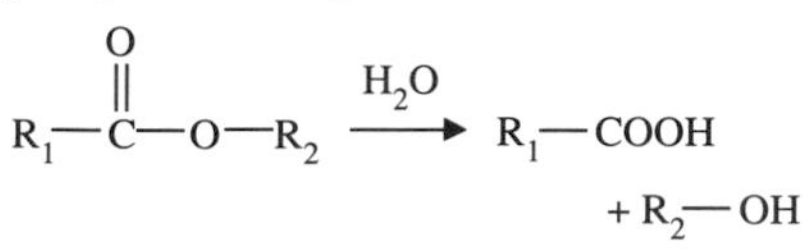 $R_1-C(=O)-O-R_2 \xrightarrow{H_2O} R_1-COOH + R_2-OH$	esters, amides, carbamates

[a] Some intermediates not shown.

molecular oxygen is unique to the microbial world. Oxidative transformations are carried out in anaerobic environments also, but here the reactions are catalyzed by hydroxylases, which apparently obtain oxygen atoms by cleavage of water. Examples of biodegradation pathways are shown in Figure 1 for aerobic degradation of cyclic alkanes, and for aerobic and anaerobic degradation of benzenoid compounds.

As we have seen, when microorganisms introduce oxygen into an organic substrate, it is usually for the purpose of initiating a pathway for complete destruction of the molecule. However, it has been known for many years that under laboratory conditions, at least, some compounds may be partially oxidized even when they cannot serve as sources of carbon and energy for the microorganisms. Leadbetter and Foster referred to this phenomenon as *cooxidation,* and the definition has since been expanded to include any transformation of a nongrowth substrate in the obligate presence of a utilizable substrate, referred to as *cometabolism*. There is little proof of the occurrence of cometabolism in nature, but it seems likely that such transformations are commonplace yet infrequently observed because the intermediate degradation products are usually degraded by other microorganisms sooner or later.

II. Effect of Chemical Structure and Properties on Biodegradation

Studies in the chemical industry, investigations in universities, and the results of environmental monitoring have shown that relatively small changes in chemical structure can appreciably alter a chemical's susceptibility to microbial degradation. These studies have resulted in a series of generalizations

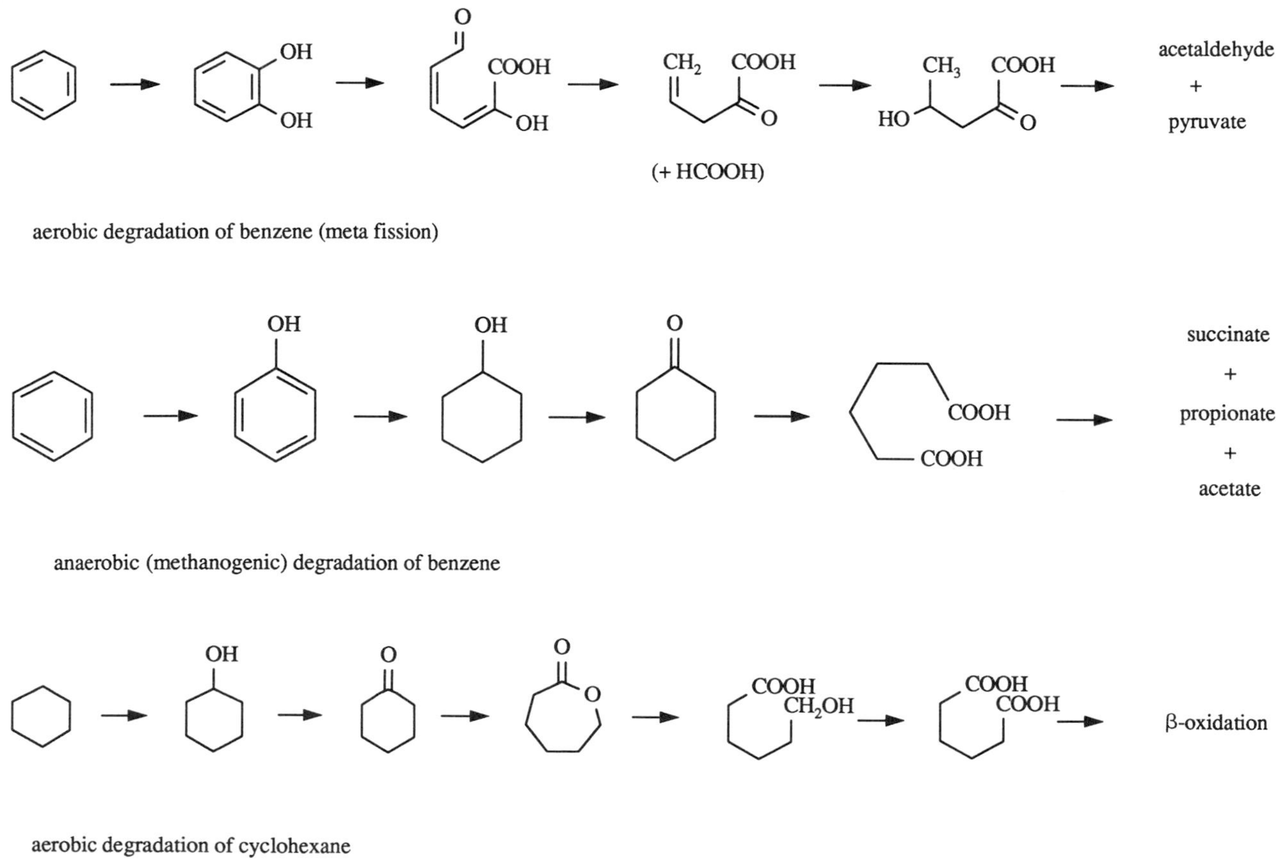

aerobic degradation of benzene (meta fission)

anaerobic (methanogenic) degradation of benzene

aerobic degradation of cyclohexane

Figure 1 Examples of biodegradation pathways. Only the major steps leading to central metabolic pathways are shown.

regarding the effects of chemical structure on biodegradability. The following summarizes the molecular features that generally confer recalcitrance.

Branching, especially tertiary (N) and quaternary (C)
Polymerization
Presence of halogen, nitro, nitroso, aryl sulfonate, aryl amino, and azo substituents (especially when there is multiple substitution)
Aliphatic ether linkages
Polycyclic residues, especially with >3 fused rings
Heterocyclic residues

For the most part, these features affect the ability of the compound to serve as an inducer or substrate, or both, of degradative enzymes and cellular transport systems.

It is well known, for example, that increasing chlorine substitution enhances recalcitrance of chlorinated biphenyls, phenols, benzoates, and many other classes of chemicals. The influence of chlorine substitution on aerobic biodegradability of PCBs is summarized in Table II. Other studies indicate that the introduction of *o*-methyl, -nitro, and -sulfonate groups, and halogens other than chlorine, may also convert a molecule that was otherwise biodegraded readily into one that is degraded only very slowly, or not at all.

As important as the type of substituent present is its position. The effect of position has been noted for many compounds including substituted biphenyls, phenols, phenoxy herbicides, benzoates, anilines, fatty acids, and aryl–alkyl compounds used as surfactants. The greater resistance to biodegradation of *meta*- than of *ortho*- or *para*-disubstituted benzenes is well known, but may apply only to degradation in soil. Other systems behave differently; differences are also observed from one chemical class to another.

Other useful generalizations are that highly branched compounds are more resistant to biodegradation than unbranched compounds, and that matter in solution is more easily degraded than insoluble material. In the latter case, the effect on biodegradability is only indirectly related to chemical structure. Despite much recent effort, the effect of solubility has proved to be difficult to characterize at the cellular and molecular levels. It is thought to involve one or more of the following:

1. the ability of the compound to reach the reaction site in the microbial cell
2. the rate of solubilization of the compound
3. the accessibility of the compound, as determined by its tendency to sorb to solids

For many xenobiotic chemicals, resistance to biodegradation is probably a function of both insolubility and molecular structure. Furthermore, it should be noted that insolubility does not necessarily imply persistence, since there are many very insoluble compounds that are easily biodegraded.

Many of the organic compounds in commerce are identical to naturally occurring substances. Of these substances, Dagley said

> . . . it is reasonable to believe at the present time that every biochemically synthesized organic compound is biodegradable.

Nevertheless, human activities have produced nu-

Table II Effect of Chlorine Substitution on Aerobic PCB Degradation

	Half-lives resulting from biodegradation			
Environment	Mono- and dichloro	Trichloro	Tetrachloro	Pentachloro and higher
Surface waters				
Fresh	2—4 days	5–40 days	1 wk.–2+ mos.	>1 year
Oceanic	←several months→		←>1 year→	
Activated sludge	1–2 days	2–3 days	3–5 days	?[a]
Soil	6–10 days	12–30 days		>1 year

[a] It is not clear how long the highly chlorinated PCBs would last under activated sludge treatment, but there appears to be no significant biodegradation during typical residence times.

merous structures never before seen or at least infrequently encountered in nature. It is worth considering, then, how it is possible that many such structures can be attacked by microorganisms, some very readily. The answer probably lies in a phenomenon referred to as "fortuitous" or "gratuitous" metabolism. This phenomenon is attributable to the fact that degradative enzymes are generally not absolutely specific for their natural substrates. Some enzymes, such as those catalyzing hydrolytic transformations, actually tend to be quite tolerant of substrate modification. From this perspective, cometabolism can be viewed simply as a consequence of the acceptance of a novel substrate by an initial, relatively tolerant enzyme, followed by the failure of a less-tolerant enzyme to accept the new product in a subsequent step.

III. Effect of Environment on Biodegradation

The susceptibility of an organic chemical to microbial degradation in nature is determined not only by chemical structure but also by environment. Biodegradability under environmental conditions can never be predicted from structure alone; it is not simply a property of the chemical. Even for biogenic compounds, biodegradation can occur only if environmental conditions are favorable.

It is useful to divide environmental factors affecting biodegradability into organism-related variables and physical/chemical variables. The following summarizes the variables thought to be most important.

Organism-related
- Concentration of viable microorganisms
- Acclimation
- Intra- and interspecies interactions (predation, mutualism, etc.)

Physical/chemical
- Concentration of xenobiotic chemical
- Temperature
- Nutrients
- Dissolved oxygen/redox potential (E_h)
- Sorption of xenobiotic chemical and/or microorganisms
- Salinity/ionic strength (pI)
- Hydrostatic pressure
- pH

It is obvious at the outset that a genetically capable microorganism must be present before biodegradation of the compound can occur. On the other hand, few studies have addressed the factors that determine the extent and maintenance of acclimation to degradation of specific organic compounds in natural populations, and almost nothing is known about the influence of interspecies and intraspecies interactions on biodegradation rates *in situ*.

Recent studies have shown that, for some chemicals, rates of biotransformation in aquatic systems are directly related to total microbial biomass. For example, microbial degradation of chemicals for which biological hydrolysis is the first step may be relatively consistent from site to site, if the biodegradation rates (more correctly, the rate constants) are corrected for the total microbial concentration. But other studies suggest that the ability of the microbial community to become acclimated to the chemical may be more important in determining biodegradation rates of many chemicals. Continuous release of a chemical should be conducive to acclimation, whereas in a spill situation or with intermittent exposure, a lower initial rate of biodegradation would be expected. Therefore, in environmental situations in which exposure to a chemical is chronic, biodegradation rates may be higher than predicted from measurements at other sites.

Although little is known in a mechanistic sense about microbial interactions, what we can say with certainty is that the complete ultimate degradation of a compound may require the participation of several microbial species or strains, since the genetic capability to mineralize the compound may not reside in a single organism. This observation has important consequences for biodegradation testing, for it suggests that testing should be performed with mixed microbial inocula or environmental samples if the purpose is environmental fate assessment. Pure culture studies are extremely useful for elucidating biodegradation pathways that *might* be environmentally important, but the biodegradative capacity of microbial communities as a general rule is much greater than that of pure cultures. Toxicity is also less likely to influence test results, because there is a greater chance that a microorganism capable of detoxifying the chemical will be present in mixed culture.

In addition to these organism-related factors, biodegradation rates may be affected by a wide variety of physical/chemical variables. Some of these, such as temperature and concentration of the test compound, have a direct influence on the reaction rate, whereas others, such as sorption of the test compound and availability of essential nutrients, exert their influence indirectly, by regulating the bioavailability of the test compound or activity of the degrading population. In still other cases, such as the effects of salinity and hydrostatic pressure, neither the mechanism by which biodegradation rates are affected nor the overall environmental significance is well understood.

Biodegradation rates are strongly affected by temperature, and the Arrhenius equation has been shown to be useful for predicting temperature effects. But deviations from the Arrhenius equation do occur; for example, evidence exists that psychrophilic ("cold-loving") microbial populations may degrade a compound more rapidly at low temperatures than would be predicted from thermodynamic considerations alone. Chemical concentrations in the environment can vary over several orders of magnitude, not only from one environment to another, but also at a given site over time. The rates of biological (enzymatic) reactions tend to be hyperbolic functions of substrate concentration, and biodegradation is no exception. Thus concentration effects similarly tend to obey a relatively simple rate law, the Michaelis–Menten relationship, with biodegradation rates being proportional to initial concentration at low concentrations, and independent of initial concentration at high levels. But there are many exceptions to this rule as well. For example, at the very low substrate concentrations typical of many aquatic environments, threshold concentrations may sometimes exist below which either degradation does not occur, or more likely, occurs more slowly than expected due to failure to induce sufficient levels of enzyme activity, or other factors. At the other end of the scale, the high concentrations of xenobiotic chemicals that can occur after spills or in waste treatment facilities may be toxic to microbial populations.

Microrganisms need a variety of nutrients in addition to a source of carbon to survive. These include the macronutrients nitrogen, phosphorus, and sulfur, and micronutrients such as essential trace elements and sometimes growth factors. If the environment is deficient in one or more of these, the biodegradation rate of a chemical may be affected. Marine environments tend to be deficient in nitrogen and phosphorus, and this deficiency is probably an important factor limiting rates of petroleum biodegradation following oil spills at sea.

As we have seen, oxygen is especially important because of its dual role as electron acceptor and cosubstrate in degradative reactions. Dissolved O_2 levels are unlikely to limit biodegradation rates in the water column in aquatic environments except under certain conditions, such as in stratified lakes. This is primarily a result of the high affinity of aerobic microorganisms for O_2. But there can be little doubt as to the importance of oxygen tension in determining the rates and pathways of biotransformation in aquatic sediments. Bottom sediments are typically stratified, with oxygen tension and redox potential decreasing with depth, often to a redox potential of -300 mV or less. Although some biotransformation pathways, such as reductive dechlorination, proceed more rapidly under anaerobic conditions, ultimate degradation of organic matter is generally much slower. Biodegradation of chemical contaminants in ground water aquifers is probably also limited by low levels of dissolved O_2 in most cases.

Sorption, salinity (ionic strength), and hydrostatic pressure are additional variables that may affect biodegradation rates under certain conditions, but their significance is poorly understood. Salinity *per se* is unlikely to be a barrier to biodegradation in marine environments, but this does not seem to be the case in estuaries, where a major portion of the microbial population is of nonmarine origin. Hydrostatic pressure may be an important factor limiting biotransformation rates in the world's oceans, where 73% of the total volume of water is found 1,000 to 6,000 m below the surface. There are several reports of the isolation of "barophilic" marine bacteria, but the limited information available on the effects of pressure on natural populations indicates that, even if barophilic microorganisms are present, microbial activity under pressure is lower than in decompressed samples. More research has been done on the effects of sorption on biodegradation, but the subject remains controversial. Solid surfaces can lower biodegradability by protecting the chemical from microbial attack, but they can also enhance degradability by increasing the microbial density or nutrient concentration at the surface, or by enhancing the buffering capacity of the medium. An example of the former effect is the protection from attack of the herbicides paraquat and diquat by intercalation into the interlamellar spaces of clay particles. In contrast, suspended sediments appear to enhance biodegradation of some chemicals in aquatic environments. This effect may be more directly attributable to a higher microbial population density in the presence of sediment.

IV. Biodegradation Testing

It is usually necessary to conduct laboratory experiments to determine biodegradability, since biodegradation is difficult to study in the field. An immense number of test methods have been developed, but in general, the type of method chosen depends upon the investigator's purpose. Most methods attempt to simulate to some degree one or more of three microbiological habitats: water (aquatic environments), soil, and wastewater (sewage) treatment. All three environments potentially include anaerobic as well as aerobic compartments. Aerobic and anaerobic biodegradation are usually quite different in terms of rates and products, and the investigator may wish to focus on one or both processes. The degree of complexity of test systems varies, depending upon the degree of realism being sought. The spectrum ranges from pure cultures to microcosms, in which the environment is effectively moved into the laboratory, as in studies using authentic sediment-water core samples.

Regardless of the purpose of testing, there are at least three important considerations in selecting or designing a biodegradation test. These are the concentration of test chemical in the test system, the source of the microbial inoculum, and the analytical method employed to detect biodegradation. For most studies designed to simulate soil or aquatic environments, concentrations should be low, since concentrations of xenobiotic chemicals normally are in the parts-per-million range or lower in the environment. Exceptions to this rule arise, of course; examples are chemical spills and some hazardous waste sites, where much higher levels may be encountered. In practice, the physical/chemical properties of the chemical, especially its water solubility and sorption capacity, and the analytical method that is selected will also influence the choice of test chemical concentration.

A wide variety of compound-specific and nonspecific analytical methods are available for monitoring the progress of biodegradation. If the investigator's interest is in determining the extent of primary degradation (loss of parent compound), chromatographic methods such as thin-layer chromatography (TLC), gas chromatography (GC), or GC coupled with mass spectrometry (GC/MS), and high-performance liquid chromatography (HPLC) are normally the methods of choice. Ultraviolet (UV), visible, and infrared (IR) absorption spectrophotometry are also frequently used, although they tend to be less sensitive and subject to greater interference than chromatographic methods.

Most nonspecific analytical methods monitor one of three parameters. These are O_2 uptake during the course of oxidation of the test chemical (biological oxygen demand or BOD), CO_2 production resulting from mineralization of the chemical, and loss of dissolved organic carbon (DOC) resulting from conversion of the organic, carbonaceous test substance to inorganic products. All three reflect ultimate degradation of the test chemical. Nonspecific methods are generally less sensitive than the chemical-specific methods listed above, and they do not yield any data on the rate of chemical disappearance *per se*. On the other hand, they have come into widespread use in screening studies because they are simple and relatively inexpensive, and yield environmentally conservative data in the sense that biodegradation is detected only if the chemical is completely degraded to inorganic products.

As we have seen, tests that employ mixed microbial populations are preferable to pure cultures if the purpose is to gain some sense of environmental biodegradation rates. Such tests may be divided broadly into two types: screening tests, in which the test chemical is incubated in a nutrient medium inoculated with some source of a mixed microbial population such as soil or activated sludge, and ultimate degradation is followed using a nonspecific analytical method; and grab sample tests, in which a compound-specific analytical method is used to monitor biodegradation in some natural matrix (e.g., sediment or soil). Grab sample methods generally can also be thought of as "simulation" tests, although the extent to which they actually simulate environmental conditions varies substantially.

Test guidelines, primarily for screening tests, have been published by various national and international organizations. The guidelines of the Organization for Economic Cooperation and Development (OECD) have been adopted by EPA's Office of Pollution Prevention and Toxics (OPPT) for use in testing new and existing chemicals under the Toxic Substances Control Act (TSCA). These and related OPPT guidelines are listed in Table III. A large body of

Table III Biodegradation Test Guidelines

Title	Code of Federal Regulations[a]	OECD guideline number
Aerobic aquatic biodegradation	40 CFR 796.3100	—
Ready biodegradability		
Closed bottle test	40 CFR 796.3200	301D
Modified AFNOR test	40 CFR 796.3180	301A
Modified MITI test	40 CFR 796.3220	301C
Modified OECD screening test	40 CFR 796.3240	301E
Modified Sturm test	40 CFR 796.3260	301B
Anaerobic biodegradability of organic chemicals	40 CFR 796.3140	—
Inherent biodegradability		
Modified SCAS test	40 CFR 796.3340	302A
Modified Zahn-Wellens test	40 CFR 796.3360	302B
Inherent biodegradability in soil	40 CFR 796.3400	304A
Simulation test-aerobic sewage treatment: coupled units test	40 CFR 796.3300	303A

[a] CFR Citation July 1, 1990.

data from screening studies has accumulated over the last 20 years, and these studies have included "round robin" evaluations of several methods; that is, series of tests using a fixed set of test chemicals and methods but conducted simultaneously in several different laboratories. In general, it can be stated that these tests yield fairly reproducible qualitative results for chemicals that are either very biodegradable or very resistant to biodegradation, but they often give variable and inconsistent results for chemicals of intermediate biodegradability.

V. Kinetics of Biodegradation

Given that biodegradation must ordinarily be studied in the laboratory, a critical aspect of environmental fate assessment is the extrapolation from laboratory data to the environment. To accomplish this quantitatively, test data *per se* are not sufficient, no matter how closely the test method approximates environmental conditions. It is also necessary to have a mathematical description or model of the biodegradation process.

The starting point most often used to derive such models is the well-known Michaelis–Menten equation, which describes the kinetics of enzyme catalysis. In its simplest form,

$$\frac{-d[S]}{dt} = \frac{V_{max}[S]}{K_m + [S]}, \tag{1}$$

where $[S]$ is the substrate concentration, V_{max} is the maximum velocity of the reaction, and K_m is $[S]$ when the reaction velocity is 1/2 of V_{max}. Equation 1 is valid for all enzymatic reactions, provided other variables that may affect the reaction velocity, including the catalyst concentration, are held constant. When substrate concentrations are much lower than K_m, which is frequently (but not always) the case in aquatic and terrestrial environments, Eq. 1 can be approximated by a first-order decay model,

$$\frac{-d[S]}{dt} = k[S], \tag{2}$$

where k is a constant of proportionality, the first-order rate constant. In fact, the biodegradation of many xenobiotic chemicals seems to follow first-order kinetics in both soil and water grab samples. Therefore, for these cases Eq. 1 appears to be a valid model of the relationship between reaction velocity and substrate concentration in mixed culture, even through it is derived from the theory of enzyme catalysis.

First-order biodegradation rate constants are typically obtained from "die-away" experiments, in which disappearance of the test chemical is followed over two or more half-lives. The rate constant is the slope, after changing the sign from negative to positive, of a log normal plot of substrate concentration versus time. The rate constant can also be obtained from V_{max} and K_m, which may be derived from heterotrophic activity measurements or by other means.

In the strictest sense, use of these kinetic parameters to predict biodegradation rates under field conditions or at untested sites requires a good understanding of the physical, chemical, and biological factors that control biodegradation *in situ*. If sufficient information were available, the effects of important environmental variables could then be quantitatively incorporated into mathematical models commonly used to predict biodegradation rates. In this way, rate constants could be "normalized" for significant variables. Of course, it is seldom possible to achieve such an understanding, much less derive algorithms. Nevertheless, some progress has been made in this area, as the following two examples illustrate. Perhaps the best known is the so-called "second-order" approach, discussed briefly in Section IV. As originally proposed, second-order rate constants, k_2, are calculated from first-order rate constants (k) obtained in die-away tests, simply by dividing k by a measure of the total bacterial population ($[B]$) in the sample:

$$k_2 = k/[B]. \quad (3)$$

As indicated earlier, this works for aquatic biodegradation of some xenobiotic chemicals but not for many others. The other example is a model developed by Walker for predicting herbicide persistence in field soil, based on laboratory studies in which herbicide loss was first order with respect to concentration, but depended upon temperature and soil moisture content. The familiar Arrhenius equation adequately describes the effect of temperature on herbicide loss. The general relationship between herbicide loss and soil moisture content is as follows:

$$t_{1/2} = aMC^{-b}, \quad (4)$$

where a and b are constants, MC is the soil moisture content and $t_{1/2}$ is the half-life for substrate disappearance, which is related to the pseudo first-order rate constant as follows:

$$t_{1/2} = 0.693/k. \quad (5)$$

VI. Biodegradability Assessment

Biodegradability is often a key issue in environmental risk assessments. Examples of where such assessments are performed include the review of industrial chemicals under TSCA, pesticide reviews under the Federal Insecticide, Fungicide, and Rodenticide Act (FIFRA), and site remediation activities under Superfund legislation. The extent to which biodegradation data are used varies, but in general, the quantitative use of such data has been limited. The review processes for new and existing chemicals under TSCA provide a good illustration of how biodegradability is assessed when data are scarce, and how this information is used in federal regulatory activities.

A. Biodegradability Assessment under TSCA

Under Section 5 of TSCA, manufacturers or importers of chemicals not on the TSCA inventory of existing chemicals—"new" chemicals—must notify EPA 90 days prior to the initiation of import or manufacture. The 90-day review period gives OPPT the opportunity to assess each chemical for potential ecological or human health effects and exposure, and to take regulatory action if necessary. OPPT receives about 2,000 Premanufacture Notices (PMNs) each year. Most of these undergo at least a screening-level review, and a significant number (perhaps 10%) proceed to in-depth review. The assessment of a new chemical's likely biodegradability in both the environment and wastewater treatment is an important part of the review process for all chemicals at the screening level, and for many undergoing detailed review.

Since TSCA does not require the development of environmental fate data as a precondition for approval, few PMNs contain biodegradation data. The result is that biodegradability must be estimated for most new chemicals. Generally, this involves a semi-quantitative analysis of a chemical's likely rates and pathways of degradation in various systems, based on structural analogy with chemicals that have been studied. New chemicals in screening-level review are scored for likely biodegradability under aerobic conditions in wastewater treatment (generally activated sludge secondary treatment) and receiving waters, and under anaerobic conditions. Both ultimate and primary degradation are considered. If sufficient toxicity concerns are expressed in the initial review, a PMN chemical may enter a detailed review. One outcome of detailed review may be rulemaking to require the development of test data.

Section 4 of TSCA is EPA's tool for gathering data on chemicals already in commerce. Under Section 4, if OPPT finds that a chemical may present an unreasonable risk or that there may be substantial exposure, it may require industry to develop the data needed to make a more reasoned evaluation of its probable effects and environmental fate. If persistence in the environment is an important consideration, test rules may incorporate a requirement for biodegradation testing. Test rules must then incorporate the methods by which data are to be developed. The OECD/OPPT test guidelines listed in Table III are frequently used to determine biodegradability of existing chemicals. In addition, some recent rules have required microcosm testing. Table IV compares the general characteristics of screening tests such as those in Table III with those of the microcosm test system most commonly specified by OPPT, as well as a natural waters die-away test that is intermediate in complexity.

B. Structure/Biodegradability Relationships

The paucity of submitted biodegradation and other fate data for new (PMN) chemicals has created substantial pressure to find better ways of estimating biodegradability. Biodegradability estimation is also an integral part of the review process for existing chemicals, since they must be screened to identify good candidates for testing. Structure/biodegradability relationships (SBR) provide a means to this end.

Quantitative SBR (QSBR), analogous to the quantitative structure/activity relationships (QSAR) that have long been important in drug design, have been developed for numerous classes of chemicals. Unfortunately, this approach has not been very useful in fate assessment under TSCA. The principal reasons are that the QSBR that exist apply only to a few very narrowly defined classes of chemicals, and that the biodegradability endpoints are very specific, and usually offer little insight into biodegradability under

Table IV Comparison of Selected Test Methods

Feature	Test method		
	Ready biodegradability tests[a]	Fate screen[b]	Eco-core[c]
OECD guideline no.	301A through E	—	—
Design principle	Artificial, mineral nutrient medium die-away	Natural waters, shake flask die-away; with & without suspended sediment	Simulation of water column plus intact bottom sediment
Quality of simulation of receiving waters environment	Poor	Fair	Good
Provides quantitative rate data?	No	Yes	Yes
Analytical method	Nonspecific (DOC, COD, CO_2)	Chemical-specific	Chemical-specific
Restrictions on solubility and volatility of test chemical?	Yes for some guidelines, no for others	No highly volatile chemicals	None
Cost	Perhaps $4,000–$8,000	Intermediate	Perhaps $50,000–$70,000

[a] Also includes Aerobic Aquatic Biodegradation, 40 CFR 796.3100.
[b] ASTM guideline E1279-89: shake flask die-away method.
[c] Sediment/water core system developed by EPA's Gulf Breeze, Florida, laboratory.

other conditions. Strictly speaking, QSBR can only be used to predict biodegradability by interpolation, and the chemicals subject to review under TSCA rarely fall into any of these classes. PMN chemicals, especially, tend to be much more complex structurally.

Another approach to SBR has seen increasing emphasis in the past several years, partly in response to the need for methods applicable to a wider range of chemical structures. This involves the application of statistical methods such as a discriminant function analysis and cluster analysis for the purpose of extracting from data on structurally diverse chemicals the factors that best account for biodegradability or persistence. For example, models have been developed for predicting whether biodegradation of untested chemicals is likely to be fast or slow, using chemical substructures and other variables as predictors. These efforts have been fairly successful by statistical criteria, but in general they are hampered by a lack of large and consistent datasets. The models are expected to be useful only for screening-level prediction of biodegradability.

VII. Emerging Technologies in Biotreatment

The central role of biodegradation in familiar processes such as agricultural composting and activated sludge secondary treatment is well established. The potential advantages of biological treatment in remediation of hazardous wastes are only now being recognized. Current treatment processes generally either relocate waste materials or confine their further spread (but leave the wastes essentially intact), or resort to expensive solutions such as incineration, which may further impact environmental quality. In contrast, biodegradation offers the potential to degrade many hazardous chemicals either completely or to harmless products that can be recycled naturally, and at relatively low cost. Growing awareness of the shortcomings of conventional treatment practices coupled with recent advances in our understanding of biodegradation in general have combined to encourage development of bioremediation alternatives.

Two of the most promising areas for near-term application involve combined and sequential treatment methods. Both approaches acknowledge the need to apply more than one treatment technology to successfully remediate many complex wastes. Combined treatment employs complementary physical/chemical and biological treatment methods. Such work has emphasized the use of KPEG (potassium polyethylene glycol) chemical treatment followed by various biological treatment processes, for the purpose of treating soils contaminated with PCBs, dioxins, and other halogenated organics. Sequential treatment attempts to take advantage of the observation that for some chemicals, the parent compound and intermediate degradation products are more rapidly degraded under different redox conditions. Examples include DDT and methoxychlor, which can be degraded much more effectively under alternating aerobic and anaerobic conditions than under one or the other alone.

Bioremediation has moved from laboratory to field in efforts to clean up contaminated ground water. Contaminated ground water is commonly anaerobic, and the rate of aerobic biodegradation tends to be limited by the rate at which oxygen is supplied to the system. Therefore, *in-situ* bioremediation of ground water can take two approaches: (i) overcoming mass-transfer limitations by supplying oxygen (for example, as hydrogen peroxide) to the water; and (ii) enhancing anaerobic processes for degradation of the contaminants (for example, by supplying alternative electron acceptors such as nitrate, and nutrients for microbial growth). In either case, effective bioremediation is limited by the geology, hydrology, and geochemistry of the subsurface environment, and a thorough site assessment is a prerequisite for selection of appropriate technology.

A substantial effort has also been mounted to develop innovative ways of exploiting conventional engineering solutions to wastewater treatment. For example, CERCLA leachates can be treated in conventional activated sludge secondary treatment, but the treatment process is often inefficient or simply involves intermedia transfer such as air stripping of volatiles. This problem may be particularly acute where anaerobic digestion is not employed and wastewater includes halogenated organics, since dehalogenation is more efficient anaerobically than aerobically. Solutions may include technology such as pretreatment in fixed-film anaerobic reactors and anaerobic fluidized beds. All such methods are designed to take advantage of several characteristics of biological treatment systems that increase the likelihood that biodegradation will occur: (i) metabolically diverse biomass; (ii) high microbial density; and (iii) presence of essential nutrients and potential cosubstrates.

The concept of using controlled environmental conditions to promote the activity of specific types of microorganisms could actually be taken much further than even the examples above would suggest. Other microorganisms such as algae, fungi, actinomycetes, and photosynthetic bacteria could be useful in removing certain types of xenobiotic chemicals from waste materials. As Kobayashi and Rittmann pointed out in 1982,

> The main concept needed for the biological removal of manmade organic compounds is that the types of microorganisms that can be useful, and their selective conditions, are diverse, extending beyond current practice.

The use of genetically engineered microorganisms to biodegrade specific wastes also holds substantial, but as yet unfulfilled promise.

Related Articles: ECOTOXICOLOGICAL TESTING; HERBICIDES; PLANTS AS DETECTORS OF ATMOSPHERIC MUTAGENS; POLYCHLORINATED BIPHENYLS (PCBs), EFFECTS ON HUMANS AND THE ENVIRONMENT; SOIL DECONTAMINATION.

Bibliography

Alexander, M. (1981). Biodegradation of chemicals of environmental concern. *Science* **211,** 132–138.

Alexander, M. (1985). Biodegradation of organic chemicals. *Environ. Sci. Technol.* **18,** 106–111.

Dagley, S. (1987). Lessons from biodegradation. *Ann. Rev. Microbiol.* **41,** 1–23.

Fewson, C. A. (1988). Biodegradation of xenobiotic and other persistent compounds: the causes of recalcitrance. *TIBTECH* **6,** 148–153.

Gibson, D. T., ed. (1984). "Microbial Degradation of Organic Compounds." Marcel Dekker, New York.

Goring, C. A. I., Laskowski, D. A., Hamaker, J. W., and Meikle, R. W. (1975). Principles of pesticide degradation in soil. *In* "Environmental Dynamics of Pesticides" (R. Haque and V. H. Freed, eds.). Plenum, New York.

Grady, Jr., C. P. L. (1985). Biodegradation: Its measurement and microbiological basis. *Biotechnol. Bioengr.* **27,** 660–674.

Howard, P. H., and Banerjee, S. (1984). Interpreting results from biodegradability tests of chemicals in water and soil. *Environ. Toxicol. Chem.* **3,** 551–562.

Klečka, G. M. (1985). Biodegradation. *In* "Environmental Exposure from Chemicals," Vol. 1 (W. B. Neely and G. E. Blau, eds.). pp. 109–155. CRC Press, Boca Raton, FL.

Kobayashi, H., and Rittmann, B. E. (1982). Microbial removal of hazardous organic compounds. *Environ Sci. Technol.* **16,** 170A–182A.

Biological Monitors of Pollution

David J. Schaeffer and Edwin E. Herricks
University of Illinois at Urbana–Champaign

Glossary

Assay Discrete measurement or analysis unit employed to examine or assess a trait. Assays are specific procedures for the examination or evaluation of a change or response, typically using a single measure or metric to assess a response.

Ecological system Subset of an ecosystem that is amenable to manipulation or experimentation, which retains the dynamic characteristics of a natural ecosystem.

Test battery Two or more assays that have similar cost, simplicity, and measurement traits.

Test system Defined measurement or analysis unit, which integrates a complex response, and is employed to examine or assess. Test systems range from simple biochemical assays to experimental manipulation of ecological systems.

Trait Particular type of measurement made on or with a living system.

Xenobiotic Chemical, compound, or material foreign to a biological system.

BIOLOGICAL MONITORING *(often contracted as "biomonitoring") is defined as the analysis of the performance of living systems structured to provide essential information for decision making. We typically distinguish between bioassays that are laboratory based tests that incorporate rigorous experimental protocols and bioassessments that are field-based analyses, ranging from the description of organisms present to the manipulation of ecosystem properties and processes, that lack strict experimental controls. Effective biological monitoring programs used for the assessment of the hazard of chemicals released to the environment requires organization of assays, test systems, and monitoring and assessment procedures in test batteries that meet specific criteria for data sensitivity, reproducibility, and validity using both experimental and descriptive analyses.*

I. Scope of Monitoring Problem

It is well known that chemicals produce some type of biological effect if concentration or duration of exposure exceed an organism's capacity to acclimate or adapt. Sometimes, as with drugs and pesticides, these biological effects are considered beneficial if the appropriate target function or organism is affected. However, any chemical has the potential of producing adverse effects. Adverse effects range from induction of enzymes (e.g., aryl hydrocarbon hydroxylase, *P*-450), tumor induction (e.g., cigarette smoke), altered behavior (e.g., lead), and pulmonary function (e.g., particulates, nitrogen and sulfur oxides), to acute mortality and chronic toxicity caused by chemicals or chemical mixtures in wastewater effluents, eventually resulting in ecosystem damage from air and water pollutants. Adverse effects can progress from a mild stress that has no lasting effect to extremely hazardous stress levels that cause mortality and degrade ecosystem integrity. Under mild stress conditions, the exposure may produce an acclimatory response that improves an organism's survival to future exposure. Under chronic exposure conditions, organism's can adapt, and may actually come to depend on low levels of a chemical in the environment. Hazardous conditions are produced when a chemical concentration is too high, causing high mortality levels, or when the release of a chemical affects nontarget organisms causing the ecosystem to sustain long-term or irreparable damage. This review is concerned with the use of biological responses at all levels of biological or ecological organization to assess the relative hazard from the use or inadvertent release of chemicals that affect life (xenobiotics). Because hazard assessments are usually performed to provide a foundation for environmental protection, we will emphasize ecological or ecological system monitoring as the most effective means of environmental protection.

An ecological system is complex. This complexity forces an emphasis on observation rather than experimentation. For example, experimentation requires reduction of system complexity to achieve control of test conditions. Controlled experiments cannot assess all of the possible conditions in a natural environment. An obvious solution is to test contaminant effects under "realistic" conditions that provide an accurate assessment of the hazard to ecosystem state, condition, and risk. This can be accomplished only if existing observational information provides a basis for experimental design.

Testing under realistic conditions or the corollary, realistic testing or evaluation of a given condition or state, is the basis for development of a hazard assessment procedure that will assess the effects of contaminant on the environment. Testing under realistic conditions can be approached in a number of ways. One approach attempts realism by quantifying and controlling multiple environmental variables in a laboratory experiment. This approach increases realism but falls short of providing realistic conditions. It is simply impossible to conduct the number of experiments necessary much less select species that adequately reflect environmental effect. A field experimental approach leaves environmental variation uncontrolled but controls the setting of the experiment. Field experimentation allows testing under realistic conditions, but study complexity, dollar costs, and possible environmental damage limit this approach. A solution is to select a hazard assessment procedure that provides different levels of realism under conditions which are progressively more realistic. This article will identify assays, test systems, and monitoring or assessment procedures that meet requirements for different levels of environmental or ecological realism in the assessment of hazards posed to the environment from the planned or inadvertent release of chemicals.

II. Test Battery Identification for Hazard Assessment Monitoring

Any chemical, physiological, or behavioral response can be measured experimentally; or a range of structural and functional parameters can be monitored for change, to assess the effect of a xenobiotic. Assays, test systems, and monitoring or assessment procedures may operate at any level of the biological/ecological hierarchy (Table I) and can be analyzed together or sequentially to provide a test battery that will effectively assess hazard. To identify a test battery, it is necessary to select assays, test systems, and monitoring or assessment procedures that provide diagnostic measurements of environmental state or condition. We distinguish between a biological organism that is used to evaluate the toxicity of an environmental sample without regard to the role of that organism in the specified ecosystem and biological tests that consider the test species' role(s) in that ecological system as the primary objective of environmental monitoring. The primary focus of test battery selection is the identification of two or more measurements with defined (or definable) relevance to the specified environmental system.

The assessment of potential hazard requires selection of a test battery that considers test realism and realistic testing. The selection of test batteries for lower levels of the hierarchy is based on criteria that emphasize experimental control while test batteries at higher levels are based on criteria that emphasize ecological or environmental realism and relevance. Thousands of assays and test systems are available to determine xenobiotic effect. However, few test systems, whether laboratory- or ecosystem oriented, have received extensive testing and validation: most have been used only a few times with a single, or small number of chemicals.

Developing a battery of test systems couples multiple measures that may range over one or more levels of the biological/ecological hierarchy. The test battery can provide a metric that integrates multiple responses produced by assays and test systems. Test batteries can also include monitoring and assessment procedures that combine assays and test systems with well-designed environmental parameter analysis to determine the state or condition of the environment. Where experimental control is possible, such as in laboratory-based toxicity testing, it is possible to assess xenobiotic effect with high ac-

Table I Biological and Ecological Organization Hierarchies

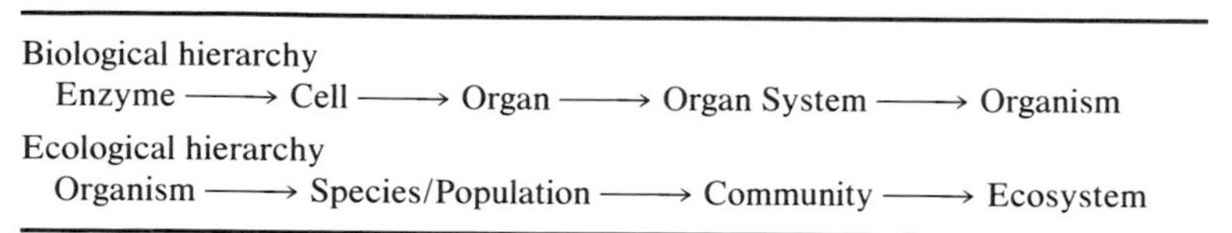

Biological hierarchy
Enzyme ⟶ Cell ⟶ Organ ⟶ Organ System ⟶ Organism
Ecological hierarchy
Organism ⟶ Species/Population ⟶ Community ⟶ Ecosystem

curacy and precision. Unfortunately, accuracy and precision may be unrelated to the application of results from a lower level to a higher level of the hierarchy. As the hierarchical complexity increases, precise and accurate relationships between cause and effect are lost. Experimentation is limited in higher levels of the hierarchy by the lack of environmental control (thus reducing the accuracy and precision of cause and effect determinations), but the application of results from a lower to a higher level in the hierarchy is improved. The determination of the environmental effect of a xenobiotic using a test battery is a compromise between experimental verification of an effect and a realistic determination of the consequence of that effect on the environment.

III. Test System Selection Criteria for Hazardous Materials

To help focus the selection of assays and test systems in a test battery, we have developed a series of descriptions of test system characteristics and suggested criteria for selecting test systems. Test systems can be based on: the media, the methods of exposure, the types of responses (chemical, physiological, or organism), the ways of measuring those responses, the effects on the organisms (such as genetic, toxic, and bioaccumulation effects), the exposure method (extracts, gases, water, or solid phase), and the exposure technique (contact, inhalation, ingestion). Test system evaluation criteria include accuracy, precision, sensitivity, variability, realism, and the related considerations of convenience, cost, and technical requirements (personnel and facilities). An assay or test system can be evaluated to determine whether it is a consensus and standard, or is tentative or developing. Assessments can follow standard practice, be designed to meet site-specific requirements, or depend on nonstandard procedures that provide essential information. Faced with this array of possible assays and test systems and the possibilities of test and test-system categorization, we have organized a method of selection of test systems based on experimental efficiency and test realism. This method of selection follows a standard tier testing protocol, but recognizes that within any tier different levels of ecological or environmental realism and relevance is possible. Hazard evaluation, presently used in environmental management and regulation, depends on tier testing using different batteries of tests in each tier. Hazard evaluation tiers include a *Screening Tier,* which employs simple and relatively inexpensive tests to make a preliminary determination of hazard/risk. Screening is followed by the *Predictive Tier,* which employs more complex, longer-term testing to predict environmental or ecological effects produced by contaminant release. *Confirmatory Tier* testing involves environmental/ecological analysis to determine actual effect, and assess prediction accuracy. Monitoring tier analyses require long term surveillance to identify low level or unanticipated long term effects.

When testing for environmental realism or relevance, the analysis requires an expansion of tier testing. There are three levels of analysis within each tier: At the highest, or *Definitive Level,* is quantification of properties or processes that exactly define the state or condition of an ecosystem or the environment. Although ecosystem characteristics can be quantified, the exact definition and indicators of state and condition are still being defined by ecologists. Definitive Level analyses are seldom obtained for any tier. The next level of analysis, the *Classification Level,* requires quantifying properties that are being, or that could be, used to classify state, condition of an ecosystem. Classification may be based on criteria developed to meet regulatory requirements or more arbitrarily on perceptions of environmental quality. Classification Level analyses may include measurements made through time, or the use of test systems which have appropriate experimental controls. The lowest and most common level, Predictive, involves analyses intended to predict Classification or Definitive Level responses in an ecosystem. *Predictive Level* testing is usually comprised of laboratory and field-oriented test systems which have unknown or uncertain relationships to ecosystem indicators, or are not used in developing relationships between trophic levels in an ecosystem. Most currently available test systems operate at the Predictive Level.

It is reasonable to organize ecological system diagnostic tests into batteries, and batteries into tiers. We suggest that the tiers be the response variables found to be changed in stressed ecosystems: nutrient pool, primary productivity, size distribution, species diversity, and system retrogression are critical to ecosystem level hazard assessment. Batteries within tiers are organized around characterization of exposure, response, and recovery of ecosystems from anthropogenic stress, and "uncertainties." Exposure considers fate, transport, and environmental modification; duration, frequency, intensity, and novelty of exposure; and differential exposure regime within an ecosystem. Response considers effects on components of ecosystems; effects on processes of ecosystems; relevant scales and characterization of response; endpoints for response characterization; and relevant indicators for endpoints. Recovery considers indicators for components and processes; irreparable harm and/or the ability to adapt; resilience and homeorhesis; scales of physical and biotic renewal processes. Uncertainties include variability in exposure; variability between ecosystems and between ecological systems within an ecosystem; extrapolation across types of stresses; and extrapolation across types of ecosystems.

The recognition of levels of analysis within tiers simplifies the organization of data collection activities to structure a hazard assessment using biological monitoring procedures. Test systems can be selected to provide experimental validation of effect. Further, the applicability of laboratory experimentation to ecosystems can be clearly defined through use of tier testing and nested levels of analysis. It is also possible to develop biomonitoring programs employing bioassessment approaches that take advantage of natural or human-caused manipulations of ecosystems. Careful analysis of bioassessment data from either planned or unplanned manipulations of ecosystems can transform a simple descriptive study into a definitive analysis of ecological consequence. Experimentation on a small scale or manipulation on a large scale is, in turn, the basis for hypothesis development. Although hypotheses can not always be rigorously tested through bioassessment, the outcome of similar experimental manipulations can be compared if the data collection effort adheres to defined quality assurance procedures.

Tier testing, which incorporates levels of analysis in each tier, provides an organized approach to data collection in ecosystems. Tier testing also makes it possible to identify scientifically valid experimental or manipulative procedures for ecosystems. Tests of data sensitivity, reproducibility, and validity, using both experimental and descriptive analysis, are possible in properly organized and planned biomonitoring programs.

The selection of any test or test system is based on criteria of accuracy, precision, sensitivity, and variability. The most common means of meeting these criteria is selection of standard methods, or following a standardized methodology. When selecting test systems for hazard assessment or analysis, however, it may not be possible to use standard methods. Because any biological or ecological response can be used to measure the effect of a xenobiotic, the best or most appropriate measure may not be standardized. In fact, the likelihood of encountering a standardized test system declines with level in the hierarchy. For example, it is possible to conduct repetitive analyses at multiple laboratories to standardize an enzyme assay or even a toxicity test with a species from a defined stock. In these tests, environmental conditions are controlled and an expected set of responses can be identified for measurement. When conducting a hazard assessment at higher levels of ecological organization it is not possible to rigorously control experiments and repetitive analyses are either fiscally or practically impossible. Further, it is often extremely difficult to predict the exact response. Flexibility must be maintained in sampling procedures and analysis methods selected to ensure that all effects are identified and quantified. When effects may change from season to season, in some instances from day to day, standardization may lead to misinterpretation if not a misunderstanding of cause, effect, and realistic consequence. At higher levels of the hierarchy, an emphasis should be placed on standardization of methodology with an emphasis on reproducibility rather than precision to develop data with an acceptable variability. Further, sufficient flexibility must be maintained in sampling and analysis methodologies to take advantage of naturally changing conditions that can reveal relationships between xenobiotic exposure and an identified effect.

IV. Test System Selection Decision Trees

The selection of assays, test systems, or monitoring assessment procedures follows an organized procedure that uses a set of decision trees we have developed for use with any environmental contaminant. The selection protocol assumes the user has a well-developed background in environmental toxicology, ecotoxicology, biomonitoring, or ecology. Protocol application will be most effective when sufficient existing data is available to support decision tree analysis and identification of important or critical issues on the site to be tested. The emphasis in the following protocol is to lead the user to test systems specific for the ecosystem being analyzed.

The test system selection process begins with the use of a general decision tree. This decision tree was developed to identify the basic data and information requirements for test system selection. Two general approaches to test system selection are then possible. The first follows from hazard assessment procedures developed over the past ten years. It focuses on a characterization of the chemical being released and its environmental fate. A knowledge of environmental fate coupled with screening or predictive analysis of concentration/effect relationships provides the basis for hazard assessment. In this application, this approach is used to identify fate, then determine what organisms are at risk. Once organisms at risk have been identified, test systems for analysis are selected. Using relationships between trophic levels, specifically identified connections between organisms such as competition or predator–prey relationships, ecosystem effects are predicted. The major limitation to this hazard assessment approach is that it is most powerful when only a single chemical is being assessed. In addition, the reliability of the approach declines rapidly when confounding physical, chemical, or biological factors are important.

The second general approach to test system selection is based on a comprehensive analysis of the ecosystem in question. We term this the bioassessment approach. The major requirement is that a reference area, unaffected by contaminant release, is available for comparison of descriptive studies or experimental manipulations. The primary objective of initial bioassessments is to identify ecosystem properties and processes, or resident species/populations or communities that are affected by contaminant. This approach depends on observation of an effect and then uses an understanding of the function and structure of the ecosystem to select appropriate measurements and trait/metrics for analysis. The test systems identified through this process are evaluated in light of demonstrated interlevel or connectance relationships. The use of test systems will typically involve experimental manipulation using unaffected areas as references.

A. General Decision Tree

1. Does contaminant release involve a high magnitude stress?
   NO—Continue with item 2.
   YES—Are there confounding physical or chemical impacts?
      NO—Proceed with selection of test systems suitable for measurement of acute contaminant effects if stress duration is short, otherwise assess chronic contaminant effects using Decision Tree 2—evaluate more than one trophic level (go to Decision Tree 1).
      YES—Continue with item 3.

2. Does contaminant release involve a low magnitude stress?
   NO—Validate lack of stress on ecological system by comparison with areas not subject to contaminant release (follow steps in Decision Tree 2 to complete validation).
   YES—Are there confounding physical or chemical impacts?
      NO—(a) If duration is short term, conduct general bioassessment for a period not shorter than the life cycle of the longest lived organism in the ecological system—compare results with reference system. (Go to Decision Tree 2.)
      NO—(b) If duration is long term proceed with bioassessment and compare results with reference site to determine changes in ecological system state or condition and perform a damage assessment.
      YES—Continue with item 3.

3. Are confounding effects known?
   NO—Compare a range of measures. See item 5 or go to Decision Tree 2
   YES—Are confounding physical or chemical effects natural or anthropogenic in origin?
      If natural—Are confounding effects episodic or random?

Episodic—Design bioassessment to assess cumulative effects and conduct specific analysis to determine organism adaptive or acclimatory capacity to natural stress.
Random—see item 4.
If anthropogenic—Are confounding effects subject to management or control?
NO—see item 5.
YES—Exercise management and control and perform a series of field manipulations to assess contaminant effect and effects of other physical or chemical stress.

4. Do random confounding events occur frequently?
NO—Are there characteristics of the ecological system that provide insight into the probable frequency of confounding events?
NO—see item 5.
YES—Based on study of probable frequency, design monitoring program to measure confounding events to establish natural variability.
YES—Based on study of probable frequency, design a monitoring program to measure confounding events and establish natural variability

5. Establish analysis program that independently defines identified confounding effect impact; factor known impact or effect into analysis of contaminant effect.

B. Decision Tree 1

1. Initial Data Requirements

Contaminant characterization: general characteristics (water solubility, vapor pressure, rates of volatilization, hydrolysis, photolysis, sorption–desorption, partition coefficients, boiling and melting points, biodegradation rate, conjugation rate, leaching, dissipation characteristics, molecular structure); deposition rate (vegetative surfaces, soil, water); media translocation/accumulation; expected transformation products (due to photolysis, microbial degradation, polymerization, conjugation, oligomerization, dimerization). Contaminant fate, contaminant release quantity, and expected mode of toxicity should also be determined.

1. Is sufficient data available to determine the fate and exposure concentration of the contaminant?
NO—Develop information.
YES—Identify the primary receptor (vegetation, soil, water) of the comtaminant—go to step 2!

2. Is the material accumulating?
NO—Determine the half-life of the contaminant and its degradation or transformation products.
YES—Determine accumulation and degradation data using *2a* through *2e*.
2a. Is half-life on the order of months to years?
NO—Confirm degradation rate under use conditions of site of release.
YES—Verify no bioaccumulation in organisms typical of release site.
2b. Is degradation due to chemical or physical alteration?
NO—Select test system to determine biodegradation requirements and determine contaminant assimilation capacity.
YES—Determine half-life and mechanisms of transformation; determine if half life is altered by physiographic conditions.
2c. Are degradation or transformation products known?
NO—Determine primary and secondary degradation and transformation products and begin analysis with step 1 (above).
YES—Continue analysis starting with step 1 (above) for each end product.
2d. Is the abiotic accumulation site likely to produce exposure to living systems?
NO—Conduct periodic monitoring to assess contaminant levels.
YES—Select test system and determine effect concentration.
2e. Can biotic accumulation be traced through a food chain?
NO—Conduct testing of species accumulating contaminant to determine effect and consequence of accumulation.
YES—Conduct testing of species in food chain accumulating contaminant to determine effect and consequence of accumulation.

Primary decision point—In the process of developing data necessary for completion of Decision Tree 1, test systems are required in several steps. The test systems for use in this analysis should be regularly used aquatic or terrestrial test systems that have a well-developed comparative database to maximize interpretative strength (see Tables II and III). When data are required for species from site ecosystems, those organisms should be selected based on their ecological or management importance. If this decision tree is completed to this point sufficient data on chemical characteristics, fate, and environmental effect will have been collected at this point to determine species-specific effects. It is now necessary to conduct further analysis with more complex systems and/or select test systems that allow connection between hierarchical trophic levels. The protocol for developing hierarchical relationships is presented following Decision Tree 2.

C. Decision Tree 2

1. Initial Data Requirements

General ecological system characterization (go to Note 1 for general procedures); determine system state or condition variability, expected oscillations in system components, expected structure, expected function. Identify contaminant and determine primary and secondary degradation or transformation products.

1. Is the contaminant being assimilated by the ecological system?

NO—Assess stress levels and determine stress effects on system structure and function (go to Note 2 for stress level determination procedures).

1a. Is stress evident in either structure or function?

NO—Monitor system state and/or condition to assess potential for accumulation to a threshold or onset of long-term effect.

YES—dentify critical ecosystem characteristics subject to stress and prioritize if possible.

1aa. Will physiographic factors alter the prioritization of critical factor effect listing?

NO—Go to Note 3; select traits and or metric(s) to assess stress and response for critical ecosystem characteristics as prioritized.

YES—Reprioritize critical ecosystem characteristic list then review critical characteristic/measurement table and select traits or metric(s) to assess stress and response for critical factors as prioritized.

YES—Monitor assimilation rate to determine if contaminant is exceeding assimilative capacity and accumulating in the system.

2a. Is the half life of the contaminant greater than the generating time for key species in the ecological system?

NO—Monitor contaminant levels to determine if trends indicate change.

YES—Determine effect of contaminant levels on key species—emphasis should be placed on measures of function.

Note 1: Methods for the critical assessment of ecosystem state and condition are readily available and the design of bioassessment programs should at a minimum assess ecosystem organization, function, state descriptors and habitat, and should meet the following criteria:

Criterion 1—The measure must be biological or have proven relationships to biological/ecological effect.

Criterion 2—Measure must be amenable to application at other trophic levels or reflect effects at other levels of the biological/ecological hierarchy.

Criterion 3—Measure must be a sensitive monitor, appropriate to the issue at hand.

Criterion 4—Range of measure must be suitable for the application.

Criterion 5—The measure must be reproducible and precise.

Criterion 6—Variability of the measure must be low. If variability is high the distributional characteristics of the data must be known.

Note 2: Stress assessment is conducted through comparison of quantitative measures of ecosystem properties and process with a reference, unaffected ecosystem.

Note 3: The proposed methodology for selection of traits to be measured uses Tables II and III to identify high priority critical characteristics.

As a companion to the decision trees, we have prepared Tables II and III to assist in the development of hazard assessment procedures. Table II provides a compilation of the measures commonly used to quantify environmental characteristics, grouped by application to different levels of the ecological hierarchy. For each measurement, we have identified the measurement potential and accuracy that can be expected in field or laboratory applications. Measurement reproducibility, the incorporation of estimates of expected variability and capacity for exact result replication, is provided, as is our assessment of the ecological relevance of the measure. Table III provides a summary of the availability of test systems for the assay of important ecosystem, community, and individual or population level metrics. When selecting a test battery, the users of the decision trees should consider both the availabil-

Table II Measures Used to Quantify Environmental Characteristics for Pollution Monitoring

		Meas. potential		Meas. accuracy			
	Units	Field	Lab	Field	Lab	Repro.	Ecol. rel
Individual level measures							
1. Fitness—the individual's relative contribution of progeny to the population	N	M	M	NBC	M	L	L
2. Disease—a destructive process in an organism with a specific cause	N	H	H	H	H	H	L
3. Mutation—a change in the genetic make-up of an organism	B/G	M	H	M	H	M-H	L
4. Reproduction—the capacity of an organism to produce progeny	N	M	H	L	H	L	M
5. Physiology—the organic processes and phenomena of an organism	B/P	M	H	L	H	L	L
6. Acclimation—capacity of an individual to modify life functions within the range of fixed genetic capability	B/P	M	H	L	H	M	M
7. Individual behavior—the response system of an organism	O/D	M	H	L	M	L	M
Measures for populations and below							
8. Intraspecific behavior—social response systems between organisms of the same species							
8.1 territoriality—social pattern of space utilization	O/D	H	H	M	H	H	M
8.2 dominance—rank-ordered systems which determine individual access and priority to natural resources	O/D	M	H	M	H	M	M
8.3 reproductive—specific sets of behavior associated with reproduction	O/D	M	H	M	H	M	M
9. Epidemiology—the pattern of disease in a population	O/N	L	H	L	H	L	H
10. Genotypic variation—the characteristic variation of genetic make-up in a population	B/P	L	H	L	H	M	TR-L
11. Phenotypic variation—the characteristic variation in the physical characteristics of a population	O/P	M	H	L	H	M	TR-S
12. Reproduction—capacity of a population to sustain itself through the production of progeny by individual members of the population	N						
12.1 age/size specific mortality	N	L	H	L	H	M	H
12.2 age/size specific survivorship	N	L	H	L	H	M	H
12.3 age/size specific fecundity	N	L	H	L	H	M	H

13. Physiology—the processes, activities, and phenomena of living systems							
13.1 metabolism	B/P	M	H	L	H	M	M-H
13.2 bioaccumulation	A	M	H	M	H	M	H
13.3 translocation	A/P	L	H	L	H	M	H
14. Adaption—modification of genetic scope of a population through change in genetic make-up of individuals	B/P	L	H	L	M	M-L	TR
System level factors							
15. Interspecific behavior—response systems between species (limited in this analysis to competitive interactions)							
15.1 predation—capture of live animals for food	O/N	M	M	M	H	M	H
15.2 competition—mutual utilization of a limited resource	O/N	M	M	L	M	M	H
16. Decomposition—the mineralization of fixed carbon compounds							
16.1 biomass loss	N	M	H	M	H	M	H
16.2 total gas production	A/N	L	H	L	H	L	H
16.3 nutrient and mineral residuals	A/N	M	H	M	H	M	M
17. Production—the fixation of carbon							
17.1 gross productivity—total production/area/time	A/N	M	H	M	H	M	H
17.2 biomass accumulation	N	H	H	M	H	M	H
17.3 total O_2 release	A/N	M	H	M	H	M	M
18. Recovery—the process where an ecological system or system characteristic returns to a normal or usual state following disturbance or displacement	O/QI	M	H	L	H	L	H
19. Resilience—the degree, manner, and phase of restoration of initial structure and function of an ecological system following disturbance	O/QI	L	H	L	H	L	H
20. Resistance—the capacity of an ecological system or system characteristic to resist disturbance or displacement of structure and function	O/QI	L	H	L	H	L	H
21. Connectivity—description of all linkages between elements of an ecological system; quantification of permanent or continuing relationships between species	O/N	M	H	M	H	M	H
22. Featured/Indicator species—a species indicative of community or ecological system condition or a species judged important based on economic or political criteria	O/QI	H	H	M	H	M	M
23. Keystone species—a species judged to play a role in maintenance of ecological system structure and/or function greater than would be expected from its numerical abundance	O/QI	L	H	L	H	L	H
24. Patterns of succession—the sequential establishment of species in an ecological system	O/QI	M	H	M	H	L	H

(*continues*)

Table II (*Continued*)

	Units	Meas. potential		Meas. accuracy		Repro.	Ecol. rel
		Field	Lab	Field	Lab		
25. Guild theory—the organization of species into groups of similar function, habitat, or process							
25.1 composition	O/N	H	H	M	H	M	M
25.2 relationships	O/Ql	M	M	L	M	L	H
25.3 interactions	O/Ql	L	M	L	M	L	H
26. Species diversity—the distribution of individuals among a series of identifiably discrete units	N	M	H	M	H	M	M
27. Vegetative structure—the physical dimensions of plant species asssessed for area and height	N	H	NA	M	NA	M	H
Major abiotic elements							
28. Soil mediation of chemical effects							
28.1 nutrient and mineral retention	A/N	M	H	H	H	M	M
28.2 leaching processes	A/N	M	H	H	H	M	M
29. Cover/habitat definition							
29.1 physiography	D/N	H	NA	M	NA	M	L
29.2 structural diversity	D/N	M	H	M	H	M	H
29.3 chemical composition	D	M	H	M	H	M	M

Measurement units: N, numerical; B, biochemical; G, genetic; O, observational; D, descriptive; A, analytical; Ql, qualitative.

Measurement potential/accuracy: L, low; M, medium; H, high; TR-L, time related—long; TR-S, time related—short; TR-V, time related—variable; NA, not applicable.

Ecological relevance: L, low; M, medium, H, high.

Table III Availability of Assays and Test Systems to Determine the Response of the Environment to Pollutants

	Bacteria	Fungi	Bryophytes and algae	Vascular plants	Protozoa	Mollusca	Crustacians & insects	Earthworms	Nematoda	Turbellaria	Amphibians	Reptiles	Birds	Fish	Mammals
Ecosystem level[a]															
Biomass production	X	X	X	X	X	X	X	X	X	X	X	X	X	X	X
Nitrogen fixation	X	X	U	U	NS	NS	NS	X	U		NS	NS	NS	NS	NS
Soil retention	NS	NS	NS	NS	NS	NS	NS	NS	NS	NS	NS	NS	NS	NS	NS
Community level[b]															
Production/ decomposition	X	X	X	X	X	X	X	X	X	X	X	X	X	X	X
Guild structure	X	X	X	X	X	X	X	X	X	X	X	X	X	X	X
Predator/prey	NS	NS	NS	NS	X	X	X	X	X	X	X	X	X	X	X
Vegetative analysis	NS	NS	X	X	NS	NS	NS	NS	NS	NS	NS	NS	NS	NS	NS
Diversity	X	X	X	X	X	X	X	X	X	X	X	X	X	X	X
Genetic alteration	U	U	U	X	X	U	U	U	U	X	U	U	U	U	X
Bioaccumulation	X	X	X	X	X	X	X	X	X	X	X	X	X	X	X
Individual/population level															
Reproduction	X	X	X	X	X	X	X	X	X	X	X	X	X	X	X
Life table/production	NS	NS	X	X	X	X	X	X	X	X	X	X	X	X	X
Disease	X	X	X	X	X	X	X	X	X	X	X	X	X	X	X
Mutation	X			X				X			X	X		X	X
Carcinogenesis								X		X				X	X
Teratogenesis										X		X	X	X	X
Behavior	NS	NS	NS	NS	NS	U	X	U	X	X	X	X	X	X	X

Abbreviations: X, indicates a testing or analysis method exists for one or more species in a taxonomic group; U, indicates uncertain availability of testing or analysis method; NS, indicates lack of direct ecological relevance for specified measurement.

[a] Assessments for biomass production, nitrogen fixation, and predator/prey relationships were developed by an expert group.

[b] Recovery is also appropriate at this level. It is possible to measure colonization following stress; measurement of recovery is problematic because adequate definition for recovery does not exist.

ity and the likelihood that an assay or test system has sufficient accuracy and reproducibility to meet environmental relevance criteria.

V. Biological Monitors of Pollution

Because all living systems respond to changed conditions of the environment by some biochemical, physiological, or other alteration in function, it is possible to identify a virtually unlimited number of candidate assays, test systems, or monitoring and assessment procedures for use as biological monitors of pollution. Advances in technology provide the means for the development of bioprobes that use a living system response to detect both the presence and magnitude of pollution. Advances in analytical and statistical techniques also make it possible to identify and interpret cause and effect at all levels of the biological/hierarchy. Providing a comprehensive review of assays, test systems, and monitoring and assessment procedures is beyond the scope of this article. Providing an organized procedure to select and use assays, test systems, and monitoring and assessment procedures is possible. Biological monitors of pollution are identified and selected based on their suitability for the measurement desired and evaluation criteria that include accuracy, precision, sensitivity, variability, realism, and related considerations of convenience, cost, and technical demands (personnel and facilities). The use of biological monitors of pollution can include controlled laboratory experimentation or uncontrolled field assessments, each meeting different selection and realism criteria. Through use of the decision trees we have included with this paper, it should be possible to optimize the selection and application of biological monitors to provide efficient and effective means of assessing the hazard of the planned or inadvertent release of chemicals to the environment.

Related Articles: ENVIRONMENTAL MONITORING; MONITORING INDICATORS OF PLANTS AND HUMAN HEALTH FOR AIR POLLUTION; PLANTS AS DETECTORS OF ATMOSPHERIC MUTAGENS; PROTISTS AS INDICATORS OF WATER QUALITY IN MARINE ENVIRONMENTS.

Bibliography

American Public Health Association (1990). "Standard Methods for the Examination of Water and Wastewater" 17th Ed. American Public Health Association, Washington, D. C.

Hellawell, J. M. (1986). "Biological Indicators of Freshwater Pollution and Environmental Management." Elsevier Applied Science Publishers, London.

Herricks, E. E., and Schaeffer, D. J. (1985). Can we optimize biomonitoring? *Environ. Manage.* **9**(6), 487–492.

Herricks, E. E., and Schaeffer, D. J. (1987). Selection of test systems for ecological analysis. *Water Sci. Technol.* **19**(11), 47–54.

Kelly, J. R., and Harwell, M. A. (1989). Indicators of ecosystem response and recovery. *In* "Ecotoxicology: Problems and Approaches," (S. A. Levin, M. A. Harwell, J. R. Kelly, and K. D. Kimball, eds.) pp. 9–41, Springer-Verlag, New York.

Novak, E. W., Porcella, D. B., Johnson, K. M., Herricks, E. E., and Schaeffer, D. J. (1985). Selection of test methods to assess ecological effects of mixed aerosols. *Ecotoxicol. Environ. Saf.* **10,** 361–381.

Rapport, D. J., Rieger, H. A., and Hutchinson, T. C. (1985). Ecosystem behavior under stress. *Am. Nat.* **125,** 617–640.

Schaeffer, D. J., Herricks, E. E., and Kerster, H. W. (1988). Ecosystem Health. I. Measuring ecosystem health. *Environ. Manage.* **12(4)**, 445–456.

Schaeffer, D. J., Cox, D. K., and Deem, R. A. (1987). Variability of biological test systems. *Water Sci. Technol.* **19,** 39–45.

Sheehan, P. J., Miller, D. R., Butler, G. C., and Bourdeau, P. (1984). "Effects of Pollutants at the Ecosystem Level." John Wiley and Sons, New York.

Biotechnology: Occupational Health Issues

Alan M. Ducatman
West Virginia University

Glossary

Biotechnology Nonspecific term for the application of molecular biological techniques to industrial processes. Genetic engineering is often implied.

Electrophoresis Movement of particles in an electric field toward one or the other electric pole, dependent on charge and polarity. Biologists use this to separate materials such as proteins.

ELISA (enzyme-linked immunosorbent assay) Binding assay using a ligand, often an antibody linked to an enzyme. Substrate is added and the amount of enzyme bound can be quantified. The process is often for the detection of infectious diseases. The process works when there is an antibody of high avidity and low cross-reactivity.

Erythropoieten Human protein containing sialic acid that stimulates red cell formation in bone marrow.

High pressure (high performance) liquid chromatography Technique for separating solute compounds in which the mobile phase is a liquid and the stationary phase is a polar absorbent to which an array of functional groups may be bonded. Retention (position) of solutes in the process depends on their relative affinity for the liquid and stationary phases.

Recombinant DNA Describes the deliberate joining of two or more pieces of DNA in a new pattern. The usual intention is the creation or repression of a specific gene product.

Transgenic Insertion of a foreign gene into a reproducing cell line, often within a live organism.

L-Tryptophan 2-amino-3-indoletropionic acid. Tryptophan is an essential amino acid; the body is not capable of synthesizing it from other amino acids for the creation of proteins and physiologically active amines such as serotonin, which influences the sleep cycle.

INDUSTRIAL MICROBIOLOGY began with fermentation technology and is older than recorded history. Modern industrial microbiology, or biotechnology, is an outgrowth of several discoveries of the past two decades. Biotechnology relies on DNA techniques which manipulate the molecular genetics of organisms for human benefit. Biotechnology ventures feature recombinant and hybridoma technology, as well as enzyme engineering or other protein engineering. These new technologies have accelerated the pace and inroads of industrial microbiology in health care diagnosis and treatment, in agriculture and food production, and even in the petrochemical industry. Research and industrial investments amount to billions of dollars and are spread over more than one thousand entities around the world. Production involves a strategic marriage between molecular genetics and process engineering.

I. Introduction

Recombinant biology is the assembly of DNA from selected organisms using restriction enzymes. Desired DNA sequences are introduced into cell genomes for the purpose of making specific products. It is now possible to create quantities of materials such as immunogenic viral subunits and rare mammalian polypeptides that were unavailable or previously constrained by impurity and cost. For example, somatostatin was previously available in milligram quantities from millions of sheep brains; it is presently available from a few liters of recombinant *E. coli* broth. Similarly, amounts of erythropoieten purified from human urine were inadequate to create a commercial pharmaceutical; sales of the recombinant product are already greater than $300 million.

Central to the recombinant effort is DNA sequencing, or genetic mapping. This effort often involves reproducing gene segments in great quantities, now done through polymerase chain reactions. In addition, common chemical extractions are required to make specific identifications. These include column and gel electrophoresis, as well as fluorescent technologies involving exposure to ultraviolet light and mutagenic chemicals.

Hybridoma technology descends from a technique for fusing mouse myeloma with antibody-secreting cells. Somatically fused cells can also have fused nuclei, so that genetic material from both organisms is reproduced. Fused cell lines, including human cells if desired, that are relatively hardy, rapidly reproducing, and engineered to produce specific products in high volumes can be selected. The promise of hybridoma technology is obvious. The less obvious risk is that long-lived cell lines are a target for infections which may change their characteristics in undesirable ways.

The impacts of biotechnology on health care illustrate the scope and speed of this industrial revolution. Four types of biotechnical advances are altering the practice of medicine: diagnostic tests for infectious diseases, diagnostic tests for genetic diseases, mass production of therapeutically important proteins, and, ultimately, gene therapy. Specific examples include an array of monoclonal antibodies for diagnostic work, subunit vaccines such as hepatitis B (produced in yeast for immunization of humans), and physiologically active peptides such as human insulin, human and bovine growth hormones, tissue plasminogen activator (TPA), α-interferon, erythropoieten, fertility-related hormones, and a variety of growth- and colony-stimulating factors. These products are often unique and are typically easier to manufacture, safer to use, and far less expensive than competing products of conventional technologies. Similar production advances will soon alter the agricultural and chemical industries.

The purpose of this article is to demonstrate that these advances are not without potential problems for workers. A variety of biological, chemical, and physical hazards are attendant upon the daily activities of biotechnologists. This article describes the biotechnology facility, its activities, health and safety hazards, as well as appropriate medical and industrial hygiene responses.

II. Facility and Containment Considerations

Design considerations begin with site selection. Adequacy of water and sewer hook-ups, consequences of sewer and stack emissions, and neighborhood responses to the presence of toxic waste surface haulers must be understood by those wishing to avoid altercations, inconvenience, or even irretrievable investment losses.

There are five space functions in a biotechnology research facility: laboratory, laboratory support, building support, administrative, and meeting. Strategic safety planning must account for adequate operations space. For example, protein chemistry requires more bench and aisle space than tissue culture. Fume hood density will also be affected by operational requirements for protein chemistry. Architectural plans must account for integrated life safety designs. These range from emergency egress to emergency power and shutoffs. They include surface finishes which allow for pest control, spill control, decontamination, and cleanliness. Adequate spaces for animal handling and waste handling are critical.

The following problems are noted consistently. Biologists want office work spaces within their laboratories but fail to realize the health and regulatory consequences of work spaces which encourage inappropriate attire and even eating in laboratories. Good separation between functions is critical, even at the cost of space occupied by walls. Some architects are inappropriately concerned with the visual imprint of support spaces, such as waste handling areas and accessible hood chases. Design planners must be realistic about unavoidable space and operational costs associated with safety aspects of support services, even over aesthetic desires. Cost concerns can cause frequent and potentially unsafe modifications in building plans. It is important that users (rather than owners or architects) clearly describe their probable future activities and collaborate directly with safety professionals so that design shortcomings do not preclude necessary functions.

Facilities should accommodate a high degree of containment even if the present work is relatively safe. Investigators can be adamant that they will never need a containment facility, until a new research or production opportunity suggests one is needed—tomorrow. Levels of biological containment are based on microorganism hazard and the selection of appropriate control techniques.

Biosafety level 1 work with nonhazardous organisms can be done on the open bench. Even so, there are decontamination requirements and standard prohibitions against eating and mouth-pipetting in the laboratory. Liquid wastes containing recombinant material from level 1 facilities are decontaminated before disposal, although the rationale for this activity relies more on custom than hazard. Biosafety level 2 facilities may contain some oncogenic viral work regarded as moderate risk. Efforts are made to contain aerosols within biological safety cabinets or laboratory hoods, and supervisory personnel must have specific training or experience in proper handling of pathogens.

Biosafety level 3 implies work with organisms that threaten human health or the environment. Level 3 facilities have controlled access corridors, air locks, or some form of double-door entry. Technical manipulations are done in biosafety cabinets rather than on the open bench. Wall penetrations for plumbing, gas lines, electricity, and other fixtures are caulked or otherwise sealed with cleanable materials. Air flow is directional to the outdoors, filtered, and not recirculated. Biosafety level 4 design is most relevant to military applications. Level 4 facilities are contained in their own building and feature a variety of barriers such as monolithic walls, air locks, contiguous shower and change rooms, double-door autoclaves, and completely separate ventilation systems.

Most research and production activity occurs at levels 1 and 2. Outbreaks of laboratory-acquired infectious disease have not yet been reported under these conditions. Incidents involving chemical toxicity and allergy have been more common.

III. Regulatory Framework

Health and safety activities of the biotechnology community have related principally to environmental and product approval regulations.

The Department of Agriculture (primarily under the Plant Pest Act) and the Food and Drug Administration (under the Food, Drug, and Cosmetic Act) have line authority over products which can be introduced for field/human testing and for commercial sale in the United States. The U.S. Environmental Protection Agency (EPA) also has the ability to consider each "new organism" a potentially toxic chemical under the Toxic Substances Control Act. EPA could therefore require a manufacturing entity to seek a premanufacturing permit. This has not been a major regulatory hurdle for the industry to date. EPA also maintains its normal regulatory control over planned and accidental releases to water or air.

Biotechnology worker health has received less consideration than environmental planning or product licensing in the United States. Several overlapping, nonvertically integrated regulations apply to worker protection, and custom or even grant-funding requirements may be the most relevant standards for some hazards. The Occupational Safety and Health Administration (OSHA) Hazard Communication Standard applies to chemical aspects of large-scale manufacturing. For small-scale research and development, chemical use is regulated under the OSHA Laboratory Standard. This is described in law as a "performance-based" standard, but it is actually a behavioral standard more reminiscent of the Nuclear Regulatory Commission's approach to low-level ionizing radiolabels used in research laboratories. Small-scale biotechnology entities should have already created their Chemical Hygiene Plan with associated training and operating procedural requirements in order to comply with the OSHA Laboratory Standard.

There is no enforceable regulation of most biological hazards to workers, although OSHA has announced intentions to expand the Laboratory Standard to include biological agents. Work with biological agents may engage safety requirements of grant-funding agencies, such as the National Institutes of Health (NIH) Recombinant Guidelines, or medical surveillance requirements of local governments. These are informal mechanisms. Enforcement at the level of worker health has not been an issue for the biotechnology industry to date. The OSHA standard concerning blood-borne pathogens is an exception. It was applied initially in health care settings only; enforcement is now broadened to include independent research entities handling human tissues. Biotechnology leaders have not uniformly recognized the health and regulatory importance of this standard. Among biotechnologists who work with blood or other tissue, accessibility to handling policies, to emergency response policies, and to hepatitis B immunization has been uneven.

IV. Biological Hazards—Infection

The biological hazards of the biotechnology industry are infection and allergy, analogous to similar problems in the conventional pharmaceutical industry. Fewer cells (and fewer wild-type cells) are required for production; the size of spills and releases and risk of wild-type exposure should be diminished.

Organisms commonly used for biotechnology are recombinant varieties of nonpathogenic *E. coli* and asporogenic bacilli, including *Bacillus subtilis, Neospora,* and *Saccharomyces* species. These organisms are not infectious for healthy immune systems and are designed for properties which do not have host advantages for the organism under wild conditions. Worker infection should generally be less of an issue than for traditional biological work.

Nevertheless, there are infectious disease risks. Biotechnologists may handle animals and are thus susceptible to bites and scratches, as are all animal handlers. Vaccine research may be carried out with potentially infectious viruses used as either vectors or hosts. These include polio viruses, retroviruses, herpes viruses, and especially vaccinia. *Mycobacteria* and *Vibrio* species are potential host bacterial cells. Primary infection, secondary infection, and misleading or "false" seroconversion are concerns of exposed workers and their families.

There are more complex, theoretical hazards as well. For example, hybrid cells are not immune to infection with viruses. Murine cell lines may harbor C-type leukemia viruses. These viruses may survive after the host murine cell is hybridized with a human cell line. An unanswered question is whether habitual residence in human/mouse hybrid cells will eventually select for resident murine C-type leukemia viruses that can extend their host range to humans alone. Extension of host range is among the most troubling questions posed by critics of the biotechnology industry.

Amphotropic (engineered) retroviruses are used to carry specific DNA information into host cells. Under invasive conditions such as needlestick puncture, whole intact viruses can transfer information to exposed workers. The appearance of unusual cancers among researchers at Pasteur Institute led to speculation that this may have occurred among biotechnologists, but no data have been presented to support this idea.

Interventions must be considered carefully. When the risk is small, recommendations for prevention of infectious diseases in laboratories may do more harm than good. Immunization research often utilizes vaccinia (cowpox) virus because it can carry large amounts of DNA. Vaccinia is usually harmless and was used with great success to prevent smallpox infection when natural reservoirs of smallpox still existed. The Centers for Disease Control recommend live vaccinia immunization prophylaxis for scientists doing recombinant vector research with the identical vaccinia strain.

The primary risk of vaccinia infection is to immunosuppressed individuals who encounter the virus through direct immunization or exposure to recently immunized individuals. Encephalitis afflicts approximately one in ten thousand vaccinated individuals. Disseminated infection can also follow immunization when the host is immunosuppressed or when an immunosuppressed family or community contact is exposed to an infected (immunized) host. There are three problems with a policy of immunizations for scientists. First, few scientists will become infected through their research. The condition for which the immunization is designed, laboratory-acquired vaccinia, is rare. Second, should laboratory accidents occur, exposures are likely to deliver less of an immune challenge than the vaccine anyway. Most important, immunization research suggests that vaccinia-naive subjects will mount more complete responses to recombinant vaccinia-based immunizations against other agents, such as human immunodeficiency virus (HIV). Unnecessarily immunized scientists (and their secondarily exposed families) could be less able to respond to the vaccinia-based recombinant vaccines they will develop. In balance, we believe that the recommendation for vaccinia immunization carries more risk than benefit.

Research and production facilities employing human tissue must have in place the same workplace protection against blood-borne pathogens that is now well-recognized in the health care industry.

V. Biological Hazards—Physiologically Active and Immunogenic Proteins

Important products of the biotechnology industry include proteases, hormones, growth factors, and other complex proteins. Products and intermediate products without direct, intended physiologic activity may still be immunogenic, capable of inducing annoying or life-threatening allergy. Some, such as proteolytic enzymes used in detergents and now made from engineered organisms, have well-known allergenic hazards. Others are novel and of unknown allergic property. For example, proteins similar to those found in spider webs are a production goal of the specialty plastics industry seeking high-strength, low-weight products. Humans have not previously been exposed to production quantities of these materials, and allergic responses are unpredictable.

A common step in the production process is the disruption of engineered cells by application of intense pressure gradients. This step is completely enclosed in order to preserve the product and protect production workers. Nevertheless, containment failures are possible in the presence of pressure gradients, exposing workers to natural cell products, including endotoxins, and to unrefined proteins of the manufacturing process. If containment failures persist over time, workers may mount immune responses in the same way that bakers become allergic to proteins in their flour. Since *Saccharomyces* species (yeast) are among the common host organisms for engineered protein production, the analogy to bakers' allergies may be quite appropriate. Several allergic responses are said (but not yet reported in peer review literature) to have required hospitalization among biotechnology workers. When there are production stages with large amounts of potential allergen or endotoxin, we have encouraged workers who manipulate the system to use respirators to guard against the eventuality of containment failure, even though the process is enclosed.

Products with intended physiologic activity add a second level of concern that parallels historic concerns of the conventional pharmaceutical industry. Development of conventional pharmaceutical products such as estrogens resulted in some untoward consequences, including breast enlargement of both sexes and withdrawal bleeding in exposed female workers. The products of biotechnology are often potent and specific, creating opportunities for physiologic responses (and for targeted medical surveillance of exposed workers). A number of the early products are cell-stimulating factors, for example. Blood counts may therefore be investigated as simple markers for exposure. Specific subsets of blood counts may be pertinent.

Physiologic response to intentionally immunogenic particles may create interesting dilemmas. Human immunodeficiency virus workers became concerned with the possibility of "false" seroconversion to noninfectious subunit particles, particularly in vaccine research. Noninfectious protein and nucleotide subunits are frequently designed to elicit an immunogenic response. The implications of false seroconversion to a socially important test, such as an ELISA for HIV, has considerable significance for workers, even in the absence of any possibility of laboratory-acquired infection. So far, an acceptable response to this problem has been to educate workers that alternative diagnostic tests can sort out the difference between true and false seroconversions, should follow-up of baseline testing ever be necessary. However, it should be anticipated that such simple solutions will not always be applicable or adequate.

The recent appearance of the eosinophilia myalgia syndrome reminds us that pharmaceutical contaminants may have disastrous consequences for consumers and implies that workers may have some physiologic risks as well. The syndrome consists of peripheral eosinophilia with a variety of muscle and other organ problems, arising from a peculiar mononuclear/eosinophilic exudate into skin, lungs, muscle, and other organs. It was apparently caused by a dimer form of L-tryptophan contaminating an over-the-counter tryptophan product sold in health food stores. The syndrome is debilitating, irreversible, and potentially fatal. The reason for contamination now appears to be inadequate carbon filter purification of the tryptophan product made by a newly engineered bacillus. The bacillus produced more tryptophan than preceding strains and also produced a previously unknown toxin, the dimer thought to cause eosinophilia myalgia.

Perhaps because of consistent and occasionally frivolous criticism of biotechnology critics, an industry theme has been that the nature of biotechnology virtually precluded environmental or product safety problems. The eosinophilia myalgia episode brings home the reality that all microbiologic industries, including this one, face health and safety hazards.

VI. Chemical Hazards

Biotechnical procedures use toxic chemicals. Typical uses are as cell culture additives, for protein chemistry including extractions, for identification procedures such as high pressure liquid chromatography and gel electrophoresis, and as sterilants. Although the public associates the industry with biohazards, chemical exposures have caused the most visible and consistent problems for workers to date.

Cell culture additives range in hazard level from merely dusty to potent mutagens. Broth mixtures may contain substantial quantities of urea and buffers. Urea has low acute toxicity, yet it is a skin and mucous membrane irritant and a weak mutagen. Toxicants such as azides, with their cyanide-like metabolism, may be added to phage cultures in order to suppress bacterial overgrowths. Antibiotics are frequently added to culture media for suppression of contamination or for linked gene selection. For example, methotrexate is a folic acid antagonist lethal to cells unable to express the enzyme dihydrofolate reductase. Linkage of a second, desirable gene to the gene expressing dihydrofolate reductase means that surviving cells will have an increased chance of expressing the "linked" trait.

Biologists regard methotrexate as a routine reagent to be added to stock solutions. Unaided, they take no special precautions. For occupational physicians, methotrexate is an important cytotoxin, capable of causing aplastic anemia and immunosuppression at high doses. More importantly, methotrexate is a teratogen and exposures should be minimized. Techniques for weighing, mixing, and handling should be related directly to the preparation and disposal of stock solutions containing even small amounts of hazardous additives. (Biologists can be taught effectively that if the additives were not toxic in low concentrations, they would not be used in the first place.)

Extractions from whole tissue or tissue culture are essential for biotechnology research and production. Extractions are chemically intensive operations, employing a variety of solvents, acids, and protein cleavage agents. Some agents used by biotechnologists have clear fire safety and explosion hazards, such as perchloric acid and hydrazine. Protein cleavage agents such as cyanogen bromide are highly corrosive to mucous membranes. Use of these substances requires chemical safety training more commonly found in chemistry departments, including advance planning for spills and leaks.

High pressure liquid chromatography (HPLC) is a technique for extracting and identifying nucleic acid components. Although the technique employs a variety of chemicals, equipment ventilation is rare in biology research settings. A chemical of particular interest is acetonitrile (CH_3CN), one of the most common bulk extraction chemicals. High dose human exposures to acetonitrile give delayed cyanide-like (possibly thiocyanate) toxicity. We have seen several subacute, reversible illnesses following inadequately protected acetonitrile spill cleanup procedures. Experience with acetonitrile spills reinforces the importance of preventing untrained custodial staff from participating in toxic spill cleanups.

A highly publicized goal of biotechnology research is to completely sequence the DNA-base pairs of the genome of species, including humans. Sequencing operations use radionuclides and toxic chemicals. Methylating agents, such as dimethyl sulfide, and cleavage reagents, such as hydrofluoric acid (HF), are used for sequencing and are all highly corrosive to human tissue. Biotechnologists are even using anhydrous HF, a highly dangerous acid previously used only in specialty microelectronic and material science laboratories. Special handling and emergency response training is required for all anhydrous HF users, and the medical department or emergency room providing facility coverage should be aware of the special treatment requirements of HF exposures.

Nucleic acid components are identified through gel electrophoresis techniques, which employ plates covered with polyacrylamide gel. Although premanufactured gel is available commercially, many scientists prefer to mix their own from monomeric acrylamide, which is a powder. The powder is too light to permit all operations to be performed under the air velocity conditions associated with functioning laboratory hoods. Acrylamide is an important cumulative neurotoxin and genotoxin. Skin contact is unacceptable. Unfortunately, the accumulation of white powder around balances and other equipment in many biotechnology laboratories attests to inadequate recognition of the industrial hygiene requirements of work with cumulative toxins.

Ethidium bromide is used in water or buffer solutions as a fluorescent stain that intercalates into DNA, allowing visualization of nucleic acids. Of interest, biologists recognize readily that their use of this unique agent relates to its probable genotoxicity. Of all the toxins they use, biologists are most likely to be appropriately cautious with this exotic chemical, possibly because they already think of it in terms of molecular interactions with DNA. Reproductive toxins will be the focus of chemical safety in this industry for the foreseeable future, and ethidium bromide is potentially a most important reproductive toxin.

Workplace sterility is a persistent need of both research and production biology. Biotechnologists use a variety of chemical (and physical) sterilants. Problems have been encountered from skin contact with irritants such as quaternary ammonium compounds and from inappropriate handling of potential carcinogens such as ethylene oxide and formaldehyde. Ethylene oxide is still used for sterilization of expensive biotechnology equipment which can be steam intolerant.

Treated steam may be introduced into recombinant microbiological and agricultural facilities in order to achieve required levels of humidity. Diethylaminoethanol (DEAE), cyclohexylamine, and a variety of other potentially irritating chemicals are handled in quantity by facility personnel and introduced into human breathing zones at very low concentrations. It is not known with certainty whether amines and other low concentration additives in ambient steam represent a human health hazard, but anecdotal reports from the microelectronics industry suggest that they may lead to increased respiratory symptoms.

VII. Physical Hazards

Laboratory achievements move inexorably to production facilities. Several types of physical hazards have been associated with rapid scale-up from research to production. Lacerations and punctures, notably from glassware handling or animal bites and scratches, have been the most common type of injury. A great deal of glassware is used in all phases of biotechnical operations, including some large flasks for production cultures. Back injuries, related in part to the handling of awkwardly designed equipment, represent the largest source of lost-time injury. Gas cylinders weighing up to 100 lb. are particularly awkward for personnel more accustomed to laboratory-scale equipment. Ergonomic planning for the larger equipment associated with production efforts should be part of all scale-up operations.

Cryogenic and high-temperature thermal burns are physical hazards of both laboratory and production activities. Liquid nitrogen, liquid oxygen, and liquid carbon dioxide may be piped into production sites or used to cool sensitive equipment. Workers must be taught to use a variety of personal protective devices, including gloves and gauntlets. An interesting insight into the youth of this industry is that tube ruptures and thermal burns are said to have occurred in several new operations which simply moved their laboratory-grade tubing across to production facilities. Electrical shock and fire are obvious risks in endeavors where both research and production involve large, sophisticated, electrically powered equipment. All equipment should be grounded. Running wires should be shielded from tripping hazards.

There are several types of ionizing and nonionizing radiation hazards in biotechnology. Ultraviolet light is used for visualization of nucleotides in DNA research and for sterility in several types of operations. Personnel working in ultraviolet areas require skin protection and specific eye protection. We have recently seen a case of keratitis attributable to incorrect selection of eyewear. It is wise to remember that janitors who change germicidal light sources may have particular ocular risks if they get close to the bulb during operation.

Biotechnology facilities typically use several radioisotopes for labeling studies, including tritium, carbon-14, sulfur-35, calcium-45, phosphorous-32, chromium-51, and iodine-125. Although equal opportunity considerations make broad recommendations difficult, it is wise to build enough flexibility into any program so that pregnant workers are not expected to perform iodinations. Facilities using any radioisotopes need a complete radiation protection program with licensure. This requires specific administrative procedures, training (and documentation), safe-handling techniques and engineering containment, routine facility radiation surveys, worker exposure monitoring, emergency plans including contamination response, and management audits for each function. The competition among molecular biologists is intense and there is pressure for time-saving short cuts. More than any other group of scientists, biologists have to be "reminded" not to eat in their radiation facilities.

Shiftwork in biotechnology has some unique aspects because the work cycle is timed to less than completely predictable outcomes of cell growth. In addition, an economic advantage of biotechnology production will be the relatively small size of the production crew relative to the amount of product. This economic advantage will not be helpful in the planning of realistic, healthy shiftwork schedules. In addition, the pressure to reach research and pilot plant goals ahead of competing groups is intense, and scientific and production staff may push themselves through very long work schedules (without complaint to health care personnel). As the industry matures, it will have to examine both individual health and plant safety aspects of work schedules.

VIII. Medical Surveillance and Epidemiology of Biotechnologists

Laboratory-acquired epidemics have not been reported among biotechnologists in any peer review format. The "Pasteur cluster" is not even a coherent disease cluster (diagnoses are neither a single cancer nor a group of logically related cancers). Therefore, the only reported disease outbreak among workers appears most likely to have been due to chance. The absence of serious workplace disease outbreaks speaks well for the efforts of both line workers and health and safety personnel who are struggling to deal with the hazards of a new industry and the handling of toxic chemicals by biologists. Alternatively, serious consequences may be waiting only for sufficient latency. Serious levels of chemical and biological exposure are most likely to be associated with the scale-up from research to production. The production phase of this industry is still in its infancy.

Is the situation analogous to the early days of microelectronics when disease-free "clean industry" still seemed a realistic concept? It will take time, but eventually we will have an overview of the most relevant human health outcomes from biotechnology work. A review of the risks suggests that mutagenic chemical exposure and immunologic response are the most likely outcome areas.

A problem of medical surveillance as it pertains to the biotechnology industry is that logistically practical testing often fails to address the hazards of concern. There are consensus medical surveillance standards enshrined in the local regulations of certain locales where biotechnology is done, that call for periodic blood counts, "liver functions," spirometry, and other similarly available routine tests of workers. This surveillance is based on early committee recommendations of a panel designed to evaluate environmental health risks of the industry. Years of such testing have not revealed problems attributable to biotechnology work. Unfortunately, the outcomes of greatest interest are not addressed. How much ethidium bromide is incorporated into the DNA of biologists, and is it now causing reproductive harm? Do biotechnology workers mount immune responses to proteins and cellular components of their work, and if so, are they of any significance to human health? Surveillance tests that respond to these questions are achievable but not presently available or economical.

It would be nice to report that the scientists leading the biotechnical revolution have perceived the need for designed, prospective surveys of health outcomes in biotechnology workers. The reality is that leading scientists have been opposed to investing in this kind of effort. When the story of the Pasteur cluster broke, the letter sections of leading journals were full of reasons why epidemiologic efforts should not be mounted. Disease outcomes will be reported in the future for this industry as for all others, but only when they are so obvious that detection requires no coherent epidemiologic effort.

These observations lead to three suggestions for the future protection of biotechnology workers. First, health care providers to the industry should identify specific surveillance needs based on plausible outcomes of exposure. Targeted surveillance ideas might be turned over to the scientists at risk, who have the technical competence to explore the feasibility of developing specific surveillance techniques. Interventions already known to be worthwhile, such as hepatitis B immunization for all human tissue research, should be implemented immediately. Second, the industry must begin to look after its own epidemiology. Most individual biotechnology enterprises are still too small to do any kind of meaningful internal population monitoring for chronic disease outcomes, but successful enterprises are beginning to grow rapidly. Leadership enterprises should encourage industry-wide collaborative population epidemiology efforts to insure that the many potential health consequences of this industry are adequately surveyed. Most important, each start-up enterprise needs thorough and ongoing assessments of biological, chemical, and radiation safety. The potential for toxic exposures to workers of this industry is high. Engineering containment must be planned from the start of each new start-up enterprise.

Related Article: GENETICALLY ENGINEERED MICROORGANISMS, MONITORING AND CONTAINING.

Bibliography

Department of Defense. (1991). Biological Defense Safety Program (AR 385-69) 32 CFR Part 626. 56 FR 18, pp. 3186–3193. January 28.

Department of Labor, Occupational Safety and Health Administration (OSHA) (1990). Occupational Exposure to Hazardous Chemicals in Laboratories; Final Rule. 29 CFR 1910.1450: 55 FR 21:3327–3335, January 31.

Doolittle, R. F. (1991). Biotechnology—The enormous cost of success. *N. Engl. J. Med.* **324,** 1360–1362.

Ducatman, A. M., and Liberman, D., eds. (1991). "The Biotechnology Technology Industry: Occupational Medicine State of the Art Reviews" Vol. 6, pp. 157–168. Hanley & Belfus, Philadelphia.

Hyer, W. C., Jr., ed. (1990). "Bioprocessing Safety: Worker and Community Safety and Health Considerations." American Society for Testing and Materials, Philadelphia.

Liberman, D. F., and Gordon, J., eds. (1989). "Biohazards Management Handbook." Marcel Dekker, New York.

Omenn, G. S., and Teich, A. H., eds. (1986). "Biotechnology and the Environment." Noyes Data, Park Ridge, New Jersey.

Upfal, M., and Doyle, C. (1990). Medical management of hydrofluoric acid exposure. *J. Occup. Med.* **30,** 726–731.

Botulinum Toxin

Lance L. Simpson
Jefferson Medical College

Glossary

Acetylcholine Small molecule that is stored and released by certain nerves to act as a chemical message at synaptic and neuromuscular junctions.

Antigen Any substance that evokes antibody formation when added to the appropriate tissues or administered to living creatures.

Antitoxin Antibody that reacts with a specific toxin (i.e., the toxin that was used as an antigen).

Binary toxin Poison that is composed of two separate and independent components, both of which are needed to alter cell shape or function.

Neuromuscular Junction Site at which voluntary nerves make functional contact with muscle; one site at which acetylcholine is released as a chemical mediator.

Serotype One member of a group of substances each of which is characterized by its antigenic properties.

Toxin Noxious or poisonous substance of biological (e.g., microbiological) origin.

BOTULINUM TOXIN *(derived from the Latin word for sausage) is a term used to describe three different classes of proteins of microbial origin. These include botulinum neurotoxin (serotypes A, B, C, D, E, F, and G), which acts selectively on the peripheral cholinergic nervous system; the botulinum binary toxin (also known as C2 toxin), which acts somewhat ubiquitously on eukaryotic cells; and the botulinum C3 exoenzyme (not truly a toxin), which does not yet have a well-characterized spectrum of activity. The neurotoxin, the binary toxin, and the exoenzyme are synthesized by the organism* Clostridium botulinum, *an anaerobic, rod-shaped bacterium that is distributed widely in nature. Only botulinum neurotoxin is known to be an etiologic agent in human disease. Poisoning can be due either to introduction of the microorganism with subsequent* in situ *production of the toxin or to direct introduction of the toxin. In either case, severe poisoning is marked by paralysis of voluntary muscles, including muscles of movement and respiration.*

I. Origin of the Toxins

Botulinum neurotoxin is synthesized mainly by the organism *Clostridium botulinum*, and it exists in at least seven different serotypes designated A, B, C, D, E, F, and G (Fig. 1). In most cases, individual strains of *Clostridium botulinum* synthesize only one serotype of botulinum neurotoxin. The various serotypes are similar in origin, macrostructure, and mechanism of action, but differ in antigenicity. Generally speaking, antibodies against one serotype do not neutralize the biological activity of other serotypes. An exception to this rule involves serotypes C and D, which are closely related.

Although *Clostridium botulinum* is the major source of the neurotoxin, other clostridia have been shown to produce these substances. *Clostridium barati* and *Clostridium butyricum* have been implicated as sources of toxin that can cause human illness. *Clostridium novyi* has been induced to make the toxin in a laboratory setting.

The nucleic acid that encodes the toxin has been isolated for most but not all of the serotypes. The genetic message for types A, B, and E is found in the host genome; the message for type G is found in a

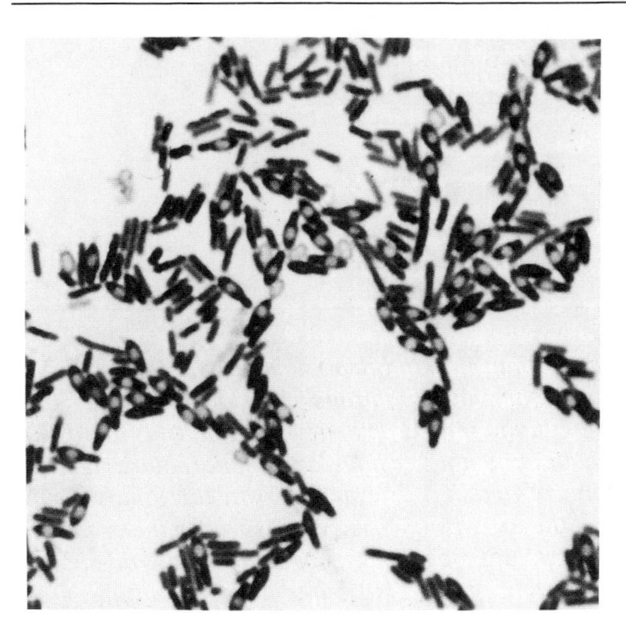

Figure 1 Characteristics of clostridia. Clostridia are anaerobic bacteria that are widely distributed, of rod-like morphology, spore-forming, gram-positive, and may or may not be proteolytic.

plasmid; and the message for types C and D is carried by a virus that infects *Clostridium botulinum*. The location of the message that encodes type F has not been determined. The fact that serotypes C and D are encoded by messages carried by a virus has been the basis for two notable findings. First, organisms that are cured of their viral infection cease to produce toxin, but they continue to grow and multiply normally. This finding suggests that the toxin plays little or no role in the physiology of the organism. Second, the viruses that encode serotypes C and D have been used to infect other clostridia (e.g., *Clostridium novyi*), and these organisms have been induced to synthesize botulinum neurotoxin. This finding shows that *Clostridium botulinum* is not unique in its ability to make and release toxin.

The botulinum binary toxin and the botulinum exoenzyme have thus far been shown to originate only from *Clostridium botulinum* types C and D. A slight historical digression will help to put this matter into context. Until the relatively recent past, botulinum neurotoxin was believed to exist in eight serotypes designated A, B, C1, C2, etc. As the nomenclature indicates, types C1 and C2 toxins were thought to have a common source and a common mechanism of action, but this belief has been shown to be partially in error. Type C2 toxin, now known as the botulinum binary toxin, is synthesized by organisms that make type C and type D botulinum neurotoxin, but it is fundamentally different in structure, immunology, and biological activity. The binary toxin is also different in genetic origin. When clostridia that make types C and D neurotoxin are cured of their viral infection, they cease to make the neurotoxin, but continue to produce the binary toxin. The latter is presumably encoded by the host genome.

The botulinum exoenzyme was also discovered as part of a mistaken effort to characterize botulinum neurotoxins. Reports were published stating that botulinum neurotoxins C and D possessed the enzymatic activity of a mono(ADP-ribosyl)transferase. Later work showed that the enzymatic activity was due to a contaminant that was initially called type C3 toxin but is now called type C3 exoenzyme or the botulinum exoenzyme. Like the botulinum neurotoxin, the botulinum exoenzyme ceases to be produced by organisms that have been cured of viral infection.

II. Structure of the Molecules

Botulinum neurotoxin is synthesized as a 150,000 dalton protein that possesses only a fraction of the biological activity of the mature toxin (Table I). To become fully active, the toxin must undergo posttranslational modification by a protease. Most clostridia possess proteolytic enzymes and are able to process the toxin, but some organisms release the nonactivated toxin. The latter material can be converted to fully activated toxin by exposure to exogenous proteases such as trypsin.

At least one of the mechanisms for proteolytic activation of the toxin has been determined. The toxin is synthesized as a single-chain molecule that has one critically important disulfide bond. When this molecule is cleaved by protease, it is converted to a dichain molecule in which a heavy chain polypeptide ($M_r \sim 100{,}000$) is linked by a disulfide bond to a light chain polypeptide ($M_r \sim 50{,}000$). Proteolytic cleavage to yield a dichain molecule appears to be necessary but not sufficient for full activation. Additional changes, involving other sites of cleavage and/or changes in conformation, are needed before the toxin becomes fully potent. Neither the isolated light chain nor the isolated heavy chain possesses inherent toxicity. Furthermore, reduction of the disulfide bond between the chains causes loss of toxicity. The data indicate that the dichain molecule must remain intact in order to poison vulnerable

Table I Characteristics of the Biological Products from *Clostridium botulinum*

	Botulinum neurotoxin	Botulinum binary toxin	Botulinum exoenzyme
Number of serotypes	7	?	?
Molecular weight	~150,000	~150,000	~26,000
Number of chains	2	2	1
Site of action	Cholinergic nerve endings	Various cell types	Various cell types
Mechanism of action	Endoprotease	ADP-ribosyltransferase	ADP-ribosyltransferase
Substrate	?	Actin	Regulatory proteins

cells. The role that the various parts of the toxin play in expressing biological activity will be discussed shortly.

The amino acid composition of the various serotypes of botulinum neurotoxin have been determined, and the complete primary structures of several of the serotypes have been published. As expected, the results show that there is some homology in primary structure, but there are also areas of disparity. The similarities are a reflection of the common origin, macrostructure, and biological activity; the dissimilarities are a reflection of evolutionary dispersal and of the resulting immunologic properties.

A limited amount of study has been done on secondary and tertiary structure of the toxin. None of this work has helped to clarify the mechanism of toxin action, but—as more is learned about the molecule—a detailed structure–function analysis will emerge.

There are less data available on the botulinum binary toxin and the exoenzyme, although the literature is growing rapidly. The binary toxin is a relatively unique substance because it is composed of two separate and independent polypeptide chains. The two exist freely in solution and are not linked by covalent or noncovalent forces. The heavy chain is synthesized as a polypeptide with a molecular weight of ~100,000 and it must undergo proteolysis to yield a mature product that has a molecular weight of ~85,000. The light chain is synthesized as a mature protein that has a molecular weight of 45,000 to 50,000. The amino acid compositions of the heavy and light chains are known, but structural information is not available. The heavy and light chains of the binary toxin are antigenically distinct.

The botulinum exoenzyme is the smallest of the products produced by *Clostridium botulinum*, with a molecular weight of ~26,000. Unlike the larger neurotoxin and binary toxin, it is composed of a single polypeptide chain. The exoenzyme has been cloned and a complete primary structure deduced. The results show that the types C and D neurotoxin are different from the exoenzyme, even though they are encoded by the same viruses.

III. Human Exposure

Clostridia are ubiquitous organisms, which means that there is the possibility for widespread exposure. However, there are special conditions that govern the interaction between clostridia and human hosts, and these conditions tend to minimize the potential for adverse outcomes.

There are three general types of mechanism by which the microorganism or the toxin might enter a patient (Table II). The most prevalent is ingestion of a food (e.g., honey) contaminated with the bacteria. The organism is so widespread that this is likely to be a common occurrence, but the disease known as "botulinum poisoning" is quite rare. This can be attributed to the fact that *Clostridium botulinum* does not ordinarily colonize the human gut. More precisely, it does not compete well with the normal flora in the human gut, and thus it is unlikely to find a niche in anyone who is beyond infancy.

The latter point is one of some importance, be-

Table II Botulism

Classification	Source of problem	Likely victim
Infection	Food (such as honey) contaminated with the bacteria	Infants>>adults
Intoxication	Improperly prepared food contaminated with the toxin	Adults>>infants
Trauma	Needles or other objects or matter contaminated with bacteria	Illicit drug users

cause infants from birth until approximately 9 to 12 months of age are susceptible to infection by clostridia. Spores can germinate in the bowel, after which the organisms will grow and produce toxin inside the infant. Although not widely appreciated, infectious botulism leading to illness in infants is the most common cause of the disease. Infectious botulism can also occur in adults, although the incidence is very low. There is no explanation for this phenomenon, but it seems reasonable to assume that afflicted adults have a naturally occurring or drug-induced change in their intestinal flora.

Another cause of botulism is ingestion of food that is tainted with toxin. This source of illness has been referred to for decades as "food poisoning," and it is often wrongly assumed to be the major cause of botulism. There is a rather common scenario for the disease, and it proceeds as follows. A food product that is contaminated with the organism is packaged (e.g., canned, bottled) in an anaerobic environment that is ideal for the bacteria. Over time the organisms grow and multiply, and in the process they release toxin into the packaged food. An unsuspecting person might then consume the food and its poisonous contents.

In order for this sequence of events to occur, there must be at least three lapses in the preparation of food. To begin with, foods are normally washed before processing, and this removes a significant amount of surface bacterial contamination. Next, packaged foods are often treated with chemicals or with heat to retard bacterial growth. And finally, unpackaged foods are usually cooked before being served, and the toxin molecule is very heat labile.

Of these three potential problems, the one that is most often the culprit in human botulism is the failure to package correctly. Properly cleaned and preserved foods should not provide an environment for clostridial growth, and hence any additional protective measures should be unnecessary (though they may be judicious). This information serves as a backdrop to a historic/economic observation. It has been noted that during times of economic downturn, there is an increased amount of home packaging of foods. This has been accompanied by an increase in the incidence of botulism, due to a failure to adhere to safe procedures for food processing.

The final defined cause of botulism is that due to trauma. This is quite rare, and perhaps the best recognized example is the illicit drug user. Persons who use nonsterile needles to inject drugs are in danger of introducing bacteria. If injections are made into traumatized tissue that is necrotic or poorly perfused, there is the possibility for bacterial growth. As with the introduction of the organism into the gut, there can be local synthesis and release of toxin that ultimately causes disease. The process would appear to be much like that seen with inadvertent injection of *Clostridium tetani* and the subsequent development of tetanus, but the incidence of the problem is much lower.

IV. Signs and Symptoms of Poisoning

Of the three substances produced by *Clostridium botulinum,* only the neurotoxin has been linked to human disease. The illness, known as botulism, is potentially devastating, but fortunately it is not a commonly occurring disease.

Patients who present with botulism may have a wide array of problems, and these can be confusing to the unsuspecting physician, but a detailed examination would reveal that there is an underlying pattern. The classic patient will complain of nausea and vomiting, of blurred vision and difficulty with accommodation, of a dry mouth with difficulty in speaking and swallowing, of general muscle weakness, of constipation and perhaps urinary retention, and most disturbingly, of difficulty with breathing.

These problems can be sorted out in the following way. Vomiting is not a consequence of botulism; botulinum neurotoxin is not an emetic agent. Patients who experience nausea and vomiting are either reacting to contaminants in the tainted food, or their response is secondary to an authentic toxin-induced problem (i.e., blurred vision and unsteady gait due to muscle weakness can produce dizziness and its sequelae). On the other hand, the difficulties that range from blurred vision to shortness of breath are all truly related to the toxin, and they are all explainable on the basis of peripheral nervous system dysfunction.

The peripheral nervous system is divided into two components, known as voluntary and autonomic. The voluntary nervous system governs such actions as movement and respiration; the autonomic nervous system regulates body functions such as blood pressure and heart rate, intestinal motility, and urination. The voluntary nervous system is exquisitely sensitive to botulinum toxin. In severe cases of poisoning, a patient will be totally paralyzed, unable to move and not capable of breathing. The autonomic nervous system is less sensitive to the toxin, but it too may be disrupted in serious cases of poisoning. Thus the cardinal signs of poisoning can all be seen as evidence of loss of voluntary or autonomic nervous system function.

It is important to note that one component of the nervous system is not directly affected by the toxin. The central nervous system, including the brain and spinal cord, is not poisoned, because botulinum neurotoxin does not cross the blood–brain barrier. This means, on the one hand, that the brain is free of toxin-induced problems; but on the other hand, it means that patients are conscious and fully aware of their predicament.

V. Mechanism of Neurotoxin Action

A. Macroscopic Actions

The peripheral nervous system, composed of an extraordinary number of nerves, networks, and synaptic connections, is extremely complex. However, botulinum toxin exerts its effects by acting at only one type of site in this complex system. The toxin acts only on those nerve endings that store and release the transmitter substance acetylcholine. Nerve endings that use acetylcholine as a transmitter are called cholinergic.

There are three locations in the peripheral nervous system where transmission is cholinergic in nature; one of these is in the voluntary nervous system and two are in the autonomic system. In the voluntary system, it is the junction between motor nerves and voluntary muscle that becomes paralyzed by the toxin. The autonomic system is somewhat more intricate, due to the fact that it is divided into two components, and each component is divided into two halves. The two components of the autonomic system are called parasympathetic and sympathetic. Each of these two components sends nerves out of the central nervous system and into a small bundle called a ganglion. Here the primary nerve, called preganglionic, makes contact with a secondary nerve called postganglionic, and the latter travels to and innervates the various organs of the body.

All ganglionic synapses are cholinergic; that is, all preganglionic nerves synthesize and release acetylcholine to stimulate postganglionic nerves. Hence, all nerve endings in the ganglia are potentially susceptible to the toxin. The postganglionic, parasympathetic nervous system also uses acetylcholine, and it too can be poisoned, but postganglionic sympathetic nerves release transmitters such as norepinephrine and are thus largely resistant. The existence of two cholinergic sites in the parasympathetic system might seem to indicate that it should be more vulnerable to poisoning than the voluntary system, but this is not the case. As noted above, nerves in the voluntary system are more sensitive to toxin (i.e., are poisoned at lower doses), and thus it is the voluntary system that is most likely to be impaired.

Regardless of the component of the nervous system involved or the location of the nerve ending, the outcome is the same. Botulinum neurotoxin acts on all cholinergic cells to block the release of acetylcholine. Blockade of transmission between motor nerves and voluntary muscles causes paralysis (i.e., loss of respiration); blockade of transmission between autonomic nerves and body organs causes disruptions in normal physiology (i.e., loss of intestinal motility).

B. Microscopic Actions

The overwhelming majority of studies on botulinum neurotoxin have been conducted on voluntary nerves such as those that innervate the muscles of respiration. This work has resulted in a nearly, but not completely, clear picture of how the toxin acts.

Botulinum neurotoxin possesses a tissue-targeting domain that directs the molecule to cholinergic nerve endings. The receptor for the toxin has not been isolated and purified, but available evidence suggests that it is a glycoprotien that has sialic acid residues. In order to exert its poisoning effect, the toxin must enter the cytosol of the nerve ending, and it does this by a process known as receptor-mediated endocytosis. The toxin–receptor complexes are captured by an inward budding of the membrane, and these buds break off from the cell membrane to form small sacs called endosomes. As the endosomes travel from the surface to the interior of the nerve ending, the toxin escapes and reaches the cytosol.

The precise action of the toxin in the cytosol is still unknown, but there is a strong basis for believing that it must be enzymatic. The toxin is so potent that a multiplicative action (i.e., catalytic) seems the only feasible explanation. In fact, recent work suggests that botulinum neurotoxin is an endoprotease that cleaves one or more polypeptides that are essential for release of acetylcholine.

Some progress has been made in defining the structure–function relationships in the toxin molecule. The 100,000-dalton heavy chain appears to mediate binding. Indeed, there is evidence that the tissue-targeting domain may be localized to the carboxy terminus of the heavy chain. There is also evidence that the amino terminus of the heavy chain plays a key role, perhaps guiding the toxin as it escapes from the endosome. The light chain is known to produce a blockade of acetylcholine release; it is this portion of the molecule that is believed to act as an enzyme to modify an intracellular substrate that governs transmission.

In keeping with presumptions about structure–function relationships, neither the isolated heavy chain nor the isolated light chain can paralyze transmission when added to the outside of the cells. Only the intact toxin can block acetylcholine release from intact nerve endings. However, when the light chain is experimentally introduced into cells, in a manner that bypasses binding and endocytosis, it can block transmitter release.

VI. Actions of the Binary Toxin and Exoenzyme

A. Binary Toxin

In addition to the neurotoxin, types C and D *Clostridium botulinum* synthesize a potent binary toxin. This toxin is composed of two separate and independent polypeptide chains, with a heavy chain of ~100,000 daltons and light chain of ~45,000 to 50,000 daltons. The binary toxin is different from the neurotoxin in that the two chains are not linked by a disulfide bond. Indeed, there are no known covalent or noncovalent forces that link the two polypeptides. However, the neurotoxin and the binary toxin are similar in the sense that both must undergo proteolytic processing to become fully active. In the case of the binary toxin, the heavy chain is cleaved by proteases to give an active product of ~85,000 daltons.

The mechanism of action of the binary toxin is relatively well understood. The heavy chain is a tissue-targeting moiety that binds to receptors on eukaryotic cells. The identity of the receptor remains a mystery, but it must be a molecule that is widely distributed in nature because virtually all eukaryotic cells are susceptible to poisoning by the binary toxin.

The light chain will not by itself associate with cells, but it will absorb to cells that have previously been exposed to the heavy chain. This suggests that the heavy chain, either by itself or in concert with the membrane, forms a docking site for the light chain. The two then enter the cell by receptor-mediated endocytosis.

The subcellular and even molecular actions of the binary toxin have been established. The light chain of the toxin is an enzyme that possesses mono(ADP-ribosyl)transferase activity; the substrate is monomeric or G-actin. Monomeric actin is the building block of the cytoskeleton of cells. By virtue of enzymatically modifying actin, the toxin produces several adverse outcomes. At a molecular level, the toxin blocks the formation of actin filaments, and thus a poisoned cell loses the ability to form extensions. The toxin also hastens the normal turnover and disaggregation of actin filaments, and this too causes a loss of cytoskeleton. At a cellular level, the toxin causes cells to collapse into small bags. In the absence of a cytoskeleton, the cell membrane and cell contents shrink to occupy the smallest possible physical volume.

B. Exoenzyme

The botulinum exoenzyme is also produced by types C and D *Clostridium botulinum*. It is not properly referred to as a toxin because at meaningful concentrations it does not produce adverse effects on cell function. In fact, the exoenzyme does not have a binding domain and has not been shown to associate with cells. The exoenzyme will not express its effects unless added to a broken cell preparation or artificially introduced into the cell interior.

The exoenzyme, like the binary toxin, has been shown to possess mono(ADP-ribosyl)transferase activity, but in this case the substrate is one of a small group of proteins that is known to regulate cell growth and differentiation. Exoenzyme that is injected into cells can produce marked changes in morphology and function due to alteration of regulatory proteins.

VII. Preventive Medicine and Therapeutic Intervention

The most desirable approach to dealing with botulism is to prevent its occurrence, and at least in theory this is achievable. The toxin is a large protein molecule that is reasonably antigenic, and several types of vaccines have been prepared. In the United States, the vaccine is generated by isolating a complex of neurotoxin and associated proteins, mainly hemagglutinins. This mixture is treated with formalin under conditions that cause the mixture to lose its neurotoxicity while retaining its antigenicity. This is the same approach that is used in preparing the vaccine against tetanus. The formalin-treated preparation is administered to patients in a series of three subcutaneous injections. This produces resistance to the toxin, although the extent of resistance varies widely from individual to individual.

The current practice is to obtain blood samples from injected persons at regular intervals, but no more than seven years after the final of the three injections. If the titer of circulating antibody is low, the patient in question is given a booster. Otherwise, the practice of assaying specimens at regular intervals continues until the titer of antibody falls. It is noteworthy that the current procedure for tetanus immunization is to administer boosters at approximately ten-year intervals, unless there is clinical indication for more frequent injections.

As the foregoing comments indicate, there is a toxoid and a standard procedure for its administration, but this has little impact on the general population. The incidence of botulism in the United States is rather low, and because of this there has never been a concentrated effort to prepare and fully evaluate a toxoid according to the current and stringent guidelines of the Food and Drug Administration. The toxoid that is now available was prepared according to techniques that were developed during the 1940s, when there was the fear that the toxin would be used as an agent in biological warfare. This toxoid is now, however, administered almost exclusively to physicians and scientists who do research on clostridial toxins. Although a considerable body of information has accumulated from observations on injected persons, the toxoid continues to be regarded as an investigational drug and is not produced by any pharmaceutical company.

Investigators who wish to have access to the toxoid should contact the Centers for Disease Control, which dispenses the material. Injections must be given under the supervision of a physician, and the response to the vaccine must be documented. The Centers for Disease Control serves as a repository for all reports of adverse reactions. Because the toxoid is investigational, and because its potential for causing developmental disorders has not been determined, the material should not be administered to women who are pregnant.

That there is not widespread immunization against botulism means that the disease can occur, albeit with a low incidence. When botulism does occur, the types of therapeutic intervention that are considered are antitoxin administration (i.e., type-specific antibody), pharmacologic antagonists, and supportive treatment. Antitoxin can play an important role, depending on the age of the patient and the cause of the disease, as can supportive treatment. On the other hand, pharmacologic intervention is of little or no value; no broad spectrum antagonist of the several botulinum serotypes has yet been identified.

When an outbreak of botulism occurs, and if adult patients are diagnosed early in the course of the disease, the administration of antitoxin may be rational. Type-specific antibody can reach and neutralize toxin in the general circulation. However, there are several *provisos* to bear in mind when considering administration of the antibody. First, the material available in the United States is of equine origin, and therefore a physician should test for sensitivity reactions before administering the antitoxin when practical. Second, the antibody can reach toxin in blood, interstitial fluid and other extracellular fluids, but it cannot enter the nerve ending to neutralize toxin that has already been endocytosed. And finally, if the patient is not diagnosed until late in the course of the disease—and especially if body specimens have been assayed for toxin and the titer is low—there may be little advantage to administering antibody. Supportive therapy may suffice.

The comments above apply to patients who are adults, because experience has led to a different strategy for infants. It is generally not advisable to administer foreign proteins to infants whose immune system is still in the developmental stage. Therefore, the accepted approach to dealing with a young patient is merely to provide supportive therapy. Experience indicates that this is an acceptable course, because infant botulism tends to be self-resolving.

The next of the three forms of therapeutic intervention is the use of pharmacologic agents, but this has not proved to be a very effective means of intervention. The two classes of drugs one might consider are antibiotics for those patients who have infectious botulism, and drugs that promote acetylcholine release for all patients with botulism. Regrettably, neither has been demonstrated to have a beneficial effect in most patients. Antibiotics have not been thoroughly tested, and in some quarters there is even fear of testing them. There is speculation that antibiotic-induced lysis of bacteria could lead to a sudden burst of toxin being released into the general circulation. Furthermore, certain antibiotics have the ability to cause muscle weakness, and this can exacerbate botulism.

A variety of drugs has been tested as potential therapeutic agents in botulism, all on the principle that they promote transmitter release, but results have been disappointing. In certain cases (e.g., guanidine), the ability of the drug to enhance transmitter release is marginal under physiological conditions, and thus there is little basis for expecting a notable therapeutic effect. In other cases (e.g., 3,4-diaminopyridine) the drug has a dramatic effect on transmitter release, but for some as yet undetermined reason the drug is beneficial in only selected patients (e.g., those with type A botulism). Although the results in the past have not been encouraging, there is still reason to press the search for pharmacologic antagonists.

The tactic of using supportive therapy continues to be the mainstay in dealing with victims of botulism. This includes the use of artificial respiration for patients who cannot breathe on their own, as well as methods for feeding and voiding patients. These supportive measures must remain in place until patients regain the function of paralyzed muscles. The recovery process can be protracted, requiring weeks to months, but in most cases recovery is complete.

Acknowledments

Supported in part by NIH grant NS-22153 and by DOA contracts DAMD17-86-C-6161 and DAMD17-90-C-0048.

Bibliography

Habermann, E., and Dreyer, F. (1986). Clostridial neurotoxins: Handling and action at the cellular and molecular levels. *Curr. Topics Microbiol. Toxin Rev.* **5**, 177–190.

Melling, J., Hambleton, P., and Shone, C. C. (1988). *Clostridium botulinum* toxins: Nature and preparation for clinical use. *Eye* **2**, 16–23.

Middlebrook, J. L. (1986). Cellular mechanism of action of botulinum neurotoxin. *J. Toxicol. Toxin Rev.* **5**, 177–190.

Simpson, L. L. (1986). Molecular pharmacology of botulinum toxin and tetanus toxin. *Ann. Rev. Pharmacol.* **26**, 427–453.

Simpson, L. L., ed. (1989). "Botulinum Neurotoxin and Tetanus Toxin." Academic Press, San Diego.

Smith, L. D. S., Sugiyama, H., eds. (1988). "Botulism: The Organism, Its Toxins, the Disease," 2nd edition. C. C. Thomas, Springfield, Illinois.

Bronchoalveolar Lavage: Detecting Markers of Lung Injury

Daniel Dziedzic, Candace S. Wheeler, and Kenneth B. Gross
General Motors Research Laboratories

Glossary

Alveolus Sac-like gas exchange unit of the lung.
Bronchoscope Instrument used to access and examine the lungs' airways.
Epithelium Cellular layer that lines the surface of an organ.
Inflammation Reaction to injury involving infiltration of cells and mediators.
Interstitium Spaces between the cells and tissue of an organ.
Lung segment Portion of the lung supplied by a major airway and blood vessels.

***BRONCHOALVEOLAR LAVAGE (BAL)** refers to the irrigation or washing out (Fr.* laver*—to wash) of the conducting airways (G.* bronchos*—windpipe) and gas exchange regions (L.* alveus*—hollow sac, cavity) of the lung (Fig. 1). This technique allows for quantitative sampling of both cellular and humoral components from the epithelial surfaces of the lung. BAL is widely used to study the response of the lung to a variety of inhaled toxicants and disease states, as well as a means of harvesting cells, lung lipids, particles, enzymes, and other soluble factors. The importance of bronchoalveolar lavage in evaluating the physiology and pathophysiology of the lung increases as we improve our ability to analyze and understand the significance of the cells and substances that are captured by BAL.*

I. Historic Development and Current Status

A. Historic Experience

1. Animal Studies

BAL was first used in 1919 to wash the lungs of dogs instilled with virulent pneumococci in an attempt to prevent the development of pneumonia. Additional studies followed as early as 1928 directed at the potential therapeutic uses of BAL. For example, studies using BAL to remove radioactive materials from the lungs of animals suggested the use of this technique to treat humans exposed to radioactive aerosols (1972 and 1975). BAL was first used as a method to harvest alveolar macrophages (AMs, Fig. 1) and other free cells from the surface of the lung in 1955. Many studies followed, especially in the late 1960s and early 1970s, that attempted to identify and regulate the many factors that influence cell yield. The use of multiple washings, various washing solutions, and changes in wash temperature and pH have been studied in an attempt to make the technique more sensitive and reproducible as well as to address particular experimental concerns. Addition of a local anesthetic, such as lidocaine, to the wash solution was shown in 1979 to significantly increase the cell yield when compared to washings obtained using normal saline. By 1982, practical techniques for performing BAL on all types of small laboratory rodents had been designed and the use of repeated lavages on small laboratory animals had also been addressed. BAL is now used extensively to assess lung injury initiated by a variety of insults and toxic agents.

2. Human Studies

While BAL in animals was developed as an experimental tool for research and could be used as a

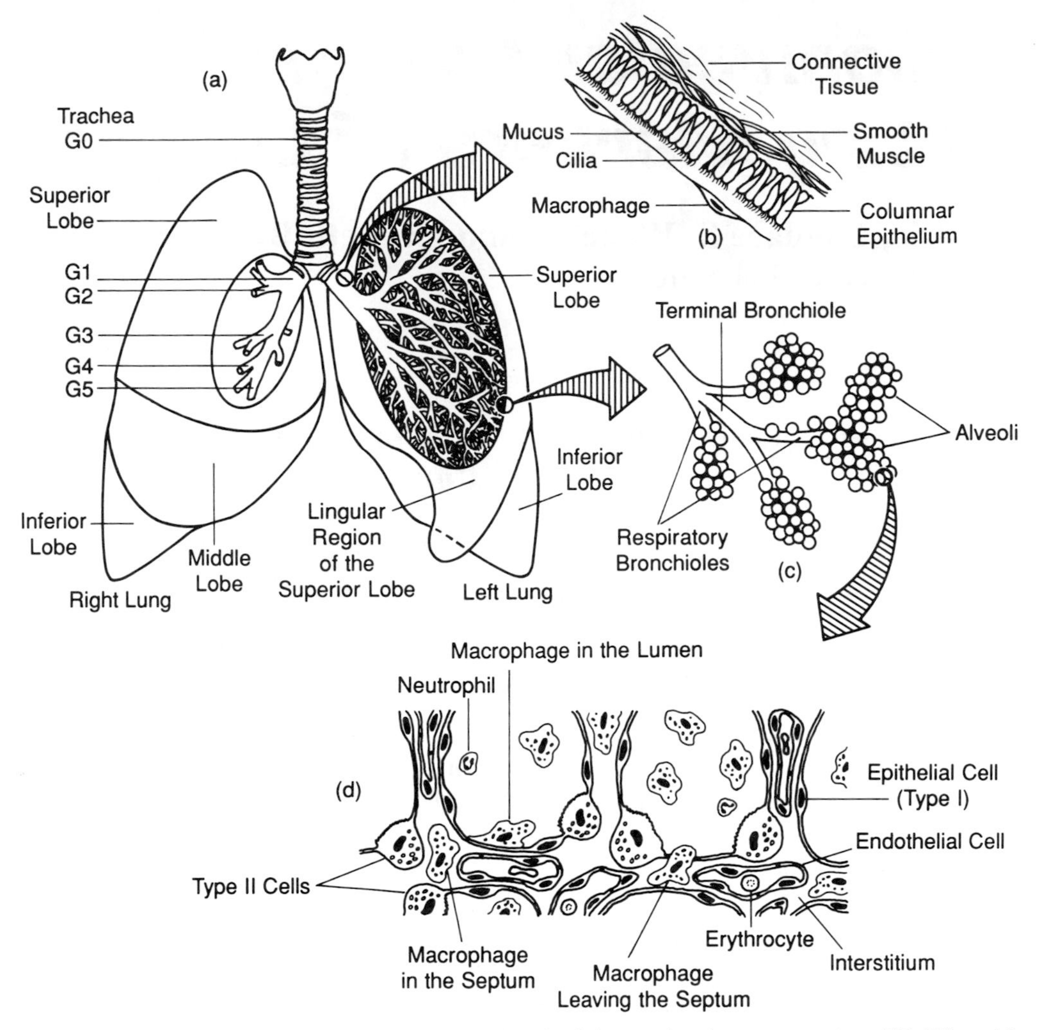

Figure 1 Schematic drawing of the lung. (a) lung, showing lobes, major airway generations (G0-G5), and fine branching pattern; (b) magnification of airway epithelium; (c) terminal airways with associated gas exchange units (alveoli); (d) fine structure of alveoli and associated cell types.

terminal procedure, the use of BAL in humans followed the development of bronchoscopy as a safe and effective diagnostic tool in the clinical setting. The rigid bronchoscope was designed in the early 1900s and was used as an instrument for visual diagnosis and for performing therapeutic lavages for certain clinical conditions. In 1964, a rigid bronchoscope was used for BAL as a technique to collect fluids in the respiratory tract. Balloon catheters also became popular as a tool for performing BAL because of their flexibility, size, and ability to lavage one lung or section of a lung discretely. With the development of fiber optics, the flexible bronchoscope was introduced in the late 1960s. The first use of human lavaging as a technique for harvesting macrophages for research purposes using a flexible fiberoptic bronchoscope was reported in 1973. Flexible fiberoptic bronchoscopes have improved in design over the years regarding flexibility, compactness, and image resolution. Bronchoscopes with outer diameters smaller than 4 mm are currently available. They can be inserted into the trachea either through the nasal passages or the mouth and can be wedged into a segmental bronchus as far down as 3 to 4 generations. More recently, techniques have been developed, such as the use of double balloon catheters, that can effectively isolate a small section of a mainstem bronchus for lavaging of this area.

B. Current Uses

Currently, the most common use for BAL in both animal and human studies is as a research technique. Significant insights have been gained into lung cell physiology and pathophysiology as well as immune defense and inflammatory changes in the lung. Some common applications in experimental studies of pulmonary responses to inhalation of hazardous materials include identifying thresholds that initiate lung injury, assessing the magnitude of injury produced, and determining the extent of reactive inflammatory and host-defense changes. BAL has also been used experimentally to study cellular and biochemical mechanisms of lung injury, the time course of changes, dosimetry, and long-term consequences of pulmonary damage, including whether changes are reversible or irreversible. In animal studies, BAL can be used to compare and rank series of compounds for their ability to cause lung injury.

In addition to research applications, BAL has also been used in patients to diagnose and stage disease, and to treat certain clinical conditions. BAL succeeds in diagnosis of opportunistic infection with *Pneumocystis carnii* in the immunocompromised (AIDS) patient greater than 90% of the time. Increasing success has also been reported for diagnosis of other infectious processes such as viral disease and for diagnosis of primary and metastatic tumors. Therapeutically, BAL has been used to remove excessive accumulations of fluids, cells, and debris in patients with pulmonary alveolar proteinosis, a lung condition characterized by the presence of proteinaceous fluids in the alveoli. It is well established that multiple washings of the whole lung produce symptomatic relief and functional restoration. BAL has also been tested for therapeutic benefits in other conditions such as asthma, pneumoconiosis, and, to a lesser extent, removal of damaging radioactive or irritating materials with varying success. Currently, while BAL is clearly useful in treatment of alveolar proteinosis, therapeutic use in other lung conditions is still being examined.

C. Comparison with Other Techniques

BAL offers specific advantages compared with other techniques in analyzing changes in the lung. BAL is uniquely capable of obtaining viable cells from surfaces of peripheral portions of normal and injured lungs without introducing added, excessive damage. These cells can be analyzed directly using cellular and biochemical approaches, or can be cultured and analyzed under controlled conditions. Likewise, BAL can be employed to sample the biochemical environment of the lung. Characterizing the appearance and disappearance of biochemical molecules allows inferences about the cellular physiology in normal and pathologic conditions. The lavage process can be coupled with lavage of other parts of the respiratory tract including the nasal cavity and large airways to provide a complete picture of surface changes at various levels of the respiratory system. Because BAL can be performed more than once in the same subject, it can be used to monitor long-term consequences in the lung as a function of time of exposure or recovery.

The use of BAL can be compared with histologic analysis from biopsy or experimentally harvested lung specimens. Histologic examination provides information from fixative-treated tissue about structure of the various cell types in and on the airway and alveoli. In contrast, BAL washes the internal surface of the lung; therefore, the cells that are retrieved represent populations that reside on this surface. Generally, lavage samples do not include significant numbers of cells of the lung wall itself. Another approach to obtaining lung cells is to sample or biopsy the lung tissue, mince it, and then purify the desired cell type. This is more traumatic and requires more steps to obtain the desired cell populations. Macrophages and other lung cells obtained from minced biopsy specimens have been compared with those obtained from the same subjects using BAL. Generally, more cells can be obtained from lung minces than from lavage. Nonetheless, the presence of surface markers, degree of cellular maturation, and functional capacity are identical or very similar. This indicates that BAL produces a population of cell types that is very similar, but not necessarily identical, to those found in the whole lung.

II. Methodology

A. Animal Models

1. Lavage Techniques

The entire lungs of smaller animals are usually removed or lavaged in the body as part of a terminal procedure. Larger animals can be kept alive providing for the possibility of performing the procedure at several different times. While repeated lavages are performed in larger animals and humans, serial lavages of small animals is complicated by the need to instill the lavage fluid while maintaining anesthesia and ventilation. Techniques for serial lavage have been worked out for a number of different species of small laboratory animals although all species do not tolerate the procedure equally well. In addition, these techniques are more difficult, requiring more skill and time to perform. The actual wash procedure is, however, very similar in either case.

Animals are generally anesthetized using sodium pentobarbital injected intraperitoneally. The abdominal aorta is severed, reducing blood volume, and preventing it from entering the lungs. In the case of excised lungs, air is introduced into the thorax by puncturing the diaphragm and the chest cavity is opened. The lungs are dissected free of other tissue and suspended in physiological saline to eliminate hydrostatic gradients that may lead to uneven filling during the wash procedure. The trachea is then cannulated by inserting a catheter usually made of small diameter polypropylene tubing. The catheter should be filled with saline to avoid injecting air into the lungs. A small negative pressure is produced in the lungs to minimize the amount of trapped air at the beginning of the procedure.

The wash fluid used in the lavage procedure can be physiologic saline or a balanced salt solution, generally warmed to 37°C. Some studies have shown that elimination of the divalent cations, Ca^{2+} and Mg^{2+}, from the wash solution will increase the yield of cells from the alveolar surface. The Ca^{2+} and Mg^{2+} ions greatly affect the adhesive forces that exist between AMs and the alveolar surface. The wash solution is instilled into the lungs using a syringe attached to the catheter by a three-way valve, and then withdrawn from the lungs by negative pressure using a second collection syringe. This process is usually repeated six or more times. The volume of fluid instilled depends on the lung volume. Generally, the wash volume is calculated as 80% of the total lung volume based on body weight or as five times the tissue volume of the lung. For small animals, usually 5 ml per gram lung provides a complete uniform washing of the lung. We have found that using less fluid on the first washing and increasing to approximately 5 ml per gram lung on subsequent washings increases return and prevents damage to the lung (Table I).

Studies looking at protein recovery with each successive wash have shown that the majority of material is collected in the first two washes and that yield decreases with each successive wash. In contrast, primarily because of their adherent properties, relatively few cells are recovered in the initial lavage, but yield increases in the next several washes. Six repetitions are generally adequate to remove the majority of free cells and lining constituents without causing excessive damage to the lungs themselves. Care should be taken, especially when serial lavages are required, not to wash the lungs excessively as this may damage the lungs and compromise further study.

Table I Comparison of Typical BAL Parameters and Cell Yields from Eight Different Species, Including Humans

Species	Body weight (g)	Wash volume (ml)	Total cells recovered ($\times 10^6$)	Cell differential (% total)				
				AM	PMN	EOS	LYMPH	OTHER
Mouse[a]	25–40	0.7–1	0.3–1	90–96	2	1–2	1–2	0–1
Hamster[a]	100–150	3–5	2–5	95–97	1–3	1–2	1–3	0–1
Rat[a]	200–325	5–8	4–10	96–99	1–2	1–2	1–2	0–1
Guinea Pig[a]	300–500	6–10	5–15	70–75	2	20–25	3–5	1
Rabbit[a]	2–3 kg	25–40		96–98	1–2	1–2	1–2	0–1
Monkey[b]	8–10 kg	10–25	12–20	89–92	1–2	1–2	8–9	0–1
Dog[b]	15–25 kg	10–30	10–40	70–75	5–10	7–8	8–14	1–2
Human[b]	70 kg	20–60	20–45	84–92	1–2	1	7–15	0–1

Abbreviations: AM, alveolar macrophage; PMN, polymorphonuclear leukocyte; EOS, eosinophil; LYMPH, lymphocyte.
[a] Denotes whole lung lavage.
[b] Denotes segmental lavage.

During the lavage process, the instilled saline washes away much of the free material including debris present in the lungs. The recovered fluid is often frothy due to the presence of surfactant, the lipids that line the alveolar surface of the lung. Therefore, once the specimen has been obtained, filtering through a single layer of loose gauze removes much of the excess debris and mucus. In addition, while proteins, primarily albumin, and glucose are present in the recovered fluid in sufficient quantity to provide nutrients temporarily to the respiratory cells until they can be further processed, the lavage fluid should be kept on ice during the lavage procedure to minimize cell stress and enzymatic degradation. Finally, because AMs adhere readily to glass surfaces, keeping the recovered fluid on ice and using a polypropylene or siliconized collection vessel instead of glass or polystyrene will minimize cell loss.

The advantage of excising the lung is to allow visual inspection of the lavage process and the lungs for evidence of disease. More commonly, however, the lavage is performed without taking the lungs out of the thorax. This saves time and avoids the possibility of damaging the lungs. Damaging the lungs causes leakage of the lavage fluid from the lungs and decreases the lavage efficiency. BAL performed in this way is accomplished by first severing the abdominal aorta and puncturing the diaphragm to allow the lungs' elastic recoil to force as much air as possible out of the lungs. The neck is then opened and the trachea is intubated as before and the washing procedure is carried out as previously described.

Techniques for lavage of living animals have been developed so that serial lavages can be performed if required. This is best done on larger animals where segmental lavages are possible but may be done on smaller animals with special care. Following anesthesia, a cuffed endotracheal tube is introduced through the larynx and placed in one of the major bronchi. The cuff is inflated to create a tight seal thus allowing one side of the lung to be lavaged while the other continues to meet the ventilatory needs of the animal. Since the right lung comprises 60% of the total lung tissue, lavage of the left lobe is preferred and tolerated better by the animal. Smaller segmental lavages may also be performed by using smaller caliber endotracheal tubes wedged in appropriately sized airways. Sterile saline is instilled and recovered through the endotracheal tube. As is the case with each of the lavages described, not all of the instilled fluid is recovered from the lungs and return may be less in animals possessing extensive collateral ventilation. Return is generally lowest on the first wash with subsequent returns closely approximating the amount of fluid instilled. All saline not recovered during the lavage process is absorbed into the capillaries and lymphatics of the lung. The lavage procedure may produce a mild inflammatory response in the lung characterized by some interstitial swelling, an influx of polymorphonuclear neutrophils (PMNs), and a subsequent increase in AMs. The response is generally temporary lasting only a few days.

2. Procedural Variables

Control of all aspects of the lavage process is important if lung lavage is to be used to quantify changes in the number or type of cells recovered in the lavage or to assess lung injury. To obtain quantitatively consistent recoveries of free cells, all variables affecting lavage efficiency must be controlled. Such factors include the type of anesthesia used, the time between animal death and lavage, the emptying of air from the lung prior to instillation of the wash fluid, the type and volume of wash fluid used, leakage and lung damage, pathological changes induced in the lung, and the number of washes performed. Other factors such as age, sex, and, of course, species of animal used, as well as body and lung weights can influence the number of free cells recovered by the lavage. More subtle effects may be due to wash temperature and the duration of the wash cycle. Warming the solution to 37°C has been shown to increase the yield of respiratory cells slightly. We, as well as others, have observed that gentle massaging of the lung during the wash cycle, and, in particular, between the instillation and withdrawal of the wash solution can more than double the number of cells recovered by the lavage.

Usually the lavage fluid recovered from each of the washes is pooled for each animal or individual. However, studies have been done where each fraction is kept separate for analysis. These studies indicate that the initial fraction may be somewhat more representative of the airways than subsequent fractions. In addition, there is generally a progressive decrease in the materials lavaged from the lung with subsequent fractions indicating a progressive removal of the material available for lavage. By pooling the fractions, much of this variation and any variations in the lavage procedure are reduced.

In the case of segmental lavages where only a small fraction of the total alveolar surface is lavaged, different results may be obtained depending on the segment studied if the response of the lung to insult or injury is not evenly distributed over the entire lung. Therefore, where possible, sampling from more than one segment or targeting the appropriate area of the lung is important. BAL of the entire lung or lung lobe may dilute the responses observed but also serves to assure sampling of the effects in isolated affected areas.

B. Human Studies

1. Lavage Techniques

BAL, as a nonterminal procedure in animals, is usually performed under general anesthesia. In contrast, BAL in humans is accomplished in conscious subjects. The details for performing a BAL on a human subject will vary from researcher to researcher and from institution to institution. The techniques presented are those we have successfully employed at the General Motors Research Laboratories and are adequate for healthy volunteer subjects. Patients with disease may require other considerations. Most importantly, human BAL should be carried out only by a licensed physician fully trained in the technique.

Subjects should not have ingested food for several hours prior to the procedure because of the potential for regurgitation and aspiration of the stomach contents. Generally, premedications are administered. Codeine may be administered intramuscularly (IM) as an antitussive and to calm the subject. Atropine (IM) is administered to alleviate the parasympathetic reflexes and bronchospasm that may result from the physical contact of the bronchoscope with the epithelium. The subject gargles with several ounces of lidocaine in water to begin the process of local anesthesia. A 2% lidocaine solution is then sprayed onto the posterior region of the pharynx until the gag reflex is eliminated. The subject then lies down. Supplemental oxygen may be given via nasal cannula. If the bronchoscope is inserted through the nasal passages the oxygen cannula can be placed in the mouth. To allow smooth insertion of the bronchoscope via the transnasal route, an aerosol nasal decongestant (e.g., 0.5% phenylephrine) can be administered to shrink the mucous membranes followed by viscous lidocaine or cocaine (4%) to anesthetize the nasal passages. For transoral bronchoscope insertion, a bite block should be used to protect the bronchoscope. As a safety precaution, the oxygen saturation of blood hemoglobin and pulse rate are monitored by pulse oximetry. A 2 or 4% lidocaine solution is injected through the bronchoscope to anesthetize the vocal cords and the airways as the scope is inserted. Suction of mucus and excess lidocaine is achieved with a regulated vacuum source set at a negative pressure of about 100 mm Hg. The bronchoscope is wedged in a third or fourth generation bronchial segment (see Fig. 1). Frequently, the right middle lobe and the lingula are chosen.

Once the bronchoscope is wedged in place, the regulated suction can be reduced to 60 mm Hg. Sterile saline is used as the lavage fluid and is usually warmed to 37°C. Aliquots of the saline have been used ranging from 20 to 60 ml. Data suggest that smaller aliquots preferentially sample the airway contents, while larger aliquots sample relatively more of the alveolar contents. Often five washings of a particular lung area are performed with the first washing being more representative of bronchial components of the epithelium than are the subsequent four washings. Washings are collected in a vacuum trap made of a material such as polypropylene to prevent cells from adhering to the sides. Keeping the trap immersed in ice will help extend the viability of the recovered cells. Return of the lavage fluid for five washings is approximately 60–70% with somewhat less return than the overall average occurring on the first washing. If the return is significantly less than this, it is possible that the bronchoscope is wedged forming a seal leading to collapse of the distal airways and diminished return. Slight movement and repositioning of the bronchoscope will often break the seal, opening the airways and allowing much of the previously trapped fluid to be recovered.

Once the lavage is complete, the bronchoscope is withdrawn and the subject is monitored for a period of time. Rales may be heard upon auscultation. Often the unrecovered fluid will stimulate the subject to cough. The cough will disappear as the liquid is adsorbed into the lymphatics and capillaries. The subject should not eat or drink until pharyngeal anesthesia has worn off so that aspiration into the lungs does not occur.

2. Ethical Considerations

Bronchoscopy and BAL are generally safe procedures when performed by a qualified, competent physician. The rate of complications in therapeutic and diagnostic lavage has been reported to be in the

0–3% range. In BAL for research purposes involving the use of healthy volunteers without underlying pulmonary disease, the complication rate is much lower. Complications that could occur are rare and include minimal hemorrhaging, laryngospasm, bronchospasm, and reactions to the medications.

In order to perform experimental research protocols involving BAL on human subjects, ethics dictate that a several step procedure be followed. First, approval must be granted by the human research committee or review board of the investigator's institution. Then subjects must be recruited, and informed consent obtained. Finally, a health history and a clinical exam should be performed on each subject that may uncover underlying pulmonary conditions that could complicate the BAL. During the BAL procedure, emergency medications, such as those routinely found in hospital emergency carts, should be readily available.

III. Lung Changes and Injury Responses

A. Lung Injury

The lung is often a major target organ for the toxic effects of many atmospheric pollutants, both gaseous and particulate in nature. Development of conditions in the lung that clearly lead to a compromise of the individual's ability to exchange respiratory gases with the environment to the point that it affects well-being and ability to function can be defined as lung damage. Lung damage can result in frank lung disease involving debilitating changes to the lung like emphysema and pulmonary fibrosis. In other instances, lung injury may be temporary and reversible as may occur with some infections. And in still other instances, a cascade of injurious changes in the lung may be initiated by an agent, and continue to evolve long after the toxic agent has disappeared, as occurs in the animal model of elastase-induced emphysema.

However, many responses and changes in the lung are not clearly definable as lung disease or damage, but yet can be discerned by BAL. This is particularly true with the inhalation of low levels of pollutants. Not every measurable change is necessarily a pathologic response. Toxicity refers to a state of being poisonous. Poison, in turn, refers to a substance that is injurious to health. Not all responses evoked by substances are injurious. The lung, like other organs, is a dynamic entity and continually responds to chemical and physical stimuli. Therefore, a discernible response of the lung does not in itself necessarily indicate lung damage. Often, the line between normal physiologic responses and adverse pathophysiologic responses is not clear. This issue is a focal point of debate in the regulation of atmospheric pollutants.

B. Cellular Changes

1. Cellular Collection and Preparation

In many cases, the cellular constituents obtained in the lavage provide a good indication of lung injury. Both the number and type of cells obtained as well as their viability and state of activation provide insight into the potentially injurious effects of an inhaled agent. Cell counts are performed either by a hemocytometer or automated cell counter. A differential cell count is usually made from cytospin (cytocentrifuged) preparations stained with Wright–Giemsa stain. For the most representative sample, it is best to perform the initial cell counts and differentials on raw, unconcentrated lavage fluid as some cell types are preferentially lost during the centrifugation process. Cell counts are generally expressed as cells per ml of lavage fluid, per gram lung, or per kilogram body weight. In smaller animals where the entire lung is lavaged, quantitation is not as big a problem as it is in larger animals or humans where only a portion of the lung is lavaged. Cell numbers are directly affected by the volume of fluid instilled and the total volume of fluid recovered from the lung. For segmental or single lobe lavages, if a standard wash volume is used, cell counts are assumed to be representative of a standard area of the lung. However, because the site of injury and area of response may not be homogenous throughout the lung, where possible, lavages of the entire lung or more than one site should be performed to reduce sampling error and assure that the response will be observed. Addition of external markers such as methylene blue or analysis of internal markers such as albumin have also been used as references to standardize various samples.

Both differential cell counts and an estimate of total cell number are important when reporting the results of cell analysis based on BAL. Each gives important, unique, and complementary data. Neither alone can give the complete information necessary to assess inflammation and possible lung injury.

A change in the normal differential cell count may signal an increase in inflammatory cells, in particular PMNs and lymphocytes. However, a significant reduction in the absolute number of macrophages lavagable from the lung could also account for the apparent proportional increase in inflammatory cells. Knowing both an estimate of the actual number of cells recovered and the differential profile markedly increases our understanding of the cellular responses of the lung.

Identifying and analyzing the cellular profile of the lung lavage aids in distinguishing inflammatory processes in which lymphocytes are primarily involved from those in which PMNs or macrophages play a primary role. Also, further study of the cells isolated from lung lavage can reveal information on their biochemical and functional nature. Cells can be analyzed for their morphology, structural integrity, viability, activational state, specific immune properties, and presence of specific surface markers. Functional properties such as the ability to migrate, respond to chemotactic stimuli, phagocytize, and synthesize DNA provide an understanding of the mechanism of response.

2. Macrophages

AMs are the predominant cell type present in lung lavage. These cells make up 80–95% of the cells harvested from the lung and are important in the defense of the lung against a variety of insults. These cells phagocytize foreign materials, bacteria, and cellular debris. They also regulate other cell types in the lung through the release of various mediators and chemotactic factors. Changes in macrophage numbers or function are a factor in determining lung injury and in characterizing the pathogenesis of such a response. While a decrease in macrophage number or phagocytic capacity may result in a reduction in clearance of inhaled materials and thus an increase in the effective dose of the potentially injurious agent, an increase in AM number may also have negative effects on the lung through the release of proteases, active oxygen species, and fibroblast-regulating mediators.

3. Polymorphonuclear Leukocytes

PMNs are a class of granulated cells, including neutrophils, eosinophils, and basophils, that are involved in defense responses in the body's tissues. Under normal conditions, PMNs comprise a relatively small percentage (less than 5%) of the cells in the lung lavage. Inhalation or instillation of particulate materials in the lung have been characterized by an early and marked influx of PMNs. Although the ability of the lung to recruit and mobilize PMNs is important for its defense against infection and disease, a large and prolonged increase in PMNs may also be detrimental to the lung. This occurs through an increase in vascular permeability, the release of lysosomal enzymes and proteases, and the release of the active oxygen species produced by these cells. For example, PMNs may cause lung damage by secreting neutral proteases and oxygen radicals. An increase in reactive oxygen species that may initiate lipid peroxidation can be detected using BAL following inhalation of materials such as asbestos, ozone, hyperoxia, and nitrogen dioxide. However, an increase in eosinophils in the lung may provide an even greater potential for damage by secreting not only neutral proteases and oxygen radicals but also other factors such as eosinophil basic protein. This factor may injure the epithelial lining of the lung and has been detected in the lavage fluid of patients with idiopathic pulmonary fibrosis. The presence of high numbers of eosinophils in BAL fluid has also been useful in the diagnosis and characterization of eosinophilic pneumonia.

4. Lymphocytes

Lymphocytes are immune cells that play diverse roles in the defense against invading microorganisms, foreign materials, and cancer cells. Lymphocytes usually comprise less than 10% of the total cell population recovered from the lung by BAL. Changes in lymphocyte cell numbers in the BAL fluid may reflect immunologic changes in the lung itself. For example, an increase in lymphocytes may be involved in a variety of hypersensitivity reactions. Like AMs, lymphocytes may also contribute to the fibrotic response of the lung through the release of factors that stimulate fibrogenesis and activate macrophages. Increases of lymphocytes in the lavage fluid have been observed from patients with hypersensitivity pneumonitis, berylliosis, and tuberculosis where an immunologic mechanism is thought to be primarily responsible, as well as other disorders, such as sarcoidosis, where a specific immunological response has not been identified.

5. Red Blood Cells

Because red blood cells are rarely found in BAL of normal animals or humans, their presence may suggest hemorrhage and serious lung injury. A small amount of blood in the lavage fluid may be produced during the lavage procedure itself. However, any significant presence of blood in the lavage may reflect endothelial damage, a rupture of lung capillaries, and bleeding into the alveolar space.

C. Humoral Components

1. Proteins

Changes in the type and quantity of proteins found in BAL have been described in characterizing lung injury. Depending on the sensitivity of the assay, protein measurements can be made on the unconcentrated lavage fluid. A great number of proteins have been identified in the lavage fluid. Larger proteins (150,000–300,000 daltons) are usually present in only barely detectable quantities due to the selective permeability of the capillary endothelium. The presence of appreciable quantities of these proteins is a very sensitive indicator of endothelial damage and leakage of these serum proteins into the alveolar space.

The most widely assayed protein in the BAL fluid is albumin. While it is also usually the most abundant protein in BAL fluid, it is primarily a serum protein. Elevated levels of this protein are generally due to endothelial damage and leakage of this protein across the endothelial and epithelial barriers. Increased levels of albumin in the lavage fluid indicate pulmonary edema, a common result produced by acute lung damage. Measurements of albumin have also been used to standardize the presence of other serum proteins that may diffuse across the air–blood interface.

BAL has also been used to study processes occurring in the interstitium by characterizing the proteins from the extracellular matrix. Elevated hydroxyproline levels have been studied as a marker for collagen synthesis and fibrosis. Fibronectin is another cell surface protein involved in collagen synthesis and the activation of fibroblasts. Because of the high molecular weight of this molecule, it cannot easily cross the alveolar epithelium. Therefore, increases in fibronectin in the lavage fluid reflect local production of the molecule and may serve as a good marker for fibrosis or similar lung injuries.

2. Enzymes

A variety of enzymes washed from the lungs in the bronchoalveolar fluid have been studied as markers of lung injury and disease. The presence of cytoplasmic enzymes in the extracellular supernatant of the BAL fluid is often used as a reflection of cell damage and cell lysis. An increase in extracellular lactate dehydrogenase (LDH) is a good indication of acute general cellular injury (damage). However, studies that identify the specific isoenzyme patterns produced by the enzyme have been used to show that the LDH present in different cell types produces a different isoenzyme pattern. Therefore, LDH provides a way to distinguish between different types of injury and to identify the target damage produced by various lung insults.

Alkaline phosphatase activity also rises following cell damage and is, therefore, a good indicator of lung damage. Alkaline phosphatase is an enzyme associated with the plasma membrane. Isoelectric focusing has been used to determine the isoenzyme pattern of alkaline phosphatase activity giving some indication of the source of the enzyme and, therefore, the possible site of lung injury. Alkaline phosphatase may be associated with epithelial type II cells and present in much lower concentrations in AMs, making it a good marker for type II cell injury.

A variety of lysosomal enzymes have also been used to characterize lung damage. Acid hydrolases are lysosomal enzymes present in AMs, epithelial type II cells, and PMNs. They are released from the cell during phagocytosis or after cell injury or cell death. Although their primary function is to kill and digest phagocytized bacteria and to promote wound healing through the degradation of damaged tissue, prolonged and excessive release of these enzymes may produce destruction of normal tissue in the lung.

Similarly, proteases are responsible for the turnover and remodeling of the tissues of the lung. If left uninhibited, proteolytic activity may lead to an enlargement of the alveoli and the development of emphysema. Two such enzymes, elastase and collagenase are secreted by both PMNs and AMs and are responsible for the degradation of elastin and the loss of lung elasticity as well as the degradation of collagen and the turnover of connective tissue in lung.

Angiotensin-converting enzyme is primarily located in the pulmonary endothelium, and increases of this enzyme in the lavage fluid have been used to detect damage to endothelial cells. Also synthesized by AMs, increases in this enzyme may be produced by increases in macrophage cell numbers and activation of these cells in the lung. Comparing increases in angiotensin-converting enzyme to relative increases in albumin helps determine the source of enzyme and the mode of injury.

3. Lipids

In addition to proteins, a significant portion of BAL is composed of lipids. Secreted by epithelial type II cells, pulmonary surfactant represents most of the lipids present in BAL. Changes in the amount of surfactant present may be due to changes in either the synthesis or removal of lipids in the lung. A decrease in BAL lipids may reflect a decrease in lipid synthesis, an increase in transport both in the alveolus or the broncho-mucocilliary escalator, an increase in lipid degradation, or enhanced ingestion by alveolar macrophages. An increase in the phospholipid content of BAL has been observed experimentally in several diseases, including silicosis and radiation injury in animals and proteinosis in man. In addition to changes in the amount of lipid present in the lavage, changes in the chemical nature of the lipids themselves, such as the degree of saturation of phospholipids, have been observed and result in altered surface tension and functional properties.

4. Immunoglobulins

Immunoglobulins are glycoproteins that are synthesized and released into organ and body fluids by plasma cells, those B lymphocytes that are activated to respond to foreign material. Immunoglobulins can be produced continually in the body or they can be produced on demand to neutralize specific foreign molecules. The latter comprise a subset that are referred to as antibodies and serve a variety of functions in host defense. They directly kill or lyse foreign material, or they can work by activating complement, another native protein group that facilitates cell destruction and uptake by phagocytic cells. Five major classes of immunoglobulin have been identified in human fluids: IgA, IgG, IgM, IgE, and IgD. The effect of hazardous material on immune responses involving immunoglobulins is being extensively investigated.

Lavageable immunoglobulins of IgA, IgG, IgM, and IgE classes reside in the pulmonary system. IgD has not been reported in BAL. BAL has shown that immunoglobulins are distributed throughout the pulmonary system in a pattern that reflects compartmentalization of immune function. For example, the lavage fluid from the upper airways, including the nasal passages, is characterized by the presence of IgA, an antibody that is synthesized by cells and then secreted onto tissue surfaces. This antibody protects the mucous surface of the airway structure. Because IgA is secreted, it is particularly amenable to analysis using BAL. In some protocols, the first wash of the lavage fluid is separated from subsequent washes because it is rich in IgA and airway surface components. IgM is the largest of the immunoglobulins with a molecular weight of 900,000, which prevents it from being transported across blood/tissue barriers in the lung. Because of this, it is present in low concentrations in normal lavage fluid. However, both IgA and IgM can be identified in lavage fluid in larger quantities after immunologic stimulation.

In contrast to the upper airways, IgG predominates in lavage fluid from the lower airways. IgG subclasses have been found, including IgG_{1-4} in humans, in the lavage fluid of the peripheral lung. IgG is normally present in blood and the presence of IgG in lavage fluid in some cases relates to breakdown of the air/blood barrier. In other cases, lymphocytes synthesize antibody when immunologically activated and antibody secreting cells appear in the deep lung. Correlation of immunoglobulins to albumin levels in lavage fluid and serum substantiates that lung immunoglobulin can be locally synthesized.

IgE is present in very low concentrations in lavage fluid from the normal lung. BAL has been used to show that IgE levels are increased in asthma, hypersensitivity states, and infection with parasites. IgE plays a key role in interactions with basophils, mast cells, and macrophages in stimulating the release of powerful vasoactive and smooth muscle mediators. Lavage fluid from allergic asthmatics shows elevated levels of specific and total IgE compared to nonallergic subjects; the IgE levels in BAL are correlated to levels in blood. Likewise, BAL has proven that allergic sensitization with ovalbumin in guinea pigs can be used as a model of IgE-mediated immediate and late bronchoconstriction. Finally, BAL has helped establish the relationships between IgE secretion, degranulation of mobile and resident cells (including mast cells and basophils), and secretion of vasoactive substances.

5. Complement

Complement is a name for a system of up to 20 serum proteins that plays several key functions in the inflammatory response. Like several other classes of inflammatory mediators, complement can stimulate inflammation by increasing vascular permeability and through the production of mediators that act as potent chemoattractants for inflammatory cells like PMNs. It can also facilitate destruction and degradation of foreign cells (e.g., bacteria) by interaction with humoral and cellular constituents of the lung, and it can down-regulate inflammation through a number of components that inhibit activated forms of the complement system.

Many of the components of complement have been identified in BAL from normal and damaged lung tissue in both their native and activated forms. Lavagable complement resides on the lung surface and has been implicated as a mediator of the inflammatory response to fibrous materials such as asbestos and to bacteria. Activated complement components (such as C5) have been lavaged from patients with active lung damage produced by sarcoidosis and adult respiratory distress syndrome. In addition, C3b has been observed in idiopathic pulmonary fibrosis. These results suggest that complement activity is present and accounts for some of the pathophysiologic responses that can occur during lung damage. However, its role requires further clarification and explanation.

6. Cytokines

Cytokines are polypeptide mediators of cellular homeostasis that have been implicated in the response to chemically mediated lung damage. Cytokines can activate and deactivate cellular functions and are important in remodeling of interstitial matrix after injury. They are important in normal, adaptive, and pathologic function, and they regulate the response to other classes of mediators including eicosanoids, hormones, and neurotransmitters. Cytokines can regulate the function of the cell that produces it (autocrine function) and affect nearby cells as well (paracrine function).

BAL is often used to harvest cells that can then be studied for their ability to secrete cytokines. Because of the ease of obtaining macrophages via BAL, the production of cytokines by these cells has been extensively studied. Interleukin 1 (Il-1) is an important initiator of inflammation, which can be produced by cells of the monocyte/macrophage series as well as other cells. Lavageable macrophages from the lung produce both Il-1α and Il-1β but to a lesser extent than blood monocytes. Il-1 enhances leukocyte attraction, induces arachidonic acid metabolism, and stimulates tissue type procoagulant activity. It changes cell morphology, alters adhesion properties, and serves as a growth stimulant for lymphocytes, fibroblasts, epithelial, and endothelial cells. Silica, asbestos, coal dust, and bleomycin exposure increase levels of Il-1 production.

Like Il-1, tumor necrosis factor-α (TNF) can initiate inflammatory changes in the lung. Many of the cellular changes that are produced by Il-1 can be produced by TNF; TNF combined with Il-1 functions are likely additive. TNF potentiates injury produced by foreign substances, including endotoxin, and can produce hemorrhage, leakage of lung fluids, and epithelial damage. In contrast to Il-1, TNF can be produced from stimulation of AMs to a greater extent than blood monocytes. Increased secretion occurs in stimulated macrophages lavaged from sarcoid patients. The degree of Il-1 and TNF secretion has been cited as one correlate that can differentiate relatively inert from more fibrosis-causing dusts.

Amplification of immune responsiveness can be achieved in the lung by lymphocytes that secrete interleukin 2 (Il-2). This mediator amplifies the lymphocyte response by inducing its own receptor (Il-2R) and expanding lymphocyte populations. Human T lymphocytes from BAL specimens respond to Il-2 to the same extent as T cells obtained from peripheral blood. Il-2 has been seen in asbestos-exposed animals.

Cytokines with regulatory and effector functions, such as transforming growth factor-β (TGF-β) and interferons can be identified using BAL. TGF-β remodels the intracellular matrix of the lung and can increase collagen and fibronectin secretion. Interferon-γ (INF-γ) is a potent immunoregulator that is important in antiviral defense mechanisms and in the regulation of the immunologic activation state of macrophages.

We have previously indicated that the presence of a given substance does not necessarily indicate damage but may reflect other types of physiologic responses instead. BAL has been used to demonstrate that protective and adaptive effects can be produced against toxic agents by cytokines. In the rat, oxygen exposure increases capillary permeability, which can be detected in increased protein levels of BAL fluid. After exposure to TNF and/or TNF–Il-1 combinations, the increased protein levels after oxygen exposure are less apparent. This protective effect is associated with induction of antioxidant enzymes, such as SOD, catalase, and glutathione peroxidase, and protects against oxygen-induced mortality.

7. Eicosanoids

The term eicosanoids refers to the myriad of 20-carbon fatty acid metabolites of arachidonic acid.

This includes the prostaglandins, leukotrienes, and thromboxanes. Arachidonic acid and the eicosanoids are found in essentially every tissue in the body. They have been implicated as the mediators of numerous physiological events and are likely involved in the response of the lung to injury or disease. Generally, eicosanoids appear to be secreted by one cell and produce an effect on another cell. Some of the actions of the eicosanoids may be due to a change in the amount produced or in the way the target cell responds to the substance. Some eicosanoids are relatively unstable or are metabolized rapidly and, therefore, have very short half-lives. Consequently, they may be present in the subject but cannot be measured quickly enough in BAL fluid to be properly quantitated. In some cases, this analytical problem can be addressed because a stable metabolite or breakdown product may be formed that can be measured. Some eicosanoids are continually produced and secreted by cells under normal conditions, while others may be secreted only in response to a particular stimulus or challenge. This means that a challenge could, for some eicosanoids, result in an increase (up-regulation) or decrease (down-regulation) in their production.

Leukotrienes C_4 and D_4 (LTC_4, LTD_4) are potent pulmonary vasoconstrictors and, therefore, increases in vascular permeability may occur and lead to the development of edema in the lung. Other eicosanoids such as the prostaglandins D_2, E_2, and I_2 (PGD_2, PGE_2, PGI_2), decrease pulmonary vascular tone while increasing permeability, indicating that the mechanism by which the increase in vascular permeability is achieved is different from that for LTC_4 and LTD_4. Leukotriene B_4 (LTB_4) is believed to be strongly involved in the inflammatory process by virtue of its powerful chemoattractant and activating properties for PMNs. Prostaglandin $F_{2\alpha}$ ($PGF_{2\alpha}$), thromboxane A_2 (TXA_2) as well as all of the leukotrienes cited above can induce constriction of the smooth muscle of the airways (see Fig. 1), indicating their potential for participating in processes that lead to bronchoconstriction.

Eicosanoids have been implicated in the acute response to various insults by virtue of their altered concentrations found in BAL fluid after the challenge. Ozone inhalation in humans increases the amount of PGE_2 in lavage fluid, but not the amount of LTB_4. In rabbits, exposure to high levels of ozone results in lavageable increases in both PGE_2 and $PGF_{2\alpha}$. Administration of ibuprofen, an inhibitor of cyclooxygenase (responsible for the first step in the synthesis of prostaglandins and thromboxanes from arachadonic acid), inhibits the degree of sepsis-induced lung injury as measured by increases in BAL fluid total protein. Many eicosanoids are produced by macrophages harvested by BAL. Macrophages collected from smokers produce much lower amounts of the eicosanoids than do those from nonsmokers. Although much data have accumulated, the possible linkage of alterations in eicosanoid metabolism to the etiology of chronic structural and functional lung disease is still tenuous, with research demonstrating associations rather than cause-and-effect relationships.

IV. Conclusions and Future Directions

We have examined herein the use of bronchoalveolar lavage by reviewing the historical development of the technique, discussing the methods employed in animals and humans, and describing the cellular and humoral changes that may occur in the lung during exposure to hazardous materials that could produce lung injury and inflammation. The approaches described have already become well-established and provide a means to study physiology and pathophysiology of the respiratory system at the cellular and subcellular level. Using this approach, detailed "fingerprints" of the multiple effects that initiate pulmonary damage are being characterized.

In the future, further refinement of the technique will provide for a more discrete and reproducible sampling of specific portions of the airway. In addition, increased use of molecular biology techniques will allow studies of the state of gene activation, gene-level dysfunction in disease, and latent or subclinical forms of genetic and cellular injury. Given the long latency periods of severe forms of lung disease including lung cancer, identification of markers early in the disease process can provide useful tools in understanding individual susceptibility to disease development. Some have proposed that BAL could be used as a tool to introduce genetic materials directly into lung cells or into harvested cells that can be modified and returned to the host, representing a new form of therapeutic intervention. Therefore, even though the process of BAL has been developing for decades, the power of BAL as a diagnostic, research, and therapeutic technique is only beginning to be exploited.

Acknowledgment

We would like to thank Carol Zawacki for her artistic contributions in composing and executing the schematic drawings of the lung.

Bibliography

Linder, J., and Rennard, S. (1988). "Bronchoalveolar Lavage." ASCP Press, Chicago.

Reynolds, H. Y. (1988). Bronchoalveolar Lavage. *In* "Textbook of Respiratory Medicine" (J. F. Murray, and J. A. Nadel, eds.) W.B. Saunders, Philadelphia.

Subcommittee on Pulmonary Toxicology, National Research Council (1989). "Biological Markers in Pulmonary Toxicology" National Academy Press, Washington, D.C.

Cadmium

Anna Steinberger
University of Texas Medical School at Houston

Glossary

Biohazard Potential risk in living organisms.

Oxide Oxidation product of an element after it combines with oxygen.

Salt Compound formed by the interaction of an acid and a base.

Spermatogenesis Process by which the male gametes, spermatozoa, are produced in the testes.

Xenobiotics Substances foreign to a biological system, from the Greek *xenos* (stranger).

CADMIUM is a heavy metal occurring widely in nature as a contaminant of another heavy metal, zinc. Because of many desirable physical and chemical properties, it is commercially produced and used in such diverse industries as paint, electrical, and automobile. It has no known physiologic function. However, trace amounts of cadmium are naturally present in water as well as in plants and animals used for food; thus the human population is inevitably exposed to it via the digestive system and also by inhaling contaminated air. Excessive exposure to cadmium or its compounds may represent a health-threatening biohazard and should be avoided. The precise mechanism(s) by which cadmium exerts toxic effects remains to be clarified, and a considerable amount of research is being conducted toward this goal.

I. Physical and Chemical Properties of Cadmium

Cadmium (usually abbreviated as Cd) is a silver-white metallic element which is chemically related to zinc and mercury. It is slightly harder than tin but softer than zinc and emits a crackling sound when bent. Cadmium forms alloys with various heavy metals, many of which are very useful in industry because of low melting points. Some of the physical and chemical properties of cadmium are listed in Table I.

II. Occurrence in Nature and Commercial Production

Cadmium was first discovered in 1817 by Fredrich Strohmeyer in a sample of zinc carbonate. Cadmium does not occur in nature as an uncombined element and is nearly always found associated in small amounts (not exceeding 3%) with zinc. Commercially, most cadmium is recovered as a by product during various steps of zinc purification; thus its output is directly linked to the output of zinc. Only one mineral contains cadmium to any appreciable quantity; that is greenockite or cadmium sulfate found in Greenock, Scotland, and several other areas including the state of Pennsylvania. The United States is the world's largest producer of cadmium (over 5,000 tons annually).

III. Major Uses of Cadmium

A large amount of cadmium produced worldwide is used in electroplating of steel, iron, copper, brass, and other metals to protect them from corrosion. Cadmium is also often used in electrical parts because of its relatively low electrical conductivity and the ease of soldering to it. A small portion is used to

Table I Physical and Chemical Characteristics of Cadmium

Atomic weight	112.41
Atomic number	48
Melting point	320.9°C
Boiling point	767°C
Electrical resistance at 20°C	7.14×10^{-6} ohm cm^2
Density at 20°C	8.65 g/ml
Most common salts:	Cadmium hydroxide Cd $(OH)_2$ Cadmium chloride $CdCl_2$[a] Cadmium bromide $CdBr_2$[a] Cadmium iodide CdI_2[a, b] Cadmium sulfate $CdSO_4$[c] Cadmium sulfide CdS[d]

[a] Soluble in water.
[b] Soluble in alcohol.
[c] Used in electric batteries.
[d] Occurs naturally in greenockite.

produce various pigments. The sulfides and sulfoselenides provide the best yellow and red colors seen in many finishes, including industrial enamels and lacquers used for automobiles. Cadmium elements used in electrical batteries have a much longer life than lead elements, and, because of its high neutron absorptive qualities, cadmium is often used in the control rods of nuclear reactors. Cadmium chloride ($CdCl_2$), bromide ($CdBr_2$), and iodide (CdI_2) are all soluble salts, and CdI_2 is sometimes used in photography.

IV. Biotoxicity of Cadmium

Many cadmium compounds are toxic to biological systems and must be handled with caution. The United States Environmental Protection Agency considers cadmium to be a potential human carcinogen. Cadmium can enter the organism via the small intestine with food and water intake or through the respiratory system as aerosol particles. Certain cadmium compounds are very poisonous, while others seem to have a lesser effect. This is probably due to differences in absorption efficiency. Systemically, cadmium leads to irreversible kidney damage, whereas the metal vapor itself and oxides which form from its combustion in air can cause violent reactions and severe lung damage when inhaled in large quantities. Occupational exposure symptoms are often delayed 24 hours followed by edema, often leading to pulmonary insufficiency. Fortunately, warning symptoms such as nausea and diarrhea often occur before more permanent injury is caused by additional exposure. Thus, exposure to mists and aerosols of cadmium solutions should be considered hazardous and should be avoided. Insoluble cadmium compounds, such as cadmium sulfide and cadmium oxide, can be tolerated in relatively larger amounts.

Cadmium is an environmental pollutant to which exposure has increased directly proportional to the progress of industrialization. It reaches the general population through water and food supplies and, to a lesser extent, by air. Certain professionals, for example painters and workers in electrical battery industries, are by the nature of their occupation exposed to relatively higher doses. The term *environmental pollutants* is often used for chemical substances which have no known physiological function. One of the main reasons for the inevitable exposure of the general population to cadmium is its association with zinc, an obligatory trace element for most organisms, plants, and animals used as food, particularly since certain organs, organisms, and plants (e.g., liver, kidney, clams, and mushrooms) tend to accumulate cadmium. This implies that goods consumed by the general population are daily cadmium sources. Moreover, the increasing use or combustion of fossil fuels in cars and industries significantly contributes cadmium to air pollution. In addition, heavy smoking of tobacco can increase the daily cadmium intake of individuals by about 50%.

In contrast to zinc, cadmium is a nonphysiological metal; no essential biological functions have been ascribed to it. Its chemical similarity to the biologically important zinc suggests that cadmium may replace zinc in many zinc-dependent enzymes which are crucial for body metabolism (e.g., carbonic anhydrase, thymidine kinase, DNA-dependent RNA polymerase, DNA polymerase, and many others). Another interesting sequela of cadmium's similarity to zinc is that metallothioneine, a protein which acts as a zinc and copper deposit in most organisms, is induced by and binds cadmium, thus creating to some degree protection from its toxic effects.

Most environmental pollutants to which human beings are inevitably exposed circulate in the blood and can cause damage in a variety of cells and organs in the body. Cadmium has been related to cell and organ damage particularly in the kidney, lung, endothelium, and testis. Stricter governmental regulations concerning permissible levels of cadmium in food and water have considerably reduced the hazards of toxic damage. However, the total elimination of its toxic effects will only be achieved after its precise cellular site and molecular mechanisms of action are fully understood. More research is needed to gain a better understanding of how the damage occurs and what preventive means could be taken to avoid it. Many studies addressing these issues are currently conducted both *in vivo* and *in vitro,* using modern techniques of toxicology, biochemistry, and molecular and cellular biology.

Biochemical methods can detect changes in proteins, lipids, carbohydrates, the activity of specific enzymes, and so on. Cellular biology, on the other hand, can reveal changes in specific cell types or their organelles. Both methods can be effectively combined for cytochemical or histochemical analysis of localized cellular changes. Changes occurring in a few specialized cells can be detected using light or electron microscopy or by histochemical methods. These determinations are usually conducted on properly preserved tissue sections in which the normal architecture and structural relationships have been retained.

Dynamic cellular effects of toxic substances, particularly the reversibility of such effects, are often investigated in cell culture where changes in cell metabolism and/or viability can be investigated over a period of time under strictly controlled conditions. Although the results obtained from *in vitro* experiments cannot (and should not!) be extrapolated directly to a whole living organism, they can provide valuable clues as to the site and mechanism(s) of action. Cell and tissue culture can provide valuable models for experimental approaches which often are not possible or practical *in vivo*. The identification of specific sites affected by environmental pollutants (or other xenobiotics) and the mechanism(s) involved in the expression of their toxicity could lead to more rational approaches in the development of effective preventive measures.

Many examples of toxic effects induced by cadmium have been described in the scientific literature. Cadmium binds easily with sulfhydryl groups, suggesting that cell membrane proteins such as glutathione and other sulfhydryl-containing cellular proteins and enzymes may be the initial targets for cadmium in the cell. At high doses (5 mg/kg body weight) cadmium induces irreversible chromatin condensation in cultured rat liver cells, which prevents RNA transcription and thus leads to cell death. The effect is very rapid; it is noticeable after approximately 15 min. Similar changes in nuclear euchromatin also occur in thymocytes (precursors of peripheral T lymphocytes). In these cells cadmium also inhibits decondensation of chromatin, thus preventing cell division and formation of the immunologically important circulating T lymphocytes. In the spermatozoa, the nucleus contains highly condensed chromatin, which becomes decondensed in the female oocyte at the time of fertilization. By interfering with the decondensation process, cadmium may prevent fertilization from occurring. The extent of chromatin changes induced by cadmium appears to be related to the dose and cell structure. A cell with a relatively small amount of cytoplasm (e.g., thymocyte or spermatozoan) is affected by lower doses of cadmium than large cells (liver parenchymal cells). Increased permeability of vascular endothelium is another example of cadmium toxicity. Blood vessels in the testis are particularly sensitive to this effect and are believed to be the primary reason for testicular damage following exposure to cadmium. There is also evidence that cadmium may directly affect the Sertoli cells—somatic cells of the seminiferous tubule which play a key role in the process of spermatogenesis.

With low doses of cadmium, cellular defense mechanisms were found to operate in the liver and other organs. There is a dramatic increase of high-molecular-weight (HMW) proteins which for the most part have not been identified. In addition, there is an increase of metallothionein, a protein high (30%) in cysteine content which strongly binds cadmium in the form of cadmium thiolate. An increase in both these proteins has a protective function against cadmium toxicity, but only to a limited extent. Some of the HMW proteins induced by cadmium have been related to the microtubular system or the heat shock proteins (sometimes called "stress" proteins) which increase following exposure to various environmental stimuli such as heat, metals, hypoxia–ischemia, and viruses. The increase in HMW proteins can be detected biochemically or microscopically as an increase in cellular and nuclear size (hypertrophy). The second protective mechanism involves an induction of metallothioneine, a low-molecular-weight (MW = 10,000) cytosolic protein occurring in most cells and believed to regulate zinc homeostasis. It is normally not found in body fluids, and its presence in urine is generally used as a parameter of kidney damage. In this context, it is important to realize that metallothioneine can also be induced by other metals (zinc, mercury, or copper) as well as a number of nonmetallic stimuli such as heat, cold, consumption of alcohol, strenuous exercise, and X-irradiation. The genetics of the HMW proteins and metallothioneine are well known, and their phenotypic expression provides the basis for differences in susceptibility of various cell types and organisms to the adverse effects of cadmium. Disturbance in either of these proteins by environmental pollutants has serious implications for cell function and organ pathology.

The protective effect offered by increased synthesis of HMW proteins and metallothioneine is limited. With chronic exposure to cadmium (even at low doses), synthesis of these proteins increases, leading to cell hypertrophy and eventually to cell death. Burst cells have been observed in the liver. Bursting of the cell membrane is facilitated by a reduced content of 5′-nucleotidase, an enzyme necessary for membrane integrity which declines following exposure to cadmium in a dose-dependent manner. A reduction in other critical enzymes was also reported. Normally, liver cells do not divide and cadmium can accumulate in them for many years. As mentioned earlier, cadmium can also cause damage in other organs—notably in the kidney, testis, cardiac muscle, lung alveoli, and capillary endothelium. The reasons for higher sensitivity of these organs to cadmium are not clearly understood.

After ingestion, cadmium is rapidly absorbed as a free ion by the epithelial layer of the small intestines. The absorption rate of cadmium is influenced by many factors, such as the amount of protein, iron, and calcium in the food and genetic makeup. With industrial development and increasing use of fossil fuels, the air we breathe contains a certain amount of cadmium in the form of oxides, chlorides, and other compounds. The alveolar type I epithelial cell in the lung involved in the exchange of oxygen and carbon dioxide is most sensitive to the toxic effects of cadmium. After cadmium gets into the blood circulation, either through the lungs or intestine, it can cause damage in other organs as well. A secondary mechanism for cadmium-induced changes following damage of liver cells is the development of autoantibodies to various membrane components. For example, antilaminin antibodies have been found in industrial workers chronically exposed to even low doses of cadmium; however, the impact of such antibodies on the total health of an individual has not been assessed.

There are several important points to be considered concerning the effects of cadmium (or other xenobiotics) on human population:

1. Much of our knowledge about the site and mechanism of action has been derived from experiments conducted in tissue/cell cultures or in animal models. The effects observed in such studies are not always predictive of similar effects occurring in human beings. Scientists are continuously searching for better predictive models of human toxicity. In the case of cadmium, there is considerable clinical evidence that renal and pulmonary effects can result from occupational exposure.
2. Often the doses of toxicant used in experimental protocols far exceed the levels of normal exposure in nature; however, humans have been known to be exposed to toxic doses of cadmium and its effects are cumulative.
3. Besides genetic differences in sensitivity, many factors in the diet and lifestyle can significantly affect the degree of individual damage. In some cases, environmental pollutants which were found to be toxic in animal models and/or cell cultures had little or no reported effects in the human, and vice versa.
4. Besides its direct effects on public health, cadmium can contribute to the effects of other environmental pollutants and factors (both external and intrinsic) in causing health hazards. Although the placenta presents an effective barrier to cadmium, thus preventing fetal exposure to cadmium which may be present in the maternal circulation, one should not assume that *no* risk to the fetus is involved.

Related Article: METAL–METAL TOXIC INTERACTIONS.

Bibliography

Copius Peereboom-Stegeman, J. H. J. (1987). Toxic trace elements and reproduction. *Toxicol. Environ. Chem.* **15,** 273–292.

Copius Peereboom-Stegeman, J. H. J. (1988). Cadmium effects on the female reproductive tract. *Toxicol. Environ. Chem.* **23,** 91–99.

Hecht, N. B. (1987). Detecting the effects of toxic agents on spermatogenesis using DNA probes. *Environ. Health Perspect.* **74,** 31–40.

Janecki, A., Jakubowiak, A., and Steinberger, A. (1992). Effect of cadmium chloride on transepithelial electrical resistance of Sertoli cell monolayers in two-compartment cultures—A new model for toxicological investigations of the blood–testis barrier in vitro. *Toxicol. Appl. Pharmacol.* **112,** 51–57.

Morselt, A. F. W. (1991). "Environmental Pollutants and Diseases" (H. P. Witschi and K. J. Netter, eds.), special issue of *Toxicology,* Vol. 70, No. 1, pp. 1–128. Elsevier Scientific Publishers, Limerick, Ireland.

Chloroform

Luciano Vittozzi and Emanuela Testai
Istituto Superiore di Sanità, Rome, Italy

Glossary

LC_{50} (lethal concentration$_{50}$) Concentration of a chemical causing death in 50% of exposed animals.

LD_{50} (lethal dose$_{50}$) Dose of a chemical causing death in 50% of treated animals.

Microsomal monooxygenase system Superfamily of cytochrome P450 and related enzymes, localized in the endoplasmic reticulum, that catalyze the transfer of electrons to molecular oxygen or to a reducible substrate, resulting in the reduction of the substrate or monooxygenation of the substrate by reduced oxygen species.

Reduced glutathione (GSH) γ-Glutamylcysteinyl glycine, a tripeptide present at high concentrations in the cysotol of hepatocytes (8–10 μg/g wet tissue). It reacts with electrophilic molecules, thus protecting cellular structures.

***CHLOROFORM** was used as a surgical anesthetic until the 1960s. It was replaced by safer agents because of its toxicity for the heart and the liver. Its use as an ingredient in drug and cosmetic products was banned later when its carcinogenicity in rodents was confirmed. Chloroform is now used mainly in the chemical and pharmaceutical industry as a solvent and a reaction intermediate.*

I. Chemical Properties

A. Physicochemical and Organoleptic Properties

Chloroform is a volatile, nonflammable, colorless liquid. It is miscible with most organic solvents and oils. Its major physicochemical properties and partition coefficients are summarized in Tables IA and IB. Its threshold odor concentration is 250 ppm (parts per million, mg/kg).

B. Major Impurities

Major impurities found in the technical grade product are CCl_4, CH_2ClBr, and CH_2Cl_2. Ethanol (0.5–1%), tymol, *t*-butylphenol, or *n*-octylphenol (<0.01%) can be used as a stabilizer.

C. Chemical Incompatibility and Dangerous Reactions

Chloroform explodes when in contact with aluminum or magnesium powder. Mixtures with potassium, sodium, or lithium or with N_2O_4 explode on hitting. Violent reactions occur on adding chloroform to methanolic sodium hydroxide or to acetone in the presence of a base [KOH or $Ca(OH)_2$], or on mixing it with gaseous fluorine or with tri-isopropyl phosphine. On contact with disilane, it reacts with incandescence. Potassium *tert*-butilane catches fire on contact with chloroform.

Chloroform exposed to light and air slowly produces chlorine and phosgene; phosgene and hydrochloric acid are formed in the presence of water and iron.

Table IA Physicochemical and Other Properties of Chloroform

Molecular weight	119.39
Relative density (at 20°C)	1.48
Vapor pressure (Pa) at 20°C	2.138×10^4
Melting point (°C)	−63.5
Boiling point (°C)	61.7
Solubility (g/liter) in water at 20°C	7.95
Log partition coefficient, *n*-octanol/water	1.97
Specific conversion factor	1 ppm = 4.8822 mg/m^3

Table IB Partition Coefficients for Human Tissues at 37°C

Blood/air	8–10.3
Fat tissue/air	280
Liver/air	17
Brain/air	16–24
Kidney/air	11
Muscle/air	8–11
Lung/air	7

Table II LD(C)$_{50}$'s of Chloroform in Different Animals

Animal	LD_{50} (oral), mg/kg b.w.	LD_{50} (i.p.), mg/kg b.w.	LD_{50} (s.c.), mg/kg b.w.	LC_{50} (inh.), mg/liter
Rat	300–2180	894–1930		
Mice	120–1120	623–1862	704–3280	28–42.9
Rabbit			800–3000	58.7–70.9
Dog		1000		100

Abbreviations: i.p., intraperitoneal; s.c., subcutaneous; inh., inhalatory; b.w., body weight.

II. Toxicological Properties

A. Acute Toxicity

The acute toxic doses and concentrations of chloroform are reported in Table II. Chloroform-induced death is usually due to liver damage, with the exception of male mice of very sensitive strains, whose death is caused by kidney damage. In these strains of mice (such as DBA, C3H, C3Hf, CBA, Balb/c, and C3H/He), the higher susceptibility to $CHCl_3$ acute toxicity with respect to other strains (such as C57Bl, C57/6JN, C57Br/cd, Albino, and St) is genetically controlled. An absolute sex-related difference with respect to kidney damage, but not to liver damage, has been described in mice. Independent of the strain, female mice do not develop renal lesions; their death is due to $CHCl_3$-induced hepatic damage. This feature is regulated by the androgen-dependent renal metabolism of mice. Some influence of the age of both rats and mice on $CHCl_3$ acute toxicity has also been described.

Liver and kidney are the main targets for $CHCl_3$-induced toxicity; however, effects on other organs may be also present.

After acute chloroform exposure, early fatty infiltration throughout the liver lobule, balloon cells, dilation, and fragmentation of the granular endoplasmic reticulum of the hepatocytes with loss of ribosomes and centrilobular glycogen depletion are detectable [0.05–0.2 ml/kg body weight (b.w.)]. Liver damage may progress to severe centrilobular necrosis (0.3–1 ml/kg b.w.). The necrotic mass, which may extend throughout the hepatic lobule, is usually infiltrated with leukocytes.

Kidney lesions, when present, range from nuclear pycnosis, hydropic degeneration, the presence of hyaline droplets, and increased eosinophilia (0.025–0.050 ml/kg b.w.), to necrosis and loss of epithelium of the proximal convoluted tubules (0.1–0.2 ml/kg b.w.).

B. Potentiation of the Acute Toxicity

Pretreatment with phenobarbital (PB) increases the severity of steatosis and centrilobular hepatic necrosis in rats and in mice of both sexes that are already at chloroform doses of 0.2 ml/kg b.w. Lipid peroxidation of membranes and loss of cytochrome P450 are additional effects observed only in PB-induced animals. Renal toxicity seems to be unaffected or even decreased by PB-pretreatment. Dietary intake of polybrominated biphenyls greatly enhances both nephrotoxicity and hepatotoxicity of $CHCl_3$ in some strains of male mice; in the more susceptible strains, renal toxicity results remain unchanged. Pretreatment of rats and mice with 3-methylcholanthrene or β-naphthoflavone only moderately increases $CHCl_3$-induced hepatotoxicity. Various aliphatic alcohols (from methanol to decanol) as well as linear secondary ketones and ketogenic compounds potentiate the hepatotoxic effects of $CHCl_3$ in rats, significantly decreasing the LD_{50}'s even if administered as a single oral dose 18–24 hours prior to a challenge dose of the halomethane. Ketones also potentiate the nephrotoxicity of $CHCl_3$. Chlordecone (a nonlinear ketone) is able to increase $CHCl_3$ hepatotoxicity and lethality in rats and mice approximately 10 times. Nonpretreated Mongolian gerbils are more sensitive to chloroform toxicity than gerbils pretreated with phenobarbital, chlordecone, mirex, or 3-methylcholanthrene.

C. Chronic Toxicity, Carcinogenicity, and Cocarcinogenicity

Since 1945, some long-term studies on $CHCl_3$ toxicity have been carried out using mice, rats, and beagle dogs. In all cases, $CHCl_3$ was administered by the oral route; the two studies in which $CHCl_3$ was tested by subcutaneous or intraperitoneal injection in mice have been considered inadequate.

As shown in Table III, the carcinogenic potential of chloroform is species-, strain-, sex-, and organ-specific.

Chloroform administered by gavage dissolved in corn oil produced hepatocellular carcinomas in B6C3F1 mice. Both males and females were susceptible, although females gave a greater response. Nodular hyperplasia and focal necrosis occurred in the livers of many tested mice; metastases were also present (in the lungs of some animals). However, the administration of $CHCl_3$ in drinking water to female B6C3F1 mice failed to produce liver tumors, suggesting the crucial role of the vehicle used for the treatment. When administered by gavage dissolved in a toothpaste, chloroform produced kidney epithelial tumors and non-neoplastic renal disease in male ICI mice but not in females nor in other strains of mice of either sex.

The oral administration of chloroform, regardless of the vehicle, produced renal tubular adenomas and adenocarcinomas in male Osborne Mendel rats. No kidney tumors were found in treated females. In both sexes, but especially in females, neoplastic nodules of the liver (cholangiofibromas and cholangiosarcomas) or thyroid tumors were also observed. Another strain of rats [Sprague–Dawley (SD)] was resistant to tumor induction following administration by gavage of $CHCl_3$ dissolved in a toothpaste.

A study with beagle dogs resulted only in an increased incidence of hepatic nodular hyperplasia.

Contrasting results were obtained on the tumor-promoting activity of chloroform. Oral administration of $CHCl_3$ promoted the growth and spread of tumors *in situ* in mice inoculated with carcinoma cells; it also increased the number of liver preneoplastic foci induced in female SD rats with a single oral dose of diethylnitrosamine. Other studies

Table III Carcinogenicity Tests on Chloroform

Animal species	Dose (mg/kg b.w.)	Time (weeks)	Vehicle	Result
B6C3F1 mice				
m	138, 277	78	Corn oil	Hepatomas
f	238, 477	78	Corn oil	Hepatomas
m	34, 65, 130, 263	104	Drinking water	Negative
ICI mice				
m	17, 60	80	Toothpaste	Kidney tumors
f	17, 60	80	Toothpaste	Negative
C57 Bl mice, m, f	60	80	Toothpaste	Negative
CBA mice, m, f	60	80	Toothpaste	Negative
CF/1 mice, m, f	60	80	Toothpaste	Negative
OM rats				
m	90, 180	78	Corn oil	Kidney tumors
f	100, 200	78	Corn oil	Negative
m	34, 65, 130, 263	104	Drinking water	Kidney tumors
SD rats, m, f	60	80	Toothpaste	Negative
Beagle dogs, m, f	16, 30	7 years	Toothpaste	Negative

Abbreviations: OM, Osborne Mendel; SD, Sprague–Dawley; m, male; f, female.

showed that after $CHCl_3$ administration (either orally or subcutaneously) no tumor initiating or promoting activity was present in male SD rat liver or in C57 B16 mice irradiated with fast neutrons. No effects were detected on the incidence of liver and lung tumors in Swiss mice treated with $CHCl_3$ after a single intraperitoneal dose of ethylnitrosourea.

D. Teratogenicity

Chloroform was found embryotoxic, fetotoxic, and teratogenic in two studies in which rats and mice were treated by inhalation from day 6 through day 15 of gestation (100 or 300 ppm; 7 hr/day). Retarded fetal development, delayed ossification, acaudia, and imperforate anus were the most frequent effects observed.

When the duration of daily inhalation exposure was shorter (1 hr/day), even at high concentrations (2,200–4,100 ppm), $CHCl_3$ was fetotoxic but not teratogenic in rats. Similar results were obtained when $CHCl_3$ was administered by the oral route to rats, mice, and rabbits (20–400 mg/kg b.w. daily): maternal toxicity, embryotoxicity, and fetotoxicity were the only observed effects.

It has been suggested that modes and duration of exposure are critical for the maintenance of damaging levels of $CHCl_3$ in the dam and in the fetus, leading to the expression of its teratogenic potential.

E. Toxicokinetics

1. Absorption

Due to its lipophilic nature and relatively limited water solubility, chloroform is easily absorbed by passive diffusion in the lungs, the gastrointestinal tract, and the skin.

Administration of chloroform in corn oil or gelatine capsules results in virtually complete absorption, and peak blood levels occur 1 hour after ingestion in mice, monkeys, and humans. During inhalatory exposure, the equilibrium between total body and inspired air is reached in 2 to 3 hours in humans at rest, with a retention value of about 65% at equilibrium.

Blood concentrations of 7 to 16.5 mg/dl are found during exposure to anesthetic concentrations of $CHCl_3$ in the air (8,000–10,000 ppm). This relationship between air and blood concentration is kept down to very low exposure levels.

The permeability constant for the percutaneous absorption of chloroform in dilute aqueous solutions is 0.13 ml/cm^2 hr for hairless guinea pigs. A similar value was calculated for humans showering with chlorinated tap water. The necrotizing effects and the dehydration of the stratum corneum dramatically limit dermal uptake of pure chloroform.

2. Distribution

Chloroform accumulates in tissues according to its affinity for them (see Table IB). The adipose tissue is by far the most important site of $CHCl_3$ storage. In the liver, extractable $CHCl_3$ also attains concentrations greater than in the blood, while in all other tissues its concentration is about equal to that in the blood. Chloroform also crosses the placental barrier and reaches the fetal tissues. After dosing animals with radiolabeled chloroform, the radioactivity levels in the liver and kidney increase with time, due to the accumulation of bound metabolites in these organs. The accumulation of bound metabolites in the renal tissue is higher in male mice than in female mice.

Tissue chloroform concentration in patients who died during surgical anesthesia were in the order brain > lung > liver; these levels were similar to those found in the blood during surgical anesthesia. In humans who lived in nonindustrial areas, chloroform was found in the fat, liver, kidney, and brain at 5–68, 1–10, 2–5, and 2–4 μg/kg wet tissue, respectively.

3. Excretion

Unchanged chloroform is excreted essentially through the lungs. In squirrel monkeys, some chloroform seems to be excreted with the bile but very little is found in the feces, possibly because of enterohepatic circulation. The dose fraction which is expired unchanged varies widely among different species and appears to be related to the extent of its biotransformation to CO_2.

The kinetics of the pulmonary excretion of chloroform is made up of three phases representing passive diffusion from (1) a richly perfused tissue compartment (liver, kidney, gastrointestinal tract, brain, and heart), (2) muscles and skin, and (3) the adipose tissue. The half-times of these processes in humans are 15–30 min, 90 min, and 30–40 hr, respectively.

The long half-life of chloroform in the adipose tissue and its high affinity for this tissue suggest that some accumulation of the compound may occur during repeated exposure. This hypothesis is also supported by the detection of the compound in autoptic samples of adipose tissue and in the breath of persons living in air usually not contaminated with $CHCl_3$.

F. Metabolism

The liver and the kidney are the major sites of chloroform metabolism, although other organs and tissues, such as striated muscle, may be competent. The biotransformation of chloroform is accomplished by microsomes of the liver and kidney cortex and comprises oxidative and reductive pathways. In the liver, both pathways are supported by cytochrome P450 isozymes. Cytochromes P450 IIE1 and IIB1/2 are likely to be involved in at least the oxidative biotransformation. The enzymatic oxidation of chloroform yields trichloromethanol, which spontaneously decomposes to phosgene and hydrogen and chloride ions. Phosgene, in turn, can react with water and other cell constituents to form CO_2, chloride ions and covalent adducts with protein, phospholipid polar heads, and DNA. Reduced glutathione reacts easily with phosgene to produce stable diglutathionyl adducts, thus preventing the damage of cellular structures. The enzymatic reduction of chloroform yields free dichloromethyl radicals and possibly the corresponding carbanion. Unlike the reduction of carbon tetrachloride, reduction of chloroform cannot proceed further to the carbene. The damages to the cell structures associated with this pathway are not yet characterized but are typical of the radical: binding to fatty acyl chains of phospholipids and, to a much lesser degree, hydrogen abstraction to give dichloromethane. The damage of cellular structures by the radical is only partially prevented by GSH.

The extents of the oxidative and the reductive *in vitro* biotransformations by liver microsomes vary broadly across the species and depend strongly on the oxygen and $CHCl_3$ concentrations. In the range of hepatic physiological oxygen tensions, the reductive metabolism is active only in some of the species tested.

The major metabolites of $CHCl_3$ detected *in vivo* are CO_2 and chloride ions. CO_2 is mainly expired, but a small proportion may be incorporated into normal tissue constituents such as urea and methionine. Expired $CHCl_3$ and CO_2 together account for about 90% of the administered dose in several species tested (Table IV). Among humans, overweight people excrete much less chloroform than lean persons. The fraction of chloroform expired unchanged is also dependent on the dose. In humans it varies from 0 to 65%, with increasing doses from 0.1 to 1 g per person. It appears therefore that a saturation of the metabolic pathways may occur in animals and humans. In agreement with the *in vitro* findings, phosgene, chloroform-derived radicals, and covalent adducts to tissue proteins and lipids have also been detected *in vivo*. Additionally, low amounts of CO, derived from the adducts of phosgene with reduced glutathione, have been detected as carboxyhemoglobin in the blood of treated animals.

G. Mechanism of Action

It has been shown under different experimental conditions that the levels of protein covalent binding or of the adduct of phosgene with cysteine correlate with different indices of necrosis in the liver and the

Table IV Species Differences in $CHCl_3$ Metabolism[a]

Species	Expired $^{14}CHCl_3$	Expired $^{14}CO_2$	Total ^{14}C recovery[b]
C57 B1, CF/PL CBA mice	6.1	85.1	95.6
SD rats	19.7	65.9	93.2
Squirrel monkey	78.7	17.6	98.3
Man[c]	40	50	—

[a] Results are expressed as percentage dose (mean values).
[b] Total ^{14}C radioactivity measured in animals 48 hr after an oral dose of $^{14}CHCl_3$ of 60 mg/kg b.w.
[c] Data on man refer to an oral dose of 500 mg per person.

kidney. Therefore, the early biochemical alterations produced by chloroform have been generally attributed to the *in situ* formation of phosgene. However, available data are not suitable to establish any correlation between the formation of radicals through the reductive pathway and $CHCl_3$-induced acute toxicity. Following chloroform intoxication, an early formation of covalent adducts with proteins, lipids, and other cytosolic components occurs due to the chemical reactivity of the unstable intermediates. Only minor levels of adducts with nucleic acids have been observed. The result is a general impairment of the structure and function of the cell membranes, especially of the endoplasmic reticulum. Among the most relevant early events are the detachment of ribosomes from the endoplasmic reticulum and the activation of phospholipase A_2. Inhibition of calcium pump and increase of intracellular calcium seem to occur at late stages of the sequence of the toxic events. Peroxidation of the membrane lipids is present in some cases.

The potentiation of chloroform toxicity has been related to the increase of its metabolic activation due to cytochrome P450 inducers. However, inhibition of tissue regeneration and possibly other mechanisms are involved.

The mechanisms by which chloroform induces cancer is still unknown. No direct correlation of the two chloroform activation pathways with the species-specific carcinogenicity of the compound has been established. Moreover, the nearly absent genotoxic potential of $CHCl_3$ suggests an epigenetic mechanism. It has been proposed that chloroform-induced necrosis (both in the liver and in the kidney) is a necessary prerequisite for tumor formation.

III. Human Exposure and Toxicity

A. Modes and Levels of Human Exposure

Besides accidental or intentional exposure, humans may be exposed to $CHCl_3$ in a number of ways. Indeed, chloroform is present (1) in drinking water supplies as a product of the haloform reaction during chlorination processes [level range in some U.S. cities: 0.1–490 ppb (parts per billion, μg/liter)]; (2) as residue in food, especially fatty items (level range 0.4–1,100 ppb, depending on the foodstuff nature and mode of storage), and in pharmaceutical preparations; (3) in the atmosphere as a common air contaminant, deriving from industrial discharges and from photochemical reactions in the troposphere (level range from 0.009 up to 11 ppb in industrialized areas); (4) in the work place (level of occupational exposure range: 2–205 ppm in different plants); (5) in the water and air of indoor swimming pools (level ranges 17–47 ppb and 0.3–3.2 ppm, respectively). Usually oral exposure (through diet and drinking water) is by far the most relevant, but inhalatory exposure in more polluted areas, as well as absorption due to the domestic and recreational use of chlorinated waters, may contribute significantly.

B. Chloroform Toxicity

Chloroform is a central nervous system depressant and a gastrointestinal irritant. Prolonged chloroform anesthesia may depress the cardiovascular system and stimulate the vagus nerve, leading to cardiac arrest. A phenomenon known as "late chloroform poisoning" may appear one to three days after $CHCl_3$ anesthesia. Patients develop severe prostation, profuse vomiting, jaundice, unconsciousness, and coma leading to death within a few hours. Postmortem histopathologic analysis of the liver shows necrotic areas extending from centrilobular regions into periportal zones. In the same patients, degenerative changes in kidney, heart, and brain were also present.

Redness of conjunctival tissue and sometimes damage to the corneal epithelium have been described after exposure to concentrated chloroform vapors.

C. Epidemiological Studies

A retrospective mortality study on anesthesiologists and other physicians exposed to chloroform and other volatile compounds suggested an association between $CHCl_3$ exposure and carcinogenicity. Several ecological studies have shown some definite association between trihalomethane (and particularly $CHCl_3$) levels in finished drinking water supplies and cancer mortality for specific sites, such as lower gastrointestinal tract and bladder cancer. Similar results were indicated by some case-control studies, while one further study found no evidence of a relationship between $CHCl_3$ in drinking water and colorectal cancer in white females.

The results of all these studies are weakened by the study design and by confounding factors, including the general quality of drinking water. Therefore, the evidence of chloroform carcinogenicity in humans is still not conclusive.

IV. Risk Evaluations

The official evaluations of carcinogenic risk for humans are listed below.

1. International Agency for Research on Cancer (IARC): 2B (the agent is possibly carcinogenic to humans).
2. U.S. Environmental Protection Agency (EPA): B2 (the compound is a probable human carcinogen, with sufficient evidences of carcinogenicity in experimental animals and inadequate human carcinogenicity data). Moreover, EPA calculated that a lifetime consumption of drinking water containing 60 ppb chloroform or inhalation of air containing 0.08 ppb causes an increased carcinogenic risk of 1/100,000.
3. The World Health Organization (WHO) established a guideline level of 30 ppb in drinking water, corresponding to an increased carcinogenic risk of less than 1/100,000 tumor incidence for a lifetime consumption of 2 liters of water per day per person.

A recent estimation of the risk of liver cancer in humans by means of a physiologically based pharmacokinetic model suggests that continuous exposure of humans to 2,200 or 13,100 ppb in contaminated air or water, respectively, is associated with a 1/100,000 risk of liver cancer.

The threshold limit values (TLV) for occupational exposures are listed below.

1. U.S. American Conference of Governmental Industrial Hygienists (ACGIH): 10 ppm as TLV–TWA (time weighted average).
2. U.S. Occupational Safety and Health Administration (OSHA): 2 ppm as PEL (permissive exposure levels)–TWA. The OSHA safety standards establish that occupational exposure to $CHCl_3$ must not exceed 50 ppm.
3. U.S. National Institute for Occupational Safety and Health (NIOSH): 2 ppm as REL (recommended exposure level) ceiling (60 min).

Related Article: Volatile Organic Chemicals.

Bibliography

Davidson, I. W. F., Sumner, D. D., and Parker, J. C. (1982). Chloroform: A review of its metabolism, teratogenic, mutagenic and carcinogenic potential. *Drug Chem. Toxicol.* **5,** 1–87.

Howard, P. H., ed. (1991). "Handbook of Environmental Fate and Exposure Data for Organic Chemicals," Vol. II, Solvents, pp. 100–109. Lewis Publishers, Chelsea, MI.

International Agency for Research on Cancer (1979, 1987). "IARC Monographs on the Evaluation of the Carcinogenic Risk for Chemicals to Humans," Vol. 20 (1979) and Suppl. N.7, Volumes 1–42 (1987).

Reitz, R. H., Mendrala, D. A., Corley, R. A., Quast, J. F., Gargas, M. L., Andersen, M. E., Staats, D. A., and Conolly, R. B. (1990). Estimating the risk of liver cancer associated with human exposures to chloroform using physiologically based pharmacokinetic modeling. *Toxicol. Appl. Pharmacol.* **105,** 443–459.

Testai, E., Gemma, S., and Vittozzi, L. (1992). Bioactivation of chloroform in hepatic microsomes from rodent strains susceptible or resistant to $CHCl_3$ carcinogenicity. *Toxicol. Appl. Pharmacol.* **14,** 197–203.

Chromium

Katherine S. Squibb
University of Maryland School of Medicine

Elizabeth T. Snow
New York University Medical Center

Glossary

Allergen/antigen A substance capable of producing an immune response.
Allergic dermatitis An inflammatory skin disorder caused by exposure to an allergen.
Anaphylactic response A life-threatening response to exposure to an antigen in a sensitized individual.
Anthropogenic Related to humans and human activities.
Bioavailability The extent to which an element or a compound is capable of crossing cell membranes and interacting with living matter.
Bioconcentration Accumulation of an element or a compound directly from water.
Carcinogenic Capable of producing malignant growth.
Chromate (CrO_4^{2-}) The salt form of chromic acid.
Chromite ($FeCr_2O_4$) A mineral composed of iron, chromium, and oxygen.
Fibroblast A connective tissue cell type.
Genotoxicity Adverse effects of an element or a compound on an organism's genome (DNA).
Hypersensitivity A state of altered reactivity in which the body reacts with an exaggerated immune response to a foreign substance.
Matrix Basic material of origin to which other substances are adsorbed or in which they are embedded.
Mutation Permanent, heritable change in a chromosome or gene.

***CHROMIUM (Cr)** is an early Group VI transition element that occurs naturally in the environment primarily as Cr (III). It is the twenty-first most abundant element in the earth's crust but occurs in much greater concentrations in the earth's core and mantle. There are over 30 known chromium-containing minerals, including oxides, hydroxides, carbonates, sulfides, nitrides, silicates, and chromates. The most economically important chromium ore is chromite ($FeCr_2O_4$). Chromium is a natural constituent of living matter and an essential trace element for both plants and animals. At higher doses, however, chromium is toxic to living systems, particularly in its oxidized form [Cr(VI)].*

I. Production and Uses

Chromium is widely used in the production of steel and other metal alloys. The high resistance of chromium steels to corrosion and their intense hardness and strength stimulated their development and use in the mid to late 1800s. Since the mid 1930s, production and use of chromite has increased dramatically in direct proportion to the development of technology. Presently, about 10 million tons of chromite ore are consumed annually on a global basis. Of this, approximately 76% is used for metallurgical purposes, 13% for refractory uses, and 11% for chemical applications. Projection studies indicate that refractory uses for chromium will decline in years to come; however, an overall 3.4% annual growth in chromium usage is expected in the United States due to increases in its use by metallurgical and chemical industries.

A. Metallurgy

The most important metallurgical use of chromium is the making of various types of steel. Production of low-chromium steels (less than 3% Cr) accounts for 20–25% of chromium consumption in the United States; these steels are used in the manufacture of vehicles, industrial equipment, and specialized machinery (Table I). Steels containing median range amounts of chromium have a greater resistance to corrosion and oxidation and vary with respect to carbon content. Higher carbon containing steels are used in the production of small tools such as wrenches, screw drivers, and pliers and in magnetic steels.

The use of chromium in stainless steels accounts for approximately 75% of the chromium used in steel production in the United States. There are basically three types of stainless steel (ferritic, martensitic, and austensitic), which differ with respect to their content of chromium, carbon, and other elements (e.g., Ni and Mn). Each of these stainless steels possesses characteristic properties that make them ideal for specific types of applications (Table I).

Small amounts of chromium (<5% of the annual consumption of chromite ore) are also used for the production of nonferrous alloys with elements such as nickel and cobalt. These superalloys find specialized uses in nuclear reactors, jet engines, gas turbines, and cutting tools.

B. Refractory Uses

Two unique properties of chromite, its very high melting point (2,040°C), and its resistance to corrosion by acids and bases at high temperatures, make it ideal for refractory uses. Chromite and chrome–magnesite are used in bricks and mortars for furnace linings in paper, glass making, and metal refining industries.

C. Chemical Uses

Sodium chromate and sodium dichromate are the two principle compounds of chromium produced from chromite ore for use by chemical industries. Other chromium compounds include potassium chromate and potassium dichromate, ammonium dichromate, chromic acid, and chromic sulfate.

One of the primary industrial uses of chromium (as chromic acid) is electroplating, in which metal surfaces are finished with a thin layer of chromium either for decorative purposes or to obtain a corrosion-resistant surface. This use of chromium accounts for approximately 20% of the amount of chromium used by chemical industries (Table II). Production of chromium-containing pigments accounts for another large portion of chromium consumption within this category. Color pigments utilize lead chromate and Cr(III) oxide greens; while

Table I Metallurgical Uses of Chromium

	Special properties	Composition	Uses
Low Cr steels	Increased hardness and strength	Cr < 3%	Manufacture of automobiles, locomotives, aircraft, ships, oil and gas equipment, and machinery
Tool steels	Enhanced corrosion and oxidation resistance	Cr 3–12%	Castings and small tools, magnetic steels, and valves
Stainless steels			
Ferritic	Resistance to oxidation at high temperatures	Cr 15–30% Carbon 0.08–0.2%	Architecture; car trim; high temperature equipment in furnaces, and heat exchangers
Martensitic	Strength and hardness	Cr 10–18% Carbon 0.08–1.1%	Cutlery, valves, shafts, bearings, steam turbines, furnance parts
Austensitic	Chemical inertness, shock resistance, superior formability	Cr 16–26% Ni or Mn 6–22%	Automobile industry, kitchen equipment manufacturing, aircraft industry, petrochemical industry

Table II Chemical Uses of Chromium

Industrial application	Percentage of total Cr utilized by chemical industries
Pigments	25.7
Color	
Corrosion inhibition	
Metal finishing (electroplating)	20.9
Decorative plating	
Hard plating for corrosion resistance	
Leather tanning	15.7
Chemical manufacture	8.1
Wood preservation	4.3
Water treatment	4.3
Drilling muds	4.3
Textiles	3.3
Catalysts	1.4
Other	8.1

Source: "Health Assessment Document for Chromium," U.S. EPA, 1984.

corrosion-inhibiting pigments contain zinc chromate as a primary ingredient.

The tanning of leather using chromic sulfate is another important chemical use of chromium. The formation of stable complexes between chromium and components of rawhide enhances its stability and produces a leather that is resistant to bacteria. In the United States, an estimated 18,000 metric tons of chromic sulfate was used by the leather tanning industry in 1979, approximately 16% of the chromium used by chemical industries (Table II).

Other chemical uses of chromium include the incorporation of chromium lignosulfonates into drilling muds to decrease corrosion of drill strings; the inclusion of chromated copper arsenate as a preservative agent in wood treatments; the use of sodium dichromate and other chromic salts by textile industries to improve colorfastness and to oxidize dyed textiles; and the use of various chromium compounds as catalysts in chemical reactions.

D. Uses in Medicine and Research

One of the primary uses of chromium in medicine and biomedical research is the utilization of radioactive chromium (^{51}Cr) as a means of tracing specific cell populations. Because the chromate anion is readily accumulated and retained by living cells, cells can be labeled with ^{51}Cr *in vivo* and their half-life or distribution within the body followed by gamma ray detection or autoradioagraphy. Parameters such as total blood volume, cell turnover, splenic function, and gastric bleeding can be measured by labeling red blood cells. This technique has also been extended to other cell populations such as white blood cells, which can be labeled to study the distribution and functions of these cells *in vivo*. Laboratory techniques have also been developed that provide a means of measuring immune cell functions *in vitro* by measuring the release of ^{51}Cr from prelabeled target cells.

II. Chemical and Physical Properties

Chromium (Cr) is a Group VI transition element with an atomic weight of 51.996. Although inorganic chromium can exist in a range of valence states from −2 to +6 (Fig. 1), Cr(III) and Cr(VI) are the two most common oxidation states. Cr (III) is thermodynamically the most stable species of chromium in acid solution. Metallic Cr(0) is produced through industrial processes such as electrolytic or aluminum reduction.

In nature, chromium occurs primarily as Cr(III) in a variety of mineral types. Trivalent chromium compounds such as chromium chloride, chromium oxide, and chromium sulfate are generally insoluble in water; however, chromium chloride hexahydrate and the acetate and nitrate salts are readily soluble. In aqueous solutions, Cr(III) forms hexaaquo complexes $[Cr(H_2O)_6]^{+3}$, which readily polymerize under basic conditions or when heated to form hydroxy- or oxo-bridged compounds

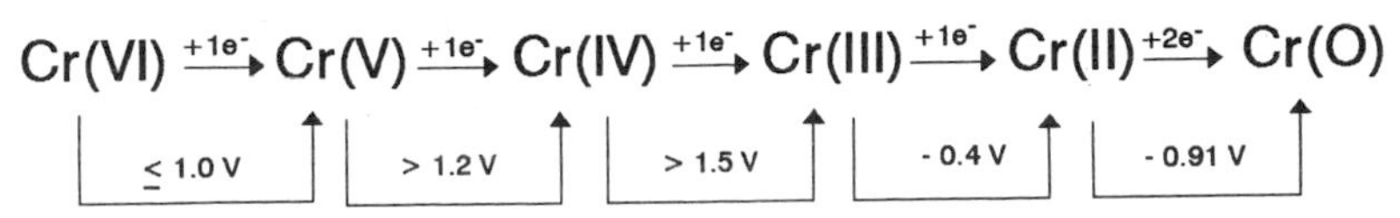

Figure 1 Reduction potential (E°) diagram for chromium (derived from Nriagu and Nieboer, 1988).

$[(H_2O)_5—Cr—O—Cr—(H_2O)_5]^{+4}$ (Fig. 2). Cr(III) also forms relatively kinetically stable hexacoordinated octahedral complexes with ligands such as water, ammonia, urea, ethylenediamine, halides, sulfates, and organic acids.

Hexavalent chromium salts vary greatly with respect to their water solubility. Sodium and potassium salts of Cr(VI) are very soluble in water, while lead and zinc salts are highly insoluble. Calcium salts of hexavalent chromium are sparingly soluble in water. Cr(VI) is a strong oxidizing agent, and thus it is rapidly reduced in nature in the presence of oxidizable organic matter. In solution Cr(VI) exists as the oxo ionic species, hydrogenchromate, chromate, and dichromate (Fig. 3), with the chromate form predominating at basic and neutral pHs. At lower pHs, the hydrogenchromate and dichromate species increase in concentration. Thus the oxidizing ability of Cr(VI) is greatest under acidic conditions.

III. Methods of Analysis

A. Analytical Methods

There are numerous methods available for chromium analysis that vary primarily with respect to their ability to differentiate between Cr(III) and Cr(VI) and in their limits of detection (Table III). Visible spectrophotometry and atomic absorption spectrophotometery (AAS) are the two most commonly used detection techniques, although others are useful for specific types of samples and low levels of detection.

In the colorimetric assay, Cr(VI) is determined by reaction with diphenylcarbazide. Total Cr [and Cr(III) by difference] can also be determined using this technique if the samples are completely oxidized prior to analysis. Atomic absorption analysis measures total chromium concentrations, but this method can be combined with pretreatment techniques that separate Cr(III) from Cr(VI) if concentrations of the individual oxidation states are desired. Flame AAS provides a detection limit of about 50 μg/l while the graphite furnace AAS technique allows detection of chromium concentrations as low as 0.1 μg/l. Inductively coupled plasma (ICP) emission is also a useful method that allows determination of total chromium in the part-per-billion range.

Potential interference problems can occur with many of these techniques due to matrix effects or the presence of other elements or anions. These can be alleviated by pretreatment of the samples.

B. Sample Pretreatment Techniques

Prior to analysis, most sample types require some form of pretreatment for solubilization, removal of interfering materials, preconcentration of the chromium or to separate Cr(VI) from Cr(III) if information on speciation is needed. Pretreatment methodologies can be grouped into four main categories: wet and dry ashing, precipitation, solvent extraction, and chromatography (Table IV). The choice of techniques is determined primarily by sample type and the selected subsequent method of analysis.

Wet or dry ashing techniques are used to solubilize biological tissues and environmental samples. Although these procedures do not allow for separation of Cr(III) and Cr(VI), they can reduce undesirable matrix effects that often occur with biological samples. Problems with high ionic strength are still encountered with some samples, however.

Unless highly contaminated, most natural (drink-

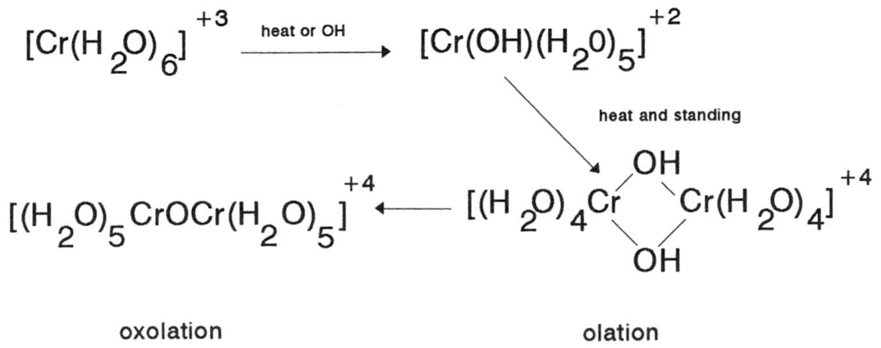

$$[Cr(H_2O)_6]^{+3} \xrightarrow{\text{heat or OH}} [Cr(OH)(H_2O)_5]^{+2}$$

heat and standing

$$[(H_2O)_5CrOCr(H_2O)_5]^{+4} \longleftarrow [(H_2O)_4Cr(OH)_2Cr(H_2O)_4]^{+4}$$

oxolation olation

Figure 2 Formation of oxobridged compounds by Cr(III).

$$CrO_4^{-2} \underset{-H^+}{\overset{+H^+}{\rightleftarrows}} HCrO_4^{-1} \underset{-H^+}{\overset{+H^+}{\rightleftarrows}} H_2CrO_4 \qquad K_{a1} = 10^{-0.6} \quad K_{a2} = 10^{-5.9}$$

chromate hydrogenchromate

$$2HCrO_4^{-1} \rightleftarrows Cr_2O_7^{-2} + H_2O \qquad K = 10^{-2.2}$$

dichromate

Figure 3 Cr(VI) species in solution.

ing or surface) water samples require some method of preconcentration prior to chromium analysis. Preconcentration can be achieved by precipitation, solvent extraction, or chromatographic techniques (Table IV). These techniques also provide a means of separating Cr(III) from Cr(VI).

C. Precautions

Care must be taken in the collection, storage, and analysis of samples to guard against contamination and loss of chromium as well as changes in oxidation state, particularly if the sample contains manganese (Mn IV). Contamination can occur in the collection of samples from stainless steel utensils and trays. High purity reagents must also be used to prevent high background levels of chromium. Loss of chromium can occur through volatilization during wet and dry digestion procedures if temperature guidelines are not carefully followed.

IV. Environmental Chemistry

A. Atmospheric Chromium

1. Distribution and Sources

Airborne concentrations of chromium in remote areas of the world range from 0.005 to 1.2 ng/m^3, with the lowest concentrations occurring in regions near the South Pole. Due to its high crustal abun-

Table III Analytical Methods of Chromium Analysis

Method	Selectivity	Detection limit	Interferences
Visible spectrophotometry (540 nm) Diphenyl carbazide reaction	Cr(VI)	100 μg/l	Fe, Ni, PO_4^{-3}
Atomic absorption spectrophotometry (AAS)	Total Cr		
Acetylene flame		50 μg/l	Fe
Nitrous oxide/acetylene flame		50 μg/l	
Graphite furnace		0.1–1 μg/l	Possible matrix effects with biological samples
Direct current plasma emission spectrophotometry (DCP)	Total Cr	10 μg/l	
Inductively coupled plasma emission (ICP)	Total Cr	1.8–7 μg/l	
Electrothermal atomic absorption spectrophotometry (EAAS)	Total Cr	0.02–0.1 μg/l	
X-ray fluorescence (XRF)	Total Cr	0.017 μg/m^3 (air) 1.5 μg/g (dried solution deposit)	Possible matrix effects
Proton induced x-ray emission spectrophotometry (PIXE)	Total Cr	0.3 μg/l	Possibly Fe
Gas chromatography	Cr(III)		

Table IV Sample Preparation for Chromium Analysis

Pretreatment	Methodology	Selectivity	Sample type
Wet and dry ashing	Wet ashing: Digestion with concentrated acid or mixture of acids. Dry ashing: graphite furnace ashing or low temperature oxygen plasma ashing	Total Cr	Air particulates, biological samples, food, soils, and sediments
Precipitation	Coprecipitation of Cr(III) with $Fe(OH)_3$	Cr(III); Cr(VI) can be reduced to Cr(III) by Fe(II)	Surface and drinking water samples
	Precipitation of Cr(VI) with $AlCl_3$	Cr(VI)	Wastewater and industrial effluents
Solvent extraction	Cr(VI) complexation with APDC followed by extraction into methylisobutyl ketone (MIBK).	Cr(VI): Cr(III) can be oxidized to Cr(VI)	Drinking water, surface water, industrial effluents, and biological samples after wet or dry ashing
	Chelation with trifluoro-acetylacetone (HFTA) followed by extraction into benzene and analysis by GC/EC, GC/MS, or GC/AAS	Cr(III); Cr(VI) can be reduced to Cr(III) with sodium sulfite	Drinking or surface water samples and biological samples after wet or dry ashing
Chromatography	Alumina adsorption	Cr(VI) or Cr(III)	Drinking or surface water samples
	Anion exchange	Cr(VI)	Drinking or surface water samples
	Chelating ion exchange	Cr(III); Cr(VI) can be reduced to Cr(III) with sodium bisulfite	Drinking or surface water samples
	HPLC anion exchange column eluted with Na_2CO_3 after reaction with Na_2EDTA	Simultaneous determination of Cr(III) and Cr(VI)	Water extracts

Source: From "Toxicological Profile for Chromium," ATSDR, 1989 and "Health Assessment Document for Chromium," U.S. EPA, 1984.

dance, chromium occurs naturally in dust; however volcanic fallout can also contribute significantly to ambient levels of this element

Worldwide urban air concentrations of chromium average about 30 ng/m^3; however, concentrations ranging from 100 to 200 ng/m^3 have been reported in some cities. Elevated concentrations are generally due to anthropogenic sources of chromium, which include emissions from iron and steel industries, chemical manufacturing plants that utilize chromium, the combustion of fossil fuels by power plants, and the dispersion of chromium-contaminated soils from landfills and hazardous waste sites. Cooling towers that use chromate salts as corrosion inhibitors have also been found to release significant quantities of chromium to the atmosphere.

Chromium is removed from the atmosphere by wet and dry deposition. Depositional patterns in sediment and soil cores provide supporting evidence that a large portion of present day atmospheric chromium is anthropogenic in origin and also suggest that recent controls on industrial emissions of chromium have been effective. Although chromium deposition rates are higher at present than during prehistoric and precolonial times, they appear to have been decreasing in many areas since the mid 1970s.

2. Chemical and Physical Properties

The physical and chemical characteristics of airborne chromium are variable and complex due to the large number of sources that contribute chromium to the air. Chromium originating from fossil fuel combustion, metallurgical production, and cement production is released primarily as Cr(III) and Cr(0), while chromium in fumes from welding shops consists of soluble Cr(VI) and insoluble Cr(VI), Cr(III), and Cr(0) complexes in different proportions and different matrices depending upon the type of welding. Other sources of Cr(VI) compounds include chrome production, chrome plating, and cooling towers. Particles from tanning and leatherwork factories usually contain 2–6% Cr(III) in an organically bound form, while particles from primary ferrochrome smelters consist of Cr(VI) in the form of chromate and dichromate and Cr(III) in the form of chromite.

Particle size plays an important role in the airborne transport and potential human health hazard of atmospheric chromium. Particles with aerodynamic diameters of <20 μm can remain airborne for extended periods of time and can be transported considerable distances from emission sites. The deposition of particles in the lung is also size-dependent, with larger particles depositing in the tracheobronchial and nasopharyngeal regions while mid-sized particles (1–5 μm) deposit in the lower lung, where they are potentially more dangerous.

The solubility of the chromium present in airborne particles is also an important determinant of its bioavailability and potential toxicity. Emissions with greater concentrations of soluble Cr(VI) compounds are thought to present a greater health hazard to humans. Generally, the solubility of chromium in particulate emissions is low in water and dilute acids, but increases with increasing acidity. Chromium associated with the matrix of particles is less soluble than chromium adhered to particle surfaces; studies have shown that chromium associated with coal fly ash is generally fairly evenly distributed between the matrix material and surface microlayer.

B. Chromium in Soils

1. Chemical and Physical Behavior

Natural concentrations of chromium in soils vary with respect to the origin of the soil and can range from trace amounts up to 20,000 mg/kg; average concentrations, however, range from 10 to 100 mg/kg. Most of the chromium present in noncontaminated soils is insoluble. Extraction with dilute acids normally releases only 0.01 to 4% of the total chromium present in a soil sample.

The chemistry of chromium in chromium-contaminated soils is very complex and only partially understood. It has long been presumed that in the presence of organic matter or Fe(II), soluble Cr(VI) is rapidly converted to insoluble Cr(III) in the form of chromic hydroxide [$Cr_2O_3 \cdot nH_2O$], which is relatively nonreactive and immobile. Recent studies have demonstrated, however, that Cr(III) can react with organic compounds in soil to form soluble Cr(III) complexes, and under certain conditions, such as in the presence of Mn(IV), Cr(III) may be converted to Cr(VI). Studies of landfills have shown that Cr(VI) can reach the groundwater through the leaching process, although most of the Cr(VI) emerging from landfills is removed by adsorption, reduction, and precipitation. The balance between these processes depends upon the nature of the soil—its pH, clay mineral content, and reducing capacity.

2. Bioavailability

The bioavailability of chromium in soil to plants depends on soil conditions and many factors that control the oxidation and reduction of the chromium ion. As a soluble anion, Cr(VI) easily crosses cell membranes and can be reduced and trapped intracellularly. The cation Cr(III), on the other hand, is normally soluble only when bound to organic complexes and thus does not readily penetrate into cells. Plant uptake studies have demonstrated that concentrations of chromium in plant tissues correlate with soil Cr(VI) concentrations rather than total chromium or Cr(III) concentrations.

C. Chromium in Freshwater and Marine Ecosystems

1. Chemistry

Chromium enters rivers, lakes, and estuaries through atmospheric deposition, surface runoff, industrial and sewage plant discharge, and landfill leachates. Chromium entering the water column as Cr(III) rapidly associates with sediment particles and is deposited near the source. The fate of chromium entering as Cr(VI) depends to a greater extent on conditions present in the receiving water. In the presence of organic and particulate matter, Cr(VI) may be rapidly reduced to Cr(III) with subsequent deposition of the Cr(III) in local sediments. If little or no organic or particulate matter is present, however, Cr(VI) may have a fairly long half-life in the receiving water body and can be carried some distance from the source as dissolved chromium in the water column. Small amounts of organically bound Cr(III) may also remain in the soluble phase. Reported concentrations of dissolved chromium in surface seawater around the world range from 0.04 to 0.5 ppb for Cr(III) and 0.02 to 0.21 ppb for Cr(VI). Concentrations in sediments range from 4 to 1000 ppm total chromium in coastal and offshore marine ecosystems.

2. Bioavailability

Although studies have shown that a large portion of sediment-bound chromium is biologically inactive, organisms living in areas with elevated sediment Cr concentrations do accumulate excess chromium. Bioconcentration factors for chromium relative to seawater range from 10^2 to 10^3 for fish and 10^4 to 10^5 for plankton and annelids inhabiting chromium-contaminated areas. Higher concentrations in biota often occur in epithelial tissues such as skin, gills, and digestive tracts, which suggests that surface adsorption phenomena may play a role in chromium accumulation.

Sediment bacteria are also important in determining chromium bioavailability. Studies have shown that facultative anaerobes release soluble chromium from sediments, thereby increasing its waterborne concentration. Also, bacterial accumulation of chromium can serve as a mechanism by which increased amounts of chromium can enter the aquatic foodchain.

V. Biology of Chromium

A. Biological Chemistry

Although there are six possible oxidation states for chromium (Fig. 1), only Cr(VI) and Cr(III) possess the stability to exist for extended periods of time in cells. Cr(V) species are sufficiently long-lived to be observed experimentally, and may play an important role as active intermediates in the toxicity of chromium; however, this species of chromium does not form stable biological compounds. Cr(IV) is highly reactive and has not been observed directly in experimental systems, but it exists as an intermediate in the reduction of Cr(V). Cr(II) is also very short-lived due to its strong reducing capacity and rapid oxidation to Cr(III).

Cr(VI) and Cr(III) have very distinct chemistries in biological systems because Cr(VI) exists as an anion (CrO_4^{-2}) in aqueous solution, while Cr(III) acts as a cation. Cr(VI) is able to cross cell membranes using existing anion (PO_4^{-2} and SO_4^{-2}) transport systems, while Cr(III) with its positive charge does not readily enter cells. The greater toxicity of Cr(VI) versus Cr(III) in biological systems is in large part due to this difference in cellular uptake. The mechanism of toxicity of Cr(VI) in cells, however, is thought to involve the reduction of Cr(VI) to Cr(III) by intracellular reducing agents such as ascorbate, glutathione, and microsomal enzymes.

Cr(III) prefers to form coordination complexes with a coordination number of 6. It has a strong affinity for charged oxygen groups and at physiological pH forms hydroxo and inert polynuclear complexes with bridging oxo groups. Cr(III) can also bind to nitrogen and sulfhydryl groups and is known to react with amino acids, proteins, and nucleic acids. The rate of ligand exchange for Cr(III) is very slow (10^{-6}/sec with water, even slower with other ligands) which suggests that Cr(III) complexes are relatively nonlabile and explains chromium accumulation and retention in cells.

B. Role of Chromium as an Essential Element

Chromium is required for normal glucose tolerance in animals and humans. Supplementation of chromium-deficient diets with Cr(III) or a Cr-containing organic compound isolated from brewer's yeast [called the glucose tolerance factor (GTF)] has been shown to increase insulin activity and glucose utilization. The mechanism by which this occurs is not well known, however, and the identity of the CrGTF(s) is still controversial. Cr(III), or a CrGFT complex, may act as a cofactor with insulin or it may act independently at the membrane level. Signs of chromium deficiency also include elevated serum cholesterol and triglyceride levels, peripheral neuropathy, and metabolic encephalopathy. The role played by Cr in these processes is also poorly understood.

No Recommended Daily Allowance (RDA) has been developed for chromium. Due to the lack of signs of chromium deficiency in the general population, average daily intake levels for chromium appear to be sufficient. These range from 5 to 100 μg/day. Normal blood and urine concentrations for chromium range from 0.5 to 5 μg/l and 5 to 10 μg/l, respectively.

C. Chromium Metabolism

1. Absorption

a. Inhalation

Inhalation is an important route of exposure to chromium compounds. Deposition and retention patterns in the lung are dependent upon the size and solubility of the particles, with larger particles depositing in the upper airways and smaller particles (<5 μm) depositing in the small airways and alveolar regions. Absorption is also regulated by the oxidation state of the chromium. In experimental animals, water soluble Cr(III) compounds have been shown to be absorbed more slowly than Cr(VI) compounds [55% versus 89% after 24 hr in rats for Cr(III) versus Cr(VI), respectively]. The greater retention of Cr(III) in lung tissue is thought to be due to its binding to extracellular macromolecules and/or its precipitation within the lung as hydrous chromic oxide.

b. Gastrointestinal Absorption

The absorption of ingested chromium is also dependent upon valence state and the form of chromium administered. Inorganic forms of Cr(III) are minimally absorbed (0.4–3%), while Cr(III) bound to the organic glucose tolerance factor (GTF) present in yeast is more readily absorbed (10–15%). Reported absorption values for Cr(VI) compounds range from 2 to 11%.

The importance of the interaction of chromium with gastric juices in determining intestinal absorption has been demonstrated in experiments comparing intragastric versus intraduodenal administration of chromium. Absorption of Cr(VI) increased 5–10-fold when doses were administered directly in the intestine; absorption of Cr(III) was affected to a smaller extent. Both Cr(VI) and Cr(III) bind to components present in gastric juice. In addition, acidic gastric juices possess the capacity to reduce Cr(VI) to Cr(III).

c. Dermal

Some absorption of both Cr(VI) and Cr(III) compounds can occur through the skin. The rate of absorption is dependent upon many factors, including valence state, pH, and concentration. In animals, there is no significant difference in the dermal absorption of sodium chromate (CrVI) compared with chromic (CrIII) chloride at concentrations up to 0.239 *M* Cr, but at higher concentrations (0.261–0.398 *M*) sodium chromate absorption is higher. pH has also been shown to alter sodium chromate absorption, with higher absorption occurring at pH values greater than 6.5.

2. Tissue Distribution and Elimination

The tissue distribution and elimination of absorbed chromium depends upon the valence state of the chromium as it enters the bloodstream and the reducing capacity of the serum. Cr(III) binds to serum proteins such as transferrin and is distributed primarily to liver, kidney, and spleen. Very little Cr(III) enters red blood cells.

Cr(VI), however, rapidly crosses red blood cell membranes and accumulates in the blood to a much greater extent than Cr(III), particularly in the rat. Trapping of Cr(VI) in red blood cells occurs through an intracellular reduction of Cr(VI) to Cr(III), with subsequent binding of the chromium to hemoglobin. Species differences in chromium blood concentrations occur and may be due, in part, to differences in ability of chromium to bind to the hemoglobin molecule. Significant accumulation of chromium also occurs in the liver, spleen, lung, kidney, and bones of animals, particularly following chronic exposure to elevated concentrations of the chromate ion.

Whole body elimination of a single intravenous dose of Cr(III) in rats occurs in three phases, with half-life values of 0.5 days, 5.9 days, and 83.4 days. Urine is the primary route of excretion for Cr(III), however some fecal excretion does occur. Studies in rats indicate that Cr(VI) compounds are more rapidly excreted than Cr(III), with whole body half-lives of 22 and 92 days for sodium chromate (CrVI) and chromic (CrIII) chloride, respectively. Urinary elimination also predominates for Cr(VI) compounds, however biliary excretion may play a somewhat larger role, particularly following exposure by inhalation.

The elimination of chromium from body tissues has been determined in the rat following a single subcutaneous dose of Cr(VI). Lung, kidney, brain, heart, and testes elimination curves consisted of two phases, with terminal half-life values of 20.9, 10.5, 9.6, 13.9, and 12.5 days, respectively. The biologic half-life of chromium in blood (after reaching a peak at 5 days) was 13.9 days. Liver tissue elimination of chromium occurred in three phases, with half-life values of 2.4 hours, 2.2 days, and 15.7 days.

VI. Toxicology of Chromium

Although chromium compounds are both ubiquitous and essential, high concentrations of chromium are toxic to humans, animals, plants, and lower organisms. Very high doses of chromium may be lethal or produce severe systemic toxicity. Chromium compounds have been found to be toxic in cell free systems (Table V), in bacteria and nonmammalian eukaryotes (Table VI), in mammalian cells in

Table V Genotoxicity of Chromium Compounds in Subcellular Systems

Assay/endpoint	Very sol. Cr^{+6} (e.g., $K_2Cr_2O_7$, Na_2CrO_4)	Sl. sol. Cr^{+6} (e.g., $CaCrO_4$)	Insol. Cr^{+6} (e.g., $PbCrO_4$)	Sol. Cr^{+3} (e.g., $CrCl_3 \cdot 6H_2O$)	Insol. Cr^{+3} (e.g., Cr_2O_3)
DNA interactions					
DNA binding					
purified DNA	−			++	
intracellular	+			+	
Strand breaks					
purified DNA	−, + (with H_2O_2 or GSH)			−, +(with —SH compounds)	
nuclei/nucleoids	+	−		+	
depurination	+			−	
Crosslinks					
DNA–DNA	−			++	
DNA–protein	−			++	
DNA replication					
Inhibition	+ (0.5–5 m*M*)[a]			−/+, ↑ then ↓ (1–10 μ*M*)[a]	
Infidelity	−/+			++	
Mutagenesis (DNA treated *in vitro*)	+/−			+ (2–3×)[b]	

Source: Data included in this table have been compiled from De Flora *et al.*, 1990; ATSDR, 1989.
[a] Effective concentration: m*M*, 10^{-3} moles/liter; μ*M*, 10^{-6} moles/liter.
[b] Increase in mutagenesis over control values.

culture (Table VII), and in animals (Table VIII and Table IX). These tables summarize available data on the toxic effects of various chromium compounds. The data for water soluble Cr(VI) ["sol. Cr^{+6}"], slightly soluble Cr(VI) ("sl. sol. Cr^{+6}"], water insoluble Cr(VI) ["insol. Cr^{+6}"], soluble Cr(III) ["sol. Cr^{+3}"], and insoluble Cr(III) compounds ["insol. Cr^{+3}"] have been tabulated separately to emphasize the role of chromium valence and solubility in the relative toxicity of these compounds *in vivo* and *in vitro*. Representative compounds in each category have been noted at the top of each table.

The relative toxicity of different chromium compounds *in vivo* is dependent on three properties: the ability of the compound to enter the cell, the intracellular redox metabolism of the compound, and the reactivity of the compound or its metabolites. Thus, Cr(III), the most stable form of chromium available in the environment, can bind to DNA and produce effects on DNA replication in cell free systems (Table V); but it does not readily enter into cells and, therefore, is less toxic than Cr(VI) *in vivo*. Likewise the bioavailability of chromium compounds may depend on the route of uptake. Soluble Cr(VI) is taken up by all types of cells via cellular anion transport channels; thus, these compounds are bioavailable (and toxic) to bacteria and lower eukaryotes as well as to mammalian cells in culture and *in vivo*. However, particulate forms of chromium cannot be taken up by the cells of bacteria and many lower eukaryotes, such as yeast. Thus insoluble and sparingly soluble chromium compounds are not bioavailable or toxic to these species unless they are first solubilized (Table VI). In contrast, many types of mammalian cells, including human lung cells, can take up insoluble Cr(VI) and Cr(III) compounds by the process of phagocytosis (i.e., small insoluble particles of chromium are engulfed by the cell membrane and taken up in membrane-bound vacuoles). The inside of these vacuoles is acidic; thus, once the metal particles are taken up they are slowly dissolved and metal ions are released inside the cell where they can easily reach the DNA and other cellular constituents. When tested, insoluble chromium compounds have often given a positive genotoxic response in mammalian cells, both *in vitro* (Table VII) and *in vivo* (Table VIII). This is consistent with the observation that insoluble chromium species [e.g., $ZnCrO_4$ and Cr_2O_3], have been shown

Table VI Toxicity of Chromium Compounds in Bacteria and Nonmammalian Eukaryotes

Assay/endpoint	Very sol. Cr^{+6} (e.g., $K_2Cr_2O_7$, Na_2CrO_4)	Sl. sol. Cr^{+6} (e.g., $CaCrO_4$)	Insol. Cr^{+6} (e.g., $PbCrO_4$, $ZnCrO_4$)	Sol. Cr^{+3} (e.g., $CrCl_3 \cdot 6H_2O$, $Cr(CH_3COO)_3$)	Insol. Cr^{+3} (e.g., Cr_2O_3)
Bacteria					
E. coli					
DNA strand breaks	+				
SOS induction	++	−	−/+ (+ when dissolved)	−−/+	
Differential killing repair⁻ versus repair⁺)	++	+	−	−, + (with 0.2 *M* phosphate)	−
Mutagenesis	+++ (2–100×)[a]	+ (3–4×)[a]	−/+ (+ when dissolved)	+++ (6–75×)[a]	
B. subtilis					
Differential killing	+				
Salmonella typhimurium					
SOS induction	+/−				
Differential killing (repair⁻ versus repair⁺)	+			−	
Mutagenesis					
frameshift (GC)	+ (3–4×)[a]	+	−/+ (+ when dissolved)	−	
bp subs (GC)	++ (3–8×)[a] (requires pKM101)	++ (requires pKM101)	−/+ (+ when dissolved)	−	
bp subs (AT)	++ (3–8×)[a]	++ (4–8×)[a]	−		
Yeast					
Saccharomyces					
Recombination (gene conversion)	++		+ (when dissolved)	+ (with 0.1 *M* phosphate)	
Mutagenesis	++		−	+ (with 0.1 *M* phosphate)	
Aneuploidy	+				
Petite induction		+			
Plants					
Micronuclei	−/+				
Chromosome aberrations				+	
DNA damage (UDS)				+ (weak)	
Insects					
Drosophila melanogaster					
Mutagenesis	+	+	−/+ (+ when dissolved)		
Chromosome loss	−/+				
Birds					
Chick embryo cells					
DNA strand breaks	++			−	
DNA crosslinks	+			−	

Source: Data included in this table have been compiled from De Flora *et al.*, 1990; ATSDR, 1989.
[a] Increase in mutagenesis relative to untreated controls.

Table VII Genotoxicity of Chromium Compounds in Mammalian Cells in Culture

Assay/endpoint	Very sol. Cr^{+6} (e.g., $K_2Cr_2O_7$, Na_2CrO_4)	Sl. sol. Cr^{+6} (e.g., $CaCrO_4$)	Insol. Cr^{+6} (e.g., $PbCrO_4$, $ZnCrO_4$)	Sol. Cr^{+3} (e.g., $CrCl_3 \cdot 6H_2O$, $Cr(CH_3COO)_3$)	Insol. Cr^{+3} (e.g., Cr_2O_3)
Cell cycle					
Inhibition	++	+	+	+/− (weak)	+
Clastogenicity					
Rodent cells					
Micronuclei	++				
SCEs	++	++	+ (↑ when dissolved)	+/−	+/−
Chromosome aberrations	++	++	+	+	+ (?)
DNA strand breaks	++	++	−	−	+ (?)
DNA crosslinks	++	+		−	
Human cells					
Aneuploidy		+		+/−	
Micronuclei	++			+/−	
SCEs	++	+		+/−	
Chromosome aberrations	++	+	+ (when dissolved)	+/−	
DNA strand breaks	++	+		−	
DNA crosslinks	++			−	
Mutagenicity					
Chinese hamster					
V79–$6TG^R/8AG^R$	++ (2–12×)[a]	+	++ (2–17×);[a] ↑ if dissolved	− (0–2×)[a]	+ (2–6×)[a] (18-hr treatment)
CHO–$6TG^R/8AG^R$	++ (3–20×)[a]	+++ (8–100×)[a]			
V79–OUA^R	+/−	+/− (0–6×)[a]			
Mouse					
L5178Y/$tk^{+/-}$–TFT^R	+/+++ (2–65×)[a]	−	−		
$C_3H_{10}T\frac{1}{2}$–OUA^R		−			
Human, HFC					
$6TG^R$	++ (3–17×)[a]	+++ (26–43×)[a]	+ (2–7×)[a]	++ (2–17×)[a]	++ (5–18×)[a]
OUA^R		+/− (1–6×)[a]		−	
Cell transformation					
SHE cells	+/−	+	+	−	+
BHK cells	++	++	+	−	+/−
$C_3H_{10}T\frac{1}{2}$	−	−	+/−		
Balb/3T3		+			

Source: Data compiled from De Flora *et al.*, 1990; IARC, 1980; ATSDR, 1989; Biedermann and Landolph, 1990.
[a] Increase in mutagenic response relative to controls.

Table VIII Genotoxicity of Chromium Compounds *in Vivo*

Species	Endpoint	V. sol. Cr^{+6} (e.g., $K_2Cr_2O_7$, Na_2CrO_4)	Sl. sol. Cr^{+6} (e.g., $CaCrO_4$)	Insol. Cr^{+6} (e.g., $PbCrO_4$, $ZnCrO_4$)	Sol. Cr^{+3} (e.g., $CrCl_3 \cdot 6H_2O$, $Cr(CH_3COO)_3$	Insol. Cr^{+3} (e.g., Cr_2O_3)
Nonmammals						
Chick embryos	DNA strand breaks	+				
	DNA crosslinks	+				
Fish	Chromosome aberrations	+				
Mammals						
Rats	DNA damage	+			−	
	Chromosome aberrations	+			+/−	
	SCEs				+	
Mice	Micronuclei	++	−	+	−	
	Chromosome aberrations	+/−	+			
	SCEs	+	+			
	Mutations	+/−	−			
Chinese hamsters	Micronuclei	+	−			
	SCEs		+			
Humans	SCEs	+/−				

Source: Data compiled from De Flora *et al.*, 1990; IARC, 1980; ATSDR, 1989.

to induce tumors in rodents (Table X) as well as in humans (Table XI).

A. Effects of Chromium on Cell-Free Systems

The toxicity of various chromium compounds has been investigated in many different ways. Table V summarizes the genotoxic effects of chromium in subcellular or cell free systems. Chromium, especially Cr(III), can interact directly with DNA to produce DNA crosslinks, strand breaks, and other conformational changes. Cr(VI) compounds do not directly interact with DNA (except at very high concentrations) unless they are incubated in the presence of either biological or chemical reducing agents. However, within a cell or in the presence of glutathione (GSH), hydrogen peroxide (H_2O_2), or other cellular reductants, the reduced chromium [Cr(V) or Cr(III)] can also interact with DNA and/or proteins to form stable Cr(III) complexes. Once formed, these complexes are slow to exchange, and intracellular Cr(III) complexes are poorly repaired. Chromium compounds also alter the kinetics and fidelity of DNA replication. The importance of these effects in the overall genetic toxicity of chromium *in vivo* remains to be determined.

B. Bacteria and Nonmammalian Species

Soluble chromium compounds, especially Cr(VI) compounds (which mimic biologically active phosphate and sulfate anions and are easily transported across cell membranes through normal anion transport channels) are toxic to living organisms and to cells in culture. Cr(III) compounds are not as biologically active as Cr(VI) in whole cell systems. Table VI summarizes the various types of genotoxic damage induced by chromium compounds in bacteria and nonmammalian eukaryotes. Chromium induces DNA damage as measured by DNA strand breaks, SOS induction (an inducible response to DNA damage in bacterial cells), and differential killing of cells that are deficient in DNA repair. Cr(VI) also induces various types of DNA mutations in bacteria and higher organisms. In the Ames assay, using different strains of *Salmonella typhimurium,* salts of Cr(VI) can induce frameshifts (small insertions or deletions) and basepair substitution mutations (bp subs) at GC target sites in the presence of the

Table IX Systemic Effects of Chromium Compounds in Humans and Animals

Exposure	Species	Dose	Observed effects
Inhalation—acute	Rodent (various)	LC_{50} Cr(VI) = 30–80 mg/m^3 $\geq$0.6 mg Cr(III)/m^3; $\geq$0.025 mg Cr(VI)/m^3	Respiratory distress, irritation, body weight loss, altered immune response, response of lung macrophages and increased lung weight, abnormal histopathology of the lung; female rats were more sensitive than male rats.
	Human	$\geq$35 mg aerosolized Cr(VI)/ml	Asthma, respiratory distress, allergic sensitivity of Cr(VI) > Cr(III)
Inhalation—chronic	Rodent (various)	$\geq$0.2 mg Cr(VI)/m^3; $\geq$[0.6 mg Cr(III) + 0.9 mg Cr(VI)]/m^3; 4.5 mg Cr(VI)/m^3 as $CaCrO_4$	Increased lung macrophage activity, altered lung morphology and function; some hematological changes and effects on the immune system.
	Human	$\geq$0.002 mg Cr(VI)/m^2	Decreased lung function, upper respiratory distress, nasal ulcerations and soreness, nose bleeds, some evidence of kidney damage. Chromium sensitive individuals may experience an anaphalactic response upon exposure to Cr(VI).
Oral—chronic (in drinking water)	Rat	LD_{50}♀ sol. Cr(VI) ≈ 16 mg/kg; LD_{50}♂ ≈ 25 mg/kg; LD_{50}♀ $CaCrO_4$ = 108 mg/kg; LD_{50}♂ = 249 mg/kg; LD_{50} sol. Cr(III) = 200–2400 mg/kg; LD_{50} Cr_2O_3 < 3 g/kg	Hypoactivity, tearing, diahhrea, weight loss, death. Soluble Cr(VI) in drinking water of pregnant mice resulted in developmental fetal abnormalities. Reduced sperm count and morphologically altered sperm have been noted in exposed males.
Oral—acute	Human males	LD_{100} $\geq$ 4 mg/kg sol. Cr(VI)	Ingestion of either 7.5 mg/kg $K_2Cr_2O_7$ or 4 mg/kg CrO_3 produced gastrointestinal ulceration, liver and kidney damage, and death.
Dermal—acute	Rabbit and guinea pig	LD_{50} sol. Cr(VI) $\geq$ 300 mg/kg	Necrosis, erythema, diahhrea, hypoactivity
	Human males	?	Treatment of a facial carcinoma with Cr_2O_3 resulted in severe nephritis and death.
Dermal—chronic	Guinea pig	$\geq$0.004 mg Cr(VI); $\geq$0.008 mg Cr(III)	Necrosis and skin sensitization followed by erythematic reaction upon further exposure; response to Cr(III) similar to Cr(VI).
	Humans	?	Chromium sensitivity and dermal ulcerations
Intraperitoneal injection	Rodent (various)	$\geq$0.03 mg Cr(VI)/kg/day; $\geq$0.3 mg Cr(III)/kg/day	Kidney damage, lipid peroxidation, developmental abnormalities, decreased spermatogenesis.

Source: Data compiled from IARC, 1980; ATSDR, 1989.

plasmid pKM101 (i.e., including SOS processing) and can induce basepair substitutions at AT target sites. Chromium salts also induce chromosomal changes and smaller mutations in yeast cells, plants, and insects (e.g., the fruit fly, *Drosophila melanogaster*).

C. Chromium Toxicity to Mammals and Mammalian Cells in Culture

In contrast to the lack of effect of Cr(VI) in subcellular systems and a low degree of bioavailability of insoluble chromium salts in bacteria and lower eukaryotes, both soluble and insoluble chromium salts are toxic *in vivo* and to mammalian cells in culture. All forms of chromium induce some degree of chromosomal damage [including aneuploidy and micronuclei formation due to aberrant chromosomal segregation, chromosomal aberrations, and sister chromatid exchanges (SCEs)], DNA damage, and/or mutagenesis in mammalian cells in culture (Table VII). Chromium is a moderate to strong mutagen at the *hprt* locus [inducing resistance to 6-thioguanine ($6TG^R$) or 8-azaguanine ($8AG^R$)] and at the *tk* locus (TFT^R) in mouse lymphoma cells. It is, at best, a weak mutagen at the *Na,K-ATPase* locus (OUA^R);

Table X Experimental Carcinogenesis by Chromium Compounds

Exposure	Strain	Cr Compounds: V. sol. Cr^{+6} $K_2Cr_2O_7$, Na_2CrO_4	Sl. sol. Cr^{+6} $CaCrO_4$	Insol. Cr^{+6} $PbCrO_4$, $ZnCrO_4$	Sol. Cr^{+3} $Cr(NO_3)_3$	Insol. Cr^{+3} Cr_2O_3
	Rats					
Oral	Long Evans/BD				–	
Inhalation	Wistar	–				+
Intratracheal instillation	Wistar/Sprague Dawley	+/–	++			+
Bronchial/pleural implant	Bethesda Bl	–	++	–/+	–	+
Intramuscular	Strangeways/Fischer		++	+/+		
Intramuscular implant	Bethesda Bl	+	++	++	?	+
Subcutaneous	Sprague Dawley			++		
	Mice					
Oral	Swiss				–	
Inhalation	A/Swiss/C57Bl	–	+			
Intratracheal instillation	A/Swiss/C57Bl	–		–		
Intramuscular implant	C57B1	–	+	–	–	–
Subcutaneous	Swiss Webster		+			
	Others					
Inhalation/instillation	Guinea pig	–?		–		
Inhalation/instillation	Rabbit	–		–		

Source: Data compiled from IARC, 1980; IARC, 1987; ATSDR, 1989.

indicating that chromium produces frameshift or deletion mutations, but not a high frequency of base pair substitution mutations, in mammalian cells (Table VII). Human fibroblast cells (HFC) are especially sensitive to mutagenesis by all forms of chromium. All forms of chromium that have been tested also induce genotoxicity *in vivo* (Table VIII) and both insoluble and weakly soluble chromium compounds, including Cr(III) oxide, have been found to be carcinogenic to rodents (Table X). While higher

Table XI Chromium-Induced Carcinogenesis in Humans

Occupation	Study population	Cancer type	Estimated relative risk	Estimated exposure	Chromium compounds
Chromate manufacture	10 worldwide populations; ≥8000 workers	Lung	2- to 80-fold increase in cancer among workers	25–7000 μg/m³ Cr(VI), some Cr(III); inhalation exposure	Mostly Na_2CrO_4 and $CaCrO_4$
Chromate pigment production	7 worldwide populations; ≥4000 workers	Lung	1- to 44-fold increase in risk	0.01–2 mg/m³ insoluble Cr(VI); inhalation exposure	Insoluble $ZnCrO_4$ and $PbCrO_4$
Chromium electroplating	6 worldwide populations, ≥4000 workers	Lung	1- to 9-fold increase in risk	0.1–50 mg/m³ soluble Cr(VI); dermal and inhalation exposure	Soluble CrO_3
Ferrochromium workers	3 populations in the Soviet Union, Norway, and Sweden; ≥4000 workers	Lung	<1- to 6.6-fold increase in risk	0.01–2.5 mg/m³ Cr(III) and Cr(VI); inhalation exposure	CrO_3 ?

Source: Data compiled from IARC, 1980; ATSDR, 1989; Langård, 1990.

concentrations of Cr(III) may be needed to achieve the same biological effectiveness as Cr(VI), there is little evidence to support the assumption that Cr(III) is not and cannot be genotoxic *in vivo*.

D. Systemic Effects of Exposure to Chromium Compounds *in Vivo*

As detailed in Table IX, chronic exposure to chromic acid and chromium salts leads to ulcers of the skin and nasal mucosa, perforation of the nasal septum, and irritation of the gastrointestinal tract. Chronic inhalation or dermal exposure to chromium compounds may lead to the development of asthma and other symptoms of respiratory distress. Acute exposure to chromium is toxic to the gastrointestinal tract and kidneys. There is, however, little evidence for liver toxicity from chromium except after acute exposures to lethal or near-lethal concentrations. Dermal exposure to chromic acid and chromium salts may produce ulcerations and allergic dermatitis. Chromium compounds, especially compounds that are insoluble or slightly soluble in water, also produce cancer in animals (Table X).

VII. Human Exposure and Toxicity

A. Routes of Exposure

The primary routes of normal human exposure to chromium compounds include ingestion of chromium in food and drinking water and inhalation of chromium in cigarette smoke and air. Occupational exposure to higher concentrations of chromium may occur via ingestion, dermal exposure, or inhalation. The highest occupational exposures to chromium have been seen in the stainless steel industry and welding, chromate and chromeplating industries, in the manufacture of chrome pigments, and in leather tanning. The respiratory tract is the most important route for both acute and chronic exposure to excess chromium. Soluble Cr(VI) complexes tend to be highly oxidizing and corrosive, producing ulcers at sites of chronic exposure. Cr(III) complexes, because they are not as readily absorbed as Cr(VI), have not been as well studied.

B. Indices of Exposure

Because chromium is an essential trace element and nearly ubiquitous in the environment, assessment of exposure to chromium must take into account background levels of this element found in tissues of unexposed populations. Acute, recent exposure to excess chromium is best measured by determination of urine or serum chromium concentrations. Under chronic exposure conditions, concentrations of chromium in urine appear to be related to Cr(VI) exposure, but do not provide a good measure of Cr(III) exposure. Serum chromium levels, however, appear to parallel exposure to both Cr(VI) and Cr(III) under occupational exposure conditions (including dermal exposure) and may thus be the best measure of overall exposure. Chromium concentrations in hair can also provide a measure of chronic exposure, primarily to Cr(III). At high levels of exposure to Cr(VI), one may also measure chromium accumulated by red blood cells and the production of DNA–protein crosslinks in peripheral blood lymphocytes.

C. Chromium Hypersensitivity

Chronic chromium exposure has been shown to result in the production of either a systemic or dermal allergic response. Skin allergies are common with chromium and other metals, such as nickel. Chromium effects on the immune system have also been documented in various animal species; however, these studies have shown both enhancements and depressions, depending on the dose and experimental system used. Exposure of sensitive individuals to soluble Cr(VI) may precipitate an anaphalactic response.

D. Carcinogenicity and Genotoxicity

Numerous reports of increased incidence of lung cancers among occupationally exposed chromium workers have convincingly shown a strong correlation between exposure to insoluble and weakly soluble Cr(VI) and human cancer (Table XI). Chromium compounds are also carcinogenic in animal models and thus have been determined to be probable human carcinogens.

VIII. Regulatory Issues and Disposal Guidelines

Due to the carcinogenicity and toxicity of chromium, regulations and advisories exist regarding acceptable concentrations of both Cr(III) and Cr(VI) compounds in air and water. Ambient and drinking water standards and guidance values for chromium in water range between 0.05 and 1.1 mg/l. Air quality standards range from 0.05 mg/m^3 and 50 mg/m^3 of chromium (Tables XII and XIII).

Although in most instances, regulatory values and criteria are similar for Cr(III) and Cr(VI) compounds, the greater bioavailability and carcinogenicity of the hexavalent form has dictated greater attention to Cr(VI) compounds. The EPA oral Reference Dose (RfD) for insoluble Cr(III) salts is 1 mg/kg/day; while the chronic oral RfD for Cr(VI) soluble salts is 0.005 mg/kg/day. These RfD values were calculated from NOAEL (No Observable Adverse Effect Level) doses determined in animal studies.

The industrial discharge of Cr compounds is regulated by EPA under the Clean Water Act Effluent Guidelines. The following industrial point sources fall under this regulation: textile, electroplating, and organic and inorganic chemical industries; petroleum refining; iron and steel manufacturing; nonferrous metal and ferroalloy production; leather tanning and finishing; steam electric, asbestos and rubber industries; timber products processing; metal finishing; mineral mining; paving and roofing; paint and ink formulating; gum, wood, carbon black, and batter manufacturing; coil coating; porcelain enameling; aluminum, copper, and nonferrous metal forming; and electrical and electronic component manufacturing.

Table XII Regulatory Values for Chromium Compounds

Agency	Regulatory value	Comments[a]
		Air
OSHA	0.5 mg/m^3	PEL for soluble chromic (CrIII) and chromous (CrII) salts
	1.0 mg/m^3	PEL for soluble chromium salts or chromium metal
	1 mg/10 m^3	ACC for chromic acid (CrVI) and chromates (CrVI)
		Water
EPA	0.05 mg/l	MCL for total Cr in drinking water
		Nonspecific media
EPA	1 lb	Reportable quantity (RQ) for release of metallic chromium (dia. $<100\ \mu m$) (federal law CERCLA 103a and b)
	1000 lb	RQ for ammonium dichromate, chromic acid, calcium chromate, potassium chromate, potassium dichromate, sodium chromate, sodium dichromate, strontium chromate, chromic acetate, chromic sulfate, and chromous chloride (federal law CERCLA 103a and b)
EPA	1 lb	Threshold planning quantity (TPQ) for chromic acid if it exists in a powdered form with a particle size $<100\ \mu m$, is in solution or molten form, or has a National Fire Protection Association rating of 2, 3, or 4 for reactivity (federal law Section 304 of SARA)
	1000 lb	TPQ for chromic acid under conditions other than those listed above (federal law Section 304 of SARA)
	1 lb	Required immediate reporting of release of chromic chloride to local emergency planning committees and state emergency planning commissions (federal law Section 304 of SARA)

Source: Data compiled from ATSDR, 1989.

[a] PEL, Permissible exposure limit; ACC, acceptable ceiling concentration; MCL, maximum contaminant level.

Table XIII Guidance Values for Chromium Compounds

Agency	Regulatory value	Comments[a]
		Air
ACGIH	0.5 mg/m^3	TLV–TWA value for chromium metal and Cr(II) and Cr(III) compounds
	0.05 mg/m^3	TLV–TWA value for water soluble Cr(VI) compounds, certain water insoluble Cr(VI) compounds, and chromite ore processing for an 8-hr workday, 40-hr work week
NIOSH	0.05 mg/m^3	TWA criterion for chromic acid for an 8-hr workday, 40-hr work week
	0.2 mg/m^3	Ceiling limit (15 min) for chromic acid
	25 μg/m^3	TWA criterion for noncarcinogenic Cr(VI) chromate and dichromate compounds
	50 μg/m^3	Ceiling limit (15 min) for Cr(VI) chromate and dichromate compounds
		Water
USPHS	0.05 mg/l	Drinking water critera for Cr(VI)
EPA	0.12 mg/l	MCLG for total Cr for drinking water
	0.98 mg/l (fw, acute)	Ambient water quality criteria for Cr(III)
	0.12 mg/l (fw, chronic)	
	1.0 mg/l (sw, acute)	
	0.1 mg/l (sw/chronic)	
	0.016 mg/l (fw, acute)	Ambient water quality criteria for Cr(VI)
	0.011 mg/l (fw, chronic)	
	1.1 mg/l (sw, acute)	
	0.05 mg/l (sw, chronic)	
	1.4 mg/l	Ten-day health advisory for a 10-kg child for Cr(VI)
	0.24 mg/l	Longer term health advisory for a 10-kg child for Cr(VI)
	0.84 mg/l	Longer term health advisory for a 70-kg adult for Cr(VI)
	0.12 mg/l	Lifetime health advisory for a 70-kg adult for Cr(VI)
WHO	0.05 mg/l	Recommended European drinking water standard for Cr(VI)

Source: Data compiled from EPA, 1984.
[a] TLV, threshold limit value; TWA, time weighted average; MCLG, maximum contaminant level goal; fw, freshwater; sw, seawater.

Related Articles: Inorganic Mineral Particulates in the Lung; Metal–Metal Toxic Interactions; Mutagenicity Tests with Cultured Mammalian Cells: Cytogenetic Assays; Organic Micropollutants in Lake Sediments.

Bibliography

ATSDR (Agency for Toxic Substances and Disease Registry) U.S. Public Health Service (1989). "Toxicological Profile for Chromium." ATSDR/RP-88/10, Public Health Service Center for Disease Control, Atlanta, GA.

Biedermann, K. A., and Landolph, J. R. (1990). Role of valence state and solubility of chromium compounds on induction of cytotoxicity, mutagenesis, and anchorage independence in diploid human fibroblasts. *Cancer Res* **50,** 7835–7842.

De Flora, S., Bagnasco, M., Serra, D., and Zanacchi, P. (1990). Genotoxicity of chromium compounds: A review. *Mutat. Res.* **238,** 99–172.

Environmental Protection Agency (1984). "Health Assessment Document for Chromium." Environmental Criteria and Assessment Office. Office of Research and Development. U.S. Environmental Protection Agency, Research Triangle Park, N.C. EPA-600/8-83-014F.

IARC (International Agency for Research on Cancer) (1980). "Some Metals and Metallic Compounds." International Agency for Research on Cancer, Lyon.

IARC (International Agency for Research on Cancer) (1987). "Overall Evaluations of Carcinogenicity: An Updating of IARC Monographs Volumes 1 to 42." International Agency for Research on Cancer, Lyon.

Langard, S. (1990). One hundred years of chromium and cancer: A review of epidemiological evidence and selected case reports. *Am. J. Indus. Med.* **17,** 189–215.

Nriagu, J. O., and Nieboer, E. (1988). "Chromium in the Natural and Human Environments." John Wiley and Sons, New York.

Cleaning and Laundry Products, Human Exposure Assessments

P. J. (Bert) Hakkinen
The Procter & Gamble Company

Glossary

Cleaning and laundry products Household detergent products in liquid, granular, solid, or aerosol form. Some are used solely for the cleaning of hard surfaces (e.g., countertops and dishes), and some are used solely for clothes washing. Others are used for both the cleaning of hard surfaces and clothes washing.

Diary study Involves having consumers keep a written diary to record how they use a product.

Exposure Contact with a product and its components, by way of the skin, eyes, gastrointestinal tract, and respiratory tract, usually expressed in terms of the quantity coming in contact with the human system (e.g., "there is __ mg of exposure to chemical __ per kilogram human body weight per day"). Further, if available, a permeability constant for a chemical can be used to calculate flux rate through the skin, and/or percent absorption values for a chemical can be used to calculate the dose absorbed from skin, gastrointestinal, or respiratory tract exposures. Also, physiologically based pharmacokinetic and pharmacodynamic modeling, or actual measurement of possible organ or tissue levels can be performed for a chemical.

Flux rate [Skin permeability constant (or "permeation coefficient") for a chemical in cm/hour or cm/minute] × (concentration of chemical in mg/cm^3 in exposure solution).

In-home observation A way to obtain first-hand knowledge about how consumers use and store a product, and how the package is disposed of, by observing behavior in the home.

Recall study Involves asking consumers about what products are used for various tasks, and how those tasks are performed. May be severely biased data due to lack of memory reliability.

Respiratory minute volume Volume of air breathed in (or out) per minute. Referred to as the "inhalation rate" in many publications; thus, it could be confused with respiratory rate or frequency (i.e., breaths per minute). Varies with an individual's age, weight, activity, and other factors.

Respiratory tract Can be divided into three anatomic regions: the nasopharyngeal, tracheobronchial, and pulmonary ("deep lung") regions.

CAREFUL CONSIDERATION of the types and extent of potential employee and consumer exposures, combined with "hazard identification" and "dose–response" information, is needed for well-founded human risk assessments for cleaning, laundry, and other consumer products. "Hazard identification" for a consumer product involves assessing whether the product or any particular component is associated with any human health effect(s), while "dose–response" assessments involve examination of the relationship between the degree of exposure to a product or component and the magnitude of any particular adverse effect(s). Good exposure assessments require realistic data to assess the extent of possible skin, inhalation, and ingestion exposures to products and components. This article describes the types and sources of data for cleaning and laundry product exposure assessments, provides comments about how to help ensure appropriate use of the data, and provides examples of exposure calculations.

I. Introduction

Based on how any similar products are used, and from a review of the premarket consumer testing and any test marketing of a particular product, a toxicologist or other exposure assessor can determine whether a product and its components are likely to be inhaled, ingested, or to contact the skin during consumer use. That is, the exposure assessor needs to gain an understanding about how consumers actually use a product, and if there are any uses different from those recommended by the manufacturer.

Key data for quantifying consumer exposures to cleaning and laundry products include surveys of how the products are used by consumers (e.g., the concentrations of product used for a particular type of task, how long a task takes, and how many times a day or week a task is performed).

Examples of the key exposure-related data for cleaning and laundry product ingredients (and possible impurities and contaminants) include: (1) in laundry products, analytical determinations of the levels of fabric deposition and subsequent transfer to skin, (2) in dishwashing products, analytical determinations of the residue levels on dishes, utensils, and glasses, (3) skin permeability constants and/or skin absorption data, (4) oral absorption and lung deposition/retention data, and (5) physical/chemical property data. The physical/chemical properties of the product's components must be evaluated to understand whether a particular component is so volatile in practice (e.g., in the task solution) that it would be expected to be released into the air of the room during use. This expectation could lead to a need for analytical work to quantify the concentrations in room air.

Although much of what follows applies to consumer products in general, the focus is on cleaning and laundry product examples. To provide a detailed understanding of the types of exposures associated with one particular type of cleaning or laundry product, dishwashing products are used as the example. Dishwashing products are considered a good example of products used in a wide variety of tasks, since they are often used for non-dishwashing tasks (e.g., the washing of clothing and the cleaning of various hard surfaces).

II. Types of Potential Employee Exposures

As would apply to the manufacture of most other consumer products, the types of potential manufacturing employee exposures to dishwashing products and their components include skin, eye, and oral contact during manufacture, and possible inhalation of dusts, aerosols, and/or vapors.

Industrial hygienists and other exposure assessors have developed procedures to quantify the air levels of materials, including products and their components. Also, procedures are available to assess skin exposures. Possible skin exposure approaches include the use of fluorescent tracers to assess the areas of skin deposition, and the analysis for material collected on absorbent patches (i.e., surgical gauze pads or cellulose patches attached to various body locations) or absorbent gloves, or in solvents or solutions from hand washes.

If judged to be needed, protective clothing and gloves, and protective creams can be used to provide a barrier to skin exposures (and oral exposure to material deposited on skin and transferred to food being eaten). Also, protective eyeglasses, goggles, and face shields can be used when indicated to protect against eye exposures. To protect against inhalation and/or skin exposures (and oral exposures from the swallowing of materials deposited in the respiratory tract), changes may be necessary in a manufacturing plant's design (e.g., conversion of a particular manufacturing step from an "open" to a "closed" system, thus preventing normal employee exposures). Changes in a manufacturing plant's ventilation system (e.g., increasing the air changeover rate in a particular work location), and/or respirators can also be used when indicated to protect against inhalation exposures.

III. Types of Potential Consumer Exposures

The types of potential consumer exposures to dishwashing products and their components include the following.

A. Normal Uses

1. Skin exposure from use of hand dishwashing liquids for dishwashing, hand laundering, and hard surface cleaning; and from the wearing of hand-laundered clothing.
2. Possible oral exposure from dinnerware, utensil, and glass deposition.
3. Inhalation exposure to volatile components of hand dishwashing liquids before, during, and after hand dishwashing, hand laundering, and hard surface cleaning.
4. Inhalation exposure to volatile components of granular or liquid automatic dishwashing products.
5. Inhalation exposure to particulate components of granular automatic dishwashing products from pouring the product into the automatic dishwasher's dispenser cup.

B. Accidental Exposures and Nonrecommended Uses

1. Accidental ingestion from the product bottle or package, task solution, pouring cup, or automatic dishwasher dispenser cup.
2. Accidental eye contact from the product bottle or package, or task solution.
3. Skin contact from nonrecommended uses (e.g., handwashing).
4. Inhalation from possible exposure to volatile components during and after nonrecommended uses.

IV. Sources of Exposure Data

The sources of data for cleaning and laundry product (and other consumer product) exposure assessments include the following:

In-home observation data
Diary studies
Recall studies
Toll free "800-line" telephone calls
Poison control center studies
Objective measurements
Technical literature
Trade organizations
Government agencies

The following provide cleaning and laundry product examples of data obtained from the above sources.

An in-home observation study of a granular laundry product found that some consumers poured the product into a measuring cup, and then into the washing machine. Other consumers dipped the measuring cup directly into the detergent box, while others poured the product directly into the washing machine from the box. The laboratory studies that followed this study measured the amount of detergent dust that could be generated during pouring of the product, and incorporated knowledge about such factors as how the task was performed, the typical distance from the box to the measuring cup, and so forth.

Diary and recall studies of a laundry detergent can include questions such as the make of washing machine used, the numbers of laundry loads washed in the past seven days, the typical amount of product used, and the typical machine settings (e.g., wash time and water fill level). Likewise, diary and recall studies of a hand dishwashing liquid can include questions such as the amount of product used to wash a load of dishes, how much water was added to the sink, how long the task took, and whether gloves were worn.

Toll free "800 line" telephone numbers are listed on the packages of many household products in the United States. Consumers use these toll free lines to seek product information, report problems, and to express satisfaction or concerns. From such telephone calls, detailed data can be collected on the types of tasks for which a product is used, accidental ingestion volumes, age and weight of the exposed person, the source of the exposure (e.g., from product package or from a measuring cup), and the exposure outcomes. For example, two Procter & Gamble studies of accidental ingestion reports received via 800-line telephone calls indicate that (1) lemon aesthetics in hand dishwashing products are *not* associated with an increase in reported accidental ingestion frequency or volume in young children, and that (2) most accidental ingestions of laundry and dishwashing products involve children weighing 20 kg or less, and one teaspoon of product or less.

Poison control centers can provide valuable information regarding accidental exposures to consumer products. One example is a multi-year Procter & Gamble–sponsored study using data about automatic dishwashing detergents gathered from poison control centers. Information was obtained on the source of exposure (e.g., product package or dishwasher dispenser cup), the amount ingested, and the reported symptoms from the accident. Another example of poison control center data is the "Annual Report of the American Association of Poison Control Centers National Data Collection System," published yearly in the *American Journal of Emergency Medicine*. Included are data on human exposure cases reported by numerous United States poison control centers during the preceding year. The data include the distribution by age for exposures associated with various types of products, and the clinical outcome of the exposures.

Objective measurement studies are those specifically designed to experimentally evaluate a particular aspect of consumer exposure to a product and its components. Consumers can perform all of, or portions of, an objective study themselves (e.g., by timing the duration of a task, or placing a task solution into a container for later concentration analysis). Also, laboratory studies simulating in-home exposures can be conducted. For example, a 20-m^3 (room-size) environmental chamber was used to simulate consumer hand dishwashing and laundry machine usage. This study found average values of one part per million or less of ethanol in the chamber's air due to volatilization from ordinary usage of the prototype dishwashing liquid and laundry detergent. Other examples of objective measurements include Procter & Gamble data obtained from studies with numerous children indicating that adding a bitter-tasting substance to a liquid detergent would be expected to significantly reduce the probability of an accidental ingestion involving multiple swallows. Another example is a study in which data obtained from hundreds of children indicated that reducing the orifice diameter of a package closure reduced the volume of a swallow for a low-viscosity liquid, but did not appear to affect swallow volume for an intermediate- or high-viscosity liquid. Also, swallow volume was found to be inversely related to the viscosity of the test liquid.

Consumer exposure data that might be available from the technical literature, trade organizations, and government agencies include trends in product use and accidental exposure, amount used per task, and task frequency and duration, with data obtained from diary and recall studies, objective measurements, etc. An example of a government agency-sponsored source of exposure information is a series of United States Environmental Protection Agency-sponsored telephone and diary surveys of the extent of consumer exposure to various household products, including cleaning and laundry products.

V. Ensuring Appropriate Use of Exposure Data

It is important to understand the statistical distribution of consumer exposure data since not all consumer data follow a normal distribution. For example, the number of machine laundry loads washed per week follows a log-normal distribution. Neither normal nor log-normal models should be used to estimate percentile values from the mean and standard deviation, unless the data are shown to reasonably fit the model used. As an example, using a normal model when the log-normal model really applies to some data could significantly *under*estimate the 99th percentile value and significantly *over*estimate the 50th percentile value.

Further, exposure assessments for consumer products are often presented in terms of average (i.e., mean) or median anticipated use, and "worst case" use. In the latter case, data pertaining to the extremes of product usage parameters are combined to provide a conservative assessment of the highest possible consumer exposure. Before exposure calculations are finalized, the data should be examined for potential correlative interactions, especially those that are inversely related. If this is not done, the final estimate of consumer exposure may be unduly conservative and totally unrealistic. For example, it would be inappropriate to use the highest recorded task duration value, the highest recorded daily task frequency, and the highest recorded product concentration for a task all in the same exposure calculation for a cleaning product (e.g., when used to clean countertops), if the task duration is known to decrease as (a) the number of times the task is performed increases, and/or (b) the concentration of product used for the task increases.

Also, rather than arbitrarily including every task for which consumers are known to use a particular product in a "worst case" exposure assessment, exposure assessors should examine whether any one particular consumer is likely to perform all the tasks during a given day or other time period. A true "maximally exposed" consumer can be established by careful collection and analysis of exposure data for individual consumers using a particular product (e.g., by looking at the particular tasks, task frequencies, or task durations associated with individual consumers).

VI. Examples of Exposure Calculations

The following equations are examples of the types of calculations used for performing exposure assessments for cleaning and laundry products. Note the need for many different types of consumer exposure data (e.g., task frequency and duration and concentration, and area of exposed skin), and information specific to a chemical (e.g., dermal permeability constant, amount deposited on fabric, particle size, etc.). The values shown for various factors are either from Procter & Gamble consumer and analytical studies, or from the technical literature.

A. Consumer Skin Exposure Assessments

1. Direct skin contact to a product or a task solution (e.g., from hand dishwashing, hand laundering, and hard surface cleaning) can be assessed as follows:

$$\left[\begin{pmatrix}\text{flux rate}\\ \text{of material}\end{pmatrix} \times \begin{pmatrix}\text{duration}\\ \text{of exposure}\end{pmatrix} \times \begin{pmatrix}\text{area of}\\ \text{exposed skin}\end{pmatrix}\right] \Big/ \begin{pmatrix}\text{body}\\ \text{weight}\end{pmatrix}$$

where flux rate = (permeability constant) × (concentration of material).

2. The assessment of the daily exposure from residue on clothing would be calculated by:

$$\left[\begin{pmatrix}\text{amount deposited}\\ \text{on fabric}\end{pmatrix} \times \begin{pmatrix}\text{\% transfer from}\\ \text{fabric to skin}\end{pmatrix}\Big/100^{*} \times \begin{pmatrix}\text{surface area}\\ \text{of exposed skin}\end{pmatrix} \times \begin{pmatrix}\text{expected or measured}\\ \text{\% absorption}\end{pmatrix}\Big/100\right] \Big/ \begin{pmatrix}\text{body}\\ \text{weight}\end{pmatrix}$$

3. Key information and considerations for skin exposures:

(a) For hand dishwashing, a 0.20% task solution is typically used (=15 ml of a liquid dishwashing product diluted in 7.57 liters of sink water).

(b) Hand dishwashing typically takes 10 min or less per task and occurs one to two times a day.

(c) If hand dishwashing is performed without gloves, the skin area exposed to the task solution is 720 cm^2 if the hands of a 70-kg adult are immersed, and 1,680 cm^2 if the hands and forearms are immersed.

* Divided by 100 to convert from a percent value to the appropriate fraction (e.g., 88% becomes 0.88).

B. Consumer Inhalation Exposure Assessments

1. The assessment of daily exposure from inhalation of a volatile material would be calculated in the following manner:

$$\left[\begin{pmatrix}\text{vapor phase}\\ \text{concentration}\end{pmatrix} \times \begin{pmatrix}\text{duration of}\\ \text{exposure}\end{pmatrix} \times \begin{pmatrix}\text{respiratory minute}\\ \text{volume}\end{pmatrix} \times \begin{pmatrix}\text{\% retained in}\\ \text{lungs (if known)}\end{pmatrix}\Big/100\right] \Big/ \begin{pmatrix}\text{body}\\ \text{weight}\end{pmatrix}$$

2. The assessment of daily exposure to a particulate material (i.e., deep lung burden method) may be calculated by

$$\left[\begin{pmatrix}\text{total aerosol}\\ \text{concentration}\end{pmatrix} \times \begin{pmatrix}\text{\% material in respirable}\\ \text{size range (if known)}\end{pmatrix}\Big/100 \times \begin{pmatrix}\text{\% deposition in}\\ \text{deep lung (if known)}\end{pmatrix}\Big/100 \times \begin{pmatrix}\text{duration of}\\ \text{exposure}\end{pmatrix} \times \begin{pmatrix}\text{respiratory minute}\\ \text{volume}\end{pmatrix}\right] \Big/ \begin{pmatrix}\text{body}\\ \text{weight}\end{pmatrix}$$

3. Key information and considerations for inhalation exposures:

(a) The degree of physical activity influences how much air is inhaled; the respiratory minute volume is 11.25 liters/minute for an adult at rest, and 21.75 liters/minute for an adult performing moderate activity.

(b) The "air exchange rate" of the room or house influences the concentration of a material released into the air. A representative air exchange rate for a recently constructed, well-sealed U.S. house has been reported to be about one-half changeover per hour, while rates as great as 4.0 and lower than 0.25 have been reported.

(c) The size of a particle influences where the particle may be deposited in the respiratory tract. The percent deposition in the deep lung (i.e., alveolar region) is ~22% for a particle having 0.75-μm aerodynamic diameter; ~30% for a particle having a 1.5-μm aerodynamic diameter; ~25% for a particle having a 3.0-μm aerodynamic diameter, and ~5%

for a particle having a 6.0-μm aerodynamic diameter.

C. Consumer Oral Exposure Assessments

1. The assessment of oral exposure from dishes, utensils, and glassware can be calculated as follows:

(a) Unrefined exposure estimate:

$$\left[\begin{pmatrix}\text{deposition of material on dishes,}\\ \text{utensils, and glassware in } \mu\text{g/cm}^2\end{pmatrix} \times \begin{pmatrix}\text{cm}^2\text{/person/day of dishes, utensils, and glassware}\\ \text{in contact with food or consumed fluid}\end{pmatrix} \times \left(\frac{\%\text{ absorption (if known)}}{100}\right)\right] \Big/ \begin{pmatrix}\text{body}\\ \text{weight}\end{pmatrix}$$

(b) Refined exposure estimate:

$$\left[\begin{pmatrix}\text{amount of material extractable from dishes, utensils, and}\\ \text{glassware into food simulating solvent in } \mu\text{g/cm}^2\end{pmatrix} \times \begin{pmatrix}\text{cm}^2\text{/person/day of dishes, utensils, and glassware}\\ \text{in contact with food or consumed fluid}\end{pmatrix} \times \left(\frac{\%\text{ absorption (if known)}}{100}\right)\right] \Big/ \begin{pmatrix}\text{body}\\ \text{weight}\end{pmatrix}$$

2. Key information and consideration for oral exposures:

(a) Only about 2% of homemakers do not always rinse every dish, utensil, and glass with clear water after washing by hand. A consumer would be exposed to only about 0.1 mg detergent residue/kg body weight/day even if *unrinsed* dishes, utensils, and glasses were used.

(b) The volatility of a material initially deposited on dinnerware may result in evaporation of the material before the dinnerware is used.

VII. Summary

Exposure assessments are a key part of the overall risk assessment of cleaning and laundry products, and other consumer products. Exposure data are obtained from a number of sources, including the published literature (e.g., body surface areas), surveys of how consumers use products (e.g., typical tasks, number of times a day or week a product is used for a particular task, how long each task takes, and the concentration of the product used for a task), toll free 800-line telephone calls, poison center surveys, trade organizations, analytical determinations (e.g., possible levels of dinnerware deposition and subsequent extraction into food), and toxicokinetic (e.g., skin permeability constants) and physical/chemical property data (e.g., vapor pressure at a particular temperature).

Further, it is important to understand what the best sources of each type of data are, and whether the data follow a normal, log-normal, or other type of statistical distribution. Finally, it is important to consider whether the separate values used in an exposure calculation are correlated with each other, and whether the combinations of tasks and task values (e.g., concentrations, durations, and frequencies) used in a "worst case" assessment truly represent what a maximally exposed consumer is likely to do with a product. Not doing any of the above can result in an unduly conservative and unrealistic exposure assessment.

Related Articles: INDUSTRIAL SOLVENTS; ORGANIC SOLVENTS, HEALTH EFFECTS.

Bibliography

Brown, S. L. (1987). Exposure assessment. *In* "Toxic Substances and Human Risk: Principles of Data Interpretation" (R. G. Tardiff, and J. V. Rodricks, eds.), Plenum Press, New York.

Brown, S. L., and Rossi, J. E. (1989). A simple method for estimating dermal absorption of chemicals in water. *Chemosphere* **19,** 1989–2001.

Bus, J. S., and Gibson, J. E. (1985). Body defense mechanisms to toxicant exposure. *In* "Patty's Industrial Hygiene & Toxicology, second edition, Vol. 3B, Biological Responses" (L. Cralley, and L. Cralley, eds.), John Wiley & Sons, New York.

Calvin, G. (1992). Risk management case history—Detergents. *In* "Risk Management of Chemicals" (M. L. Richardson, ed.), The Royal Society of Chemistry, United Kingdom.

Franklin, C. A., Somers, D. A., and Chu, I. (1989). Use of percutaneous absorption data in risk assessment. *J. Amer. Coll. Toxicol.* **8,** 815–827.

Gibson, W. B., Keller, P. R., Foltz, D. J., and Harvey, G. J. (1991). Diethylene glycol mono butyl ether concentrations in room air from application of cleaner formulations to hard surfaces. *J. Expos. Analys. Environ. Epidem.* **1,** 369–383.

Hakkinen, P. J., Kelling, C. K., and Callender, J. C. (1991). Exposure assessment of consumer products: Human body weights and total body surface areas to use, and sources of data for specific products. *Veter. Hum. Toxicol.* **33,** 61–65.

McKone, T. E. (1989). Household exposure models. *Toxicol. Lett.* **49,** 321–339.

Temple, A. R., and Spoerke, D. G. (1989). "Cleaning Products and Their Accidental Exposure" Sixth Edition. The Soap and Detergent Association, New York.

U.S. Environmental Protection Agency, Exposure Assessment Group, Office of Health and Environmental Assessment (1989). *Exposure Factors Handbook*. EPA/600/8-89/043, Washington, District of Columbia.

Wooley, J., Nazaroff, W. W., and Hodgson, A. T. (1990). Release of ethanol to the atmosphere during use of consumer cleaning products. *J. Air & Waste Mgmt. Assoc.* **40,** 1114–1120.

Cobalt Dusts

Dominique Lison
Catholic University of Louvain, Belgium

Glossary

Alveolitis Pathological syndrome in which alveolar walls are infiltrated by edema and inflammatory cells, accompanied by enlargement and desquamation of alveolar epithelial cells.

Hard metals Cemented alloys of tungsten carbide and cobalt to which various other carbides and metals may also be added. These alloys are almost as hard as diamond and are remarkably resistant to high temperature; they therefore have multiple applications for the manufacture of a wide variety of components such as drill tips and cutting tools.

Lung fibrosis Morphologic term used to describe the excessive appearance of connective tissue (chiefly collagen fibers) within alveolar structures.

Obstructive syndrome Functional alteration characterized by an obstruction to the airflow due to narrowing of large or small airways. The disorder may be reversible as in the case of asthma, or irreversible as in chronic bronchitis and emphysema.

Pneumoconiosis Generic term refering to a non-neoplastic lung reaction caused by the inhalation of any mineral or organic dust, excluding asthma, bronchitis, and emphysema.

Restrictive syndrome Functional alteration of the lungs, which cannot expand as fully as they should as a result of any cause such as impairment of full movement of the chest walls or diffuse interstitial fibrosis.

COBALT is a hard, magnetic, silvery metal closely resembling iron and nickel in appearance, with a high melting point, and is relatively unreactive and insoluble in water. Because of these desirable properties, it is widely used alone or in association with other metallic compounds for a number of industrial applications. Various types of cobalt (from the word Kobold *meaning goblin, given by the ancient German miners) containing dusts (metal, oxides, alloys) may be produced during these operations and the inhalation of these dusts may give rise to a number of toxic manifestations.*

I. Introduction

There are health hazards associated with excessive exposure to cobalt containing dust. The major source of exposure is occupational, arising principally from the handling of cobalt metal and oxides but also from the production, processing, and use of hard metals and alloys. Lung diseases, including alveolitis, diffuse interstitial fibrosis, and bronchial asthma constitute the principal adverse effects. There is no indication that cobalt compounds represent a health risk for the general population.

Cobalt is the 33rd element in abundance in the earth's crust with an average concentration of 20 $\mu g/g$, but much higher concentrations are found in ore deposits from which approximately 25,000 tons of cobalt metal is produced annually. Cobalt is mostly obtained as a secondary product during the processing of other metals (chiefly copper, nickel, and lead). The most important minerals are cobaltite (CoAsS), smaltite ($CoAs_2$) and erythrite $[C0_3(AsO_4)_2]$; the major sources are Zaire, Zambia, Canada, and the former USSR. The principal consumer of cobalt is the United States, which uses about half of the world production.

Cobalt is a hard silvery or greyish-white metal

(atomic weight, 58.93; melting point, 1495°C; boiling point, 2900°C; and density, 8.84). It is practically insoluble in water but dissolves at a rate of 0.003 mg/l in physiological solution and 152.5 mg/l in blood plasma (37°C). Cobalt is easily solubilized in HCl, H_2SO_4, and HNO_3 but not in HF. Cobalt is magnetic, shiny, ductile, and brittle; it is a relatively unreactive material that does not oxidize in dry or moist air at ordinary temperature; at higher temperatures, oxidation is only marginal.

Ionic cobalt can exist in either a bivalent or a trivalent form. Different oxides are used in industry: (1) cobaltous oxide (CoO) varies in color from olive green to red; usually, however, it appears as a dark-grey powder (melting point, 1935°C); (2) black cobaltic oxide (Co_2O_3) is formed when cobalt is heated in an excess of air at low temperature; higher temperatures convert it into cabaltocobaltic oxide (Co_3O_4), the most stable oxide of cobalt (III). The latter decomposes in CoO at 900–950°C. Cobalt oxides are insoluble in water but are soluble in acids. Cobalt chlorides and sulfate and a number of complexes with amines, nitrites, and cyanides are other important industrial compounds.

Cobalt is an essential nutrient to humans; hydroxycobalamin (vitamin B_{12}) is the major physiological form in which cobalt functions in the human body. Vitamin B_{12} deficiency leads to the development of pernicious anemia. Mammals are unable to synthesize this molecule although the bacterial flora in the digestive tract of ruminants can form vitamin B_{12} molecules from cobalt ions. In humans, vitamin B_{12} must be provided in the diet; the minimum recommended daily intake for an adult is 3 μg, corresponding to 0.012 μg of cobalt.

Cobalt is also associated with the regulation of a number of enzymes as well as (probably in interaction with copper and iron) the production of erythropoietin, the red cell stimulating factor.

While dietary cobalt deficiency in humans is usually associated with vitamin B_{12} deficiency (and hence with anemia), dietary cobalt toxicity is rare. Adverse effects due to the inhalation of cobalt-containing dusts and allergic reactions are, however, more common.

II. Sources and Levels of Exposure

Cobalt has been used for more than 2000 years for coloring glass, ceramics, and jewelry, but it has only been since the beginning of this century (1927) that cobalt was used for the production of alloys. Because of the practical advantages of a high melting point, strength, and resistance to oxidation, about 70% of the cobalt consumed in the United States is to produce magnetic and hard alloys including:

- aluminium–nickel–iron–cobalt alloys for the production of permanent magnets used in the electronic and electrical industries
- vanadium–iron–cobalt alloys used for permanent ductile telephone components
- vitallium, an alloy of cobalt, chromium, nickel, and molybdenum used in bone replacement prostheses such as artificial hips and knee joints
- cobalt alloyed with chromium and nickel used in heat-resistant gas turbines and jet engine components

The most important application of cobalt is, however, in the manufacture of cemented tungsten carbide, where it is used as binder of the different powder components. The tungsten carbide content of hard metal usually exceeds 80%, and the cobalt content is often less than 10%, but it may be as high as 25%. The tungsten metal powder is reacted with carbon under an atmosphere of hydrogen to form tungsten carbide, which is, in some cases, mixed with nickel or the carbides of niobium, molybdenum, chromium, titanium, tantalum, or vanadium. Cobalt is added to the mixed powder, which is then filtered; paraffin is usually added to provide cohesiveness for processing; organic solvents such as acetone and *n*-hexane may be included for mixing but are later driven off by a drying process. The material is then shaped and presintered and deparaffinized (at 1000°C) before grinding into final form; finally, it is sintered progressively up to 1500°C. Workers are exposed to cobalt-containing dust during the mixing, filtering, shaping, and grinding operations.

These metals possess extraordinary properties of hardness (about 90% that of diamond) and remarkably resist heat and wear. Moreover, unlike other metals, the "toughness" properties of hard metals increase with temperature. This characteristic is used in the manufacturing of hard metal-tipped tools that remain sharp up to operating temperatures of 1700°C; the normal maximum temperature of steel-made cutting tools is approximately 1000°C. The tips of saws, cutters, drill bits, and other special devices often consist of hard metal which is cemented to the body of the tool. These metals are also used in armaments for the nosecones of armor-piercing ammunitions and for armor plating. High-speed dental drills are made from hard metals as well. Flame plating of certain components with hard metal protects from wear; tire studs, and some ballpoint pens are other applications.

Cobalt exposure has also been reported among diamond polishers; certain modern polishing disks have a frame of half-hard steel (more than 98% iron) and a polishing surface consisting of 10–20% in microdiamonds cemented into ultrafine cobalt metal powder (80–90%). Other components such as tungsten carbide, iron, nickel, manganese, sulfur, and silicon may also be present (depending on the disk type). Typical manufacturing of these disks includes the following operations: cold pressing of microdiamonds and cobalt, mild sintering to 600–700°C (since higher temperatures cause graphitization of diamond), and hot pressing of the diamond–cobalt mixture onto the steel frame. After a number of grinding and polishing sessions, microdiamonds and cobalt are abrased and the cutting surface is resmoothed; this operation can be repeated several times whereupon the disk normally has to be replaced. In an attempt to recondition these costly tools, diamond powder (sometimes titanium or zirconium particles) may be glued (with arabic gum or dextran) onto the polishing surface of abrased disks.

Cobalt compounds also have important uses as catalysts in the chemical industry. Cobalt promotes oxidation and desulfuration processes in the refining of crude oils. Cobalt naphtenate, octanoate, oleate, resinate, tallate, and linoleate are used as drying agents for oil-based paints, varnishes, and printing inks.

The exposure level of the workers in these different industries varies widely. Large amounts of cobalt are handled in a cobalt-producing plant, in cobalt salt manufacture, and in the cemented tungsten carbide industry and cobalt-containing alloy manufacture. The potential for exposure to cobalt dusts is also great in workers who perform cutting, grinding, boring, sawing, or polishing with the aid of cobalt-tipped devices. Wet operations are not helpful in reducing exposure as it has been shown that cobalt may dissolve and accumulate in the recycled coolant that is finely aerosolized during the grinding process. Exposure of painters to cobalt driers and pigments is thought to be minimal.

Typical concentrations in the working environment range from 0.01 to 1.7 mg/m^3, but concentrations reaching 10 and even 100 mg/m^3 have been reported. Studies conducted in the cemented tungsten carbide industry have recorded average airborne cobalt concentrations ranging from 0.16 to 0.21 mg/m^3. In one of these studies, the most exposed workers were found to be those handling the powders (0.048 mg/m^3), followed by persons employed at pressing operations (0.033 mg/m^3), grinders (0.016 mg/m^3); general inspection personnel received the lowest exposure level (0.015 mg/m^3). A survey conducted in a nickel refinery reported airborne cobalt levels ranging from 0.02 to 0.099 mg/m^3. In a plant producing cobalt salts, values of 0.001 to 0.002 mg/m^3 were reported at the drying operations, whereas values as high as 0.2–0.5 mg/m^3 were found in the powdering installations. A study of the diamond industry recorded cobalt air concentrations ranging from 0.006 to 0.045 mg/m^3. Mean airborne cobalt concentrations of up to 0.135 mg/m^3 were observed in a plant producing diamond–cobalt tipped tools.

Cobalt is present in very low concentrations in ambient air; concentrations of 5 and 1 ng/m^3 have been measured in urban and rural areas, respectively.

It is important to emphasize that not only quantitative but also qualitative conditions must be taken into account to adequately characterize exposure levels and hence the inherent health risk. It has indeed been shown that the biological reactivity of cobalt metal dust can differ depending on, among other things, particle size or the presence of other particles (e.g., carbides).

III. Metabolism

The human body contains about 1 to 2 mg of cobalt; the highest concentrations have been found in kidney, liver, heart, bone, and skin.

In subjects not occupationally exposed to cobalt, the blood concentration is less than 2.0 μg/l; in urine it does not exceed 2 μg/g creatinine. In subjects occupationally exposed to cobalt dust, the measurement of cobalt concentration in urine can be proposed for the assessment of current exposure levels since a number of studies have demonstrated a linear relationship between airborne exposure and urinary levels. For blood cobalt concentration, a linear relationship is mainly observed at high exposure levels, making it less useful in industrial hygiene.

No human data are available to assess the degree of respiratory absorption of cobalt. From animal studies, it has been estimated to represent about one-third of the deposited dose. Observations on occupationally exposed workers have confirmed that cobalt is taken up to a significant extent from the lung since dramatic increases in both blood and urine cobalt levels are immediately observed after a short exposure period. In a study of workers from the powder-preparation unit of a hard metal manufacturer, the Friday afternoon cobalt content in urine was found to be increased up to 700 times control values. In another survey on workers employed in the mixing and weighing of cobalt-diamond powders, values of 25 to 40 μg of cobalt/g creatinine were recorded in spot urine samples collected at the end of work shifts; nonexposed employees excreted less than 0.2 μg/g creatinine. The gastrointestinal absorption of soluble cobalt compounds such as cobalt chloride amounts to approximately 25%, with large variations depending on the amount of cobalt absorbed, the fasting state, or the addition of albumin or iron to the diet.

After absorption from the lung or the gastrointestinal tract, cobalt is distributed mainly to the liver and kidneys. There is no indication that cobalt accumulates with age in the human body.

Excretion of parenterally administered cobalt takes place mainly via urine but also via feces 50 and 10% of the dose excreted after 8 days, respectively. Most inhaled cobalt is rapidly excreted in urine (within a few days), but a compartment is cleared more slowly (months) since urinary excretion levels remain ten times above normal values even after four weeks of vacation. Accidental contaminations of workers with radioactive cobalt have confirmed the latter observations; a fast elimination phase is followed by a period of protracted cobalt retention.

IV. Health Effects

Workers can be exposed to cobalt by several routes:

- inhalation represents the principal route of exposure primarily when cobalt-containing materials are heated, handled, or subjected to operations such as grinding;
- dermal exposure can occur when these materials are handled manually especially in solution; and
- ingestion is more often the consequence of lack of observance of good work hygiene practices.

A. Respiratory System

Workers who have inhaled cobalt-containing particles may develop respiratory diseases. However, exposure to a single substance in the occupational environment is rare and, in the case of cobalt, most reported respiratory effects concern mixed exposures such as in the cemented tungsten carbide. So far, information concerning the influence of concomitant exposure conditions is only fragmentary.

The term "hard metal disease" is usually used to characterize the respiratory pathology associated with exposure to cobalt dusts; this term however may be confusing since it often includes different types of lesions affecting either the airways (obstructive disorder) or the parenchymal tissue (restrictive disorder); moreover, these lesions are unlikely to result from a unique pathogenic mechanism. Different clinical presentations can be distinguished in several disease manifestations.

1. Upper Respiratory Tract

Aspecific irritative manifestations including chronic rhinitis and rhinopharyngitis, sinusitis, sore throat, and a deterioration of the sense of smell have been reported primarily in workers in the hard metal industry.

2. Asthma

Exposure to cobalt dust alone or in association with other metals may cause bronchial asthma in a small proportion of workers (<5%). Usually, it develops after the worker has been exposed for periods between 1 and 18 months and presents the classical features of occupational asthma, that is in occurring toward the end of the day or in the evening, disappearing during the weekend or the holidays, and recurring on the first day back at work. These manifestations are generally accompanied by wheezing, chest tightness, and cough. The chest radiograph usually shows no abnormality, and a reduction of peak flow rates or forced expiratory volume (FEV_1) in the course of the working week is typical. The worker may voluntarily change from the offending job because of discomfort. After the change, the symptoms improve remarkably and complete remission is often observed in the absence of further exposure. If exposure is continued, the worker may develop permanent bronchoconstriction.

The exact mechanism of cobalt asthma remains unknown; a type I allergic reaction may play a role, while a nonreaginic or nonimmunologic mechanism cannot be ruled out. A possible link between cobalt-induced asthma and skin allergic reactions has been suggested but this remains a matter of controversy. A specific bronchial challenge test consisting of exposure to controlled concentrations of nebulized cobalt sulfate or chloride has been described; it can help to confirm the etiology of the disease; immediate, late or dual reactions can be observed. Skin and RAST tests have also been proposed but appear less useful than inhalational tests.

There is substantial evidence that cobalt alone is the causative agent of the asthma syndrome since, (1) it has been observed not only in hard metal workers but also in workers exposed to cobalt metal, cobalt salts, or oxides and in the diamond polishing industry; and, moreover, (2) it can be reproduced by inhalational challenge of cobalt salts alone.

3. Parenchymal Lung Diseases

The most severe type of lung change observed among workers exposed to cobalt-containing particles is characterized by the presence of interstitial fibrosis and restrictive pulmonary impairment leading to gas exchange alterations and a possibly fatal outcome. Lung fibrosis was first described in 1940 in Germany in hard metal workers; since then, different types of lesions, with distinct clinical and morphological presentations have been described.

In some cases, the clinical picture is that of an acute or subacute alveolitis characterized morphologically by the presence of numerous giant multinuclear cells occupying the alveolar spaces, while the alveolar walls are slightly affected by fibrosis. This lesion is usually distributed uniformally in the lungs. The presence of giant multinucleated cells is a typical feature of cobalt-induced alveolitis; electron microscopic studies have shown that these cells can originate both from alveolar macrophages and type II pneumocytes, but the biological significance of their presence remains to be clarified. Clinically, the worker rapidly develops fever, cough, and dyspnea very often after a brief period of exposure. Radiography shows wedge shaped or ground glass reticuloacinar opacities in the basal portions of the lungs. The subject is obliged to take time off work and the symptoms improve spontaneously or under glucocorticoid therapy; recurrence, however, of the symptoms is usual when the patient returns to work.

In other cases, the clinical picture gradually evolves to a diffuse fibrotic pattern: alveolar walls are thickened by fibrotic tissues and intraalveolar cells are few; perivascular and peribronchial fibrosis can also be observed. This lesion is termed "diffuse interstitial pulmonary fibrosis" and most probably represents the end-stage of the pathological process (mural form). This phenomenon is often more insidious and progressive and occurs typically in older workers with long exposure histories (up to 20–25 years). The subject develops cough, labored breathing on exertion, and tachypnea; clubbing and a substantial weight loss are also common. Chest radiographs reveal linear striations and diffuse reticulonodular opacities in the middle and lower lung zones. Lung function studies typically show a restrictive syndrome accompanied by a reduction in the diffusing capacity of carbon monoxide (DL_{CO}) and a drop of PaO_2 during exercise. In the final stages, corpulmonale and cardiorespiratory failure lead to death.

The transition from alveolitis to permanent fibrosis is probably gradual; these two conditions are the extremes of a continuous process. In between, there exist varying degrees of alveolitis and fibrosis depending on the duration and severity of exposure and also individual susceptibility factors. It is indeed noteworthy that a majority of exposed workers remain unaffected, whereas only a small percentage (less than 5%) develop interstitial fibrosis.

Modern techniques such as bronchoalveolar lavage, microprobe, and neutron activation analysis can provide some additional information on the presence of metals in the lung; such information may help in confirming the etiology of the disease. In most cases, cobalt content is found to be very low in fibrotic lungs by comparison with other inert particles such as tungsten and tantalum carbides. This phenomenon is usually explained by the high solubility of cobalt in biological fluids causing its rapid disparition from lung tissue.

Most researchers attribute the cause of fibrotic disease to cobalt, since animal studies have shown that cobalt inhalation can cause major pulmonary lesions, whereas tungsten carbide or other components of hard metals do not produce significant toxic change. However, it must be emphasized that in the majority of these studies, very high doses of cobalt are administered resulting in lesions morphologically different from those observed in humans. Moreover, it must be pointed out that most cases of interstitial lung diseases have been observed in the hard metal and diamond polishing industries where workers are usually exposed to a mixture of particles, whereas only two isolated cases have been recorded after exposure to pure cobalt powders. These observations suggest that the role of other constituents and their possible interaction with cobalt particles must be considered in the pathogenesis of interstitial lung lesions. This view is supported by experimental studies indicating that the toxicity of cobalt is enhanced in the presence of so-called inert particles such as carbides; the mechanism of this interactive toxicity, however, remains to be clarified.

B. Skin Effects

Industrial exposure to cobalt-containing dusts can cause allergic contact dermatitis characterized by a positive patch test for cobalt; only minute traces of the metal are necessary to cause sensitization. The allergy usually presents as an urticaria with erythemous papules mostly of the hands. About 10% of workers in the cemented tungsten carbide industry may develop a contact allergy to cobalt, which is often associated with an allergy to nickel (cross-reaction). The differentiation with an irritant reaction is difficult; both can even coexist. The irritating action of cutting oils and fluids used in grinding operations performed with hard metal tools must be stressed since appreciable amounts of cobalt can solubilize in these fluids. This creates a favorable environment for the development of allergy on an irritated skin. Similar situations can be encountered in the paint industry and during cosmetic, cement, ceramic, and pottery production. Jobs associated with frequent traumatic lesions or abrasions of the skin, a previous contact allergy, or atopy represent predispository factors for the development of cobalt allergy. Spanish investigators have also reported a photocontact dermatitis induced by cobalt. The diagnosis of allergic cobalt dermatitis is based on the demonstration of contact with cobalt associated with a positive patch test (e.g., with 1–2% cobalt chloride) and possibly a prick test with 1:1000 cobalt chloride giving a delayed tuberculin reaction in a nonatopic subject. The latter test may, however, cause a nonspecific urticarial reaction. The clinical value of the lymphocyte transformation test seems limited in the case of cobalt allergy.

C. Other Organs

1. Effects on Blood

Cobalt has an erythropoietic effect resulting in an increase in blood volume and in total erythrocyte mass. This action has been used in the treatment of refractory anemia and in anephretic patients. Polycythemia has also been observed in heavy drinkers of cobalt-contaminated beer. The possibility that polycythemia may result from industrial exposure has been raised in a few studies conducted in the hard metal and diamond polishing industries. Some evidence for a role of cobalt in retarding clot formation has been presented but is not well documented.

2. Myocardial Effects

Cobalt has been shown to be the cause of a cardiomyopathy diagnosed in heavy beer drinkers who consumed beer containing high cobalt concentrations; poor nutrition and alcohol consumption were probably also important favorizing factors. The classical syndrome is characterized by left and then right heart failure accompanied by cardiomegaly, gallop rhythm, cyanosis, low cardiac output, pericardial effusions, and hypotension. Lactic acidosis and high levels of serum glutamic oxaloacetic transaminase, lactic dehydrogenase, and creatine phosphokinase are found in laboratory studies. A few cases of cardiomyopathy have also been recorded after heavy industrial exposure to cobalt dust.

3. Thyroid Gland Effects

The goitrogenic effect of cobalt is well-known, particularly in patients treated for hematological disorders. It has been suggested (on the basis of animal studies) that alterations of thyroid function might occur following occupational exposure; there exists, however, no further information supporting this hypothesis. An infraclinical hypothyroid state (decreased T3, T4, and increased TSH) has been observed among workers from a cobalt refinery plant.

4. Other Effects

Adverse reactions in various organs such as the liver, kidneys, testes, and pancreas have been observed in animals injected with high doses of cobalt; the relevance of these observations for industrial exposure has not been documented. Insufficient information is available to allow definitive conclusions concerning the teratogenic activity of cobalt in humans.

5. Mutagenic and Carcinogenic Properties

Limited mutagenic properties have been demonstrated for cobalt chloride in some *in vitro* tests on yeast and plant cells. In mammalian cells, however, cobalt appears to be devoid of mutagenic activity.

A number of studies carried out on rats have induced rhabdomyosarcomas at the site of injection of cobalt metal, cobalt chloride, and sulfide. Mice injected with twice the rat dose do not develop tumors. Intratracheal administration of cobalt produces lung tumors in rats but not in hamsters. There have been few reports of cancer developing in cobalt dust workers so that insufficient evidence is available to conclude whether cobalt is carcinogenic to humans. A possible relationship between progressive lung fibrosis and the development of lung cancer in hard metal workers remains to be documented. The fact that cobalt can induce tumors at the injection site in rats has resulted in a recommendation to thoroughly rinse a wound contaminated with cobalt.

V. Prevention

A. Engineering Control

Handling cobalt in a closed system and reduction of dust generation represent the most effective methods of limiting worker exposure. Well-designed ventilation, including local exhaust systems, can reduce the accumulation of airborne dust. General plant maintenance must be conducted regularly to prevent cobalt-containing dust from accumulating in work areas.

Most countries have adopted a limit of 0.05 mg/m^3 for occupational exposure to airborne cobalt (TLV–TWA) as recommended by the American Conference of Governmental Industrial Hygienists. In Germany, a maximum permissible exposure concentration of cobalt has not been established since it is considered as a possible carcinogen; a TRK (Technische Richtkonzentration) of 0.5 mg/m^3 has been proposed. It should be pointed out, however, that current recommendations do not take into account whether exposure is to cobalt alone or to cobalt associated with other compounds (e.g., carbides). A number of epidemiological and experimental studies indeed suggest that the critical exposure level might be lower when cobalt-containing mixtures are handled.

Personal hygiene measures can also contribute to prevent the development of adverse effects:

- eating, smoking, and drinking in the work area should be prohibited
- workers should be encouraged to wash or shower after each work period and before eating

Various protective equipment such as gloves and protective clothes are useful. Respirators may be needed but do not represent a substitute to proper engineering control.

B. Medical Surveillance

Workers exposed to cobalt-containing dusts should be submitted periodically (at least once a year) to careful medical examination. Special attention should be given to the skin and the respiratory system; lung function tests including spirometry and possibly diffusion capacity for CO (DL_{CO}) should be carried out annually; chest radiography appears less sensitive for early detection. Palpation of the thyroid gland for enlargement is a simple measure that should be included in the medical examination. Other tests such as thyroid function studies and electrocardiogram do not seem necessary unless specific signs or symptoms are present. End-of-shift cobalturia is a good indicator of overall exposure levels.

Related Articles: Inorganic Mineral Particulates in the Lung; Metal–Metal Toxic Interactions.

Bibliography

Criteria for controlling occupational exposure to cobalt. *NIOSH Occupational Hazard Assessment.* DHHS (NIOSH) publication No. 82–107 U.S. Government Printing Office, Washington, D.C., Oct., 1981.

Gennart, J. Ph., and Lauwerys, R. (1990). Ventilatory function of workers exposed to cobalt and diamond containing dust. *Int. Arch. Occup. Environ. Health* **62,** 333–336.

Lauwerys, R., and Lison, D. (1993). Health risks associated with cobalt exposure: An overview. *Sc. Tob. Environ.* (in press).

Léonard, A., and Lauwerys, R. (1990). Mutagenicity, carcinogenicity, and teratogenicity of cobalt metal and cobalt compounds. *Mutation Res.* **239,** 17–27.

Lison, D., and Lauwerys, R. (1990). In vitro cytotoxic effects of cobalt-containing dusts on mouse peritoneal and rat alveolar macrophages. *Environ. Res.* **52,** 187–198.

Rüttner, J. R., Spicher, M. A., and Stolkin, I. (1987). Inorganic particulates in pneumoconiotic lungs of hard metal grinders. *Brit. J. Ind. Med.* **44,** 657–660.

Scansetti, G., Lamar, S., Talarico, S., Botta, G., Spirelli, P., Sulotto, F., and Funtari, F. (1985). Urinary cobalt as a measure of exposure in the hard metal industry. *Int. Arch. Occup. Environ. Health* **57,** 19–26.

Swennen, B., Buchet, J.-P., Stanescu, D., Lison, D., and Lauwerys, R. (1993). Epidemiologic survey on workers exposed to cobalt oxides, cobalt salts, and cobalt metal. *Br. J. Ind. Med.* (*in press*).

Cyanide

Gary E. Isom
Purdue University

Glossary

Alkaline chlorination A process whereby cyanides are oxidized to less toxic products that can be disposed of by standard procedures.

Cyanides Group of chemicals that readily dissociate to or are metabolized in the body to the cyanide (CN^-) ion.

Cyanogenic glycosides Class of compounds produced by plants that can be metabolized to free cyanide in the body.

Cytochrome oxidase Terminal oxidase enzyme in the electron transport chain that utilizes oxygen in the production of cellular energy.

Histotoxic hypoxia Cellular asphyxiant state in which utilization of oxygen by the cell is inhibited, but the availability or delivery of oxygen to the cell is normal.

Rhodanese A sulfur-transferase enzyme that converts cyanide to the nontoxic metabolite thiocyanate.

***CYANIDE** is a term used to designate a group of toxic compounds including hydrogen cyanide, potassium cyanide, and sodium cyanide in which the toxicity is attributed to the cyanide (CN^-) anion. The primary biological action of cyanide is to block tissue utilization of oxygen by inhibiting the enzyme, cytochrome oxidase. Cyanide is one of the more rapid-acting toxic compounds; in acute toxicity consciousness can be lost within seconds to minutes. Since cyanides are so rapid acting, the storage, use, and disposal of these compounds must be closely controlled and monitored. Education of employees on the chemical and toxicological properties of cyanides is thus necessary for safe handling.*

I. History and Occurrence

The toxicity of cyanide to humans has been known for over 200 years. Scheele, the Swedish chemist who first synthesized cyanide in 1782, was one of its first victims; he died from acute cyanide poisoning in 1786. Since that time cyanide has been responsible for an extensive number of poisonings in a variety of settings, including occupational exposures, murders, judicial executions, and chemical warfare. Use and misuse of this compound has justifiably given it the reputation as one of the more toxic agents known.

Cyanide has a long history of use as a toxic compound. Napoléon III was the first to employ hydrogen cyanide as a chemical warfare agent. During World Wars I and II, the gas was used on the battlefield by both sides. Hydrogen cyanide gas was used by the Germans during World War II for genocidal mass killings. More recently, potassium cyanide was used in the 1978 Jonestown mass suicide in which over 900 people died.

Since cyanide salts are readily available, it has been used in intentional adulteration of consumer products, including nine deaths resulting from consumption of a cyanide-tainted over-the-counter pain reliever in 1982 and a decongestant in 1991. Cyanide compounds continue to be of significance toxicologically as a result of their use in industrial processes, agricultural applications, and environmental pollution. Additionally, a significant number of poisonings occur from dietary ingestion of cyanogenic plant products that break down to cyanide.

The number of deaths from accidental cyanide poisonings appears to be declining, perhaps as a result of public awareness of potential toxicity and of more appropriate industrial hygiene. The United Kingdom National Poisons Unit reported 134 cases of cyanide poisoning with 16 deaths in the period of 1963 to 1984. Potassium and sodium cyanide were the most frequent forms of cyanide involved in the intoxications. In 1983, the American Association of Poison Control Centers reported 105 cyanide exposures with 3 deaths. The reported incidence of acute toxicity is low; nevertheless its rapid action and potential for severe toxicity mandate that it be handled with caution and that exposures be treated appropriately.

II. Chemical and Physical Properties

The toxic cyanide ion can be produced from hydrogen cyanide, the alkali metal salts (potassium and sodium), and metal complexes. A number of compounds including cyanohydrins, aliphatic nitriles and cyanogenic glycosides can produce the cyanide ion in the body following metabolism.

The cyanide compounds of primary concern are hydrogen cyanide, potassium cyanide, and sodium cyanide (see Table I). Hydrogen cyanide is a highly volatile colorless liquid with a boiling point of 25.7°C. The vapor is the more commonly encountered form; it is a colorless gas that has an odor of bitter almonds. The density of gaseous hydrogen cyanide is about the same as air, and it will readily penetrate air spaces and is transported in air currents. At ambient temperatures hydrogen cyanide represents a fire and explosion hazard and upon storage under basic conditions, it can undergo exothermic polymerization to form an explosive product.

The potassium and sodium salts of cyanide are white solids that are highly soluble in water and in solution readily dissociate to the cyanide anion. In damp air, both salts have a slight odor of bitter almonds. Aqueous solutions of the salts are strongly alkaline and ingestion of aqueous salt solutions can produce a strong caustic effect in the stomach in addition to the characteristic systemic toxicity. Caution should be used in handling, storage, or use of cyanide salts near acids or acid salts. Addition of the salts to acid will result in the rapid liberation of hydrogen cyanide gas. Hydrogen cyanide can be generated from exposure of the salts to high concentrations of carbon dioxide such as those that could be produced with a carbon dioxide fire extinguisher.

The heavy metal salt complexes of cyanide do not readily dissociate in aqueous solution (Table II). However, these compounds should always be handled with caution since they have a high potential for producing systemic toxicity. Additionally, they can liberate hydrogen cyanide following exposure to acids. Among other cyanide compounds of toxicological interest is ammonium cyanide, which is a white solid. Ammonium cyanide is unstable and ammonia and hydrogen cyanide are produced following decomposition. Cyanogen and chlorine cyanide (cyanogen chloride) are colorless gases with highly irritant properties and can produce systemic toxicity following inhalation.

III. Detection

A. Odor

The bitter almond odor of hydrogen cyanide can be detected by 60–70% of the population. The threshold for those sensitive to the odor is estimated to be 1–5 ppm concentration in air. It should be stressed that even at high toxic concentrations some individuals cannot smell hydrogen cyanide. Also, the action of cyanide is so rapid that detection of the odor may be too late to prevent severe intoxication. Hence, detection of cyanide should not be dependent upon the characteristic odor.

B. Analysis and Monitoring

A number of analyses have been developed for quantitation of cyanides in environmental samples. The procedures vary in simplicity, cost, and sensitivity. A listing of some of the more commonly used techniques are listed in Table III.

When selecting an analysis technique, factors to consider include sample stability, assay reliability, and the cyanide species present. Also when storing samples, it should be kept in mind that hydrogen cyanide will volatilize from acidic samples.

In water samples, cyanides can exist as free cyanide (hydrogen cyanide, potassium cyanide, or sodium cyanide) and metal–cyanide complexes. Most methods readily measure free cyanides in water, and the chloramine-T method will measure both

Table I Chemical and Physical Properties of Common Cyanides

Substance	Chemical formula	Physical properties	Molec. wt.	Melting pt.	Boiling pt.	Vapor density (air = 1)	Solubility with water	Synonyms
Hydrogen cyanide	HCN	Colorless gas or liquid	27.03	−13.2°C	25.7°C	0.94	Miscible	Prussic acid Hydrocyanide Acid formonitrile
Potassium cyanide	KCN	White solid	65.11	636°C	NA[a]	NA	71.6 g/100 ml	Cyanide of potassium
Sodium cyanide	$NaCN$	White solid	49.02	564°C	1500°C	NA	48 g/100 ml	Cyanide of sodium
Calcium cyanide	$Ca(CN)_2$	Amorphous white solid	92.12	>350°C	NA	NA	Readily soluble	Black cyanide (crude form)
Sodium nitroprusside	$Na_2Fe(NO)(CN)_5$	Red crystals	261.93	NA	NA	NA	40 g/100 ml	Sodium nitroferric cyanide Nitride
Copper cyanide	$Cu(CN)_2$	White solid	89.56	473°C	Decomposes	NA	Insoluble	

[a] NA, not applicable.

Table II Heavy Metal Cyanides

Cobalt cyanide
Cupric cyanide
Cuprous cyanide
Silver cyanide
Ferricyanide
Ferrocyanide
Zinc cyanide

free cyanide and metal–cyanide complexes. For sampling of air, hydrogen cyanide in the sample is trapped on either a basic solid sorbent or a cellulose ester membrane on which particulate cyanide is trapped and hydrogen cyanide is collected in a basic solution (Table IV). The cyanide content is then quantitated by an accepted analytical technique. For selection of an analysis procedure and details on performing the assays, see the NIOSH Manual of Analytical Methods.

In most procedures a number of substances can interfere with accurate cyanide quantitation. For ion-selective electrodes, many metal ions and halides can interfere. In measurement of cyanide by colorimetric techniques, several classes of compounds interfere with the assays, including oxidizing agents, nitrates, nitrites, thiocyanate, and sulfides. The presence of these substances in a sample must be considered when selecting a method for quantitation of cyanides.

C. Biological Analysis

Several methods have been used for the diagnostic or forensic detection of cyanide in biological samples. The simplest qualitative procedure is the Schoenbein test, which consists of acidifying the sample by adding tartaric acid. The sample is slowly warmed and a filter paper strip freshly treated with 10 parts of 0.2% gum guaiac in ethanol; 3 parts of 0.1% copper sulfate in water is held above the sample. The filter paper will turn blue if a toxic level of cyanide is present.

For quantitation of cyanide in biological samples a number of sensitive methods have been employed. In selection of the procedure, several factors should be considered, including type of tissue, storage duration, and temperature. Cyanide can be lost from tissue samples upon storage for extended durations at room temperature. Whole blood, plasma, or serum is usually analyzed; however, any tissue sample can be used. Detection of cyanide in urine is difficult, and thiocyanate, a metabolite of cyanide, is more readily detected and can be used as a marker of cyanide exposure.

Some of the more frequently used methods to analyze biological samples are summarized in Table V. When assaying biological samples, many interfering substances such as the cyanide antidote sodium thiosulfate can be encountered. This may necessitate removal of cyanide from the sample by microdiffusion and trapping the cyanide in an alkaline media prior to analysis. The Conway microdiffusion approach involves placing the sample in the outer well of the microdiffusion chamber followed

Table III Representative Cyanide Assay Methods

Method	Technique	Detection limits	Reference
Colorimetric	Sample is acidified, reflux-distilled and HCN absorbed by NaOH solution treated with chloramine-T and pyridine-pyrazole or pyridine barbituric acid. Absorption at 580 nm.	0.02 ppm	EPA method 335.2
Ion-specific electrode	mV Reading of CN electrode vs. reference electrode.	0.01 ppm (for 200 l of air)	McAnalley *et al.*, *J. Anal. Toxicol.* **3,** 111, 1979
Ion chromatography/ amperometric detection	NaOH solution containing CN is converted to formate by electrochemical detection.	0.04 ppm (for 2.66 l of air)	Dolzine *et al.*, *Anal. Chem.* **54,** 470, 1982.
Titrimetric	Sample acidified, reflux-distilled and HCN absorbed by NaOH trap. NaOH solution is titrated with $AgNO_3$/with *p*-dimethylaminobenzalrhodamine as indicator.	1 ppm	Ryan and Culshaw, *Analyst* **69,** 370, 1944.

Table IV Monitoring Techniques for Hydrogen Cyanide[a]

Samplier	Sampling flow rate	Measurement	Range	Accuracy
Solid sorbent tube (soda lime, 600 mg/200 mg)	0.05–0.2 l/min	Colorimetric analysis with chloramine-T technique—absorption at 580 nm.	3–300 μg CN^-/sample	2–15 mg/m³ HCN
Filter and bubbler (cellulase ester membrane + 0.1 *N* NaOH)	0.5–1.0 l/min	mV reading of CN^- electrode vs. reference electrode.	0.05–2.0 mg CN	5–21 mg/m³ HCN 2.6–10 mg/m³ KCN

[a] For details of methods see NIOSH Manual of Analytical Methods.

by acidification. The evolved hydrogen cyanide is trapped in a solution of sodium hydroxide in the center well. An aliquot of the center well is then analyzed by a detection method.

IV. Sources of Exposure

Exposure to cyanide can occur under a variety of settings and in some cases from unexpected sources. It is important to recognize the potential for toxicity from these sources, both for prevention and diagnosis of acute intoxication.

A. Dietary

The primary source of cyanide in the diet is from cyanogenic plants. Hundreds of plants have been reported to produce cyanide-containing compounds or cyanogenic glycosides. Following ingestion, the glycosides are hydrolyzed to generate cyanide in the gut, which is then absorbed to produce toxicity. Some of the more important plant sources include the kernels (pits or seeds) of the chokecherry, bitter almond, peach, apple, and apricot. Livestock poisonings have been reported after consumption of a variety of grasses and shrubs, including sorghum, Johnson grass, and Sudan grass. The cyanogenic glycoside content of these plants can vary depending upon growing conditions (soil conditions and weather).

In some tropical areas, the tuberous root of Cassava (*Manihot esculenta*) is a source of carbohydrates and a mainstay of the diet. High levels of cyanide can be detected in improperly prepared cassava and its products. Topical ataxic neuropathy (a motor disability characterized by partial leg paralysis) and visual failure are related to chronic cyanide exposure in Cassava diets.

B. Industrial

Hydrogen cyanide and its salts are used for a wide variety of industrial processes (Table VI). Cyanide salts or metal complexes are used in the metal industry for case hardening and metal plating. Sodium cyanide is used extensively in the extraction and recovery of metals from ore. Cyanides are also used as chemical intermediates in the synthesis of some

Table V Quantitation of Cyanide in Biological Samples

Method	Procedure	Sensitivity	Reference
Colorimetric	Oxidation of CN to cyanogen which reacts with pyridine-pyrazolone to form a chromophore.	0.004 μmol/ml	Epstein, *Anal. Chem.* **19,** 272, 1947.
Potentiometric	Ion-selective electrodes containing silver sulfide membrane.	10^{-3}–$10^{-6}M$	Frant *et al., Anal. Chem.* **44,** 2227, 1972.
Spectrophotofluorometry	Reaction of cyanide with *p*-benzoquinone to form a fluorescent derivative.	0.001 μmol/ml	Guilbault *et al., Anal. Chem.* **37,** 1395, 1965.
Gas chromatography	Use of chloramine-T to oxidize cyanide to cyanogen chloride, followed by GC detection.	0.25 μg/ml	Valentour *et al., Anal. Chem.* **46,** 924, 1974.

Table VI Industrial and Agricultural Applications of Cyanide

Synthesis of nitrites, carbylamides, and cyanohydrins
Preparation of select resin monomers
Metallurgy extraction of gold and silver ores
Electroplating of steel
Case hardening of steel
Cleaning and coating silver
Silver polishes
Photography fixatives
Process engraving
Rotenticide
Insecticide

polymers and nitriles, as well as in rotenticides and insecticides.

Hydrogen cyanide is applied directly as a gas for vermicidal fumigation of greenhouses, ships, and buildings. Nut meal and seed can be fumigated with hydrogen cyanide in vacuum chambers. In agricultural operations, hydrogen cyanide has been employed as a scale insect fumigant for citrus trees under tents.

Nitriles and cyanohyrin compounds contain organically bound cyanide. Following systemic absorption of aliphatic nitriles, many of these compounds can be metabolized by liver enzymes to generate free cyanide (Table VII). Delayed acute cyanide intoxication can occur following exposure to these agents. The toxicity is delayed (the time varies depending on the compound and amount of exposure) for the generation of cyanide in the body by metabolism. These agents are used as chemical intermediates and acrylonitrile has been used as a fumigant gas of grains. A number of cyanide poisonings have been reported following ingestion of cosmetic nail remover containing acetonitrile. Symptoms of toxicity develop over a 3–12 hour latency period, which is required to metabolize the acetonitrile to cyanide.

Table VII Aliphate Nitriles wherin Acute Toxicity Is Attributed to Cyanide Generation by Liver Metabolism

Acetone cyanohydrin
Acetonitrile
Acrylonitrile
n-Butyronitrile
Malononitrile
Propionitrile
Succinonitrile

C. Fires

Incomplete combustion of carbonaceous compounds can generate hydrogen cyanide. Many deaths in fires are not directly attributable to thermal injury, but to the poisonous gases generated. Carbon monoxide and hydrogen cyanide can be liberated by burning of synthetic polymers such as polyurethane, polyacrylonitrile, and other structural materials. The gases can be generated rapidly in flash fires and accumulate to lethal levels in enclosed areas such as buildings, aircraft cabins, and ship fires. Carbon monoxide and hydrogen cyanide can produce an additive toxicity in which each compound contributes to the toxic effects. Toxic levels of cyanide have been detected in the blood and tissues of fire victims and firemen.

Tobacco smoke, which may contain as high as 1,600 ppm, is the most common source of exposure to hydrogen cyanide. The cyanide may arise from combustion of tobacco proteins and nitrate. Use of absorbent filters will lower the amount of hydrogen cyanide delivered to the lungs by as much as 50–80% compared to nonfiltered cigarettes. The use of low-tar, low-nicotine cigarettes does not reduce the amount of hydrogen cyanide in smoke. Tobacco amblyopia is a syndrome of visual failure that is attributed to chronic cyanide exposure in heavy smoking.

V. Environmental Contamination and Ecotoxicity

Cyanides that contaminate the environment may originate from different anthropogenic sources such as industrial venting of gases, discharge of cyanide-containing wastes, and migration of cyanide from waste dumps. Low levels of cyanide from natural sources is rarely detected in the environment since it is highly reactive and rapidly degrades. Vehicle exhaust is the largest source of cyanide in the air, but the use of catalytic converters reduces the formation of hydrogen cyanide. Emission of hydrogen cyanide may occur under inadequate emission control of industrial processes, or its production by coke-ovens and blast furnaces. Atmospheric hydrogen cyanide can undergo photodecomposition, react with hydroxyl radicals, or be removed by wet deposition.

Surface water and groundwater can be contaminated with cyanide following discharge of wastes containing cyanide salts or metal complexes. This has been reported with mining residues and wastes of metal finishing processes. The free cyanide ion is decomposed with alkaline chlorine oxidation treatment, and the metal complexes (iron and nickel) require more vigorous chemical processing. Road salts used for de-icing of highways in the winter are also a source of cyanide. Cyanide-containing salts are added to road salt as anticaking compounds. Following the spreading of salts on the highways, they can wash off into streams and storm sewers. In acidic water, the cyanide salts and metal complexes form HCN, which volatilizes. Volatilization is the major route for loss of cyanide from water. Cyanides are also removed from water by biodegradation and sorption onto particulate material followed by sedimentation.

Soil contamination with cyanide salts or metal complexes can occur following disposal of cyanide-containing wastes in dumps and the use of road salts containing cyanides. The fate of cyanide in the soil is complex and is pH dependent. In acidic surface soil, cyanides volatilize and are lost as hydrogen cyanide. In subsurface soils, cyanides can decompose slowly or be degraded by microorganisms. Cyanides can also leach into groundwater or directly contaminate surface runoff water. In basic soils, cyanides are more stable and less mobile. Also, cyanide can react with soil constituents to form other chemical species. Cyanide reacts with sulfur from pyrite and other mineral constituents to form thiocyanate. Additionally, free cyanides will complex with trace metals to form metal complexes.

Contamination of the environment with cyanide represents a toxicological risk. In humans, the potential for toxicity is dependent upon the amount and duration of exposure. Human toxicity is reviewed in Section VII. In wildlife, toxicity to free cyanide is well documented. Mammalian species are extremely sensitive to cyanide and deaths can occur following inhalation of volatilized hydrogen cyanide or ingestion of contaminated food sources. Accumulation of cyanide in water can produce death in freshwater and marine invertebrates. The lethal level of hydrogen cyanide to various species of fish is in the range of 30 to 150 μg/l. The degree of toxicity is dependent upon a number of factors, including water pH, oxygen content, temperature, species of fish, and body size. Long-term exposure of fish to sublethal concentrations as low as 5 μg/l has been reported to reduce weight gain, swimming performance, and reproduction. The toxicological effects on other aquatic species is not known.

VI. Disposal of Cyanides

Cyanides from industrial wastes should be treated by an acceptable method before release into aquatic ecosystems or disposal in a secured sanitary landfill. A variety of methods involving the decomposition of cyanides to less toxic compounds by physical or chemical processes have been used for waste management (Table VIII). Among the difficulties encountered with many methods is the delineation of the cyanide species present in the waste and the use of an efficient treatment process that meets discharge criteria and does not generate degradation products that are ecotoxic.

Alkaline chlorination (pH 9–11) with chlorine plus sodium hydroxide or hypochlorite is one of the most frequently employed and best developed of the commercial methods for treating cyanides. In this process, cyanide solutions are oxidized to less toxic cyanate. The reaction is summarized as:

$$CN^- + Cl_2 \rightarrow CNCl + Cl^-$$

$$CNCl + NaOH \rightarrow NaCNO + NaCl + H_2O$$

In natural degradation, waste solutions containing cyanides are placed in holding ponds and allowed to degrade. After prolonged retention (several months)

Table VIII Methods for Waste Management of Cyanides

Lagooning for natural degradation
Ozonization processes
Acidification–volatilization–reneutralization
Ion exchange
Alkaline chlorination
Activated carbon absorption
Electrolytic processes
Catalytic oxidation
Biological treatment (CN-metabolizing microorganisms)
Alkalinization–ferrous sulfate treatment

in mildly acidic or neutral solutions, over 90% of the cyanide is lost by volatilization of molecular hydrogen cyanide and the remainder is oxidized or photodecomposed.

Commercial biodegradation processes use microorganisms that metabolize cyanate to less toxic products. Typically, dilute cyanide solutions (10 ppm) containing biological nutrients are filtered through biological contactors colonized by microorganisms. A drawback to this procedure is that high free cyanide concentrations can kill the microorganisms; in most cases the waste solutions must be diluted. With this process, the cyanide content of a 10 ppm cyanide waste solution can be reduced 100-fold.

VII. Toxic Actions

A. Mechanism of Toxicity

Cyanide is a rapid-acting toxicant whose primary mechanism of action is inhibition of cellular metabolism. Cyanide inhibits cellular aerobic metabolism by inhibiting cytochrome oxidase, the terminal enzyme in the respiratory chain. The cyanide anion (CN^-) is the species that binds to the ferric iron (Fe^{2+}) component of the enzyme; thus any cyanide compound that will dissociate to the anion has the potential to inhibit cytochrome oxidase. The result is inadequate cellular utilization of oxygen resulting in histotoxic hypoxia. All cellular processes dependent upon energy are rapidly inhibited. Tissues such as the brain and heart that are highly dependent upon aerobic metabolism are most susceptible to the toxic actions of cyanide. The uptake and transport of oxygen by the blood is not impaired; only the tissue utilization of oxygen is blocked. Hence the blood is saturated with oxygen, producing a cherry-red color of venous blood that is characteristic of acute cyanide intoxication.

In addition to cytochrome oxidase, over 40 enzymes are known to be inhibited by cyanide. Potentially any metalloenzyme can be inhibited by cyanide, including superoxide dismutase, catalase, and peroxidases. The cyanide anion forms complexes with the metal component of these enzymes, resulting in impairment or complete inhibition of enzyme activity. Cyanide inhibition of these enzymes probably does not play an important role in the rapid onset of acute toxicity, but may underlie some of the symptoms of chronic exposures and delayed or residual symptoms of acute intoxications.

B. Routes of Exposure and Dosimetry

Hydrogen cyanide can be readily absorbed from most sites of exposure, including the lungs, skin, and mucous membranes. Potentially, acute intoxication could result from exposure of any portion of the body to hydrogen cyanide gas or its aqueous solutions. Cyanide salts more typically produce toxicity by ingestion; however, skin absorption of cyanide from solutions of cyanide salts has been reported. Occupational exposures to hydrogen cyanide or the salts are most common by inhalation, oral, or skin routes. In industrial poisonings it is not uncommon that exposure is through multiple routes. The dose required to produce toxicity is dependent upon the form of cyanide (gas or salt), the duration of exposure, and the route. Blood levels greater than 0.2 μg/ml suggest toxicity.

1. Inhalation

Hydrogen cyanide is readily and rapidly absorbed from the lungs; following inhalation, symptoms occur within seconds and death within minutes. Inhalation of dusts containing cyanide compounds can produce toxicity. Hydrogen cyanide is an irritant to the lungs and can produce localized damage which may manifest as pulmonary congestion, dryness, and burning in the throat and cough.

Inhalation of 0.3 mg/l (270 ppm) of hydrogen cyanide is considered immediately lethal. In cases where death has occurred following inhalation of high concentrations, the total absorbed dose of hydrogen cyanide may be as low as 0.7 mg/kg. The estimated lethal dose for humans after 10 minutes of exposure to the gas is 0.2 mg/l (181 ppm). Concentrations of hydrogen cyanide as low as 90 ppm should be considered dangerous to life.

2. Oral

Solutions of hydrogen cyanide are rapidly absorbed from the stomach, whereas cyanide salts may undergo slower absorption. High doses of the salts may produce severe intoxication within minutes. Lower doses may exhibit symptoms within a few minutes following ingestion and death may be delayed for 1–2 hours. Strong solutions of hydrogen cyanide and the salts are corrosive and will produce local damage, including ulceration and inflammation of the stomach.

Death has been reported following ingestion of an estimated 30 mg of hydrogen cyanide. For humans, the average lethal dose (LD_{50}) of sodium cyanide is projected to be 2.86 mg/kg. The lethal dose for potassium cyanide would parallel that of the sodium salt.

3. Dermal

Cyanide toxicity following skin exposure is rare, but should be considered a possibility following exposure to hydrogen cyanide gas or direct contact with cyanide solutions or compounds. For this reason, adequate skin protection must be used in combination with respiratory protection when exposure to hydrogen cyanide gas or dust particles containing cyanides occurs. In addition to the skin, cyanides can be absorbed by the conjunctiva of the eye. Dry dusts containing cyanides and the salts are not rapidly absorbed, but in the presence of moisture or when in solution, they can be absorbed across the skin to produce toxicity. Injury or abrasion of the skin can increase the absorption.

Skin contact with cyanide-containing dusts or solutions can be irritating and localized skin damage can be produced. Reports of "cyanide burns" have been noted after exposure to hydrogen cyanide gas and dermatitis and ulceration may occur after contact with solutions of cyanides, including potassium and sodium cyanide and metal complexes. Repeated contact with cyanide solutions can produce slow healing ulcerations that increase the absorption of cyanide.

Deaths have been reported after dermal exposure; the LD_{50} is estimated at 100 mg/kg for hydrogen cyanide solutions. Prolonged (greater than 30 minutes) skin contact with solutions of concentrations as low as 0.035% may produce symptoms.

C. Toxicokinetics

After absorption cyanide exists in equilibrium as the anion and undissociated HCN. The pK_a of HCN is 9.21; therefore at physiological pH 7.4, the compound exists primarily as hydrogen cyanide. Nonionized hydrogen cyanide can readily cross biological membranes, whereas the distribution of the cyanide anion across membranes is more limited. Cyanide is rapidly distributed throughout the body with the highest concentrations occurring in the liver following oral administration. Regardless of the route of exposure, the concentrations of cyanide are consistently high in the heart and brain—the major target organs of toxicity. Blood cyanide levels correlate well with symptoms. The minimal lethal whole blood levels have been reported in the range of 250 μg/dl to 300 μg/dl.

Cyanide is eliminated from the body by multiple pathways. A small portion (1–2%) is lost through the lungs by exhalation. A number of minor pathways of metabolism (less than 15% of total) accounts for cyanide elimination. This includes conversion to 2-aminothiazoline-4-carboxylic acid, incorporation into the one-carbon metabolic pool, or the combination with hydroxycobalamin to form cyanocobalamin.

The major route for detoxication of cyanide is the enzymatic conversion to the nontoxic metabolite thiocyanate. This reaction is catalyzed by two sulfur transferase enzymes, rhodanese (thiosulfate-cyanide sulfurtransferase) and β-mercaptopyruvate-cyanide sulfurtransferase. The primary pathway for metabolism is thought to be rhodanese, which is widely distributed throughout the body with the highest concentration in the liver. This enzyme catalyzes the transfer of a sulphane sulfur from a sulfur donor (such as thiosulfate) to cyanide to form thiocyanate. In acute intoxication, the limiting factor in cyanide detoxication by rhodanese is the availability of adequate quantities of sulfur donors. The endogenous stores of sulfur donors are rapidly depleted and cyanide metabolism is slowed. Hence, sodium thiosulfate serves as antidote that accelerates the metabolic inactivation of cyanide (see Section VIII).

The metabolites of cyanide are eliminated from the body in urine, with thiocyanate being the primary compound. Urinary levels of thiocyanate have been used as a marker of exposure to cyanide as they increase after exposure to low levels of cyanide. However, caution should be used when interpreting urinary thiocyanate levels since heavy smoking will increase the urinary concentrations and can be mistaken for cyanide exposure from nonsmoking sources.

D. Symptoms of Toxicity

1. Acute Intoxication

Inhalation of hydrogen cyanide produces symptoms within seconds and death within minutes. A transient period of central nervous system stimulation may occur and is characterized by excitement and hyperventilation accompanied by flushing, tachycardia, headache, and dizziness. This is rapidly followed by unconsciousness, asphyxial convulsions, and death. The cause of death is respiratory arrest accompanied by cardiovascular collapse. Ingestion of salt solutions may produce a slower onset of symptoms that will progress in a similar pattern to that observed with hydrogen cyanide.

Victims may present either in an excited state similar to severe anxiety or in a stuporous, depressed condition. Many poisoned individuals are found unconscious. The diagnosis of cyanide intoxication can be difficult and in many cases diagnosis is dependent upon a history of possible exposure (cyanide in the work place or salts found at the site of poisoning). Venous blood may be cherry red and the victim's breath may smell of bitter almonds. Hydrogen cyanide may produce pulmonary edema perhaps as a result of its corrosive action. It is common that the cyanide poisoned patient exhibits systemic acidosis.

Following appropriate antidotal and supportive treatment, most nonfatal cases have a complete recovery without sequelae. In severe poisoning by the oral route, the victims may remain in a stuporous state over a period of hours and finally slip into a coma and die of respiratory depression.

In rare cases, the victim may develop a postexposure sequelae which may present permanent disabilities. In these cases the victim appears to undergo complete recovery, followed within days or weeks by a delayed onset of the postexposure syndrome. The postexposure sequelae may be characterized by brain damage presenting as memory deficits, personality changes, and a Parkinsonian-like condition with permanent motor disability.

2. Chronic Toxicity

Since cyanide does not accumulate in the body, a toxic syndrome may develop as a result of low level recurring exposure. It has been reported that chronic industrial exposure to hydrogen cyanide produces dizziness, headache, pulmonary irritation, nausea, and weakness. These symptoms may linger for weeks after discontinuance of exposure. Chronic symptoms have been observed in electroplaters and silver polishers exposed for years to cyanide. The symptoms include motor muscle weakness with arms and legs predominantly afflicted and headaches. There have been reports of thyroid disease in individuals exposed to low levels of cyanides over extended periods. This appears to be more prevalent in individuals with iodine deficiency. These symptoms may be attributed to the thiocyanate metabolites as opposed to cyanide directly.

Several clinical states are associated with chronic low level exposure to cyanide from either dietary sources or tobacco smoke (see Table IX). The role of cyanide in these diseases has not been conclusively proven and the syndromes may result from a combination of cyanide, slow cyanide metabolism, and dietary deficiencies. Rare inherited defects of cyanide metabolism may predispose some individuals to toxicity including cases of tobacco amblyopia. Also dietary deficiencies may decrease the rate of cyanide metabolism.

Nigerian nutritional ataxic neuropathy is attributed to repeated ingestion of Cassava, which may contain cyanide as a result of improper processing of the tuber. The syndrome occurs more frequently in adult males and is characterized by a spastic paralysis, visual deficits, and deafness. The role of cyanide as the primary toxicant in this syndrome is questioned; nevertheless, the combination of dietary deficiency and cyanide appear to be etiological factors.

Cyanide has not been shown to be carcinogenic in laboratory studies. Chronic feeding does not produce cancer in laboratory animals. Teratogenic effects have been observed in laboratory animals, but the potential for *teratogenesis* in humans has not been established. Exposure of pregnant animals to exceedingly high levels of cyanide have produced neurological defects among the offspring.

VIII. Treatment and Prevention of Toxicity

A. Treatment of Intoxication

Since cyanide is a rapid-acting compound, proper treatment of an intoxication must be initiated imme-

Table IX Clinical States Associated with Chronic Cyanide Exposure

Tobacco amblyopia
Retrobulbar neuritis
Optic atrophy of Leber
Nigerian nutritional ataxic neuropathy

diately. In industrial settings where cyanide salts are used or hydrogen cyanide may be produced, all personnel should be familiar with the potential for toxicity and emergency personnel should be well versed in therapy. Following a severe exposure to hydrogen cyanide, the victim should be removed from the environment and respiratory assistance given if respiration is depressed and the individual is unconscious. It is recommended that the victim be decontaminated by removal of clothing and, if dermal exposure has occurred, the skin washed. Supportive and symptomatic therapy is important throughout the treatment. This includes monitoring for cardiac arrhythmias and prompt treatment of acidosis.

Specific antidotal therapy can be highly effective in decreasing the morbidity and mortality of severe intoxications. The Lilly cyanide antidote kit is available for specific treatment of acute poisonings. All emergency personnel serving areas where cyanide is used should be trained in the use of the kit. The conventional therapy used in the United States is a two-step process in which amyl nitrite and sodium nitrite are used to convert hemoglobin (Fe^{2+}) to methemoglobin (Fe^{3+}). The ferric ion (Fe^{3+}) of methemoglobin then complexes the cyanide anion. Methemoglobin competes with cytochrome oxidase for the anion, thus either preventing or reversing inhibition of the respiratory enzyme. The binding of the cyanide anion by methemoglobin in essence inactivates the cyanide. The second step of antidotal treatment is the use of sodium thiosulfate as a substrate for rhodanese, the enzyme which catalyzes the biotransformation of cyanide to the nontoxic metabolite thiocyanate, which is then eliminated in urine.

Successful treatment of severe acute cyanide intoxication is dependent upon prompt action. Amyl nitrite pearls are available for inhalation therapy and can be used on site. If the victim is breathing, a crushed amyl nitrite pearl is held close to the victim's nose for 15–30 seconds. This can be repeated no more than five times. In severe intoxications, respiration may be markedly or completely depressed and artificial respiration should be instituted immediately, preferably with a resuscitator delivering oxygen. Mouth-to-mouth respiration should be used with caution since one case has been reported in which an individual applying mouth-to-mouth respiration developed cyanide poisoning from inhaling the breath of the victim. In severe poisonings, intravenous injection of 0.3 g sodium nitrite is given as a 3% solution (10 ml) at a rate not in excess of 2.5–5 ml/min. Caution should be used since excess generation of methemoglobin can be toxic. The sodium nitrite is followed immediately by intravenous injection of 12.5 g of sodium thiosulfate (50 ml of a 25% solution) at the same rate as sodium nitrite. In children, the dose of sodium nitrite is 6–8 ml/m^2 of body surface area and sodium thiosulfate at 1.65 ml/kg. Although cyanide blocks the cellular utilization of oxygen, studies demonstrate that oxygen combined with sodium nitrite and sodium thiosulfate can be beneficial. It is recommended that oxygen be administered concurrently with the antidotes. The patient should be monitored closely for at least 24–48 hours and if symptoms reappear or persist, then one-half the original doses should be administered.

If cyanide salts have been ingested, the compound should be removed from the stomach by gastric lavage. Due to cyanide's rapid action, gastric lavage should follow specific antidotal therapy. In severe poisonings, proper antidotal therapy must be initiated without delay.

B. Prevention of Poisoning

1. General Safety

Employees working in an environment that has the potential for exposure should be thoroughly familiar with the chemical and toxic properties of the cyanides. Training should include proper emergency procedures and contingency plans. For prevention, hazard labeling of storage containers and posting of use instructions are necessary in an industrial setting. Proper impervious protective clothing, hand protection, and suitable respiratory protective equipment should be used. In the laboratory, use of hydrogen cyanide should be conducted in a mechanically vented fume hood.

2. Exposure Limits

The various guidelines for maximal permissible workplace exposure to cyanides are included in Table X. For dietary exposure, the World Health Organization recommends an acceptable daily food intake of 0.05 mg cyanide. In drinking water, the lifetime acceptable daily intake for drinking water is 1.5 mg/kg (equivalent to 0.02 mg/kg/day based on a 70-kg adult).

Table X Exposure Standards for Cyanides

Compounds	OSHA[a]	IDLH[b]	TLV ACGIH[c]
Hydrogen cyanide	4.7 ppm ceiling (5 mg/m^3)-STEL[d] (with skin notation)	50 ppm	10-ppm ceiling values 11.0 mg/m^3 (with skin notation)
Sodium cyanide	5 mg CN/m^3 (eq. to 9.4 mg/m^3 NaCN) 8-hr TWA[e]	50 mg/m^3	5 mg CN/m^3 (eq. to 9.4 mg/m^3 NaCN) 8-hr TWA
Potassium cyanide	5 mg CN/m^7 (eq. to 12.5 mg/m^3 KCN) 8-hr TWA	50 mg/m^3	5 mg CN/m^7 (eq. to 12.5 mg/m^3 KCN) 8-hr TWA

[a] NIOSH recommended exposure limits or the same as OSHA.
[b] IDLH, immediately dangerous to life or health.
[c] TLV, threshold limit value.
[d] STEL, short-term exposure limit.
[e] TWA, time-weighted average.

3. Respirator Usage

Guidelines for use of respiratory equipment in hydrogen cyanide contaminated areas have been established by the National Institute for Occupational Safety and Health. In conjunction with respirator usage, protective clothing should be worn if there is a possibility of skin contact with liquid hydrogen cyanide. For atmospheres containing hydrogen cyanide concentrations less than 90 ppm, it is recommended that one use a Type C supplied-air respirator with a full facepiece operated in a pressure demand or continuous-flow mode. For concentrations greater than 90 ppm, the respirator should be a self-contained breathing apparatus with a full facepiece operated in pressure demand (positive pressure) worn under a gas-tight suit providing whole body protection.

Bibliography

Ballantyne, G., and Marrs, T. C., eds. (1987). "Clinical and Experimental Toxicology of Cyanides," Wright, Bristol.

Hardy, H. L., and Boylen, G. W., Jr. (1983). Cyanogen, hydrocyanic acid and cyanides. "Encyclopedia of Occupational Health and Safety," vol. 1, p. 574. International Labor Corp., Geneva.

Hartung, R. (1982). Cyanides and nitriles. *In* "Patty's Industrial Hygiene and Toxicology," 3rd ed., vol IIC. Wiley, New York.

Kirk, R., and Othmer, D. (1979). "Encyclopedia of Chemical Technology," 3rd ed., vol. 7. Wiley, New York.

"Toxicological Profile for Cyanide." (1988). Agency for Toxic Substances and Disease Registry, U.S. Public Health Service.

Vernnesland, B., Conn, E. E., Knownles, C. J., Westley, J., and Wissing, F., eds. (1981). "Cyanide in Biology," Academic Press, New York.

Way, J. L. (1984). Cyanide intoxication and its mechanism of antagonism. *Ann. Rev. Pharmacol. Toxicol.* **24,** 451.

Cytotoxicity of Mycotoxins

Raghubir P. Sharma
Utah State University

Glossary

Aflatoxins Group of mycotoxins produced by molds (e.g., *Aspergillus flavus*) particularly in stored agricultural crops (e.g., peanuts). Some aflatoxins are potent carcinogens.

Citrinin Mycotoxin produced by molds of genus *Penicillium* (*P. citrinum*) and *Aspergillus* (*A. niveus*). Reported to be bactericidal to some gram-positive organisms.

Cytotoxicity Toxic effect produced by a chemical in cells. This may include both morphologic and functional changes, either *in vivo* or in cells in culture.

MDBK cells Madin–Darby bovine kidney cells. Cell line derived from a kidney of an apparently normal, adult steer by S. H. Madin and N. B. Darby (February 18, 1957). Probably the first permanent cell line from a large domestic animal to be established.

Mycotoxin Poisonous substance produced by fungi. These chemicals are considered as secondary metabolites of fungal organisms with no known physiological function in the organism producing them.

PFBK cells Primary fetal bovine kidney cells, derived from fresh tissue; somewhat less differentiated compared with MDBK cells.

Rubratoxins Mycotoxins produced by selected species of *Penicillium* (*P. rubrum*).

Trichothecenes Mycotoxins produced by various molds, especially those of *Fusarium* species. So named because of their chemical structure.

***MYCOTOXINS** are secondary metabolites of various fungi. These chemicals comprise a diverse group of compounds that differ widely in their chemical properties and toxic effects. The primary health risks of mycotoxins are via contaminated food and feed since many fungi grow in foods, mainly during storage or in the field under stress conditions. The occurrence of various fungal organisms in food commodities, and hence the possible presence of mycotoxins, depends on factors such as region, season, and harvesting and storage practices. Crops growing in warm and humid climates, (e.g., in the tropics and subtropical countries) are more likely to have fungal contamination than those grown in temperate climates. More than 100 different fungal species are known to produce mycotoxins that have been associated with various health problems in humans and animals worldwide.*

I. Mycotoxins and Health Risks

One of the first health problems associated with mycotoxins was that produced by ergot of rye in the 1850s in the United States. Rye and other cereals infested with *Claviceps purpurea* caused a variety of diseases ranging from neurological disorders to gangrenous degeneration of appendages in both people and livestock. In the middle of the twentieth century, various human diseases were associated with other mycotoxins (e.g., the human stachybotryotoxicosis caused by moldy bread infected with *Fusarium graminearum,* and alimentary toxic aleukia caused by ingestion of overwintered grains infested with *F. poae* and *F. sporotrichioides*). A great interest in mycotoxin research followed the outbreak of turkey X disease in England, later found to be caused by aflatoxins from *Aspergillus flavus*.

Not all metabolites produced by fungi are necessarily highly toxic. Useful drugs such as penicillin are produced by fungi but are not termed mycotoxins. Similarly, the presence of fungal spores in organic matter is not always related to the presence of mycotoxins; the toxins are produced only under certain conditions which are perhaps different than those required for optimal growth of the fungi themselves. Factors such as drought or extremes in temperature are presumed to be favorable for mycotoxins to be produced by their respective fungi. Other stress conditions such as mechanical or insect damage, soil temperature, and excess moisture during storage also favor the formation of mycotoxins in food commodities contaminated with fungi.

The presence of mycotoxins in food may be considered an important health risk. Although a variety of mycotoxins is highly toxic after an acute exposure, such exposures are highly unlikely in natural situations. Most food products will contain a very small amount of mycotoxin unless they have perished and are unpalatable or unsightly. Human exposure to mycotoxins will therefore be to small amounts but for prolonged periods, if contaminated foods are ingested. Several mycotoxins are again chronic toxicants, whose effects may not be apparent for a long time. For instance, the aflatoxins produced by *Aspergillus* species are considered to be one of the most potent carcinogens known.

The mechanism of action of only a few mycotoxins is well understood. In some cases only the effects are known on a target organ or cell in the biological organism. Some mycotoxins may have selective toxicity on specific organs or functions, whereas others may cause generalized effects (Table I). The target tissue may also depend on the mode of entry of mycotoxins in the host; it has only recently been speculated that the presence of aflatoxins in grain dust can be a cause of pulmonary disease.

A variety of mycotoxins have cytotoxic properties. These mycotoxins can therefore cause alterations in both structure and function of the cells with whom they come in contact. Certain mycotoxins, such as aflatoxin B_1, need to be metabolized to active products *in vivo;* accordingly, the organ (e.g., liver) that is involved in its primary metabolism will be most likely to show cytotoxic effects. Some mycotoxins have selective effects (e.g., citrinin is primarily nephrotoxic).

Because of the large number of mycotoxins identified thus far, only the major ones will be discussed here. These have been selected primarily because of their prevalence in foods and feeds and their potential for health hazards in humans and animals. Although this article deals with the cytotoxic effects of these mycotoxins, a relevant description of their general toxic effects is also included. The major mycotoxins discussed here include aflatoxin B_1, fusarium toxins (such as T-2 toxin and zearalenone), rubratoxin, citrinin, and a brief description of other miscellaneous compounds.

II. Various Mycotoxins, Sources, and Chemistry

A. Aflatoxins

The aflatoxins, particularly aflatoxin B_1, have been studied more than any other mycotoxin. They are produced by the fungi of *Aspergillus* genus, particularly by *A. flavus* and *A. parasiticus*. Common commodities associated with aflatoxins include peanuts, corn, cottonseed, millet, sorghum, pistachio nuts,

Table I Various Mycotoxins–Their Sources and Target Organs for Toxicity

Mycotoxin	Fungal source	Target organ(s)
Aflatoxins	*Aspergillus flavus*	Liver, kidney
	Aspergillus parasiticus	Upper respiratory tract?
Fusarium toxins		
T-2 toxin	*Fusarium tricinctum*	Mucous membranes, lymphatic organs
Zeralenone	*Fusarium roseum*	
Macrocyclic trichothecenes	*Myrothesium roridum*, etc.	Dividing cells, lymphatic organs
Rubratoxin	*Penicillium rubrum*	Liver
	P. purpurogenum	
Citrinin	*Penicillium citrinin*	Kidney
	Penicillium viridicatum	
Ochratoxin A	*Aspergillus ochraceous*	Kidney
	Penicillium viridicatum	

and Brazil nuts. The mycotoxin has also been found in barley, beans, cassava, peas, sesame, soybean, sweet potato, and wheat in other parts of the world.

Aflatoxin B_1 is the most toxic in this class and is mainly associated with *A. flavus*. The toxicology of various aflatoxins is described in detail elsewhere in this handbook and will only be briefly discussed here. Aflatoxin B_1 is highly toxic as it is capable of producing ring epoxides *in vivo* that form adducts with a variety of macromolecules such as DNA, RNA, proteins, and glutathione. Chemical structures of various mycotoxins are depicted in Figure 1.

B. Trichothecenes

Simple trichothecenes, of which T-2 toxin is a prototype, are produced by various species of *Fusarium, Trichothecium, Myrothecium, Cephalosporidium, Stachybotrys, Trichoderma, Cylindrocarpon,* and *Verticimonosporium*. The fungus isolated with T-2 toxin-infected corn in the United States is *F. sporotrichioides*. These fungi and associated toxins have been associated with alimentary toxic aleukia of humans and animals in the Soviet Union, bean-hull poisoning of horses in Japan, and fungal corn toxicosis of various livestock species in the United States.

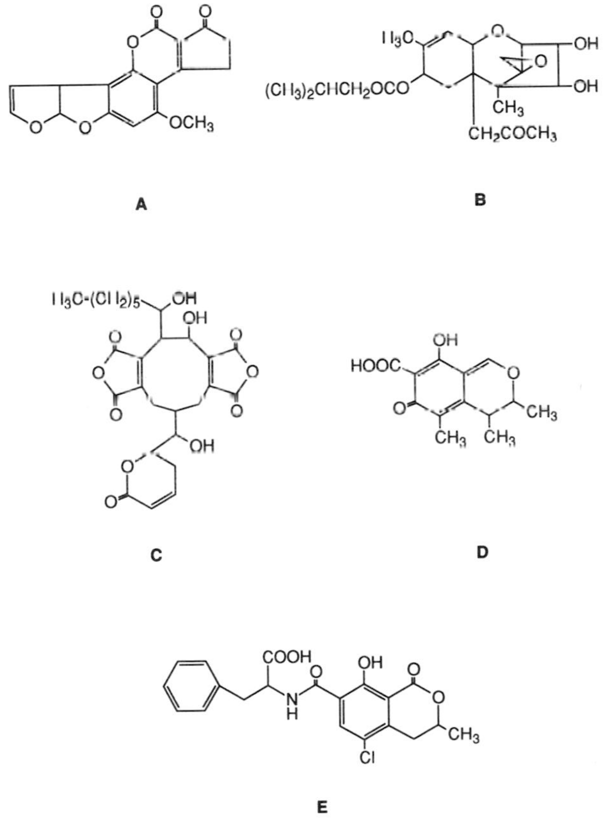

These mycotoxins have a common trichothecene skeleton (see Fig. 1), which consists of cyclopentane, cyclohexane, and a six-membered oxyrane ring with four methyl groups. In addition to the T-2 toxin, other simple trichothecenes include neosolaniol, HT-2 toxin, nivalenol, fusarenon-X, deoxynivalenol, and crotocin.

The macrocyclic trichothecenes are di- or trilactone esters of verucarol with various macrolide rings. These chemicals vary widely both in structure and toxicity parameters. In addition to being produced by a number of fungi, they are also produced by certain Brazilian species of plants belonging to the genus *Baccharis*. The fungi producing these chemicals include *Myrothecium roridum, M. verucana, Stachybotrys atra, Cylindrocarpon* sp., *Verticimonosporium diffractum, Cryptomola acutispora,* and *Phomopsis leptostromiformis*. The molds often contaminate hay and straw that, when consumed by farm animals due to shortage of normal feed, may give rise to problems like stachybotryotoxicosis.

Figure 1 Chemical structures of selected mycotoxins. (A) Aflatoxin B_1, (B) T-2 toxin, (C) Rubratoxin B, (D) Citrinin, and (E) Ochratoxin A.

C. Rubratoxins

Various rubratoxins are produced by *Penicillium rubrum* and *P. purpurogenum;* the two forms isolated thus far are rubratoxin A and rubratoxin B. The B form is a cyclic bisanhydride (see Fig. 1) and rubratoxin A is a dihydro derivative of one of the anhydride groups. The fungi producing rubratoxins are often found along with *A. flavus* in moldy corn.

D. Citrinin

Citrinin is produced by *Penicillium citrinum* and other Penicillium species. This mycotoxin has been found in wheat, oats, barley, and rye. Citrinin has also been observed in ground nuts infected with *Aspergillus flavus, Penicillium citrinum,* and *A. terreus*. It has been reported in rice and corn flour. The most frequent food commodity contaminated is fruit juices. Compared to other mycotoxins, citrinin is a relatively simple molecule (see Fig. 1), which chemically is (3R-*trans*)-4,6-dihydro-3,4,5-trimethyl-6-ox-D-3H-2-benzopyran-7-carboxylic acid.

E. Ochratoxins

Several ochratoxins have been isolated from the genera *Aspergillus* and *Penicillium*. The fungi are prevalent throughout the world and hence the potential of contamination by ochratoxins is considerable. Structurally these mycotoxins are isocoumarin derivatives. The chemical name for ochratoxin A (Fig. 1) is *N*-((5-chloro-8-hydroxy-3-methyl-1-oxo-7-isochromanyl)carbonyl)-3-phenyl alanine.

III. Comparative Toxicity

A. Aflatoxins

Aflatoxin B_1 is activated by microsomal cytochrome oxidases into highly reactive metabolites, particularly in liver. To a lesser extent this mycotoxin can also be metabolized via cooxidation involving prostaglandin-H synthetase in various other tissues. Liver is by far the primary target organ although biochemical changes have been elucidated in various extrahepatic tissues. The transformed chemical includes metabolites such as aflatoxin B_1-2,3-epoxide, a highly carcinogenic substance. There is a wide variation in the lethal toxic doses of this chemical in different species. As indicated in Table II, ducks and rainbow trout are highly susceptible to the toxic effects of aflatoxin B_1 and are also sensitive to its carcinogenic potential. Various experimental species show a comparable variation in the binding of aflatoxin B_1 to hepatic DNA; rats are relatively resistant. Some of the DNA adducts of aflatoxin B_1 that may lead to carcinogenicity include the N^7-guanine and formamidopyrimidine adducts. Alkylation of aflatoxin B_1 with DNA could result in a loss of purine or pyrimidine bases resulting in apurinic sites. The active aflatoxins are also potent mutagens in both microbial and eukaryotic cell systems.

The exposure to aflatoxin B_1 leads to malaise, loss of appetite, and lower growth rate and productivity. The pathologic effects are most dramatic in liver; in ducklings and adult rats the periportal zone is affected, whereas in pigs, guinea pigs, and dogs the lesions are predominantly in the centrilobular zone. Hemorrhagic necrosis of kidney, heart, spleen, and pancreas is also noted. In chickens, hemorrhagic gastroenteritis has been observed. In humans, various syndromes including jaundice, ascites, portal hypertension, gastrointestinal bleeding, and possibly encephalopathy and fatty degeneration of viscera (Reye's Syndrome) have been implicated with the aflatoxin B_1 exposures. Although there is a likelihood that exposure to aflatoxin B_1 results in hepatocellular carcinoma in people exposed to dietary mycotoxin and perhaps cancer of the upper airways in workers exposed to contaminated grain dust, convincing evidence is as yet unavailable.

Table II Relative Toxicity of Various Mycotoxins

Mycotoxin	Species employed	LD_{50} [mg/kg (route)]
Aflatoxin B_1	Duck	0.34–0.56 (oral)
	Rainbow trout	0.81 (oral)
	Monkey	3 (oral)
	Rat	7 (oral)
T-2 toxin	Mouse	9 (oral)
	Hamster	10.2 (oral)
	Rat	17.9 (oral)
Rubratoxin B	Rat	6.4 (oral)
	Hamster	0.4 (ip)
	Chicken	83.2 (oral)
Citrinin	Rat	67 (ip or sc)
	Guinea pigs	37 (sc)
	Guinea pigs	19 (iv)
	Guinea pigs	110 (oral)
	Hamster	75 (ora)
Ochratoxin A	Rat	20 (oral)
	Chicken	3.3 (oral)

Abbreviations: ip, intraperitoneal; iv, intravenous; sc, subcutaneous.

B. Trichothecenes

The acute poisoning by T-2 toxin is characterized by vomiting, inflammation, diarrhea, damage to bone marrow, thymus, spleen, and mucous membrane of intestines and a depression in circulating white blood cells. In humans, the alimentary toxic aleukia has various stages such as (1) a burning sensation of mouth, tongue, throat, palate, esophagus and stomach, followed by vomiting, diarrhea, and abdominal pain; (2) progressive leukopenia and granulopenia then sets in, often accompanied by bacterial infection. In the progressive stage (3), there are petechial hemorrhages of skin and necrosis, and finally (4), severe leukopenia and necrosis of pharyngeal tonsils and throat leading to angina. Symptoms in exposed animals are somewhat similar and depend on the amount of toxin given and the duration of exposure.

Macrocyclic trichothecenes are highly toxic and cytotoxic, although the toxic potential may vary widely. The lethal oral dose may range from 0.6 mg/kg for roritoxin B to about 6 mg/kg for baccharinoid B4. The toxicity is believed to be caused by the potential to be inhibitors of protein synthesis. Indeed, because of this property, various macrocyclic trichothecenes have been proposed and tested as anticancer agents. Although death and emaciation is generally reported as a result of the toxicity of these mycotoxins, no specific organ has been described as the target organ other than dividing cells. Because they affect actively proliferating cells, many of these chemicals have considerable immunotoxic properties.

C. Rubratoxin B

Rubratoxin B is a hepatotoxin in rat and mouse and perhaps in large livestock species. The toxin affects hepatic protein synthesis and polyribosome integrity; hepatocellular electron transport and oxygen consumption in mitochondria; and causes inhibition of adenosine triphosphatase and drug metabolizing enzymes, and reduces glutathione content in liver. The primary organ of insult appears to be liver although biochemical effects have been noticed in other organs. Livestock exposed experimentally to rubratoxin B exhibit hemorrhage and congestion in most tissues, a mottling of the liver, hepatocellular necrosis, and degeneration. Experimental administration of rubratoxin B in pregnant mice causes embryonic death, malformation, and resorption. It is also embryocidal and teratogenic in the chicken.

D. Citrinin

Citrinin is acutely toxic to laboratory animals. The LD_{50} for most species ranges between 19 and 200 mg/kg by various exposure routes (Table II). This mycotoxin is primarily nephrotoxic and has been implicated in porcine nephropathy, which is characterized by polyuria, glucosuria, and proteinuria. A chronic interstitial renal disease called endemic Balkan nephropathy is considered to be caused by mycotoxins including citrinin (in addition to ochratoxin A). The disease is characterized by normocytic, normochromic anemia, azotemia and persistent proteinuria, followed by renal failure over several years. Renal damage is the primary lesion in experimentally induced citrinin toxicity in animals. The action of citrinin is to disrupt renal tubular function; focal areas of necrosis are observed in the proximal convoluted segments. The necrotic areas are accompanied by extensive mineral deposits. Experimental nephropathy after citrinin administration is, however, considered reversible. Citrinin has been shown to be an embryocidal and is a fetotoxic chemical.

E. Ochratoxins

Ochratoxin A is the most toxic of this class. In exposed animals it produces toxic nephritis, characterized by tubular degeneration, atrophy, and interstitial fibrosis. In laboratory studies, microhemorrhagic enteritis of cecum, colon, and rectum; enlargements of lymph nodes; and liver damage have been observed accompanied by growth retardation and decreased food intake. The presence of ochratoxin A was implicated in Balkan endemic nephropathy, the exact etiology of which remains unclear.

IV. Cytotoxicity

A. Aflatoxins

The cytotoxic effects of aflatoxins, particularly that of aflatoxin B_1, have been investigated in primary rat hepatocytes. The critical macromolecular target of aflatoxin B_1 is considered to be cellular deoxyribonucleic acid (DNA), which can form adducts with the chemically reactive metabolites of this mycotoxin. A number of cell systems have been evaluated for aflatoxin B_1-induced cytotoxicity. These include rat, mouse, or hamster fibroblasts; human lung cells, and primary as well as established renal cell lines. In both MDBK and PFBK cell cultures, the effects of aflatoxin B_1 can be shown to occur at millimolar concentrations (Table III). In all cases, the presence of this mycotoxin decreases cell survival, growth rate, attachment of cells to the culture surface, and cellular degeneration. Morphologically, a shrinkage of cytoplasmic material, loss of microvilli, and an apparent cleavage of nuclear membranes becomes evident (Fig. 2). The cells become somewhat elongated and spindle shaped. Extensive fibrous materials and a complete loss of cell boundary architecture become apparent in severe toxicity in culture. When observed with a transmission electron microscope there is cytoplasmic vacuolization and degeneration of mitochondria, along with a disruption of nuclear architecture (Fig. 3).

The ability of aflatoxin B_1 to cause cytotoxic ef-

Table III Cytotoxicity Parameters of Selected Mycotoxins

Mycotoxin	Tissue or cell	Cytotoxicity (M)
Aflatoxin B_1	Rat primary hepatocytes	LC_{50} 8 × 10^{-7}
		LC_{Lo} 3 × 10^{-7}
	Rat hepatocytes	LC_{50} 3 × 10^{-8}
	Mouse hepatocytes	LC_{50} 3 × 10^{-5}
	Mouse primary hepatocytes	LC_{50} 0.03 × 10^{-6}
	MDBK cells	LC_{50} 1 × 10^{-4}
	PFBK cells	LC_{50} 2.5 × $10-^{5}$
T-2 toxin	MDBK cells	LC_{Lo} 1.5 × 10^{-6}
	Human fibroblasts	LC_{50} 1.5 × 10^{-6}
	PFBK cells	LC_{Lo} 1 × 10^{-9} (growth inhibition)
	HL-60	9 × 10^{-9} (cellular differentiation)
Rubratoxin B	Corneal cells	>1.4 × 10^{-4} (cytotoxic)
	HeLa cells	2 × 10^{-4}
	Hepatocytes	6 × 10^{-5}
Citrinin	MDBK cells	LC_{50} 4.7 × 10^{-5}
	PFBK cells	LC_{50} 3.8 × 10^{-4}
Ochratoxin A	GBK cells	0.05–0.1 × 10^{-3} (thymidine incorporation)

Abbreviations: LC_{Lo}, lowest concentration of mycotoxin to show a cytotoxic effect; LC_{50}, concentration of the mycotoxin to show a half-maximal effect; in most cases the parameter evaluated was cell viability or attachment to culture surface.

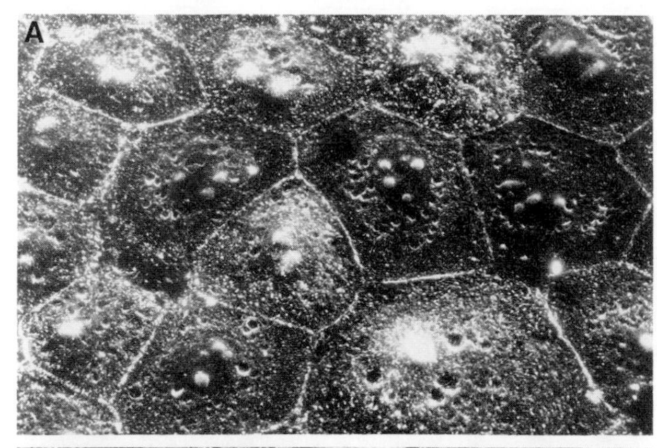

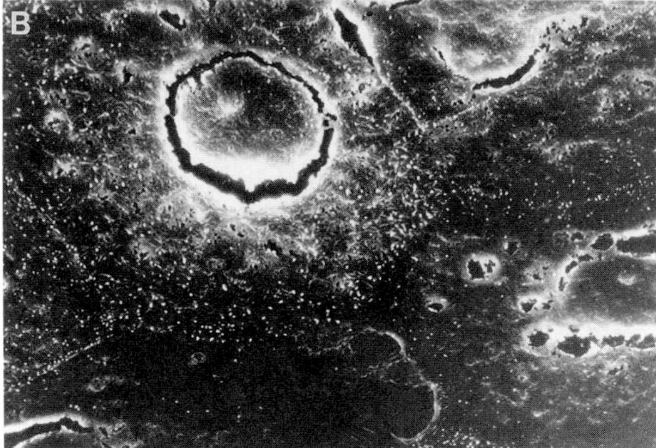

Figure 2 Scanning electron micrograph of MDBK cells in culture in the presence of aflatoxin B_1. (A) Control; ×1500. Note the hexagonal architecture of cells with cell–cell boundaries in a monolayer. (B) Similar cultures in the presence of 10^{-4} M aflatoxin B_1 for 24 hours; ×3600. Shrinkage of cytoplasmic material, presence of debris, loss of microvilli, and cleaved nuclear membranes are evident.

fects is directly dependent on the metabolism of this mycotoxin in various cell systems. Since both hepatic and renal cells are abundant in microsomal mixed function oxidases, the systems involved in activating aflatoxin B_1 have been observed even in fibroblasts derived from various species. In cells of the upper respiratory tract, nonciliated cells are particularly capable of converting aflatoxin B_1 (because of the presence of microsomal metabolizing systems) and are affected preferentially by aflatoxin B_1. Ciliated cells (which lack these microsomal enzymes) are somewhat less vulnerable (Fig. 4).

B. Trichothecenes

The cytotoxicity of T-2 toxin and related trichothecenes has been evaluated in a number of culture systems. T-2 toxin decreases protein synthesis in MDBK and Chinese hamster ovary (CHO) cells at 1.4 and 5.6 μM, respectively. In PFBK cells, the reduction of growth is observed at 1 nM and is preceded by accumulation of electron-dense vesicles in the cytoplasm (Fig. 5). The LC_{50} of T-2 and T-2-tetrol in normal human fibroblasts is 1.5 and 30 μM, respectively. The cellular damage formed are in nuclei, rough endoplasmic reticulum, and ribosomes with little or no effect on lysosomes and plasma membranes.

In the HL-60 cell line established from peripheral blood of a patient with acute promyelocytic leukemia, trichothecenes caused differentiation of the cells along with cytotoxic effects. The cells were differentiated to macrophage-like granulocytes at 2–5 ng/ml of T-2 toxin; the effects were similar at higher levels of T-2 triol; acetyl T-2 tetrol was not effective in differentiating HL-60 cells.

The macrocyclic trichothecenes are highly cytotoxic compounds. In general the order of toxicity

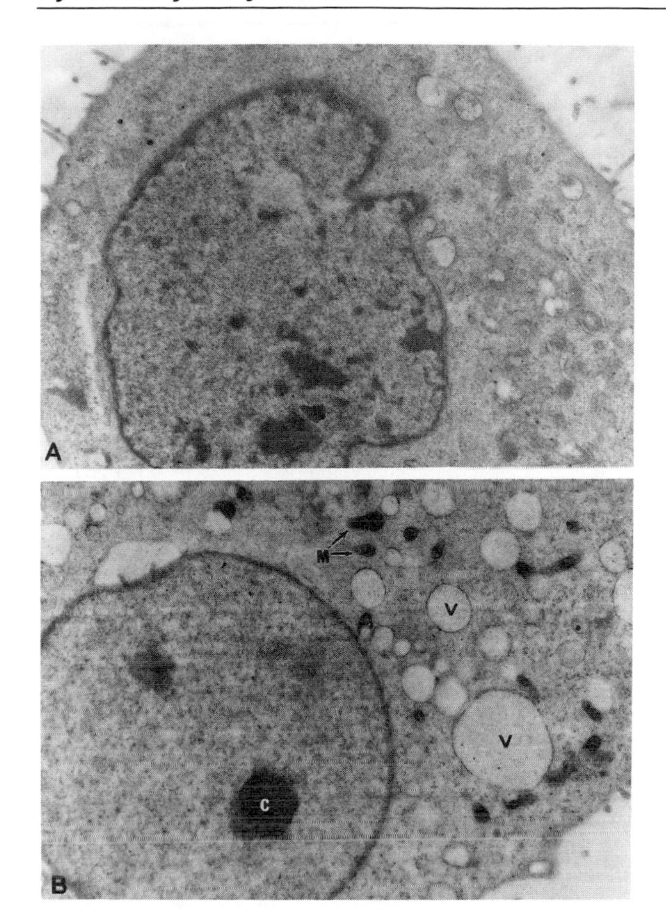

Figure 3 Cytotoxicity of aflatoxin B_1 in PFBK cells. (A) Control; ×7100. (B) Aflatoxin B_1 at 10^{-4} *M* for 24 hours; ×7100. Note cytoplasmic vacuolization (v), degenerating mitochondria (M), and condensed nuclear material (C). (Reproduced with permission of Academic Press, Inc.)

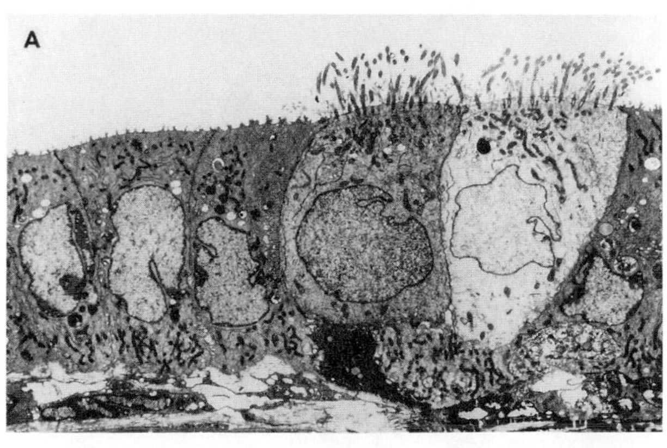

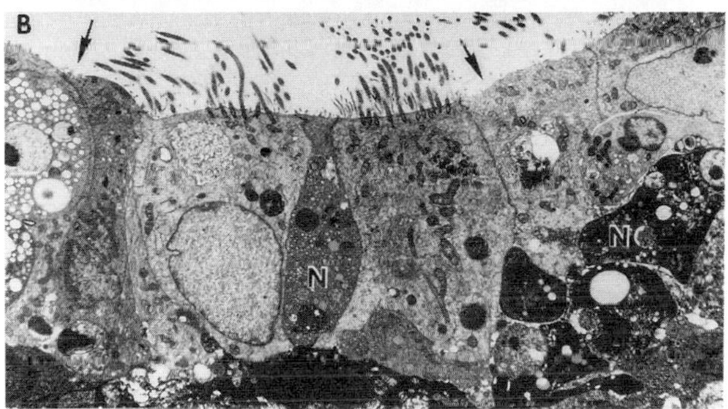

Figure 4 Hamster tracheal epithelial cells in culture (A) Control; ×2100. (B) After incubation with 10^{-6} *M* aflatoxin B_1 for 24 hours; ×2100. The nonciliated cells (indicated by arrows) are vacuolated and necrotic (N), or contain electron-dense cytoplasmic droplets. The ciliated cells have occasional electron-dense cytoplasmic materials but appear otherwise normal. (Photograph courtesy of R. A. Coulombe and reproduced with permission of Elsevier Scientific Publishers.)

has been described as roritoxins ~ myrotoxins > roridins ~ verrucarins > baccharinoids. The cytotoxicity is generally produced around 10 ng/ml and varies widely in different cell lines (Table IV). The cytotoxicity relates well to the *in vivo* toxicity of these compounds.

C. Rubratoxin B

The cytotoxicity of rubratoxins is highly variable depending on the cell type. A concentration of 100 μg/ml of rubratoxin B (but not 30 μg/ml) inhibited growth of HeLa cells, elongated the M phase of mitosis, and caused accumulation of abnormal mitotic cells and polynuclear cells. Pulmonary and renal cells are more sensitive than HeLa cells. Cultured hepatocytes also show sensitivity to rubratoxin B at 30 μg/ml.

D. Citrinin

Cytotoxic effects of citrinin have been evaluated in kidney cultures. The presence of 10^{-4} *M* citrinin causes extensive detachment of cells and decreases cell viability in MDBK and PFBK cells. The MDBK cells appear flattened or shrunken and devoid of cell-to-cell contact, and are often unable to form a characteristic monolayer. The projecting nucleoli, often observed in normal cells in scanning electron microscopy, are often absent after citrinin treatment (Fig. 6). The effects are similar in PFBK cells. The synthesis of nucleic acids is inhibited, perhaps due

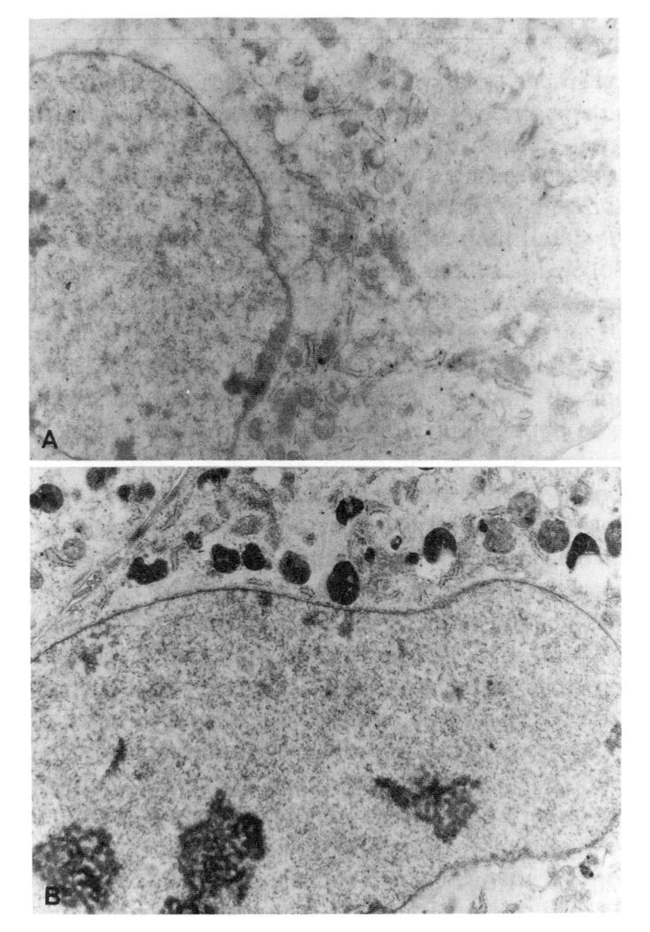

Figure 5 Effects of T-2 toxin in PFBK cells. (A) Control culture; ×7000. (B) Cell incubated with 10^{-8} *M* T-2 toxin for 24 hours; ×7000. Note accumulation of electron-dense particles in the cytoplasm and disorientation of chromosomal material. The mitochondria, endoplasmic reticulum, and nuclear membranes appear normal and the cell membrane and microvilli are also intact.

Table IV Effects of Selected Mycotoxins on Splenic Lymphocytes

Mycotoxin	Parameter observed	Cytotoxicity LC_{50} (*M*)
Aflatoxin B_1	RNA synthesis[a]	$0.2-1 \times 10^{-4}$
T-2 toxin	Viability[b]	1×10^{-8}
Macrocyclic trichothecenes		
Roritoxin B	Viability	0.029×10^{-6}
Myrotoxin B	Viability	0.087×10^{-6}
Roridin A	Viability	0.6×10^{-8}
Verrucarin A	Viability	20.1×10^{-6}
16-Hydroxyverrucarin A	Viability	6.2×10^{-6}
Verrucarin J	Viability	0.8×10^{-6}
Baccharinoid B_{12}	Viability	1.3×10^{-6}
Roridin E	Viability	0.9×10^{-6}
Baccharinoid B_4	Viability	2.9×10^{-6}
Baccharinoid B_5	Viability	3.7×10^{-6}
Rubratoxin B	Growth inhibition	2×10^{-4}
Citrinin	DNA synthesis[c]	1×10^{-4}
Ochratoxin A	DNA synthesis[c]	3.5×10^{-5}

[a] Measured as radiolabeled uridine incorporation.
[b] Dye exclusion assay.
[c] Measured as labeled thymidine incorporation with phytohemagglutinin stimulation.

to inhibition of transport of precursor compounds into cells. Activities of certain marker enzymes (e.g., K^+-dependent phosphatase, alkaline phosphatase, and succinic dehydrogenase) are reduced, indicating a general depression of cellular metabolic processes.

E. Miscellaneous

1. Cytochalasins

Although not an appreciable health risk in either humans or animals, a group of highly cytotoxic mycotoxins has been isolated from cultures of *Phoma exigua, Helminthosporium dematioideum, Metarrhizium anisophdae,* and *Zygosporium masonii,* that has a specific effect on cells in culture. These chemicals of related structure called cytochalasins bind to filamentous actin and inhibit crosslinking between actin and proteins. These mycotoxins, therefore, prevent division in dividing cells. Certain cytochalasins such as cytochalasin A and cytochalasin B inhibit glucose uptake in human erythrocytes; this property is, however, not shared by cytochalasins C, D, and E. Cytochalasins allow cells to undergo mitosis and nuclear division; cytokinesis is inhibited. Formation of multinuclear cells may thus result in the presence of appropriate concentrations of cytochalasins. The cells may undergo further division to produce unicellular cells if cytochalasins are removed. Treatment of cultured cells with cytochalasin B may cause segregation of the nuclei into smaller pockets causing a bulging from the top of cells. Experimental treatment of laboratory animals causes accumulation of large amounts of fluids at the injection site and congestive degenerative changes and necrosis of liver, kidney, spleen, pancreas, small intestines, and also pulmonary hemorrhages, brain edema, and injury to vascular walls. It appears that the effect is primarily on capillary walls causing

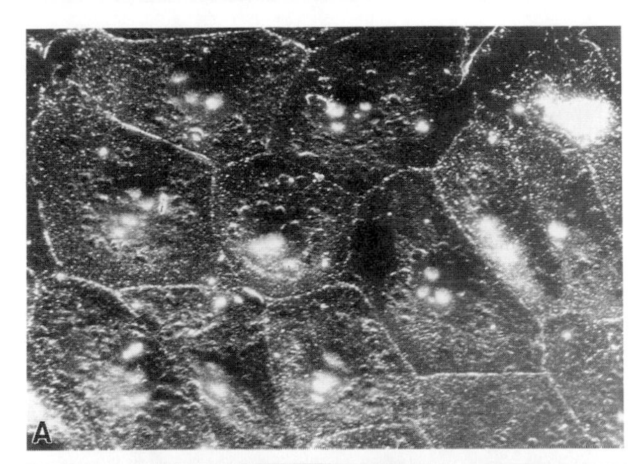

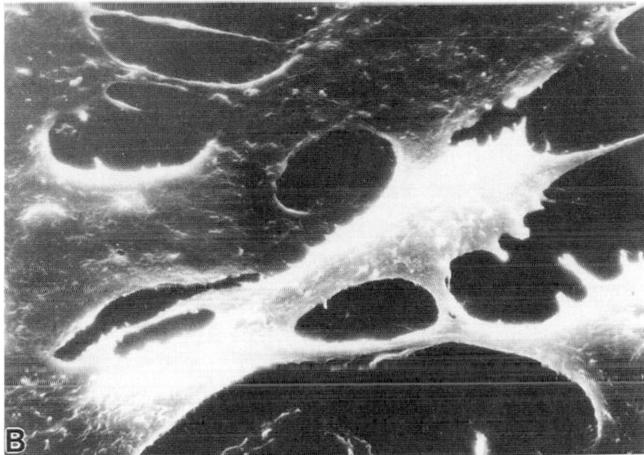

Figure 6 Cytotoxicity of citrinin in MDBK cultures. (A) Control cultures; ×1500. (B) Cells in presence of 10^{-4} *M* citrinin for 24 hours; ×1500. The cells lose their characteristic form and are connected to each other by filamentous structure and are unable to form a monolayer.

extravascular effusion, loss of plasma proteins, and shock resulting in death.

2. Patulin

Patulin is a toxic metabolite produced by the genera *Penicillium* and *Aspergillus*. This mycotoxin is most commonly associated with the common storage rot fungus of apples, *P. expansum*. Although no outbreaks of disease have been reported in people consuming apples contaminated with patulin, the natural occurrence of patulin suggests that it may be a possible contaminant of human and animal food. The ranges of LD_{50} in mice (mg/kg) are 25–46 (oral), 10–33 (subcutaneous), 5–15 (intraperitoneal), and 15–25 (intravenous). Gross lesions in laboratory species are gastrointestinal hyperemia, fluid distention, gastritis, and enteritis. Patulin is embryotoxic but not teratogenic in mice and rats. It has been shown to cause lesions of the extremities in chicken embryos. This mycotoxin has been observed to be mutagenic in *Bacillus subtilis* and *Saccharomyces cerevisiae* and causes increased frequency of sister chromatid exchange; however, it was negative in Ames test, SOS chromotest (in *E. coli)*, and DNA repair test in primary hepatocytes.

Patulin is a mitotic poison affecting dividing cells. Mouse fibroblasts are highly sensitive to this mycotoxin, 154 μg/ml of patulin causing a 50% reduction of multiplication in suspension cultures. HeLa cells treated with patulin (32 μg/ml) showed single and double strand DNA breaks and slowed all phases of cell division. Presence of patulin may produce an increased incidence of polyploid cells.

V. Effects of Selected Mycotoxins on Lymphocytes

Because of their biological reactivity, a number of mycotoxins have been evaluated for their potential effects on the immune system. With a few exceptions (e.g., citrinin), most mycotoxins are depressants of immune response. In some cases the cytotoxicity of mycotoxins has been evaluated on immunocompetent cells, particularly the splenic mononucleolar cells (lymphocytes). Even citrinin, which caused immunostimulation after treatment of laboratory animals, causes inhibition of various cellular functions when the splenocytes were treated with this mycotoxin *in vitro*. The parameters observed on the splenic cells after a direct addition of mycotoxins vary widely and range from cell viability (dye exclusion) to synthesis of macromolecules (DNA or RNA) in either resting or mitogen-induced proliferating cells. Actions of selected mycotoxins in cultures are listed in Table IV.

It is apparent that the cytotoxicity of mycotoxins on splenic cells is highly variable. T-2 toxin is perhaps one of the most cytotoxic of several mycotoxins that have been evaluated; the effect is related to its high toxicity and potent immunosuppressive potential. Other simple trichothecenes will perhaps be cytotoxic to lymphocytes as well, but there are limited data available on unstimulated lymphocyte cultures. Both aflatoxin B_1 and citrinin showed cytotoxic effects on lymphocytes in much higher concentrations. The relative lack of aflatoxin B_1 cytotoxicity on lymphocytes can be explained on the basis of the limited activation of this mycotoxin by these cells.

Related Article: FUNGICIDES.

Bibliography

Joffe, A. J. (1986). "Fusarium Species: Their Biology and Toxicology." John Wiley, Toronto.

Krogh, P. ed. (1987). "Mycotoxins in Food." Academic Press, San Diego.

Salunkhe, D. K., Wu, M. T., Do, J. Y., and Maas, M. R. (1980). Mycotoxins in food and feeds. *In* "Safety of Foods" (H. D. Graham, ed). AVI Publishing, Westport, CT.

Sharma, R. P., and Salunkhe, D. K., eds. (1991). "Mycotoxins and Phytoalexins." CRC Press, Boca Raton, FL.

Steyn, P. S., and Vleggaar, R., eds. (1986). "Mycotoxins and Phytotoxins. Elsevier, Amsterdam.

Ueno, Y. (1983). "Trichothecenes: Chemical, Biological and Toxicological Aspects." Elsevier, New York.

Diesel Exhaust, Effects on the Respiratory System

John F. Gamble
Exxon Biomedical Sciences, Inc.

Glossary

Bias Some aspect of a study that has introduced a systematic error into the results, thereby producing spurious results.

Case-control study Subjects are selected on the basis of whether they do (cases) or do not (controls) have the disease. Cases and controls are then compared on the proportion of who were exposed. This comparison is the odds ratio.

Cohort study Individuals free of disease are selected on the basis of exposure and followed up to determine the incidence of disease over a specified time interval. The measure of risk is the ratio of the incidence of disease among exposed divided by the incidence of disease among nonexposed. If an external population is used as nonexposed, this is the standardized mortality ratio (SMR).

Confounding Factor associated with exposure that can independently affect the risk of disease. If a confounder is more common among the exposed group (or cases) than among nonexposed (or controls) then a spurious relationship of health outcome and exposure will be observed.

Cross-sectional study Individuals are selected at some point irrespective of either exposure or disease. The prevalence (or mean score) of the health outcome is examined in relation to exposure while adjusting for known and measured risk factors, and then compared to some nonexposed comparison or control population.

Relative risk (RR) Measure of the risk of disease calculated by dividing the (observed) disease occurrence in the exposed group by the (expected) disease occurrence in the nonexposed or comparison group. This measure varies with epidemiologic study design. (1) Standardized mortality ratio (SMR) is the observed divided by expected deaths in a cohort mortality study. SMRs > 100 indicate increased risk. (2) Odds ratio (OR) is the occurrence of exposure in cases compared with occurrence of exposure in controls in a case-control study. The reference group is set at 1, an OR > 1 indicates increased risk. (3) In cross-sectional studies RR may be simply the ratio of symptom prevalence in exposed population to prevalence in nonexposed; or the ratio of mean ventilatory function in exposed and nonexposed.

DIESEL EXHAUST contains a broad range of individual chemicals that present a variety of possible respiratory health effects beyond simple asphyxiation including increased symptoms of both an acute and chronic nature; reduced pulmonary function both over a workshift (acute) and baseline (chronic); and increased pneumoconiosis. There is also concern over the potential for increased risk of lung cancer. Products of combustion included in the exhaust are carbon dioxide and monoxide, nitrogen oxides, sulfur oxide compounds, uncombusted hydrocarbons, partially oxidized hydrocarbons (such as aldehydes, ketones, phenols), and submicron carbonaceous particulates. Like all products of combustion, diesel exhaust contains variable amounts of chemical species that at some concentration are odoriferous, irritating, and mutagenic.

I. Introduction

The evidence for assessing respiratory health effects caused by diesel exhaust is from epidemiologic studies of diesel exposed workers. Morbidity is evaluated mostly by cross-sectional surveys and shift studies. Mortality is evaluated mostly by retrospective cohort studies and less successfully by population-based case-control studies. In this article, the results of these studies will be presented in brief. Conclusions as to causative relationships are on the degree of risk, presence of a biologic gradient (exposure–response relationship), consistency of the effects in different studies, adequacy of the studies, and plausibility. Exposure estimates are based on surrogates for diesel exhaust exposure, most commonly NO_2, dust, and/or tenure. The adequacy of the studies relates to factors such as the absence or control of confounding exposures such as smoking and nondiesel dust (e.g., mine dust, asbestos); assessment of exposure; and adequate latency and exposure for the disease (lung cancer) to develop.

II. Morbidity

A. Acute Health Effects

Odor and irritation are obvious features of diesel exhaust and have perhaps been personally experienced by some. Several groups exposed to diesel exhaust have been studied to estimate the occurrence of acute symptoms or changes in ventilatory function over the shift (Tables I and II).

1. Symptoms

In an experimental study, volunteers were exposed to variable levels of diesel exhaust gases (particulates were removed) for up to 10 minutes. Exposure to an average NO_2 concentration of 4.2 ppm (1.5–7.0) produced intolerable eye irritation in over 50% of the subjects in less than 10 minutes. Exposure to about half this concentration reduced the eye irritation to a tolerable level. The lowest exposure of 1.3 ppm NO_2 (0.2–2) produced little irritation. These exposures were comparable to those found in railroad shops in the 1960s and are higher than the average exposures measured in the subsequent epidemiologic studies of miners, bus garage workers, and stevedores.

Symptoms observed in the subsequent studies record the prevalence of symptoms, but not their intensity. Eye irritation was the most common in all of the four populations of diesel exposed workers with about 50% prevalence on average and as high as 80% in highly exposed groups. Several other symptoms are above expected or show an increasing prevalence with increasing exposure. These include nose irritation, headache, difficult or labored breathing, chest tightness, and wheeze.

Symptom rates over the shift are higher in diesel-exposed workers than in other exposed populations, and there is some evidence of an exposure–response relationship (see Table II).

2. Acute Change in Ventilatory Function

Ventilatory function (ΔFEV_1, ΔFVC, ΔFEF_{50}) measured before and after work was not reduced in any of the studies of diesel-exposed workers. Exposure to salt mine dust plus diesel exhaust, where NO_2 concentrations were about 5 ppm, reduced FEV_1 to a level approaching clinical significance. In an experimental study, exposure for one hour to an average NO_2 exposure of 4.2 ppm produced no observable increase in pulmonary resistance.

3. Summary

The prevalence of acute symptoms (eye and nose irritation, chest tightness, wheeze) is elevated in diesel-exposed populations, and the prevalence increases as exposure increases. These increased symptoms occur at exposure levels that do not reduce ventilatory function over a workshift. Eye irritation is the most sensitive symptom.

B. Chronic Health Effects

Measures to assess the chronic effects (Table III) of exposure to diesel exhaust include respiratory symptoms as determined by standardized respiratory questionnaires, baseline ventilatory function as measured by spirometry, and radiographic evidence of pneumoconiosis. There is no evidence that pneumoconiosis increases due to diesel exposure in any of the studies where evaluated.

The studies are mostly cross-sectional in design (two prospective reports). The exposed populations include bus garage workers, coal miners, salt miners, trona miners, potash miners, iron ore miners, railroad workers, and stevedores. Surrogate measures of diesel exposure include NO_2, respirable or total dust, and tenure. The amount of diesel use in the mining populations was quite variable and could be as low as one or two vehicles in a mine. In gen-

Table I Acute Health Effects and Diesel Exhaust Exposure

Study group	Exposure (time)	Exposure (ppm) NO_2	Exposure (ppm) SO_2	Exposure (ppm) HCHO	Exposure (mg/m^3) RP	Acute health effects		
						Eye irritation score[a]		
Volunteers—exp't. study (*Arch. Env. Hlth.* **10,** 165, 1965)	≤10 min	1.3	0.2	<0.1	0	avg. = 0.86	peak = 2	
		2.8	0.5	<0.1	0	= 1.47	= 3	
		4.2	1	<0.1	0	= 2.65	= 3	
	60 min	4.2	1	<0.1	0	No effect on pulmonary resistance		
						Eye irritation	Headaches	Throat irritation
Egyptian bus garage workers (*Ind. Hlth.* **4,** 1, 1966)	~ 8 h	0.4–0.7				42%	37%	19%
						Symptom prevalences (%)		Change in ventilatory function
U.S. bus garage workers (*Env. Res.* **42,** 201, 1987)	~8 h	0.23		—	0.24	Eye irritation	54/82 ($p<.0005$)	ΔFVC: −12 mL
		Symptom prevalence in avg. exposure group/ high exposure groups (>0.3 ppm NO_2)				Nose irritation	40/46 (NS)	ΔFEV_1: −19 mL
						Sneeze	35/41 (NS)	ΔFEF_{50}: −102 mL/sec
						Cough	49/61 (NS)	
						Headache	32/39 (NS)	
						Difficult breathing	17/27 ($p<.05$)	
						Chest tightness	14/25 ($p<.05$)	
						Wheeze	18/32 ($p<.005$)	

(*continues*)

Table I (*Continued*)

Study group	Exposure (time)	(ppm)			(mg/m³)	Acute health effects	
						ΔPFT per ppm change in NO_2	
Salt miners in diesel and nondiesel mines (*ACGIH Symp.*, 1978)	~8 h	1.5	<1	<0.5	0.65–0.71	ΔFEV_1: −14 ml ΔFVC: −9 ml ΔFEV_{50}: −99 ml/sec	
						Symptom prevalences (exposed/controls; %)	Decrease in ventilatory function (%)
Stevedores (diesel and gas exhaust) vs. controls (*Appl. Ind. Hyg.* **21**, 133, 1987)	~8 h	<limit of detection (HCHO)				Eye irritation 59/27 Nasal complaints 47/0	ΔFEV_1 1.5/1.1 ΔFVC 0.3/2.6
Coal miners diesel mines electric mines (*Am. Rev. Resp. Dis.* **125**, 39, 1982)	~8 h	 0.3 NO_2 <0.1 NO_2				shift change in FEV_1 −32 mL −33 mL	
		NO_x[a]		HCHO	RP	ΔFVC (%)	ΔFEV_1 (%)
Stevedores (*Am. J. Ind. Med.* **17**, 341, 1990)	~8 h	1.08 1.12 Controls		0.27 0.03 —	0.12 0.23 —	−2% (−109 ml) −5% (−291 ml) +1% (+55 ml)	−1 (−44 mL) −2 (−92 mL) −1 (−45 mL)

Abbreviations: HCHO, formaldehyde; RP, respirable particulate; ns, not statistically significant ($p>0.5$); NO_x, nitrogen oxides; NO_2, nitrogen dioxide.
[a] 0 = none, 1 = same, 2 = tolerable, 3 = intolerable.

Table II Irritation and Acute Symptoms[a, b]

Study populations	Prevalence (%)				
Diesel-exposed	Irritation		Symptoms		
	Eye	Nose	Headache	Chest tightness	Wheeze
U.S. bus garage workers					
all	54	40	32	14	18
high exposed	82	46	39	25	32
dose response	NO_2, dust	no	dust	NO_2, dust	NO_2, dust
Egyptian bus garage workers	42		37		
Ferry stevedores	59	47		29	24
controls	27	0		9	9
Non-diesel-exposed workers					
Battery acid	28	33	22	17	9
Rubber workers	33				
Printers	49	63	54		
controls	17	15	38		

[a] Comparisons are among diesel-exposed workers during a work shift; non-diesel-exposed persons provided for comparison.
[b] Conclusion: Exposure to diesel exhaust results in eye irritation in over 50% of exposed workers and is related to degree of exposure. Other less common acute symptoms associated with diesel exhaust exposure are nose irritation, headache, chest tightness and wheeze.

eral, the degree of diesel exposure is not well characterized, and the length of diesel exposure is generally not provided.

Dust measurements in the mine studies are of total mine dust and sometimes respirable (<10 μm) dust. Diesel particulate is submicron in size, while only a small proportion of mine dust is this small. NO_2 sources include both diesel exhaust and explosives. In both cases there is no way to distinguish the source, and there are no retrospective estimates of exposure. A major confounding exposure in these studies is mine dust which can potentially cause symptoms, reductions in ventilatory function, and pneumoconiosis. Because symptoms and ventilatory function are nonspecific affects, it is difficult to distinguish between mine dust effects and diesel exhaust effects. Smoking is also a potential confounder, causing both symptoms and decreased ventilatory function but is generally adjusted for in the analyses.

Comparisons are either internal (an exposure–response analysis comparing high versus low exposed workers) and/or external (mean ventilatory function values or prevalences compared to nonexposed external control populations). The various studies and results are summarized in Table III for symptoms and ventilatory function.

1. Symptoms

Chronic cough, chronic phlegm, and shortness of breath (dyspnea) are the symptoms most commonly studied and are already elevated among miners in nondiesel mines. The studies of miners to date show no consistent or convincing effects on the prevalence or incidence of these symptoms that can be conclusively attributed to diesel exhaust; that is, there are no consistent exposure–response relationships.

Among coal miners, the lack of an observable effect may be due to low exposure levels ($NO_2 <$ 1 ppm). In some cases the length of diesel exposure is short. Introduction of diesels into noncoal mines began about a decade before introduction in coal mines and they have been used more extensively in the former. Therefore, exposure in noncoal mines is higher than in coal mines, as well as higher than in bus garages.

There are no consistent or convincing relationships between symptom prevalence and surrogate exposures to diesel exhaust among noncoal miners. Although overall symptom prevalences are higher than in nonmining populations, the lack of exposure–response relationships makes it difficult to differentiate between mine dust effects and diesel exhaust effects.

Table III Chronic Health Effects among Diesel- and Non-Diesel-Exposed Controls

Population	No. of subjects	Comparison	Symptoms	Ventilatory function	Exposure	Comment
			% (exp/control)			
Railroad engine house (*Ind. Med. Surg.* **33**, 121, 1964)	210	154 yardmen	Cough (7.6/15.4) Phlegm (9/18.2) Dyspnea (7.6/16.9)	No difference between exp and control	median: winter/summer NO_2: 0.5/0.1 Acrolein: 0.2/0.1 SO_2: 1/0.1	No observed effects of diesel exposure
Egyptian bus garages ($n = 2$) (*Ind. Hlth.* **4**, 1, 1966)	161 (72% smokers)	~workers w/o diesel exposure	Chronic bronchitis (8.7/5.6)		NO_2: 0.3–0.6 SO_2:0.13–0.15 HCHO: 0.5–0.9 Smoke: 2.19–2.46	Throat and nasal irritation and tearing of eyes experienced but no increased respiratory symptoms
			% (exp/control)	% Predicted		
U.S. bus garages ($n = 4$) (*Env. Res.* **44**, 6, 1987)	283	external blue collar control pop.	Cough: 24/10 no trend Phlegm: 26/14 no trend Dyspnea: 8/5 trend	FVC: 104 trend FEV_1: 103 trend	Tenure used as surrogate for exposure	Increased cough and phlegm
			% (exp/control)			
Stevedores (*Appl. Ind. Hyg.* **2**, 133, 1987)	17	11 administrators	Cough (18/27) Chronic bronchitis (29/36) Dyspnea (6/6)	FVC: 111 FEV_1: 98	Exposure low; exposure to both gasoline and diesel exhaust	No apparent effect of diesel exhaust exposure
			% (obs/control)		Average	
Salt mines w/ diesel use ($n = 5$) (*Am. J. Ind. Med.* **4**, 435, 1983)	NS: 50 SM: 150	external blue collar workers; expected no. from internal comparison	Cough: NS 10(23) no trend SM 32(24) Phlegm: NS 16(27) trend SM 33(27) Dyspnea: NS 6(7) no trend SM 8(7)	FVC: 98 no trend FEV_1: 98 no trend	NO_2: 1.33 RP: 0.57	No apparent effect of diesel exhaust exposure
			% (exp/controls)	% Predicted		
Potash mines (Canada) (*J. Occup. Med.* **26**, 209, 1984)	NS: 478 SM: 799	394 community controls (<50% participation); trends (tenure)	Morning cough: NS 3.6/4 no trend SM 27.3/18.5 no trend Chronic bronchitis: NS 8.4/1 no trend SM 27.9/10.7 trend Dyspnea: NS 13.2/6.1 trend SM 29.9/15.2 trend	FVC: 104 105 no trend FEV: 106 104 no trend	No measure of diesel use or exposure; mean tenure = 6 yr.	Possible increased symptoms but confounded by mine dust

Potash mines (U.S.) (n = 6) (*Ann. Occup. Hyg.* **26,** 817, 1982)	NS: 176 SM: 454	exp/blue collar internal E–R (NO_2, diesel use)	Cough: NS 12/6 no trends SM 37/20 Phlegm: NS 21/8 no trends SM 38/23 Dyspnea: NS 4/1 no trends SM 11/9 Productive cough: NS 7/3 SM 28/7	mean (liters) FVC: NS 5.10/4.88 SM 4.93/4.80 no trends SM 3.70/3.69 no trends	Range of mean values in 5 mines was: NO_2 (0.1–3.3) Aldehydes (0.1–4) Resp dust (1.6–7.7)	Possible increased symptoms but confounded by mine dust
Iron mine (*J.O.M.* **12,** 348, 1970)	NS: 60 SM: 60 (in both exposed & controls)	UG vs. S miners matched for age and smoking	Productive cough: NS 7/3 SM 28/7	FVC: 4.25/4.17 FEV_1: 3.45/3.46	More UG workers considered air bad than surface workers; NO_2 (0.5–1.5 ppm) mine dust (3–9 mg/m^3)	High symptom rate may obscure exposure–response high mine dust; may cause high symptom rates
			% Diesel/electric (controls)			
Coal mines (U.S.) (*Ann. Occup. Hyg.* **26,** 799, 1982; *ibid* **32,** 635, 1988)	823 (52% smokers)	UG diesel vs. UG electric; surface diesel vs. electric mines	Cough: UG 23.6/16.5 S 20.1/17.6 no trends	FVC & FEV_1 adj values less than controls for both UG and S workers at diesel mines	Range of mine avg was: NO_2 (.13–.28 ppm) mine dust respirable) (.93–2.7 mg/m^3) very low diesel exposure	No obvious effect attributable to diesel exposure
		Trends (by tenure)	Phlegm UG (25.5/22.8) S (23.6/21.8) no trends Dyspnea UG (9.3/23.8) S (6.3/6.6)	—	—	—
Coal mines (U.S.) *Arch. Env. Hlth.* **39,** 389, 1984)	280	5-yr prospective change from group of 828	—	—	—	No support for the hypothesis that exposure to diesel emissions leads to chronic respiratory effects as measured by respiratory symptoms and ventilatory function
Coal mines (U.K.) (*Brit. J. Ind. Med.* **41,** 214, 1984)	560	Compare w/self; 4-yr prospective	—	—	Range of mine averages were: NO_2 (.02–.08)	NO_2 = surrogate for diesel and shotfire exposure; no relationship between NO_2 and respiratory symptoms or ventilatory function)

Abbreviations: NS, nonsmokers; SM, smokers; FVC and FEV_1, in liters; UG, underground miners; S, surface workers.

Studies of railroad workers in engine houses, bus garage workers, and stevedores do not present as great a problem with confounding exposure from nondiesel sources as the studies of miners. Exposures to NO_2 and respirable particulate around the time of the surveys were <1 ppm for NO_2 and <1 mg/m^3 for respirable particulate. These exposures are somewhat lower than in the noncoal mining environment; NO_2 exposures are similar to coal mine diesel exposures. The stevedore population had symptom prevalences equivalent or less to a comparison group of office workers. The diesel-exposed railroad engine house workers had symptom prevalences about one-half that of nondiesel–exposed railroad workers. Bus garage workers had about twice the prevalence of cough and phlegm of an external comparison group of nonexposed blue collar workers; the prevalence of shortness of breath was equivalent. There were trends for the prevalence of cough and shortness of breath to increase with longer tenure, but only in the bus garage workers where exposure may have been higher in previous years.

These data suggest the possibility of a causative association between diesel exhaust exposure and the symptoms of cough, phlegm, and shortness of breath based on the slight (<twofold) risk ratios in about half the populations studied. The lack of a consistent exposure–response gradient argues against a causal association, but the exposure estimates are imprecise and overall rates so high as to preclude any confident assertions. The slight effect of diesel exhaust on symptoms is difficult to detect in the presence of large effects from smoking cigarettes and mine dust.

2. Baseline Ventilatory Function

The assessment of baseline ventilatory function was done in the same populations as symptom prevalences, and therefore, has the same problems of confounding exposures and imprecise estimates of exposure. The Forced Expiratory Volume in 1 second (FEV_1) is a measure of obstruction in the airways and is commonly reduced in smokers. Forced expiratory flow at 50% of vital capacity (FEF_{50}) measures obstruction of small airways. Forced Vital Capacity (FVC) is a measure of the capacity of the lung, commonly reduced in restrictive conditions such as pneumoconiosis. The more sensitive measures for evaluating causal association with diesel exhaust are FEV_1 and FEF_{50}.

Most of the study groups have baseline FEV_1 and FVC values similar to those of the comparison groups. There is generally no exposure–response relationship. Thus, there is no evidence that baseline ventilatory function is compromised by diesel exposure.

However, any conclusions remain problematic because of the short exposure times, low exposure (particularly among U.S. coal miners), large effects of smoking, and relatively young age of the workers. Chronic decrements in ventilatory function probably require more time to occur than the length of exposure occurring in many of those studies.

III. Mortality (Lung Cancer)

The assessment of the risk of lung cancer due to diesel exhaust exposure is difficult for several reasons. Smoking is the best known and major cause of lung cancer and should be controlled for in evaluating the effects of diesel exhaust, particularly when the risk is low. The time between first exposure and diagnosis of lung cancer is generally considered to be twenty or more years. The latency for smoking is close to forty years. Mortality that increases by latency may constitute a diminishing of the "healthy worker effect" and the increasing effect of smoking. Thus a simple analysis by just latency or just tenure does not adequately control for the interrelated exposure variables of smoking, diesel exhaust, and sometimes asbestos. Exposure to both cigarette smoke and diesel exhaust may be highly correlated, and their separate effects difficult to extricate with few nonsmoking diesel-exposed cases.

Exposure to diesel exhaust has yet to be adequately quantified. Even qualitative exposure estimates are difficult because of inadequate work histories and because of the gradual introduction of diesel engines into the workplace. Dieselization in the railroad industry was complete around 1959. Diesel trucks were in the majority by about 1961. Without adequate work histories (a phenomenon more common than not in these studies), even the length of time exposed to diesel exhaust is problematic. Asbestos (from steam locomotives) is a possible confounding exposure among railroad workers and possibly among truck mechanics and bus garage workers.

Evaluation of the risk of lung cancer due to diesel exhaust should attempt to account for these diverse effects. The effects of short latency and misclassification of exposure are to underestimate risk. The effect of confounding (by smoking and asbestos) is (most likely) to overestimate risk.

Two kinds of study designs are useful in evaluating the risk of lung cancer from diesel exhaust, namely cohort studies and case-control studies. The measure of risk in the cohort studies is the standardized mortality ratio (SMR) and is the ratio of observed deaths in the cohort to expected deaths in the comparison population. The measure of risk in the case-control study is the odds ratio (OR). A value greater than 100 for SMRs and greater than 1 for ORs indicates an increased risk. Although all of the epidemiologic studies have some of the problems noted above, the studies included in the tables have fewer biases and/or have been given more weight by governmental agencies in their assessments of carcinogenic effects of diesel exhaust.

A. Cohort Studies

The occupational groups in cohort studies (Table IV) were London transport workers, Canadian and U.S. railroad workers, heavy construction workers, and Swedish bus garage workers. The SMR for the London transport workers was 101, but follow-up (and therefore latency) was too short to reach any conclusion as to risk.

There were three cohorts of railroad workers. In the cohort of Canadian railroad workers, the relative risk was increased slightly for "possibly" and "probably" exposed workers compared to nonexposed. However, exposure is based on inadequate work histories and is confounded with asbestos exposure. Latency and smoking histories are unknown. No conclusions can be drawn from this study.

The following year a pilot study of American railroad workers was reported. SMRs were less than 100 overall and in all tenure groups. The risk ratio when comparing exposed to nonexposed workers was 1.42. The lack of an exposure–response relationship, potential asbestos exposure among the exposed, incomplete work histories, lack of smoking information, and the modest increase in risk do not permit a conclusion.

Investigators from the same laboratory reported on a larger cohort of American railroad workers who were 40–64 years old in 1959 and with 10–20 years of tenure. Follow-up was until 1980, so latency could be as long as 41 years. Age-adjusted relative risk comparisons were with nonexposed railroad workers. After 1959 any workers in jobs having asbestos exposure were excluded. The group with the most exposure to diesel exhaust (and probably least exposure to asbestos) was the group 40–44 in 1959. The relative risk was 1.57, which was the highest risk of any of the age groups. Since jobs prior to 1959 were included in the analysis, prior asbestos exposure is possible. However, the risk of lung cancer based on the 1959 job was 0.8, and the stability of jobs may be such that asbestos is probably not a major confounding factor. No smoking information was available. A current sampling of railroad jobs, however, showed significant environmental tobacco smoke present; levels did not appear to be higher in diesel-exposed jobs than nondiesel jobs. If the current smoking pattern is consistent with that of the study population, smoking might increase the risk of lung cancer in the nonexposed more than exposed group. Two exposure–response analyses were conducted. An association of tenure with increased hazard of lung cancer was observed when exposure in the year of death and four years before were disregarded; no association was observed when recent exposure was included. The reasons for this inconsistency illustrate the imprecise nature of years of exposure as a surrogate for dose, but other reasons for the inconsistency are otherwise not obvious. The cohort study results are contrary to the results of the case-control study of railroad workers by the same authors and summarized in Table V. The authors found no source of systematic bias in this study, and recognized the potentially important effects of unknown biases or confounding when risks are this low. They concluded that a modest excess risk of lung cancer was observed.

The cohort of heavy construction operators showed no apparent increased risk even among workers with long tenure and long latency. The lack of smoking information would appear to bias results only if smoking is reduced among the exposed workers, which is unlikely. The lack of any observed risk could be due to a low diesel exposure, but the intensity of exposure is not known. Results are inconclusive because of possible confounding from reduced smoking-exposed workers, limited information on intensity of exposure, and the 40% loss of work histories.

The cohort of Swedish bus garage workers (with seventeen cases) is the smallest of the cohort studies. Exposure was ranked qualitatively based on: (1) the number of types of buses and engine-running time, (2) size and ventilation of garages, and (3) job types and work practices for which detailed historical data were available. Intensity was

Table IV Retrospective Cohort Studies of Lung Cancer among Diesel-Exposed Workers

Source	Cohort	SMR (95% C.I.)	Exposure–response	Comments
Rushton *et al.* (1983). *Brit. J. Ind. Med.* **40,** 340.	London bus garage workers ≥1-yr tenure	101(82, 110) obs = 102	Not done	No smoking information; short latency (follow-up 1967–1975); tenure and exposure unknown
Howe *et al.* (1983). *J. Nat. Cancer. Inst.* **70,** 1015.	Canadian RR retirees dying 1965–1977	106(99, 113) obs = 933	No exposure: 1.0 Possibly: 1.20 Probably: 1.35	No smoking; inadequate work history; confounding by asbestos; short latency; intensity of exposure unknown
			Tenure (yrs)	
Schenker *et al.* (1984). *Brit. J. Ind. Med.* **41,** 320.	American RR workers 45–64 ≥10-yr service by 1967	82(59, 111) obs = 43	10–19: 83(51, 122) 20–29: 84(74, 95) ≥30: 91 (80, 103) (internal comparison ratio of exposed/nonexposed = 1.42 (.45, 2.39)	Diesel exposure based on 1959 job; RR of exposed vs. unexposed = 1.42; inadequate exposure histories; no smoking information; potential asbestos exposure; "consistent with a small increase in the risk of respiratory cancer"
Wong *et al.* (1985). *Brit. J. Ind. Med.* **42,** 435.	Heavy construction operators union; ≥1-yr 1964–1978	99(88, 110) obs = 309	<5: 45(21,78) 5–9: 75(48, 108) 10—14: 107(80, 139) 15–19: 102(77, 131) ≥20: 107(91, 124)	Four high-exposure jobs SMR ≤100 high exposure: 94 low exposure: 86 unknown: 67 no history: 119 (no smoking info; tenure may be overestimated; missing work history on ~40%; intensity of exposure unknown)

			Latency (yrs)	
			<10: 66(43, 93) 10–19: 89(72, 109) ≥20: 112(97, 129)	
Garshick *et. al.* (1988). *Am. Rev. Resp. Dis.* **137,** 820.	American RR workers 40–64 yrs old in 1959 with ≥10-yr tenure; follow-up to 1980 study, started work 1939–19[illegible]0	RR = 1.57 (1.19, 2.06) in 40–44 yr olds in 1959; jobs with asbestos excluded	tenure—yrs 1–4: 1.34(1.08, 1.65) 5–9: 1.33(1.12, 1.58) 10–14: 1.33(1.10, 1.60) ≥15: 1.82(1.3, 2.55) (last 4-yr tenure excluded; only workers with diesel exposure)	Internal comparison group: 95% diesels by 1959; excluded jobs ≥1959 with asbestos exposure; no smoking info; 12% missing death cert.; "exposure to diesel exhaust is associated with a modest excess risk of lung cancer"; current diesel exposure = 71–141$\mu g/m^3$, nonexp = 1–69$\mu g/m^3$; no exposure–response if exposure to year of death included
			Intensity × yrs	
Gustavvson *et. al.* (1990). *Scand. J. Work. Env. Hlth.* **16,** 348.	Stockholm bus garage workers ≥6 months 1945–1970; follow-up 1952 to 1986	122(71, 196) obs = 17	0–10: 97(30, 194) 10–30: 152(46, 317) >30: 127(49, 242)	Reference group: from Stockholm adjusted for occupational activity; no apparent exposure–response or increased SMR; no analysis by latency. (See nested case-control study in Table V)

Table V Selected Case-Control Studies of Lung Cancer and Diesel Exposure

Source	Study group	Controls	Exposure	Odds ratio (95% C.I.) (smoking adjusted)		Comments
Hall and Wynder (1984) *Env. Res.* **34,** 77	Male hospital patients 20–60 yr	Patients w/o tobacco-related diseases matched on age, race, hospital, room-status	Usual lifetime occupation	Not exposed	1.0	No information on latency, tenure; much higher proportion of exposed smoked; when stratified by smoking "there was no significant risk associated with an occupation which has postulated diesel exposure"
				Exposed (usual-work)	1.4(0.8, 2.4)	
				Heavy equip. operators & repairmen	1.9(0.6, 5.5)	
				Diesel exp—yrs (adjusted)		
Garshick *et al.* (1987) *Am. Rev. Resp. Dis.* **135,** 1242	U.S. railroad workers ≥10-yr tenure, ≤64 yrs old	RR workers matched on DOB and DOD	Years in diesel-exposed jobs after 1959	0–4	1.0	Adjusted for smoking and asbestos exposure; unadjusted OR by diesel—yrs = 0 1.0; 1–10: 1.32; 11–19: 0.98; ≥20: 1.60
				5–19	1.02(0.72, 1.45)	
				≥20	1.64(1.18, 2.29)	
				Est. at 20 yr	1.41(1.06, 1.88)	
Boffetta *et al.* (1988) *Am. J. Ind. Med.* **14,** 403	Volunteers in ACS study; men 40–79 yrs old	Noncases from same ACS cohort	Self-reported diesel exposure (Y/N)	Overall	1.18(.97, 1.44)	OR adjusted for age, smoking, other occup. exposure
				0	1.0	
				1–15	1.05(.80, 1.39)	
				>15	1.21(.94, 1.56)	
			Diesel truck drivers (controls = nondiesel truck drivers)	0	1.0	Presumably adj. for smoking but not stated in text
				1–15	0.87(.33, 2.25)	
				>15	1.33(.64, 2.75)	
			Diesel truck drivers		1.22(.77, 1.95)	Ever employed as truck driver
			Nondiesel truck drivers		1.19(.74, 1.89)	
			Probable	Truck driver	Self-reported	
Boffetta *et al.* (1990) *Am. J. Med.* **17,** 577	Same as Hall and Wynder (1984) except: 2 controls/case and added exposure data	1–15 yr	0.52(0.15, 1.86)	1.83(.31, 10.7)	0.90(.4,2.0)	No exposure data; OR adjusted for race, yrs education, age, smoking, asbestos
		16–30	0.70(0.34, 1.44)	0.94(.4, 2.2)	1.04(0.4, 2.5)	
		>30	1.49(0.72, 3.11)	1.17(.4, 3.4)	2.39(0.9, 6.6)	
		Total	0.95(.78, 1.16)	0.88(.67, 1.15)	1.21(.7, 2.0)	
			Diesel truck driver	Truck mechanic		
Steenland *et al.* (1990). *AJPH 80,* 670	Teamsters Union ≥ 20 yr, died 1982–1983	Every 6th death excluding lung and bladder cancer, motor vehicle accident; 1–24 yr	1.27(.7, 23.3)	1.69(.6, 4.7)		Overall OR from longest job (from next of kin pension application; diesel or gas truck unknown); E-R adjusted for age, smoking, asbestos; next-of-kin data; no exposure data
		25–34	1.26(.7, 2.2)	1.39(.6, 3.1)		
		≥35	1.89(1.0, 3.4)	1.09(.4, 2.7)		
		overall	1.42(0.9, 2.3)	1.35(0.7, 2.5)		

Abbreviations: DOB, date of birth; DOD, date of death.

ranked on a six-point scale (0, nonexposed; 1, low exposed; 1.5; 2.25; 3.38; 5.06 for highest exposure). Cumulative exposure was Σ(intensity $\times$ time worked), and ranged from 0 to 112 (median = 13). Several groups were used for comparison; the most appropriate was a local reference group of workers. The overall SMR for lung cancer was 122, and there was no apparent evidence of an association with cumulative diesel-exhaust exposure. No apparent association was observed between cumulative asbestos exposure and lung cancer, so confounding from asbestos exposure seems unlikely, although the occurrence of two mesotheliomas in such a small population is suggestive of potentially significant exposure to asbestos. This is not true, however, if exposure occurred prior to working at the bus garage. The potential minimum latency is 16 years, so the lack of a latency analysis should not bias the results. Exposure estimates are qualitative, details are skimpy, and few measures of exposure were available. Diesel fuel had been used in buses since 1945 so the long time for exposure to occur and the presence of adequate historical information should reduce exposure misclassification to a minimum. The cohort study does not suggest an increased risk of lung cancer attributable to diesel exhaust. However, the results of a nested case-control within this cohort are contradictory and suggest an increased risk among the highly exposed (see Section III, B).

B. Case-Control Studies

There are five case-control studies (Table V) that attempt to control for smoking and other confounders, and one nested case-control study within the Swedish bus garage cohort where smoking was not known. The first study used usual lifetime occupation as the criteria for exposure. Both cases and controls were hospital patients and controls were males with non-tobacco related diseases. This study highlights the importance of controlling for smoking in studies of occupational exposure related to lung cancer. The smoking adjusted risk ratio for usual diesel exposed employment was 1.4, down 30% from the unadjusted risk ratio of 2. The risk ratio for heavy equipment operators and repairmen was 1.9, down 40% from the unadjusted risk ratio of 3.5. When adjustments are made for smoking, there was no apparent association between diesel exhaust exposure and risk of lung cancer. This study is inconclusive because of the lack of information on latency and tenure (or other measure of cumulative exposure).

A modification of this ongoing case-control study was subsequently published in 1990. The modifications included more cases, two controls/case, and a more complete occupational history (usual employment plus up to five other jobs that included duration and self-reported exposure to diesel exhaust). After adjustment for smoking and other confounders, the risk ratio for probable exposure was 0.95, down 27% from an unadjusted risk ratio of 1.31; for truck drivers the risk ratio was 0.88, down 33% from 1.31. There were no obvious trends with increasing tenure, although the risk of lung cancer after more than 30 years tenure was modestly increased for those with probable exposure and self-reported exposure. The authors suggest the evidence does not support the hypothesis that occupational exposure to diesel exhaust increases the risk of lung cancer. This study shows that controlling for confounding exposures (particularly smoking) brings the risk to the null value, and the risk may be further confounded by other lifestyle factors as well as recall bias (for self-reported exposure).

An earlier study reported on a case-control analysis drawn from active and retired male U.S. railroad workers with $\geq$10 years railroad service. The most exposed group was workers $\leq$64 years of age at death. Diesel-years exposure was counted beginning in 1959 when 95% of the locomotives were diesels. Job histories were not available prior to 1959; so all employment prior to 1959 was based on the 1959 job. Odds ratios were adjusted for both smoking and asbestos exposure. These are both confounding exposure as more of the cases than controls smoked and had asbestos exposure, and cases also were heavier smokers than controls. The conditional logistic regression (OR) for 20 years diesel exposure was 1.41 and was increased only in the >20-year exposure group. These results are confusing because the unadjusted OR is lower (1.39) than the adjusted OR (1.41), when it should be higher because of the confounding exposures. Second, calculation of unadjusted OR stratified by a different set of diesel-year exposures shows elevated risks in the low-exposure and high-exposure group, and no elevated risk in the intermediate exposure group. The authors suggest possible misclassification of diesel exposure based on job classification, and point out that the effects of confounding can never be eliminated at this modest a risk. The seeming internal contradictions in the data are also troublesome.

The American Cancer Society (ACS) study of male volunteers included three occupation-related questions in the work history questionnaire: current job if employed; last job if retired; usual jobs if longer than the other two. The overall adjusted risk ratio for those exposed to diesel exhaust was 1.18, down 16% from the unadjusted rate of 1.40. When stratified by years exposure, the risk ratio was 1.21 for those with >16 years and 1.05 for those with 1–15 years, a modest exposure–response. The risk (adjusted for smoking) of lung cancer due to any employment in a diesel job was reported for selected occupations: railroad worker = 1.59 (0.9, 2.7); truck driver = 1.24 (0.9, 1.7); heavy equipment operator = 2.60 (1.1, 6.1); and miner = 2.67 (1.6, 4.4). When truck drivers were classified into either diesel or gasoline engines, there was no difference in the risk ratio (1.22 vs. 1.19). There was, however, an apparent exposure–response relationship: risk ratio = 0.87 for 1–15 years diesel exposure and 1.33 for >15 years diesel exposure. This study is suggestive of a possible modest increase in lung cancer after long exposure. However, these results are not conclusive because of a lack of information on latency, incomplete job history, and no quantitative estimates of diesel exhaust exposures.

A study of lung cancer deaths within the Teamsters Union compared the risk of different jobs while controlling for smoking and other confounders. The diesel truck driver (history from next of kin) with ≥35 years employment showed elevated risk (1.89). Truck mechanics on the other hand had the highest risk (1.69) in the shortest tenure group (1–24 years), and the risk decreased as exposure duration increased. The principal limitation (as in all the other studies) is a lack of data regarding level of diesel exhaust exposure for different jobs. Mechanics are likely to have high exposure in the winter and potential confounding from low levels of exposure to asbestos. The authors suggest another limitation is too short a latency. However, the two longest employment groups have 25 or more years latency which should be adequate for some lung carcinogens. If the ≥35 years tenure necessary for elevated risk in truck drivers is valid, then a 20 or more year latency may not be adequate to assess risk. The results for diesel truck drivers and mechanics are contradictory. The truck drivers risk increases and the mechanics risk decreases with tenure. Most of the controls were said to be from dairies. Farmers consistently show a lower incidence of lung cancer; so the control group may not be appropriate and may be spuriously elevating risk. The authors suggest an increased risk of lung cancer for diesel truck drivers, and suggest the results should be considered cautiously because of their inconsistency and limitations.

The nested case-control study of bus garage workers added three more cases (n = 20) of lung cancer compared to the cohort study. Smoking histories were not available, but since the controls were from the same cohort of bus garage workers, smoking habits are unlikely to be different between cases and controls. Smoking is therefore unlikely to bias the results. Again no trend for an increased risk with asbestos exposure was observed. Three different exposure–response analyses were performed. While all showed a trend for the risk to increase as cumulative exposure increased, only the weighted regression analysis was statistically significant. The odds ratios and 95% CIs for the highest exposure group were as follows: crude OR =2.14 (0.68, 6.8); logistic regression OR = 2.63 (0.74, 9.4); weighted regression OR = 2.43 (1.3, 4.5). This logistic regression makes the fewest assumptions and is the more popular model. Both of the regression models are affected by the rating scale used and the small number of data points. The crude OR is not as affected by the qualitative nature of the exposure rating scale but does not take into account all the data as when estimating a trend. The authors conclude that the risk of lung cancer is increased in the high-exposure group.

In 1988 NIOSH issued a Current Intelligence Bulletin recommending that diesel exhaust be regarded as a potential occupational carcinogen. This recommendation was based on carcinogenic and tumorigenic responses in rats and mice exposed to whole diesel exhaust, and not on the epidemiology studies.

In 1989 IARC concluded there was "limited evidence" for the carcinogenicity in humans of diesel engine exhaust. Limited evidence means a positive association has been observed and a causal interpretation is credible, but chance, bias, or confounding could not be ruled out. This conclusion as it relates to lung cancer was based largely on the two cohort studies of railroad workers and a case-control study of U.S. railroad workers. IARC concluded there was sufficient evidence of diesels causing carcinogenicity in experimental animals.

Since these recommendations, there was published a population-based case-control study of ASC volunteers; a hospital-based case-control and the nested case-control studies of Teamsters Union members and bus garage workers. The major occupation with enough persons for analysis was truck drivers. The results are inconsistent as indicated by the summary exposure (duration)–response relationships from these studies and are shown below.

All but the Swedish study have been adjusted for smoking but have no quantitative measure of exposure. The first two studies (above) are small and show either no exposure–response or a suggestion of one. The study of Teamsters shows almost twofold increased risk after 35 or more years. The bus garage workers in the highest exposed group are at about a 2.5-fold increased risk. All of the union members had applied for pension benefits and all had been in the Teamsters Union for at least 20 years. The inverse trend with exposure among truck mechanics confuses the causation issue if one assumes similar exposures of truck drivers and mechanics. During the winter, exposure among mechanics may be higher than among truck drivers. The Teamster's study has the largest number of cases, but exposures are problematic because there are no quantitative measures and no good verification of whether the trucks had gasoline or diesel engines. However, the estimates of diesel exposure are an improvement over those in the population and hospital-based studies. The effect of cessation of diesel exposure after retiring would suggest an underestimate of risk, but no analysis was done relating risk to time after retirement.

The Swedish bus garage workers study provides the most reliable estimates of exposure, and exposure may have been fairly high. Personal time-weighted average exposures in the early 1980s in some instances approached 0.7 mg/m^3 of diesel exhaust. The authors estimated that exposure for 20 years to total dust levels experienced in the 1980s might double the lung cancer risk after 20 years of exposure. The confusing aspects of this study are the somewhat contradictory results between the cohort and case-control studies, although statistically there is no difference because of the very wide confidence intervals.

Because of the inconsistent and somewhat confusing data from human studies and the emphasis on animal data by NIOSH and IARC, experimental exposures of rats will be briefly reviewed.

C. Animal Studies and Plausibility of Lung Cancer

Table VI summarizes the results of lifetime inhalation studies of rats to experimental exposures of diesel exhaust. Rats develop increased tumors at exposures as low as 1 mg/m^3 in one experiment and no increased tumors when exposures were as high as 3.5 mg/m^3 in another. Tumor incidence is exposure-related. The lowest concentration at which tumors are significantly increased is 2.2 mg/m^3.

Exposure to concentrations as low as 0.35 mg/m^3 also produces impaired clearance, and therefore a progressive increase in the particle mass retained in the lung. Chronic exposure to low concentrations (0.35 mg/m^3) did not lead to similar accumulations of diesel particulate, nor to tumors. Exposure to high concentrations of normally non-tumor-producing particles (coal dust, titanium dioxide) also produces lung tumors (presumably a tumorigenic effect of particles per se). These findings raise the possibility that the tumor response in the rat may be a nonspecific response to the accumulation of particles per se from the impaired clearance.

Finally, because of the smaller lung plus higher ventilation rate, the rat deposits about three times the amount of diesel particulate on the unciliated portion of the airways as humans. Thus when the exposure of rats is adjusted to approximate human exposure and deposition, the lowest dose at which tumors develop is about 13 mg/m^3.

In short, the rat is a suspect model on which to make extrapolations to human risk. The actual exposure at which tumors occur appears to be at least an order of magnitude higher than occupational exposures, and perhaps two orders of magnitude higher than outside air exposure, and may be a nonspecific response to particle overload.

Population-based			Hospital-based			Teamsters union (nested)		
Exposure (yrs)	*n* cases	Risk ratio	Exposure (yrs)	*n* cases	Risk ratio	Exposure (yrs)	*n* cases	Risk ratio
1–15	6	0.87	1–15	4	1.83	1–24	48	1.27
>15	12	1.33	16–30	12	0.94	25–34	72	1.26
			>30	7	1.17	≥35	56	1.89

Table VI Lung Tumor Incidence in Lifetime Exposure Studies of Rats Exposed to Diesel Exhaust

Source	Species–exposure	n	NO_2 (ppm)	RP (mg/m^3)	RP adjusted	Malignant tumors	
						n	%
Takemoto *et al.* (1986)	Female rats 4h/d;4d/wk	15	2–4	2–4	4.5	0	—
		12	0	0	0	0	—
Ishinishi *et al.* (1986)	Male and female rats; 16h/d;6d/wk	124	1.41	2(LD)	12	2	1.6
		123	0.70	1(LD)	6	5	4.1
		125	0.26	0.4 (LD)	2.4	0	0
		123	0.08	0.1 (LD)	0.6	2	1.6
		123	0	0(LD)	0	3	2.4
		124	3	3.72(HD)	22.3	8	6.5[a]
		123	1.68	1.84(HD)	19.8	4	3.3
		125	1.02	0.96(HD)	5.76	0	0
		123	0.46	0.46(HD)	2.76	1	0.8
		123	0	0(HD)	0	1	0.8
Iwai *et al.* (1986)	Female rats 8h/d/;7d/wk	19	1.8	4.9	14.7	5	26.3[a]
		22	0	0	0	0	0
Lewis *et al.* (1986)	Male and female rats; 7h/d/;5d/wk	183	1.5	1.95	5.85	0	0
		180	0	0	0	3	1.7
Mauderly *et al.* (1986)	Male and female rats; 7h/d;5d/wk	143	0.7	7	21	14	9.8[a]
		131	0.3	3.5	10.5	1	0.8
		138	0.1	0.35	1.05	1	0.7
		141	0	0	0	2	1.4
Stöber *et al.* (1986)	Female rats 19h/d;5d/wk	96	1.5	4	28.5	1	1.0
		96	0	0	0	0	0
Brightwell *et al.* (1986)	Male and female rats; 16h/d;5d/wk	143	2	6.6	39.6	55	38.5[a]
		144	0.6	2.2	13.2	14	9.7[a]
		143	0.2	0.7	4.2	1	0.7
		260	0	0	0	3	1.2

Abbreviations: RP, respirable particulate from diesels; LD, light duty diesel engine; HD, heavy duty diesel engine; adjusted RP = (hr/d/8) × 3RP; all references are in Ishinishi *et al.*, 1986.
[a] $p<.05$ greater than controls.

Tables VII and VIII summarize the observations for classifying diesel exhaust as a carcinogen based on a weight of the evidence approach. The bioassay observations clearly show lung carcinogenicity occurrences are not the cause of death and are accompanied by toxicity and disrupted homeostasis. The bioassay evidence is considered sufficient to show carcinogenicity, but is weak because the effects were seen only at very high exposures and even then did not contribute to mortality. The corroborative evidence is considered to weigh against the relevancy of the rat response under conditions of exposure that humans experience in the workplace.

IV. Summary

Diesel exhaust causes acute symptoms and irritation in the workplace at concentration levels of $NO_2 < 1$ ppm. Eye irritation is the most sensitive indicator of exposure to diesel exhaust.

Ventilatory function (FEV, FEF_{50}) is significantly reduced over the workshift when NO_2 levels approach 5 ppm, but the magnitude of the reductions are minimal below exposures of 5 ppm NO_2.

Diesel-exposed workers generally show an increased prevalence of chronic symptoms of cough, phlegm, and dyspnea. However, many of these

Table VII Observations from Animal Studies Relevant to Ascertained Human Lung Cancer Risk from Diesel Exhaust[a]

Supportive	Non-supportive
(+) Inhalation is the relevant route of administration.	• Different route of administration.
(+) Activity in several species (rats, mice).	• Activity in one species only (not in hamsters).
(+) Lung cancers in both rats and mice.	• No site correspondence across species.
• Activity at only one site (one report of splenic malignant lymphoma with frequent complications of leukemia in rats).	(−) Tumors in lung only.
(+) Activity in tissues and analogous to humans (lung).	• Activity in animal tissues not found in humans.
• No evidence of cellullar toxicity at target site (not in hamsters).	(−) Toxicity in lung of rats and mice including decreased clearance and severe inflammation.
• Tumors appear early.	(+) Most tumors in rats appear after two years.
• Tumors progress rapidly.	(−) Benign tumors only at low exposures, benign plus malignant at highest exposure.
• Tumors usually fatal.	(−) Tumors were not the primary cause of death.
• Activity at several exposures.	(−) Malignant tumors significantly increased at highest exposure levels only. Overall tumor rate dose-related.
(+) Background lung tumor rates low.	• Background rates high at target site.

[a] (+) Factor supports predictivity of the animal repsonse for human risk; (−) factor detracts from predictivity of the animal response for human risk; • generic description of the factor in the weight-of-evidence approach to evaluating carcinogenicity.

Table VIII Corroborative Nonbioassay Observations for the Evaluation of the Relevancy of Animal Bioassays in Ascertaining Risk of Lung Cancer in Humans from Exposure to Diesel Exhaust[a]

Supportive	Non-supportive
(+) Diesel particulates are mutagenic.	• Nongenotoxic.
(+) Metabolites of diesel particulate extracts bind covalently with DNA.	• Not reactive with DNA.
(+) PAH activity expressed via alkylating metabolite (bay region dihydrodiol epoxides) and metabolism probably similar in humans and animals.	• Metabolic pathways differ.
• Mechanisms of action known to occur in humans. (+) release of organics from particles and interaction with lung macromolecules (i.e., DNA adducts); degree of release dependent on residence time.	• Mechanisms not relevant to humans. (−) possible nonspecific particle effects (mitogenesis) in rats but not humans at current exposures (−) long residence and clearance time (due to overload) occurs in rats, but presumably not humans (−) prolonged retention and slow release in high doses to animals (not humans) is (probably) correlated to tumor incidence (−) induction of enzymes to increase metabolism of PAH (and therefore tumorigenicity) requires high concentrations (7 mg/m^3)
• No evidence of disruption of homeostasis.	(−) Evidence that homeostasis is disrupted (reduced body weight, inc BUN, dec urinary protein, impaired liver function possible, hypoxia due to restrictive and obstructive airway disease, inc heart wt and BP).

[a] (+) Factor supports the idea of a relevant human risk; (−) factor operates differently in humans and is therefore not predictive of human cancer risk; • generic description of the factor.

studies are confounded by mine dust, and there is a general lack of an exposure–response gradient.

Baseline ventilatory function is often higher than predicted in the diesel-exposed workers, and there is an inconsistent association with exposure. There is little evidence supporting a causal association between diesel exhaust exposure and reduced ventilatory function.

The epidemiological studies of diesel exposed workers show a modest and inconsistent risk of lung cancer. The modest risk observed could be due to a number of factors including chance, confounding, bias, lack of adequate latency, too low an exposure, or misclassification of exposure. Animal experiments show a risk of lung cancer at adjusted exposures 10 or more times greater than workers are likely to have been exposed. These high exposures impair clearance, producing particulate accumulation, which may produce tumors by a mechanism unlikely to function in humans and unrelated to the mutagenic substances in diesel exhaust. Thus diesel exhaust is considered a "possible human carcinogen" using a weight-of-the-evidence methodology of classification. That is, there is inadequate-to-limited evidence of carcinogenicity in humans in occupational exposure settings, and sufficient evidence of carcinogenicity in animals but with inadequate information regarding relevance of the animal response to humans.

Related Article: Dust Particles: Occupational Considerations.

Bibliography

Ashby, J. T. *et al.* (1990). A scheme for classifying carcinogens. *Reg. Toxicol. Pharmacol.* **12,** 270–95.

Higgins, I. T. T. (1984). Air pollution and lung cancer: Diesel exhaust, coal combustion. *Prev. Med.* **13,** 207–218.

IARC Monographs on the Evaluation of Carcinogenic Risks to Humans, Volume 46. Diesel and Gasoline Engine Exhaust and Some Nitroarenes, 1989.

Ishinishi, N., Koizumi, A., McClellan, R. O., and Stöber, W. eds. (1986). "Carcinogenic and Mutagenic Effects of Diesel Engine Exhaust," Elsevier, Amsterdam.

McClellan, R. O. (1986). Health effects of diesel exhaust—A case study in risk assessment. *Am. Ind. Hyg. Assoc. J.* **47,** 1–13.

McClellan, R. O. (1987). Health effects of exposure to diesel exhaust particulates. *Ann. Rev. Pharmacol. Toxicol.* **27,** 279–300.

NIOSH Current Intelligence Bulletin No. 50 (1988). Carcinogenic Effects of Exposure to Diesel Exhaust, US DHHS/PHS/CDC/NIOSH/DSDTT.

Steenland, K. (1986). Lung cancer and diesel exhaust: A review. *Am. J. Ind. Med.* **10,** 177–189.

Dioxin, Health Effects

Han K. Kang
U.S. Department of Veterans Affairs

Glossary

Agent Orange A mixture of phenoxyherbicides that was used in military operations in Vietnam from 1965 to 1971. It contains two active ingredients, 2,4-D and 2,4,5-T.

Chloracne Acne-like skin condition known to result from exposure to a group of compounds in which chlorine atoms are bound to an aromatic hydrocarbons.

NHL Non-Hodgkin's lymphoma; a heterogeneous group of malignant neoplasms of the lymphoreticular system.

STS Soft tissue sarcoma; extraskeletal, nonepithelial sarcomas that arises mainly in muscle, fat, or fibrous connective tissue.

TCDD 2,3,7,8-Tetrachlorodibenzo-*p*-dioxin.

***DIOXIN,*[1] *or 2,3,7,8,-tetrachlorodibenzo-p-dioxin (TCDD), is often referred to as one of the most toxic manmade chemicals known. Many animal toxicity studies support this claim although there is a wide variation in biological response to the chemical among different species. In humans, however, scientific consensus on the magnitude and extent of health hazards from dioxin exposure has yet to be reached. In the meantime public concern over dioxin exposure continues to mount as evidenced by the publicity surrounding an industrial accident in Sevaso, Italy; the well-known incidents that developed at Times Beach, Missouri, and Love Canal, New York; and the continuing debate concerning the military use of herbicides such as Agent Orange in Vietnam.*

I. Introduction

TCDD is not commercially manufactured, but rather is formed as an unwanted by-product during the manufacture of certain chemicals such as the herbicide 2,4,5-trichlorophenoxyacetic acid (2,4,5-T), the fungicide pentachlorophenol, and bactericide hexachlorophene. In recent years it has been suggested that TCDD can be produced in certain combustion processes of carbonaceous fuels. Trace amounts of chlorinated dioxins have been found in samples collected from incinerator stacks and fossil-fueled power plants. There are 75 possible dioxin isomers that differ only in the number and position of attached chlorine atoms. TCDD is considered to be the most toxic among these isomers; it is lipophilic and extremely stable and persistent in the environment.

The primary sources of dioxin in the environment are the industrial manufacture of chlorophenol or its derivatives, and the chemical disposal sites containing the wastes from these industries. Small amounts of chlorinated dioxin contamination have been found in fish and wildlife in areas around chlorophenol manufacturing industries and chemical waste disposal sites. Analyses of adipose tissue from the general population of industrialized countries have indicated the presence of a number of 2,3,7,8-substituted dioxins and dibenzofurans at part per trillion (ppt) levels; TCDD levels in human adipose tissue seldom exceed 20 ppt. Among individuals with known histories of heavy dioxin exposure, the levels of TCDD are found to be much higher [e.g., chemical workers, 246 ppt (42 to 750 ppt); U.S. Air Force Ranch Hand Veterans, 115 ppt (17 to 423 ppt); Sevaso Italy residents, 10,400 ppt (828-27,800 ppt)].

[1] The opinions expressed in the article are those of the author and do not necessarily reflect official positions or policies of the Department of Veterans Affairs.

II. Short-Term Health Effects

Much of current knowledge of the effects of dioxins on humans comes from studies of industrial accidents. Symptoms of acute exposure to chemicals that contain dioxin are fatigue, nausea and vomiting, headache, irritation to the eyes, skin, and respiratory tract, and weight loss. Chloracne has been the most frequently reported effect. Chloracne as its name suggests, is similar in appearance to the common forms of acne. Irrespective of the route of exposure, lesions generally develop on the face within a few weeks of exposure. In mild cases, chloracne cysts may be confined to the area around the eyes extending along the temples to the ears. In more severe cases it may also appear in many places on the body, for example, the arms, chest, back, abdomen, thighs, and genitalia. The chloracne frequently disappears completely but some lesions may persist for many years in extreme cases. Overexposure to other chlorinated aromatic hydrocarbons such as chloronaphthalenes, chlorobiphenyls, chloroazobenzenes, chlorodibenzofurans, and chlorodibenzodioxins are also known to cause an acnegenic response.

Other responses less consistently reported are porphyria cutanea tarda, hepatomegaly, hyperpigmentation, and peripheral neuropathy. At the cellular level, enzyme induction, and altered liver function and lipid metabolism have been reported. A summary of various symptoms and clinical findings reported for individuals exposed to dioxin is presented in Table I.

Table I Various Symptoms and Clinical Findings Reported for Industrial Workers Exposed to TCDD

Type of exposure	Effects
1949 Explosion at Monsanto plant, Nitro; 117 workers	Skin, eye, respiratory tract irritation, headache, nausea, dizziness, chloracne, muscle pain, fatigue, nervousness, irritability, hepatomegaly, peripheral neuropathy, increased serum lipid levels.
1968 Explosion in UK; 41 workers	Chloracne, elevated triglyceride and GGT, elevated urinary D-glucaric acid.
1953 Explosion at BSF plant in FRG; 53 workers	Chloracne, polyneuritis, sensory impairment, fatigue, drowsiness.
1976 Accident at ICMESA plant in Italy	Chloracne, elevated GGT and ALT, elevated urinary D-glucaric acid.

Abbreviations: GGT, gamma glutamyltransferase; ALT, alanine aminotransferase.

III. Long-Term Health Effects

Because of the evidence presented in some of the animal studies and the several cases of soft tissue sarcoma reported among workers involved in the manufacture of phenoxyherbicides, studies of long-term health effects in humans have been focused primarily on cancer and adverse reproductive outcome. In animals, TCDD is a potent carcinogen and causes fetal death, birth defects, or spontaneous abortion after maternal exposure. To date, however, paternal exposure to TCDD failed to produce any unfavorable pregnancy responses in laboratory animals.

In this review, the selected epidemiologic studies of populations potentially exposed to phenoxyherbicides or dioxins will be summarized with respect to cancer and adverse reproductive outcome. All of these studies are either studies of specific diseases (e.g., case-control studies of soft tissue sarcoma) or studies of specific populations (e.g., chemical workers, Vietnam veterans).

In a case-control study, persons with a given medical condition are compared to persons without the condition for certain risk factors. For the study of rare diseases, the case-control design is usually the only practical or efficient approach to identifying risk factors. One major limitation of this design is that the exposure information is often collected from interviews with study subjects or their next-of-kin, contributing to inaccurate recall of exposure by study subjects and observational bias by interviewers.

In a study of a specific population, a cohort of individuals is selected based on common characteristics in the past and disease outcome up to some defined point in time is followed and evaluated. The cohort study is therefore suitable for studying multiple health outcomes. The evidence of exposure is usually documented in the study group minimizing misclassification of exposure status. On the other hand, for many cohort mortality studies that rely on death certificates for outcome, misclassification of causes of death can be an important limitation.

A. Studies of Cancer

1. Soft Tissue Sarcomas

Soft tissue sarcomas (STS) are defined as extraskeletal, nonepithelial sarcomas that arise chiefly in muscle, fat, or fibrous connective tissue. They are relatively rare tumors accounting for about 1% of all cancer and about 2% of all cancer deaths. Between 5000 and 7500 new cases are diagnosed each year in the U.S.

The first report of a relationship between STS and exposure to phenoxyherbicide appeared in northern Sweden in 1977. Hardell described seven STS patients who had been exposed to phenoxyherbicides. Prompted by this observation, he and his colleagues conducted two case-control studies in which STS cases were compared to individuals without STS for history of phenoxyherbicide exposure. The first study consisted of 52 male adult STS cases admitted to the University Hospital in Umea, (northern) Sweden, during 1970–77. Four controls were selected for each STS case who were matched for age, sex, place of residence, and vital status (alive or dead). Occupational and other exposure histories were obtained from the subjects or their next-of-kin through mail questionnaires supplemented by telephone interviews as necessary. The estimated relative risk for phenoxyherbicides was 5.3 (95% confidence interval, 2.4–11.5). The authors reported that phenoxyherbicides were used to control unwanted hardwood in Swedish forests and for other agricultural purposes since the early 1950s.

The second study was carried out in southern Sweden where herbicides were mainly used for agricultural purposes. A total of 110 STS cases and 220 controls were included in the study. Assessment of exposure was also carried out by mail questionnaires supplemented when necessary by telephone interviews. The association was as strong as the first study. The estimated relative risk was 6.8 (95% confidence interval 2.6–17.3).

The first study was strongly criticized for including 7 STS cases that had been reported earlier. At least 6 of 13 STS cases classified as exposed to phenoxyacetic acid were cases already reported in 1977. Other criticisms included the possible recall bias as a result of the prevailing publicity concerning the carcinogenicity of phenoxyherbicides, observation bias by interviewers, and lack of control for confounding variables. The second study was also criticized for similar reasons, except for the fact that no hypothesis-generating STS cases were included in the second study.

In an attempt to address these criticisms, Hardell conducted another case-control study: a study of colon cancer and phenoxyherbicide exposure. His reported rationale was that if there was a recall bias by STS cases, one might expect similar recall bias by colon cancer cases. He reported that no significant association was found between exposure and colon cancer (odd ratio, 1.3; 95% confidence interval, 0.6–2.8). In addition, analysis of the 52 STS cases using the 154 colon cancer patients as controls continued to show a strong positive association. The estimated relative risk was 5.5 with 95% confidence interval of 2.2–13.8. He concluded, therefore, that recall bias could not have accounted for the earlier findings.

Many case-control studies have been conducted since the publication of these two studies. None of the subsequent studies provides convincing evidence that there is an association between the risk of STS and exposure to phenoxyherbicides, between STS and those occupations with potential phenoxyherbicide exposure, or between STS and military service in Vietnam. As presented in Table II, all studies published subsequent to the Swedish reports failed to support the strong association between STS and herbicide exposure. Several possible reasons for the conflicting results among the Swedish studies and the subsequent studies have been offered by several investigators. Smith *et al.* (1984) suggested that the difference in the level of dioxin contamination in Swedish and New Zealand phenoxyherbicides could explain the difference in findings. But, in their later paper in 1986 they dismissed the possibility because it was found that the levels of dioxin contamination in new Zealand herbicides were similar to the Swedish samples. Wood *et al.* (1987) postulated several possibilities: differences in the intensity or duration of exposure and variations in the distribution of specific inherited, life-style and/or environmental factors. It is unknown whether in fact there is a substantial difference between the Swedish and the Washington State study subjects with respect to any of these factors. Because little is known about the etiology of STS, even if there is a measurable difference in population characteristics between the two groups, its effects on the outcome is only speculative.

Kang *et al.* (1986, 1987) offered two alternate explanations for the lack of a positive association between STS and Vietnam service or surrogate mea-

Table II Selected Case-Control Studies of Soft Tissue Sarcomas

Reference	Risk factor	OR (95% C.I.)
Hardell and Sandstrom (1979)[1]	PH	5.3 (2.4–11.5)
Eriksson *et al.* (1981)[2]	PH	6.8 (2.6–17.3)
Hoar *et al.* (1986)[2]	PH	0.9 (0.5–1.6)
Woods *et al.* (1987)[4]	PH	0.89 (0.4–1.9)
Smith *et al.* (1986)[5]	PH	1.1 (0.7–1.8)
Balarajan and Acheson (1984)[6]	Farming	1.15 (0.83–1.59)
Smith *et al.* (1982)[7]	Agriculture & forestry	1.03 (0.9–1.81)
Greenwald *et al.* (1984)[8]	v	0.53 (0.21–1.31)
Kang *et al.* (1986)[9]	v	0.83 (0.63–1.09)
Kang *et al.* (1987)[10]	v	0.82 (0.55–1.21)
CDC (1990)[11]	v	0.74 (0.39–1.41)

Abbreviations: OR (95% CI), odds ratio (95% confidence interval); PH, phenoxyherbicides; v, military service in Vietnam.

[a] 90% confidence interval.

1. *Br. J. Cancer* **39,** 711; 2. *Br. J. Ind. Med.* **38,** 27; 3. *JAMA* **256,** 1141; 4. *JNCI* **78,** 899; 5. *Chemosphere* **15,** 1795; 6. *J. Epidemiol. Comm. Health* **38,** 113; 7. *Comm. Health Studies* **6,** 114; 8. *JNCI* **73,** 1107; 9. *JOM* **28,** 1215; 10. *JNCI* **79,** 693; 11. *Arch. Int. Med.* **150,** 2485.

sures of Agent Orange exposure. First, they stated that observation time might have been too short. Approximately 80% of STS cases in their studies were observed less than 10 years after the last use of Agent Orange in Vietnam. Second, Vietnam veterans in general might not have been exposed to sufficient amounts of Agent Orange. Notwithstanding the perceptions of Vietnam veterans, there is increasing evidence that U.S. troops as a group were not heavily exposed to dioxin or Agent Orange in Vietnam. The CDC (1988) reported that the serum levels of 2,3,7,8-TCDD for 646 ground combat troops who served in heavily sprayed areas of Vietnam were not significantly different from the levels of 97 veterans who did not serve in Vietnam. In a recent study, Kang *et al.* (1991) reported that the mean dioxin level in adipose tissue of 36 Vietnam veterans was similar to that of 79 non-Vietnam veterans or 80 civilian controls. Furthermore, none of the surrogate measures of Agent Orange exposure was associated with the dioxin levels in the adipose tissue specimens of Vietnam veterans.

2. Non-Hodgkin's Lymphoma (NHL)

Non-Hodgkin's lymphoma is a heterogeneous group of malignant neoplasms of the lymphoreticular system. NHL is relatively uncommon in the U.S. with about 15,000 new cases estimated each year.

The origin of scientific investigations for the relationship between NHL and phenoxyherbicides closely parallels that of the soft tissue saracoma investigation. As was the case for STS, Hardell (1979) first reported a number of men with malignant lymphoma who had previous exposure to phenoxyacids or chlorophenols. Hardell *et al.* (1981) then followed up their clinical observations with a case-control study of malignant lymphoma. The study consisted of 169 malignant lymphoma cases (60 Hodgkin's disease, 4 unclassifiable lymphomas, and 105 NHL) and 338 controls. The study used the same design that they used in the soft tissue sarcoma studies. They reported that exposure to phenoxyacid may be a causative factor in malignant lymphoma. The estimated relative risk was 4.8 (95% C.I., 2.9–8.1). The relative risk was not separated for HD and NHL. Unlike the STS studies, the NHL studies have presented mixed results (see Table III).

Hoar *et al.* (1986) conducted a population based case-control study in Kansas to determine whether agricultural use of herbicides contributed to the risk of soft tissue sarcoma, Hodgkin's disease, and NHL. For NHL cases, a random sample of 200 men was selected from 297 men diagnosed with NHL from 1978 through 1981. Of those, selected,

Table III Selected Case-Control Studies of Non-Hodgkin's Lymphoma

Reference	Risk factor	OR (95% C.I.)
Hardell *et al.* (1981)[1]	PH	4.8 (2.9–8.1)[a]
Hoar *et al.* (1986)[2]	PH	2.2 (1.2–4.1)
Pearce *et al.* (1986)[3]	PH	1.4 (0.7–2.5)[b]
Woods *et al.* (1987)[4]	PH	1.7 (1.04–2.8)
CDC (1990)[5]	v	1.5 (1.0–2.3)
Dalager *et al.* (1991)[6]	v	0.9 (0.64–1.28)

Abbreviations: OR (95% C.I.), odds ratio (95% confidence interval); PH, phenoxyherbicide; v, military service in Vietnam.

[a] Odds ratio for developing malignant lymphomas, which include non-Hodgkin's lymphoma and Hodgkins's disease.

[b] 90% confidence interval.

1. *Br. J. Cancer* **43,** 169; 2. *JAMA* **256,** 1141; 3. *Br. J. Ind. Med.* **43,** 75; 4. *JNCI* **78,** 899; 5. *Arch. Int. Med.* **150,** 2473; 6. *JOM* **33,** 774.

172 NHL cases were histologically confirmed by a panel of three pathologists. The controls were selected from the general population of Kansas. Information on the exposure to chemicals was obtained through telephone interviews with subjects or their next-of-kin. An attempt was made to verify self-reported agricultural chemical exposure by way of seeking collaborative evidence from suppliers. Men exposed to phenoxyherbicides had a 2.2-fold increased risk of NHL (95% C.I., 1.2–4.1). In this study 2,4-dichlorophenoxyacetic acid (2,4-D) was virtually synonymous with phenoxyherbicides. Only 3 cases and 18 controls used 2,4,5-T. 2,4-D has been the most commonly used herbicide in Kansas and it is not likely to be contaminated with dioxin, especially 2,3,7,8,-TCDD. 2,4-D and 2,4,5-T were the two main components of the herbicide Agent Orange sprayed in South Vietnam by the U.S. Air Force. No association was found by this study between the agricultural use of herbicides and development of STS or HD. The authors believed their findings on NHL were consistent with the Swedish study but did not speculate on the reasons for the inconsistency concerning STS and HD.

Pearce *et al.* (1986), after studying 83 NHL cases and 168 cancer controls and 228 general population controls in New Zealand, reported no statistically significant association between exposure to phenoxyherbicide and NHL. The odds ratio was 1.4 (90% confidence limits 0.7–2.5, $p = 0.26$). Exposure data were collected from telephone interviews.

Woods *et al.* (1987), on the other hand, reported a positive association between NHL and farming (O.R. = 1.33, 95% C.I. = 1.03–1.7) and forestry herbicide application (O.R. = 4.8, 95% C.I. = 1.2–19.4). For those potentially exposed to phenoxyherbicides in any occupation for 15 years or more during the period prior to 15 years before cancer diagnosis, the odds ratio was 1.71 (95% C.I., 1.04–2.8). This was a population based case-control study conducted in western Washington State to determine phenoxyherbicide exposure and the risk of STS and NHL. A total of 128 STS cases, 576 NHL cases, and 694 controls without cancer were interviewed personally for occupational histories and other data. As was the finding by Hoar *et al.* (1986) the study failed to find a positive association between STS and phenoxyherbicide exposure.

The CDC initiated several case-control studies of cancers in order to examine their associations with military service in Vietnam. The CDC analyzed 1157 men with pathologically confirmed NHL and 1776 controls for their veteran status and other military characteristics. 99 NHL cases and 133 controls had served in Vietnam whereas 94 cases and 203 controls served in the military at anytime from 1964–1972 but not in Vietnam. This resulted in an odds ratio of 1.52 (95% C.I., 1.0–2.32). The excess NHL risk was statistically significant only among sea-based blue-water Navy veterans. The risk was not associated with surrogate measures of Agent Orange exposure such as dates of service, type of unit, and military region. The authors ruled out the possibility of Agent Orange being responsible for the NHL risk. But, they didn't suggest any other risk factor that may be associated with Vietnam service.

Dalager *et al.* (1991) conducted a hospital-based case-control study to examine the relationship between NHL and military service in Vietnam. The cases consisted of 201 Vietnam era veteran patients who were treated in one of 172 VA hospitals from 1968–1985 with a diagnosis of NHL. 358 Vietnam era veteran patients with diagnoses other than malignant lymphoma served as a comparison group. Military service information was obtained from reviews of military personnel records. In contrast to the CDC findings, service in Vietnam did not increase the risk of NHL. The branch adjusted odds ratio was 1.03 (95% C.I., 0.7–1.5). A total of 100 NHL cases and 187 controls had served in Vietnam, whereas 101 cases and 171 controls had never served in Vietnam. Surrogate measures of potential Agent Orange exposure such as service in a specific military branch, in a certain region within Vietnam, or in a combat role as determined by military occupational specialty were not associated with any increased risk of NHL.

3. Other Cancers

A few case-control studies have been published that examine the relationship between other cancers and potential phenoxyherbicide exposure. Hardell *et al.* (1984) studied 102 liver cancer cases for phenoxyherbicide exposure and concluded the relative risk was 1.7, which was not statistically significant. The CDC (1990) reported no significant association between primary liver cancer and Vietnam service.

Hardell *et al.* (1982) also reported on nasal and nasopharyngeal cancer and their relationship to phenoxyherbicide exposure. There were 44 nasal cancer cases, 27 nasopharyngeal cancer cases, and 541 controls in the study. They reported a twofold increased risk of these cancers associated with phenoxyherbicide exposure, but it was not of statistical significance. As described earlier, Hardell (1979) and Hardell and Sandstrom (1981) suggested the possibility of a Hodgkin's disease and phenoxyherbicide relationship. Hardell and Bengtsson (1983) later analyzed 60 HD cases and found 32% of cases and 10% of controls were exposed to phenoxyherbicide or chlorophenols. However, Hoar *et al.* (1986) failed to find the relationship between the risk of HD and farming, or phenoxyherbicide use. The CDC reported that Vietnam veterans are not at higher risk for HD compared with non-Vietnam veterans. Of the 310 HD cases, 28 had served in Vietnam whereas of the 1766 controls, 133 had served in Vietnam. This resulted in the odds ratio of 1.23 (95% C.I., 0.65–2.36). The risk did not vary by branch, calendar year of service, location in Vietnam, rank, or self-reported assessment of exposure to Agent Orange.

B. Studies of Specific Populations at Risk

The risk of cancer associated with potential phenoxyherbicide exposure has been studied among groups of industrial workers involved in the manufacture or use of chemicals, among professional herbicide applicators, among residents of specific areas, and among Vietnam veterans. Because of the rarity of the cancers of interest, almost all of these studies lack adequate statistical power to detect even a moderate increase in risk. Furthermore, because mortality outcomes were evaluated primarily based on death certificate data, the probable misclassification of underlying causes of death is of concern.

1. Industrial Workers

Workers involved in the manufacture of phenoxyherbicides had a significant potential for exposure to dioxin. A recent study showed that the mean serum TCDD levels in a sample of 253 workers from two chemical plants that manufactured the herbicide was 233 pg per gram of lipid (ppt), whereas a mean level of 7 ppt was found among 79 unexposed workers. All the workers' exposure to TCDD had stopped 15 to 37 years earlier. As shown in Table IV, results from the studies of industrial cohorts are mostly nonpositive and inconsistent.

One of the most comprehensive mortality studies on industrial workers who had significant potential for TCDD exposure was conducted by the National Institute for Occupational Safety and Health (Fingerhut, 1991). They identified 5172 workers from all U.S. chemical companies that had made TCDD contaminated products between 1942 and 1984. Occupational exposure to substances contaminated with TCDD was confirmed by measuring serum TCDD levels of workers from two chemical plants. Vital status was determined as of December 31, 1987. Cause of death was obtained from death certificates and classified by two nosologists according to the rules of the International Classification of Diseases. The mortality experience of these workers was evaluated by life-table analysis. Standardized mortality ratios (SMRs) were calculated by dividing the observed deaths by the expected deaths after stratifi-

Table IV Selected Mortality Studies of Individuals with Potential Phenoxyherbicide Exposure

Reference study population	Results
Thiess *et al.* (1982)[1]	Increase in stomach cancer (3 observed vs. 0.7 expected, $p = 0.03$).
Richimaki *et al.* (1982)[2]	No significant increase in any cancer; no death from STS or lymphoma.
Zack and Suskind (1980)[3] Monsanto workers	All causes of death (32 observed vs. 46.6 expected); all cancer (9 observed vs. 9.04 expected); no stomach or liver cancer.
Zach and Gaffey (1983)[4] Mosanto workers	Increased risk in bladder cancer (9 observed vs. 0.91 expected).
Lynge (1985)[5] Danish workers	Increased risk in soft tissue sarcoma (5 observed vs. 1.84 expected), lung cancer (11 observed vs. 5.33 expected).
Wiklund *et al.* (1989)[6] Swedish pesticide applicators	Significant decrease in cancers of the liver, pancreas, lung, and kidney as well as in overall cancer. (SIR, 0.86; 95% C.I., 0.79–0.93).
Bond *et al.* (1988)[7] Dow Chemical workers	No significant increase in any cancer.
Fingerhut *et al.* (1991)[8] U.S. Dixon Registry	Increased risk in all cancer (114 observed vs. 78 expected) and soft tissue sarcoma (3 observed vs. 0.3 expected).

1. *Am. J. Med.* **3,** 179; 2. *Scan. J. Work Environ. Health* **8,** 37; 3. *JOM* **22,** 11; 4. *Environ. Sci. Res.* **26,** 575; 5. *Br. J. Cancer* **52,** 259; 6. *Br. J. Ind. Med.* **46,** 809; 7. *Br. J. Ind. Med.* **45,** 98; 8. *NEJM* **324,** 212.

cation to adjust for age, race, and calendar year of death. The U.S. general population was used as the comparison group. This study presumably included workers from Monsanto and Dow plants, which were studied earlier. Among the overall cohort, the authors found no significant increase in mortality from cancers of prior interest in this cohort: buccal and pharynx, 5 observed vs. 7 expected; stomach, 10 observed vs. 9.7 expected; liver and biliary, 6 observed vs. 5.2 expected; trachea, bronchus and lung, 89 observed vs. 80.1 expected; Hodgkin's disease, 3 observed vs. 2.5 expected; non-Hodgkin's lymphoma, 10 observed vs. 7.3 expected; soft tissue sarcoma, 4 observed vs. 1.2 expected. However, among the 1520 workers who were exposed for 1 year or more and who had at least 20 years of latency, mortality was significantly increased for soft tissue sarcoma (3 observed vs. 0.3 expected, SMR = 922, 95% C.I. = 190–2695) and for cancers of the respiratory system (43 observed vs. 30.2 expected, SMR = 142, 95% C.I. = 103–192).

As the authors cautioned, conclusions about increases in the risk of STS are limited by the small number of STS deaths (four from the overall group and three from the subgroup) and misclassification of underlying cause of death recorded on death certificates. Two of the four STS deaths and one of the three STS deaths recorded as such on death certificates were found to be deaths due to other cancers upon review of tissue specimens. Misclassification of STS based on death certificates has been occasionally observed. It was reported earlier that only about 56% of soft tissue sarcoma deaths coded on the death certificates were confirmed by hospital records.

Although information on smoking was not available, the authors believed that the risk of respiratory cancer was probably not due to confounding by smoking for the following reasons. First, other deaths related to smoking were no higher than expected. Second, the excess risk was greater among the subgroup with presumed higher exposure. Third, confounding by smoking was unlikely to account for an excess risk of more than 10 to 20%. Fourth, actual adjustment based on the smoking prevalence of surviving workers at two plants did not substantially change the results.

It is also possible that the excess mortality risk reported by the study may have been due to exposure to chemicals other than TCDD. First, the median years of exposure to TCDD contaminated processes for all workers in the study was much less than one year, whereas the median total employment years for the workers was approximately 10 years. In other words, TCDD was one of many chemicals to which these workers were exposed, and exposure was for a relatively brief period. Second, there was no significant linear trend of increasing mortality with increasing duration of exposure to products contaminated with TCDD. Third, mortality from all cancers and from respiratory cancers were also significantly higher among those with over 20 years of employment and with over 20 years of latency. Because of equally plausible alternate explanations for the observed excess risk among the subgroup of workers, one should consider this study as only suggestive.

2. Herbicide Applicators

Riikimaki *et al.* (1982) studied a cohort of 1926 Finnish men who had sprayed 2,4-D and 2,4,5-T for brushwood control during 1955–1971. Analysis of historic samples of herbicides showed that TCDD content ranged between 0.1 and 0.9 mg/kg. The duration of exposure to the herbicide had been brief (no more than 2 months) for most of the workers. It was noted, however, that no special precautions were taken to avoid inhalation of the aerosols or skin contact. As of 1980, the total number of deaths (n = 144) and deaths from cancer (n = 26) were significantly less than expected based on Finnish national death rates. No death from soft tissue sarcoma or malignant lymphoma was observed. The authors cautioned that because of the limitations of the small cohort, low exposure, and brief follow-up period, one should not consider the study definitely negative.

Blair *et al.* (1983) evaluated the mortality experience of a cohort of 3827 men licensed to spray pesticides in Florida. Organochlorine and organophosphate chemicals were the principal chemicals used; some phenoxyherbicides (2,4-D,2,4,5-T) may have been handled by these applicators as well. Among those licensed for 20 or more years, the risk of death from lung cancer was significantly elevated (8 observed vs. 2.8 expected). Mortality from lung cancer was greater among those who were first licensed before age 40 than among those after age 40. Information on smoking was not available.

Wiklund *et al.* (1989) analyzed a cohort of 20,245 licensed Swedish pesticide applicators for the risk of cancer. Their cancer experience was ascertained by the Swedish Cancer Register as of December 31, 1982, which provided a mean follow-up time of 12.2 years. There was a statistically significant decrease in the incidence of all cancers (liver, pancreatic, lung, and kidney) (558 observed vs. 649.8 expected). No excess risk of STS, NHL, or HD was observed.

3. Vietnam Veterans

From 1965 to 1970, the U.S. Air Force sprayed more than 11 million gallons of Agent Orange in South Vietnam for defoliation and crop destruction purposes. Approximately 2 million U.S. military personnel served 1-year tours during the same period. Concerns persist that as a result of exposure to Agent Orange and other chemicals in Vietnam, Vietnam veterans may be at increased risk for soft tissue sarcoma and other cancers. Results of mortality studies of Vietnam veterans with respect to cancer are not consistent with each other (Table V).

Data from Wisconsin and Massachusetts indicate that veterans who served in Vietnam may have an increased risk of soft tissue sarcoma. No such findings were reported by the CDC, U.S. Air Force, or Breslin *et al.*, however U.S. Marine Corps Vietnam veterans were seen to have a statistically significant excess of non-Hodgkin's lymphoma when compared with Marines who did not serve in Vietnam (PMR = 2.10; 95% C.I. = 1.17–3.79). NHL has been associated with exposure to phenoxyherbicides, arsenicals, dapsone, and certain viruses.

Table V Selected Mortality Studies of Vietnam Veterans

Reference study population	Results
Fett *et al.* (1987)[1] Australian Vietnam veterans	Increased overall mortality rate, no difference in death rate from neoplasms.
CDC (1987)[2] U.S. Vietnam veterans	Increased risk in traumatic deaths (RR = 1.25).
Breslin *et al.* (1988)[3] U.S. Vietnam veterans	Increased risk in traumatic deaths, increased mortality from lung cancer (PMR = 1.58) and NHL (PMR = 2.1) among marines.
Kogan and Clapp (1988)[4] Massachusetts Vietnam veterans	Increased risk in STS deaths (PMR = 8.8).
U.S. Air Force (1991)[5] U.S. Vietnam veterans	No significant differences regarding accidental, cancer, and circulatory deaths.

Abbreviations: RR, relative risk; PMR, proportionate mortality ratio.

1. *Am. J. Epidemiol.* **125,** 869; 2. *JAMA* **257,** 790; 3. *JOM* **30,** 412; 4. *Int. J. Epidemiol.* **17,** 39; 5. *JAMA* **164,** 1832.

The Vietnam veterans had a potential for exposure to all of these agents. Agent Blue was an organic arsenical compound used in Vietnam and dapsone was used as an antimalarial drug by some of the troops.

Lung cancer was also significantly elevated (PMR = 1.58; 95% C.I. = 1.09–2.29) among Marines who served in Vietnam relative to Marines who did not serve in Vietnam. The veterans from New York with service in Vietnam also had relatively more cases of lung cancer than other Vietnam veterans, but the excess was not statistically significant.

Among the Vietnam veterans who were most likely have had a heavy exposure to Agent Orange were 1261 "Ranch Hand" veterans. These were Air Force veterans who were responsible for spraying Agent Orange in Vietnam. There have been no significant differences regarding accidental, malignant neoplasm, or circulatory deaths observed in this group compared with 19101 other Air Force veteran controls. Twelve cancer deaths were observed among Ranch Hands, whereas 14.6 deaths were expected. One Ranch Hand and one control subject have died of STS. No Ranch Hands but 10 control subjects have died of malignant lymphoma.

It is not obvious whether the inconsistent findings are the result of the relatively small number of deaths analyzed in each study or whether they suggest an underlying difference in the mortality experience of the different Vietnam veteran populations. Except for the Breslin *et al.* study, which analyzed the patterns of mortality among 24,235 Vietnam veteran deaths and 26,685 non-Vietnam veteran deaths, the number of deaths analyzed in these studies was relatively small resulting in inadequate statistical power to detect differences.

C. Adverse Reproductive Outcomes

In laboratory animals, maternal exposure to dioxin at very low doses (0.1 mg/kg/day) can cause an excess of birth defects and fetal loss. But paternal exposure of animals to dioxin, even at a high dose level has failed to show any adverse reproductive outcomes.

A series of ecological (correlation) studies was undertaken, which suggested potential teratogenic and other adverse reproductive effects of dioxin on humans. Studies of this type evaluate reproductive outcomes of residents living in an area presumed to be contaminated with dioxin. Outcome data are often based on self-reported events. Adjustment for confounding variables is seldom feasible. In addition to these inherent problems of studies of this type, many studies conducted by Vietnamese investigators have not been published in peer-reviewed scientific journals making them inaccessible for critical review. Although many of these studies generate public interest in potential reproductive hazards of dioxin, results of these studies should be viewed only as hypothesis generating.

Published studies of reproductive outcome among well-defined cohorts with potential exposure to dioxin have been largely negative (Table VI). Smith *et al.* (1982) evaluated professional New Zealand 2,4,5-T sprayers and a comparison group of agricultural contractors for pregnancy outcome of their spouses. Information on the number of births, congenital defects, and miscarriages was obtained by mail questionnaire. The response rates of the applicators and agricultural contractors were 89% and 83%, respectively. There was no association between potential exposure to dioxin and either congenital defects (RR = 1.19; 90% C.I. = 0.58–2.45) or miscarriage (RR = 0.89; 90% C.I. = 0.61–1.30).

Townsend *et al.* (1982) studied reproductive events of wives of Dow Chemical plant employees exposed to chlorinated dioxins. A control group consisted of wives of men who had no dioxin exposure and whose employment entry dates were similar to those of men in the exposed group. Exposure classification was based on company employment records and historic area monitoring data. Reproductive outcome data were obtained from personal interviews with wives of employees. There were no statistically significant differences in estimates of risk for adverse pregnancy outcome among wives in different exposure categories. Nine variables were considered and adjusted for in the analyses: mother's age, birth control method, complication during labor and delivery, medical conditions during pregnancy, medication or treatment, alcohol consumption, gravidity, high risk occupation, smoking status. The main limitations of this study were that outcome data were obtained from interviews and that the refusal rates for interviews were somewhat high (34% of study group and 35% of control group).

Suskind and Hertzberg (1984) conducted a study of Monsanto chemical company workers involved in an industrial accident in 1948. As a part of the study, 204 "clearly exposed" workers and 163 control workers were asked about their wive's pregnancy outcomes. Among 189 exposed workers who reported reproductive events, there were 655 pregnancies, 69 miscarriages, 11 still births, 17 neonatal deaths, 18 children with birth defects. These rates were not significantly different from what were observed in the control group.

Stockbauer *et al.* (1988) reported on the reproductive outcome of mothers who resided in areas contaminated with waste oil and dioxin. Maximum TCDD soil levels at the area varied from 241 to 2,200 ppb. 410 births were identified from the "exposed" mothers. 820 births were selected from "un-

Table VI Reproductive Outcomes among People with Potential Exposure to Dioxin

Reference	Study description	Results
Smith *et al.* (1982)[1]	548 male 2,4,5-T sprayers; 441 male agricultural workers; mail questionnaires.	No significant difference in congenital defects (RR = 1.19, 90% C.I. = 0.58–2.45) and miscarriage (RR = 0.89, 90% C.I. = 0.61–1.30).
Townsend *et al.* (1982)[2]	370 wives of employees with potential dioxin exposure; 345 wives of control male workers, personal interview.	No association between adverse pregnancy outcomes (spontaneous abortions, still-births, infant deaths, birth defects) and paternal dioxin exposure.
Suskind and Hertzberg (1984)[3]	189 exposed male chemical workers; 155 male controls; self-reported reproductive outcomes data from male workers.	No association between adverse pregnancy (miscarriage, still births, neonatal deaths, birth defects) and paternal dioxin exposure.
Stockbayer *et al.* (1988)[4]	410 births from mothers residing in TCDD contaminated area; 820 control births; based on medical records.	No significant difference in birth outcomes.

Abbreviations: RR, relative risk; CI, confidence interval.
1. *Arch. Environ. Health* **37,** 197; 2. *Am. J. Epidemiol.* **115,** 695; 3. *JAMA* **251,** 2372; 4. *Am. J. Epidemiol.* **128,** 410.

exposed" mothers as a comparison group after matching for maternal age and race, year of birth, hospital of birth, and plurality. Data on reproductive outcome were obtained from hospital records and vital statistics records. The authors reported increased risk ratios for infant, fetal, and perinatal death, low birth weight, and several subcategories of birth defects among the exposed group. But none of these increased risk ratios was statistically significant. The authors concluded that substantial adverse reproductive effects in humans were unlikely for the reproductive outcome studies at the environmental dioxin concentrations measured in contaminated Missouri residential sites.

The possibility of fathering a child with birth defects as a result of exposure to Agent Orange has been one of the major concern of Vietnam veterans. There have been numerous instances when Vietnam veterans reported fathering children with congenital malformation. As much as the veterans' concern is genuine, there is little or no evidence to support an association between military service in Vietnam and the risk of fathering children with birth defects (Table VII).

Studies based on self-reported reproductive outcome tend to associate Vietnam service and adverse reproductive outcome. Stellman *et al.* (1988) reported an increased risk of miscarriage among spouses of Vietnam veterans who were members of the American Legion. Similarly, in the CDC Vietnam Experience Study, Vietnam veterans reported more reproductive and child health outcomes than did non-Vietnam veterans. However, the rates of total birth defects recorded on hospital birth records were similar to the two groups of children. This seems consistent with Vietnam veterans' observed tendency to over report adverse events concerning their own health status. U.S. Air Force Ranch Hands also reported more children with birth defects than controls. Hospital records of all children born to Ranch Hands and their controls are now being collected to verify the results obtained from self-reporting. Several case-control studies have failed to show an association between paternal military service in Vietnam and birth defects, still births, neonatal deaths, or spontaneous abortions. Studies in the Atlanta area, Australia, and the Boston area all failed to substantiate these veterans' concerns.

Table VII Paternal Military Service in Vietnam and Adverse Pregnancy Outcome

Reference	Study description	Results
Erickson *et al.* (1984)[1]	7133 babies born with serious structural congenital malformation, 4246 control babies matched to the cases by race, year of birth, hospital of birth	No association with father's Vietnam service or Agent Orange exposure likelihood.
Donovan *et al.* (1984)[2]	8517 babies born with congenital abnormalities; 8,517 control babies matched to cases by hospital, birth year, age of mother and hospital payment category.	OR = 1.02, 95% C.I. = 0.078–1.32 self reported birth defect data:
Aschengrau and Monson (1989)[3]	201 women having spontaneous abortion, 1,119 women having full-term live-born infants.	OR = 0.88, 95% C.I. = 0.42–1.86
CDC Vietnam Experience Study (1988)[4]	7924 Vietnam veterans, 7364 non-Vietnam veterans; self-reported reproductive outcomes, hospital record review.	OR = 1.3 (95% C.I. = 1.2–1.4) hospital birth records: OR = 1.0 (95% = C.I. = 0.8–1.4).
Stellman *et al.* (1988)	6810 American Legionnaires; Mail questionnaire.	Increased miscarriages: Vietnam veterans' spouses, 7.6%; non-Vietnam veterans' spouses, 5.5%.
Aschengrau and Monson (1990)[6]	857 babies with congenital abnormalities, 61 still birth cases and 48 neonatal death cases, 998 normal controls.	No statistically significant association with all congenital anomalies, major malformation, minor malformation, stillbirths, or neonatal deaths.
U.S. Air Force (1988)[7]	1208 Ranch Hands, 1668 controls, post-Southeast Asia service live birth outcomes.	Ranch Hand, 76 children with birth defects; controls 44 children with birth defects based on self-reported data.

1. *JAMA* **252,** 903; 2. *Med. J. Aust.* **140,** 394; 3. *JOM* **31,** 618; 4. *JAMA* **259,** 2715; 5. *Environ. Res.* **47,** 150; 6. *Am. J. Public Health* **80,** 1218; 7. USAF, School of Aerospace Medicine, Brooks Air Force Base.

IV. Conclusion

The risk of soft tissue sarcoma related to phenoxyherbicide exposure has been overstated. Almost all studies published subsequent to the initial Swedish studies have failed to substantiate the 5- to 6-fold excess risk of STS reported by the Swedish studies. No significant positive association has been observed in any study conducted in countries outside Sweden. Cohort mortality study results are too inconsistent and the number of deaths evaluated has been too small to draw firm conclusions. The weight of the evidence does not support an etiologic role of phenoxyherbicides (or dioxin) for the risk of soft tissue sarcoma in humans.

Evidence for an unusual risk of non-Hodgkin's lymphoma among people exposed to phenoxyherbicides (or dioxin) in much stronger. The Swedish study findings have been replicated in two other case-control studies. Two Vietnam veteran studies also reported a significant elevation of NHL risk. An association which persists in several studies conducted under different circumstances, with other study populations and with different study methods, is less likely to be spurious in nature than isolated observations from single studies. For other cancer risks in humans, the results are too inconsistent and too equivocal to suggest any relationship to phenoxyherbicide exposure. More studies, especially well designed case-control studies, are necessary to definitively answer this question.

The combined weight of evidence from industrial workers and residents living in contaminated areas suggests that exposure to dioxin does not appear to constitute substantial reproductive hazards. No significant association was found between adverse pregnancy outcome and paternal dioxin exposure. Compared to non-Vietnam veterans, Vietnam veterans are not at higher risk of fathering children with birth defects. It should be noted, however, almost all of the studies concern paternal exposure to dioxin. Because maternal exposure to dioxin can cause adverse reproductive events in animals, but not paternal exposure, these conclusions remain tenuous.

Bibliography

Aschengrau, A., and Monson, R. R. (1989). Paternal military service in Vietnam and risk of spontaneous abortion. *J. Occup. Med.* **31,** 618–623.

Aschengrau, A., and Monson, R. R. (1990). Paternal military service in Vietnam and the risk of late adverse pregnancy outcomes. *AJPH* **80,** 1218–1224.

Bond, G. G., Bonder, K. M., and Cook, R. R. (1989). Phenoxy herbicides and cancer; insufficient epidemiologic evidence for a causal relationship. *Fund. Appl. Toxicol.* **12,** 172–188.

Boyle, C. A., Decoufle, P., and O'Brien, T. R. (1989). Long-term health consequences of military service in Vietnam. *Epidemiol. Rev.* **11,** 1–27.

The Centers for Disease Control. (1990). The association of selected cancers with service in the U.S. military in Vietnam. I. Non-Hodgkin's lymphoma. *Arch. Intern. Med.* **150,** 2473–2483.

The Centers for Disease Control. (1990). The association of selected cancers with service in the U.S. military in Vietnam. II. Soft-tissue and other sarcomas. *Arch. Intern. Med.* **150,** 2485–2492.

Dalager, N. A., Kang, H. K., and Burt, V. L. (1991). Non-Hodgkin's lymphoma among Vietnam veterans. *J. Occup. Med.* **33,** 774–779.

Fingerhut, M. A., Halperin, W. E., and Marlow, D. A. (1991). Cancer mortality in workers exposed to 2,3,7,8-tetrachlorodibenzo-*p*-dioxin. *NEJM* **324,** 212–218.

Lilienfeld, D. E., and Gallo, M. A. (1989). 2,4-D, 2,4,5-T, and 2,3,7,8-TCDD; an overview. *Epidemiol. Rev.* **11,** 28–58.

Michalek, J. E., Wolfe, W. H., and Miner, J. C. (1990). Health status of Air Force veterans occupationally exposed to herbicides in Vietnam. II. Mortality. *JAMA* **264,** 1832–1836.

Stellman, S. D., Stellman, J. M., and Sommer, J. F. (1988). Health and reproductive outcomes among American Legionnaires in relation to combat and herbicide exposure in Vietnam. *Env. Res.* **47,** 150–174.

Dust Particles: Occupational Considerations

William Jones, Jane Y. C. Ma, and Vincent Castranova
National Institute for Occupational Safety and Health

Joseph K. H. Ma
West Virginia University

Glossary

Aerosol Solid or liquid particles suspended in a gas.

Birefringence Difference in refractive indices for a substance.

Organic dusts Particles originating from plant, animal, or microbiologic sources.

Pneumoconiosis Reaction of the lungs to inhalation of dust.

Sign of elongation Elongation of a substance in relation to refractive indices.

OCCUPATIONAL RESPIRATORY DISEASES resulting from the inhalation of dusts generated in working environments have been recognized for centuries. The effects of asbestos, silica, coal, and cotton dusts, for instance, are well documented, although mechanisms for disease initiation and progression are not yet fully understood. At present, despite the availability of improved air sampling techniques and dust control methods, disease persists.

I. Health Effects

Pneumoconiosis is the reaction of the lungs to inspired dust. The pathogenicity of a dust is determined by the effective or retained dose within the lungs and the biological reactivity of the dust with lung tissue. The amount of dust retained in the lungs is the difference between the amount inhaled and deposited in the airways and the quantity cleared from the lung by dissolution, phagocytosis, or the action of the mucociliary escalator (i.e., the movement of airway mucous up the respiratory tree and out of the lungs by the action of airway ciliary cells). Important characteristics of exposure include: the airborne concentration as well as the physical and chemical properties of the dust (aerodynamic diameter, shape, surface area, surface charge, surface reactive groups, contaminants adhered to the surface, etc.). These characteristics determine the acute and chronic reactions of the lungs to dust exposure.

The text that follows discusses "fibrous", "nonfibrous," and "organic" dusts as distinct entities. This distinction is not absolute and was made for ease of presentation. For example, there are obviously fibrous components of many organic dusts. In addition, dusts described in the nonfibrous section may contain particles having a ratio of length to width of ≥ 3 (a morphological distinction made by some groups for defining certain types of fibers).

A. Fibrous Dusts

1. Asbestos

Asbestos refers to a group of naturally occurring fibrous minerals. These silicate minerals can be divided mineralogically into serpentines and amphiboles. Chrysotile is in the serpentine group. It is a sheet silicate mineral rolled into a hollow tube structure. Amphiboles include crocidolite, actinolite, tremolite, amosite, and anthophyllite. These minerals consist of double chains of linked silicon oxygen tetrahedra with trace metals found between these chains.

Asbestos is resistant to thermal and chemical degradation. Therefore, it has many commercial uses as an insulator or construction material. Occupations exposed to asbestos include: asbestos mining, asbestos milling, auto repair, shipbuilding, construction, and demolition.

Asbestosis is the pneumoconiosis caused by inhalation of asbestos. It is characterized by diffuse interstitial fibrosis and thickening of the visceral pleura (i.e., the external lining of the lungs). Symptoms associated with asbestosis include labored breathing upon exertion and cough. Pulmonary function tests indicate restrictive lung disease with a decrease in vital capacity and gas exchange. X-rays often show small irregular opacities in the lower regions of the lungs and pleural thickening. Gross examination of diseased lungs shows the presence of plaques on the lung surface. These pleural plaques are a cellular, fibrous material composed of collagen arranged in a basket weave pattern. In severe cases, this fibrous material can surround the lungs and restrict their full expansion. Histologically, asbestosis is characterized by fibrosis (i.e., increased collagen in the interstitial spaces). Fibrotic lesions most often occur in the lower lobes of the lungs near the bifurcation of the respiratory bronchioles where asbestos fibers lodge. These fibrous lesions surround asbestos bodies (i.e., asbestos fibers coated with an iron–protein material deposited by alveolar macrophages). In severe disease, this diffuse fibrosis may advance to conglomerate fibrosis where alveolar spaces are obliterated.

In addition to pulmonary fibrosis, inhalation exposure is also associated with mesothelioma and lung cancer. Mesothelioma is a malignant tumor of mesothelial cells lining the thoracic or more rarely the abdominal cavity. It is very rare in individuals who have not been exposed to asbestos. Those exposed to crocidolite or amosite appear to have the greatest incidence of mesothelioma.

Asbestos exposure can also lead to lung cancer. These tumors are bronchial carcinomas. Since tumors are associated with the fibrotic lesions of asbestosis, they are most common in the periphery and lower lobes of the lungs. In asbestos workers, smoking and asbestos are multiplicative risk factors for lung cancer (i.e., the risk of lung cancer in a nonsmoking asbestos worker is $\approx$5-fold, that of a nonexposed smoker is $\approx$10-fold, and that of a smoking asbestos worker is $\approx$60-fold higher than normal).

Stanton and co-workers proposed that the pathogenic properties of asbestos are due to its dimensions with long thin fibers being more toxic. The needle-like shape of asbestos acts to lodge it in airways and prevents phagocytosis by alveolar macrophages. Indeed, it is believed that reactive products secreted from macrophages during "frustrated phagocytosis" cause lung damage and induce collagen secretion by interstitial fibroblasts. The durability and needle-like shape of asbestos are also thought to play a role in the migration of asbestos fibers through lung tissue to the thoracic (pleural) cavity where plaque formation and mesothelioma occur.

2. Man-Made Mineral Fibers

Due to the pathogenicity of asbestos, man-made mineral fibers have been introduced that are noncrystalline and differ chemically from asbestos. These fibers are made from glass, natural rock, or slag and include fibrous glass, rock wool, and ceramic fibers.

These fibers are resistant to high temperature and chemicals and exhibit a high tensile strength. They are, therefore, used as asbestos substitutes for thermal and acoustical insulation and reinforcement in plastics. Workers exposed are those involved in the production, cutting, and packing of these fibers; those in the construction industry involved in installing or removing insulation; and those involved in refitting refractory linings of furnaces in the iron and steel industry.

Since the use of man-made mineral fibers is relatively recent, little data exists concerning the fibrogenicity of these fibers in humans. However, some epidemiological evidence has been obtained linking fibrous glass with a moderate elevation in lung cancer risk.

Cellular and animal studies are more numerous. Cellular studies indicate that man-made mineral fibers can be cytotoxic to lung cells. Injection of these fibers into the abdominal cavity results in tumor formation (mesothelioma). Intratracheal instillation of this material has been associated with fibrosis but not with lung tumors. To date neither fibrosis nor lung cancer has been found in animals exposed to fibrous glass, rock wool, or slag wool by inhalation. However, recent inhalation experiments indicate that ceramic fibers induce fibrosis and mesothelioma in hamsters and fibrosis in rats.

Some man-made mineral fibers would be judged as pathogenic according to the "Stanton Hypothesis" which states that long thin particles are associated with disease. However, fibrous glass may be less durable than asbestos, and the ability of fibrous glass to be bent may lessen its ability to migrate into the pleural space. At present, the question of the toxicity of man-made mineral fibers is an area of extensive research.

B. Nonfibrous Dusts

1. Silica

Mineralogically, silica can exist in either a crystalline or amorphous form. Quartz, cristobalite, and tridymite are silicon dioxide (SiO_2) crystals. Silica is used in glass manufacturing, pottery making, and sandblasting. In addition, workers in foundries, mining, tunneling, quarrying, and road working can be exposed to crystalline silica.

Silicosis is a fibrotic disease produced by inhalation of silica-containing dusts. High exposures to crystalline silica can result in acute silicosis. Acute silicosis develops rapidly (1–3 yr) and is characterized by labored breathing (dyspnea), fatigue, cough, and weight loss. Histologically, acute silicosis is characterized by alveolar proteinosis (i.e., edema, increased number of macrophages in the alveolar walls, and positive PAS staining of protein-like material in the alveolar airspaces). Acute silicosis may also be associated with fibrosis. However, these lesions are diffuse rather than nodular in appearance and are located in the middle and lower lobes of the lung. As the disease progresses, dramatic decreases in pulmonary function, lung volumes, and gas exchange are noted.

Chronic silicosis occurs 20–40 yr after initial exposure to crystalline silica. The disease progresses in degree from simple silicosis to complicated silicosis (progressive massive fibrosis). In simple silicosis, few symptoms are noted and pulmonary function is relatively normal. However, x-rays reveal small rounded opacities in the upper lobes of the lungs. These opacities are indicative of silicotic nodules which are made up of collagen arranged in a whirled or circular pattern. As the disease becomes more severe, it is classified as "complicated silicosis." At this time, shortness of breath upon exercise is common. Pulmonary function tests reveal restrictive lung diseases (decreased vital capacity) and x-rays reveal opacities that are more numerous, larger, and may be associated with contraction of lung tissue in the upper lobes resulting in expanded airspaces (emphysema).

Several theories have been proposed to explain the importance of surface properties to the cytotoxicity of silica. One theory suggests that SiOH groups on the crystal surface form hydrogen bonds with lung tissue resulting in damage. A second theory suggests that SiO^- groups are critical to cytotoxicity. A third proposal states that when silica is cleaved, radicals ($\dot{S}i$ or $Si\dot{O}$) are formed on the surface. These radicals can oxidize and damage cell membranes (lipid peroxidation). In addition to the lung damage caused directly by silica, this dust can cause inflammation (i.e., migration of phagocytes into the airspaces) and activate these cells to release reactive products (superoxide anion, hydrogen peroxide, hydroxyl radical, and lysosomal enzymes) that can further damage lung tissue. Silica also stimulates alveolar macrophages to produce cytokines such as interleukin 1, platelet activating factor, tumor necrosis factor, fibronectin, macrophage-derived growth factor, and platelet-derived growth factor. Some of the mediators are inflammatory while others are fibrogenic (i.e., causing proliferation of fibroblasts and increased collagen production by these cells).

Available animal and epidemiological data suggest that amorphous silica is far less toxic than crystalline silica.

2. Coal

It has long been known that coal miners can be stricken by a chronic lung disease commonly called "black lung disease." Disability is associated with chronic airway obstruction and thus labored exhalation. The more precise term to describe the lungs' reaction to inhalation of coal mine dust is "coal workers' pneumoconiosis."

Coal workers' pneumoconiosis (CWP) is categorized according to disease severity as simple CWP and progressive massive fibrosis. In simple CWP, there may be few symptoms and only small declines in forced expiratory lung volumes. However, x-rays show small rounded opacities in the upper lobes of the lungs. Histologically these opacities are associated with coal macules (i.e., black areas of the lung 1–4 mm in diameter). These macules are located near the respiratory bronchioles and are concentrated in the upper lung lobes. Adjacent to these macules may be areas of focal emphysema and fibrosis containing reticulin but little collagen. In progressive massive fibrosis, breathing may be labored especially upon expiration. Pulmonary function indicates increased airway resistance and obstructive lung disease characterized by decreased forced expiratory volumes. A decrease in gas exchange may also be present. X-rays demonstrate multiple irregular opacities which are larger and more numerous than with simple CWP. Histologically, those opacities are associated with large coal macules (2 cm in diameter). Lesions of collagen arranged in bundles rather than in a circular pattern are present and emphysema often occurs around these lesions.

CWP is characterized radiologically by the size and number of opacities as category 0, 1, 2, and 3. There is a direct relationship between cumulative coal mine dust exposure (i.e., the product of dust level and exposure duration) and the probability of progressing to category 2 or greater. There is also an increased risk of progressive massive fibrosis in miners with category 2 and 3 disease. In addition, the risk of CWP and progressive massive fibrosis at any given dust level increases with coal rank (i.e., disease risk is greater in anthracite miners than in bituminous miners).

There is no association between inhalation of coal mine dust and lung cancer. However, stomach cancer is elevated in coal miners. It has been proposed that coal dust is cleared from the lungs by the mucociliary escalator, swallowed, and nitrosated under the acidic conditions in the stomach. Nitrosated coal dust has been shown to be highly mutagenic and genotoxic.

In animal models, inhalation of coal dust results in activation of alveolar macrophages and increased release of reactive products that may cause lung damage. Such activated macrophages have been obtained from lungs of coal miners with CWP. In addition, these cells secrete fibrogenic cytokines such as fibronectin and alveolar macrophage-derived growth factor. Such results correlate with the presence of macrophages in coal macules characteristic of CWP. Later studies suggest that surface radicals can be generated when coal is crushed. These radicals could result in direct damage to lung tissue.

3. Diesel Particulate

During operation, diesel engines produce particulate material that is contaminated with a variety of organic chemicals adsorbed to the particulate surface. Diesels are extensively used in trucks, buses, and railroad engines. They are now being introduced within mines to power equipment. Workers who service these engines and some miners are exposed to this particulate.

Little information concerning disease in workers is available. However, cellular studies indicate that the organic chemicals adsorbed onto diesel particulate is highly mutagenic. Exposure of animals to diesel particulate has resulted in tumor formation at doses above 4 mg/m^3. Diesel particulate has been shown to decrease the phagocytotic potential of alveolar macrophages, to decrease particle clearance by the lungs, and to decrease the ability of the lungs to prevent the spread of pulmonary viral infection. Epidemiology studies on workers exposed to diesel particulate are less conclusive. However, some studies show increased risk of lung cancer while others report a mild elevation in reports of respiratory infections.

4. Other Silicates

Silicate minerals such as kaolin, beryl, kyanite, mica, talc, and vermiculite are generally considered to possess relatively low toxicity and are classified as nuisance dusts by OSHA. However, toxicity may occur when these dusts are contaminated with silica or asbestos or when pulmonary deposition of dust is high enough to compromise particulate clearance.

C. Organic Dusts

1. Cotton Dust

Exposure to cotton, flax, and hemp dust is associated with a set of respiratory symptoms referred to as byssinosis. These symptoms are characterized by chest tightness and respiratory tract irritation on the first day of the work week following a break of two or more days. This is often referred to as the "Monday response." Symptoms begin 2–3 hr after exposure and increase in severity throughout the workday. Therefore, the time course is distinct from asthma which would occur immediately upon exposure. Workers in the cotton or linen textile industry are exposed to cotton or flax dust. Disease is more common in the early stages of the textile process where the cotton fiber is less pure and contaminated with other plant parts such as the bract, leaf, and stem. Occupational exposure to hemp dust is associated with byssinosis in workers involved in producing rope from these plant fibers.

Schilling introduced a system for categorizing the severity of byssinotic symptoms. With grade 0 byssinosis, there are no symptoms. Grade 1/2 displays occasional chest tightness and respiratory tract irritation on Monday, while in grade 1 byssinosis these symptoms occur regularly on Monday. In grade 2 byssinosis symptoms occur on other work days as well as Mondays.

In acute byssinosis, chest tightness and a fall in forced expiratory volumes occurs over the working shift on Mondays. With chronic byssinosis, obstructive lung disease (i.e., decreased forced expiratory volumes) becomes apparent even on nonworkdays.

Byssinosis is associated with increased rates of chronic cough, phlegm, and chronic bronchitis (defined as production of phlegm on most days for at least three months over a year). In late stages of the disease, dyspnea may occur upon exercise.

Both the occurrence of byssinosis and workshift decreases in forced expiratory volumes are related to the dust levels in the workplace. Although decreases in forced expiratory volume correlate with byssinotic symptoms, they do not seem to be the cause of perceived chest tightness.

No x-ray changes are associated with byssinosis and no fibrosis has been identified histologically. The occurrence of bronchitis in byssinotics is associated with hyperplasia (increased number) and hypertrophy (increased size) of mucous glands in the bronchi.

Several mechanisms have been proposed to explain the onset of byssinosis. Some evidence suggests that cotton dust stimulates histamine release from airway mast cells. This histamine would induce constriction of airway smooth muscle and result in the observed decrease in forced expiratory volume. Evidence also indicates that cotton dust causes an inflammatory response dominated by the influx of leukocytes into the airways. These activated leukocytes could produce mediators (leukotriene D_4) which cause airway smooth muscle contraction. Still other evidence suggests that cotton dust directly causes airway constriction.

The etiologic agent of byssinosis is still undefined. Tannin, gossypol, and other plant chemicals in the cotton bract, leaves, and stems are biologically active and have been proposed as causitive agents. In addition, cotton dust contains significant microbial contamination, and bacterial products such as endotoxin can cause biological effects similar to cotton dust itself. Such effects include infiltration of leukocytes and acceleration of breathing rate. It may be that the syndrome of byssinosis results from the combined effects of several agents in cotton dust.

2. Other Agricultural Dust

Vegetable dusts generated by handling or cutting rice, corn, oats, wheat, hay, straw, wood, and compost can cause adverse lung reactions. Similar reactions occur upon exposure to dusts in swine or dairy confinement buildings and to dusts associated with poultry housing and processing. Therefore, farmers, foresters, woodworkers, grain handlers, food processors, and gardeners can be exposed.

The response to agricultural dust exposure is complex and not fully understood. It depends on factors such as exposure level and sensitization. Organic dust toxic syndrome (ODTS) refers to an adverse reaction to relatively high dust levels. No prior exposure sensitization is required for this reaction to occur. Symptoms appear 4–6 hr after exposure and include pulmonary inflammation, fever, chills, fatigue, headache, and chest tightness. Chest x-rays are usually normal and bronchoalveolar lavage samples are characterized by an increase in polymorphonuclear leukocytes. Serum antibodies to the dust in question need not be present.

Hypersensitivity pneumonitis (i.e., allergic alveolitis) comprises a group of allergic lung diseases resulting from sensitization and recurrent exposure to organic dusts. It includes such diseases as farmers' lung, mushroom workers' disease, pigeon breeders' disease, and bark strippers' disease. Symptoms include inflammation of the alveoli characterized by monocytes and lymphocytes with few polymorphonuclear leukocytes. The acute response includes fever, chills, shortness of breath, malaise, decreases in pulmonary function showing both restrictive and obstructive changes, a decrease in diffusion capacity, and a decrease in arterial oxygen levels. X-rays show lung infiltration characteristic of alveolitis and serum antibodies to the inhaled antigen are always present. Acute symptoms appear 4–10 hr after exposure and subside in 18–24 hr. Upon repeated exposure, the disease may progress to a chronic state where granulomatous lesions and diffuse fibrosis can be identified by x-ray. In chronic disease, pulmonary function is compromised even between exposures. Symptoms include those of restrictive lung disease with decreased gas exchange and lower arterial oxygen levels.

The etiologic agents in ODTS and hypersensitivity pneumonitis most likely originate from the microbial contaminants found in organic dusts. ODTS may be in part a reaction to endotoxin which is known to cause fever and leukocytosis. Hypersensitivity pneumonitis is most like an allergic reaction to bacteria or fungi.

II. Sampling and Analysis

A. Fibrous Dust

Sampling and analysis of fibrous dust including asbestos can be done using several techniques that can be grouped into the two general categories of bulk and air analysis.

Bulk sampling and analysis involve the identification of components present in some bulk material, typically a building product. Usually the concern is whether or not asbestos is present. Polarized light microscopy is the most common method used although x-ray diffraction, infrared spectroscopy, and electron microscopy are also useful.

Bulk samples are usually collected with some type of coring device since many samples are multilayered. At the laboratory, samples are typically examined first with a stereomicroscope so that observation of particle morphology, color, and homogeneity can be made. Next, subsamples are transferred into a drop of refractive index liquid and examined under the polarized light microscope. Identifying particles microscopically requires the determination of a variety of optical properties including morphology, refractive index, birefringence, color, extinction characteristics, and sign of elongation. Figure 1 shows how some fibrous materials appear under the polarized light microscope.

For air sampling of fibrous dust, the most common method used is the collection of samples on some type of filter followed by fiber counting using microscopy. For light microscopy, samples are collected on cellulose ester filters that are cleared with acetone. Fiber counts are made using phase contrast illumination. Fibers collected from air samples are typically too small to identify with light microscopy so this technique is not qualitative beyond simple morphological considerations. When fiber identification is required, electron microscopy analysis of the filter samples is required. Fibers are identified by examination of morphological features, electron diffraction patterns, and x-ray analysis data. There is also a direct reading meter available that responds preferentially to elongated particles by exploiting the tendency for elongated particles to align themselves within a magnetic field.

B. Nonfibrous Dust

Particle size is a critical factor in estimating health risk since it is predominately particle size (and specifically aerodynamic size of a particle) that governs where deposition will occur in the respiratory system. In dealing with the problem of dust sampling, scientists have sought to design samplers that simulate the human respiratory system in terms of size selectivity. For instance, certain dusts cause pneumoconiosis that is related mainly to the fraction of dust that deposits in the alveolar region of the lung. Silica and coal dust are prime examples. For these dusts "respirable" samplers are used. These are samplers that preferentially sample that fraction of the dust that enters the alveolar region of the lung. In this country the most common means for making this measurement is to use a small battery-operated pump to first draw air through a miniature cyclone to remove the nonrespirable particles and then through a filter to capture the respirable portion. Samples are either collected on the workers themselves (personal sampling) or are collected from the workers' environment (area sampling). For most other dusts, "total" dust sampling is done. This is done by first drawing air through some arbitrary inlet and then through a filter. It is important to note that all inlets will exhibit size-selective characteristics especially for large particles where inertial forces of the particles tend to resist the drag forces of the fluid as the air enters the inlets. The inlets that are used for "total" dust sampling are not, however, linked to human lung deposition. Recent research has been directed at refining this concept of total dust. New definitions for particle-size-selective sampling have been proposed that include, in addition to respirable dust, thoracic and inhalable dust. For each of these criteria there is defined a distinct penetration curve that was selected by review of the best available experimental data. Thoracic dust includes that dust which would penetrate past the larynx. Since this includes both large and small airway penetration, it would be reasonable to apply this criterion to dusts which cause bronchitis. Inhalable particulate is dust that would penetrate anywhere in the respiratory tract from the head to the alveoli. This would logically be applied to dusts that can cause effects throughout the entire respiratory system. Certain agricultural dusts would seem to fit this category. Since these proposals were made, samplers have been developed for both thoracic and inhalable dust.

An alternative solution to the problem of size-

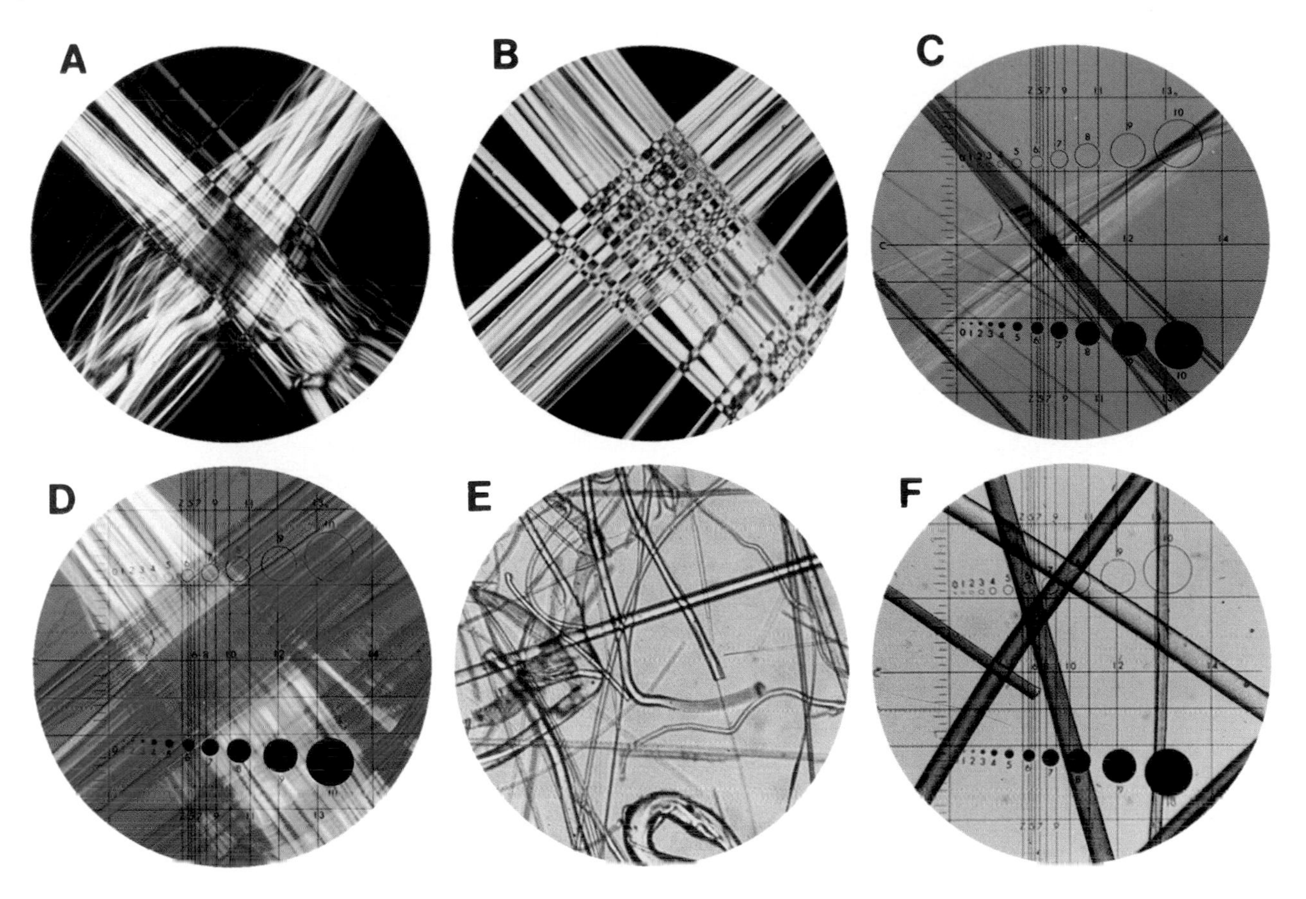

Figure 1 Photomicrographs of "fibrous" materials under various conditions of illumination. (A) Chrysotile asbestos (crossed polars); fiber bundles are ~250 µm in diameter. (B) Amosite asbestos (crossed polars); fiber bundles are ~250 µm in diameter. (C) Crocidolite asbestos (crossed polars/compensator); circle 10 = 80 µm. (D) Chrysotile asbestos (crossed polars/compensator); circle 10 = 80 µm. (E) Glass wool (brightfield); large fibers are ~10 µm in diameter. (F) Human hair (slightly uncrossed polars); circle 10 = 200 µm.

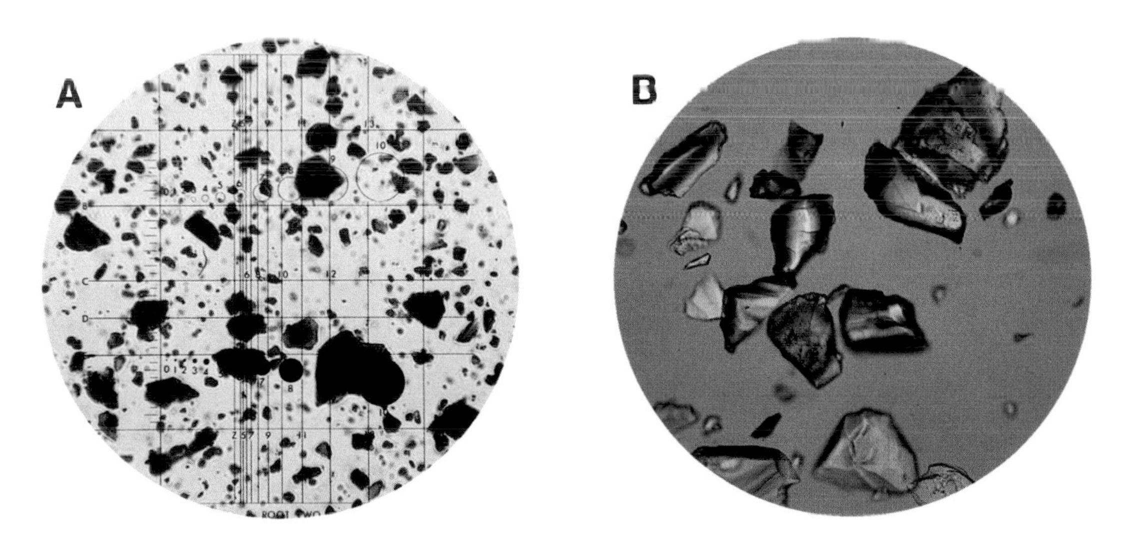

Figure 2 Photomicrographs of "nonfibrous" particles under various conditions of illumination. (A) Coal dust (brightfield); circle 10 = 20 µm. (B) Silica (slightly uncrossed polars); large particles are 100–200 µm. (C) Corundum (slightly uncrossed polars); circle 10 = 80 µm. (D) Calcite (crossed polars); particles are ~200 µm. *(Figure continues.)*

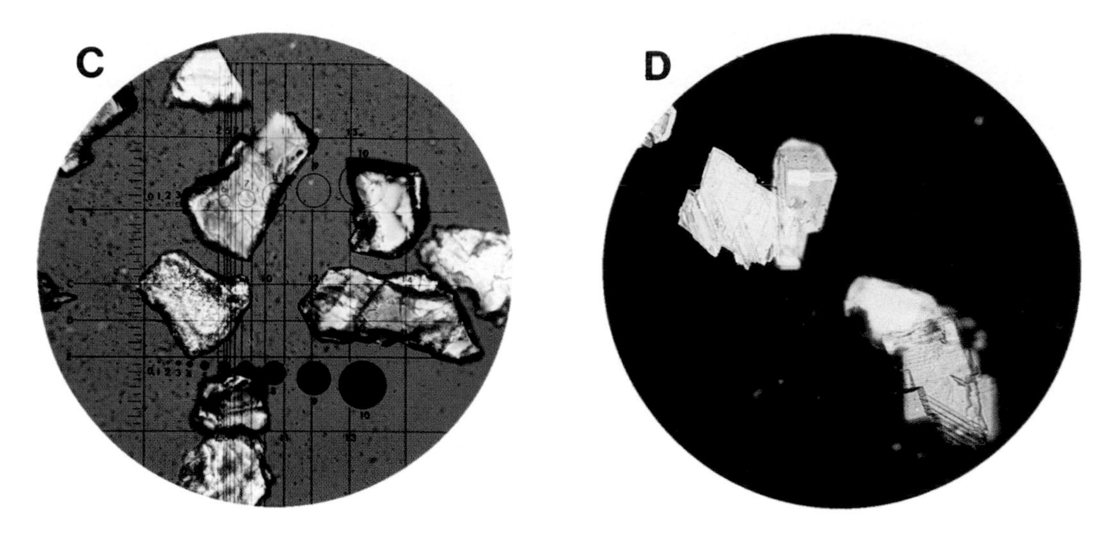

Figure 2 *(Continued)*

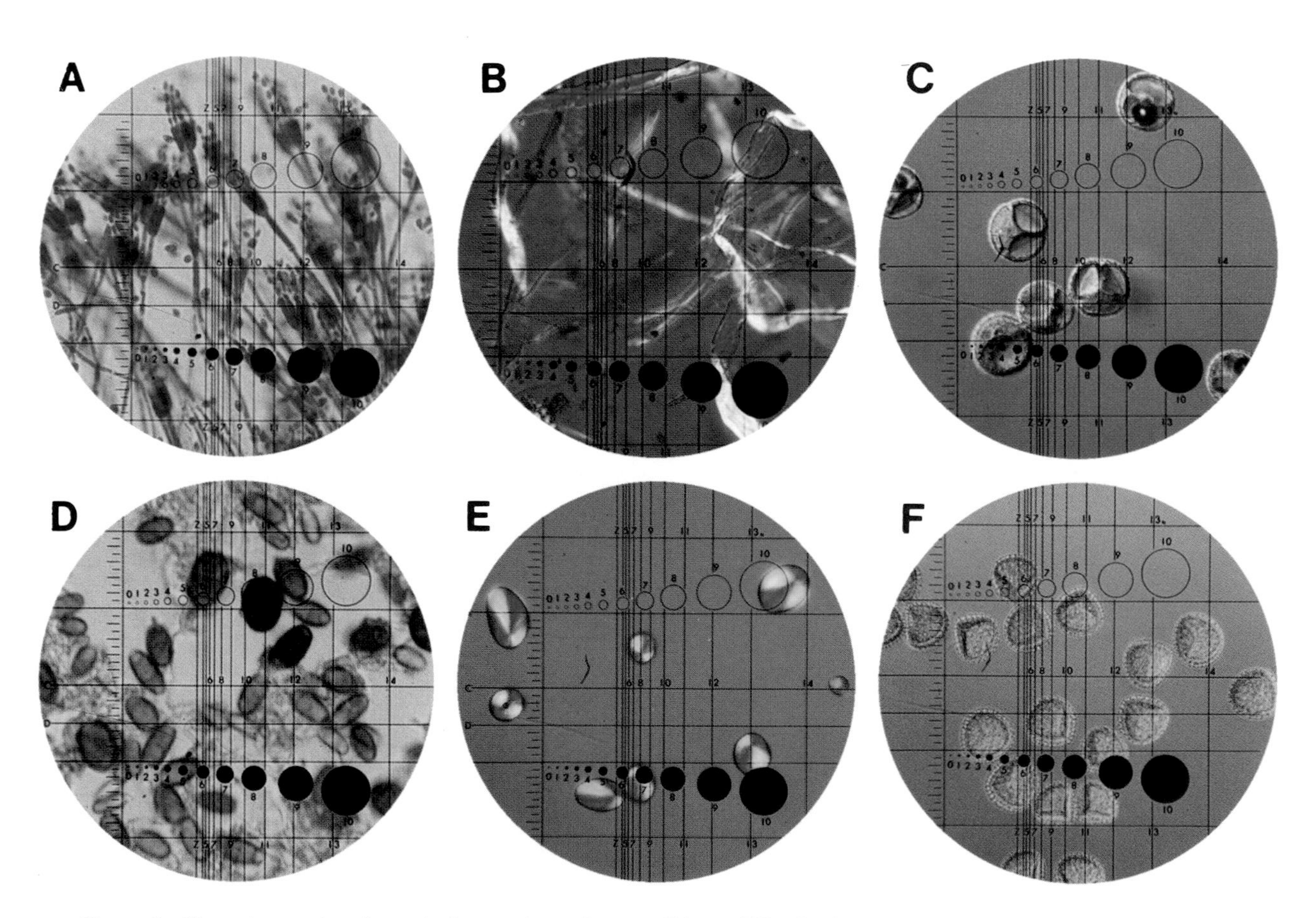

Figure 3 Photomicrographs of organic dusts under various conditions of illumination. (A) Fungal colony, stained (brightfield); circle 10 = 20 μm. (B) Cotton dust (crossed polars/compensator); circle 10 = 40 μm. (C) Ragweed pollen (differential interference contrast); circle 10 = 20 μm. (D) Fungal spores (brightfield); circle 10 = 20 μm. (E) Starch particles (crossed polars/compensator); circle 10 = 20 μm. (F) Lycopodium spores (Rheinburg illumination); circle 10 = 30 μm.

selective sampling is to measure the complete size distribution of the aerosol. Once one knows this, estimates of exposure based on any criteria can be made. A common sampler for making such measurements is the impactor. In this device, dust-laden air emerges from an orifice and is directed to a surface referred to as an "impactor plate." The air is forced to make an abrupt change in direction as it moves around the plate and those particles that have gained sufficient momentum will cross the airstream lines and hit or "impact" the plate while others will follow the airstream around the plate. Impactors are commonly operated in multistage configurations and both series and parallel versions have been described. Particles are sized aerodynamically so that data from these units can be used to predict human lung deposition.

Several techniques are available for the analysis of dust samples. Gravimetric analysis is considered routine for general hazard evaluation. In many cases, though, gravimetric assay must be complemented by other methods to determine dust composition.

For crystalline materials, x-ray diffraction is useful since crystalline compounds produce unique x-ray diffraction patterns that can be matched with patterns from known standards. Infrared spectroscopy is also applied to the analysis of some dust samples. These two techniques are commonly used for analysis of crystalline silica. Light microscopy, and especially polarized light microscopy, is very useful for identifying small particles. Figure 2 shows a number of different dusts as they appear under the polarized light microscope. Electron microscopy offers a variety of analytical techniques that are also useful for particle identification.

Direct reading meters are available also for measuring airborne dust levels and a number of them are portable. In one type, particles are deposited on a grease-coated film. The mass concentration of dust is measured by determining attenuation of beta radiation through the dust deposit. In another design, dust levels are measured by the change in vibration of a piezoelectric quartz crystal as particles are collected onto its surface. Several instruments are available that measure aerosols by light scattering. Size distribution measurements may also be made with direct reading instruments. In one type of sampler, aerodynamic size is recorded by measuring the velocity of particles in an accelerating airstream.

C. Organic Dust

Sampling organic dusts offers additional challenges for environmental scientists due to the typical complex nature of these aerosols and the added problem of maintaining viability of microorganisms. The area is further complicated by the lack of dose–response data on individual components that are found in these samples.

For measuring levels of bacteria and fungi, two samplers have historically been considered more or less as "standard" samplers. These are the Andersen viable sampler and the AGI-30 all-glass impinger although consideration of these samplers as "standard" is probably as much due to tradition as it is to aerodynamic considerations or some notion of absolute accuracy.

In the Andersen sampler, dust is impacted directly into culture plates. The plates are brought back to the laboratory and kept under controlled conditions. After a period of time, the colony-forming units are counted. A variety of collection media (some general, some specific) have been reported for both bacteria and fungi. One version of the sampler has six stages arranged in series, each with progressively lower cut-off diameters. The result is that the dust particles are partitioned throughout the sampler as a function of aerodynamic size thus enabling one to estimate human lung deposition. There is also a two-stage version that provides less size resolution but is more convenient since there are fewer culture plates to prepare and count. There is also a single-stage version that has been shown to be useful in indoor air quality surveys where concentrations of microorganisms are relatively low. Another version of this type of sampler is the slit-to-agar sampler. In this device, a rectangular jet stream is directed to the surface of a culture plate that slowly rotates. These samplers provide data on the relationship between concentration of microorganisms and time.

An advantage of these types of samplers is that the microbes are collected directly in the media and no further handling is required. They also tend to have reasonable collection efficiencies for smaller microorganisms. A disadvantage is that they are prone to overloading (especially the one- and two-stage versions) in areas of high concentration such as agricultural environments.

In the AGI-30 sampler, particles are impacted onto or into a liquid. At the laboratory, the solution is diluted, plated out, and after a period of time, colony-forming units are counted. These samplers are especially useful in highly contaminated environments since they afford infinite dilution potential. A disadvantage is that the collection efficiency drops off for smaller (submicrometer) organisms.

A third method of viable sampling is to collect samples on filters. The filters are then plated directly or the dust is washed off, diluted, and plated. A distinct advantage of filter sampling is that most filters are highly efficient for small particles. A problem is that viability may not be maintained as well as with impinger samples or samples collected directly onto media. This is probably less of a problem with fungal spores and endospore-producing bacteria.

There is another class of samplers available for estimating concentrations of bacteria, fungi, and pollen. These samplers collect particles in a form that enables one to do counts under the microscope. There is a trend these days for doing microscopic counts rather than viable sampling in occupational environments. The idea is that the nonviable organisms can also cause health effects and these microbes are not considered with viable techniques. Also missed are those organisms that will not grow with a particular medium or set of environmental conditions. With the microscope, one can count them all. The size of fungal spores and pollen is such that they are easily recognized using light microscopy. Bacteria are more difficult to resolve even with various contrast enhancements (phase contrast, interference contrast, darkfield, etc.); thus they usually require some staining technique. Acridine orange stain followed by epifluorescence microscopy is one alternative. A problem with microscopic counting is that, although an analyst can learn fairly quickly to resolve between a bacteria, a fungal spore, or a pollen grain, it takes much more skill to identify organisms even to the genus level. In some cases, electron microscopy reveals more detail that aids in identification. However, confident identifications can often only be made by observing colony characteristics. In an attempt to combine approaches, methods have been proposed where one takes a single filter sample and from that does (1) acridine orange counts, (2) electron microscope examination, and (3) culturing for identification.

In addition to bacteria, fungi, and pollen, there are also methods for sampling and analysis of viruses and protozoa, as well as for some of the toxic products of microorganisms (endotoxin and mycotoxins).

Many organic dusts (especially agricultural dusts) also contain a variety of other materials such as mineral components, insect parts, starch particles, and various other botanical components. With such complex exposures, gravimetric analysis is of limited value. It is more appropriate to try to identify and quantify the individual components of the sample. Polarized light microscopy again can be quite useful here since many of the materials can be readily identified. Figure 3 shows some organic dusts as seen under the polarized light microscope.

III. Control Methods

A. Air Standards

In the United States there are three groups that deal with exposure limits for airborne particles in the general occupational environment. The American Conference of Governmental Industrial Hygienists (ACGIH) was organized in 1938 with a mission to promote standards and techniques in occupational health. Each year ACGIH publishes an updated version of threshold limit values (TLVs) for chemical and physical agents. Documentation for the TLV list is also available, and this provides the background information on how individual limits are selected.

The National Institute for Occupational Safety and Health (NIOSH), which promotes education and conducts research in the area of occupational safety and health, also makes recommendations for safe exposure limits for various contaminants. Many of these recommendations are contained within criteria documents. These are reports that also include information on sampling and analysis, surveillance, control methods, labeling procedures, and so forth.

The Occupational Safety and Health Administration (OSHA) is the agency that enforces health and safety regulations in the general workplace. Part of these regulations are the permissible exposure limits (PELs) which were initially adopted from the TLV list. For many contaminants the NIOSH recommendation, TLV, and permissible exposure limits are identical. Table I lists the exposure limits for selected dusts. For some agents such as asbestos, OSHA has in addition to the exposure limit, regulations that cover other aspects of exposure including surveillance, personal protective equipment, recordkeeping, training, and monitoring schedules.

The Mine Safety and Health Administration (MSHA) is the agency that enforces health and safety regulations specifically in mining environments. The MSHA air quality standards have also traditionally been linked to TLV lists.

Table I TLVs, OSHA Standards, and NIOSH Recommendations for Selected Dusts

Dust	TLV	OSHA	NIOSH	Sampler
Silica (crystalline quartz)	0.1 mg/m^3	0.1 mg/m^3	0.05 mg/m^3	Respirable
Asbestos	Varies for individual forms	0.2 fibers/cc	0.1 fibers/cc	Total
Coal dust	2.0 mg/m^3	2.0 mg/m^3		Respirable
Cotton dust	0.2 mg/m^3	1.0 mg/m^3	0.2 mg/m^3	Verticle elutriator
Nuisance particulates	10.0 mg/m^3	15.0 mg/m^3		Total
		5.0 mg/m^3		Respirable
Cellulose	10.0 mg/m^3	15.0 mg/m^3		Total
		5.0 mg/m^3		Respirable
Fiberglass	10.0 mg/m^3		5.0 mg/m^3	Total
			3 fibers/cc	Total

B. Engineering Controls

There are a variety of control methods available that can be used to reduce dust exposures in occupational environments. These include ventilation, isolation, substitution, and dust suppression by wetting.

Ventilation systems can be divided into general categories: local exhaust ventilation and general ventilation. In local ventilation systems, the contaminant is captured near the point of generation. In general ventilation systems, the entire room is supplied with intake and exhaust air to dilute concentrations within the area. Local exhaust has the advantage of lower air flow requirements and the contaminant is captured before it enters the general workroom air. General ventilation is usually restricted to the removal of low levels of relatively nontoxic contaminants from decentralized sources.

Substitution of hazardous materials with less harmful ones is also a means for controlling occupational exposures. There are materials for instance which may be substituted for silica in several applications. There are also insulation products that can be used in place of asbestos. Process changes and isolation of hazardous areas can also be effective control measures.

Dust suppression can often be accomplished by the application of water or other suitable liquid. This technique is used extensively in mines to reduce the concentration of coal and other dusts. Also, water that is forced through drill bits in rock drilling operations can be effective in dust reduction.

C. Personal Protective Equipment

Personal protective equipment, and in the case of dust exposure, specifically respiratory protection, should only be considered for operations where it is not possible to control exposure by other means. They should never be considered as an alternative to engineering controls.

There are two general categories of respirators: air purifying and atmosphere-supplying. The latter can be further divided into self-contained breathing apparatus (SCBA) and supplied air respirators.

Air purifying devices for use against dust exposures consist of a facepiece along with some type of mechanical filter. Facepieces come in three types. Quarter masks cover the mouth and nose, half masks fit over the chin, and full-facepieces cover from above the eyes to below the chin. There are some types of single-use respirators where the facepiece itself serves as the filter. Dust filters usually are made of a fibrous material and collection mechanisms include impaction, sedimentation, interception, diffusion, and electrostatic attraction. There are a variety of filter types designed to provide protection against different classes of airborne particles. There is usually some trade-off between collection efficiency, resistance to breathing, and clogging time. A special class of air purifying respirator is the powered air-purifying respirator (PAPR). With this device a blower is used to pass contaminated air through a filter to provide purified air to a facepiece, helmet, or hood. An advantage is that it provides an airstream that can have a cooling effect in warm temperatures. Some are also designed with loose fitting hoods or helmets that enable them to be worn by some individuals with beards or facial scars as long as the beard or scar does not lie along any sealing surface of the hood/helmet (many helmets have cheek and/or temple seals).

When a high level of protection is required or when there is low oxygen, atmosphere-supplying respirators are appropriate. One type, the airline respirator, uses compressed air from a stationary source to supply air through a hose to a facepiece. In the SCBA design, the wearer carries the breathing gas source, thus eliminating the need for connection to some stationary source. These units are either open circuit (where exhaled air is exhausted to the atmosphere) or closed circuit (where exhaled gas is recirculated).

NIOSH has a program of certification and evaluation for respirators. The certifications are done in collaboration with the Mine Safety and Health Administration. OSHA has requirements for respirator programs that contain provisions for training, fit testing, cleaning, inspection, selection, and storage.

IV. Summary

Despite well-documented effects of dust exposure in occupational environments, health problems persist. Exposure standards exist, however, for many dusts and a variety of engineering controls has been described. When controls are not possible, personal protective equipment is available. Sampling and analysis systems have been developed for many dusts with sampler designs motivated mainly by the size-selective characteristics of the human respiratory system. Although lung deposition is principally a function of particle size, the type and extent of biological response are also affected by other chemical and physical features of the particles as well as the concentration and duration of exposure. Organic dust exposures are especially complex in terms of both exposure and response. However, it is just this complexity that makes the study of such exposures essential to improving control, diagnosis, and treatment. Although precise dose–response data are difficult to achieve even for well-studied dusts, occupational settings often provide the best data available for estimating risks to community populations where environmental dose is much more difficult to estimate.

Related Articles: DIESEL EXHAUST, EFFECTS ON THE RESPIRATORY SYSTEM; INORGANIC MINERAL PARTICULATES IN THE LUNG; SILICA AND LUNG INFLAMMATION.

Bibliography

Bollinger, N. J., Schutz, R. H. (1987). "NIOSH Guide to Industrial Respiratory Protection." NIOSH Publication No. 87–116, Cincinnati, OH.

Davis, J. M. G. (1986). A review of experimental evidence for the carcinogenicity of man-made vitreous fibers. *Scand. J. Work Environ. Health* **12,** 12–17.

McCrone, W. C., McCrone, L. B., and Delly, J. G. (1984). "Polarized Light Microscopy." Microscope Publications, Chicago, Illinois.

Merchant, J. A., Boehlecke, B. A., Taylor, G., and Pickett-Harner, M., eds. (1986). "Occupational Respiratory Diseases." DHHS (NIOSH) Publications No. 86–102.

Parkes, W. R. (1982). "Occupational Lung Disorders." 2nd ed. Butterworths, Boston.

Plog, B. A., Benjamin, G. S., and Kerwin, M. A. (1988). "Fundamentals of Industrial Hygiene." National Safety Council, Chicago, Illinois.

Rylander, R. (1990). Introduction: Organic dusts and disease. *Am. J. Ind. Med.* **17,** 1–2.

Rylander, R., Schilling, R. S. F., Pickering, C. A. C., Rooke, G. B., Dempsey, A. N., and Jacobs, R. R. (1987). Effects of acute and chronic exposure to cotton dust: The Manchester criteria. *Br. J. Ind. Med.* **44,** 577–579.

Schilling, R. S. F. (1950). Byssinosis in cotton and other textile workers. *Lancet* **2,** 261–265.

Stanton, M. F., Layard, M., Tegeris, H., Miller, A., May, M., Morgan, E., and Smith, H. (1981). Relation of particle dimension to carcinogenicity in amphibole asbestos and other fibrous minerals. *J. Nat. Cancer Inst.* **67,** 965–975.

Ecotoxicological Testing

Irene Scheunert
GSF-Institut für Bodenökologie, Munich

Glossary

Biocenosis Community of plants and animals that is in steady interaction by mutual influences and dependencies.

Chromatography Separation and detection methods to analyze chemical mixtures qualitatively and quantitatively.

Convection Passive movement of solutes with the moving water.

Diffusion Physical process by which molecules, atoms, and ions, because of thermic mobility, move from sites of higher concentration to those of lower concentration.

Dispersion Distribution or mixing of solutes in the moving pore water; results from different flow velocities of individual water volumes.

Henry's law constant (*H*) The dimensionless H is the ratio between concentration of a chemical in gas and that in water at equilibrium. The "approximated" H is the quotient of saturation vapor pressure and maximal water solubility.

Lysimeter Equipment to determine water evaporation, transpiration, evapotranspiration, chemical mobility, and leaching by boxes or columns containing soil.

***n*-Octanol/water partition coefficient** Ratio between the concentration of a chemical in n-octanol and that in water after equilibration; measure of lipophilicity.

Partition Distribution of a chemical between two partly miscible liquids after equilibration.

Substrate Material subjected to biodegradation.

Xenobiotic Man-made chemical compound "foreign" in nature.

***ECOTOXICOLOGICAL TESTING** comprises experiments carried out to describe qualitatively and quantitatively toxic effects on cell, organ, organism, population, and ecosystem levels, and if possible, the dependence on chemical concentrations and mode of action. The aim is the prediction of potential adverse effects in the environment. Tests for effects must be accompanied by tests for exposure estimation.*

I. Introduction

Ecotoxicology is concerned with the toxic effects of chemical and physical agents on living organisms, especially on populations and communities within defined ecosystems; it includes the transfer pathways of those agents and their interactions with the environment. The effect of a chemical in an organism may be defined as the time integral of the sum of chemical concentrations minus the sum of elimination processes.

In testing effects, dosis–effect relationship curves are established, demonstrating a "no-observed-effect" level, a sublethal range comprising for example, EC_{10}, EC_{50} concentrations, and a lethal range comprising LC_{10}, LC_{50} concentrations. EC_{10} means the concentration at which 10% of the organisms tested exhibit a statistically significant effect of the chemical. LC_{10} means the concentration at which 10% of the organisms tested are killed. In all cases, control organisms without the chemical are required.

The term "exposure" is not well defined. In a general sense, exposure means contact of a physical or chemical agent with a boundary of the target or

Handbook of Hazardous Materials

receptor. It is measured in terms of concentration of the agent at the outer or inner boundary, and the frequency and duration of contact. Another definition is that exposure is the amount of a particular physical or chemical agent that reaches the target. Exposure analysis comprises tests on the transport of the chemical in soil, water or air, uptake by organisms, conversion reactions in soil, water and organisms, and biotic and abiotic mineralization. Thus, both the influences of biota and their environment on the chemical and the influences of the chemical on biota are a subject of ecotoxicological testing.

In direct measurements of processes in exposure analysis, the amount of chemical that has undergone the respective process is quantified. By contrast, in indirect measurements, the amount is concluded as a difference between initial amounts and those left after the respective process. In this case, processes other than the one to be measured, that may cause a decrease in the quantity of test chemicals, have to be excluded. Since the process is quantified as a difference, standard deviations of the initial and the remaining chemical amounts have to be considered very carefully.

Two kinds of information are usually needed: relative information that allows comparison between chemicals tested under similar conditions, and absolute information on process rates and quantities of products and effects under natural environmental conditions. For the first type, standardized experimental conditions often require significant deviations from those of the natural environment in order to obtain relevant results. These experiments include the determination of physicochemical data and the more simple laboratory tests that provide qualitative or relative information on the transport, transformation and accumulation of chemicals and their effects.

Absolute information requires more complex laboratory tests under simulated environmental conditions and direct field tests. These tests give results that are representative of the general environment, but are only confirmatory within the specific environmental condition of the test. Since a multitude of factors affects experimental results, comparisons between chemicals should be undertaken only with caution. Both groups of tests are necessary and have specific utility, nevertheless greater insight into environmental behavior of chemicals is achieved if results are properly synthesized.

In the following text, both laboratory and field tests for exposure to chemicals and effects of chemicals in terrestrial and aquatic ecosystems will be presented and discussed. This article refers to well-established tests only. Recent developments, such as biomarker research, are not considered here.

II. Ecotoxicological Testing in Terrestrial Systems

A. Tests for Exposure Prediction

1. Adsorption

Adsorption–desorption in soil plays a dominant part in all physical processes affecting the exposure of biota to chemicals, such as volatilization from soil into the air, soil–water transfer resulting in mobility in soil, and leaching of pesticides to surface- or groundwater, or the uptake by plants or soil fauna. Additionally, biological activity and degradation are also affected by adsorption. In general, adsorption is defined as the adhesion or attraction of one or more ionic or molecular layers to a surface. In soil, the sum of all kinds of fixation of ions or molecules on or within the solid phase is called adsorption. There is a continuum of possible adsorption interactions, starting with fixed site adsorption and ending with partition between three-dimension phases such as aqueous solution and soil organic matter. Bonds vary between those that are reversible and those resulting in so-called "bound residues."

Adsorption is quantified by the adsorption coefficient, which defines the ratio between the concentration adsorbed and that in the water phase. Adsorption coefficients may be estimated with the aid of correlations with more simple physicochemical parameters that are easier to determine, the most important being the *n*-octanol/water partition coefficient. However, since this parameter necessarily comprehends only the three-dimensional "partition" of the chemical between soil–water and hydrophobic soil organic matter and neglects other adsorption mechanisms including that on soil mineral surfaces, such correlations always imply various "outliers."

The dependence of adsorption upon concentration in the solution is described by empirical equations among which the Freundlich isotherm equation fits best in most cases:

$$x/m = K \cdot C_e^{1/n} \tag{1}$$

where x/m is the amount adsorbed per unit amount of adsorbent, K the Freundlich adsorption coefficient, C_e the equilibrium concentration in solution, and $1/n$ a constant.

The experimental determination of adsorption coefficients is achieved by equilibration of the chemical dissolved in water with soil during agitation. The shortcoming of this method is its deviation from real conditions where the soil–water ratio is much higher and where no agitation occurs. However, tests with higher soil–water ratios and without agitation in the case of organic compounds fail to yield reproducible results, since equilibration needs a very long time during which decomposition of the test compound may occur.

2. Volatilization

Volatilization of chemicals from soil is the transfer of the chemical as a gas through the soil–air interface under environmental conditions; volatilization from plants is the corresponding process. Its ecotoxicological importance is due, on the one hand, to the fact that it is the main pathway of exposure reduction of the soil ecosystem to not-readily biodegradable chemicals.

On the other hand, volatilization of chemicals may contribute significantly to air pollution and, thus, to the long-range transport of chemicals by air.

Volatilization may be estimated by physicochemical data of chemicals, such as vapor pressure, water solubility, and soil adsorption coefficients, as well as by the gas phase transfer velocity, which is partly dependent upon chemical substance properties such as molecular mass and atomic diffusion volume, and partly on environmental conditions such as air velocity.

Experimental methods to test volatilization of chemicals from soils and plant surfaces are not yet standardized in official guidelines. Some methods described in the literature are listed in Table I. It should be mentioned that in general only direct measurement of the volatilized amounts after trapping gives satisfactory results, since indirect methods estimating volatilization as a difference between initial and remaining soil concentrations are mostly inaccurate and may, additionally, include other processes, such as degradation or irreversible adsorption in the measurement.

3. Mobility and Leaching

The mobility of chemicals in soil is of paramount importance for ecotoxicological evaluation. This phenomenon is related to adsorption on the one hand and the mass flux of dissolved fractions on the other. Adsorption retards the mass flux which consists of diffusion, convection, and dispersion and which is also a function of the removal of chemicals from the solution by biological and chemical reactions. Mobility is the cause for chemicals leaving their original application site and threatening other environmental compartments (especially groundwater) by leaching.

The assessment of mobility and leaching of chemicals by the determination of adsorption coefficients only underestimates the influence of soil texture affecting the mass flux of dissolved chemicals. Experimental simulation of the complete migration and leaching process, therefore, is indispensable. Some methods are listed in Table II.

The advantages of large-scale experiments compared to the use of small columns are a reduction of wall effects falsifying the results. Disturbed soil filled into columns or lysimeters is a good representative of a certain soil area, but neglects natural macropores such as earthworm ducts or soil fractures by which preferential flow occurs. This is probably the main reason for the rapid appearance of, for example, herbicides in groundwater a short

Table I Tests for Assessment of Volatilization of Chemicals from Soil to Air

Test type	Principles	Examples
Laboratory tests	Indirect measurement	Decrease in chemical concentration on the treated material
	Direct measurement	Trapping of chemicals volatilized from soil or plant surfaces; trapping of volatilized chemicals after incorporation into soil; trapping of chemicals volatilized from model ecosystems
Open air tests	Indirect measurement	Mass balance studies with ^{14}C-labeled chemicals in lysimeters; concentration decrease in the field
	Direct measurement	Trapping of volatilized chemicals in chambers on the soil; trapping of volatilized chemicals over fields in different heights

Table II Tests for Assessment of Mobility of Chemicals in Soil and for Leaching from Soil

Test type	Principles	Examples
Laboratory tests	Measurements in various soil depths and in percolation water	Columns with disturbed soil; columns with soil cores
	Measurements in leachate and run-off water	Small model ecosystems
Open air tests	Measurements in various soil depths and in percolation water	Outdoor lysimeters
	Measurements in various soil depths, drainage water, and groundwater	Open field tests

time after application. Undisturbed soil cores, on the other hand, include this factor in the evaluation; however, they represent only a small ecosystem section and disregard natural spatial variability in fields. Therefore, replicates may exhibit enormous differences in the leaching behavior of xenobiotics.

4. Degradation

Degradation of xenobiotic organic substances in soil comprises both their chemical transformation to conversion products and their total mineralization to carbon dioxide and other small molecular fragments.

In soil, organic xenobiotics undergo various abiotic and biotic transformation reactions. Alterations in the chemical structure of a chemical, as initiated either by biotic or abiotic factors, result in the formation of new xenobiotic compounds which may be more or less ecotoxic than the parent compounds and which, in any case, represent unwanted chemical residues in the ecological system. Furthermore, due to the changes in chemical structure, conversion products may differ significantly from the parent compounds in physicochemical properties and, thus, also in their distribution behavior in the environment. Therefore, it is important both to identify their chemical structure and to quantitate their formation and degradation kinetics.

Mineralization of organic xenobiotics in soil is a desirable process since it is the only pathway for a final elimination of the xenobiotic from the environment. This process may also proceed both abiotically and biotically. Its kinetics govern the time-course of decline of total residues—both of parent compounds and of all conversion products—in soil.

A survey of methods to test degradation of organic xenobiotics in soil is given in Table III. In this table, indirect measurement means the determination of the non-degraded substance amount that remains after a certain incubation period, or the oxygen consumed for degradation. In direct methods, the final products of degradation are determined. Indirect methods based on analytical determination of the original foreign compound measure both mineralization and the formation of organic conversion products; however, they also include disappearance processes such as volatilization or, in the case of field tests, leaching into deeper soil layers, that are not assayed. Therefore, in many cases they do not offer a real description of xenobiotic chemical degradation.

The isolation, chemical structure elucidation, and quantification of transformation products in soil is of high importance for the ecotoxicological evaluation of foreign compounds in terrestrial ecosystems. However, this may be neither performed in short-term tests, nor standardized in any way since the procedure is dependent on the chemical structure

Table III Tests for Assessment of Degradation of Organic Chemicals in Solid Materials

Test type	Principles	Examples
Laboratory tests	Indirect measurement	Decrease in chemical concentration in treated materials; O_2 consumption
	Direct measurement	CO_2 production; trapping of $^{14}CO_2$ from ^{14}C-labeled xenobiotics; CO_2 and CH_4 in anaerobic systems; CO_2 formed abiotically after UV-irradiation
	Direct and indirect measurement	Parent compound and conversion products
Open air tests	Indirect measurement	Decrease in chemical concentration in samples from treated fields
	Direct measurement	Trapping of $^{14}CO_2$ in chambers on lysimeters

both of parent compounds and of conversion products and sometimes needs complex separation and purification schemes including various chromatographic techniques. The use of ^{14}C-labeled compounds facilitates the detection and isolation of as many conversion products as possible. It also enables the detection of so-called "bound residues" in soil, which are unextractable by solvents and thus normally escape analytical detection.

The determination of mineralization of foreign compounds in soil is best feasible by using ^{14}C-labeled compounds and determining $^{14}CO_2$. Because of the large amount of CO_2 produced by normal soil respiration, the determination of the portion of CO_2 originating from degradation of the xenobiotic is problematic without labeling the xenobiotic.

5. Uptake by Plants and Soil Fauna

Uptake of soil pollutants in plants comprises a sequence of processes. The uptake implies two major pathways: uptake by roots and foliar uptake. Both pathways must be preceded by processes outside the plant, the first one by the desorption from soil solid phase into soil water, the second one by volatilization from soil into the air. Since all of these processes are correlated with physicochemical substance properties of the chemicals in different ways, uptake of chemicals by terrestrial plants—in contrast to bioconcentration in aquatic organisms—cannot be predicted by simple correlations with physicochemical properties. More complex models are needed that also include numerous environmental conditions.

Experimental tests designed to determine the uptake of xenobiotic compounds by plants are listed in Table IV. For organic compounds, the separate determination of root and foliar uptake in closed systems is important since both pathways strongly depend on environmental conditions in different ways and have to be considered separately in prediction models. On the other hand, a closed laboratory system deviates from natural conditions. Therefore, field tests have to be carried out also, and the results of both test types have to be evaluated in combination.

In contrast to testing of biological effects, the uptake of soil pollutants by soil fauna has not been standardized. Some earthworm species are the preferred test organisms. Molluscs have also been used sporadically.

The bioconcentration factor, defined as the quotient of the concentration in the organism divided by the concentration in the surrounding medium, can be estimated for lipophilic chemicals by the *n*-octanol/water partition coefficient which is a measure of lipophilicity of the chemical. The formation of three-phase equilibria between soil solids, earthworms, and soil water has to be regarded.

B. Tests for Biological Effects

1. Effects on Soil Microflora

Test methods to assess the effects of chemicals on the microflora of soils are based on sum parameters either taking into account more or less the total microflora, or activities performed by special microbial groups important for soil fertility. Single species tests are not usual in soil microflora testing. Because of the high complexity of soil microflora, they would not be relevant for the environment. Methods based on sum parameters should be able to measure life activity and to quantify differences in life activity between untreated soils and soils treated with chemicals. Table V gives a review of some methods used to cover the total microflora.

The determination of ATP is a good measure of life activity since ATP is present only in living cells and is decomposed quickly when the cell dies. The respiration method comprises the measurement of CO_2 released from soil by the activity of living or-

Table IV Tests for Assessment of Uptake of Chemicals from Soil by Plants

Test type	Principles	Examples
Laboratory tests	Measurement of total uptake	Open plant–soil tests; open terrestrial ecosystems; aerated closed terrestrial ecosystems
	Separate measurement of root and foliar uptake	Special aerated closed plant–soil systems
Open air tests	Measurement of total uptake	Open outdoor lysimeters; open field trials

Table V Sum Parameters to Determine Effects of Chemicals on Total Soil Microflora

Method	Principle	Organisms detectable	Problems
ATP method	ATP in living cells → extraction → determination by luminescence (luciferin–luciferase) method	All living cells	Extractability from soil
Respiration	Living cells → CO_2 → determination by IR measurement	Preferably aerobic organisms	Abiotic CO_2 formation
FDA method	Fluoresceine diacetate → hydrolysis by enzymes → fluoresceine → determination by spectral photometry	Preferably fungi	Limited information
Microcalorimetry	Living cells → heat output → determination	All living cells	Abiotic heat output
Fe(III) reduction	Insoluble Fe(III) → reduction by bacteria → soluble Fe(II) → determination by atomic absorption spectrometry	Preferably anaerobic bacteria	Limited information

ganisms. Preferably, aerobic microorganisms are detected; however, anaerobic microorganisms also form CO_2. In order to include "dormant" microorganisms in the determination, glucose is added to the soil before CO_2 is measured. Instead of CO_2 determination, O_2 consumption may be used as an indicator of soil respiration.

Fluoresceine diacetate is a colorless substance which is hydrolyzed (by a group of microbial enzymes such as proteases, lipases and esterases) to colored fluoresceine, which can be determined photometrically. Fungi are preferentially detected.

In microcalorimetry, the output of heat as a result of life activity is determined in a bioactivity monitor. The extent of reduction of the low soluble Fe(III) oxides to highly soluble Fe^{2+} can be used as a test parameter for the activity of microorganisms such as *Clostridium* species and *Bacillus polymyxa,* which are widely distributed in soils and can thus be regarded as representative of total soil microflora.

Most of the tests determining the activity of special groups of soil microorganisms are related to nitrogen turnover. Nitrogen transformation, as performed by nitrifying bacteria, is the subject of tests for which guidelines are available. Test criteria are nitrite and nitrate formation rates. Arginine ammonification, glutamic acid degradation, and nitrogen fixation are other microbial activities related to nitrogen turnover that may serve as indicators of ecotoxicological effects of chemicals in soil. The latter activity is performed by a small, highly specialized group of microorganisms, which is very sensitive to such chemicals as fungicides. The ability to reduce molecular nitrogen to ammonia is assayed by quantifying the reduction of added acetylene to ethylene. This test is very sensitive but is not indicative of the total soil microflora.

Soil enzyme activity also may be an indicator of toxic effects of chemicals on soil. One example is the activity of soil dehydrogenase.

Microbial community structure is not covered by any of the tests described above. However, due to the high importance of this soil property, not only for soil fertility but also for the capacity of the soil to degrade xenobiotics, development of such tests is needed.

2. Effects on Terrestrial Fauna

For assaying the effects of foreign chemicals on soil fauna, thus far only monospecies tests have been elaborated. Earthworms are the preferred organisms. Indicators of effects are acute toxicity as indicated by lethality, body weight reduction, and other visible changes after exposure, and reproduction toxicity as indicated by cocoon production and cocoon viability. Exposure is performed by injection, feeding, immersion, or application on contact filter paper in natural or artificial soil, or artisol, a synthetic silica substrate mixed with water.

In standardized guidelines such as the OECD-guideline, the compost-inhabiting species *Eisenia foetida* is used because this species is relatively easy to maintain. An indicator of acute toxicity is the median lethal concentration LC_{50}. The use of both species and substrates differing from those in natural terrestrial environments imposes problems for the extrapolation of test results to normal field conditions.

Soil-dwelling invertebrates proposed for testing the toxicity of chemicals include the springtail and other beneficial arthropods, nematodes, spiders, and predatory mites. Other terrestrial organisms for which toxicity tests exist are honey bees, birds, and mammals. In the case of honey bees, guidelines for both acute toxicity and a semi-field test are available. For birds, dietary acute toxicity and avian reproduction are assayed. For mammals, dietary acute and chronic toxicity are the subjects of test guidelines.

3. Effects on Flora

Effects of hazardous chemicals on flora are studied in higher plants (e.g., oat, radish, tomato, cress, turnip, and others) by assaying such indicators as seed germination, growth (as indicated by plant weight), root elongation, and ethylene stress.

In OECD- and EG-tests, concentration–effect relationships are established, and LC_{50} values for emergence and EC_{50} values for growth are calculated. Visible phytotoxic effects are also recorded.

In the root-elongation test, statistically significant changes in the root or in the shoot length of seedlings grown in any test medium (as compared to controls) indicate a phytotoxic effect by some cause.

The plant stress ethylene bioassay uses the formation of ethylene by plants under stress conditions as a measure for toxic effects of chemicals. The degree of chemical stress is estimated by ethylene and ethane analysis.

For gaseous hazardous substances, testing of toxic effects on plants is performed in closed chambers. However, it should be considered that nearly all organic pollutants undergo measurable volatility from soil. The exposure of plants by this pathway is not controlled by the open tests discussed above. Thus, in the case of organic chemicals with high phytotoxic potentials, in simple "open" tests, the treated samples may exert toxic effects on controls or on lower-dosed samples via the air, leading to test results of dubious value. Therefore, the gas phase should be controlled by using closed laboratory systems in any testing of chemicals having measurable volatility.

III. Ecotoxicological Testing in Aquatic Systems

A. Tests for Exposure Prediction

1. Volatilization

Apart from adsorption of chemicals in sediments, the volatilization from water is the main pathway for the decrease in exposure of aquatic organisms to chemicals. Test methods to determine adsorption in sediments are similar to those in soils and will therefore not be discussed again.

As in the case of volatilization from soils, volatilization of chemicals from aquatic systems also results in air contamination and subsequent transport of the contaminants to remote ecosystems.

Volatilization of chemicals from aqueous solution is a function of the concentration difference between air and water and the interfacial mass transfer velocity (usually expressed in m sec^{-1} or m h^{-1}). If the interface between air and water is considered to consist of a water and an air film, which have to be penetrated by the chemical by diffusion, the overall mass transfer velocity K has the following magnitude:

$$K = \left[\frac{1}{k_l} + \frac{R \cdot T}{k_g \cdot H}\right]^{-1} \qquad (2)$$

where k_l is the liquid film mass transfer velocity, k_g the gas film mass transfer velocity, H the "approximated" Henry's law constant (the quotient of saturation vapor pressure and saturation water solubility), R the gas constant, and T the absolute temperature. Under outdoor conditions where the chemical concentration in the air is nearly zero, the volatilization process follows first-order kinetics and may be quantified by the half-life of the chemical in aqueous solution.

Experimental methods to determine the volatilization of chemicals from aqueous solution are shown in Table VI.

Indirect measurement by determining the concentration decrease in water may be accurate if other processes causing decrease of concentration (e.g., degradation or adsorption) can be excluded. The determination of dimensionless air water Henry's law constants under equilibrium conditions by measuring concentrations in air and water phases may replace the more complicated determination of volatility under wind conditions when [according to Eq. (2)] the volatilization rate and transfer velocities k_l and k_g are negligible. This is the case if Henry's constants are very low (i.e., for chemicals with low vapor pressure and/or high water solubility).

2. Degradation

Degradation of chemicals in aquatic systems occurs both abiotically and enzymatically. Whereas in soils (with high microbial content) biotic degradation processes prevail and photochemical decomposition occurs only at the soil surface, in aqueous solution biotic and abiotic reactions have the same importance. For both reaction types, experimental tests have been elaborated. Some of them are compiled in

Table VI Tests for Assessment of the Volatilization of Chemicals from Water to Air

Test type	Principles	Examples
Laboratory tests	Indirect measurement	Decrease in chemical concentration in open vessels; decrease in chemical concentration in wind tunnels; decrease in chemical concentration in closed aerated systems
	Direct measurement	Air–water Henry's law constants; trapping of volatilized chemicals in closed aerated systems
Open air tests	Indirect measurement	Decrease in chemical concentration in flowing waters; decrease in chemical concentration in ponds
	Direct measurement	Trapping of volatilized chemicals over flooded fields at different heights

Table VII. Degradation in the sediment may be assessed by methods listed in Table III for soils. However, a major difference between terrestrial soils and aquatic sediments has to be considered: Whereas soils in most cases are aerobic (i.e., contain free oxygen and a corresponding adapted microflora), sediments often are anaerobic; besides carbon dioxide, other gases such as CH_4, N_2, N_2O, and H_2S may be evolved upon biodegradation of chemicals. In testing anaerobic degradation, anaerobic conditions must be maintained (e.g., by using a nitrogen atmosphere). Besides CO_2, the other gases should be determined also.

Nearly all of the degradation tests in use today are laboratory tests. In contrast to degradation experiments in soil where sufficient microorganisms are available to perform the degradation, most degradation experiments in aqueous solutions require the addition of microbial inoccula. In indirect measurements, substance losses by processes other than degradation can be avoided by the use of closed vessels and by the control of adsorption losses. If, besides the test chemical, no other organic substrate is present in the solution, the remaining organic carbon is a good measure of the nondegraded portion of the test substance. The determination of dissolved or total organic carbon avoids the more complicated determination of parent compound. However, if other carbon sources have been added (e.g., activated sludge), blanks must be analyzed in the evaluation of the remaining organic carbon, or the parent compound and metabolites must be determined by residue analytical procedures.

In the open air, degradation is assayed only by indirect measurement (i.e., measurement of concentration decrease of the test substance in water). This method is even more questionable than in terrestrial fields since volatilization as a competitive process is higher from water than from soil and interferes with degradation measurement. Adsorption in sediments also feigns degradation. Therefore, field methods can be used only for substances that are neither volatile nor adsorbable in sediment to a significant extent (e.g., fully water-soluble xenobiotics).

3. Bioaccumulation

Bioconcentration of chemicals by organisms is the enrichment of chemicals in the organisms by direct uptake from the surrounding medium, resulting in a concentration which is higher than that of the medium; the uptake by contaminated food is not considered.

Biomagnification of chemicals by organisms is the enrichment of chemicals in the organisms by direct

Table VII Tests for Assessment of Degradation of Organic Chemicals in Aqueous Solution

Test type	Principles	Examples
Laboratory tests	Indirect measurement	Decrease in chemical concentration; decrease in dissolved or total organic carbon
		Turbidity and dissolved organic carbon; decrease in chemical concentration occurring abiotically after irradiation; O_2 demand
	Direct measurement	CO_2 production; CO_2 and CH_4 in anaerobic systems
	Direct and indirect measurement	Parent compound and conversion products
Open air tests	Indirect measurement	Decrease in chemical concentration in water bodies

uptake by food. In natural aqueous systems, this process takes place in parallel to bioconcentration.

Bioaccumulation of chemicals by organisms is the enrichment of chemicals in the organisms by direct uptake from the surrounding medium and by food.

Bioconcentration is quantified by the bioconcentration factor. It is the quotient of concentration in the organism divided by that in the medium under equilibrium conditions. Bioconcentration factors of lipophilic chemicals may be estimated from *n*-octanol/water partition coefficients of the respective chemicals.

Most available ecotoxicological tests consider only bioconcentration; if the fish are fed during the experiment, biomagnification may be part of the measured uptake process. Algae, mussels, and fish (e.g., catfish, zebrafish, carp, guppy, rainbow trout, bluegill, fathead minnow, golden ide, and several marine fish) are used as test organisms. Bioconcentration factors are determined after equilibrium is reached. If this is not possible within a reasonable time span, bioconcentration factors can also be calculated as the quotient between uptake and elimination rates.

A static test is a test with aquatic organisms in which no flow of test solution occurs. Solutions remain unchanged throughout the duration of the test. The decrease in chemical concentration in the solution is a measure of bioconcentration. At the end of the test, fish are also analyzed.

A semi-static test is a test without flow of solution, but with occasional batchwise renewal of the test solution after prolonged periods (e.g., 24 hr). A flow-through test is a test in which water is renewed continuously in the test chambers, the test substance being transported with the water used to renew the test medium.

In field tests, enclosures with defined added amounts of chemicals may be used to determine bioconcentration factors. In free-living organism samples, mean concentrations of chemicals may be determined and divided by those in water. The results are bioaccumulation factors.

Some tests for bioconcentration and bioaccumulation are listed in Table VIII. This list does not portend to be exhaustive.

B. Tests for Biological Effects

1. Effects on Invertebrates

Tests to determine effects of hazardous chemicals on aquatic invertebrates comprise short-term and long-term tests that investigate acute and chronic toxicity of chemicals. They are carried out as single species tests with sensitive species, as multispecies tests assaying effects on populations and biocenoses in the laboratory, and as field tests and field observations. Table IX lists some examples.

In single species tests, single species are used in laboratory systems, and the effects of chemicals are assessed by the measurement of certain life functions such as growth, mobility, or reproduction. Sum parameters are used as indicators for effects on certain organism groups, such as the chlorophyll fluorescence for photosynthetically active aquatic organisms, or respiration for microorganisms in activated sludge. Multispecies tests regarding, for example, the complete natural plankton, measure population dynamics comprising the changes in dominance structure, species abundance and biomass, species diversity, local organism distributions, biological fluctuations, and reproduction rates. Experiments that investigate these factors and processes are complex and must not be classified as simple tests. Under field conditions, enclosures have to be used when a comparison with controls is utilized. However, it must be considered that enclosures themselves sometimes develop differing populations even without the influence of a chemical.

Table VIII Tests for Assessment of Bioconcentration or Bioaccumulation of Chemicals from Aqueous Solution

Test type	Principles	Examples
Laboratory tests	Indirect measurement of bioconcentration	Mussel tests; algae tests; static fish tests
	Direct and indirect measurement of bioconcentration	Static fish tests; semi-static fish tests; flow-through fish tests
Open air tests	Direct and indirect measurement of bioconcentration	Tests in enclosures in open water bodies
	Direct and indirect measurement of bioaccumulation	Measurements in open natural water bodies

Table IX Methods to Determine Effects of Chemicals on Aquatic Invertebrates

Test type	Principles	Examples
Laboratory tests	Measurement of effects on single species	Algae growth inhibition test; *Daphnia* acute immobilization and reproduction test; rotifer tests determining individual densities, egg production, and egg development
	Measurement of sum parameters	Chlorophyll fluorescence method; activated sludge respiration inhibition test; bacterial bioluminescence inhibition assay
	Measurement of population dynamics	Aquaria with plankton populations
Open air tests	Measurement of sum parameters	Chlorophyll fluorescence method
	Measurement of population dynamics	Plankton assessments in enclosures
	Observations supporting test results	Plankton assessments in open water bodies

2. Effects on Fish

In fish, acute and prolonged toxicity may be assessed. The tests may be carried out as static tests, semi-static tests, and flow-through tests, as in the case of bioconcentration studies. In acute toxicity tests, mortalities are recorded at different concentrations of the chemical in water. The LC_{50} values, the maximum concentration tested producing no mortality, and the minimum concentration tested producing total mortality are recorded. Recommended species are the zebra-fish, fathead minnow, common carp, red killifish, guppy, bluegill, and rainbow trout.

Prolonged toxicity tests also cover toxic effects other than lethality. The threshold level of observed effects is the lowest concentration of the test substance in the test solution at which the substance is observed to have an effect other than a lethal one on a significant number of fish. NOEC (no observed effect concentration) is the highest-tested concentration of a test substance at which no statistically significant lethal or other effect is observed.

Effects other than lethal effects include all effects observed on the appearance, size, and behavior of the fish that make them clearly distinguishable from control animals (e.g., different swimming behavior, different reaction to external stimulation, changes in appearance, reduction or cessation of food intake, or changes in length or body weight).

IV. Conclusions

Laboratory and field tests yield good partial information on single factors affecting the transport and toxic effects of chemicals in the environment. The interaction between these factors in ecosystems is complex and sometimes not well understood. Field tests give results that are relevant for a specific environment, but which are difficult to interpret. Basic research is required, to develop tests that are designed to elucidate the fundamental relations that permit laboratory observations to be interpreted in terms of field situations.

Related Articles: Biodegradation of Xenobiotic Chemicals; Bioconcentration of Persistent Chemicals.

Bibliography

Bear, J., ed. (1972). "Dynamics of Fluids in Porous Media." Elsevier, New York.

Butler, G. C., ed. (1978). "Principles of Ecotoxicology," SCOPE 12. Wiley, Chichester.

Ebing, W., ed. (1992). "Chemistry of Plant Protection, " vol. 8. Springer Verlag, Berlin.

Kamely, D., Chakrabardy, A., and Omenn, G. S., eds. (1990). "Biotechnology and Biodegradation." Portfolio Publishing Company of Texas, Woodlands.

Korte, F., ed. (1987). "Lehrbuch der Ökologischen Chemie." Thieme, Stuttgart.

Lyman, W. J., Rechl, W. F., and Rosenblatt, D. H., eds. (1982). "Handbook of Chemical Property Estimation Methods." McGraw-Hill, New York.

"OECD-Guidelines for Testing of Chemicals." (1981). OECD, Paris.

Persoone, G., Jaspers, E., and Claus, C., eds. (1984). "Ecotoxicological Testing for the Marine Environment," vols. 1 and 2. The State University of Ghent, Belgium.

Reid, R. C., Prausnitz, J. M., and Sherwood, T. K. (1977). "The Properties of Gases and Liquids." McGraw-Hill, New York.

Sheehan, P., Korte, F., Klein, W., and Bourdeau, P., eds. (1985). "Appraisal of Tests to Predict the Environmental Behaviour of Chemicals," SCOPE 25. Wiley, Chichester.

"U.S. EPA-Environmental Effects Test Guidelines." (1982). U.S. Environmental Protection Agency, Washington, D.C.

Effects of Environmental Toxicants on the Cytoskeleton

Monique Cadrin and David L. Brown
University of Ottawa

Kenneth R. Reuhl
Rutgers University

Glossary

Actin-binding proteins Diverse group of proteins that physically associate with microfilaments and modulate their dynamics and functions.

Cytoskeleton Filamentous system of polymerized proteins consisting of microtubules, intermediate filaments, and microfilaments that regulate structural and functional organization of the cell.

Intermediate filaments 10-nm-diameter protein filaments expressed in a cell- and tissue-specific manner.

Microfilaments 6 nm filaments composed of actin monomers.

Microtubule-associated proteins (MAPs) Diverse group of proteins that physically associate with microtubules and modulate their dynamics and functions.

Microtubules 25-nm-diameter tubular polymers of $\alpha\beta$ tubulin dimers.

***THE CYTOSKELETON** plays an essential role in many cellular processes, including cell movement along a substratum, intracellular movement of particles, development and maintenance of cellular shape, cell–cell interactions, organization of organelles in the cytoplasm, and movement of chromosomes during cell division. The cytoskeleton is composed of three basic types of filamentous structures: microtubules, intermediate filaments, and actin microfilaments (Table I).*

I. Interpretation of Cytoskeletal Changes

The central roles served by the cytoskeleton in cellular structure and function make it a likely target for toxic injury. Nevertheless, alterations in cytoskeleton following exposure to toxins and toxicants must be interpreted with caution, and the temptation to prematurely assign mechanistic importance to cytoskeletal changes without fully understanding their origin must be avoided. There are several reasons why studies of cytoskeletal responses to toxicants are particularly prone to misinterpretation.

Much of the work describing cytoskeletal damage by environmental agents has been done *in vitro* or with cultured cells. While valuable, these approaches cannot replicate the biological complexity of intact tissues. Intracellular homeostasis, hormonal or other circulating factors, and tissue-specific cell modifications of cytoskeleton (e.g., posttranslational modifications) influence cytoskeletal pattern function and stability; such factors can not be fully reproduced in culture.The investigator must always ask whether the effect produced in the *in vitro* or cell culture model represents biological reality.

Most toxicants have many molecular targets. Under conditions of toxic cell injury, cell responses can lead to changes in cellular homeostasis that result in

Handbook of Hazardous Materials

Table I Components of the Cytoskeleton

Components	Composition	Diameter
Microtubules	Tubulin and MAPs	~25 nm
Intermediate filaments	Family of cell- and tissue-specific proteins	~10 nm
Microfilaments	Actin and actin-binding proteins	~6 nm

secondary changes in cytoskeleton. For example, alterations in membrane permeability can result in ionic or fluid imbalances which would favor microtubule disassembly or inhibit microtubule assembly. Similarly, compounds that induce extensive redox cycling, such as quinones, will affect the cytoskeleton indirectly. In addition, toxicants having an effect on one component of the cytoskeleton may indirectly alter another cytoskeletal component. For example, the disassembly of microtubules frequently results in the collapse and aggregation of intermediate filaments. Differentiating primary from secondary cytoskeletal effects is crucial to the proper interpretation of a toxicant's mechanism of action.

Many elements of the cytoskeleton are extremely labile and can be rapidly altered by relatively modest cellular stress. Changes in cytoskeleton in response to exposure to a toxicant may not be pathological but rather a reflection of the cell's normal response to injury or irritation. Most cells possess a capacity to change shape in response to alterations in local environment, and many cytoskeletal changes represent such adaptations.

II. Microtubules

A. Structure and Function

Microtubules are ubiquitous cytoskeletal elements in eukaryotic cells that are critical to cellular functions such as cell motility, mitotic chromosome movement, cell shape, and vesicular transport. Microtubules consist of polymerized $\alpha\beta$ tubulin dimers and a variable number of microtubule-associated proteins (MAPs). The microtubules have a rod-shaped structure that is formed by 13 parallel rows of polymerized $\alpha\beta$ tubulin dimers, or protofilaments, that run parallel to the long axis of the tubule. Microtubules are polarized structures; the addition and loss of tubulin occurs preferentially at one end, designated the plus end. In most animal cells microtubules are primarily assembled from a single microtubule organizing center, the centrosome, with the plus end distal to this site (Fig. 1A). Microtubules are dynamic structures in a constant state of assembly and disassembly. Studies of microtubule dynamics in cells in tissue culture have shown that most interphase cytoplasmic microtubules have a short life measured in minutes and that microtubules at the mitotic spindle may be disassembled and replaced within seconds. In contrast, microtubules in some specialized cells such as neurons are believed to have much longer lives.

The lifetime of microtubules in cells is regulated by a large number of factors including ionic microenvironment, energy status of the cell, and availability and posttranslational modifications of microtubule structural proteins. Of particular importance are the set of proteins referred to as MAPs (microtubule associated proteins). These proteins nucleate microtubule assembly and stabilize the assembled polymer. There is also evidence that they participate in interactions between microtubules and other organelles. MAPs can be differentially distributed within a cell; for example, in neurons MAP1 is located in the axon and cell body while MAP2 is restricted to the dendrite. It is likely that this confers differential stability to microtubules in different cellular compartments.

B. Toxic Effects on Microtubules

A large number of toxic compounds affect microtubules (Table II). These compounds are generally divided into two classes: those that have affinity for specific binding sites on tubulin or MAPs and those that alter the microenvironment necessary for microtubular integrity.

Toxins that bind specifically to tubulin include a number of plant natural products such as colchicine, the vinca alkaloids vinblastine and vincristine, and taxol. Colchicine and the vinca alkaloids result in disassembly of the normal microtubule array, whereas taxol acts to stabilize microtubules and can make them resistant to disassembly. Toxic metals such as methylmercury and a variety of complex organic agents are less specific, but also directly interact with tubulin and can result in microtubule disassembly (Fig. 1B). Sulfhydryl residues on tubulin appear to be particularly vulnerable to binding with toxicants. As many as fifteen sulfhydryl groups are present per tubulin dimer, and binding by a toxi-

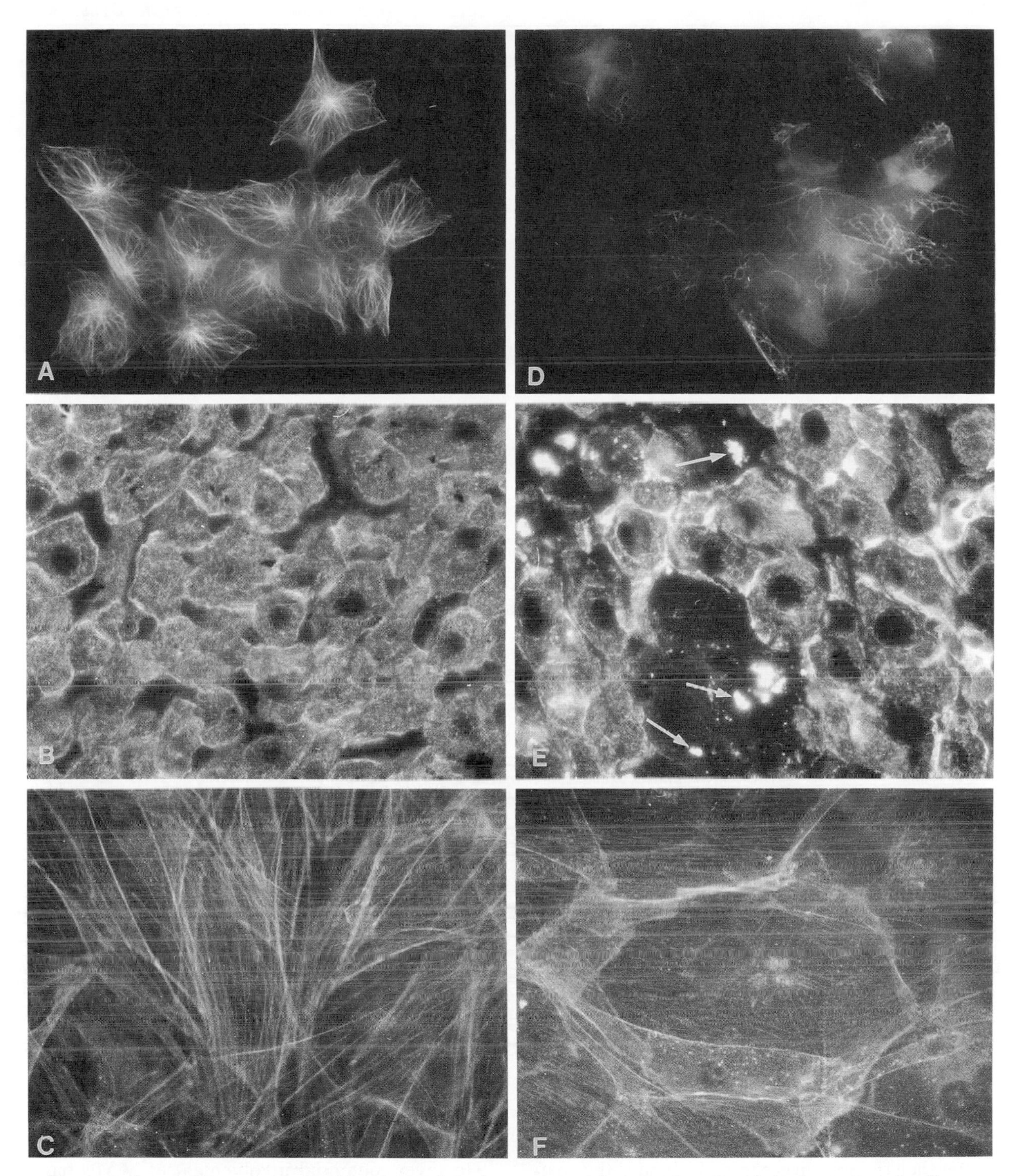

Figure 1 Immunofluorescence staining of cytoskeletal components in control and toxin/toxicant treated cells. (A,B) Microtubules in embryonal carcinoma cells grown in tissue culture. Control cells (A) show a dense network of microtubules extending from the single microtubule organizing center, the centrosome. Methylmercury treatment (B) (5.0 μM for 1 hr) results in a drastic reduction in the number of microtubules present in the cells. (C,D) Immunofluorescence staining of mouse liver frozen sections with an antibody to cytokeratin. In control hepatocytes (C), the intermediate filaments form a complex filamentous network that is more dense at the cell periphery. In mice maintained on a diet containing 2.5% griseofulvin (D) there is a collapse of intermediate filaments and formation of Mallory bodies (*arrows*) in some hepatocytes. (E,F) Distribution of actin in preadipocytes stained with rhodamine–phalloidin. This toxin binds specifically to polymerized F actin. In control cells (E) actin is detected near the plasma membrane and in the form of stress fibers. Cytochalasin D treatment (F) (4 μM for 4 hr) results in actin filament disassembly and a reduction in the number of stress fibers.

Table II Toxins/Toxicants Affecting the Cytoskeleton[a]

Compound	Microtubules	Intermediate filaments	Microfilaments
Acrylamide		X	
Aluminum		X	
Cadmium			X
Carbon disulfide		X	
Cisplatin	X	X	
Cytochalasin			X
Elatone	X		
Ethanol	X	X	X
Griseofulvin	X		
2,5-Hexanedione		X	
IDPN		X	
Lead	X		
Methyl carbamate	X		
Methylmercury	X		
Nickel	X	X	
Phalloidin			X
Phenytoin			X
Porphyrin	X		
Pyridoxine	X	X	
Taxol	X		
Vinca alkaloids	X		
Zinc			X

[a] Representative examples of toxicants and toxins believed to exert primary effects on cytoskeleton. The list is not intended to be exhaustive.

cant to as few as two of these sulfhydryl groups can inhibit microtubule polymerization. These compounds may act at the growing end of the microtubule to inhibit microtubule elongation or to accelerate depolymerization. Some compounds such as methylmercury appear to do both. The result is a loss of normal microtubule network and disturbance of microtubule-dependent functions.

Microtubules within cells are heterogeneous in nature, and display distinctly different sensitivities to toxic agents. This heterogeneity is a consequence of differences in protein composition and in the post-translational modifications to tubulin and MAPs. The difference in sensitivity of the different microtubules has toxicological significance. Some cells contain a substantial number of highly modified microtubules that are resistant to disassembly by traditional microtubule toxins such as colchicine. Consequently, a toxicant may have profound effects upon a subpopulation of microtubules within a cell. In the case of neurons, the perikaryal microtubules are labile and prone to toxicant-induced disassembly, while microtubules in mature axons are more resistant to disassembly. Cellular functions subserved by these resistant microtubules are more likely to be preserved.

Some toxic compounds, including various toxic metals and the chemotherapeutic drug, estramustine, may exert their effects upon the MAPs. However, potential involvement of MAPs has not been rigorously investigated for most toxicants.

The list of compounds that affect microtubule integrity secondarily is much larger than direct microtubule perturbing agents, and includes compounds that alter intracellular calcium homeostasis or energy status. This would potentially include any compound that causes cell damage. The degree of secondary microtubule damage depends on the type and severity of the cell injury. If cellular homeostasis is regained, it is likely that the microtubule network will be completely rebuilt. Consequently, distinguishing primary from secondary microtubules damage represents a particular challenge in the interpretation of toxic mechanisms.

III. Intermediate Filaments

A. Structure and Function

Intermediate filaments constitute a complex family of polypeptides that are classified into six groups differentially expressed in various tissues. Types I and II intermediate filament proteins, the cytokeratins, are found in epithelial cells. Type I corresponds to the acidic keratins and type II to the neutral basic keratins. Equimolar amounts of type I and II are necessary to form the cytokeratin intermediate filament. The type III group is composed of vimentin of mesenchymal cells, desmin of muscle cells, and glial fibrillary acidic proteins of astrocytes. The type IV group is formed by the three neurofilaments proteins of nervous system. The type V group is represented by the nuclear lamins, nestin present in the neuroepithelial stem cells corresponds to type VI.

The intermediate filaments form a complex filamentous network in the cells (Fig. 1C). The specific role of intermediate filaments in cells is poorly understood. Their close association with the nuclear envelope and cell membrane suggest involvement in signal transduction. It is also believed that they play a major role in cytoplasmic organization. Some intermediate filament types (e.g., cytokeratins) may strengthen epithelial cells against mechanical stress. There is good evidence that the nuclear lamins function to maintain the structural integrity of the cell nucleus.

B. Toxic Effects on Intermediate Filaments

Because the diverse nature of different intermediate filament classes, no generalizations can be made regarding their vulnerability to specific toxicants. *In vivo* as well as *in vitro* studies suggest that intermediate filament proteins are primary targets for a number of toxic agents (Table II). In most cases examined, toxic injury to intermediate filaments is manifested by alterations in structural organization, particularly by accumulation of intermediate filaments in the form of dense bodies, tangles, or cytoplasmic swelling packed with intermediate filaments. Neurofilament aggregation has been particularly well studied. *In vivo* studies have shown that neurofilament aggregations are induced by many compounds including aluminum, acrylamide, β,β'-iminodipropionitrile (IDPN), solvents such as carbon disulfide 2,5-hexanedione and hexacarbon and by triethyl lead. Other intermediate filaments such as vimentin and cytokeratins also show aggregation and bundling following exposure to toxicants.

The mechanisms by which toxicants affect intermediate filament organization are unclear. Studies indicate that some toxicants react directly with intermediate filaments to induce cross-linking of adjacent filaments. Cross-linking may result from cationic binding between the toxicant and the negatively charged amino acids on the intermediate filament proteins, as has been postulated for aluminum. Alternatively, the toxicant may chemically alter components of the intermediate filament. For example, γ-diketones such as 2,5-hexanedione may react with lysine residues to form amines which cyclize to pyrroles. Auto-oxidation of the pyrroles could result in cross-linking of intermediate filaments. Cross-linking may also be induced indirectly by compounds that induce enzymes, such as transglutaminase, which are also known to cross-link proteins. This mechanism has been proposed to explain the formation of Mallory bodies, intermediate filament aggregates observed in hepatocytes following ethanol intoxication in humans or griseofulvin intoxication in mice (Fig. 1D).

Toxicants may also change intermediate filament organization by altering the level of filament phosphorylation. For example, treatment of epithelial cells in culture with acrylamide results in cytokeratin aggregates concomitant with a decrease in cytokeratin phosphorylation. Altered intermediate filament phosphorylation is also observed in longstanding neurofilamentous aggregates following γ-diketone exposure and in neurodegenerative diseases such as Alzheimer's. It is not yet clear whether or not changes in intermediate filament phosphorylation are primary or secondary effects of toxicity.

IV. Microfilaments

A. Structure and Function

Microfilaments represent a class of thin filamentous protein polymers present in all eukaryotic cells (see Table I). Actin exists in cells either in a globular monomeric, or G form, or as thin, tight helical pairs of polymerized filaments (F actin). Approximately 50% of cellular actin exists as F actin while the remainder exists either as free actin monomers or as small complexes with proteins. The exact ratio of free to polymeric actin varies according to cell type and the physiological state of the cell. A dynamic equilibrium exists between the polymerized filaments and the unpolymerized actin. The actin filaments are polarized structures; one end is referred to as "barbed" and the other end as "pointed." Elongation of the filament can occur from either end, although addition at the barbed end is favored. The actin filaments can be readily disassembled and reassembled in response to changes in the cellular microenvironment.

A large number of actin-binding proteins have been identified that participate in the regulation of filament assembly–disassembly, facilitate interactions between F actin filaments resulting in actin cross-linking and bundling, and control actin interactions with the plasma membrane or other organelles.

Actin participates in many critical cellular functions, including cytokinesis, motility, maintenance of cell shape, and modification of cell shape during injury. Actin filaments are particularly prevalent immediately beneath the plasma membrane and in bundles (stress fibers) extending through the cytoplasm (Fig. 1E), where they reinforce the general cytoarchitecture and anchor transmembrane proteins such as receptors and adhesion molecules to the cell. In this manner, intercellular and cell-substrate adhesions may be stabilized. Actin filaments are also attached to microtubules and intermediate filaments and participate in the transduction of extracellular information and rearrangement of cytoskeletal structure.

B. Toxic Effects on Actin Microfilaments

Relatively few environmental compounds are known to exert their primary effects upon the microfilament component of the cytoskeleton. Examples include heavy metals such as zinc and cadmium, complex organic molecules such as phorbol esters, and toxins, such as cytochalasin, phalloidin, and Toxin B of *Clostridium difficile* (Table II). The mechanisms by which cytochalasin and phalloidin affect actin have been characterized at the molecular level, and involve shifting the dynamic equilibrium toward disassembly (cytochalasin, Fig. 1F) or assembly (phalloidin). The mechanisms by which metals or the phorbol esters affect actin are more obscure, and may involve interaction of the toxicant with a variety of calcium-dependent molecules, such as protein kinase C, or with crucial actin-binding proteins. Thus far, direct interactions between toxicants and actin-binding proteins have not been examined.

Alterations of the actin cytoskeleton as a secondary event would appear to be common. Disturbance of the subplasmalemmal actin network results in the cytoplasmic "blebbing observed in cultured cells exposed to a wide variety of toxic compounds. Exactly how these blebs form is unclear; however, the process appears to involve local loss of actin support of the plasma membrane, leading to evagination of the membrane and cytoplasm.

V. Concluding Remarks

The cytoskeleton is a highly complex, delicately regulated system. While there is considerable evidence for direct toxic effects on specific components of the cytoskeleton, the complex interrelationships of these components makes distinguishing primary events from secondary events elusive.

Recent evidence suggests that changes associated with oxidative stress can perturb intracellular homeostasis sufficiently to alter cytoskeleton. These changes may involve direct oxidation or alkylation of cytoskeletal protein thiols, an increase in intracellular calcium, activation of calcium-dependent proteases, or other as yet uncharacterized pathways. Since oxidative stress is an extremely common response of cells to a variety of toxic compounds, damage to the cytoskeleton is a likely element in the toxic response to many agents.

Acknowledgment

Preparation of this work was supported by MRC (MC) and NSERC (DLB) and NIH ES-04976 (KRR).

Bibliography

Bretscher, A. (1991). Microfilament structure and function in the cortical cytoskeleton. *Ann. Rev. Cell Biol.* **7,** 337–374.

Burgoyne, R. D., ed. (1991). "The Neuronal Cytoskeleton." Wiley-Liss, New York.

Clarkson, T. W., Sager, P. R., and Syversen, T. L. M., eds. (1986). "The Cytoskeleton: A Target for Toxic Agents." Plenum Press, New York.

Gelfand, V. I. (1991). Microtubule dynamics: Mechanism, regulation, and function. *Ann. Rev. Cell Biol.* **7,** 93–116.

Pollard, T. D., and Goldman, R. D., eds. (1993). "Cytoplasm and Cell Motility." Current Opinion in Cell Biology, Vol. 5.

Sager, P. R., and Matheson, D. W. (1989). Mechanisms of neurotoxicity related to selective disruption of microtubules and intermediate filaments. *Toxicology,* **49,** 479–492.

Environmental Cancer Risks

William Lijinsky
NCI-Frederick Cancer Research and Development Center

Glossary

Carcinogen Agent that causes cancer.

Dose–response Relationship of size of dose to physiologic response; a larger dose of a carcinogen produces more cancers or reduces the time for cancer to appear.

NDMA Nitrosodimethylamine; the simplest and most widespread nitrosamine.

Nitrosamine Dialkylamine in which the third substituent is a nitroso group; formed by reaction of a secondary or tertiary amine with a nitrosating agent.

PAC Polycyclic aromatic compound; contains three or more aromatic rings fused together.

Pyrolysis Heating without combustion.

***THE INVESTIGATION** of agents that cause cancer began early in this century and has continued steadily; the most important findings have been the connection between cigarette smoking and lung cancer, and the identification of occupational cancers related to exposure to carcinogenic substances in factories. The most important types of cancer in industrialized countries are lung, breast, colon, prostate and bladder, while the most common types in less developed countries are liver, stomach, cervix, and esophagus. Immigrants tend to develop cancers typical of their new countries, indicating that genetics may play a minor role in humans, a phenomenon also observed in species of experimental animals. These disparities indicate that differences in exposure to environmental cancer-causing agents are responsible for many cancers. These include sunlight and other radiation, infectious agents (viruses), habits (tobacco use), diet (including products of preparation and cooking, and residues of agricultural chemicals), drugs, and medicines. Exposure to these agents is usually multiple and over a lifetime an individual is exposed to various combinations, the combined effects of which are difficult to calculate. Most exposures are to low concentrations of the carcinogens which, nevertheless, result in a cancer incidence of one in four in populations of industrialized countries (approximately 4 million per year, of which more than 2 million are fatal). Stomach cancer, however, appears at high incidence in both industrialized and less developed countries, causing the yearly deaths of one million people worldwide.*

I. Specific Cancer Incidence in Various Regions and Societies

The large difference in incidence of most common human cancers among the regions of the world indicate that among humans there are many factors other than genetics that contribute to cancer occurrence. For example, the high incidence of colon and breast cancer in Europe and the United States contrasts with the low incidence of liver cancer in those countries, whereas colon and breast cancer are relatively rare in less developed countries (e.g., Africa and Asia) in which liver cancer is common. Lung cancer is common where tobacco smoking is common. Stomach cancer is common in Eastern Europe, Russia, Japan, and South America, but has declined in Western Europe and the United States. Emigrants from Africa and Japan have tended to assume the cancer patterns of the new country by the second or third generation. These differences suggest the importance of diet, daily habits, and occupation in determining the common cancers that appear in different societies and geographical areas. Exposure to chemical carcinogenic agents seems to be an important contributory factor; it is protection from such exposure that offers the greatest promise for reducing cancer risks, as the numbers of people with even relatively uncommon cancers is large and the overall incidence of cancer (25% of the population) is great.

II. Main Types of Carcinogen

Until recently, the discovery of new carcinogens was sporadic, and most carcinogens were members of large groups of highly carcinogenic compounds. These include polynuclear aromatics (PACs) which are both homocyclic and heterocyclic, aromatic amines, *N*-nitroso compounds, hydrazines, nitrohydrocarbons, carbolines, metal compounds and some mycotoxins. However, a later basis for testing compounds for carcinogenicity (by the National Toxicology Program) has changed to a primary concern with widely used compounds of economic importance, and is not directly related to chemical structural characteristics. A high proportion of more recently studied compounds are not mutagenic, thus weakening the earlier link between mutagenicity and carcinogenicity. Activity of compounds in short-term assays is no longer considered a reliable guide to carcinogenic effectiveness nor to mechanisms of carcinogenesis. Chronic toxicity tests in animals are the only reliable source of information about the carcinogenic activity of chemicals. The relative differences in response to a carcinogen among species provides an additional measure of carcinogenic effectiveness, although there is no means of predicting from the results in experimental animals which tumor would be induced in humans. Table I shows the primary types of environmental carcinogens and their use or origin.

This list is not comprehensive; some carcinogens made by living organisms are described elsewhere in this volume. In the absence of evidence to the contrary it must be assumed that exposure to even small amounts of carcinogens contribute to an increased cancer risk. Calculation of the increased risk from information derived from experiments in animals is not straightforward, because the relative carcinogenic potency in experimental animals and in humans is not known, nor is the shape of the human dose–response curve at low doses. Calculation of carcinogenic risks to humans from the results of short-term, primarily mutagenesis assays is not possible.

Table I Types of Carcinogen

Class	Example	Source
Polycyclic aromatic	Benzo(*a*)pyrene	Burning organic matter, engines, fuel, air, cooking, tobacco smoke
Nitroarene	1,4-Dinitropyrene	Air, diesel engines
Aromatic amine	Benzidine	Dyes and chemical manufacture
	Methyl IQ	Frying, broiling of meat/fish
Nitrosamine	Nitrosodimethylamine Nitrosomorpholine	Food, air, toiletries, tobacco, factories
Vinyl compound	Vinyl chloride	Plastic manufacture
Azo-dye	4-Dimethylaminoazobenzene	Dyes, food colorings
Halogenated hydrocarbon	Carbon tetrachloride	Solvents, pesticides, agricultural chemicals
Hydrazine	1,1-Dimethylhydazine	Chemical factories, rocket testing, some plants
Inorganic	Lead chromate	Metal smelting and processing, paints, sludge
Hydrocarbons	Benzene	Fuels, solvents, chemical manufacture

III. Measurement of Carcinogenic Potency

Measurement of relative carcinogenic potency among carcinogens of the same chemical class or relative to the same target organ was in earlier days considered simple. This precipitated the introduction of the Iball Index for assessing the carcinogenic potency of mouse skin carcinogens (e.g., polycyclic aromatic hydrocarbons, PAHs)—the higher the response, the more potent the carcinogen. The calculations are not, however, directly transferable to other species, partly because the carcinogenic response of PAHs is lower in rats, for example, then in mice. As a group, PAHs appear to be quite effective skin carcinogens in humans, and much of the early work by Kennaway, Clar, Shear, Andervont, and their associates dealt with these compounds precisely for that reason. PACs are also present in tobacco smoke, in cooked foods, and as an air pollutant, but their potency by inhalation or ingestion in experimental animals is quite low, compared with their effect in mouse skin. Mixtures in which they occur are very complex and contain multitudes of PACs, so that extrapolation of risk for all compounds would be a daunting task.

There have been few dose–response studies of carcinogens sufficiently extensive to justify mathematical analysis of the results and potential extrapolation to humans. However, one large study in rats with the aromatic amine acetylaminofluorene (albeit a compound in which human exposure is restricted to a few scientists), showed the usual sigmoid curve relating tumor incidence to dose. Similar curves are seen for dibenz (*a,h*) anthracene in mouse skin, by nitrosomorpholine in rats (Fig. 1), and by a number of other nitrosamines to which humans are frequently exposed. Among nitrosamines, and even more so for other types of carcinogen, the size of the minimally effective dose in experimental animals varies widely. The smallest cumulative effective dose over the lifetime of rats is approximately 2 to 5 mg/kg body weight of nitrosodimethylamine (NDMA), nitrosodiethylamine, or nitrosomorpholine; it is considerably larger in mice. Other carcinogens are less potent in rodents (i.e., larger doses are required to produce the same tumorigenic effect). Levels for benzo(*a*)pyrene in hamster lung are 50–100 mg/kg, vinyl chloride 4000 mg/kg, azo dyes 2000 mg/kg, carbon tetrachloride 5000 mg/kg in rats, and 2-naphthylamine or benzidine 30,000 mg/kg in dogs. In the cases of substances known to induce cancer in humans—a small number, usually related to occupational exposures—it is not possible to measure carcinogenic potency, nor the minimum effective dose over a lifetime. In the past, before a substance was known to be carcinogenic, exposures were not known; after it was known to be carcinogenic, exposures were reduced or eliminated.

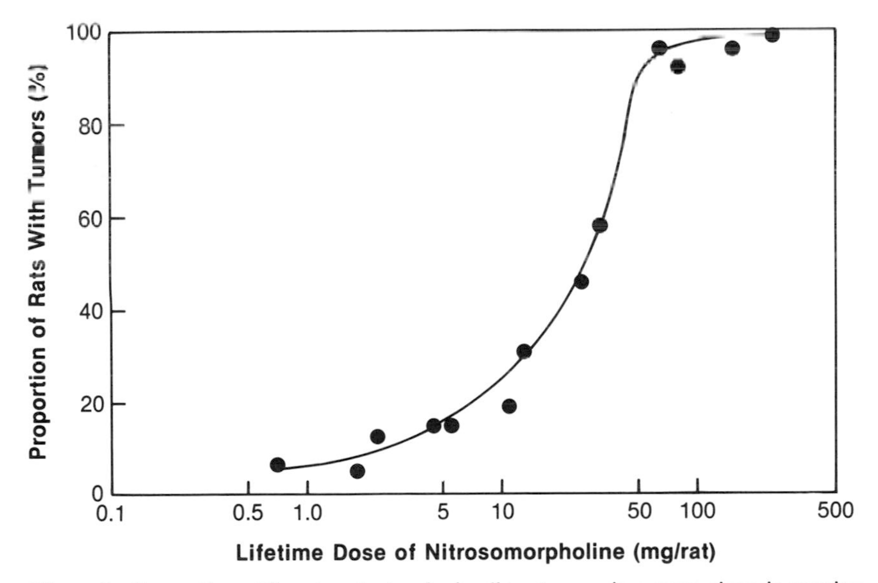

Figure 1 Proportion of female rats developing liver tumors in groups given increasing doses of nitrosomorpholine in drinking water. Total lifetime doses administered ranged from 0.7 to 250 mg per rat.

IV. Extrapolation of Risks

As cancer is a result of chronic toxicity, the usual process of translating the results of experiments in animals to predictions of human response is extremely difficult. Not only are there differences between experimental animals and humans in metabolism of a carcinogenic agent, but the multitude and timing of intervening steps before an identifiable tumor mass emerges might be quite different between species, (e.g., humans outlive rodents by 25 or 30 times). On the other hand, it would be unjustified to ignore the results of experiments showing the induction of cancer in animals after application of a substance. The simplest use of the information would be a direct relationship between dose and carcinogenic effect, represented as the proportion of animals with tumors. Carcinogenic effect can also be calculated from the decreased time it takes for larger doses of the agent to cause lethal tumors. Such calculations of increased cancer risk from exposure to a given dose of a particular carcinogen might occur on the order of 1 in 10^5, for example, and might vary by one or more orders of magnitude. However, in safeguarding the public health, the most conservative number showing the higher, rather than the lower, potential increased risk must be used. When there is exposure to several or many cancer-causing agents in the assessment, the several risks must be at least added, although there is a possibility of inhibition of the action of one carcinogen by another, or—more likely—a possibility of synergism of action among some of them. The experimental basis for assessing the action of mixtures of carcinogens is unfortunately inadequate, because of the complexity and expense of suitable experiments.

Although the observation that a substance that induces tumors in experimental animals does not prove that it would have the same effect in humans, it is more likely to pose an increased carcinogenic risk to humans than a substance that does not induce tumors in experimental animals. It should be treated accordingly: observed differences in potency among carcinogens suggests that a very potent carcinogen, such as tetrachloro-*p*-dioxin, aflatoxin B_1, benzo(*a*)pyrene or nitrosomorpholine should be treated with considerably more concern than a weak carcinogen, such as DDT, dioxane, or dioctylphthalate. Similarly, volatile carcinogens such as nitrosodimethylamine, vinyl chloride, and acrylonitrile should be contained more securely than the nonvolatile nitrosodiethanolamine, nickel compounds, and chromates. In the case of "daily habits" and food additives it is possible to virtually eliminate exposure to associated carcinogens by avoiding their use, but in occupational settings, or in air and water exposures, concentrations of carcinogens must be reduced to the lowest possible levels.

V. Mechanisms of Action of Carcinogens

The precise mechanism by which any carcinogen gives rise to tumors is unclear. Considerable effort during the past 50 years or more has been devoted to investigating the metabolism and activation of carcinogens to reactive intermediates that interact with cellular macromolecules (e.g., DNA). Exposure induces mutations, which are then believed to bring about transformation of some cells to the state of uncontrolled cell division referred to as neoplasia or cancer. Cancers typically arise months or years after the beginning of exposure, even in cases where exposure is continuous; little is known of the intermediate stages, or why so long a time elapses before manifestation of the mutational changes results in a diagnosable tumor.

Many carcinogens are mutagens when tested *in vitro* in systems; (e.g., those using specially engineered bacteria or mammalian cells in culture). There are also many carcinogens that are not mutagens in any of the test systems, yet the tumors they induce in animals are not different in character or behavior from those produced by mutagenic carcinogens. There have been attempts to draw distinctions between mutagenic and "nongenotoxic" carcinogens, and to regard the latter as less important or more tolerable than the former. There is no foundation for this distinction, because evidence is weak that the mutagenic DNA adducts formed with the carcinogen are prime causes of the tumor. Adducts frequently occur in both target and nontarget organs in large numbers of cells, only a few of which eventually form tumors. Other factors (biological, biochemical, and physiological) thus play a predominant role in conversion of normal cells by the carcinogen into tumor cells.

It seems clear that properties associated with particular chemical structures enable those molecules to cause cancer, but it is not known why certain organs in some species, but not in others, are targets. Changes induced by the carcinogen in the cell surface as well as the interior of the cell and the nucleus participate in the complex sequence of events that produce altered cells that progressively change to autonomous, malignant cancer cells.

VI. Methods of Risk Control

Carcinogens of many different types and potencies are found in a large number of sources, and control of human exposure to them depends on their nature (chemical and physical), concentration, the milieu in which they are found, the mode of exposure, and "carcinogenic effectiveness" or potency. Although the risks posed by small amounts of potent carcinogens and large amounts of weak carcinogens might be similar, different approaches are likely to be used to protect the public from the two. The simplest method, of course, is to avoid exposure entirely, (e.g., by ceasing production of particular carcinogenic chemicals). This occurred in the case of benzidine, an intermediate in dye manufacture that causes bladder cancer in dogs and humans, but liver tumors in mice and rats. Its manufacture and use was phased out decades ago. The insecticide DDT is a weak carcinogen that used to be manufactured in vast quantities, but is no longer sold in industrialized countries, although it is sold in less developed countries. Other carcinogens are inevitably byproducts of manufacture, for example nitrosamines in the rubber industry, so that human exposure can be reduced, but not eliminated. Some natural products are carcinogens, or are closely related to carcinogens (e.g., hydrazine derivatives in certain mushrooms), and are difficult to avoid.

It is fortunate that few chemical compounds are carcinogenic, and that few of those are potent carcinogens. Of the several thousand compounds tested thus far in animals, 10% or less are carcinogenic, which makes the problem more manageable. Nevertheless, most cancers in humans have not been linked to specific environmental exposures, other than the carcinogens in tobacco smoke, and there are undoubtedly important contributors to human cancer that are yet undetected; some of them might be formed within the body, thereby complicating protective measures.

A. Carcinogens in Food and Drinks

The most notorious carcinogens in food are probably the aflatoxins, which are natural products of molds and were detected as the liver toxins in contaminated peanut meal (which killed turkeys) because they are strongly fluorescent in ultraviolet light. They are the most potent carcinogens known to rats, but very weak carcinogens in mice and monkeys. Their activity in humans is not clear, but it is prudent to assume that humans are no less susceptible than rats to these carcinogens. The same is true of other naturally occurring plant toxins. Exposure to those carcinogens produced by molds can be reduced by improving methods of storage so that molds do not grow, or at least where their growth is limited.

An important source of carcinogens in food is methods of processing and preservation, which includes cooking, a process by which chemical reactions produce attractive (appetizing) flavors and colors in food, which otherwise might be bland and uninteresting (pity our distant ancestors). Food additives, including spices and condiments, accomplish the same end. Preservation and pickling extend the time for which food may be kept without bacterial spoilage, beyond that achieved by cooking. All of these processes, however, have the potential for exposure to a variety of carcinogens (albeit usually at low concentration). These exposures cannot be ignored, however, because the carcinogens include some of the most potent we know.

Products of cooking that might be carcinogenic are of two kinds, those produced at moderate temperature and those produced at very high temperatures. The latter primarily comprise end-products of pyrolysis (e.g., PACs which include hydrocarbons and heterocyclic compounds). Among these, PAHs are the most extensively studied and are produced when the source of heat is below the food (as in charcoal or wood broiling). Here, melted fat drips on to the hot fuel, is pyrolyzed, and PAHs rise in the smoke to be deposited on the food. Fatty meat, therefore, contains the highest concentrations of these carcinogens—as much as 50 μg per kg—while chicken, fish, or lean meat contain much less. The simplest way to reduce this exposure to carcinogens is to heat the food from the top or sides, or to segregate the food from the heat source.

Food cooked at lower temperatures can contain carcinogens of a different type, namely heterocylic polyaromatic amines (carbolines) formed by heating amino acids, as extensively studied in Japan. A series of amines formed from tryptophan, glutamic acid, and other natural nitrogenous compounds in food have been isolated (using their strong mutagenicity as an index) and characterized chemically. Many of them are powerful bacterial mutagens, but they are carcinogens of medium potency. Oral ingestion produces tumors of the liver, intestine, and other organs in rats. Although present at low concentrations in cooked food, they nevertheless contribute to an increased carcinogenic risk in humans. Avoiding them completely would require abandoning cooking and eating food raw, as may have been Nature's intent in our design.

Pyrolysis of foods, particularly of fats, produces a great variety of organic compounds, many of which are simple carbonyl compounds (such as acrolein) and unsaturated compounds. It is probable that some of these compounds are carcinogenic, but few have been identified and even fewer have been tested for carcinogenic activity. Since there are no established causative agents for such common human cancers as those of the colon, breast, prostate, and stomach, it is wise to reduce exposure to those carcinogens associated with diet.

Apart from cooking, the most common source of non-naturally occurring carcinogens in food is preservation and flavoring/coloring. Since the discovery a century ago that certain azo dyes are carcinogenic (usually inducing liver tumors in rats), close attention has been paid to the possible carcinogenicity of all synthetic food colorings and use of many of them has been discontinued. Not only are several of them carcinogenic, but a number of those based on 2-naphthylamine, benzidine, and other carcinogenic aromatic amines can be degraded by intestinal bacteria to form those amines.

Among many preservatives, the antioxidants butylated hydroxyanisole (BHA) and butylated hydroxytoluene (BHT) have shown some carcinogenic activity, particularly BHA which induced tumors of the forestomach in rats. It would seem that alternatives to these widely used compounds should be found. In relation to carcinogens, the most important food additive may be sodium nitrite and its precursor, sodium nitrate. Sodium nitrate occurs naturally in all plants, is present in human body fluids, and is a normal constituent of urine. Nitrate is reduced by bacteria in the mouth to nitrite which is consequently swallowed to become part of stomach fluids in low concentrations. Nitrite used to "cure" (color, flavor, and preserve) meat and fish is a prominent component of bacon, ham, sausages, corned beef, and other cured meats, which are consumed in very large quantities in industrialized countries, where they represent perhaps a quarter of all meat eaten. Nitrites react with secondary amines, and more slowly with tertiary amines, to form nitrosamines which, as a group, are the most potent and most broadly acting carcinogens known. Because nitrosamines—or more comprehensively—*N*-nitroso compounds, have induced virtually every kind of human tumor in some species of animal, they can be considered to contribute to the formation of any cancer found in humans who are exposed to them. Among those nitrosamines that have been subject to dose–response studies, even the lowest dose rates, amounting to a total of a few milligrams exposure during the lifetime of a rat, have given rise to tumors.

No "safe" level of exposure to carcinogenic *N*-nitroso compounds has been established, and they have induced tumors in every species to which they have been administered, more than 40. Exposure in foods might be reduced by eliminating or reducing the amount of nitrite used in meat curing. However, this would not entirely eliminate exposure because of the ease with which they can be formed endogenously by reaction of nitrites or other nitrosating agents with secondary or tertiary amines. This occurs in mildly acid conditions (pH 3 to 5), which exist in the stomach during digestion of a meal. The nitrite can come from residues in cured meats or fish, or in smaller amounts from saliva. The reaction is faster with less basic amines (such as morpholine) and slower with strongly basic amines (such as dimethylamine). Other structural characteristics also influence the rate of formation of nitrosamines. The reaction is catalyzed by nucleophilic anions and by a number of carbonyl compounds, so that nitrosamines can be formed in neutral or slightly basic conditions as well. Among amines that form *N*-nitroso compounds are naturally occurring plant components, residues of agricultural chemicals, and a host of drugs and medicines. The latter include some antihistamines and analgesics that have been fed to rats together with sodium nitrite and have given rise to large numbers of tumors caused by the nitrosamines formed in the stomach. A method has been developed for estimating the amount of *N*-nitroso compounds that could be formed endogenously under particular conditions, but the estimates are subject to considerable error. To reduce exposure to this particular carcinogenic risk it is wise to avoid large doses of nitrosatable amines as well as large amounts of vegetables with high nitrate content (such as beets and spinach); and, to take ascorbic acid or vitamin E; they compete with amines for reaction with nitrite and therefore lower the amount of *N*-nitroso compound formed in the stomach.

The formation of NDMA by reaction of amine alkaloids in barley with nitrogen oxides in burning fuel has led to the presence of that nitrosamine in malt used to make beer. A decade ago this caused some consternation because, although the concentration of the nitrosamine in beer (in Germany) was 50 ppb or less, beer is drunk in such vast quantities in Germany. Only certain types of beer, however, had high concentrations of NDMA; a worldwide examination of beer showed that concentrations of 5 to 10 ppb were common. A rapid response of the brewers led to redesign of malt-houses so that the barley was segregated from the hot fuel containing the nitrosating gases. Present-day beers contain 5 ppb of NDMA or less. There are low concentrations of NDMA in whiskey, from the same source.

The synthetic estrogenic compound diethylstilbestrol has been used as an implant to hasten maturation of chickens and fatten cattle and has thus sometimes been found as a residue in meats. It is carcinogenic in humans, causing gynecomastia in industrially exposed male workers and adenocarcinoma of the uterus in girls whose mothers took it pharmaceutically while they were in the womb. The effect was discovered because that tumor is usually very rare in humans.

Other carcinogenic risks in food are from residues of solvents (methylene chloride) used to extract, for example, caffeine from coffee. Residues of agricultural chemicals, including a number of nitrosatable amino compounds might be precursors of nitrosamines when they enter the stomach. Included are carbaryl and other *N*-methylcarbamate insecticides, herbicides such as fenuron and monuron, and thiram. Many other pesticides, (e.g., chlordane and heptachlor, aldrin and dieldrin, atrazine, and daminozide) can be residues on plants, and when consumed those that are carcinogenic represent an increased cancer risk. More stringent control of the use of such carcinogenic, but industrially useful substances will minimize their harmful effects. Though present at low concentrations, exposure can be frequent and long continuing, and effects can persist over decades. Another source of carcinogenic residues in food is migration of substances from packaging materials. These include dyes from labels, carcinogenic monomers from plastics (e.g., vinyl chloride, acrylonitrile), urethane from plastic drink bottles, and plasticizers, including dioctyl phthalate and dioctyl adipate. Although individual exposures to each of these may seem trivial, their widespread use in vast numbers of products can result in a considerable lifetime exposure. As discussed earlier, combinations of carcinogens can act additively or possibly even synergistically, thereby magnifying apparently small individual effects.

B. Pesticides

In addition to pesticide residues in foods, the problem of carcinogenic pesticides is magnified in the heavy exposure of workers during pesticide manufacture and application in the field. Cancer-causing properties are, however, far from the most important attributes of pesticides, many of which are powerful poisons, particularly the organophosphorus compounds, including parathion and malathion. Those that are carcinogenic are mostly less potent than nitrosamines or PAHs.

The carcinogenic pesticides are primarily chlorinated hydrocarbons, such as chlordane, heptachlor, toxaphene, aldrin, and dieldrin. These compounds usually induce liver tumors in mice and often also in rats. This result is often dismissed as unimportant, because liver cancer is not common in Western countries. However, it must be remembered that aflatoxin B_1 is almost exclusively an inducer of liver tumors in rats, and that by no means is the target organ of a carcinogen in a rodent always indicative of the target organ of that carcinogen in humans.

Other carcinogenic agricultural chemicals include aminotriazole (used on cranberries), daminozide (used on apples), and ethylenethiourea (a degradation product of bis-diethylthiocarbamate fungicides) which induces thyroid tumors in rats. Dichloroethane and dibromochloropropane are widely used as fumigants for grain and soil. The pesticide trifluralin and its analogs contain small amounts of nitrosamines, such as nitrosodi-*n*-propylamine as by-products of manufacture involving nitration. The content of nitrosamines was, however, previously much higher, but the manufacturing procedure was modified when the nitrosamine product was identified. Few agricultural chemicals are potent carcinogens, yet exposure to them can occur frequently in consumers and workers. They can, cumulatively, constitute a considerably increased potential cancer risk, as in the case of other mixtures of carcinogens where an additive effect can be expected.

C. Occupational Exposure

Occupational exposure to carcinogens represents the greatest risk of cancer in humans. Indeed, workers in certain occupations can attribute development of particular cancers almost entirely to exposure to identifiable carcinogens. Most knowledge of causation by specific agents is due to clinical observation in groups of people in particular occupations who develop particular (and unusual) types of cancer. An early example was the observation of scrotal skin cancer in young men who were chimney sweeps by the surgeon Percival Pott in 1775. Similarly, bladder cancer was observed in workers in dye factories (originally called by Rehn "aniline cancer") as a result of exposure to aromatic amines. Observations by Volkmann in the 1870s of skin cancer in coal tar and shale oil workers, were attributed to PAHs in materials (e.g., Creosote) coupled with the later finding (as early as 1915) by Yamagiwa and Ichikawa that painting coal tar or similar hydrocarbons on the skin of mice gave rise to skin cancer. This was the first connection between a cancer in humans from exposure to chemical agents and induction of the same type of cancer in animals by the same agents. There have also been many instances of a carcinogen inducing different tumors in experimental animals and in humans. Radiation has been identified as carcinogenic for a long time, following the early discovery that X-rays cause several types of cancer, (e.g., skin and thyroid). Excessive exposure to sunlight has been linked to skin cancer, including melanoma, and fair-skinned people are particularly susceptible. Early workers who painted watch and clock dials with luminous paint containing radium salts frequently licked the brush and were early victims of cancer (especially of the bones) caused by the alpha particles emitted by the radium.

Another physical agent, asbestos, has been long associated with cancer of the lung. Yet it was only later through the devoted efforts of Selikoff and other physicians that attention was drawn to the widespread effects of asbestos in the working population, including those installing insulation, plumbers, and construction workers. Cigarette smokers exposed to asbestos have an enormously increased risk of developing lung cancer—almost 100%—much greater than among those who only smoked, or who only worked with asbestos. This is one of the clearest cases of synergism of action of two carcinogenic agents of quite different chemical nature.

Among industrial application of inorganic materials, nickel refining (through exposure to nickel carbonyl) and chromate production have long been associated with lung cancer in workers. More recently other compounds of nickel and beryllium have been studied and found to be carcinogenic. Cadmium compounds have been implicated in tumors of the prostate in humans, while arsenic compounds have been considered carcinogenic for centuries, since the arsenic derivatives used to control molds in European vineyards were associated with cancer in workers who tended the vines; workers in metal smelters who were exposed to arsenic also were reported to develop cancer. Arsenic compounds, which will be discussed elsewhere, represent the only group of compounds known to induce cancer in humans for which there is no adequate model in experimental animals. Some lead salts have induced kidney tumors in animals, but evidence in humans is inadequate. Although chromates (hexavalent chromium) are carcinogenic, chromium salts (trivalent chromium) appear not to be; nor is there evidence that mercury compounds are carcinogens.

In 1974, epidemiological observations showed a connection between a rare cancer, angiosarcoma of the liver, in workers in a polymer factory, and heavy exposure to vinyl chloride found in autoclaves. A few years earlier an Italian investigator, Viola, had reported that inhaled vinyl chloride gave rise to tumors in rats and mice; a later experiment showed the induction of angiosarcomas of the liver in those animals, in addition to other tumors. Drastic reductions (over some protest) in the allowable concentrations of vinyl chloride in factory air were imposed (to about 1 ppm from 50 or more) the effect of which has been a greatly reduced risk from this carcinogen to large numbers of workers. Vinyl bromide, much less commonly used, is an equivalent hazard, but the search for other vinyl compounds that might be carcinogenic has revealed only acrylonitrile in the same class of carcinogenic potency as vinyl chloride. Acrolein is extremely toxic (which makes chronic testing of it difficult) and is only weakly carcinogenic, if at all. Vinyl acetate is not highly toxic and is weakly carcinogenic. Formaldehyde and acetaldehyde, which are large bulk industrial chemicals, induce tumors very effectively in rats at relatively low concentrations, but only by inhalation and only in the nasal mucosa; in humans there is not yet epidemiological evidence of the carcinogenicity of formaldehyde, which is a common metabolite produced within the body.

As a large group of industrial chemicals, solvents are manufactured in huge quantities and thus human exposure is considerable. Exposure occurs not only in factories (where it can be controlled) but in small workshops (dry-cleaning establishments, metal working and mechanical shops) in which controls are impractical. Benzene, one of the largest volume chemicals made, is used in many industrial processes. There is epidemiological evidence of its causing cancer in humans (myelogenous leukemia, and possibly other tumors). Evidence in laboratory animals was until recently unsatisfactory, possibly in part because the pure chemical is tested in animals but human exposure is to the technical grade compound, which contains thiophenes and many other compounds with unknown carcinogenic properties. Benzene is a common constituent of gasoline (petrol) in substantial concentrations, so that people who pump fuel for cars and trucks (often the drivers of the vehicles) are frequently exposed. Toluene and the xylenes, which are also heavily used solvents, are not carcinogenic.

Among the chlorinated solvents, many are used in large quantities and several are carcinogenic, including carbon tetrachloride, chloroform, methylene chloride, perchlorethylene, trichlorethylene, and several isomeric tri- and tetra-chlorethanes. The first four compounds provided adequate evidence of carcinogenicity in animals (liver tumors in rats and mice), but results with the others are equivocal. There is a lack of epidemiological evidence for the carcinogenicity of these compounds, individually or as a group. There seems to be sufficient grounds for choosing one of these chlorinated solvents over others on the basis of their likelihood to lessen the cancer risk of people exposed.

There is long-standing epidemiological evidence that certain industries carry with them the burden of increased cancer incidences among their employees. Among them is the dyestuffs industry, discussed earlier in connection with human exposure to aromatic amines. Another is coal gas and coal tar, and their connection with PAHs. There is current concern with coke oven workers and foundry workers (i.e., lung cancer increase), furniture and cabinet makers (cancer of the nasal passages), rubber workers (bladder, lung, and kidney cancer, and leukemia) while in leather tanning there seems to be an overall increase in cancer but not markedly of any particular type. Studies of the types of carcinogen prevalent in these factories have been undertaken. There is great interest in the results of chronic studies in animals associated with compounds used in these industries.

In the vicinity of coke ovens there are high concentrations of PAHs and other PACs in the atmosphere, as well as what are classified as "coal tar fumes," which comprise a mixture of compounds of enormous complexity. What might be responsible for the elevated risk of lung cancer and kidney cancer in the workers is not specifically known. In foundries, exposure to PACs also occurs, together with metal fumes, formaldehyde, and many simple compounds derived from the liquids used for quenching hot castings.

Among furniture and cabinet makers there is a considerable increase in incidence of nasal tumors but not in other tumors of the respiratory tract. These tumors seem to be due to exposure to wood dust, particularly the dust from hard woods, since other chemicals or treatments can be ruled out because of the uniformity of response of workers at different stages of manufacture. That said, there is only speculation but no specific information about the nature of the carcinogenic compounds that might be present in wood. Experiments in animals with wood, wood dust, or extracts have been uniformly unsuccessful in eliciting tumors.

Employment in the rubber industry has been associated traditionally with an increased incidence of bladder cancer. This was formerly ascribed to exposure to aromatic amines, which were extensively used in formulating rubber (notably 4-aminobiphenyl, which is an acknowledged cancer causing agent in humans). However, the discovery in the 1970s of the prevalence of nitrosamines in rubber factories indicated that these powerful carcinogens are major contributing factors to the many cancers that are associated with working in rubber factories. Volatile nitrosamines present include NDMA, nitrosodiethylamine, nitrosodi-*n*-butylamine, nitrosomorpholine, nitrosopyrrolidine, nitrosopiperidine, and methylnitrosoaniline. They are formed by reaction of a large variety of amino compounds, including dialkyldithiocarbamates and dialkylbenzothiazoles with nitrogen oxides or with the powerful nitrosating agent nitrosodiphenylamine (a retarder of vulcanization) at the elevated temperatures used in rubber manufacture. Concentrations of these nitrosamines measured, ranged from 1 to 240 μg per cubic meter of air, providing an exposure of an average worker of from 10 μg to almost 2.5 mg per day of some nitrosamines. This could be many times more carcinogenic than all other exposures to carcinogens. Concentrations of nitrosamines were particularly high in areas where rubber formulation takes place or where tires are cured; due to ready diffusion of these volatile compounds, exposure to nitrosamines anywhere in the factory can be considerable. As has been discussed, nitrosamines are very broadly acting carcinogens and can potentially contribute to an increased risk of any cancer. Butadiene is a widely used compound in the manufacture of synthetic rubber, and is also a broadly acting carcinogen, inducing a variety of tumors in rats and mice by inhalation. Epidemiological studies of the effects of butadiene have begun.

NDMA is a common by-product of the tanning of hides and processing of leather, so that workers in those industries have considerable exposure to this potently carcinogenic nitrosamine. Concentrations as high as 47 μg per cubic meter of air have been measured in some areas. A number of pesticides are organic acids (di- and trichlorophenoxyacetic acids, for example) and these are often sold as neutral salts with organic amines such as dimethylamine. As a consequence, NDMA formed by reaction with nitrogen oxides or sodium nitrite (used as a corrosion inhibitor in cans) is a common contaminant of pesticide formulations; exposures occur in workers during manufacturing.

NDMA is commonly detected in the rocket manufacture and rocket propellant industries, because of its close association with unsymmetrical dimethylhydrazine. Dimethylhydrazine is usually manufactured by reduction of NDMA, and thus contains traces of the latter. When dimethylhydrazine burns (oxidizes), NDMA is one of the products. Thus, workers in areas in which rockets are test-fired, or used in the space program or militarily, are exposed to NDMA if dimethylhydrazine is used; this exposure, however, might be of limited duration for most workers. The hydrazines used in rocket fuels are themselves carcinogenic, although much less potent than nitrosamines. Hydrazine, monomethylhydrazine, and unsymmetrical dimethylhydrazine are all used in various proportions; all are volatile. In rats, hydrazine has produced tumors of the liver, lung and nasal cavity; methylhydrazine is very weakly carcinogenic in hamsters; and dimethylhydrazine induces tumors of kidney, liver, lung, and blood vessels in rats and mice.

A major source of exposure to nitrosamines is modern synthetic cutting oils. These are considered a superior replacement for the traditional petroleum-based cutting oils which contained (unless highly refined) carcinogenic polycyclic aromatic hydrocarbons. For the past 15 or 20 years, cutting oils have contained some mineral oil together with water and triethanolamine as an emulsifier and sodium nitrite as a corrosion inhibitor. Early in this period, investigators in Sweden and the United States discovered nitrosodiethanolamine (NDELA) at high concentrations in some commercial cutting oils. This carcinogen arises from the interaction of diethanolamine (present in commercial triethanolamine) with nitrite. The process is facilitated in the presence of carbonyl compounds but is slow in neutral or basic conditions. There is usually ample storage or process time, however, for NDELA to form and concentrations as high as 3% have been measured, although they are usually lower. The emulsifier, triethanolamine, is also the source of NDELA found at part per million levels in many cosmetics. Nitrosamines (including NDELA) readily penetrate the skin, and human exposure from this source can thus be considerable. It is probably premature for epidemiological observations to reveal a connection between cancer incidence and exposure of workers to NDELA. NDELA is a less potent carcinogen than NDMA, but has induced tumors of a variety of organs in rats after an exposure for 2 years to concentrations similar to those encountered by many workers.

There are many industrial alkylating agents and precursors of polymers that are mainly used "captive," but are manufactured in large quantities and to which there can thus be exposure of limited numbers of workers. Some of these are potent carcinogens and deserve mention. Among them are epichlorhydrin, diepoxybutane, styrene oxide (but not styrene), ethylene oxide (widely used as a sterilant in hospitals), benzyl chloride, and benzotrichloride. All of these have produced significant incidences of tumors in one or more species of experimental animals when appropriately tested, and must therefore be considered more likely to increase the risk of cancer in humans exposed than equivalent substances that do not produce tumors in animals. Chloromethyl ether and bis-chloromethyl ether are by-products of manufacturing that are formed in the air from formaldehyde and hydrogen chloride gas; they are also chemical industrial intermediates. They induce, among other types, lung tumors in rats and mice and there is epidemiological evidence of a production of lung cancer in exposed workers.

D. Daily Habits

The most prominent of the daily habits associated with an increased risk of cancer is tobacco smoking (more particularly, cigarette smoking). This habit has been associated worldwide with more than a million deaths a year: 300,000 of these in the United States (from heart attacks, strokes, and cancer). The cancers that have been linked to cigarette smoking are those of lung, larynx, bladder, oral cavity, and pancreas. Other exposures to carcinogens probably contribute to the risk of these cancers and, similarly, the carcinogens in cigarette smoke contribute to the increased risk of many other cancers in which smoking is not the principal cause. One of the reasons for these interrelations is that the most important carcinogens in cigarette smoke are the broadly carcinogenic nitrosamines. There are, in fact, nitrosamines in cigarette smoke that are peculiar to tobacco products (called by Hoffman and Hecht "tobacco-specific nitrosamines") in addition to nitrosamines that are common in other environmental sources. There are additionally carcinogens other than nitrosamines present in cigarette smoke, including polynuclear aromatic compounds, aromatic amines, radionuclides, and others unidentified, but these occur in lesser concentrations and are less potent.

Nicotine is the prevalent alkaloid in tobacco and is a tertiary amine that is nitrosatively dealkylated by reaction with a nitrosating agent, such as nitrogen oxides in burning tobacco, to form nitrosonornicotine, which is a relatively weak carcinogen. However, another product of the nitrosation of nicotine is 4-(methylnitrosamino)-1-(3-pyridyl)-1-butanone, usually abbreviated NNK. This is a very potent carcinogen (on a par with NDMA and similar nitrosodialkylamines) and gives rise at quite low doses to tumors of the lung and liver in rats. There are other minor tobacco-specific nitrosamines present, but these are of considerably lower potency. Other nitrosamines present (usually at lower concentrations) include NDMA, nitrosopyrrolidine, methylnitrosoethylamine, and nitrosodiethylamine. The mainstream smoke of one cigarette has been variously reported to contain a total of 0.4 to 0.8 μg of nitrosamines. The sidestream smoke of cigarettes contains larger amounts of the nitrosamines, but these are largely dissipated into the air, so that their effect is diminished.

Burning tobacco is not the only source of tobacco-specific nitrosamines, since they are also formed during the curing of tobacco, notably in the fermentation step when nitrate is reduced to nitrite, which then reacts with nicotine. Thus, there is present in unburnt tobacco, particularly that prepared for chewing and snuff-dipping, high concentrations of nitrosonornicotine and NNK (but not NDMA and other volatile nitrosamines that are formed by burning tobacco). The concentration of nitrosamines in snuff and chewing tobacco is often tens of parts per million and it is extracted by the saliva and swallowed in this form. In addition to a probable systemic carcinogenic effect, which has not been quantified, there is certainly an association between the oral use of tobacco and cancer of the mouth and associated organs, including the tongue, buccal cavity, larynx, and pharynx. In parts of India, tobacco is chewed with lime (calcium hydroxide) which liberates the basic nitrosamines NNN and NNK that are associated with a high incidence of oral cancer. Cancer of the mouth has also been observed in the United States (particularly in southern states) and linked with the use of snuff which is kept for hours between the gum and cheek. Chewing of betel nut with tobacco and lime is common in many areas of the world. This exposes people to nitrosamines formed from the amines arecoline and guvacoline (in betel nut), which react with nitrite always present at low concentrations in saliva. Nitrosoguvacoline is not carcinogenic, but methylnitrosaminopropionitrile (formed from arecoline) is a potent carcinogen.

Alcoholic beverages are associated with cancer of the liver and of the esophagus, especially in some parts of the world. However, it does not seem that ethanol itself is carcinogenic (a number of animal studies have been entirely negative), which leads to the conclusion that some of the hundreds of compounds that have been identified in alcoholic beverages are responsible for the carcinogenic effects. Furfural is of low carcinogenic potency but is a common constituent of alcoholic beverages and of many processed carbohydrate foods. As discussed previously, there is NDMA in beer and liquors made from malt. These are not, however, the beverages most strongly associated with cancer.

Other habits that entail exposure to carcinogens are those involved with hygiene and appearance. Investigations many years ago showed that many components of hair dyes were mutagenic and some of them, on subsequent testing, proved to be carcinogenic (mainly aromatic amines). There is no good epidemiological evidence relating use of hair dyes with particular cancers, but those that are carcinogenic must be considered likely to increase the cancer risk of users. The use of cosmetics entails exposure not only to various dyes and colorings which might be carcinogens, but there are nitrosamines derived from reaction of amines with nitrosating agents. They include NDELA and methylnitrosododecylamine and -tetradecylamine, which are bladder carcinogens. Nitrosamines are readily absorbed through the skin.

E. Drugs and Medicines

Although medicinal drugs are not strictly "environmental" some of them are used by so many people and in such large quantities that those that are carcinogenic must be considered part of environmental exposure to carcinogens. As pointed out by Schmähl, many drugs used in cancer therapy are carcinogens that cause cancers in people who were cured of one cancer by the drug; frequently these secondary cancers do not take long to develop, perhaps a few years. The usefulness of such drugs is not disputed, even if it carries considerable risk.

Of much greater concern are the large number of drugs and medicines that are carcinogenic and that are used to treat less serious complaints—often trivial ones—and for which noncarcinogenic replacements are usually available. They include readily nitrosatable drugs which can form nitrosamines in the material itself during storage (by reaction with oxides of nitrogen) or in the stomach when nitrite is present from cured meats and, less importantly, from saliva. Among these nitrosatable drugs are pyramidone, several tetracyclines, antihistamines (chlorpheniramine, diphenhydramine), cyclizine, thorazine, and many others; none of these has given evidence of carcinogenicity to experimental animals in the absence of nitrite, but many have induced tumors when fed together with nitrite.

Phenacetin has been largely withdrawn from the market because of indications that it induces kidney tumors in humans. The once commonly used sleep-aid methapyrilene was also withdrawn after demonstrations that it caused liver cancer in rats. The antischistosomal agents lucanthone, hycanthone, and niridazole have all been carcinogenic in experimental animals, and millions of people have been treated with them in less developed countries. 8-Methoxypsoralen has been used with ultraviolet radiation to treat the skin disease psoriasis and the agents have been cocarcinogenic in mice. Several hypolipidemic drugs, including clofibrate, nafenopin, ciprofibrate, and others have induced liver tumors in varying degrees in rats.

Some of the drugs that have been tested, both naturally occurring and totally synthetic, have provided weak or limited evidence of carcinogenicity in animals.

F. Air and Water

The carcinogens that are in air and water have been alluded to in other places. Some of them arise from natural activities, such as the products of pyrolysis in the smoke of volcanoes and of natural fires, and products of decomposing organic matter leached into water. However, the carcinogens of greatest concern are products of industrial activity and related to human inventiveness. Incomplete combustion of fuels, for example, puts thousands of tons of benzo(*a*)pyrene and other carcinogenic PACs into the air each year; nitropyrenes and other nitroarenes are formed in diesel engines. These compounds are enormously diluted except at the source at which they are produced. Carcinogenic pesticides (of which DDT is the most studied example) run off fields into streams, rivers, and lakes, and would probably be of little concern if they were not concentrated through the food chain (and often stored in fat) where they may appear at relatively high concentrations.

Contaminated air exists in and around factories in which carcinogenic chemicals are produced as major products or as by-products. For example the nitrosamines and solvents in tire factories will affect ambient air. Nitrosamines are transported in the air when amines and nitrogen oxides (e.g., NDMA) mix. Many solvents and halogenated compounds (for example, ethylene dichloride and dibromide, which are carcinogenic) are present in air because they are used in factories or in agriculture. Benzene and other solvents enter the atmosphere whenever fuel tanks (large or small) are filled; benzene is an established carcinogen in humans. Very large numbers of people are exposed to benzene through association with the petroleum industry. Other hazards in the category of carcinogenic air pollution include exposure to tobacco smoke, exhaled or in secondary side-stream smoke from cigarettes. This is now considered a much more serious carcinogenic risk to workers in restaurants and bars than was formerly the case.

The contamination of water with carcinogens is most serious when the water is used for subsequent human consumption. All of the agricultural and industrial wastes that enter the air also find their way into water, occasionally large bodies of water where their carcinogenic properties are reduced by dilution. However, when the run-off or industrial effluents reach streams, rivers, or small lakes and drinking water is drawn, the carcinogens can become deleterious to public health, because of the large volumes of water consumed. Drinking water is settled and filtered, sometimes through activated charcoal, but in many water supplies it is not practical to detect and remove carcinogens at the parts per billion level which might still, cumulatively, be toxicologically important. Among the carcinogens that might be present are the extremely potent tetrachloro-*p*-dioxins and chlorinated dibenzofurans, which are formed by heating the almost ubiquitous polychlorinated biphenyls (PCBs) and polychlorophenols (the latter are used as herbicides and wood preservatives). Ironically, our attempts to ensure a bacteriologically safe water supply by chlorination often leads to formation of chloroform and other carcinogenic halogenated organic compounds, through reaction of chlorine with the complex organic material "humic acid" in the water.

VII. Summary

Carcinogens occur in the human environment in great variety, but an individual's exposure to a particular substance is usually small, except in a few occupations. Some of the more prevalent carcinogens have been mentioned here, but they vary greatly in potency. Their potency in humans cannot be measured, and cannot be estimated precisely. Some are present at much lower concentrations than others, yet seem to have a large carcinogenic effect (for example the high incidence of cancer in smokers exposed to a few micrograms of nitrosamines a day). It seems wise to identify and quantify as many carcinogens as possible to which people in large numbers might be exposed, and prudent to reduce our exposure to as many of these as can be avoided. There will be others which cannot be avoided, or avoided only with difficulty but, since the effects of the exposures are cumulative over a lifetime, the fewer exposures to carcinogens the more likely a life free of cancer can be achieved.

Related Articles: CHLOROFORM; CHROMIUM; COBALT DUSTS; DIESEL EXHAUST, EFFECTS ON THE RESPIRATORY SYSTEM; DIOXIN, HEALTH EFFECTS; FORMALDEHYDE: EXPOSURE EFFECTS ON HUMAN HEALTH; FUNGICIDES; INDUSTRIAL SOLVENTS; LEAD POISONING; MUTAGENICITY TESTS WITH CULTURED MAMMALIAN CELLS: CYTOGENETIC ASSAYS; NATURAL ANTICARCINOGENS AND MECHANISMS OF CANCER; ORGANIC SOLVENTS, HEALTH EFFECTS; PLANTS AS DETECTORS OF ATMOSPHERIC MUTAGENS; OZONE LUNG CARCINOGENESIS; PEROXISOMES; POLYCHLORINATED BIPHENYLS (PCBs), EFFECTS ON HUMANS AND THE ENVIRONMENT; RUBBER INDUSTRY, TOXICITY OF WORK ENVIRONMENT; SULFUR MUSTARD; VOLATILE ORGANIC CHEMICALS; XYLENES.

Acknowledgment

Research sponsored by the National Cancer Institute, DHHS, under contract No. NO1-CO-74101 with ABL. The contents of this publication do not necessarily reflect the view or policies of the Department of Health and Human Services, nor does mention of trade names, commercial products, or organizations imply endorsement by the U.S. Government.

Bibliography

"Chemical Carcinogens," 2nd edition. (1984). (C. E. Searle, ed.), American Chemical Society, Monograph No. 182, Washington, D.C.

"Chemical Carcinogenesis, Models and Mechanisms," (1988). (F. Feo, P. Pani, A. Columbano, and R. Farcea, eds). Plenum Press, New York.

"Environmental and Occupational Cancer: Scientific Update" (1990). (M. M. Mehlman, ed.) Princeton Scientific, Princeton, New Jersey.

"Monographs on the Evaluation of Carcinogenic Risk to Humans," Supplement 7, (1987). International Agency for Research on Cancer, Lyon, France.

Environmental Monitoring

David L. Swift
The Johns Hopkins University

Glossary

Absorption Process by which gases or vapors are dissolved into liquids and thus removed from the gas phase.

Adsorption Process by which gases or vapors are chemically or physically bonded to solid surfaces and thus removed from the gas phase.

Ambient monitoring Process of contaminant monitoring in a medium away from either the source or the receptor of that substance.

Extractive sampling Removal of a contaminant from its medium, by some physical or chemical process, onto a collection medium suitable for subsequent analysis.

Grab sampling Removal of a single sample in a particular location over a short time interval for the purpose of a "snapshot" determination of contaminant concentration.

Inventory monitoring Release of contaminants based on a material accounting of the source substances and their process consumption.

Personal sampling Sampling contaminant in the immediate vicinity of a potentially exposed individual to estimate exposure.

Proportional sampling Sampling in a moving medium in a manner that weighs the sampling contaminant release or transport according to the medium flow or volume.

Source monitoring Monitoring in the immediate vicinity of a contaminant source in order to evaluate its release rate.

Source strength Amount of a contaminant released from a source per unit time.

***ENVIRONMENTAL MONITORING** is the process by which purposeful measurements are made to evaluate the sources, occurrence, transport, transformation, removal, exposure, effects, or control of substances or agents (e.g., sound or electromagnetic radiation) in the physical environment. The concept of monitoring goes beyond measurement in that a specific objective guides the process; the word* monitor *is derived from the Latin word* monere *(to warn), implying that the recognition, evaluation, and control of some potentially harmful effect is involved. Another word that is often used in its place is* surveillance; *here the word is from the French* surveiller *(to watch over) and implies additionally that there is a degree of continuous observation. The concepts of warning and overseeing imply that the process of monitoring is not used simply to satisfy curiosity but is guided by a particular end.*

I. Purposes of Environmental Monitoring

Environmental monitoring of hazardous materials can be undertaken for any of a number of purposes, and it is important to identify the purpose at the outset, as this will influence the design and implementation of the process. For materials whose environmental transport or transformation in one or more media (air, water, soil) is not known, monitoring can be carried out to determine this unknown.

This information will provide the basis for choosing which medium to sample and the time scale for repeated sampling. For some well-characterized hazardous materials the partition among the media is well known, as are the transport rates. These rates vary markedly for different media, generally fastest in air, slower in water, and slowest in soil. For a given medium this will also depend on the medium properties, such as convective motion in air or water, and the physicochemical properties of the hazardous material, such as solubility, vapor pressure, absorbability, and chemical reactivity with other environmental substances.

In some cases the partition of hazardous materials between the media can be predicted readily on the basis of known physicochemical characteristics, but in other instances it may be necessary to monitor a substance in a specific location to determine partition, transport, and transformation. Partition coefficients for many substances have been compiled, but these values are usually for a single component in two media (e.g., air and water) and do not take account of the normal situation where several components are present and the partition is influenced by the other components.

Closely related to monitoring for media transport and transformation is monitoring for the purpose of determining the temporal and spatial distribution of hazardous material in the physical environment. It can generally be taken as axiomatic that hazardous materials are distributed unevenly both spatially and temporally when the length and time scales are appropriate to the patterns of potential exposure. In most instances, an insufficient knowledge of the distribution of sources, mechanisms of transport, and interactions with other materials makes the mathematical modeling of the spatial and temporal distribution of a hazardous material infeasible. Thus one must monitor the material to develop a picture of its distribution. If such monitoring reveals a particular spatial and temporal pattern that implies the controlling influence of certain mechanisms of transport, it is then often possible to model the expected changes in distribution when certain physical parameters are altered, such as the convective motion of air or water or the porosity of soil.

Monitoring is often undertaken specifically for the purpose of developing a mathematical model for the movement and fate of environmental substances. The model thus developed will have a number of input parameters from which some prediction of the concentration, uptake, exposure, or ultimate fate of the substance in question can be obtained. Models are clearly of greater benefit if they cover a large variety of possible situations; the more restricted the model, the less useful it is in terms of new situations. Very simple models may only predict concentrations within a very wide range, but in many instances this may be adequate to make a preliminary decision about a necessary action.

Knowledge of the long-term temporal trend of hazardous material concentration in a location of interest is another appropriate purpose of environmental monitoring. This only differs from the above temporal information in that the time is greater; many hazardous materials are very persistent as well as being toxic or otherwise harmful. The determination of a material's persistence in a given situation is important in the choice of remedial action (i.e., whether to opt for removal or wait for degradation to more benign substances).

The evaluation of potential or actual exposure of a target host to a hazardous material may be the purpose for environmental monitoring. The first two purposes described above do not require the assumption of an exposure situation but rather assume that a distribution of the material may be used for some other purpose. Monitoring for host exposure accepts the idea that there are a number of factors beyond the spatial and temporal patterns of the material that influence exposure, including the movement of the host from one region of concentration to another and the temporal pattern of exposure (such as during working hours or other defining periods of exposure). The physical placement of monitoring instruments and their operation during specific periods of time is important if exposure is not continuous.

Of somewhat greater importance in many instances is monitoring for the purpose of evaluating a health effect. Here the time scale depends on the temporal pattern of the effect, which may range from acute irritation to chronic disease with a long latent period and a long period of disease progression.

When a control measure is put into effect, it is important to know what change has been realized in the environmental concentration of a hazardous material. Thus another major motivation for environmental monitoring is to evaluate the benefit or lack thereof associated with a control measure. This could take several forms, including exposure assessment, one point ambient concentration, or spatial distribution, but in every case the environment is appropriately characterized before and after the control measure is put into effect. This form of monitoring requires planning so that a representative baseline value can be established before control measures are introduced.

Monitoring may also be undertaken to compare the environmental level of a hazardous material to a regulatory level established by law or practice. This could be done to provide information necessary to decide on a compliance strategy or to anticipate what effects might be observed if no control is elected. Regulatory levels are often expressed in terms of either a concentration averaged over a specified period not to be exceeded more than a certain frequency or as peak concentration. Monitoring must comply with this definition in terms of sample frequency, averaging time, and minimum length of the monitoring program.

Monitoring may be performed in an emergency situation to evaluate the need for action and to determine when levels of hazardous materials have decreased enough to involve personnel in cleanup. Initially, in such a situation, monitoring is undertaken to establish the temporal concentration trends, and if a particular concentration rises above some predetermined level, emergency action, such as evacuation or cessation of use of drinking water, may be undertaken. Monitoring will provide information about when emergency actions are no longer warranted. In the past, emergency response monitoring capabilities were seldom available except in instances where routine monitoring was in place and could be utilized for the situation. In the future, it will be important to have the capability to respond quickly with appropriate mobile monitoring systems and instruments and a method to design a monitoring strategy rapidly.

A particularly important subset of health effect environmental monitoring is that in support of epidemiological investigations. Measuring effects over a sizeable population without the comparable exposure data as determined by some types of monitoring renders the potential relationship between agent and effect very poorly supported. Of course, in some retrospective studies no exposure monitoring was performed and recourse must be taken otherwise. In occupational instances, it is common to characterize exposure according to job classifications in spite of the variability within such groupings and the assumptions needed to reconstruct a past exposure. Prospective studies must include in their protocol an adequate monitoring design based on the objectives of the study.

Monitoring is an appropriate step when any decision is made that may ultimately result in hazardous material release to the environment. It is important to establish a background or baseline value of an appropriate surrogate material or the substance in question before the process of possible release is begun, particularly if there is suspicion that the material may already be present from past activity or may occur naturally. In this way the impact of the activity on the environmental concentration can be separated from preexisting sources.

It is clear from the above discussion that environmental monitoring may be carried out for any of a number of purposes. It is important that the purpose be identified clearly, as this will influence the design of the monitoring program and the methods employed. Too often monitoring has been carried out with emphasis on the capabilities and limitations of specific instruments and with relatively little thought as to how instruments are employed and how many measurements are required. Making too few measurements in a monitoring program will fail to answer the basic question being asked or to meet the specific objective; too much monitoring is a waste of resources and leads to the overaccumulation of data, most of which will probably never be analyzed. Of course, if one does not have a basis for decision, the tendency is to err on the side of too much; but there has been an accumulation of studies over the last twenty years that should suffice as a guide to efficient design of monitoring schemes.

In the United States in recent years, the phenomenon of "the pollutant of the month" has occurred, wherein a new environmental pollutant with potentially significant effects is revealed, attended with a great deal of publicity in the national media, and accompanied by calls from legislators and other officials to "get control" of the problem quickly. Some responsible agency is then called upon to hastily enact a monitoring program to define the magnitude of the problem. While monitoring to respond to public pressure is a stated purpose of such instances, these actions frequently are characterized by lack of careful design and forethought and can result in wasteful "wheel spinning" until a new pollutant appears and interest in the present effort wanes.

II. Types of Environmental Monitoring

Exposure to and subsequent effects from hazardous materials comprise a process of several steps regardless of the receptor of interest, be it humans, other animal species, other biota, or specific parts of the physical environment. These steps may include material generation, transport through one or more environmental media, physical and chemical transformation, exposure, uptake, and receptor processing (which may consist of intrareceptor transport, transformation, and elimination or storage). At each stage of this process, the substance(s) of interest, or a suitable surrogate, can be monitored in order to meet one of the above purposes. At what point in the process one chooses to monitor may depend on the purpose; for example, if monitoring is performed to assess and control source emission, then obviously monitoring is best performed at the source. It is seldom that monitoring is performed at every step simultaneously or even at several steps along the path of exposure.

Source monitoring is a common approach which may be carried out in several ways. Sources may be classified as point sources, distributed stationary sources, or distributed mobile sources. Sources are defined as point sources if their characteristic dimension is small with respect to the length scale between the source and any potential receptor of interest; the concept is useful in that mathematical models of transport from point sources are common and readily predict contaminant concentration given the source "strength" (quantity released per unit time).

Distributed sources, whether fixed or mobile, can also be simulated by mathematical models whose form depends on whether the source extent is best approximated by linear, areal, or volumetric. Source strength in these cases is defined as the quantity released per unit time per unit length, area, or volume, respectively. For both types of models, the strength can vary in time, be continuous, or be an impulse of short duration. Such models require that the temporal pattern of release quantity is known, and this is the usual objective of source monitoring.

The simplest approach to source monitoring is that of inventory monitoring which does not even involve a direct measurement. In some cases, if the amount of a substance utilized for some purpose can be reliably and directly linked to its emission or release, this suffices to specify the source strength. All that is then necessary is to determine how much of the substance is used, hence the name *inventory monitoring*.

Even though one should never overlook the possibility that source strength may be suitably estimated by inventory, one is rarely so fortunate for this to be the case and frequently must resort to actual measurement of the source strength by sampling at the "point of discharge." The interpretation of such sampling to derive the source strength will be described below. A third method for determining source strength over some period of time is a material balance in which material loss is measured by some physical means rather than by inventory.

Source monitoring measurements may be done at short range or remotely. A useful feature of source monitoring is that concentrations of contaminants are normally at their peak just at discharge and progressively become more dilute as they are dispersed throughout their media of transport. Thus, if instrument sensitivity is a limiting factor in the characterization of release, it may be desirable to monitor at the source, particularly if it is a point source.

The most widely used monitoring approach is *ambient monitoring*. On the line from source to receptor, ambient monitoring gives more information about the receptor exposure than does source monitoring and is often a much more practical approach for several reasons. If there are unknown transformation steps between the source and the receptor, ambient monitoring gives a better indication of exposure to the material of interest than does source monitoring. An *ambient monitoring network* is capable of providing temporal and spatial profiles of contaminant concentration that satisfy the requirements of several monitoring purposes. The influence of various factors such as ambient temperature, water table, wind profiles, and seasonal soil permeability may be determined by ambient monitoring networks.

Ambient monitoring is useful for the detection of release of contaminants when source monitoring is not employed, but if the distance to an ambient sampling site is great, the ability to provide emergency response is limited. For example, when radioactivity was released from Chernobyl, the first indication in Western Europe of its transport was an ambient airborne monitoring station in Sweden. However, by the time it was detected, the cloud had already traveled more than 1000 km and had passed over several other countries as well as the Soviet

Union. Steps taken to limit exposure were not as effective as if the release had been reported immediately, accompanied by whatever source monitoring information was available.

The third broad area of monitoring is at the site of effect. It is appropriate first to consider human monitoring, since most investigations and concerns regarding hazardous material focus on human health. We may consider human monitoring from several viewpoints beginning with the "interface" between the ambient medium and the human, namely the immediate vicinity of the human, in order to estimate exposure. In the occupational health field this is referred to as *personal sampling,* in which a sampling device is located on the person in some fashion to detect exposure or its appropriate surrogate.

A monitoring effort built around personal sampling is an effective way to estimate the time and intersubject averaged exposure to a contaminant if it is feasible and practical. The technique to do so for airborne substances accessible via the respiratory tract has been employed for years in the occupational health field and is under development for community air pollutants. An analogous method for substances whose exposure route is dermal absorption is not well developed, other than simple wipe tests on suspect surfaces. Perhaps more research effort will yield a methodology for simulation of dermal exposure.

In some cases, the tissues and fluids of the body can be monitored to detect exposure to hazardous substances. An excellent example of this is the analogous situation of ionizing radiation, where routine monitoring or urine, feces, blood, and (occasionally) other tissues is undertaken to detect exposure to radioactive materials. With gamma emitting isotopes it is also possible to monitor using a whole body counter, a less invasive procedure than blood or tissue samples. Long-lived isotopes such as plutonium are appropriate candidates for long-term whole body external detector monitoring to evaluate the biological storage properties and calculate tissue dosimetry, especially for radiosensitive tissues or organs.

Human monitoring may also be undertaken by measurements of physiological or biochemical parameters indicative of exposure to specific contaminants. For example, exposure to irritant airborne particles or vapors results in increased airway resistance, which can be monitored by standard pulmonary function tests. Studies of exposure to air pollutants have often employed these sensitive tests as indicators of conditions leading to more serious chronic diseases or the exacerbation of existing conditions such as asthma or respiratory allergy. Many contaminants are metabolized in the body, but if the primary metabolite is known and not present endogenously at significant levels, this metabolite can be monitored as an indicator of exposure.

Monitoring of contaminant concentrations in other animal species or in plants can provide useful indications of transport, uptake, and effects that may occur in humans. Commercially valuable species or cultivated crops that suffer damage or become unusable because of exposure to contaminants are of concern in themselves, besides providing a flag for possible human exposure. Under most circumstances it is easier and less restrictive to monitor tissue or fluid samples from animals and plants rather than from humans to detect hazardous contamination concentration. Much experience has been gained in the extrapolation from animal species to man of contaminant uptake and effects from controlled animal toxicology studies, and their use in establishing regulatory standards for human exposure.

Monitoring of certain segments of the physical environment in order to track harmful effects, broadly defined as "quality of the environment" or ecological effects, has been widely practiced. In community air pollution such effects are known as "welfare effects," since they do not directly and immediately impact human health. Such things as atmospheric visibility, water turbidity, dustfall, and odors are among the items in this category, and monitoring of such conditions can provide data indicative of improving or deteriorating conditions.

Although all of the above types of environmental monitoring have been employed to investigate environmental effects, trends, and the influence of controls, they are not all equally applicable to the special case of hazardous materials. Most hazardous material environmental issues center around emission, transformation, and human health, although this may involve intermediate biotic species such as plant and animal food sources. The choice of a particular type of monitoring is determined by the purpose being pursued and the available resources, including time, manpower, and equipment. In some cases a more complete picture can be obtained by using more than one type of monitoring, usually simultaneously. These issues suggest that consideration be given to the strategy and design of environmental monitoring.

III. Strategy and Design of Environmental Monitoring

The idea of strategy and design for environmental monitoring implies that, before monitoring is undertaken, some time for reflective thought should be taken to decide the purpose of monitoring and how that purpose can best be realized. Too often in the past, monitoring has been undertaken simply on the basis of "like-to-know" or under hurried conditions when decisions are made simply on the basis of the available instrumentation. Even for "emergency" conditions, it seems reasonable that anticipatory plans can be developed so that a particular program can be lifted off the shelf when it is deemed to meet the requirements of a specific type of situation. Environmental agencies at all levels of government have been criticized in the past for their lack of planning and the absence of coordination among monitoring programs carried out by different groups.

The opposite problem exists in the case of some environmental regulation in which the monitoring plan is spelled out in the legislation in a detailed fashion and is a requirement of compliance with the law. It is quite easy to meet these criteria by simply following the requirements in a cookbook fashion. This meets the legal requirement but may not be adequate to answer the additional questions that may arise in a specific situation, and it leaves no room for scientific judgment based on professional experience. These legislative mandates usually center on source or ambient monitoring as the basis for action and give no attention to the question of exposure likelihood or effects monitoring in humans or other segments of the biota. Even in such situations, questions of the quality, reproducibility, and handling of environmental monitoring data must be addressed by the program director.

The strategy of monitoring must first deal with the type of monitoring to be employed; this is, to a large extent, as discussed above, determined by the purpose(s) of the monitoring effort. If the purpose of monitoring is centered on the source release evaluation, source monitoring will suffice, perhaps augmented with a modest effort to measure ambient concentrations. The strategy adopted will depend also on the degree of knowledge of the properties of the source and media of transport. Are the contaminants already identified by other means or does a screening phase have to be included to enumerate the materials of interest? In either case the strategy must include a decision about how many and which contaminants to monitor. Is there a single surrogate material that can serve to assess the release or exposure of all contaminants? Again, this may depend on a screening effort to enumerate the contaminants and on a consideration of their properties.

If ambient monitoring is called for, does the monitoring effort require a small data collection to provide parameters for a well-accepted and applicable model? Are the transport properties of the medium (or media) well known so that a small number of spatial samples will suffice reasonably to predict the isopleths of concentration? Are temporal variations of source strength small, permitting a small sample in time to predict long-term averages, or does this have to be the objective of a screening pilot study? Can the results of other monitoring studies provide guidance in the decisions about the characteristics of contaminant release, transport, ambient concentration, and exposure?

It is much less straightforward to decide on a strategic approach for human monitoring than for source and ambient monitoring, because the physical factors provide only some help in deciding how to monitor in these instances. Can the entire exposed population be monitored, or does one have to choose a representative sample? What sort of profile of exposure is needed to reach a conclusion about exposure? What effect(s) should be chosen to define the degree of health effect or to act as a surrogate for a possible chronic effect? Such decisions must be undertaken on the basis of appropriate past studies and professional judgment.

Likewise, ecological monitoring leaves much room for judgment in deciding a strategy of approach and offers little applicable experience in past studies. There must be a base of data to establish a causal relationship between the ecological effect and the presence of a hazardous material. Some ecological effects may not occur as a result of the presence of a hazardous material, but both may be due to a third factor. In other instances, ecological changes are simply reflections of secular variation, not directly linked to the hazardous substance.

The design of monitoring efforts requires the combination of a strategy decision and the practical realities of taking environmental samples. The general decision of what and where to sample based on the generation and transport of hazardous substances must be fleshed out with further decisions on how many samples are adequate to reach the desired objective, how frequently they are taken, and what choice of instruments shall be employed in the effort. This is sometimes dictated by the properties of the substances and available instrumentation, but often there are a range of choices. For example, is it desired to make the measurements of substances at the source or in the ambient environment *in situ*, or will samples be taken for subsequent analysis? These issues will be discussed in Section VI.

Design must also take into consideration that technical failures will occur. Many air monitoring networks for community air pollutants have a significant fraction of invalid data because of instrument down time, lack of calibration, failure of data transmission or recording, mislabeling of samples, or faulty analytical procedures. Such inevitable events must be allowed for even with the most careful attention to sampling procedures.

Air and water monitoring networks for contaminants have been operated in many countries for more than twenty years, and this experience is useful in considering how to design monitoring for hazardous materials. Although monitoring networks are ideally designed on the basis of known principles of medium transport, the issues of cost, security, availability of convenient public sites, and public acceptance often weigh heavily in the final decisions about monitoring locations. Additionally, the choice of whether to employ fixed sites or mobile monitoring techniques may depend on strategy, purpose, and the above practical considerations.

IV. Monitoring in Different Media

Source or ambient monitoring may be undertaken in air, water, or soil, or in more than one medium. Design considerations of sampling number and frequency depend on the medium properties as well as instrument and analytical issues. When contaminant substances are released into air, water, or soil, they are subject to dispersion by convective and diffusive processes. These two processes differ in that convective processes depend on the bulk movement of the medium or a constituent thereof, while diffusive processes involve contaminants traveling from regions of high concentration to those of low concentration without the aid of bulk motion. Generally speaking, convective motion in air and water is greater by several orders of magnitude than that in soil, so that greater times are required for convective transport in the latter. Convective motion in soil is primarily by the movement of soil moisture or groundwater responding to gravitation, thermal, pressure, or capillary forces.

Convective magnitudes in air and water cover the same range, but diffusive transport, whether due to molecular motion or to the random fluctuations in convective velocities, is normally greater in air than in water. Both air and water are subject to wide variation in the magnitude of convective velocity; air movement in the atmosphere results from winds driven by weather factors, while water movement derives from gravitation and buoyant forces, the latter resulting from density gradients. These features suggest that temporal fluctuations in contaminant concentrations in air and water generally have much shorter characteristic times than those in soil, and these principles should govern the design of monitoring programs.

Because of the large interfacial areas between the media, transport across these boundaries may play a significant role in the overall fate of released hazardous contaminants. Transport of gases, vapors, or particles from air to water or soil takes place by settling, adsorption, absorption, chemical reaction, or condensation. Transport from water to air may result from evaporation, desorption, reaction, or atomization. Transport from soil to air is primarily by wind driven dust formation, reaction, desorption, or evaporation. Transport of contaminants from soil to water is primarily by dissolution, desorption, solid suspension, or reaction. Transport of contaminants from water to soil can result from sedimentation, adsorption, pore inclusion, reaction, or filtration. Thus a number of physiochemical processes can be involved in the intermedia transport of hazardous materials.

V. Determinants of Monitoring Samples

Environmental samples taken as a part of a monitoring activity are normally measured to determine the quantity of a contaminant, but the value is expressed as a concentration (a quantity per unit of the medium). The concentration is usually the property that expresses a contaminant's capacity to produce a particular effect. In many measurement methods, only the quantity of contaminant is determined by the analytical method, and the measurement of the medium is left to the person taking the sample. This is the first determinant of a monitoring sample.

For liquid or solid medium samples, the determination of volume or mass is fairly straightforward. Air contaminants present a more difficult problem, particularly since air volumes change more markedly with temperature, pressure, and relative humidity than do liquids or solids. If the vapor, gas, or particulate contaminant is removed from air by filtration, reaction, absorption, or adsorption, the volume of air must be measured to express air concentration. There are several common methods to make this measurement, depending on the volume required and the flow rate. These include rotameters, mass flow meters, and constant flow devices employing a critical orifice. Such a flow device is part of a "sampling train" that includes the sample collector, the flow measuring device, a controller (if necessary), and a pump to move air through the collector.

In order to make valid comparisons of air concentration under different environmental conditions, it is common to specify a standard condition of volume, often taken as 25°C and 760 mm Hg pressure. It is then necessary to measure (or assume) the conditions during sampling and convert the volume to that at standard conditions to express the concentration.

Contaminants can be separated from their medium at the site of the sampling by the process of extractive sampling. This general process is illustrated by filtration of air or water to remove suspended particles or by adsorption or absorption of contaminants from air or water onto or into a suitable solid or liquid collection medium. The contaminant thus separated from the medium may then be subject to an appropriate analytical procedure, such as gravimetric determination, spectrophotometry, or gas chromatography. The alternative to extractive sampling is determination of the contaminant quantity in the medium either at the site of measurement or remotely. The quantity of sample in this instance is determined by the specific measurement technique.

The other major determinant of a monitoring sample is the sample frequency and length of sample averaging or integrating. In this context, the methods are known as grab sampling, intermittant sampling, continuous sampling, and proportional sampling. Grab samples are one time samples for the purpose of determining a sample property at a specific moment in time. Intermittant samples are taken for some specified time at a given frequency, with suitable intervals of down time based on a knowledge or hypothesis of the time of change of contaminant concentration. Continuous sampling is only possible with certain kinds of instruments, and it is usually done with a specified averaging or integrating time. Proportional sampling adjusts the sampling time or volume to the medium conditions (such as water flow rate or volume) in order to weigh the samples with respect to the flow or volume contribution.

VI. Measuring Instruments: The Tools of Monitoring

Instruments for environmental monitoring may be classified in general as field or laboratory instruments. Field instruments make their measurement on site; the sampling and measurement process are not generally separated in time with field instruments, unless the instrument contains a series of collectors that are measured after all the samples are taken. Generally speaking, field instruments are portable, rugged, and able to operate under a variety of conditions, and they do not require highly regulated and controlled power sources.

With the advent of microelectronics, it is possible for many new field instruments to be operated relatively unattended, with measurement values being stored in a chip memory for a period of time. Some field instruments are designed to be passive; they do not require imposed medium flow but rather depend on natural medium movement. There has been considerable progress of late in miniaturizing some field instruments to be used for source, ambient, or personal monitoring.

Laboratory instruments belong to the domains of analytical physics, chemistry, and biology, and the major issues of monitoring have to do with appropriate sample collection and transport to the laboratory. Sampling methodology depends both on the desired sample frequency and duration and on the basic accuracy and sensitivity of the measuring instrument. However, it is a mistake to insist on an instrument of extremely high sensitivity when one of acceptably lower sensitivity will suffice. Limitations on sample accuracy may depend as much on the sampling procedure and the possible changes before measurement as they do on the measuring instrument. Contamination, sample degredation, and conversion of contaminants during sampling are all problems that must be addressed in monitoring.

Monitoring instruments cover a very wide range of sample sizes, flow rates, detection principles, measurement accuracies, and measurement precisions. It is important to choose an instrument that meets the criteria based on the monitoring purpose. For example, in the area of gas and vapor sampling, grab samples can be obtained very easily and inexpensively with detector tubes if one is satisfied to accept an accuracy of ±50%. Similarly, personal monitors consisting of passive diffusion collectors provide a relatively inexpensive method to estimate exposure, with similar accuracies. Alternatively, there are much more accurate methods available if needed, primarily employing laboratory analytical instruments. The handbooks listed in the bibliography contain descriptions of many monitoring instruments for environmental sampling.

VII. Processing and Interpreting Monitoring Data

Monitoring efforts that entail a very limited number of samples present few problems in data processing, even though the validity of the measurement should be determined by well-established techniques. These techniques include measurement calibration, blank checks, checks of known spiked samples, and procedures for outlier values. Larger data sets from monitoring efforts require more attention to assure that the percentage of invalid measurements is minimized. When examining samples to detect trends over time or with imposed controls, it is particularly important to use appropriate statistical methods to establish the significance of these trends.

When spatial distributions of contaminants are being determined, there are a number of simple models of convection, dispersion, and decay that can be employed to test the reasonableness of the measured values. Professional judgment is important to the interpretation of monitoring data in order to avoid drawing improper conclusions from faulty data. Too often in past monitoring efforts, data of very marginal quality has been taken at face value without subjecting it to physical reasoning based on transport principles.

VIII. Conclusions

In this article, the general principles of environmental monitoring have been discussed, placing emphasis on the importance of both design and measurement. The critical role of monitoring purpose in choosing a particular type of monitoring and a monitoring strategy has been outlined. Monitoring design based on strategy, practical limitations of monitoring sites, and sampling choice is an important component of the monitoring process.

The determinants of monitoring, such as flow measurements, sample size, and frequency, have as much to do with correct monitoring as do the measuring instrument properties. A wide variety of instrumental choice are available, and the decision of a method and instrument needs to be matched to the monitoring requirements. Some methods of monitoring are unique to the medium of interest, and monitoring in more than one medium may be necessary. Data collection, processing, and interpretation are important components of the monitoring effort and must receive their share of careful attention.

Related Articles: BIOLOGICAL MONITORS OF POLLUTION; MONITORING INDICATORS OF PLANTS AND HUMAN HEALTH FOR AIR POLLUTION; PLANTS AS DETECTORS OF ATMOSPHERIC MUTAGENS; PROTISTS AS INDICATORS OF WATER QUALITY IN MARINE ENVIRONMENTS; ZINC.

Bibliography

Hering, S. V., ed. (1989). "Air Sampling Instruments," 7th ed. ACGIH, Cincinnati, Ohio.

Lippmann, M., and Schlesinger, R. B. (1979). "Chemical Contamination in the Human Environment." Oxford University Press, New York.

National Academy of Sciences (1977). "Environmental Monitoring," Vol. IV, Analytical Studies for U.S. E.P.A. National Academy of Sciences, Washington, D.C.

Ethanol Fuel Toxicity

Eduardo Massad, György M. Böhm, and Paulo H. N. Saldiva
The University of Sao Paulo

Glossary

Alcohol Group of chemical compounds with a hydroxyl group, that is a C = O in the end of a carbon atom chain.

Alcoolita Dinitro triethylen glycol; chemical compound used to additivate ethanol in order to be used in diesel engines.

CO Carbon monoxide; poison to the respiratory system resulting from an incomplete oxygenation reaction.

HC Total hydrocarbons.

NO_x Nitrogen oxides.

IN CITIES with heavy traffic like Los Angeles, Mexico City, Tokyo, or Sao Paulo, the main source of air pollution is certainly automobile exhaust. Fuel composition is thus a determinant of the atmospheric spectrum of gaseous and particulate substances. In general this is a very sensitive system; slight changes in fuel composition may have a tremendous effect on air pollution and consequently on the health of the population as a whole.

When a new fuel is introduced (e.g., the Brazilian partial substitution of gasoline by ethanol in spark ignition engines), a significant change in the resulting air pollution profile of the affected area is expected. This has indeed been gradually observed in the ten years subsequent to ethanol introduction in Brazil.

I. Introduction

Since the petroleum crisis of the early 1970s, Brazil is the only country that has attempted a large scale program to substitute a non renewable fuel source by an entirely renewable one (i.e., hydrated ethanol). More than 4 million cars are now fuelled exclusively by ethanol. Additionally, gasoline sold in petrol stations contains 22% ethanol and is consequently lead free.

The use of ethanol as fuel is not a new idea. It was was initially introduced in 1832 in the preparation of illumination gas. A few years later, N. A. Otto used it in a spark ignition engine that he patented in 1861. Hence, ethanol has historical priorities over gasoline as engine fuel. However, despite these first successful attempts the price of ethanol was not competitive with that of petroleum, and its use as a fuel was only revived by Germany, Austria, and Italy during World War II, when it was extracted from potatoes.

In Brazil, there were several reasons for choosing ethanol as a good substitute for gasoline after the petroleum crisis; the principle one was its production capacity. Less than 2% of the Brazilian geographic area was required to produce enough fuel to replace crude oil importation. Also, from the 1360 W/m^2 of solar radiation that reach the earth, the Brazilian area gets 200 W/m^2. Although several other regions have higher solar radiation, they do not have the other essential component for ethanol production—water—which is extremely abundant in the Brazilian territory.

Like any other large-scale energy program, the Brazilian National Ethanol Program has advantages and disadvantages. The disadvantages include the potentially drastic consequences of the by-products of large-scale ethanol production and the monoculture of sugar cane. The advantages include the lower levels of toxic substances expelled by automobile exhaust into the atmosphere. Of course, this is a simplified view; it is not in the scope of the present article to explore this subject in detail.

Whether or not this ethanol program is changing air pollution for the better has been the concern of the Laboratory of Experimental Air Pollution of the University of Sao Paulo since 1980. It has been the aim to assess the biological effects of the new fuel as compared to the traditional one (namely ethanol versus gasoline) in a comparative nature.

II. Chemical Composition and Properties

A. Fuel

Ethanol, or ethylic alcohol (or simply alcohol), is a chemical substance belonging, together with the phenols, to organic compounds with a hydroxyl, or —OH, group; that is, the oxygen is joined to a hydrogen atom as well as to a carbon atom, —C—OH. Its formula is H_5C_2—OH, and it is usually extracted for fuel purposes from sugar cane or beet root in tropical and temperate countries, respectively.

The word "alcohol" is derived from the Arabic *kuhl* (also *kohl* or *kohol*). It originally was used to mean a "very fine powder" but gradually came to connote essence. Later, the term was applied to wine spirits, which were referred to as *alcohol vini* and, eventually, simply as alcohol. Its discovery is credited to the alchemists.

In Brazil, the alcoholic content chosen for fuel is 91.1 to 93.9% by volume (i.e., a hydrated form of ethylic alcohol). The so-called "P.A." (proper for analysis) could also be used as a fuel, but it has not been because of its high price. Commercial ethanol should not be used as a fuel because its high degree of impurity, primarily due to the presence of solid residuals and acidity, which can cause engine damage.

The basic properties of ethanol fuel and the physicochemical properties of ethanol, gasoline, and various mixtures of ethanol–gasoline are summarized in Tables I and II, respectively.

Table I Characteristics of Hydrated Ethanol Fuel

Characteristic	Values
Specific mass at 20°C	0.8073–0.8150
Alcoholic degree (% by vol.)	91.1–93.9
Maximum residual (mg/100 ml)	5.0
Maximum total acidity (mg/100 ml)	3.0
Maximum quantity of esters (mg/100 ml)	6.0
Alcohols of higher chains (mg/100 ml)	8.0
Maximum quantity of aldehydes (mg/100 ml)	6.0
Alkalinity	Negative
General aspects	Limpid and without residual materials

B. Engine Exhaust

Due to its characteristics as a fuel, ethanol requires high compression rates. This causes an increase in the emission of NO_x, unburned fuel, and aldehydes.

Table IIA Physicochemical Characteristics of Ethanol–Gasoline Mixtures

Volume percentage		Density at 20°C	Vapor pressure at	Octane number	Caloric content
Gasoline	Ethanol	(g/m^2)	37.3°C (kg/cm^2)		(cal/g)
100	0	0.7333	0.6749	73.0	10.626
95	5	0.7350	0.6749	74.0	10.326
90	10	0.7378	0.6749	76.5	10.071
85	15	0.7410	0.6538	78.9	9.825
80	20	0.7440	0.6046	80.3	9.680
75	25	0.7464	0.5976	82.0	9.307
70	30	0.7502	0.5976	84.0	8.977
0	100	0.7950		89–92	6.299

Table IIB Physicochemical Characteristics of Ethanol–Gasoline Mixtures

Volume percentage		Compression rate	Thermal yield	Increase in consumption
Gasoline	Ethanol	(max.)	(%)	(%)
100	0	7.0 : 1	37.3	0
95	5	7.2 : 1	37.7	2
90	10	7.5 : 1	38.3	2.8
85	15	8.0 : 1	39.3	3.3
80	20	8.3 : 1	39.8	4.0
75	25	8.5 : 1	40.2	6.3
70	30	8.7 : 1	40.5	9.3
0	100	9.0 : 1	41.7	Not tested

On the other hand, the emission of carbon monoxide (CO) is substantially diminished in the exhaust of ethanol engines when compared with gasoline. This is primarily due to differences in the air/fuel ratio for the compression rates used by the ethanol engine. If the air/fuel ratio is changed in order to lower the emissions of aldehyde and unburned fuel, the exhaust CO concentration is proportionately increased.

The actual differences between the exhaust emission of ethanol and gasoline engines are not well characterized. Some controversial values have been presented in the literature, but as a general rule it may be said that there is a substantial reduction in the levels of CO and total hydrocarbons (HC) and an increase in the levels of NO_x and aldehydes in the ethanol exhaust emissions. Table III summarizes the average differences between emissions of gasoline and ethanol as described in the literature.

In our laboratory, we found an emission spectrum for gasoline and ethanol engines shown in Figures 1A and 1B, respectively, as a function of the air/fuel ratio represented as CO emission in percentages by volume.

Table III Relative Emissions of Gasoline and Ethanol Engines

Fuel	CO	HC	NO_x	Total aldehydes
Gasoline	100	100	100	100
Ethanol	42.8	61.9	101	215

III. Biological Effects of Fuels

The manipulation of fuel by employees and customers involves several degrees of risk attributed to cutaneous absorption, accidental inhalation, and ingestion. In this section, we describe results in our laboratory of experiments with animal models in order to assess the potential toxicity of these practices.

Experiments were undertaken to assess the biological effects of ethanol additives, a blend of ethanol with other chemicals developed for diesel engines. As mentioned earlier, the successful substitution of gasoline by ethanol in spark ignition engines could not be reproduced in diesel engines. Diesel oil in fact explodes at pressures found inside engine chambers whereas ethanol does not. A possible technical solution for this problem is to introduce additives into ethanol by a number of substances, the most frequently used of which were the nitrated compounds. A mixture that has been applied with success in diesel truck engines comprises "Alcoolita" (4.5%), castor oil (1.0%), and "Max Lub 8027" (0.025%).

Alcoolita is the nitrated compound (dinitro triethylene glycol) responsible for the ethanol detonation inside the combustion chambers. The other components of the mixture are a lubricant (castor oil) and an anticorrosive buffer (Max Lub).

A common practice among customers in Brazil is to transfer fuel from one vehicle to another by a process of siphoning, which consists of aspirating the fuel through a plastic or rubber tube. Once filled, the tube is positioned into a recipient tank below the level of the fuel container. During this process a

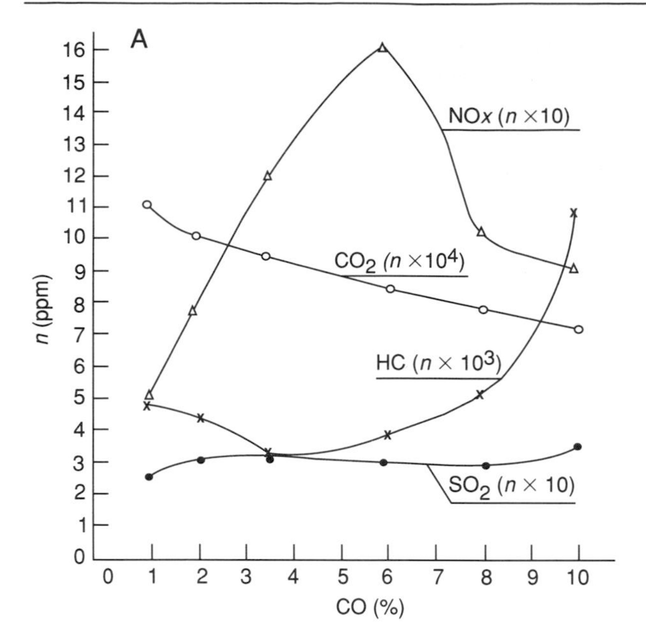

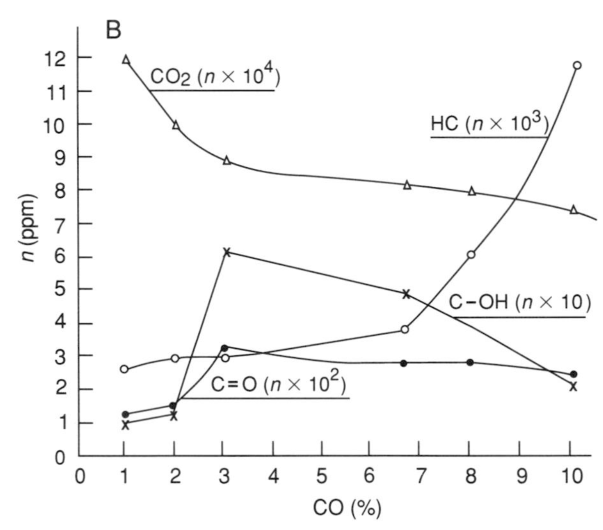

Figure 1 (A) Emission spectrum of the gasoline engine expressed in terms of the air/fuel ratio, represented by carbon monoxide concentrations. NO_x, nitrogen oxides; CO_2, carbon dioxide; HC, total hydrocarbon compounds; SO_2, sulfur dioxide. (B) Ethanol engine under identical conditions. C—OH, Hydroxyl radical; C═O, carbonyl radical.

variable amount of fuel is inevitably inhaled and ingested. It was this dangerous practice that caused much concern with the potential risks in the new fuel mixture called "additivated alcohol."

The general metabolic effects of nitrated compounds are well characterized and include an acute lowering of arterial pressure, followed by seizure, shock, and eventually death. The chronic effects include the biosynthesis of such carcinogenic compounds as nitrosamines.

A. Ingestion

The acute and chronic effects of ethanol ingestion are well known and can be easily found in any medical textbook. We shall limit our analysis to the mixture employed in diesel engines, namely additivated ethanol.

In order to assess the biological effects attributed to long-term ingestion of high quantities of activated ethanol, we submitted rats to an average daily intake of 0.4 ml for a period of 51 weeks. This group was compared with controls (that received only water) for body weight gain, mutagenic tests, hematologic, and anatomo pathologic examinations. The rats that received additivated ethanol showed an average weight gain of 25.7%, compared to a 39.7% average weight gain in the control group. This difference was statistically significant. No evidence was found that the additivated ethanol provoked any kind of tumor, despite the extremely high doses received by the animals. The mutagenic tests were also negative. Hematologic examinations showed that the rat that ingested additivated ethanol had a lower number of red blood cells. These alterations as well as the pulmonary infections found in the test groups are the same as those described in chronic ingestion of pure ethanol.

B. Inhalation

When filling tanks, fuel station employees very often spill fuel on the hot chassis while maintaining an ear close to the tank to detect fuel levels. During these maneuvers, they inhale significant amounts of vapors. In order to assess the toxicity of the several types of fuels by inhalation, we carried out a set of experiments in which rats were subjected to test atmospheres comprising vapors of ethanol, additivated ethanol and each of its components, gasoline, and diesel. Control animals were subjected to water vapor. Only acute intoxication analysis was performed; the main biological effect tested was the survival time as a function of exposure to high concentrations of fuel vapors. The rats were placed in a special chamber that maintained identical physical conditions throughout the experiments. They were separated from the fuel to be tested by a convex perforated floor that avoided the mixing of their defects with the liquid and permitted the passage of the vapors. The evaporation was optimized by bubbling synthetic air through the fuel at normal and controlled temperatures. Table IV shows the result of the acute intoxication tests for the several fuel vapors analyzed.

It can be noted that the survival times of rats submitted to pure or additivated ethanol are essentially the same but differ from the rest of the test substances. It should be noted that the differences found in the mean survival rates, when compared with the evaporation velocity of each of the test substances, point to the fact that the fuel vapor pressure influences inhalation toxicity. This explains why the rats exposed to diesel survived for more than 24 hours, while those subjected to gasoline died in such a short period of time. However, this does not invalidate our results. On the contrary, they point to the high toxicity of gasoline vapors, in part due to its intrinsic toxicity and in part due to its extremely high vapor pressure. Also, the absence of differences found between the pure and the additivated ethanol points to the relatively safe use of the additivated ethanol in diesel engines, at least from the point of view of inhalation toxicity.

C. Cutaneous Contact

We standardized a cutaneous test in which guinea pigs were submitted to cutaneous contact with ethanol, additivated ethanol and each of its components, gasoline, and diesel for the period of 15 days.

These skin tests showed no positive correlation with either additivated ethanol or Alcoolita. The component Max Lub displayed the severest reaction with hemorragic and necrotic lesions, very similar to those of diesel oil.

The vasoactivity of additivated alcohol and its components was investigated with the colloidal carbon test. Vessels were labeled only in the presence of Alcoolita, and these changes were slight and reversible. Thus, the conclusion is that the handling risks of additivated ethanol itself are similar to ethanol and much lower than those of gasoline. The problem, however, lies in the toxicity of its components (mainly Alcoolita); therefore the manipulation of this substance—or similar additives—should be carefully monitored and the preparation of additivated alcohol should be performed only by specialized personnel.

Table IV Survival Times in Rats Exposed to Fuel Vapor Atmospheres

Fuel	Mean survival time (min)
Max Lub	603
Additivated ethanol	227
Ethanol	225
Alcoolita	10.3
Gasoline	6
Diesel	No deaths in 24 hr

IV. Biological Effects of Exhaust Fumes

Among the potential toxic effects of ethanol when used as an engine fuel are those caused by the composition of exhaust fumes that differs significantly from that yielded by traditional fuels such as gasoline and diesel. We thus carried out a set of experiments that aimed to assess these effects in the exhaust of ethanol fuels in spark ignition engines. However, we know of no experiment similarly performed to analyze the effects of additivated alcohol in diesel engines.

We classify the toxic effects of exhaust fumes into short- and long-term effects. The former are principally caused by acute situations like traffic jams or exposure in confined garages and car workshops. The latter represent the most realistic situation in which an entire region is submitted to the exhaust expelled to the atmosphere. We carried out experiments intended to compare the biological effects of the exhaust of a new fuel with the traditional one (namely, ethanol versus gasoline)

A. Short-Term Effects

Because the stationary situation is the most acutely toxic, the lethal concentration for 50% of a sample of animals (LC_{50}) was determined. We also determined the mutagenicity of the exhaust fumes to bacteria by the Ames test and are currently performing a set of experiments to determine the comparative effects of exhaust fumes of gasoline and ethanol engines on the athletic performance of trained rats. The expectation has been that the acute exposure of biological models to a wide range of concentrations can provide some insights into the potential human response to acute hazards associated with the new fuel.

1. LC_{50} Experiments

We tested three atmospheres: ethanol exhaust gas, gasoline exhaust gas, and carbon monoxide (CO) controls, all diluted with clean air. During the 3-hr exposure period, the rats were housed in aluminum inhalation chambers specially designed for engine exhaust fume experiments.

The ethanol and gasoline exhaust fumes were generated by two previously unused Fiat engines of 1300-cm^3 displacement operating unloaded in the stationary mode at a constant speed of 1000 RPM.

The emission mode was fixed at 2.0% CO for both engines and we used two CO analyzers, one monitoring the CO concentration in the exhaust pipe and the other the CO inside the inhalation chambers. The ratio of the concentrations read by the two CO analyzers gave a measure of the dilution rate (Table V).

We adopted the following statistical model in these experiments:

$$Y_{ij} = \alpha_j + \beta X_{ij}, \quad (1)$$

where Y_{ij} is the probit of the expected P_{ij}; α and β are the unknown parameters; and X_{ij} is the $\log_{10}$ of the ith dose of the jth atmosphere ($i = i, \ldots, K_j$; $j = 1,2,3$).

The parameters of the Eq (1) were estimated by the probit analysis method; the results can be seen in Table VI.

We then estimated the log LC_{50} and constructed the confidence intervals for this parameter with a coefficient of 0.95. Calculating the antilogarithms of these estimations, we obtained estimates and confidence intervals for the LC_{50} of the test atmospheres (Table VII).

These experiments demonstrate that the acute toxicity of the exhaust of the gasoline-fueled engine is greater than that of the ethanol-fueled engine; the higher LC_{50} value of ethanol exhaust fumes indicates a lower toxicity.

The superimposition of the confidence intervals of the CO and ethanol LC_{50} values shows an equality of both gaseous mixtures and suggests that the acute effects of ethanol exhaust gases depend mainly on their CO content.

The exhaust fumes of gasoline and ethanol are complex gaseous mixtures containing many substances. By comparing the acute toxicity based on CO levels, we have attempted to indirectly appraise the other components that may or may not be common to both types of exhaust gases. Therefore, one of the main conclusions of these experiments is that gasoline exhaust fumes contain noxious substances other than CO, which are presumably responsible for its greater toxicity. This does not seem to be true

Table VI Estimates of the Parameters of the Statistical Model Used to Calculate LC_{50} Values in Rats Exposed to Exhaust Fumes

Test atmosphere	Estimated equation
Gasoline	$Y = -103.5505 + 32.954X$
Ethanol	$Y = -104.3183 + 32.954X$
CO	$Y = -104.4346 + 32.954X$

Table V Dose Response of Exhaust Fume Atmospheres

Atmosphere	CO dose (ppm)	Dilution rate	Log dose	Response (%)
CO	2025		3.30643	20
	2050		3.31275	45
	2100		3.32222	55
	2162		3.38486	70
Gasoline exhaust	1838	10.88	3.26435	10
	1995	10.03	3.27761	60
	2000	10.00	3.30103	55
	2097	9.54	3.32160	90
Ethanol exhaust	2007	9.97	3.30255	40
	2087	9.58	3.31952	50
	2192	9.12	3.34084	90
	2275	8.97	3.35698	75

Table VII LC_{50} and Confidence Intervals of Test Exhaust Fume Atmospheres

Test atmosphere	LC_{50}	Confidence interval at 95%
Gasoline	1968	1940–1992
Ethanol	2076	2042–2102
CO	2093	2068–2121

for ethanol exhaust fumes, at least in the case of acute toxicity.

2. Mutagenicity Tests

In this set of experiments strains of *Salmonella typhimurium* were exposed, in the presence or absence of rat-liver microsomes, for a 1-hr period to exhaust fumes of ethanol- and gasoline-fueled engines. These engines operated in the same experimental design described above but with emission concentrations of CO of 3.0% by volume (ethanol) and 3.5% by volume (gasoline).

The results of these tests can be seen in Table VIII. The highest mutagenic induction was obtained only with the TA100 strain, at CO concentrations of 13,000 and 14,500 ppm inside the chamber, in the presence or absence of S9; clearly CO concentrations higher than 14,500 ppm could be too toxic for the test cells and could thus nullify the mutagenic activity.

The results of the test with ethanol exhaust can be seen in Table IX. Mutagenic activity of the ethanol exhaust was observed only with the TA102 strain. The highest inductions obtained with S9 were associated with 8,500 and 10,500 ppm of CO, whereas those in the absence of S9 were obtained with 11,000 and 12,500 ppm of CO.

In conclusion, these experiments determined the mutagenic activity of gasoline exhaust fumes, revealing mutagenic activity of base-pair substitution without the need for metabolic activation, indicating the presence of direct-action mutagens. The experiments with ethanol exhaust suggest an indirect mutagenic activity of the oxidant type, probably due to the presence of aldehydes in these fumes.

Table VIII Mutagenicity Test with *Salmonella* Strain TA100 and Gasoline Exhaust Fumes

Dose interval investigated (CO ppm)	Induce revertants	
	+S9	−S9
1,000–3,000	31 ± 15	47 ± 31
5,000–8,000	37 ± 16	45 ± 19
9,000–11,500	97 ± 11	97 ± 11
13,000–14,500	249 ± 44	264 ± 78
15,000–16,000	191 ± 40	137 ± 44
16,500–17,500	86 ± 8	—[b]

[a] Spontaneous revertants (+S9 = 184 ± 44, −S9 = 174 ± 32) substracted.
[b] Toxic.

Table IX Mutagenicity Test with *Salmonella* TA102 and Ethanol Exhaust Fumes[a]

Dose interval investigated (CO ppm)	Induce revertants	
	+S9	−S9
1,000–2,500	82 ± 44	82 ± 36
3,000–5,500	154 ± 74	126 ± 68
6,000–8,500	204 ± 76	296 ± 78
8,500–10,500	451 ± 125	323 ± 175
11,000–12,500	342 ± 116	347 ± 204
13,500–15,500	149 ± 80	175 ± 42
16,000–17,500	—[b]	—[b]

[a] Spontaneous revertants (+S9 = 371 ± 74; −S9 = 352 ± 92) substracted.
[b] Toxic.

3. Physical Performance Tests

The preliminary results of a set of experiments intended to assess the comparative effects of gasoline and ethanol exhaust fumes on the aerobic and anaerobic physical performances of trained rats are described here.

The animals were trained on aerobic and anaerobic schedules, three times weekly, for a period of 5 months. The aerobic training consisted of 45 minutes of running in a conveyor belt at the speed of 16 m/min, and the anaerobic training consisted of pushing a trolley with a variable load in a specially designed sloped track.

After the training period, the rats were subjected to exhaust fumes from ethanol and gasoline engines operating in the previously described experimental regime at dilution rates that resulted in 100 ppm of CO. The aerobic group was exposed to the exhaust fumes simultaneously with the performance test in the conveyor belt. The anaerobic group was maintained inside the exposure chambers for a period of time equivalent to the average time the aerobic group was able to run at 100 ppm of CO. After exposure to the test atmospheres, the rats were then evaluated in the performance test on the sloped track. Table X and XI show the preliminary results of the experiments with the aerobic and anaerobic groups, respectively. As can be seen, we found a statistically significant difference in the average running time between the test performed before exposure and during the exposure of the aerobic group. This points to a greater acute toxicity of gasoline exhaust fumes on aerobic performance of trained rats. The ethanol group showed no significant difference between the tests before and during exposure.

Of note in Table XI is the absence of significant differences in both the gasoline and ethanol emissions in the aerobic experiments. This can be credited in part to a reduced influence of pollution in anaerobic physical activity and in part to our performance testing techniques. The ideal situation would be to test the anaerobic performance simultaneously with the presence of the exhaust atmospheres as in the aerobic experiments. However, technical problems at the time the preliminary experiments were carried out forced us to adopt the procedures described. We are attempting to solve these problems, and we will repeat the tests to verify these preliminary results. In any case, the gasoline emission gases have once again indicated a greater acute toxicity to animals when compared with ethanol.

B. Long-Term Effects

Long-term effects of ethanol exhaust fumes (once again compared with gasoline) were investigated through a set of experiments in which three groups of 12 rats and 10 mice were submitted to gasoline exhaust, ethanol exhaust, and to clean air (control groups). A fourth batch of mice was used as a control in the mutagenicity test and received cyclophosphamide. Table XII summarizes the physical parameters and gas concentrations inside exposure chambers. The three groups of rats were subjected to the test atmospheres 8 hours daily, 5 days weekly for 5 weeks. The mice (only used for mutagenicity tests) were exposed to the same atmospheres 8 hours each day for 10 days.

Assessments were made of the pulmonary function, hematological and biochemical parameters, pathological alterations, and the mutagenicity potential of the exhaust gases. Tables XIII and XIV summarize the results of the pulmonary function tests (with the parameters that showed statistically significant differences) and the micronucleus mutagenicity test, respectively.

Relevant pathological alterations were observed only in the respiratory system. The morphological findings were similar in the three groups of rats:

Table X Post-Training Physical Performance Test with the Aerobic Group

Test atmosphere (100 ppm of CO)	Average running time (min)	
	Before exposure	During exposure
Gasoline	108.33	74.16[a]
Ethanol	112.22[b]	114.89

[a] Statistically significant ($p = 0.0076$).
[b] No statistically significant difference between ethanol and gasoline groups before exposure.

Table XI Post-Training Physical Performance Test with the Anaerobic Group

Test atmosphere (100 ppm of CO)	Average load units (100 g)	
	Before exposure	After exposure
Gasoline	24.83	25.00
Ethanol	23.36	23.91

Table XII Physical Parameters and Gas Concentrations in Exposure Chambers

Parameter	Test atmosphere		
	Gasoline	Ethanol	Control
CO (exhaust pipe, %)	3.5	3.0	—
CO (inside chamber, ppm)	509	509	*a*
Dilution rate	68.8	58.8	—
NO_x (inside chamber, ppm)	0.50	0.66	*a*
SO_2 (inside chamber, ppm)	1.0	*a*	*a*
HC (inside chamber, ppm)	50.0	50.9	*a*
C = O (inside chamber, ppm)	*a*	0.97	*a*
Temperature (°C)	26	27	27
Pressure (atm)	1	1	1
Relative humidity (%)	45	48	43

[a] Below sensitivity.

Table XIII Pulmonary Function after Long-Term Exposure to Exhaust Fumes

Parameter	Gasoline		Ethanol		Control	
Mean expiratory flow (ml/sec)	Pre	Post	Pre	Post	Pre	Post
0–25% vital capacity	27.8	24.7	20.5	24.8	21.8	28.3
25–50% vital capacity	28.3	25.3	24.1	27.2	22.7	25.8
50–75% vital capacity	19.0	17.5	17.3	16.6	13.6	17.7

chronic bronchitis, focal interstitial pneumonitis, subpleural emphysema, and, in some animals, lobar condensation with great accumulation of neutrophils in the alveolar lumen. The pathological findings were qualitatively similar in the three groups of rats but rather different in intensity. Both the gasoline and ethanol groups exhibited more intense lesion controls, but the alterations were most striking in rats exposed to gasoline exhaust. Hematological examinations resulted in alterations typical of chronic hypoxia, both for the gasoline and ethanol groups but not for controls, which was probably due to the effects of carbon monoxide. Biochemical alterations were credited to chronic infection and pulmonary tissue destruction.

These results again point to an increased toxicity of gasoline exhaust fumes. As can be seen in Table XII, the primary difference observed between the two exhaust tests is the presence of 1 ppm of SO_2 inside the gasoline chamber. This gas was absent in the ethanol emission mixture. The concentration found is more than sufficient to cause the observed effects. The total hydrocarbons were present in practically the same concentration, but it should be mentioned that their individual compositions should be expected to be rather different from a qualitative point of view. Thus the SO_2 levels found in our experiments added to the particulate matter (not measured but certainly present in the gasoline exhaust) and the HC could be the responsible factor for the pulmonary lesions observed in rats exposed to gasoline exhaust fumes.

Table XIV Micronucleus Mutagenicity Tests after Long-Term Exposure to Exhaust Fumes

Parameter	Gasoline	Ethanol	Control
Normochromatic erytrocytes (NCE) frequency	15.47	18.15	18.58
Polychromatic erytrocytes (PCE) frequency	85.60	50.10	82.60
Micronucleus (MN) in PCE	0.80	0.18	0.25

V. Possible Indirect Effects

As mentioned previously the main difference between gasoline and ethanol exhaust fumes is the presence of higher concentrations of aldehydes in the latter. The primary aldehydes produced by ethanol engines are acetaldehyde, formaldehyde, and acrolein. These aldehydes, when released in the atmosphere, are quickly taken up in the photochemical chain of reactions and therefore have a very short half-life. Hence, all the secondary effects in the atmosphere due to ethanol exhaust are probably due to the by-products of the reaction of aldehydes with sunlight.

Extremely reactive radicals are formed (e.g., hydroperoxyl, HO_2) from the photolysis of formaldehyde:

$$HCHO + h\lambda_{370} \xrightarrow{\kappa = 2.2 \times 10^{-3}\ \mathrm{min}^{-1}} H_2 + CO \quad (1)$$

$$HCHO + h\lambda_{370} \xrightarrow{\kappa = 2.0 \times 10^{-3}\ \mathrm{min}^{-1}} H + HCO \quad (2)$$

where h is Planck's constant, κ the equilibrium rate constant, and λ the wavelength in nanometers.

The radicals produced in reaction (2) interact with the molecular oxygen to produce hydroperoxyl groups:

$$HCO + O_2 \longrightarrow HO_2 + CO \quad (3)$$

$$H + O_2 \longrightarrow HO_2 \quad (4)$$

Besides the photodissociation of formaldehyde, which leads to the hydroxiperoxyl HO_2, the radical

OH can by itself produce the formyl radical from formaldehyde:

$$HCHO + OH \longrightarrow HCO + H_2O \quad (5)$$

The transformation to HO_2 occurs through reactions (3) and (4).

Another possibility is

$$OH + H_2 \longrightarrow H + H_2O \quad (6)$$

followed by reaction (4), and

$$OH + H_2O_2 \longrightarrow HO_2 + H_2O \quad (7)$$

although this occurs less significantly.

The peroxide of reaction (7) is generated by the following reactions:

$$M + HO + HO \longrightarrow H_2O_2 + M \quad (8)$$
$$HO_2 + HO_2 \longrightarrow H_2O_2 + M \quad (9)$$

Acetaldehyde is known to react with the HO_2 radical according to the following reaction:

$$HO_2 + CH_3CHO \longrightarrow H_2O_2 + CH_3CO \quad (10)$$

All these reactions take place in the presence of solar radiation.

It can thus be concluded that the primary secondary effect of the exhaust fumes of ethanol engines in the atmosphere is to increase the production of reactive radicals, which can have potentially drastic consequences to the atmospheric pollution panorama. However, these are only theoretical speculations; to the best of our knowledge there has been no systematic research on the actual effects of these substances released by ethanol engines. What is clear is that animals submitted to long-term exposure to the air of Sao Paulo show alterations that cannot be credited to an increase of photochemical oxidants but rather to SO_2, CO, and, to a lesser extent, hydrocarbons and particulate matter. However, the direct effects of aldehydes may be drastic to the health of the population.

To estimate these effects, we carried out a set of comparative experiments in which animals were exposed to test atmospheres comprising the vapors of ethanol, methanol, acetaldehyde, and formaldehyde.

The effects of the inhalation of ethanol are quite the same as those caused by its ingestion. However, due to reasonably high exhaustion of unburned fuel by engines, and also to the great concern caused by the possibility of the secondary synthesis of aldehydes in the combustion chamber of ethanol engines, we decided to analyze the effects of exposure to ethanol as compared with other potential substitutes of gasoline in spark ignition engines (i.e., methanol) and to acetaldehyde and formaldehyde.

Wistar rats were subjected for short periods of time to the vapors of the above chemicals to assess their acute toxicity, and also for 8 hours of daily exposure, 5 days weekly, for a period of 5 weeks, to assess chronic effects. The test atmospheres were generated by a simple device in which the chemical is heated to its boiling temperature and then infused into the inhalation chambers. The principle biological parameters assessed were clinical observation and histopathological examination.

The ethanol and methanol vapors caused similar reactions, both with clinical and histopathological observations. After 5 weeks of exposure the main findings were peribronchial inflammatory reactions and interstitial pneumonitis.

As for aldehydes, acetaldehyde caused a strong irritant reaction of upper airways, chronic inflammatory reactions of the lungs with interstitial pneumonitis and squamous metaplasia, and central nervous system damage; formaldehyde caused similar respiratory reactions to the acetaldehyde, but these were of more severe intensity in the form of upper airway irritation of the nose and larynx characterized by acute inflammatory reaction with areas showing ulcerations. However, there were almost no central nervous lesions with formaldehyde exposure.

Figure 2 summarizes the effects of the substances described in this section. It indicates that all the vapors analyzed cause an irritation/inflammation of the respiratory system and a depression of the central nervous system that culminate in death.

The morphologic lesions observed in the study animals exposed to aldehyde vapors were induced at concentrations around 100 ppm and 6 ppm for acetaldehyde and formaldehyde, respectively. Because the concentration of acetaldehyde (60 ppm) and formaldehyde (below detection sensitivity) in ethanol exhaust are lower than (although very close to) the damaging limit, these concentrations are still presumably safe, at least in experimental animals.

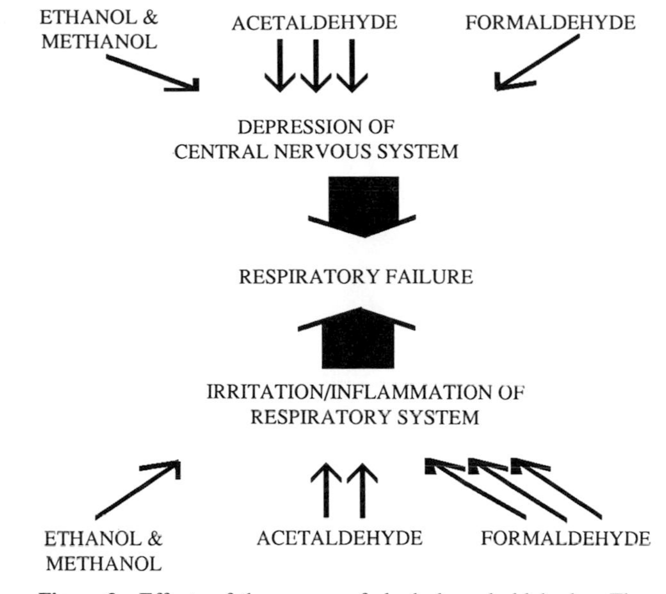

Figure 2 Effects of the vapors of alcohols and aldehydes. The number of arrows represents the intensity of each vapor effect. Respiratory failure is always the final event.

Related Article: INORGANIC MINERAL PARTICULATES IN THE LUNG.

Bibliography

Böhm, G. M., Saldiva, P. H. N., Massad, E., Pasqualucci, C. A. G., Muñoz, D. R., Gouveia, M. A., Cardoso, L. M. N., Caldeira, M. P. R., and Silva, R. (1988). Health effects of hydrated ethanol used as automobile fuel. *Proc. VIIIth Intl. Symp. Alcohol Fuels,* pp 1201–1025, Tokyo.

Lofti, C. F. P., Brentani, M. M., and Böhm, G. M. (1990). Assessment of the mutagenic potential of ethanol auto engine exhaust gases by the *Salmonella typhimurium* microsomal mutagenesis assay, using a direct exposure method. *Environ. Res.* **52,** 225–230.

Massad, E., Saldiva, C. D., Cardoso, L. M. N., Silva, R., Saldiva, P. H. N., and Böhm, G. M. (1985). Acute toxicity of gasoline and ethanol automobile engine exhaust gases. *Toxicol. Lett.* **26,** 187–192.

Massad, E., Saldiva, P. H. N., Saldiva, C. D., Caldeira, M. P. R., Cardoso, L. M. N., Morais, A. M. S., Calheiros, D. F., Silva, R., and Böhm, G. M. (1986). Toxicity of prolonged exposure to ethanol and gasoline autoengine exhaust gases. *Environ. Res.* **40,** 479–486.

Saldiva, P. H. N., Massad, E., Caldeira, M. P. R., Calheiros, D. F., Saldiva, C. D., Nicolelis, M. A. L., and Böhm, G. M. (1985). Pulmonary function of rats exposed to ethanol and gasoline fumes. *Brazil. J. Med. Biol. Res.* **18,** 573–577.

Fluoride Toxicity

Frank A. Smith
University of Rochester School of Medicine and Dentistry

Glossary

Carpopedal spasm Muscle spasms of the hands, fingers, feet, or toes.

Chemosis Edema of the conjunctiva forming a swelling around the cornea.

Hyperkalemia Elevated blood potassium level, reflected in an increased peak of T wave of the electrocardiogram.

Hypocalcemia Reduced blood calcium level, reflected in prolonged QT interval of the electrocardiogram.

Hypomagnesemia Reduced blood magnesium level.

Hypovolemic Shock State of shock in which the blood volume has been reduced.

Hypoxia Oxygen deficiency.

Malaise General discomfort or uneasiness.

Na^+–K^+ ATPase Enzyme involved in the transport of Na^+ and K^+ across cell membranes.

Nephrotoxicity Injury to the kidney cells.

Polyuria Excessive excretion of urine.

Proteinuria Protein in the urine.

Tachycardia Rapid beating of the heart.

Tetany Intermittant muscular contraction.

Ventricular fibrillation Rapid twitching of the ventricular muscle of the heart.

***FLUORIDES** (viewed here as inorganic compounds containing fluorine) are ubiquitous in nature. This is not surprising, inasmuch as fluorine is the most electronegative and reactive of the elements. It occurs naturally in more than 50 minerals. The toxicology of fluoride as it pertains to humans has been a part of the medical literature since at least 1873, when a fatality following the ingestion of a solution of hydrofluoric acid was reported. Our appreciation and understanding of the biological effects associated with fluorides have subsequently increased enormously, stimulated by their increasing uses in industry, dentistry, and medicine.*

The fluoride-containing mineral cryolite (Na_3AlF_6) is an important raw material in the production of aluminum by the Hall electrolytic process. Following exhaustion some years ago of the cryolite deposits in Greenland, the aluminum industry has turned to the use of synthetic cryolite. Rock phosphate (apatite) of which there are extensive deposits in Tennessee and Florida, is an important source of phosphoric acid and of phosphate fertilizers. Fluorsilicic acid recovered from the production of phosphoric acid is the principal source of fluoride added to domestic water supplies as a means of reducing the incidence of dental caries. More than half of the U.S. population lives in areas where the water supply is fluoridated up to levels recommended for optimal dental benefit. The apatite also contains some fluorapatite $[Ca_{10}(PO_4)_6F_2]$. During processing, fluoride containing dusts may be deposited on the surrounding territory, where it may become a hazard to vegetation and to grazing livestock. Consequently, much research has been conducted into the effects of fluoride on these important economic assets.

Domestic production of fluorspar or fluorite (CaF_2) is limited chiefly to Illinois. More than 20% of the fluorspar produced in this country in 1990 was used as a flux in steelmaking and in iron and steel foundries. Indeed, fluorite (from the Latin, "to flow") has been used by metallurgists for centuries. It is also used in the ceramic industry, in the manufacture of glass, enamels, and welding rod coatings. Over 60% of the fluorspar was used in the production of hydrofluoric acid (HF). Major uses of HF include the manufacture of synthetic cryolite and aluminum fluoride required for the electrolytic production of aluminum, in petroleum alkylation, in the processing of uranium and of rare metals, etching of glass, manufacture of semiconductors, manufacture of herbicides, and a variety of fluoride salts. Hydrogen fluoride solutions and gels are readily available for home use as rust removers. The largest use of HF is in the manufacture of various fluorocarbons (e.g., fluoropolymers and chlorofluorocarbons). Production of the latter compounds is expected to decrease over the next ten years, however, as efforts to protect the stratospheric ozone layer are implemented. More than 100 occupations, encompassing over 350,000 workers, have been identified as having a potential exposure to fluoride and/or hydrogen fluoride. Moreover, fluorides are readily accessible in the form of fluoridated toothpastes, mouthwashes, and vitamin preparation.

From the foregoing, it is evident that there is a significant potential for exposure to fluorides. Moreover, exposures may be acute or chronic in nature, and routes of entry include absorption through the skin, inhalation, and ingestion, singly and in combination. However, when handled properly, and treated with caution, fluorides do not constitute an unusual or unique hazard and exposures can be made to conform to acceptable limits. As was stated by Paracelsus more than 400 years ago, "All substances are poisons; there is none which is not a poison. The right dose differentiates a poison and a remedy."

I. Toxicity in Humans

Acute poisoning with fluoride results from the intentional or accidental ingestion of fluoride solutions or salts (the latter have been mistaken for confectioner's sugar and for powdered eggs) and from absorption from the respiratory tract and/or skin following releases of gaseous hydrogen fluoride or splashes from hydrofluoric acid solution. Regardless of the route, the effects of the fluoride are the same and fatalities may occur. Exposures to hydrogen fluoride gas or solution may be further complicated by the chemical burn these agents can cause. During chronic exposures (e.g., occupational exposures to fluoride-containing dusts, gaseous hydrogen fluoride, prolonged use of water supplies containing excessive levels of fluoride) osteosclerosis may develop over a period of time. The severity of this problem will depend upon the magnitude and duration of the exposure.

A. Lethal Dose

For the human subjects fatally poisoned by the ingestion of fluoride, we rarely have good information relating to the amount of fluoride taken. Based on a limited number of cases reported in the literature, it has been estimated that a dosage of 32–64 mg F/kg would be certainly lethal to a 70-kg person. The probably toxic dosage, defined as the threshold dosage for which the victim should receive immediate emergency treatment, has been estimated to be 5 mg F/kg. This observation should be useful from a practical viewpoint, albeit that it is based on limited data.

B. Symptomatology

Upon ingestion, inhalation, or skin penetration, fluoride is absorbed into the blood. Approximately half of this fluoride is excreted into the urine over the next 24 hours. Most of the remainder is deposited in calcified tissue, chiefly the skeleton. These two processes, elimination in the urine and deposition in bone, constitute the body's means of detoxifying fluoride. Of the two, skeletal deposition is the faster. If the dose is sufficiently large to overwhelm these mechanisms, acute toxicity ensues. If death is to be the final outcome, it often occurs in the first 2–4 hours and usually within the first 12 hours. It may, however, be delayed as long as 24 hours. If the victim can be effectively supported through the first 24 hours, the prognosis is good. Following ingestion, gastrointestinal symptoms of nausea, vomiting, abdominal pain, and diarrhea may be evident before signs of systemic poisoning are seen. Reported signs of acute systemic poisoning include

General: malaise, weakness, pallor
Cardiopulmonary: tachycardia, hypotension, prolonged QT interval on the electrocardiogram, ventricular fibrillation, pulmonary edema
Neurologic/neuromuscular: central nervous system depression, respiratory depression and paralysis, seizures, carpopedal spasm, tetany
Metabolic: hypocalcemia, hyperkalemia, hypomagnesemia

Not all of these signs and symptoms are necessarily observed in every victim. Serum or plasma fluoride concentrations ranging between 2 and 30 μg/ml have been reported in fatal cases of poisoning. The dose, however, is rarely known. Moreover, because fluoride is rapidly absorbed into and removed from the blood, concentrations may have been higher at some time prior to death. Normal plasma fluoride concentrations are generally less than 0.05 μg/ml.

C. Mechanism of Action

Upon entry into the bloodstream, a series of events is initiated, which may well terminate fatally. These are shown in Figure 1, where it is assumed that a fluoride salt (e.g., sodium fluoride) has been ingested. Upon entering the acid milieu of the stomach, the fluoride is converted to poorly dissociated hydrogen fluoride. This uncharged molecule readily passes across the gastric mucosa and the intestinal wall by passive diffusion. The irritation of these membranes, gastroenteritis, gives rise to nausea, vomiting, hemorrhage, and diarrhea. The attendant loss of fluid contributes to an electrolyte imbalance, a state of hypovolemic shock, and decreased blood pressure. Myocardial hypoxia ensues, accompanied by a state of acidosis, which in turn favors the presence of poorly dissociated hydrogen fluoride over ionic fluoride. At the cell surface, fluoride inhibits Na^+–K^+ATPase, leading to an increase in intracellular sodium, an increased sodium–calcium exchange, and increased intracellular calcium and a state of hypocalcemia. This latter state is also contributed to by an increase in calcium uptake by bone as fluoride is deposited there as fluorapatite. Some calcium fluoride may be formed and precipitated in renal tubular fluid, urine, and possibly elsewhere. The hypocalcemia may induce painful involuntary muscle contractions (tetany) evidenced in carpopedal spasm, twitching of limb muscles, laryngospasm, and cardiospasm. These responses may be due to a fluoride-facilitated neuromuscular transmission by increasing the sensitivity of cholinergic receptors to acetylcholine. The increase in intracellular calcium results in a loss of intracellular potassium to blood and a state of hyperkalemia. The hyperkalemia and hypocalcemia are reflected in the electrocardiogram by a peaking of the T wave and a

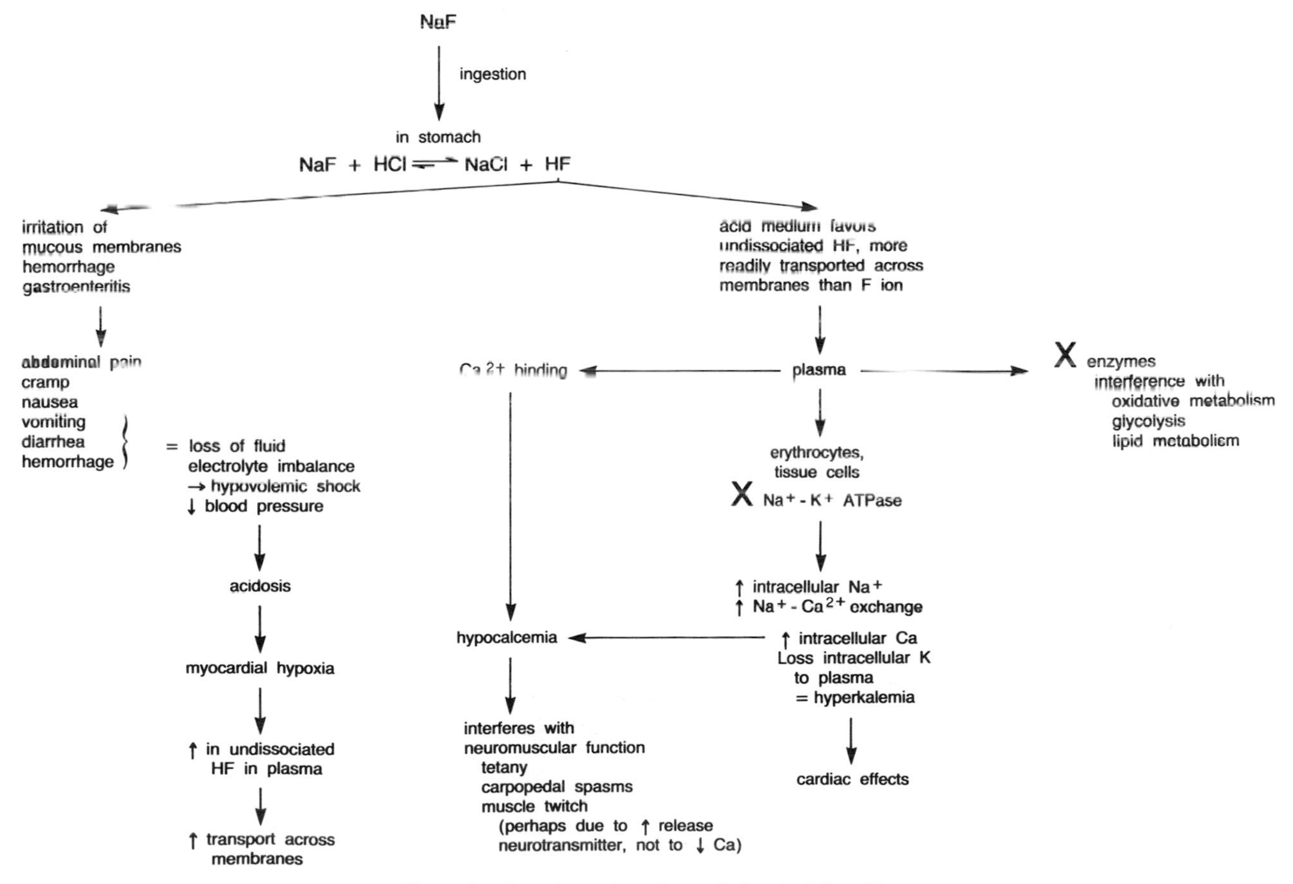

Figure 1 Acutely toxic actions of absorbed fluoride.

prolongation of the QT interval, respectively, though there may be hypocalcemia without a longer QT interval. Ventricular fibrillation can follow, sometimes preceded by ventricular trachycardia. The ventricular fibrillation eventually resists treatment. Sinus tachycardia is the most common cardiac finding. Cardiovascular collapse is probably the most common proximal cause of death. Effects on the brain are commonly seen as lethargy, stupor, and coma, and as respiratory arrest just before terminal cardiac arrest. Renal affects (e.g., proteinuria and polyuria) may be seen in some instances. The lungs may show congestion and hemorrhagic edema consistent with terminal congestive heart failure. Injury to the myocardium, thought to be due to fluoride in the tissue with associated hypocalcemia, has been reported.

Management of acute fluoride poisoning following ingestion is chiefly supportive in nature. As in any poisoning emergency, the physician must strive to maintain cardiac and respiratory function, prevent further absorption of the poison, and immobilize, remove, or otherwise inactivate that which has already been absorbed. Shock and the declining blood pressure are combated by infusion of glucose in saline or of plasma or whole blood. This also helps correct the dehydration, alleviate thirst, and maintain a mild diuresis to aid in excreting fluoride. Gastric lavage with calcium gluconate or with lime water, or the ingestion of milk, retards further absorption of fluoride. Inclusion of calcium gluconate in the infusion fluid also helps to offset hypocalcemia; considerable amounts of calcium may be required. Hypomagnesemia, if present, should be corrected. Acidosis should be corrected by infusion of sodium bicarbonate to reduce the presence of more readily absorbed undissociated hydrogen fluoride, and to improve the renal excretion of fluoride. Hemodialysis may be necessary to help remove fluoride and potassium from the blood. Cardiac monitoring by electrocardiogram should be instituted to warn of possible arrhythmias; monitoring should continue for up to 72 hours inasmuch as appearance of the hyperkalemia is sometimes delayed. The urine should be monitored for osmolality and volume to warn of possible polyuria.

II. Acute Effects

Hydrofluoric acid burns may be caused by either gaseous HF or by aqueous solutions of HF. The solutions encountered industrially range between 0.5 and 70% or higher. Concentrations commonly used are 10 and 70%. Rust remover preparations, readily available to anyone, contain up to 12% HF in a gel form or in solution. Most reported burns are to the fingers or other areas of the extremities and are generally due to dilute aqueous solutions. Accidents involving spills or splashes of more concentrated solutions may be more life threatening. In these accidents, the physician must be aware that there may be damage to the respiratory tract, and eyes from fumes and droplets on the victim. The physician may also be subject to injury from these sources.

A. Skin

Injuries may range from burns of the fingers from pinhole leaks in protective gloves to splashes on the torso, arms, neck, or face, to one instance of total immersion in a tank of 10–12% HF and anhydrous ammonia (the victim survived). Skin burns are the most common form of HF injury, and are rarely fatal. The injury proceeds in a characteristic two-step fashion. Initially, there is a corrosive skin burn similar to that caused by other mineral acids, but less severe because of the limited dissociation of HF. When the acid solution is dilute, the reaction may be delayed up to several hours, and the victim may not realize that he or she has indeed been injured until a characteristic severe, throbbing pain sets in. In the second phase, a secondary chemical burn is produced as the acid penetrates into the deep subdermal layers, with associated severe destruction and liquefaction necrosis. The severe pain accompanying the burn is thought to be due to the release of potassium into the extracellular fluid and the intense stimulation of the nerve endings.

Hydrofluoric acid burns have been classified with respect to acid concentration as follows:

<20% HF: Pain; erythema may be delayed up to 24 hours
20–50%: Burn usually apparent in 1–8 hours
>50%: Immediate, intense pain with tissue destruction

Fatal systemic poisoning can occur from burns of the trunk, neck, face, or respiratory tract following absorption of lethal amounts of fluoride from the injured surface. The site of the burn is important. Absorption is more rapid from thin or highly vascular tissue than from, for example, the palms of the hand or soles of the feet. Death due to systemic fluoride poisoning has been reported following a facial burn involving only 2.5% of the body surface area. It has been recommended that persons with burn areas greater than 50–100 cm^2 be hospitalized immediately; if the burn area exceeds 65 cm^2, the QT interval on the electrocardiogram should be monitored for indications of impending hypocalcemia. For burn areas greater than 100–150 cm^2, the patient may need to be placed in an intensive care unit.

Speed in treating HF burns is of the essence, because of the rapidity with which the acid penetrates deep into the underlying tissue. Essentially, treatment consists first of stopping further exposure, and then inactivating that fluoride present in the tissue. The first objective is accomplished by copious lavage with water under a faucet or in a shower as appropriate. At this time, all contaminated clothing is removed. The current agent of choice for inactivating fluoride is calcium gluconate. A 2.5% calcium gluconate gel is applied liberally and frequently to the affected area. A 5 or 10% solution of calcium gluconate may be infiltrated into the burn area at the rate of 0.5 ml/cm^2 of surface; if injected into a finger, no more than 0.5 ml per digit should be administered. Burns beneath the finger nail can be extremely painful; these can be treated with intraarterial injections, eliminating the possible need to remove the nail. The treatment does require hospitalization.

A second frequently used treatment is the application (as soaks or compresses) of iced aqueous solutions of Zephiran (benzalkonium chloride, 0.13%) or Hyamine (benzethonium chloride, 0.2%). It is thought that the chloride atom in these quaternary amine compounds exchanges with the fluoride in the tissues. These solutions tend to irritate facial skin; therefore, calcium gluconate treatment is the better choice for head and neck burns.

Following the initial treatment to minimize the specific fluoride effect, further treatment should follow the usual guidelines for chemical or thermal burns.

B. Eyes

The eye is highly susceptible to damage by HF gas or hydrofluoric acid solution. Damage by the latter is far more extensive than that caused by other acids in similar concentrations and is worse than that caused by the other halogen acids. The acidity per se of the HF solution is of less importance than is the ability of the undissociated HF to penetrate rapidly and deeply, thus carrying fluoride well into the tissue. The cornea and conjunctiva can be extensively damaged by exposure to HF vapor alone. The signs and symptoms of ocular injury by HF include the following:

Rapid onset of pain
Edema of the eyelid
Tearing
Conjunctival inflammation, chemosis, corneal opacification, and decrease in visual acuity

Conjunctival hemorrhage may be evident. Chronic conjunctival and corneal inflammation, scarring, and perforation may develop later. The corneal epithelium may be severely denuded, with reepithelialization taking place over a period of several weeks. There may be recurrent breakdown and erosion of the corneal epithelium. Generally, the ocular effects of hydrogen fluoride exposure are noted within the first day. The prognosis is not good if treatment is delayed or inadequate; permanent eye injury can occur associated with permanent loss of vision.

Hydrofluoric acid burns of the eye are to be considered as ophthalmic emergencies, and the attention of an ophthalmologist should be sought promptly. Immediately upon injury, the eye should be irrigated with water or saline until lavage is possible with at least two liters of 1% calcium gluconate in saline. Isotonic magnesium chloride solution has also been used, though the former solution is preferable. Following washing, drops of 1% calcium gluconate may be used every 2–3 hours for 48–72 hours. Quaternary amine compounds or 10% calcium gluconate can cause extreme irritation and damage and should not be used.

C. Respiratory System

Patients with hydrofluoric acid burns of the face and neck (indeed all patients with extensive burns) should be closely monitored for signs of pulmonary edema on the assumption that burns of the respiratory tract may have occurred. Airborne concentrations of 5 ppm (4.1 mg/m^3) are irritating to the nose and eye. Heavily contaminated clothing, especially in the chest area, may produce concentrations of 10^4–10^5 ppm in the victim's breathing zone [the limit for occupational exposures is 3 ppm (2.5 mg/m^3)]. Effects of exposure may range from mild upper airway and eye irritation to airway obstruction to pulmonary edema to death. The edema may be rapid or delayed in onset. The swelling of the oral or pharyngeal mucosa may obstruct the airway and require tracheostomy or endotracheal intubation for relief. Pulmonary reactions may persist for several weeks.

Immediate first aid consists of administering 100% oxygen, followed by a 2.5% solution of calcium gluconate given by a nebulizer. The edema is treated by the usual standard methods.

D. Kidney

Acute fluoride poisoning may induce in some individuals a polyuria resembling diabetes insipidus, which may persist for days to months. In a few instances, the acute polyuric renal failure has terminated fatally. Interest was sharply focused on the problem when it was realized that an unexpected polyuria was being encountered in some surgical patients who has been anesthetized with methoxyflurane (2,2-dichloro-1,1-difluorethyl methyl ether). It was shown that fluoride ion is released when the parent compound is metabolized *in vivo*, and that the nephrotoxicity is directly related to the level of serum fluoride. Clinical experience has shown this relationship to be as follows:

Peak serum F associated with	
No nephrotoxicity:	<0.76 (μg F/ml)
Subclinical toxicity:	0.95–1.5
Mild clinical toxicity:	1.7–2.3
Clinical toxicity:	1.5–3.3

It is believed that fluoride brings about the diuresis by inhibiting the resorption of salt (NaCl) and therefore water in the ascending limb of the loop of Henle and by interfering with the action of the antidiuretic hormone in regulating water absorption in the collecting ducts of the kidney.

Means should be instituted to reduce the blood levels of fluoride, (e.g., by establishing a state of alkalosis and by increasing fluid intake). Hemodialysis may be necessary.

III. Chronic Effects

A. Bone

As stated earlier, the deposition of fluoride in the skeleton is one of the two major mechanisms by which the body detoxifies absorbed fluoride. Fluoride brought into contact with the bone mineral surface by the extracellular fluid penetrates the hydration shell of the bone crystal. These crystals are essentially hydroxyapatite and upon reaching the mineral surface, the fluoride exchanges with the hydroxyl group to form fluorapatite. By this means, the level of fluoride in the circulating plasma can be rapidly lowered. The deposited fluoride is released gradually to the plasma over a prolonged period of time and the plasma concentrations do not rise to dangerous levels.

There are no significant deleterious effects on the bone of such brief encounters as might occur in an acute fluoride poisoning. However, ongoing cycles of rapid deposition and slow release, as occur for example in occupational exposure to fluoride or in the use of water supplies containing dangerously high concentrations of fluoride, can lead to a progressive building of bone fluoride concentrations with attendant hypermineralization, and potentially harmful consequences. An awareness of this problem came about when it was realized that long-term Danish cryolite workers were afflicted with a condition characterized by excessive bone formation, fusion of vertebrae of the spinal column, and calcification of ligaments of the pelvic floor, leading to a crippling loss of skeletal mobility. Bending over to pick up a tool from the floor became impossible, and the condition was referred to as "poker back." Extensive studies of these workers established that the causative agent was the fluoride in cryolite. Crippling fluorosis has not been seen in American workers. An osteosclerosis, defined as an increased density of bones to x-rays (radiopacity) has been seen in some employees in various industries associated with fluoride exposure. The increased density is attributable to increased bone formation brought about by fluoride. Osteosclerosis is detectable in bone with fluoride concentrations as low as 4000–5000 ppm in bone ash. Severe osteosclerosis is associated with concentrations greater than 10,000 ppm. Normal concentrations for persons of comparable age are approximately 1000 ppm.

As yet, the only practical way to achieve a lowering of skeletal fluoride is to reduce fluoride intake. It has been estimated that 8–9 years are required to reduce bone fluoride levels by one-half. Lower fluoride intake by the worker can be achieved by ensuring that airborne concentrations in the work place do not exceed the mandated standard of 2.5 mg F as dust/m^3 or 3 ppm F as HF. Community water supplies containing more than the concentrations recommended for optimal dental benefits should be replaced if possible or the water treated to reduce the fluoride concentration.

B. Cancer

The ubiquitous distribution of fluoride in nature makes it virtually impossible to avoid exposure to this element, and it is not surprising, therefore, that concern surfaces again and again about its possible carcinogenicity. It suffices to say that the bulk of the published literature supports the view that fluoride is not associated with an increase in cancer risk in humans. Recent studies in New York, the Midwest (Iowa and Nebraska), and in Alberta, Canada have led to the conclusion that there is no link between fluoridation of public water supplies and osteosarcoma. A recent rodent study in which rats were fed diets containing 1.8, 4.5, or 11.3 mg F (as NaF/kg of body weight for up to 99 weeks) found no evidence that fluoride altered the natural incidence of preneoplastic and neoplastic lesions at any site (including bone) in animals of either sex.

Bibliography

Caravati, E. M. (1988). Acute hydrofluoric acid exposure. *Am. J. Emerg. Med.* **6,** 143–150.

Gosselin, R. E., Smith, R. P., Hodge, H. C., and Braddock, J. D. (1984). "Clinical Toxicology of Commercial Products," 5th ed. pp. III-185–III-193, Williams and Wilkins, Baltimore, MD.

Hrudey, S. E., Soskolne, C. L., Berkel, J., and Fincham, S. (1990). Drinking water fluoridation and osteosarcoma. *Canad. J. Publ. Hlth.* **81,** 415–416.

McCulley, J. P., Whiting, D. W., Petitt, M. G., and Lauber, S. E. (1983). Hydrofluoric acid burns of the eye. *J. Occup. Med.* **25,** 447–450.

McGuire, S. M., Vanable, E. D., McGuire, M. H., Buckwalter, J. A., and Douglass, C. W. (1991). Is there a link between fluoridated water and osteosarcoma? *J. Am. Dent. Assoc.* **122,** 39–45.

McIvor, M. E. (1990). Acute fluoride toxicity: Pathophysiology and management. *Drug Safety* **5,** 79–85.

MacKinnon, M. A. (1988). Hydrofluoric acid burns. *Dermatol. Clinics* **6,** 67–74.

Mahoney, M. C., Nasca, P. C., Burnett, W. S., and Melius, J. M. (1991). Bone cancer incidence rates in New York State: Time trends and fluoridated drinking water. *Am. J. Publ. Hlth.* **81,** 475–479.

Maurer, J. K., Cheng, M. C., Boysen, B. G., and Anderson, R. L. (1990). Two-year carcinogenicity study of sodium fluoride in rats. *J. Natl. Cancer Inst.* **82,** 1118–1126.

Mayer, T. G. (1985). Fatal systemic fluorosis due to hydrofluoric acid burns. *Annals. Emerg. Med.* **14,** 149–153.

Shupe, J. L., Peterson, H. B., and Leone, N. C. eds. (1983). "Fluorides: Effects on Vegetation, Animals, and Humans." Paragon Press, Salt Lake City.

Smith, F. A., and Hodge, H. C. (1979). Airborne fluorides and man. Part I. *CRC Critical Reviews in Environmental Control* **8,** 293–371. Part II, **9,** 1–25.

Upfal, M., and Doyle, C. (1990). Medical management of hydrofluoric acid exposure. *J. Occup. Med.* **32,** 726–731.

Formaldehyde: Exposure Effects on Human Health

Michael D. Lebowitz
University of Arizona

James J. Quackenboss
U.S. Environmental Protection Agency

Glossary

Asthma Variable airflow obstructive disease that occurs as a reaction to the chemical (or to other stimuli). It is characterized (and tested) by bronchial hyperreactivity in this case to HCHO.

Conversion factor mg/m^3 = 1.2 ppm, at 20–25°C at sea level pressure.

Detection threshold The concentration at which one can detect the presence of HCHO (recognition of its identity may have a different threshold).

Metabolism The biochemical change of HCHO in the body.

Respiratory system The organ system responsible for breathing (i.e., oxygen and carbon dioxide exchange); it includes the nose, the pharyngeal passage, the mouth and throat, the trachea, other airways (such as bronchi), alveoli/air sacs.

Sensitization The production of an allergic state, in this case for skin or lungs.

FORMALDEHYDE is the most common aldehyde found in the environment, and is the most important one produced commercially, primarily as resins and formalin (37–50% aqueous solution). The chemical formula is HCHO. The gas is colorless, but has a pungent odor. There is potential for exposure in numerous occupational and indoor environmental settings. HCHO is highly soluble in water and oxidizes readily. About 95% is captured by the nose during nasal breathing. It is metabolized quickly. Formaldehyde is also a natural biochemical involved in synthesis of methyl groups and it is involved in lipid metabolism (i.e., in the decomposition of peroxide).

I. Nature of Formaldehyde: Its Sources and Distribution

The natural background levels of formaldehyde are very low (5–10 $\mu g/m^3$). Synthetic sources include direct emissions, especially from the production and use of formaldehyde gas, resin, or formalin, and secondary reactions of oxidized hydrocarbons from stationary and mobile sources (including incomplete combustion). The major source of exposure for humans is in the indoor environment, where products containing formaldehyde are common. Concentrations indoors are influenced by temperature, humidity, ventilation, and reactive gases such as sulfur dioxide. These sources include particle (or press-) board and plywood, insulation, fabrics, paper products, leather products, cosmetics, pesticides, fertilizers, cigarette smoke, and heating and cooking. Very little enters from outdoors (see Table I). Short-term outdoor peaks (e.g., with peak traffic in street canyons or under conditions of photochemical

Table I Contribution of Various Atmospheric Environments to Average Exposure[a]

Source	Exposure (mg/d)
Air	
Ambient (10% of time)	0.02
Indoor	
Home (65% of time)	
Conventional	0.5–2
Prefabricated (particle board)	1–10[b]
Workplace (25% of time)	
No occupational exposure[c]	0.2–0.8
With 1 mg/m^3 occupational exposure	4–5
ETS	0.1–1
Smoking (20 cigarettes/day)	1

Source: WHO/EURO, 1987. "Air Quality Guidelines for Europe," (EURO Series #23). pp. 91–104. Reprinted with permission from WHO/EURO, Copenhagen.

[a] Contribution of food and water is low so they are ignored here.

[b] Currently unusual.

[c] Assuming the normal formaldehyde concentration in conventional buildings.

smog) are about 50–100 $\mu g/m^3$. Formaldehyde resins (e.g., phenyl formaldehyde) and insulation (e.g., urea formaldehyde in foam insulation) are particle forms of the compound. Solvents and other liquids with HCHO can form aerosolized HCHO. HCHO gas can also adsorb onto particles.

Formaldehyde gas can be actively collected using the chromotropic acid or pararosanoline methods, or passively using Drager tubes or LBL sodium bisulfite containing tubes. The passive samplers require about 5–7 days of exposure in most indoor environments. Thus, they yield long-term integrated average concentrations. Screening questionnaires can be used to assess likely concentrations indoors.

II. Detection Thresholds and Range of Sensory Effects

The possible routes of exposure to formaldehyde are by ingestion, inhalation, dermal absorption, and occasionally through blood exchange (as in dialysis).

Inhalation is the major route whereby formaldehyde enters the body. The WHO/EURO approach can be used to obtain daily intake estimates (Table I). If one assumes that the average adult breathes 15 m^3/day, with the exposures in Table I and different estimates of time spent in various environments, one can calculate inhalation exposure per day. Average time estimates are: 60–70% of time in the home, 25% at work, and 10% outdoors. If normal work exposures are assumed to be similar to home exposures, the daily exposure resulting from breathing is about 0.7 mg/day. A few individuals may inhale more than 3 and possibly as much as 5 mg/day. Occupational exposure can contribute significantly to total exposure. An example of a high exposure of 1 mg/m^3 for a 25% time-weighted period (with 65% time in the home and 10% outdoors) would give a daily respiratory intake of about 4–5 mg/day.

Skin contact occurs during the use of cosmetics, household products, disinfectants, textiles (especially synthetic), orthopedic casts, and work exposure to HCHO gas or aqueous solutions. Skin exposure from most of these sources is likely to remain localized. Absorption is estimated to be negligible. Contact with liquid barriers, as in the eyes, does not appear to lead to absorption either.

Concentrations in drinking water can be expected to be less than 0.2 mg/day and are thus negligible. (Some formaldehyde occurs naturally in raw food and is produced in cooking or smoking of food.) Daily ingestion is estimated to be about 1–14 mg/day for an average adult. However, most of ingested HCHO is in a bound and unavailable form.

In certain medical treatments, formaldehyde in water solution can enter the bloodstream directly. Specifically, this can occur during blood dialysis or in circulation-assisted surgery, as machines and tubes are sometimes disinfected with formaldehyde.

Formaldehyde gas is quickly absorbed in the respiratory and gastrointestinal tract. It is quickly metabolized to formic acid by oxidation in the liver and by erythrocytes in the blood. It can subsequently undergo oxidation to carbon dioxide (CO_2) and water, be eliminated as a sodium salt in the urine, or be entrained in the carbon pool. The half-life of HCHO or its metabolites in blood is 1–2 minutes. (Studies show no differences in blood levels of formaldehyde itself between exposed and unexposed humans.) Metabolites can circulate to other organ systems. About 15% is excreted quickly as formic acid or as bound formaldehyde. The lungs exhale the CO_2 and water (about 85%). Formaldehyde-bound particles, however, may be inhaled and reside for periods of days/weeks/months in the airways of the lung, depending on their chemical form and size.

Formaldehyde is produced in small quantities as a normal metabolite and may be found in the liver.

Formaldehyde dehydrogenase converts methanol to formic acid, which has a major role in the acidosis of methanol poisoning. However, formate poisoning, also occurring with methanol ingestion, does not lead to high HCHO levels in the organism. Condensation products from formaldehyde and amines are the possible metabolic compounds found in the brain. Formaldehyde may affect uptake of folate by cells. (Folate deficiency can result in neurological defects in infants and hematologic abnormalities.)

Biological mechanisms of irritation and sensitization have been reviewed more thoroughly elsewhere (referenced). Briefly, formaldehyde gas has significant effects on retinal oxidative phosphorylation. Some adverse effects of HCHO may be related to its high reactivity with amines and formation of methylol adducts. Formaldehyde inhibits defense mechanisms in the airways by increasing mucolytic (and decreasing ciliary) activity. Primary irritant effects may include inflammation. Formalin has produced edema as well as inflammation. It produces a proliferation of lymphocytes and histiocytes; in dialysis-related exposure, it produces eosinophilia and protein conjugates that act as sensitizers. (Systemic sensitization is related to formation of eosinophilia after blood exchange contamination.) Cholinesterase activity is also reduced by exposure to HCHO. Other effects of HCHO-bound proteins on epithelial cells and receptors appear critical in sensitization. Formaldehyde can produce coagulation of proteins upon contact with membranes, possibly leading to an immunological reactive complex. Stimulation of macrophages by HCHO lead to generation of hydrogen peroxide. Particulates will carry formaldehyde gas, and some HCHO is in particle form. Thus formaldehyde may sometimes have a major effect in the airways and alveoli air sacs where HCHO–particle interactions are likely to be involved in the various processes discussed.

An important characteristic of a chemical such as formaldehyde is the concentration at which it can be detected and recognized. Odor detection of formaldehyde gas occurs with concentrations as low as 0.06 mg/m^3. Both odor recognition and other sensory/skin detections are more likely to occur at about 0.1 mg/m^3 (see Table II). Sensory irritation can be recognized by subjective symptoms or by objective indicators (e.g., eye blink rate, tearing, amount of inflammation/swelling, changes in breathing patterns) including triggering of hypersensitivity patterns.

III. Irritation: Exposures and Responses

Acute irritation in humans has been studied epidemiologically and in controlled human exposure studies. The irritation response is mostly a reflex phenomenon, which can be a direct cellular response or a nerve fiber (i.e., ANS) response. Symptoms after

Table II Effects of Formaldehyde on Humans after Short-Term Exposure

Concentration of formaldehyde (mg/m^3)		
Estimated median	Reported range	Effect
0.1	0.06–1.2	Odor threshold in 50% of people (including repeated exposure)
0.5	0.1–1.9	Eye irritation threshold
0.6	0.1–3.1	Throat irritation threshold
3.1	2.5–3.7	Biting sensation in nose, eyes
5.6	5–6.2	Tolerable for 30 min (tearing)
17.8	12–25	Strong flow of tears, lasting for 1 hr
37.5	37–60	Danger to life, edema, inflammation, pneumonia
125	60–125	Death

Source: WHO/EURO, 1987. "Air Quality Guidelines for Europe," (EURO Series #23). pp. 91–104. Reprinted with permission from WHO/EURO, Copenhagen.

short-term exposure to formaldehyde are irritation of the eyes, nose, and throat, sneezing, coughing, chest tightness, and shortness of breath. There have also been reports of nausea and discomfort. Exposure to formaldehyde vapor causes direct irritation of the skin. (A single application of 1% formalin in water on human skin will produce an irritant response in about 5% of the population.)

Table II shows the thresholds of human responses. Eye and throat irritation starts at 0.05 mg/m^3 with 50% of individuals showing irritation at 0.1 mg/m^3. Due to absorption in the upper respiratory tract, higher concentrations of formaldehyde gas are required to stimulate bronchial receptors in the lower airways; these concentrations are greater than those causing sensory irritation. However, HCHO–particle interactions can affect the lower airways at about the same concentrations as gas-produced sensory irritation. Lung function studies show increased airway resistance and decreased compliance. At higher concentrations, or if formaldehyde is adsorbed onto particles, peripheral lung tissue receptors might be stimulated by the effects of formaldehyde. At concentrations between 12 and 25 mg/m^3, symptoms are severe and normal breathing becomes difficult. Effects on pulmonary tissue and the lower airways are likely to occur at gaseous concentrations of 3–6 mg/m^3, and HCHO–particle concentrations at or below 1 mg/m^3.

There appears to be a possible development of tolerance, as symptoms are often more severe at the start of exposure and then may diminish. Improvements in irritant symptoms are reported after removal of/from sources. Children have been reported to be more sensitive to irritant effects. Also, formaldehyde appears to be a predisposing factor to respiratory tract infections because of its effect on immune defense mechanisms, especially in children.

A. Epidemiological Studies

Studies have been made on the effects of formaldehyde on chronic disease in occupationally exposed populations. In a study at wood-processing plants, workers exposed to formaldehyde had a higher incidence of chronic upper respiratory disease. Particle-bound formaldehyde was not measured in this study, and the air concentrations of formaldehyde gas in all locations were below 0.4 mg/m^3.

Workers in an acrylic wool filter department also were studied. The work environment had phenol levels of 7–10 mg/m^3 and formaldehyde levels of 0.5–1 mg/m^3. Exposed workers had lower lung function values. (They also had exposure to particles and fibers that were not monitored.)

Studies in embalmers, pathology technicians, and batt makers (those involved in phenol–formaldehyde–plastic foam matrix embedding of fiberglass) have demonstrated irritant effects in the upper and lower respiratory tracts, and at the same time, effects on the skin, nervous, and gastrointestinal systems. Kilburn reported a summary of studies that showed increases in irritant symptoms and decreases in lung function. In studies of batt makers and histology technicians, symptoms were greater for batt makers, especially those involved in the hot (high-exposure) areas of the process; exposure for the latter group was about 6 mg/m^3, while exposure for the techs ranged from 0.5–2.3 mg/m^3. All exposed groups had 2–3 times more symptoms than hospital controls (matched by age, ethnicity, and smoking). Respiratory symptoms evaluated included cough, breathlessness, and wheezing. Kilburn's group also evaluated chronic bronchitis and emphysema, but these didn't differ much between groups (emphysema only occurred in the most highly exposed).

Studies of symptoms and lung function in several community populations have shown similar results of exposures, though often at lower concentrations (possibly due to the existence of other, concomitant exposures). Those with sensitized airways or skin have a greater reaction to the irritant. Chronic airway obstructive disease, (e.g., chronic bronchitis and emphysema) due to formaldehyde has been observed in various epidemiological studies. Several other studies, in various settings in Eastern Europe and the Soviet Union, have found relationships between formaldehyde exposure and chronic lung disease.

The community population study reported by Krzyzanowski and collegues showed higher rates of diagnosed chronic bronchitis in children that correlated to monitored formaldehyde exposure. Specifically, the rates went from about 3.5 to 28.6% (and were related to kitchen concentrations). The higher rate occurred with current concentrations above 72 μg/m^3. (The latter did not reflect the higher exposures experienced during the development of the disease.)

It has been concluded by several committees that chronic obstructive lung disease is unlikely to occur in people exposed to less than 1.8 mg/m^3. According to a WHO/EURO report, it is quite possible that some chronic lung disease will occur in those exposed to concentrations above 5 mg/m^3.

Nervous system (neural and behavioral) symptoms have been a subject of contention, in terms of finding them specifically related to formaldehyde exposure alone, and because they often mimic the nonspecific complaints of "sick/tight building syndrome." The symptoms have been found in pathology technicians who use various solvents and fixatives and in plastic workers using phenol–formaldehyde resins, where appropriate controls were used. These symptoms included the usual gastrointestinal symptoms (e.g., nausea) as well as problems with sleep, memory, equilibrium, and mood. In the Kilburn studies, batt workers had more "vegetative" symptoms (headache and fatigue, as well as loss of appetite, indigestion, and nausea) than male histotechs and controls, but female histotechs reported as much as the batt workers (and female controls reported as much as male histotechs). In comparing histotechs by length of formaldehyde exposure and by concomitant xylene exposure, trends of other symptoms with amount of exposure were also observed. These included lightheadedness, dizziness, unstable moods, irritability, loss of appetite, and insomnia. With concomitant increased xylene exposure, higher rates of these symptoms occurred, as well as headache (but not fatigue), and both indigestion and nausea. (Such symptoms have been found also in conjunction with skin and breathing problems. These types of symptoms, and others, have also been found in those smoking marijuana soaked in formaldehyde (called AMP) a prominently abused substance. Some of these human neurological findings are supported by animal studies with injected formalin.

Several studies have investigated possible effects of formaldehyde on pregnancy outcome, frequency of menstrual disorders, pregnancy complications, and low-birth-weight babies. It is not, however, possible to draw definite conclusions because of the scarcity and limitations of available data.

IV. HCHO-Specific Asthma and Skin Sensitization

A. Asthma

The first published case of formalin asthma was in an occupational setting. Large numbers of case studies have been reported in Europe on formaldehyde-induced asthma, claiming that an allergic mechanism was responsible. Immunological data have not usually been presented, although one study showed an IgG increase in one case.

In formaldehyde–resin makers and other formaldehyde workers, it has been shown that formaldehyde, including formalin (40% formaldehyde by volume in aqueous solution) will produce primary irritational or allergic hypersensitivity. Further, formaldehyde reaction products, for example, dimethylurea, can be sensitizers. Animal species, especially guinea pigs, have shown sensitization with repeated inhalation (or dermal contact).

Other studies have lead an NAS committee to conclude that formaldehyde has been shown to cause bronchial asthma in humans. The committee stated that asthmatic attacks related to exposures to formaldehyde at low concentrations are specifically due to formaldehyde sensitization in some cases. Controlled inhalation studies with formaldehyde (usually HCHO-particles) are positive in these instances.

However, in an evaluation of 230 individuals with explicit suspicion of formaldehyde asthma, it was found that only 12 actually reacted to bronchial-provocation tests with formaldehyde gas. These patients had significant dose–response reactions between 0.4 and 2.5 mg/m^3 HCHO. Of the remaining cases (218), 71 reacted to histamine bronchial-provocation challenge, indicating actual asthma; the actual diagnosis in the remaining 147 is unclear. (If the 147 didn't have asthma, the rate of HCHO-asthma was 12/83, or 14%; it could be higher, however, as others could have reacted to HCHO-particles.)

Two cases of occupational asthma caused by urea-formaldehyde particle board were reported to have positive histamine bronchial challenge tests, but did not exhibit increased RAST-specific IgE antibodies directed against formaldehyde–human-serum-albumin conjugates. A third asthmatic didn't have a response with formaldehyde exposure, indicating (to the investigators) a specific sensitization to the agent in the first two instances. (Formaldehyde also seems to act as a direct airway irritant in persons who have bronchial-asthmatic attacks from other causes.)

In the study of batt workers and histotechs, Kilburn reported an increase in diagnosed asthma (16%, about two-fold more) only in the low-exposure batt workers and more diagnosed allergy (39%, about 1.5 times more) in the histotech workers. Airway obstructive disease (including asthma) due to formaldehyde has also been observed in various other occupational studies. In all likelihood, based on a variety of recent studies, the potency of formaldehyde occupationally is probably less than 5%. In addition, some individuals with allergic contact dermatitis may have mild-to-severe asthmatic reactions with continued exposure to HCHO.

Acute asthma-like symptoms of chest tightness and dyspnea have been reported in dialysis patients when formaldehyde-treated membranes were used. The symptoms were correlated with complement activation (C3a). Asthma-like symptoms and bronchospasm were also reported with similar exposure; this was associated with increased eosinophilia, IgE, elevations, and complement activation (C3a, C5a). Formaldehyde-conjugated specific IgE (RAST) elevations were found in some; other agents were found to have similar effects (e.g., ethylene oxide, phthalates, isocyanate).

Kryzanowski and collegues have reported higher rates of diagnosed asthma in children in a community population study, related to currently monitored formaldehyde exposure, with a synergistic contribution of environmental tobacco smoke (ETS). The prevalence rates were 23.8% in homes with current HCHO levels in kitchens of over 72 $\mu g/m^3$, a doubling compared to other homes. It was doubled again if there was concomitant ETS exposure. (It is assumed that current levels are much lower than those experienced during the development of asthma.) Formaldehyde-induced persistent bronchial responsiveness occurred primarily in the lower socioeconomic status (SES) group in this population. (The monitoring used a modified Lawrence Berkeley Laboratory (LBL)-validated passive tube method with no temperature or humidity interferences in the range experienced in this locale.) They also found acute changes in children, especially asthmatics, related specifically to overnight exposures in their bedrooms; morning peak expiratory air flow rates (PEFR) had decreased significantly with a demonstrable exposure–response relationship (Fig. 1).

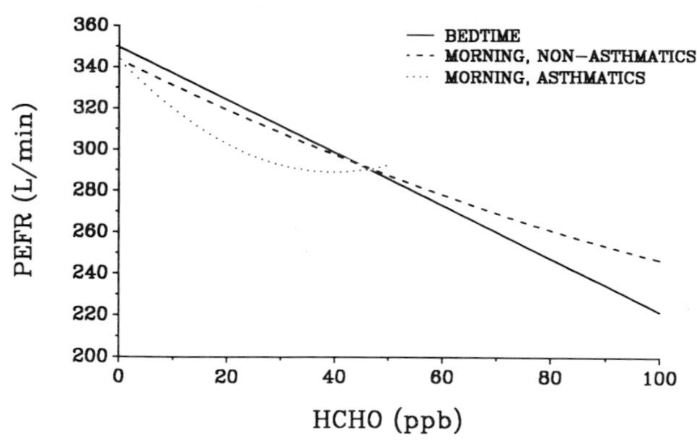

Figure 1 PEFR estimated by random effects model in children. (Reprinted from Krzyzanowski *et al.*, 1990, with permission from Academic Press.)

B. Skin Sensitization

Skin contact with formaldehyde has been reported to cause a variety of acute and chronic problems in humans, including irritation, allergic contact dermatitis (Type I allergy), and urticaria (Type IV allergy). Dermal exposure to formaldehyde may cause an acute inflammatory reaction of the affected region at concentrations of 1.2 to 3.6 mg/m^3 over periods of minutes to hours. These irritant effects may be tolerated by normal adults even though they produce noticeable skin irritation from direct exposure.

There is a chronic form of allergic contact dermatitis. In a sensitized individual, as low as 0.24 mg/m^3 has produced a dermal sensitization response under controlled conditions. In tests conducted on supposedly normal subjects, using an induction dose (0.37–3.7%) followed by a 3.7% challenge, the follow-up challenge indicates that 4.5–7.8% of the subjects showed evidence of skin sensitization. Based on the variety of data available, it is estimated that between 1 and 6% of individuals in the United States are probably sensitive to low concentrations of formaldehyde.

V. Cancer

Formaldehyde has been found to produce a high incidence of squamous cell carcinomas in the nose of two strains of rats in both sexes but only at the highest dose (18.7 mg/m^3). The dose–response relationship was clearly nonlinear, with a disproportionate increase in tumors at higher concentrations. In mice, no statistically significant increase in nasal tumors was found, even at 18.7 mg/m^3. In more limited studies, formaldehyde has not been carcinogenic in hamsters.

Formaldehyde reacts with amino groups in RNA, breaking the hydrogen bonds forming the coiled

RNA. It reacts with DNA less frequently, because the DNA hydrogen bonds are more stable. One study of hemodialysis patients showed chromosome abnormalities.

Data concerning the potential carcinogenicity of formaldehyde in humans are available only from studies of occupationally exposed populations. The occupations studied consist of those using formalin to preserve biological specimens, and workers involved in the production and use of formaldehyde. A mortality study of British pathologists found only an excess of brain tumors (more likely due to solvents), no excess of lymphatic or hematopoietic neoplasms, and no nasal or nasal–sinus tumors. A study of American anatomists also yielded a significant excess of brain tumors only. Limitations of all the studies include lack of exposure assessment. Further, smoking or exposures to other chemicals could not be controlled in most of the studies. In summary, none of the studies has provided strong, conclusive evidence of the carcinogenicity of HCHO in humans.

VI. Summary and Conclusions

Formaldehyde concentrations in the indoor environment are many times higher than those outdoors. Formaldehyde has an odor, the threshold of which is a level below the concentrations that are of little concern in human discomfort. Formaldehyde is primarily an irritant to eyes, nose, throat, skin, airways, and nervous system. Some sensitive individuals likely react to lower levels. Skin and inhalation sensitization occurs. Long-term pulmonary effects are possible, probably at concentrations above 5 mg/m^3. Formaldehyde poses a different degree of problems in indoor environments.

It has been recommended by WHO/EURO that in an effort to avoid complaints about the odor of formaldehyde indoors, concentrations should be below 0.12 mg/m^3. (Some sensitive individuals, however, can sense much lower formaldehyde concentrations.) Further, indoor air usually contains other organic compounds that, alone or in combination with formaldehyde, may have odorous and irritating properties.

WHO/EURO and NRC/NAS have evaluated control measures with the following conclusions. In existing buildings, replacement of materials and furnishings is the preferred control measure, but sealing of materials alone can be very helpful. Increased ventilation, possibly combined with excess heating, can be helpful in some cases to accelerate vaporization from materials and subsequent removal. (It has also been reported that fumigation of problem houses with ammonia has reduced levels considerably over long periods of time.) The use of appropriate materials in new or remodeled interiors is a good prevention method, assuming that appropriate ventilation is also present. Product standards can be formulated to restrict maximum permitted values of formaldehyde released. Also, building regulations can regulate the use of formaldehyde-releasing materials, such as particle board, paint, and glue, and can prescribe the ventilating rates.

Related Article: Plants as Detectors of Atmospheric Mutagens.

Acknowledgment and Disclaimer

Although the research described in this article has been funded wholly or in part by the U.S. Environmental Protection Agency, and James Quackenboss now works for the Agency, it has not been subjected to the Agency's required peer and policy review and therefore does not necessarily reflect the views of the Agency, and no other official endorsement should be inferred.

Bibliography

Allegra, L., Fabbri, L. M., Picotti, G., and Mattoli, S. (1989). Bronchial epithelium and asthma. *Eur. Respir. J.* **2**(Suppl 6), 460S–468S.

Gardner, D. E., Crapo, J. D., and Massaro, E. J., eds. (1988). "Toxicology of the Lung." Raven Press, New York.

Hendrick, D. J. (1983). The formaldehyde problem: A clinical appraisal. *Immunol. Allergy Practice* **5(4)**, 21–32.

Kilburn, K. H., Warshaw, R., Boylen, C. T., Johnson, S. J. S., Seidman, B., Sinclair, R., and Takaro, Jr., T. (1985). Pulmonary and neurobehavioral effects of formaldehyde exposure. *Arch. Environ. Hlth.* **40(5)**, 254–260.

Krzyzanowski, M., Quackenboss, J. J., and Lebowitz, M. D. (1990). Chronic respiratory effects of indoor formaldehyde exposure. *Environ. Res.* **52,** 117–125.

NRC/NAS (National Research Council) (1981). "Formaldehyde and Other Aldehydes." National Academy Press, Washington, D.C.

Quackenboss, J. J., Lebowitz, M. D., Michaud, J. P., and Bronnimann, D. (1989). Formaldehyde exposure and acute health effects study. *Environ. Intl.* **15,** 169–176.

"Report on the Consensus Workshop on Formaldehyde." (1983). *Environ. Hlth. Perspect.* **58,** 323–81.

WHO. (1987). "Air Quality Guidelines for Europe." (European Series #23). pp. 91–104, Copenhagen.

WHO. (1989). "Organic Pollutants." (Euro Reports and Studies #111), Copenhagen.

Fungicides

Robert A. Frakes
Maine Department of Human Services

Lebelle R. Hicks
Maine Department of Agriculture

Glossary

Active ingredient Component of a fungicide product that is specifically intended to kill or retard fungal growth.

Dermatitis Irritation or inflammation of the skin.

Inert ingredients Ingredients in a fungicide product other than the active ingredient which are ineffective toward the target organism.

LC_{50} Median lethal concentration. The concentration of a chemical in air or water which causes death in 50% of test animals living in that air or water for a specified period of time.

LD_{50} Median lethal dose. A single dose of a chemical, usually given orally or dermally, which causes death in 50% of test animals.

LOEL Lowest observed effect level. The lowest experimental dose or concentration at which a statistically or biologically significant effect is observed.

NOEL No observed effect level. The highest experimental dose or concentration at which there are no statistically or biologically significant increases in any effect observed in the exposed population when compared to the appropriate control population.

Nontarget Organisms other than the intended pest species which may be affected by the application of a fungicide. These include nonpathogenic soil fungi and bacteria, beneficial insects, fish and aquatic invertebrates, and wild birds.

Sensitization An allergic reaction of the skin or lungs to exposure to a chemical substance. Once sensitized, an individual may react adversely to even very low levels of exposure to the substance.

Systemic A fungicide which exerts its protective effect through being taken up by the plant and distributed to its tissues, rather than as a surface coating.

Tolerance The maximum residue concentration of a fungicide allowed to remain in a crop after harvest.

***FUNGICIDES** are pesticides whose target pests are the nonchlorophyll bearing plants, fungi. Fungi include rusts, mildew, smuts, rots, and blights. Fungicides in the United States must be registered for use by the Environmental Protection Agency (EPA) under provisions of the Federal Insecticide Fungicide and Rodenticide Act (FIFRA). Chemicals used as fungicides include organic compounds such as heterocyclic nitrogen compounds, dithiocarbamates, substituted benzenes, and thiophthalimides, as well as inorganics and organometals such as sulfur, copper, and mercury compounds. While antibiotics, fumigants, antimicrobials, wood preservatives, and antifouling paints have fungicidal properties, this article will be limited to the organic agricultural fungicides.*

Fungicide products comprise both the active ingredients and the "inert" ingredients. The "inerts" are not necessarily toxicologically inert, but rather are not active toward the target pests. The physical forms of the products can be liquids, emulsifiable concentrates, soluble concentrates, wettable powders, or granulars. The physical formulation and the percentage of active ingredients are important in determining usage and toxicity characteristics. Registration of fungicides requires scientific tests of both the Technical Grade Active Ingredient (TGAI) and the end products. Chemical characteristics, doses, and concentrations discussed below are expressed in terms of the active ingredients rather than the formulated products.

I. Heterocyclic Nitrogen Compounds

A. Introduction

Compounds are classified as heterocyclic nitrogen fungicides based on the presence of a ring containing one or more nitrogen atoms. This is a strictly chemical classification and is not meant to reflect biological activity. Also considered in this group are compounds whose metabolically active form is a heterocyclic nitrogen, for example thiophanate-methyl.

Benomyl (methyl-1-(butylcarbamoyl)-2-benzimidazole carbamate) has been chosen as the representative compound for this class of chemicals as it is metabolized to another heterocyclic nitrogen, carbendazim [methyl-2-benzimidazole carbamate (MBC)]. This is the same active metabolite as seen with thiophanate-methyl. Benomyl has tolerances for over 100 commodities including livestock (cattle, hogs, goats, poultry), fruits (e.g., apples, bananas, grapes, and citrus) and many vegetables. The tolerances range from 0.1 ppm in livestock byproducts and products to 125 ppm on dried grape pomace and raisin waste.

Benomyl is a systemic fungicide first registered for use in 1969. While benomyl is a nitrogen heterocycle with a carbamate functional group, it has no anticholinesterase activity. The biologically active derivative of benomyl is MBC. MBC is also the active metabolite of thiophanate-methyl and is included in the tolerances for both benomyl and thiophanate-methyl. It is a fungicide in its own right, but is not yet registered for use in the United States. The toxicity profile of MBC, as expected, is similar to that of benomyl with regard to effects on the testes, developmental effects, and oncogenic potential.

The physical characteristics of benomyl, as well as the other heterocyclic nitrogen compounds, are shown in Table I. Technical benomyl is a colorless crystalline solid, with a negligible vapor pressure at room temperature and decomposes without melting. The solubility at 25°C in water is very pH dependent. Benomyl is very soluble at pH 1, has a solubility of 4 mg/kg at pH 3 to 10 and decomposes at pH 13. The solubility of benomyl in organic solvents varies from 4 g/kg in ethanol to 400 g/kg in heptane. Formulations containing benomyl include wettable powders, flowable liquids, and oil dispersible products.

B. Animal Toxicology

1. Metabolism

Absorption of benomyl through skin in rats shows a nonlinear decrease with increasing dose and a nonlinear increase with increasing time. After 10 hours of exposure, the percent of benomyl absorbed ranges from 0.031 in the high dose group to 3.5 in the low dose group. By 10 hours, 96 to 99% of the absorbed dose is excreted in the urine in all treated groups.

Limited metabolic studies in rats with benomyl suggest that the urine is the major route of excretion (86%) and that the major metabolites are the sulfate and glucuronide conjugates of 5-OH MBC. Metabolic studies (with doses of 0.1 mg/kg) were done in mice, rabbits, and sheep via oral or intraperitoneal dosing. The metabolite distribution was similar in all three species regardless of route of administration. As with the rats, the sulfate and glucuronide conjugates of the hydroxylated metabolites were identified in the urine of mice. No parent compound was detected in either the urine or feces.

2. Acute Studies

Heterocyclic nitrogen fungicides are of low acute toxicity (Table II). Most of these compounds have oral and dermal LD_{50} values in the range of 1000 to 10,000 mg/kg. Dazomet, imazalil, and triademefon are slightly more toxic than the others. Acute inhalation data are available for benomyl. The inhalation LC_{50} is greater than 4.01 mg/l, and a no observed effect level (NOEL) for aspermatogenesis of 0.20 mg/l in air was determined for this compound. Benomyl dry flowable formulation causes reversible corneal opacities and mild iritis in eyes. Benomyl also causes irritation of the skin and moderate sensitization.

3. Subchronic and Chronic Toxicity

In the subchronic studies, the effects observed were hematological changes and changes in the testes. In

Table I Physical and Chemical Characteristics of Fungicides

Common name	Class[a]	CAS no.	Physical form	Molecular weight	Vapor pressure at room temp.[b]	Melting point (°C)	Solubility in water at room temp.[b]
Metham-sodium	DC	137-42-8	white crystalline solid	129.2	—	—	722 g/l
Ferbam	DC	14484-64-1	black powder	416.5	negligible	>180 decomposes	130 mg/l
Thiram	DC	137-26-8	colorless crystals	240.4	negligible	146	30 mg/l
Ziram	DC	137-30-4	white solid	305.8	negligible	240	65 mg/l (25°C)
Mancozeb	EBDC	8018-01-7	yellowish powder	polymer	negligible	decomposes	practically insoluble
Maneb	EBDC	12427-38-2	yellow crystalline solid	265.3	negligible	decomposes	virtually insoluble
Metiram	EBDC	9006-42-2	light yellow solid	polymer	<0.1 mPa	>140 decomposes	2 mg/l
Nabam	EBDC	142-59-6	colorless crystals	256.3	—	—	200 mg/l
Zineb	EBDC	12122-67-7	light colored powder	275.8	negligible	decomposes	10 mg/l
Captan	TP	133-06-2	white solid	300.6	<1.3 mPa (25°C)	175–178	3.3 mg/l (25°C)
Captafol	TP	2425-06-1	colorless crystalline solid	349.1	negligible	159–161	1.4 mg/l
Folpet	TP	133-07-3	white crystals	296.6	1.3 mPa	177–180	1 mg/l
Hexachlorobenzene	SB	118-74-1	colorless crystals	284.8	1.45 mPa	226	practically insoluble
Pentachlorophenol	SB	87-86-5	buff color crystals	266.3	16 Pa (100°C)	191	20 mg/l (30°C)
Pentachloronitrobenzene	SB	82-68-8	pale yellow crystals	295.3	1.8 Pa (25°C)	142–146	0.87 mg/l
Chlorothalonil	SB	1897-45-6	white crystalline solid	265.3	1.3 Pa (40°C)	250–251	0.6 mg/l (25°C)
Dichloran	SB	99-30-9	yellow crystalline solid	207.0	0.16 mPa	195	7 mg/l
Chloroneb	SB	2675-77-6	white crystals	207.1	400 mPa (25°C)	133–135	8 mg/l (25°C)
Anilazine	HN	101-05-3	white–tan crystalline solid	275.5	—	159–160	practically insoluble
Benomyl	HN	17804-35-2	white crystalline solid	290.3	negligible	decomposes	4 mg/l (25°C)
Carbendazim	HN	83601-81-4	nearly white solid	191.2	<100 nPa	310 decomposes	7–28 mg/l
Dazomet	HN	533-74-4	white–slightly gray crystalline solid	162.3	0.37 mPa	104–105	3.0 g/l (decomposes in water)
Etridiazole	HN	2593-15-9	reddish brown liquid	247.5	13 mPa	—	50 mg/l (25°C)
Fenarimol	HN	60168-88-9	white crystalline solid	331.2	0.013 mPa (25°C)	117–119	13.7 mg/l (25°C)
Imazalil	HN	35554-44-0	slightly yellowish to brown oil	297.2	0.0093 mPa	50	293 mg/l
Iprodione	HN	36734-19-7	white crystals	330.2	<0.133 mPa	136	13 mg/l
Myclobutanil	HN	88671-89-0	light yellow solid	288.8	213 mPa (25°C)	63–68	142 mg/l (25°C)
Propiconazole	HN	60207-90-1	pale liquid	342.2	0.133 mPa	bp 180	110 mg/l
Thiabendazole	HN	148-79-8	white powder	201.2	nonvolatile	304–305	<50 mg/l (25°C)
Thiophanate	HN	23564-06-9	colorless crystalline solid	370.4	—	195 decomposes	almost insoluble
Thiophanate-methyl	HN	23564-05-8	colorless crystalline solid	342.4	—	172 decomposes	slightly soluble
Triadimefon	HN	43121-43-3	white–tan crystals	293.8	<0.1 mPa	82.3	70–260 mg/l
Triadimenol	HN	55219-65-3	white–tan crystals	295.8	<1 mPa	121–127	95 mg/l
Triforine	HN	26644-46-2	colorless crystals	435.0	0.027 mPa (25°C)	155	6 mg/l
Vinclozolin	HN	50471-44-8	white solid	286.1	<10 mPa	108	1.0 g/l
Carboxin	O	5234-68-4	off-white crystals	235.3	<133 Pa	91.5–92.5	170 mg/l (25°C)
Dinocap	O	39300-45-3	dark brown liquid	364.3	—	bp 138–140	practically insoluble
Dodine	O	2439-10-3	white crystalline	287.4	—	136	63 mg/l (25°C)
Fosetyl-AL	O	39148-24-8	colorless powder	110	negligible	>200 decomposes	122 g/l
Metalaxyl	O	57837-19-1	colorless crystals	279.3	0.293 mPa	71.8–72.3	7.1 g/l

[a] DC, dithiocarbamate; EBDC, ethylenebisdithiocarbamate; TP, thiophthalimide; SB, substituted benzene; HN, heterocyclic nitrogen; O, other.

[b] Unless another temp. is specified.

Table II Acute Toxicity of Fungicides

Compound	Species	Oral LD_{50} (mg/kg)	Dermal LD_{50} (mg/kg)	Inhalation LC_{50} (mg/l)
Dithiocarbamate				
Metham-sodium	rat	1700–1800		
	rabbit		1300	
Ferbam	rat	>4000–>17,000		
Thiram	rat	780–865		
	mouse	1500–2000		
	rabbit	210		
Ziram	rat	1400	>6000	
Ethylenebisdithiocarbamate				
Mancozeb	rat	8000–11,200	>15,000	
Maneb	rat	6750–8000		
Metiram	rat	>10,000		
	mouse	>5400		
	guinea pig	2400–4800		
Nabam	rat	395		
Zineb	rat	4400–8900	>2500	
	mouse	7000–7600		
Thiophthalimide				
Captan	rat	9000–17,000		
	mouse	7000–7840		
Captafol	rat	5000–6200		
	rabbit		>15,400	
Folpet	rat	>10,000		
	rabbit		>22,600	
Substituted benzene				
Hexachlorobenzene	rat	10,000		
Pentachlorophenol	rat	50–210		
Pentachloronitrobenzene	rat	1200–1600	>5000	1.4
	mouse	1400		2.0
Chlorothalonil	rat	>10,000		
	rabbit		>10,000	
Dichloran	rat	4000		
	mouse	1500–2500	>5000	
Chloroneb	rat	>11,000		
Heterocyclic nitrogen				
Anilazine	rat	2710		
	rabbit	460	>9400 (intact)	
			>2500 (abraded)	
Benomyl	rat	>10,000	>2000	>4.01
	rabbit		>10,000	
Carbendazim	rat	>1500		
	rabbit		>10,000	
	dog	>2500		
Dazomet	rat	520	>2000	
Etridiazole	rat	1077		
	mouse	2000		
	rabbit		1700	

(continues)

Table II (*Continued*)

Compound	Species	Oral LD_{50} (mg/kg)	Dermal LD_{50} (mg/kg)	Inhalation LC_{50} (mg/l)
Fenarimol	rat	2500		
	mouse	4500		
	rabbit		>2000	
Imazalil	rat	320		
	rabbit		4200–4880	
Iprodione	rat	3500		
	mouse	4000		
Myclobutanel	rat	1600–2290	>5000	
Propiconazole	rat	1517	>4000	
Thiabendazole	rat	3000		
	mouse	3810		
	rabbit	3850		
Thiophanate	rat	>15,000	>15,000	
Thiophanate-methyl	rat	6640–7500	>10,000	
Triademefon	rat	363–568		
	mouse	989–1071		
	rabbit		>2000	
Triadimenol	rat	700–1500	>50,000	
	mouse	1300		
Triforine	rat	>16,000	>10,000	>4.5
	mouse	>6000		
Vinclozolin	rat	10,000		
	guinea pig	8000		
Other				
Carboxin	rat	3820		>20
	rabbit		>8000	
Dinocap	rat	980–1190		
Dodine	rat	1000		
	rabbit		>1500	
Fosetyl-AL	rat	5800	>3200	
	mouse	3700		
	rabbit	2680		
Metalaxyl	rat	669	>3100	
	mouse	788		
	hamster	7120		
	rabbit		>3170	

90-day rat and dog dietary studies, benomyl had a NOEL of 500 ppm. Effects observed at higher doses included liver and hematological effects. A decrease in testicular weight was observed in rabbits receiving dermal doses of 1000 mg/kg/day and higher, 5 days a week for 3 weeks. The NOEL in this study was 500 mg/kg/day. Similar effects were observed in the chronic studies. The effects of benomyl on the testes include epidermal degeneration and decrease in weight. In a two year chronic bioassay in rats, the NOEL for all toxicological effects was 2500 ppm (125 mg/kg/day). In chronic studies in dogs, liver cirrhosis was observed. There was no evidence of delayed neurotoxicity in hens treated with benomyl.

MBC has a toxicological profile similar to that of its parent compound, benomyl. For similar studies, the NOEL's for MBC are either the same as those for benomyl or lower, indicating a greater potency.

4. Carcinogenicity

Benomyl has been tested for oncogenicity in lifetime dietary studies in rats and mice. Liver tumors (hepatocellular carcinomas and adenomas) were observed in male and female mice, but not in rats. Lung tumors observed in male mice at the low and medium doses were not considered to be compound related.

MBC has also been associated with oncogenic effects. It produced liver tumors in two strains of genetically related male and female mice. However, MBC was negative in rats and in a third strain of mice. EPA has classified benomyl and MBC as Class C (possible human) oncogens.

5. Mutagenicity

A battery of mutagenicity studies have been done using benomyl. Benomyl and MBC are both spindle poisons and as such may result in nondisjunction and aneuploidy. They have also been associated with mutational events and chromosome damage, as indicated by positive results obtained in the Ames reverse mutation assay in the absence of metabolic activation, the mouse lymphoma mutation assay, the mouse micronucleus test and the sister chromatid exchange test. In the mouse lymphoma test, the assay was positive at doses of 12–25 mg/ml (metabolic activation) and 50 mg/ml (no metabolic activation). Lower concentrations were negative and higher concentrations were cytotoxic. The Chinese hamster ovary cell assay and DNA repair assays (rat and mouse) were negative.

In addition to these studies, there is a body of literature addressing the antitubulin effects of benomyl as a tool in basic research. There are also a number of studies that have investigated the mechanism and the specific nature of the developmental effects on the testes. More information on these can be found in the bibliography provided at the end of this article.

6. Reproductive/Developmental

Benomyl is teratogenic at doses greater than 30 mg/kg/day in rats and 50 mg/kg/day in mice. The effects observed in the rat include ocular anomalies and those observed in mice include tail anomalies, fused ribs and vertebrae, and cleft palate. At doses where the teratogenic effects are observed, fetotoxicity in the form of decreased body weight and delays in development are also observed. There was no maternal toxicity found in any of these studies.

C. Effects on Human Health

For many of the fungicides in this class, there have been no reports of effects on humans related to their use. Others such as imazalil, thiabendazole, and thiophanate have been used as antiparasitic agents in animals and humans. Symptoms and signs of thiabendazole exposure following ingestion as a drug include dizziness, nausea, vomiting, diarrhea, epigastric distress, lethargy, fever, flushing, chills, rash, and local edema, headache, tinnitus, paresthesia, and hypotension. Persons with kidney and liver problems are vulnerable to these effects. No adverse effects from the use of thiabendazole as a fungicide have been reported.

The major toxic effects of anilazine, benomyl, thiophenate-methyl, and triademefon are dermal, including contact dermatitis and sensitization.

D. Environmental Concerns

1. Nontarget Species

As expected with a group of compounds as diverse as the heterocyclic nitrogen fungicides, the acute toxicity to fish shows a wide range of variation. LC_{50} values range from less than 0.5 ppm for benomyl and MBC in rainbow trout to greater than 1000 ppm for triforine in bluegill sunfish and rainbow trout. Avian toxicity is low. Iprodione has an acute LD_{50} of 930 mg/kg in bobwhite quail and 10,400 mg/kg in mallard ducks. Fenarimol, MBC, thiophanate-methyl, and triforine have acute LD_{50} values greater than 2000 mg/kg.

2. Persistence and Environmental Fate

Environmental fate studies indicate that benomyl in soil is moderately persistent, relatively immobile, and readily hydrolyzed to nonactive compounds by bacteria and fungi. The soil studies included a range of organic matter content from 0.7 to 83.5%. The soil half-life is between 3 and 6 months (turf application) or between 6 and 12 months (bare soil). There is no evidence of season-to-season carryover in soil. In water, benomyl is converted to MBC within 4 days. The MBC is further degraded to 2-aminobenzimidazole.

E. Storage, Handling, and Disposal

Given the low dermal absorption of benomyl, EPA has concluded that proper personal protective clothing, especially for respiratory exposure for mixers, loaders, and applicators, will result in an adequate margin of safety. Benomyl products should be stored in a dry place to avoid loss of fungicidal activity. As with all pesticides, do not store or dispose where contamination of water supplies, food, or feed may result.

II. Dithiocarbamates

A. Introduction

The dithiocarbamate class of fungicides includes the methyldithiocarbamates, dimethyldithiocarbamates, diethyldithiocarbamates, and ethylenebisdithiocarbamates (EBDCs). Of these, the EBDCs are the largest and most important group. They are also the only members of this class capable of being metabolized to ethylene thiourea (ETU). The occurrence of ETU as a contaminant, environmental decomposition product, and urinary metabolite common to all EBDCs has been a cause for concern because of its mutagenic, teratogenic, carcinogenic and goitrogenic properties. Thus, most of the following discussion of toxicity of EBDCs will be with regard to this compound.

The EBDCs were first used as fungicides in 1935, when nabam was used to control fungal diseases on tomatoes. Zineb was developed in 1943, when zinc sulfate was added to nabam suspensions to improve the fungicidal activity. Zineb was the reaction product and later became widely used as a fungicide. Since then, other salts and metal complexes of EBDC have been developed, including maneb (manganese salt), mancozeb (zinc–manganese complex), amobam (diammonium salt), and metiram (zinc ammoniate). Each contains an organic and a metallic or inorganic moiety. The organic moiety is the same for each compound, although metiram and mancozeb are linear polymers based on the same organic structure. The name is generally based on the metallic moiety, for example nabam contains sodium, maneb contains manganese, and zineb contains zinc.

Maneb, mancozeb, and metiram are registered for use to prevent damage by fungi to a wide variety of food and ornamental crops, as well as being used as seed and seed piece treatments, as soil treatments, and for home gardens. Nabam is registered as a fungicide on ornamental plants and as a biocide in industrial cooling tower water systems, oil well drilling fluids, pulp and paper mills, and sugar mills. All uses of amobam and zineb have been cancelled since 1986 and 1988, respectively. Approximately 12 to 18 million pounds of all EBDCs are used in the United States each year, primarily on potatoes, apples, curcurbits, tomatoes, onions, sweet corn, and small grains.

B. Animal Toxicology

1. Metabolism

The metal moiety of each EBDC is lost in metabolism, so that nonmetal metabolites of the different compounds are identical. Those include ethylenethiuram disulfide and monosulfide, ethylene-bis-isothiocyanate sulfide (EBIS), thiourea, ethylene thiourea (ETU), ethylene urea (EU), ethylenediamine (EDA), and carbon disulfide. The toxicity of parent EBDC compounds is relatively low; most of the toxicity is due to the metabolites, particularly ETU.

ETU itself is further partially metabolized followed by excretion (primarily in the urine) as both unchanged ETU and its metabolites. In rats, about 80% of an administered dose of ETU was excreted in the urine within 24 hours. Besides ETU itself, imidazoline, inidazoline, EU, and other metabolites were found. In contrast, cats excreted primarily *S*-methyl ETU in the urine, along with lesser amounts of ETU, EU, and other metabolites. Rats and mice apparently metabolize ETU by somewhat different pathways, and mice are able to eliminate ETU more rapidly than rats. Thus, although the general metabolic pathways of EBDCs and ETU are known, there are many unidentified metabolites and large interspecific differences in the metabolism and disposition of these compounds.

2. Acute Toxicity

The EBDC parent compounds are of low or very low acute oral toxicity, with LD_{50} values ranging from 2400 to 11,200 mg/kg in rats and mice. The exception to this rule is nabam, which is more than an order of magnitude more toxic than the others (see Table II). Dermal LD_{50} values are too high to measure in this group. Signs of acute poisoning include ataxia, hyperactivity followed by inactivity, loss of muscular tone, and weakness of the hind legs. Low acute toxicity is in part due to relatively low absorption of the parent compound and elimination unchanged in the feces. A single sublethal dose of EBDC can cause changes in thyroid function, and thyroid hyperplasia can result from repeated doses.

3. Subchronic Toxicity

The most important target organ in EBDC and ETU subchronic and chronic studies is the thyroid. Maneb, zineb, and mancozeb have been shown to cause reduced uptake of radioactive iodine by the thyroid and/or hypertrophy of the thyroid gland. These effects occur promptly after administration of the compound, appear to be reversible in the early stages, and occur at dosages lower than those affecting other processes such as reproduction. These effects may be due, at least in part, to metabolism of EBDCs to ETU, which is clearly toxic to the thyroid.

In a 13-week subchronic study, thyroid follicular hyperplasia was seen in all rats exposed to 750 or 500 ppm ETU in the diet and in 80% of rats ingesting feed containing 250 ppm. Thyroid adenomas occurred in 65%, 45%, and 35% of the rats in these groups, respectively. In the same study, mice fed 2000 ppm or 1000 ppm ETU in the diet had 100% and 95%, respectively, incidences of thyroid follicular hyperplasia.

Rats fed diets containing 75 to 150 ppm ETU for 7 weeks showed decreased body weight and feed consumption, while thyroid weight increased linearly with dose. Blood levels of T4 were significantly reduced in animals fed 150 ppm ETU. The changes were partly reversed by removing ETU from the diet. In another study, Sprague-Dawley rats showed altered thyroid function when fed ETU in the diet at concentrations of 125 or 625 ppm. Thyroid follicular hyperplasia was noted at levels of 25 ppm and above, with a no-effect level of 5 ppm.

In subchronic studies in Rhesus monkeys, dietary levels of 50 ppm ETU and above caused pituitary hypertrophy, thyroid hypertrophy and hyperplasia, and changes in pituitary and thyroid hormone levels. The NOEL for these effects in monkeys was apparently 10 ppm.

4. Chronic Toxicity

Long-term dietary studies of EBDCs have produced a variety of nonthyroid effects in rats, mice, and dogs at high doses. Rats, for example, tolerated 250 ppm of maneb in the diet for 2 years. However, dietary levels of 1250 ppm or above produced growth depression and increased liver weight. Dogs fed maneb at 75 mg/kg/day for 1 year showed anorexia, weight loss, tremors, weakness, hind limb neurological effects, and impaired kidney function. In contrast, a higher dose of zineb in the diet of dogs produced only thyroid effects. Mancozeb produced no toxicological effects in a long-term study in rats fed at a dietary level of 100 ppm.

In a long-term study in hamsters and rats, ETU caused hypercholesterolemia in both species at the 5 ppm dietary level, the lowest dose tested. This effect was attributed mainly to thyroid impairment in the rat and hepatic impairment in the hamster.

Five groups of 68 male and 68 female Charles River rats were fed ETU at levels of 5, 25, 125, 250, or 500 ppm in the diet for up to two years. Increased thyroid weights were seen in rats fed ETU at 250 and 500 ppm in the diets. Uptake of I-131 was significantly decreased in both sexes fed 500 ppm. Moderate pituitary hyperplasia was also present at 500 ppm. Slight, irreversible thyroid hyperplasia was found in the rat even at the lowest dietary levels of 5 and 25 ppm ETU. As 5 ppm was the lowest dose tested, a NOEL for this effect was not determined. In another study in rats, ETU administered in the diet at levels of 175 or 350 ppm for 18 months produced hyperplastic goiter with extreme enlargement of the thyroid.

5. Carcinogenicity

A few carcinogenicity studies using nonstandard protocols have been conducted on EBDC parent compounds. In one study, increased numbers of lung adenomas were observed in mice fed weekly doses of maneb by stomach tube for 6 weeks. Similarly, another study found an increased incidence of lung adenomas in mice fed weekly doses of zineb for 6 weeks. Maneb has also induced lung adenomas in rats when given orally once a week for six months. A number of other studies of zineb and maneb in rats and mice have not produced a tumorigenic response. These studies of EBDC parent compounds are generally considered to be inadequate for evaluating the carcinogenic potential of EBDCs.

No such uncertainty exists regarding the common EBDC metabolite, ETU. The U.S. EPA classifies ETU as a probable human carcinogen (Group B2) based on four studies. In the first study, two hybrid strains of mice were dosed with 215 mg/kg/day ETU by stomach tube from days 7 to 28 of life, then fed 646 ppm ETU in the diet until termination of the experiment at 82–83 weeks. These doses were considered to be the maximal tolerated dose. The incidence of hepatomas was significantly increased in males and females of both strains when compared with controls. Lymphomas were also observed in some animals. Thyroids were not examined.

In the second study, administration of 175 or 350 ppm technical grade ETU in the diet of Charles River CD rats for 18 months, followed by control diet for 6 months, produced thyroid carcinomas in males and females at both dose levels. None was observed in the controls. Two male rats with thyroid carcinoma exhibited pulmonary metastases.

Charles River rats were fed ETU at levels of 0, 5, 25, 125, 250, or 500 ppm in the diet for two years. After 12 months, rats fed 125 ppm or more displayed nodular hyperplasia of the thyroid, while thyroid carcinomas were found in some rats fed 250 or 500 ppm. After 24 months, the incidence of thyroid carcinomas was 89% in the high-dose group and 23% in the 250 ppm group, compared with 3% in the controls. The authors classified ETU as a thyroid carcinogen in the rat at 250 and 500 ppm in the diet, and a weak thyroid tumorigen at 125 ppm. They did not consider 25 and 5 ppm to have been tumorigenic, although both levels increased the incidence of histological thyroid hyperplasia.

Finally, in a study recently completed by the National Toxicology Program, ETU fed in the diet for up to two years caused significantly increased incidences of liver adenomas and carcinomas in male and female mice at both the high and low doses. Increased incidences of thyroid follicular cell adenomas and carcinomas were also observed in both sexes at the high dose only. In rats, dietary ETU was associated with higher rates of thyroid adenocarcinomas in both sexes at the high dose.

It has been suggested that ETU is not a direct acting hepatocarcinogen, but that it increases liver tumor incidence by an indirect mechanism, possibly involving its antithyroid properties.

The U.S. EPA recently cancelled the use of EBDCs on 11 food crops due to concerns about human cancer risk. The cancellations are based on EPA's determination that unacceptable human exposure to ETU may result from ingestion of some foods containing EBDC residues.

6. Mutagenicity

The EBDCs and ETU have been tested for genotoxicity in a wide variety of mutagenicity assays, with both positive and negative results. Results have been inconsistent among different investigators using the same assay, and there is no single assay where all the EBDCs provide consistently positive results. ETU and EBDCs have produced positive, if not striking, results for sister chromatid exchanges, unscheduled DNA synthesis, and gene mutation in Chinese hamster ovary (CHO) cells. On the other hand, the Ames *Salmonella* assay and most others have generally been negative. The overall body of evidence suggests that, while ETU and the EBDCs are capable of inducing a variety of genotoxic effects, they do not appear to be potent genotoxic agents.

Another concern is that ETU may be nitrosated when it occurs together with nitrates in the acidic environment of the human stomach. Nitroso-ETU and ETU given in combination with sodium nitrite have been shown to induce potent genotoxic effects *in vitro* and *in vivo*.

7. Reproductive/Developmental

ETU has demonstrated potent teratogenic activity in a large number of studies in the rat. Developmental effects have been observed at dose levels that produced no apparent maternal toxicity. A serious teratologic effect observed in rats has been the development of hydrocephalus and other CNS defects (e.g., motor impairment with a hopping gait) in the offspring *postnatally*, resulting in a high mortality rate among the pups. In one study, pregnant Wistar rats were given a single oral dose of 15, 30, or 45 mg/kg on day 15 of gestation. Dams receiving 15 mg/kg gave birth to normal litters and raised normal offspring. The dose of 30 mg/kg produced hydrocephalus and microphthalmia in 90% of the offspring (0% in controls) and these died during the first nine weeks after birth. The anomaly was recognized, grossly, on days 6–9 of postnatal life, increased thereafter in severity, and was always fatal. Some of the offspring that survived 9 weeks had motor impairment, a hopping gait, and lesions in the neuraxis. All pups from dams given 45 mg/kg died within 4 weeks of birth.

In another study with similar results, postnatal survival was impaired in a dose-related manner following maternal treatment with doses of 10 mg/kg and higher on day 17, 18, 19, or 20 of gestation. The reduced survival was attributable to the development of hydrocephalus in the pups, starting about 4 days after birth. In addition, of surviving offspring from dams treated with 10 or 20 mg/kg, a high percentage (19–26%) showed hydrocephalus at 6 months of age. Doses of 1, 3, or 5 mg/kg did not adversely affect postnatal development. The NOEL for this effect was 5 mg/kg in a single oral dose. No maternal toxicity was reported.

Multiple malformations were produced *prenatally* when ETU was administered orally once a day to female rats from day 21 before conception to day 15 of gestation or from days 6–15 or 7–20 of gestation. Dose levels used were 0, 5, 10, 20, 40, or 80 mg/kg/day. Gross abnormalities occurred at dose levels of 10 mg/kg/day and above in a dose-dependent manner, including neural tube defects, abnormal pelvic limb posture, and short or kinky tail. Retarded parietal ossification occurred at all dose levels and was also dose related. Therefore, 5 mg/kg/day is a LOEL for retarted fetal development.

Other studies in rats have produced a wide variety of anomalies involving the nervous, urogenital, and ocular systems, and skeletal anomalies affecting the axial and appendicular skeletons. The types of anomalies produced were dependent upon the time of treatment.

Other laboratory species appear to be less sensitive than the rat to the developmental toxicity induced by ETU. Studies in mice, rabbits, and cats have found that ETU is not teratogenic or only weakly so, at maternally toxic doses. The difference between rats and other species may be accounted for by interspecific differences in metabolism of ETU.

Parent EBDCs produce developmental effects similar to those of ETU but usually at much higher doses. Maneb has produced a diversity of severe anomalies in rats at high doses (>770 mg/kg), but not in mice. In a standard three-generation test, maneb had no effect on fertility, viability, or litter size and produced no anomalies in the F_3 generation at a dietary level of about 12.5 mg/kg/day. However, maneb apparently caused reduced fertility in rats given a dose of 10 mg/kg/day or higher for 11 or 12 months. Zineb has caused reduced fertility, increased fetal death, and various anomalies in rats, and reduced fetal survival in mice. Mancozeb has also caused reduced fertility in rats, but has no teratogenic or embryotoxic effects. In a recent teratology study in rabbits, nabam induced major malformations and other effects including hydrocephaly, incomplete ossification, and resorptions at doses that were not maternally toxic.

C. Effects on Human Health

All EBDCs can cause irritation of the skin, respiratory tract, and eyes. In addition, maneb and zineb have caused cases of chronic skin disease (sensitization) in occupationally exposed workers. Skin disease has been observed in 38–61% of citrus growers in Japan. In patch tests, maneb caused the greatest primary irritation of all compounds used by these citrus growers. In a study of 54 employees of a mancozeb manufacturing plant, none showed any abnormalities that could be related to mancozeb exposure, except for a few cases of dermatitis. Mancozeb has also produced dermatitis in a few vineyard workers and potato planters. In an experimental study of zineb in humans, 50 subjects were patch-tested on the forearm for zineb sensitivity. Forty-nine had no reaction, while one subject developed primary irritation. In contrast, 86 cases of dermatitis were documented in workers who had entered tobacco fields sprayed with zineb 2–15 days earlier. The rash lasted 5–15 days, and teenagers and young women were affected more than other groups. Contact dermatitis was also observed in a zineb factory (21.4% of workers), but the incidence was reduced significantly by taking steps to minimize worker exposure to the compound.

In addition to its dermal effects, zineb has been associated with other effects in exposed workers, including changes in liver MFO function, aberrant metaphases in lymphocyte cultures, moderate anemia and other blood changes in female workers, and changes in respiratory parameters in male workers.

Cases of acute poisonings in humans have been extremely rare, probably due to the poor oral and dermal absorption of EBDCs. However, in one case a man experienced behavioral changes, loss of consciousness, convulsions, and an abnormal electroencephalogram (EEG) after spraying a combination of maneb and zineb. In another case, a farmer with hereditary enzymatic defects (low glucose-6-phosphate dehydrogenase and catalase activities) developed acute hemolytic anemia after a single exposure to a zineb-sprayed field. Another woman working in a field sprayed with zineb developed severe allergic symptoms. While the body of evidence for EBDCs suggests that the primary effect seen in humans is dermal (irritation and allergic sensitization), severe systemic effects may be observed in rare cases involving sensitive individuals.

ETU has been shown to affect thyroid function in humans. Five workers who mixed ETU into master batch rubber at a factory in the United Kingdom were studied for three years and compared to matched controls. The mixers had significantly lower levels of total thyroxine (T4) than the controls. One mixer had an appreciably raised level of thyroid stimulation hormone (TSH). Exposure levels were not estimated.

D. Environmental Concerns

1. Nontarget Species

Maneb and mancozeb are highly toxic to fish. LC_{50} values for maneb reported in the older literature for bluegill, rainbow trout, carp, and harlequin fish, ranged from about 0.5 to 1.9 ppm. However, more recent 96-hour LC_{50} values are about one-fifth of these (0.20 ppm for maneb and 0.46 ppm for mancozeb in rainbow trout, and 0.27 ppm for maneb in bluegill). Metiram is only slightly toxic to fish (LC_{50} = 46 to 100 ppm). Maneb is also teratogenic to developing frog embryos and inhibits regeneration of limbs in salamanders at low concentrations (1 to 5 ppm). Toxicity to aquatic invertebrates varies. Maneb is only mildly toxic to the crayfish but is highly toxic to the water flea (*Daphnia*). Estuarine invertebrates are especially sensitive to maneb and nabam, and may be at risk from industrial uses of the latter.

Mancozeb may also represent a concern with regard to impacts on reproduction in wild birds. Dietary concentrations of 250 ppm and higher have caused adverse reproductive effects in mallards. Bobwhite reproduction was effected at 500 ppm. The NOEL for reproductive effects in the mallard may be as low as 50 ppm. EPA recently concluded that repeated application of mancozeb, maneb, or metiram on apples, potatoes, or cranberries may pose a reproductive risk to birds. Acute oral toxicity of EBDCs to birds is low (LC_{50} >10,000 ppm).

2. Persistence and Environmental Fate

EBDCs and their degradation products are eliminated relatively rapidly from plants and animals, and would not be expected to bioaccumulate in the environment. Metabolism of EBDCs in mammals is characterized by rapid excretion of parent compounds and metabolites in urine and feces. In long-term bioaccumulation studies, no residues of maneb were detected in the tissues of rats or dogs fed maneb for up to two years.

Persistence of EBDCs and ETU in plants has been studied extensively due to concerns that residues of these compounds may persist on fresh fruits and vegetables after harvest, or be present in processed foods. The half-life of maneb ranges from 6.4 to 14 days in beans and tomato leaves, respectively. Maneb is detectable in tomatoes as long as 10 weeks after application. Mancozeb has a half-life of 14 days in apples. The half-life of ETU in crops ranges from 1.7 to 28 days (soybeans and grapes, respectively). EBDCs and their metabolites are apparently not taken up from soil by plants to any significant extent. In one study, only traces of ETU or maneb were found in tomato plants grown in soil in which these compounds had been applied.

EBDCs and ETU apparently do not persist in soils. Soil half-lives of 4 to 8 weeks for maneb, and less than one week for ETU, have been reported. Leachability studies have not shown significant downward leaching in soil of low-solubility EBDCs such as maneb. However, ETU is highly water soluble and would be expected to be mobile in the environment. One study has shown that ETU does move downward in soil with leachate water. ETU rapidly photodegrades in surface water, but may be more persistent in groundwater.

E. Storage, Handling, and Disposal

EBDCs should not be stored where they will contaminate water, food, or feed. Store in a dry, well-aired location. Care should be taken not to inhale dust or spray mist. Appropriate protective clothing should be worn when handling these compounds. Where no disposal procedures are recommended by federal, state, or local authorities, bury unused product or empty containers in a safe place away from water supplies. Do not reuse the empty containers. Open dumping is prohibited.

III. Substituted Benzene Compounds

A. Introduction

Fungicides classified as substituted benzenes have a single benzene ring with at least two chlorinated positions. Hexachlorobenzene (HCB) has been included, in spite of the fact that it no longer has U.S. registration. In addition to at one time being a fungicide in its own right, HCB is a common contaminant of other compounds in this class. Similarly, pentachlorophenol (PCP) has been included for historical purposes. The only U.S. registrations remaining for PCP and its sodium salts are the wood preservative uses. The other substituted benzenes, pentachloronitrobenzene (PCNB), chlorothalonil, chloroneb, and dichloronitroaniline (DCNA), are currently registered for food uses. Chemical and physical properties of these compounds are found in Table I.

PCNB has been selected as representative of this class as it shares characteristics with others in the class. It is a nonsystemic fungicide used to treat soil, transplanted seedlings, and turf. Substituted benzene compounds, including PCNB, are commonly contaminated by hexachlorobenzene, and pentachlorophenol is a metabolite common to this group. Where appropriate, the toxicity information on these compounds will be discussed.

B. Animal Toxicology

1. Metabolism

Blood levels of radiolabelled PCNB peaked 12 hours after administration of an oral dose of 5 mg/kg in rats. The half-life in the blood was calculated to be 21.8 hours. The results of a distribution study in which rats were fed PCNB for 7 months, indicated that while the HCB, pentachlorobenzene, and PCP contaminants and metabolites are accumulated in the fat, PCNB itself is not.

The major route of excretion for orally administered PCNB is the feces (85 to 88%). There is evidence of biliary excretion. Reported metabolites in rats and dogs are pentachloronitroaniline, PCP, *N*-acetyl-*S*-pentachlorophenyl cysteine, and their conjugates. In monkeys, blood levels of PCNB peaked twice, first at 1.5 hours and again at 7 hours. Approximately equal amounts were excreted in the feces and urine. Pentachlorothioanisole and bismethylmercaptotetrachlorobenzene were identified as metabolites. Several of the other substituted benzenes (e.g., HCB and PCP) share the glutathione conjugation pathway followed by formation of the mercapturic acid.

Dermal absorption of PCNB is low. Only 1–2% was absorbed following a 4-hour dermal exposure in rats.

2. Acute Toxicity

The most acutely toxic of the substituted benzene fungicides is PCP, with oral LD_{50} values between 50 and 210 mg/kg. This acute toxicity is evident in humans, as PCP has been responsible for several deaths in cases of misuse and the absence of adequate protective equipment. PCNB has reported oral LD_{50} values in the range of 1200 to 1600 mg/kg in rats and mice, and the dermal LD_{50} is greater than 5000 mg/kg. The other members of the class have very low acute toxicity.

3. Chronic Toxicity

Contamination of PCNB with HCB must be considered when evaluating the chronic database for PCNB. Changes in the liver, including decrease in liver to body-weight ratio, increase in serum alkaline phosphatase, cholestatic hepatosis, and secondary bile nephrosis, were observed in chronic studies in dogs. A NOEL in dogs of 30 ppm in the diet for liver effects was determined. In a rat study with doses of 0, 100, 400, and 1200 ppm, survival to the termination of the study was poor in all dose groups, and therefore a NOEL could not be determined. The PCNB is this case was contaminated with 2.7% HCB.

4. Carcinogenicity

Oncogenicity studies that have been done with PCNB have been both positive and negative for increases in tumors. This discrepancy is probably related to the level of HCB contaminant present in the PCNB. For instance, in mouse studies with greater than 2.5% HCB, there were increases in hepatomas, whereas none were observed in the mouse study where HCB concentration was known to be 0.07% HCB. This tumor type is consistent with those observed in HCB chronic bioassays. EPA stated in their 1987 reregistration guidance document that in the positive mouse study for PCNB, the concentration of HCB may have been as high as 11%.

HCB is classified as a Group B2 (probable human) carcinogen by EPA, and PCNB is classified in Group D (inadequate evidence of carcinogenicity in animals and humans). Historically, the levels of HCB in PCNB (technical grade) were between 1.5 and 11%. In an effort to reduce exposure to HCB from PCNB use, U.S. EPA required that the level of HCB contaminant be reduced to 0.1% as of April, 1988.

5. Mutagenicity

The majority of mutagenicity tests for PCNB have been negative. These tests included forward and reverse mutation assays, recombination assays, recessive and dominant lethal assays, and unscheduled DNA synthesis *in vitro*. The only positive assay reported was the *Escherichia coli* assay.

6. Reproductive/Developmental

As was the case in the oncogenicity tests, HCB contamination appears to confound the developmental studies. At the 500 mg/kg/day dose level, when the PCNB had a contamination level of 11% HCB, there was a decrease in litter size and an increase in malformations in rats. No such result was observed when the HCB concentration was 1%. These studies are insufficient to determine the teratogenic potential of PCNB.

C. Effects on Human Health

The most serious health effects reported for substituted benzenes are related to exposure to HCB and the use of pentachlorophenol. PCP increases metabolic rate and uncouples oxidative phosphorylation and is readily absorbed through skin. In infants, two cases of fatal poisoning and 7 cases of intoxication occurred in a hospital where PCP had been misused in the laundry and diapers were contaminated. The level in diapers ranged from 26 to 172 ppm.

Occupational exposure to PCP has resulted in death in several instances. Lack of personal protective clothing, prolonged skin contact, or bad personal hygiene practices (lack of bathing and not changing clothes) were documented in these cases. Symptoms observed included high temperature, sweating, dehydration, pain in the chest or abdomen, and rapid onset of coma. In the 20 cases documented, time between onset of symptoms and death ranged from 3 to 30 hours. In nonfatal poisoning, symptoms may include anorexia, weight loss, constriction in the chest, dyspnea, headache, nausea, and vomiting.

Chloracne has been reported in production workers, associated with the hexa, hepta, and octachlorodibenzo-*p*-dioxin and dibenzofurans that occur as contaminants of pentachlorophenol.

More than 3000 Turkish farm dwellers who consumed HCB-treated wheat that had been meant for seed developed hepatic porphyria cutanea tarda. Most adults recovered, but some nursing infants did not. HCB is no longer registered and PCP is a restricted use pesticide used as a wood preservative only.

For most of the other substituted benzenes, exposure to skin can result in irritation with or without sensitization.

D. Environmental Concerns

1. Nontarget Species

PCNB is highly toxic to fish, having an LC_{50} of 0.88 ppm for bluegill sunfish and an LC_{50} of 0.50 ppm for rainbow trout. Chlorothalonil has a similar order of toxicity, with LC_{50} values ranging from 0.25 ppm to 0.43 ppm for rainbow trout, bluefill sunfish, and channel catfish. EPA requires label precautions to prevent contamination of surface water by products containing these compounds as the active ingredient.

Substituted benzene fungicides are of low acute toxicity to avian species such as mallard ducks and bobwhite quail. Eight-day dietary LC_{50} values for PCNB, chlorothalonil, and chloroneb in these species were greater than 5000 ppm.

2. Persistence and Environmental Fate

The substituted benzenes are insoluble or only slightly soluble in water, and are not likely to contaminate groundwater, with the possible exception of PCP. The half-life of PCNB is water has been estimated at 1.8 days. Processes responsible for its disappearance include volatilization and sorption to particulate matter. Substituted benzenes in surface water are likely to bioaccumulate in fish. The bioconcentration factors (BCFs) reported for PCNB are 950 to 1130 in the golden orfe and 260 to 590 in rainbow trout. BCFs for chlorothalonil and PCP are similar (264 and 475, respectively). HCB, the common contaminant in this group, has the highest BCF at 1160 to 3740.

There is evidence of carryover in root crops rotated with crops that have been treated with PCNB. Therefore, a 12-month rotation restriction is required with regard to root crops. Similarly, the chlorothalonil label does not permit the planting of certain crops within 12 months of application.

E. Storage, Handling, and Disposal

The personal protective equipment (PPE) requirements vary with the product. One product containing chloroneb has no PPE requirements, while another containing chlorothalonil requires long-sleeved shirts, long pants, and gloves. Due to the toxicity profile of PCNB, personal protective clothing is required for applicators, but not for workers entering fields after treatment. Reentry intervals have not been required for PCNB. Storage recommendations for substituted benzenes range from "store in a cool, dry place" for PCNB to "not for use or storage in or around the home, vapors will cause damage, do not leave in sunshine, and do not use, pour, spill, or store near heat or open flame" for PCP.

IV. Thiophthalimides

A. Introduction

Captan is representative of the thiophthalimide class of fungicides, which are sometimes called chloroalkylthio fungicides or dicarboximides. The chemical name for captan is *N*-trichloromethylthio-4-cyclohexane-1,2-dicarboximide. This class also includes folpet and captafol, which have very similar structures. Since its introduction in 1949, captan has become widely used as a nonsystemic fungicide to protect food crops and plant seeds from fungal attack. Registered uses on food include apples, cherries, peaches, nectarines, almonds, California grapes, Florida strawberries, blueberries, blackberries, and raspberries. At least 75% of all vegetable seeds are treated with captan. The largest uses are on apples and as a seed treatment, which account for 30 and 28% of the total, respectively. Postharvest use of captan is also significant, especially on apples, pears, and cherries. Captan is also applied to packing boxes used for storage and shipping of fruits and vegetables, and is registered for use on ornamentals, including house plants. About 10 million pounds of captan active ingredient is sold in the United States each year. Formulations include dusts, wettable powders, aqueous suspensions, and granules.

B. Animal Toxicology

1. Metabolism

Orally administered radiolabelled captan was metabolized and excreted rapidly by the rat, with most of the radioactivity excreted in the urine and feces within 24 hours. Some radioactivity was also excreted in expired air. Metabolites include tetrahydrophthalimide, tetrahydrophthalic acid, thiazolidine-2-thione-4-carboxylic acid, and derivatives of dithiobis(methanesulfonic acid). Apparently, thiophosgene is an intermediate in the production of the two sulfur-containing metabolites. There is some evidence that significant degradation of captan occurs in the gastrointestinal tract.

2. Acute Toxicity

Captan has a low acute oral toxicity to laboratory animals. Oral LD_{50} values ranging from 480 to 17,000 mg/kg have been reported in the rat. Acute oral LD_{50} values in male and female mice were 7840 and 7000 mg/kg, respectively. Sheep and cattle appear to be more sensitive. A single oral dose of 250 or 500 mg/kg was lethal to some sheep. When applied dermally, captan produced little or no sensitization or irritation in rabbits and moderate sensitization in guinea pigs.

3. Subchronic and Chronic Toxicity

Subchronic and chronic studies of captan in rats, mice, guinea pigs, pigs, dogs, sheep, and cattle have been unremarkable. The most commonly reported effects of feeding high levels of captan in the diet have been decreased food intake and reduced body weight. Dogs dosed orally for over a year showed slightly increased liver and kidney weights, but had normal gross appearance, histopathology, and clinical chemistry. When rats were fed diets containing from 0 to 10,000 ppm captan for up to two years, growth retardation was observed at concentrations of 1000 ppm and above. Testicular atrophy was observed at a concentration of 10,000 ppm, but other parameters were not significantly different from controls at any concentration.

4. Carcinogenicity

Captan is considered to be carcinogenic in male and female mice, and shows some evidence of carcinogenicity in rats. In one study conducted by the National Cancer Institute (NCI), mice fed dietary levels of 8000 or 16,000 ppm for 80 weeks showed a significant increase of duodenal adenocarcinomas in both males and females. In another study, neither of two strains of mice placed on a lower dietary level of captan (500 ppm) showed an increased tumorigenic response. However, the ability of this study to detect tumors was poor due to the small group size (18 mice per sex) and the single dose level used.

Oncogenicity studies of captan in rats have produced equivocal results. In female rats, NCI found a small increase in adrenal gland and thyroid gland tumors, but the statistical signficance of this was questionable and it was not considered to be convincing evidence of captan carcinogenicity. Another study detected a small increase in combined benign and malignant kidney tumors in male rats at the highest dose tested. A marginally significant increase in uterine tumors was observed in a third study. EPA concluded that captan may be weakly oncogenic in the rat based on the male kidney tumors. EPA has classified captan as a Group B2 (probable human) carcinogen. This classification is supported by captan's structural similarity to folpet and captafol, both of which have demonstrated oncogenic effects, and by captan's mutagenic effects in a variety of *in vitro* test systems.

5. Mutagenicity

Captan is considered to be mutagenic, based on extensive testing in a wide variety of genotoxicity assay systems. It has produced positive results in over twenty bacterial strains, 3 strains of fungi, and one strain of yeast. It was negative in a few strains of *Salmonella typhimurium*. In mammalian cell culture systems, captan has been found to inhibit mitosis, increase chromosomal and chromatid breaks, and induce unscheduled DNA synthesis. Test methods using intact organisms have generally been less responsive to captan. Chromosomal aberrations were not observed in the mouse micronucleus test, and captan was not mutagenic in fruit flies. Dominant lethal tests in several strains of mice were generally negative, although one dominant lethal study in rats and one in mice showed increased early fetal deaths. It has been suggested that this difference in mutagenic response between *in vivo* and *in vitro* systems may be due to detoxification of the compound by the intact organism.

6. Reproductive/Developmental

Because of their structural similarity to thalidomide, the thiophthalimide fungicides have been a concern with regard to teratogenicity. Captan has been tested for teratogenic potential in more than 20 separate studies in rats, mice, hamsters, dogs, rabbits, and monkeys and has not shown convincing evidence of teratogenicity in any of these species. Some malformations were observed in one study each in rats, hamsters, and dogs. However, the results were difficult to interpret because the effects were not dose related, the anomalies were not characteristic of thalidomide, and the incidence of malformations was low. In two three-generation rat studies, captan fed at 1000 ppm in the diet had no effect on reproduction or fertility.

Perhaps the most convincing evidence of captan's lack of developmental toxicity comes from a study in which pregnant rhesus monkeys and stump-tailed macaques were administered oral doses of 0 to 75 mg/kg/day on days 21 to 34 of gestation. Thalidomide was given at a dose of 5 to 10 mg/kg/day as a positive control. No fetal anomalies, abortions, or maternal toxicity were observed in either species given captan. Thalidomide produced typical deformities and numerous abortions in both species of primate.

Like captan, captafol and folpet are not teratogenic.

C. Effects on Human Health

All of the thiophthalimide fungicides are moderately irritating to skin, eyes, and mucous membranes, and may produce dermal sensitization. Sensitivity to captan has been demonstrated by applying the material to the backs of volunteers. Conversely, captan has been used in the treatment of fungal dermatitis, and was well tolerated by the patients, especially when applied as a water suspension. Captafol has been associated with irritation and sensitization in several occupational settings. For example, a high incidence of skin irritation was observed among workers using the fungicide in tangerine orchards in Japan. The irritation occurred in 25 to 40% of exposed persons, appeared 1 to 3 days after exposure, and usually disappeared within a week. Reaction to captafol may vary from slight to severe, and may include dermatitis, conjunctivitis, stomatitis, painful bronchitis, and phototoxicity. The irritation is often localized to the eyelids or conjunctiva. Captafol induced dermatitis may be associated with other effects such as hypertension, proteinuria, anemia, and depressed cholinesterase activity. No cases of systemic poisoning in humans by thiophthalimides have been reported.

D. Environmental Concerns

1. Nontarget Species

As with target microorganisms, application of captan results in toxic effects on nontarget soil microorganisms and aquatic algae. Captan has been shown to significantly reduce the biomass of fungi and bacteria when applied to soil. However, the effect is short-lived, probably due to rapid degradation of captan in the field. In water, captan inhibits the growth of marine algae (10 to 50 ppm), blue-green algae (1 ppm), and freshwater phytoplankton (0.01 to 8 ppm).

Toxicity of captan to invertebrates has been studied extensively. LC_{50} values determined for aquatic invertebrates range from 1.5 ppm in water fleas to 15,631 ppm in crayfish. Captan is considered to have low toxicity to beneficial insects such as honey bees, predatory beetles, and parasitic wasps. Captan was determined to be very toxic to earthworms in a laboratory study; however, application to fields did not result in significant reduction in the earthworm population.

Captan is highly toxic to fish. Acute toxicity tests in 9 species of fish have yielded 24- and 96-hour LC_{50} values ranging from 0.026 to 0.200 ppm. Captan has low acute and subchronic toxicity in birds by the oral route. Redwinged blackbirds and starlings were not affected by a single oral dose of 100 mg/kg. In chickens, dietary levels of 1000 ppm and higher had no adverse effects on adult birds or on fertility or hatchability of eggs. However, captan injected into the eggs of snapping turtles and chickens was teratogenic and embryotoxic.

2. Persistence and Environmental Fate

Captan is rapidly degraded in the environment and does not persist or bioaccumulate. The primary metabolite in animals is tetrahydrophthalimide (THPI) which is rapidly excreted in the urine (see Section IV,B,I, "Metabolism"). The compound was not found in tissues of hogs or hens fed captan in the diet, and it did not accumulate in eggs. Half-life values for captan in plants were not available. However, residues on crops generally decline to below detectable levels within 40 days after application. Captan's persistence in soil depends upon moisture content, having a half-life of one to three days in moist soil but over 50 days in dry soil. Mobility in soil is very limited, primarily due to its low solubility in water and its rapid hydrolysis in moist soil environments. For those reasons, captan would not be expected to contaminate groundwater or surface water bodies, unless by direct runoff or drift during application. Captan which did enter water bodies would be short-lived. It's half-life in sea water is one to two days, and one-half day or less in fresh water.

E. Storage, Handling, and Disposal

Captan should be stored in a cool, dry place. Because of possible dermal irritation, contact with skin or clothing should be avoided. A respirator, goggles, and protective clothing should be worn when handling products containing this material.

V. Other Fungicides

There are a number of important fungicides that do not fit into the chemical classes as defined. For reference, some of these compounds are included in Tables I and II. Of the fungicides in this category, dinocap and dodine are irritants in humans. Dinocap also causes allergic sensitization.

Related Articles: CYTOTOXICITY OF MYCOTOXINS; PESTICIDES AND FOOD SAFETY.

Bibliography

"Captan Position Document 2/3 (PD 2/3)." U.S. Environmental Protection Agency, Washington, D.C., 1985.

Choudhury, H. (1987). Health and environmental effects profile for pentachloronitrobenzene. *Toxicol. Industr. Health* **3**(1), 5–69.

"EBDC Special Review Technical Support Document." U.S. Environmental Protection Agency, Washington, D.C., 1989.

"Farm Chemicals Handbook '91." Meister Publishing Co., 1991.

"Guidance for the Reregistration of Pesticide Products Containing Benomyl as the Active Ingredient." U.S. Environmental Protection Agency, Washington, D.C., 1987.

Hayes, W. J., and Laws, E. R., eds. (1991). "Handbook of Pesticide Toxicology," vol 3. Academic Press, San Diego.

Morgan, D. P. (1989). "Recognition and Management of Pesticide Poisonings," 4th ed. Office of Pesticide Programs, U.S. Environmental Protection Agency, Washington, D.C.

"Pesticide Background Statements, Vol. II. Fungicides and Fumigants." Forest Service, U.S. Department of Agriculture, Washington, D.C., 1986.

Worthing, C. R., and Walker, S. B., eds. (1987). "The Pesticide Manual: A World Compendium, " 8th ed. British Crop Protection Council. Lavenham Press, Lavenham, Suffolk.

Genetically Engineered Microorganisms: Monitoring and Containing

Asim K. Bej and Meena H. Mahbubani
University of Alabama at Birmingham

Glossary

Antibody Protein produced by B lymphocytes cells that recognizes a particular foreign antigen and induces the immune response.

Antisense RNA RNA transcribed from the nonsense DNA strand of a gene; it is complementary to the messenger RNA of that gene.

Attenuation Regulation of transcription termination involved in controlling the expression of some bacterial operons.

Auxotroph Nutritional mutant that requires growth factors not needed by the parental strain.

DNA fingerprint Characterization of one or more features of an individual's genome by developing a DNA fragment band (allele) pattern. If a sufficient number of different size bands are analyzed, the resultant bar code profile will be unique for each individual except identical twins.

Polymerase chain reaction (PCR) Method for amplifying DNA *in vitro* involving the use of oligonucleotide primers complementary to nucleotide sequences in a target gene and copying the target sequences by the action of DNA polymerases.

Promoter Binding site for RNA polymerase to initiate transcription.

Replicon Unit of the genome or a plasmid in which DNA is replicated; contains an origin for initiation of replication.

Restriction endonuclease Enzyme that cleaves specific double-stranded DNA sequences generating blunt or single-stranded (sticky) ends.

Transposon Genetic element that is capable of inserting itself at a new location in the genome (without any sequence homology with the target locus).

GENETICALLY ENGINEERED MICROORGANISMS (GEMs) *are microorganisms that have been designed to contain modified genetic properties either through the selection of desirable traits or the introduction of new genetic information on DNA or both. The rapid advancement of biotechnology and the application of recombinant DNA technology to engineer novel microorganisms for human benefit have caused concerns in both the scientific community and the general public over potential risks. Primarily, these concerns arise from the potential harm that can be caused to human health and to the surrounding environment by these engineered microbes upon deliberate or unintentional release. Several cultivation-dependent and cultivation-independent genetic-based monitoring methods have been developed to track the GEMs, should they escape from the laboratory or be deliberately released to the environment. The most important problem of recalling the released GEMs from the environment has been addressed by engineering conditional or unconditional death in the GEMs by incorporating a "built-in self-destruct" mechanism, which, upon completion of its task,* *would eliminate the host microbe. Although none of the systems created so far are "fail-safe," the future direction in this area of research for creating safe GEMs for human benefit shows great promise.*

I. Introduction

In the early 1970s, genetic engineering of DNA led to the possibility of creating novel microorganisms. This manipulation of the genetic material of the microorganisms required a vast understanding of the impact that the GEMs might have upon the environment. Scientists and naturalists predicted that if the GEMs were released into the environment it would disrupt the food chain and possibly cause disease in plants, animals, and humans. It was also predicted that if the recombinant DNA (rDNA) moved to another organism it could cause massive mortalities (caused by a hypothetical "Andromeda Strain" pathogen produced by recombinant DNA technology) or alter the global climate. These concerns led to the Asilomar Conference in February 1975, which proposed the use of physical containment measures to reduce the likelihood that GEMs might escape from the laboratory, and biological containment measures to reduce the possibility of the survival of the rDNA in the environment if the GEMs were to escape from the laboratory. The Recombinant Advisory Committee (RAC) at the National Institute of Health (NIH) has established guidelines for the safe handling of GEMs. During the last few years, concerns have shifted away from the risks of constructing GEMs in laboratories, to the deliberate release of GEMs into the environment.

To be able to deal with potential adverse and undesirable effects of GEMs, it is necessary to be able to monitor them and the rDNA in the environment, should the rDNA move out of the original host microorganism into another host organism. Suitable monitoring techniques are needed to track the fate of the GEMs and its rDNA in the environment.

II. Monitoring

Several different methodologies have been developed for monitoring GEMs each of which has advantages and disadvantages when dealing with diverse environments and microorganisms (Table I). A monitoring technique must be applicable under a wide variety of environmental conditions; technically simple to apply; able to detect, identify, and enumerate GEMs; sensitive and specific enough to detect small microbial populations; capable of differentiating specific GEMs from other organisms in the environment; able to discriminate GEMs from other strains of the same species; and is expected to be efficient, cost-effective, and time-economic.

A. Plate Count

The plate count method for bacterial enumeration is considered to be the gold standard in microbiology. This method is highly applicable to the enumeration of bacteria from environmental samples. The growth of the GEM is favored over nonrecombinant strains by using selective media formulated to take advantage of the nutritional and/or antibiotic resistance characteristics of the GEM. The GEM can be engineered to contain certain markers such as chromogenic markers, antibiotic resistance markers, heavy metal resistance markers, and markers of metabolic traits. A marker gene must be expressed in the host and the gene product should be stable in the host, transported to its final location in the cell, and readily detected or selected. At least two markers should be introduced in the GEM to ensure high specificity. Inclusion of marker genes linked to the rDNA sequence permits the direct monitoring of the rDNA.

Plating requires the growth of the organism under the specific conditions used in the procedure and may thus underestimate populations. Due to the lack of a universal medium that permits the growth of all host microorganisms in the sample, significant populations may be missed due to the fact that the fate of the rDNA may be independent of that of the original host organism. Also false negatives may result if organisms are stressed and thus viable but not culturable at time of sampling. Other problems that may be encountered with plating are: poor or slow growth of the organisms; no growth due to a requirement for a cometabolic substrate or auxotrophic nutrient requirement; the need for another population for cross-feeding; colony–colony inhibition; and the failure to detect poorly selectable phenotypes or genotypes that are not expressed in the specific media.

Despite these limitations, the plate count method has relatively high sensitivity and reliability and is easy to perform. It allows the selection of specific populations and the multiplication of populations present in low numbers.

Table I Various Methods for Detection of GEMs in the Environment

Detection method	Advantages	Disadvantages
Culturing on liquid media or on solid agar media	Relatively simple; selectable when used with chromogenic indicator or antibiotics; acceptable level of sensitivity; diverse number of microbes can be done by enrichment and using nonselective conditions	Can take days to weeks to culture certain fastidious microbes; microbes must be in viable culturable condition; loss of the selectable marker may prevent detection; media, temperature (etc.) dependent
Direct immunofluorescence microscopic examination	Simple and requires less time to prepare; highly specific when monoclonal antibody is used; direct detection of the target microbes in environmental samples; acceptable sensitivity when the samples are concentrated; both viable culturable and viable nonculturable microbes can be detected	Nonspecific when polyclonal antibody is used; expression of an antigen gene may be unstable in a given environment; does not discriminate between dead and live microbes; may cause eye fatigue; environmental contaminants may interfere with the antibody binding reaction
Protein product colony ELISA blot	Specific; can discriminate live from dead microbial population	Low sensitivity; the gene for specific target protein may be unstable in a given environment
Gene probes	Highly specific; time efficient	Variable sensitivity; nonradioactive methods are not as good as the radioactive detection; trained technician may be required; cannot discriminate live from dead microbial population
Polymerase chain reaction (PCR)	Highly specific; sensitivity level can be one microbe detection from variable amounts of environmental samples; extremely time efficient	Environmental contaminants may inhibit the reaction; unique target gene information is required; gene probe may be required for confirmation; cannot distinguish host; trained technician may be required
DNA fingerprint	Highly specific; good sensitivity	Trained technician may be required; time consuming

The first deliberate release of a GEM to the environment, an ice minus strain of *Pseudomonas syringae* was monitored by five different types of sampling and selective plating methods. The different sampling methods were 6-stage Andersen samplers, Reynier slit samplers, all glass impinger samplers, gravity plates, and sentinel potted plants. It was found that all glass impingers and gravity plates (exposed petri dishes) were most desirable due to such attributes such as reliability, ease of use, and cost-effectiveness. The plating medium used was Kings B agar containing cycloheximide (100 μg/ml), Dupont benolate (100 μg/ml), and rifampicin (50 μg/ml). Colonies with appropriate colony morphology, and resistance to rifampicin and which were fluorescent under long UV, were considered presumptively to be ice minus bacteria.

A genetically engineered pseudomonad strain has been created that contains *lac*Z and *lac*Y genes of *Escherichia coli*. Primary selection is done on a chromogenic substrate 5-bromo-4-chloro-3-indolyl-β-D-galactoside (X-gal), on which the recombinant strain produces blue colonies, while the wild type gives white colonies. Thus, conventional plating together with recombinant DNA methods can be used for monitoring GEMs in the environment.

B. Most Probable Number (MPN) Method

This method is similar to the plating method in that it allows multiplication of the organisms. MPN is usually done in liquid media which allows the easy selection of populations that metabolize a single carbon source. With MPN it is necessary to establish the appropriate conditions for culturing the organism of interest. MPN has many of the advantages and disadvantages of plating.

C. Colony ELISA Blot

In this method, the environmental sample is filtered through a hydrophobic grid membrane. Cells are collected and developed on the membrane. A replica is made onto nitrocellulose paper, cells are lysed with chloroform, and the proteins are immobilized on the nitrocellulose paper. The rDNA sequences are detected indirectly by using rabbit antibody against an antigen which is the gene product of a recombinant gene. This method is limited by the requirement that the organisms be able to grow on the grid membrane and may not allow direct quantitation of microorganisms.

D. Direct Immunofluorescence Microscopic Detection

The use of fluorescent antibodies for enumerating GEMs in the environment does not require culturing. The antibody is specific to the organism of interest. In the ideal case, the antibody reacts specifically with the gene product coded exclusively by the GEM. In most cases however, cross-reactivity occurs with nontarget organisms. The detection limit is 10^3 to 10^4 organisms per gram, a level of sensitivity that may not be sufficient for monitoring purposes.

E. DNA Probes

DNA probe hybridization involves the detection of target nuclei acid sequences by the reannealing of two complementary denatured DNA strands. One DNA strand, the probe, is labeled radioactively or nonisotopically. The probe binds with the complementary strand and the hybrid DNA carrying the label is detected. This reveals the presence of the complementary strand. Radioactively labeled probes may be labeled with ^{32}P by nick translation or by the end-labeling method using T4 polynucleotide kinase. Nonisotopic probes may be labeled with various hapten molecules such as biotin or detected by chemiluminescence.

Factors affecting the sensitivity and specificity of DNA probe hybridization are the length of the probe sequence, the type of label, the specific activity of the isotopic label, the stringency of hybridization, the abundance of the target, and the degree of mismatch between the base pairs. DNA probes have been used for the detection of GEMs in several different experimental designs (Table II).

1. Colony Hybridization

In this procedure, microorganisms are first cultivated on a solid medium and colonies are transferred to nitrocellulose or nylon membranes. The colonies in the membrane are lysed and the single-stranded DNA is immobilized on the hybridization nitrocellulose or nylon membrane. The membrane is prehybridized to block nonspecific nucleic acid binding sites and then incubated with a labeled probe to allow hybridization of complementary sequences. In the case of a radiolabeled probe, autoradiography is performed and the GEM colonies are located on the membrane.

The number of microorganisms detected by colony hybridization is four to five times higher than

Table II Application of Gene-Probe Methods for Detection of Genetically Engineered Microorganisms

Genetically engineered microorganism	DNA sequence in rDNA for gene-probe target	Method applied for gene-probe hybridization analysis
Escherichia coli	2,4-dichlorophenoxy acetic acid degrading gene	Colony hybridization
E. coli	Thymidine kinase gene	Colony hybridization
E. coli	Napier grass DNA	Direct extraction of total genomic DNA followed by PCR amplification
Pseudomonas cepacia AC1100	Highly repeated 1.3 kilobase DNA sequence	Direct extraction of total genomic DNA followed by PCR amplification
P. putida	Transposon 5 (Tn5)	Most probable number followed by colony hybridization
Pseudomonas spp.	Neomycin phosphotransferase II gene (*npt*II)	Direct extraction of nucleic acids followed by colony hybridization
Rhizobium spp.	Tn5	Most probable number followed by lysis and hybridization
Azospirillium lipoferum	Tn5	Most probable number followed by lysis and hybridization
R. leguminoserum b.v. *phaseoli*	Tn5	Direct extraction of genomic DNA followed by PCR amplification

that detected by plating, and the sensitivity is one target colony in one million background colonies. However, it may fail to detect target populations that are below 0.1% of the bacterial population in fresh water microcosms. Colony hybridization requires that the target organism must comprise a relatively high proportion of the total enumerated colonies. The major disadvantage of colony hybridization is that organisms must be cultured. Hence, colony hybridization has limitations such as the selectivity of the cultivation medium, the inability of the cultivation medium to isolate more than a few percent of the total microorganisms at the time of sampling, the abundance of the organism in the environment, and the amenability of the colony growth form to the hybridization protocol. The sensitivity of the assay can be increased by plating the isolated bacteria onto selective agar before colony hybridization.

Colony hybridization has been used to enumerate naphthalene and toluene degradative microorganisms with a sensitivity of one colony in 10^6 colonies from the environment. Populations of degrading polychlorinated biphenyls and mercury resistance genes have been identified in environmental isolates by colony hybridization.

2. MPN Gene Probe Method

The MPN gene probe method is used to increase the sensitivity of detection by amplifying the bacteria carrying the engineered DNA sequence. The method is based on the dilution-to-extinction probability theory. Multiple serial dilutions of a sample are placed in a microtiter plate, enrichment medium is added, and the plate is incubated. After allowing time for the growth of the organisms, cells are lysed and the DNA transferred and immobilized on a nitrocellulose membrane. A labeled gene probe is used for hybridization to detect the rDNA sequence. The MPN is calculated from the pattern of positive and negative hybridizations at the various dilutions. The sensitivity level achieved is 10–100 cells per gram of soil. This method was used to monitor GEM transport and colonization of plant roots by transposon 5 (Tn5) mutants of *Azosprillium*.

3. Direct Probing of DNA

DNA can be recovered from environmental samples and used for hybridization with a labeled probe in two ways. Cells may be lysed, followed by DNA extraction; or microbes may be first separated by differential centrifugation, followed by cell lysis and DNA extraction. In the direct lysis method, the cells are lysed in the soil by ballistic disintegration followed by alkali extraction of the DNA by phenol, cesium chloride density gradients, and/or hydroxyapatite column chromatography. When the microbes are first extracted from the soil, approximately 30% of the bacteria are recovered. After cell extraction, the bacteria are concentrated, and the DNA is extracted by enzymatic and detergent lysis of the cells. The DNA may be purified by cesium chloride density gradients or hydroxyapatite column chromatography. Direct lysis–DNA extraction gives a greater total yield of DNA. Humic contaminants may be removed by using polyvinylpolypyrrolidone (PVPP). DNA extracted by these methods is pure enough for DNA–DNA hybridization but may not allow some restriction enzymes to cut target sequences.

Direct probing of DNA does not require cultivation, and therefore the spread of the rDNA sequence in the environment can be tracked and quantified from both culturable and nonculturable organisms. This is important in some cases, such as when assaying for mercury resistance genes where both gram-positive and gram-negative mercury resistant bacteria are found, but gram-negative bacteria are more easily cultured. The sensitivity of direct probing is limited by the efficiency of DNA recovery.

Both filter-immobilized and solution DNA hybridizations have been used to study genetic diversity in environmental samples. Filter hybridization has a detection limit of 10^4 organisms per gram sample while solution hybidization can achieve a sensitivity of as few as twenty-five organisms per gram sample. A solution hybridization method was developed for detecting 2,4,5-T-degrading GEMs in sediment. DNA was recovered from the sample followed by hybridization in solution with a radiolabeled RNA gene probe. After nuclease digestion of nonhybridized probe RNA, the DNA–RNA hybrids formed in the solution hybridization reaction were separated by Sephadex or hydroxyapatite column chromatography and detected in a liquid scintillation counter. The detection level achieved was 100–1000 target cells per gram of sediment. Solution hybridization is useful when detecting a small amount of target DNA against a large background of nonspecific DNA as in environmental samples. Unlike a dot blot assay, which requires dilution of the DNA sample, solution hybridization can screen the entire DNA sample in one assay.

4. Polymerase Chain Reaction (PCR)–Gene Probe Detection

PCR has been used to directly amplify target sequences from environmental samples. The sensitivity of detection can be increased 10^3 fold or more by PCR amplification of the target sequence in the extracted DNA followed by DNA hybridization. This approach was successfully used for the detection of the 2,4,5-trichlorophenoxy acetic acid-herbicide-degrading bacterium *Pseudomonas cepacia* AC1100 at a sensitivity level of 1 cell per gram sediment. A 1.3-kilobase (kb)-specific region of DNA from the GEMs was amplified by PCR and hybridized with a radiolabeled probe. The presence of 20 μg nonspecific DNA did not hinder the amplification and subsequent detection of 0.3 pg of target DNA sequence. PCR–gene probe detection was also used to detect a 0.3-kb *Pennisetum purpureum* DNA fragment cloned in a 2,4-dichlorophenoxyacetic acid degradative plasmid, pRC10 in an *E. coli* host. The GEMs were incubated for 10 to 14 days in filter-sterilized lake and sewage water samples prior to PCR amplification and gene probe detection. This method is able to detect picogram quantities of DNA in a complex environment.

5. DNA Fingerprinting for Detection of Gene Migration

In this method, DNA is extracted from the environmental sample, digested with appropriate restriction enzymes, and a Southern blot DNA–DNA hybridization is performed. The size of the labeled fragment is compared with the size of the fragment produced from the original GEMs. A fragment of a different size would indicate an alteration or deletion. Point mutations, however, will not be detected. This approach was used to detect the migration of the rDNA of genetically engineered *P. fluorescens* and *P. syringae,* which were the first GEMs to be deliberately released to the environment.

III. Containment

Historically biological containment was designed to reduce the potential for survival of rDNA if the GEMs escape from the laboratory. This consisted of creating and using "safe" plasmids as cloning vectors and debilitated host bacteria. Obviously these are not practical for deliberate environmental releases. GEMs must be able to compete successfully with the indigenous community in order to survive and perform their desired function. Thus, a containment system for an environmental release must be "conditionally lethal" to the host microorganism. It would be viable only under permissive growth conditions and would kill itself once it had performed its function in the environment.

A. Safe Cloning Vectors

The Charon cosmid vectors based on derivatives of the bacteriophage lambda, carry mutations that decrease the host range of recombinant molecules or reduce the possibility of producing a nonlytic recombinant phage. Plasmids have also been constructed that lack transfer function and are not mobilizable in triparental matings. Other "safe" plasmid vectors have features such as temperature-sensitive replicons and the replacement of antibiotic resistance genes with marker genes complementing auxotrophic mutations for the selection of transformed cells.

B. Debilitated Host Bacteria

These are extremely labile bacteria which presumably would not survive outside the laboratory. The *Escherichia coli* X^{1776} contains mutations such that it has an obligate requirement for exogenous diaminopimelic acid and thymine, and is very sensitive to bile, antibiotics, detergents, and DNA-damaging agents. Other mutations that could be introduced to reduce the likelihood for long-term survival are the ability to synthesize cyclic AMP (cAMP) and the cAMP receptor protein (CRP). Another approach would be to introduce a gene such as *hok,* which mediates a lethal collapse of the transmembrane potential. A certain fraction of cells die in each generation and this reduces the possibility of the GEMs establishing a population larger than the wild-type species in the same ecological niche.

C. Restriction of the Vector to the GEM

This may be achieved by using a plasmid vector with a nonsense mutation in a gene indispensible for plasmid replication and/or maintenance and a suppressor mutation in the chromosome that would allow translational read-through of the message of the gene.

An alternate strategy would be to endow the vector with a "kill" function such that if transferred to another microorganism, it would cause death. The "kill" function could be a restriction enzyme where the ability to synthesize the modification methylase would be specified by a chromosomal gene. When the vector transferred into a new microorganism, the restriction enzyme would be synthesized and would digest the recipient's DNA and cause cell death. Unfortunately, neither of the two above strategies have ever been put into practice.

D. Insertion of Gene in the Bacterial Chromosome

rDNA may be integrated in the host chromosome for stable maintenance of the gene in the host and to prevent its escape from the cell as may happen when present on a plasmid. An impaired Tn5 vector has been developed which inserts the *Bacillus thuringiensis* toxin gene in the chromosome of *Pseudomonas fluorescens,* which is then unable to undergo further transposition.

The limitation of this approach is that, since there would be only one copy of the gene of interest transferred to the chromosome, high expression of the gene would not be achieved as would be possible on a high copy number of plasmid vector.

E. Balanced Lethal Systems

A balanced lethal system has been developed involving the gene for diaminopimelic acid (DAP) biosynthesis. The chromosome of the bacterial cell has a deletion mutant for DAP while the cloning vector contains the wild-type DAP gene to complement the chromosomal gene defect. The bacterial cell is unable to survive without the vector. Chromosomal deletions of the aspartate β-semialdehyde dehydrogenase (*asd*) gene and DAP were made. The plasmid cloning vector contained the asd^+ gene and DAP. To limit the vector to the host, one could endow the asd^+ vector with a "kill" function such as "*hok*" (*ho*st *k*illing). The *hok* may be fused to promoter P_{lac} that would be repressed in a bacterial host with a chromosomal $lacI^q$ (controlling) gene. If the plasmid is transferred to a wild-type strain with the $lacI^+$ gene, depression of the P_{lac}–*hok* fusion would occur resulting in cell death. A restriction enzyme gene or other "killer genes" may be used in place of the *hok* gene.

F. Conditional Lethal Systems

A conditional lethal system contains a gene that kills the released cell at a time chosen by the investigator. A lethal gene is placed under control of a promoter which is expressed under certain environmental conditions such as in the absence of a pollutant (Fig. 1). Lethal bacterial genes include bacteriocins, restriction endonucleases, and lethal genes in high and low copy number plasmids. These genes are lethal to the host bacterium if expressed without concomitant expression of "protecting" genes, or if expressed at altered levels or times in the cell life cycle. Lethal genes found on low copy number plasmids are useful for biological containment. Several suicide systems using such lethal genes have been constructed (Table III).

1. Conditional Lethal Systems Using Bacterial Lethal Genes

The *hok* gene is found on plasmid R1 (antibiotic resistance plasmid), which encodes a protein consisting of 52 amino acids. The Hok protein disrupts the electrical (ionic) potential of the bacterial cell

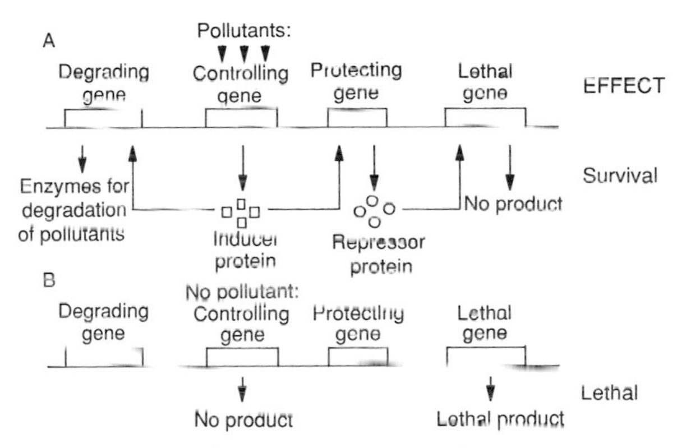

Figure 1 Schematic representation of a "built-in self-destruct" (conditional lethal) system for biological containment of genetically engineered microorganisms (GEMs). All genes are present either in the chromosome of the host microbe or on a plasmid which is harbored in the host cell. (A) In the presence of pollutants, the controlling gene will produce an inducer protein which will cause induction of the degrading gene and the protecting gene. As a result, the pollutants will be degraded to nontoxic products and the protecting gene products will repress the lethal gene, thus guarding the GEM from "self-destruction." (B) When there is no pollutant left in the environment, or should the GEM escape from the site of release to a place where there is no pollutant, the controlling gene will not produce inducer protein, thus the protecting gene will be turned-off due to the lack of induction. As a result, the lethal gene will be derepressed and the production of lethal protein will eliminate the GEM from the environment.

Table III Biological Containment of Genetically Engineered Microorganisms Utilizing Lethal Genes from Bacteria

Lethal gene	Controlling gene/ element for lethality	Signal for lethality	Type of promoter	Plasmid designation	Source[a]
hok	*lac*I[q]	IPTG	P_{lac}	pBAP19*h*	1
gef	*lac*I[q]	IPTG	P_{tac}	pSM970	2
hok or *gef*	High conc. of inorganic phosphate (hi P_i)	Low P_i	P_{ugp}	pAB66	1[b]
hok	*sok*	Low P_i	P_{ugp}	pAB660	1
hok	*fim*E, *fim*B	Random inversion (flip-flop) of the fim_A promoter	P_{fimA}	pPKL100	2
gef	*xyl*S, *Pm::lac*I	3-methyl benzoate (3MB)	P_{tac}	pCC100 pCC102 pHB101	3
hok	*c*I857	Nonpermissive growth temperature	lambda P_R	pKG345	2
*kil*A	*xyl*S	3MB	OP2	pEPA70	4[b]
hok	*xyl*S	3MB	OP2	pEPA112	4[b]
*rel*F	*lac*I[q]	IPTG	P_{lacUV5}	pDW205	5

[a] The plasmid constructs are from the following laboratories: 1, Asim K. Bej and Ronald M. Atlas, Department of Biology, University of Louisville, USA; 1[b], Asim K. Bej, Department of Biology, University of Alabama at Birmingham, USA; 2, Søren Mølin and his colleagues, The Technical University of Denmark, Copenhagen, Denmark; 3, Juan-Luis Ramos, C.S.I.C., Estacion Experimental del Zaidin, Spain; 4, Environmental Research Laboratory, U.S. Environmental Protection Agency, Gulf Breeze, Florida, USA; 5, S. M. Knudsen and O. H. Karlstrom, Institute of Microbiology, University of Copenhagen, Denmark.

[b] Plasmid constructs that have not yet been published.

membrane and causes cell death. Also located within the *hok* gene is the *sok* (*s*uppression *o*f *k*illing) gene which produces antisense RNA against the *hok* messenger RNA (mRNA) and prevents synthesis of Hok protein. Thus the bacterial cells carrying the R1 plasmid containing the *hok–sok* system protect themselves from death. Since *hok* mRNA is more stable than *sok* antisense RNA, a daughter cell that does not receive the R1 plasmid after cell division, dies due to Hok expression. The *hok* gene has a broad host range of both gram-negative and gram-positive bacteria and is suitable for biological containment.

Several lethal systems have been constructed with the *hok* gene for both gram-negative as well as gram-positive bacteria. The plasmid pKG345 contains the *hok* gene downstream from the lambda P_R promoter and the *c*I857 temperature sensitive repressor gene. While cells containing the plasmid grow at 30°C, at 42°C viability is reduced due to inactivation of the repressor. A model conditional suicide system established for biological containment of bacteria grown in fermenters carries the *hok* gene regulated by the *E. coli* tryptophan (*trp*) promoter on plasmid pNWL7. In the presence of tryptophan, the *trp* promoter is repressed as tryptophan forms an active repressor complex with *E. coli trp*R repressor. In the absence of tryptophan, the cells stop growing and rapid killing of the cells occurs. Approximately 0.1% of the cells survive and a more efficient system may be developed by removal of the *trp* attenuator sequence in the plasmid pNWL7 or by substituting the *trp* promoter with a stronger promoter.

The *hok* gene has also been used in a model for containment of bacteria to be released. Plasmid pPKL100 contains the *hok* gene downstream from the *E. coli fim*A promoter. The *fim*A promoter randomly "flip-flops" and is active in only one orientation. This systems responds to the ratio of two regulatory proteins encoded by the *fim*E and *fim*B genes. The *hok* gene is thus randomly activated through inversion of the promoter DNA, causing random cell killing and an increase in the doubling time of the host cells.

A model system, utilizing the *hok* gene, has been developed for conditional death rather than conditional maintenance of a strain as in the two previous cases. The *hok* gene has been fused with the *lac* promoter in a plasmid pBAP19h. The *hok* gene is turned on upon addition of an inducer, called isopropyl-β-D-thiogalactopyranoside (IPTG), which binds with the repressor molecules produced by the *lac*I^q gene, thus killing the host bacterial cells released in the environment. The model suicide vector has been shown to be functional in soil as well as in laboratory cultures. However, the method does not provide a fail-safe system for the containment of GEMs.

A homolog of the *hok* gene, which was isolated from the *E. coli* chromosome, called *gef* (*g*ene *e*ffect *f*atal), was used in a conditional lethal construct and was shown to have a similar killing efficiency of the host microbe as the *hok* gene.

The *hok* and *gef* have been fused with a more efficient P_{tac} promoter, which contains elements from both *lac* and the *trp* promoters. These plasmid constructs, though efficient suicide systems in *E. coli,* are not as effective in pseudomonads. Similar constructs with the *hok* and *gef* genes using the P_{ugp} promoter, which is induced by a high concentration of inorganic phosphate (P_i) and glucose have been tested in *E. coli,* in which they show high killing efficiencies.

Another lethal gene called *rel*F, which is homologous to *hok* and *gef* genes, was used as a model suicide system under a tightly controlled promoter to estimate the frequency of mutation of such a system upon induction. These plasmid constructs were used to identify the factors required for the construction of an efficient conditional lethal system for GEMs.

The *kil*A gene is lethal to *E. coli,* which are normally protected from its action by the expression of the *kor*A (*k*ill *o*ve*r*ride) gene. In organisms designated to degrade toxic wastes, the *kil*A gene can be manipulated to be constitutively expressed at a low level and the *kor* gene can be controlled by a promoter that is activated by the presence of the toxic compound. In the presence of the compound, the organism will be protected and will perform degradation of the compound. In the absence of the toxic compound, the organism will die. The OP2 promoter from the TOL plasmid pWWO, which is activated by benzoate and related aromatic acids in the presence of the *xyl*S regulatory gene, has been used to construct containment plasmids pEPA88 and pEPA86. While pEPA88 contains the OP2 promoter upstream to the *kil*A gene, pEPA86 contains the OP2 promoter upstream of the *kor*A gene. Cells containing pEPA86 are killed in the presence of benzoate, while cells containing pEPA88 require benzoate for survival. In a study with *P. aeruginosa,* it was found that pEPA88 causes a nonlethal cessation of growth that may be reversed if OP2 promoter activity is repressed.

2. Cold-Sensitive Mutants to DNA-Damaging Agents

A conditional lethal system that might be suitable for agricultural products (i.e., needed only during the warm weather) utilizes mutants with cold sensitivity to DNA-damaging agents. Mutant bacterial strains ACG3 and ACG4 of *P. aeruginosa* strain PAO1 grow normally at 37°C in the presence of methylmethane sulfonate, but not at 25°C. The disadvantages of this system are that mutants can easily revert to wild-type and that the mutation is not easily transferable to other strains, so a new mutation must be made for each released strain.

3. Supersensitivity to Mercury

Another type of conditional lethal construct consists of a partial metabolic pathway such that a toxic compound accumulates in the cell and causes cell death. Plasmid pEPA19 contains the *mer*R and the *mer*I genes, which are a regulatory gene and the gene required for mercury transport, respectively. Plasmid pEPA19 does not, however, contain the *mer*A gene for murcuric reductase. Tests conducted with *P. aeruginosa* PAO1 showed that transformed cell populations of 10^3 cells/ml were completely killed in the presence of 0.5 to 1.0 μg/l mercuric chloride, while nontransformed cells were not affected. The major drawbacks of this system are that the system is ineffective in higher cell concentrations and that mercury is an extremely toxic compound.

IV. Problems with Biological Containment and Possible Solutions

A. Stability of the Construct and Incomplete Killing

This seems to be the most important problem. There are several different possibilities regarding the fate

Table IV Predicted Fates of Intentionally or Unintentionally Released Genetically Engineered Microorganisms, and Their Engineered Gene(s) in the Environment

Scenario	Detection/level of the GEM	Detection/level of the engineered gene(s)
The GEMs with the engineered gene(s) find the environmental conditions favorable for them; thus they survive and multiply.	High	High
The host microbes lose the engineered gene(s) along with the plasmid and multiply without engineered gene(s).	High	Low
The engineered gene(s) are transferred to other microbes in the environment from the host; the recipient microbes multiply although the original host microbes do not exist in the environment any more.	Low	High
The GEMs and the engineered genes in them are retained in the environment; neither is eliminated or greatly increased.	Same	Same
The plasmid containing the engineered gene(s) is lost from the host microbe, but the microbe is retained in the environment.	Same	Low
The engineered gene(s) are transferred to other microbes in the environment and eventually the original host microbes carrying those genes die due to competition or adverse environmental conditions.	Low	Same

of the GEMs and its rDNA in the environment, which have been described in Table IV. In all of the constructs developed so far, population killing was incomplete. Survivors were isolated that had lost the containment plasmid or which had deleted, mutated, or rearranged plasmids. Experiments with cells transformed with pBAP19*h* plasmid containing the *hok* gene under the control of *lac*I promoter and the *lac*I regulatory gene, showed that survivors contained mutations that were resistant to the *hok* gene. In an experiment to determine the mutation rate of the conditional lethal system, it was found that a mean of 3.4 colonies (with a variance of 123) appeared per IPTG-containing plate. The frequency of mutation was calculated from this experiment to be 1×10^{-6} per cell per generation. The stability of a conditional lethal system may be increased by including two or more conditional lethal systems in the organism and by placing these in the chromosome rather than on a plasmid. One or more repressor genes should be included in the chromosome or in the plasmid. It has also been shown that use of a tightly controlled promoter can reduce the mutation rate by selecting a lesser number of mutants upon induction. The inclusion of two or more conditional lethal systems with tightly controlled promoter and regulatory systems may overcome the problem of incomplete killing.

B. Requirement for the Addition of an Innocuous Chemical to the Growth Medium for Control

This may be a problem if the released organisms are transported beyond the release site and do not receive the lethal signal. This problem may be solved by using the invertable *fim*A "flip-flop" promoter from *E. coli* or by using the lethal and "protective" genes such as the *hok–sok* and *kil–kor* systems in a containment strategy.

V. Concluding Remarks

The application of genetic engineering for creating novel microbes holds promise for solving some of the problems facing the world today. The deliberate release of GEMs to the environment can benefit humans by increasing agricultural productivity and removing chemical contaminants such as herbicides, polychlorinated biphenyls (PCBs), dioxin, etc. GEMs with "killer" genes may also be used as safe live vaccines for livestock and even for humans. There is however, the concern that if the rDNA transfers to another organism a new pathogen or environmentally disastrous organism may be unintentionally created. We have yet to design a "fail safe'' system for controlling the deliberately released GEMs in the environment. The conditional lethal systems seem to be the way of the future.

Related Article: BIOTECHNOLOGY: OCCUPATIONAL HEALTH ISSUES.

Bibliography

Atlas, R. M., Bej, A. K., Steffan, R. J., and Perlin, M. H. (1989). Approaches for monitoring and containing genetically engineered microorganisms released into the environment. *Hazard. Waste Hazard. Mater.* **6**(2), 135–144.

Bej, A. K., Mølin, S., Perlin, M., and Atlas, R. M. (1992). Maintenance and killing efficiency of conditional lethal constructs in *Pseudomonas putida. J. Industrial Microbiol.* **10,** 79–85.

Bej, A. K., Perlin, M. H., and Atlas, R. M. (1988). Model suicide vector for containment of genetically engineered microorganisms, *Appl. Environ. Microbiol.* **54,** 2427–2477.

Contreras, A., Mølin, S., and Ramos, J. L. (1991). Conditional-suicide containment system for bacteria which mineralize aromatics. *Appl. Environ. Microbiol.* **57,** 1504–1508.

Federal Register, Department of Health and Human Services (1986). "Guidelines for Research Involving Recombinant DNA molecules, Part III," National Institute of Health, Washington, D.C.

Fiksel, J., and Covello, U. T. (1986). "Biotechnology Risk Assessment: Issues and Methods for Environmental Introductions," Pergammon Press, New York.

Halvorson, H. O., D. Pramer, and M. Rogul, eds. (1985). "Engineered Organisms in the Environment: Scientific Issues," American Society for Microbiology, Washington, D.C.

Jain, R. K., Burlage, R. S., and Sayler, G. S. (1988). Methods for detecting recombinant DNA in the environment. *CRC Crit. Rev. Biotechnol.* **8,** 33–84.

Knudsen, S. M., and Karlstrom, O. H. (1991). Development of efficient suicide mechanisms for biological containment of bacteria. *Appl. Environ. Microbiol.* **57,** 85–92.

Mølin, S., Boe, L., Jensen, L. B., Cristensen, C. S., Givskov, M., Ramos, J. L., and Bej, A. K. (1993). Suicide genetic elements and their use in biological containment of bacteria. *Ann. Rev. Microbiol.* **47,** in Press.

Mølin, S., Klemm, P., Poulsen, L. K., Biehl, H., Gerdes, K., and Andersson, P. (1987). Conditional suicide system for containment of bacteria and plasmids. Bio/Technol. **5,** 1315–1318.

Omenn, G., ed., (1988). "Environmental Biotechnology: Reducing Risks From Environmental Chemicals Through Biotechnology," Plenum Press, New York.

Sussman, M., Collins, C. H., Skinner, F. A., and Stewart-Tull, D. E., eds. (1988). "The release of Genetically-Engineered Microorganisms," Academic Press, New York.

Steffan, R., J., and Atlas, R. M. (1991). Polymerase chain reaction: Applications in environmental microbiology. *Ann. Rev. Microbiol.* **45,** 137–61.

Hazardous Substances, Regulation by Government Agencies

Troyce D. Jones
Oak Ridge National Laboratory[1]

Glossary

Act A law, decree, or edict (e.g., an act of Congress).

ANPRM (ANPR) Advance Notice of Proposed Rule Making. A preliminary notice (published in the *Federal Register*) that an agency is considering a regulatory action. An ANPRM is issued before a detailed, proposed rule is developed—it describes areas subject to regulation, alternatives under consideration, and solicits public comment.

CFR Code of Federal Regulations. A codification of the general and permanent rules published in the *Federal Register* by the departments and agencies of the federal government. The Code contains 50 titles that identify broad areas of regulation; issued quarterly, it is revised annually.

Corrosivity A characteristic that identifies waste that must be segregated because of its capacity to extract and solubilize toxic contaminants.

Environment Defined by the Environmental Protection Agency (EPA) as: (1) "Water, air, land, and all plants and humans and other animals living therein, and the interrelationships which exist among these"; (2) (a) "The navigable waters, the waters of the contiguous zone, and the ocean waters of which the natural resources are under the exclusive management authority of the United States under the Fishery Conservation and Management Act of 1976"; and (b) "Any other surface water, groundwater, drinking water supply, land surface or subsurface strata, or ambient air within the United States or under the jurisdiction of the United States"; or (3) "The sum of all external conditions affecting the life, development, and survival of an organism."

Federal Register A daily publication that announces all proposed and final federal regulations. The *Register* also contains notices of public meetings and other agency-related events.

Hazard A forseeable but unavoidable danger. Defined by EPA as: "The probability that a given pollutant will have an adverse effect on man or the environment in a given situation, the relative likelihood of danger or ill effect being dependent on a number of interrelated factors present at any given time."

Ignitability A characteristic of hazardous waste causing it to become a fire hazard under routine disposal or storage conditions.

Law A system or collection of rules.

[1] This article has been authored by a contractor of the U.S. Government (Oak Ridge National Laboratory managed by Martin Marietta Energy Systems, Inc.) under contract No. DE-AC05-84OR21400 with the U.S. Department of Energy. Accordingly, the U.S. Government retains a nonexclusive, royalty-free license to publish or reproduce the published form of this contribution, or allow others to do so, for U.S. Government purposes.

Material Matter of any kind or description including, but not limited to, dredged material; solid waste; incinerator residue; garbage; sewage; sewage sludge; munitions; radiological, chemical, and biological warfare agents; radioactive materials; chemicals; biological and laboratory waste; wreck or discarded equipment; rock; sand; excavation debris; and industrial, municipal, agricultural, and other waste. The term does not, however, refer to sewage from vessels or oil except to the extent that such oil is taken on board a vessel or aircraft for the purpose of dumping.

NPRM Notice of Proposed Rule Making. Issued by an agency and published in the *Federal Register* to solicit comment on a proposed regulation. An NPRM must include a statement of the time, place, and nature of the public rulemaking proceedings; reference to the legal authority under which the rule is proposed; and either the terms or substance of the regulation or a description of the subject and issues.

PL Public law.

Protect Prevent any unreasonable adverse effects.

Reactivity A characteristic that identifies hazardous waste that is unstable or extremely reactive under routine management.

Regulation A rule or order prescribed by authority to regulate conduct.

Risk Exposure to the probability of injury or loss.

Risk assessment The core of the EPA's "protection of human health"; intended to characterize the probability of adverse effects from human exposure to environmental hazards. Risk assessments are quantitative and chemical-specific—they use statistical and biologically based models to calculate numerical probabilities of risk.

Rule A principle or regulation governing conduct, procedure, etc.

Title Used in the U.S. Code of Federal Regulations to identify one of fifty broad areas of regulation.

Toxic Defined by the EPA as: (a) "A chemical with a median lethal dose (LD_{50}) of more than 50 mg/kg but not more than 500 mg/kg of body weight when administered orally to albino rats weighing between 200 and 300 g each"; (b) "A chemical that has a median lethal dose (LD_{50}) of more than 1000 mg/kg of body weight when administered by continuous contact for 24 hr (or less if death occurs within 24 hr) with the bare skin of albino rabbits weighing between 2 and 3 kg each"; or (c) "A chemical that has a median lethal concentration (LC_{50}) in air of 200 parts per million or less by volume of gas or vapor, or 2 mg/liter or less of mist, fume, or dust, when administered by continuous inhalation for 1 hr (or less if death occurs within 1 hr) to albino rats weighing between 200 and 300 g each."

USC *United States Code*. Contains a consolidation and codification of all laws of the country; includes 50 titles subject to federal law.

THE ENVIRONMENTAL PROTECTION AGENCY *has used different definitions to characterize hazardous substances—they include: (1) "Any material that poses a threat to human health and/or the environment" [hazardous substances may be toxic, corrosive, ignitable, explosive, or reactive; any substance designated by the EPA to be reported (if a designated quantity of the substance is spilled in the waters of the United States or if otherwise emitted to the environment) is hazardous]; (2) (a)"any substance designated pursuant to Section 311(b)(2)(A) of the Federal Water Pollution Control Act," (b) "any element, compound, mixture, solution, or substance designated pursuant to Section 102 of the Act," (c) "any hazardous waste having the characteristics identified under or listed pursuant to Section 3001 of the Solid Waste Disposal Act (but not including any waste the regulation of which under the Solid Waste Disposal Act has been suspended by act of Congress)," (d) "any toxic pollutant listed under section 307(a) of the Federal Water Pollution Control Act," (e) "any hazardous air pollutant listed under Section 112 of the Clean Air Act," and (f) "any imminently hazardous chemical substance or mixture with respect to which the administrator has taken action pursuant to Section 7 of the Toxic Substances Control Act." The term does not include petroleum, including crude oil that is not otherwise specifically listed or designated as a hazardous substance under* a *through* e *above. Similarly, it does not include natural gas, natural gas liquids, liquified natural gas, or synthetic gas usable for fuel (or mixtures of natural gas and synthetic gas). These definitions determine the substances covered by the Comprehensive Environmental Response Compensation and Liability Act (i.e., "Superfund").*

I. Historical Perspective

Although statutory control of dangerous materials and activities is not new, environmental concerns have continued to increase in an exponential manner over the past three decades. The thus expanding regulatory manifesto has been illustrated by a multitude of individual statutes, each with an associated complexity. For example, the CFR documents span 50 broad areas of regulation, contain well in excess of 100,000 pages, and list an almost indeterminable number of substances, mixtures, activities, and actions. Two events may have amplified or potentiated the regulation of hazardous substances by government agencies—they include a popular hypothetical account of a pesticide-killed environment published by Rachel Carson in 1962 (*Silent Spring*) and the problems caused in the late seventies by deposits of underground solid and liquid wastes at Love Canal near Niagara Falls, New York. The Clean Air Act (CAA) of 1963 is arguably the practical beginning of many environmental laws concerned with hazardous substances. While portions of Carson's monograph give remarkable insight into environmental problems, other passages may seem in retrospect to be exaggerated somewhat unrealistically from a current perspective. Nevertheless, the work seeded enduring environmental concern. *Silent Spring* should, however, also be remembered for statements such as, "It is not my contention that chemical insecticides must never be used. . . . We have put poisonous and biologically potent chemicals indiscriminately into the hands of persons largely or wholly ignorant of their potential for harm. . . . This is an era of specialists, each of whom sees his own problem and is unaware of or intolerant of the larger frame into which it fits." Indeed, Carson's observation about specialists being largely "unaware of or intolerant of the larger frame" could be said to describe modern science itself in microcosm.

II. Public, Political, Scientific, and Regulatory Agency Views

Since the CAA was enacted, environmental statutes have been promulgated at unbelievable rates—rates that are either too fast or too slow depending upon one's perspective. In spite of the seemingly large numbers of federal laws and the great bulk of text needed to codify them, there remains widespread frustration from a perceived inadequacy and lack of progress in control of environmental pollutants. This concern is reflected by members of the House of Representatives, senators, some presidents, and large segments of the general public alike. For example, in 1986 Representative Dennis Eckart from Ohio stated (*Congressional Record*—House; May 13, 1986; p.2632) that "Over 700 contaminants have been identified in our water. Yet EPA since 1974 has been able to regulate fewer than 25 . . . you can lead EPA to water, but you cannot necessarily make them regulate it," and Senator Dave Durenberger from Minnesota stated (*Congressional Record*—Senate; May 21, 1986; p.6285) that "The Safe Drinking Water Act is indeed simple in theory. In fact, Congress expected the program to fall quickly into place. The original statute is replete with quick deadlines stated as mere days from enactment—60 days, 90 days, 180 days. It is now 12 years later and the Safe Drinking Water Act once again comes to the floor of the Senate with most of the original promise unfulfilled. The act has failed miserably. In all of the time that has elapsed since enactment, the Environmental Protection Agency has set standards for only a handful of contaminants—six pesticides, all of which have been banned for use—a half a standard for trihalomethanes, one of the principal contaminants of concern in the early 1970s, and radionuclides."

The EPA's declared goal of the "protection of human health and the environment" and its codified means for reaching those objectives depend delicately upon a functional interface between science and policy. The diverse objectives, personal values, and technical expertise involved on both sides of the science–policy interface make the process complex and controversial within federal agencies, all branches of the government, science, and segments of the general population. The frustrations described by members of Congress are equalled in magnitude but often reversed in direction from the concerns of scientists. A characteristic comment representative of a toxicological perspective has been articulated by Richard A. Merrill as "the regulatory process sometimes mistreats the work of the scientific community. Perhaps no field has been victimized as often as toxicology. . . . A central difference [exists] between the objectives of science and the role of government—Science investigates and attempts to explain natural phenomen[a]: it is cautious, incremental, and truthseeking. Government regulation seeks to affect human behavior and settle human disputes . . . to solve problems . . . A regulator often cannot withhold a decision . . . while averting an unknown, but possibly trivial, risk . . . the regulator is invariably forced to intervene (i.e., to decide, before knowledge is complete)."

P. R. Portney, from the perspective of policy analysis, has taken a somewhat more balanced view that envelops the science–policy interface by noting that "the clean air and clean water laws promise 'safe' air and water quality, call for the establishment of literally tens of thousands of discharge standards, mandate the creation of comprehensive monitoring networks, and impose numerous other important tasks on the administrator of the EPA. Yet the laws allotted just 180 days for completion of many of these responsibilities. Today, more than seventeen years after passage of the laws, many of those assignments have yet to be carried out."

III. How Federal Agencies Coordinate Control

If a substance is classified as hazardous because of carcinogenic properties, there are several federal organizations that may be involved as stated in §305.82-5 of *Federal Regulation of Cancer Causing Chemicals*—". . . [B]ecause agencies differ in their functions and [the fact that] their selection of candidates for evaluation and regulation must be coordinated with other agency programs, each agency should establish its own priorities. The several agencies responsible for regulating carcinogens should, however, periodically compare their rankings and, where feasible, coordinate their testing, evaluation, and regulatory efforts. Consultation with scientific, industrial, and public interest organizations should be encouraged." Additionally, regulatory agencies are expected to use existing voluntary consensus standards and to collaborate with government scientific bodies such as the National Toxicology Program, National Institute of Environmental Health Sciences, and National Center for Toxicological Research, and to avoid inconsistencies in mixed scientific–policy issues.

The primary federal agencies responsible for regulating hazardous materials are the Food and Drug Administrations (FDA), the Environmental Protection Agency (EPA), the Occupational Safety and Health Administration (OSHA), the Nuclear Regulatory Commission (NRC), and the Consumer Products Safety Commission (CPSC).

The FDA was formed in 1938 to control foods for humans and animals; human and veterinary drugs; medical devices; and cosmetics. As mandated by the well-known Delaney (Food Additives) Amendment of 1958, the FDA requires that any food additive be found "safe" before it can approve its use.

OSHA was formed in 1970 and regulates workplace conditions in diverse areas ranging from chemicals to agriculture. By charter, OSHA is directed to select workplace standards "which most adequately assure[s] . . . that no employee will suffer material impairment of health or functional capacity."

The EPA, formed in 1970, regulates emissions of pollutants concerned with exposure of the general population from various industrial sources. EPA became responsible for functions previously under the jurisdiction of the Departments of Agriculture; Education; Health; Interior; and Welfare and for issues acquired from the Atomic Energy Commission, the Federal Radiation Council, and the Council on Environmental Quality (i.e., the FD&C Act of 1958; CPSC Act of 1972; FFD&C Act of 1938; and OSH Act of 1970).

Under Section 104(i)6 of the Comprehensive Environmental Response, Compensation, and Liability Act of 1980 (CERCLA), the Agency for Toxic Substances and Disease Registry (ATSDR) responsibilities include assessment of populations with current or potential exposure to waste sites, development of health advisories, and conducting follow-up health studies to evaluate future health effects. The ATSDR health assessment is usually qualitative, site specific, and focuses on medical and public health perspectives in contrast to the EPA's functions related to risk assessment and risk management. Where ATSDR is involved, the EPA and ATSDR should coordinate health-related studies during the remedial process. EPA is expected to use the risk assessments developed as part of the public health evaluation process and include ATSDR analyses into risk mangement decisions. Under the Superfund Amendments Reauthorization Act of 1986 (SARA), ATSDR is required to conduct health assessments for all sites on (or nominated for) the *national priorities list* for site remediations.

Toxicity, especially carcinogenicity, is one of several considerations that determine if a substance is hazardous. Because ignitable, corrosive, reactive, and explosive properties can be defined in terms of unambiguous chemical and physical constants, most of the variability within the statutory process derives from toxicological considerations.

Although agencies concerned with carcinogenic hazards have attempted both individually and collectively to coordinate criteria for evaluating scientific data, much incoherence has resulted. Some can be attributed to the inherent variability in methods of data selection, biological evaluation, and mathematical assumptions chosen to analyze test results in animals and subsequent extrapolation of those measures of dose and response to hypothetical human populations assumed to be exposed (for a lifetime of 70 years) to low concentrations of pollutants in water or air.

The efforts of individual agencies to establish criteria for identification and evaluation of potentially carcinogenic hazards have been criticized widely. Because a particular compound may have large commercial value, industrial laboratories frequently sponsor their own research conducted by highly trained scientific staff members and state-of-the-art research techniques. Frequently the industrial perspective has been to question the logic of allowing agency-specific standardized codifications to take precedence over industrial compound-specific determinations based on considerations of carcinogenic mechanisms. Because of such controversy, the Administrative Conference of the United States has published some recommendations for rulemaking. A summary is given in Table I.

IV. How Rules Are Promulgated

Even though agency-specific philosophies differ widely, all are moderated by recommendations of the Administrative Conference Act. The Administrative Conference Act was established to study the efficiency, adequacy, and fairness of the administrative procedures used by federal agencies in carrying out administrative programs. The Conference makes recommendations for improvements to agencies and to the President, Congress, and the Judicial Conference of the United States. Such a recommendation ("Procedures for Negotiating Proposed Regulations") provides a method of developing regulatory programs—negotiations between substantially affected interests, with a view to minimizing protracted adversary proceedings and litigation. This recommendation suggests procedures for conducting such negotiations, and commends their use to agencies when particular circumstances suggest a likelihood of success. "Procedures for Negotiating Proposed Regulations" helps to provide safeguards against arbitrary or capricious decisions by agencies and to ensure that agencies develop sound factual bases for rulemaking. A summary for federal agencies is given in Table II.

V. Philosophical Templates Used for Rulemaking

It is of interest that laws promulgated in support of the FDA require that users of "new substances" prove their safety before humans may be exposed, whereas, laws to support objectives of EPA and OSHA require regulators to show that a substance is hazardous before exposures can be restricted.

Similarly, definitive evidence of carcinogenicity is actually required for a compound to be controlled as a carcinogenic agent. A far more reasonable policy would be to give first-order considerations to "protection of human health" and second-order considerations to factual representation of biochemical and biological processes actually underlying cancer disease mechanisms.

As described above, environmental laws reflect manifold objectives: for example, several laws attempt to obviate risk completely by mandating that ambient standards be set to provide an "adequate margin of safety"; the Clean Water Act of 1972 (CWA) transfers the ambient water quality standards to the states but additionally requires an implied margin of safety for aquatic life. Other standards are based on technological feasibility; yet others such as The Toxic Substances Control Act (TSCA) and The Federal Insecticide, Fungicide, and Rodenticide Act of 1972 (FIFRA) are based on cost–benefit type considerations.

Since 1974, executive branch agencies have been required to prepare impact analyses for major rulemaking proposals. Cost against benefit analysis may be part of the impact analysis except when Congress has specified otherwise. Regulations were intended to reduce the risks of accidents and illness and usually impose compliance costs on the regulated activities and on rulemaking agencies. Given this situation, the Administrative Conference of the United States has noted that "agencies cannot avoid placing a value—either explicitly or implicity—on the societal benefits of risk reduction . . . (even

Table I Outline of Considerations for Carcinogenic Hazards as Recommended by the Administrative Conference of the United States

Generic rulemaking

1. Generic rules may be used provided they do not foreclose reexamination of scientific conclusions. Agencies should proceed cautiously because of complexity, uncertainty, and continued advances in science.
2. Agencies should develop principles to identify, rank, and evaluate carcinogenic chemicals. The process should involve opportunities for persons from outside the agency. Whenever debate is limited, compliance with statutory requirements is essential.
3. Agency statements should attempt to distinguish between elements summarizing current consensus and representations of policy judgments reached in the absence of consensus.
4. Issues in identifying, evaluating, and regulating carcinogenic hazards may be common to all agencies and should be resolved consistently.

Quantitative assessment of risk

1. To the extent regulatory statutes and data permit, agencies should estimate and describe the magnitude of risk posed by prevailing levels of exposure. Within the same constraints, the health benefits from mitigation of exposure should be estimated.
2. The weight accorded risk estimates should reflect (a) governing statutes, (b) adequacy of available data on carcinogenic potency and human exposure, and (c) the acceptance of methods used.
3. Estimates of risk should include (a) toxicological, epidemiological, and exposure data used, (b) assessments underlying extrapolations from animals to man or from high to low concentrations, (c) other assumptions about the behavior of the substance or human exposures, and (d) uncertainty of the estimates.
4. Statements should explain imprecision and uncertainty about disease, cancer, and death because of the public response. When available data are inadequate, the risk of misinterpretations may justify a decision not to assess risk or benefits.

Public participation

1. Agencies should follow procedures that provide opportunities for all affected interests to provide insight.
2. Agencies should encourage and facilitate the participation of independent experts in toxicology, epidemiology, risk and exposure estimation, and other relevant disciplines.
3. Congress should refrain from procedural requirements which are burdensome.

Additional procedures for comment

1. Agencies should enlarge public participation and increase effectiveness by
   (a) providing for the possibility of two cycles of notice and comment
   (b) providing for additional comment when new, important, or serious conflicts of data are involved
   (c) incorporating a summary of the agency's attitude, description of data and where available for inspection
   (d) explain[ing] procedures and open[ing] comment
   (e) open[ing] conferences to all interested groups
   (f) providing for agency questions including those from interested persons
2. Provide opportunities for issues of fact and cross-examination.
3. An agency should use (a)–(f) and cross-examination only to improve quality or fairness of the process.

though) . . . Placement of a dollar value on human life is controversial and complex.'' In practice, some agencies reject all such explicit considerations, while other agencies routinely include such estimates into proposed rules. Some financial considerations for regulatory decisionmaking as recommended by the Administrative Conference of the United States are summarized in Table III.

A California initiative usually referred to as ''Proposition 65'' has shifted the ''burden of proof'' to the regulated industry for continuing the use of substances listed by the state. Previously, no actions were taken until the regulators had met the burden of proof. Proposition 65 has strongly affected regulations of several states as well as federal agencies. For example, the EPA uses the ''California List'' for land disposal restrictions regarding metal-bearing and cyanide-containing wastes.

Like Proposition 65, litigation specific to vinyl chloride (under the Clean Air Act of 1987) has affected the regulation of other carcinogenic hazards. The vinyl chloride decision set a precedent for a two-stage regulatory process. In the first stage, the EPA is required to establish the emission level that is acceptable or safe based on health considerations; cost must not be considered. In a second stage, EPA must apply an ''ample margin of safety'' to derive national emission standards for hazardous air pollutants (NESHAPs). Here, economics, feasibility, and other factors may be considered. Because, EPA's analytical approach involves data selection and subjective modeling assumptions, it is quite difficult to

Table II Summary of "Procedures for Negotiating Proposed Regulations" by Federal Agencies

1. Agencies should consider using regulatory negotiations as described below for drafting the text of a proposed regulation. A proposal to establish a regulatory negotiating group could be made either by the agency (e.g., in an ANPRM) or by an interested person.
2. Congress should pass legislation to conduct rulemaking according to the following recommendations. The legislation should provide substantial flexibility for agencies to adapt negotiation techniques to individual proceedings.
3. In legislation authorizing negotiation, Congress should authorize use of a "convenor" to organize negotiations. The convenor should be an individual, government agency, or private organization that is neutral with respect to policy under consideration. The convenor would (i) advise the agency as to whether negotiation is feasible and practical, and (ii) collaborate with the agency as to who should participate in negotiations.
4. An agency should select and consult with a convenor at the earliest practicable time. The convenor should determine whether a negotiating group should develop a proposed rule. The convenor should consider risks that negotiation would increase the likelihood of a consensus proposal that would limit output, raise prices, restrict entry, or otherwise contribute to unreasonable restraints on competition. Other factors include: (a) issues should be mature and ripe for decision—there should be deadlines for negotiations and for issuing the rule, (b) resolution should not require a compromise of participants' fundamental tenets, (c) usually negotiations cannot be conducted with a large number—usually no more than 15 participants, (d) there should be diverse issues that participants can rank, (e) no single interest should dominate, (f) participants should negotiate in good faith to draft a proposed rule, and (g) the agency should designate a representative, but he or she cannot bind the agency.
5. If negotiation is appropriate, the convenor should determine the interests that may be affected substantially, individuals to represent those interests, scope of issues, and a schedule. The negotiating group should be balanced.
6. The proposed regulation may be within the scope of a nongovernmental standards writing organization. The convenor may recommend that negotiations be conducted under that committee's auspices.
7. The agency should publish in the *Federal Register* advanced notice of proposed rulemaking (ANPRM) and indicate issues and participants already identified.
8. The agency should identify a senior representative in the ANPRM.
9. Unlike intervenors, the negotiating group will be performing a function normally performed within the agency, and the agency should reimburse direct expenses.
10. The convenor and the agency may consider the use of a mediator.
11. The goal of the negotiating group should be consensus on a proposed rule. Following either consensus or lack-of-consensus, the group should prepare a report that states findings and recommendations.
12. The group may close meetings only to protect confidential data or enhance likelihood of reaching consensus.
13. The agency should publish the proposed rule in its notice of proposed rule making (NPRM).
14. The group should review any comments in response to the NPRM. The final responsibility of issuing the rule resides with the agency.

Table III Financial Considerations for Decisionmaking as Recommended by the Administrative Conference of the United States

Valuation of human life in regulatory decisionmaking
1. When an agency adopts a regulation to reduce risk to human life, it should disclose the dollar value per statistical life used.
2. Agencies should recognize limitations to incorporate all the variables that affect societal valuations of human life.
3. Whenever agencies discount costs and benefits, they should disclose rates used, methodology applied, and sensitivity of conclusions to rates used.
4. Office of Management and Budget should serve as a central clearinghouse on research and information of life valuation issues.

satisfy the intent of the vinyl chloride ruling in a realistic, practical sense.

VI. Some Federal Laws and Regulations

Historically, there have been many attempts to regulate substances and/or actions perceived as hazardous. These attempts have been somewhat sparse until about 1963. Since then, agency regulation of hazardous substances has been in "early bloom" as summarized in Table IV. "Mature bloom" may occur in a decade or two and could extend to all substances and activities as described in the section on "future control by government agencies."

Such broad-initiative control may be a surprise in some technical areas, but should not be unexpected to those trained in toxicology or oncology. Indeed, Paracelsus (1493–1541) noted that "all substances are poisons: there is none which is not."

VII. Toxicological Unknowns and Attainability Are Primary Weaknesses

The nontoxic considerations of hazardous substances are defined unambiguously in terms of simple chemical and physical constants. Thus, any changes are class-wide and not compound-specific. Because toxicants are evaluated individually, future changes can be either compound-specific or broad in scope. There has not yet been a goal to provide uniform, comprehensive control of carcinogenic substances. The scope for control of all potentially carcinogenic hazards has been described (in Federal Regulations of Cancer Causing Chemicals" as: ". . . because the number of chemicals that may be carcinogenic is larger than government agencies can evaluate or regulate, attention should be concentrated on those that pose the greatest risks and can be controlled economically." This somewhat relaxed attitude (about the protection of human health) is a possible outgrowth of an early belief that few chemicals were actually carcinogenic and consequently exposure to them possibly did not (or may not) add significantly to the natural or spontaneous cancer rates in normal human populations living under individually variable conditions.

Such expectations are based mostly on faith and historical conditioning—both of which can be misleading. The literature is rich with accounts of unexpected toxicity. One interesting example can be learned from Xenophon's account of *The Persian Expedition*—"[viz] there were great numbers of bee hives in these parts, and all the soldiers who ate the honey went off their heads and suffered from vomiting and diarrhea and were unable to stand upright. Those who had only eaten a little behaved as though they were drunk, and those who had eaten a lot were like mad people. Some actually died."

VIII. Future Control by Government Agencies

In the future, it should almost be obligatory that "greatest risks" and "controlled economically" will be redefined because, according to Federal Regulations, ". . . criteria for selecting candidates for regulation should include the extent of the hazard posed by a chemical—a function of its potency, the conditions of exposure, and the number of people exposed . . . selection of candidates for regulation will often be an iterative process."

Currently, some statutory concentrations for chemicals are based on the assumptions that a particular compound is carcinogenic only if it has been found to cause cancer by the route of exposure being regulated. A safer policy could be to assume that any carcinogenic insult can, in theory, potentiate cancer in any organ tissue with which it comes in contact. Each nucleated cell—in any organ—contains the full genetic code for the entire human body. Typically, only phenotypic expression of different proteins translated from the same genomic DNA distinguishes a cell in one organ from a cell in a different organ. Therefore, lack of positive data for a

Table IV A Simplified Guide to Some Federal Laws and Regulations Hazardous Substances

The Refuse Act of 1899
Purpose: to prevent impediments to navigation—it shall not be lawful to throw, discharge, or deposit any refuse matter into any navigable water of the United States.

FDA/The Federal Food, Drug, and Cosmetic Act of 1906
Purpose: to control foods for humans and animals, human and veterinary drugs, medical devices, and cosmetics.
Includes: Amendments in 1938, 1962, and the Food Additive Amendment of 1958.

The Water Pollution Control Act of 1948
Purpose: to encourage control of water pollution by research and investigation. Federal construction loans were authorized (but none was ever approved). Actual control was left to state and local governments.

Air Pollution Control Act of 1955
Purpose: authorized federal funds to assist the states in air pollution research and in training technical and managerial personnel.

Water Pollution Control Act Amendments of 1956
Purpose: authorized states to establish water quality criteria, sponsored conferences to negotiate clean-up plans, federal discretionary enforcement for interstate waters, and authorized construction grants up to 55% of cost.

CAA/The Clean Air Act of 1963
Purpose: to protect and enhance air quality; initiate and accelerate R&D; provide technical and financial aid to state and local governments; and promote regional control programs.
Includes: PL 84-159 (1955), PL 88-206 (Clean Air Act of 1963), PL 89-220 (Motor Vehicle Air Pollution Control Act of 1965), PL 89-272 (1965), PL 89-675 (1966)/PL 91-604 (1970)/PL 95-95 (1977)—(amendments), PL 90-148 (The Air Quality Act of 1967), PL 92-157 (Comprehensive Health Manpower Training Act of 1971), PL 93-15 (1973), PL 93-319 (Energy Supply and Environmental Coordination Act of 1974), PL 95-623 (1978), PL 96-209 (1980), PL 97-300 (1982), PL 97-325 (1982), PL 98-213 (1983), PL 100-202 (1987), PL 95-190 (1977)/PL 96-300 (1980)/PL 97-23 (1981)/PL 98-45 (1983)—Safe Drinking Water Act as amended, and PL 101-549 (1990).

Water Quality Act of 1965
Purpose: to attain ambient water quality standards to be set and enforced by the states, limit discharges from individual sources by state implementation plans, provide federal oversight through approval and enforcement procedures.

NEPA/National Environmental Policy Act of 1969
Purpose: to encourage productive and enjoyable harmony between humans and the environment; moderate environmental damage and stimulate health and welfare of humans; learn about ecosystems; and establish a Council on Environmental Quality.
Includes: PL 91-190 (1970), PL 94-52 (1975), PL 94-83 (1975), and PL 99-160 (1985), PL 100-202 (1987), PL 100-404 (1988), PL 100-144 (1989).

OSHA/Occupational Safety and Health Act of 1970
Purpose: to regulate commerce and to provide for safe and healthful working conditions by: reducing hazards and providing safety programs; providing that employers and employees have responsibilities and rights; authorizing the Secretary of Labor to mandate standards affecting interstate commerce and creating a Health Review Commission, providing for research; studying occupationally linked diseases; providing medical protection criteria; providing training programs; providing for the development and promulgation of standards; providing an enforcement program; involving the States; reporting procedures; and unite labor and management about safety.
Includes: PL 91-190 (1970), PL 94-52 (1975), PL 94-83 (1975), and PL 99-160 (1985), PL 100-202 (1987), PL 100-404 (1988), PL 100-144 (1989).

CWA/Federal Water Pollution Control Act of 1972 (Clean Water Act as amended)
Purpose: to restore and maintain water quality by eliminating discharged pollutants into navigable waters by 1985; protect fish, shellfish, and wildlife by 1985; prohibit toxic discharges; finance waste treatment; control pollutants in states; and support R&D to eliminate discharges into navigable waters, waters of contiguous zone, and oceans.
Includes: PL 92-500 (1972), PL 93-207 (1973), PL 93-243 (1974), PL 93-592 (1975), PL 94-238 (1976), PL 94-273 (1976), PL 94-558 (1976), PL 95-217 (1977), PL 95-576 (1978), PL 96-148 (1979), PL 96-478 and PL 96-483 (1980), PL 96-561 (1980), PL 97-35 (1981), PL 96-510 (1981), PL 97-117 (1981), PL 97-164 (1982), PL 97-216 (1982), PL 97-272 (1982), PL 97-440 (1983), PL 98-45 (1983), PL 100-4 (1987), PL100-202 (1987), PL 100-404 (1988), PL 100-688 (1988), PL 101-144 (1989), and PL 101-380 (1990).

FIFRA/The Federal Insecticide, Fungicide, and Rodenticide Act of 1972 (amended to Federal Environmental Pesticide Control Act)
Purpose: The Administrator of EPA and the Secretary of Agriculture, shall identify pests that must be controlled. R&D programs will promote the safe use and effectiveness of chemical, biological, and alternative methods to control agricultural pests.
Includes: PL 92-516 (1972), PL 93-86 (1973), PL 93-629 (1975), PL 94-51 (1975), PL 94-109 (1975), PL 94-104 (1975), PL 95-113 (1977), PL 95-251 (1978), PL 95-396 (1978), PL 96-539 (1980), PL 97-98 (1981), PL 98-201 (1983), PL 98-620 (1984), PL 100-202 (1987), PL 100-418 (1988), PL 100-532 (1988), PL 101-624 (1990), and PL 102-237 (1991).

(*continues*)

Table IV *(Continued)*

CPSC/The Consumer Product Safety Act of 1972

Purpose: to regulate products that pose an unreasonable risk of injury or illness to consumers. Consumer products "include all articles sold for general or personal use in the home, school, or recreational settings, except products regulated by EPA and FDA."

Includes: FHSA/The Federal Hazardous Substances Act of 1960

MPRSA/Marine Protection, Research, and Sanctuaries Act of 1972

Purpose: to regulate the transportation of material for the purpose of dumping into oceans, and the dumping from outside the United States if the dumping occurs in the territorial sea or contiguous zone of the United States.

Includes: PL 92-532 (1972), PL 93-254 (1974), PL 93-472 (1974), PL 94-62 (1975), PL 94-326 (1976), PL 95-153 (1977), PL 96-332 (1980), PL 96-381 (1980), PL 96-470 (1980), PL 96-572 (1980), PL 97-16 (1981), PL 97-109 (1981), PL 97-375 (1982), PL 97-434 (1983), PL 98-498 (1984), PL 99-272 (1986), PL 99-499 (1986), PL 100-4 (1987), and PL 100-688 (1988).

SDWA/Safe Drinking Water Act of 1974

Purpose: develop national drinking water regulations; provide for enforcement of regulations; restrict materials used to install or repair water systems; protect underground sources; prevent tampering with public water systems; ensure availability of treatment supplies; support research; assistance, and training; provide grants for states; and require records and inspections.

Includes: PL 93-523 (1974), PL 94-317 (1976), PL 94-484 (1976), PL 95-100 (1977), PL 96-63 (1979), PL 96-502 (1980), PL 98-620 (1984), PL 99-339 (1986), PL 100-572 (1988).

RCRA/Resource Conservation and Recovery Act of 1976

Purpose: objectives are to protect human health environment and conserve resources by: assisting states and local governments and interstate agencies with solid waste management; provide training grants for solid waste disposal systems; prohibit open dumping and remediate existing sites; manage hazardous waste to protect human health and the environment; require that hazardous waste be managed properly in the first instance; minimize hazardous waste by process substitution, recovery, and treatment; establish Federal–State partnerships; promulgate guidelines; promote R&D; promote solid waste management to preserve quality of air, water, and land; and establish cooperation to recover resources from solid waste.

Includes: PL 94-580 (1976), PL 95-609 (1978), PL 96-482 (Solid Waste Disposal Act of 1980), PL 96-463 (Used Oil Recycling Act of 1980), PL 96-510 (Comprehensive Environmental Response, Compensation, and Liability Act of 1980), PL 97-272 (1982), PL 97-375 (1982), PL 98-45 (1983), PL 98-371 (1984), PL 98-616 (Hazardous and Solid Waste Amendments of 1984), PL 99-160 (1985), PL 99-339 (1986), PL 99-499 (1986), PL 100-202 (1987), and PL 100-582 (1988).

TSCA/Toxic Substances Control Act of 1976

Purpose: to regulate commerce and protect human health and the environment by requiring testing and necessary use restrictions on certain chemical substances, and for other purposes. Manufacturers and processors of chemical substances and mixtures should develop data on effects to health and the environment. Chemicals which present unreasonable risk should be regulated and action taken on imminent hazards. Authority should not impede unduly or create unnecessary economic barriers to technological innovation.

Includes: PL 94-469 (1976), PL 96-510 (1980), PL 97-98 (1981), PL 97-129 (1981), PL 98-80 (1983), PL 98-377 (1984), PL 98-620 (1984), PL 99-499 (1986), PL 99-519 (1986), PL 100-11 (1987), PL 100-202 (1987), PL 100-368 (1988), PL 100-418 (1988), PL 100-551 (1988), PL 100-577 (1988), PL 101-144 (1989), PL 101-508 (1990), PL 101-624 (1990), and PL 101-637 (1990).

CERCLA/Comprehensive Environmental Response, Compensation, and Liability Act of 1980 (SARA as amended/Superfund Amendments Reauthorization Act of 1986)

Purpose: to provide for liability, compensation, cleanup, and emergency response for hazardous substances released into the environment and the cleanup of inactive waste disposal sites.

Includes: PL 96-510 (1980), PL 97-98 (1981), PL 97-216 (1982), PL 97-272 (1982), PL 98-45 (1983), PL 99-160 (1985), PL 99-272 (1985), PL 99-499 (SARA of 1986), PL 100-202 (1987), PL 100-144 (1989), PL 100-203 (1987), PL 100-551 (1988) TSCA, PL 101-221 (1989), PL 101-239 (1989), PL 101-380 (1990), PL 101-508 (1990), PL 101-584 (1990), and 101-624 (1990).

HMTA/Hazardous Materials Transportation Act

Federal Hazardous Substance Act of 1927 (amended in 1976).

Purpose: for control of flammable, corrosive, allergenic, or toxic materials in consumer products.

particular route of exposure seems to be a specious and potentially unsafe way to identify carcinogenic risks. Of course, pharmacological studies could be used to help fill data gaps.

Often efforts in regulatory toxicology are directed at extrapolations of test results from one subjectively selected animal study to man. To draw conclusions of such cosmic importance according to principles of "good science" would require: a theoretical and/or operational understanding of all involved response mechanisms both in humans and in the biological test model selected as a proxy for humans, and unquestionable confidence in the validity and relevance both of the test data and analytical methods used to derive the conclusions. It is commonly acknowledged that risk assessments suffer immense deficienceis in all of these aspects. Thus, it would appear that the product from such a logic represents neither good science nor good policy.

Historically, noncarcinogens have been controlled individually and without considering concomitant interactions with carcinogens. If a chemical is lacking one intrinsic trait needed to be operationally defined as a carcinogen, it has not been permitted to "borrow" that trait from other pollutants or endogenous biochemical processes. However, simple cell killing or wounding of tissue either by chemicals or by mechanical insults is highly effective at promoting cancers initiated by other insults. A safer logic would be to allow chemicals to "borrow and lend" properties of toxicity because, in contrast to test animals, humans are always exposed simultaneously and serially to multiple physiological stresses.

In 1986, the EPA proposed to regulate certain mixtures. Previously, only products in which a listed chemical was the sole active ingredient were regulated. Formulations comprised of two or more active ingredients (one or more of which could be acutely toxic) were priorly unregulated. (See the references by Jones *et al.* and Owen and Jones as to how broader considerations may be used to define iso-effect concentrations for human exposures to environmental mixtures.)

Without doubt, the number of regulations dealing with hazardous substances will continue to increase. Some substances currently treated as carcinogenic hazards are probably over-regulated by orders of magnitude in a relative sense. But, in contrast, the current intent to treat substances as carcinogenic if and only if they are carcinogenic by their own individual, intrinsic properties will likely weaken and lead to more realistic regulations of all exposures that contribute significantly to chronic toxicity in human environments.

Control of hazardous substances is a slowly unfolding process. The efforts cannot wait for decades simply because the environment has a unique role in the support of life on this planet. As Jack London had Wolf Larsen articulate in *The Sea Wolf*—"Why, if there is anything in supply and demand, life is the cheapest thing in the world. There is only so much water, so much earth, so much air; but the life that is demanding to be born is limitless." Larsen's perspective is unusual compared with a more common persuasion that life itself is unique, precious, and generally superior to all lifeless considerations.

Wolf Larsen's insight should be remembered even though he callously depreciates the value of life. The long-term regulation of hazardous substances by government agencies can be previewed reasonably well by other examples from Plato (*The Laws*)—"Anyone convicted of fouling water by magic poisons should, in addition to his fine, purify the spring or reservoir, using whatever method of purification the regulators of the Expounders prescribe as appropriate to the circumstances and the individuals involved" and from More (*Utopia*)—"Nor do they allow anything filthy and impure to be brought into the city, whose corruption could pollute the air and carry in some disease."

Related Articles: BIODEGRADATION OF XENOBIOTIC CHEMICALS; BIOTECHNOLOGY: OCCUPATIONAL HEALTH ISSUES; CHROMIUM; CYANIDE; HAZARDOUS WASTE REDUCTION; PESTICIDES AND FOOD SAFETY; POLYCHLORINATED BIPHENYLS (PCBS), EFFECTS ON HUMANS AND THE ENVIRONMENT.

Bibliography

Administrative Conference of The United States 1 CFR, Chapter III.

Dower, R. C., Freeman III, A. M., Russell, C. S., and Shapiro, M. (1990). *In* "Public Policies for Environmental Protection" (P. R. Portney, ed.), Resources for the Future, Washington, D. C.

Efron, E. (1984). "The Apocalyptics." Simon and Schuster, New York.

Environmental Protection Agency (1986). Hazardous Waste Management System; Identification and Listing of Hazardous Waste; Commercial Chemical Products. *Federal Register* **51**(30), 5472–5481.

Government Industries, Inc. (1990). "Environmental Regulatory Glossary" (G. W. Frick and T. F. P. Sullivan, eds.). Fifth Edition, Rockville, Maryland.

Jones, T. D. (1991). On the future of regulating chemicals. *Environ. Auditor* **2**(3), 101–102.

Jones, T. D., Owen, B. A., and Trabalka, J. R. (1991). Protection of human health from mixtures of radionuclides and chemicals in drinking water. *Arch. Environ. Contam. Toxicol.* **20,** 143–150.

Jones, T. D., Owen, B. A., Trabalka, J. R., Barnthouse, L. W., Easterly, C. E., and Walsh, P. J. (1991). Chemical pollutants: A characterized logos for future planning. *Environ. Auditor* **2**(2), 71–88.

Merrill, R. A. (1986). Regulatory toxicology. *In* "Casarett and Doull's Toxicology." Macmillan, New York.

Owen, B. A., and Jones, T. D. (1990). Hazard evaluation for complex mixtures: Relative comparisons to improve regulatory consistency. *Reg. Toxicol. Pharmacol.* **11,** 132–148.

Hazardous Waste Reduction

A. J. Englande, Jr.
Tulane University

Glossary

Hazardous waste Nonproductive hazardous output from an industrial operation into all environmental media. Hazardous refers to harm to human health or the environment and includes wastes which exhibit the characteristics of ignitability; reactivity; corrosivity and/or toxicity. It also includes specific listed waste.

Multimedia All environmental media including air, water, and soil to which contaminants may be discharged.

Non-waste and low-waste technologies Application of technological modifications that lead to the creation of less waste through greater physical efficiency of energy and materials use, and processes that convert waste products from conventional processes into new secondary products (by-products) that can be reused.

Recycling Processes constituting "use or reuse" and "reclamation." "Use or reuse" is the procedure whereby a residual is employed as an ingredient in an industrial process to make a product or as an effective substitute for a commercial product. "Reclamation" is a procedure whereby a material is treated to recover a useable product, or where a material is regenerated.

Small quantity generators SQGs are those generators that produce between 100 kg and 1000 kg of hazardous waste per month.

Source reduction Any method or technique applied at or before the point of generation, the application of which reduces or eliminates the use or generation of hazardous substances so as to reduce the risk to public health and the environment.

Standard industrial classification The SIC System identifies industries and establishments by types of activity. Establishments are grouped together if they exhibit similar industrial characteristics.

Treatment Any method, technique, or process, designed to change the physical, chemical, or biological character or composition of a material to (1) recycle energy or material resources from the material; (2) render such material nonhazardous, or less hazardous; (3) render the material safer to dispose of; or (4) render the material more amendable for recycling or storage.

Waste exchange Transfer of wastes through clearinghouses which match potential waste users with generators of the wastes.

Waste minimization Reduction, to the extent feasible, of any solid or hazardous waste that is generated or subsequently treated, stored, or disposed of.

Waste reduction Source reduction and recycling.

Waste streams Residuals resulting from industrial processes which require recycling, treatment, or disposal.

***HAZARDOUS WASTE REDUCTION** refers to inplant process modifications that reduce the volume or degree of hazard of waste generated as well as promote the reuse and reclamation of hazardous residuals. Hazardous waste minimization means the reduction, to the extent feasible, of any hazardous waste that is generated or subsequently treated, stored, or disposed of. Waste minimization techniques focus on source reduction or recycling activities that reduce either the volume or the toxicity of* *waste generated. Consequently, waste reduction and waste minimization are used interchangeably. Waste reduction/minimization can save significant amounts of money through more efficient use of valuable resources and reduced waste treatment and disposal costs. Waste reduction can reduce a generator's hazardous-waste-related financial liabilities: the less waste generated, the lower the potential for negative public health and environmental impact. Taking the initiative to reduce hazardous waste is also good policy. Polls show that reducing toxic chemical risk is the public's primary environmental concern. Waste reduction will pay off tangibly when local residents are confident that industry is making every effort to handle its waste responsibly.*

I. Introduction

Waste reduction has long been recognized as the preferred option for the management of hazardous wastes. As early as 1976, the United States Environmental Protection Agency (EPA) adopted a hierarchy as the guide for hazardous waste management options that identified reducing the amount of waste at the source through changes in the industrial processes as the most desirable option. Treatment, destruction, and disposing of waste on land was deemed the least desirable option. Similar hierarchies have been adopted by several international environmental and development agencies, as well as many of the advanced industrial countries of the world.

Managing hazardous industrial waste continues to be an important national issue in the United States. In 1984, Hazardous and Solid Waste Amendments (HSWA) to the federal Resource Conservation and Recovery Act (RCRA) of 1976 made waste minimization (i.e., source reduction and recycling) national policy. Congress stated that, wherever feasible, the generation of hazardous waste is to be reduced or eliminated as expeditiously as possible. Whatever waste is generated, however, is to be treated, stored, or disposed of so as to minimize the present and future threat to the environment. The Pollution Prevention Act of 1990 continued the focus on waste reduction and includes a cross-media Toxics Release Inventory (TRI) aimed at reducing a targeted set of 17 toxic chemicals by 50% in five years.

Several terms have been used to describe preferred options for waste management; these include waste reduction, waste minimization, waste abatement, waste prevention, waste avoidance, pollution prevention, and source reduction. These terms mean different things to different people, and thus, constitute an obstacle to comparing and assessing effectiveness claims and reported costs and benefits.

A narrow definition of waste reduction, as used by some, considers only activities that prevent the waste from being generated in the first place. This definition excludes follow-up activities such as waste recycling, reuse, or treatments that reduce the volume or toxicity of the waste destined for ultimate disposal. A broader definition of waste reduction which includes source reduction as well as reclamation and reuse (waste recycling) is used herein. Definition of terms identified herein are somewhat different therefore than those defined by the 1990 act.

II. Potential for Waste Reduction

To date, EPA has classified 207 separate hazardous waste streams, originating from 33 industrial sectors. Estimates for the total amount of hazardous waste generated annually range between 158 and 266 million metric tons, depending on the study used.

Disagreement exists on the identity of major generators. However, all agree that the chemical industry (Standard Industrial Classification, SIC, 28: Chemicals and Allied Products) alone contributes far more hazardous waste than any other industrial sector (roughly 50% or more). General agreement is also reached on the fact that small quantity generators (SQG) collectively contribute only a small amount of hazardous waste—approximately 0.5% of hazardous waste by volume, annually. However, small quantity generators represent by far the largest number of hazardous waste producing establishments in the United States.

Despite the amount of waste generated, the SQGs are of particular interest because they may be less capable of dealing with the waste generated than larger volume generators. They also are likely to have more difficulty in complying with the new regulations and may also be less capable of implementing waste minimization techniques.

Estimates of possible waste reductions vary widely, but all indicate that a substantial, untapped potential for waste reduction within American industry exists. EPA estimates that waste reduction of 50% of target chemicals is possible over the next five years; the Congressional Office of Technology Assessment (OTA) also estimates that a 50% reduction is possible over the next five years. Meanwhile, the Chemical Manufacturers' Association reports that total waste generation declined 11% between 1981 and 1986, with the generation of hazardous solid waste dropping 56% during this period. Differences in calculated percentiles vary due to definitions employed, changes in productivity, and many other factors.

Most current waste minimization estimates are based on case studies of waste minimization techniques applied at individual facilities. EPA has evaluated more than 100 case studies, about 30% of which involved chemical plants. Other major industry groups included were plating, textiles, electronics, and printing. Waste minimization estimates developed are shown in Table I. The greatest potential for waste minimization occurs in the metal products cleaning industry, where reductions of 30–48% are possible. The lowest occurs in the petroleum products industry with a potential for waste reduction at 12–30%.

The percentages shown in Table I are minimum ranges, since they are based only on source reduction techniques. Other techniques such as offsite recycling or use of a waste exchange further reduce the need for disposal capacity.

III. Waste Reduction Methods

As defined by the hierarchy of hazardous waste management options (Table II), elimination or reduction of the waste at its source is more desirable than end-of-pipe treatment. Source reduction followed by recycling, therefore, are the options of choice. Added regulations, increasing treatment and disposal expenses, and increased liability costs have stimulated the following: (1) initiation of waste reduction audits which are designed to identify waste elimination, reduction, and recycling opportunities; assess and select appropriate techniques; implement and monitor a waste reduction program; (2) initiation of research and development (R&D) programs on non-waste and low waste technologies; and (3) expansion of recycling and recovery waste exchanges.

Table I Estimates of Waste Reduction Potential for Various Industries

Industry	Potential waste reduction (%)
Electroplating	20–48
Metal finishing	18–33
Metal products cleaning	30–48
Paint application	28–43
Paint manufacturing	18–33
Petroleum products	12–30
Printed circuit boards	18–48
Wood processing	13–40

Source: U.S. EPA, "Waste Minimization Issues and Options," EPA/530-SW-86-041, October 1986.

Table II Waste Management Hierarchy

Source reduction	
Waste avoidance	Don't make it.
Waste abatement	If you have to make it, minimize its volume and toxicity.
Waste Recycling	
Waste reuse	See if you or someone else can use it.
Waste recovery	If it can't be used as is, reclaim as much as possible that is useful.
Waste treatment	Treat what can't be reclaimed to render it safe.
Waste Disposal	Dispose of residues to air, water, or land.

A. Source Reduction

Source reduction is any activity that reduces or eliminates the generation of hazardous waste within a process. Source reduction approaches can be summarized as follows:

Process technology and equipment changes: This involves changes in the basic technology and equipment of production. Such reduction may be achieved by making changes in equipment, adopting a different way to make a chemical, or modernizing plant production processes to be more efficient. Plastic media paint stripping as a substitute for solvent stripping is an example.

Changes in plant operation: This involves better plant management or housekeeping to significantly reduce waste. Examples of operation changes include better handling of materials to reduce fugitive releases and improvements in preventive maintenance. Other examples include employee training, management initiatives, inventory control, waste stream segregation, and spill and leak prevention.

Process input changes: This involves changes in

raw materials used in the manufacturing process. Examples include using water instead of organic solvents or using materials with lower levels of contaminants. Replacement of cyanide in electroplating solutions is another example.

End product changes: This involves changes in design, composition, or specifications of end products. This approach is dependent on the restrictions imposed on the product by the customer or by performance specifications. Reformulating a product in the form of pellets rather than as a dust can reduce waste generation during packaging.

B. Waste Recycling

Waste recycling capitalizes on the fact that contaminants in many industrial process waste streams are misplaced raw materials, products, or by-products that can be recovered and reused. The value of using the waste as feedstock for the same or other processes has been recognized with increasing frequency by waste generators and independent reclaimers. Recycling reduces waste generation and can provide revenues to offset costs of waste disposal and plant operation.

Recycling is defined as those practices in which wastes are reclaimed or reused. Reclamation involves the processing or enhancement of waste to make it suitable for a subsequent use or to recover specific constituents or other marketable products (e.g., thermal energy). Waste reuse involves the use of untreated waste as feedstock or for other beneficial purposes. Waste recycling can take place onsite or offsite.

1. Onsite Recycling

It is most effective to recover waste at the point of generation where risks due to contamination and transportation and handling are minimized. In-process reclamation is the most widely used onsite recycling practice and is used extensively in such industrial operations as metal finishing, photographic processing, and dry cleaning. In many cases, a feedback treatment/recovery loop is installed where recycling takes place automatically.

Generic examples of direct use of waste include countercurrent washing and cleaning, use of alkaline waste streams to neutralize acidic wastes or precipitate heavy metals from metal-bearing wastes, and incineration of solvent sludges and organic wastes for heat recovery.

Onsite recovery of silver from hospital and photographic wastes is now commonplace with the recovered silver being sold to commercial reclaimers. Solvents are the most often recycled major class of industrial wastes. Solvents are commonly used in many industrial applications including dry cleaning, cold cleaning, solvent extraction, and vapor degreasing, and in the production of coatings, stains, wood treatment chemicals, printing inks, and agricultural chemicals. Most applications of solvents result in generation of solvent wastes. Solvent wastes include contaminated or spent solvents from process applications as well as off-specification batches of products containing solvents. Among the solvents that are recycled in significant volumes are xylene, toluene, acetone, methyl ethyl ketone, 1,1,1-trichloroethane, methylene chloride, methanol, perchloroethylene, trichloroethylene, methyl isobutyl ketone, and isobutanol.

2. Offsite Recycling

Offsite recycling may occur when recovery equipment is not available onsite, quantities generated are too low for effective recovery, or if waste materials generated cannot be reused in the production process. Recovery methods used offsite are similar to those employed onsite. Offsite recycling involves transportation of the waste to commercial recyclers or to an end user for reclamation or reuse. A commercial recycler processes the waste and returns it to the generator or sells the constituents recovered from it on the open market. Recycling of oils, solvents, electroplating sludges, lead-acid batteries, drums, and mercury represents the major waste sent offsite for recovery.

Incineration with heat recovery is a form of waste reuse suitable for large volume wastes with a high heat content. This can be affected either on or offsite. The heat released in incineration can be captured for electricity generation or for industrial use.

Because many types of waste oils and solvents exhibit properties of potentially good fuels, combustion of these wastes provides the double incentive of reduced steam generation costs, and solving an otherwise expensive waste disposal problem. Waste oil and spent solvent combustion in boilers also supports the philosophy of resource recovery. The term "waste oil" refers to used motor vehicle crankcase oils and spent machinery lubricating oils; whereas, "spent solvent" refers to a broad classification of waste liquid hydrocarbons used by a large group of industries. Generally only nonhalogenated solvents are suitable as waste fuels for industrial boilers.

3. Waste Exchanges

The major role of waste exchanges is to facilitate waste reuse by matching potential waste users with generators of the wastes. Hence, transfer and reuse of a waste product not usable by the generating industry or business but economically feasible for use by another is affected. As the costs of raw materials and landfill disposal rise, the diversion of potentially valuable waste items from solid and liquid waste streams becomes an economically viable alternative. The exchange and reuse of waste materials also saves energy in the processing of raw materials and in the reuse of energy-rich oil, solvent, and wood wastes.

Waste exchanges in North America play an important role in assisting waste generators in the identification of recycling and waste minimization opportunities as a means to comply with regulatory waste management regulations. There are two basic types of waste exchanges: information exchanges and material exchanges. Information exchanges act as clearinghouses for information to inform and contact waste generators and waste users through the use of publications (e.g., catalogs) and through other means of communication such as telephone, FAX, and computer networks. These exchanges will typically function as nonprofit operations. The material exchange differs from the clearinghouse in that it either operates as a brokerage service between waste generators and waste users or may take physical possession of the material and perform whatever functions (e.g., reprocessing, marketing, etc.) required for material marketing. Material exchanges will typically function as for-profit operations.

The waste exchange catalogs include waste materials that are "available" from and "wanted" by both privately owned firms and public agencies. Waste exchanges throughout North America have cooperatively established eleven standard categories for both "materials available" and "materials wanted." These eleven categories are: acids; alkalis; other inorganic chemicals; solvents; other organic chemicals; oils and waxes; plastics and rubber; textiles and leather; wood and paper; metal and metal sludges; and miscellaneous.

Information in the catalog is generally presented by a code containing only the quantity, description, availability, and general location of the waste. Use of a coding system maintains the confidentiality of the lister. Those interested in the waste may contact the waste exchange service which forwards inquiries to the lister. The lister, in turn, responds to the inquiries. If the potential for an exchange exists, subsequent negotiations between the two parties to complete the transfer take place without the involvement of the waste exchange service. This represents a passive exchange (compared to active) involving only the exchange of information.

A 1988 survey listed twenty-two waste information exchanges operating in North America. Twenty were operated on a nonprofit basis and only two for-profit. The survey assessed the effectiveness of six nonprofit waste exchanges operating and publishing waste exchange catalogs. These included: Northeast Industrial Waste Exchange, Inc. (NIWE); Canadian Waste Materials Exchange (CWME); Industrial Materials Waste Service (IMES); Great Lakes/Midwest Waste Exchange; Southeast Waste Exchange (SEWE); and Southern Waste Information Exchange (SWIX).

Results of the survey indicated a typical rate of transfer of "materials available" and "materials wanted" for a major clearinghouse type of waste exchange is conservatively estimated at between 10 and 30%. Successful transfer for North American exchanges of waste materials listed as available (not including those wanted) were therefore estimated between 800,000 to 2,400,000 metric tons. Besides the direct economic benefits incurred, equally important are the environmental benefits made possible by waste exchange operations which offer alternatives to the landfilling of wastes. It should be noted that European waste exchange organizations, established as early as 1972, typically have a 30–40% success rate in exchanging materials contained in their listings.

As the price of waste disposal has increased in recent years, so too have rewards for waste exchange participants. The Canadian Waste Materials Exchange claimed $1.4 million (Canadian dollars) in cost savings in fiscal year 1986; IMES reported $7 million during fiscal year 1985; NIWE claimed $1.7 million in fiscal year 1985; and the Southeast Waste Exchange indicated a $1 million saving for clients during fiscal year 1985–1986. Transfers were valued in terms of energy savings, decreased raw material requirements, and waste disposal costs.

Constraints on successful waste exchange include possible long transport distances between the generation and reuse points, and cost of waste purification prior to reuse. Factors which enhance waste exchange include the inherent value of the material, high concentration and purity, quantity and reliability of availability, and high offsetting costs for ultimate disposal.

Factors which must be considered before transfer is consummated include:

Technical feasibility—the matching of the chemical and physical properties of available waste streams with the specifications of raw materials to be replaced.
Economic feasibility—balancing of disposal costs foregone and raw material costs saved against the administrative and transport costs of implementing a waste transfer.
Institutional and marketing feasibility—guarantees of supply and anonymity; and mutual confidence among generator, user, and transfer agent.
Legal and regulatory feasibility—protection of confidentiality, legality, and unlikelihood of liability suits.

A major concern is potential liability under RCRA or the Comprehensive Environmental Response, Compensation, and Liability Act of 1980 (CERCLE) and the relatively vague and changing definition of solid wastes versus hazardous wastes. Active exchanges which tend to serve as middlemen appear to be most vulnerable.

IV. Incentives and Disincentives

A. Waste Reduction Incentives

There exist three primary incentives for source reduction or recycling opportunities: economics, legislation (current and future), and corporate image. The greatest incentive for generators to reduce their hazardous waste volume is the high and escalating cost of other forms of hazardous waste management. Land disposal, which once cost as little as $10 per ton of waste, now costs at least $240 per ton. Disposal sites with adequate capacity are in short supply, regulatory restrictions are becoming increasingly stringent, and prices continue to rise. Implementation of the "land ban" required by RCRA amendments will require many untreated wastes previously sent to landfills to be incinerated or otherwise treated at costs many times higher than those for land disposal. Other costs include waste storage expenses, transportation fees, administrative and reporting burdens, potential financial liabilities from accidental releases, and insurance (which may not even be available).

An added incentive becoming increasingly important is the role and weight of public opinion and pressure to reduce hazardous waste generation. This and public sensitivity for a safe and clean environment underscores the need for an environmentally conscientious good corporate image.

Federal and state economic incentives include the following options: awarding grants for information dissemination, waste audits, implementation of waste reduction measures and R&D projects; loan guarantees for approved waste reduction technologies; tax credits to encourage development of specific technologies; and state-issued loans as a possible substitute for loan guarantees. A more detailed delineation of waste minimization incentives follow.

Increases in the costs of hazardous waste management: EPA and state regulations have been the primary cause of increased costs in treatment, storage, and disposal of hazardous wastes, especially in relation to landfills, surface impoundments, and storage and accumulation tanks. The current series of land disposal restrictions under HSWA of RCRA will limit the number of untreated wastes that can be disposed of on land and thus are likely to increase the cost of disposal. HSWA also imposes more stringent standards on surface impoundments and hazardous waste storage tanks.

Difficulties in siting hazardous waste treatment, storage, and disposal capacity: Intense public opposition to the siting of many types of hazardous waste facilities including NIMBY, the "not in my backyard" syndrome, may cause shortages to persist even when market demand is strong. A generator's only alternative in many cases may therefore be a reliance on source reduction and onsite recycling to reduce the amount of waste that would otherwise be sent to offsite management facilities.

Permitting burdens and corrective action requirements: Even though the demand for new treatment and disposal capacity is high, permitting procedures will tend to delay and thereby increase the costs of all forms of treatment and disposal. No new hazardous waste management facility may be constructed until it has acquired a RCRA permit—a costly process that usually takes several years to complete.

Financial liability of hazardous waste generators: Generators using offsite treatment, storage, or disposal face financial liability for two reasons: (1) a potential for mismanagement of wastes by facility operators, and (2) the possibility of improper design of the disposal facility itself. When less waste is generated, it reduces potential liability for future disposal and thus is an incentive for both source reduction and onsite recycling.

Shortages of liability insurance: The traditional means for obtaining coverage for potential hazardous waste management liabilities is through insurance, but, for many generators and owners or operators of treatment, storage, and disposal facilities (TSDFs), liability insurance is no longer available, or is available only at an extremely high cost. In recent years, premiums have increased 50 to 300%. Increases in insurance costs or an inability to obtain insurance will inevitably result in higher treatment and disposal costs or the loss of available treatment or disposal capacity.

Public perception of company responsibilities: While the strongest incentives for implementing waste minimization techniques are probably economic, many companies are establishing waste minimization programs out of sensitivity to public concern over toxic chemicals. The desire to achieve a "good citizen" image is good public relations and makes good business sense.

B. Waste Reduction Disincentives

Disincentives for waste reduction and reuse include technical barriers, economic barriers, regulatory disincentives, traditional attitudes, and lack of information about developments.

Economic barriers: Even though waste minimization practices often lead to cost savings, availability of capital for plant modernization is often a significant obstacle to implementation. Although major companies may have sufficient access to upgrade inefficient processes, small and medium-sized companies often do not. Offsite recycling may be curtailed due to potential financial liability under CERCLA.

Technical barriers: By implementing waste reduction programs, some plants and companies have been able to significantly lower their waste generation. Achieving further waste reduction, however, becomes increasingly more difficult and less cost-effective as the more obvious and resource-effective possibilities are exhausted. Certain products simply cannot be manufactured without hazardous waste generation. Offsite recycling is often technically limited by process realities and administrative logistics. Off-specification chemicals for example offer little onsite or offsite recycling potential. Also industries producing high volume or high toxicity wastes often operate largely through batch processes which can present problems for certain onsite recycling techniques.

Regulatory disincentives: Some of the provisions of current environmental statutes, including RCRA, tend to discourage waste minimization. Stringent regulations governing hazardous material handling, transportation, and use and delays in permitting new technologies tend to promote end-of-pipe treatment and disposal and discourage resource recovery through reclamation and reuse. For example, once a material is classified as a hazardous waste, a plant may not be able to sell it to a reclaimer who may prefer not to get involved in the complex paperwork and handling procedures required to handle a "hazardous waste." Vagueness and changing definitions such as solid/hazardous waste and toxicity can profoundly hamper waste reduction and recycling efforts due to perceived liability concerns. Regulations need to be administered consistently and predictably and be flexible enough to encourage the use of methods that reduce waste generation.

Traditional attitudes: Reduction in generation of hazardous waste may be impeded because of a tendency in industry to select proven production technologies rather than alternatives that may generate less waste. Once a manufacturing process is operating, there may be an even greater reluctance to make major modifications that could affect reliability. There is a natural tendency to resist change, new concepts, and untried methods, and instead to modify existing processes, equipment, and procedures to which the industry is accustomed.

Lack of information on waste reduction measures or technologies: Perhaps the most significant technical barrier to waste minimization may often be a lack of suitable engineering information on source reduction and recycling techniques. Available information suggests that this is frequently the case with small- and medium-sized companies. Reducing the generation of hazardous waste gives a company a competitive advantage if waste management costs are a significant fraction of production costs. Thus, many firms are reluctant to release information about their waste reduction practices because doing so may help competitors.

A summary comparison of incentives and disincentives for source reduction, offsite recycling, and onsite recycling is provided in Table III.

Table III Incentives and Disincentives for Various Waste Minimization Techniques

	Source reduction	Offsite recycling	Onsite recycling
Incentives			
Increased cost of waste management	X	X	X
Difficulties in siting new HW management facilities	X		X
Permitting burdens and corrective action requirements	X	X	X
Financial liability of HW generators	X		X
Shortages of liability insurance	X		X
Public perception	X	X	X
Disincentives			
Economic barriers			
Lack of capital	X		X
Financial liability		X	
Technical barriers			
Attitudes toward unfamiliar methods	X	X	X
Batch processes			X
Lack of information	X	X	X
Technical limits of process	X		
Technical quality concerns	X		
Regulatory barriers			
Need to obtain TSD permit	X	X	X
Perceived stigma of managing HW		X	X
Revisions to other environmental permits	X	X	X

Source: "Report to Congress: Minimization of Hazardous Waste," prepared by the Office of Solid Waste, U.S. EPA, October, 1986.

V. International Waste Reduction Activities

A. United Nations

Some of the earliest initiatives in waste reduction came from international organizations. The United Nations Economic Commission for Europe (ECE), a UN subsidiary organization with headquarters in Geneva, sponsored the first International Conference on Non-Waste Technology in Paris in 1976. In 1979, the ECE adopted a detailed "Declaration on Low- and Non-Waste Technology and Reutilization and Recycling of Wastes," which recommended action on both the national and international levels to develop and promote low- and non-waste technologies

Low- and non-waste technology philosophy promotes the "do-it-cheaper-with-less-energy-and-with-less-waste" concept in the manufacturing process rather than the end-of-pipe abatement methods for protecting the environment. Emphasis is placed on both technological modifications that lead to the creation of less waste through greater physical efficiency of energy and materials use, and on those processes that convert waste products from conventional processes into new secondary projects (by-products) that can be reused.

Examples of activities include: design of products to minimize energy use and/or raw materials, either during manufacture, use, or disposal/recovery; design and operation of production processes to reduce energy and raw materials usage; design and location of manufacturing plants to facilitate the reuse of waste materials and energy from one plant by another; design and operation of waste handling processes to recover "residuals"; and design and operation of central "waste exchanges" to permit the wastes from one industry to be converted into raw materials for another or the same industry.

Resultant international ECE activities include: publication of a four-volume compendium on low- and non-waste technologies in 1982, listing over 80 examples of successful pollution prevention efforts by European industrial firms; publication of a compendium of lectures by experts in low- and non-waste technology in 1983; organization of a European Seminar on Clean Technologies at the Hague in 1980; forming a Working Party on Low- and Non-Waste Technology and Reutilization and Recycling of Wastes which has met annually since 1980; and the establishment of an Environmental Fund for demonstration of innovative technologies that are broadly applicable to reducing pollution. A sum of 6.5 million in European Currency Units (about 6.1 million U.S. dollars) was set aside for this purpose in 1985.

Congruent with these activities, the governing council of the United Nations Environment Programme (UNEP) set forth a firm direction for UNEP and its Industry and Environment Office (IEO) to serve as the brokers of a global information network on clean production technologies (also called low- and non-waste technologies). To promote cleaner production worldwide, a seminar on cleaner production was held in September 1990 in Canterbury, England. Some 150 invited speakers and participants were on hand from both developed and developing countries, to share pertinent experiences, policy issues, and technology transfer opportunities.

B. Western Europe

In Europe, much of the impetus for waste management comes from the efforts of the European Economic Community, which has frequently characterized waste minimization as a goal of member countries. The Third Environmental Action Program of 1981 emphasized the inclusion of environmental protection into all aspects of economic and social development. The program directed that the greatest emphasis be placed on prevention of waste generation, on designing products to facilitate recycling, and on waste management that progressively substitutes reuse for disposal.

European programs usually are concerned with a broad range of wastes. Wastes include both toxic and conventional pollutants—as well as nonhazardous solid wastes. Waste management authorities may have responsibility for only certain subsets of wastes, but agencies specifically directed to promote clean technologies deal with a wide variety of wastes. For example, the French National Agency for the Recovery and Disposal of Waste (ANRED) deals only with solid and RCRA-type hazardous wastes, but the French Mission for Clean Technologies deals with all types of pollution. Similarly, the Danish National Agency for Environmental Protection is divided into a large number of waste-specific units, but the Clean Technology Office researches reduction of all kinds of pollution. The Norwegians have taken a rather unique course of regulating by industrial sector, rather than by environmental medium. Thus, the Norwegian environmental regulations are multimedia in an attempt to avoid shifting of waste among environmental media.

To promote waste minimization, European governments have not relied on regulatory requirements to stimulate waste minimization but rather on economic measures. Their efforts have mainly taken the form of grants or loans to fund research on new low-waste technologies and tax incentives and disincentives to influence the actions of hazardous waste generators. Grant and loan programs for clean technology R&D, which have not been widely used in the United States, are a particularly common feature of European waste reduction efforts. Every Western European country active in waste reduction has had such a program in place at the national level for at least 15 years. Goals of the EEC for 1987–1992 focus on "clean technologies," multimedia action, economic incentives, product life-cycle considerations, education, and new regulatory measures.

C. Japan

Japan can probably be considered the world's premiere recycling country. This is due to its limited space and resources. The population of Japan is about half that of the United States, but its land area is approximately that of California. Forty-five percent of Japan's 119 million citizens live in three cities: Tokyo, Osaka, and Nagoya. Japan is also heavily dependent on imported raw materials including 98.8% of its oil and 99% of its iron ore. These factors combine to shape waste management policies in Japan and result in the implementation of many creative and sophisticated techniques.

Recycling is the "most desirable" waste management method in Japan. The Waste Disposal and Public Cleansing Law of 1970 specifically identifies recycling and reuse as the means to reduce wastes in Japan by emphasizing that the enterpriser must endeavor to lessen the amount of wastes by regeneration or reuse of wastes.

Extraordinary success has been experienced by this approach. it should particularly be noted that the Japanese define municipal solid waste as that remaining after recycling which requires treatment and disposal and does not include industrial waste. Hence recycled materials are considered as resources and not wastes. It is estimated that approximately 50% of all categories of municipal solid waste (by Western definition) is recycled. Data indicate that: 66% of bottles are used an average of 3 times;

42% of glass bottles are made from cullet (crushed glass); 51% of all paper is recovered; and 55% of steel is recycled. The importance and success of recycling in Japan may be attributable to: (1) a long history of recycling; (2) the country's enormous reliance on imported primary raw materials; (3) the need to control pollutants from landfills and incinerators; (4) government support; (5) a wide-ranging public education program; and (6) social factors.

In Japan, the market is the driving force for waste reduction although much attention is given to governmental/legislative initiatives and to the public image of the firm. In no case have performance standards been required by legislation. Rather, the Japanese rely on working out standards with firms at the local, or prefecture, level for the management of wastes in the specific area. The environmental and health and welfare agencies at the national level set general standards but then work through the prefectures for specific development and application of requirements.

The law makes firms responsible for whatever happens to their wastes. As in the United States, firms are responsible for the disposition of wastes through transportation and disposal. The Japanese make little use of litigation. Liability and insurance issues, therefore, are not nearly as important as in the United States. On the other hand, there is a state-financed victim compensation system for those who are injured through exposure to environmental contaminants. There exist far fewer personnel responsible for enforcement in Japan than in the United States since there is a far stronger impetus in the business culture for self-enforcement.

Japan does not use tax incentives for encouragement of waste handling practices. Such a system was attempted, but abandoned due to excessive red tape for large firms and technical complexities and management limitation for small firms. Financial assistance, however, is available for environmental projects through loans from the federal and prefectural levels.

In Japan, the Mnistry of International Trade and Industry (MITI) plays an important role in providing subsidies, financial assistance, and advice; while local governments are integrally involved in operating specific projects.

One of the largest waste reduction/recycling efforts is the Clean Japan Center (CJC), a research and development corporation formed in 1975 through joint government/private funding. Since its inception CJC has been building one new demonstration plant each year. Local governments typically participate in the building and running of a plant, and then buy it after the demonstration period. The CJC also has a large public relations program that works with local governments and citizens, collects and disseminates information about waste disposal and recycling, provides technical consultation services, and exchanges information with similar associations in other countries.

Recycling in Japan is also facilitated by "passive" waste exchanges in each prefecture. These handle both solid and hazardous wastes and claim a 40–60% success rate in some of the more industrialized areas.

D. Study of International Waste Minimization

Results of an EPA study focusing on government laws and programs affecting waste minimization practices and requirements in other countries including Japan, Canada, Germany, Sweden, the Netherlands, and Denmark are summarized in Table IV.

Different social milieus, infrastructures and geographical relationships, and business patterns make some European and Japanese policies and practices difficult to apply directly in the United States.

As in the United States, however, market factors are the principal force driving waste minimization decisions. These include cost of materials, cost of disposal, and cost of alternatives to disposal. Within the context of these market factors, the relative importance of other influences (i.e., legislation; regulation and government administrative activity; technical and financial support; and corporate public image) varies from country to country.

Commonalities in waste minimization strategy and efforts exist. All of the countries rely on cooperative, voluntary efforts. All of them stress the importance of low-pollution source reduction and recycling technologies, waste exchange, and information sharing. As in the United States, these countries operate on a two tier system: states, provinces, or prefectures deal directly with waste generators, while central governments provide direction and support. All countries surveyed in an EPA study of foreign waste reduction practices have rejected the notion of mandatory performance standards. Several countries have committed significant resources toward working with generators to reduce waste volumes.

Table IV Waste Minimization Practices in Other Countries

Practice	Japan	Canada	Germany	Sweden	Netherlands	Denmark
Tax incentives						
Waste end taxes			X		X	X
Tax incentives	X	X	X			
Economics						
Price support system for recycling			X		X	
Government grant as subsidy	X	X	X	X	X	X
Low interest loans	X		X			
Technical assistance						
Information and referral system	X	X	X		X	X
Site consultation	X	X				
Training seminars		X	X			X
R&D assistance						
Technical development labs				X	X	X
Demonstration projects	X	X	X	X		
Industrial research			X	X	X	
Permits and plans						
National waste management plans					X	
Waste reduction agreements	X					
Waste reduction as a part of permits				X		
Waste exchange						
Regional waste exchanges	X	X	X		X	X
Public information						
Focus on corporate image	X					
Focus on consumer practices			X			

Source: "Foreign Practices in Hazardous Waste Minimization" (Medford, MA: Center for Environmental Management, Tufts University, 1986).

VI. Waste Reduction Activities of the U.S. Government

A. Congress

Managing hazardous waste is an important national issue that Congress has addressed most recently by passage of the 1984 Hazardous and Solid Waste Amendments (HWSA) to the federal Resource Conversation and Recovery Act (RCRA) of 1976. Congress has declared waste minimization (i.e., source reduction and recycling) to be national policy. Supportive studies directed to this policy have been affected by the Office of Technology Assessment (OTA). OTA is an analytical arm of the U.S. Congress whose basic function is to help legislators anticipate and plan for the positive and negative impacts of technological changes.

OTA's first report on hazardous waste in 1983, *Technologies and Management Strategies for Hazardous Waste Control,* was used by Congress to examine the environmental problems and high long-term cost of land disposal practices and the benefits and availability of alternative waste treatment technologies. Congress made substantial use of that analysis in its 1984 amendments to RCRA. By request of numerous committees of Congress, OTA undertook and successfully completed an evaluation of waste reduction in 1986. *Serious Reduction of Hazardous Wastes* examines what is meant by hazardous waste, waste reduction, and waste reduction technology. The report explores the meaning and consequences of giving primacy to waste reduction over waste management, and puts waste reduction into the context of industrial production and efficiency, recognizing the current constraints of the American economy. The range of policy options examined is intended to assist an extensive policy debate—similar in extent and importance to the energy efficiency debates of the past 15 years. Congress reconfirmed pollution prevention as national policy with passage of the Pollution Prevention Act of 1990.

B. Environmental Protection Agency

The U.S. EPA, in response to the HWSA mandate, has announced plans for a series of actions to facilitate source reduction and recycling. Activities involve a division of labor betwen EPA headquarters, the EPA regional offices, and state programs. Some of the initiatives are described here.

The EPA has established the Pollution Prevention Division (PPD) in the Office of Pollution Prevention and Toxics (OPPT). Recognizing that an effective response would require a broad multimedia approach, a Pollution Prevention Advisory Committee comprising senior agency managers was also created. The objective of PPD is to move the agency, states, industry, and society to incorporate pollution prevention into their regulatory programs, manufacturing activities, and attitudes.

The agency also created a Waste Minimization Branch within the Office of Research and Development to assure that the technological tools needed to make prevention happen were identified and made available.

Current EPA strategy is resultant from the 1990 Pollution Prevention Act. The EPA has implemented a "33/50 Program" which seeks voluntary reductions of 17 targeted high-risk industrial chemicals by 50% by the end of 1995. The Toxics Release Inventory (TRI) tracks toxic chemicals from 28,000 industrial facilities by public reporting. Future issues faced by EPA include promoting a lifecycle perspective, creating a pollution prevention ethic, and measuring pollution prevention.

Pollution Prevention Division (PPD): Four mechanisms are being employed by PPD to promote a crossmedia perspective and to integrate pollution prevention into the fabric of all aspects of EPA's programs. These include: (1) Publication of a Pollution Prevention Policy Statement articulating EPA's hierarchy for waste and emission management; (2) Review of existing statutes, regulations, guidelines, and policies to determine areas where waste reduction and recycling can be incorporated and to ensure that future regulations and statutes conform with EPA's Pollution Prevention Policy; (3) Examination of initiatives including regulatory flexibility and financial incentives which may be included as the basis of statutory provisions; and (4) Cooperation with the Office of Research and development (ORD), the Office of Solid Waste (OSW), and the Office of Federal Activities to assure that the federal procurement process and product specifications are geared toward fostering products and processes that incorporate pollution prevention when possible.

The Pollution Prevention Division also has a number of activities which are designed to foster development of new state and local programs which already exist. Chief among these are the development of regional pilot programs and administration of state multimedia grants.

Resource awards to seven EPA regions have been made to establish pilot multimedia pollution prevention programs. Also, awards of over 15.5 million dollars to more than 56 state organizations have been made to support the development of multimedia pollution prevention programs capable of working directly with the private sector, local governments, and other constituencies.

Office of Research and Development: The Office of Research and Development (ORD) has a number of activities underway to support EPA's pollution prevention program. A comprehensive research plan which focused on needed research has been completed. On-going initiatives are summarized here.

Waste Reduction Innovative Technology Evaluation (WRITE) This program is designed to identify, evaluate, and demonstrate new pollution prevention technologies. Cooperative agreements are being established with industry and academia to develop and test new technologies. ORD is currently focusing on the food, textile, lumber, printing, chemical, petroleum, fabricated metals, machinery, electronics, and transportation industries.

Waste Reduction Assessment Program (WRAP) This program is geared to encourage the use of waste minimization assessments by the industrial community. The office has developed a general assessment manual and is establishing training seminars and workshops. The manual will be field tested in several industries, and plans are underway to develop industry-specific manuals.

Waste Reductions Evaluations at Federal Sites (WREAFS) This program consists of a series of cooperative demonstration and evaluation projects at federal facilities. The office is conducting demonstrations and evaluations of pollution prevention technologies at several government sites and will apply the lessons learned to other federal sites as well as to industry.

Waste Reduction Institute for Scientists and Engineers (WRISE): The institute includes technical representatives from academia, industry, and government. These representatives will help further technology transfer, assist in citing and selecting demonstration projects, and provide assistance in identifying areas for further research among other activities.

Pollution Prevention Information Clearinghouse (PPIC) A national waste minimization clearinghouse, PPIC has been established to foster the transfer of policy, program, technical, and economic information. Elements include an electronic bulletin board, a newsletter, a hotline, and other outreach mechanisms. This activity is currently under PPD.

Office of Solid Waste (OSW): During 1986–87 the OSW conducted a survey of hazardous waste facilities in which questionnaires were sent to a sample of commercial and private facilities that store, treat, recycle, or dispose of hazardous wastes. One major objective of the survey was to develop a categorization of the types of treatment and recycling practices that hazardous waste facilities use based on their specific capacity and operating costs. The questionnaire was designed to elicit information on specific types of treatment and recycling practices, for example, solvent recovery systems including fractionation, batch still distillation, solvent extraction, and thin film evaporation. Results update and amplify the existing 1984 Directory of Commercial Hazardous Wastes Treatment and Recycling Facilities.

C. Department of Defense

In 1980, 1987, and 1989 the U.S. Department of Defense (DOD) issued policy directives instructing the armed services to significantly reduce hazardous waste generation. Major waste minimization programs are currently underway and/or planned by the services. The major waste generating commands within DOD have established 50% reduction goals in hazardous waste generation by 1992.

Army: The Army currently generates over 100,000 metric tons of hazardous waste annually, the majority of it coming from the industrial activities under the Army Material Command. Hazardous wastes include solvents, batteries, industrial sludge, and conventional ordinance wastes. Most installations have a hazardous waste minimization plan (and as of October 1988) some 30 Army installations have been audited for hazardous waste minimization. The Army HAZMin program requires a 50% reduction in hazardous wastes generated by the end of 1992.

Navy: Generating over 4 million metric tons of hazardous waste annually, the Navy has developed a waste minimization program that focuses on material substitution, process changes, recycling and treatment, and proper hazardous material control. In 1985, the Navy determined that 90% of its hazardous waste came from 20 major generators and decided to concentrate resources in these areas. Numerical reduction goals were established for these and other installations.

Air Force: The goal of the Air Force has been to reduce hazardous waste disposal by 10% each year for 5 years, beginning in fiscal year 1988. The Air Forces's hazardous waste minimization program, known as PACER REDUCE, studies and tracks minimization initiatives and the procedures for material acquisition. The Air Force also has been active in research and has developed an innovative method for stripping paints from aircraft using plastic beads rather than chemicals. The new technology has been adopted by private industry and is used by both the Navy and the Army for non-aircraft applications.

DOD's Depot Maintenance Community, which accounts for 80% of waste generated by DOD, has formed a joint environmental panel which is examining different possibilities for waste reduction, particularly in the areas of plating, coating, and stripping. Efforts focus on improving information transfer in pollution prevention technologies, establishing joint testing parameters for evaluating new technologies, and coordinating with EPA on a variety of initiatives.

DOD and EPA representatives continue to meet to discuss joint projects and information transfer within the federal community. DOD is developing a reporting system for waste minimization projects that will make use of existing EPA forms and will move forward to address the more difficult problem of developing simple but representative statistics of accomplishments that measure the value of toxicity as well as volume reductions.

D. Department of Energy

In 1989, the Department of Energy (DOE) committed itself to a 10-point multibillion dollar initiative designed to chart a new course for the department toward full accountability in the areas of environment, safety, and health. It is designed to demonstrate that DOE is committed to complying with the environmental laws of the United States and is capable of discharging its many responsibilities which include protecting public health and safety. DOE cites waste minimization and recycling as important components of this endeavor. The department indicates that waste minimization (the reduction in the generation of radioactive, hazardous, and mixed waste before treatment, storage, or disposal) is a legal requirement, an ethical responsibility, and often a financial benefit. DOE's intent is to make waste minimization a key factor, not only in process and facility modification, but also in the procurement of goods and services. The major new modernization goal of minimizing waste generation entails a significant research, development, demonstration, testing, and evaluation (RDDT&E) component within the Office of Environmental Restoration and Waste Management. The Office of Technology Development will manage the Office of Environmental Restoration and Waste Management's contribution to the design and demonstration of new processes to avoid the generation of waste-containing hazardous constituents. Equipment used in waste processing will be designed for cleaning with nonhazardous substances and/or to yield a nonhazardous product.

Recycling is another major initiative. Contaminated metals from decontamination and decommissioning operations and decommissioned facilities may be used in noncritical structural or shielding applications for DOE high level radioactive waste (HLW) disposal and monitored retrievable storage facilities. Facilities will be designed to recycle the large volume of waste process water they generate rather than use it "once through."

Complex-wide strategies will address issues such as the management of mixed (radioactive and chemically hazardous) wastes, regional treatment facilities, and intersite recycling and processing of recovered materials. DOE's strategy for managing mixed wastes is to minimize their generation, treat the hazardous constituents of those that must be generated in accordance with the EPA's Land Disposal Restrictions, and dispose of only those that have been treated to attain the least hazard possible in a manner appropriate for the radioactive constituents.

Latest DOE policy affirms the environmental management hierarchy and promotes use of life-cycle cost analyses. A Waste Minimization (WMin) Crosscut Plan has been initiated, which establishes a Waste Minimization and Pollution Prevention Executive Board.

VII. State Efforts to Promote Waste Reduction

State programs which actively and directly promote waste reduction and recycling increased from two in the 1981 to thirty-six in 1988 with an additional six programs conducted by universities and private organizations. Most activities fall under the following general categories:

Technical assistance, including waste audits, seminars, process design consultation, and information clearinghouses;
Education projects, which publicize and promote waste minimization with pamphlets, conferences, and awards;
Economic incentives, which provide industry with grants, tax benefits, and loans, or which levy fees and taxes;
Waste exchange programs, which bring together waste generators with possible users of their wastes;
Research and development efforts, which develop new waste minimization techniques and industrial processes; and
Regulatory requirements, which mandate waste minimization or make other options less attractive.

Table V summarizes a survey of state waste minimization programs conducted for the National Governor's Association (NGA). The most common components of state waste minimization programs are technical assistance and education activities, with twenty-nine states engaged in each of these activities. On the average, most programs are budgeted at $150,000 and involve one to two staff people. It should be noted that state waste minimization programs continue to change, reflecting the influence of state legislation, federal programs, and individual experiences. Thus, any summary is quickly

Table V State Waste Minimization Programs (as of October 1988)[a]

States	Program	Technical assistance	Education	Economic incentives	Waste exchanges	Research/ development	Regulatory requirements
Alabama	Y	Y	Y	N	N	N	N
Alaska	Y	Y	Y	N	N	N	N
Arizona	N	B	Y	N	Y	Y	Y
Arkansas	Y	Y	Y	Y	N	N	N
California	Y	Y	Y	Y	Y	Y	N
Colorado	N	—	—	—	—	—	—
Connecticut	Y	Y	Y	Y	Y	Y	Y
Delaware	Y	N	N	Y	N	N	Y
Florida	Y	Y	Y	N	Y	N	N
Georgia	Y	Y	N	N	N	N	N
Hawaii	N	—	—	—	—	—	—
Idaho	N	—	—	—	—	—	—
Illinois	Y	Y	Y	Y	Y	Y	B
Indiana	Y	Y	Y	N	Y	N	N
Iowa	Y	Y	Y	N	N	N	N
Kansas	Y	N	Y	Y	N	N	N
Kentucky	Y	Y	Y	Y	N	Y	Y
Louisiana	Y	N	N	Y	N	Y	Y
Maine	N	—	—	—	—	—	—
Maryland	Y	Y	Y	Y	Y	N	Y
Massachusetts	Y	Y	Y	N	N	N	N
Michigan	Y	Y	Y	Y	Y	N	N
Minnesota	Y	Y	Y	Y	Y	Y	N
Mississippi	N	—	—	—	—	—	—
Missouri	Y	Y	Y	Y	Y	N	N
Montana	Y	Y	Y	B	Y	B	B
Nebraska	N	—	—	—	—	—	—
Nevada	N	—	—	—	—	—	—
New Hampshire	Y	B	B	Y	Y	B	B
New Jersey	Y	O	Y	O	Y	O	O
New Mexico	N	—	—	—	—	—	—
New York	Y	Y	Y	Y	Y	N	Y
North Carolina	Y	Y	Y	Y	Y	Y	N
North Dakota	N	—	—	—	—	—	—
Ohio	Y	Y	Y	B	Y	Y	Y
Oklahoma	Y	Y	Y	Y	N	N	N
Oregon	Y	Y	Y	Y	N	N	N
Pennsylvania	Y	Y	Y	Y	Y	Y	Y
Rhode Island	Y	Y	O	Y	Y	Y	N
South Carolina	Y	Y	N	N	Y	N	N
South Dakota	N	—	—	—	—	—	—
Tennessee	Y	Y	Y	Y	Y	Y	N
Texas	Y	O	O	Y	Y	N	N
Utah	N	—	—	—	—	—	—
Vermont	N	—	—	—	—	—	—
Virginia	Y	Y	Y	N	O	N	N
Washington	Y	Y	Y	N	Y	Y	N
West Virginia	N	—	—	—	—	—	—
Wisconsin	Y	Y	Y	Y	Y	Y	N
Wyoming	N	—	—	—	—	—	—

Source: National Governors' Association, compiled from "NGA Hazardous Waste Capacity Assurance Project: Summary of State Hazardous Waste Planning Activities," prepared by DPRA, May 1988; "State Hazardous Waste Minimization Programs," prepared for NGA by ICF, Incorporated, October, 1988; and "Survey of State Hazardous Waste Minimization Programs" for the Delaware Department of Natural Resources by Kathy Stiller, August, 1988.

Note: Component figures supplied only for the 36 states that have adopted a waste minimization program.

[a] Y = Yes; N = No; O = Pending; B = Blank.

outdated. The National Roundtable of State Reduction Programs has been instrumental in providing an effective forum for allowing an annual update of state programs and congressional and federal activities dealing with waste minimization. A summary of several of the more established state programs is included here.

The Pollution Prevention Pays Program in North Carolina is a nonregulatory state-funded program that began in 1983. The program is the lead agency for waste minimization in the state, and works cooperatively with the Division of Environmental Management, Hazardous Waste Management Branch, Department of Human Resources and the Governor's Waste Management Board. This multimedia program has been the model for many other states. The program includes an information clearinghouse, on-site technical assistance, informational technical assistance in the form of waste-stream-specific waste minimization reports, public education and outreach, challenge grants, research and education, and a cooperative Governor's Award. The program operates on a budget of $550,000 and is staffed with three professionals.

The clearinghouse lists more than 2,500 documents, personal contacts, and case studies on waste reduction. In addition to providing documents, clearinghouse staff conduct customized literature searches and prepare individualized reports for business and communities detailing cost-effective waste reduction options. The program also publishes more than fifty waste reduction reports.

The Minnesota Technical Assistance Program (MnTAP) offers assistance in hazardous waste compliance and reduction through the University of Minnesota. The staff provides direct assistance to generators and an outreach program that includes seminars and workshops. MnTAP also sponsors an engineering intern program through which engineering students are placed in industry to assist in waste management and provide technical expertise. EPA provided $100,000 to MnTAP in 1987–1988 for research projects dealing with small quantity generators and waste reduction. Other cooperative programs for matching grants and a Governor's Award are implemented by the Minnesota Waste Management Board. The MnTAP Program was operated on a fiscal 1988 budget of $260,000 and has four professional staff members with approximately six or seven engineering interns per year.

The New York waste minimization program is administered by both the New York State Department of Environmental Conservation and the Environmental Facilities Corporation (EFC). The EFC waste minimization program is one of the oldest in the nation (founded in 1983). Its activities include a waste exchange, on-site technical assistance to industry, an information clearinghouse, a quarterly newsletter, and low-cost loans to industry for improvements in waste management. The waste minimization program is part of the EFC's Industrial Materials Recycling Program, which has a budget of $495,000 and supports a staff of six.

VIII. Public Interest Organizations

Public interest organizations, which are essential for the functioning of a democracy, play a leading role in promoting the institutionalization of waste and toxics reduction. They introduce citizens' interests and concerns of citizens not included in the policymaking and program planning methodology of established corporate and political structures. Through incorporation of values outside of traditional technical analysis, public interest groups inform and expand public debate. Public interest groups continue to be largely responsible for exploring public and private policy prospects for waste reduction and have been instrumental in promoting its adoption as policy and practice. These groups play important and diverse roles in promoting a more balanced and comprehensive environmental and socioeconomic evaluation of the costs and benefits of waste reduction. This is evidenced by the work performed by organization such as the League of Women Voters, INFORM, Pollution Probe Foundation (Canada), The Environmental Defense Fund, The Citizen Clearinghouse for Hazardous Wastes, The Institute for Local Self-Reliance, and others. A description of several of the more prominent public interest groups promoting waste reduction are briefly described in the following.

The Environmental Defense Fund (EDF) assumes an active role in promoting waste reduction strategy and technology transfer through education. Of all the large environmental groups, EDF alone produces a publication *Approaches to Source Reduction: Practical Guidance from Existing Policies and Programs* which details benefits from wastes reduction and provides a source of information on strategies that states have taken to provide incentives or actively promote wastes reduction.

The National Campaign against Toxic Hazards (NCATH) provides technical assistance, organized training, media outreach, and legal assistance to state organizations, most notably the citizen action affiliates. It also undertakes research and prepares model legislation. The distinctive nature of NCATH is that expertise is directed toward grassroots citizen organizations. NCATH, for example, provided effective leadership in winning passage of the $9 billion Superfund reauthorization. NCATH orchestrated the "Superdrive for Superfund," where samples were collected from 200 toxic waste sites and 2 million signatures were obtained on petitions from citizens demanding a strong Superfund cleanup program.

The Environmental and Energy Study Institute (EESI) is a well-known and respected public interest group on Capitol Hill. EESI provides information to legislators and engages in drafting environmental and energy-oriented legislation. In contributing to the implementation of waste reduction and clean technology, it works to facilitate the identification of concrete programs, specific tasks, and details arising from policy debate. Legislators use this information to evaluate what legislative provisions are necessary to implement waste reduction and clean technology transfer.

INFORM is a public interest organization based in New York that has played an analytical role in informing the debate on waste reduction. It also performs an educational role in the preparation of waste reduction literature and research tools in several states. INFORM's most noted activity in promoting waste reduction are its publication of *Cutting Chemical Wastes: What 29 Organic Chemical Plants Are Doing to Reduce Hazardous Wastes* and *Garbage Management in Japan—Leading the Way.*

The Local Government Commission (LGC) is based in California and is organized to assist local governments in protecting the health, safety, and welfare of their constituents. LGC has adopted waste reduction as one of the issues by which they motivate local governments to assume a leadership role. They provide educational and outreach services, technical assistance, and regulatory advice. The LGC presents workshops to local government officials interested in encouraging waste reduction and identifies the incentives they can provide to assist local industries to adopt waste reduction strategies. The workshops provide experienced instruction on how to design local waste reduction programs and where to obtain resources to provide technical assistance.

Bibliography

Forcella, Dominic. (1989). "Volume I: The Role of Waste Minimization." Center for Policy Research, National Governors' Association, Washington, D.C.

Freeman, H. M., ed. (1988). "Standard Handbook of Hazardous Waste Treatment and Disposal." McGraw-Hill, NY.

Ghassemi, M. Waste reduction: An overview. *The Environmental Professional,* **11,** 2.

Hershkowitz, A., and Salerni, E. (1987). "Garbage Management in Japan." INFORM, Inc., NY.

ICF Consulting Associates, Inc. (1985). "Economic Incentives for the Reduction of Hazardous Wastes." Prepared for the California Department of Health Services.

Jones, E. B., Herndon, R. C., and Moerlins, J. E. (1989). "An Assessment of the Effectiveness of North American Waste Exchanges in 1988." Presented to North American Conference on Waste Exchange, San Antonio, Texas.

Martin, L. R. (1989). Demanding waste reduction: The roles of public interest organizations in promoting the institutionalization of waste and toxics reductions. *The Environmental Professional* **11,** 2.

National Research Council. (1985). "Reducing Hazardous Waste Generation: An Evaluation and Call for Action." National Academy Press, Washington, D.C.

Sarokin, D. J., Muir, W. R., Miller, C. G., and Sperber, S. R. (1985). "Cutting Chemical Wastes: What 29 Organic Chemical Plants Are Doing to Reduce Hazardous Wastes." INFORM, Inc., NY.

U.S. Congress, Office of Technology Assessment. (1986). "Serious Reduction of Hazardous Waste for Pollution Prevention and Industrial Efficiency." OTA-ITE-317, Government Printing Office, Washington, D.C.

U.S. Environmental Protection Agency. (1986). "Report to Congress: Minimization of Hazardous Waste." U.S. EPA, Washington, D.C.

Herbicides

Lennart Torstensson
Swedish University of Agricultural Sciences

Glossary

Active ingredient Active part of a formulated herbicide product.

ADI Acceptable daily intake of an herbicide during a lifetime which appears to be without appreciable risk on the basis of all facts known at the time. It is usually expressed as mg of the chemical per kg of body weight.

Formulated product Liquid concentrate or solid particles containing the active ingredient of an herbicide as well as other substances which enhance its effect.

LD_{50} Dose (given as mg per kg body weight) that kills 50% of the test animals.

Persistence Resistence to deomposition.

Pesticide Includes different types of pest control products (e.g., herbicides, fungicides, insecticides, rodenticides).

Potentiation The possibility that the toxic effect of an herbicide could be enhanced by another component in the formulation, by another pesticide with which it is used, or by other chemicals in the environment.

***HERBICIDES** are chemical substances used to control unwanted vegetation (e.g., weeds) in agricultural and horticultural crops, shrubs in reforestations, or all plant growth on areas like industrial sites, railway embankments, and garden pathways. The herbicides constitute a large and very diverse group of organic chemicals. It is estimated that 150 or more substances with herbicide activity, appearing in still more formulated products, are commercially available. The intention is that the herbicide only affects the unwanted plant species (i.e., by design interferes with plant systems) and should be nontoxic to mammals or other organisms. However, herbicides are not always that selective in their actions and may occasionally cause damage to nontarget organisms. They may also contribute to the general increase in the pollution of air, soil, and water to our planet.*

I. History and Classification

A. Use of Herbicides

The problem of weeds and their control has been present since the early days of agriculture. The simple methods of the preagricultural revolution era were replaced by the concept of crop rotation and the prophylactic measures embodied in "good husbandry." The development of fertilizers, tractor power, and mechanization further increased the ability to reduce crop/weed competition. It was the discovery of the organic herbicides, however, which truly presented the farmer, horticulturist, and forester with the power to control weeds in crop or noncrop situations. In early stages of development, herbicides had a broad range of effects. Later, it become possible to create substances with selective phytotoxicity. Today, the sophistication is such that it is possible to control broad-leaved weeds in narrow-leaved crops, broad-leaved weeds in broad-leaved crops, narrow-leaved weeds in narrow-leaved crops, or narrow-leaved weeds in broad-leaved crops.

B. Classification

Herbicides may be classified in a variety of ways, including by chemical group, selectivity, nature of action, or application characteristics. A number of herbicide formulations are common.

1. Chemical Groups

Herbicides can be classified on the basis of their chemical affinities (Table I). Herbicides of the same chemical group tend to have similar physiological characteristics and it may be possible to predict how new compounds of the group may act. However, minor differences in chemical structure may lead to considerable differences in selectivity.

2. Selectivity

Total or nonselective herbicides are those which kill all vegetation, whereas selective compounds will control some plants while leaving other species unharmed. Nonselective herbicides (e.g., sodium chlorate) are used to eliminate weeds from industrial sites, railway tracks, paths, etc. Selective compounds (e.g., 2,4-D and MCPA) control weeds without adversely affecting the growth of the crop. Selectivity may be due to differences in the retention, uptake, movement, metabolism, or biochemical action of the herbicide in crops and weeds.

3. Nature of Action

Contact herbicides affect only those parts of the plant to which they are applied, whereas translocated or systemic substances can move within the plant to regions remote from the point of application. Contact substances, such as paraquat, tend to be relatively phytotoxic, acting on the membrane systems of the leaf tissues and inhibiting photosynthetic and respiratory metabolism. Conversely, phloem-translocated herbicides (e.g., 2,4-D and MCPA) tend to act relatively slowly and are translocated to the regions of metabolic activity. Normally, root-absorbed compounds are transported in the xylem to the shoots.

4. Application Characteristics

Herbicides may be classed as soil- or foliage-applied. Soil-applied compounds are normally absorbed by the roots or emerging shoots and transported to the shoot in the transpiration flow. Foliage-applied compounds penetrate the outer waxy cuticule and are absorbed into the leaf tissue, where they may or may not be translocated basipetally in the phloem. Soil-applied herbicides normally are of relatively low water solubility and ideally remain in a discrete band at the soil surface. They are normally absorbed by the root or emerging shoots of weeds and may be translocated in the xylem to the shoots; many of them, such as urea and triazine compounds, act on the light reactions of photosynthesis. Soil-applied herbicides may be subject to a number of degradation mechanisms in the soil and only a portion of the applied dose may be available for absorption by the plant.

Crops and weeds have usually both emerged at the time of spraying a foliage-applied herbicide and its efficiency and selectivity thus depend on the efficiency of spray retention, cuticule penetration, tissue absorption, translocation, and metabolism or fixation/binding at nonactive sites.

The timing of application is normally described in relation to the stage of crop development, such as presowing, preemergence, or postemergence treatment. Selectivity of an herbicide may depend partly on differences in the rooting depth of the crop and weed. It may also involve differences in herbicide retention, leaf absorption, translocation, or metabolism; differences in the suceptibility of the ultimate enzyme target sites may also play a role.

5. Herbicide Formulations

The active part of an herbicide, known as the active ingredient (a.i.), is generally formulated by a manufacturer as a liquid concentrate or in solid particles. The former has to be diluted by the user, generally with water, and many herbicides are formulated in such a way as to be readily dispersible in water. Compounds which are water-insoluble are generally formulated as emulsifiable concentrates or wetable or flowable powders (Table II).

Many herbicide formulations contain adjuvants which enhance their leaf retention and cuticule penetration. Surfactants or surface-active agents are used extensively for this purpose. There are three classes of surfactants—cationic, anionic, and nonionic, compounds belonging to the last two groups are most commonly used as herbicide adjuvants.

II. Appearance in the Environment

To use herbicides in the most efficient manner (e.g., in effective economic weed removal), it is necessary to have detailed knowledge of the appearance and presistence of the substances in the environment. This knowledge is also urgently needed to avoid the spread of pollution outside the treated areas, as well as to prevent damage to the environment. The con-

Table I Classification of Herbicides According to Chemical Group

Chemical group	Examples
Amides	Diphenamide (*N,N*-dimethyldiphenylacetamide) Propyzamide (3,5-dichloro-*N*-(1,1-dimethyl-propynyl)benzamide)
Anilides	Flamprop-isopropyl (isopropyl-*N*-benzoyl-*N*-(3-chloro-4-fluorophenyl)-D,L-alaninate) Metazachlor (2-chloro-*N*-(pyrazol-1-ylmethyl) acet-2′,6′-xylidide) Propachlor (2-chloro-*N*-isopropylacetanilide) Propanil (3′,4′-dichloropropionanilide)
Aromatic acids	Dicamba (3,6-dichloro-*o*-anisic acid) 2,3,6-TBA (2,3,6-trichlorobenzoic acid)
Carbamates	Asulam (methyl-4-aminophenylsulfonyl-carbamate) Phenmedipham (methyl-3-(3-methylcarbaniloyloxy)carbanilate) Propham (isopropyl phenylcarbamate)
Haloalkanoic acids	Dalupon (2,2-dichloropropionic acid) TCA (CCl_3COONa)
Heterocyclic nitrogen compounds	
Triazines	Atrazine (2-chloro-4-methylamino-6-isopropylamino-1,3,5-triazine) Metamitron (4-amino-4,5-dihydro-3-methyl-6-phenyl-1,2,4-triazine-5-one) Metribuzin (4-amino-6-*t*-butyl-4,5-dihydro-3-methylthio-1,2,4-triazine-5-one) Simazine (2-chloro-4,6-bis(ethylamino)-1,3,5-triazine) Terbutylazine (2-*t*-butylamino-4-chloro-6-ethylamino-1,3,5-triazine)
Pyridazines	Chloridazone (5-amino-4-chloro-2-phenyl-pyridazin-3(2*H*)one)
Pyridines	Diquat (1,1′-ethylene-2,2′-bipyridyldiylium dibromide) Paraquat (1,1′-dimethyl-4-4′-bipyridylium dichloride) Picloram (4-amino-3,5,6-trichloropyridine-2-carboxylic acid) Triclopyr (3,5,6-trichloro-2-pyridyloxyacetic acid)
Pyrimidines (uracils)	Bromacil (5-bromo-3-*t*-butyl-6-methyl-uracil)
Unclassified	Aminotriazole (1*H*-1,2,4-triazol-3-ylamine) Benazolin (4-chloro-2-oxobenzothiazolin-3-ylacetic acid) Bentazon (3-isopropyl-1*H*-benzo-2,1,3-thiadiazin-4-one 2,2-dioxide) Endothal (7-oxabicyclo(2.2.1)heptane-2,3-dicarboxylic acid) Ethofumesate ((±)-2-ethoxy-2,3-dihydro-3,3-dimethylbenzofuran-5-yl methanesulfanate
Inorganic compounds	Sodium chlorate ($NaC10_3$)
Nitriles	Bromoxynil (3,5-dibromo-4-hydroxybenzonitrile) Dichlobenil (2,6-dichlorobenzonitrile) Ioxynil (4-hydroxy-3,5-iodobenzonitrile)
Nitroanilines	Trifluralin (α,α,α-trifluoromethyl-2,6-dinitro-*N,N*-dipropyl-*p*-toluidine) Pendimethalin (N-(1-ethylpropyl)-2,6-dinitro-3,4-xylidine))
Nitrophenols	Dinoseb (2-sec-butyl-4,6-dinitrophenol) DNOC (4,6-dinitro-*o*-cresol)
Nitrophenyl ethers	Nitrofen (2,4-dichlorophenyl-4-nitrophenyl ether) Oxyfluorfen (2-chloro-α,α,α-trifluro-*p*-toluyl-3-ethoxy-4-nitrophenyl ether)
Organoarsenic compounds	Cacodylic acid (dimethylarsinic acid)
Organophosphorus compounds	Fosamin-ammonium (ammonium ethylcarbamoylphosphonate) Glufosinate (ammonium D,L-homoalamin-4-yl methylphosphinate) Glyphosate (*N*-(phosphonomethyl)glycine)
Phenoxyalkanoic acids	
Phenoxyacetic acids	2,4-D ((2,4-dichlorophenoxy)acetic acid) MCPA ((4-chloro-2-methylphenoxy)acetic acid) 2,4,5-T ((2,4,5-trichlorophenoxy)acetic acid)
Phenoxybutyric acids	2,4-DB (4-(2,4-dichlorphenoxy)butyric acid) MCPB (4-(4-chloro-*o*-toluyloxy)butyric acid)

(*continues*)

Table I (*Continued*)

Chemical group	Examples
Phenoxypropionic acid	Dichlorprop (2-(2,4-dichlorophenoxy)propionic acid) Fluazifop-butyl (butyl 2-(4-(5-trifluoromethyl-2-pyridyloxy)phenoxy)propionate) Mecoprop (2-(4-chloro-*o*-tolyloxy)propionic acid) Napropamid (*N,N*-diethyl-2-(1-napthyloxy) propionamide)
Phenyl ureas	Diuron (3-(3,4-dichlorophenyl)-1,1-dimethylurea Isoproturon (3-(4-isopropylphenyl)-1,1-dimethylurea Linuron (3-(3,4-dichlorophenyl)-1 methoxy-1-methylurea)
Sulfonyl ureas	Chlorsulfuron (1-(2-chlorophenylsulfonyl)-3-(4-methoxy-6-methyl-1,3,5-triazin-2-yl) urea) Sulfometuron methyl (methyl 2-(3-(4,6-dimethylpyrimidin-2-yl)ureidosulfonyl) benzoate)
Thiocarbamates	Diallate (*S*-2,3-dichloroally diisopropylthiocarbamate) EPTC (*S*-ethyl dipropylthiocarbamate) Triallate (S-2,3,3-trichloroallyl diisopropylthiocarbamate)

tamination by herbicides and other pesticides of surface and groundwaters is discussed below as is the transport routes of herbicides in the environment.

A. Surface and Groundwaters

A number of different pesticides has been found as contaminants in both surface and groundwaters. In surface waters, like creeks and streams, a total of 18 pesticides—11 herbicides, 2 fungicides, and 5 insecticides, have been identified in agricultural areas of Sweden. The most frequently found substances were the herbicides atrazine, bentazone, dichlorprop, MCPA, and mecoprop. The largest amounts of the phenoxy acids were found during the spraying season in amounts peaking at 25 μg/l.

Around 50 compounds have been found in groundwater in more than 20 states of the United States. Of these, about one-fifth are herbicides like atrazine, bentazon, bromacil, cyanazine, 2,4-D, dinoseb, metolachlor, metribuzine, and simazine.

Contamination of waters by herbicides and other pesticides has been frequently reported from many countries during recent years. The amounts found have generally not exceeded toxic levels, but are an illustration of the serious problem caused by general chemical contamination of our environment.

B. Transport Routes

A number of mechanisms and factors influence the transport of herbicides in the environment. From the place of application, a substance may move above the ground, in the growth zone and further into the atmosphere, on the soil surface, or within the ground, both in the nonsaturated and the saturated zone. In all zones, the mobility of water is of great importance for the transport of chemicals. In addi-

Table II Classification of Herbicide Formulation

Formulation	Characteristics
Dry preparation	Pelleted, granulated, or microgranulated goods, powder, smoke, and gaseous preparations, etc.
Fluid preparation	Solvent, emulsifiable and flowable concentrate, aerosols, etc.
Emulsifiable concentrate (EC)	A fluent oil solution giving a stable milky homogeneous fluid, emulsion.
Soluble powder (SP)	Easily gives a clear solution.
Wettable powder (WP)	Holds the herbicide in a "filler" such as clay and a dispersing agent which produces a fine suspension of solid particles when water is added.
Flowable powder (FP)	Gives a fine-grained, turbid suspension of particles.
Flowable or suspension concentrate (FWC, SC)	Homogeneous, fluent dredge of a very fine (micronized) flowable powder.
Water dispersible granule (GW)	Granules dredged and dispersed in water.
Aerosol	A solvent in a pressurized pack, which gives a cloud of very small droplets.

tion to transport, loss of herbicides from a certain zone is also dependent on decomposition processes as well as incorportion into soil organic matter.

Transport of herbicides in the environment occurs through diffusion, through the influence of wind and water, or through mobility of solid material. During transport, the herbicide may occur as a gas or in solid phase, be dissolved in water, or absorbed to particles of different kinds. The rate of the transport, as well as that of different decomposition processes, is influenced by a great number of factors (Fig. 1). A survey of possible transport routes that may result in the appearance of herbicides in surface and groundwater is given in Figure 2.

1. Transport by Wind

a. Evaporation

Evaporation of an herbicide from the soil surface depends on the vapor pressure of the substance and varies with temperature, water solubility, and adsorption. As the vapor pressure is an equilibrium property yet an equilibrium with the atmosphere is impossible, the rate of evaporation is determined by the wind and the surface area where the substance is distributed. At rising temperatures, the vapor pressure increases.

After a spraying operation, at least part of the herbicide is dissolved in available water. Because of this, the distribution of the compound between water and air is also of interest. It is particularly important for substances with both a low vapor pressure and a low water solubility.

Examples of herbicides known to evaporate after application on bare soil are trifluralin (80% within 2–3 days) and chlorpropham (49% within 50 days).

b. Wind Drift

Wind drift of herbicides in connection with spraying operations is a well-known problem. It has caused damage to sensitive plants on other fields, in greenhouses, and in gardens. It also contributes to the general pollution of the agricultural landscape.

Wind drift (even slight winds of 1–3 m/s) may lead to minor amounts of the herbicide (often 0.1% or less) detected hundreds of meters from the sprayed field. In stronger winds, the wind drift may become more obvious.

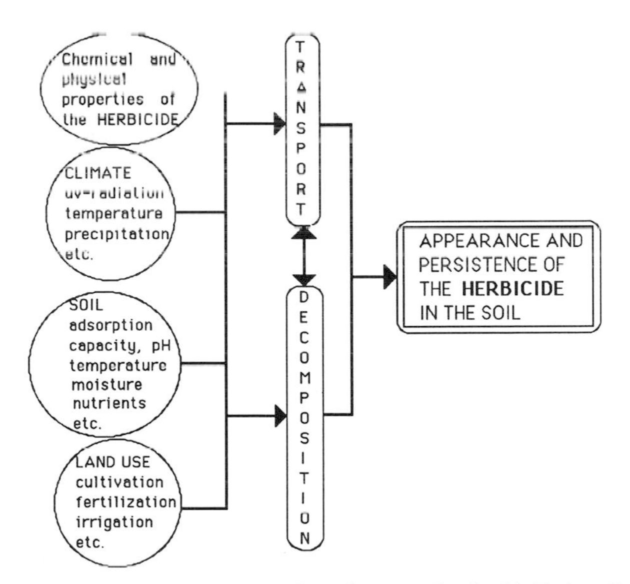

Figure 1 The appearance and persistence of an herbicide in soil depends on its transport and decomposition, mechanisms influenced by a great number of factors.

2. Transport on the Ground Surface

Transport of herbicides on the ground surface may occur in flowing water or by the wind. In water, the slope and ground porosity, as well as the intensity and/or continuity of the water supply through precipitation or irrigation, is of importance. The adsorption properties of the herbicide are also of great importance. With low adsorption, the substance easily penetrates the ground and the risk of loss by surface run-off is little. With strong adsorption, a major part becomes bound to particles at the soil surface. Those particles may then be transported by wind or water over short or long distances. The losses by surface water run-off vary between less than 0.1 to 10% or more, depending upon the slope of the area and precipitation.

Soil particles with a diameter of more than 0.06 mm are normally transported only a few meters by the wind. Indeed, even particles in the range of 0.002 mm fall down relatively soon after the wind has enveloped them. However, smaller particles, including adsorbed herbicides, may be transported by wind over very long distances.

3. Transport within the Soil Profile

a. Diffusion

Diffusion is a process by which a chemical substance (i.e., an herbicide) tends to obtain an equal distribution in a given space. It implies a transport from areas of high concentrations to those of lower concentration. The rate of diffusion depends on the herbicide's concentration and adsorbtion properties

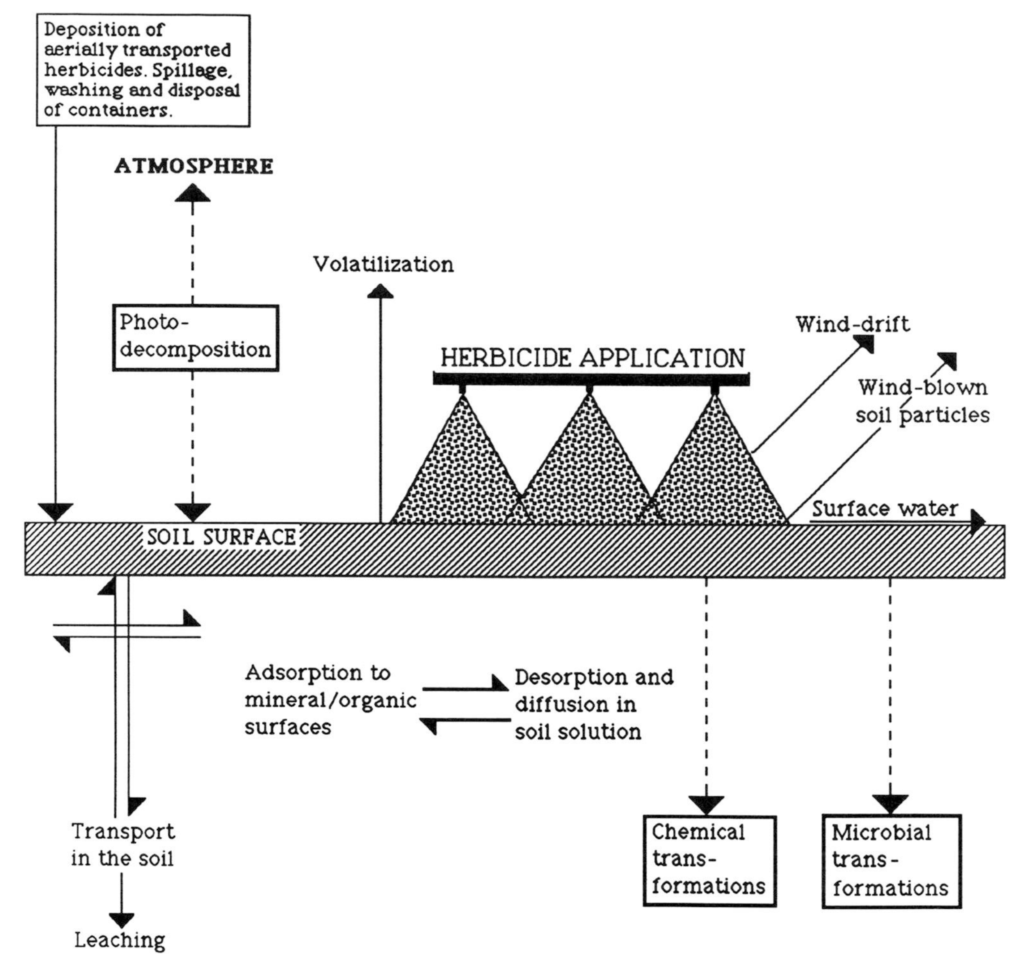

Figure 2 Entry of herbicides into the soil environment. Movements of herbicides, →; agencies affecting the herbicides - - >. [From Torstensson, L. (1988). Microbial decomposition of herbicides in the soil. *In* "Outlook on Agriculture," Vol. 17. pp. 120–124. Pergamon, Oxford. Reproduced with permission.]

as well as the soil's temperature, porosity, and water content. Diffusion in the gaseous phase is 10,000 times faster that in the water phase.

b. Transport of Particles

Herbicides adsorbed to soil particles move with them and may thus very rapidly penetrate cracks and channels in the soil. As a result, immobile herbicides may thus, in minute amounts, penetrate deep into the soil profile.

c. Root Exudates

Herbicides taken up by plant foliage can be transported through the roots and thereby be exudated. This transport is quite rapid and exudation may occur within hours of application. Herbicides are transported through the topsoil (that is normally a barrier to immobile substances) down to the subsoil. Exudated herbicides can, therefore (in some cases) be taken up by "untreated" plants which have their roots in the vicinity.

d. Transport by Water

The most important route of herbicide transport in the soil is with the soil water. The water solubility and adsorption properties of the herbicide, together with the amounts of mobile water, determine the extent of the transport. The transport is generally downward or in some cases occurs horizontally, during dry conditions it can be upward.

It is of particular interest to be able to predict the mobility in soil of an herbicide as a function of adsorption and water flow through the soil. Both TCA and glyphosate are soluble in water. TCA is weakly adsorbed to soil while glyphosate is strongly adsorbed. Therefire, TCA is very mobile in the soil, while glyphosate is nearly immobile.

Thus, the potential mobility of herbicides can be compared by determining their adsorption to soil. Soil, water, and the herbicide are mixed in a test tube and allowed to equilibrate. Then the amounts of herbicide adsorbed to the soil and in solution are determined. The results are evaluated by the Freundlich's equation:

$$\log x/m = \log K_f + n \log c,$$

where x/m is the amount of herbicide adsorbed to the soil (μg/g soil), c is the amount in water (μg/ml water), K_f is the adsorption coefficient (the amount of herbicide adsorbed per gram of soil in equilibration with 1 μg per ml water), and n is the slope of the line. Examples of K_f values for linuron in some soils are given in Table III. The different K_f values indicate that linuron is adsorbed differently; if the values are ≤ 2, there is a great risk of mobility of linuron in that soil.

The main adsorbents of herbicides in soil are clay minerals, organic matter, iron, and aluminum hydroxides. For many herbicides, there is a strong correlation between K_f and content of organic carbon. This has been used to compare the adsorbtion strength of different herbicides. The K_f value is divided by the content of organic carbon in each soil and multiplied by 100. The adsorbtion constant (K_{oc}) for linuron on the basis of data presented in Table III is 540 ± 21%.

Table III K_f and K_{oc} Values for Linuron in Some Soils

Organic C (%)	Clay (%)	K_f	K_{oc}
1.39	4	9.0	648
1.86	10	11	591
2.49	25	16	643
3.65	32	25	685
11.6	7	73	629
0.83	5	2.9	349
1.28	7	4.9	381
1.97	9	6.3	320
4.64	5	21	453
8.62	33	50	580
6.35	15	36	567
0.59	ND[a]	3.3	559
0.31	ND	1.8	581
0.24	ND	1.3	542
0.07	ND	0.4	571
			$\overline{X}$ = 540 ± 21%

[a] ND, not determined.

Table IV Expected Mobility for Some Herbicides Based on Their K_{oc} Values

K_{oc} value	Expected mobility	Examples
0–50	Very high	Dalapon, dicamba, TBA, TCA, picloram, hexazinone, chloramben, chlorsulfuron, sulfometuron methyl
50–150	High	Atrazine, chlorthiamide, 2,4-D, MCPA
150–500	Medium	EPTC, simazine
500–2000	Low	Diuron, linuron
2000–5000	Very low	Chloroxuron, trifluralin
>5000	Immobile	Paraquat

Table IV gives examples of herbicides within different K_{oc} intervals that are expressions for expected mobility. However, as with the linuron example, if the substance is applied on soils with a low content of adsorbing materials, the risk of mobility may nevertheless be high. This is also the case for other normally strongly adsorbed substances used on areas low in adsorbents.

III. Persistence

Persistence of an herbicide is a reflection of its resistance to decomposition. The persistence time is the time required for the compound to be degraded to below the detection limit of chemical analysis; the "chemical persistence" time. Persistence time can also be defined as the time required for a sensitive organism to no longer react to the chemical. Often, cultivated plants are used to determine the persistence time for herbicides used in agriculture (i.e., the phytotoxic persistence time). This is of practical importance in farming where it defines the answers the period of time after usage of a certain herbicide before a new sensitive crop can be sown. For many herbicides, there is a considerable difference between the chemical persistence time and the phytotoxic persistence time (see Table V).

Table V Chemical and Phytotoxic Persistences for Some Herbicides

Herbicide	Chemical persistence	Phytotoxic persistence
MCPA	1–4 weeks	1–4 weeks
TCA	6–12 months	6–12 months
Glyphosate	6–18 months	1–2 weeks
Diquat	Years	<1 week in a mineral soil

A number of factors influences the persistence time for a particular herbicide (see Fig. 1). Transport discussed previously. Decomposition may be of abiotic and/or biotic nature.

A. Abiotic Decomposition

1. Photochemical

Sunlight consists of several types of radiation, among them ultra violet radiation or UV light. Photochemical decomposition of the majority of herbicides is mediated by UV light under experimental conditions, the energy of UV radiation is capable of breaking the bonds in herbicide molecules.

There is sufficient evidence to suggest that photodecomposition also occurs under field conditions and that a considerable proportion of some herbicides may be transformed by this mechanism. Since UV light is incapable of penetrating soil, this type of decomposition only occurs when sunlight directly hits the molecules, (e.g., on the soil surface, on plant surfaces, and in the atmosphere).

A number of herbicides are known to be partly degraded by photochemical reactions (e.g., phenoxy acids, triazines and phenyl ureas). However, the decomposition is not complete but gives products similar to those occurring in other types of decomposition.

2. Chemical Decomposition

In soil and water, chemical reactions may change herbicide molecules. Such reactions have been proposed to be the main mechanism for the degradation of only a few herbicides. However, the importance of chemical reactions should not be underestimated, as they may also contribute essential stages to the principal biological routes of degradation.

Probably the two most important factors governing chemical transformation in soil are moisture and pH, although other reagents may also be involved. The rates of the reactions are highly influenced by pH. This is demonstrated for the sulfonyl ureas that hydrolyze rapidly at pH <6 but very slowly at pH >7. The rate of the reactions may sometimes increase if the herbicide is adsorbed to clay (e.g., for the triazines).

Free radical reactions may also degrade herbicides. The biota in the soil may partly be responsible for the generation of these free radicals (e.g., via hydrogen peroxide produced by microbial extracellular oxidase enzymes). Chemical decomposition, as well as photochemical decomposition, thus lead to products that may be further degraded by biological reactions.

B. Biotic Decomposition

1. Organisms Involved

Decomposition of the main part of applied herbicides is carried out by the soil microorganisms. Decomposition also occurs in plants and animals. Apart from decomposition in soil mediated by cell-bound enzymes, there are also enzymes bound to soil particles. These enzymes are able to catalyze the breakage of bonds in herbicide molecules. The enzymes have been exudated from living microorganisms or plant roots or are released from dead microorganisms, soil animals, or plant roots.

Microorganisms are able to degrade a wide variety of chemicals, ranging from simple polysaccharides, amino acids, proteins, lipids, etc. to more complex materials such as plant residues, waxes, and rubbers. Without the degradative capacity of microorganisms, such organic compounds would accumulate and pollute the environment. Microorganisms are also capable of degrading synthetic chemical compounds. A variety of mechanisms underlies these activities and a number of factors influences rates and routes of the transformations.

2. Mechanisms of Microbial Decomposition

Microbial metabolic activities require energy. Most organic materials can serve as a source of energy for at least some microorganisms. Another environmental attribute of microbial metabolism is their ability to adapt through induction or mutation, particularly toward chemicals that are initially toxic to them.

An important criterion used in classifying the enzymatic reactions involved in microbial transformations of herbicides (Table VI) is whether or not

Table VI Classification of Microbial Activities in Connection with Decomposition of Herbicides

Mechanism	Consequences
Enzymatic reactions	
Direct degradation of herbicides in central metabolism of microorganisms in which the herbicides serve as energy sources to supply growth (catabolism) and where adaptation phenomena appear.	Repeated application of an herbicide to the same field results in faster decomposition.
Incidental transformation of herbicides by microorganisms via peripheral metabolic processes in the absence of the perfect coordination of the process which is characteristic in central metabolism (cometabolism).	All herbicides may be decomposed by this mechanism.
Incidental transformation of herbicides by extracellular enzymes.	All herbicides may be decomposed by this mechanism.
Nonenzymatic reactions	
Normal activities of soil-living microorganisms such as change of pH or generation of free radicals or different reactive substances	Influence on biological and nonbiological reactions resulting in transformation of herbicides in soil.

the microorganisms derive energy from the process. This consideration is significant from a practical viewpoint, since it is of importance in predicting an herbicide's persistence. Thus, catabolic metabolism requires a favorable chemical structure of the herbicide which allows it to be utilized as a carbon and energy source.

If the microorganisms do not derive energy from the transformation of an herbicide (Table VI, cometabolism), they depend upon other more accessible carbon sources, which increase their general metabolic activities. Thus, the rate of microbial metabolism can be controlled by changing either the amount of herbicide or other added carbon source, depending upon the type of microbial degradation activity. In general, cometabolism is the prevalent form of microbial metabolism when the amount of herbicide is low in comparison with other carbon sources.

In addition to enzymatic reactions, microorganisms may contribute to the transformation of herbicides via nonenzymatic reactions (Table VI). Large pH changes (1 to 2 pH units or more) are often associated with such microbial activities together with changes in nutritional sources. There are ways in which microbial products can promote photochemical reactions. First, microbial products can act as photosensitizer's by absorbing energy from light and transmitting it to the herbicidal molecule. Second, microbial products can facilitate photochemical reactions by serving as the frequently essential donors or acceptors (e.g., of hydrogen and OH^-) for photochemical reactions.

a. Metabolism and Cometabolism

The microbial degradation associated with a given herbicide can be enhanced by the rate and frequency of its application (Fig. 3). This phenomenon is observed and has been found to have practical consequences for the persistence of herbicides such as 2,4-D, dalapon, EPTC, MCPA, and TCA. A loss of

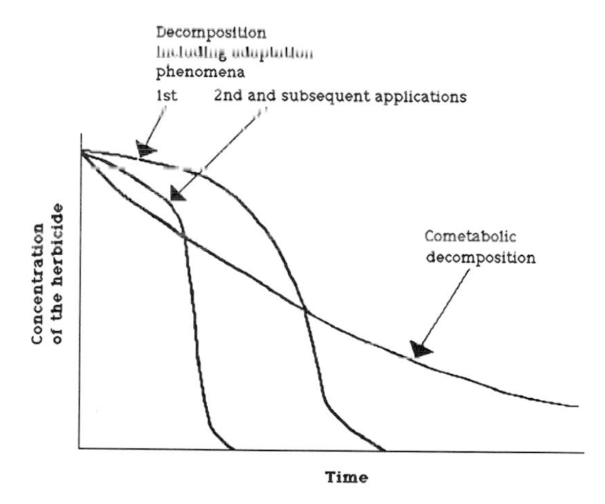

Figure 3 Principal differences between microbial cometabolic decomposition of herbicides and decomposition where adaptation to the herbicide occurs. [From Torstensson, L. (1988). Microbial decomposition of herbicides in the soil. *In* "Outlook on Agriculture," Vol. 17, pp. 120–124. Pergamon, Oxford. Reproduced with permission.]

application efficiency of certain soil-applied herbicides may result through the development of microbial populations capable of rapidly degrading the substances. The reduced pesticide efficiency differs from the separate problem that arises when the pests themselves develop resistance to chemical treatment.

The term *cometabolism* has no succint definition and has been used in several different ways. In studies of herbicide decomposition, it is often necessary to distinguish between degradation by catabolism in connection with adaptation phenomena and that caused by other metabolic transformations. The use of the term *cometabolism* for these transformations has resulted from the focus of many investigators on the microbial interactions with several substances and on the interrelation between their metabolism. Therefore, the use of the term might be justified.

From a practical viewpoint it is of interest that adaptation phenomena do not appear during cometabolic decomposition of herbicides. This means that repeated applications of the substance have no influence on the rate of its disappearance from the soil (Fig. 3).

b. Rate of Degradation

The rate of microbial decomposition of herbicides in soil is a function of three variables: (i) the availability of the chemical to the microorganisms or enzyme systems which can degrade it; (ii) the quantity of these microorganisms or enzyme systems; and (iii) the relative activities of these organisms or enzyme systems. Edaphic factors, such as contents of organic matter and clay, moisture level, temperature, pH, aeration, and nutrient status, are of importance as moderators and driving factors.

The degree to which an herbicide is decomposed in soil is determined by the adsorption/desorption characteristics of the substance and the simultaneous occurrence of the substance and microorganisms at the same site. The distribution of soil microorganisms is not uniform throughout the soil profile. The organism density is highest in the uppermost part of the soil and declines with depth. This means that if an herbicide is mobile and passes through the topsoil layer into the subsoil, the chance of its becoming microbially degraded is considerably reduced. Additionally, within the topsoil there are differences in microorganism distribution with higher densities occurring in the rhizosphere.

The role of microbial biomass in the transformation of organic matter in soil is crucial and the rate of turnover and mineralization of organic substrate are largely governed by the activity of the soil biomass. Inhibition of microbial activity by a low or high temperature, drought, waterlogging, extremes of pH, or xenobiotic substances may result in the persistence in soil of potentially decomposable and mineralizable compounds such as herbicides.

From a number of studies, we know that herbicides may be transformed by a single organism, but we also know that the normal degradation of native organic matter proceeds in a sequence of metabolic steps carried out be a wide variety of microorganisms. With an increasing biomass in soil, the chances also increase that the microorganisms synthesizing the "right" enzymes are present in a number that forms the basis for a high rate of decomposition of herbicides and other chemicals in the soil (Fig. 4).

IV. Toxicology

A. Toxicity of Herbicides

The toxicity to mammals of any particular herbicide can readily be determined and compared with that of other herbicides in appropriate tests on laboratory animals. However, a proper evaluation of any toxic hazards that a compound might present to humans requires that additional factors be taken into account. There is a need to investigate all potentially adverse effects found during routine toxicological investigations and to undertake such additional studies as are appropriate to assess the significance of these effects.

Since every herbicide is on some occasion likely to come into contact with the skin or be ingested, toxicity data are required on oral ingestion and skin application. Similarly, inhalation data must be furnished when there is danger of uptake by the respiratory tract. Repeated use may also be adequately covered in the toxicological study.

To define acute and short-term hazards resulting from use, to classify, and to assist in the design of longer-term tests, all herbicides must be assessed by certain primary toxicological studies. The single-dose toxicity by appropriate routes of administration is established with an LD_{50} value for at least two common laboratory species, preferably including a rodent and a non rodent. Examples of typical LD_{50} values are given in Table VII.

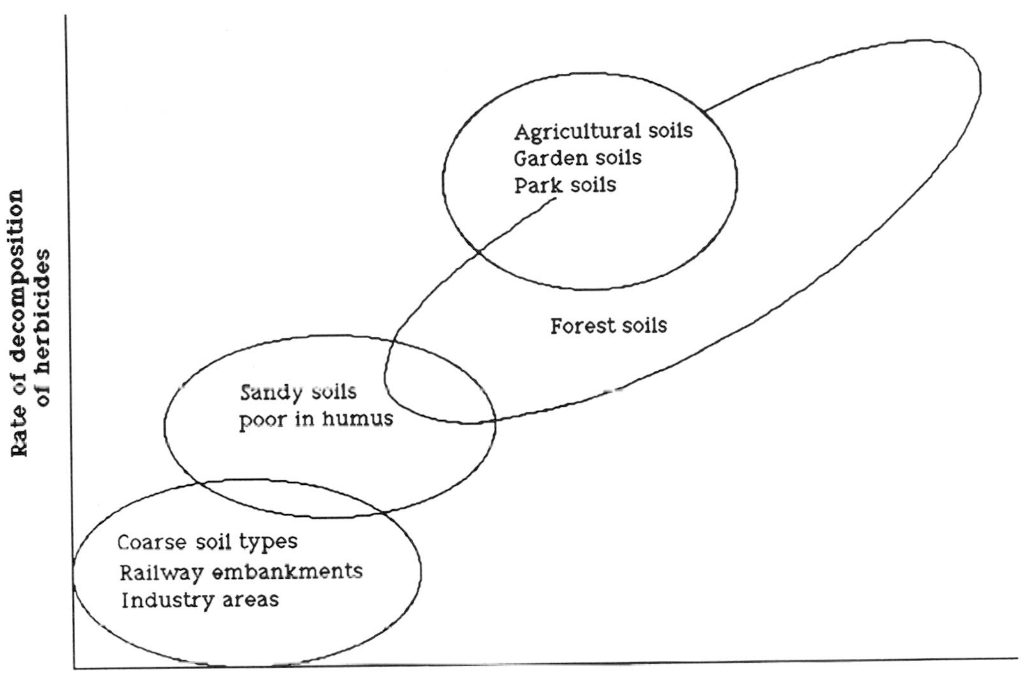

Figure 4 The importance of soil biological activity for rate of decomposition of herbicides in the same soil. [From Torstensson, L. (1988). Microbial decomposition of herbicides in the soil. *In* "Outlook on Agriculture," Vol. 17, pp. 120–124. Pergamon, Oxford. Reproduced with permission.]

Table VII LD_{50} Values for Some Herbicides Established in Rat or Rabbit

Herbicide	LD_{50} Value (mg/kg body weight)	
	Oral	Dermal[a]
Chlorsulfuron	5550–6300	>3400 (R)
Dicamba	1700	>2000 (R)
Dichlobenil	>3160	1350 (R)
Dinoseb	58	80–200
EPTC	2525	>5000 (R)
Glyphosate	4050	>5000 (R)
Hexazinone	1690	>5280 (R)
Linuron	4000	>5000
MCPA	700	>1000
Metribuzin	2200–2345	>20000
Phenmedipham	>8000	>4000
Simazine	>5000	>3100
Sulfometuron methyl	>5000	>2000 (R)
Terbuthylazine	2000	>3000
Triallate	1675–2165	8200
Trifluralin	>10000	>2000 (R)

[a] (R), rabbit; all others are for rat.

When primary toxicological studies are completed, the future studies needed for the toxicological assessment of the particular compound is planned. They include metabolic studies in animals, neurotoxicology, risk of potentiation, long-term (chronic) toxicity and carcinogenecity studies, mutagenicity tests for reproductive toxicity and embryotoxicity (including teratogenesis) as well as tests on immunological response and the enzyme system.

Finally, the results of the animal experiments must be extrapolated to humans and any information which gives an indication of human sensitivity to the herbicide under consideration is of value. This additional information is derived from epidemiological studies and health controls of workers in the pesticide chemical factories and in the field. It may be obtained from the investigations of incidents of accidental or deliberate overdosage. Volunteers have sometimes submitted themselves to trials in which accurately measured doses have been administered under conditions where strict biochemical monitoring of the effects is possible.

B. Exposure to Herbicides

1. Application

The potential hazard for occupational exposure to pesticides has been a concern for many years. In particular, acute toxicity has been the focus of occupational health programs. However, since most herbicides have low acute toxicity in comparison with many other pesticides, only a few occupational health studies have been made on them.

An increased awareness of long-term effects has reinforced the aim of reducing the uptake of any chemical. This does not mean that full protection is sought in every case. The degree of protection should be balanced against known suspected effects of the chemical used. Exposure studies provide a basis for the use of relevant protective equipment and alterations to working procedures.

Observations of actual work procedures must include all the various steps involved (e.g., handling the herbicide container, mixing, loading the tank, spraying, adjusting booms and cleaning nozzles, cleaning the spraying equipment, changing clothes, personal hygiene, eating and smoking). The subject must be followed continuously for the entire session to cover unexpected events. For instance, in an interview study of 240 farmers and spraymen, one-third said that they would clean the nozzles in the field by blowing through them with their mouth.

Since many herbicides are applied as sprays, it is popularly believed that the greatest risk of exposure is by inhalation. Herbicides can be airborne as vapors, aerosols, or dusts. Measurements of exposure in the breathing zone give the sum of what is inhaled and what is swallowed. In general, airborne exposure is small in comparison with dermal exposure. On the other hand, respiratory and oral exposure lead in most cases to a more efficient uptake than dermal exposure.

Determination of skin contamination is not in itself a way of assessing dermal absorption. However, controlled experimental volunteer studies in humans with phenoxy herbicides have shown that the dermal absorption is the main route of uptake.

It is obvious that people working with pesticides in the field will be exposed. In many cases the same worker mixes, loads, and sprays the herbicide. In other cases there are special mixers, loaders, spraymen, and flagmen, who are exposed to different extents. Another group of potentially exposed people are those entering the field shortly after spraying for thinning, picking, etc. Different methods of application give rise to different degrees of exposure. Thus tractor spraying usually exposes the driver more than does airplane spraying. Knapsack application frequently gives rise to leakage and an increased risk of dermal contamination. Other workers who are at risk of exposure are those involved in the manufacture of herbicides (i.e., production, formulating, and packaging staff).

2. Food and Water

Many publications and recommendations have been prepared by both national and international committees relating to the levels at which residues of herbicides may persist on crops to which they have been applied. Since herbicides are used to kill or restrict the growth of plants, problems of residues in the harvested crop will not often arise. Residues may sometimes persist in straw or fodder crops, and stock may have access to treated crops.

The maximum residue limits should not exceed those resulting from good agricultural practice and should not give rise to intakes exceeding the acceptable daily intake (ADI). On these criteria, good agricultural practice is defined as the officially recommended or authorized usage of herbicides under practical conditions at any stage of the production, storage, transport, distribution, and processing of food or other commodities. Also bearing in mind that variation in requirements within and between regions, and taking into account the minimum quantities necessary to achieve adequate control, herbicides must be applied in a manner that leaves a residue which is the smallest amount practicable and which is toxicologically acceptable.

Groundwater was once thought protected from chemicals applied on the soil's surface, but current evidence shows the presence of a great number of different chemicals, including herbicides, in the groundwater of many countries. The ability to detect increasingly lower concentrations of pesticides, increased knowledge of and concern about potential chronic effects, and the finding of low levels of pesticides in drinking water has made the public generally more suspicious of the beneficial aspects of pes-pesticide usage. However, from a toxicological point-of-view, the chemical residues can be handled in the same way as those in food.

3. Third Parties

The use of poisons for suicidal intent has led to use of the herbicide paraquat for this purpose. As a group, however, herbicides are not sufficiently toxic to be lethal, although the solvent may have lethal effects in some instances.

V. Side Effects on Flora and Fauna

To use herbicides effectively and economically, and to avoid unacceptable side effects in the environment, we have to know their appearance and persistence in the environment as discussed in Sections II and III. Most herbicides today have a good selectivity in their action against weed species. However, despite this selectivity, effects on nontarget organisms have been observed. This may involve direct toxic effects on sensitive species (e.g., certain plants and soil microorganisms) or indirect effects caused as a consequence of the weed treatment, (e.g., on insects and wildlife).

A. Soil Organisms

1. Microorganisms

Microorganisms usually occupy a volume of less than 0.1% of the soil, but are responsible for numerous transformations that cycle elements and energy in nature. However, population densities of bacteria may be quite high (10^9/g of soil is not uncommon) and the length of fungal hyphae may be some thousands of meters per gram of soil. Therefore, the biomass of microorganisms per hectare may reach several tons.

The diversity of species of microorganisms in soil is great. The microbial population exists in a dynamic equilibrium formed by interactions of abiotic and biotic factors that can be altered by modifying environmental conditions. If this equilibrium is disturbed by an herbicide, it may lead to consequences for the turnover of carbon and nutrients in the soil resulting in changes to primary production.

It can generally be stated that herbicides affects growth and other activities of soil microorganisms. This leads to changes in the composition of species of microflora and in interactions between different species. However, in the majority of cases where herbicides have been tested, the influence has been concluded to not result in any effect of appreciable importance for the function of the soil ecosystem.

In some cases, herbicides have contributed a more or less potent effect on several microorganisms or microbial processes in soil (e.g., organisms involved in nitrogen fixation, or nitrification, as well as mycorrhiza, and plant pathogens). Examples of such substances are amitrol, atrazine, diallate, dicamba, dinoseb, paraquat, simazine, and sodium chlorate.

2. Animals

About 10% of energy flow in soil is caused by soil animals. The microbial decomposition of organic matter is, in most cases, dependent on interactions with soil animals.

It is difficult to identify herbicide effects on the soil fauna in practical agriculture because herbicide application is only one component of a system of mangement. Some herbicides like atrazine, dalapon, dinoseb, linuron, and simazine may have a significant effect on the soil fauna, although several studies suggest that the effects are largely indirect. A decrease in the overall number of soil fauna species could primarily be caused by the destruction of the plant cover so that the soil fauna would be secondarily exposed to greater physical stress from extreme temperature and drought.

B. Insects

Insects in agricultural areas may carry out many activities such as causing crop damage (or conversely being "natural enemies" to the pest insects), pollinating a wide range of plants, or being food for many bird species. Herbicides generally have minor toxic effects on insects. The indirect effects are probably more important. Removal of weeds greatly decreases insect densities in small grains and other crops, hence some weed growth is essential if high insect densities are to be maintained.

C. Plants

The regular use of herbicides in agriculture since the end of 1940s has caused a decrease in the occurrence of weeds in fields, both in the number and diversity of species. Weed species that are sensitive to herbicides now represent a smaller portion of the weed population, while the numbers of insensitive species have increased, particularly grass weeds. Within certain weed species there has been observed a change toward more tolerant and/or resistant specimens/populations (Table VIII).

In the field corridors (e.g., the unplanted areas of the field) and in the immediate surroundings of the field, the use of herbicides and fertilizers has caused

Table VIII Increased Tolerance/Resistance in Weeds after Long Continuous Use of Herbicides

Herbicide	Weed	Tolerant/resistant
Haloalkanoic acids		
Dalapon	Quackgrass (*Elymus repens*)	tolerant
Phenoxyalkanoic acids		
2,4-D	Dandelion (*Taraxacum vulgare*)	tolerant
MCPA	False camomile (*Matricaria maritima* ssp. *inodora*)	tolerant
	Canada thistle (*Cirsium arvense*)	tolerant
Phenyl ureas		
Metoxuron	Annual bluegrass (*Poa annua*)	tolerant
Pyridines		
Paraquat	Annual bluegrass (*Poa annua*)	resistant
Triazines		
Atrazine	Allseed (*Chenopodium polyspermum*)	resistant
	Galinsoga (*Galinsoga ciliata*)	resistant
	Ragwort (*Senecio vulgaris*)	resistant
	Common lambsquarters (*Chenopodium album*)	resistant
	Speedwell (*Veronica persica*)	tolerant
	Common chickweed (*Stellaria media*)	resistant
	Spreading orache (*Atriplex patula*)	resistant
Simazine	Ragwort (*Senecio vulgaris*)	resistant
	Shepherd's purse (*Capsella bursa-pastoris*)	tolerant
	Common lambsquarters (*Chenopodium album*)	tolerant
	Annual bluegrass (*Poa annua*)	resistant

a trivialization of wild flora. These areas often have been islands with the few plant remnants from an older type of landscape.

Quite another problem involves what may happen if certain weeds are not removed from a grain crop. Different species of the fungus *Fusarium* may grow in the crop (including its weeds) on the field. Under certain conditions, the *Fusarium* spp. produce mycotoxins such as trichothecenes of different kinds. The trichothecenes are highly toxic and are a threat to major grain consumers like poultry and swine. A positive correlation has been found between the occurrence of quackgrass (*Elymus repens*) and concentrations of deoxynivalenol, one of the trichothecenes, in grain.

D. Wildlife

Herbicides have seldom been reported as directly toxic to wildlife. However, substances used for defoliation or other drastic changes of the vegetation cover may alter the possibilities for both mammals and birds to find food and cover. The structure of forest bird communities is strongly influenced by the successional stages of the vegetation. Herbicides can alter the successional stage of a given habitat and thus affect nesting birds. Several studies, on the effect of phenoxy acids (2,4-D, and 2,4,5-T) and glyphosate on bird communities have reported densities altered by spray treatment.

Field margins, defined as the unplanted areas of the field and its boundary, and the outermost edges of the crops themselves, are important habitats for many bird species, especially species of gamebirds. Partridges (*Perdix perdix* and *Alectoris rufa*) nest almost entirely in field boundaries. After hatching, the chicks of these species move into crops to feed, as do chicks of the pheasant (*Phasianus colchicus*). Chicks of grey partridges (*P. perdix*) in particular, feed in cereal fields and during the first 2–3 weeks of life their digestive systems are not well adapted to plant food. Therefore the presence of large numbers of insects, which at this crucial age form their staple diet, are vital to their survival. Increased herbicide use over the last four decades has removed the host plants of these insects and, more recently, the use of insecticides to control aphids has caused direct mortality of other insect species. The consequent reduction in the numbers of chick-food insects caused by the disruption of food chains has been a major factor in the decline of the wild grey partridge.

E. Aquatic Organisms

Certain herbicides are used against aquatic weeds. However, as was discussed in Section II, many other herbicides can also appear in water systems where they may cause effects on the aquatic flora and fauna. Herbicides with effects on photosynthetic systems have been toxic at the parts per billion level. Other types of herbicides may be toxic at the parts per million level.

VI. Regulations

As in so many factors which may be on the surface beneficial to humans (such as the use of herbicides to protect crops) hazards to human and animal health and the environment may also occur if their use is not properly controlled. This type of problem is not restricted to herbicides, but arises with all other chemicals encountered or used by humans in daily life.

Consequently, before an herbicide can be allowed to pass into general use, sufficient information must be available for a reasonable estimate to be made of its potential for good and harm. The correct measures can then be devised for its control and recommendations made on how surplus herbicides and containers can be safely disposed of with the aim of reducing the risks to the community and the environment.

Most countries regulate the sale and supply of herbicides and other pesticides and, for that purpose, call upon pesticide manufacturers to submit information on the effect and safe use of their products. Differences exist between countries on the extent and scope of these requirements. However, the type of information needed includes:

1. the formulation of the proprietary product;
2. identity of the active ingredient;
3. chemical properties of the active, pure, and technical ingredients;
4. physical properties of the active, pure, and technical ingredients;
5. intended uses and methods of application; and
6. experimental data on efficiency.

Data on herbicide residue including statements on principal residues (parent compound, breakdown products, and metabolites), and methods of residue analysis are also important particularly in edible crops, food, or feedstuffs. These must be supplemented by experimental data on toxicity in animals, including acute, short-term and long-term toxicity, carcinogenecity, neurotoxicity and reproductive studies, including teratogenicity and mutagenicity. Observations in humans and information on diagnosis and treatment as well as environmental and wildlife data must analyze degradation in soil and water, adsorption and mobility in soil, effects on soil flora and fauna, and toxicity to birds, fish and bees. Finally, conditions of safe disposal of surplus herbicides and herbicide containers are essential.

Related Articles: Dioxin, Health Effects; Organic Micropollutants in Lake Sediments; Peroxisomes; Pesticides and Food Safety.

Bibliography

Garner, W. Y., Honeycutt, R. C., and Nigg, H. N., eds. (1986). "Evaluation of Pesticides in Groundwater." ACS Symp. Ser. 315. Washington, D.C.

Grover, R., ed. (1988). "Environmental Chemistry of Herbicides." Vol. I. CRC Press, Boca Raton, Florida.

Hutson, D. H., and Roberts, T. R., eds. (1987). "Progress in Pesticide Biochemistry and Toxicology." Vol. 6. "Herbicides." Wiley, New York.

Racke, K. D., and Coats, J. R., eds. (1990). "Enhanced Biodegradation of Pesticides in the Environment." ACS Symp. Ser. 426. Washington, D.C.

Richardson, M., ed. (1986). "Toxic Hazard Assessment of Chemicals." The Royal Society of Chemistry, Burlington House, London.

Worthing, C. R., ed. (1987). "The Pesticide Manual: A World Compendium." 8th ed. The British Crop Protection Council. Lavenham, Suffolk.

Hydrogen Sulfide

Sheldon H. Roth
University of Calgary

Glossary

Anosmia Loss of the sense of smell due to damage of olfactory mucosa or nerve fibers.

Anoxia Lack of oxygen in circulating blood or tissues.

Asphyxiant Agent that prevents oxygen transportation to tissues using oxygen.

Conjunctivitis Inflammation of the connective coat or membrane continuous over the inside of the eyelid and outside of the eyeball.

Keratoconjunctivitis Inflammation of both the cornea and membrane lining the eyelids (conjunctiva).

Olfactory paralysis Transient or permanent loss of the sense of smell.

Toxicity Capacity of a substance to produce adverse effects in biologic systems.

HYDROGEN SULFIDE (H_2S) is an extremely toxic gas to humans and animals. It is both an environmental and industrial pollutant, colorless, flammable, heavier than air, and has a characteristic odor of "rotten eggs" at very low concentrations. At high concentrations the sense of smell is lost, and acute exposures can often be fatal. The gas occurs in the environment as a result of anaerobic decomposition of organic-rich material, and is a contaminant of natural gas, volcanic gases, and sulfur springs. It is produced as a by-product or utilized in a number of industrial processes and remains a major occupational health hazard, reported to be one of the leading causes of death in the workplace. There is a paucity of toxicological data, and the majority of studies have been conducted at lethal or sublethal concentrations. Information regarding chronic low-dose exposures is severely lacking. Although hydrogen sulfide is referred to as a broad spectrum toxicant, the nervous system is regarded as the primary target organ. The risks due to repeated exposures and the effects following exposure have been greatly underestimated. Current measures of treatment of poisoning are not adequate, and it is essential that the health effects of chronic low level exposures be determined using modern scientific methods.

I. Historical Background

The first published account of hydrogen sulfide poisoning was described by the Italian physician Bernardino Ramazzini in 1713. In his book "De Morbis Artificum," he described the painful irritation and inflammation of the eyes of sewer cleaners. Ramazzini hypothesized that these symptoms were caused by exposure to a "volatile acid" that was also responsible for the blackening of silver and copper coins. It is interesting that this first documented report of eye irritation due to hydrogen sulfide (H_2S) gas later became the basis for establishing a threshold limit value for exposure in the workplace. In 1775, a chemist, Carl Wilhelm Scheele, was credited with discovering the toxic gas that could be produced by combining acid with metal (poly)sulfides or heating sulfur in hydrogen gas. The gas was later analyzed in 1796 by Berthollet. In 1862, a very explicit description of the effects of the toxic fumes generated in the ancient sewer systems of Paris provided the basis for a dramatic scene of bravery for one of Victor Hugo's main characters in the novel

"Les Miserables." In 1934, R. R. Sayers published a paper describing the acute and subacute effects of H_2S in sewers and sewage treatment plants.

II. Physical and Chemical Properties

Hydrogen sulfide is a flammable, colorless gas with a characteristic and often nauseating odor of rotten eggs at low concentrations. It is heavier than air ($d = 1.19$) at standard temperature and pressure (STP). The gas can burn in air with a pale blue flame forming water, SO_2, and elemental sulfur (see Eqs. 1 and 2). The explosive limits of H_2S are 4.3 to 46% by volume in air, and it will autoignite at 260°C. H_2S can be oxidized readily by a variety of oxidizing agents to the primary products SO_2, sulfate, and elemental sulfur. It is soluble in both water and organic solutions (see Table I). In aqueous solution (Eq. 3), H_2S dissociates into a hydrosulfide ion (HS^-) and sulfide ion (S^{2-}). The pK_a values for H_2S are 7.04 for the HS^- ion and 11.96 for the S^{2-} ion; therefore at physiological pH 7.4, approximately one-third of the total sulfide will exist as the undissociated form and two-thirds as the hydrosulfide ion. Aqueous solutions of chlorine, bromine, and iodine may react with H_2S to form elemental sulfur. The relative lipid solubility of the undissociated form of H_2S in solution permits the gas to easily penetrate biological membranes.

$$2H_2S + 3\,O_2 \longrightarrow 2H_2O + SO_2 \quad (1)$$
$$2H_2S + O_2 \longrightarrow 2H_2O + 2S \quad (2)$$
$$H_2S \rightleftharpoons H^+ + HS^- \rightleftharpoons H^+ + S^{2-} \quad (3)$$

III. Atmospheric Chemistry

There have been relatively few measurements of atmospheric concentrations of H_2S in ambient air in either urban or rural areas. In the atmosphere, H_2S can undergo chemical and photochemical oxidation reactions yielding sulfuric acid and/or sulfate ion. The atmospheric lifetime of H_2S is influenced by many factors including temperature, humidity, light,

Table I Physical and Chemical Properties of Hydrogen Sulfide (H_2S)

Properties:	colorless, flammable, gas, highly toxic, characteristic offensive odor
Synonyms:	hydrogen sulfuric acid; sulfur hydride; sulfureted hydrogen; stink damp
Molecular weight:	34.08
Melting point:	−85.49°C
Boiling point:	−60.33°C
Density (gas):	1.192 g/L (air = 1.00 g/L)
Autoignition temperature:	260°C
Explosive limits in air:	4.3 to 46% by volume
Solubility of 1 g H_2S:	187 ml H_2O at 10°C 242 ml H_2O at 20°C 314 ml H_2O at 30°C 405 ml H_2O at 40°C 94.3 ml alcohol at 20°C 48.5 ml ether at 20°C Also soluble in gasoline, kerosine, crude oil, CS_2, glycerol, aqueous solutions of amines, alkali, carbonates, bicarbonates, hydrosulfides
Vapor pressure:	18.5 atm (18.75×10^5 Pa) at 20°C 23.9 atm (23.9×10^5 Pa) at 30°C
Odor threshold:	0.0008 to 0.20 mg/m^3 (0.0005 to 0.13 ppm)
pK_a of 0.01–0.1 M:	7.04 for HS^- at 18°C
Aqueous solutions:	11.96 for S^{2-}
Conversion factors:	1% by volume = 10,000 ppm 1 mg/L = 717 ppm (STP) 1.4 mg/m^3 = 1 ppm

wind, pressure, release characteristics from the source, dispersion patterns, and geophysical features of the land. Photolysis or photo-oxidation can occur in the presence of oxygen, ozone, and SO_2. It has been estimated that H_2S can reside in a relatively clean atmosphere (area remote from emission sources) for approximately 1–2 days, but this residence time will be greatly reduced to a few hours in a polluted atmosphere.

IV. Analytical Methods

Historically, the detection of H_2S was based on the blackening of common metal objects such as coins, keys, as well as lead-based paints and lead-acetate papers. More sophisticated measurement techniques have been developed in the past years which utilize colorimetric, photometric, photoionization, fluorescence, gas chromatography, and electrochemical methods. The colorimetric reaction of H_2S with *N,N*-dimethyl-*p*-phenylenediamine and ferric chloride to form methylene blue has been recognized as a standard analytical method. It is both specific and accurate for H_2S; detection levels are reported to be 0.0012 to 0.1 mg/m^3 (0.0008 to 0.07 ppm). Reducing agents may interfere with the reaction but do not present a major problem. Gas chromatography coupled with flame photometric detection is also an accurate and sensitive method for determining levels of H_2S at concentrations of 0.005 to 0.013 mg/m^3 (0.0035 to 0.009 ppm). This method is useful for quantitative analysis of individual sulfide gases. Both of these methods can be used for analyzing higher concentrations using appropriate dilution procedures. Lead acetate impregnated paper has proven to be a very useful method for measurement in the field. It is sensitive to low concentrations, but will cross-react with other oxidizing agents. The most sensitive methods appear to involve fluorescent detectors capable of measuring levels as low as 0.002 $\mu g/m^3$ (0.0014 ppb) and ion-selective electrodes useful for determining concentrations in biological fluids. Several types of direct reading instruments have been developed that utilize a variety of detection systems including the methylene blue reaction, lead acetate paper, and electrochemical sensors. The majority of these instruments are semiquantitative and sensitive to threshold levels of $\sim$1 mg/m^3 (0.7 ppm). Long-duration detector tubes are now commonly used and will respond to a broad range of H_2S concentrations (1 to 84 mg/m^3 or 0.7 to 56 ppm). Electrochemical detectors have become popular for monitoring occupational environments. These units are generally reliable, accurate, and provide direct real-time concentrations in air for H_2S in the range 0.7 to $>$140 mg/m^3 (0.5 to $>$100 ppm). They are subject to interference with other substances and must be routinely tested and calibrated. Although there is still some argument whether air samples should be made on an intermittent or continuous basis, there are many advantages to real-time analysis.

V. Sources

Hydrogen sulfide is both an environmental and industrial pollutant or contaminant. It is now recognized that H_2S is an extremely toxic substance to humans, animals, and vegetation. The increased awareness of the potential widespread exposure and severe toxicity of this gas has resulted in the distribution of numerous brochures and documents related to the hazards and regulations of H_2S.

A. Natural Sources

H_2S occurs widely in nature as a constituent of natural gas, petroleum, sulfur deposits, volcanic gases, and sulfur springs. It is one of the principal compounds in the environmental sulfur cycle. H_2S is produced as a result of decomposition of organic protein by bacteria, fungi, and actinomycetes. The natural production of H_2S is balanced by the oxidation of H_2S to elemental sulfur and sulfate by anaerobic bacteria in soil and water; these constitute other pathways of nature's sulfur chain. It has been reported that production of H_2S by natural sources accounts for approximately 90% of the atmospheric H_2S. This has been estimated to be on the order of 90 to 100 million tons; 60–80% from terrestrial and 20–30% from aquatic sources. The combination of organic material, sulfate, and relatively low oxygen concentrations that are usually found in swamps, bogs, slurries, and still ponds are ideal for the production of H_2S.

The excessive release of naturally occurring H_2S is most often the result of human activities such as exploration and extraction of petroleum deposits, coal mining, and development of geothermal wells for generation of electrical power. Natural gas that is contaminated with sulfurated compounds, including H_2S, is termed 'sour gas'. Depending on the geological formation, H_2S may be present in natural gas in concentrations ranging from <1% to over 90% by volume.

B. Industrial Sources

Large quantities of H_2S are utilized in various industrial processes (see Table II) as primary or intermediate reagents. Many commercial processes also produce H_2S as a by-product. In 1977, the U.S. National Institute of Occupational Safety and Health listed 73 industries and/or occupations that present a potential risk to workers due to excessive or accidental release of the gas. Hydrogen sulfide still remains a potential hazard in industrial activities such as petroleum and gas refining, manufacturing of pulp and paper (using the Kraft process), production of viscose rayon, heavy water, some metal alloys, and certain food products, tanning of animal hides, incomplete combustion of fuels, decomposition of organic waste material in sewers and agricultural slurries, and spoilage in the holds of fishing ships. It has been described as one of the leading causes of sudden death in the workplace.

Table II Examples of Occupations and Industries Associated with Potential for Hydrogen Sulfide Exposures

Animal fat and oil processing	Livestock farmers
Animal manure slurries	Metallurgists
Artificial-flavor processing	Mining
Barium salt makers	Natural gas production/processing
Blast furnace workers	Papermakers
Brewery workers	Petroleum and production/refinery workers
Cable splicers	Phosphate purifiers
Caisson workers	Photoengravers
Carbon disulfide production	Pipeline maintenance workers
Cellophane production	Pyrite burners
Chemical laboratory personnel	Rayon makers
Cistern cleaners	Refrigerant process
Coal gasification	Rubber and plastics processing
Coke oven workers	Septic tank cleaners
Copper-ore sulfidizers	Sewage treatment plants
Dyemakers	Sewer workers
Felt making	Sheepdippers
Fermentation process workers	Silk makers
Fertilizer makers	Slaughterhouse workers
Fishing and fish-processing workers	Smelting workers
Geothermal wells	Soapmakers
Gluemaking	Sugar beet and cane processers
Gold-ore workers	Sulfur spa workers
Heavy-metal workers	Sulfur products processing
Heavy-water manufacturing	Sulfuric acid purification
Hydrochloric acid purifiers	Synthetic-fiber makers
Hydrogen sulfide production	Tannery workers
Landfill workers	Textile printers
Lead ore sulfidizers	Tunnel workers
Lithographers	Wool processing
Lithopone makers	

VI. Levels

A. Ambient Levels

Measurement of ambient levels of H_2S in either urban or rural centers is not routinely performed. The limited data available in the literature are often lacking details regarding the sampling times and methods. Earlier results, which account for the majority of data, are not reliable due to the methodologies and instrumentation used, which were relatively low in sensitivity and/or accuracy. However, based on this available data, it has been proposed that the general population in urban centers is usually not exposed to H_2S concentrations exceeding 1 $\mu g/m^3$ (0.7 ppb). Average ambient air levels have been estimated to be of the order of 0.0003 mg/m^3 (0.0002 ppm). Data averaged from various urban centers in the United States range from 0.0001 mg/m^3 (0.00007 ppm) to 0.006 mg/m^3 (0.0042 ppm) with maximum values on the order of 0.01–1.4 mg/m^3 (0.007–1.0 ppm). It must be emphasized that apparent ambient levels are very dependent on climatic and geographical parameters. Time-averaged values calculated over relatively long time intervals may not reflect occurrences of short periods of very high levels that exceed the average by several orders of magnitude, and consequently present potential health risk to the general population. It has not been possible to define a NOEL (No Observed Effect Level) or LOAEL (Lowest Observed Adverse Effect Level) for H_2S exposures.

In the past, there has been relatively little attention to the potential dangers of agricultural activities related to H_2S. Animal waste materials are often stored in manure pits or slurries, and with the appropriate conditions, very large quantities of H_2S can be produced. It has been well documented that with agitation, H_2S can be released at concentrations of 196 to over 700 mg/m^3 (140 to $>$ 500 ppm), and on occasion as high as 1540 mg/m^3 (1100 ppm).

B. Industrial/Occupational Levels

The exposure of H_2S is a potential occupational hazard for many thousands of workers (estimated to be approximately 125,000 in the United States). It is generally accepted that under normal conditions, workers are not exposed to levels exceeding recommended limits (i.e., time-weighted averages of 10–15 mg/m^3 or 7–10 ppm) established by most governmental agencies (see Table III). It can be assumed that the effects of H_2S are generally depen-

Table III Exposure Limits and Guidelines

Report	Concentration (mg/m³)	(ppm)	Exposure time
Clean Air Act (Alberta, Canada)	0.014	0.01	1 hr
WHO Odor Nuisance Guideline	0.007	0.005	30 min
WHO Health Effect Guideline	0.150	0.107	24 hr
Alberta OHS Occupation Exposure Limit (OEL)	14	10	8 hr
OSHA Permissible Exposure Limit (PEL)	14	10	TWA
	21	15	STEL
	28	20	CL
		20	TWA
		50	10 min[a]
ACGIH	14	10	TLV:TWA
	21	15	STEL
NIOSH Recommended Exposure Limit (REL)	15	10.7	10 min
	14	10	CL 10 min

Abbreviations: TWA, time weighted average for 8-hr workday; STEL, short-term exposure limit (defined as a 15-min TWA); CL, ceiling limit; TLV:TWA, threshold limit value—time weighted average (time weighted concentration for 8-hr workday).

[a]10-min maximum peak in 8-hr shift.

dent on the concentration and duration of exposure that justifies the use of time-weighted averages for setting occupational limits; however, short-term high levels or "spikes" may not be evident and could prove to be more toxic than proportionally shorter exposures to higher levels.

There are, however, a number of reported incidents (and many unreported) where exposures in excess of the limits have occurred as a result of accidental release. The effects of H_2S on humans, animals, and vegetation have been documented usually as individual case reports. Many of these have been reviewed extensively in recent years. In most instances, the concentrations have not been accurately determined, but have been estimated based on observed clinical symptoms or calculated release quantities. A number of acute high dose exposures have provided estimates of lethal concentrations for humans. Although there are few well-documented reports on the levels of H_2S of accidental releases, values of 150 mg/m^3 (107 ppm) to 18,000 mg/m^3 (12,000 ppm) have been published.

C. Odor Threshold

The human olfactory system is a very sensitive detector for H_2S. The characteristic odor of H_2S at low concentrations has been the cause of numerous complaints by people living in the vicinity of emission sources. This response has been used as a physiological endpoint (see Fig. 1), and under laboratory conditions, it has been shown that the human can

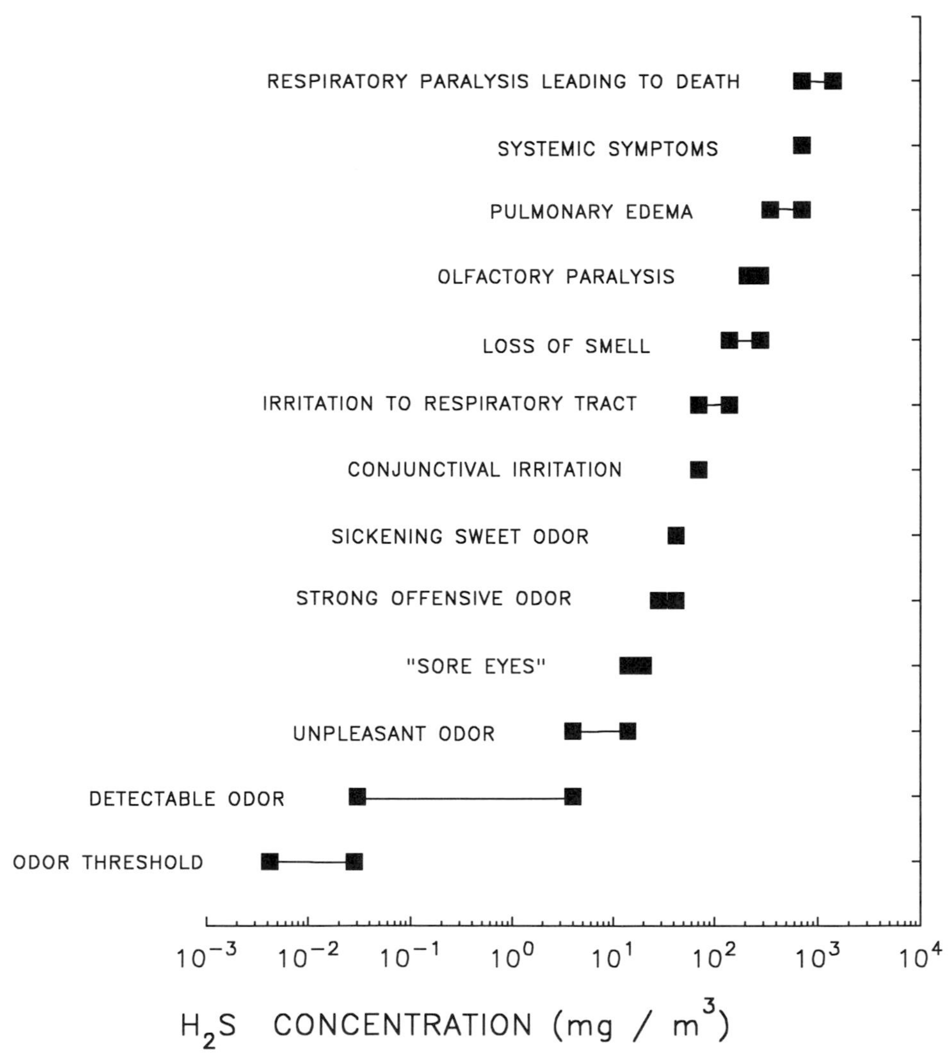

Figure 1 Physiologic responses of humans to hydrogen sulfide.

detect levels as low as 0.0008 to 0.20 mg/m^3 (0.0005 to 0.13 ppm). The broad range of threshold concentrations for odor perception clearly exhibits the biological differences in sensitivity of individuals. At concentrations of 4.2 to 42 mg/m^3 (3 to 30 ppm), the smell of H_2S has been described as offensive, resembling "rotten eggs." Above 42 mg/m^3 (30 ppm) to approximately 140 mg/m^3 (100 ppm), the odor becomes sickeningly sweet. At concentrations greater than 140 mg/m^3 (100 ppm), anosmia or loss of sense of smell is often reported as a cause of olfactory fatigue. Thus smell may not be a reliable warning of potential danger to toxic levels of H_2S. There is some concern, however, that the odor is often not reported following exposure to high levels of H_2S because individuals have recovered with some memory loss. Thus the effects of olfactory fatigue and paralysis still require further evaluation.

Guidelines for ambient air concentrations appear to be related to the odor thresholds and are orders of magnitude lower than those assumed to be a risk for human health. The World Health Organization in 1987 recommended upper levels of 7 μg/m^3 (5 ppb) over 30 minutes for odor nuisance, and 150 μg/m^3 (107 ppb) for a 24-hour period for health effects. The Clean Air Act of Alberta, Canada recommends 0.014 mg/m^3 (0.01 ppm) for a one-hour average. These levels can be compared to occupational exposure limits (see Table III).

VII. Toxicology

Hydrogen sulfide is a broad spectrum toxicant that affects most organ systems in a variable manner. The degree or intensity of effect is dependent on concentration of gas, rate and duration of exposure, and tissue type. It is now accepted that the nervous system is the primary target in regards to overall toxicity.

In general, the actions of H_2S are similar at high concentrations on human subjects and experimental animals. The majority of animal studies have been conducted at lethal or sublethal concentrations. The lethality data are quite consistent for most animal species with some minor differences in susceptibility. The avian (bird) species appears to be the most sensitive; a 100% lethal dose (LD_{100}) for canaries is approximately 140 mg/m^3 (100 ppm). The rank order of sensitivity to H_2S for different species is as follows: canaries > dogs > guinea pigs > goats > humans > white rats, although some reports suggest that guinea pigs are as sensitive as birds. For most animal species local irritation of eyes and throat will occur after several hours at 150–225 mg/m^3 (100–150 ppm) and at 300–450 mg/m^3 (200–300 ppm) in one hour. Systemic symptoms will begin to occur within one hour at 750–1050 mg/m^3 (500–700 ppm) and become severe at 1350–2250 mg/m^3 (900–1500 ppm). Animals exposed to 2700 mg/m^3 (1800 ppm) will experience immediate collapse, respiratory paralysis, and death. The reported LC_{50} (concentration that is lethal for 50% of the population) for Sprague-Dawley rats is 662 mg/m^3 (444 ppm), mice 140 mg/m^3 (100 ppm at 7.5 hr), and cats 0.025 mM/kg.

In humans H_2S is both an irritant and asphyxiant. As a gas it is rapidly absorbed via the lung, and sufficient exposure time and/or concentration will result in systemic effects shown in Figure 1.

A. Target Organ Systems

Tissues with exposed mucous membranes or with high oxygen (O_2) demands are most susceptible to the effects of H_2S at low concentrations. Of the possible target organs affected by H_2S gas, the nervous system, lung, heart, eyes, and respiratory tract appear to be most affected. The nervous system is likely the most important in relation to overall toxicity, and is regarded as the primary target organ. Table IV lists the effects of H_2S on each target organ system. There are very few controlled studies of the effects of H_2S exposure on human subjects; thus most of the published data have been derived from case reports following industrial and/or accidental exposures or from studies on experimental animals. There is a lack of scientific data on chronic low level exposures.

B. Toxicokinetics

Since hydrogen sulfide is a gaseous substance, it is rapidly absorbed via inhalation; absorption through the skin is negligible. Inorganic sulfides are normally present in the body in very small amounts. H_2S is well distributed to the brain, liver, kidney, pancreas, and small intestine. Although the toxicokinetics of H_2S have not been studied in man, animal studies have identified three primary metabolic (detoxification) pathways occurring primarily in red blood cells and liver mitochondria: oxidation to sulfate, methylation, and reaction with metalloproteins and disulfide-containing proteins. Disulfide bonds are

Table IV Effects of H_2S on Target Organ Systems

Target organ	Symptoms
Nervous system	Acute intoxication may produce fatigue, vertigo, intense anxiety, olfactory paralysis, convulsions, collapse, unconsciousness, respiratory arrest, cardiac failure, memory loss (amnesia), depression, headache, morphological and neurochemical alterations in brain tissue; threshold for systemic effects 700 mg/m³ (500 ppm)
Respiratory system	Dyspnea, pulmonary tract irritation, bronchitis, rhinitis, pharyngitis, laryngitis, sore throat, cough, chest pain, pulmonary edema, cyanosis, hemoptysis, pneumonia, increased mucosal permeability, breathlessness, wheezing
Cardiovascular system	No significant effects at <28 mg/m³ (<20 ppm); transient changes in electrocardiograms at high concentrations; decrease in blood pressure, ventricular extrasystoles, arrhythmias, migraine
Reproductive system	Reported effects are ambiguous; no significant effects at <140 mg/m³ (100 ppm); high levels may result in dystocia (increase in delivery time)
Development	No significant effects in growth parameters; significant, subtle alterations in architecture, growth characteristics and neurochemistry of developing (rat) brain; increased incidence of fetal toxicity and embryonic development
Eye	Irritation and inflammation of conjunctival and corneal tissue (keratoconjunctivitis); blurred vision, halo around lights, 'gritty' sensation or pain
Olfactory system	Malodorous response, anosmia (loss of smell), olfactory fatigue, olfactory nerve paralysis, decreased sensitivity with age
Skin	Reversible discoloration, spots, rash
Hepatic tissue (liver)	No significant data
Kidney	No significant data
Gastrointestinal tract	Nausea, vomiting, diarrhea, stomach cramps
Hematopoietic (blood forming) system	Variable results; does not form sulfhemoglobin, may inhibit heme synthesis, and disturb iron metabolism
Immunological system	Depression of macrophage function
Endocrine	Decrease milk production (cows); increase plasma cortisol levels; dose-dependent lesions of thyroid gland; reduction or blockade of oxytoxin receptor
Psychological	Impairment of cognitive function, neuropsychological and emotional changes
Carcinogenesis	No significant data
Mutagenesis	No significant data

susceptible to cleavage by aqueous sulfide. Rapid detoxification has often been interpreted incorrectly suggesting that H_2S is not a cumulative poison. In addition, some of the metabolic products may be responsible for toxic reactions.

C. Mechanisms of Action

Hydrogen sulfide has been shown to interact with a large number of enzyme systems (e.g., alkaline phosphatase, adenosine phosphatase, and those involved in protein synthesis), but cytochrome oxidase has been identified as the critical enzyme target for toxicity. This is similar to the action of cyanide, but somewhat more potent. Cytochrome oxidase is involved in the electron transport system in oxidative phosphorylation that generates cellular energy. Enzyme inhibition would thus result in reduced oxygen availability and cellular anoxia. A functional histotoxic (tissue toxicity) hypoxia has been suggested to be a primary mechanism of toxicity in tissues with a high O_2 demand. H_2S does not impair transport of O_2 by hemoglobin, nor does it cause formation of sulfhemoglobin as reported in earlier studies. At low concentrations, H_2S may also inhibit heme synthesis, and perhaps disturb iron metabolism. In addition to the biochemical effects, the direct irritant actions on mucous membranes and stabilization (i.e., anesthesia) of excitable tissues may also play a role in the overall toxic reactions.

D. Cumulative Effects

There has been considerable controversy whether certain symptoms would continue following acute (or chronic) exposure to H_2S. In animals exposed repeatedly to H_2S at concentrations of ~150 mg/m^3 (~100 ppm), inhibition of cytochrome oxidase and a decrease in cerebral RNA synthesis have been shown to be cumulative. Recent preliminary studies in the author's laboratory have shown cumulative effects on central neuronal activity. The "sequelae" and cumulative adverse effects after H_2S poisoning or intoxication have been documented, but the risks have been largely underestimated. It has been reported that following poisoning, workers have developed an increase in sensitivity plus an aversion to the odor of H_2S, in addition to neuropsychological, cardiovascular, and cognitive disorders. Tolerance may also occur in some subjects, but this remains to be evaluated. It is essential that the health effects on the human population of chronic low level exposures to H_2S be determined using scientific and epidemiological methods.

VIII. Clinical Symptoms

Clinical symptoms are frequently classified in terms of acute, subacute, and chronic exposure. The World Health Group on Environmental Health Criteria for hydrogen sulfide suggested that acute intoxication (exposure) be defined as the effects of a single exposure of H_2S within seconds or minutes at concentrations of approximately 1400 mg/m^3 (1000 ppm); exposure for several hours at 140–1400 mg/m^3 (100–1000 ppm) is termed "subacute intoxication." Chronic intoxication would involve intermittent exposures of 70–140 mg/m^3 (50–100 ppm). Subacute intoxication is associated with eye irritation and perhaps pulmonary edema with prolonged exposure and/or exposure to higher concentrations. Chronic intoxication does not refer to prolonged exposure to low concentrations associated with ambient or environmental levels. The symptoms associated with exposure are shown in Table V.

Table V Clinical Symptoms following Hydrogen Sulfide Exposure

Visual "fogging"	Loss of appetite (anorexia)
Conjunctivitis	Headache
Intolerance to light (photophobia)	Abnormal peripheral reflexes
Tearing	Weakness of extremities
Eye pain	Depression
Rhinitis	Lethargy
Bronchitis	Fatigue
Sore throat/cough	Anxiety
Chest pain	Decreased libido
Shortness of breath (dyspnea)	Insomnia
Expectoration of blood (hemoptysis)	Nystagmus
Pulmonary edema	Irritability
Cyanosis	Dizziness
Bradycardia	Amnesia
Nausea	Disequilibrium
Vomiting	Unconsciousness
	Convulsions

IX. Treatment

It is necessary to treat H_2S exposure (poisoning) to prevent death and the development of adverse sequelae. The current therapeutic measures are of questionable benefit, and new approaches are required. It is essential that the exposed victim be removed from the contaminated area immediately and be provided with supportive care. If spontaneous breathing is not restored, then artificial methods must be employed at once. If a heartbeat is not evident, then cardiopulmonary resuscitation (CPR) should be initiated. It is also recommended that rescuers be advised to enter areas of contamination with caution, to avoid being overcome themselves. There have been several antidotes described, but all remain controversial. Two approaches are often described: administration of nitrites, which is believed to scavenge sulfide from cytochrome oxidase, and hyperbaric oxygen (HBO). Many recent reports suggest that nitrites offer little benefit and may prove to be harmful. Oxygen at atmospheric pressure may not be useful but administration at elevated pressures (HBO) has proven to be effective in some cases. It has been proposed that persulfide reagents may be effective in lowering sulfide levels in tissues, however the clinical benefit will require further research.

Bibliography

Beauchamp, R. O., Bus, J. S., Popp, J. A., Boreiko, C. J., Andjelkovich, D. A. (1984). "A Critical Review of the Literature on Hydrogen Sulfide Toxicity." CRC Critical Reviews in Toxicology, vol. 13. pp 25–97. CRC Press, Boca Raton, Florida.

Ellenhorn, M. J., and Barceloux, D. G. (1988). Medical Toxicology pp. 836–840. Elsevier, New York.

Glass, D. C. (1990). A review of the health effects of hydrogen sulfide exposure. *Ann. Occup. Hyg.* **34,** 323–327.

International Programme on Chemical Safety (1981). "Hydrogen Sulfide." Environmental Health. Criteria 19. World Health Organization, Finland.

Prior, M. G., Roth, S. H., Green, F. H. Y., Hulbert, W. C., Reiffenstein, R. J., eds. (1989). Proceedings of International Conference on Hydrogen Sulphide Toxicity, Banff. pp. 231. Sulphide Research Network, Edmonton, Alberta.

Reiffenstein, R. J., Hulbert, W. C., Roth, S. H. (1992). Toxicology of hydrogen sulfide. *Ann. Rev. Pharm. Toxicol.* **32,** 109–134.

Report on H_2S Toxicity (1988). Edmonton, Alberta Health.

Immune Response to Environmental Agents

Noel R. Rose
The Johns Hopkins University

Glossary

Adjuvant Compound capable of potentiating an immune response.

Antibody Protein that is produced as a result of the introduction of an antigen and that has the ability to combine with the antigen that stimulated its production.

Antigen Substance that can induce a detectable immune response when introduced into an animal.

Antigenic determinant Smallest fragment of an antigen that determines the specificity of the antigen–antibody reaction.

Antigen processing Series of events that occurs following antigen administration and preceding antibody production.

Autoimmunity Immunity to self-antigens (autoantigens).

B cell In a strict sense, a bursa-derived lymphocyte in avian species and, by analogy, a bursa-equivalent lymphocyte in nonavian species. B cells are the precursors of the plasma cells that produce antibody.

Cell-mediated immunity Immunity brought about primarily by lymphocytes and macrophages.

Complement Series of interacting serum proteins that combine with antigen–antibody complexes to cause lysis of cells.

Delayed-type hypersensitivity Cell-mediated reaction measured by means of a skin test.

Gamma globulins Serum proteins showing gamma mobility in electrophoresis and comprising the majority of antibodies.

Helper T cell Subtype of T lymphocytes that cooperates with B cells in antibody formation.

Humoral immunity Immunity brought about primarily by molecules in body fluids, particularly antibody and complement.

Hybridoma Transformed cell line grown *in vivo* or *in vitro* that is a somatic hybrid of two parent cell lines and contains genetic material from both.

Immunoglobulin Glycoprotein composed of heavy and light chains that functions as antibody.

Lymphocyte Mononuclear cell, 7–12 μm in diameter, containing a nucleus with densely packed chromatin and a small rim of cytoplasm.

Macrophages Phagocytic mononuclear cells that derive from bone marrow precursors and serve accessory roles in cellular and humoral immunity.

MHC Major histocompatibility complex designated as H-2 in mouse and HLA in humans.

NK cell Natural killer cell capable of killing tumors and virus-infected cells without prior sensitization.

Phagocytes Cells capable of ingesting particulate matter.

Plasma cells Fully differentiated antibody-synthesizing cells derived from B lymphocytes.

T cell Thymus-derived lymphocyte that participates in a variety of immune reactions.

Variable (V) region Amino terminal portion of the heavy or light chain of an immunoglobulin molecule, containing considerable heterogeneity in the amino residues compared to the constant (C) region.

***THE BODY** normally maintains its individual integrity by splitting substances absorbed from the environment into their constituent units. Thus, proteins from foods ingested through the gastrointestinal tract are broken down to their constituent amino acids, while carbohydrates are cleaved to their monosaccharide units. After absorption, these peptide and monosaccharide units are synthesized into new proteins and polysaccharides unique for the host. Individuality, which is genetically determined, is expressed through the enzymes that catalyze the formation of functional and structural molecules of the host's composition. The immune system is called into action when foreign substances evade the normal method of assimilation. It then represents the body's reaction to environmental agents. If an undigested foreign molecule enters the body via an abnormal route, it is recognized as such by the immune system, which then takes steps to eliminate it. The normally functioning immune system, therefore, has two primary capabilities: the ability to recognize all foreign molecules and the ability to distinguish self from nonself.*

I. Introduction

Immune recognition is the task assigned to a specialized population of cells, the lymphocytes, which circulate in the blood stream and penetrate the tissues. Lymphocytes are small round cells with a relatively large nucleus and scant cytoplasm. On the surface, lymphocytes bear specially configured receptors with the property of binding foreign molecules. Each mature lymphocyte bears a particular receptor. During the development of lymphocytes, a sufficient diversity of receptors is generated so that any foreign molecule entering the body will encounter a lymphocyte with a complementary receptor on its surface.

Following its interaction with a foreign molecule, the lymphocyte is stimulated to enlarge and divide. Eventually, lymphocyte proliferation generates a family or clone of lymphocytes, each bearing the same receptor as its progenitor cell (Fig. 1). Thus, over a period of a few days, a large population of lymphocytes capable of binding the foreign molecule is produced. Eventually, the proliferating lymphocytes give rise to functional cells. Some of these functional cells (called B cells) secrete protein molecules that appear in the bloodstream as antibodies. Antibodies have the ability to bind the foreign molecule that instigated their production. The immune response also generates a population of lymphocytes (T cells) that bind the foreign molecule directly. They bring about a series of reactions termed "cell-mediated immunity." The foreign molecule that initiated the immune response is called an antigen. Antigens are defined by an ability to instigate an immune response and to react with the resulting antibody or effector T cell.

In this article, we describe the major properties of the immune response and the methods used to measure it. The molecular basis of antigenicity and the mechanisms by which antigenic molecules interact with immunologically active cells are described as well. The factors that regulate the immune response are examined. Finally, the practical ways in which the products of immunity can be used to assess environmental agents and the adverse effects of environmental agents on the immune response are discussed.

II. Antibody-Mediated Responses

Antibody activity takes place primarily in the gamma globulin fraction of the blood serum. Gamma globulins with antibody activity are called immunoglobulins.

Five major classes of immunoglobulins are distinguished: IgG, IgM, IgA, IgD, and IgE. Antibodies of the IgG class are the most plentiful in normal adult serum and consist of four polypeptide chains: two heavy chains with a molecular weight of approximately 53 kDa and two light chains with a molecular weight of approximately 23 kDa. Disulfide bonds unite these chains into a macromolecular complex of about 160 kDa. Starting at the amino end (Fig. 2A), the first 110 amino acids of each chain are highly variable and bring about the unique ability of the antibody molecule to bind a particular antigen. The actual binding site consists of the variable portions of the light and heavy chains. Each IgG molecule has two such binding sites. The constant portions of the light chains and heavy chains contain the class-specific properties. IgG immunoglobulins are capable of activating complement, the system of proteolytic enzymes present in normal plasma that plays an important role in the defense against invading microorganisms. They can also opsonize invading microorganisms and thereby promote their ingestion and destruction by phagocytic cells. Finally, IgG antibodies are the only immunoglobulins capable of

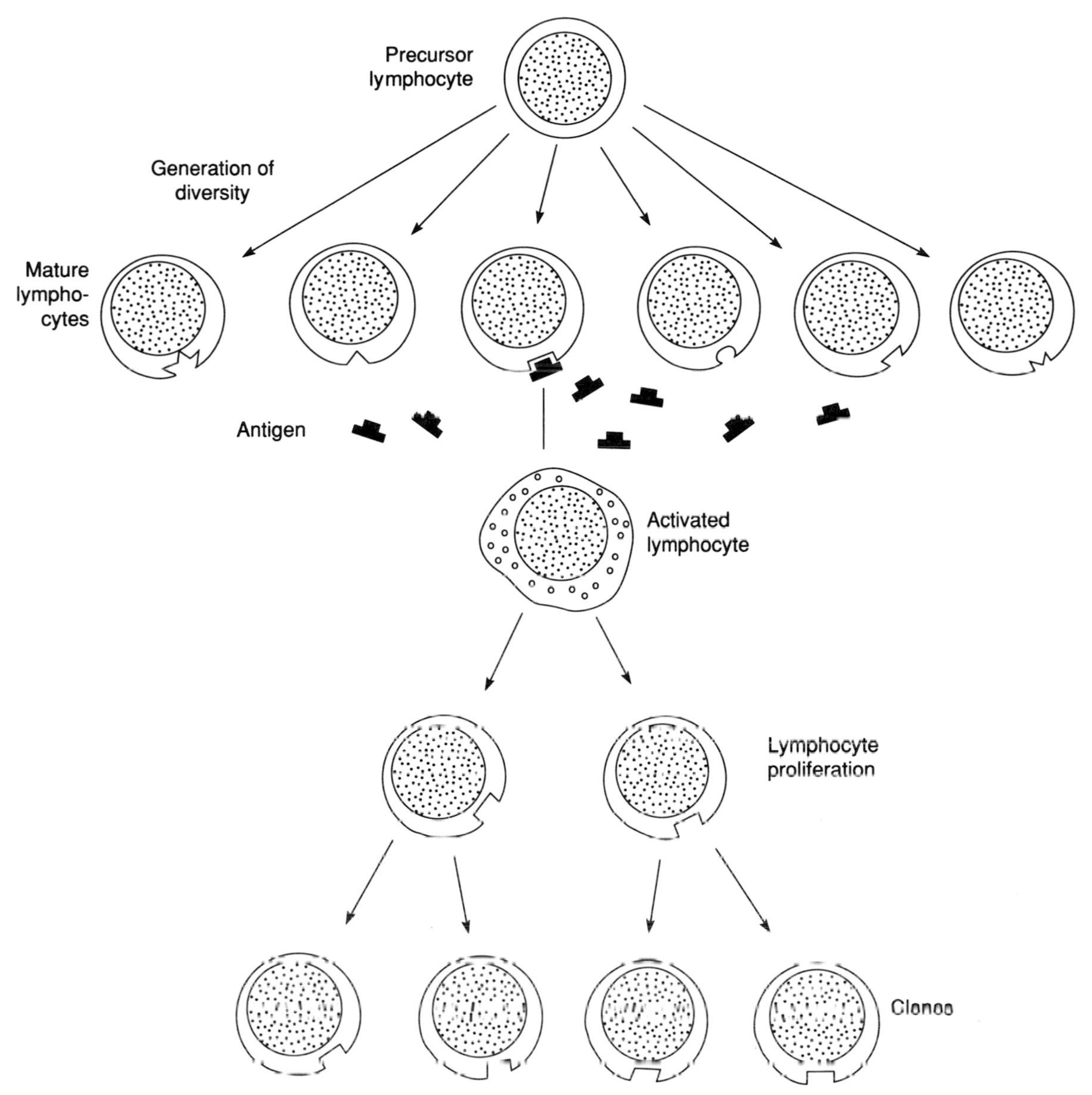

Figure 1 Clonal selection. [From N. R. Rose (1991). Autoimmune disease. *In* "Encyclopedia of Human Biology" (R. Dulbecco, ed.), vol 1, pp. 519–533. Academic Press, San Diego. Reproduced by permission.]

crossing the placenta and conferring immunity to the neonate.

IgM antibodies are usually the first immunoglobulins to appear after an antigen is encountered. They have a molecular weight of approximately 900 kDa and are found mainly in the intravascular compartment. The IgM molecule consists of 10 light chains and 10 heavy chains arranged as five subunits, each resembling the IgG molecule (Fig. 2B). Based on two binding sites per subunit, the IgM antibody has a valency of 10. The IgM molecule is capable of activating the complement system, but often has relatively low affinity for its respective antigen.

The IgA antibodies circulate in the blood stream in low amounts, but are present in relatively large amounts in internal secretions such as saliva and gastric juice. They are, therefore, the major effectors of secretory immunity. In the blood stream, IgA antibodies circulate as monomers with a molecular weight similar to that of IgG, but in bodily secretions they are more observed as dimers with molecular weights up to 390 kDa (Fig. 2C). Although they cannot activate complement by the classic pathway,

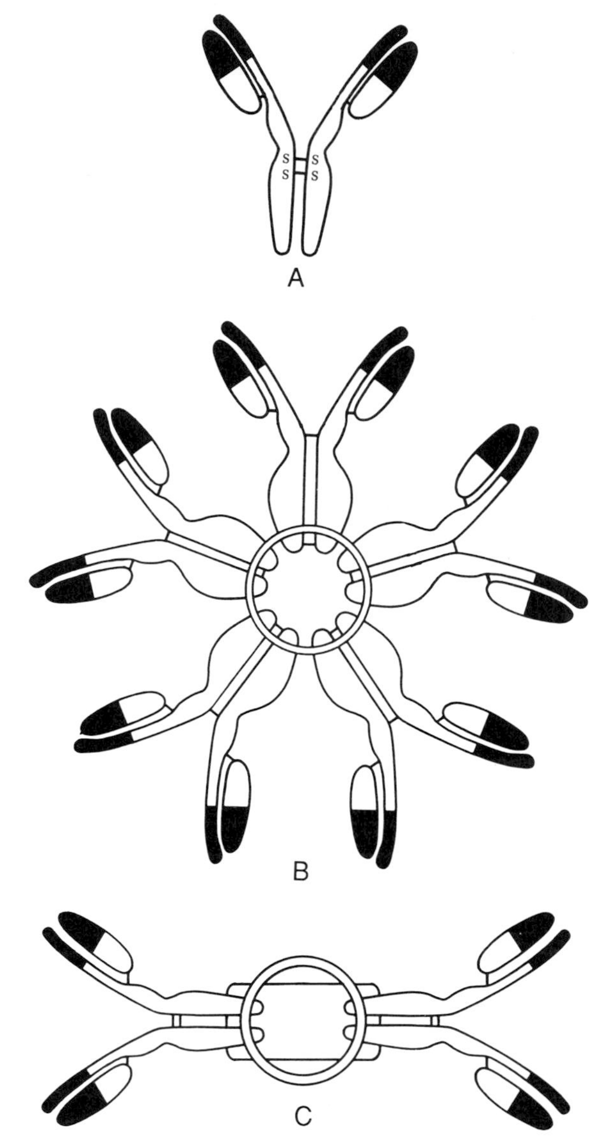

Figure 2 (A) IgG molecule; (B) IgM molecule; (C) IgA molecule.

they can produce complement activation via an alternative route. IgA molecules, therefore, are important in defense against microorganismic invaders entering through mucosal surfaces.

The constant region of the IgE antibody confers the unique ability of binding tightly to basophils and mast cells. Granules of these specialized cells contain mediators such as histamine, serotonin, and leukotrienes. When antigens encounter basophils or mast cells with IgE antibody attached to their surfaces, the granules discharge these mediators into the surrounding fluid, causing the typical symptoms of allergy. IgE has a molecular weight slightly greater than IgG, but is found only in trace amounts in the serum.

IgD, finally, is an immunoglobulin that remains part of the lymphocyte membrane and is not secreted in significant amounts into the blood stream. It is believed to play a role in regulating lymphocyte function.

All antibodies are the products of B cells. These lymphocytic cells arise from the bone marrow in adult mammals, but their maturation in birds depends upon a unique lymphoid organ of the hindgut, the bursa of Fabricius. The development of a functional B cell from its progenitor in the bone marrow is determined by a series of peptide hormones referred to as colony-stimulating factors. As a consequence of the maturational process, the B lymphocyte acquires a receptor for a specific antigenic determinant; that is, a short sequence of peptides or sugars on an antigen molecule. The B cell recognizes these antigenic determinants on the basis of their three-dimensional configuration, reflecting their tertiary or quaternary structure.

When it encounters its corresponding antigenic determinant, the B cell enlarges and divides progressively. Eventually, it differentiates into a plasma cell that specializes in secreting large amounts of immunoglobulin. The immunoglobulin produced by the plasma cell bears the antigen-specific B cell receptor and represents the circulating antibody molecule.

A plasma cell secretes antibody of a single specificity only. Yet, antibodies of many different specificities correspond to many antigenic determinants found in nature. To produce a sufficient number of different binding sites on antibody products, B cells use unique genetic mechanisms that consist of recombining a relatively large number of gene segments controlling the variable region of the immunoglobulin molecule with a very limited number of gene segments controlling the more constant portions. Different light chains and heavy chains can be paired to increase opportunities for variability. Finally, the variable-region genes of the B cell are highly mutable. Somatic mutation increases opportunities for different binding sites and, by a selective process, allows the antibody population to fit the antigenic target with progressively greater affinity during the evolution of the immune response. Generally, an antiserum (that is, serum containing antibodies of various specificities) sample taken late in the course of an immune response binds antigen with greater affinity than does a serum sample taken during the early stages of the immune response.

III. Cell-Mediated Responses

In most instances, the B lymphocyte is incapable of responding to antigen by itself, but requires the cooperation of the second major lineage of lymphocytes, the T cells. After originating in bone marrow, the T lymphocyte matures within the thymus gland through the action of simple peptide hormones produced by thymic epithelial cells. Similar to the B lymphocyte, each T lymphocyte is programmed to recognize a particular antigenic determinant. After an encounter with a corresponding antigenic determinant, the T cell may differentiate along one of several pathways. Some lymphocytes differentiate into helper T cells that cooperate with B cells in the synthesis of antibody. Other populations of T cells regulate immune responses or develop into effectors. They produce the classic delayed hypersensitivity skin test reaction observed when an appropriate antigen is introduced into the skin of mammals. Alternatively, they may function as cytotoxic T cells capable of killing target cells that bear the corresponding antigenic determinant on their surfaces. Since these reactions occur in the absence of antibody, they are referred to as cell-mediated immune responses.

After an encounter with antigen, a certain fraction of T cells (as well as B cells) reverts to a quiescent form. Because they bear the antigen-specific receptor, these lymphocytes are capable of initiating an immune response. For that reason, the next encounter with the same antigen usually brings about a more rapid and vigorous immune response. The long-lived lymphocytes responsible for this second ary, or anamnestic, response are referred to as memory cells.

In contrast to B cells, T cells do not recognize free antigenic molecules. T cell recognition depends upon presentation of antigen by a specialized population of cells referred to as antigen-presenting cells (APC). Among the cells capable of presenting the antigen to T cells are macrophages (important phagocytic cells found in tissue spaces) and dendritic cells (which create a network in lymphoid organs and skin). Antigenic peptides are taken up via endocytosis by these cells, degraded by proteolytic enzymes, and combined with class I or class II products of the major histocompatibility complex (MHC). Class I MHC products are present on all nucleated cells of the body, whereas class II products are found mainly on macrophages, dendritic cells, and other APC. A small peptide product, representing an individual antigenic determinant of engulfed and hydrolyzed antigen joined with the MHC product, is transported to the APC surface membrane (Fig. 3).

Receptors on immunocompetent T cells recognize the MHC–peptide complex. Antigenic determinants bound to class I MHC products are generally recognized by cytotoxic T cells, whereas determinants complexed with class II MHC products are recognized by helper T cells. Like immunoglobulins, T cell receptors consist of two polypeptide chains, each of which has a constant and a variable region. Their diversity is encoded by multiple gene segments and arises from the recombination of these segments.

To initiate antibody production, receptors of appropriate specificity on B cells interact with intact antigen that may be taken up and degraded within the B cell so that the resulting peptides bind to class

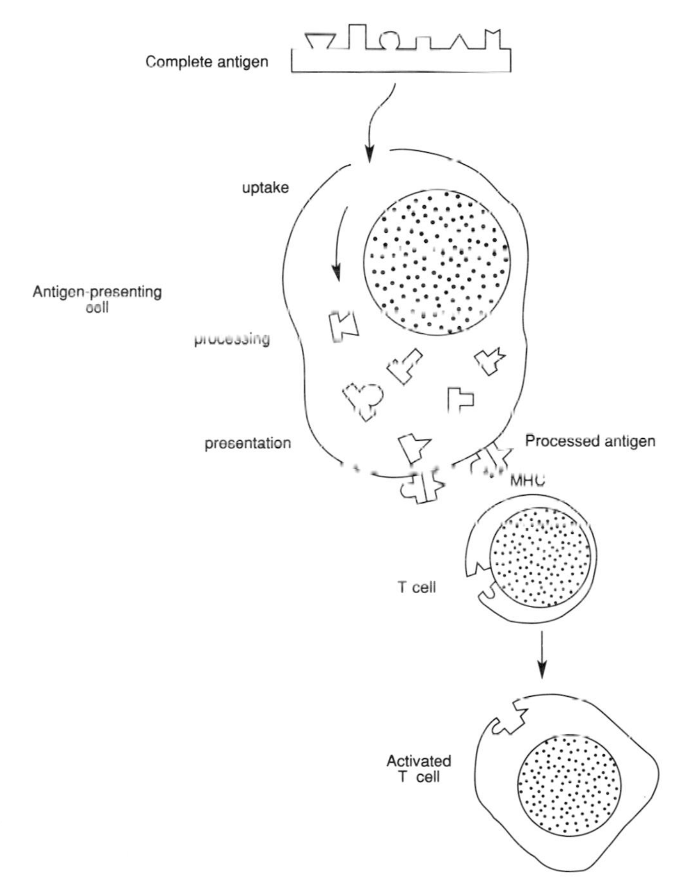

Figure 3 Antigen uptake, processing, and presentation to T cells. [Adapted from N. R. Rose (1991). Autoimmune disease. *In* "Encyclopedia of Human Biology" (R. Dulbecco, ed.), vol. 1, pp. 519–533. Reproduced by permission.]

II MHC molecules. These complexes migrate to the B-cell surface where they may be recognized by antigen-specific helper T cells. When the helper T cell is activated by a corresponding MHC peptide complex, it generates a number of protein products with specific biological activities. These are referred to as cytokines. Some cytokines activate the B cell and thereby promote the process of antibody production.

For T cell–B cell cooperation to take place, therefore, both cells must recognize the same antigen. However, they often respond to different determinants of the same antigenic molecule. It should be pointed out that the B cell can also act as an APC, so that the T cell–B cell encounter may be a two-way interaction.

Other cytokines recruit and stimulate T cells in a non-antigen-specific manner and thereby augment cell-mediated immune responses. Still other cytokines activate macrophages to turn them into even more effective phagocytes. Therefore, the production of cytokines is itself a useful indicator of an active cellular immune response.

IV. Reactions of Antibodies

Antibodies induced by the introduction of an antigen into a host will react specifically with that antigen. Under such circumstances, the antigen is said to be immunogenic. Most foreign molecules with a molecular weight of 1,000 kDa or more are antigenic. Smaller molecules may be immunogenic only if they are attached to larger carrier proteins. Such smaller molecules are referred to as haptens. Although they are not capable of initiating immunologic responses by themselves, haptens can react with preformed antibodies. Carbohydrates as well as proteins can be immunogenic. Unfortunately, polysaccharides are generally poor antigens in infants, probably because the molecules are poorly cleaved by digestive enzymes of the APC. In fact, injecting excessive amounts of a polysaccharide antigen can actually lead to unresponsiveness.

Antigen–antibody interactions are relatively strong because they result from an accretion of a number of low-energy noncovalent bonds. The major bonds are electrostatic or Coulombic forces resulting from a complement of positively and negatively charged groups and Van der Waals forces operating over short ranges. Both of these forces require a close fit between antigen and antibody and the total picture closely resembles the original ''lock and key'' concept of Paul Ehrlich.

Closeness of fit confers extraordinary specificity on antigen–antibody interactions. Antibody to *p*-azobenzene arsenate, for example, is highly dependent upon the precise configuration of the benzene ring. Placement of an orthomethyl group on the benzene ring reduces the binding energy to a very low level. A displacement of the benzene ring by merely 1.5 Å (e.g., by inserting a methylene group) significantly reduces complementarity with the antibody. Thus, the precise configuration of the antigenic determinant dictates its ability to fit into the combining site of the antibody.

All antibodies directed to a single autogenic determinant are not necessarily identical. Since they are the products of different B cell clones, each antibody combines with its corresponding antigenic determinant in a slightly different manner. It is, however, now possible to obtain a population of identical antibodies by the cell hybridization technique. Using this method, immune B cells, which have a limited lifespan in culture, are crossed with immortal plasmacyte tumor cells. The resulting hybridoma is capable of secreting large amounts of monoclonal antibody. These reagents have proved to be of great value in enhancing the specificity of immunochemical analyses.

V. Reactions of T Cells

Like B cells, T cells are clonally diverse. Their clonality is due to an antigen-specific receptor. As mentioned above, the most important distinction between immunoglobulin receptors and T-cell receptors is that the latter recognize antigenic determinants only when they are associated with MHC gene products.

During maturation in the thymus, T cells undergo two important selection processes. First, negative selection removes most of the T cells capable of reacting with antigens of the host itself. In this way, the body avoids the potentially deleterious effects of autoimmunity. Next, positive selection favors the survival of T cells capable of recognizing the MHC of self combined with some foreign antigenic determinant. At this point, an important departure occurs in the pathway to T-cell maturation. T cells that recognize antigen in the context of class II MHC products develop a surface marker referred to as CD4. These cells are destined to become initiators of cellular immune response and helpers in antibody production. Other T cells recognize antigen in the context of class I MHC products. They become the cytotoxic T cells or the regulatory T cells and bear the CD8 marker.

Unlike immunoglobulin, the T-cell receptor remains an integral part of the lymphocyte. The first step in T-cell activation following contact with an APC that possesses an appropriate MHC-peptide complex on its surface is a series of metabolic reactions involving uptake of thymidine.

T-cell proliferation is accompanied by secretion of cytokines. These factors act on other cell types and influence cell division, differentiation, metabolism, and effector function. Cytokines are responsible for the mobilization of cell-mediated immunity. Some of the most prominent cytokines produced by T cells are the interleukins (IL).

IL-2 is secreted by activated T cells. It promotes growth of T cells and initiates the development of a population of lymphocyte-like cells referred to as natural killer (NK) cells. IL-4 is also produced by stimulated T cells and, together with IL-5 and IL-6, causes B cells to undergo maturation, proliferation, and to subsequently secrete immunoglobulin. These interleukins also influence the class of immunoglobulins produced by B cells. Other cytokines, termed "interferons," were initially described as proteins that interfere with replication of pathogenic viruses. A number of interferons have been characterized among which interferon-γ plays a significant role in immune response. It increases expression of class I and class II MHC products and stimulates macrophage and NK cell activity. Finally, another cytokine, tumor necrosis factor, injures tissue cells and produces cachexia (i.e., the anorexia of chronic disease).

VI. Regulation of the Immune Response

The important role played by the MHC in T-cell maturation, antigen presentation, and T cell–B cell interactions have all been described. It is not surprising, therefore, that the MHC also takes a part in determining immune response efficiency. Indeed, class II MHC genes were originally referred to as "immunoregulatory" or Ir genes because in mice they control the ability to recognize simple peptide antigens.

Two mechanisms have been proposed to explain the function of MHC genes in the regulation of immune responses. First, MHC is critical in the negative and positive selection of T-cell populations. Therefore, the distribution of T cells with various specificities differs from one animal to another depending upon the genetic constitution (genotype) of the MHC. Second, MHC determinants must bind antigenic determinants before they can be recognized by T cells. The degree of binding affinity determines T-cell response. The fine structure of the class II MHC product on the cell surface influences the immune response.

The important role of the MHC in regulating immune response is reflected in human immunology. Autoimmune diseases are caused by a disturbance in immunoregulation resulting in the production of antibodies and/or T cells that react with antigens of the host itself. Among the diseases characterized as autoimmune are diabetes mellitus, rheumatoid arthritis, and myasthenia gravis. The risk of developing any of these diseases is related to the genotype of the host's MHC.

In addition to the MHC, other genetic factors influence immune response. They include genes that regulate the production of immunoglobulins and of the T-cell receptor. Genes that control cytokine production and hormone levels are also important. Females, for example, are generally poorer antibody producers than males. Environmental factors also have an impact. Irradiation, infection, nutrition, and stress all affect the immune response.

In addition to host factors, intrinsic characteristics of antigens are important in determining immuno response. The molecule must have a firm configuration; therefore, lipids are generally poor antigens. Even different parts of the same large molecule differ in immunogenicity. Amphophilic portions of a large protein molecule are most likely to be recognized by T cells. Sequences with a certain degree of molecular flexibility are likely to be more immunogenic. These properties probably determine the ability of the antigenic determinant to form a complex with the class II MHC product.

The intrinsic immunogenicity of substances can often be enhanced by administration of adjuvants. One function of an adjuvant is to provide a depot for antigen to be released to the immune system over a period of time. Alum precipitation, encasement of antigen in microspheres, or injection of antigen–antibody complexes produce such depots. Other adjuvants additionally induce an intense local inflammatory response. The best studied example is complete Freund's adjuvant. It consists of an emulsion of soluble antigen in mineral oil with the addition of killed acid-fast bacteria. These bacteria attract large numbers of inflammatory cells that enhance antigen uptake and subsequent production of cytokines. This type of adjuvant is particularly useful in promoting cell-mediated immunity.

VII. Assessment of Environmental Agents

One of the most important contributions of immunology to environmental studies has been the use of immunochemical methods for the demonstration of minute quantities of antigenic substances in complex mixtures. These methods are referred to as immunoassays. The specificity and sensitivity of antigen–antibody reactions have made these methods widely applicable.

As described previously, all antibody molecules are bivalent or multivalent, since they have two or more identical binding sites. This characteristic of the antibody molecule has allowed it to become an important reagent for immunochemical analysis. When added to a solution containing a mixture of molecules, an antibody will specifically crosslink only those molecules bearing a particular antigenic determinant. Depending on the ratio of antigen to antibody and on the characteristics of the antigen, the complexes will become insoluble and produce a visible precipitate. The optimal antigen-to-antibody ratio is generally the one in which neither free antibody nor free antigen can be demonstrated in the supernatant solution. (In practice the greatest amount of precipitate occurs in mixtures with a slight excess of antigen.) The precipitate can be isolated and measured quantitatively using chemical or physical methods.

Antigen–antibody interactions can also be studied in a semisolid medium such as agar gel. Based on the concentration of antigen and antibody and diffusion characteristics, bands of precipitate will appear in the gelified medium that can be visualized directly or subjected to chemical or enzymatic colorimetric reactions. Gel precipitation is especially useful in resolving mixtures of antigens. For more complex mixtures, the antigen solution can be first separated by electrophoresis or diffusion and then reacted with the antiserum. Immunoelectrophoresis and Western immunoblotting are examples of this methodology.

Agglutination differs from precipitation in that the antibody causes aggregation of insoluble particles into large clumps. It is generally more sensitive for the demonstration of small amounts of antibody. For direct agglutination, particles bearing the appropriate antigenic determinant are mixed with serial dilutions of antiserum. Because of the greater sensitivity of agglutination over precipitation, it is sometimes advantageous to convert precipitation to agglutination by fixing a soluble antigen to the surface of an appropriate particle. This method is called indirect agglutination and red blood cells, latex, or other particles are frequently used as carriers.

The cardinal principle in all immunoassays is the appropriate selection of the antibody reagent. Animals respond to complex foreign molecules by producing a number of antibodies with differing specificities. Before use in an immunoassay, an antiserum must be carefully evaluated. Unwanted antibodies must be eliminated by immunoabsorption or by enrichment by affinity chromatography. An alternative procedure uses monoclonal antibodies wherein the entire population of immunoglobulin molecules has a single specificity.

The sensitivity of immunoassays depends fundamentally upon the affinity of an antibody for the corresponding antigen. Antibody affinities vary according to the method of immunization, the sampling time, the immunogenicity of the antigen, and the genetic constitution and environmental status of the host. For example, IgG antibodies sampled late in the course of immunization generally have greater affinity for antigens than IgG antibodies taken soon after immunization.

The sensitivity of immunoassays can be increased greatly by using appropriate labels on the antigen or the antibody. The labels most commonly employed are radioisotopes since they can be readily measured with a high degree of sensitivity. More recently, nonradioisotopic labels, such as enzymes or fluorescent groups, have replaced many of the original radioimmunoassays.

VIII. Immunotoxicology

Environmental agents may have profound effects on the immune response of the host. A new area of study, referred to as "immunotoxicology," has been developed to investigate injuries to the immune system resulting from deliberate or accidental exposure to environmental agents. The mechanism whereby injury occurs is twofold: one phenomenon involves suppression of immunity resulting in enhanced susceptibility to infection or malignancy; the other concerns the excessive or dysregulated immune response seen in allergy or autoimmunity.

There is an increasing public awareness that environmental agents may suppress immune processes. This concern is substantiated by numerous data collected from studies in animals and humans treated with immunosuppressive drugs. A well-established correlation has been observed between immunosuppressive treatment of patients and an increase in the incidence of infections and neoplastic disease. It has been estimated, for example, that 50% of transplant patients receiving immunosuppressive drugs develop cancer within 10 years following transplantation. Similarly, recipients of cancer chemotherapy have an increased incidence of infection and malignancy. The types of resulting neoplastic diseases are often quite selective; for example, patients with chemical immunosuppression show a high prevalence of lymphomas, leukemias, and skin tumors. Kaposi's sarcoma, a rare form of cancer, occurs at a much increased frequency in severely immunosuppressed patients.

The type of infection in immunosuppressed individuals can often be related to the particular site of immune system injury. Patients with a defect in B-cell function and antibody synthesis are prone to develop acute pyogenic infections due to staphylococci, streptococci, pneumococci, *Pseudomonas,* and *Haemophilus*. If the complement system is injured, meningococcal infections are relatively prevalent. On the other hand, patients with defects in cell-mediated immunity tend to develop generalized infections due to intracellular pathogens. Thus, there is an increased prevalence of viral infections, such as rubeola, varicella, or generalized vaccinia; mycosis-like candidiasis and cryptococcosis; parasitic infections such as toxoplasmosis and pneumocystis pneumonia; and bacterial infections, including generalized tuberculosis and related acid-fast infections. If phagocytosis is impaired, many common, usually harmless, bacteria, including *pseudomonas* and nonpathogenic staphylococci from the skin, are able to establish themselves.

Enhancement of the immune response by environmental agents may produce (1) immediate hypersensitivity reactions associated with IgE production and mediator release, or (2) delayed hypersensitivity associated with T-cell proliferation and cytokine secretion. Immediate hypersensitivity is widely recognized as a major cause of human allergy. The signs of an allergic response are related to the route of exposure: inhalation of the inciting agent causes rhinitis or asthma, while exposure through the skin produces the rash of atopic dermatitis, and ingestion of the agent may lead to food intolerance. All of these conditions are associated with a genetic predisposition to produce unusually high levels of IgE in response to antigen. Elevated levels of IgE antibody can be demonstrated directly in the patient by means of skin tests. Following injection of a minute amount of the chemical into the skin, the individual develops an immediate wheal-and-flare response due to the local liberation of histamine and other mediators.

Delayed hypersensitivity is the cause of contact dermatitis and of certain generalized, chronic inflammatory diseases. In addition to full antigens, simple chemicals can often produce contact dermatitis. Evidence suggests that the chemicals, acting as haptens, bind to autologous proteins of the skin. A specialized dendritic cell of the skin, the Langerhans' cell, appears to be important in the induction of sensitivity to chemicals. Langerhans' cells express class II MHC products and, therefore, serve as effective antigen presenters. After contact with the sensitizing chemical, T cells enter the lymphatic circulation and localize at draining lymph nodes. Immune T cells produced in this manner are distributed throughout the body, including the site of application of the immunizing chemical. The characteristic sign of delayed hypersensitivity is the delayed skin test. It consists of a local erythema and induration (thickening) that appears 18–24 hr after intradermal injection of antigen. This local reaction usually reaches its peak at 24 to 48 hr and then gradually fades.

In the lung, gastrointestinal tract, and other tissues, delayed hypersensitivity reactions are often associated with granuloma formation. Granulomas are focal, chronic inflammatory responses that may represent manifestations of a specified cell-mediated response. The process begins by phagocytosis of the inciting agent by mononuclear cells, which secrete a variety of cytokines. More monocytes are recruited, sometimes resulting in cellular fusion and production of multicellular giant cells. Surrounding these giant cells are collections of epithelial cells formed from mature macrophages that enlarge and flatten, arranging themselves as palisades around the giant cell core. As the central cells of the granuloma lose their blood supply, they die and later calcify. The presence of these granulomas, then, suggests the presence of a cell-mediated immune response. It should be remembered, however, that many immunologically inert materials, such as silica, can also induce granuloma formation.

A blurring of the distinction between antigens of self and nonself on the part of the host leads to autoimmunity. Autoimmune disease may be the destructive consequence of self-directed reactions. The host's own immune response may produce functional impairment, inflammation, and tissue damage.

Some autoimmune diseases are due to circulating antibody; for example, autoimmune hemolytic anemia is caused by antibodies to the red blood cells, and myasthenia gravis by antibodies to the acetylcholine receptor at the neuromuscular junction. Sometimes autoimmune disease is the consequence of the union of antibody and circulating antigen in the production of immune complexes. Such complexes may be deposited in various small vessels of the kidney, lung, skin, and brain where they may activate the complement system and instigate an inflammatory response. Other autoimmune diseases are due to cell-mediated immunity. In insulin-dependent diabetes mellitus, for example, T cells specifically target the insulin-secreting beta cells of the pancreatic islets. This causes islet cell inflammation as well as a decrease in circulating insulin.

The initiation of autoimmune disease depends on both genetic and environmental factors. Genetically, the class II MHC genotype of the patient is the best predictor of many autoimmune diseases. Human MHC is referred to as HLA, and HLA typing has become a common tool in clinical immunology where it can be used as a marker for individuals most disposed to developing a particular autoimmune disease. Some autoimmune diseases are associated with a particular gene controlling the variable portion of the T-cell receptor. The presence of these genes may serve as predictors for inordinate susceptibility to that particular disease. In addition to the genes that regulate the immune response, other inheritable traits may increase the genetic predisposition to autoimmune disease. They include genes that affect the target organ, such as the thyroid gland in the case of chronic autoimmune thyroiditis. Inheriting both types of genes, those affecting immunoregulation and those involving target organ vulnerability, constitutes the greatest predisposition to autoimmune disease.

A number of environmental agents are known to induce autoimmune responses. Among them are some commonly used drugs: α-methyldopa may induce autoimmune hemolytic anemia in susceptible individuals; penicillamine can cause autoimmune myasthenia gravis, and procainamide is the cause of a drug-induced form of systemic lupus erythematosus. Chemical pollutants may also be important. Chlorinated hydrocarbons and mercuric chloride cause glomerular nephritis in susceptible experimental animals and probably in some susceptible humans. Progressive systemic sclerosis has been associated with exposure to trichloroethylene, vinyl chloride, perchloroethylene, and aromatic hydrocarbon solvents such as benzene and toluene. Silica may induce silicosis, a granulomatous reaction in the lungs. Dietary iodine is thought to be responsible for an increased incidence of chronic thyroiditis.

The mechanisms by which these environmental agents act may differ greatly from disease to disease. Sometimes the agents cause dysregulation of the immune system. Mercuric chloride is known to be a nonspecific stimulator of B cells. Excess iodine on the other hand disturbs the normal function of the thyroid gland. α-Methyldopa probably alters the surface of the erythrocyte, whereas procainamide may resemble antigens found in the cell nucleus. The latter mechanism, referred to as "molecular mimicry," is always a hazard when environmental agents happen to have a configuration similar to antigens in the body of the host.

Related Article: BIOTECHNOLOGY, OCCUPATIONAL HEALTH ISSUES.

Bibliography

Newcombe, D. S., Rose, N. R., and Bloom, J. C., eds. (1991). "Clinical Immunotoxicology," Raven Press, New York.

Paul, W. E., ed. (1989). "Fundamental Immunology," second edition, Raven Press, New York.

Roitt, I. M., Brostoff, J., and Male, D., eds. (1989). "Immunology," second edition, Gower Medical, London.

Rose, N. R., Milgrom, F., and van Oss, C. J., eds. (1979). "Principles of Immunology," second edition, Macmillan, New York.

Stites, D. P., and Terr, A. I., eds. (1991). "Basic and Clinical Immunology," seventh edition, Appleton & Lange, Norwalk.

Industrial Solvents

Ingvar Lundberg
Karolinska Hospital, Stockholm, Sweden

Carola Lidén
Karolinska Hospital, Stockholm, Sweden
National Institute of Occupational Health, Solna, Sweden

Glossary

Hippuric acid Colorless crystals obtained from the urine of domestic animals; slightly soluble in water and alcohol; melts at 188°C.

Phenol Toxic, white crystalline mass used in solvents, resins, and weed killers.

Styrene Fragrant liquid, soluble in alcohol and ether, used in making polystyrene plastics and rubbers.

Toluene Flammable, toxic, colorless liquid; used in explosives, high-octane gasoline, and organic synthesis.

***INDUSTRIAL SOLVENTS** are organic liquids commonly used to dissolve other organic materials such as oils, fats, resins, rubber, and plastics. Solvents are used in a wide variety of industrial processes and very large quantities are consumed. They are chemically diverse and may be divided into a number of categories based on their chemical structure. The main categories are shown in Table I. The partition coefficients mentioned in the table are determined as the concentration ratio of the solvent (equilibrated in fluid) and gas mixture at a specific temperature. Thus, they express the affinity of the solvent for different media.*

This article will exclusively concern data on health effects from solvent exposure obtained in occupational studies. Environmental concentrations of solvents are in general very low and the effects of the much higher occupational exposures have to be used to assess possible effects from environmental exposure.

I. Exposure and Primary Uses

NIOSH (National Institute for Occupational Safety and Health) estimates that 9.8 million people were exposed to solvents in the United States in the first part of the 1970s. In the 1980s approximately 15% of the Danish labor force was estimated to be occupationally exposed to solvents daily. Thus, exposure to organic solvents is very common and constitutes one of the most important occupational chemical hazards. Some occupations may entail particularly heavy solvent exposures. A number of such occupations are listed below.

Painting with solvent based paints, particularly spray painting (mixed solvents)
Floor-laying (mixed solvents)
Production of glass-fiber reinforced polyester plastics, particularly lamination of large areas as in the production of boats (styrene)
Paint manufacturing, particularly manual cleaning of equipment with solvents (mixed solvents)
Rotogravure printing (toluene)
Metal degreasing (commonly trichloroethylene)
Dry cleaning (commonly perchloroethylene)

Most workers are exposed to mixtures of organic solvents and the majority of these workers are exposed to solvents in paints and lacquers. The most

Table I Physicochemical Properties of Some Commonly Used Solvents and Their Odor Thresholds

Chemical class of hydrocarbon	Solvent	Boiling point (101.3 kPa, °C)	Vapor pressure (25°C, kPa)	Flash point (°C)	Water solubility (weight%)	Partition coefficients			Odor threshold (ppm)
						Water/air	Blood/air	Oil/air	
Aliphatic	*n*-Hexane	156	16	<−20	<0.1				
Aromatic	Benzene	80	10	−11	<0.1	3	8	490	
	Toluene	111	2.9	4	<0.1	2	16	1470	0.03–3
	Xylenes	138–144	0.6–1.1	23–30	<0.1	2	42	4050	20–40
	Styrene	146	0.6	32	0.3	5	40	5000	0.05–25
Halogenated	Methylene chloride	40	45	none	2	7	9	150	50–80
	Chloroform	61	21	none	0.7	4	10	400	200
	Carbon tetrachloride	77	12	none	<0.1	0	2	360	25–50
	Trichloroethylene	87	8	none	0.1	1.5	9	700	20–100
	Tetrachloroethylene	121	1.9	none	<0.1	0	13	1900	50–70
	Methyl chloroform	74	13	none	<0.1	1	2	350	75–100
Esters	Ethyl acetate	77	10	−4	9				50–200
Ketones	Acetone	56	25	−19	100	400	250	90	200–450
	Methyl ethyl ketone	80	9	−7	27	250	200	260	<25
	Methyl-*n*-butyl-ketone	127	0.4	23		110	130	1640	
Alcohols	Ethanol	78	6	12	100				350
	Isopropanol	82	4	12	100				200
	n-Butanol	118	0.6	29	8				15–25
Glycol ethers	2-Ethoxyethanol	135	0.5	40	100				
Miscellaneous	Carbon disulfide	46	40	−30	0.3	1	2	—	1
	Dimethylformamide	153	0.4	58	100				
Petroleum distillates	White spirit	150–200	<0.7	35–40	<0.1	—	—	—	variable

important occupational groups include painters (e.g., house painters, car painters) and paint manufacturing workers. The solvent mixtures often contain a petroleum distillate (brush painting) (e.g., painter's naphta or white spirit), or xylene or toluene (industrial painting) as main components.

A. Decreasing Exposure Levels in Industry

The preventive measures taken against solvent health hazards are manifested in the lists of hygienic standards that are published in most industrialized countries. Figure 1 shows the ACGIH (American Conference of Governmental Industrial Hygienists) standards and TLVs (Threshold Limit Values), for some common solvents over a period of two decades.

In general the safe-level standards have been reduced. The reduction is most pronounced for methylene chloride. The finding that this solvent was metabolized to carbon monoxide caused the dramatic reduction between 1970 and 1980 while the potential carcinogenicity of the compound, as shown in animal experiments, was the reason for the reduction during the last decade. The very low standard for benzene is due to its hematotoxicity and carcinogenicity. For most other solvents, the hygienic standards are commonly based on neurotoxic and irritative effects.

It is likely that the standards in fact give a rather accurate picture of the exposure conditions within reasonably sized operations in the industrial branches that have the heaviest exposures. Thus, the reduction of the standards probably reflects decreasing exposure levels in industry. Moreover, it is likely that there has been a concomitant reduction in the number of heavily exposed workers within the process industry due to automation which reduces the number of workers and moves the remaining workers further away from the production process.

B. Reducing Exposure by Diminishing the Solvent Content in Paints

In artisan occupations (like painters) the possibilities of reducing the exposure by closed systems and exhaust ventilation are much smaller than in industry. In fact, the efforts to reduce exposure to paint solvents have followed another route, (i.e., to reduce the solvent content in the paints by replacing the solvent soluble binders with those that are water soluble). This tendency is prevalent in all industrialized countries although the pace in the shift may vary.

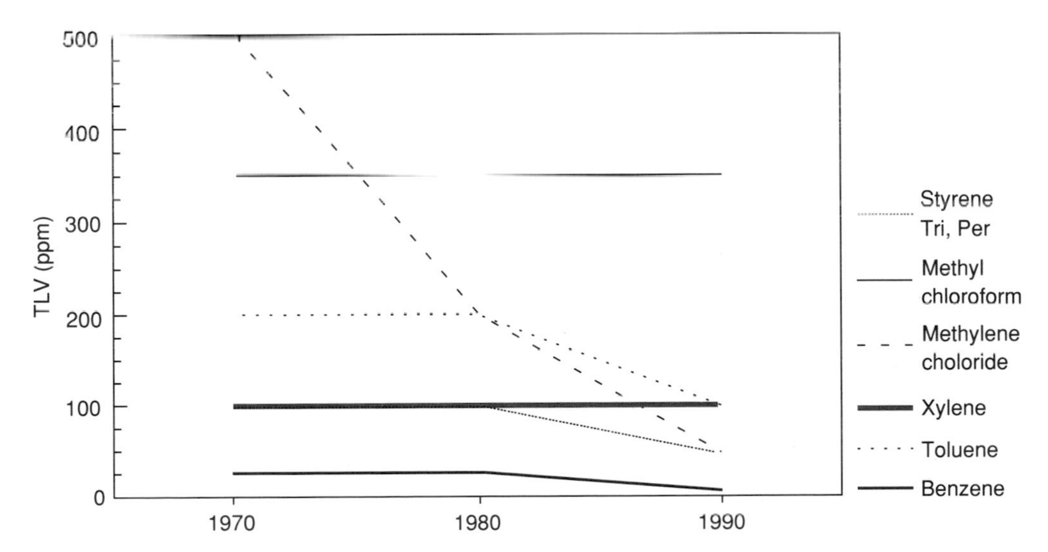

Figure 1 Threshold limit values (TLVs) for some common solvents proposed by the American Conference of Governmental Industrial Hygienists (ACGIH) 1970, 1980, and 1990.

II. Uptake, Biotransformation, Distribution, and Elimination

A. Uptake and Distribution

1. Lungs

During occupational exposure, solvents may be taken up through the lungs or through the skin while uptake through the gastrointestinal tract is negligible.

Generally the lungs by far provide the most important uptake route. The critical determinant of pulmonary uptake is the solubility of the solvent in blood. The uptake at rest for methylene chloride, trichloroethylene, toluene, xylene, and styrene is between 50 and 70% of the amount inhaled.

The amount of solvents taken up per unit time increases during excercise. Styrene, xylene, acetone, and butyl alcohol are more soluble in blood and tissues than methylene chloride, trichloroethylene, and toluene, for example, and show a linear increase in uptake with increasing workload, due to increasing lung ventilation. Thus, it is important to determine workload when considering the potential health effects of solvent exposure.

The effects that most solvents have in common (i.e., on the nervous system and skin) are to a large extent dependent on their solubility in fat. Thus, a significant part of the amount of solvents taken up is distributed to fat tissues. Five to twenty percent of the total uptake of xylene, styrene, methylene chloride, and toluene is retained in fat deposits in the human body. In studies on occupationally exposed workers it has been shown that a significant accumulation of some solvents in fact occurs in body fat, implying a continuous low-dose solvent exposure also occurs during work-free periods.

2. Skin

Skin absorption from exposure to solvent vapors is negligible for most common solvents. However, solvents which contain a lipophilic as well as a hydrophilic part (e.g., glycol ethers and dimethylformamide) are very readily absorbed through the skin, and skin exposure to solvent vapors may contribute in large part to the total uptake.

For most common solvents, rather extreme skin exposures are needed before skin absorption can reach the level of the lung uptake. For example, if both hands were kept in a xylene bath for 15 min, the uptake would be of the same as the respiratory uptake from air exposure to 100 ppm during the same period of time.

Lists of hygienic standards often contains a special notation, "skin," to mark if a substance is easily absorbed through this route. The notations are in general based on limited data and the absence of the notation should not be regarded to mean that the solvent is safe from this point of view.

B. Biotransformation and Elimination

In general, much of the absorbed solvent is biotransformed to more water-soluble metabolites that may be excreted in urine. The main site of metabolism is the liver. Up to 95% of xylene and styrene absorbed and up to 80% of the total amount of trichloroethylene and toluene absorbed is biotransformed. However, some common solvents (like methyl chloroform and perchloroethylene) are metabolized to a very small extent and largely excreted unchanged through exhalation.

Biotransformation has important implications in the evaluation of solvent toxicity and in the assessment of solvent exposure. Solvents may be biotransformed into reactive metabolites that are more toxic than the parent compound. In fact, all solvents that show a predictable liver toxicity (such as carbon tetrachloride, chloroform, and dimethylformamide) are actually toxic through reactive metabolites.

Similarly, peripheral neurotoxicity of *n*-hexane and methyl-*n*-butyl ketone is caused by metabolites of the solvents. Metabolites from other solvents, such as the epoxides produced from trichloroethylene and styrene, are less reactive and do not cause liver toxicity but may instead have carcinogenic effects, possibly in other tissues. The carcinogenicity of benzene is also linked to its metabolism, although the ultimate carcinogen has not been identified.

Nontoxic metabolites from other solvents may be used to estimate the dose of the solvent that has been absorbed, see as is discussed in Section III below.

Solvent metabolism varies substantially between different individuals leading to potential differences in susceptibility to solvent effects. Ingested ethyl alcohol (ethanol) inhibits the metabolic degradation of toluene and xylene, increases the blood levels, and decreases the rate of elimination from blood of these solvents, thus probably enhancing the central nervous system effects of these solvents. Smoking may increase the elimination of toluene from blood.

III. Biological Monitoring

For some common solvents, the solvent itself or some metabolite may be measured in end-exhaled air, blood, or urine in order to monitor the amount of the solvent taken up by the body. Examples are shown below.

Solvent	Indicator
n-Hexane	2,5-Hexanedione in urine
Benzene	Phenol in urine Benzene in exhaled air
Toluene	Hippuric acid in urine Toluene in venous blood
Xylenes	Methyl hippuric acid in urine
Styrene	Mandelic acid and phenyl glyoxylic acid in urine
Trichloroethylene	Trichloroethanol and trichloroacetic acid in urine
Methyl ethyl ketone	Methyl ethyl ketone in urine
Carbon disulfide	2-Thiothiazolidine-4-carboxylic acid (TTCA) in urine
Methyl chloroform	Methyl chloroform in end-exhaled air Trichloroethanol and trichloroacetic acid in urine
Perchloroethylene	Perchloroethylene in end-exhaled air Perchloroethylene in venous blood Trichloroacetic acid in urine

Biological monitoring is most often used as an alternative to air exposure measurements. Urinary metabolites of trichloroethylene and styrene as an estimate of the dose (and exposure) to these solvents, provide the most commonly used biological indicators of solvent exposure. The combined urinary excretion of mandelic acid and phenyl glyoxylic acid also give a good measure of preceding styrene exposure at low exposure levels; mandelic acid, however, may also be used alone as an acceptable indicator.

Trichloroethanol and trichloroacetic acid are excreted in urine after trichloroethylene exposure. Trichloroacetic acid has a long half-life (around three days) and will thus integrate the measured exposure over a longer period of time. Trichloroethanol has a much shorter half-life and may give a reasonable estimate of the exposure during a particular workshift. The combined excretion of the metabolites has also been suggested to estimate the exposure.

Hippuric acid is a toluene metabolite but also a metabolite of benzoic acid and thus it may very often be present in urine in significant amounts without any toluene exposure. Toluene exposures below 100 ppm cannot be quantified by hippuric acid excretion. Cresol excretion has more recently been suggested as an indicator of toluene exposure as well as used to measure toluene in venous blood.

Phenol in urine cannot be used for the quantification of benzene exposures around the present hygienic standards in industrialized countries as it is produced from other sources in too large quantities. Although exhaled air analysis is in general unreliable for exposure quantification, it appears that benzene in exhaled air is currently the best possibility to estimate benzene exposure.

Methyl hippuric acids give a reliable estimate of the xylene exposure. However, since xylenes are usually used in solvent mixtures and since the goal is generally to evaluate overall solvent exposure, the practical use of this indicator has been limited.

IV. Health Effects

In this dicussion, we will differentiate between short- and long-term exposures, denoting exposures of hours to days, and months to years, respectively. We will also differentiate between acute, subchronic, and chronic effects. Acute effects are readily reversible after the discontinuation of exposure. Subchronic effects are most often reversible within six months and always within one year. Chronic effects are not reversible, only partly reversible, or reversible after more than one year.

A. Nervous System

1. Peripheral Nervous System

Workers with long-term, heavy exposure to *n*-hexane (above 100 ppm) may develop symmetrical sensory dysfunction in the hands and feet with loss of superficial pain reception and sensitivity to temperature change, light, touch, and vibration in the toes. In more severe cases the motor function is also affected with muscular weakness in the limbs. Similar symptoms may occur among workers heavily exposed to methyl-*n*-butyl ketone.

The toxicity of the two solvents depends on a common metabolite, 2,5-hexanedione. In fact it has been shown that other diketones with the same distance betwen the ketone groups also share the ability to damage peripheral nerves.

If severe polyneuropathy has occurred, the most important treatment is removal from exposure. After discontinuation of exposure, recovery will occur slowly reflecting the regeneration of peripheral nerves. In severely injured individuals, residual permanent impairment has sometimes been noted.

Long-term exposure to styrene may induce a mild sensory neuropathy. The neuropathy may manifest itself as an increased pain and tingling in the limbs although the effects at time-weighted average air concentrations at around 100 ppm and below seems to be predominantly subclinical. The effects are largely reversible over a month. Concurrent alcohol consumption diminishes the peripheral nerve effects.

Toluene does not seem to induce peripheral nerve effects in exposed groups.

Mixed solvents (used for different types of painting), however, may slightly impair peripheral nerve functioning after long-term exposure. However, the effects seem to be sublclinical in the large majority of exposed workers.

2. Central Nervous System

In the middle of the 19th Century, the vulcanization of rubber was developed and carbon disulfide was found to be a suitable solvent in the process. Soon after its introduction, the French physician August-Louis Delpech described severe neuropsychiatric effects among rubber workers and was able to identify carbon disulfide as the causative agent. Some symptoms (e.g., excitement followed by collapse) could appear after only a few hours of exposure. Other symptoms (e.g., fluctuations of mood, sleeping and memory problems, loss of sensations from different parts of the body, and impotence) appeared much slower.

These symptoms as described by Delpech seem largely to fit with modern conceptions of the acute and chronic effects of solvents on the central nervous system.

Short-term high exposures to organic solvents produce narcotic effects. In fact, several solvents (e.g., trichloroethylene and chloroform) have been used as general anesthetics. The narcotic effect of a solvent is directly related to its solubility in fat. Acute symptoms include headache, dizziness, confusion, a feeling of drunkenness, and where exposure continues, unconciousness and death may follow.

Such symptoms have in fact been very common in several heavily exposed occupations such as painters, paint manufacturing workers, and floor-layers; workers often report fainting on the job due to solvent exposure. The acute symptoms are reversible after the discontinuation of exposure.

Several laboratory investigations have shown that current solvent hygienic standards in many countries are too permissive to completely protect from acute effects on the central nervous system.

3. Subchronic and Chronic Effects

a. Mixed Solvents

A large number of disability pensions attributable to neuropsychiatric disorders has been reported in solvent-exposed occupational groups. Other studies have subsequently verified this increase although the authors were reluctant to attribute the excess to solvent exposure. Several studies have also reported excess neuropsychiatric symptoms and impaired neuropsychological performance in active house painters, car painters, paint manufacturing workers, and jet-fuel exposed workers although some studies could not reproduce the findings.

b. Toluene

Toluene is a popular sniffing agent and the resulting symptoms and signs among long-term inhalers are well known. Tremor, unsteady gait, and memory impairment belong to the most commonly encountered symptoms but a decreased sense of smell, a destruction of the visual nerve, hearing impairment, and nerve damage in the limbs might also occur.

In workers with long-term occupational exposure, there seems to be an increase of the same neuropsychiatric symptoms as those found in association with mixed-solvent exposure. However, impairment in neuropsychological tests has not been found after long-term occupational exposure to air concentrations currently prevalent in industrialized countries (100 ppm and below).

c. Styrene

Sleepiness, an apathetic feeling, and equilibrium problems have been reported after very high exposures. Lower exposures may elicit more subtle effects. A slowing of the reaction time has been found in groups with 8-hr time-weighted average exposures above 50 ppm. Performance in other neuropsychological tests may also be affected. Subchronic or chronic symptoms, and effects in neuropsychological tests, have been found after exposures to roughly the same air concentrations.

d. Chlorinated Solvents (Trichloroethylene)

Trichloroethylene exposure may specifically damage one of the nerves responsible for the sense of perception from the face (i.e., the trigeminal nerve) resulting in a feeling of facial numbness. However, this condition does not seem to occur at exposure levels presently prevalent in industrialized countries.

A classical study, published in 1955, showed an increased prevalence of neuropsychiatric symptoms with increasing exposure among workers exposed to trichloroethylene in a mechanical engineering industry. A "psycho-organic syndrome" was diagnosed characterized by intellectual impairment with memory deficiencies and affective changes. The prevalence of this syndrome increased with increasing exposure levels. The authors considered that the prevalence of neuropsychiatric symptoms and signs was definitely increased at exposure levels above 40 ppm.

Other chlorinated solvents (e.g., methylene chloride and methyl chloroform) have also been shown to increase neuropsychiatric symptoms among exposed workers.

Thus, long-term exposure to different solvents may elicit similar symptoms from the central nervous system. However, some doubt remains that long-term occupational solvent exposure may contribute to chronic effects on the central nervous system. Studies performed thus far have been criticized for unsatisfactory evaluation of alternative explanations of the association, such as alcohol consumption and primary mental capacity. Thus, although further research is needed to clarify the association, most researchers now agree that solvent exposure may indeed cause chronic neuropsychiatric impairment.

e. Symptoms and Signs

The symptoms described by solvent-exposed workers include: memory problems, concentration difficulties, and affective changes such as aggressiveness and depression, fatigue, vertigo, decreased libido, sleeping problems, frequent headaches, and vegetative symptoms such as palpitations and increased sweating.

The time of appearance and severity of the symptoms depend on the exposure level. In heavily exposed occupations, the symptoms may appear after some years of exposure.

The development of the symptoms is insidious. In the beginning, symptoms may disappear over weekends and holidays, but over a period of years they will become chronic. Ultimately, they will typically be perceived by the individual as a fundamental change in personality.

Some of the symptoms, such as memory problems and concentration difficulties, may also be demonstrated as an impairment in the corresponding functions in neuropsychological tests. Such tests were originally developed to measure less severe forms of brain damage and have been widely used in the diagnosis of early stages of dementia.

f. Chronic Toxic Encephalopathy

An organic brain syndrome, chronic toxic encephalopathy, may develop after long-term, heavy exposure. This syndrome is characterized by the symptoms listed below.

- Primary symptoms
  - Reduced intellectual function
    - Impairment of memory and concentration
    - Difficulties in learning, abstract thinking (in severe cases)
  - Personality changes such as increased depressiveness or aggressiveness
- Ancillary symptoms
  - Dizziness
  - Sleeping problems
  - Palpitations, increased sweating
  - Reduced interest in sex

Diagnostic requisites include the following:

- Long and/or intense exposure to industrial solvents (usually more than 10 years in a heavily exposed occupation);
- Typical symptoms (as listed above)
- Neuropsychological test results that concur with the diganosis;
- A reasonable probability that other causes are excluded.

The symptoms, as well as the typical results in neuropsychological testing, are not specific to solvent-induced changes but may occur due to other known causes, such as alcoholism or depression, or for unknown reasons. EEGs and CAT scans do not help in the individual diagnosis. In conclusion, the evaluation of the syndrome requires expertise in occupational medicine.

g. Time Course and Prognosis

The symptoms and the neuropsychological impairment seem to progress as long as exposure continues. When exposure has stopped, the condition may be alleviated or remain stationary; progression does not seem to occur.

B. Skin

Contact dermatitis (inflammation of the skin due to contact with an external agent) represents 90% of occupational skin diseases. The most common localization is the hands. Hand eczema might be an irritant contact dermatitis, due to skin irritation, or an allergic contact dermatitis due to contact allergy.

Skin exposure to organic solvents is responsible for up to 20% of occupationally induced cases of dermatitis. The main effects of organic solvents on the skin are irritant contact dermatitis and acute chemical burns, but contact allergy occurs rarely.

The skin may be exposed to organic solvents when cleaning hands with solvents or cleaners containing solvents. Although poor work habits and lack of proper protective equipment may also contribute to the exposure, knowledge of hazards and suitable protection is nonetheless often insufficient.

1. Skin Irritation

Many solvents cause irritant contact dermatitis through their defatting action on the skin. Exposure results in dryness, erythema, and edema. Repeated exposure may easily develop into an irritant contact dermatitis. In some cases, prolonged exposure and occlusion causes chemical burns. The development of contact dermatitis also results in increased water loss and facilitated uptake of other substances.

Aromatic solvents are more irritating than aliphatic, causing irritant dermatitis, severe defatting, and burns after prolonged contact.

Halogenated solvents also cause dermatitis. Skin contact with tetrachloroethylene may cause erythema, blistering, burns or exfoliation of the skin. Trichloroethylene is not a strong skin irritant, but prolonged contact or occlusion may cause chemical burns. This has, for example, frequently been caused by dry cleaned clothes which had not been properly dried. Methylene chloride may also cause chemical burns; methyl chloroform may cause dermatitis by prolonged or repeated exposure.

Alcohols are less irritating than aldehydes or ketones. They act as drying agents and provoke irritant contact dermatitis, but the irritancy decreases as the molecular size increases.

Esters are more potent skin irritants than the corresponding alcohols. Ketones are mild skin irritants.

Carbon disulfide is one of the most irritating solvents.

2. Sensitization

As organic solvents often cause irritant contact dermatitis, they may facilitate dermal sensitization to allergens that workers may be exposed to. Contact allergy to organic solvents is rare, however.

The use of turpentine as a paint solvent has declined considerably but it is still often a component of oil-based paints and varnishes and is also used as an ingredient in other products. Turpentine was previously a rather common cause of allergic contact dermatitis among painters as it contains monoterpenes whose oxidation products are sensitizers.

Single cases of contact allergy to the alcohols methanol, ethanol, ethylene glycol, and propylene glycol have been reported, as well as a few reports on contact allergy to styrene and dioxane.

3. Protective Equipment

Protective gloves should be used to prevent dermal exposure to organic solvents. They are, however, manufactured by different formulations—according to use and thickness, by different methods, and from several polymeric materials, with variable resistance to chemicals. They therefore have different properties with regard to their protective effect against organic solvents. When selecting protective gloves, permeation test data must be considered, as well as risk of side effects (i.e., of the gloves themselves) and the risk of contamination. Barrier creams are less efficient than protective gloves in reducing percutaneous absorption of organic solvents.

4. Diagnosis and Prognosis

Hand eczema is the most common skin lesion caused by organic solvents. Dermatitis on the face might also be caused by organic solvents through hand contamination or by solvents that appear as an aerosol. Dermatitis on the neck and flexures (elbow, knee, etc) may be caused by solvents in dry-cleaned working clothes which have not been properly dried. Hand eczema is often caused by a combination of several factors such as exposure to solvents, water, detergents, and other skin irritants, as well as contact allergens.

Mild hand eczema might clear spontaneously. Hand eczema, however, often becomes chronic or recurrent, due to repeated exposure to skin irritants and contact allergens.

Contact allergies are quite persistent and contact allergy to a solvent (i.e. to turpentine) can only be prevented by avoiding further contact.

Hand eczema caused by exposure to solvents sometimes necessitates a change of job, particularly in cases where exposure is unavoidable.

C. Carcinogenic and Teratogenic Effects

1. Carcinogenicity

a. Benzene

The carcinogenicity of benzene has been established through a number of case reports and epidemiological and experimental studies. Benzene may induce different forms of leukemia and probably also lymphomas and myelomas.

Exposure–response calculations have suggested that benzene may also pose a significant cancer risk at very low exposure levels, around 1 ppm. The time from the first benzene exposure until the diagnosis of leukemia averages around 10 years, and is thus shorter than for most other occupational cancers.

There are no known mechanisms of early diagnosis of benzene-induced hematopoetic malignancies, thus no screening procedures can be recommended.

In order to monitor exposed populations, cytogenetic methods have been used. Benzene induces chromosomal aberrations in peripheral blood lymphocytes.

b. Methylene Chloride, Trichloroethylene, and Perchloroethylene

These commonly used chlorinated solvents have shown at least some carcinogenicity in animal experiments. For methylene chloride, the evidence for carcinogenicity has been considered "sufficient" by the WHO Cancer Research Institute.

The epidemiological studies carried out thus far concerning methylene chloride do not show any carcinogenic effect, but the limited numbers of studies as well as the short follow-up periods necessitate further investigation.

Despite the vast number of workers exposed to trichloroethylene, only few cohorts have been followed perhaps because only a few workers at each work-site are engaged in the degreasing operations where trichloroethylene is typically used. Thus, studies are thus far limited in number and have short follow-up periods. Nevertheless, one of these studies suggests an excess of tumors in the urinary tract as well as in the hematolymphatic system. Similarly, some studies also suggest that perchloroethylene might induce tumors of the urinary tract.

c. Styrene

Styrene is suspected to be carcinogenic, but the evidence is mainly theoretical. The main metabolite, styrene oxide, is an established carcinogen but seems to be readily degraded in humans. Cytogenetic effects have been found in workers exposed to comparatively low levels of stryene (i.e. around 25 ppm). There are a few animal carcinogenicity studies with somewhat contradictory results. The WHO has concluded that these studies provide limited evidence of carcinogenicity.

A number of epidemiologic studies have been carried out. The studies that have been performed in the glass-reinforced plastics industry, where the styrene exposures are highest, have been negative so far. However, there is a very limited number of individuals with heavy exposures and long-term follow-up in those studies. Several studies performed in the styrene–butadiene rubber industry have shown excesses of hematolymphatic tumors, but their exact cause remains elusive.

d. Chloroform, Carbon Tetrachloride

Carbon tetrachloride causes liver cancer in animal experiments. It is possible that the carcinogenicity of carbon tetrachloride is associated with its potential to induce toxic liver damage. Chloroform causes liver and kidney tumors in rodents where its acute toxicity might contribute to the observed carcinogenicity.

e. Mixed Solvents

Several studies have shown an excess risk of lung cancer and cancer of the urinary bladder among painters. Based on an evaluation of these studies, the IARC has considered painting an occupation at increased risk of cancer. However, it is not known whether solvent exposure might contribute to this observed increase in tumors.

A number of studies have also shown an increase of lymphomas (mainly non-Hodgkin lymphomas) and myelomas in occupational groups exposed to mixed solvents. Whether solvent exposure has caused these tumors remains unclear.

2. Teratogenicity

The potential toxicity of organic solvent exposure to the fetus (teratogenicity) has caused widespread concern. Studies performed in the last decades have shown that laboratory workers, women with occupational exposure to perchloroethylene, methylene chloride, styrene, toluene, mixed solvents, and painter's naphta have an increased risk of spontaneous abortions and it seems likely that several solvents may contribute to miscarriages.

Recent studies, however, have shown that neither exposure to organic solvent mixtures nor exposure to perchloroethylene at the levels prevalent in Sweden in the 1980s seem to increase the risk of spontaneous abortions. This finding may be due to decreasing exposure levels.

A number of studies have suggested that both maternal and paternal exposure to mixed solvents may increase the risk of congenital malformations. Some studies also suggest that toluene and styrene exposure of the mother may increase the frequency of malformations. However, no specific type of defect or syndrome has repeatedly been associated with solvent exposure.

Some studies have also suggested that male exposure to organic solvents may influence spermatogenesis and thereby increase the risk of abortions among the wives of exposed workers.

Glycol ethers with short alkyl groups (e.g., 2-methoxy- and 2-ethoxyethanol), are teratogenic as well as being testicular toxins at rather low doses in experimental animals. There has thus far been no epidemiological confirmation of the teratogenic potential but this observation has entailed a shift to the use of glycol ethers with longer alkyl chains in most branches of industry.

Thus a large number of studies have suggested that occupational organic solvent exposure during pregnancy may damage the fetus. Current hygienic standards are usually not based on the probability of teratogenic effects and it seems warranted to keep the solvent exposure of pregnant women significantly below these standards.

D. Liver and Kidneys

1. Liver

Some halogenated solvents are classical heptotoxins. 1,1,2,2-Tetrachloroethane was used in the aeroplane industry during World War I and caused jaundice and deaths from liver necrosis. The use of carbon tetrachloride as a solvent has been accompanied by numerous case reports of liver toxicity including death due to liver and kidney necrosis. Chloroform is also heptotoxic although at higher doses. It has been shown that the liver toxicity of these solvents is due to metabolic activation with formation of reactive metabolites. The effect of these solvents on the liver are potentiated by intake of barbiturates and alcohol.

Since liver toxicity has been a striking feature of several of the most common solvents during the first half of the century, it has also been suspected that other solvents may also be toxic to the liver. However, it seems that occupational solvent exposure (including very heavy exposures) does not affect liver enzymes in serum as measured by routine tests of liver damage. Thus, it seems unlikely that common solvents may cause serious liver damage.

There are several recent reports on the liver toxicity properties of dimethyl formamide which currently seems to be the most widely used of the liver toxic solvents. This observed toxicity depends on metabolic activation; the solvent has a very large uptake through the skin which makes it particularly dangerous. Another solvent in widespread use that is potentially hepatotoxic is tetrahydrofuran.

At least two solvents, trichloroethylene and dimethylformamide, share the ability to interefere with alcohol metabolism in the liver. A flushing of the face and neck as well as severe nausea is induced if alcohol is ingested in conjunction with high exposure to these solvents.

2. Kidney

For many decades it has been known that certain solvents, like carbon tetrachloride and ethylene glycol, may induce renal failure in humans after short-term heavy exposures. Short-term high exposures to other solvents such as toluene, xylene, trichloroethylene, methylene chloride, and methyl chloroform have been associated with renal failure in a few reports. However, it seems that acute occupational intoxications with toluene, xylene, methylene chloride, styrene, trichloroethylene, perchloroethylene, and 1,1,1-trichloroethane, do not directly damage the kidneys in most workers.

In recent years, most of the interest concerning solvent nephrotoxicity has been devoted to the possible induction of a serious inflammation in the kidneys, glomerulonephritis. Although chronic glomerulonephritis is an uncommon disease, it is the single most common cause of kidney transplantation.

Some studies have suggested a direct association but these remain unconfirmed.

Patients with glomerulonephritis show an increased excretion of proteins as well as red blood cells in urine. On a group basis, such increases have also been found among active workers exposed to xylene, toluene, and styrene. Thus it seems reasonable to consider solvent exposure in cases of proteinuria or glomerulonephritis of otherwise unspecified origin.

E. Other Effects

1. Irritative Effects

Most solvents produce irritative effects in the mucous membranes of the eyes, nose, and throat. Cough as well as a burning sensation is induced. However, all common solvents have weakly irritating properties when compared with intense irritants such as toluene diisocyanate, acrolein, or formaldehyde. The eyes are often most sensitive. Heavy exposures are commonly associated with cough as well as chest tightness and a feeling of breathlessness, although the origin of the latter symptom is not clear. At very high exposure levels, a number of common solvents (such as toluene, xylene, and methylene chloride) seem to induce pulmonary edema (collection of fluid in the lungs) or chemical pneumonitis (inflammation in the lungs due to a chemical agent).

Different solvents have varying potentials to irritate mucous membranes. Ethyl acetate, styrene, and xylene seem to be irritating at comparatively low concentrations. Methylisobutyl ketone, butyl alcohol, toluene, and methylethyl ketone show an intermediate potency while higher concentrations are needed for the irritative effects from ethanol, methanol, isopropyl alcohol, and acetone.

2. Circulatory System

People who abuse solvents for their inebriating properties may die suddenly during the intoxication probably due to solvent induced disturbances in the heart rate. These effects should in general not be expected unless severe intoxication is present.

Exposure to carbon disulfide has been shown to increase the incidence of myocardial infarction. Carbon disulfide alters fat metabolism and increases arteriosclerosis.

Methylene chloride is metabolized to carbon monoxide. Carbon monoxide binds very effectively to hemoglobin in the red blood cells, thereby diminishing the hemoglobin available for oxygen transport. The hygienic standard for carbon monoxide has been established to limit the hemoglobin occupied by carbon monoxide (COHb) to below 5% as levels above 5% have been reported to be associated with an increase in symptoms among angina pectoris patients. Exposures to 100 ppm of methylene chloride may produce COHb levels immediately above 5%. However, the biotransformation to COHb becomes saturated if the exposure exceeds 250 ppm, and COHb levels above 10% among nonsmokers are unlikely to be due to methylene chloride exposure. Epidemiological studies performed thus far have found an incidence of heart disease in methylene-chloride–exposed workers that is comparable to that in control populations.

3. Blood

Heavy benzene exposures may induce a diminished production of all types of blood cells in bone marrow. It is unclear if any type of blood cell is particularly susceptible to benzene. However, serious anemia due to bone marrow failure (aplastic anemia) is unlikely to appear until time-weighted average exposure levels have reached at least one order of magnitude above current hygienic standards in industrialized countries. It has been suspected that other aromatic solvents, with a chemical structure similar to benzene (such as toluene and xylene) would also damage bone marrow. However, this is not observed, probably due to their different routes of biotransformation.

V. Recommendations for Monitoring of Exposed Groups

Most solvents irritate mucous membranes and affect the central nervous system and skin. In general, monitoring of exposed groups should focus on these effects. Skin problems (which are clearly visible) and the exposure conditions causing the damage are often obvious. On the other hand, subtle effects on the central nervous system are not easily diagnosed. These patients present with vague and unspecific symptoms that generally lack objective correlates. Thus, questionnaires have been developed and seem currently to be the best instruments available for the monitoring of effects particularly in groups exposed to solvent mixtures.

Inorganic Mineral Particulates in the Lung

John E. Craighead
University of Vermont

Glossary

Crystallographic Pertaining to the structural arrangement of atoms comprising a mineral.

Mucociliary escalator The physiological process employing mucin-secreting and ciliated cells by which foreign material is cleared from the lungs.

Phyllosilicates Silicate minerals arranged in leaf-like or layered crystallographic patterns (e.g., talc).

Pneumoconiosis Lung disease resulting from the inhalation and retention of organic or inorganic dust.

Pulmonary fibrosis The formation in the lung of fibrous tissue as a reparative or reactive process.

Silicate Mineral dioxide form of silicon and various cations (e.g., asbestos, talc, kaolin).

Threshold Lowest limit of dust exposure that causes a pathologic response.

INORGANIC MINERAL PARTICLES are derived from naturally occurring mineral deposits or are products of industry. They differ widely in chemical and physical structure and are variable in size. Respirable particles may have pathogenic effects on humans.

I. Introduction

The respiratory tract exchanges large volumes of gases with the external environment on a continuous basis. At rest, the average adult human inhales approximately 500 ml of air with each respiratory excursion (i.e., the so-called tidal volume). At a respiratory rate ranging from 8 to 18 per minute (depending on physical activity), each of us exchanges roughly 5 to 15 $\times$ 10^3 liters of air each 24 hours. Since ambient air invariably suspends countless particles of inorganic and organic dust, tissues of the respiratory tract interact with an enormous mass of foreign material on a continuous basis. Humans are endowed with a highly efficient clearance mechanism to eliminate the bulk of this burden. Countless excellent anatomic and functional descriptions of the respiratory tract have been published. Accordingly, a detailed presentation will not be provided here, but the relevant features of several anatomic units will be emphasized.

II. Anatomy and Physiology

The nares of the nose, comprised of complex turbinates lined by mucin-secreting and ciliated cells, form an efficient barrier for large and structurally complex particles entering the respiratory tract. However, in mouth breathers, the nose is bypassed and its function is supplanted, in part, by the oral pharynx. The tracheobronchial tree is an arborescing, interlinked tubular system in which velocity of air is oriented in a laminar fashion, so that it is maximal in the center of the airway lumen, but turbulent at its interface with the mucosa. Functionally, the airways from the larynx to the terminal bronchioles are lined by a layer of mixed mucin-secreting and ciliated cells enveloped by a mucin barrier that varies in thickness and viscosity depending upon external stimuli. Beginning in the trachea, the airways undergo some 20 to 23 bifurcations into units of increasingly smaller dimension, but considerable variability exists from one anatomic location in the lung to another (Fig. 1). These bifurcations prove to be impaction sites for particles and gases that enter the airway. The most distal segment of the respiratory tract (the acinus) is comprised of the respiratory bronchiole and its branches, which lead into the alveolar ducts. Alveoli "bud" from the respiratory bronchioles and alveolar ducts. The respiratory bronchioles in healthy persons are lined by simple, low, cuboidal cells, but this layer often undergoes metaplasia and secretes mucin in smokers. The alveoli are lined by simple laminated type I pneumocytes and the type II cells which secrete surfactant, a protein that reduces surface tension in the small airway branches. Velocity of air movement appears to be negligible at the level of the respiratory bronchiole and beyond. The air spaces and airway branches are invested with an interconnecting interstitium, which is dynamically altered by physiologic influences and during disease processes. The lymphatics course through this interstitium adjacent to blood vessels and airway branches.

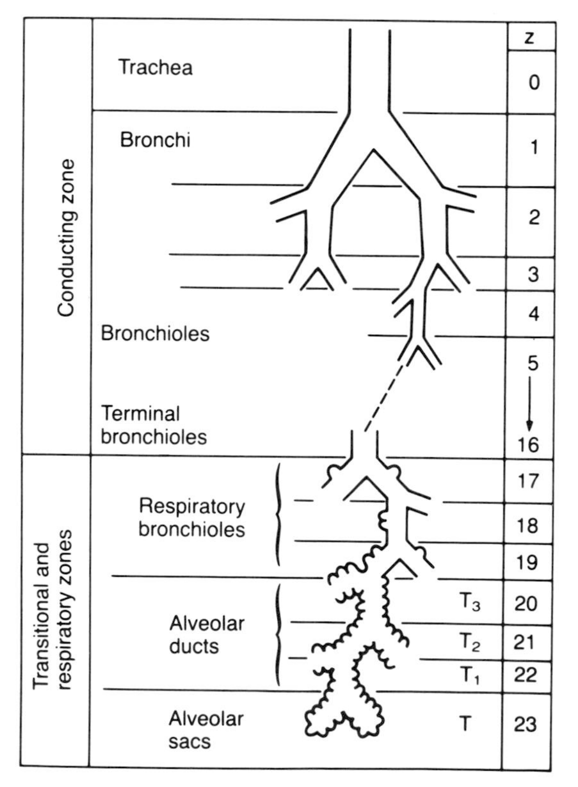

Figure 1 Branching divisions of the respiratory tract of humans depicting conducting airways and zones for respiratory gas exchange. (Reproduced with permission from Weibel, E. R. "Morphometry of the Human Lung," Academic Press, 1963.)

III. Deposition of Airborne Particulates in the Lungs

The depth of penetration of dust particles into the lungs depends on the physical and chemical characteristics of the particles and their inherent electrostatic charge, as well as host factors such as preexisting disease and smoking (Fig. 2). Large particles of high density tend to deposit in the upper respiratory tract (i.e., turbinates of the nose and oral pharynx), whereas smaller particles ($<4\ \mu m$) enter the bronchi of the lower respiratory tract where they often impact with the walls of the airways or intersect with bifurcations (Fig. 3). Movement of particulates down the airways is influenced by physiologic mechanical forces and the intrinsic properties of the particle. For example, fibers tend to orient parallel to the long axis of the airway; the depth of penetration into the lung, therefore, is inversely proportional to the diameter of the fiber, not its length. The aerodynamic properties of a particle such as a fiber are also influenced by surface configuration and charge. For example, the positively charged, flexible, irregular fibers of chrysotile asbestos tend to interact with and impact on the walls of major airways in contrast to the sleek rigid amphibole asbestos fibers, which are transported by laminar flow. Such considerations have enormous implications with regard to clearance and the health effects of the particulate.

Structural Features Of Common Dusts in the Occupational Environment

Granular
- Silica
- Feldspar
- Nonfibrous zeolite

Amorphous
- Tungsten and tantalinium

Fibrous
- Asbestos
- Fiberglass
- Fibros zeolite–eronite
- Titanium dioxide

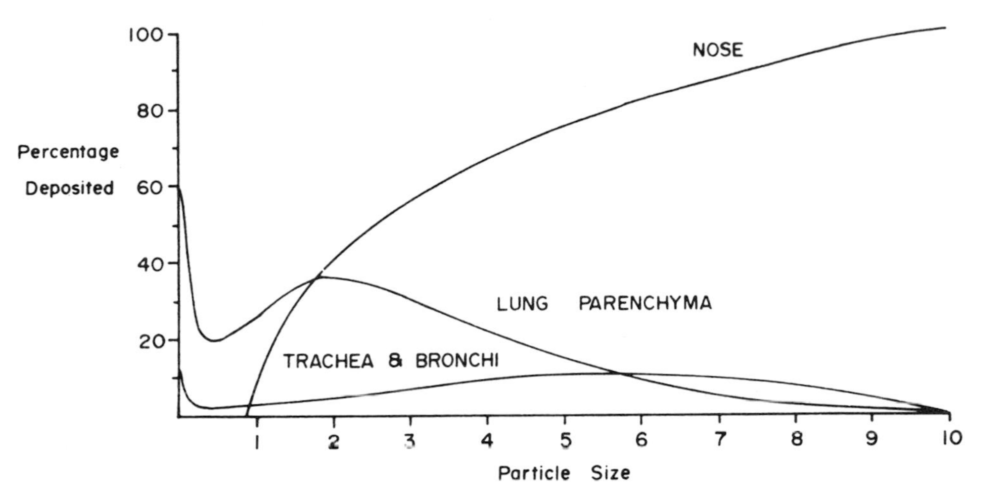

Figure 2 Deposition of particulates in respiratory tract by size. The nose plays a major role in filtering large particles of dust from the ambient air. Smaller particulates deposit in the bronchi and the acini. (Reproduced with permission from W. Keith, C. Morgan, and A. Seaton, eds. "Occupational Lung Diseases," W.B. Saunders, Philadelphia, 1984.)

Platy	*Mixture*
Talc	Slate and shale
Mica	China stone
Kaolinite	Volcanic ash
Vermiculite	Bentonite

Fine particles (0.5–4 μm) of dust are suspended by Brownian movement in the distal airways where they preferentially accumulate in the respiratory bronchioles and alveoli; diffusion of toxic gases and particulate deposition at these sites often provokes the development of lesions in these anatomic units.

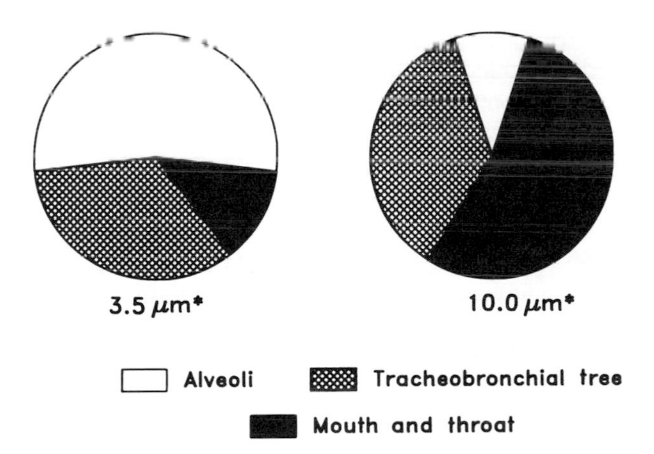

Figure 3 Sites of deposition of dust particulates in the lower respiratory tract in relation to mean particulate aerodynamic diameter.

IV. Clearance

Particles of dust accumulated in the nares and oropharynx are readily cleared and either expectorated or swallowed. Clearance of particulate matter from the lower respiratory tract is more complex and is influenced by numerous variables, only a few of which can be satisfactorily defined quantitatively. Not the least of these factors is the variability between individuals, the effects of aging, and cigarette smoke. Three elements in the lung play key roles in clearance. The mucociliary escalator system eliminates a substantial proportion of particles inhaled into the lung, particularly those >4 μm in diameter. The flow rate is claimed to be approximately 1 cm per minute in the trachea, 4 mm per minute in the segmental bronchi, and 1.6 mm per minute in more distal bronchi, although differences in anatomic location influence rate, as do variations, in the rheology of the mucus blanket. Secondly, macrophages, which reside in the smaller airspaces, scavenge small particles entering the distal airways of the lung. These cells sparsely populate the respiratory bronchioles and alveoli of the normal lung, but are present in great abundance in the lungs of smokers and those exposed to other environmental pollutants. Monocytes are recruited into airspaces from the interstitium and the circulation in response to undefined stimuli from inhaled particulates. After phagocytosis, they migrate up the mucociliary escalator or between the lining cells of the acinus to sequester in the interstitium, or are transported in the lymphatics to more central sites. Finally, the lymphatic system invests the lung with a dense network of interlinking channels. Although the complementary role of these three clearance mechanisms endows the lung with a highly efficient cleansing system, many fine particulates of dust are retained within the lungs for prolonged periods of time after inhalation.

Overall, clearance of particulates by the lungs takes place in two temporal phases; the half-life of the first is approximately 20–30 days and the second, 200 or more days. In experimental studies, about one quarter of the larger particles (diameter 3–4 μm) is removed in the first phase. The relative contribution of the mucociliary escalator and the alveolar macrophage to clearance in this initial phase is uncertain. Little quantitative experimental information exists regarding the role of the lymphatics in the dynamics of clearance of particles from the distal airspaces. As indicated above, macrophages play a cardinal role as scavengers of foreign materials in the acini, yet some particles are transported across the type I and II cell barrier of the alveoli, either by endocytosis or by intercellular routes, where they accumulate in the interstitium (Fig. 4). It is unclear to what extent these isolated particulates migrate between the loosely structured endothelial lining cells of the lymphatics and are transported by lymph flow to central locations. Macrophages laden with dust particles also migrate to the interstitium and are transported by the lymphatics. The peripheral lymphatics of the lungs form a subpleural arborescing network that ultimately drains to centrally located hilar and mediastinal lymph nodes through major lymphatics in lung septae. Centrally located lymphatics are found adjacent to the respiratory bronchioles and follow the course of airways and vessels to the hilar lymph nodes. For reasons that are unclear, dust-laden macrophages commonly deposit adjacent to the airways and vessels, where they can provoke a fibrotic response (Fig. 5). Possibly the dust burden of the macrophages impedes the migration of these cells to lymph nodes in more central locations in the chest. Nonetheless, hilar and mediastinal lymph nodes do accumulate countless particulates and thus mirror the exposure of the lungs to dust over a lifetime.

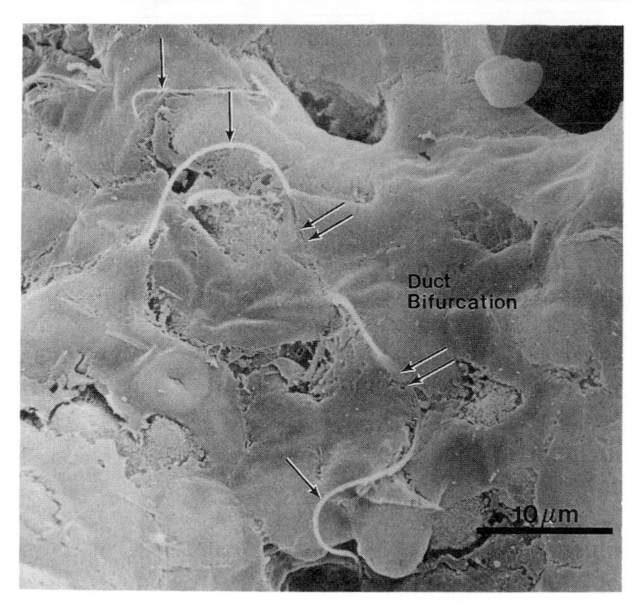

Figure 4 Alveolar duct epithelium of rat exposed experimentally to chrysotile asbestos. Note the incorporation of the long, curvy fiber through the intercellular spaces of epithelial barrier. (Courtesy of Dr. Arnold R. Brody.)

Customarily, the concentrations of dust in the ambient air do not exceed the capacity of the lungs' clearance mechanism. However, a threshold exists at which point particulate accumulation in the lung exceeds the transport capacity of the mucociliary escalator and the macrophage pool. This amount approximates 1 mg/g mouse lung, although the size and nature of the particles is a critical determinant of the outcome. When the threshold is exceeded, dust accumulates linearly in the lung in relation to ambient air concentrations. Under these circumstances, macrophages are "poisoned" and immobilized, and massive amounts of dust can accumulate. Inorganic particulates of low pathogenic potential can produce disease in this manner. For example, workers in the China clay and talc industries were regularly exposed to quantities of dust over a working lifetime that overwhelmed defense mechanisms. Under the microscope, the lungs of these men were found to have accumulated massive amounts of particulate material, associated with a fibrotic tissue reaction (Fig. 5).

V. Ambient Air Exposure and Lung Dust Burdens

The concentrations of dust in the ambient air are dramatically influenced by geographic and metero-

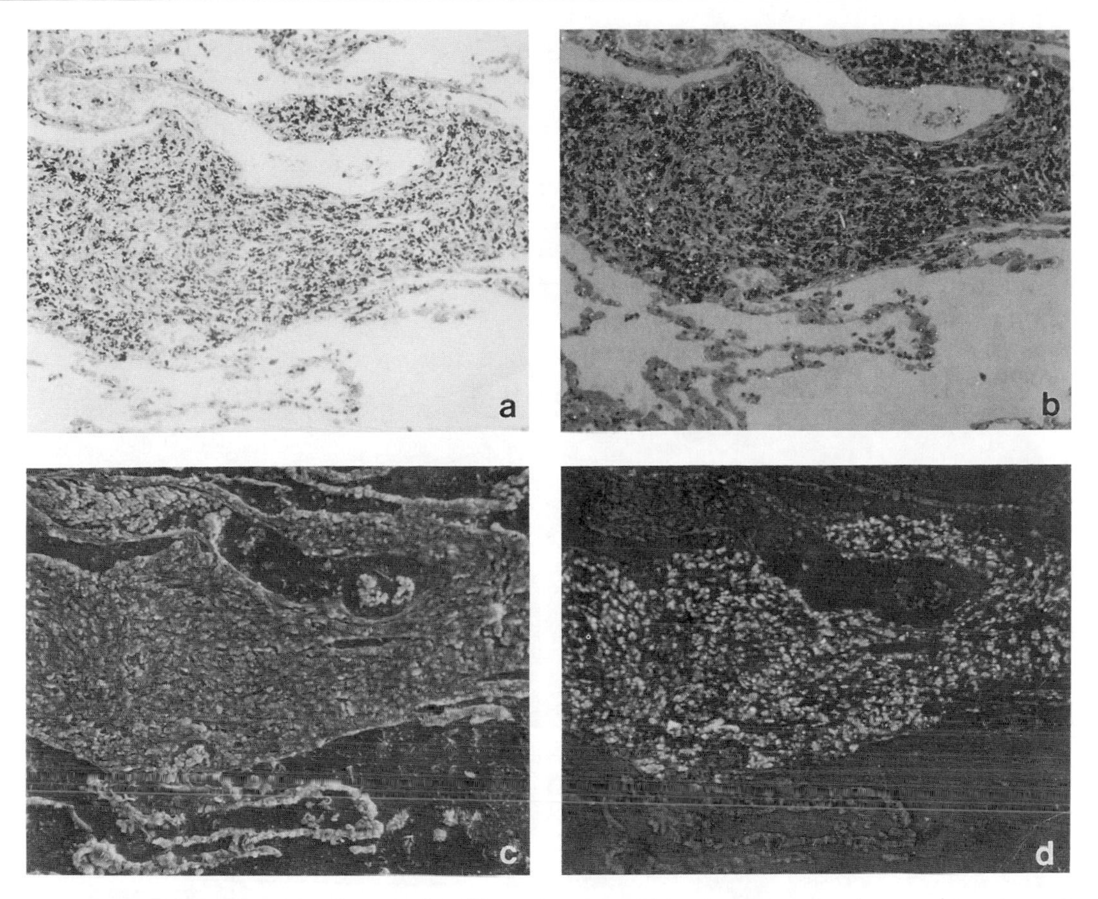

Figure 5 Accumulations of dust in the perivascular and peribronchiolar tissue of the lung in a hematite miner. (a) Light microscopic illustration of dust largely having the appearance of granular carbon. (b) Polarization photomicrograph in which birefringent particulates are found scattered in the perivascular accumulations of fibrous tissue. (c) At a higher level of resolution, one sees the same tissue by scanning microscopy. (d) Backscatter scanning electron micrograph in which the tissue is screened for iron-containing particulates.

logical considerations. Air in the urban environment usually contains high concentrations of carbon and mullite (byproducts of coal and petroleum combustion), titanium from paints and pigments, and fragmented chrysotile from friction products, particularly brake shoes. In agricultural areas, aluminum silicate particles increase in concentration, representing the dust from dry fields and unpaved roads.

Information has accumulated on the dusts in the lungs of members of the general population who are not exposed occupationally. These particles were not eliminated by the clearance mechanisms described above, but may reflect material in a dynamic equilibrium with the inhaled air in the lung. Estimates based on analytical studies indicate that approximately 1×10^8 to 2×10^9 particles of silicates are found in a gram of dried human lung tissue. The health implications of these particulates are not known with certainty but are believed to be low.

VI. Industrial Air Exposure and Lung Dust Burdens

A vast body of scientific information has accumulated concerning the evaluation of the air of the occupational environment and the assessment of lung burdens of dust accumulating as a result of these exposures. The importance of inorganic particulates in the genesis of mineral lung disease (pneumoconiosis) can hardly be overemphasized and its recognition has played a significant role in development of contemporary occupational health regulation.

A catalog of the great diversity of pneumoconiosis in humans is beyond the scope of this chapter. The general reference citations and a specific listing of key references to some of the more important disease conditions are provided in the bibliography.

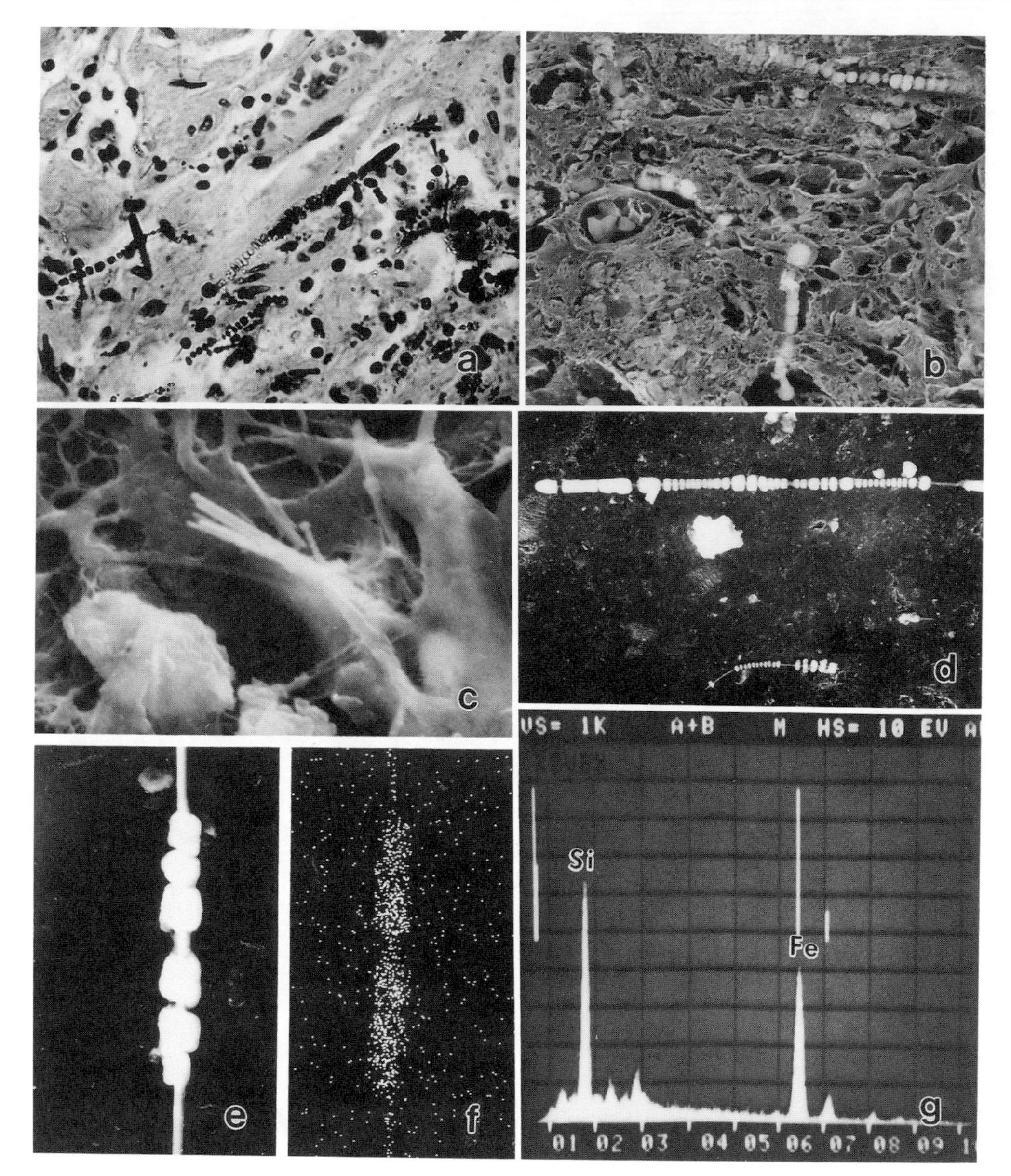

Figure 6 Definition of asbestos fibers in tissues by various morphological approaches. (a) Histological tissue section with several asbestos bodies defined by Prussian blue technique. This technique stains iron in the body. Note the beaded composition of the bodies that, on close inspection, have within them a fibrous core. (b) Scanning electron micrographic illustration of asbestos bodies at low levels of resolution in a tissue section. (c) Fibers of asbestos in lung tissue as demonstrated by scanning electron microscopy. (d) Asbestos bodies at high level of resolution. The larger body illustrates the typical appearance of a structure with an amphibole core, whereas the smaller body has a curved fiber core of chrysotile. Note the difference in size of the asbestos bodies. (e) An asbestos body defined by scanning electron microscopy. (f) The ferruginous material in the coat of the asbestos body in (e) is defined by backscatter x-ray spectroscopy. (g) An x-ray spectrographic elemental definition of the composition of a fiber. The major peak on the left represents silicon and the peak on the right indicates iron. The smaller peaks represent magnesium and various other cations. The pattern is consistent with an amphibole asbestos.

Several of the more common minerals responsible for disease are categorized as follows.

Relative Disease-Producing Capacity
of Selected Environmental and Occupational Dusts

Highly fibrogenic	*Mildly fibrogenic*
Asbestos (amphibole > chrysotile) (Fig. 6)	Talc (Fig. 9)
Silicon dioxide (crystobalite > alpha quartz) (Figs. 7,8)	Mica (Fig. 10)
	Graphite (Fig. 11)
Nonfibrogenic particles ("nuisance" dusts)	
Titanium dioxide (rutile)	
Feldspar and related granular aluminum silicates	
Kaolinite and related platy aluminum silicates (Fig. 12)	
Aluminum silicate combustion product (mullite)	
Metal (steel, titanium)	
Carbon	

Countless analytic studies have now shown that a dosage threshold exists below which exposure to dust fails to produce clinical disease. Nonetheless, the threshold must be defined in the context of the air concentration of individual mineralogical particulate types and the duration of exposure (Fig. 13), as well as the intrinsic pathogenetic potential of the mineral particle in question. For example, the SiO_2 cristobalite is far more toxic and pathogenic than the mineralogically similar alpha quartz. These differences would appear to relate to the crystallographic properties of these two variants of SiO_2. And the fibrogenic response to the amphibole asbestos types exceeds the effects of chrysotile asbestos at equivalent airborne dust concentrations. Since different dusts cause disease by differing pathogenetic mechanisms, the issues are exceptionally complex. The scientist must amalgamate the results of experimental animal studies with findings in epidemiological investigations on populations of occupationally exposed workers. These questions cannot be resolved by analyzing risk in the context of dust accumulation in the lungs. Deposition and clearance are dynamic concepts. Disease represents the outcome of a complex series of pathogenetic events that only in part reflects the mineral dust concentrations in lung tissue.

VII. Identification and Quantification of Inorganic Particles in Lung Tissue

Analysis of lung tissue to detect the presence of and to subsequently characterize foreign inorganic particulates is conducted using several different well-

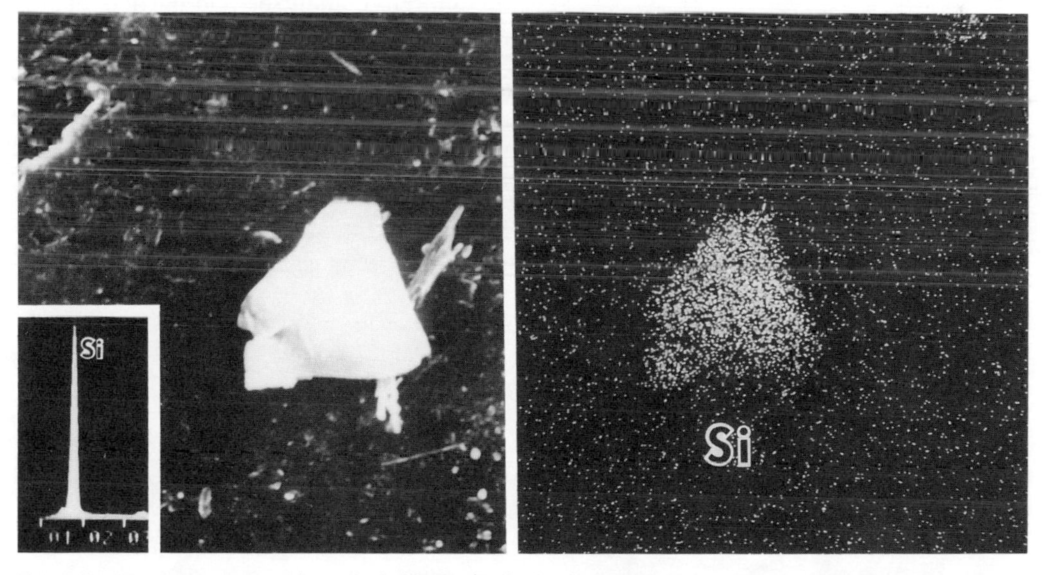

Figure 7 X-ray spectrographic analysis illustrating by backscatter imaging the silicon in a particulate in the lung. The insert illustrates the x-ray spectrographic pattern. One particle contains only silicon and is presumably quartz, whereas the second is an aluminum silicate. The backscatter technique can be used to locate individual particles *in situ* in intact lung tissue.

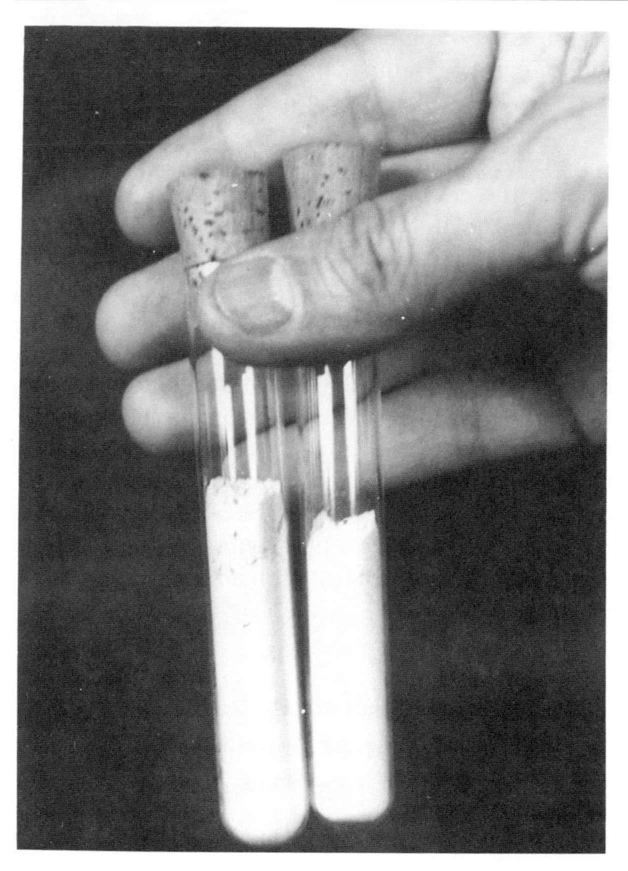

Figure 8 Dust extracted postmortem from the lung tissue of a worker with silicosis. Undoubtedly, the dust particulates are comprised of both quartz and silicates, but an analysis is not available.

Figure 10 Scanning electron micrograph of mica particulates in a digest of lung tissue from a worker exposed to mica dust. [Reproduced with permission from *Arch. Pathol. Lab. Med.* (1988). **112,** 673–720.]

established techniques. The choice of methodology is largely based on the question posed. The investigator can use either intact thin sections of lung tissues for *in situ* localization of particulates (see Fig. 6) or, alternatively, digestates of tissue for semiquantitative analyses and particle characterization.

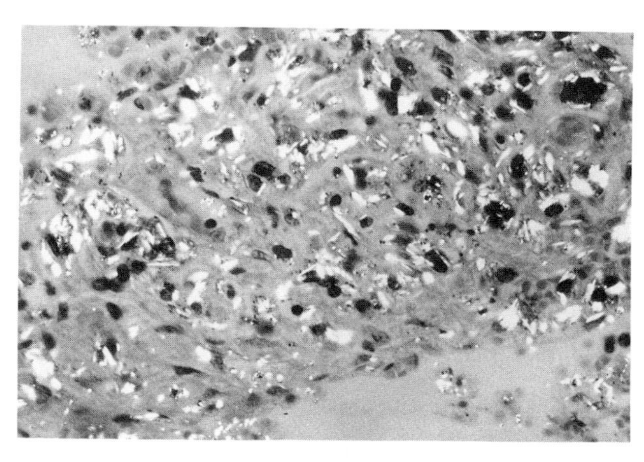

Figure 9 High resolution polarization microscopic demonstration of talc particulates in lung tissue. Note the overall lancet-shaped configuration of the particles that are variable in size.

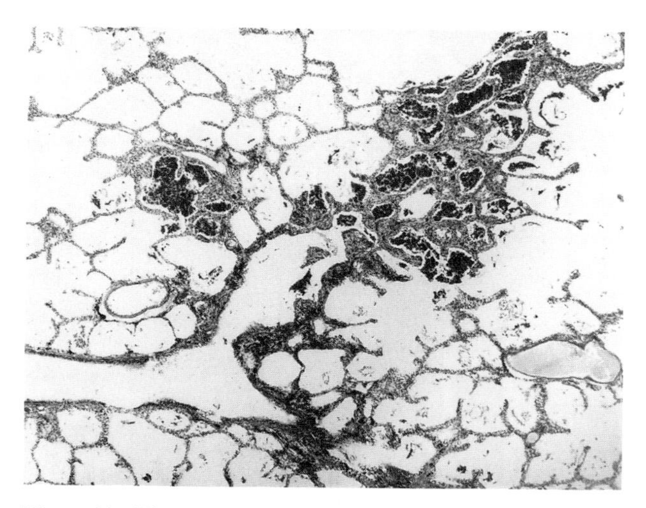

Figure 11 Histologic section of the lung of a graphite worker. Note the accumulation of carbonaceous material within pulmonary alveolar macrophages and the interstitium. Graphite consists of carbon and varying amounts of silica.

Figure 12 A whole-lung section of the lung of a China clay worker who was exposed to kaolinite dust for an extended period of time. Note the massive consolidated accumulations of gray material which is kaolinite. Elsewhere, the lungs show deposits of carbonaceous material adjacent to kaolin-containing nodules. The accumulations of carbon adjacent to the larger lesions develop as a result of lymphatic obstruction. [Reproduced with permission from *Arch. Pathol. Lab. Med.* (1988) **112,** 673–720.]

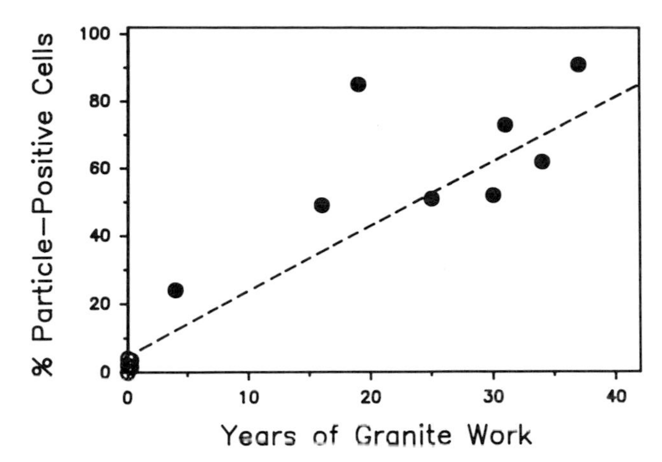

Figure 13 Relative number of pulmonary alveolar macrophages containing dust particulates in bronchoalveolar lavage specimens from granite workers in relation to duration of employment. The data suggest a cumulative burden of dust during the period of employment. The analyses of the particulate-containing cells are done by polarization microscopy or x-ray spectrometry.

In situ morphological techniques are conducted at the level of resolution of the light microscope using polarization microscopy to detect birefringent particulates or by scanning electron microscopy (see Fig. 5).

Polarization microscopic analysis of particle morphology and assessment of the intensity of birefringence can assist in the classification of particles. For example, SiO_2 particles (alpha quartz) are fine, granular, and difficult to detect by polarization microscopy. In contrast, phyllosilicates such as talc, mica, and kaolin are irregular, lancet-shaped, or platy particles of varying size that are brilliant and obvious by this technique (Figs. 9,14). The sensitivity of polarization microscopy depends upon a number of technical and optical considerations and upon the skill and experience of the examiner. However, this technique does not permit the mineralogical identification of a particle. Some inorganic particles (for example, asbestos) have limited or no refractile properties when examined with commonly used polarization microscopes.

Analytical studies requiring a higher degree of sensitivity and specificity employ scanning or transmission electron microscopy supplemented by x-ray diffraction spectrometry. Using these techniques, an investigator can screen a histological section of intact tissue for a specific element and then establish the relative elemental composition of the particle by focused x-ray spectrometry (see Figs. 6g,7). Painstaking analysis of tissue by these approaches can yield information on the relationship of inorganic particles to tissue lesions. This approach, however, yields no quantitative data.

Analysis of digestates of tissue potentially provides quantitative information on dust accumulations in the lung, assuming appropriate standards have been established. It is particularly important to establish the range of dust concentrations in the lungs of a representative sample of a population if a critical interpretation of the findings in the tissue of a specific individual is contemplated. As indicated above, most members of the general population have within their lungs a great diversity of particu-

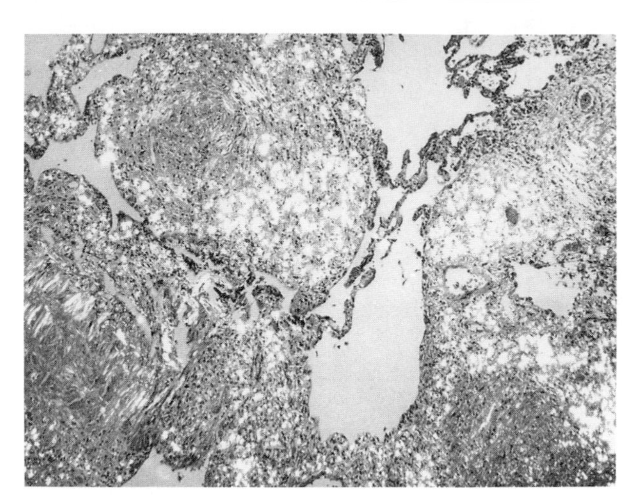

Figure 14 High-resolution polarization photomicrograph demonstrating silica and silicate particles in a perivascular proliferation of connective tissue. The rod- and lancet-shaped structures represent silicates. The granular material is SiO_2 (quartz). [Reproduced with permission from *Arch. Pathol. Lab. Med.* (1988) **112,** 673–720.]

lates, and abnormal accumulations can only be assessed in this context (Figs. 15, 16).

Digestion of known amounts of lung tissue with oxidizing agents yields sediments containing concentrates of dust appropriate for electron microscopic evaluation. After suitable preparation, the morphology of the particle can be evaluated and x-ray spectrometry used to assess the elemental

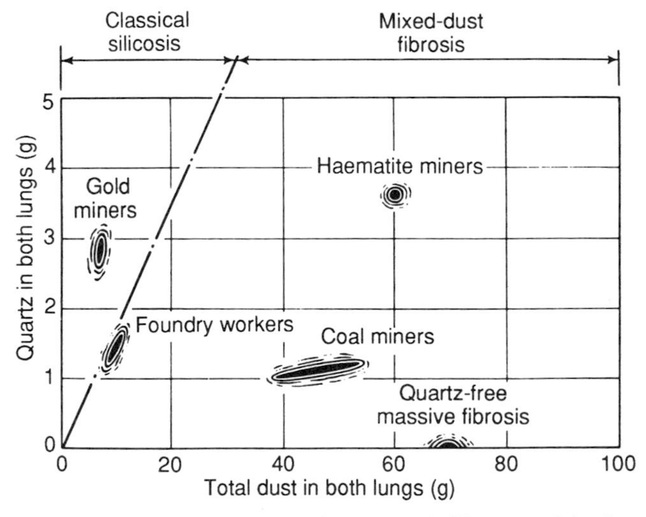

Figure 15 Relative amounts of quartz and silicate particles in relation to the total dust burden of the lungs among men employed in various industries. These then are examples of mixed-dust pneumoconiosis. [Reproduced with permission from *Br. J. Tuberc. Dis. Chest.* (1957) **4,** 297–308.]

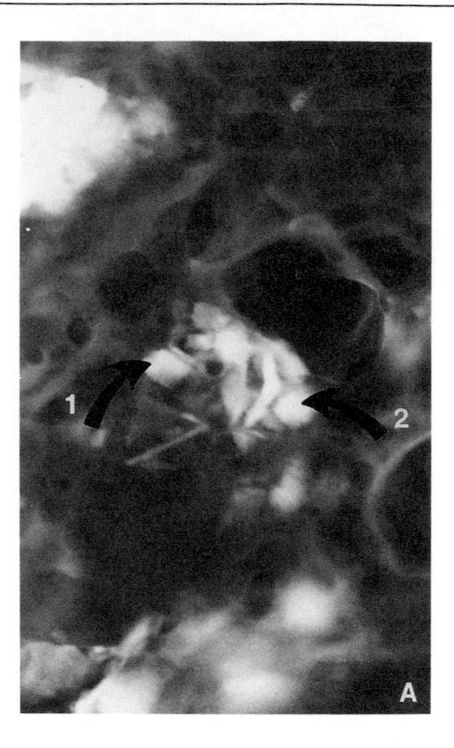

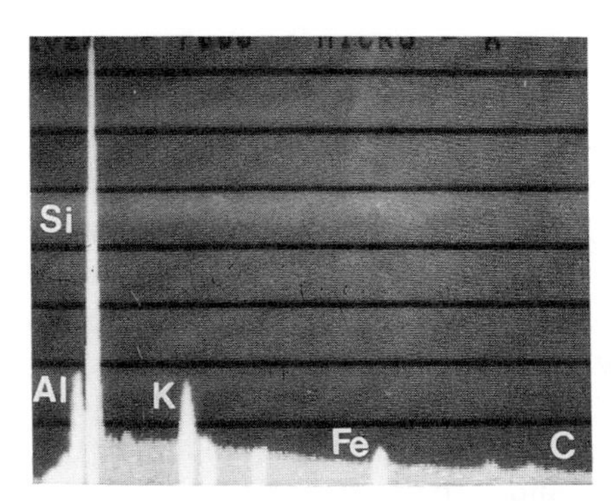

Figure 16 *In situ* demonstration of a cluster of inorganic particulates in the lung of an individual exposed to a mixture of aluminum silicates and quartz. (A) Scanning electron micrographic image; (B) and (C) demonstrate the spectrograph of particles 1 and 2, respectively.

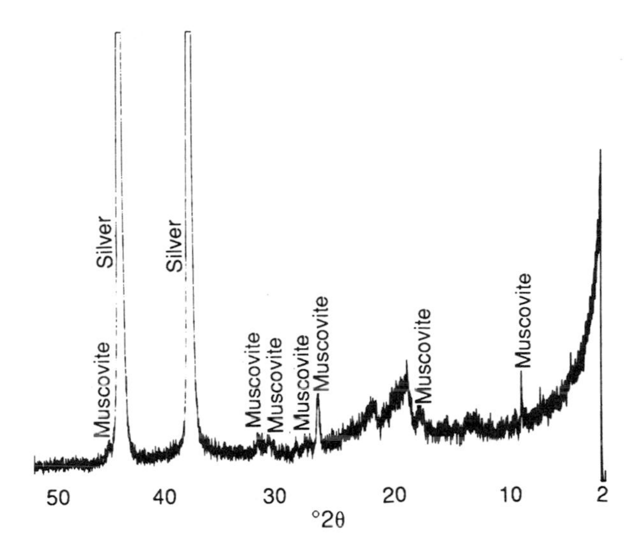

Figure 17 X-ray diffraction spectrum of dust in a lung digest containing both quartz and muscovite. Specific minerals exhibit reproducible patterns. The concentration of dust can be determined quantitatively using appropriate controls.

composition. This technique provides semiquantitative information on the relative concentrations of elements in an individual particle. Unfortunately, elements of low molecular weight are not detected by x-ray spectrometry. For example, the approach is useless in evaluating beryllium (an important cause of pulmonary fibrosis in the past) in lung tissue. Selected area diffraction spectrometry by transmission electron microscopy is used to establish the crystallographic features of an individual particle using selected area diffraction spectrometry. With these data, it should be possible to categorize an individual particle mineralogically. Obviously, the detailed evaluation of a digestate from a sample of human lung is a time-consuming and technical undertaking.

X-ray diffraction spectrometry is a highly refined approach to the analytical semiquantitative analysis of lung digestate for an overall content of inorganic materials (Fig. 17). The technique requires established reference standards to document a mineral "fingerprint." This procedure is, however, timeconsuming and largely of value in research.

Related Articles: Bronchoalveolar Lavage: Detecting Markers of Lung Injury; Chromium; Cobalt Dusts; Cyanide; Dust Particles: Occupational Considerations; Environmental Cancer Risks; Ethanol Fuel Toxicity; Fluoride Toxicity; Hydrogen Sulfide; Monitoring Indicators of Plants and Human Health for Air Pollution; Polychlorinated Biphenyls (PCBs), Effects on Humans and the Environment; Silica and Lung Inflammation; Sulfur Mustard; Zinc.

Bibliography

General

Brain, J. D., and Valberg, P. A. (1979). Deposition of aerosol in the respiratory tract. *Amer. Rev. Respi. Dis.* **120,** 1325–1373.

Craighead, J. E., Kleinerman, J., Abraham, J. L., Gibbs, A. R., Green, F. H. Y., Harley, R. A., Ruettner, J. R., Vallyathan, N. V., and Juliano, E. B., Silicosis and Silicate Disease Committee (1988). Diseases associated with exposure to silica and nonfibrous silicate minerals. *Arch. Pathol. Lab. Med.* **112,** 673–720.

Craighead, J. E., Abraham, J. L., Churg, A., Green, F. H. Y., Kleinerman, J., Pratt, P. C., Seemayer, T. A., Vallyathan, V., and Weill, H. (1982). Asbestos-associated diseases. The pathology of asbestos-associated diseases of the lungs and pleural cavities: Diagnostic criteria and proposed grading schema. *Arch. Pathol. Lab. Med.* **106,** 544–597.

Kleinerman, J. (1979). Pathology standards for coal workers' pneumoconiosis. *Arch. Pathol. Lab. Med.* **103,** 373–432.

Vallyathan, N. V., Green, F. H. Y., Craighead, J. E. (1980). Recent advances in the study of mineral pneumoconiosis. *Pathol. Ann.* **15,** 77–104.

Aluminum

Jederlinic, P. L., Abraham, J. L., Churg, A., Himmelstein, J. S., Epler, G. R., Gaensler, E. A. (1990). Pulmonary fibrosis in aluminum oxide workers. Investigation of nine workers, with pathologic examination and microanalysis in three of them. *Am. Rev. Respir. Dis.* **142,** 1179–1184.

Asbestos

Roggli, V. L., Pratt, P. C., Brody, A. R. (1986). Asbestos content of lung tissue in asbestos associated diseases: A study of 110 cases. *Br. J. Ind. Med.* **43,** 18–28.

Assorted Silicates

Vallyathan, N. V., Craighead, J. E. (1981). Pulmonary pathology in workers exposed to nonasbestiform talc. *Human. Pathol.* **12,** 28–35.

Hard Metal Particulates

Anttila, S., Sutinen, S., Paananen, M., Kreus, K-E., Sivonen, S. J., Grekula, A., and Alapieti, T. (1986). Hard metal lung disease: A clinical, histological, ultrastructural and X-ray microanalytical study. *Eur. J. Respi. Dis.* **69,** 83–94.

Siderosis

Angervall, L., Hansson, G., Rockert, H. (1960). Pulmonary siderosis in electrical welder. *Acta. Pathol. Microbiol.* **49,** 373–380.

Assorted Pathological Conditions

Craighead, J. E., Emerson, R., Stanley, D. O. (1991). Slate-workers' pneumoconiosis. *Hum. Pathol.* **23,** 1098–1105.

McLaughlin, A. I. G. (1957). Pneumoconiosis in foundry workers. *Br. J. Tuberc. Dis. Chest* **4,** 297–309.

Stettler, L. E., Groth, D. H., Mackay, G. R. (1977). Identification of stainless steel welding fume particulates in human lung and environmental samples using electron probe microanalysis. *Am. Ind. Hyg. Assoc. J.* **38,** 76–82.

Wagner, J. C., Pooley, F. D., Gibbs, A., Lyons, J., Sheers, G., Moncrieff, C. B. (1986). Inhalation of china stone and china clay dusts: relationship between the mineralogy of dust retained in the lungs and pathological changes. *Thorax* **41,** 190–196.

Lead Poisoning

Deborah A. Cory-Slechta
University of Rochester School of Medicine and Dentistry

Glossary

Blood lead Concentration of lead measured in whole blood.

Body burden Total amount of lead currently resident in the body.

Diagnostic chelation Test involving a single administration of the chelating agent $CaNa_2$-EDTA designed to evaluate whether the body burden of lead is elevated.

Therapeutic chelation Administration of chelating agents that bind lead and other metals and facilitate their urinary excretion.

ZPP accumulation Buildup of zinc protoporphyrin in blood as a consequence of the inhibition by lead of ferrochelatase in the heme biosynthetic pathway.

LEAD POISONING has traditionally referred to any of a constellation of effects on various organ systems, including the central and the peripheral nervous systems, the GI tract, kidney, and the hematopoietic system, which can prove lethal, that result from exposure to high levels of lead (Pb). In more remote times, sources of lead exposure responsible for poisoning included leaded water pipes and aqueducts, lead-lined cookware, lead-glazed ceramics, and manufacturing processes incorporating lead. More recent exposures derive from lead smelting operations and the manufacture and use of Pb-containing products such as paint and batteries and the dispersion of Pb from gasoline. Today, the incidence of frank lead poisoning per se, both in children and adults, has been generally alleviated by the imposition of stricter industrial hygiene measures and by the discontinuation of the use of lead in paint and gasoline. The problem, however, has taken on new dimensions in light of accumulating evidence over the past 10 years revealing that increasingly lower levels of lead exposure are associated with subclinical effects on several different organs and systems. The sources for these exposures are the residual environmental contamination from the previous uses of lead in paint and gasoline, and, more regionally, from smelting sources.

I. Historical Perspective

Knowledge of the potential toxicity of lead appears to extend almost as far back into history as its use. Lead is stated to have been among the metals of antiquity and beads of lead along with ornaments made of gold and copper dating possibly from as early as 7000 B.C. were found in Anatolia, the Asiatic part of modern Turkey. It has also been reported that lead was among the materials used in the construction of the Hanging Gardens of Babylon. It is thought that in pre-Roman times, however, the importance of lead actually derived from its association with silver. Silver was found in varying concentrations in the ore galena, which is primarily lead sulfide. In the process of cupellation, lead oxide is a byproduct.

The Romans first escalated the production and geographic distribution of lead through considerable expansion of its uses. Probably the best known of these were the lead aqueduct and water-pipe systems that were found throughout the Roman Empire. Evidence of this type of use can be seen today in the old Roman bath house uncovered in Bath, England, where lead linings and pipes were used to move and contain the waters of the local hot springs. The Romans also made considerable use of lead in their cooking practices, resulting in its incorporation into both food and wine. Lead linings were common in bronze cooking vessels as they prevented the bitter taste that was otherwise produced. Lead-lined vessels were likewise used in the preparation of wines. Of course, lead was also an ingredient of coins, and of the glazes used in ceramics, a practice that still occurs in many places today. These multiple uses of lead by the Romans resulted in extensive contamination of the population and have led several historians to speculate that the fall of the Roman Empire was the result of widespread lead poisoning. Indeed, some have even attributed the bizzare behavior of several noteable Roman emporers to their overexposure to lead.

In light of their far-ranging uses of lead, one might think that the Romans had been unaware of the hazards attendant to this substance. This does not appear to have been the case, however. Pliny wrote that exposures to this substance could be associated with "dangling, paralytic hands," a reference to one of the classical signs of Pb-induced peripheral neuropathy known today as "wrist-drop." Pliny also pointed out the particular dangers inherent to the inhalation of Pb vapors. These reports were preceded by those of the Greek physician Nikander, who provided in the *Alexipharmac* a description of classical lead poisoning: pallor, colic, paralysis of limbs, and ocular disturbance leading to death.

The decline of the Roman Empire was accompanied by a decrement in the production and usage of lead. But the requirement for lead production was resurrected by the events defining the Industrial Revolution, and it was in this period of history that Pb production worldwide increased markedly. This, in turn, was followed in the early nineteenth century by an increasing problem with occupational lead poisoning. Since it was commonplace for women and children to also be employed in industry at that time, the toxic effects associated with exposure to lead were not limited to adult males. This period also saw many reports of the impact of lead on various aspects of reproductive function, including observations of abortion, stillbirth, premature delivery, and increased infant mortality. Based on such episodes of toxicity, regulations were gradually implemented in industrial hygiene practices, including those that forbade employment of women in industries with considerable lead exposure. The outcome of this policy was an eventual decline in the reported incidence of reproductive failure as well as in other indices of toxicity.

Industrial regulations were by no means the end of the lead toxicity problem, however. Toxicity associated with environmental contamination resulting from the use of Pb-containing products became acutely evident in the United States by 1920 and continued well into the 1960s. This was manifest in numerous reports of childhood lead poisoning, particularly in urban areas, which were attributed primarily to the ingestion of lead-containing paints used in the residences of these children. One report documented 89 cases of lead poisoning in Boston occurring between 1924 and 1933. Almost half of these presented with lead encephalopathy, a syndrome which may include grand mal epilepsy, coma, delirium, headaches, and tremor; 11 of these patients died. Another study reported that 182 children, 28% of whom died, were diagnosed and treated for lead encephalopathy in Chicago between 1959 and 1963. Such episodes made it apparent that lead encephalopathy was the predominant form of lead poisoning in infants and children.

The prognosis for lead poisoning improved markedly with the introduction of two antidotes, British Anti-Lewisite (BAL) and calcium disodium edatate ($CaNa_2$-EDTA). Combined treatment with these agents was found to decrease mortality substantially, especially when measures to control cerebral edema were included. It was reported in one study, for example, that the mortality rate for acute Pb encephalopathy dropped from about 60% to 20–30% as a function of combined BAL–$CaNa_2$-EDTA chelation therapy; further improvements in treatment protocols led to mortality figures as low as 5%. The efficacy of $CaNa_2$-EDTA alone for occupational Pb poisoning was established as well.

However, survival by no means insured a return to normal function, since residual neurologic and behavioral effects were noted in many children who had undergone chelation therapy. In one report on the fate of 425 survivors of lead poisoning, a total of 39% were left with permanent neurological deficits. The residual effects found in those that initially presented specifically with encephalopathy were even more pronounced: 82% of those children retained permanent neurological deficits, such as cerebral palsy, mental retardation, and convulsive seizures.

II. Current Sources of Exposure

Another facet of the impact of lead exposure that became apparent during this period was that even *asymptomatic* Pb exposure could result in permanent mental impairment. In one report, for example, of a group initially considered to be asymptomatic, 9% were diagnosed at follow-up as being mentally retarded. Concern aroused by such findings led to widespread blood lead screening programs initiated in accord with the 1971 Lead-Based Paint Poisoning Prevention Act. These analyses revealed that the incidence of elevated lead absorption in children was far higher than had been initially supposed, with an incidence as high as 25 to 45% cited for some high-risk areas. These screening initiatives also made clear that the extent of the problem was widespread, and was not at all confined to urban areas or to regions surrounding lead smelters.

It was the widespread scope of the exposure problem and the potentially more insidious nature of toxicity that provoked the questions that are being actively pursued today regarding the nature and extent of toxic effects at increasingly lower levels of exposure. With respect to CNS function, for example, the fact that high level asymptomatic exposure could result in permanent mental retardation led to the speculation that even lower levels of exposure could also be associated with behavioral and/or neurological abnormalities of a much more subtle nature. It was not long thereafter that prospective studies of the effects of environmental lead exposure in children and of occupational lead exposure in adults were initiated. Results from many such studies confirmed initial suspicions that increasingly lower levels of lead produced toxicity that became apparent when more sensitive measures of effect were applied. These effects, moreover, were by no means restricted to the CNS, but were manifested in other target organ systems as well. Collectively, these studies led to reductions of the designated safe level of lead in blood by various regulatory agencies, and caused many investigators and regulators to question whether there was indeed any threshold for Pb-induced toxicity.

The Lead-Based Paint Poisoning Prevention Act passed in the United States in 1971 prohibited any further use of lead-based paint in residential structures that were financed through federal assistance programs and thus curtailed one source of Pb exposure that was particularly detrimental to infants and children. In the United States, the 1980s saw the removal of most of the lead from gasoline, a major source of environmental contamination as over 90% of the lead emissions to the atmosphere derived from this source. These measures put the United States in a leading role with respect to elimination of sources of lead exposure when compared to many other countries (e.g., the Eastern European countries, where hygiene practices are relatively more lax and where the use of leaded gasoline continues.)

The legacy of many years, however, of the persisting use of lead and its unabated distribution into the environment continues to haunt us even today and will undoubtedly do so long into the future. The decline in the use of lead resulting from its discontinuation as an antiknock additive in gasoline was paralleled by a decline both in ambient air lead concentrations and in the average blood lead concentrations in this country. However, extensive prior dispersal of lead into the environment, particularly through gasoline, had already resulted in geographically extensive contamination of air, soil, and groundwater and subsequent incorporation into food and water supplies. In fact, for nonoccupationally exposed adults, lead exposure today derives almost exclusively from the ingestion of food, water, and beverages (see Table I) that have been contaminated either directly via deposition on food materials, or indirectly by transfer from other surfaces as schematically illustrated in Figure 1.

An additional residual contamination is that which has occurred as the result of the use of lead-based paint. As that paint begins to peel and disintegrate, the lead it contains becomes incorporated into dirt and dust where it can reach extremely high concentrations. It has been estimated that lead-based paint is still found in 57 million privately owned and occupied homes in the United States, of which almost 10 million are thought to be lived in by families with children under the age of 7, a population thought to be at particular risk. Almost 11 million homes have lead in the interior surface dust which exceeds safe levels. Almost 4 million of those have peeling lead-based paint and/or excessive amounts of dust that contains lead. In fact,

Table I Relative Baseline Human Lead Exposures

Age/sex/source	Total lead consumed (μg/day)	Total lead consumed (per kg body wt, μg/kg·day)	Atmospheric lead (per kg body wt, μg/kg·day)
Child (2-yr-old)[a]			
Inhaled air	0.5	0.05	0.05
Food and beverages	25.1	2.5	1.0
Dust	21.0	2.1	1.9
Total	46.6	4.65	2.95
Adult female[b]			
Inhaled air	1.0	0.02	0.02
Food and beverages	32.0	0.64	0.25
Dust	4.5	0.09	0.06
Total	37.5	0.75	0.33
Adult male[c]			
Inhaled air	1.0	0.014	0.014
Food and beverages	45.2	0.65	0.28
Dust	4.5	0.064	0.04
Total	50.7	0.728	0.334

Source: EPA Air Quality Criteria for Lead, 1986.
[a] Body weight, 10 kg.
[b] Body weight, 50 kg.
[c] Body weight, 70 kg.

this dust and dirt constitutes a major source of lead exposure in children in addition to that taken in via food and water, as is also shown in Table I. This particular source of Pb exposure in young children is considered to be the result of the extensive amount of hand-to-mouth behavior engaged in by those age groups.

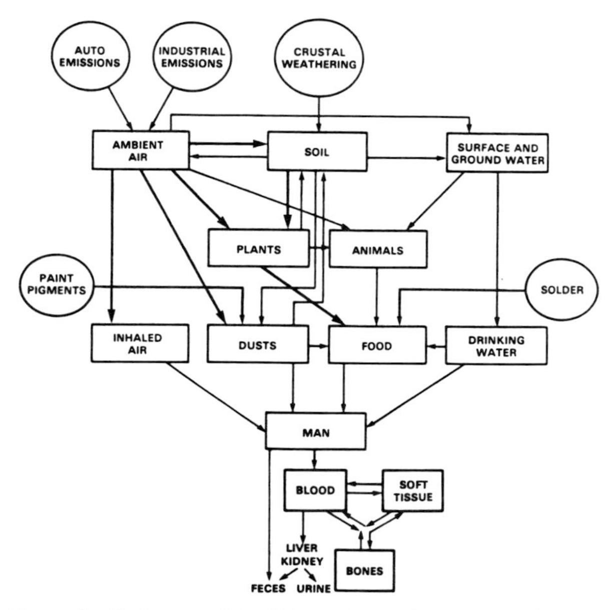

Figure 1 Pathways of lead from the environment to humans. Primary compartments involved in partitioning of internal body burden of absorbed/retained lead and those of lead excretion. Source: EPA Air Quality Criteria for Lead, 1986.

III. Target Organs

The target organs and systems impacted by lead exposure do not strictly reflect its relative distribution within the body. The greatest deposition of lead is in the skeletal system, which accumulates more than 90% of the total body burden. The half-life of the longest of the bone lead compartments is extremely long, on the order of several decades. For these reasons, the total body burden of lead is best reflected in bone lead concentrations, but this is a measure which is difficult to obtain, particularly on a widespread basis. Lead does accumulate in soft tissue as well, although at far lower concentrations than in bone. Among soft tissues examined, accumulation is generally greatest in kidney. Concentrations of lead in brain tend to be lower, even though the CNS is particularly vulnerable to lead, especially during earlier stages of development. The half-life of lead in soft tissue is considerably shorter than that in bone and is thought to be on the order of one or two months at best. Lead also accumulates in blood, but with a relatively short half-life (i.e., on the order of 30 days) reflecting the turnover rate of erythrocytes. Thus blood lead measures, while generally easily obtained and consequently relied on as some indication of lead dose, represent only recent exposure levels rather than cumulative lead exposure.

Unfortunately, although the levels of lead in the environment have declined, so have the blood lead levels at which adverse effects on various target organs are observed. Today, the levels of lead in blood resulting from unavoidable chronic environmental exposure overlap considerably with those which are now implicated in causing a variety of health concerns, including subtle changes in behavior and CNS function and elevated blood pressure. These blood lead thresholds of adults are shown in Table II and those of children in Table III.

A. Central Nervous System

Exposures to high levels of lead have been shown in both adults and children to result in encephalopathy. Disease manifestations include such features as dullness, irritability, tremor, inadequate attention span, headaches, hallucinations, and loss of memory. This syndrome can progress rapidly to delirium, mania, coma, convulsions, paralysis, and even death. In adults, this syndrome is normally observed only when blood concentrations exceed 100 μg/dl, but in children, the same syndrome can occur at levels as low as 80 μg/dl. In addition, it appears that encephalopathy in response to lead exposure tends to occur with a higher incidence in children than in adults. The blood lead concentration at which encephalopathy is observed in any individual actually varies widely both in children and adults; some individuals may experience encephalopathy at 80 μg/dl whereas others remain asymptomatic until far higher blood lead levels (e.g., over 300 μg/dl) are attained.

Neuropathological findings in adults and children succumbing to lead encephalopathy are also similar. In such cases, lead does not appear to produce any type of selective CNS lesion, but rather a diffuse pattern of effects that have been reported to include an edematous and congested brain, alterations in brain capillaries such as the activation and dilation of intracerebral capillaries, swelling of endothelial cells, and diffuse astrocytic proliferation. In the cerebellum, ischemic changes are particularly pronounced.

A variety of studies over the past twenty years have been directed at characterizing subclinical lead toxicity in adults (i.e., central nervous system effects occurring at blood lead levels below 80 μg/dl and in the absence of any signs of overt intoxication). Collectively these studies have generally revealed evidence for a variety of behavioral and neurological impairments at exposures associated with blood lead concentrations as low as 40 μg/dl. Deficits in such capacities as oculomotor performance, eye–hand coordination, reaction time, hand dexterity, and even IQ and cognitive function were first reported at blood lead levels of about 50–60 μg/dl. Later studies reported changes in learning and memory capabilities, verbal concept formation, and visual motor function at levels as low as 40 μg/dl, the level currently considered a threshold value for such impairments in adults.

Similar questions have been pursued in children in long-term prospective studies in which women were followed from the time of pregnancy and subsequent lead exposure and neurobehavioral development of the offspring tracked over protracted periods of time. Taken together, these studies have suggested that blood lead concentrations as low as 10 μg/dl may be associated with CNS deficits. Naturally, such findings have raised numerous controversies, particularly with respect to the adequacy of controlling for possible confounding variables and covariates. But the general consistency of findings across studies, particularly when innumerable possible confounders and covariates have been controlled for, argues strongly for the reliability of the effect.

The first such study, carried out in first and second grade children, reported an average decrement in IQ of about 4 points on the WISC IQ scale associated with blood lead concentrations in the 30–50 μg/dl range. This study also reported that with increasing blood lead concentration there was an increase in the frequency of maladaptive classroom behavior based on classroom teachers' reports. Although a change of four IQ points may seem like a very small effect, it was later calculated that a deficit in the mean IQ of this magnitude in a normal population would not only be associated with a threefold increase in the number of children considered severely impaired (as defined by an IQ below 80) but also by a reduction of 5% in the number of those considered to be of superior intelligence (i.e., an IQ greater than 125).

Numerous studies followed that attempted to

Table II Summary of Lowest Observed Effect Levels for Key Lead-Induced Health Effects in Adults

Lowest observed effect level[a] (PbB)	Heme synthesis and hematological effects	Neurological effects	Effects on the kidney	Reproductive function effects	Cardiovascular effects
80–120		Encephalopathic signs and symptoms	Chronic nephropathy		
80	Frank anemia				
60				Female reproductive effects	
50	Reduced hemoglobin production	Overt subencephalopathic neurological symptoms		Altered testicular function	
40	Increased urinary ALA and elevated coproporphyrins	Peripheral nerve dysfunction (slowed nerve conduction)			
30					Elevated blood pressure (white males) aged 40–59
25–30	Erythrocyte protoporphyin (EP) elevation in males				
15–20	Erythrocyte protoporphyin (EP) elevation in females				
<10	ALA-D inhibition				?

Source: EPA Air Quality Criteria for Lead, 1986.
[a] PbB, blood lead concentration ($\mu g/dl$).

Table III Summary of Lowest Observed Effect Levels for Key Lead-Induced Health Effects in Children

Lowest observed[a] effect level (PbB)	Heme synthesis and hematological effects	Neurological effects	Renal system effects	Gastrointestinal effects
80–100		Encephalopathic signs and symptoms	Chronic nephropathy (e.g., aminoaciduria)	Colic, other overt gastrointestinal symptoms ↓
70	Frank anemia			
60		Peripheral neuropathies ↓ ?		
50				
40	Reduced hemoglobin synthesis Elevated coproporphyrin Increased urinary ALA	Peripheral nerve dysfunction (slowed NCVs) CNS cognitive effects (e.g., IQ deficits) ?		
30			Vitamin D metabolism interference ↓ ?	
15	Erythrocyte protoporphyrin elevation	Altered CNS electrophysiological responses ↓ ?		
10	ALA-D inhibition Pyrimidine-5′ nucleotidase activity inhibition ↓ ?			

Source: EPA Air Quality Criteria for Lead, 1986.
[a] PbB, Blood lead concentrations, (μg/dl).

replicate and extend such findings. One of these utilized a sample population that was considered to be relatively free of the other potential covariates of IQ. It reported that infants with blood lead concentrations greater than 10 μg/dl scored lower on the Mental Development Index of the Bayley Scales. Three other studies relating aspects of cognitive function to blood lead concentration in children have likewise shown inverse linear relationships between neuropsychologic measures and blood lead levels, with no evidence of any threshold down to the level of about 6–10 μg/dl of lead in blood. These are blood lead concentrations that are widely prevalent in many areas of the United States.

Aspects of cognitive function do not appear to be the sole effects of chronic low-level lead exposure in children. Several studies have corroborated the original reports of behavioral and attentional deficits in the classroom as indicated by teacher ratings. Deficits in reaction times have been reported under some circumstances. In addition, based on the NHANES II data set, the age at which various developmental milestones occurs, such as that at which a child first speaks, walks, and sits up have been found to be directly related to lead exposure. Low-level exposures to lead have also been shown to impair postural equilibrium and to elevate hearing thresholds. At least one prospective study in Cincinnatti suggests that the effects of these low-level lead exposures on CNS development are mediated via effects on birth weight and gestational age, both of which were also reduced by Pb exposure. Effects of Pb on birth weight, however, have not been uniform in all studies.

As yet, the neurobiological basis of Pb-induced CNS toxicity remains to be determined. The diversity of effects produced suggests that there may be multiple mechanisms, or alternatively, that a single basic mechanism impacts numerous systems to produce this array of CNS deficits. Moreover, the fact that CNS effects can be produced in both children and occupationally exposed adults suggests that a mechanism based solely on developmental disturbances is not sufficient. An additional issue that has yet to be adequately resolved is the reversibility of CNS effects resulting from low-level lead exposure.

B. Peripheral Nervous System

The high dose lead exposures typically associated with occupational exposure have frequently resulted in peripheral nervous system damage, as reflected in one of the classical signs of toxicity known as wrist drop in which the hands dangle from the wrist. This syndrome is often more pronounced in the dominant hand. In addition, it can sometimes be accompanied by a similar effect at the ankle, known as ankle drop. These manifestations are the result of the impact of Pb on large myelinated nerve fibers. At the pathological level, segmental demyelination may be observed, along with axonal degeneration and endoneural edema. Frequently, by the time peripheral neuropathy is manifested, the damage can be assumed to be permanent.

Studies aimed at evaluating whether subclinical effects of lead on the peripheral nervous system exist have focused on assessment of nerve conduction velocity, a noninvasive procedure that can be applied to either sensory or motor nerves. As in the case of CNS function, utilization of these more sensitive measures has revealed that effects of Pb on peripheral nervous system function are actually evident at far lower blood Pb levels than those associated with wrist and ankle drop. Collectively, these studies reveal that Pb exposure is indeed associated with a slowing of nerve conduction velocity, and that this effect is most consistently reflected in the median motor nerve. In addition, such effects are noted at blood lead levels well below 70 μg/dl and possibly even as low as 30 μg/dl.

One prospective study tracked workers from the time of employment to four years. Nerve conduction velocities in workers of high versus low lead exposure levels were then compared after one, two and four years of employment; high-Pb-exposure workers had blood lead concentrations that ranged from about 30 to 50 μg/dl. At each of the time points of measurement, the nerve conduction velocities of the median motor nerve of the high-level Pb-exposed group were significantly slower than those of the low-level Pb-exposed group. These changes in nerve conduction velocity may be at least partially reversible. Nerve conduction velocities of the median motor nerve of a group of Pb-exposed workers increased significantly after blood lead concentrations decreased from mean values of 52 to 33 μg/dl at one month to three years after therapeutic chelation with $CaNa_2$-EDTA. Thus, studies suggest the possibility of Pb-induced changes in the PNS as assessed via nerve conduction velocity at blood lead levels as low as 30 μg/dl in adults.

C. Kidney

Detrimental effects on the kidney of exposure to high levels of lead (i.e., lead nephropathy) have long been noted both in adults and children. Many studies of occupationally-exposed populations have been carried out documenting this nephropathy and the wide range of blood lead concentrations with which it is associated (40 to over 100 μg/dl) as well as the variation among individuals in susceptibility to this effect.

In children with lead nephropathy, the full Fanconi syndrome can be seen (i.e., glycosuria, aminoaciduria, and phosphaturia), but these occur at blood lead levels in excess of 150 μg/dl, generally reflecting quite high exposure levels. Once blood Pb levels reach at least 80 μg/dl, proximal tubular transport impairment is seen. Studies carried out in a population of Australian children many years ago who had been acutely exposed to very high levels of lead suggested that severe childhood lead poisoning was associated with chronic nephritis in adulthood. Additional studies since that time have qualified these findings to suggest that chronic nephritis in adulthood may only follow childhood lead poisoning when early exposure is particularly protracted and results in an extensive body burden of lead.

The syndrome of chronic lead nephropathy in adults differs in several respects from the acute lead nephropathy that is characteristic of childhood lead poisoning (e.g., in being more difficult to identify). Cases of lead nephropathy among occupationally exposed workers are generally diagnosed on the basis of a positive $CaNa_2$-EDTA challenge or provocation test outcome and additional evidence of impaired renal function in the absence of other identifiable etiologic or renal disease. Another difference seems to be that in adults, the intranuclear inclusion bodies found in the kidney in response to Pb exposure are less common, although it should be noted that there is no necessary correspondence between the presence of intranuclear inclusion bodies and renal impairment. The full Fanconi syndrome seen in children is not prevalent in adults where, in general, proximal tubular deficits are more difficult to demonstrate. Instead, hyperuricemia is a frequent finding in the chronic lead nephropathy of adults.

It has been suggested, however, that the differences in severity of renal impairment between children and adults may reflect differences in the conditions and degrees of lead exposure, in that such episodes in children normally follow short-term high-level exposure whereas adult occupational exposure is generally of a more prolonged duration. An additional difference may also be that gout occurs in about half of the adult population with lead nephropathy. One speculated explanation for this is that the hyperuricemia produced by Pb exposure would create the appropriate conditions for development of gout. To date, the impact of chronic lead exposure of an environmental nature on the renal status of the general population remains unknown.

Pathologic changes in kidney have been reported for cases of human Pb exposure and the nature of these effects are generally corroborated by experimental animal studies. These findings indicate reversible lesions that include Pb-containing intranuclear inclusion bodies, swollen mitochondria, cytomegaly, and an elevation in iron-containing lysosomes in proximal tubular cells. Interstitial fibrosis, on the other hand, is considered to be an irreversible lesion. Functional renal changes include the increased excretion of amino acids, as well as elevations in both serum urea nitrogen and uric acid concentrations, an inhibition of membrane marker enzymes, and alterations in mitochondrial respiratory and energy functions. Less consistent are findings relating to Pb-induced changes in glomerular filtration rate, renal bloodflow, and the status of the renin–angiotensin system. Questions also remain with regard to the mechanism of Pb-induced renal injury and to the relationships between Pb-induced changes in the renin–angiotensin system and hypertension.

D. Hematopoietic System

The formation of the heme molecule, a molecule with an array of important functions, takes place via the heme biosynthetic pathway. Heme is, of course, the primary constitutent of hemoglobin and is also a component of several other hemoproteins, including the cytochromes, the *P*-450 component of the mixed-function oxygenase system, and myoglobin. Lead has effects at multiple points within the heme biosynthetic pathway that result in the net accumulation of several byproducts of the pathway. The accumulation of these byproducts have variously served both as indices of Pb exposure, as well as indices of Pb effect.

There are three primary effects of Pb on the heme biosynthesis pathway. One of these is the stimulation of the mitochondrial enzyme aminolevulinic acid synthetase (ALA-S), the enzyme involved in the formation of aminolevulinic acid (ALA). A second effect is the inhibition of the cytosolic enzyme aminolevulinic acid dehydratase (ALA-D) which normally results in the formation of porphobilinogen from two units of ALA. These two effects result in the accumulation of ALA in blood, urine, and soft tissue. The third effect is the inhibition of the enzyme ferochelatase which is the enzyme mediating the incorporation of iron into the heme molecule, a step of the pathway occurring intramitochondrially and which results in the buildup of protoporphyrins.

Numerous studies have examined the blood lead levels at which such effects occur. For increased ALA-S activity, this threshold appears to be about 40 μg/dl. Erythrocyte ALA-D is much more sensitive to inhibition by Pb exposure, having been shown to be inhibited at a blood lead concentration of only 10 μg/dl, but is even thought by many to be inhibited at virtually all blood lead levels, without any apparent threshold. More arguable, however, is the apparent biological or health relevance of ALA-D inhibition at these extremely low exposure levels. Since substrate (ALA) accumulation only begins to occur at levels of 40 μg/dl, inhibition of ALA-D at blood lead levels below this may simply reflect the extensive reserve capacity of the enzyme. The blood lead threshold for zinc protoporphyrin (ZPP) accumulation appears to be on the order of 25–30 μg/dl for males, and occurs at the slightly lower levels of 15–20 μg/dl in females and children. ZPP accumulation is thought to be a better indicator of Pb burden than is ALA, since it only occurs during the formation of the erythrocyte, thus reflecting the long-term presence of lead. In contrast, ALA-D inhibition occurs even in the mature erythrocyte. This results in a several week lag time from onset of Pb exposure to ZPP accumulation. For the same reason, ZPP accumulation can also reflect continued exposure to lead resulting from bone resorption after external lead exposure has ceased.

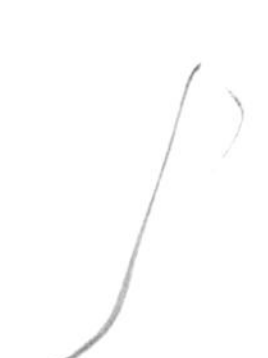

As previously pointed out, heme is a constituent of many other protein and enzyme systems, and consequently the deleterious effects of Pb on heme production are not restricted to the erythropoietic system. For example, inhibition of the activity of renal 1-hydroxylase, a cytochrome *P*-450-mediated enzyme, decreases serum 1,25-dihydroxyvitamin D levels. The reduction of the activity of the hepatic enzyme, tryptophan pyrrolase, can increase levels of tryptophan in plasma, and levels of tryptophan, serotonin and 5-hydroxyindoleacetic acid in brain. The hepatic protein, cytochrome *P*-450, a heme-containing protein, can likewise be inhibited with obvious consequences for detoxification mechanisms.

Lead exposure is also associated with anemia, an effect attributed to both reduced hemoglobin production as a result of the decline in heme formation, and to the shortening of erythrocytic lifespan. In turn, the shortened red blood cell lifespan has been attributed to an increase in cell fragility as a consequence of Na/K-ATPase inhibition leading to cell shrinkage; and to inhibition of pyrimidine-5-nucleotidase, the enzyme that dephosphorylates pyrimidine nucleotides necessary for cellular energetics in the maturing erythrocyte. The inhibition of pyrimidine-5-nucleotidase now appears to occur in children at extremely low blood levels (i.e., at about 10 μg/dl).

Some investigators have even speculated that alterations in heme metabolism may be related to the neurotoxicity of Pb, based on several lines of evidence. Among these are the findings that increases in ALA are noted in neural tissue as well as non-neural tissue; the similarities of some of the classical signs of lead encephalopathy and certain of the neurological components of acute intermittent porphyria; the demonstration *in vitro* that ALA can inhibit the release of the neurotransmitter GABA from presynaptic receptors, and the fact that the inhibition of tryptophan pyrrolase can increase brain levels of tryptophan and serotonin. Such evidence, however, is currently indirect at best and fails to take into account the possibility of Ca^{2+}-mediated effects, the numerous documented effects of Pb on other neurotransmitter systems, and the increasing evidence of neurotransmitter interactions.

E. Cardiovascular System

The possibility of a relationship between blood lead levels below 40 μg/dl and hypertension has generated almost as much controversy as that proposed between low-level Pb exposure and IQ decrements in children. The situation is further analagous in that it is well known that at high exposure levels, Pb can be associated with cardiac abnormalities. Lead poisoning has been shown to occur in conjunction with degenerative changes in heart muscle in children. Moreover, electrocardiographic abnormalities at a rate four times higher than anticipated were reported in a population of adults that had experienced chronic lead poisoning. Chelation therapy for lead intoxication has been shown to reverse electrocardiograph abnormalities.

Whether more subtle cardiac effects (as reflected in blood pressure elevations) occur at even lower Pb exposure levels has been the subject of numerous investigations. As is also the case with the Pb–IQ controversy, there are of course many potential confounding variables that can impact blood pressure that must be considered in study design. Table IV presents the results of several analyses based on two large-scale studies of the relationship between lead exposure and blood pressure. In reviewing the collective results of these studies, as well as those of numerous smaller-scale studies, the EPA in 1986 came to the conclusion that the evidence for a small but statistically significant effect of lead exposure on blood pressure was highly convincing. This effect appeared to be most pronounced in males between the ages of 40 and 59 and for systolic more than for diastolic pressure. The blood lead level associated with such effects still remains to be definitively determined, but studies show reliable associations at Pb blood levels of even 30 μg/dl with some studies even suggesting that these effects may extend to as low as 7 μg/dl. The degree to which such effects extend to other age and gender groups remains unclear.

The mechanism(s) of lead effects on blood pressure and other aspects of cardiovascular system function are not completely understood, but there is support for any of several different possibilities. In analyzing and integrating the results of both human and experimental animal studies relating to this issue, the EPA in 1986 suggested that the primary defect in the pathophysiology of hypertension is the alteration in calcium binding to plasma membranes of cells, an effect which could in fact derive from alterations in membrane sodium permeability. The alterations in calcium binding could then lead via different pathways to elevated blood pressure. One way would be to alter the sensitivity of the vascular smooth muscle to vasoactive stimuli, an

Table IV Coefficients for the Natural Log of Blood Lead Concentration (log PbB) versus Blood Pressure (BP) in Men With and Without Adjustment for Site Variables

Study group/blood pressure	Unadjusted for site	Adjusted for site	Reference
British regional heart study[a]			
Systolic (n = 7371)	1.68[f]	2.09[f]	1, 2
Diastolic (n = 7371)	0.30	1.81[g]	
NHANES II[b]			
Systolic (n = 2254)	5.23[g]	3.23[f]	3, 4
Diastolic (n = 2248)	2.96[g]	1.39[f]	
NHANES II[c]	3.43[g]	1.95[e]	5
Systolic (n = 2794)	2.02[g]	0.36	
Diastolic (n = 2789)			
NHANES II[d]	8.44[f]	5.01[e]	6, 7
Systolic (n = 543)	3.95[f]	2.74[e]	
Diastolic (n = 565)			
NHANES II[d]	6.27[f]	3.46[e]	5
Systolic (n = 553)	4.01[f]	1.93[e]	
Diastolic (n = 575)			

Source: EPA Air Quality Criteria for Lead, 1986.
[a] Subjects were white males aged 40–59.
[b] Subjects were males aged 20–74.
[c] Subjects were males aged 12–74.
[d] Subjects were males aged 40–59.
[e] $p < 0.05$
[f] $p < 0.01$
[g] $p < 0.001$

1. Pocock, S. J., Shaper, A. G., Ashby, D., Delves, T., and Whitehead, T. P. (1984). Blood lead concentration, blood pressure and renal function. *Br. Med. J.* **289,** 872–874.
2. Pocock, S. J., Shaper, A. G., Ashby, D., and Delves, T. (1985). Blood lead and blood pressure in middle-aged men. *In* "International Conference: Heavy Metals in the Environment," (T. D. Lekkos, ed.), vol. 1, pp. 303–305. CEP Consultants, Athens.
3. Schwartz, J. (1985a). Evidence for a blood lead-blood pressure relationship [memorandum to the Clean Air Science Advisory committee]. U.S. Environmental Protection Agency, Office of Policy Analysis, Washington, D.C. Available for inspection at, U.S. Environmental Protection Agency, Central Docket Section, Washington, D.C., docket no. ECAO-CD-81-2 II A.C.S.
4. Schwartz, J. (1985b). Response to Richard Royall's questions on the blood lead blood pressure relationships in NHANES II [memorandum to Dr. David Weil]. U.S. Environmental Protection Agency, Office of Policy Analysis, Washington, D.C. Available for inspection at U.S. Environmental Protection Agency, Central Dockety Section, Washington, D.C.; docket no. ECAO-CD-81-2 IIA.C.5.
5. E. I. du Pont de Nemours (1986). Du Pont comments on the February 1986 draft document. E. I. du Pont de Nemours & Co., Wilmington, DE. Available for inspection at U.S. Environmental Protection Agency, Central Docket Section, Washington, D.C.; docket no. ECAO-CD-81-2 IIA.E.C.3.6.
6. Schwartz J. (1986a). NHANES II blood pressure analysis [memorandum to Lester Grant]. U.S. Environmental Protection Agency, Office of Policy Analysis, Washington, D.C. Available for inspection at U.S. Environmental Protection Agency, Central Docket Section, Washington, D.C.; docket no. ECAO-CD-81-2 IIA.C.11.
7. Schwartz J. (1986b). Blood lead and blood pressure (again) [memorandum to Lester Grant]. U.S. Environmental Protection Agency, Office of Policy Analysis, Washington, D.C. Available for inspection at U.S. Environmental Protection Agency, Central Docket Section, Washington, D.C.; docket no. ECAO-CD-81-2 II.A.C.11.

effect which in and of itself would be sufficient to explain hypertension. Alternatively, or even in conjunction with such an effect, an alteration could occur in the neuroendocrine input (e.g., changes in renin secretion rate) to vascular smooth muscle.

F. Reproduction

Many reports of increased rates of abortion, miscarriage, and stillbirth have occurred as a result of occupational exposures of women to very high levels of lead during the 19th century. In fact, and perhaps based on this knowledge, various admixtures of lead were widely used at that time as abortifacients. However, many of those early reports of reproductive failure suffered from methodological difficulties. Later, regulations were introduced restricting employment of women in such industries, thus there was no population from which to replicate the prior studies linking high levels of lead to reproductive toxicity. Nor is there enough information from human studies that is useful in gauging the impact of Pb on the ovarian cycle or on other aspects of fertility. Experimental animal studies that address such issues have been carried out primarily in rodents. They reveal irregular or even absent estrus cycles, ovarian follicular cysts, and reductions in numbers of corpora lutea. Similar results which were found to be reversible following termination of exposure are noted in Pb-dosed monkeys.

There are reports in human males of lead effects on various aspects of testicular function (including abnormal spermatogenesis) which occur at blood lead concentrations of about 40–50 μg/dl. Again, early studies suffered from methodological difficulties but their findings appear to be corroborated by the later results of experimental animal studies. The latter also suggest that such effects may occur via direct effects of Pb on testicular enzymes.

Lead is clearly able to cross the placenta and to accumulate in the fetus although the precise kinetics are not well delineated. Early studies detailing effects in human offspring exposed to Pb prenatally yielded inconsistent findings, with some studies suggesting shortened gestation, decreased birthweights, and increased incidence of stillbirths, while others reported no evidence for such effects. Many of these studies, however, suffered the limitations of small sample sizes and inadequate determinations of Pb exposure. Later experimental animal studies have suggested that prenatal Pb exposure could indeed induce a number of teratogenic and fetotoxic effects. More recent human studies that have overcome the primary deficiencies of early studies suggest a correspondence between the outcomes of experimental animal and human findings. At least one such study provides evidence of a relationship between low-level Pb exposure and the incidence of minor congenital anomalies, an effect that requires further resolution. While there seems to be little consistent evidence that birth weight per se is affected, several studies have reported significant negative relationships between prenatal Pb exposure and gestational age (Table V). An effect of Pb on growth has also been suggested in both human and experimental animal studies that is speculated to be related to Pb-induced changes in neuroendocrine function.

G. GI Tract

Despite the fact that colic is a consistent symptom of Pb poisoning both in occupationally exposed workers and in lead-poisoned children, it has received limited experimental attention. It is known that there are wide individual differences in the blood lead levels at which such effects are experienced, and that these levels can be below 50 μg/dl. There is almost no information from which to determine the mechanistic basis of the effects.

H. Immune System

Early descriptions of Pb toxicity generally considered five major target organs, including the central nervous system, peripheral nervous system, GI tract, kidney, and hematopoietic system. Since that time, it has become clear that the immune system must also be considered a target organ for lead. Experimental animal studies assessing the impact of Pb exposure on host resistance to infectious agents have revealed increased mortality at exposure levels as low as 13 ppm administered orally to mice for 10 weeks. Few comparable studies in lead-exposed humans have as yet been carried out, but there is evidence that Pb workers with blood lead levels in the range of 22 to 89 μg/dl experience more colds and influenza infections per year. More epidemiologic work is needed both to corroborate such findings and to examine the nature of the dose–effect relationships. Other supportive evidence reveals that Pb exposure enhances transplantability of tumors, the oncogenicity of leukemia viruses, and the development of chemically induced tumors.

Even in the absence of sound epidemiological data, studies in experimental animals on lead effects on various components of the immune system and on immune function provide strong evidence that Pb is immunosuppressive at very low exposure levels. In that regard, lead has been shown to have little effect on serum immunoglobin levels, but under several different conditions to impair antibody production. These effects have been manifest to a

Table V Association of Prenatal Lead Exposure with Gestational Age and Birth Weight[a]

Pb-exposure index	n	Blood levels (μg/dl)	Gestational age effect	Birth weight effect	Reference
delivery	185	6.5	?	none	1, 2
cord	162	5.8	?	none	
delivery	216	6.5	pos.	neg.[a]	3
cord					
delivery	4354	6.5	none	none	4
cord					
prenatal	185	8.3	neg.[b]	neg.[b]	5
delivery	749	11.0	neg.[b]	pos.[c]	6
cord		10.0	neg.[b]	pos.[c]	
delivery	236	14.0	neg.[b]	none	7
cord		12.0	neg.[b]	none	

Source: EPA Air Quality Criteria for Lead, 1986.

[a] Birth weight showed no relationship, but the trend in percentage of small-for-gestational-age infants was nearly statistically significant at $p < 0.05$.

[b] $p < 0.05$.

[c] See text for possible explanation of reduced blood lead levels in mothers whose infants were low in birth weight.

1. Ernhart, C. B., Wolf, A. W., Kennard, M. J., Filipovich, H. F., Sokol, R. J., and Erhard, P. (1985). Intrauterine lead exposure and the status of the neonate. *In* "International conference: Heavy metals in the Environment." (T. D. Lekkas, ed.), vol. 1, pp. 35–37. CEP Consultants, Ltd., Athens.
2. Ernhart, C. B., Wolf, A. W., Kennard, M. J., Erhard, P., Filipovich, H. F., and Sokor, R. J. (1986). Intrauterine exposure to low levels of lead: The status of the neonate. *Arch. Environ. Health* **41,** 287–291.
3. Bellinger, D. C., Needleman, H. L., Leviton, A., Waternaux, C., Rabinowitz, M. B., and Nichols, M. L. (1984). Early sensory-motor development and prenatal exposure to lead. *Neurobehav. Toxicol. Teratol.* **6,** 387–402.
4. Needleman, H. L., Rabinowitz, M., Leviton, A., Linn, S., and Schoenbaum, S. (1984). The relationship between prenatal exposure to lead and congenital anomalies. *J. Am. Med. Assoc.* **251,** 2956–2959.
5. Dietrich, K. N., Krafft, K. M., Shukla, R., Bornschein, R. L., and Succop, P. A. (1987). The neurobehavioral effects of prenatal and early postnatal lead exposure. *In* "Toxic Substances and Mental retardation: Neurobehavioral Toxicology and Technology" (S. Schroeder, ed.), vol. 8, pp. 71–96. American Association on Mental Deficiency Monograph Series, Washington, D.C.
6. McMichael, A. J., Vimpani, G. V., Robertson, E. F., Baghurst, P. A., and Clark, P. D. (1986). The Port Pirie cohort study: Maternal blood lead and pregnancy outcome. *J. Epidemiol. Commun. Health* **40,** 18–25.
7. Moore, M. R., Goldberg, A., Pocock, S. J., Meredith, A., Stewart, I. M., Macanespie, H., Lees, R., and Low, A. (1982). Some studies of maternal and infant lead exposure in Glasgow. *Scott. Med. J.* **27,** 113–122.

variety of challenges, including typhus, pseudorabies virus, and *Salmonella* typhomurium after exposures to levels as low as 10 ppm administered orally for 10 weeks.

Evaluation of these effects at the cellular level has involved quantification of IgM and IgG plaque-forming cells and these studies further substantiate effects of Pb at quite low levels of exposure, as shown in Table VI, with indications that both IgM and IgG plaque-forming cell numbers are decreased in response to lead. *In vitro* studies suggest that these effects may actually occur at the macrophage level. Lead exposure also appears to alter cell-mediated immunity as manifested by a decrease in delayed-type hypersensitivity responses after an exposure of only 13.7 ppm for 4 weeks. Macrophage activity and reticuloendothelial system function are also altered in response to low exposure

Table VI Effect of Lead on the Development of Antibody-Producing Cells

Species	Antigen[a]	Lead dose and exposure	Effect[b]	Reference
Mouse	SRBC (*in vivo*)	13–1370 ppm; 8 wk	IgM PFC (D) IgG PFC (D)	1
Mouse	SRBC (*in vivo*)	0.5–2 ppm tetraethyl lead; 3 wk	IgM PFC (D) IgG PFC (D)	2
Mouse	SRBC (*in vivo*)	13–1370 ppm; 10 wk	IgG PFC (D)	3
Mouse	SRBC (*in vivo*)	4 mg (i.p. or orally)	IgM PFC (I) IgG PFC (D)	4
Mouse	SRBC (*in vivo*) SRBC (*in vitro* + 2-ME)	16–2000 ppm; 1–10 wk 16–80 ppm; 4 wk 2000 ppm; 4 wk	IgM PFC (N) IgM PFC (I) IgM PFC (D)	5
Rat	SRBC (*in vivo*)	25–50 ppm; pre/postnatal	IgM PFC (D)	6
Mouse	SRBC (*in vitro*) SRBC (*in vitro* + 2-ME)	50–1000 ppm; 3 wk 50–1000 ppm; 3 wk	IgM PFC (D) IgM PFC (N or I)	7
Mouse	SRBC (*in vitro* + 2-ME)	2–20 ppm (*in vitro*)	IgM PFC (I)	8, 9

Source: EPA Air Quality Criteria for Lead, 1986.

[a] The antigenic challenge with sheep red blood cells (SRBC) was *in vivo* or *in vitro* after *in vivo* exposure. The *in vitro* assays were performed in the presence (+, 2-ME) or absence of 2-mercaptoethanol.

[b] IgM/G PFC, Immunoglobulin M/G plaque-forming cells; D, decreased response; N, unaltered response; I, increased response.

1. Koller, L. D., and Kovacic, S. (1974). Decreased antibody formation in mice exposed to lead. *Nature* (Lond.) **250,** 148–150.
2. Blakely, B. R., Sisodia, C. S., and Mukkur, T. K. (1980). The effect of methylmercury, tetraethyl lead, and sodium arsenite on the humoral immune response in mice. *Toxicol. Appl. Pharmacol.* **52,** 245–254.
3. Koller, L. D., and Roan, J. G. (1980). Effects of lead, cadmium and methylmercury on immunological memory. *J. Environ. Pathol. Toxicol.* **4,** 47–52.
4. Koller, L. D., Exon, J. H., and Roan, J. G. (1976). Humoral antibody response in mice after single dose exposure to lead or dacmium. *Proc. Soc. Exp. Biol. Med.* **151,** 339–342.
5. Lawrence, D. A. (1981a). Heavy-metal modulation of lymphocyte activities. I. *In vitro* effects of heavy metals on primary humoral immune responses. *Toxicol. Appl. Pharmacol.* **57,** 439–451.
6. Luster, M. I., Faith, R. E., and Kimmel, C. A. (1978). Depression of humoral immunity in rats following chronic developmental Pb exposure. *J. Environ. Pathol. Toxicol.* **1,** 397–402.
7. Blakely, B. R., and Archer, D. L. (1981). The effect of lead acetate on the immune response in mice. *Toxicol. Appl. Pharmacol.* **61,** 18–26.
8. Lawrence, D. A. (1981b). Heavy-metal modulation of lymphocyte activities. II. Lead, an *in vitro* mediator of B-cell activation. *Int. J. Immunopharmacol.* **3,** 153–161.
9. Lawrence, D. A. (1981c). *In vivo* and *in vitro* effects of lead on humoral and cell-mediated immunity. *Infec. Immun.* **31,** 136–143.

levels of lead. Lead does not appear to achieve any of its immune system effects through an alteration of interferon production.

The exact mechanism(s) by which lead alters immune function is not yet known. Some investigators have speculated an involvement of sulfhydryl groups based on the involvement of thiols in lymphocyte function and the fact that some effects of Pb on the immune system can be reversed by exogenous thiol treatment.

I. Endocrine Function

Most of the evidence regarding any effects of lead on aspects of endocrine function indicate their manifestations only at quite high levels of exposure. There are several reports that both in humans and experimental animals Pb exposure can decrease thyroid function. Moreover, a decrease in uptake of ^{131}I noted in some occupationally exposed individuals can be offset by administration of thyroid-stimulating hormone. Pituitary and adrenal function alterations in response to Pb exposure have also been reported. Lead also seems to impact elements of the endocrine system involved in the regulation of reproductive function in males, as evidenced by changes in leutinizing hormone secretion and in reduced basal serum testosterone levels. Evidence for effects on ovarian function in response to Pb exposure are simply not available. Of those endocrine effects that have been noted, however, none is associated with blood lead levels below 40 μg/dl.

J. Hepatic Function

Reports relating Pb to changes in hepatic function are conflicting and sparse. Liver function tests have not typically been included in assessments of the health of lead-exposed workers. Changes in drug-metabolizing capabilities are not particularly compelling. It appears that any change in hepatic *P*-450 heme-associated enzymes caused by the effects of Pb on the hematopoietic system are not sufficient to result in any significant changes in enzyme function.

K. Bone

It is well known that over 90% of the total body burden of lead resides in the skeletal system. For a long time, this sequestration of Pb in bone was considered to serve a protective mechanism, preventing exposure of soft-tissue target organs. Thus, bone was considered to be an insidious sink for lead. That perspective has changed dramatically over the past 10 years as it has become increasingly clear that this vast store of lead can be mobilized under various physiological conditions, and can actually be redistributed to soft-tissue compartments and even back to bone.

One manipulation that appears to provoke bone Pb mobilization is $CaNa_2$-EDTA chelation therapy. Experimental studies with rats clearly show that $CaNa_2$-EDTA administration results in a pronounced mobilization of Pb from bone either following acute or chronic Pb exposure. Further experiments show that this mobilized Pb is then redistributed to soft-tissue compartments, noteably brain and liver. In both humans and experimental animals, the administration of $CaNa_2$-EDTA is accompanied by a marked increase in plasma Pb concentrations even while blood lead concentrations decline. Given that the plasma Pb compartment is the source of Pb delivered to soft tissue, these findings are consistent with the noted increases in brain and liver Pb concentrations. These studies have obvious implications for chelation therapy, as discussed below.

Another condition associated with movement of Pb from bone back to soft tissue is aging. During the processes of aging, bone resorption occurs and the rate of bone growth declines as well, a phenomenon more pronounced in females than in males. Together, these processes mean that Pb already stored in bone is released, and that newly absorbed lead cannot be incorporated into bone at the rates occurring earlier in life. These two events combine to make more lead available to soft tissue during advancing age. This phenomenon has been demonstrated in experimental animals, where a dose-related decline in bone Pb with age is observed, as well as an age-related increase in Pb levels in brain, liver, and kidney. Postmortem studies of humans have shown clear decreases in the Pb content of certain bones with age and this effect is of greater magnitude in females. In addition, analyses of data from the NHANES II study indicate higher blood lead levels in postmenopausal women than in premenopausal women, presumably due to the greater resorption of bone and release of stored lead after the onset of menopause. At least one study in humans to date has reported a higher rate of lead poisoning in aging. Both epidemiologic and smallscale studies are needed to more fully investigate the impact of aging processes on lead metabolism and toxicity in humans. With the exception of one study examining changes in bone Pb with age, most studies have been based upon sample sizes too restricted to allow any definitive conclusions.

A third condition associated with bone Pb mobilization and consequent soft tissue redistribution is the cessation of external Pb exposure. When environmental Pb exposure is terminated, the kinetics of Pb dictate that steady-state levels across physiological compartments be achieved. Given the higher concentrations of lead in bone relative to soft tissue, the gradient of movement of lead would be from bone into the blood stream and then to soft tissue. This phenomenon is evident in occupationally exposed workers following retirement, when blood Pb levels remain elevated above background levels for years afterwards as bone Pb levels reequilibrate.

A final set of conditions associated with Pb redistribution from bone are the physiological states of pregnancy and lactation. During both pregnancy and lactation, the Ca^{2+} requirement increases enormously and is often fulfilled, at least in part, by a resorption of Ca^{2+} from bone. This resorption may concurrently release lead stores, again making lead available to soft tissue. While this phenomenon has not yet been studied to any great extent in humans, it is supported by results of experimental animal studies. It is also supported by analyses of the NHANES II data set indicating that postmenopausal women who had previously been pregnant showed less of an increase in blood lead relative to premenopausal women, than did postmenopausal women who had never been pregnant. This differential increase in blood lead is thought to derive from the decrease in bone Pb stores that had already occurred in postmenopausal women during previous pregnancies.

IV. Genotoxicity and Carcinogenicity

Several epidemiological studies of industrial workers were conducted in the period of 1960 to 1980 to evaluate the carcinogenic potential of Pb. There were, however, numerous methodological difficulties, particularly with respect to documentation of lead exposure and evaluation of total lead body burden, that limit the strength of any conclusions that can be drawn. Taken together, however, these studies suggest increased mortality rates and an association of high-level Pb exposure with cancers of the respiratory tract, GI tract, and kidney. Cancer of the respiratory tract must be considered particularly tenuous in the absence of information on smoking histories. However, both GI tract and renal cancers have been repeatedly noted.

A number of experimental animal studies have also addressed the relationship between lead exposure and renal carcinogenicity. While such studies have not followed the National Cancer Institute Guidelines, they nonetheless reveal a fairly consistent pattern of renal carcinoma. In addition, taken together, they also suggest a relationship between kidney Pb burden and carcinogenicity. At the highest lead doses used, renal carcinogenicity was accompanied by kidney damage. What is generally unclear at this time is the extent to which lead compounds other than lead acetate may induce renal carcinogenicity and the extent of renal or other toxicity associated with carcinogenicity. Bolstering the possibility of a link between lead exposure and carcinogenesis are studies demonstrating the ability of at least one Pb compound, lead acetate, to produce neoplastic transformations in cells in a dose-related fashion, leading the EPA to suggest that Pb acetate has effects resembling those of other mental carcinogens such as nickel and chromium.

Lead has also been shown to induce genotoxicity. Studies evaluating chromosomal aberrations have been carried out both following *in vivo* exposure and in tissue culture. While the collective outcomes of such studies have been inconsistent, they do yield evidence of a positive effect, especially in the presence of a low Ca^{2+} environment. Chronic lead exposure has been shown in another study to increase the incidence of sister chromatid exchange, an effect which was also shown to be reversible after a 4-week cessation of exposure. In addition, experimental studies have demonstrated the ability of Pb to alter the synthesis of DNA and the fidelity of DNA replication, to inhibit RNA synthesis, to stimulate mRNA chain initiation, and to induce DNA single-strand breaks, all indicating that Pb can alter certain molecular processes that are involved in the normal regulation of gene expression. In considering the results of both carcinogenicity and genotoxicity studies, the EPA has suggested that sufficient evidence has been presented to indicate that lead can act both as an initiator and promotor of carcinogenicity.

V. Mechanism of Lead Effects

To date, there is no one unifying mechanism that has been identified which has been shown to account for all of the diverse effects of lead on the variety of organs and systems that are impacted. However, much of the work aimed at identifying such a basis has focused on the mitochondrion and on lead–calcium interactions. A role in mitochondria is based on three lines of evidence: that Pb exposure has been shown to produce structural changes both *in vivo* and *in vitro,* to accumulate in mitochondria, and to produce functional changes in energy metabolism and ion transport. The structural changes feature mitochondrial swelling and the distortion of cristae. Accumulation of Pb in mitochondria has been reported *in vivo* in kidney, liver, spleen, and brain. Functional changes that have been documented include inhibition of cellular respiration and alterations in intracellular Ca^{2+}, both in neural and in nonneural tissue at relatively low levels of Pb.

The movement of Pb into mitochondria involves active transport. The interaction with calcium occurs at this level; it has been shown that Pb and Ca show pronounced similarities in cellular metabolism, and this interaction may also account for its interference with the transport of calcium across membranes. In addition, Pb has been shown to have other effects both on Ca homeostasis and on Ca-mediated processes (shown in Table VII), which have been observed both *in vivo* and *in vitro* and which could underlie or be involved in a wide spectrum of toxic effects.

VI. Developmental Differences in Susceptibility

Lead toxicity has long been clinically apparent to be of greater vulnerability in children than adults. This is manifest on a more quantitative basis as a difference in blood thresholds for certain toxic effects of Pb, particularly for those on the nervous system and the hematopoietic system. For example, the threshold for Pb-induced encephalopathy is lower in children than adults, and in fact the CNS appears to be a particular target in children, whereas peripheral nervous system problems are more frequently encountered in lead-exposed adults. Moreover, the neurologic sequelae appear to be more severe in children. With respect to the hematopoietic system, decreases in production of hemoglobin and the accumulation of erythrocyte protoporphyrin have been demonstrated at lower blood lead thresholds in children than adults, suggesting that children might also be more susceptible to the array of biochemical consequences arising from Pb-induced alterations in heme production.

Table VII Spectrum of Known Effects of Lead on Ca^{2+} Homeostasis and Ca^{2+}-Mediated Processes at the Cellular and Molecular Level[a]

Interaction	Ca^{2+} signal transduction and homeostasis	Ca^{2+} receptor and response systems
Direct	Plasma membrane Ca^{2+} channels Mitochondria Ca^{2+} pump Ca^{2+}-ATPase	Calmodulin Protein kinase C Troponin C Osteocalcin Calbindin Oncomodulin
Indirect	Adenylate cyclase Na^+/K^+-ATPase	Adenylate cyclase
Secondary	Hydrolysis of ATP Decreased heme Protein–SH binding	Protein–SH binding

[a] Source: Pounds *et al.* (1992). Reproduced by permission.

A variety of factors conspire to produce this enhancement of susceptibility during early development. Some of these factors are related to differential metabolism of Pb in children as compared to adults. For example, uptake of Pb, both from the GI tract and from the respiratory tract, is greater in children. Human studies have yielded GI absorption figures of 10 to 15% for adults, as compared to 50% for children. Children also exhibit a greater retention of Pb; a slower decline in brain Pb has been postulated. Whereas in adults, over 90% of the total body burden is lodged in bone, the corresponding figure for children is only 73%. This perhaps reflects the greater metabolic activity of the developing skeletal system. The consequences in skeletal deposition is that, with relatively less lead tied up in skeleton, greater relative amounts are bioavailable to soft-tissue target organs such as brain.

An additional factor contributing to enhanced vulnerability early in development is based on the known interaction of Pb with many other essential metals such as Ca, Fe, and Zn. The nature of these interactions is such that deficiencies of these essential nutrients enhance aspects of Pb metabolism such as absorption. Nutritional deficiencies of Ca, Fe, and Zn are certainly not uncommon in children, especially among those who are economically disadvantaged. These states of nutritional deficiencies, by enhancing absorption, act functionally to increase exposure to Pb.

Typical patterns of behavior early in development likewise serve to increase exposure to Pb. Young children in particular tend to engage in a substantial amounts of hand-to-mouth activity. This means that once they become mobile, the contamination of, for example, hands and toys with lead-containing dust becomes an added bolus which is then ingested through hand-to-mouth behavior. A more extreme example of this behavior is known as "pica," or the ingestion of nonfood materials, and this behavior can actually persist well beyond early childhood. Pica is considered to be responsible for exposure of many children to lead via the ingestion of lead-based paint chips.

A final factor that may enhance susceptibility during the earlier stages of development is the immaturity of the blood–brain barrier in children. It may thus be more permeable to toxicants, again with the net result of a functional increase in the dose of Pb to brain.

VII. Antidotes to Lead Toxicity

The standard approach to treating lead toxicity is via the administration of chelating or complexing agents (i.e., compounds that bind metals such as lead and enhance urinary excretion). The principal agents used for the diagnosis and therapeutic treatment of elevated lead burden are calcium disodium edatate ($CaNa_2$-EDTA), combined with British Anti-Lewisite (BAL) in a therapeutic protocol context.

Diagnostic assessment is sometimes made on the basis of a measurement of lead in blood, but is thought to be more adequately reflected in the results of a $CaNa_2$-EDTA provocation test. This diagnostic provocation test involves the administration of a single injection of $CaNa_2$-EDTA and the collection of all urinary output over the next eight hours. If the amount of lead excreted exceeds a specified amount, the patient is suspected to have an elevated lead burden and is deemed a candidate for therapeutic chelation treatment.

Therapeutic chelation consists of a 3–5 day course of $CaNa_2$-EDTA treatment, which, depending upon the extent of toxicity, is administered with priming doses of BAL. This treatment may be reinitiated if blood Pb levels rebound to near prechelation levels within a certain period following therapeutic chelation.

More recent findings with respect to the potential adverse consequences of mobilizing Pb with $CaNa_2$-EDTA; the development of new alternative chelating agents; and the advent of x-ray fluorescence technology, however, are all currently contributing to a changing perspective and course of action with regard to both diagnostic and therapeutic chelation. Recent animal studies have shown that mobilization of lead from bone following a single injection of $CaNa_2$-EDTA, the equivalent of the diagnostic provocation test, is accompanied by an increase in brain and liver concentrations. These findings, bolstered by experiences reported in clinical situations, have led to editorial comments questioning the safety of the $CaNa_2$-EDTA provocation test even in asymptomatic children with moderately increased lead absorption. Furthermore, experimental animal studies reveal that a net loss of Pb from brain appears to be achieved only when $CaNa_2$-EDTA treatment is protracted well beyond the standard five-day course of treatment. Unfortunately, $CaNa_2$-EDTA cannot be administered safely to humans on a daily basis for more than five days.

Another agent that appears to be quite effective with respect to therapeutic chelation of Pb is 2,3-dimercaptosuccinic acid (DMSA). It enjoys several advantages over $CaNa_2$-EDTA, including its efficacy following oral administration, which could permit its prophylactic use. DMSA has been shown to be efficacious in accelerating urinary Pb excretion in humans, and unlike $CaNa_2$-EDTA, is not accompanied by the concommitant excretion of large amounts of other essential trace metals. It has been reported to alleviate clinical symptoms and biochemical alterations produced by Pb toxicity in occupationally exposed individuals, and to reduce blood lead and reverse hematological alterations in Pb-exposed children. When administered in a prophylactic capacity, DMSA was found to prevent the rebound of blood lead levels in children that ultimately accompanies cessation of chelation. Moreover, studies in experimental animals show mixed results with respect to the capacity to mobilize Pb from bone, but consistently show no evidence of redistribution to soft-tissue target organs during administration. Following completion of further studies of its efficacy and safety, DMSA may well be made widely available and become the agent of choice under certain conditions for the treatment of elevated Pb burden.

The ideal solution for diagnostic evaluation of elevated lead burden may reside in the technology of x-ray fluorescence, used as a noninvasive assay of bone levels. This is preferable to the $CaNa_2$-EDTA provocation test not only because of the potential adverse consequences of the latter, but also because the total body burden of lead is best and most directly reflected in bone Pb. Serial measurement of bone Pb levels using L-x-ray fluorescence have shown levels to be highest in children with the greatest Pb burdens, as indicated by the outcome of $CaNa_2$-EDTA provocative tests, thus establishing the validity of the procedure. Moreover, chelation treatment has been shown to be accompanied by a decline in bone Pb content as assessed by this technique. While x-ray fluorescence clearly has the capacity to replace the $CaNa_2$-EDTA provocation test, questions remain as to the sensitivity of the procedure, the pool of bone Pb it taps, and perhaps most importantly from a clinical perspective, whether the technology can be widely enough distributed to be available to populations at greatest risk.

Related Article: METAL–METAL TOXIC INTERACTIONS.

Bibliography

Agency for Toxic Substances and Disease Registry (1988). "Toxicological Profile for Lead." U.S. Public Health Service, Atlanta, Georgia.

Lippmann, M. (1990). Lead and human health: Background and recent findings. *Environ. Res.* **51,** 1–24.

Piomelli, S., Rosen, J. F., Chisolm, J. J., and Graef, J. W. (1984). Management of childhood lead poisoning. *J. Pediatr.* **105,** 523–532.

Pounds, J. G., Long, G. J., and Rosen, J. F. (1991). Cellular and molecular toxicity of lead in bone. *Env. Health Perspect.* **91,** 17–32.

U.S. EPA (1986). "Air Quality Criteria for Lead." Environmental Criteria and Assessment Office, Research Triangle Park, North Carolina.

Metal–Metal Toxic Interactions

Bernd Elsenhans, Klaus Schümann, and Wolfgang Forth
Ludwig-Maximilians-Universität München

Glossary

Erythropoiesis Formation of erythrocytes.

Hematocrit Volume percentage of erythrocytes in the whole blood.

Hyperzincuria Excessive excretion of Zn/d in the urine.

Hypozincemia Inadequate zinc concentrations in the plasma.

Parakeratotic lesions Retention of the nuclei in the cells of the keratin (or horney layer) normally seen in the mucous membrane.

Toxicokinetics Impact of the organism on toxic substances. Uses model calculations to describe the kinetics of absorption, distribution, metabolism, and excretion of toxic compounds.

IN BIOLOGICAL SYSTEMS *the toxicity of metals is based on concentration-dependent interactions of metal ions with organic structures and compounds thereby altering or impairing their optimal biological and physiological functions. Metal–metal interactions can be divided into two groups: First, direct ones which require the exchange or replacement of one metal by another (e.g., binding of Cr(III) or Al(III) ions to transferrin and replacement of Fe(III) ions). Second, indirect interactions in which one metal affects the handling of a second one without having a ligand in common (e.g., Ni, which does not bind to metallothionein (MT), can induce the synthesis of this low-molecular weight protein thereby influencing other metals which are bound by MT).*

I. Introduction

Most of the elements in the biosphere are metals. Reference to metals in biological systems principally means their various ionic states since elementary forms are of negligible importance. Some metals are termed "essential" and others "toxic." The term "essential" was based on the observation that deficiency states for certain metals led to retarded growth and malfunctions in various species which, in turn, could be reversed by administration of the lacking metals. When this definition of essentiality is applied to mammalian species, the following trace metals can certainly be considered as being essential: Ca, Cu, Fe, Mg, Mn, Mo, and Zn. For all of these metals, clear deficiency symptoms can be described with a corresponding reversal after administration of the appropriate metal. Though the interpretation of a mere growth-stimulating effect of certain elements is controversial in terms of essentiality, Cr, Ni, and V, as well as the metalloids As and Se, are considered to be essential elements for at least some mammalian species.

Since it is only the dose absorbed or the concentration of a metal compound in the organism which determines whether a metal is toxic or nontoxic, essential metals can also cause toxic effects. Natural and, to a greater extent, industrial and agricultural sources may contribute to a critical reduction or accumulation of metal compounds in the environment and in foods. When they accumulate beyond a tolerable concentration, these metals become acutely or chronically toxic. This holds true particularly for nonessential metals for which the organism has not developed regulatory mechanisms and which may use pathways present for the absorption, distribution, and excretion of essential metals.

With the exception of epidemiological and clinical observations, data and findings about metal–metal toxic interactions derive from animal studies. How can metal–metal toxic interactions be investigated? In the most simple type of an animal experiment, two metals are applied and the impact of this combination on the site, time course, and extent of a toxic effect are recorded. The dose levels applied in these experiments are often much higher than those found in the environment in order to produce measurable effects within a feasible period. Changes in the kinetics of a metal or of a metal-dependent reaction can be monitored either after the addition of a second metal or after the application of various states of deficiency or overload for another trace metal. Certainly, more than two metals can interfere with each other at one time and epidemiological observations rather suggest that a variety of metals is engaged in physiologically relevant interactions (e.g., in the Cd-induced "itai-itai" disease in Japan, interactions between Cd and the metabolism of Fe, Zn, Ca, and Cu were observed).

In order to obtain definitive results, however, most experiments on metal–metal interactions are restricted to the interactions between two metals. Thus far, a considerable amount of information has accumulated on which metals show toxic interactions, which target organs are involved, and what type of toxic lesions are produced. Less information is available on the mechanism of these interactions, particularly at the typical environmental low-dose levels in organisms with varying degrees of depletion for essential trace metals such as Cu, Fe, and Zn.

This article focuses on interactions between essential and nonessential metals. Toxic interactions between essential metals are only possible in cases of extreme imbalance and excesses which, in general, are easily recognized and thus less problematic. Toxic interactions between nonessential metals are mainly conceived as being additive or synergistic effects. Such interactions, however, represent an almost unknown field. Therefore, interactions among the nonessential metals are not considered here. In addition, metal–metal interactions occurring in biological systems mostly as anions are only occasionally described since these anions (e.g., molybdate, selenate, or arsenate) may compete with or behave as analogs of nonmetal anions (e.g., sulfate, thiosulfate, or phosphate).

II. Aluminum

A. Al–Fe Interactions

Aluminum (Al) may cause a microcytotic anemia in patients with renal failure who are subjected to chronic renal dialysis. Al can bind to serum transferrin which may cause a displacement or a different utilization and distribution of transferrin-bound iron (Fe). The effect of Al on Fe metabolism probably depends on the degree of Al accumulation. Enzymes for heme synthesis are inhibited by Al, such as ferrochelatase, uroporphyrin decarboxylase, and γ-aminolevulinic acid dehydratase (ALAD). Erythropoiesis can also be disturbed by Al as demonstrated by a transferrin-mediated inhibition of erythroid colony growth which is directly related to the number of free binding sites on transferrin and which is not observed in the presence of fully Fe-saturated transferrin. However, transferrin does not mediate the cellular uptake of Al.

Depending on its chemical form, high dietary Al may increase or decrease Fe absorption. On the one hand, Al chloride may render Fe more available during ingestion by reducing the formation of less absorbable Fe phosphates, thus leading to increased hepatic Fe levels as shown in lambs. On the other hand, Al-hydroxide containing antacids may decrease intestinal Fe absorption due to their capacity to neutralize gastric acid. This indicates that interactions between Al and nutrients, also trace metals, within the lumen of the digestive tract may be due to the adsorptive or chemically reactive features of $Al(OH)_3$. Thus, Al–Fe interactions are not necessarily those between cations.

B. Al–Ca Interactions

Al–Ca interactions are less clear. On a low calcium (Ca) diet, the Ca balance in adult males becomes significantly more negative during the administration of Al-hydroxide-containing antacids than in control periods without antacids. Urinary and fecal Ca excretions were increased. This was not seen when Ca intake was normal. Balance studies in sheep which were fed $AlCl_3$ (2 g Al/kg diet) also showed a reduction in the apparent absorption of dietary Ca which resulted in an increase in fecal Ca excretion; concentrations of Ca in plasma, tissues and bone, however, were not affected. The interactions of orally administered Al with Ca are not unequivocal as shown by studies in humans in which the apparent Ca absorption or retention was not affected. In these studies, however, Al was administered in lower doses as the lactate form and not as hydroxide.

Similar absorption characteristics of Al and Ca in rat duodenum suggest an interaction between Al and Ca at the site of intestinal absorption. In fact, Ca impairs jejunal absorption of Al demonstrating mutual Al–Ca interactions.

The main toxic effects of Al on bone metabolism may not depend only on an impaired intestinal absorption of Ca. They are also probably due to an accumulation of Al in bone tissue which impairs Ca utilization and decreases bone formation. At low doses of $AlCl_3$, however, Al stimulated uncoupled bone formation and induced a positive bone balance without affecting Ca metabolism.

C. Al–Zn Interactions

High dietary Al increased kidney zinc (Zn) levels in lambs on a diet low in phosphorus; addition of phosphate to the diet reversed this effect. When Zn was administered together with Al hydroxide in normal subjects, Zn absorption was decreased by 60%. The effect was observed in normal subjects as well as in patients with renal failure and can be attributed to the adsorbing and neutralizing properties of Al hydroxide.

High parenteral doses of Al chloride altered the distribution of metals normally associated with MT in rats. The liver and kidney MT levels significantly increased at high but not at low doses of Al. Since performance and food intake of the rats were impaired at high doses, these findings may result from metabolic disturbances due to an acute Al intoxication.

III. Cadmium

A. Cd–Fe Interactions

Intestinal Fe absorption is reduced in a dose-dependent manner by oral doses of Cd in rats (Table I). This seems to be due to an immediate interaction between Cd and Fe during the intestinal transfer process, as Fe absorption returns to normal within 12 hours after the last Cd administration. Consequently, symptoms of Fe-deficiency anemia are observed after chronic oral Cd exposure. Fe supplementation mitigates this type of toxic Cd effect.

Absorption of Fe and Cd depends on the Fe status of the organism (Fig. 1) and is increased in Fe deficiency, putting children and women of reproductive age at special risk because they are often Fe deficient due to fast growth and menstrual blood losses, respectively. However, anemia in Cd-exposed populations has also been observed without Fe deficiency, which might be due to a Cd effect on erythrocyte production or lifespan. In ad-

Table I Effect of Orally Administered Cd on Whole-Body Fe Retention in Rats[a]

	Group number			
	1	2	3	4
Cd dose, as $CdCl_2$ (μmol/kg body wt)	0	0.5	1.0	2.0
Fe dose, as $^{59}FeSO_4$ (μmol/kg body wt)	1.0	1.0	1.0	1.0
^{59}Fe retention (% of dose)	14.9 ± 1.9	9.1 ± 0.4	6.1 ± 0.5	4.1 ± 0.4

[a] The oral dose of Cd and ^{59}Fe were administered simultaneously. After three weeks, ^{59}Fe retention was measured in a whole-body counter. Data are means ± SEM (n = 8). Data from S. Schäfer and W. Forth, (1984). *J. Nutr.* **114,** 1989–1996.

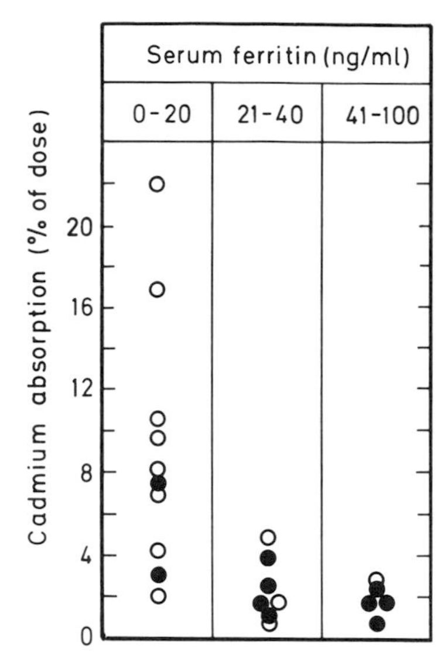

Figure 1 Cd absorption (in human volunteers) in relation to body iron status. The dose of Cd amounted to about 25 μg Cd and was administered by a test meal extrinsically labeled with 5 μCi of $^{115m}CdCl_2$. Body radioactivity was determined in a whole body counter. Absorbed Cd was determined from body ^{115}Cd counts obtained approximately 5 weeks after administration. Serum ferritin levels served as marker of the body iron status; ○ females, ● males. Data redrawn from P. R. Flanagan *et al.* (1978). *Gastroenterol.* **74,** 841–846.

dition, Cd seems to inhibit the storage of Fe by binding to catalytic sites on the ferritin molecule which inhibits the oxidation and incorporation of Fe into the protein.

B. Cd–Zn Interactions

Cd and Zn interact at the level of intestinal absorption, but the mode of interaction is more complex than that between Cd and Fe. High dietary Zn supplementation causes an inhibition of the intestinal Cd uptake. On the other hand, large intragastric Cd doses decrease Zn absorption in mice. Accordingly, exposure in early life to low dietary Cd levels has been found to reduce brain Zn and Cu levels, thus causing persistent reductions in locomotive behavior. In Fe-deficient mice, dietary Cd reduced intestinal Zn absorption, as Cd may block Zn-transferring ligands which are normally occupied by Fe. Contrary to the absorption of Cd in the Fe-deficient state, its absorption is not increased in Zn deficiency.

Many symptoms observed in Zn deficiency and after Cd application are similar, such as growth retardation, paraceratotic lesions, and impaired glucose tolerance. Consequently, Cd toxicity is increased in Zn deficiency. A mechanistic explanation of this synergism is that Cd may induce a Zn trap in the tissues: Zn disappears from the tissue when MT is degraded. Cd, however, remains in the tissue and induces synthesis of new MT which, in turn, traps Zn and decreases its availability. On the other hand, Zn and Cd pretreatment reduces the toxic effects of Cd by inducing hepatic MT synthesis and, thus, sequestration of Cd. Consequently, the toxic impacts of Cd, such as lethality, reduction of hepatic drug metabolism, and oxidative phosphorylation are decreased. As Cd persists in tissues much longer than antagonistic trace metals, toxic Cd action can be resumed after the latter have been excreted.

High Cd intake depresses the release of insulin (for which Zn is necessary). Thus, Cd may decrease glucose tolerance. Cd may also disturb Zn-containing enzymes, which are involved in nucleic acid synthesis and function. Indeed, Zn deficiency and Cd administration are found to be teratogenic in animals.

C. Cd–Ca Interactions

Osteomalacia due to Cd-mediated perturbance of Ca metabolism was ubiquitous in "itai-itai" disease. The biochemical mechanisms of this epidemiological observation remain obscure, though a variety of explicative findings have been derived from animal experiments. For example, a competitive inhibition of vitamin-D-stimulated Ca absorption by Cd was observed. Though oral Cd inhibits the synthesis of 1,25-dihydrocholecalciferol, this mechanism was not observed in long-term studies casting doubt on the relevance of this finding for "itai-itai" disease. Cd at high concentrations may occupy intestinal binding sites at the expense of Ca. Binding of Ca to the Ca-binding protein was inhibited by Cd, and was accompanied by a decreased intestinal absorption of Ca. Cd is a potential inhibitor of alkaline phosphatase, an intestinal increase of which is paralleled by increases in Ca transfer. On the other hand, Ca deficiency and cholecalciferol administration enhance Cd absorption, which is not correlated with changes of the vitamin-D-dependent Ca-binding protein. Ca ions can inhibit the saturable uptake of Cd into a nonlabile intestinal compartment.

Calmodulin serves as an interface for many Ca effects on a variety of target enzymes. Cd substitutes for Ca in calmodulin and induces conformational changes. This mechanism activates the myosin light chain kinase, which may be the reason for the observed vascular toxicity of Cd.

D. Cd–Cu Interactions

In Cd-polluted areas, increased human Cd and Zn concentrations are associated with a decrease in tissue Cu concentration. A deficit of Cu in the duodenal mucosa reduces cytochrome oxidase activity by 50%, which inhibits the growth rate in lambs. On the other hand, MT synthesis in Cu-treated mice is induced in the intestine, where it is accompanied by mucosal Cd retention. MT induction by Cu is less pronounced than by Zn and Cd. However, this is difficult to evaluate since Cu cannot be administered in the same amounts as Zn. Still, the low tolerance to Cd in Cu deficiency might be due to reduced amounts or to a different distribution of MT in the organism which, thus, may be less apt to sequester Cd.

IV. Chromium

A. Cr–Zn Interactions

Although intestinal absorption of Cr(III) appears to be a process of passive diffusion, the intestinal absorption of Cr(III) and Zn is increased in the Zn-deficient rat. In the presence of Zn, the absorption of Cr(III) is inhibited and vice versa. This shows that, at least under conditions of Zn deficiency, Cr(III) may interact with Zn-binding ligands in the small intestine.

B. Cr–Fe Interactions

Transport of Cr(III) in the blood occurs mainly in the plasma by binding to transferrin. For this transport protein, a competitive binding of Fe(III) and Cr(III) ions is established. Since Cr(III) may play a role in insulin metabolism, its displacement from transferrin in iron overload conditions, and thus its reduced body retention, suggest a contribution in the mechanism of hemochromatotic diabetes.

C. Cr–V Interactions

A toxic effect of ammonium vanadate (5 μg/g diet) in chicks is only observed when Cr(III) acetate is added to the diet. Cr or vanadate alone in the diet does not exert toxic effects. The results, however, are less reproducible and are thought to reflect a combined action of Cr and vanadium (V) on some dietary ingredients.

In chicks, V (as ammonium vanadate) at 20 ppm in the diet is extremely toxic as shown by growth depression and mortality figures. When Cr(III) chloride is added to the diet, the toxicity of V becomes less as the Cr level is increased, and (at a level of 2000 ppm Cr) the toxicity of V is virtually abolished (Table II). In addition, the presence of Cr in the diet considerably reduces the effect of dietary V on the uncoupling of oxidative phosphorylation in chick liver mitochondria, an effect attributed to the mutual inhibition of chromate and vanadate in the uptake into mitochondria. This represents an example of anionic metal–metal interactions. However, mitochondrial uptake of chromate is also decreased by sulfate and phosphate, demonstrating its analog nature with nonmetal anions.

V. Cobalt

A. Co–Fe Interactions

The main interactions of Cobalt (Co) with other metals are those with Fe, although it is not retained in the body by processes of storage or utilization as is Fe (i.e., by ferritin or heme formation). Intestinal absorption of Co is increased in Fe deficiency. There is a good correlation between Fe absorption and the urinary excretion of orally administered Co, both in normal subjects and in patients with acute or chronic blood loss (Fig. 2). These findings

Table II Effect of Dietary Cr(III) Chloride Supplementation on Vanadate-Induced Mortality of Chicks[a]

	Mortality at 3 weeks (%)	
	Dietary V (ppm)	
Dietary Cr (ppm)	0	20
0	6.7	86.6
500	6.7	66.7
1000	10.0	40.0
2000	6.7	13.3

[a] Chicks were fed a dried skim milk diet supplemented as indicated with ammonium vanadate and Cr(III) chloride. Data adapted from C. H. Hill, (1979). "Chromium in Nutrition and Metabolism" (D. Shapcott and J. Hubert, eds.). Elsevier/North Holland Biomedical Press.

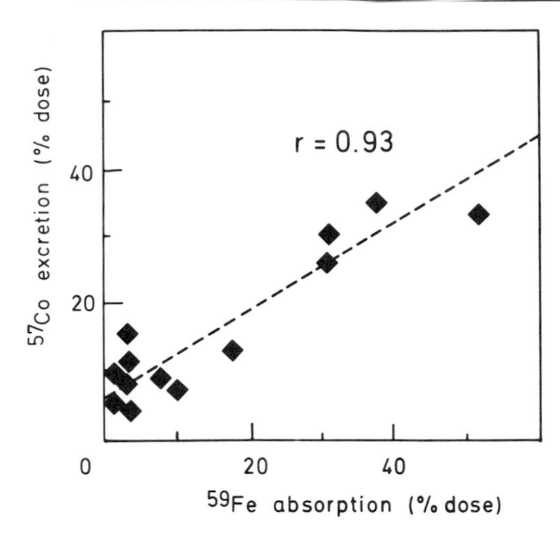

Figure 2 Relationship between Fe absorption and Co excretion in 13 patients with various disorders associated with altered Fe metabolism. The correlation is shown for the 24-hr ^{57}Co-excretion test. Redrawn from D. L. Wahner-Roedler *et al.*, (1975). *J. Lab. Clin. Med.* **85,** 253–259.

indicate that Co shares intestinal absorption mechanisms with Fe. Experimental findings suggest that Co competes for Fe transfer sites at the basolateral membrane, but not at the luminal site of the small-intestinal brush border. In addition to these similarities, however, differences also exist between the absorption of Fe and Co. In patients with exogenous Fe overload due to excessive oral or parenteral Fe treatment or transfusions (conditions that lead to a reduced intestinal absorption of Fe) no correlation is found between the absorption of Fe and Co. This indicates that Co absorption is responsive to the mechanisms that enhance Fe absorption, but not to those that inhibit it.

B. Co–Zn Interactions

Interactions between Co and Zn are less well-characterized. A competitive interaction exists between Co and Zn at the site of their intestinal absorption. This might be an analogy to corresponding Zn–Fe interactions.

VI. Lead

A. Pb–Ca Interactions

A typical design for an interactive study is to apply the highest lead (Pb) amount which had not previously produced toxic effects in combination with a normal Ca diet (200 μg Pb/ml) and to observe the effect of a Ca-deficient diet on Pb absorption, distribution, and retention. In an experiment of this type in rats, low dietary Ca enhanced intestinal Pb absorption. Fecal Pb output was reduced and urinary Pb excretion was increased. Pb concentrations in kidney and blood increased substantially in short-term experiments.

In long-term feeding studies, low dietary Ca levels also increased the bioavailability and toxicity of Pb. Pb accumulation was increased to the greatest extent in bone under these conditions. Together with the clinical observation that gastrointestinal absorption of Pb is higher in children, this puts growing organisms at special risk for Pb intoxication. However, the effect of dietary Ca intake on Pb metabolism was not linear. Increases of Pb accumulation in the bones were overproportional when the physiological requirement for Ca were not met in rats. Thus, not only dietary Ca levels, but also Ca deficiency has an impact on Pb retention in the organism. However, Ca deficiency has no direct impact on Pb absorption but rather reduces Pb excretion.

Contrary to more distal parts of the intestine, Pb is actively transported in the duodenum. When Ca and Pb are administered together intraduodenally, absorption from constant Pb doses increases in proportion to a reduction in dietary Ca. The Pb–Ca interaction does not represent a competition for common intestinal transport or binding sites, since Pb does not influence the active transport of Ca in rat intestine. Only the passive Ca transport component is affected, probably by an altered membrane permeability.

B. Pb–Fe Interactions

Nutritional Fe deficiency in animals aggravates Pb toxicity. However, Fe deprivation does not always increase Pb absorption. Under hypoxia, Fe absorption is increased whereas that of Pb remains unchanged. On the other hand, an increased absorption of Pb always seems to depend on an increased capacity of the intestinal Fe-transfer system. The percentage retained of Pb administered intragastrically at a low-dose level decreases with increasing doses of Pb administered in Fe-deficient mice. In Fe-repleated mice, the retained percentage of a dose of Pb is low and does not differ substantially with the dose administered.

In everted-sac preparations of rat duodenum *in vitro,* addition of $FeCl_2$ to the mucosal medium selectively inhibits the net transfer of Pb to the serosal side. Just before Pb administration, intragastric addition of $FeCl_2$ inhibits Pb absorption in Fe-deficient mice, and, alternatively, the addition of high doses of Pb reduces the absorption of Fe (Fig. 3). These results indicate that Pb and Fe interact at the level of the intestinal Fe-transfer system in Fe deficiency.

There is experimental evidence for a correlation between Fe status and Pb absorption and retention in humans. In one study, the severity of Pb poisoning in children was shown to increase with the prevalence of Fe deficiency. In another study, food Fe and Pb absorption were significantly correlated in 26 subjects who represented a wide-range of Fe states, ranging from Fe deficiency to Fe repletion (Fig. 4). However, in accordance with the results mentioned above, the correlation was not strong; only 50% of the subjects who hyperabsorbed Fe also hyperabsorbed Pb.

A more academic type of interaction is based on the essentiality of Pb. In Pb deficiency, which is not observed in a normal but only in a specifically modified environment, rats show reduced concentrations of Fe and increased Fe-binding capacities in the serum.

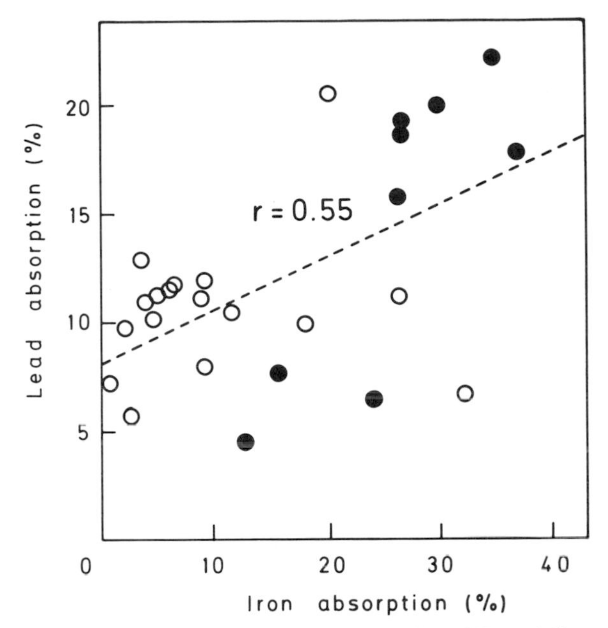

Figure 4 Correlation between the absorption of Fe and dietary Pb in human volunteers. At 7 and 12 days after ingestion of a test meal extrinsically labeled with $^{59}FeSO_4$ and $^{203}PbCl_2$, radioactivity was measured in a whole body counter to determine the retention data. ●, Fe-deficient and ○, all other subjects. Data redrawn from W. S. Watson *et al.*, (1986). *Am. J. Clin. Nutr.* **44,** 248–256.

C. Pb–Zn Interactions

Chronic Pb exposure may cause hyperzincuria and affect the Zn tissue distribution. Depending on dose and length of exposure pancreatic and hepatic zinc (Zn) concentrations can be slightly elevated.

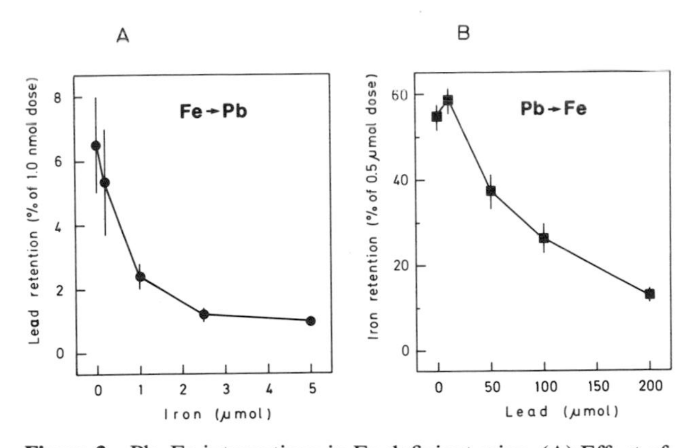

Figure 3 Pb–Fe interactions in Fe-deficient mice. (A) Effect of $FeCl_3$ dosage on the 10-day retention of a dose of 1 nmol ^{203}Pb acetate; (B) effect of Pb acetate on the 5-day retention of a dose of 0.5 μmol $^{59}FeCl_3$. Metals were administered as intragastric doses. Data are means ± SEM (n = 10). Data from D. L. Hamilton (1978). *Toxicol. Appl. Pharmacol.* **46,** 651–661.

Interactions between Pb and Zn are relevant for heme biosynthesis and Pb absorption. Heme synthesis is impaired because the Zn-dependent aminolevulinic acid dehydratase (ALAD) is inhibited by Pb. The Pb-inhibited enzyme can be reactivated by the addition of Zn. Since MT is involved in the metabolism of Zn, it also appears to affect the Pb inhibition of ALAD. Studies in rats suggest that MT exhibits a dual function in reducing Pb inhibition of ALAD by a mechanism which involves the donation of Zn to this Zn-requiring enzyme as well as the sequestration of Pb.

When the dietary supply of Zn is elevated, the severity of oral Pb toxicity decreases, as shown by decreases in blood and tissue Pb levels as well as by a reduced inhibition of kidney ALAD and a lowered Pb absorption.

D. Other Pb–Metal Interactions

Compared with a normal magnesium (Mg) status, maternal Mg deficiency results in significantly higher Pb concentrations in dam liver and offspring erythrocytes, liver, and tibia. Increasing dietary copper (Cu) concentrations aggravate Pb toxicity in rats.

VII. Mercury

A. Hg–Zn Interactions

Due to chemical similarities among mercury (Hg^{2+}), Zn^{2+} and Cd^{2+}, interactions among these ions are frequent. Zn and Hg mutually influence their tissue accumulation in the liver and the kidneys. The induction of MT and its concentration in these organs seem to be important in metal–Hg interactions. Thus, pretreatment of rats with Cd, a potent inducer of MT, increases the uptake of Hg by the liver and the kidney (Table III), but also protects the animals against the renal toxicity of Hg. On the other hand, feeding a Zn-deficient diet to rats during Hg exposure aggravates Hg toxicity, probably due to low tissue MT levels.

In rats pretreated with Cd and, thus, exhibiting high tissue levels of MT, intraperitoneal (i.p.) administration of $HgCl_2$ results in a displacement of Zn and Cd in the kidneys and hence in lower renal Zn and Cd concentrations than in Cd-pretreated controls without Hg administration. However, Hg administration increases Zn and Cd levels in the liver, an effect thought to be caused by an additional synthesis of MT due to a redistribution of Zn and Cd.

In animals without previous induction of MT synthesis, Zn–Hg interactions may be different. Zn increases the amount of Hg at the injection site and in the 48-hr feces specimins in rats when Hg and Zn are injected subcutaneously (s.c.) together in equimolar amounts. It does not change the amount of Hg in the liver, kidneys, blood, or urine which is probably due to the fact that the amount of Zn injected was small as compared to the amount of endogeneous Zn. The mechanism for the increased excretion of Hg remain obscure, however, Hg increases the amount of Zn at the injection site and in the kidneys, but reduces it in blood and 48-hr feces. This was interpreted as an interference of Hg-induced MT synthesis with the distribution of Zn. When Zn is injected (i.p.) in higher doses, Hg injected intravenously (i.v.) results in an increased accumulation of Zn in the spleen and a slight increase in the heart, but not in other organs or in the blood. Obviously, the extent and the site of such Hg–Zn interactions depend on the method of administration of the metals which reflects the importance of toxicokinetic considerations in interpreting the mechanisms of metal–metal interactions.

Hg can replace MT-bound Zn, Cu, and Cd *in vitro* when Cd-induced rat-kidney MT is treated with increasing amounts of $HgCl_2$. Hg initially replaces Zn, then Cd, and finally Cu in a sequence reflecting the metal-binding affinities of MT (Hg>Cu>Cd>Zn). These exchange reactions may also occur *in vivo* as it can be demonstrated in rats that Hg replaces Zn in hepatic MT.

Little is known about the metal interactions of methyl-Hg (MeHg) though it is converted to inorganic Hg in the organism. Thus, MeHg may possibly interact with other metals after its conversion into Hg. Repetitive s.c. doses of MeHg chloride to rats produce profound hypozincemia and an increase in hepatic Zn bound to MT by the initial 24 hr after injection. This suggests *de novo* synthesis of MT by MeHg, although MeHg does not bind

Table III Uptake and Distribution of ^{203}Hg in the Liver and Kidneys of Rats: Effect of a Cd Pretreatment[a]

	Liver		Kidney	
Fraction	Control	Cd-pretreated	Control	Cd-pretreated
Total organ	24.9	65.1	12.9	43.7
Soluble fraction	5.9	20.9	6.8	24.4
Cd–BP	1.2	18.1	1.1	6.4

[a] For pretreatment of rats with Cd, a dose of $CdCl_2$ (9.6 μmol Cd per rat) was administered subcutaneously daily for 3 days. Hg uptake and distribution were measured 1 hr after an intravenous injection of a dose of $^{203}HgCl_2$ (0.23 μmol Hg per rat). The amount of Hg found in the total organ or fraction is given in nmol. Cd–BP = low-molecular weight Cd-binding protein (metallothionein). Data from Z. A. Shaikh, R. L. Coleman, and O. J. Lucis (1973). *Trace Subst. Environ. Health* **7**, 313–321.

to MT either *in vivo* or *in vitro*. After oral (p.o.) administration of MeHg chloride to guinea pigs, tissue Zn concentrations continuously and significantly increase in the pancreas and in the cerebrum whereas in the kidneys and the liver, Zn levels increase only initially and then decline to control levels. Stress produced by MeHg administration may be involved in the induction of hepatic Zn-MT synthesis. With rapidly decreasing Hg levels, however, the kinetically different behavior of Zn in the various organs may not be solely due to a MeHg-induced MT synthesis, but also to effects of MeHg itself.

B. Hg–Cu Interactions

Hg can replace MT-bound Cu *in vitro*. However, under *in vivo* or *ex vivo* conditions, Hg is obviously not able to replace Cu in rat hepatic MT when the protein is induced by i.p. administration of $CuCl_2$. Since MT binds metals in two different clusters, the results may reflect the different affinities of these clusters for various metals.

After MT induction by Cd in rats, an i.p. administration of $HgCl_2$ decreases renal Zn and Cd levels, but increases that of Cu. This might be due to the stability of Cu–MT complexes in which Hg cannot replace Cu. Hepatic Zn, Cd, and Cu levels are increased which was suggested to result from additional MT synthesis caused by a redistribution of Zn and Cd. However, hepatic MT induction by Zn or Cd does not necessarily increase hepatic Cu levels, but may also decrease them. Since s.c. injections of $HgCl_2$ may significantly increase hepatic Cu levels in rats, the effect of Hg on hepatic Cu levels is probably more complex. This is emphasized by another study in rats which showed that i.p. administration of $HgCl_2$ increases MT and Zn levels in the liver and in the kidneys, but increases Cu levels only in the kidneys and not in the liver (Fig. 5).

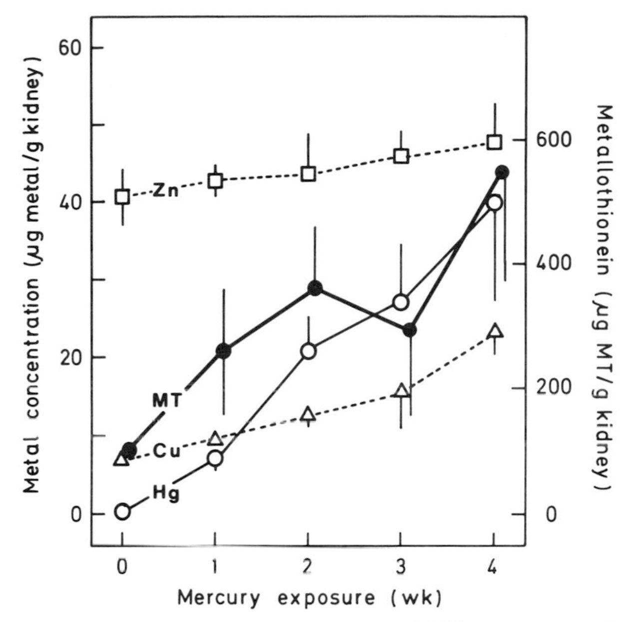

Figure 5 Effect of a 4-wk exposure to $HgCl_2$ on the Hg (○), Cu (△), Zn (□), and metallothionein (MT, ●) levels in the rat kidney. Animals received intraperitoneal injections of a solution of $HgCl_2$ (0.2 μmol/kg body wt/day) 5 days in every 7 for periods of up to 4 weeks. MT was measured by radioimmunoassay and metal concentrations were determined by atomic absorption spectrometry. $\bar{X} \pm SD$ ($n = 5$). Drawn from data from C. V. Nolan and Z. A. Shaikh. (1987). *Biol. Trace Elem. Res.* **12,** 419–428.

VIII. Nickel

A. Ni–Fe Interactions

In Fe-deficient rats, intestinal nickel (Ni) absorption is increased *in vitro* as shown by an increased transfer across the intestinal wall and also by an increased accumulation in the mucosal tissue. Accordingly, Ni toxicity is aggravated in Fe deficiency. Simultaneous administration of Ni and Fe salts into the intestine causes a reduction of Fe absorption up to 60% of control levels. These findings indicate an antagonistic behavior of Fe and Ni at the site of their intestinal absorption. Ni deprivation, however, does not necessarily facilitate intestinal absorption of Fe, but is found to impair it when the dietary form of Fe was Fe(III) sulfate instead of Fe(II) sulfate. Therefore, it was speculated that Ni affects the intestinal absorption of Fe(III) but not Fe(II) ions.

Ni deprivation in rats decreases growth and reduces hematocrits. The extent and severity of a Ni deficiency may, however, depend on Fe status. Thus, Ni deprivation can depress Fe content in liver, spleen, and kidneys in severely Fe-deficient rats, but can increase hepatic Fe levels in only marginally Fe-deficient rats. These effects are probably species dependent since Ni deprivation increases Fe concentrations in liver, spleen, and lung of sheep adequately supplied with Fe.

It is noteworthy that some of the toxic effects of Ni on Fe metabolism appear to resemble those found in Ni deprivation. At high dietary Ni-acetate levels (>500 mg Ni/kg diet) (but not at lower concentrations) Ni reduces the weight gain of rats as well as the packed cell volume and hemoglobin, but elevates the Fe level in serum, red blood cells, and inner organs. This suggests that Ni can disturb the mobilization of Fe stores when Ni-tissue levels are either too high or too low. Therefore, besides toxic and antagonistic features, a synergistic role of Ni is proposed for the optimal utilization of Fe. The stimulating effect of nontoxic, low dietary Ni concentrations on hematopoiesis is thought to be due to a Fe-sparing effect. It is speculated that Ni removes Fe from binding sites in the diet or in the organism so that Fe becomes more available. Ni–Fe interactions also appear to affect the equilibrium between Fe(II) and Fe(III) and thus the organism's utilization of Fe.

B. Ni–Cu Interactions

Ni–Cu interactions are well established. The responses of rats receiving a low copper (Cu) diet to Cu supplementation are increases in weight gain, hematocrit, hemoglobin, and Cu content of liver, heart, and kidney; Ni supplementation increases weight gains and hematocrits to the same extent as does Cu. Liver Fe levels in this study in the group with no Cu and no Ni added to the diet were high. With the addition of Ni or Cu they dropped by about 20% or 50%, respectively. To some extent, Ni seems to substitute for Cu in mobilizing Fe. Responses to Cu were not modified by simultaneous Ni supplementation, but responses to Ni were seen only in low-Cu diets. Accordingly, Ni effects were predominantly seen in Cu-deficient but not Cu-adequate rats. The results suggest that Ni does not reduce Cu absorption since Cu levels in the organs were unchanged. Thus, Ni seems to behave as an antagonist to Cu, but to act synergistically in certain Cu-mediated physiological processes.

Ni–Cu interactions depend on the dietary Fe level as indicated in experiments with rats in which high hepatic Cu concentrations of animals on a low-Fe diet were reduced by a moderate dietary Ni supplementation (Table IV).

C. Ni–Zn Interactions

Ni appears to be partially effective in preventing signs of Zn deficiency in rats because at low dietary Zn levels (<8 mg Zn/kg diet) dietary Ni (30 or 50 mg Ni/kg diet) can reverse symptoms of Zn deficiency such as increased urinary nitrogen excretion and reductions in leukocyte counts. An in-

Table IV Dependence of Rat Hepatic Cu, Fe, Ni, and Zn levels on Dietary Ni and Fe Concentrations[a]

Dietary supplement (μg/g diet)		Hepatic metal concentration (μg/g dry wt)			
Ni	Fe	Cu	Fe	Ni	Zn
0	15	65	85	0.31	96
20	15	23	105	0.37	80
100	15	129	83	0.34	92
0	100	19	737	0.07	95
20	100	17	785	0.05	103
100	100	15	933	0.16	117
Metal effect		Analysis of variance (p-values)			
Ni		0.0001	NS	NS	0.02
Fe		0.0001	0.0001	0.0001	0.0001
Ni × Fe		0.0001	0.05	NS	NS
Error mean square		558.6	9201	0.94	20299

[a] Dietary Ni was supplemented as Ni(II) chloride and Fe as Fe(III) sulfate. Rats received their respective diets for 10 wk. Hepatic metal concentrations were measured by atomic absorption spectrometry. Values are means of six rats. Data from F. H. Nielsen *et al.* (1984). *J. Nutr.* **114,** 1280–1288.

duction of MT synthesis by Ni possibly contributes to changes in the metabolism of Zn. Dietary Ni at high levels (400 mg Ni/kg diet as $NiCl_2$) causes signs of Ni toxicity in chicks fed a corn–soybean meal diet slightly deficient in Zn. The addition of Zn (40 mg/kg diet) completely prevented growth depression as Zn might have reduced the absorption of Ni.

D. Other Ni–Metal Interactions

Particularly low-soluble Ni compounds can cause cancer. In employing intramuscular (i.m.) injections to test the carcinogenicity of Ni compounds in rats, a coadministration of manganese (Mn) dust with Ni subsulfide significantly reduces the incidence of sarcomas at the injection site caused by administration of Ni subsulfide alone. Other Mn compounds do not share this rather local effect. A similar local interaction between simultaneously administered Ni and Mn chloride was found for the Mn(II)-blockade of Ni(II)-induction of heme-oxygenase activity in rat kidney.

Ca and magnesium (Mg) acetate inhibit the tumorigenic activities of Ni in the lungs of mice indicating an antagonism between Ni and Mg as well as Ca. In particular, Mg ions are likely to inhibit Ni uptake by target tissues as well as nuclear and cytosolic uptake of Ni by the target tissue cells *in vivo*. Furthermore, Mg ions may inhibit Ni binding to DNA *in vitro* and Ni-induced disturbances in DNA synthesis *in vivo*. However, the specificity of these Ni-alkaline earth metal interactions is not completely clear yet.

IX. Thallium

A. Tl–K Interactions

Thallium (Tl) can be considered to be somewhat exceptional among the heavy metals. First, Tl^{+1} is absorbed from the gastrointestinal tract to an unusually large extent (>80%) as compared with other heavy metals. Second, it is secreted into the intestinal lumen across the mucosal epithelium of the small and the large intestine. The secretion of Tl^{+1} can be described as transport against an electrochemical gradient which disappears at low temperature and, hence, is revealed to be dependent on cellular metabolism. The net transport of Tl^{+1} is abolished when Cl^- and Na^+ ions are withdrawn. It can be partly inhibited by furosemide and ouabain. The net secretion of Tl^{+1} is inhibited by K^+ ions; it has thus been concluded that Tl^{+1} shares the transport system(s) with K^+.

From the complexing capacity of Prussian Blue on the one hand and the adsorptive capacity of *Chlorella* for binding of alkali and Tl(I) ions, the decisive role of the ionic radii of these ions was derived as K^+, 133 nm $\times$ 10^{-3} and Tl^{+1}, 140 nm $\times$ 10^{-3}. Interactions between Tl^{+1} and K^+ occur in skeletal muscle, heart muscle, smooth muscle of the uterus, nervous tissue (giant axon), kidney, red blood cells, frog skin, and with purified enzymes (e.g., pyruvate kinase and K/H-ATPase). Thus, the interaction of Tl^{+1} with K^+ obviously takes place at many sites in the organism.

For the effect of Tl^{+1} and its interference with alkalimetal ions, an interference at transport sites is assumed. However, it is also plausible that interferences occur at structures of membrane systems at which an exchange of ions may take place (depending on their ionic radii) and which are not related to transport processes. Thus, the findings about Tl–metal interactions can also be interpreted by general ion exchange mechanisms at anionic sites of biological matrices.

X. Tin

A. Sn–Zn Interactions

Tin (Sn) interacts with the intestinal absorption of Zn. Rats fed dietary $SnCl_2$ (206 mg Sn/kg diet) lose significantly more Zn in feces and retain significantly lower levels of Zn in tibia and kidney than rats fed a control diet. Similar effects of Sn were observed in humans. When human subjects consumed 50 mg Sn daily (test diet) they retained significantly less zinc in the body, lost significantly more Zn in feces and significantly less Zn in urine than controls (0.1 mg Sn/d). The inhibitory effect of oral Sn on the absorption of dietary Zn was also confirmed with ^{65}Zn absorption tests in humans. Zn reduces the inhibitory effects of Sn on ALAD in rabbits indicating Sn–Zn interactions at the level of heme biosynthesis.

B. Sn–Cu and Sn–Fe Interactions

The severe growth retardation and distinct decrease in hemoglobin levels in rats fed dietary Sn

(5.3 g/kg diet) are diminished by dietary supplements of Fe and/or Cu. Rats fed >500 mg Sn/kg diet have plasma Cu levels that are only 13% of control levels and exhibit depressed Cu levels in liver and kidneys. Repeated oral intake of Sn in rats is followed by changes in the distribution pattern of Fe: significant increases in tissue Fe content were observed in liver, kidneys and spleen. Although the mechanisms of these interactions are yet unclear, the findings suggest that Sn may disturb Cu metabolism necessary for the utilization of Fe for heme synthesis. In addition, a direct effect of Sn on intestinal Fe absorption may also contribute to an impaired Fe metabolism.

C. Other Sn–Metal Interactions

The relevance of other metal–Sn interactions is not yet elucidated. Simultaneous administration of Mn chloride with Sn in rats prevents the increase in heme oxygenase induced by the latter metal. Mn and Zn administration also prevents the decline in cytochrome *P*-450 content and the changes in ALA-synthase activity associated with Sn administration. The blocking effect of these metals on heme-oxygenase induction is lost when they are administered 10 min or more after the inducer element indicating that the enzyme synthetic process triggered by Sn is initiated rapidly after administration. Sn also affects Ca metabolism. A single oral administration of $SnCl_2$ increases bile Ca content in rats in a linear fashion. Repeated Sn administration further increases biliary Ca concentrations and reduces serum Ca levels significantly.

XI. Vanadium

A. V–Fe Interactions

Vanadium (V) compounds occur in various oxidation states, mainly in the pentavalent form as the orthovanadate anion (VO_4^{3+}) and in the tetravalent form as the vanadyl cation (VO^{2+}). The chemical form of V may influence the pattern of metal–metal interactions. In general, vanadates have been used frequently in experimental work, but one has to take into account that this form of V can be reduced to vanadyl ions in the organism.

The vanadyl cation has properties that make interactions between V and Fe likely. *In vitro* studies show that ferritin can incorporate vanadium and that the metal can be exchanged between transferrin and ferritin. Since V is also bound *in vivo* by liver ferritin and plasma transferrin, it has been suggested that the Fe-binding proteins may be involved in the metabolism of V. Studies that indicate the essentiality of V in rats, show that low dietary V neither substitutes for Fe at metabolic sites nor affects Fe absorption but might facilitate the optimal utilization of Fe after its absorption. Radioisotope studies in chicks demonstrate that the absorption of V is not affected by dietary Fe, but that in Fe deficiency, more V is found in blood and liver and less in bone than in controls supplemented with dietary Fe. This points to a role of Fe saturation of transferrin and ferritin for the retention of V.

Related Articles: CADMIUM; CHROMIUM; COBALT DUSTS; LEAD POISONING; SELENIUM; ZINC.

Bibliography

Foulkes, E. C., ed. (1986). "Cadmium—Handbook of Experimental Pharmacology," vol. 80. Springer, Berlin.

Foulkes, E. C., ed. (1990). "Biological Effects of Heavy Metals," vol. I. CRC Press, Boca Raton.

Friberg, L., Nordberg, G. F., and Vouk, V. B., eds. (1986). "Handbook on the Toxicology of Metals," vols. I and II. Elsevier, Amsterdam.

Levander, O. A. and Cheng, L., eds. (1980). "Micronutrient Interactions: Vitamins, Minerals, and Hazardous Elements," Annals of the New York Academy of Sciences, vol. 355. The New York Academy of Sciences, New York.

Rowland, I. R., ed. (1991). "Nutrition, Toxicity, and Cancer." CRC Press, Boca Raton.

Sigel, H., ed. (1986). "Metal Ions in Biological Systems," vol. 20. Marcel Dekker, New York.

Singhal, R. L., and Thomas, J. A., eds. (1980). "Lead Toxicity." Urban & Schwarzenberg, Baltimore.

Monitoring Indicators of Plants and Human Health for Air Pollution

Yuchi Naruse, Terutaka Katoh, and Sadanobu Kagamimori
Toyama Medical and Pharmaceutical University

Glossary

Allergic skin test Test for screening persons with atopy who are predisposed to development of allergic disorders such as asthma and allergic bronchitis.

Annual ring growth index Ratio of the observed expected annual ring width.

Bi-band ratio Infrared color film ratios, red (near-infrared region)/blue (green region of visible light) and red/green (red region of visible light) which decrease with the advance of the degree of tree injury.

BMRC questionnaire Interview questionnaire proposed by British Medical Research Council's Committee on research into chronic bronchitis.

SPR Standardized prevalence ratio obtained by dividing the observed prevalence by the expected one adjusted for age and sex.

AIR POLLUTION caused by fuel combustion at stationary sources is induced mainly by sulfur oxides, nitrogen oxides, and suspended particulates. The combustion of unpreprocessed coal or heavy oil is likely to become a significant source of SO_2, primarily emitted into the air when exhaust gases are not scrubbed. With regard to nitrogen oxides, nitric oxide (NO) is a pollutant derived mainly from combustion. In principal, air contains 79.02% nitrogen (N_2) and 20.94% oxygen (O_2). At high temperatures (over 200°F or 1093°C) in a burning flame, N_2 and O_2 combine and form NO, a gas of little significance as an air pollutant. However, when it is vented into the air and rapidly cooled, a fraction is converted into NO_2 which has strong chemical activities. Nitrogen oxide and suspended particulates derive from all combustion processes but are present in automobile exhaust as well. The latter source has become more significant for these two air pollutants especially after the introduction of a scrubber system that eliminates SO_2 from stationary sources of fuel combustion.

I. Oil-Fired Electricity-Generating Station

When an oil-fired, electricity-generating station is constructed in rural areas where there has been no significant air pollution source, changes in the quality of ambient air are recognized. For example, a power station (350 MW) was constructed in a rural county of Japan. At that time, the registered quantity of crude oil consumed in the county was 542,552 kl. The power station used 460,611 kl (84.9%) of the total quantity. Thereafter, anti-air-pollution measures such as usage of crude oil of lower sulfur concentration and reduction of factory wastes through chimneys were introduced in the power station. The second power station (250 MW) was commissioned adjacent to the first one 6 years later. Here, the anti-air-pollution measures mentioned above were in force from the time of construction. Demographically, the study town consists of the three parts. Districts I and II are located nearer to the power stations than District III. According to data of air pollution monitoring stations in each district, among three major parameters of air pollution, the concentration of SO_2 is much higher in districts I and II. However, there was no considerable observed difference in the other two air pollution parameters between the districts (Fig. 1). Since the power stations are the major users of crude oil in study districts, they contribute substantially to the peak hourly maximum value (the highest 1-hr concentration per year) of SO_2 in the first year of operation. With the introduction of anti-air-pollution measures primarily aimed at reducing sulfur exhaust, the hourly maximum concentration of SO_2 was decreased from the first to second year when the second power station commenced operation.

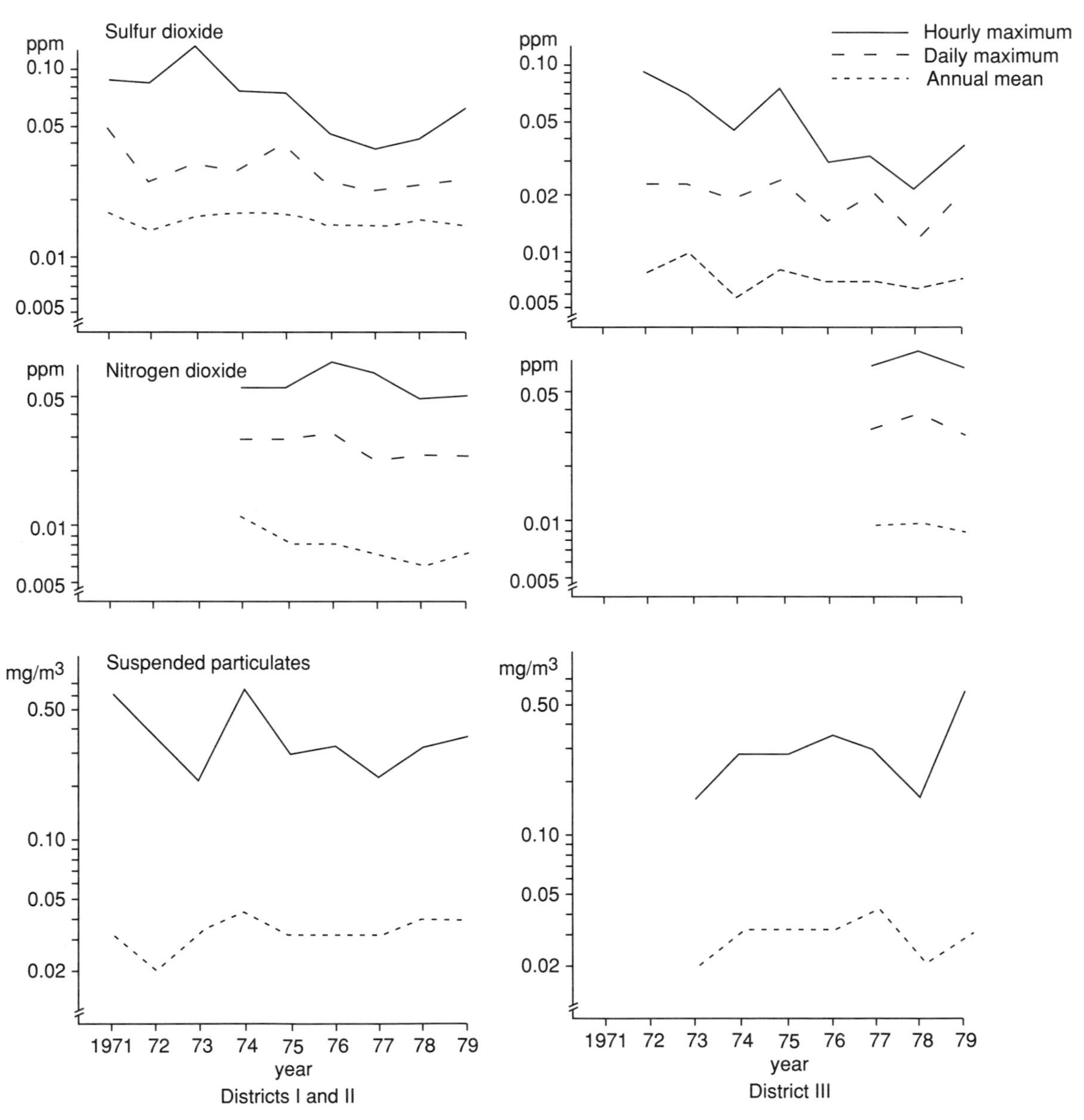

Figure 1 Changes in atmospheric concentrations of air pollutants measured at the air pollution measuring stations Districts I and II, and District III. (The daily maximum suspended particulate concentration was not available.) [From S. Kagamimori *et al.* (1990). *Environ. Res.* **52,** 47–61. Used by permission.]

II. Plant Indicators of Air Pollution

For monitoring the effects of air pollution on plants, growth reduction without visible symptoms of damage is the first obvious consideration. This effect has been studied by air-exposure experiments using greenhouses. Ambient untreated air is passed through one greenhouse, while air for the second is scrubbed first with water, thus removing 98–100% of the SO_2. As shown in Table I, there are no abnormal lesions relating to air pollution, but after 119 days there are differences in the amount of growth. The greatest reduction in growth rate occurs in plants exposed all of the time to ambient air which could thus contain pollutants that effect plant growth. In this case, the mean concentration of SO_2 was 47 $\mu g/m^3$, which was lower than that of the ambient air quality standard (80 $\mu g/m^3$ annual hourly mean in the United States). In addition, studies of exposure to SO_2 have confirmed that an increase in SO_2 concentration induces a pronounced drop in growth without visible symptoms of injury.

As a sensitive monitoring method for field surveys, infrared color film shots are available. Vigorous green leaves rich in chlorophyl reflect more abundant infrared lights compared with those from less vigorous ones (Fig. 2). Plants are identified on the film, and each of three color components, red (near-infrared region), blue (green region of visible light), and green (red region of visible light) is analyzed using a densitometer. The spectrum of the three colors is analyzed using optimal filters for red, blue, and green, respectively, and the density of each color ingredient is measured. The ratios, red/blue (R/B) and red/green (R/G) are called bi-band ratios. Due to the long distance sensitivity of infrared light analysis, this method is used in remote sensing.

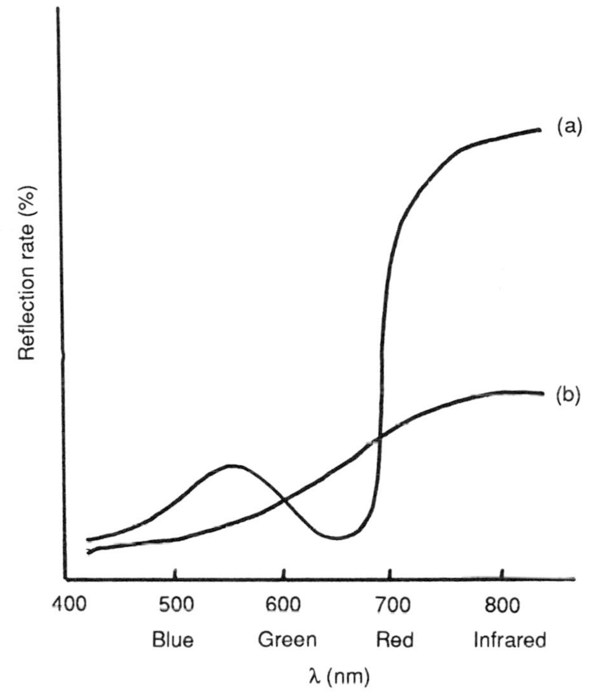

Figure 2 Reflection spectrum of plant leaves. (a) Vigorous leaves; (b) less vigorous leaves. [From S. Kagamimori *et al.* (1978). *Environ. Res.* **17,** 33–45. Used by permission.]

For monitoring visible symptoms of plant damage, changes in the shapes of trees and leaf colors are used as criteria for determining the degree of injury caused by air pollution. In Figure 3, the criteria for determining the degree of injury to the conic hope tree (Japanese cedar; *Cryptomeria japonica* D. Don) are shown. In this case, the correlation be-

Table I Growth of 72 Plants of S23 Ryegrass (*Lolium perenne*) under Each of Four Different Types of Exposure to Polluted Air Near Manchester ($\bar{X}$ ± SE)

Daily exposure (h)				
Ambient air	"Scrubbed" air	Number of tillers/plant	Number of leaves/plant	Dry weight of green shoot/plant (mg)
—	24	3.4 ± 0.25	15.5 ± 0.6	53 ± 3
24	—	3.0 ± 0.26	14.1 ± 0.6	44 ± 3
16 (16.30–08.30)	8	5.7 ± 0.32	21.0 ± 0.8	62 ± 3
8 (08.30–16.30)	16	6.8 ± 0.34	24.2 ± 0.8	62 ± 3

Source: Bleasdale, J. K. A. (1973). *Environ. Pollution* **5,** 275–285. Used by permission.

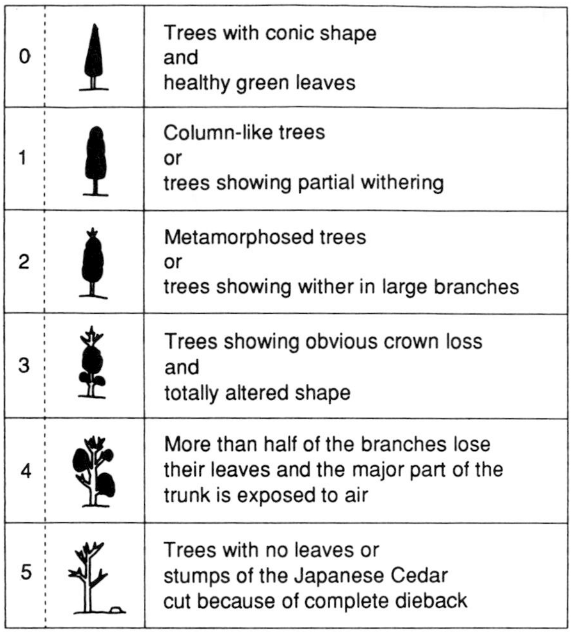

0	Trees with conic shape and healthy green leaves
1	Column-like trees or trees showing partial withering
2	Metamorphosed trees or trees showing wither in large branches
3	Trees showing obvious crown loss and totally altered shape
4	More than half of the branches lose their leaves and the major part of the trunk is exposed to air
5	Trees with no leaves or stumps of the Japanese Cedar cut because of complete dieback

Figure 3 Criteria used for determining the degree of injury to Japanese cedar. [From S. Kagamimori *et al.* (1978). *Environ. Res.* **17,** 33–45. Used by permission.]

tween the degree of injury and the bi-band ratio had been analyzed. The field survey of the cedars, which were 6 m in height or more, was performed by a trained expert throughout 60 communities. To determine the average of the injury in a community, the number of the cedars appraised at random was from 18 to 80 in each community. Additionally, aerial infrared color film shots were made of these communities (from 1100 to 1300) per day. The cedars were identified on the film, and each of three color components of the canopies of cedars was analyzed, respectively. The cedars which were less than 1 mm in diameter as measured by the densitometer's calipers were excluded from survey, even though they were identified on the film. The spectrum was analyzed randomly in each community about 20 to 30 times. Thus the average values of not only three kinds of bands but also the bi-band ratios were obtained in each community. The correlation coefficient between the degree of damage and the bi-band ratio (R/B) of each community is -0.34 ($p < 0.05$), and the correlation coefficient between the degree and the bi-band ratio (R/G) is -0.41 ($p < 0.01$). This shows a tendency of the bi-band ratio to decrease with the advancing of the degree of injury. Furthermore, the community group with a higher average score of injuries to the cedar showed lower average of bi-band ratios (see Table II). In the study area, the concentration of sulfur oxide in the air had been measured by the lead dioxide method. Therefore, it was demonstrated that the average value of each degree of injury to the cedars in communities within 1 km^2 from the station was correlated with the sulfur oxide concentration of the previous three years.

Besides these cross-sectional methods, dendrochrology is also used for plants sensitive to air pollution in order to analyze air pollution effects on plants retroprospectively. For assessing the influence of air pollution on tree plants, core samples are collected from sites representative of the study area. In general, growth inhibition is estimated by normalizing the tree age and meteorolgical conditions to obtain a standardized ring growth width (annual ring growth index). This is obtained by dividing the observed annual ring width by the expected one. The former is determined from soft x-ray films of core samples collected. The latter is calculated by fitting each measured series of annual ring widths during the period with no significant source of air pollutants to an exponential curve to standardized the trend of decrease in annual ring widths with the age. Furthermore certain techniques (e.g., the stepwise multiple regression method) is needed to adjust the influence of seasonal climatic conditions and so on.

Table II Correlation between the Averages of Injury to Japanese Cedars and the Bi-band Ratio at Each Community

The average of injuries of the cedar	Number of communities	Average bi-band and bi-band ratio ($M \pm$ SE)				
		R	B	G	R/G	R/G
0–1.9	13	2.10 (0.03)	2.13 (0.05)	1.08 (0.05)	1.00 (0.02)	1.18 (0.03)
2.0–2.4	11	2.06 (0.02)	2.06 (0.03)	1.79 (0.04)	1.00 (0.01)	1.16 (0.02)
2.5–2.9	13	2.05 (0.03)	2.12 (0.04)	1.80 (0.06)	0.97 (0.01)	1.15 (0.03)
3.0–3.4	13	2.07 (0.02)	2.11 (0.03)	1.84 (0.03)	0.98 (0.01)	1.12 (0.01)
3.5	4	2.17 (0.07)	2.30 (0.16)	2.00 (0.10)	0.95 (0.04)	1.09 (0.02)

III. Human Health Indicators of Air Pollution

Generally speaking, it is reasonable in epidemiological studies that the prevalence rate of cough as well as phlegm production are used as human health indicators of air pollution. In addition to respiratory symptoms, eye and nasal symptoms have been identified to be associated with air pollutants. Furthermore, systemic changes secondary to those in the lungs have been observed in the kidney, liver, and heart in heavy exposures. Nevertheless, since SO_2, NO_2, and suspended particulates consistent with the major pollutants in ambient air are thought to be pulmonary irritants, it is convenient and efficient to use respiratory symptoms as health indicators of air pollution. With regard to SO_2, more than 90% of inhaled gas is absorbed in the airways above the larynx as it is much more soluble in water than most other pollution gases. Under conditions of high humidity, sunlight, and in the presence of such particulate catalysts as charcoal, ferric oxide, and graphite, sulfur oxide is converted to sulfur trioxides which combine immediately with water vapor to form sulfuric acid (H_2SO_4), a strong irritant. Therefore, suspended particulates are very important in the subsequent appearance of droplets sulfuric solution in ambient air. In contrast to SO_2, NO_2 is not very soluble in water, and reaches the moisture-laden alveoli of the lungs where the very irritating gases, nitrous acid (HNO_2) and nitric acid (HNO_3) are formed.

From the standpoint of monitoring air pollution effects on human health, cigarette smoking has to be considered as a confounding factor. As shown in Table III, British Medical Research Council's (BMRC) Committee on Research into chronic bronchitis has proposed an interview questionnaire on respiratory symptoms for use in epidemiological research as early as 1966. When using this questionnaire and recording the prevalence of persistent cough and phlegm, information on smoking habits is also an essential consideration. Based on the BMRC questionnaire, the American Thoracic Society (Division of Lung Diseases; ATS–DLD) has presented another questionnaire on respiratory symptoms for children (13 years of age and older) and adults (see Table IV). In epidemiological studies, a comparison of the symptoms of respiratory impairment frequency between areas is performed by using the standardized prevalence ratio (SPR). This is obtained by dividing the observed prevalence with the expected one which is calculated from the standard population adjusted for age.

In monitoring low-grade air pollution effects on human health, the existence of hyperresponders is especially taken into account. Regarding host reactivity (to air pollution), knowledge is increasing. Not only SO_2 but also NO_2 and suspended particulates are suggested to be related to the allergic reactions associated with human respiratory symptoms. For instance, the importance of atopic predisposition to respiratory disease has been emphasized in low grade air pollution levels below ambient air quality standards. Practically, atopic subjects are screened with skin reactions to common inhaled allergens such as house-dust mites, pollen, fungi, insects, and so on. However, this predisposition could be dependent on the kind of air pollutants. Although a positive skin test group is more sensitive to the atmospheric concentration of SO_2 compared with the negative group, the positive group is not always a useful indicator for monitoring health effects of air pollution. In fact, some studies have shown a relatively poor association between respiratory symptoms and skin reactions to common airborne allergens. As subjects for monitoring health effects of air pollution, children are more easily studied because of their susceptibility to respiratory symptoms and higher positive skin reaction rates to common airborne allergens, particularly to house-dust mites. Although the elderly also have a tendency to show respiratory symptoms such as a cough and phlegm production, these are often due to other disease states which are not associated with air pollution.

IV. Correlation between Plant and Human Health Indicators

In assessing air pollution, comprehensive monitoring is desirable from an ecological standpoint. The consistent monitoring of plant and human health indicators is a good example. Furthermore, in order to speculate cause–effect relationships and evaluate either a deterioration or improvement in air pollution conditions, monitoring should be carried out for long periods of time. For plant indicators, tree rings have been demonstrated to be useful for air environment monitors in the past.

The monitored studies in Figure 4 were carried out under circumstances where SO_2, NO_2, and sus-

Table III From the BMRC Questionnaire on Respiratory Symptoms

COUGH

1. Do you usually cough first thing in the morning in the winter?
2. Do you usually cough during the day—or at night—in the winter?

If Yes to 1 or 2

3. Do you cough like this on most days for as much as three months each year?

PHLEGM

4. Do you usually bring up any phlegm from your chest first thing in the morning in the winter?
5. Do you usually bring up any phlegm from your chest during the day—or at night—in the winter?

If Yes to 4 or 5

6. Do you bring up phlegm like this on most days for as much as three months each year?

Periods of cough and phlegm

7a. In the past three years have you had a period of (increased) cough and phlegm lasting for three weeks or more?

If Yes

7b. Have you more than one such period?

BREATHLESSNESS

If the subject is disabled or unable to walk caused by any condition other than heart or lung disease, omit question 8.

8a. Are you troubled by shortness of breath when hurrying on level ground or walking up a slight hill?

If Yes

8b. Do you get short of breath walking with other people of your own age on level ground?

If Yes

8c. Do you have to stop for breath when walking at your own pace on level ground?

WHEEZING

9a. Does your chest ever sound wheezing or whistling?

If Yes

9b. Do you get this on most days—or nights?

10a. Have you ever had attacks of shortness of breath with wheezing?

If Yes

10b. Is/was your breathing absolutely normal between attacks?

CHEST ILLNESSES

11a. During the past three years have you had any chest illness that has kept you from your usual activities for as much as a week?

If Yes

11b. Did you bring up more phlegm than usual in any of these illnesses?

If Yes

11c. Have you had more than one illness like this in the past three years?

Past illnesses

Have you ever had:

12a. an injury or operation affecting your chest

12b. heart trouble

12c. bronchitits

12d. pneumonia

12e. pleurisy

12f. pulmonary tuberculosis

12g. bronchial

12h. other chest trouble

12i. hay fever?

Table IV From the ATS–DLD Questionnaire on Respiratory Symptoms

COUGH

14A. Does he/she usually have a cough with colds?
B. Does he/she usually have a cough apart from colds?
If Yes to 14A or 14B:
C. Does he/she cough on most days (4 or more days per week) for as much as 8 months of the year?
D. For how many years has he/she had this cough?

CONGESTION AND/OR PHLEGM

15A. Does this child usually seem congested in the chest or bring up phlegm with colds?
B. Does this child usually seem congested in the chest or bring up phlegm apart from colds?
If Yes to 15A or 15B:
C. Does this child seem congested or bring up phlegm, sputum, or mucus from his/her chest on most days (4 or more days per week) for as much as 8 months a year?
D. For how many years has he/she seemed congested or raised phlegm, sputum, or mucus from his/her chest?
16A. Does this child get attacks of (increased) cough, chest congestion, or phlegm lasting for 1 week or more each year?
If Yes to 16A:
B. For how many years?
C. On average, how many chest colds per years does he/she get?

WHEEZING

17. Does this child's chest ever sound wheezing or whistling
A. When (he/she) has a cold?
B. Occasionally apart from colds?
C. Most days or nights?
If Yes to 17B or 17C:
D. For how many years has wheezing or whistling in the chest been present?
18A. Has this child ever had an attack of wheezing that has caused him/her to be short of breath?
If Yes to 18A:
B. Has he/she had two or more such episodes?
C. Has he/she ever required medicine or treatment for the(se) attack(s)?
D. How old was this child when he/she had his/her first such attack?
E. Is or was his/her breathing completely normal between attacks?
19. Does this child ever get attacks of wheezing after he/she has been playing hard exercising?

CHEST ILLNESSES

20A. During the past 3 years has this child had any chest illness that has kept him/her from his/her usual activities for as much as 8 days?
If Yes to 20A:
B. Did he/she bring up more phlegm or seem more congested than usual with any of these illnesses?
C. How many illnesses like this has he/she had in past 8 years?
1. Less than 1 illness per year
2. 1 illness per year
3. 2–5 illnesses per years
4. More than 5 illnesses per year
8. Does not apply
D. How many of these illnesses have lasted for as long as 7 days?

pended particulate concentrations in air were below ambient air quality standards, even after operation of oil-fired electricity-generating stations. As Figure 4 illustrates, the annual ring growth index in Districts I and II near the power stations began to decline in the late 1960s and reached the smallest value in 1974 when the power station had already been in operation for 1 year. Due to a reduction in SO_2 exhaust after the introduction of the scrubber system in 1975, a recovery in growth commenced. On the other hand, the annual ring growth index in District III (located far from the power station) did not show such marked changes as those observed in Districts I and II and instead fluctuated around the expected value during the entire study period (Fig. 5). In monitoring respiratory symptoms of children aged 6–14 years, an interview questionnaire standardized by BMRC and the skin reaction test to house-dust mites were used. School children were interviewed about their respiratory symptoms every few years. In the

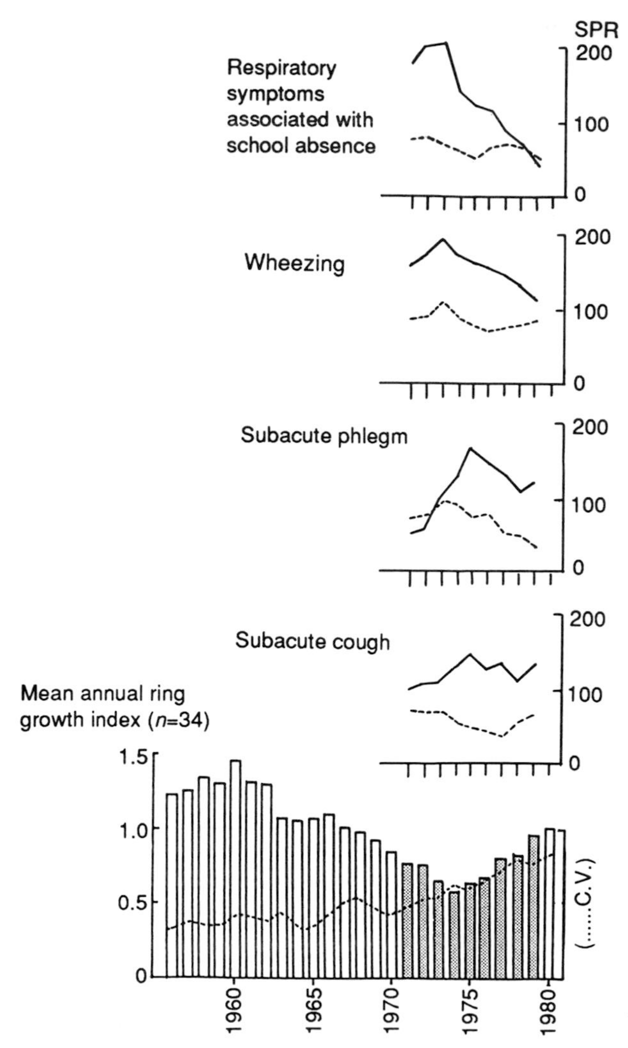

Figure 4 Changes in the mean of annual ring growth index and standardized prevalence ratio (SPR) of respiratory symptoms among school children in Districts I and II. Annual ring indexes used for analyzing the correlation with SPR are shaded. —, positive skin test group; -------, negative skin test group. [From S. Kagamimori *et al.* (1990). *Environ. Res.* **52,** 47–61. Used by permission.]

case when a positive answer was given to questions 1 and 3 (usually cough in the morning in winter) and/or questions 2 and 3 (usually cough during the day or at night in winter) of the questionnaire, the symptom was designated as subacute cough; when the answer "yes" was given to questions 4 and 6 (usually phlegm in the morning in winter) and/or questions 5 and 6 (usually phlegm during the day or at night in winter), it was designated as subacute phlegm; and when the answer "yes" was given to question 9a (wheezing or whistling), it was designated as wheezing. In addition to these definitions, any symptoms of respiratory disease which caused an absence from school of over 1 week were also taken into consideration and designated as respiratory symptoms associated with school absence. With regard to the skin reaction test, school children who had a positive reaction to house-dust mites at least once during the entire study period were incorporated into the positive skin test group, and those

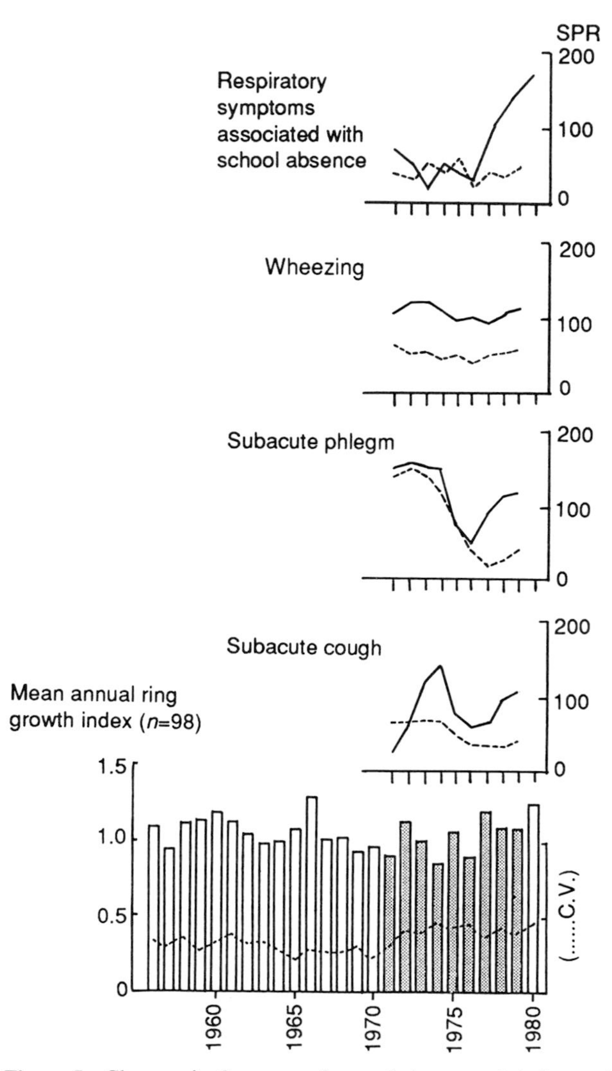

Figure 5 Changes in the mean of annual ring growth index and standardized prevalence ratio (SPR) of respiratory symptoms among school children in District III. Annual ring indexes used for analyzing the correlation with SPR are shaded. —, positive skin test group; -------, negative skin test group. [From S. Kagamimori *et al.* (1990). *Environ. Res.* **52,** 47–61. Used by permission.]

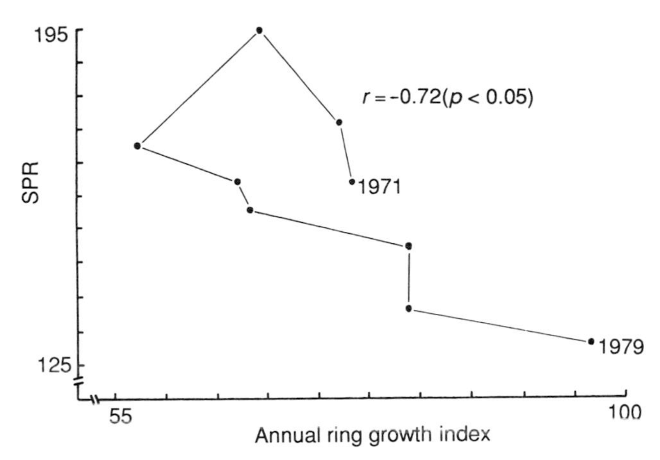

Figure 6 Scattered diagram of correlation between the annual ring growth index of Japanese cedars and the SPR for wheezing of school children with the positive skin test in Districts I and II (1971–79). [From S. Kagamimori *et al.* (1990). *Environ. Res.* **52,** 47–61. Used by permission.]

who had never had a positive reaction into the negative skin test group. In general, since lower-grade and male school children had a tendency to show the higher prevalence of respiratory symptoms adapted here, a standardized prevalence ratio (SPR) adjusted for age and sex was used for the comparison between calendar years and between districts. As shown in Figures 4 and 5, in general, SPRs of all respiratory symptoms investigated were higher in the positive skin test group than in the negative skin test group for Districts I and II, and District III. These SPRs in the positive and negative skin test groups were analyzed in terms of their correlation with the annual ring growth index for each calendar year. In District I and II, the annual ring growth index was inversely correlated with the SPRs for respiratory symptoms associated with school absence, wheezing, and subacute phlegm of the positive skin test group, respectively (See Fig. 6). With regard to the negative skin test group, the correlation was observed for wheezing and subacute phlegm. On the other hand, such a correlation between the two kinds of indicators was not demonstrated in District III, which did not have a substantial change in the atmospheric concentration of SO_2. These observations indicate an increase in the prevalence of respiratory symptoms mentioned here and a decrease in the annual ring growth index coincided following a deterioration in ambient air mainly due to SO_2 released from power stations in the early period of the study. An opposite change in response occurred following an improvement in air pollution for the symptoms and the annual ring growth index, respectively. Actually the annual ring growth index has been demonstrated to be mainly associated with SO_2 concentration in air. These findings agree with the observation mentioned above that the positive, skin test group is more sensitive to SO_2 concentration in air compared with the negative group. In particular, the positive skin test group is the suitable indicator for low-grade air pollution.

Related Articles: AMBIENT ACIDIC AEROSOLS; PLANTS AS DETECTORS OF ATMOSPHERIC MUTAGENS; RUBBER INDUSTRY, TOXICITY OF WORK ENVIRONMENT.

Bibliography

Brain, J., Beck, B. D., Warren, A. J., and Shaikh, R. A., eds. (1988). "Variations in Susceptibility to Inhaled Pollutants," The Johns Hopkins University Press, Baltimore, MD.

Moriarty, F., ed. (1988). "Ecotoxicology: The Study of Pollution in Ecosystems," Academic Press, San Diego, CA

Stern, A. C., ed. (1976). "Air Pollution," Academic Press, New York.

U.S. EPA (1981). "Air Quality Criteria for Particulate Matter and Sulfur Oxides," Environmental Criteria and Assessment Office, Office of Research and Development, Research Triangle Park, N.C.

Waldbott, G. L., ed. (1973). "Health Effects of Environmental Pollutants," C.V. Mosby Company, Saint Louis, MO.

Mutagenicity Tests with Cultured Mammalian Cells: Cytogenetic Assays

Günter Obe
University GH Essen

A. T. Natarajan
University of Leiden

Glossary

Aneuploid Cell containing less or more chromosomes than the normal complement by deletion or addition of chromosomes.

Chromosomal aberrations Structural changes in chromosomes visible under the light microscope.

DNA adducts Covalent binding of reactive electrophilic groups to different sites in the DNA.

DNA lesions Strand breaks, base damage, and DNA adducts induced by mutagenic agents.

DNA repair Cellular processes that recognize damaged DNA and correct the damage to restore the original configuration.

Fibroblasts Cells growing under *in vitro* conditions on a glass or plastic surface appearing as fiberlike structures.

Lymphocytes Mononuclear white cells in the blood that can be stimulated to divide *in vitro*.

Metaphase Phase of mitosis before separation of the chromatids in which the chromosomes are usually analyzed for the presence of chromosomal aberrations.

Mutagen Physical or chemical agent that can induce alterations in DNA leading to heritable changes.

***CYTOGENETIC ASSAYS** are used to analyze mutagenic activities of chemicals. Chemicals can induce mutations in all living systems from viruses to humans. In 1943 Charlotte Auerbach discovered that mustard gas could induce mutations in* Drosophila. *Mutations comprise a spectrum of changes in the sequence of bases in the DNA including single base pair changes (point mutations), small and large deletions, as well as rearrangements. Large deletions and rearrangements can be detected as chromosomal aberrations, whereas the smaller changes can be detected by molecular techniques. Because all mutagens induce chromosomal aberrations, a test for chromosome-breaking ability can unravel the possible mutagenic effect of a test chemical. In human populations, both point mutations and chromosomal aberrations occur. There are two types of chromosomal aberrations recognized, namely structural and numerical. About 0.6% of human newborns have either structural or numerical chromosomal aberrations. In spontaneous abortions, more than 50% are chromosomally abnormal. Most human neoplasms are associated with chromosomal aberrations. Hence testing for the ability of a chemical to induce chro-*

mosomal aberrations is of utmost relevance. Various assays are available to detect chromosome-breaking abilities or the ability to induce numerical changes (aneuploidy) using human and mammalian cells in culture.

I. Eukaryotic Chromosomes

A. Structure

Eukaryotic chromosomes generally contain one DNA molecule per chromatid (i.e., they are uninemic). DNA molecules in chromosomes are very long; e.g., the human chromosome #1 contains a DNA molecule about 7 cm in length. In the metaphase of mitosis this chromosome is about 10 μm long, indicating a very effective packing of the DNA molecule. The first level of packing is the winding of DNA on octamers of two molecules each of the histones H2A, H2B, H3, and H4 (nucleosome cores). The nucleosome cores contain about 146 nucleotide pairs and are connected by linker DNA of about 60 nucleotide pairs. In the presence of histone H1, the nucleosome cores are compacted to form a fiber about 30 nm in diameter. This fiber is further condensed to give rise to the structure found in metaphase chromosomes. This model only describes the association of DNA and histones, but there are also many nonhistone proteins, such as enzymes, hormone-receptor proteins, proteins regulating transcriptional activities, high mobility group (HMG) proteins, and scaffold proteins. The scaffold proteins remain attached to chromosomes from which the majority of the proteins, including the histones, have been eluted. The scaffold has the form of a chromosome shaped matrix from which loops of DNA emanate. It contains lamins (a component of the nuclear membrane) and topoisomerases. In mammalian (including human) chromosomes, the genome is subdivided in two subgenomes which, after special treatment and staining with Giemsa solution, are visible as reverse (R) or Giemsa (G) bands in the chromosomes. The early replicating part of the genome, visible as R bands, contains primarily the housekeeping genes, and the late replicating part of the genome, visible as G bands, contains the ontogenic or tissue-specific genes. This indicates that, apart from the complex association of DNA, histones, and nonhistones, the chromosome has a suprastructure that may also influence the intrachromosomal location of chemically induced chromosomal aberrations.

B. Cell Cycle and Chromosome Cycle

When cells multiply, there is an interphase stage between two mitoses. Within the interphase stage, the DNA replicates during the synthetic phase (S phase). The presynthetic phase is called G1 phase and the postsynthetic phase is called G2 phase. During the G1 phase, the chromosome contains one DNA molecule, while during the G2 phase it contains two identical DNA molecules. Some cells (such as human peripheral lymphocytes) when not proliferating, are in a presynthetic G0 phase.

In mammalian cells in culture, the S phase requires about 7 hours. DNA synthesis starts at multiple sites in the chromosomes and proceeds bidirectionally. The replication of a single replicating site (replicon) is finished when the replicated DNA meets the replicated DNA of the neighboring replicons. There are thousands of replicons in each chromosome and it appears that clusters of replicons initiate their DNA synthesis together. This type of multirepliconic replication of eukaryotic chromosomes is reflected by replication bands, which can be seen with the light microscope following incorporation of bromodeoxyuridine (BrdU) during parts of the S phase, and special staining of the chromosomes.

The mitotic division can be separated into discrete stages: prophase, metaphase, anaphase, and telophase. During anaphase the two chromatids separate and under the influence of the mitotic spindle move to opposite poles of the cell, thus initiating the formation of two daughter cells. Malsegregation during anaphase will lead to unequal daughter nuclei (aneuploidy due to nondisjunction).

II. Chromosomal Aberrations

A. DNA as the Target for Production

There are several pieces of evidence indicating that chromosomal DNA is the target for the production of chromosomal aberrations. These include the following observations:

1. The most effective UV wavelength for the production of chromosomal aberrations is 260 nm,

which is the wavelength maximally absorbed by DNA.

2. In cells with a photoreactivating system (e.g., chicken embryonic fibroblasts, fibroblasts from *Xenopus*), fewer chromosomal aberrations are found when UV irradiation is followed by exposure to photoreactivating light. This treatment monomerizes pyrimidine dimers in DNA, reverting them to the normal form.

3. BrdU, when incorporated in chromosomal DNA, induces chromosomal aberrations.

4. Chromosomes substituted with BrdU are more sensitive to ionizing radiation and to UV radiation than unsubstituted chromosomes.

5. There are fewer chromosomal aberrations when chromatin substituted with BrdU is treated with the restriction endonuclease *Dra*I (recognition site TTTAAA) when compared to treatment of unsubstituted chromatin.

6. Inhibitors of DNA metabolism induce chromosomal aberrations.

7. Monofunctional alkylating agents induce chromosomal aberrations whose frequencies are correlated with specific alkylating activities on chromosomal DNA.

8. The transformation of DNA single-strand breaks to DNA double-strand breaks in chromatin with single-strand-specific endonucleases leads to chromosomal aberrations.

9. Endonucleases such as restriction endonucleases and DNase I induce chromosomal aberrations.

10. Inhibition of DNA repair leads to higher frequencies of chromosomal aberrations.

11. Prolongation of DNA repair times in treated cells via liquid holding leads to fewer chromosomal aberrations.

B. DNA Lesions

Many types of DNA lesions can lead to chromosomal aberrations and sister chromatid exchanges (SCEs). The ultimate lesions for the induction of these alterations are DNA double-strand breaks. Double-strand breaks can be induced directly by treatment itself or, during subsequent cellular repair or DNA replication. Agents such as ionizing radiation and the antibiotic bleomycin induce double-strand breaks directly and efficiently induce chromosomal aberrations but not SCEs. Most of the other agents such as UV light (i.e., that induce pyrimidine dimers and 6,4-photoproducts), monofunctional alkylating agents (causing alkylation of O-6 and N-7 of guanine and of other sites), polyfunctional alkylating agents (that induce intra- and interstrand DNA cross-links or DNA-protein cross-links), aromatic amines, polycyclic hydrocarbons (bulky adducts covalently linked to DNA) induce both chromosomal aberrations and SCEs very efficiently. Some of these agents act indirectly and therefore require metabolic activation before being effective (see Section VII).

Two main types of aberrations are recognized depending on whether both or one of the chromatids in a metaphase chromosome is involved. These are called chromosome or chromatid aberrations, respectively. The type of aberration observed in metaphase depends on the stage of the cell cycle in which treatment is made and the type of mutagenic agent used. Agents that induce DNA double-strand breaks directly (e.g., x-rays, bleomycin, endonucleases) induce chromosome-type aberrations in G1 phase, chromatid-type aberrations in G2 phase, and a mixture of both types in S phase. Most chemical agents (e.g., alkylating agents, cross-linking agents, and UV light) induce chromatid-type aberrations irrespective of the cell stage in which the treatment is made. In addition, if cells in G2 phase are treated with these agents, no aberrations can be observed during oncoming mitosis but only in the next mitosis, indicating that the cells must go through an S phase before the aberrations become fixed. These agents are therefore termed "S-dependent" in contrast to those that induce DNA double-strand breaks directly and are thus termed "S-independent."

III. Cytogenetic Endpoints Used for Mutagenicity Testing

A. Chromosomal Aberrations in Metaphase

There are three types of aberrations that are generally scored in metaphase, namely gaps (unstained regions smaller than the width of the chromatid) (Fig. 1a), breaks (Fig. 1b and c) and exchanges (Fig. 1d, e, f, g, h). When both chromatids are involved in a break they are called isochromatid or chromosome breaks (Fig. 1b). Asymmetrical exchanges involving both chromatids give rise to

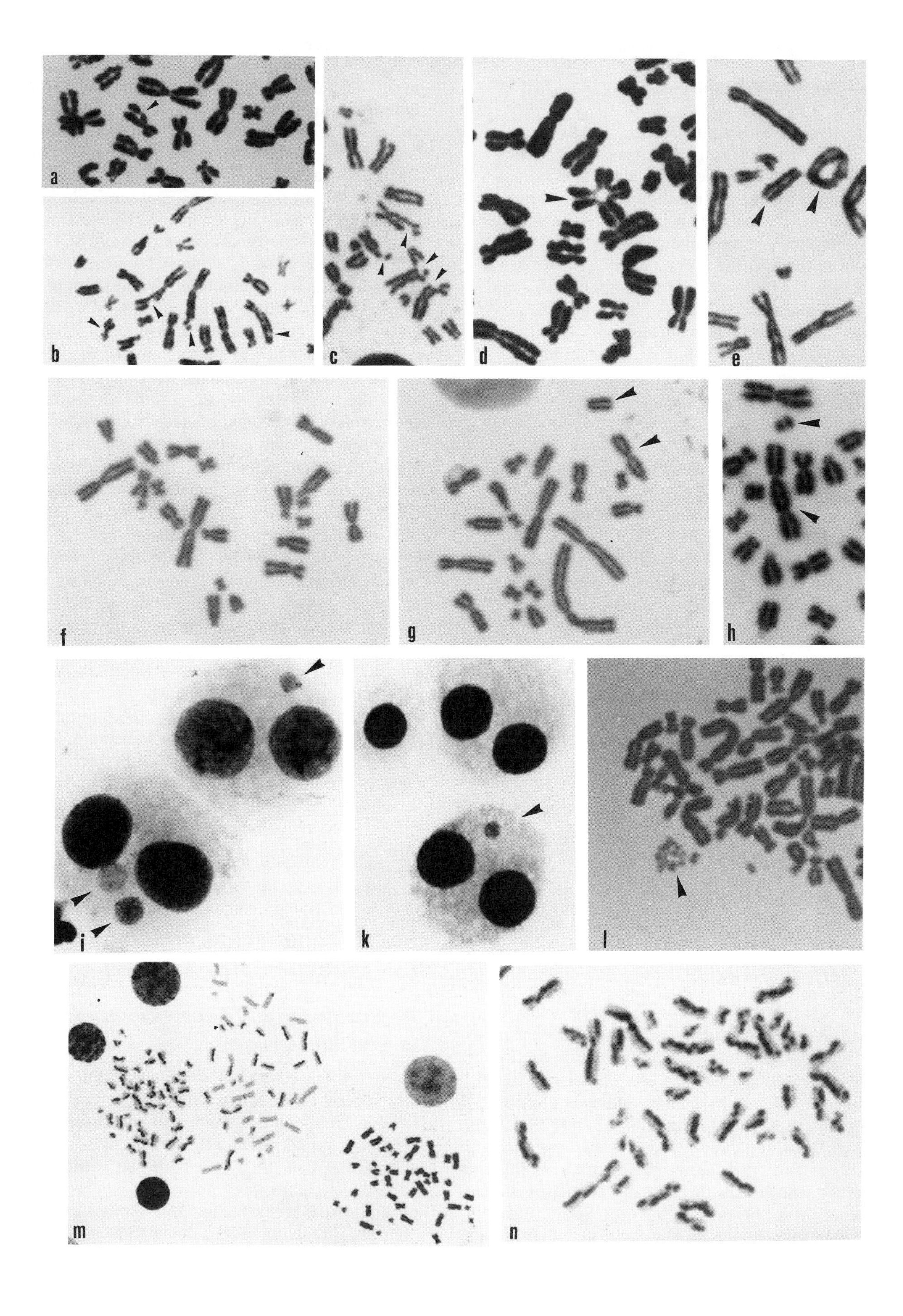

dicentrics (or polycentrics) (Fig. 1g, h). Symmetrical exchanges involving both chromatids give rise to reciprocal translocations that are generally difficult to identify and are therefore not scored. Exchanges involving single chromatids result in symmetrical or asymmetrical chromatid exchanges, which look like X-configurations and are easy to analyze (Fig. 1d).

B. Methodologies and Protocols

Established rodent cell lines are commonly used to test chemicals for an ability to induce aberrations. Chinese hamster cell lines (CHO, V79, CHL) are widely used in view of the large size of the chromosomes, which makes them amenable to easy scoring of aberrations (Fig. 1f). These cells are grown in Petri dishes or bottles in a nutrient medium containing components essential for cell proliferation. The medium is supplemented with bovine serum and antibiotics. The cell cycle duration is about 12 to 16 hours at 37°C.

Before testing for mutagenicity, a preliminary investigation of cytotoxicity of the agent is carried out. For this, exponentially growing cells are treated with different concentrations of the test agent for 1 or 2 hours and then allowed to recover. The toxicity is evaluated by (a) survival of cells, (b) determination of mitotic indices (number of mitoses in 1000 cells) after 24 or 48 hours, or (c) counting the number of cells after 48 hours recovery.

Based on the data obtained on the toxicity tests, appropriate doses of the test compounds for a chromosomal aberration test are selected. These doses are in the range below 50% cell killing or 50% inhibition of the mitotic index of the control. After initial testing, exponentially growing cells are treated for 15 minutes to 2 hours, washed, and allowed to recover for 16 to 24 hours. In such treatments, the activity of a test chemical on an early part of the cell cycle (mainly the G1 phase) can be analyzed. Multiple fixations following treatment can uncover cells treated at different stages of the cell cycle. One can also treat the cells chronically with low concentrations of the test chemical to cover the entire cell cycle. Such chronic treatments are very useful in cases where the solubility of a test chemical is very low.

Colcemid (a derivative of colchicine) is added 2 hours before fixation, in order to accumulate metaphases with contracted chromosomes (C-metaphases). The cells are suspended (by trypsination or by mitotic shake-off of CHO cells), centrifuged, and exposed to hypotonic shock before fixing in methanol–acetic acid (3:1). The fixed cells are dropped on clean slides and allowed to air dry. These preparations are stained with an aqueous Giemsa solution and scored for the frequencies of different types of aberrations. A dose-dependent increase in the frequency of aberrations is taken as an indication of a positive mutagenic effect. Several statistical tests are available to evaluate these types of data.

Some rodent cell lines can be grown in suspension (e.g., mouse lymphoma cells) and the treatment protocol for these cells has to be modified in that procedures such as trypsinization are not necessary. Human fibroblasts can also be used for testing and the protocols have to be modified to accomodate a longer cell cycle time (about 24 hours).

Human peripheral lymphocytes are employed for mutagenicity testing in many laboratories. For this purpose, whole blood minicultures (0.5 ml of blood in 5 ml culture medium) are established. The medium is supplemented with a mitogen (phytohemagglutinin or concanavalin A) in order to stimulate dormant lymphocytes to proliferate. The first round of DNA synthesis in the stimulated lymphocytes begins around 24 hours after initiation of the cultures and the first mitoses occur at about 48 hours. To achieve an optimal effect of the test chemical, the treatment should be made during the S phase of the cell cycle. The fixation protocol should take into account possible mitotic delay due to treatment. In cases of mitotic delay, the first mitosis is generally later than 48 hours.

Figure 1 Chromosomal aberrations in metaphase. (a–c) Gaps, chromatid, and isochromatid/chromosome breaks (HPL); (d) chromatid exchange (HPL); (e) ring chromosome with associated fragment (HPL); (f) metaphase of a CHO cell; (g,h) dicentric chromosomes with associated fragments (g, CHO; h, HPL); (i,k) micronuclei in cytochalasin B-induced binucleated cells (HPL); (l) prematurely condensed micronucleus (HPL); (m) metaphases grown for one (TT, *right*), two (TB-BB, *left*), and three (TB-BB and BB-BB) cell cycles in the presence of BrdU (HPL); (n) elevated frequency of SCEs (HPL). HPL, human peripheral lymphocytes; CHO, Chinese hamster ovary cells.

In all protocols, the scoring of aberrations should be confined to the cells in their first posttreatment mitosis in order to avoid selection against aberrant cells. First division mitotic cells can be easily recognized when the cells are grown in medium containing BrdU and the preparations stained with the fluorochrome–Giemsa technique (see Section V) (Fig. 1m, n). Usually 100–200 metaphases are scored for the presence of aberrations for each experimental point.

IV. Micronuclei

A. Origin and Types

A proportion of fragments or whole chromosomes, which lag in anaphase, are not included in the main nucleus and in the next interphase they can form a small nucleus, which is called a micronucleus. Depending on the amount of chromosomal material included, the sizes of micronuclei may vary. Micronuclei are easy to score (Fig. 1i, k), but can sometimes become prematurely condensed under the influence of mitosis in the main nucleus giving the appearance of pulverized or extended chromosomes (Fig. 1l).

B. Methodologies and Protocols

The simplest way to make preparations to detect micronuclei in fibroblast cultures is to grow the cells on cover glasses and fix them *in situ* with methanol. For detecting induced micronuclei, the cells should be allowed to go through at least one mitosis following treatment. For a quantitative evaluation of the frequencies of micronuclei, it is important to know how many cells have gone through mitosis. This can be achieved by growing the cells following treatment in the presence of cytochalasin B. Cytochalasin B inhibits cytokinesis generating binucleated cells. Such binucleated cells should be exclusively used for scoring micronuclei (Fig. 1i, k). The scoring of micronuclei is amenable to automation.

For studying the induction of micronuclei, essentially the same protocol is used as for metaphase analyses except that following treatment with the test chemical, cytochalasin B (3–6 μg/ml) is added to the culture and the cells are allowed to proceed through at least one cell cycle. The preparations are made with or without mild hypotonic treatment. In the case of human lymphocytes, cytochalasin B is added around 42 hours after culture initiation, and the lymphocytes are fixed at around 72 hours. To keep the cytoplasm intact, a trace of formalin can be added to the fixative (methanol–acetic acid, 3 : 1). The slides are stained with aqueous Giemsa solution or any other DNA specific stain. Usually 1000 to 2000 cells are scored for the frequencies of micronuclei per experimental point. The micronucleus frequency reflects the chromosome breaking activity of the test chemical, or its ability to induce lagging of whole chromosomes (induction of aneuploidy).

V. Sister Chromatid Exchanges

A. Origin and Types

The manner in which sister chromatid exchanges (SCEs) are formed is not known. SCEs may be cytological manifestations of DNA double-strand breakage and rejoining at homologous sites between the two chromatids of a chromosome, or they may be the result of recombinational repair. SCEs can be visualized if the sister chromatids can be distinguished either by radioactive labeling or differential staining following incorporation of BrdU (Fig. 1m, n). SCEs are induced very efficiently by several mutagenic and carcinogenic agents especially those that form covalent adducts to the DNA or interfere directly or indirectly with DNA replication. Agents that induce DNA strand breaks directly are poor inducers of SCEs, whereas agents which act in an S-dependent manner to induce chromosomal aberrations induce SCEs very efficiently. While most of the mutagens induce SCEs, not all agents that induce SCEs also induce point mutations. For example, inhibitors of DNA synthesis such as cytosine arabinoside and hydroxyurea as well as inhibitors of poly-(ADP-ribose)-polymerase induce SCEs efficiently without increasing the frequency of point mutations. SCEs are easy to score and this method appears to be of sufficient sensitivity to detect the activity of certain classes of chemical mutagens at very low doses.

B. Methodologies and Protocols

The current procedure to obtain sister chromatid differentiation is to incorporate BrdU (5–10 μM) into cells for two cycles. This will generate cells with chromosomes in metaphase in which DNA in one chromatid is bifiliarly substituted with BrdU (BB) and the other chromatid containing DNA substituted unifiliarly with BrdU (TB) (Fig. 1m). One can also employ a single round of BrdU incorporation followed by another cycle in normal medium, wherein the chromosomes will contain DNA in one chromatid unsubstituted with BrdU (TT) and the other chromatid unifiliarly substituted with BrdU (TB). To evaluate the induction of SCEs by a test chemical, the protocol employed for detection of chromosomal aberrations is used except that the cells are allowed to go through one or two divisions in the presence of BrdU. Differential BrdU content between the sister chromatids can be visualized microscopically by staining with the fluorochrome bisbenzimidazole (Hoechst 33258) and either viewing through a fluorescent microscope or by exposing the fluorochrome-stained slides to visible light or near-UV light, followed by incubation in 2 × SSC (saline sodium citrate) at 65°C for about 2 hours and staining with Giemsa solution. The chromatid containing less BrdU will fluoresce brighter or stain stronger with Giemsa stain in comparison to its sister chromatid. Any switch between segments of sister chromatids can easily be detected and evaluated (Fig. 1n). Usually 25 to 50 metaphases are scored. This method can also be used for human lymphocytes.

VI. Numerical Chromosomal Aberrations

A. Origin and Types

Malsegregation of chromosomes during anaphase can lead to daughter nuclei containing more or less than the diploid number of chromosomes (aneuploidy). Aneuploidy is generally induced by chemicals that influence cellular targets other than DNA that participate in the concerted mechanisms leading to an equal distribution of chromosome sets during mitoses in mammalian cells in culture. Mitotic division depends (1) on the spindle, which contains protein microtubules; (2) on the centrosomes which contain the microtubular structures, the centrioles; (3) on localized areas in the chromosomes, the centromeres where spindle microtubuli insert; and (4) on actin filaments, which are important for the division of the cell.

Aneuploidy can be induced in mammalian cells *in vitro* by various chemicals such as colchicine or its derivative colcemid, vinblastin, benomyl, diethylstilbestrol, and griseofulvin. Typical mutagens such as alkylating agents or ionizing radiations can also induce aneuploidy. This could be the result of damage to protein targets or of damage to chromosomal DNA (probably in the centromeric region).

B. Methodologies and Protocols

For the testing of induction of numerical aberrations in mammalian cells *in vitro*, primary cultures of embryonic fibroblasts are used. The cells are grown on cover glasses or on slides and treatment and fixation are carried out *in situ*. The number of chromosomes in metaphase is counted. Care should be taken to avoid loss of chromosomes during processing by employing only a mild hypotonic treatment. In Giemsa stained preparations, it is possible to see the cytoplasmic boundary of the mitotic cells.

In the protocol designed to detect aneuploidy in cells grown on cover glasses, chronic treatment covering one cell cycle and a recovery time covering a second cell cycle is used. Parallel preparations are made to monitor mitotic indices and cell cycle progression.

In the routine procedure adapted for making chromosomal preparations to score chromosomal aberrations or SCEs, the chromosomes are usually spread widely and this may lead to artefacts in chromosome number. Therefore these procedures cannot be employed for assessing aneuploidy.

As indicated earlier, chromosomes lagging during anaphase movement can form micronuclei in the oncoming interphase, which can be analyzed in cytochalasin B-treated binucleate cells. The presence of such micronuclei can be distinguished from those originating from lagging fragments in that they contain centromeres. The detection of centromeres within micronuclei can be achieved by immunochemical reaction with antibodies (i.e., derived from the serum of CREST-type xeroderma patients), which are specific to the centromeric proteins. By not including acetic acid in the fixative, care is taken that the structure of the chromosomal proteins is not disturbed during fixation. The other method for detecting micronuclei containing whole chromosomes is to employ *in situ* hybridization techniques using biotinylated centromere-specific DNA sequences and detecting the signal with avidine FITC. For detection of nondisjunction by micronucleus analysis, 1000–2000 binucleate cells are usually scored.

VII. Testing Indirectly Acting Mutagens

One of the problems associated with using mammalian cells in culture is the inability to activate mutagens that need metabolic activation. This can be overcome by introducing drug-activating systems such as microsomal fractions from induced rat liver (S9) during the treatment. This technique is similar to the Ames test with minor modifications. However, such a system does not represent the complete activation and deactivation system found under *in vivo* conditions.

There are, however, some permanent cell lines of human or rodent liver origin that have retained some capacity to activate mutagens. The advantage of using such cells is that both activating cells and target cells are those that allow the short-lived metabolites to react with the DNA. It is also possible to achieve metabolic activation of mutagens *in vitro* by cocultivating primary hepatocytes with target cells. Any one or more of the end-points discussed earlier (chromosomal aberrations, micronuclei, and SCEs) can be used to evaluate the ability of indirectly acting mutagens. In routine testing of chemicals, the protocol should include treatments with and without metabolic activation systems including appropriate controls.

VIII. In Vitro *versus* in Vivo Test Systems

In *in vivo* testing (usually in bone marrow) for chromosomal effects of chemicals from treated rodents is possible. Chromosomal aberrations or SCEs in dividing bone marrow cells or micronuclei in polychromatic erythrocytes are thus evaluated. Usually much higher concentrations of the test chemical are required to produce an effect *in vivo* in comparison to effective *in vitro* conditions. This insensitivity of the *in vivo* system may be due to (a) detoxification of the active metabolite of the test chemical (b) short-lived metabolites not reaching the bone marrow or (c) the relative insensitivity of bone marrow cells. Alternative end-points such as micronuclei in hepatocytes can be of relevance, because most activation occurs in the liver. Lymphocytes or splenocytes from treated animals can also be used to evaluate the mutagenic activity of test chemicals. Because most lymphocytes are dormant cells (G0), they are relatively insensitive to mutagenic damage.

The advantage of using *in vitro* systems is that higher doses can be employed for a treatment, and target cells are in direct contact with the test chemicals. Because high concentrations of test chemicals can be used, however, there is a danger of generating false positive results. High concentrations of test chemicals can lead to extreme changes in the pH and osmolality of the treatment solution, which by themselves can induce chromosomal damage. Similarly, high concentrations of solvents (e.g., DMSO) in the treatment solution or in the S9 fraction with the necessary cofactors may also influence the ultimate result. In view of the false positives generated in *in vitro* chromosomal aberration tests, there is growing scepticism about the utility of this test system in routine screening of chemicals for their mutagenic ability. However, if the criteria that have been described (i.e., use of concentrations of chemicals which are not highly cytotoxic, strict control of osmolality and pH of the treatment solution) are adhered to, then the *in vitro* mammalian cell culture system can be very valuable in mutagenicity testing.

IX. Conclusions

A dose-dependent increase in the frequency of chromosomal aberrations or micronuclei should be taken as a positive indication of the mutagenic activity of a test chemical. A dose-dependent increase of the frequencies of SCEs indicates that a test chemical induces DNA damage and may be mutagenic. This pattern holds true for all of the highly reactive genotoxic chemicals. However, in practice when chemicals or drugs with unknown properties are tested one may encounter situations where positive effects are observed only at high concentrations without a linear dose–response. In such cases, although the test chemical may appear to be positive, it is generally considered to be negative in view of the absence of a dose–response relationship.

Related Articles: BIODEGRADATION OF XENOBIOTIC CHEMICALS; ETHANOL FUEL TOXICITY; PLANTS AS DETECTORS OF ATMOSPHERIC MUTAGENS.

Bibliography

Ashby, J., de Serres, F. J., Draper, M., Ishidate, M., Jr., Margolin, B. H., Matter, B. E., and Shelby, M. D., eds. (1985). Evaluation of short-term tests for carcinogens. *Progr. Mut. Res.* **5,** Elsevier, Amsterdam.

Ishidate, M., Jr. (1988). "Data Book of Chromosomal Aberration Test *in Vitro*." Elsevier, Amsterdam.

Kilbey, B. J., Legator, M., Nichols, W., and Ramel, C., eds. (1984). "Handbook of Mutagenicity Test Procedures." Elsevier, Amsterdam.

Lewin, B. (1990). "Genes IV." Oxford University Press, Oxford, New York.

Obe, G., ed. (1990). "Advances in Mutagenesis Research," vol. 2. Springer-Verlag, Berlin.

Obe, G., and Natarajan, A. T., eds. (1990). "Chromosomal Aberrations, Basic and Applied Aspects." Springer-Verlag, Berlin.

Sumner, A. T. (1990). "Chromosome Banding." Unwin Hyman, London.

Tice, R., and Hollaender A., eds. (1984). "Sister Chromatid Exchanges." Plenum, New York.

van Holde, K. E. (1989). "Chromatin." Springer-Verlag, Berlin.

Natural Anticarcinogens and Mechanisms of Cancer

Devra Lee Davis
U.S. Department of Health and Human Services

H. Babich
Stern College, Yeshiva University

Glossary

Anticarcinogen Substance that inhibits the development of cancer, or demotes or causes the regression of initiated tumors.

Case-control (or retrospective) Human epidemiologic studies that compare cases with a given disease to those without such a disease, in order to determine likely causal factors.

Cross-sectional Human epidemiologic studies that make a "snapshot" assessment of the health and exposure status of a population.

In vitro Experimental studies involving cells in culture, bacteria, and other biological systems, but not whole animals.

In vivo Experimental studies involving whole animals.

Mixed function oxidase Microsomal enzyme system that metabolizes drugs and xenobiotics.

Prospective Human epidemiologic studies that follow large numbers of people over time in order to discern links between patterns of exposure and disease.

Tumor demotion (tumor regression) Process by which compounds directly impede or reverse initiated, neoplastically transformed cells.

Tumor initiation Process that begins neoplastic transformation of a cell and can involve binding of a carcinogen to DNA.

Tumor promotion Process by which tumors evolve from initiated cells.

***THE ROLE** which naturally occurring anticarcinogens in food may play in the development of human cancer has been studied by employing epidemiologic and toxicologic methodology. Recent trends in cancer need to be assessed in light of changes in environmental factors broadly conceived within the past two or more decades. The human diet in developed countries generally includes a number of naturally occurring, powerful anticarcinogens whose role remains elusive and inadequately assessed. Substances that occur naturally in food may influence cancer initiation, promotion, progression, and demotion, or regression, by a number of mechanisms, including (1) covalent binding with DNA of naturally occurring anticarcinogenic compounds, which effectively blocks the initiation of carcinogenesis; (2) modulation of biotransforming enzymes such as the cytochrome P450 mixed function oxidase (MFO) system, which can enhance or reduce carcinogenicity, and of detoxication enzymes, such as glutathione S-transferase; (3) inhibition of tumor promotion and progression; (4) physicochemical alteration of carcinogens by components of food or by food preparation and handling; and (5) demotion, or regression, of initiated*

This article is expanded from Davis, D. L. (1989). Natural anticarcinogens, carcinogens, and changing patterns in cancer: Some speculation. *Environ. Res.* **50,** 322–339. Copyright © 1989 by Academic Press, Inc., with permission.

and promoted tumors by compounds that reverse or suppress tumor growth factors. Common mixtures in food need to be tested both in vitro *with human cells in culture and* in vivo *with long-term rodent bioassays, focusing the studies on the potential ameolioration of the potencies of synthetic organic carcinogens by naturally occurring anticarcinogens.*

I. Introduction

This article expands on studies that evaluated experimental toxicologic and epidemiologic evidence on the impact on human cancer of synthetic and natural carcinogens. The question of the role in cancer causation of synthetic materials versus naturally occurring ones has stimulated a great deal of rhetoric and speculation, but definitive data remain scanty. In part, this is because of the complex nature of the evidence that needs to be considered. Ultimately, inferences on the role of naturally occurring anticarcinogens and carcinogens will need to be drawn from a variety of experimental and observational sources.

To estimate the impact on humans of cancer-causing agents and factors, two general types of evidence can be assessed: (1) experimental toxicological studies that administer generally high doses of test compounds to animals under controlled conditions and (2) inferential, epidemiologic studies that customarily ask subjects what they have been exposed to (*ex post facto*) or that sometimes are able to use frozen samples of tissue and blood that can be analyzed for residues which are indications of past exposure.

In addition to direct-acting carcinogens, some which come from naturally occurring sources and others which derive from added synthetic materials and contaminants, such as pesticides, the human diet includes powerful natural anticarcinogens. The role of these substances in the causation and prevention of human cancer patterns remains elusive and inadequately explored. Indeed, the relative role of dietary and other long-established causes of human cancer compared to more recent factors remains a subject of considerable debate. Throughout human evolution, the human diet has consisted of a number of naturally occurring, long-established toxins and carcinogens. Some have argued that these naturally occurring carcinogens could be responsible for the development of many types of human cancer. Others point to an array of evidence that newer factors probably explain many of the recent changes in cancer in industrial countries. This evidence includes the variety of anticarcinogens and other beneficial compounds within the human diet, the lack of studies linking food to major cancer patterns in modern countries; the exponential growth in the consumption of drugs, tobacco, and industrial chemicals; and relatively recent changes in cancer patterns.

II. Trends in Cancer in Industrial Countries

Cancer is a complex set of more than 200 diseases with multiple causes and multiple stages. To a large extent, current cancer patterns reflect genetic factors and exposures from past decades, including such environmental factors as cigarette smoking, radiation, industrial pollutants, occupational exposures, pesticides, diet, microorganisms and viruses, pharmaceutical agents, and other lifestyle factors. For a number of cancer sites, new immigrants tend to acquire the cancer rates of their adopted countries; worldwide, many cancers vary by a factor of ten or greater.

Throughout the developed world, stomach cancer continues to decline at a rapid rate, possibly reflecting improved handling and preservation of food and other widespread changes in diet, including reduced use of salts and year-round consumption of green–yellow vegetables and fruit. In much of the developing world, stomach cancer rates remain high, although rates are also beginning to decline.

In the United States since 1950, overall cancer incidence (with the exception of lung cancer) rose 29%, adjusted for changes in age distribution. Incidence for all sites rose 43.5% during this same time. Recent lung cancer increases in many industrial countries in considerable part reflect cigarette smoking practices; workplace asbestos exposures are also relevant. However, perhaps 15 to 20% of lung cancers are not known to be related to smoking or asbestos, and these have doubled in the United States, Italy, and Japan in the past two decades. Among the nonsmoking factors associated with increased lung cancer risk are increased ownership and exposure to pet birds; exposure to indoor radon; passive smoking for spouses of smokers; and, for nonsmoking females in developing countries such as China, indoor exposure to high levels of combustion products.

In the past two decades, patterns and types of cancers at a number of sites not plausibly linked with cigarette smoking practices have changed remarkably in older persons in some industrial countries. Death rates from multiple myeloma, kidney, and brain cancers have significantly increased in persons over age 54. These increases in prevalent types of cancer do not merely reflect increased longevity, as they are adjusted for the longer lifespan and increased average age of the population. Thus, while the size of the population of men ages 75 to 84 in the United States increased one-third from 1969 to 1986, their brain cancer and other nervous system death rates increased more than two times. For West Germany and Japan, the proportion of this population increased by 0.5 and 1.3, respectively, while their respective brain cancer and other nervous system death rates increased more than nine times and more than seven times. Over the same time period, women ages 75 to 84 experienced even greater rates of increase in their recorded death rates. The significance of these recorded increases must be carefully assessed in light of cultural or diagnostic factors, including improvements in the quality of ascertainment. However, diagnostic factors do not appear to be the sole explanation for these increases.

In addition to these increases in cancers in the elderly, all age groups in the United States since 1950 have experienced important changes in a number of other cancers. These changes include impressive reductions in mortality for those cancers for which treatment can be effective, including childhood, colorectal, bladder, cervix, testis cancers and Hodgkins disease, and continuing increases in mortality for multiple myeloma, malignant melanoma, and kidney, brain, and nervous system cancers (Table I).

Table I Cancer Incidence and Mortality by Site: 1950–1987[a]

Site	Incidence	Mortality
Stomach	−73	−76
Colon/rectum	11	−24
Lung and bronchus	263	240
Breast	57	4
Bladder	54	−36
Testis	96	−65
Childhood	21	−60
Melanoma	303	150
Kidney and renal pelvis	102	29
Non-Hodgkins lymphoma	154	102
All cancer	44	3
All cancer minus lung	29	19

Sources: National Cancer Institute (1991). "Annual Report on Cancer Statistics"; and American Cancer Society (1988). "Cancer Facts and Figures."

[a] Age-adjusted percentage change for all U.S. white males and females.

III. Epidemiologic Evidence on Anticarcinogens

Epidemiologic studies have identified a number of dietary factors that reduce the risk of cancer and appear to act as anticarcinogens, including antioxidants and fiber. Persons who eat diets low in β-carotene and provitamin A have an increased risk of developing cancer of the lung, according to several studies in the developed world. Cigarette smokers have lower levels of β-carotene than nonsmokers even when they consume comparable levels of dietary β-carotene; thus, cigarette smoking itself may reduce blood carotene levels. One case-control study in Hawaii found that persons consuming all vegetables (dark green vegetables, cruciferous vegetables, and tomatoes) had significantly reduced rates of lung cancer. Another case-control study found that lung cancer patients who were smokers had significantly lower levels of β-carotene in their serum, as did their family members.

Prospective and cross-sectional studies of populations with differing patterns of consumption of foods containing antioxidant vitamins have also found that consuming foods high in vitamin A protects against a number of gastrointestinal cancers, including esophageal cancer, the most prevalent cancer for much of China. A link was found between esophageal epithelial precancerous lesions and the amount of nitrosamines present in the blood of persons living in Lin-Xian, China. Residents of this area eat few fresh fruits and vegetables and are at higher risk for esophageal and stomach cancer than their compatriots who eat similar amounts of foods high in levels of NOC precursors and other compounds that are carcinogenic, but who also eat more of the protective foods. As a consequence of this finding, intervention trials are underway in China to determine whether vitamin supplementation may inhibit the development of esophageal tumors. A related intervention trial in India with quid-chewing fishermen found remarkable reductions in the frequency of oral dysplasias following a 6-month trial of vitamin A treatment. Vitamin A produced complete remission of leukoplakias in 57% of the men, and in nearly all the men it reduced micronucleated cells, which are more readily promoted to cancerous cells.

In a series of case-control studies of residents of northern Italy, a high green vegetable consumption rate was associated with a substantial reduction for epithelial cancers (e.g., esophagus, liver, larynx, and breast). Furthermore, a strong inverse relationship was noted for cancers of the upper digestive and respiratory tract and high consumption of fruit.

Cross-sectional, prospective, and case-control epidemiologic studies also indicate that individuals who consume a higher fiber diet have a reduced risk of colon cancer, and that persons with precancerous polyps can reduce their risk by consuming a high-fiber diet. Compared to omnivorous eaters, vegetarians (e.g., Seventh Day Adventists) also have lower rates of many cancers. Increased fiber intake in healthy individuals decreases the concentration of secondary bile acids and reduces the production of fecal mutagens. Fiber composition varies considerably between foods. In general, high intakes of insoluble fibers increase fecal bulk and weight, which decreases transit time and thereby dilutes bile acids that may promote colon carcinogenesis. Regarding other types of cancer, fiber also appears to be protective against breast cancer, although additional studies are underway to further test that association.

Two successful collaborations of joint U.S.–Chinese epidemiologic and biochemical research provide vital evidence linking diet, host status, and cancer patterns. A case-control study in Shanxshi Province found that persons who regularly consumed larger amounts of *Allium*-containing vegetables such as garlic and onions, which are rich in sulfhydryl groups and selenium, had markedly lower risks of contracting stomach cancer. Observing more than 30,000 person-years of men in southern Guangxi, China, F. S. Yeh and colleagues found that half of all the deaths were due to primary hepatocellular carcinoma, and that estimated exposure to naturally occurring aflatoxin-producing mold and hepatitis B virus explained much of the variation in cancer patterns in this disease.

In the developing world, differing patterns of food consumption and infectious disease are associated with differences in some cancer patterns. For instance, betel nut chewing is directly linked with a high incidence of oral cancer in central and southeast Asian countries. In the developing world, high rates of nasopharyngeal carcinoma, liver cancer, and esophageal cancer are also linked with greater multiple exposures to infectious agents, nutritional deficits, food preservatives, and other environmental factors.

IV. Toxicologic Evidence on Anticarcinogens

Before they can exert their toxic effect, many environmentally persistent carcinogens must undergo biotransformation to reactive intermediates or otherwise be activated metabolically. As a heuristic device, the stages of cancer development can be considered to include initiation of target, somatic cells; promotion of these cells; and progression and further growth. Tumor demotion or regression can also occur, wherein initiated cells have created tumors that regress following exposure to naturally occurring anticarcinogens or treatment with chemotherapy agents. Certain anticarcinogenic compounds can turn off the proliferative process of cancer—a process that can be described as demotion of initiated cells. At each step a number of complex processes occur, reflecting a balance between opposing factors such as metabolic activation and detoxication, formation of electrophilic derivatives and blocking by nucleophiles, and DNA damage and repair.

A variety of systems have been proposed for classifying the ways in which compounds can influence carcinogenicity. These include (1) covalent binding of a competing compound with DNA to block or otherwise influence the initiation of carcinogenesis by the toxic compound; (2) trapping or inactivating compounds that affect tumor promotion; (3) biotransformation of the parent compounds—either metabolized by Phase I reactions, which include oxidation, reduction, and hydrolysis and which are catalyzed by a variety of enzymes, in particular, the cytochrome *P*450–dependent monooxygenase enzymes, or by Phase II conjugation reactions, which can enhance or minimize toxicity by adding functional groups to the molecule; (4) chemical or physical conversion of metabolites by active catalysis and physical alteration of carcinogens through food preparation or digestive processes; and (5) chemically induced regression (or demotion) of tumor growth by active interference with tumor metabolism.

The number of mechanisms that may be involved in inhibiting mutagenesis and carcinogenesis is considerably greater than the few discussed here. Inhibitors may act extracellularly (by inhibiting the uptake of carcinogens or their metabolites), intracellularly (by blocking reactive molecules), or by other multiple and simultaneous mechanisms that modify tumor progression.

Table II lists a number of compounds with different anticarcinogenic properties, the commonly consumed foods that contain them, and the carcinogens against which they are effective in experimental systems. Where a cancer site is indicated, the support of human epidemiologic studies is signified.

A. Modulation of Xenobiotic Metabolism

The majority of carcinogenic chemicals do not produce their detrimental effects by themselves. Such chemicals, termed *procarcinogens*, may be involved in oxidative bioactivation reactions, principally by the microsomal cytochrome *P*450–dependent monooxygenase system, and thereby transformed to electrophilic intermediates that are reactive in adduct formation with nucleic acids and proteins. Since the manifestation of carcinogenic effects by these procarcinogens is dependent on their biotransformation to activated metabolites, dietary modifications that alter these xenobiotic oxidations would lead to a decrease of the carcinogenic effect.

For example, the ingestion of broccoli, brussels sprouts, and other cruciferous vegetables modified some *P*450 activities in small intestine, colon, and liver tissues of rodents, with such alterations presumably accounting for the decreased risk of chemically induced cancer in animals fed a broccoli-supplemented diet. H. S. Nyandieka and colleagues evaluated the influence of various nutritional factors on the modulation of aflatoxin B_1 (AFB_1)-induced liver tumors in rats. Whereas β-carotene completely inhibited the development of liver cancer, ascorbic acid (vitamin C), selenium, and uric acid reduced the percentages of tumor-bearing rats from 80% (without anticarcinogen) to 13% each. Reduced glutathione and vitamin E (α-tocopherol) decreased the percentage of tumor-bearing rats to 20 and 40%, respectively. The researchers concluded that this inhibition of AFB_1-induced liver cancer development occurred through microsomal induction and AFB1 activation. Green and yellow vegetables from Cruciferae plants contain a variety of nonvolatile indole compounds that stimulate *P*450 enzyme activity. Enhanced *P*450 mixed function oxidase (MFO) activity may be responsible for the protective effect of indole 3-carbinol (I_3C), a component of cruciferous vegetables, noted in several studies of chemical-induced carcinogenesis using the rat model system.

L. W. Wattenberg has recently summarized much of the research on those naturally occurring compounds in the diet that prevent carcinogenesis by inhibiting carcinogen activation reactions. The aromatic isothiocyanates, benzyl isothiocyanate and phenethyl isothiocyanate (PEITC), both found in cruciferous vegetables, were effective in preventing neoplasia by inhibiting the activation of benzo[*a*]pyrene (BaP), 7,12-dimethylbenz[*a*]-anthracene (DMBA), and *N*-nitrosodiethylamine (NDEA). Wattenberg noted earlier that pretreatment of mice with benzyl isothiocyanate inhibited the induction of forestomach tumors by diethylnitrosamine (DEN). DMBA activation was also inhibited by the indoles I_3C and indoleacetylnitrile, both components of cruciferous vegetables. The organosulfur compounds, diallyl disulfide, allyl mercaptan, and allyl methyl disulfide, which are found in *Allium* species, including garlic, onions, leeks, and shallots, were effective against bioactivation of NDEA. The monoterpenes, D-limonene found in citrus fruit oil and D-carvone found in caraway seed oil, inhibit bioactivation of NDEA. Induction of neoplasia by the tobacco-specific carcinogen, 4-(methylnitrosamino)-1-(3-pyridyl)-1-butanone (NNK) in mice was reduced by dipropyldisulfide and D-limonene, presumably due to their inhibition of the bioactivation of NNK. Bioactivation of NNK was also blocked by 3-phenylpropyl isothiocyanante and 4-phenylbutyl isothiocyanate. The metabolism of NNK and *N*′-nitrosonornicotine (NNN), the only tobacco constituents which induce oral cancer in an animal model, was reduced in rats fed a diet supplemented with PEITC. These findings prompted the investigators to suggest the possibility of PEITC being useful as a chemopreventive agent for cancer of the oral cavity.

Another effect of dietary supplements on procarcinogen metabolism is the induction of detoxication enzymes. For example, G. M. Alink and colleagues reported that aryl isothiocyanates, common

Table II Anticarcinogens in Foods[a]

Mechanisms and compounds	Common foods	Anticarcinogenic activity
	Inhibit covalent DNA binding	
Phenethyl isothiocyanate	Broccoli, cabbage (*Brassica*)	Tobacco-specific nitrosamine
Ellagic acid	Fruits, nuts, berries, seeds	Benzo(*a*)pyrene
	Vegetables	Aflatoxin B_1 *N*-Methyl-*N*′-nitro-*N*-nitrosourea *N*-Methyl-*N*′-nitro-*N*-nitrosoguanidine 3-Methylcholanthrene *N*-Methyl nitrosourea
Flavonoids and Polyphenolic acids (morin, myricetin, and quercetin)	Fruits and vegetables	Carcinogens in cooked foods (heterocyclic amines)
	Inhibit tumor promotion	
Retinol	Green and yellow vegetables, fruits	*N*-nitroso compounds
Tocopherol	Nuts, wheat germ	Dimethylhydrazine Nitrosamines
Ascorbic acid	Fruits, vegetables	Nitrosation
β-Carotene	Green and yellow vegetables, fruits	Lung cancer
	Vegetables	Lung cancer
	Inhibit tumor formation	
Organosulfur compounds	Garlic, onions	Stomach cancer Benzo(*a*)pyrene Dimethylhydrazine
Curcumin	Tumeric/curry	12-*O*-tetradecanoylphorbol-13 acetate
Capsaicin	Chili peppers	Benzo(*a*)pyrene constituents
	Induce biotransformation	
Indole 3-carbinol	Cabbage, brussels sprouts, cauliflower, spinach, broccoli	Benzo(*a*)pyrene, Aflatoxin B_1
Selenium	Seafood, garlic	Oxygen radical-induced peroxidation, mercury
	Reduce absorption/adsorption of carcinogens	
Fiber	Fruits, vegetables Grains, nuts	Fecal mutagens Heterocyclic amines in cooked foods
Cooking/washing	Grains, nuts, mushrooms	Cycasin, hydrazine
Riboflavin chlorophyllin	Vegetables, fruits	Cigarette smoke condensate, benzo(*a*)pyrene

[a] Data compiled from numerous studies.

in cruciferous vegetables, induce glutathione *S*-transferase, an enzyme which conjugates electrophilic metabolites of carcinogenic compounds, such as those of BaP, with glutathione. V. L. Sparnins and colleagues noted that four allyl-containing derivatives of garlic and onion (i.e., allyl methyl trisulfide, allyl methyl disulfide, diallyl trisulfide, and diallyl sulfide) inhibit BaP-induced cancer of the forestomach of mice, presumably by their induction of glutathione *S*-transferase. A. K. Maurya and S. V. Singh concluded from their studies of diallyl sulfide in mice that this organosulfur compound exerts its prime anticarcinogenic effect by modulating glutathione-dependent detoxication enzymes, such as glutathione *S*-transferase. The modulation of MFO activity, as well as the induction of glutathione *S*-transferase, by naturally occurring anticarcinogens, such as organosulfur compounds and indoles, may provide the primary mechanism by which they inhibit chemically induced carcinogenesis. Consumption of cruciferous vegetables could appreciably reduce or mitigate the adverse effects of synthetic carcinogens consumed during the same meal.

In both short- and long-term animal studies and in some human studies, a variety of antioxidant vitamins and other scavengers have been shown to modulate biotransformations of the parent compound and to thereby interfere with carcinogenesis. Ascorbic acid prevents the biotransformation of nitrosamines and other carcinogenic *N*-nitoso compounds and inhibits skin, nose, tracheal, lung, and kidney carcinogenesis. Tocopherol (vitamin E) also inhibits nitrosation and protects cells from chromate cytotoxicity. Rats exposed to both vitamin E and 2-acetylaminofluorene (AAF), a potent experimental carcinogen, developed significantly less liver microsomal enzymes and reduced rates of liver cancer. A recent comprehensive review cites 37 studies with experimental animals and 13 clinical studies with humans that show either ascorbic acid, some phenolics and polyphenolics, and tocopherol as inhibitors of endogenous *N*-nitrosation.

Dietary plant phenols, such as tannic acid, quercetin, myricetin, and anthraflavic acid, inhibit polycyclic aromatic hydrocarbon (PAH) metabolism and subsequent PAH–DNA adduct formation in the epidermis of mice. These plant phenolics also afforded protection against skin tumorigenicity, by delaying the onset and the subsequent development of DMBA-, BaP-, 3-methylcholanthrene-, and *N*-methyl-*N*-nitrosourea(MNU)-induced skin tumors.

B. DNA Binding and Tumor Initiation

As the basic framework for all living matter, DNA undergoes constant assault and repair. A number of experimental carcinogens, such as BaP and a host of nitrosamine compounds can directly adduct with the basic building-block nucleotides of DNA, forming strong covalent, carbon–carbon bonds. Vegetables, fruits, and seasonings contain a host of potent naturally occurring anticarcinogens which can inhibit this process. For example, DNA binding by AFB1 was decreased by ajoene and diallyl sulfide, both organosulfur compounds common in garlic, and by a crude garlic extract. Regardless of the route of exposure, dietary I_3C decreased DNA binding with AFB_1 in rats. With some exceptions, the anticarcinogenic effect prevailed. The inhibition of DMBA-induced mammary tumorigenesis in rats fed dietary selenium was correlated with reduced DMBA–DNA adduct formation.

Ellagic acid (EA) is a naturally occurring constituent of many fruits, vegetables, and nuts that also blocks DNA binding, and hence initiation, of several experimental carcinogens. EA may be highly biologically active and efficiently metabolized, although this has not yet been demonstrated in humans or animals *in vivo*. EA inhibits the metabolism of the carcinogen *N*-nitrosobenzylmethylamine (NBMA) and impedes DNA binding in cultured rat esophagus and reduces azoxymethane-induced colon carcinogenesis in rats. Further, EA inhibits AFB_1 mutagenesis in a bacterial assay and produces dose-dependent inhibition in the covalent binding of AFB_1 to DNA with cell cultures of both the rat trachea and human tracheobronchus and thus has the potential to act as a naturally occurring inhibitor of AFB_1.

Many studies have shown that natural anticarcinogens inhibit the clastogenic effects of carcinogens. For example, diallyl sulfide reduced MNNG-induced nuclear aberrations in cells from the glandular mucosa of rats. β-Carotene reduced the clastogenicity, as scored by chromosomal aberrations and micronuclei of erythrocytes, of cyclophosphamide in mice. DNA single-strand breaks induced by NNK were reduced in hepatocytes from rats fed a vitamin A supplemented diet as compared to hepatocytes from rats fed a vitamin A deficient diet. Analyses of cytochrome *P*450 activity suggested that the protective ability of vitamin A was related to its inhibition of the enzymes involved in the bioactivation of NNK.

For a number of carcinogens such as BaP, MNU, and NBMA, EA inhibited *in vitro* mutagenesis in human cells. It forms adducts with the carcinogenic and mutagenic metabolite of BaP, BPDE-II, thereby blocking cytotoxicity and mutagenicity of BaP. EA inhibited the formation of 6-*O*-methylguanine adducts by *N*-methyl-*N'*-nitro-*N*-nitrosoguanidine (MNNG) in human cells. Alkylation of the O-6 position of guanine in DNA correlates with mutagenesis in bacterial and mammalian cells. EA also inhibited the direct-acting mutagen MNU, by preventing methylation at the O-6 position of guanine through an EA–guanine affinity binding mechanism.

Flavonoids are ubiquitous in vascular plants higher than ferns, in which they impart a yellow or orange color. Investigators have suggested that the flavonoids in extracts of some plants have anti-inflammatory and other beneficial biological effects, including antioxidant properties, superoxide radical scavenging actions, and covalent binding with DNA. In some test systems, flavonoids inhibit the initiation by BaP and AFB_1 or otherwise produce anticarcinogenic activity. Dietary rutin and quercetin, phenolic flavonoids commonly found in many fruits and vegetables, inhibit azoxymethanol-induced colonic neoplasia in rats. These flavonoids reduced both azoxymethanol-induced hyperproliferation of colonic epithelial cells and the incidence of azoxymethanol-induced focal areas of dysplasia. In a bacterial assay, three plant flavonoids (morin, myricetin, and quercetin) inhibited the genotoxic effects of a number of potent mutagens that occur when meat or fish are cooked, such as imidazoquinoline (IQ), 2-amino-3,8-dimethylimidazo[4,5-*f*]quinosaline (MeIQx), and the tryptophan metabolites trp-*P*-1 and trp-*P*-2. In related work with rats, diets of anticarcinogen-containing fruits and vegetables (diets also included the heated meats that produce many of the above-noted heterocyclic amines) effectively blocked the action of known carcinogens such as BaP and NDMA. Capsaicin, the active principle of hot peppers, also inhibits the binding to DNA of BaP and AFB_1. Retinol (in vitamin A) can inhibit the mutagenicity of pyrolysis products of some common heterocyclic amino acid metabolites, such as trp-*P*-1, trp-*P*-2, glu-*P*-1, and glu-*P*-2, possibly by inhibiting their metabolic activation.

C. Inhibition of Tumor Promotion and Progression

Dietary calcium may play an important role in modulating carcinogenesis. There is evidence from experimental animal studies that calcium counteracts the promotion of colon cancer by dietary fat. At the basic pH of the colon, fatty acids and bile acids become highly surfactant, causing cell loss and compensatory cell hyperproliferation. Apparently, calcium reduces lipid damage in the colon by complexing with fat. Furthermore, in a study with human beings, the colonic hyperproliferation associated with increased colon cancer was reversed for short periods with administration of supplemental dietary calcium. Low amounts of dietary calcium and vitamin D increase the susceptibility of rats to DMBA-induced primary mammary neoplasia. Cell hyperproliferation, noted during the induction of colonic cancer by MNNG in rats on a low calcium diet, was reduced by a calcium-enriched diet regime.

One study of colonic tumors using mice injected with 1,2-dimethylhydrazine (DMH) found significantly more cancers in the mice receiving a low vitamin E diet, in contrast to those receiving higher supplementation.

D-Glucaric acid and D-glucaro-1,4-lactone have potent antiproliferative properties *in vivo* and are inhibitors of β-glucuronidase, whose level of activity has been positively correlated with susceptibility to chemically induced carcinogenesis. Detoxication of nucleophilic metabolites of chemical carcinogens may occur by UDP-glucuronosyltransferase. Glucuronidation is a principal conjugation pathway in vertebrate tissues; carcinogens which, after metabolic activation, are subject to glucuronidation include PAHs, some nitrosamines, aromatic amines, and fungal toxins. The elimination and/or deactivation of potentially damaging carcinogens which undergo glucuronidation is limited not only by the rate of conjugation with glucuronic acid but also by the rate of deglucuronidation by β-glucuronidase. D-Glucaric acid and D-glucaro-1,4-lactone are found in some citrus fruits and cruciferous vegetables. Dietary D-glucaric acid inhibited the chemical carcinogen-induced tumorigenesis of the mammary gland, colon, liver, skin, and kidney.

As indicated in Table II, a number of widely used condiments also inhibit tumor promotion in animals and are associated with reduced risks of some cancers in humans. Garlic and onions are *Allium* vegetables that are rich in sulfhydryl-containing compounds that inhibit *in vivo* carcinogenesis in mice. Garlic is also one of the richest sources of selenium, which has anticarcinogenic properties. Diallyl sulfide, a natural thioether in garlic, reduced the amount of colorectal adenocarcinoma in mice exposed to DMH and decreased the carcinogenic, oxidative metabolism in male rats of potent nitrosamines such as NNK. A single 5-mg dose of garlic oil (which contains diallyl sulfide) maximally inhibited skin carcinogenesis in mice exposed to the carcinogen DMBA. Allyl sulfide and I_3C significantly inhibited the development in rats of chemically induced liver cancer. E. Bresnick and colleagues observed a reduction in MNU-induced mammary cancer in rats fed a diet rich in cabbage. Similar reductions in mammary carcinogenesis were also noted in rats administered a residue obtained from cabbage. L. Santamaria and colleagues demonstrated that dietary β-carotene inhibited the progression to early and infiltrating carcinoma in the stomach of rats treated with MNNG.

Experimental work has identified a number of other condiments with potential protective effects against tumor promotion. Turmeric, a common spice in Indian cooking, contains curcumin, which has antimutagenic properties in a number of *in vitro* test systems. Glandular stomach carcinogenesis in rats, induced by MNNG, was suppressed by the administration of wasabi, a pungent spice used in Japanese meals.

Research on food technology also indicates that essential oils derived from thyme, cumin, clove, caraway, rosemary, and sage effectively inhibit the growth of *Aspergillus parasiticus* and the production of aflatoxin.

Dithiolethiones are abundant in cruciferous vegetables, the consumption of which has been associated with reducing the incidence of human cancers. The incidence of azoxymethane-induced adenocarcinomas in colon and small intestine and the multiplicity of colon adenomas and small intestinal adenocarcinomas were significantly inhibited in rats administered dietary olitpraz, a substituted dithiolethione.

Consumption of miso (Japanese soybean paste) may be a factor contributing to the lower incidence of breast cancer noted in Japanese women. A diet supplemented with miso reduced the incidence and delayed the appearance of DMBA-induced mammary tumors in rats. Rats fed the miso-supplemented diet showed trends toward a lower number of cancers per animal, a higher number of benign tumors per animal, and a lower growth rate of cancers compared to controls.

Rats fed a diet containing brassica vegetables developed enzymes in their intestines [benzopyrene hydroxylase (BPOH)] that inhibited the carcinogenic metabolism of BaP. This suggests that the inhibition of BaP carcinogenesis in the rat stomach is associated with enzyme-inducing compounds in brassica (Wattenberg, 1971). The hydrolysis products of glucobrassicin, including I_3C, indole 3-acetonitrile, diindolys 1-methane, and ascorigen (the *in situ* reaction production of I_3C and vitamin C), all induce BPOH activity in rat liver and intestine, with I_3C being the most potent inducer.

G. K. Johansson and colleagues evaluated the effects of a shift from a mixed diet to a lactovegetarian diet on some cancer-associated bacterial enzymes in human feces. Three months after the shift to the lactovegetarian diet, there was a decrease in certain bacterial enzyme activities proposed to be risk factors for colon cancer.

In a number of animal studies, I_3C directly inhibited carcinogenesis from naturally occurring carcinogens such as polynuclear aromatic hydrocarbons and AFB_1. Staging of exposure may be critical. A high dose of I_3C given before and during AFB_1 exposure reduced the tumor incidence in trout fingerlings started on diets 19 weeks post hatch. However, postinitiation exposure to high doses of I_3C increased tumors, although the numbers studied were not sufficient to confirm this effect. Several researchers are pursuing studies of other dosing regimens and exposure scenarios to investigate these effects further.

D. Physicochemical Alteration of Carcinogens

Methods of food preparation can alter the toxicity of ingested materials. Some toxins are destroyed through cooking or washing. For instance, properly washed cycad nuts do not contain the carcinogenic component cycasin. Pyrrolizidine alkaloids are also detoxified by hydrolysis. Agaritine, a naturally occurring hydrazine in mushrooms (*Agaricus bisporus*) is rather unstable and can be removed nearly completely by cooking. Two of its metabolites are carcinogenic but, like agaritine, are destroyed by proper preparation before eating. The potato, *Solanum tuberosum* L., generally contains insignificant amounts of toxic steroidal alkaloids (a mixture of α-solanine and α-chaconine). Exposure to light, length of storage, and mechanical damage will increase toxicity. Potatoes also contain an endogenous enzyme capable of hydrolyzing the glycoalkaloids in a specific manner, although their metabolism is not well understood. Cooking alters and often decreases or eliminates many of the toxins (e.g., anti-enzymes, glycosides, hemagglutins, tannins, and alkaloids), bacteria, molds, and yeasts in such common classes of foods as grains, legumes, and many buried foods such as tubers.

Not all food preparation reduces toxins. Charcoal broiling foods produces BaP and other polynuclear hydrocarbon residues on the foods. Cooking can also produce endogenous quantities of potent bacterial mutagens in some foods, such as heterocyclic amines in cooked animal or fish muscles. However, mixing 10% soy protein with ground meat prior to frying prevents the formation of these mutagenic heterocyclic amines.

Physical properties of food also affect toxicity. For instance, vegetable fibers may influence toxicity through a number of different mechanisms. Ingested fibers physically adsorb heterocyclic amines and inhibit their mutagenicity. Also, diets high in fiber may move through the body more rapidly with a reduced transit time and thereby less exposure and opportunity for uptake of fecal mutagens. In addition, fibers may dilute the overall dose of ingested carcinogens. Since 1980, some 32 studies have assessed the role of fiber-containing foods in relation to colon cancer; among these studies, 25 showed an inverse relationship.

Finally, some constituents of food may physically intervene with the process of carcinogenesis. Chlorophyllin and riboflavin may physically convert electrophilic metabolites by acid catalysis or some other process that disrupts carcinogenesis of cigarette smoke condensate in bacterial test systems.

E. Tumor Regression or Demotion

As noted, the transformation of a normal cell into a malignant cell is thought to occur through a multistage process, involving initiation, promotion, and progression. The promotion stage is a very slow process, with the latency period between initiation and the appearance of a malignant tumor spanning decades. It is during this latency period that intervention with anticarcinogens may be most effective. Such chemopreventive agents may cause a premalignant lesion to regress (i.e., tumor demotion) or may slow the neoplastic process so that the malignant tumor does not develop in the lifetime of the individual. Over the past decade, efforts to evaluate naturally occurring anticarcinogens in cancer treatment have expanded; they include trials using β-carotene, folic acid, 13-*cis*-retinoic acid, 4-hydroxyphenyl retinamide, vitamins A, C, and E, and minerals. The premise of these studies is that anticarcinogens can work both to prevent cancer from forming and to regress or demote precarcinogenic lesions that have already formed.

Precancerous lesions are broadly defined as abnormal tissue growth with altered nuclei. Important premalignant lesions known to be associated with increased cancer incidence include oral leukoplakia and erythroplakia associated with cancer of the oral cavity; cervical dysplasia and carcinoma *in situ* associated with cervical carcinoma; colonic polyps associated with colon cancer; and actinic keratosis associated with skin cancer. Many epidemiological studies have utilized remission of precancerous lesions after treatment with anticarcinogens as an indication of tumor regression. For example, H. F. Stich and colleagues evaluated the effects of β-carotene and vitamin A on precancerous lesions in the oral cavity of individuals who chewed tobacco containing betel quids daily, both before and during the study period. Administration of vitamin A for 6 months resulted in complete remission of leukoplakias in 57% and a reduction in micronucleated cells in 96% of the tobacco chewers. β-Carotene induced remission of leukoplakia in 14.8% and reduction of micronucleated cells in 98% of the tobacco chewers. After withdrawal of vitamin A or β-carotene, oral leukoplakias reappeared and the frequency of micronuclei in the oral mucosa increased. V. N. Singh and S. K. Gaby (1991) have recently reviewed the role of antioxidants and β-carotene in the remission of precancerous lesions.

Such epidemiological studies complement the data from toxicological studies with laboratory animals. For example, a series of elegant investigations showed PEITC not only inhibited DNA binding but also caused existing precarcinogenic lesions to regress, effectively producing tumor demotion. PEITC reduced DNA binding to 7-methylguanine by NNK, one of the most potent carcinogenic nitrosamines in tobacco. This naturally occurring isothiocyanate decreased DNA methylation and pyridyloxobutylation by NNK in the lung. PEITC also produced 50% fewer adducts following exposure to NNK in rats. Ultimately, PEITC reduced the number of lung tumors formed in rats exposed over a two-year period to both NNK and PEITC. Rats exposed only to NNK exhibited dose-related amounts of lung tumors, with all those in the group that received the largest exposure developing tumors. Animals that were exposed to both NNK and PEITC showed a dose-related reduction in the rate of tumors found at autopsy. PEITC functioned to demote tumors that would otherwise have developed.

Y. Kise and colleagues observed that dietary administration of selenium to hamsters reduced the induction of cholangiocarcinomas and related precancerous lesions caused by *N′*-nitrosobis (2-oxopropyl) amine.

More recent illustration of the phenomenon of tumor demotion or regression is provided by *in vivo* toxicologic studies of hamster cheek pouch and rat esophagus. Each of these two model test systems provides a well-established method for experimentally evaluating a variety of carcinogens and anticarcinogens. Using the experimental carcinogen DMBA, it has been shown that a combination of liposomes and β-carotene injected locally in oral squamous cell carcinoma of the hamster produces dose-dependent tumor regression or demotion. Studies of the experimental carcinogen NBMA, administered to rats along with the anticarcinogen EA, similarly found a dose-related decrease in the number of tumors that developed in the exposed animals.

Clinical trials of tumor demotion and regression are underway that explore these findings from a toxicological viewpoint. Several trials have found that retinoids reverse oral leukoplakia in humans, but toxicity was associated with the doses used. β-Carotene is being evaluated as an alternative agent, insofar as it also suppresses micronuclei in exfoliated oral mucosal cells.

V. Concluding Remarks

The study of foodborne carcinogens and anticarcinogens provides an important field for continued work. Future studies in this area need to take into account fully the panoply of the human diet, to approximate the mixtures in which food occurs in any toxicologic testing of these compounds, and to include *in vitro* testing using human cell cultures.

As to the relative impact of foodborne and synthetic carcinogens, recent trends in cancer patterns in older persons in industrial countries require a careful search for potential changes in the environment in the past decades that may explain these changes in prevalent types of cancer. Among factors to be explored are changes in the following: exposure to infectious disease; cigarette smoking habits; occupational exposures; diagnostic procedures; food consumption, preparation, and handling, such as refrigeration and preservation; and environmental pollution patterns, including alterations in air pollution, energy production, transportation, electromagnetic radiation, radon, and pet domestication.

Whether changes in food consumption prove to be a major factor accounting for recent cancer patterns remains to be investigated, but appears unlikely in the United States and other industrial countries given the fact that major cancers, such as stomach cancer, are continuing to decline. In some developing countries, such as China, naturally occurring carcinogens play a key role in explaining cancer patterns.

Most experts in paleolithic nutrition agree that natural selection has provided us with nutritional adaptability—the range of early diets extending from protein-rich diets of far northern peoples to the vegetable-laden diets of the African Kalahari. Lactose tolerance provides one illustration of this phenomenon of nutritional adaptation: those societies that have consumed dairy products for a considerable time exhibit higher rates of lactose tolerance. One can also imagine that cooking of food served a number of critical functions in evolution. It allowed people to survive in colder climates and increased their ability to eat animal protein. Living in dark, smokey caves through the harsh winters may also have selected for resistance in future generations to exposures to those hydrocarbons generated by pyrrolysis of foods and cooking fuel. Certainly, the experimental results show that a number of naturally occurring constituents of foods inhibit neoplasia from BaP, one of the most commonly generated products of combustion.

The impact of natural substances, other than tobacco, on current disease patterns in the United States and other industrial countries is unclear. At present, several epidemiological research studies are underway to ascertain the impact of diet on carcinogenesis in humans. Although pure extracts of some dietary constituents have been tested, well-designed *in vitro* assays, using animal and human cell cultures, of common mixtures in food need to be conducted. Whether synthetic organic carcinogens may also be inhibited by naturally occurring anticarcinogens constitutes a critical and complex subject for research.

Related Articles: Diesel Exhaust: Effects on the Respiratory System; Environmental Cancer Risks; Nitric Oxides and Nitrogen Dioxide Toxicology; Oxides of Nitrogen; Ozone Exposure, Respiratory Health Effects; Plants as Detectors of Atmospheric Mutagens.

Acknowledgments

Support for the research reported here came from a sabbatical granted to D. L. Davis by the National Research Council (NRC) of the National Academy of Sciences, with funding from the NRC, National Institute of Environmental Health Sciences, the Mobil Foundation, the Agency for Toxic Substances and Disease Registry, the Bundesgesundheitsamtes, the World Health Organization, the National Center for Toxicologic Research, and the American Petroleum Institute. This paper represents solely the views of the authors and not necessarily those of the sponsoring organizations.

Bibliography

Adami, H. O., Bergstrom, R., Sparen, P., and Baron, J. (1993). Increasing cancer risk in younger birth cohorts in Sweden. *Lancet* **341**, 773–777.

Ames, B. N., Magaw, R., and Gold, L. S. (1987). Ranking possible carcinogenic hazards. *Science* **236**, 271–280.

Ayala, F. J., and Valentine, J. W. (1979). "Evolving: The Theory and Processes of Organic Evolution." Benjamin Cummings, Menlo Park, California.

Hirona, I., ed. (1987). Naturally Occurring Carcinogens of Plant Origin." Kodansha, Tokyo.

Hoel, D. G., Davis, D. L., Miller, A. B., Sondik, E. J., and Swerdlow, A. J. (1992). Trends in cancer mortality in 15 industrialized countries, 1969–86. *J. Natl. Cancer Inst.* **84**, 313–320.

Magnus, K., ed. (1982). "Trends in Cancer Incidence." Hemisphere, Washington, D.C.

Miller, B. A., Ries, L. A. G., Hankey, C. L., and Edwards, B. K. (eds.) (1992). Cancer statistics review. National Cancer Institute, NIH Pub. No. 92–2789.

National Research Council (1982). "Diet, Nutrition, and Cancer." National Academy Press, Washington, D.C.

National Research Council (1986). "Environmental Tobacco Smoke: Measuring Exposures and Assessing Health Effects." National Academy Press, Washington, D.C.

National Research Council (1989). "Diet and Health: Implications for Reducing Chronic Disease Risk." National Academy Press, Washington, D.C.

Shankel, D. M., Hartman, P. E., Kadal, T., and Hollander, A., eds. (1986). "Antimutagenesis and Anticarcinogenesis Mechanisms." Plenum, New York.

U.S. Department of Agriculture (1986). "Composition of Foods: Raw, Processed, Prepared." Handbook No. AH-8. U.S. Department of Agriculture, Washington, D.C.

World Health Organization (1988). "Cancer Incidence in Five Continents." World Health Organization, Geneva.

Nitric Oxide and Nitrogen Dioxide Toxicology

Bruce E. Lehnert
Los Alamos National Laboratory

Glossary

Alveoli Air sacs in the lung where gas exchange occurs.

Bronchiolitis fibrosa obliterans Narrowing of the conducting airways due to the abnormal build-up of granulation tissue.

Methemoglobinemia Condition in which the hemoglobin in erythrocytes is altered by NO_X in a manner that reduces the blood's ability to carry and deliver oxygen to tissue sites.

NO_X Designation used to describe exposure atmospheres that contain mixtures of oxides of nitrogen, especially nitric oxide and nitrogen dioxide.

ppm Parts per million, a unit used to denote the concentration of a gas contaminant in air. At 25°C and 1 atmosphere pressure, ppm NO = (μg NO/m^3)/1880 and ppm NO_2 = (μg NO_2/m^3)/1230.

Pulmonary edema Lung disorder in which fluid moves from the lung's blood vessels into extravascular sites, including the alveoli.

NITRIC OXIDE (NO) AND NITROGEN DIOXIDE (NO_2) *are two gaseous oxides of nitrogen that are toxic to humans when inhaled at excessive mass concentrations. Both NO and NO_2 are generated by natural mechanisms and by a variety of processes associated with human activities. The most hazardous exposures to high concentrations of NO and NO_2 generally occur in occupational settings. The main toxicologic effect of inhaling NO is the formation of methemoglobin in the blood's erythrocytes, which reduces the ability of the blood to carry and deliver oxygen to body tissues. The inhalation of NO_2, on the other hand, causes damage to the respiratory tract. Depending on exposure concentration and duration, such injury can range from mild compromises in pulmonary function to life-threatening pulmonary edema. Bronchiolitis fibrosa obliterans is a further outcome of NO_2 exposure that can occur in the lungs of individuals who survive the pulmonary edematous response.*

I. Characteristic Properties and Occurrence of Oxides of Nitrogen

A. NO_x, the Oxides of Nitrogen

The oxides of nitrogen consist of a large family of interrelated and generally interconvertible compounds (Table I and Fig. 1), with usually varying physicochemical properties (Tables II and III). Some of the oxides of nitrogen, especially the gases nitric oxide (NO) and nitrogen dioxide (NO_2) (Table III), are of particular concern from a toxicological perspective because of their biological reactivity. The total amount of these gases, and

Table I Chemical Forms of Oxides of Nitrogen

Name	Form
Nitric oxide	NO
Nitrogen dioxide	NO_2
Nitrogen trioxide	NO_3
Nitrous oxide	N_2O
Nitrogen peroxide	N_2O_2
Dinitrogen trioxide	N_2O_3
Dinitrogen tetroxide	N_2O_4
Dinitrogen pentoxide	N_2O_5
Nitrous acid	HNO_2
Nitric acid	HNO_3
Metallic (M) inorganic nitrites	MNO_2
Metallic (M) inorganic nitrates	MNO_3
Nitroso and nitroxy compounds	R-NO
Organic nitrates	R-NO_2
Peroxyacl nitrates	R-C-OONO$_2$ ‖ O

sometimes other oxides of nitrogen such as peroxyacetyl nitrates, present in contaminated air is oftentimes summarily referred to as NO_X (pronounced "N-O-X" or "nox") in the parlance of the environmental health sciences, with the usual understanding being that the bulk of NO_X present in typical exposure atmospheres of concern is mainly due to NO_2 and NO.

B. Natural Sources of Atmospheric NO_x

Natural sources of atmospheric NO_X are numerous. Nitrous oxide (N_2O), the most abundant oxide of nitrogen in unpolluted air, and, to a lesser extent, NO and NO_2, arise from the decomposition of nitrogenous compounds (e.g., amino acids, nitrate salts, ammonia) by bacteria during the decay of organic material. In otherwise unpolluted atmospheres, N_2O generated by this process can result in harmless ambient concentrations of up to 0.5 parts per million (ppm). Other natural sources of NO_X production, particularly NO and NO_2 production, include volcanic eruptions, forest fires, nitrogen fixation by lightning, and the breakdown of mineral nitrates and nitrites. Overall, these natural sources have been estimated to account for an average background level of <5 parts per billion (ppb) NO_2 and even lesser concentrations of NO in ambient air.

C. Human Sources of Atmospheric NO_x

The main source of atmospheric NO and NO_2 due to human activities is the thermal combustion of fossil fuels, with the primary combustion product in this case initially being NO (Eq. 1). In general, higher combustion temperatures are associated with higher yields of nitrogen oxides. Most of the thermally generated atmospheric NO in urban areas comes from internal combustion engines, in-

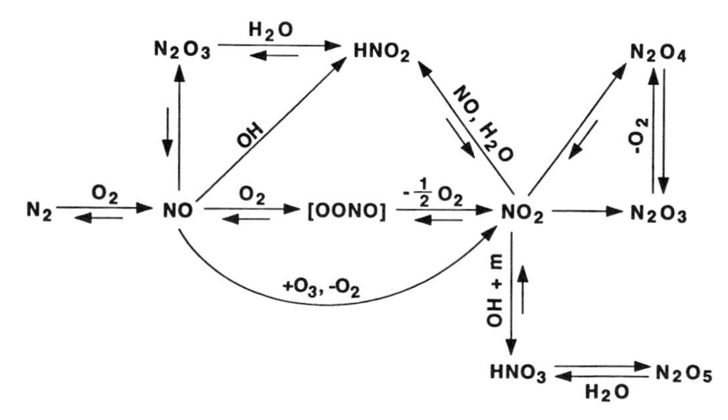

Figure 1 Interconvertibility of the oxides of nitrogen.

Table II Properties of Several Oxides of Nitrogen

Name	Molecular weight	Density (g/c^3)	Melting point (°C)	Boiling point (°C)	Other
Nitric oxide			See Table III		
Nitrogen dioxide			See Table III		
Nitrous oxide	44.01	1.977	−90.8	−88.5	Sweetish odor; supports combustion
Dinitrogen trioxide	76.01	—	−102	Unstable	
Dinitrogen tetroxide	92.02	—	−11.2	21.15	Liquid contains 0.03% NO_2 by volume; at boiling point, vapor contains 16% NO_2; the dissociation of gaseous N_2O_2 to NO_2 increases from 20% at 27°C to ~20% at 100°C
Dinitrogen pentoxide	108.01	1.642[a]	30	47	Colorless hexagonal crystals; sublimes at 32.4°

[a] At 18°C.

cluding diesel engines, although NO and NO_2 can occur locally from industrial settings where they may also be generated (e.g., fertilizer and explosive factories) and where nitric acid is used or made. As shown in Eq. 2, atmospheric NO is subsequently converted to NO_2 by complex photochemical reaction(s) involving oxidized hydrocarbons, especially substituted aromatics and olefins. Another chemical pathway by which NO is converted to NO_2 is by reaction with ozone (O_3) derived from the photolysis of atmospheric oxygen (Eq. 3). Still another means by which NO_2 is formed from NO involves reaction of NO with HO_2, a product formed by reactions involving carbon monoxide, oxygen, and the hydroxyl radical (HO•) created upon the photolysis of molecular water (Eq. 4).

$$N_2 + O_2 \rightarrow 2NO \tag{1}$$

$$NO + RO_2 \xrightarrow{h\nu} NO_2 + RO \tag{2}$$

$$NO + O_3 \xrightarrow{h\nu} NO_2 + O_2 \tag{3}$$

$$NO + HO_2 \rightarrow HO\bullet + NO_2 \tag{4}$$

Table III Properties of Nitric Oxide and Nitrogen Dioxide

Property	Nitric oxide	Nitrogen dioxide
Color as liquid	Deep blue	Yellow-brown
Color as gas	Colorless	Reddish-brown
Molecular weight	30.01	46.01
Melting point	−163.6°C	−11.2°C
Boiling point	−151.8°C	+21.2°C
Density (air = 1)	1.04	1.58
1 ppm equivalent[a]	1.23 mg/m^3	1.88 mg/m^3
1 mg/m^3 equivalent[a]	0.813 ppm	0.532 ppm
Water solubility (ml/l)	73.4	Slight

[a] At 25°C and 760 Torr. For NO, ppm = (μg/m^3)/1880; for NO_2, ppm = (μg/m^3)/1230.

The relative importance that each of the above three mechanisms may play in producing atmospheric NO_2 from NO remains controversial. Even so, it is clear that the overall process is relatively slow unless promoted by the driving energy of sunlight and the presence of ambient oxidants and hydrocarbons. Typical of some urban areas is a diurnal peak in NO concentration at the times of maximal use of automobiles (e.g., morning traffic) which is followed one to two hours thereafter by a diurnal peak in NO_2. Other cities, however, do not show this pattern of sequential NO and NO_2 peaks. The brown haze observed in photochemical smog is due to N_2O_4, the dimeric form of NO_2; NO_2 and N_2O_4 are always present together in equilibrium at normal environmental temperatures. The Environmental Protection Agency (EPA) estimated that over 8 million people had been exposed to NO_2 at concentrations in 1989 that exceeded the National Ambient Air Quality Standard for this toxic gas

(~0.05 ppm annual arithmetic mean). In southern California where photochemical smog is a particular problem, concentrations of NO_2 may exceed 0.5 ppm. Other, less well understood atmospheric reactions involving NO_X that are of environmental concern result in the formation of peroxyacetyl nitrates. These compounds are toxic to plants and are strong lacrimators, i.e., they cause tearing.

D. Commercial Production and Use

Hazardous exposures to NO_X can potentially occur during the commercial production of oxides of nitrogen and during their use in various manufacturing processes. Industrial exposures to NO_X may occur, for example, where nitric acid (HNO_3) is produced and/or used. Historically, such exposures have taken place especially during manufacturing processes that involve the dipping of metals such as copper or brass into HNO_3 or during the nitration of organic materials such as cotton. The brownish gases generated from these processes have been incorrectly referred to in the older literature as "nitrous fumes"; the gaseous mixtures do not contain nitrous oxide (N_2O), nor are they generated as a fume. Commercially, NO is produced by the oxidation of ammonia or by passing air through an electric arc. It is usually contained in steel gas cylinders after production. The use of commercial NO includes the production of nitric acid, the bleaching of rayon, and its use as an oxidant in several organic chemical reactions. NO_2 is produced commercially by the oxidation of NO, and it is stored in steel cylinders in the gaseous form or in the liquid form primarily as its dimer (N_2O_4). NO_2 is also a byproduct of the manufacture of several types of chemical materials, including some dyes, explosives, celluloid, lacquers, photographic films, synthetic cloths, and fertilizers. Because it is a strong oxidizing agent, this oxide of nitrogen has been used as jet and missile fuel components. Fortunately, governmental regulatory standards, modern precautions, and care for the health and safety of industrial workers who produce or use oxides of nitrogen have made exposure to high concentrations of NO_X a less frequent occurrence in the United States.

E. Occupational Sources

In addition to the potential exposure hazards associated with the intentional production and use of oxides of nitrogen, several other occupational sources of NO_X are associated with some job activities in which NO_X can be unintentionally generated at hazardous concentrations, especially when they are performed in confined spaces. Such activities include arc welding, electric and gas welding, electroplating and engraving, oxyacetylene cutting, and brazing. Most fatal exposures associated with NO_X generated during welding and flame cutting have occurred in the hulls of ships, in tanks, and in boxcars. Other occupational activities that introduce risk to NO_X exposure include the detonation of explosives such as dynamite and ammunitions, and firefighting. In the latter case, NO_X concentrations generated by a fire can reach lethal concentrations, as illustrated by the Cleveland Clinic fire in 1929 in which massive clouds of NO_X generated from the burning of nitrocellulose X-ray film resulted in the tragic deaths of 129 individuals. Concentrations of NO_2 as high as 4000 ppm have been associated with occupational accidents.

Farming is another occupation in which hazardous and even lethal concentrations of NO_X can be encountered. Such exposures result from the fermentation of silage in silos and cause what has been termed "silo filler's disease" or "silage gas poisoning." The first report of four deaths due to silo gas was in 1914, which was attributed to carbon dioxide at that time. Recognition that fatal concentrations of NO_2 in freshly filled silos was the actual cause of silo filler's disease was not established until 1965. In the confines of a warm, moist silo, clouds consisting mainly of NO_2 and N_2O_4, which typically occur in a ratio of 30 : 70, are produced by anaerobic fermentation. In this process, plant nitrates are first converted into nitrites and oxygen. The nitrites subsequently react with organic acids to form nitrous acid (HNO_2), which eventually decomposes to NO, N_2O_3, and NO_2 and its dimer, N_2O_4. The color of the resulting gaseous mixture is yellow to reddish-brown. The odor of household bleach is another identifying characteristic of toxic silo gas. The concentrations of these oxides of nitrogen can become exceedingly high within a week of ensilage, and the maximum concentrations achieved can depend on the amount of fixed nitrates in certain crops, especially corn. Up to 1900 ppm NO_X can occur in poorly ventilated silos. Because concentrations of NO_2 that can cause significant injury to the lower respiratory tract upon prolonged breathing initially may be only mildly to moderately irritating to the eyes or upper respiratory tract, serious injury can occur in farmers who persistently work in silos in spite of these latter, sometimes tolerable symptoms of exposure. While most people can detect the odor of NO_2 when present at 0.22 ppm, human volunteers exposed to 25 ppm NO_2 for 5 min in a test study found such an exposure to be unpleasant but tolerable. It is of importance to note that the odor of NO_2 may not be perceived when its concentration is gradually increased, and that adaptation to the pungent odor of NO_2 can occur within several minutes after the onset of exposure.

Silo filler's disease should not be confused with "farmer's lung," a hypersensitivity lung disease caused by exposure to thermophylic *Actinomycetes* and not by exposure to NO_2. A history of exposure can usually provide a means to distinguish between the two disorders. Whereas silo filler's disease occurs during the harvesting season and is associated with a recently filled silo, "farmer's lung" most often occurs when moldy forage is handled in the winter months.

Sometimes occupational exposures to NO_X can occur in even the most unexpected setting. For example, three American astronauts who participated in the Apollo–Soyuz joint mission with the Soviet Union were accidentally exposed to significant concentrations of NO_X during their reentry into the earth's atmosphere. In this incident, NO_2 entered the command module through an open cabin pressure relief valve during the inadvertent firing of the Apollo's Reaction Control System. As a result of this error, the crew members were exposed to an estimated peak concentration of 750 ppm NO_2 and to an average concentration of 250 ppm NO_2 for nearly 5 min.

Occupational exposure limits to NO and NO_2 established by several agencies are given in Table IV.

Table IV Exposure Limits

	NO	NO_2
ACGIH/TLVs[a]	25 ppm; 31 mg/m^3	3 ppm; 5.6 mg/m^3
OSHA/PELs[b]	25 ppm; 30 mg/m^3	1 ppm; 1.8 mg/m^3
NIOSH/RELs[c]	25 ppm; 30 mg/m^3	1 ppm; 1.8 mg/m^3

[a] American conference of Governmental Industrial Hygienists' Threshold Limit Values (TLV). TLVs are defined by the ACGIH as air concentrations that represent conditions under which it is believed that nearly all workers may be exposed day after day without adverse effects. The TLV values shown above are time-weighted averages (i.e., values for a normal 8-hr workday and a 40-hr work week).

[b] Occupational Safety and Health Administrations' Permissible Exposure Limits (PELs). The values given are short-term exposure limits (STELs), which are concentrations that should not be exceeded even for a brief period of time (i.e., 15 min).

[c] National Institute for Occupational Safety and Health Recommended Exposure Limits (RELs), which are considered the "least detectable effect" levels. The values given are STELs.

F. Other Sources of Exposure to NO_x

NO_x is a common indoor air pollution problem. Indoor exposure to NO_2 has been identified as a health problem in poorly ventilated apartments in which unvented gas stoves are used for cooking. Such conditions, which usually occur in inner-city dwellings, can lead to short term peak exposures ranging from 0.2 ppm to as high as 1.5 ppm NO_2. As well, peak concentrations of NO_2 as high as 2.0 ppm have been measured in homes with unventilated combustion sources. Indoor ice rinks apparently are another setting where exposure to NO_2 can occur due to emissions from ice resurfacers. Such exposures have been associated with the development of acute respiratory symptoms in hockey players and enthusiasts. Both an indoor and outdoor problem for smokers is the presence of high concentrations of NO_X in undiluted cigarette smoke, which may range from 150 to 650 ppm. Up to 250 ppm of the NO_2 species has been measured in main stream cigarette smoke. It has been estimated that a one pack per day cigarette smoker absorbs ~3 mg of NO_X daily. Even higher concentrations of NO_X are produced by cigars and pipes (e.g., 1200 ppm) but lesser concentrations are usually inhaled by users because of the dilution of the smoke as it is inhaled.

II. Pathologic Responses to NO and Related Mechanisms

A. Pulmonary Effects

The oxidation of NO to NO_2 by atmospheric O_2 makes it nearly impossible for human exposures to air contaminated with pure NO to occur in the absence of NO_2. The known oxidation of NO to NO_2 also complicates the results of several animal studies in which rigorous attempts were not made to determine the pulmonary toxicity of NO exclusive of other nitrogen oxide species. Such studies have frequently indicated that the inhalation of NO-containing atmospheres produce profiles of lung injury similar if not identical to that observed following exposure to NO_2 (to be described). As examples, the lungs of rabbits exposed for 14 days to 5 ppm of nominal NO showed evidence of endothelial cell damage, and dogs exposed to 5000 ppm concentration of nominal NO for 24 min or longer succumbed to pulmonary edema after exposure. In other animal studies in which attempts were made to remove or substantially diminish NO_2 from NO-containing atmospheres, acute or subchronic exposures to NO produced no significant lung injury as evidenced by histopathologic changes. In a recent study, for example, no lung injury was detected by lung histopathologic criteria or lung weight increases as an indicator of pulmonary edema following a 30 min exposure of rats to 1000 ppm NO, whereas detectable injury was observed in other rats exposed to 25 ppm NO_2 for the same duration. Accordingly, existing information indicates that NO is substantially less toxic as an injurious agent in the respiratory tract than is NO_2.

B. Extrapulmonary Effects and Associated Mechanisms

The main toxicologic effect of the inhalation of NO is the formation of methemoglobin (MetHb) in erythrocytes. The rate at which MetHb is formed under resting conditions is directly related to the contaminating concentration of NO in inhaled air. Though incompletely understood, the formation of MetHb is probably initiated by the binding of NO to erythrocytic hemoglobin (Hb) to form nitrosylhemoglobin (Hb–NO); NO binds to hemoglobin $\sim 3 \times 10^5$ times more strongly than does O_2 and $\sim 10^3$ times more strongly than does carbon monoxide (CO). Subsequently, ferrous Hb–NO is rapidly oxidized in the presence of oxygen to form ferric MetHb, which gives the blood a peculiar chocolate coloration. During the inhalation of NO, the steady-state concentration of Hb–NO is very low while MetHb progressively increases. Another proposed mechanism for the development of MetHb is that NO dissolved in blood is converted to nitrite, which is well recognized for its ability to cause methemoglobinemia. Regardless of the underlying mechanism(s), one outcome of MetHb formation is a decrease in the oxygen-carrying capacity of the red cells due to a reduction in the availability of normal Hb to bind O_2. Because the oxidized form of the heme iron (Fe^{3+}) causes a higher affinity for O_2 by the remaining heme iron in Hb, a decrease in the dissociation of O_2 from Hb, or a "shift-to-the-left" of the O_2–Hb dissociation curve, further compromises tissue delivery of O_2. Overall, MetHb produces a state that is effectively comparable to anemic hypoxia.

Slate-gray cyanosis due to methemoglobinemia usually becomes apparent when MetHb levels of ~30–40% are reached. Oxygen supplementation is often employed as a mainstay form of therapy in managing significant methemoglobinemia, even though the O_2-carrying and delivering capacities of Hb are limited by the presence of MetHb. In this regard, several clinical investigators advocate the use of hyperbaric oxygenation in severe cases of methemoglobinemia, but few clinical settings have this capability. In the rat, NO-induced methemoglobinemia results in deaths when blood saturation levels exceed ~65% of available Hb. Similarly, MetHb saturation levels exceeding 60% of available Hb are considered to be fatal in the human. One factor that has been shown (in animal studies) to affect the level of MetHb formed during exposure to NO is minute ventilation. For a given concentration of inhaled NO, resulting MetHb saturation levels directly scale with increases in minute ventilation.

Under normal circumstances, ~1% of the blood's Hb is in the form of MetHb. The origin of this low background level of MetHb in normal erythrocytes is not completely understood. Probable mechanisms for its occurrence include the spontaneous or auto-oxidation of intracellular Hb, the production of NO by vascular endothelial cells as well as other cell types, and the metabolic production of nitrates and nitrites from various ingested nitrogenous compounds. NO that is normally produced in the body is thought to be involved in a variety of physiologic functions, including vascular tone and blood pressure regulation, gastrointestinal contractions, and the killing of various microorganisms by some inflammatory cell types. Most recently, NO has also been implicated as being a rather novel neurotransmitter.

Symptoms associated with methemoglobinemia are similar to those observed with equivalent degrees of carboxyhemoglobin produced by the inhalation of CO. High MetHb saturation levels in the blood (e.g., 25–40%) are associated with headache, weakness, and shortness of breath. Blood gas analyses usually reveal normal arterial O_2 tension. Lower saturation levels (5–20%), however, appear to be relatively well tolerated, although individuals with methemoglobinemia may experience vision changes in dark adaptation, especially after exercising. Recently studies with rats have demonstrated a direct correlation between the percentage of MetHb saturation and reductions in work performance capacity as indexed by maximal O_2 consumption during exercise. A component of diminished tolerance to exercise following NO exposure includes the oxidation of myoglobin (an intracellular protein that both transports and stores O_2 in red skeletal muscle fibers) to metmyoglobin (MetMb). Little information is currently available as to the rate(s) at which MetHb and MetMb may differentially accumulate in the human relative to inhaled NO concentration and minute ventilation. From the perspective of occupational exposures, MetHb has been especially associated with the inhalation of NO_X generated during electric welding processes. However, MetHb levels achieved in welders usually do not exceed ~3%. More serious methemoglobinemia with associated signs of cyanosis mainly occurs following the ingestion of excessive quantities of nitrite in some cured foods, after the ingestion of nitrates in well water or in fertilized vegetables, or following the ingestion of nitrogenous compounds that are metabolically converted to nitrite *in vivo*.

Surprisingly limited information presently exists as to the rate by which MetHb is removed from the blood following NO exposure. Unlike carboxyhemoglobin, which dissociates spontaneously as a function of the partial pressures of O_2 and CO, the reduction of MetHb to normal Hb requires the expenditure of metabolic energy by erythrocytes. The major mechanism for the reduction of MetHb in the red cell is via an enzyme called methemoglobin reductase, which requires reduced diphosphopyridine nucleotide (NADH) as a cofactor. Greater than tenfold differences have been reported in the methemoglobin reductase activities of different mammalian species. In normal human erythrocytes, very high levels of MetHb are reduced *in vitro* with a half-life of 6–24 hr. In severe cases, total reconversion of MetHb to normal Hb *in vivo* may require 4–5 days. Some lines of evidence have suggested that the *in vivo* rate of MetHb reduction is complexly related to the percent MetHb saturation level achieved during the inhalation of NO, with higher percentage saturations being associated with more prolonged half-lives for conversion of MetHb to normal Hb. Another pathway by which MetHb can be reduced is by way of the so-called "dormant NADPH-reductase system." This mechanism is activated in the presence of electron acceptors such as methylene blue (tetramethylthionine chloride). In this case, the leuco or colorless form of enzymatically reduced methylene blue reduces MetHb nonenzymatically. The intravenous administration of methylene blue (1–2 mg/kg) has been shown to have dramatic effects on reducing MetHb. Ascorbic acid has also been shown in *in vivo* studies to hasten the reconversion of MetHb to Hb, but vitamin C has not received wide therapeutic used for this purpose. Other, more minor pathways in erythrocytes that reduce MetHb include the slow reduction of MetHb by reduced glutathione and by vitamin C.

Another potentially detrimental action of inhaled NO is excessive vasodilation. This effect is mediated by the binding of NO to a heme group associated with guanylate cyclase or to an enzyme involved in the activation of guanylate cyclase. Activation of guanylate cyclase leads to the conversion of guanosine 5′-triphosphate to cyclic guanosine monophosphate, the activation of a series of protein kinases, and finally the relaxation of vascular smooth muscle cells. Some investigators have suggested that the hypotension observed in endotoxin-induced shock may be due to the production of NO by NO-producing cells. Recently, the vasodilatory effects of NO have been exploited on a trial basis by using inhaled NO as an approach for the therapeutic management of pulmonary hypertension. Wider therapeutic use of NO for the treatment of this disorder awaits more extensive investigation of the toxic effects of this oxide of nitrogen.

III. Pathologic Responses to Acute, High-Level Exposures to NO_2

Aside from a condition of sudden death due to bronchospasm, laryngospasm, reflex respiratory arrest, and asphyxiation, the onset of symptoms following brief exposures to very high, life-threatening concentrations of NO_2 (e.g., 250 ppm) occurs within minutes. Hallmark features of such an exposure include pronounced lacrimation, mental confusion, laryngospasm, paroxymal coughing, irritation of the upper respiratory tract, increased difficulty in breathing due to progressive increases in airway resistance that result from bronchoconstriction and airway edema, and the onset of potentially fatal pulmonary edema, which is due to a change in the permeability status of the lung's air–blood barrier. With this disorder, fluid from the lung's vasculature progressively floods the alveoli, or air sacs, where gas exchange otherwise normally occurs. Symptoms of pulmonary edema are the frequent expectoration of blood-tinged sputum, increasing hyperpnea or shortness of breath, tachypnea or an increase in breathing frequency, tachycardia or an increase in heart rate, and, depending on the severity, cyanosis due to failure of the lungs to adequately oxygenate the blood. Several complex factors can underlie arterial hypoxemia and inadequate saturation of hemoglobin with oxygen, including an increase in alveolar interstitial thickening due to edema fluid and swollen cells, which impairs gas diffusion between the alveolar surface and alveolar capillaries, the flooding of the alveoli with blood-derived fluid and cellular constituents, and ventilation-perfusion mismatching. Chest X-rays in such cases show fluffy transudates throughout both lung fields. Chest auscultation typically reveals abnormal mid- and late breath sounds over both lungs.

At lesser but still severe concentrations, the exposed individual may cease coughing, generally "feel better," and have relatively clear chest radiographs, but this relative "calm before the storm" can be followed 4–10 hr later by the development of frank pulmonary edema. A "normal" X-ray taken shortly after NO_2 exposure, accordingly, does not rule out the possibility that pulmonary edema will develop sometime thereafter.

Therapeutic management of NO_2-induced pulmonary edema includes oxygen therapy, ventilatory support, corticosteroids (e.g., prednisone, 1 mg/kg/day), airway dilators, and blood oncotic pressure modulation. In that the bronchconstricting response to NO_2 may be mediated by the release of histamine from mast cells, antihistaminic therapy may also be useful in managing patients experiencing bronchospasm and/or excessive increases in airway resistance. The clinical presentation and the general management of NO_2-induced pulmonary edema generally follows that of adult respiratory distress syndrome. Persons exposed to high concentrations of NO_2 should be provided immediate bed rest and be maintained under close medical supervision for at least 72 hr to detect the earliest signs of pulmonary edema. The importance of bed rest cannot be overly emphasized inasmuch as experiments with laboratory animals have demonstrated that the severity of NO_2-induced pulmonary edema can be substantially potentiated by post-exposure exercise.

Histopathologic changes that occur in the respiratory tract following the inhalation of high concentrations of NO_2 have been best studied at the light and electron microscopic level in experimental animals. Such investigations have shown that the NO_2 causes the loss of cilia along the conducting airways and the sloughing of necrotic, ciliated epithelial cells that line their luminal surfaces. Ciliary action normally is involved in the cephalad propulsion of mucus out of the lower respiratory tract, an important lung defense mechanism. An expected consequence of ciliary loss, therefore, is a compromise in the ability of the lung to mobilize secretions and the failure to remove inhaled pathogenic microorganisms. More peripherally, damage to the lung occurs mainly in the terminal bronchiolar–alveolar duct region with the more proximal alveolar structures being most affected. The sparing of more distal alveolar structures is thought to be due to the more proximal uptake of the inhaled NO_2 as it courses through the respiratory tract and not due to the more peripheral alveoli being less susceptible to the injurious effects of NO_2. Indeed, the extent of alveolar damage following acute inhalation of high levels of NO_2 is a function of exposure concentration and duration, with exposure concentration being the more dominant factor. Thus, under some exposure conditions, even more distal alveolar structures can undergo extensive injury. The breach in the alveolar–capillary barrier that results in pulmonary edema is due to damage to endothelial cells lining alveolar capillaries as well as the destruction and exfoliation of type I pneumocytes that normally line ~90% of the alveolar surface. Abnormal fluid movements increase the thickness of the interstitium in the alveolar region, which, as previously indicated, impairs gas diffusion. In many forms of permeability pulmonary edema induced by toxic agents, the accumulation of fluids and proteins follows an orderly pattern of an initial interstitial filling followed by alveolar flooding. Some experimental animal data, however, have suggested that alveolar flooding may preferentially occur first and interstitial edema second in NO_2-induced pulmonary edema. Regardless, milder damage to the alveoli may result primarily in the passage of vascular fluid phase constituents into the air sacs and the formation of fibrin, whereas more extensive damage to the alveolar–capillary barrier can result in the extravasation of blood-borne cells (i.e., erythrocytes) into the alveoli as well. The passage of blood compartment constituents into the alveoli alters the function of alveolar surfactant that normally prevents alveolar collapse and thereby diminishes lung compliance. Histopathologic assessments of lung tissue obtained from fatal cases of human exposure to high NO_2 have generally been consistent with what has been observed in the animal studies. Typical observations have included the presence of intra-alveolar edema fluid and exudates, alveolar wall thickening, and leukocytic cell infiltrates. The pulmonary edematous response to NO_2 persists until these latter deleterious effects are reversed by the repair and proliferation of capillary endothelial cells and the repopulation of alveolar epithelial cells by proliferating type II pneumocytes, which subsequently further differentiate into the type I alveolar epithelial cell phenotype. The complete resolution of the pulmonary edematous response requires the reconstitution of an intact alveolar–capillary barrier. These reparative processes usually begin within 24 hr after exposure. However, some lines of recent evidence suggest that the repair mechanisms can be delayed when alveolar damage is very severe. Another outcome of NO_2-induced alveolar damage is the recruitment of inflammatory cells into the lung's alveoli (i.e., blood mononuclear phagocytes and polymorphonuclear leukocytes). These cells may further amplify the injurious response by releasing reactive oxygen species, including superoxide anion and hydrogen peroxide, as well as proteolytic enzymes as they phagocytize or engulf debris. A host of other mediators that can have detrimental effects on pulmonary vascular resistance and airway smooth muscle tone may also be elaborated by the inflammatory cells. On the other hand, these cells are known to release a variety of cytokines that are growth factors for other lung cell types, and, in this regard, the recruited inflammatory cells may serve to augment lung repair.

The acute symptoms of NO_2 toxicity in survivors can essentially resolve completely within days after the exposure. However, other symptoms may linger, including shortness of breath, malaise, weight loss, fever, and the radiographic appearance of extensive nodulation throughout both lungs. Such nodulation and the radiographic appearance of overinflation due to what is called bronchiolitis fibrosa obliterans can subsequently develop up to six weeks after an acute exposure. This airway-occluding disorder, the second component of the so-called "biphasic response pattern to NO_2," evidently emerges from the earlier damage to the airway epithelial cells and from their loss along the conducting airways, although some evidence suggests this later response may have an alveolar origin as well. In this disease process, accumulated airway exudates and necrotic cellular debris become infiltrated and organized by fibroblasts originating from the airway submucosa and/or from the alveolar region, with the progressive growth of granulation tissue within airway lumens. Occlusion of the airways with granulation tissue underlies the severe airway obstruction and lung hyperinflation. The administration of steroid therapy up to eight weeks after exposure to high concentrations of NO_2 is thought by many clinicians to be of use for preventing bronchiolitis fibrosa obliterans, but the efficacy of this therapeutic approach has been mainly antidotal. Bronchiolitis fibrosa obliterans resolves in most exposed patients, even when steroid therapy is administered at the time the disorder has already developed. Even so, functional disorders can persist in some individuals for years after acute exposures to high concentrations of NO_2. These include mild resting hypoxemia, or abnormally low arterial blood oxygen saturation at rest, diminished exercise tolerance, and small airways obstructive lung disease due to a narrowing of the airways.

Concentration–exposure time relationships in the context of NO_2 causing pulmonary edema or bronchiolitis fibrosa obliterans have not been established for the human. Thus, it should be noted that while relatively brief exposures to high concentrations of NO_2 can result in significant pulmonary injury, such exposures do not necessarily lead to the development of serious, life-threatening pulmonary edema or bronchiolitis fibrosa obliterans, as indicated by the experience of the astronauts involved in the previously mentioned reentry accident that occurred during the Apollo–Soyus mission. During the first hour after splashdown, the exposed crew members complained of tightness in the chest, reported a retrosternal burning sensation, were unable to inhale deeply, coughed nonproductively, and requested oxygen supplementation. Chest roentgenograms taken as of 4 hr after splashdown revealed no evidence of significant lung injury. On the following day, when steroid therapy was initiated, the astronauts were unable to hold their breaths or forcefully exhale sufficiently to perform pulmonary function tests, and their chest X-rays showed diffuse, nodular infiltrates consistent with a diffuse chemical pneumonitis. Arterial blood gas and pH analyses indicated hypoxia, and alkalosis due to hyperventilation. By day three, the crew members became asymptomatic, and, as of day 5 following reentry, their chest roentgenograms returned to normal. Of interest, some evidence obtained from studies of these NO_2-exposed astronauts has suggested that NO_2 may, in addition to injuring cells in the respiratory tract, bring about the breakdown of collagen that contributes to the extracellular matrix in the lung's interstitium. Information obtained from the study of hamsters and guinea pigs exposed to NO_2 supports this possibility.

IV. Pathologic Responses to NO_2 Inhaled at Lower Concentrations

Limited information about the acute, subchronic, and especially chronic effects caused by lower level exposures to NO_2 (e.g., $\leq$50 ppm) is available for the human. Indeed, many of the effects that are frequently presumed to occur in people have been observed as effects in animal studies that have involved diverse exposure protocols involving a wide range of NO_2 concentrations and exposure durations. For the sake of brevity, many of the main effects found in several of these investigations are summarized in Table V. Of primary concern are the observations that acute or subacute low-level exposure to NO_2 can impair the normal defensive functions of alveolar macrophages, depress mucociliary clearance of the conducting airways, cause hyperplasia of the bronchial epithelium, cause fibrosis in the terminal bronchioles, and reversibly increase susceptibility to lung infection. Additionally, numerous studies with laboratory animals have shown that chronic low-level exposure to NO_2 can result in changes resembling human emphysema, including alveolar duct distension, destruction of alveolar walls, and an enlargement of the distal air spaces. Extrapolating all of the effects noted in Table V, however, to the human condition may not be appropriate because of the known response differences that can occur among different mammalian species. For example, while rats and mice have been reported to develop emphysema following exposure to NO_2, the hamster appears to be resistant to this form of injury. Another confounding problem with some animal investigations of NO_2 toxicity, particularly in the older literature, that limit their interpretation are that the exposure conditions were not well characterized in terms of exactly which NO_X species may have been present in the exposure atmospheres. Moreover, it has not always been clear that the animals used in a given study were free of preexisting lung disease. The potential significance of this latter shortcoming is illustrated by a study in which emphysematous changes occurred in the lungs of influenza virus-infected squirrel monkeys after a 16-month exposure to 1 ppm NO_2, whereas no such changes were found in the lungs of uninfected animals. Yet another frequent problem with the existing database obtained from animal studies is a paucity of information concerning the toxicity of NO_2 at ambient or near ambient concentrations. Hence, animal information concerning threshold effect levels of NO_2 is limited. Nevertheless, some basic principles that are likely applicable to the human condition and lower level NO_2-induced toxicity can be gleaned from the animal literature: (1) as would be expected, the severity of pulmonary injury caused from inhaling NO_2 increases with exposure concentration; (2) for a given concentration of NO_2, pulmonary damage becomes progressively greater as the exposure duration becomes more prolonged; (3) with both acute and chronic exposures, exposure concentration is the more important variable in producing lung damage. That is to say, NO_2 toxicity does not follow Haber's Law (exposure concentration $\times$ time $= k$, where k is the magnitude of an observed effect) in that for a given constant concentration $\times$ time level, shorter term exposures to higher concentrations produce greater effects than do lower concentrations inhaled over a longer period; (4) both younger and older animals may be more sensitive to NO_2 than their more mid-aged counterparts; (5) the pulmonary response to NO_2 can be significantly altered by preexisting disease.

Information about how humans may be affected by acute, lower level exposure to NO_2 is conflicting. Some individuals experience severe substernal pain after being exposed to 50 ppm NO_2 for only

Table V Examples of Effects in Experimental Animals Caused by Exposure to Relatively Low Concentrations of NO_2

Effect	Concentration (ppm)	Exposure	Species
Destruction of cilia	15–25	Acute	Rat
Airway epithelial cell hypertrophy and hyperplasia	15–25	Acute	Rat
Pulmonary edema	7	Acute	Dog
Bronchiolitis obliterans	15–25	Acute	Rat
Alveolar macrophage accumulations	1–5	Chronic	Dog
Increase in interstitial thickness	37	Acute	Dog
Alveolar endothelial cell blebbing	3	Acute	Dog
Alveolar capillary engorgement, type I pneumocyte damage, cuboidal (type II) cell metaplasia	5–20	Acute	Rat
Alveolar septal fibrosis	12–25	Chronic	Rat
Centrilobular emphysema	12	Chronic	Rat
Decreased alveolar macrophage phagocytic activity	15	Acute	Rabbit
Decreased mucociliary clearance	6	Subchronic	Rat
Decreased resistance to bacterial infection	5	Acute	Mouse
Decreased resistance to viral infection	15	Acute	Rabbit
Impaired immune responses	20	Acute	Mouse
Increased airway reactivity	7.5	Acute	Sheep
Increased lavageable protein	0.4	Subchronic	Guinea pig
Increased respiratory rate	0.8	Chronic	Rats
Hematologic disturbances	10	Acute	Mice

1 min while others do not. In terms of NO_2-induced changes in pulmonary function, one group of investigators found that exposure of normal subjects to as little as 0.24 ppm for 20 min caused an increase in airway resistance, whereas in another study, no effect on pulmonary function was observed when healthy subjects inhaled up to 4 ppm NO_2 for 75 min with intermittent exercise. Overall, though, most human studies have indicated that the pulmonary function of healthy individuals is not significantly altered when NO_2 is inhaled below 1.5 ppm for 2 hr or less, whereas brief exposures on the order of 10–15 min of 5 ppm NO_2 can result in a variety of changes including decreases in lung capacity, increases in airway resistance, and decreases in gas diffusion capacity. Asthmatics, or more likely a subpopulation of asthmatic responders, may represent a group of individuals with a particular susceptibility to the effects of inhaling low concentrations of NO_2 (e.g., increased airway resistance, increased airway reactivity to a second bronchoconstrictor, and the potentiation of exercise-induced bronchospasm) although this possibility remains somewhat equivocal due to the conflicting results of different studies. Nevertheless, it is known that the prevalence of asthma is about four times the national average in the Black and Hispanic communities in the Harlem/Washington Heights areas of New York City. Epidemiologic analyses have revealed that the incidences of asthmatic attacks in these communities, where most dwellings are poorly ventilated and contain gas stoves (a source of NO_2 which, as previously indicated, can achieve concentrations exceeding 1.5 ppm) are more frequent when people spend more time indoors. Normal individuals with no history of asthma have also been shown to have an NO_2-induced increase in airway hyperresponsiveness to a second bronchoconstricting agent (methacholine). This effect has been found to be completely prevented by pretreatment with ascorbic acid (500 mg, 4 times per day for 3 days).

Because animal studies have indicated that NO_2 exposure to concentrations as low as 0.5 ppm can increase the susceptibility to respiratory tract infections, some attempts have been made to determine if such an effect also occurs in humans. To date, clinical investigations have produced ambiguous results and they have as yet to clarify the role that low-level exposure to NO_2 may play in causing respiratory infections. Such studies, as well as the aforementioned investigations of low-level NO_2 exposure on asthmatic symptoms, have been generally restricted to concentrations below 3 ppm for ethical reasons. Thus, existing studies do not provide information about the toxicity of NO_2 in individuals who may inhale higher concentrations during occupational exposures. A firm, direct linkage between NO_2 exposure and an enhanced susceptibility to respiratory infections in children and adults has not been possible from epidemiologic studies because of the usual concurrent exposure to other potentially detrimental air co-contaminants, including cigarette smoke.

Subchronic or chronic effects of lower level NO_2 exposure are even less well-established in humans. While numerous investigators have speculated that chronic NO_2 exposure can produce emphysema, as has been shown in numerous animal studies, epidemiologic studies of various occupationally exposed groups have failed to produce an unequivocal association. Based on some *in vitro* investigations and from a hypothetical perspective, the inhalation of an oxidant like NO_2 could alter the functional activity of alpha-1-protease inhibitor (α_1-PI) in the lung, which could diminish protection against neutrophil elastase and cause emphysema. The existing database in support of this possibility, however, remains inconsistent. In one study, an ~45% decrease in α_1-PI activity was found in bronchoalveolar lavage fluids obtained from humans 3 hr after exposure to 3 or 4 ppm NO_2 with intermittent exercise. In a more recent study, healthy humans were either exposed to three 2-ppm peaks with a 0.05-ppm background of NO_2 over a 3-hr period with intermittent exercise, or to a continuous 3-hr exposure to 1.5 ppm NO_2 with intermittent exercise. No evidence was found in this later study that indicated the low-level exposures to NO_2 caused a deficiency in α_1-PI activity in lavage fluids. It should also be noted that in a separate investigation, the inhalation of 1.5 ppm NO_2 was also found not to inhibit α_2-macroglobulin, another protease inhibitor. In addition to being a potential cause of emphysema, animal studies have also attempted to address the question as to whether or not subchronic or chronic NO_2 exposure can increase the severity of experimentally induced or preexisting emphysema. Again, the results from these studies have frequently conflicted. Hence, the role low-level NO_2 exposure may play in the development and progression of chronic obstructive airway disease remains to be determined. Some investigators have also maintained that low concentrations of NO_2 can cause pulmonary fibrosis in humans, as observed in the lungs of some welders, for example, but evidence in direct support of this contention has yet to be obtained because individuals employed in this occupation are exposed to a wide variety of inhalable materials (i.e., potentially toxic aerosols).

Another pulmonary disorder which low-level NO_2 exposure has been postulated to induce is lung cancer. NO_2 has been reported to be mutagenic in *Salmonella typhimurium* and causes chromosomal aberrations in rat lung cells. It has been proposed that NO_2, and NO for that matter, may be indirectly carcinogenic by forming nitrosamines upon reaction with endogenous amines; the generation of nitrosamines following NO_x exposure has been demonstrated in homogenates of lung tissue. Nitrite, a potent *in vitro* mutagen, is also formed in the lungs following NO_2 exposure. Nitrous acid, a potential product of NO_2 deposition in the lung and a weak mutagen, has been observed to cause base pair substitutions and/or to react with amines to form nitrosamines. However, there is presently no firm evidence that NO_2 alone is a carcinogen, cocarcinogen, a promoter of carcinogenesis, a mutagen, or a teratogen in the human. To date, epidemiologic studies have been unable to uncover or rule out a carcinogenic effect due to NO_x because individuals exposed to NO and NO_2 generally experience concurrent exposures to a variety of potentially cancer-causing particulate and gaseous agents. Nevertheless, findings that the chronic exposure of rodents to relatively high concentrations of NO_2 (40 ppm for 16 months) did not increase tumor rate argues against a direct or even indirect carcinogenic effect by this oxidant gas. On the other hand, mice predisposed to developing pulmonary adenomas developed these benign tumors earlier when exposed to 5 ppm NO_2 for 1 year. It is also evident from the animal literature that NO_2 exposure can result in cell proliferation in the lung, and such a mitogenic effect in itself is thought by many investigators to be a risk factor for the development of cancer inasmuch as mitogenesis predictably increases the likelihood of mutational events that can occur during cell division. Increased mitogenesis presumably can also lead to an expansion of cells in which a mutational step may have already occurred. It remains possible that the reaction of NO_2 with other airborne constituents may lead to the formation of cancerous agents as opposed to more direct or indirect carcinogenic effects of inhaled NO_2 *per se*. For example, nitrated polyaromatic hydrocarbons formed by reactions of polyaromatic hydrocarbons with NO_2 may be the source of diesel exhaust mutagenicity observed *in vitro*, or perhaps, such reactions may underlie the *in vivo* carcinogenicity of diesel exhaust. Regardless, some animal data have suggested that the inhalation of low concentrations of NO_2 can increase cancer cell metastasis.

V. Extrapulmonary Effects of NO_2

Irritation to the eyes, lacrimation, and conjunctivitis can result from exposure to NO_2 with the severity of these responses generally increasing with increasing concentration. Flushing the eyes with water as soon as possible after exposure is recommended as a means to limit these effects. NO_2 exposure has also been reported to cause an essentially immediately reversible disturbance in dark adaptation, or the ability to perceive dim light. This effect has been observed to occur following exposure to as little as 0.074 ppm NO_2. NO_2 exposure has also been associated with the formation of MetHb. Mechanisms involved in this effect appear to be complex, and they have as yet to be satisfactorily elucidated. Nevertheless, NO_2 has been shown *in vitro* to convert oxyhemoglobin to MetHb in a mole for mole ratio to NO_2 consumed. Nitrite and nitrate, which are also known to oxidize Hb to MetHb, are readily formed when NO_2 reacts with water ($2NO_2 + H_2O \rightarrow HNO_3 + HNO_2$). (*Note:* such reactivity with water accounts for the fact that NO_2 does not obey Henry's Law.) Regardless of how NO_2 or its products ultimately lead to the formation of MetHb, this effect is generally considered to be of far lesser importance than are the injurious effects of the NO_2 in the respiratory tract.

VI. Mechanisms of Action and Factors Influencing NO_2-Induced Pulmonary Toxicity

A. Mechanisms of Action

The actual mechanism(s) by which the toxicity of NO_2 in the lung is mediated has not been firmly established. NO_2 is a nitrogen-centered free radical with limited solubility in water. These physical properties suggest that the uptake of NO_2 in the respiratory tract primarily occurs and is governed by chemical reactions as opposed to simple physical dissolution only. Consistent with a limited solubility relative to a much more soluble gas like sulfur dioxide, the uptake or deposition of NO_2 in the respiratory tract is incomplete. In studies with rhesus monkeys, for example, only about 50–60% of inhaled NO_2 was found to be retained in the respiratory tract during quiet breathing. In ventilated rat lungs, $<40\%$ of inhaled NO_2 is actually deposited. When breathed via the oral route by humans at a low concentration (0.3 ppm) under resting conditions, ~72% of the inhaled NO_2 deposits in the lower respiratory tract. Consistent with theoretical predictions that increases in minute ventilation which accompany exercise results in a more peripheral penetration and enhanced deposition of NO_2, ~87% of the NO_2 deposited in the lower respiratory tracts of the same subjects when they breathed the gas while exercising on an ergometer.

The dominant mechanism by which NO_2 is thought to cause tissue injury is via the peroxidation of lipids in cell membranes, which results in cellular damage and death. Circumstantial evidence that this process occurs in the lower respiratory tract includes: (1) findings of lipid peroxidation products (e.g., conjugated dienes) in the lung or in exhaled air, (e.g., ethane) after NO_2 exposure; (2) the *in vitro* ability of NO_2 to initiate the autooxidation of unsaturated fatty acids in lipids via addition to the ethylene group of fatty acids; and (3) a deficiency in vitamin E, a lipophilic antioxidant in cell membranes, significantly enhances NO_2-induced lipid peroxidation *in vivo*. A second mechanism of NO_2-induced injury is that NO_2 may directly oxidize low molecular weight reducing substances and proteins, and thereby causes metabolic dysfunctions in cells and ultimately cell death.

B. Factors Affecting the Toxicity of NO_2

1. Physical Activity

As demonstrated in studies of animals exposed to NO_2 while exercising, one factor that can significantly increase the severity of NO_2-induced lung toxicity is the level of physical activity that occurs during exposure. Such increases in the level of injury by exercise during exposure are most likely due to increases in minute ventilation, and, hence, the amount of NO_2 inhaled and deposited. As previously indicated, exercise performed after exposure to NO_2 concentrations resulting in pulmonary edema can also significantly potentiate the expression of this response. The mechanism(s) involved in the potentiation of No_2-induced lung injury by postexposure exercise have not been elucidated, but one mechanism may involve the imposition of physiologic events common to exercise (e.g., increased heart rate, increased pulmonary vascular pressures and blood flow) on an already hyperpermeable lung.

2. Nutrition

Diet is another factor that may influence an individuals' susceptibility to the toxic effects of NO_2. As previously noted, experimental animal studies have shown that a deficiency in vitamin E can result in a enhancement of NO_2-induced pulmonary effects. In other studies, mortalities resulting from exposures to high concentrations of No_2 have been shown to be reduced by pretreatment with high levels of vitamin E. The integration of vitamin E into biomembranes makes it especially well positioned to block the propagation of lipid peroxidation by the ability to scavage polyunsaturated peroxy free radicals. Vitamin C (ascorbic acid), a water-soluble antioxidant, is also thought to play a role against the effects of No_2. Deficiency of this vitamin in NO_2-exposed animals has also been shown to increase the severity of lung damage. Like vitamin E, vitamin C can also donate hydrogen to free radicals and thus confine propagating reactions. Of particular relevance is the possibility that vitamin C and vitamin E act synergistically, with vitamin E serving as the primary antioxidant and vitamin C serving to regenerate the reduced active form of vitamin E from the resulting vitamin E radical. In this way, vitamin C cooperatively plays an important function in providing protection against oxidative stress by maintaining vitamin E levels in lung tissue. Vitamin C has been shown in animal experiments to be substantially decreased in lung tissue after NO_2 exposure. As previously indicated, an increase in ascorbic acid intake has been found to prevent NO_2-induced airway hyperresponsiveness in normal human subjects. Another vitamin that may play a role in the response to No_2 inhalation is vitamin A (retinol). While this vitamin does not serve as a free radical scavenger, some evidence indicates that it may be important following No_2-induced injury by influencing the regeneration and differentiation of epithelial cells as the lower respiratory tract undergoes repair. On the other hand, β-carotene, a provitamin A with antioxidant activity, may provide some protection against NO_2 toxicity because of its ability to inhibit lipid peroxidation. Some (although limited) evidence has suggested that the mineral selenium may also protect type I alveolar epithelial cells from the damaging effects of NO_2. Selenium is incorporated in the active center of the enzyme glutathione peroxidase, which utilizes glutathione to reduce organic hydroperoxides and thereby protects membrane lipids and perhaps proteins and nucleic acids against oxidant damage. Yet another dietary substance that is gaining increasing interest as a protector against NO_2 toxicity is taurine (2-aminoethanesulfonic acid). Following the treatment of hamsters for 14 days with 0.5% taurine in their drinking water, a 24-hr exposure to 30 ppm NO_2 caused no light- or electron-microscopic evidence of lung injury, whereas pathologic changes including inflammatory cell infiltrates and morphologic damage to epithelial cells in the bronchiolar and alveolar duct regions were observed in the lungs of untreated animals. The mechanism(s) by which taurine provides such a dramatic protective effect against NO_2 remains to be determined, but taurine may act by scavenging free radicals and preventing peroxidative injury and/or by stabilizing the functional status of cell membranes.

3. Age

One variable that may influence the response to inhaled NO_2 is the age of the exposed individual, but the limited evidence in support of this possibility has been obtained mainly from animal studies. Such studies have indicated that neonatal guinea pigs and rats are less susceptible to the lung-damaging effects of NO_2 than are adult animals. Repair of NO_2-induced damage to the lower respiratory tracts of older rats has also been found to be slower than that in younger rats. On the other hand, other evidence obtained with newborn rats suggests that exposure to No_2 can cause a transient delay in lung maturation. In adult humans, no significant changes in pulmonary function have been observed in older men and women (51–76 years of age) and younger adults exposed for 2 hr to 0.6 ppm NO_2.

4. Adaptation

Adaptation or tolerance is a phenomenon that has been observed to occur with a variety of oxidant gases including NO_2, ozone (O_3), and O_2. Sublethal exposure to No_2, or the other aforementioned gases, leads to an increase in the lung's resistance to the damaging effects of a subsequent bout of exposure. While such a "protective" effect in laboratory animals has been repeatedly demonstrated to prevent death upon reexposure to otherwise lethal mass concentrations of NO_2, tolerance is only limited to scope inasmuch as cell injury and death is not prevented. Mechanisms involved in the development of tolerance are incompletely understood. However, it has been demonstrated that the activities of several enzymes that influence the increase in glutathione content in the lung have been shown to be increased in rats by exposure to NO_2, and, at least for O_3, vitamin C increases in the lung have been associated with adaptation. The significance of the development of tolerance in response to an oxidant gas has been well illustrated in human studies in which subjects have been given daily exposures to O_3. In these investigations, decrements in pulmonary function have been found on the first and second days of exposure, whereas the O_3-induced changes became attenuated with succeeding daily exposures.

5. Coinhalation with Other Air Contaminants

It should be noted that human exposure to NO_2 in the absence of other air contaminants is a rare occurrence. Welders, for instance, can be exposed to a host of other substances in welding fumes along with NO_2, including cadmium, nickel, zinc oxide, iron oxide, chromium, and O_3. Clearly, the inhalation of other potentially toxic materials may alter the response profile expected with NO_2 exposure alone. As examples, animal studies have indicated that the morphological and biochemical changes observed following NO_2 exposure are enhanced by the concurrent inhalation of respirable aerosols, and the combination of NO_2 and O_3 acts synergistically in producing biochemical changes in lung tissue and in enhancing susceptibility to bacterial infection.

Related Articles: DIESEL EXHAUST, EFFECTS ON THE RESPIRATORY SYSTEM; MONITORING INDICATORS OF PLANTS AND HUMAN HEALTH FOR AIR POLLUTION; OXIDES OF NITROGEN; OZONE EXPOSURE, RESPIRATORY HEALTH EFFECTS; PLANTS AS DETECTORS OF ATMOSPHERIC MUTAGENS.

Bibliography

Guidotti, T. L. (1978). The higher oxides of nitrogen: Inhalation toxicology. *Environ. Res.* **15,** 443–472.

Lindvall, T. (1985). Health effects of nitrogen dioxide and oxidants. *Scand. J. Work Environ. Health* **11**(Suppl. 3) 10–28.

Morrow, P. E. (1984). Toxicological data on NO_X: An overview. *J. Toxicol.* and *Environ. Health* **13,** 205–227.

Mustafa, M. G., and Tierney, D. F. (1978). Biochemical and metabolic changes in the lung with oxygen, ozone, and nitrogen dioxide toxicity. *Amer. Rev. Resp. Dis.* **118,** 1061–1090.

National Institute for Occupational Safety and Health. (1976). "Criteria for a Recommended Standard: Occupational Exposure to Oxides of Nitrogen," HEW Publication No. (NIOSH) 76–149.

Nitrogen oxides. *In* "Environmental Health Perspectives" (G. W. Lucier and G. E. R. Hook, eds.), vol. 73, 1987. U.S. Department of Health and Human Services, Washington, D. C.

"Nitrogen Oxides and Their Effects on Health," (1980). (S. D. Lee, ed.). Ann Arbor Science, Ann Arbor, MI.

Overton, J. H., and Miller, F. J. (1988). Absorption of inhaled reactive gases. *In* Toxicology of the Lung," (D. E. Gardner, J. D. Crapo, and E. J. Massaro, eds.). pp. 477–507. Raven Press, New York.

Stavert, D. M., and Lehnert, B. E. (1989). Potentiation of the expression of nitrogen dioxide-induced lung injury by post-exposure exercise. *Environ. Res.* **48,** 87–99.

Stavert, D. M., and Lehnert, B. E. (1990). Nitric oxide and nitrogen dioxide as inducers of acute pulmonary injury when inhaled at relatively high concentrations for brief periods. *Inhalat. Toxicol.* **2,** 53–67.

"Textbook of Pulmonary Diseases," (1989). vols. I and II. (G. L. Baum and E. Wolinsky, eds.) 4th Ed. Little Brown, Boston.

Organic Micropollutants in Lake Sediments

Alfredo Provini and Silvana Galassi
Water Research Institute
Brugherio, Italy
and
University of Milan, Italy

Glossary

Adsorption Surface retention of compounds or ions by a solid or a liquid.

Bioaccumulation Ratio between the concentration in an organism and in the water, air, or food.

Biotic zone Surficial zone of sediment comprising living organisms.

Organic micropollutants Organic compounds mainly of anthropic origin; dangerous for living organisms even at very low concentrations.

Partition coefficient Constant ratio of the solute's concentration in one phase to its concentration in another phase.

ORGANIC MICROPOLLUTANTS, *especially those not very soluble in water, tend to accumulate in sediments. Since lacustrine sediments of the anthropized areas are the storage compartment of these compounds, sediment analysis has been successfully used to follow their geographical and temporal distribution. Sediment-bound chemicals may represent a risk for aquatic life and human health both because of remobilization and bioaccumulation. Chemical analysis is not always adequate to cover all the aspects of the organic contamination completely (because of the complexities involved), and biological testing is presently being used to assess the hazard of sediment-bound pollutants.*

I. Introduction

Organic compounds of anthropogenic origin are delivered to lacustrine systems by atmospheric transport and deposition, direct discharges, and riverine inputs. When they enter the aquatic environment, their potential for accumulation in sediment is governed by their persistence and hydrophobicity and by the sorptive properties of suspended matter and sediment. Aquatic sediments in lentic systems are the main accumulation site of very persistent and lipophilic pollutants of domestic, urban, industrial and agricultural origin. Organic micropollutants adbsorbed to sediments are usually scarcely bioavailable, but sediments from very contaminated areas might represent an important source of pollution for freshwater organisms. Thus, sediment monitoring has been recognized to be very important both to localize "hot point" pollution areas and to study the pollution of the surrounding aquatic environment.

Sediments are frequently used as indicators of pollution in aquatic environments for a twofold reason: first, sediments accumulate very hydrophobic compounds that occur in water at undetectable

Handbook of Hazardous Materials

concentrations; second, sediment analyses allow integration of the fluctuations of the concentrations of the pollutant in water over a certain period of time.

Analysis of surficial sediment provides, therefore, an average estimate of recent pollution. On the other hand, the analysis of different layers of sediment corresponding to previous deposition periods enables one to reconstruct the historical trend of pollution. Furthermore, since many hydrophobic chemicals are known to partition between water and sediment and this phenomenon can be described by simple physicochemical models, sediment monitoring can be successfully used to evaluate the hazard of some classes of accumulable pollutants for an entire aquatic environment.

II. Major Classes of Chemicals Associated with Sediments

The range of compounds exhibiting substantial accumulation in sediments includes hydrocarbons, chlorinated hydrocarbons, phthalates, and other chemicals with a limited range of bond types. In most cases the predominant bonds are C—C aliphatic, C—C aromatic, C—H, and C—Cl although other bonds are occasionally present. These compounds are stable to degradation in aquatic environments and are likely to accumulate since this process requires a significant period of time.

The potential for accumulation, that is the concentration ratio of the compound in sediment and in water at the equilibrium between the two phases, is then ascribable to the lipophilic nature of the compound. Nonpolar compounds with low solubility in water are the most lipophilic. Lipophilicity is usually defined by the *n*-octanol/water partition coefficient, K_{ow}. Chemicals with a very high accumulation potential have log K_{ow} values ranging between 5 and 7. They represent about 1090 of the approximately 19,000 chemicals currently produced commercially in the United States. K_{ow} can be calculated from water solubility to which it is related through an inverse logarithmic relationship.

K_{oc}, the partition coefficient on soil and sediment, normalized for the carbon fraction content of the sorbent, is also related to water solubility and K_{ow}. Equations that describe such relationships have been experimentally determinated for a number of chemicals and are shown in Table I. The relationships linking K_{oc} with water solubility hold in the case of the finer fraction of sediment (<60 μm); a K_{oc} reduction of 50–90% is expected for the coarse fraction.

III. Sources of Pollution

A. Aerial Transport

Very persistent and accumulable compounds, such as chlorinated pesticides and polychlorinated biphenyls (PCBs), used in huge amounts on a worldwide scale for many years (starting from the 1930s) are by now ubiquitous and appear in most remote areas of the planet. Other compounds, such as polycyclic aromatic hydrocarbons (PAHs), are not delivered purposefully into the environment but originate from anthropogenic activities.

Aerial transport is considered to be the major pathway for global contaminants. This is governed by the extent of gas phase/particle phase distribution, since the deposition mechanism and the range of transport depend on it. In the atmosphere, organic micropollutants occur in gaseous aerosol or particulate form. The extent of transport depends on a variety of factors, including particle size, air movement, washout, and vapor pressure. However, most of the environmental xenobiotic burden remains relatively close to the source of contamination, and concentrations decrease approximately logarithmically with distance from the source.

Thus, the majority of organic micropollutants entering the aquatic environment are located in lakes, rivers, and coastal waters. Benzo(*a*)pyrene concentrations measured by EPA since 1966 in urban air, for example, were higher by one order of magnitude than in nonurban areas. Most of PAHs in fact derive from urban and industrial activities.

In the case of pesticides, the atmosphere is contaminated from drifting spray or airborne par-

Table I Relationships between K_{oc}, K_{ow}, and Water Solubility

$\log K_{ow} = 4.158 - 0.8 \log WS$	($R^2 = 0.74$)
$\log K_{oc} = 3.64 - 0.55 \log WS$	($R^2 = 0.71$)
$\log K_{oc} = -0.54 \log S + 0.44$	($R^2 = 0.94$)
$\log K_{oc} = \log K_{ow} - 0.21$	($R^2 = 1.00$)
$K_{oc} = 0.63 K_{ow}$	($R^2 = 0.96$)

Abbreviations: *WS*, water solubility in ppm, *S*, water solubility as mole fraction.

ticulate matter and through volatilization from agricultural soils. During the sixties, levels of dichlorodiphenyltrichloroethane (DDT) ranged from 0.1 ng/m^3 of air to 1560 ng/m^3, with higher levels in agricultural areas. Present ambient concentrations in areas where DDT has been severely limited are lower than 0.1 ng/m^3.

Fallout from air and rainfall are considered the main source of pollution by organic micropollutants into lakes not subjected to local inputs. The concentrations of pesticides, PCBs, and PAHs in rain and snow has been measured in many studies in different areas of the world. In Table II, some examples of the levels of micropollutants in rain and snow from different geographical areas are given. Other pesticides not included in the table have later been found in rainfall in the United States. Concentrations up to about 6000 ng/l and 2700 ng/l were detected for alachlor and metolachlor, respectively. Not enough is known about the pathways into and transport in the atmosphere of moderately stable pollutants such as these, but many of them are probably more important than has been realized to date, at least in areas of intense agriculture.

Seasonal variations are observed. The higher pesticide levels measured in summer can be related to an increase in usage: the application of lindane (γ-HCH) as a soil pesticide and of herbicides for weed control is mainly done in these areas during spring and at the beginning of summer. It is interesting to note that more persistent pesticides such as lindane and terbutylazine are detected even in winter. On the contrary, the other more degradable herbicides disappear some months after the application period. In the case of PAH, levels in wet depositions are higher in winter than in summer. The behavior of PAHs has been explained both by the increase of pollution sources in winter due to domestic heating and by a longer retention time of these compounds in the atmosphere at lower temperatures. Since aerial pollution decreases logarithmically with the increase in distance from the source, it is obvious that its impact on lakes is greater in the vicinity of anthropic activities, as can be inferred by the concentrations of micropollutants in air and rain.

In the Great Lakes region, the atmospheric input of PCBs derives mainly from wet depositions with concentrations ranging between 50 and 215 ng/l, higher by far than those of remote regions (see Table II). No significant differences in PAH concentration can be found in rain and water from different geographical areas, since these compounds are mainly associated with particles and reach the aquatic environment through dry deposition.

B. Land Runoff

Runoff from land may also contribute substantial quantities of PAHs to the aquatic environment. Runoff waters from a British motorway contained

Table II Organic Micropollutants in Rain and Snow (ng/l)

	α-HCH	γ-HCH	pp'DDE	Atr	Sim	Tba	PCBs	BaP	F	B(ghi)P
Canada, 1986										
artic snow	0.43–8.70	0.22–5.31	nd–0.07		—	—	0.02–1.76	—	—	—
United States, 1985										
winter rain	—	—	—	nd	nd	—	—	—	—	—
summer rain	—	—	—	nd–1500	nd–500	—	—	—	—	—
Switzerland										
winter rain	—	—	—	nd[a]	nd–1.1[a]	nd–9.6[a]	—	25[b]	400[b]	25[c]
summer rain	—	—	—	2.4–600[a]	1.2–121[a]	0.4–198[a]	—	4[b]	20[b]	4[b]
snow	—	—	—	nd[a]	nd[a]	nd[a]	—	35[b]	160[b]	20[b]
Italy, 1984–1986										
winter rain	—	4.4–6.9	—	—	—	—	—	—	20–36	—
summer rain	—	17–28	—	—	—	—	—	—	227–204	—
The Netherlands, 1983										
rain	—	—	—	—	—	—	—	15–37	138–204	39–68

Abbreviations: α-HCH, α-hexachlorocyclohexane; γ-HCH, lindane; Bap, benzo(*a*) pyrene; F, fluoranthene; B(*g,h,i*)P, benzo(*g,h,i* perylene; Atr, atrazine; Sim, simazine; Tba, terbutylazine; nd, not detectable.

[a] 1987.

[b] 1988–89.

up to 3.1 μg/l of total PAHs including 0.57 μg/l of benzo(*a*)pyrene (BaP). PAH land runoff to the aquatic environment is estimated at 1.2% of the total, while BaP land runoff is 17% of the total from all BaP sources.

As far as PCBs are concerned, it is well recognized that a large amount of wastes have been deposited in lagoons, landfills, and open dumps. Even if statistics are not highly accurate, an estimate of the distribution and fate of PCBs in the United States shows that 27% of the total production since 1929 is in landfills and 16% either is in the environment or has been destroyed. PCB disposal represents a potential source of pollution that can last over a long period of time.

Runoff from cultivated land is the major source of pesticide pollution to inland water bodies: the extent of this phenomenon depends on the amount applied per unit of area (yearly turnover), the composition of the soil, and the physicochemical properties of the pesticide. Most of the data on agricultural runoff refer to herbicides because they are used in large amounts on soil and include a series of compounds easily leached from soil because of their relatively high water solubility. Mobility, in fact, is proportional to water solubility and inversely proportional to K_{ow} and K_{oc}, and is exactly the opposite of adsorption capacity. Runoff losses of herbicides are highest if the runoff occurs soon after herbicide application. For atrazine, one of the most used herbicides in corn and maize growing areas, runoff losses of 0.1–2.9% of the amount applied have been calculated based on concentrations measured in rivers draining some corn-growing basins in Québec.

After the application, pesticide leaching depends on its persistence in soil. Persistent pesticides are also released from treated soils during winter thaws. In northern agricultural regions, there is extensive flooding of fields because the drainage ditches are blocked with ice. After snow melting and runoff, the pesticides released from soils are discharged from the watershed into lakes.

C. Point Sources

Thousands of chemicals enter the environment in this technological era due to chemical manufacturing activities, product utilization, and energy plants. On a local scale, the spectrum of potential pollutants is much wider than that of ubiquitous contaminants. Less persistent and accumulable chemicals may reach lake sediments if they are delivered in amounts considerably larger than the receiving capacity of the drainage basin.

The identification of local contaminants requires a detailed knowledge of human activities within a watershed, which is generally lacking; information on sources and fate are therefore mainly available for widespread pollutants. A large number of industries produce wastewaters containing PAHs. Commonly cited as industrial sources are oil refineries, the plastic and dyestuff industries, high temperature furnaces, and the lime industry. BaP concentrations in the mentioned wastewaters vary from undetectable to 1000 μg/l.

Domestic wastewaters including raw sewage and storm sewer runoff may contain significant quantities of PAHs. PAH concentrations in sewage increase substantially following heavy rains presumably because of sewer runoff from roadways (see Table III). Hydrocarbons (*n*-alkanes) and chlorinated hydrocarbons (e.g., HCHs, DDTs, PCBs) have also been determined at significant levels in urban sewages (see Table IV). Wastewater treatment is not 100% efficient and residual chemicals, including intermediates from treatment processes, may be discharged into the effluent reaching the receiving surface waters (see Table IV).

Wastewaters from tanneries may contain large amounts of chlorinated pesticides. In a tannery district in Northern Italy, unusual concentrations of

Table III Concentration of PAH (μg/l) in Domestic Sewage from Hegne (GFR)

Compound	Dry weather	During heavy rain
Fluoranthene	0.352	16.35
Pyrene	0.254	16.05
Benzo(*a*) antrhacene	0.025	10.36
Benzo(*b*) fluoranthene	0.039	10.79
Benzo(*j*) fluoranthene	0.057	9.91
Benzo(*k*) fluoranthene	0.022	1.84
Benzo(*a*) pyrene	0.001	3.84
Benzo(*g,h,i*) perylene	0.004	4.18
Indeno(1,2,3-*c,d*) pyrene	0.017	4.98
Total unidentified PAH	0.075	9.20
Total PAH	0.846	87.50

Source: From Borneff and Kunte (1965). *In* Neiff, J. M. (1979), "Polycyclic Aromatic Hydrocarbons in the Aquatic Environment." Applied Science Publishers, London. Reproduced with permission.

Table IV Chlorinated and Aliphatic Hydrocarbons (μg/l) Identified in Urban Sewage

Location	PCB	DDTs	γ-HCH	*n*-Alkanes
Toulon-east, France				
(I)	0.141±0.052	0.239±0.327	0.220±0.050	274
(E)	0.089±0.054	0.190±0.168	0.215±0.031	26
Morlaix, France				
(I)	0.046±0.034	0.013±0.009	0.045±0.026	—
(E)	0.25±0.011	0.003±0.001	0.021±0.021	—
Nantes-north, France				
(I)	0.14±0.31	<0.02±0.21	<0.02	130–453
(E)	0.08	<0.02	<0.02	74–94
Nantes-south, France				
(I)	0.17–0.44	<0.02	<0.02–0.24	80–132
(E)	<0.02	<0.02	<0.02–0.05	<0.5–3.4
Marseille, France	0.11	—	0.12	—
Oxford, U.K.	0.05–0.08	0.03	—	—
New York	0.02–0.9	—	—	—
Los Angeles	0.94	0.94	—	—
Kalamazoo River WWTP, USA	0.4	—	—	—
Various sewages, USA	0.54–3.05	0.02–1.05	0.11–0.05	—

Source: Modified from Marchand, M., *et al.* (1989). Organic pollutants in urban sewage and pollutant inputs to the marine environment. Application to the French shoreline. *Wat. Res.* **23,** 461. Reproduced with permission.
Abbreviations: I, treatment plant influent; E, treatment plant effluent.

HCH isomers and DDTs were found in waters containing one-half civil and one-half industrial discharges (see Table V). This was due to the leather conservation treatment in African countries from which the leather was imported. According to the adsorption properties of these insecticides, a considerable part of this contamination is transferred to the activated sludge (see Table V).

Rivers receive most of the urban, industrial, and agricultural discharges. In many situations, lake tributaries represent point sources of pollution for a number of organic compounds. The total quantity of dissolved pollutants in lakes increases depending upon the morphological characteristic of the lakes themselves (i.e., renewal time, volume). The pollutants vehiculated by suspended solids are depos-

Table V Concentration of Chlorinated Pesticides (μg/l) in a Treatment Plant Receiving Tannery Effluents[a]

Sample	α-HCH	γ-HCH	pp'DDT	op'DDT	pp'DDE	pp'DDD	ΣDDT
Sewer	132	53	20.2	4.6	0.5	1.5	26.8
WWTP tank[a]	72	0.9	—	—	1.0	3.2	4.2
WWTP effluent	38	0.9	—	—	0.3	0.5	0.8
WWTP mud[b]							
fresh	59	—	—	—	142	351	493
2 months	—	—	—	—	286	1182	1468
5 months	52	8	—	—	3030	9361	12391

Source: Modified from Galassi *et al.* (1983). Presenza di pesticidi clorurati in acque di scarico di conceria. *Acqua & Aria* **6,** 601. Reproduced with permission.
[a] Waste water treatment plant, homogenizing tank.
[b] In μg/kg dry weight.

ited in lake sediments according to the pattern of the current. Several examples of "hot pollution" areas have been identified in deposition zones of incoming tributaries. This fact partly accounts for the wide range of concentrations measured within a single lake (see Section V).

IV. Analysis

A. Sampling and Sample Handling

Two main categories of sediment samplers are usually employed: grabs and corers. Grabs produce large samples of known surface from the top layers of sediment, but the degree of penetration into the sediment is not predictable. Furthermore, grab sampling causes a perturbation of the water/sediment interface and a sediment mixing. Grab samplers, therefore, do not allow the attribution of the collected sample to the corresponding sedimentation period. Corers, on the contrary, give a lower disturbance of the sediment and also allow the collection of deep layer sediments. Historic patterns of pollution may therefore be reconstructed, analyzing the sediment layers obtained by cutting the core in slices corresponding to a certain period of time, provided that the sedimentation rate of the lake is known. The sedimentation rate is usually calculated measuring the ^{137}Cs decay, taking into account the effect of density and porosity of the sediment.

B. Extraction

Sediment samples are usually freeze-dried before extraction. However, organic micropollutants can also be recovered from wet sediment, providing that extraction solvents of decreasing polarity are used. Dehydrating chemicals, such as anhydrous sodium sulfate, can also be added to wet sediment instead of drying it or using intermediate solvents.

Numerous solvents have been used, in conjunction with a solid extraction technique, to recover organic micropollutants bound to sediments. Due to their low polarity, hydrocarbons such as pentane, hexane, or cyclohexane are employed for lypophilic compounds such as *n*-alkanes, PCBs, and DDTs. More polar solvents such as benzene or methylene chloride are usually used for the extraction of PAHs, phthalates, and herbicides.

Continuous solvent extraction is generally performed with a Soxhlet apparatus, but good recoveries are also obtained by sonication. This last technique is much faster than Soxhlet extraction and requires very small amounts of solvent. Since many solvents may contain some impurities that interfere with subsequent analytical determinations, in many cases it is very convenient to limit the volume of the extractant. This is the case with phthalates that are ubiquitous contaminants of solvents, glassware, and laboratory equipment.

C. Cleanup

Many natural compounds of biogenic origin may occur in lake sediments. In productive lakes, much of the organic detritus accumulating in surface sediments is derived from biota occurring within the water column. The lipid fraction of these sediments includes *n*-alkanoic acids, branched/cyclic alkanoic acids, alkenoic acids, and *n*-alkanes.

Diagenetic compounds are coextracted with organic micropollutants, with which they may interfere during the analytical determinations. For this reason, cleanup procedures are usually required before analysis in order to prevent contaminations of the analytical apparatus and to render analysis easier.

Column chromatography has been extensively used to separate organic micropollutants from other organic compounds, as well as one class from another among the former. Thin-layer chromatography (TLC) may also be used as a separation technique.

Alumina, silica, Sephadex, and Florisil column chromatography have been widely applied for the cleaning-up of sediment extracts. However, the cleanup procedure should be chosen as a function of the subsequent analysis.

D. Analytical Procedures

Gas-liquid chromatography (GC) is the primary technique for the separation and quantitative determination of organic micropollutants. Due to particularly difficult separation problems, high-resolution gas chromatography is required for the analysis of environmental samples. Flame ionization detection is usually coupled to GC for *n*-alkane, PAH, and phthalate determinations. Electron capture detection is used for PCBs and DDTs and selective nitrogen/phosphorus detection for herbicides.

High-performance liquid chromatography (HPLC) has also been applied to the analysis of PAHs. Fluorescence spectrometry coupled with HPLC allows a very sensitive analysis of PAHs without the need for complicated cleaning-up procedures, due to the high selectivity of the detector.

One drawback of GC and HPLC analysis is the lack of positive identification of the compound which gives rise to a certain peak. GC–mass spectrometry (GC–MS) allows unequivocal quantitative analysis. However, since the use of the mass spectrometer as a detector often gives lower sensitivities than selective detectors, the use of GC–MS can not always be applied. GC analyses making use of capillary columns of different polarities, HPLC, and TLC may be used as confirmatory techniques.

V. Environmental Data

A number of data are available for those compounds that have a worldwide distribution (such as PAHs and chlorinated hydrocarbons), but only scattered informations are reported for the other organic micropollutants. An overview of concentration levels from different lacustrine environments and within a single lake may give an insight of the pollution status, provided that the data are obtained by similar analytical procedures. For this reason, the most recent data have been selected, and only in a few cases have the older ones been reported for comparison.

A. PAHs

As PAHs derive from various sources, the resulting PAH compounds in sediments are complex and contain a great number of individual PAHs that can not be resolved even by the modern analytical techniques. Characterization is limited to the detection of some individual species, primarily benzo(*a*)pyrene (BaP), anthracene (A), and fluoranthene (F) (see Table VI), although in a few cases up to 64 compounds have been identified. A selection of these data is presented in Table VII.

Fluoranthene is one of the main anthropogenic PAH compounds and is an excellent representative of this class since it is abundant in sediment, is associated with a combustion origin, and is correlated with the majority of the other PAHs with the exception of perylene, dibenzo(*a,c*/*a,h*) and indeno(*c,d*)pyrene. It is generally agreed that fluoranthene concentrations in sediment lower than about 400 μg/kg are typical of rural or remote areas while higher values are found in industrial or urban areas. If there are no local sources and street runoff is negligible, PAH concentration is in good agreement with the atmospheric load.

Since PAHs are also formed by biogenesis, a background level is to be expected even in unpolluted sediments. Preindustrial values of F are generally in the order of 10 μg/kg or less. Since then, concentrations have increased with different patterns in the various geographical areas, according to usage of fossil fuels. Examples are presented in Figure 1.

The concentration of BaP has been measured in a large number of freshwater sediments, primarily from European lakes (see Table VI). Sediment BaP concentrations may vary by a factor of 1000 within a single lake, possibly because of different sedimentation rates or degrees of biogenesis in different regions of the lake. When sedimentation rates are high, a seasonal cycle may be detected, not so much because of a changing equilibrium between biosynthesis by sediment anaerobes and degradation by sediment aerobes, but mainly because of the possible seasonal variation in the rate of BaP input.

A high value of BaP in Lake Cayuga sediments was measured in a sampling station located below a highway bridge. This behavior is frequently observed in other lentic environments.

B. Chlorinated Hydrocarbons

Chlorinated pesticides and PCBs have been recognized as a very relevant ecological problem since the sixties and for this reason they have been banned or limited in most European and North American countries. In spite of these restrictions, surficial lake sediments from these areas remain polluted by these contaminants (see Table VIII). Among the reported data, only in the most recent sediments (1986) from Lake Superior PCB concentrations are significantly lower than those determined several years earlier.

The positive consequences of the adopted regulation of PCB and DDT usage are well exemplified by the historical trend of their concentration in

Table VI Benzo(*a*)pyrene (BaP), Anthracene (A), and Fluoranthene (F) in Lake Sediments (μg/kg)

Location	BaP	A	F
Rubleskoye Reservoir, USSR	44	—	—
Khiminskoe Reservoir, USSR	390–500	—	—
Grosser Ploner See, FRG	1610	—	560–300
Lake Constance, FRG	443	—	400–1000
Greifensee, Switzerland	—	—	420
Lake Varese, Italy	69	18	82–192
Lake Comabbio, Italy	37–118	5–17	89–253
Lake Monate, Italy	25–41	3–8	62–128
Lake Superior, USA	—	nd–10	36–153
Lake Sagamore, USA	128	21	80–463
Woods Lake, USA	690	32	320–1250
Lake Cayuga, USA	nd–2000	nd–620	19–3900
Mirror Lake, USA	nd	nd	600–1000
Northern Lake George, USA	—	—	160
Marion County Lake, USA	—	—	140
Lake Luise, USA	—	—	375
Lake Haiyaha, USA	—	—	96
Lake Husted, USA	—	—	110
Lake Loch, USA	—	—	80
Lake Meade, USA	—	—	10
Mono Lake, USA	—	—	13–74
Utah Lake, USA	—	—	70

Abbreviation: nd, not detectable.

Lake Ontario sediments (see Fig. 2). A continuous increase was observed with the maximum peak occurring in the late 1950s for DDTs and in the mid-1960s for PCBs. In both cases, there is a good agreement between the core record and the production or history of use. The occurrence of DDT in the surficial sediment of African lakes (see Table VIII) is probably due to its present use in developing countries.

C. Phthalate Esters

Phthalate esters were introduced into the high polymer materials market in the 1920s and bis(2-ethylexyl)phthalate (DEHP), the most highly produced phthalate ester, was first synthesized in 1933. Since the 1940s, their production rates have shown a rapid increase. The levels in sediment chronologically follow the increasing commercial use of these compounds (see Fig. 3).

As previously pointed out, serious contamination problems may be associated with the analysis of these substances, as phthalate esters are ubiquitous contaminants and are present in some common laboratory items. For this reason only a few data on sediment are available (see Table IX).

D. Phosphate Esters

Phosphate esters have been widely used since the 1940s as fire retardant plasticizers in synthetic polymers, hydraulic fluids resistant to high temperatures, and lubricant additives. This class of compounds is very large, including akyl, aryl, aryl/alkyl, and haloalkyl phosphates. The most frequently detected in sediments are the less water soluble aryl/alkyl phosphates (see Table X). However, even the more soluble phosphates are detected in lake sediments when direct discharges are present. A typical example is given by tris-

Table VII Individual PAHs (μg/kg) Found in Lake Sediments

	Lake Superior	Lake Sagamore	Woods Lake	Lake Cayuga
Phenanthrene	62–98	154	324	49–660
Pyrene	21–88	320	934	<22–3500
Benzo(*a*)fluorene	—	91	294	—
Benzo(*a*)anthracene	29–65	78	362	<9–1700
Chrysene/triphenylene	73–172	191	888	14–2000
Benzo(*b*)fluoranthene	104–319	358	1784	nd–3500[a]
Benzo(*k*)fluoranthene	49–197	115	558	—
Benzo(*e*)pyrene	51–103	165	768	nd–1500
Perylene	—	605	280	nd–540
Indeno(*c,d*)pyrene	91–404	315	1294	7–1500
Dibenzo(*a,c/a,h*)anthracene	15–125	30	92	nd–870
Benzo(*g,h,i,*)perylene	41–216	303	1356	nd–1600
Coronene	—	188	801	—
Dibenzo(*a,e*)pyrene	—	15	394	—
Retene	—	191	270	—
7,12α-dimethyl-1,2,3,4,4α 11,12,12α-octahydrochrysene	—	90	100	—
3,4,7-trimethyl-1,2,3, 4-tetrahydrochrysene	—	20	200	—
Benzo(*g,h,i*)fluoranthene	—	—		nd–380
Dibenzothiophene	—	—	—	nd–260
Fluorene	4–8	—	—	—

Source: Modified from Heit, M. (1985). The relationship of a coal-fired power plant to the levels of polycyclic aromatic hydrocarbons (PAH) in the sediment of Cayuga Lake. *Water Air Soil Pollution* **24,** 41. Reproduced with permission.

Abbreviation: nd, not detectable.

[a] Sum of benzo(*b*) and benzo(*k*) fluoranthene.

monochloroisopropyl phosphate (TCIPP) in Lake Varese and Comabbio (see Table X), where concentrations in water up to 18 μg/l were also measured. The rapid increase of the synthetic polymer industry corresponds to the levels in sediment during the last years (see Fig. 4).

E. Aliphatic Hydrocarbons

Petroleum hydrocarbons from oil spills are able to persist in sediments for a long time. The odd-carbon predominance reflects the composition of organic matter of biogenic origin. *n*-Heptadecane (C_{17}) in particular is very abundant in surficial sediments of productive lakes because it is produced by algae decomposition. The isoprenoid hydrocarbon pristane (C_{20}), on the contrary, is one of the major constituents of many crude oils.

The concentration of total aliphatic hydrocarbons in surficial sediments of polluted lakes is higher than that in polluted sediments at least by a factor of ten (see Table XI). *n*-Alkanes represent a small fraction of total hydrocarbons. The *n*-C_{17}/pristane ratio gives an idea of the occurrence of natural hydrocarbons in comparison with those that are anthropogenic.

VI. Bioavailability and Environmental Hazards

A. Sorption/Desorption

The partitioning of a chemical between sediment and water can be described by

$$K_p = C_s/C_w,$$

Figure 1 BaP concentrations in the annual layers of sediments from the northern (built-up area, upper line) and southern (forest, lower line) regions of the Grosser Ploner See (GFR). [From Grimmer, G., and Bohnke, H. (1975). Profile analysis of polycyclic aromatic hydrocarbons and metal content in sediment layers of a lake. *Cancer Lett.* **1,** 75. Reproduced with permission.]

Table VIII Concentrations of Chlorinated Hydrocarbons (μg/kg) in Lake Sediments[a]

Lake	Year	PCB	ΣDDT	pp'DDT	pp'DDE	α-HCH	γ-HCH	HCB
Siskiwit Lake, Canada	1983	48	—	—	—	—	—	—
Lake Superior, USA	1973	<2.5–57 [9]	—	— (0.5)	<0.25–23	—	—	—
	1977	5–390 (170)	—	—	—	—	—	—
	1986	5.3–11.7 [8.6]	—	—	—	—	—	—
Lake Ontario, USA	1968	<5–280 (57)	0.4–218 (42.8)	—	0.4–70 (12.7)	—	—	—
	1981	(570)	(141)	(18)	(51)	(1.5)	(1)	(100)
Lake Michigan, USA	1977–80	1–2370	—	—	—	—	—	—
Lake Huron, USA	—	9–33	—	—	—	—	—	—
Lake Erie, USA	—	74–252	—	—	—	—	—	—
Lake Geneva, Switzerland	1979	20–540 (43)	0.5–49 (9.8)	— (1.5)	— (3.7)	— (1)	— (1)	— (1)
Lake Garda, Italy	1989	(89)	(14)	(0.9)	(9)	(0.3)	(0.6)	(0.4)
Lake Varese, Italy	1988	165	—	—	3.7	—	—	—
Lake Comabbio, Italy	1988	87	—	—	11.8	—	—	—
Lake Monate, Italy	1988	43	—	—	0.6	—	—	—
Lake McIlwaine, Zimbabwe	1988	—	—	32–146 (76)	—	—	—	—

Abbreviation: HCB, hexachlorobenzene.
[a] Values in brackets represent means.

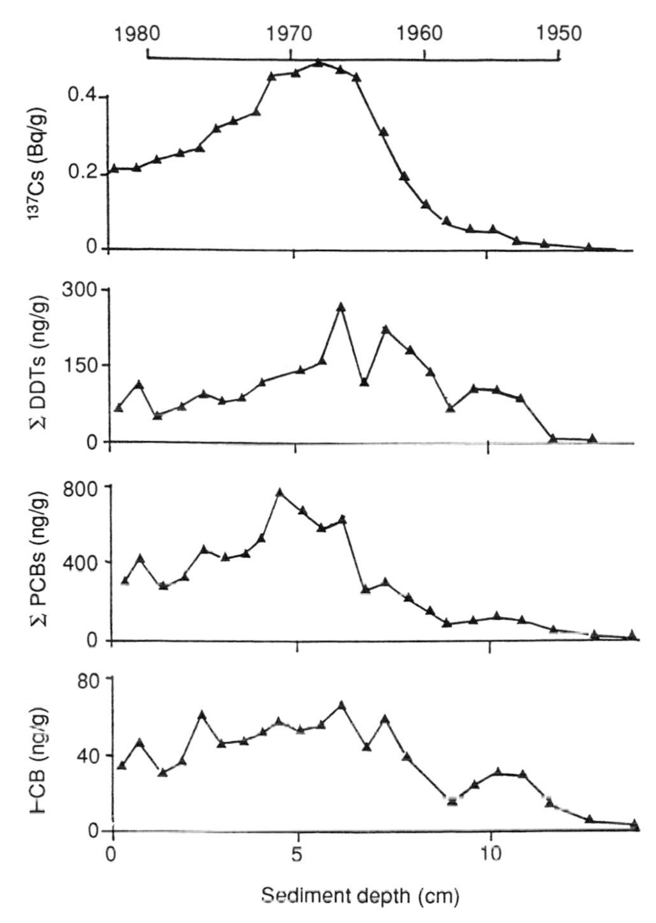

Figure 2 Chlorinated hydrocarbon concentrations versus sediment depth and age. [Modified from Oliver, B. G. *et al.* (1989). Distribution, redistribution, and geochronology of polychlorinated biphenyl congeners and other hydrocarbons in Lake Ontario sediments. *Environ. Sci. Technol.* **23**, 200. Reproduced with permission.]

where K_p is the sediment/water partition coefficient, C_s is the amount of chemical sorbed per unit mass of sediment, and C_w is the amount of chemical in aqueous phase per unit mass of solution.

Equilibrium K_p values for a given hydrophobic solute sorbed on a series of sediments are proportional to the fraction of organic carbon (f_{oc}) of the sediment, being

$$K_{oc} = K_p/f_{oc},$$

where K_{oc} is the partition coefficient on sediment as defined in Section II.

The availability of sorbed chemicals to aquatic organisms depends on the rate of desorption relative to the exposure time. The rate of desorption appears to depend on the magnitude of K_p for hydrophobic solutes. As a first approximation, the characteristic time (in minutes) for slow desorption seems similar to the numerical value of K_{oc}. In a lake, the possibility of reaching the equilibrium condition between water, suspended solids, and sediments depends not only on the lipophilic nature

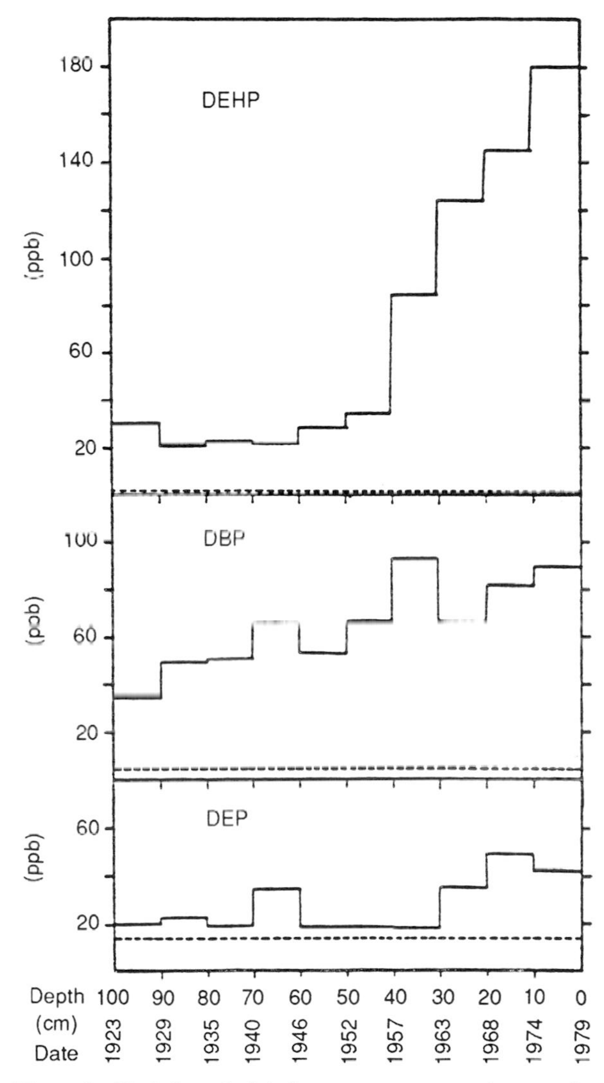

Figure 3 Variation of phthalate ester concentration as a function of sediment depth and time. The dotted lines represent background levels. [From Peterson, J. C., and Freeman, D. H. (1982). Phthalate ester concentration variations in dated sediment cores from the Chesapeake Bay. *Envir. Sci. Technol.* **16**, 464. Reproduced with permission.]

Table IX Concentrations of Phthalate Esters (μg/kg) in Lake Sediments

Lakes	DEHP	DBP	DEP	DPP	DAP	DIBP	BBP
Chesapeake Bay, USA	12–180	27–89	22–42	5.9–13	<1	5.6	—
American lakes (several)	—	—	—	—	—	—	428
Lake Constance, Switzerland	500	200	—	—	—	—	—
Lake Superior, USA	200	100	—	—	—	—	—

Abbreviations: DEHP, bis(2-ethylhexyl)phthalate; DBP, dibutylphthalate; DEP, diethylphthalate; DPP, dipropylphthalate; DAP, di(2-propenyl)phthalate; DIBP, bis(2-methylpropyl) phthalate, BBP, butylphenylmethyl phthalate.

of the chemical but also on certain characteristics of the water body (residence time, water stratification, sedimentation rate, etc.) and on the type of the immission of the contaminant (continuous, discontinuous, or occasional) into the lake.

As a general guideline, under equilibrium conditions, all the water from the top of the water column to the bottom of the biotic zone in a settled sediment can be considered as sediment interstitial water. In this case, the exposure concentrations or bioaccumulability can be calculated on the basis of partitioning.

Some exceptions to this general behavior may be observed for polar and ionizable organic compounds. In these cases, the equilibrium condition between water and sediment is not governed by the organic carbon content because of specific interaction with sediment surface functional groupings. The same seems to be true for PAH, since some reported data show that water PAH concentrations are between two and five orders of magnitude lower than those calculated from sediment concentrations, assuming equilibrium partitioning.

Table X Concentrations of Phosphate Esters (μg/kg) in Lake Sediments

Lakes	TPP	NPDPP	CPDPP	TCIPP
Lake Michigan, USA				
Waukegan Harbor	0.01	0.2–0.6	0.05–0.2	—
Waukegan Bay	nd	nd	nd	—
Lake Varese, Italy	—	—	—	0.30
Lake Comabbio, Italy	—	—	—	0.86
Lake Monate, Italy	—	—	—	nd

Abbreviations: TPP, triphenylphosphate; NPDPP, nonylphenyldiphenylphosphate; CPDPP, cumylphenyldiphenylphosphate; TCIPP, trismonochloroisopropylphosphate; nd, not detectable.

B. Biological Effects

The main route for bioaccumulation at the lower trophic levels of the food web is the direct assumption of the chemicals from water. However, in freshwater environments, some organisms, such as oligochaetes and chironomids, are sediment ingesters. Benthic invertebrates could therefore represent an important dietary source of contaminants for predators. Furthermore, these organisms are also responsible for bioturbation of settled sediments causing sediment mixing. Contaminated sediments of the biotic zone could be resuspended and become available for exchanges with the overlying surface water. Deep sediments can be more contaminated than surficial ones because of previous emmissions of pollutants into the water body or to sedimentation of contaminated material at higher rates than those needed to establish equilibrium exchanges between suspended matter and water.

Besides the risk for bioaccumulation, the con-

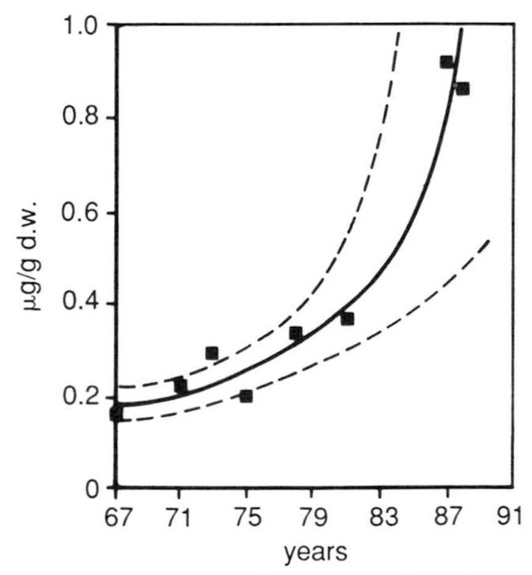

Figure 4 Temporal trend of trismonochloroisopropyl phosphate concentration in Lake Comabbio sediments.

Table XI Concentrations of Aliphatic Hydrocarbons (μg/kg) in Lake Sediments

Lakes	Total	*n*-alkanes	*n*-$C_{17}H_{36}$/pristane
Lake Constance, Switzerland	320	—	—
Lake Zug, Switzerland	50–900	—	1–2.8
Lake Washington, USA	1400	10	—
Various unpolluted lakes, U.K.	37	—	—

taminants in sediments might represent a potential source of toxicity to the organisms. An understanding of sources and effects of complex mixtures requires the complete chemical characterization of sediments. However, residue analyses are time-consuming and expensive, and they enable one to determine toxicant concentrations only as a function of the efficiency of the extraction and not of their availability to biota. Therefore, bulk chemical analysis of sediment cannot be used directly to predict biological responses. In addition, the concentrations of single chemicals are usually compared with known dose–response relationships, while the global effects of toxicants in the real environment are also the result of the interactions among toxicants.

Because of these problems, biological assays have been proposed to screen the sediment quality and the potential risk to benthic organisms. Two procedures are usually employed to perform ecotoxicological tests on sediments. The first uses aqueous extracts (elutriate), giving an idea of the potential toxicity of resuspended sediments in the water column. The second uses the whole sediment, diluting it with reference noncontaminated sediment, in order to obtain a dose–response relationship.

An example of toxicity of harbor sediment in the Lake Ontario is shown in Figure 5. In this case the

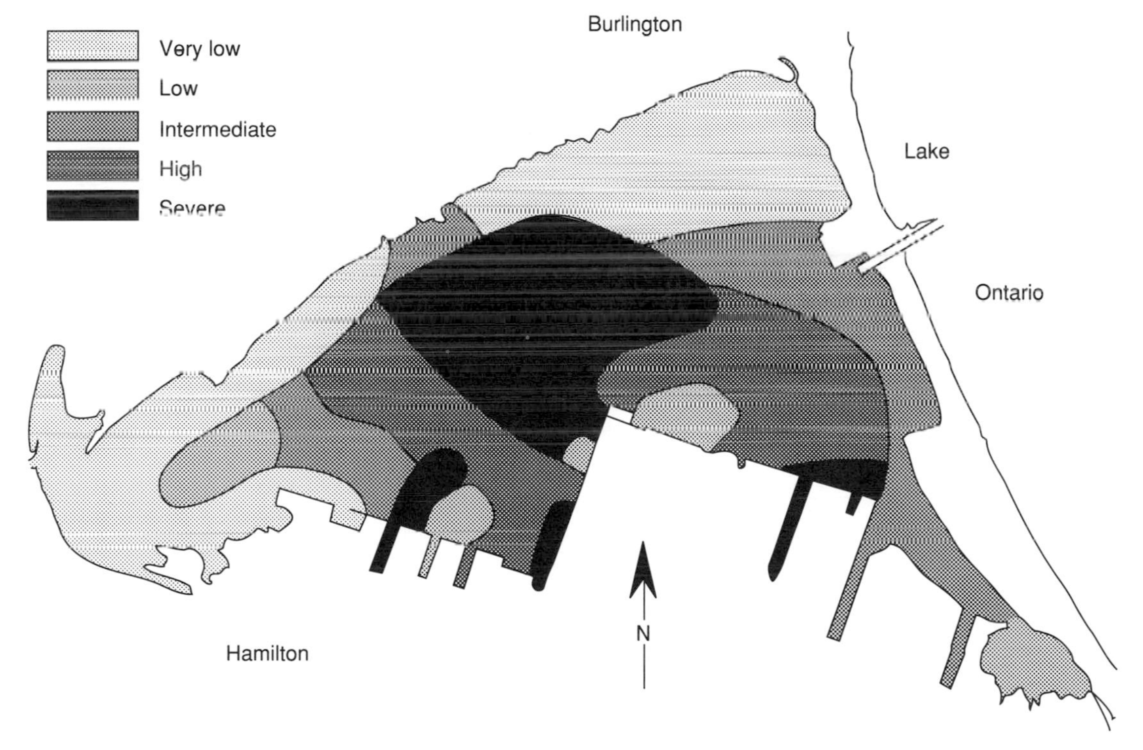

Figure 5 Toxicity of Hamilton Harbor sediments to *Photobacterium*. [From Brouwer, H., *et al.* (1990). A sediment-contact bioassay with *Photobacterium phosphoreum*. *Environ. Toxicol. Chem.* **9,** 1353. Reproduced with permission.]

toxicity was determined on *Photobacterium* using whole sediment. The relative importance of the interstitial water and the whole sediment depends on organism and toxicant type. It is therefore generally agreed that both elutriate and sediment should be tested with a battery of tests, including a series of species representative of different trophic levels.

In conclusion, even if there is a general consensus for developing and promulgating sediment water criteria, it is recognized that at this time it is not feasible to establish numerical sediment quality criteria, because of the lack of a unified approach to assess the potential hazard of sediment-bound chemicals.

Related Articles: AQUATIC TOXICOLOGY, ANALYSIS OF COMBINATION EFFECTS; CHROMIUM; SOIL DECONTAMINATION.

Bibliography

Baker, J. E., and Eisenreich, S. J. (1989). PCBs and PAHs as tracers of particulate dynamics in large lakes. *J. Great Lakes Res.* **15,** 84–103.

Baudo, R., Giesy, J. P., and Muntau, H., eds. (1990). "Sediments: Chemistry and Toxicity of In-Place Pollutants." Lewis Publishers, Ann Arbor, MI.

Dickson, K. L., Maki, A. W., and Brungs, W. A., eds. (1987). "Fate and Effects of Sediment-Bound Chemicals in Aquatic Systems." Pergamon Press, New York.

Donze, M., Nieuwendijk, C., and Van Boxtel, A. (1990). "Aquatic Pollution and Dredging in the European Community." Delwel Publishers, The Hague.

Galassi, S., Provini, A., and De Paolis, A. (1990). Organic micropollutants in lakes: a sedimentological approach. *Ecotoxicol. Environ. Safety* **19,** 150.

Kenaga, E. E., and Goring, C. A. I. (1980). Relationship between water solubility, soil sorption, octanol–water partitioning, and concentration of chemicals in biota. *In* "Aquatic Toxicology," (J. G. Eaton, P. R. Parrish, and A. C. Hendricks, eds.). American Society for Testing and Materials, STP 707, 78–115.

Oliver, B. G., Charlton, M. N., and Durham, R. W. (1989). Distribution, redistribution, and geochronology of polychlorinated biphenyl congeners and other chlorinated hydrocarbons in Lake Ontario sediments. *Environ. Sci. Technol.* **23,** 200–208.

Rand, G. M., and Petrocelli, S. R., eds. (1985). "Fundamentals of Aquatic Toxicology." Hemisphere Publishing Corporation, Washington, D.C.

Waid, J. S., ed. (1987). "PCBs and the Environment," vol. III. CRC Press, Boca Raton, FL.

Organic Solvents, Health Effects

Laura S. Welch, Katherine Hunting, and Brenda Cuccerini
George Washington University School of Medicine

Glossary

Biologic monitoring Direct measurement of levels of agents or their metabolites in biological materials.

Biotransformation Metabolism of chemicals by tissue enzymes as a first step in clearance from the body.

Dermatitis Any inflammation of the skin.

Detoxification Process which makes a substance less toxic and prepares it for excretion from the body.

Gastroschisis Failure of closure of the embryonic anterior wall at the midventral line.

Hemolysis Lysis or breaking of circulating red blood cells.

Hydrocarbon Compound consisting of the elements hydrogen and carbon.

Idiosyncratic Pertaining to a characteristic peculiar to an individual's physical or mental constitution.

Induction Production of a specific morphogenetic effect or determination of the developmental fate of a cell or tissue.

Inhibition Suppression of a process or enzyme activity by the reversible and usually noncovalent binding of a substance to an enzyme.

Lipophilic Having an affinity for lipids or being soluble in fat-like materials.

Metabolites Any of the compounds produced by enzyme activity or breakdown of a parent compound by a metabolic process.

Potentiation Making one substance more toxic by another which by itself is not toxic.

Promoter Agent which enhances tumor development when administered after a carcinogen.

Psychometric Pertaining to the measurement of mental ability, aptitudes, interests, manual ability, or special ability.

Surveillance Program whose objective is monitoring to identify and prevent injuries or illness.

Volatile Characterized by rapid evaporation.

A SOLVENT is a substance (usually a liquid) that dissolves another substance. Solvents are either water based or hydrocarbon based; an organic solvent, functionally defined, is a hydrocarbon-based liquid compound which can dissolve other substances to form a solution. It is on this group that we will focus in this article. This class of chemicals is very broad and includes aliphatic hydrocarbons, cyclic hydrocarbons, aromatic hydrocarbons, ketones, amines, esters, ethers, glycols, alcohols, aldehydes, and carboxylic acids.

I. Introduction

Solvents are classified into groups based on their chemical structure: aliphatic, alicyclic, and aromatic. They can then be classified by the presence of certain functional groups: glycols, ketones, alcohols, esters, ethers, and others. Understanding these basic structures may allow us to extrapolate data from one solvent to another, if the two have similar structures. Table I describes some major classes of organic solvents and lists examples within each class. Exposures to solvent mixtures are also common. Organic solvents as a class are relatively volatile; thus inhalation exposures are very important. Another property of solvents that influences exposure is their lipophilic (lipid-loving) nature, which means that they can be absorbed through the skin.

Table I Basic Chemical Structure and Examples of Some Common Organic Solvents

Chemical classification	Sample structure	Examples
Aromatic hydrocarbons: Characterized by the presence of hydrogen and carbon atoms arranged in a benzene ring structure.	H H C – C H – C C – H C = C H H Benzene (C_6H_6)	Benzene Toluene Xylene
Halogenated hydrocarbons: Hydrogen and carbon atoms joined by one or more halogen atom (fluorine, chlorine, bromine, or iodine).	Cl Cl – C – Cl Cl Carbon tetrachloride (CCl_4)	Chlorodifluoromethane Monochloromonofluoromethane Carbon tetrachloride Methylene chloride Perchloroethylene
Aldehydes: Characterized by a double bonded carbonyl (C = O) group joined by at least one hydrogen atom.	O C H H Formaldehyde (HCHO)	Formaldehyde
Glycol ethers: One of the two hydroxyl (OH) groups of a glycol is replaced by an ether. Ethers contain the C—O—C linkage.	H H H H – C – O – C – C – O – H H H H Ethylene Glycol Monomethyl Ether ($C_2H_3OCH_2OH$)	Ethylene glycol monomethyl ether (EGME), Ethylene glycol monethyl ether (EGEE), and their acetates Propylene glycol ether
Ketones: Contains a double-bonded carbonyl (C = O) group joined by two hydrocarbon groups.	H O H – C – C – H H H C – C – H H H Methyl Ethyl Ketone ($CH_3C(O)C_2H_5$)	Methyl ethyl ketone Acteone

Source: From Paul, M., ed. (1993). "Occupational and Environmental Reproductive Hazards: A Guide for Clinicians." Williams and Wilkins.

Solvents are used for cleaning, degreasing, thinning paints and other coatings, and other uses. Exposures can occur in the workplace or in the community; many household products contain organic solvents; and solvents are often found in community air or water from industrial releases. There are over 30,000 industrial solvents, which can be classified by common properties into subclasses. Workers in thousands of manufacturing, construction, service, and retail occupations—from the service station attendant pumping gasoline, to the electronics worker cleaning metal components, to the plastics manufacturing worker mixing a myriad of solvents—are exposed to solvents. The production of dyes, polymers, plastics, textiles, printing inks, agricultural products, and pharmaceuticals all involve solvents. Solvents are also present in many paints, adhesives, glues, coatings, and degreasing or cleaning agents. NIOSH estimated from data collected in the early 1970s that approximately 9.8 million U.S. workers were potentially exposed to organic solvents. Table II gives recent estimates from NIOSH on the number of workers exposed to a selection of common organic solvents.

Organic solvents have a broad range of human health effects. They can cause skin and gastroin-

Table II Exposure to Original Solvents in the United States

Solvent exposure	Number of workers
Acetone	1,740,181
Benzene	272,286
Dichloromethane/Methylene chloride	1,438,207
Epichlorohydrin	80,167
Ethyl alcohol	2,069,131
Ethylene glycol	1,511,030
N-Hexane	643,104
Isopropyl alcohol	4,665,549
Methyl chloroform (ethane,1,1,1-trichloroethane)	—
Methyl ethyl ketone (2-butanone)	1,447,465
Methyl chloride	—
Styrene	333,210
Toluene	2,015,883
Trichloroethane	2,528,268
Trichlorethylene	401,366
Tetrachloroethylene	668,099
Xylene	2,145,035

Source: Unpublished provisional data as of 7/1/90. NIOSH, National Exposure Survey (1981–83), Cincinnati, Ohio. USDHHS/PHS/CDC/NIOSH surveillance Branch.

testinal disorders, and impair hematologic, reproductive, and central and peripheral nervous system function. These effects will be summarized in the sections that follow.

II. Absorption of Solvents

Certain properties of solvents affect exposure of the workers using them. Solvents vary in lipid and water solubility. Solvents that are both lipid and water soluble pass through intact skin most easily, for skin has both water and lipid compartments. For those solvents that are readily absorbed through the skin, this route can represent the major route of exposure.

The property of a solvent affecting inhalation is its volatility, its tendency to evaporate into a gas or a vapor. As a general rule, the higher the volatility of a solvent, the more will be present in the breathing zone of a worker, and hence the higher the exposure. Solvents are readily absorbed across the alveolar capillary membrane of the lung. From 40 to 80% of the inhaled dose is absorbed at rest, and the total amount absorbed increases with exercise and pregnancy as both bloodflow to the lung and alveolar ventilation increase.

Once absorbed, solvents are distributed through the body, and in particular to lipid rich tissues. They may be excreted unchanged by exhalation from the lungs, or metabolized in the liver and excreted in the urine. The half-life of a solvent in the body varies greatly from compound to compound, with some as long as several days; solvents with a half-life longer than 12 hr will accumulate over the work week, resulting in a higher body burden at the end of the week from the same exposure conditions.

Exposure to some solvents can be determined by biological monitoring (i.e., measuring the solvent or a metabolite in urine or blood). These measurements can be very useful in monitoring exposure to a solvent that is absorbed through the skin; industrial hygiene monitoring in the workplace can only approximate inhalation exposure. In addition, biological monitoring allows determination of exposure in an individual. The usefulness of such monitoring depends on the solvent—ones that are excreted rapidly may not be "captured" by a test performed some time after work, and in solvents with short half-lives, the levels of metabolite may fluctuate so much from hour to hour as to make the tests an imperfect measure of overall exposure.

III. Effects of Solvents on the Skin

Solvents dissolve oils and fats, and through this mechanism cause drying of the skin. Repeated or prolonged exposure to organic solvents can cause cracking of the skin, and in some cases can lead to the development of a chronic dermatitis (inflammation of the skin). The type of dermatitis seen from solvents is called an irritant contact dermatitis; it is not due to an allergy, but rather to chronic skin irritation from the drying effects of solvents. Irritant contact dermatitis can become a chronic condition that does not reverse even though solvent exposure is reduced or stopped. It can be prevented by reduction in solvent use, for example by substituting waterless hand cleaners for solvents for cleaning, and by use of system design or gloves to reduce skin contact with the solvent. Solvent manufacturers can recommend the appropriate glove for each solvent; the glove materials vary in their resistance to specific solvents.

IV. Effects of Solvents on the Blood and Bone Marrow

Two solvents are known to have significant toxic effects on the blood producing system: benzene and ethylene glycol ethers. Benzene causes both bone marrow suppression and leukemia; the ethylene glycol ethers cause bone marrow suppression, but have not been shown to cause leukemia. NIOSH estimates that 272,000 workers in the United States are exposed to benzene.

Ethylene glycol ethers are widely used in industry in paints, varnishes, thinners, resins, textile printing, and a variety of coating operations. There are now over 100 glycol ethers available commercially, and not all carry the same reproductive risk. NIOSH estimates that 850,000 workers in the United States are potentially exposed to 2-ethoxyethanol, 2-methoxyethanol, and their acetates, the ones most clearly associated with bone marrow suppression.

Exposure to either of these solvents in sufficient quantity causes decreased production of red cells, white cells, and platelets from the bone marrow, resulting in a range of medical problems such as bleeding, increased risk of infection, and fatigue. Severe bone marrow suppression, called aplastic anemia, can be fatal. Bone marrow suppression has been found in human populations exposed to either solvent at exposure levels present in the current industrial environment.

2-Butoxyethanol, one of the ethylene glycol ethers, causes hemolysis (lysis of circulating red blood cells) in animal experiments; this has not been seen in human populations, possibly because this effect only occurs at a relatively high exposure.

V. Reproductive Effects of Solvents

There have been many studies of solvent exposure and adverse reproductive outcome in both humans and animals. Maternal exposure to organic solvents as a group is strongly associated with an increased risk for spontaneous abortion and birth defects. Some studies suggest an association with menstrual disorders and preeclampsia (severe high blood pressure in pregnancy), but the data here are more limited. One family of solvents, ethylene glycol ethers, has been shown to cause testicular dysfunction in exposed men. In addition, there are some data that paternal exposure to organic solvents is associated with spontaneous abortion and childhood cancer.

Animal studies demonstrate that solvents readily cross the placenta. Factors known to affect the rate of transfer of drugs and chemicals from the mother to the embryo include the degree of lipid solubility, placental blood flow, placental function, the strength to which the compound is protein bound, the pK_a of the compound, and the degree to which active transport occurs. Active transport is not thought to be of major importance. Generally, compounds that have a high affinity for lipid, those with a low degree of ionization, and those with a molecular weight of less that 1000 are rapidly transferred across the placenta; many solvents fit this description, for by definition they are lipid soluble and nonpolar.

A. Infertility

Ethylene glycol ethers and their acetates, discussed above under blood and bone marrow, have been shown in animal and human studies to have an effect on sperm count and fertility.

With chronic oral or inhalation exposure, ethylene glycol ethers cause focal testicular atrophy and disruption of the seminiferous tubules in mice, rats, and rabbits. Two studies of men exposed to EGME or ethylene glycol monoethyl ether (EGEE) have found a decreased total sperm count as well. In evaluating all the data, one should conclude that ethylene glycol ethers present a reproductive risk to male workers. The animal data are strong and reproducible, and an effect is seen near levels achieved in workplaces; in addition, two human studies have found an effect. NIOSH recently established a recommended exposure limit of 0.1 ppm EGME and 0.5 ppm EGEE. OSHA is currently preparing an updated assessment that reflects these data, and the permissible exposure limits will be reduced in the future.

It is likely that the toxicity of the ethoxyethanol series of glycol ethers is due to active metabolites, rather than to the parent compound. 2-Ethoxyethanol is oxidized by alcohol dehydrogenase and aldehyde dehydrogenase to ethoxyacetic acid; 2-methoxyethanol is similarly metabolized to methoxyacetic acid. Inhibition of this metabolism protects against toxicity, and administration of the metabolites causes the toxicity seen with the parent compound *in vivo*. As a result of this hypothesis and set of experiments, glycol ethers that are not metabolized by these enzymes and do not produce an aloxyacetic acid metabolite are thought not to have reproductive effects.

B. Menstrual Disorders

Several studies report that women who work with benzene, toluene, xylene, styrene, carbon disulfide, and formaldehyde have an increased incidence of menstrual disorders. This group of studies comes from Eastern Europe, but the data reported make an assessment of exposure or of case ascertainment difficult; one cannot determine the dose of solvent nor the other exposures present. These results have not been replicated in other studies to date.

A well-conducted study in the United States of over 1500 workers found no association between styrene and menstrual disorders; this degree of detail is not available on the other solvents suggested to have a problem from the Eastern European studies. At this point, it is premature to conclude that solvents in general are the cause of menstrual disorders, but further study is warranted.

C. Maternal Morbidity during Pregnancy

A case report described two cases of preeclampsia in women working in a chemical laboratory, and results from one large prospective study suggest the same effect. In this study, solvent-exposed women developed hypertension during pregnancy. The methods used were rigorous, and the result biologically plausible. Solvents have been associated with glomerulonephritis in a case series, and with proteinuria in some studies, supporting the concept that solvents could cause kidney dysfunction and so preeclampsia with exposure during pregnancy.

D. Spontaneous Abortion

Increased rates of spontaneous abortion have been reported in women exposed to styrene in the production of reinforced plastics, in women working with glues, and in women with exposure to methylene chloride or to four or more organic solvents in an industrial setting. Paternal exposure to organic solvents, toluene, or miscellaneous solvents in a range classified as high or frequent (daily use or exposure above established thresholds in biological monitoring) was associated with spontaneous abortion as well. Several solvents cause fetal loss in animal experiments; these include chloroform, xylene, styrene, and toluene. Overall, these studies show a consistent association between maternal solvent exposure and spontaneous abortion, and suggest that paternal exposure may be important as well. The data suggest that this effect occurs primarily in the high-exposure groups, but further study is warranted to characterize the dose–response.

E. Neurodevelopmental Effects

Solvents are by nature lipophilic, can cross the blood–brain barrier, and are known to affect central nervous system functioning in exposed adults. It would be expected that the developing fetal CNS would be exposed to solvents during maternal exposure.

Exposure of rats to dichloromethane, 2-ethoxyethanol, and 2-methoxyethanol both pre pregnancy and during pregnancy caused neurodevelopmental effects, including a decrease in habituation to a new environment in the pups, delayed learning in avoidance conditioning, and decreased activity in a running wheel.

One study compared the neurodevelopment of 41 children whose mothers worked with organic solvents during pregnancy with a group of children matched for maternal age, race, and child's age at testing, and whose mothers were unexposed, and found no difference between the groups in a range of developmental scales. At this time, no solvent has been shown to have a neurodevelopmental effect in humans, other than a fetal-alcohol like syndrome from high exposure to toluene.

F. Birth Defects

Several studies have investigated the link between parental exposure to solvents and birth defects in offspring. For many of these, exposure to specific solvents was not determined, and men or women were simply classified as exposed or unexposed. In others, a high rate of birth defects appears in certain industries, and because of solvent use in that industry, solvents are considered a potential cause of the defects.

Some studies have focused specifically on central nervous system defects and solvent exposure. The most susceptible period of the CNS to effects of solvents is from 10 to 18 weeks of gestation, a period of rapid growth of the CNS. Two case control studies suggest a link between parental exposure to solvents and CNS defects. One reported that malformations of the CNS were higher in children of painters and in children of men exposed to solvents, and the other, that mothers of children with CNS defects were five times more likely to have had exposure to organic solvents in the first trimester of pregnancy. In a large cohort study, exposure to aromatic solvents was more frequent among the mothers of infants with an important congenital defect. More specifically, another study found an increase in exposure to organic solvents among mothers of infants with ventricular septal defects, but not with all cardiovascular defects.

In another industry with solvent exposure, an association has been reported between gastroschisis and work in the printing industry.

Some solvents have been tested for fetotoxicity in animal systems. Those that cause malformations include chloroform, dichloromethane, chlorodifluoromethane, and methyl ethyl ketone. At levels of 3000 ppm, these solvents caused an increase in major malformations in rats. Other solvents have been studied and in well-conducted studies have been shown not to be fetotoxic or teratogenic; these include benzene, carbon tetrachloride (if given at less than the maternally toxic dose), and dichloroethane.

2-Ethoxyethanol, 2-methoxyethanol, and their acetates induce birth defects in exposed animals, including cardiovascular and skeletal malformations. Exposure of rats and rabbits by inhalation or dermal exposure caused a significant increase in cardiovascular abnormalities.

Looking at this group of studies as a whole, there are data to support an association between solvent exposure and congenital defects. In the human studies, the specific solvents responsible could not be identified, but the data suggest a general risk from organic solvents rather than a risk due to a specific one. At least two solvents cause birth defects in animal experiments; in the human studies, the exposures were generally to a mix of solvents. We do not know the level of exposure to solvents in these human studies, so cannot determine clearly which solvent groups might present a risk.

G. Childhood Cancer of the Central Nervous System

In adults, there is thought to be a lag of ten or more years between initial exposure to a carcinogen and the subsequent development of cancer. Nervous system cancer is the most common solid tumor of childhood, and the occurrence of cancer in the first decade suggests that prenatal events or exposures could contribute to development of the disease. Brain tumors have been reported in some studies to occur in excess in children whose parents had exposure to solvents, while in other studies no such association was found.

An increased rate of brain tumors was found among the children of women factory workers, and children of men who worked as painters. Some authors reported that work in the aircraft industry is associated with a higher incidence of childhood brain cancer, while another case-control study of brain cancer cases in Washington State did not find an elevated rate of parental employment in the aerospace industry. Work in the printing, chemical, or petroleum industry has been associated with an increased risk of brain tumors in children. Another study found an increased risk of CNS defects in children of printing workers, graphic artists, and workers in the chemical and petroleum industry, but not in all workers with exposures to hydrocarbons.

Work in these industries entails exposure to solvents as well as other substances; although the specific solvent present is not described in these reports. At the present time, these data suggest a risk of brain tumors from solvent exposure, but further study is needed.

VI. Neurotoxic Effects

Solvent exposure can impair both central nervous system (CNS) and peripheral nervous system (PNS) function. The types of impairment vary, depending on the specific solvent compounds and on exposure characteristics such as intensity and duration. CNS impairment is generally of greater concern and may occur either as an acute response to exposure or as a result of chronic solvent exposure. The symptoms and functional changes which are manifest as consequences of acute and chronic exposure are somewhat distinct from one another, but not completely separable.

Researchers have used a number of different methods to evaluate neurotoxic effects, including assessment of subjective symptoms, clinical examination, neurophysiological testing, and neurobehavioral performance (psychometric) testing. Psychometric tests have been the most widely used in epidemiological research and in some diagnostic settings to objectively evaluate central nervous system function. A broad variety of neurobehavioral performance tests has been utilized to assess memory, reasoning, visuo-spatial abilities, perception, reaction time, manual dexterity, and other CNS functions. These tests have proven more sensitive than traditional medical exams in the evaluation of subclinical CNS disturbances.

A. Acute Toxicity

Acute, transient CNS symptoms which have been commonly associated with occupational solvent exposure include dizziness, nausea, and vomiting, headache, drowsiness, lightheadedness, a feeling of intoxication, and incoordination. These are sometimes referred to as "narcotic" or "prenarcotic" symptoms and are due to the pharmacologic depression of central nervous system function.

It has been thought that these acute symptoms generally appear during the workday and disappear rapidly or within several hours after exposure ends, paralleling clearance of the compounds from the body. Recent evidence, however, suggests that the effects may not be as transient as commonly believed, and that some solvents and their metabolites may take longer to clear from the body.

A number of researchers have evaluated transient neurologic impairment among occupationally exposed individuals. These tests have found various CNS functions to be impaired. For acute exposures within the current permissible exposure limits, reaction time (one test of psychomotor function) is the most consistently affected function.

B. Chronic Toxicity

1. Central Nervous System Effects

Many cross-sectional epidemiologic studies have been carried out to investigate the effects of regular, low-level solvent exposure. The most frequent persistent CNS symptoms reported in association with such exposures are unusual fatigue, concentration difficulties, short-term memory problems, irritability, and depression, or mild mood disturbance.

Despite the subtle and nonspecific nature of these neurologic symptoms, and their gradual development, they have been recognized in many different populations. An excess of such chronic symptoms has been reported in epidemiologic studies of painters, paint manufacturing workers, styrene-exposed populations, and in a variety of other solvent-exposed groups. These symptoms have been termed "neuroasthenic" in much of the literature, and unlike transient prenarcotic symptoms, have been found to persist for several days or weeks following cessation of solvent exposure. The pathophysiology of these persistent CNS symptoms is not understood.

More severe and sometimes irreversible neuropsychiatric conditions have been less frequently reported in association with long-term solvent exposure. The symptoms include loss of intellectual and cognitive abilities, profound memory disturbances, and personality changes. Symptoms are accompanied by overt clinical signs and by deficits in neurophysiological parameters and psychological functioning. Suggestive evidence for these severe conditions comes from numerous case series reports and also from some studies which included comparison populations.

Psychometric test batteries have been widely used to objectively evaluate the relationship between solvent exposure and subclinical CNS functional deficits. Almost every epidemiologic investigation of workers with chronic, low-level solvent exposure has found one or more aspects of behavioral performance—either memory functions, other intellectual functions (besides memory), perceptual ability, reaction time, or other psychomotor functions (besides reaction time)—to be impaired among exposed workers. Notably, however, studies have generally found one or more CNS function to be *unaffected* when exposed workers were compared with nonexposed workers.

The large number of positive findings strongly suggests that solvents, in addition to inducing symptoms, can impair neurobehavioral functioning. No predictable pattern has emerged, however, with regard to which functions will discriminate between exposed and nonexposed workers. It is possible that specific neurobehavioral functions are not selectively affected by organic solvents. It may also be that the psychometric methods which have been used to evaluate occupationally exposed groups are not adequate for the predictable identification of specific functional changes, or have not been consistently applied and interpreted.

Identification of specific neurobehavioral functions which are selectively impaired by organic solvents may also have been hampered by a lack of distinction between acute and chronic effects of exposure. Most of the epidemiologic studies designed to assess the chronic effects of solvent exposure allowed an exposure-free interval, ranging from fifteen hours to a few days, before psychometric testing. This allows that chronic effects be assessed, as acute effects would presumably have disappeared. Recent evidence indicates, however, that some solvents and their metabolites may take days or weeks to clear from the body. Thus the exposure-free interval for many study participants may not have been long enough to allow for the complete reversibility of acute effects.

It has been suggested that some of the neurasthenic symptoms and functional impairment found in many studies may actually be due to acute or subacute, rather than chronic, effects. The heterogeneity of the exposure-free interval, both between studies, and between individuals from within the same group, may account for some of the variability in psychometric findings. Some researchers have argued that there is no clear evidence that behavioral performance impairments attributed to long-term exposure are distinct from the effects which have been observed in acute response to exposure. It is possible that acute and chronic effects are both due to similar underlying pathology. This is an as yet unresolved issue. Some overlap of acute and chronic CNS effects probably exists in many working populations.

The available evidence from these studies indicates that central nervous system toxicity can occur at exposures at or below current occupational standards. Although several studies did identify dose–response relationships, the literature is inconsistent in its indication of dose–response relationships between acute symptoms, neurasthenic symptoms, or neurobehavioral test impairment, and intensity of exposure. It is of interest that there does not appear to be a relationship between chronic symptoms and functional impairment. Although a few studies showed a relationship between behavioral performance and neurasthenic symptoms, most studies did not demonstrate this relationship.

2. Peripheral Nervous System Effects

Some solvents, including *n*-hexane, methyl *n*-butyl ketone, and carbon disulfide, cause a mixed sensorimotor neuropathy that affects distal locations first (feet and hands) and ascends to more proximal parts of the body. Symptoms include numbness, tingling, and weakness. This type of peripheral neuropathy is the result of axonal degeneration. Trichloroethylene has a specific neurotoxic effect on the trigeminal nerve, resulting in sensory loss to the side of the face.

Case reports and epidemiologic studies from workers exposed to other mixed solvents have sometimes found evidence of impaired sensory or motor nerve function, but the patterns are less clear than for the specific solvents mentioned above.

3. Summary

A considerable range of solvent-associated CNS impairment has been demonstrated in occupationally exposed populations. Although acute effects have been characterized fairly consistently, the results of epidemiologic studies vary widely with regard to the extent and severity of chronic neurological impairment, particularly at the upper range of severity. The effects of chronic solvent exposure, as currently understood, have been described as qualitatively similar syndromes of CNS dysfunction which vary in severity and reversibility. This variability has not been shown to be correlated with variations in exposure, although exposure differences may play some role. The relative contribution of transient and persistent effects to the impairment seen in working populations needs further characterization.

The gradual, subtle, and sometimes progressive action of solvents on the central nervous system has been difficult to characterize because many epidemiological studies have relied on subjective symptoms or on nonstandard batteries of neurobehavioral function tests. To achieve some diagnostic uniformity in epidemiologic research, a categorization scheme for CNS conditions caused by workplace exposure to toxic solvents, metals, and pesticides was proposed by a World Health Organization (WHO) working group and modified at a subsequent international workshop. These characterizations of both acute and chronic CNS disorders are presented in Table III and serve as a useful summary of the solvent-related neurologic changes which have been demonstrated in occupational groups. Acute effects and chronic effects are indicated for the conditions which are most likely to be present in an actively working population, whereas the most severe chronic conditions would be present only in disabled workers.

VII. Solvents and the Liver

A. Introduction

Liver injuries from solvents such as carbon tetrachloride and chloroform, have been recognized for over 50 years. Researchers have subsequently continued to identify the effects of solvents on the liver and define the role of the liver in both producing toxicity under some circumstances and preventing it under others. The liver plays a major role in the detoxification of solvents and their elimination from the body. Detoxification, also known as biotransformation, occurs in a number of sites including the kidney, lung, and gastrointestinal tract but the primary site of biotransformation and detoxification is the liver.

During the biotransformation process, metabolites that are formed may be either less toxic or more toxic than the parent compound. These metabolites are then either excreted by way of the kidney or the gastrointestinal tract. If not excreted, they may act on a target organ such as the liver or an extrahepatic organ system.

The majority of all hepatic injuries are caused by viruses, metastatic cancers or the voluntary consumption of alcohol rather than by occupational or environmental exposure to chemical or specific solvents. Frequently, the clinical or morphologic picture created by occupational solvent exposure may be similar to that due to drugs, alcohol, or viral injury.

Much of the information obtained on hepatotoxicity in humans is based on clinical reports, retrospective studies, or extrapolation from animal data. Many of these reports or studies do not address the role of multiple chemical exposures but rather address the effects of exposure to one chemical.

Acute hepatotoxicity caused by solvents is now rare. More often, the acute toxicity from solvents involves multiple systems where the liver is not the primary concern or the target organ most affected. Acute, large exposures to solvents are rarely seen because many of the most hepatotoxic agents are either no longer manufactured or are carefully controlled. When acute hepatotoxicity does occur, it is fairly easy to detect by obtaining a detailed history, physical examination, and laboratory studies. What is not always easy is the clear identification that the toxicity is due to a solvent rather than to more common causes.

The solvents that are toxic to organs other than the liver are biotransformed by the liver enzyme systems, such as the cytochrome *P*-450 system, into toxic metabolites that are then carried to the target organ. Examples of this phenomenon include benzene which causes bone marrow suppression and leukemia; *n*-hexane which causes neurotoxi-

Table III Acute and Chronic CNS Disorders Attributable to Occupational Exposures

Acute central nervous system conditions
Acute intoxication[a]
  Pathophysiology: pharmacologic effect
  Duration: minutes or hours; no sequelae
  Clinical: acute CNS depression, psychomotor impairment
Acute toxic encephalopathy
  Not clearly documented with occupational solvent exposure
  Pathophysiology: cerebral edema, CNS capillary damage
  Duration: hours or days; may cause permanent deficits
  Clinical: coma, seizures
Chronic central nervous system conditions
Organic affective syndrome (neurasthenic symptoms)[b]
  Pathophysiology: unclear
  Duration: days or weeks; no sequelae
  Clinical: depression, irritability, loss of interest in daily activities
Mild chronic toxic encephalopathy (CNS symptoms and objective functional impairment)[b]
  Pathophysiology: unclear
  Course: insidious onset; duration: weeks or months; reversibility: variable
  Clinical: fatigue, mood disturbances, memory complaints, attentional complaints
  Reduced CNS function
    Psychomotor function (speed, attention, dexterity)
    Short-term memory
    Other abnormalities common
Severe chronic toxic encephalopathy (dementia)
  Pathophysiology: unclear, often associated with structural CNS damage
  Course: insidious onset; duration: indefinite, usually irreversible
  Clinical manifestations
    Loss of intellectural abilities of sufficient severity to interfere with social or occupational functioning
    Memory impairment
    Other
      impairment of abstract thinking
      impaired judgment
      other disturbances of cortical function
      personality change
  Reduced CNS function
    Types of abnormalities similar to mild chronic toxic encephalopathy; deficits more pronounced and pervasive
    Some neurophysiologic and neuroradiologic tests abnormal

Source: Baker, E. L., and Fine, L. (1986). *J. Occup. Med.* **28,** 126–129.
[a] Acute effects for conditions most likely to be present in an active working population.
[b] Chronic effects for conditions most likely to be present in a working population.

city, and ethylene glycol ethers which cause reproductive toxicity.

B. Hepatotoxicity

The hepatotoxicity of a particular substance may be classified in different manners. The first scheme classifies substances (chemicals and drugs) by the probability and predictability of its causing liver toxicity; the second scheme classifies the substance by the morphology or type of damage caused to the liver.

In the first classification scheme, the substance is further classified as Type I or Type II on the basis of its mechanism of toxicity. Type I injuries are said to occur if the toxic lesions are caused by those chemicals or substances that always cause hepatotoxicity and in which the toxicity is dependent on the dose and duration of exposure. These lesions are both predictable and reproducible in animals and in humans. Type II responses are dose- and time-independent, and are not predictable, nor can they be reproduced in animal models. These reactions are idiosyncratic and occur in susceptible persons. Type II lesions can also be subclassified into either hypersensitivity reactions or metabolic aberration reactions. Hypersensitivity reactions usually require a period of exposure of 2 to 5 weeks before a reaction occurs. When rechallenged, the sensitized person would immediately develop a reaction. Metabolic aberration occurs when the individual has an apparent defect in a metabolic pathway. Both Type I and Type II lesions may then be classified by the specific morphology or type of damage that is done to the liver.

Whether the agent causes a Type I or Type II lesion, the pathologic responses of the liver to the toxic agent may be divided into five categories. These categories may coexist in a single case and one chemical may produce more than one type of injury.

- **Steatosis** This is fat accumulation within the liver cells (hepatocytes). It may be associated with a decrease in plasma lipids and lipoproteins. Carbon tetrachloride and ethanol are examples of solvents that cause steatosis.
- **Hepatocellular necrosis** This is damage to or death of the hepatocyte. Damage to the cell may involve various organelles including the plasma membrane, mitochondria, lysosomes, and nucleus. This necrosis then leads to hepatic dysfunction resulting in disruption of cellular activity or cell death. Solvents causing hepatocellular necrosis include carbon tetrachloride, allyl alcohol, bromobenzene, bromotrichloromethane, ethanol, trichloroethane, and trichloroethylene.
- **Cholestasis** Cholestasis is produced when there is interference with the hepatic mechanisms of bile excretion. With the increased retention of bile salts and bilirubin, jaundice and itching occurs. In animals, it has been shown that ketone and ketogenic chemicals enhance some types of cholestatic reactions.
- **Fibrosis and cirrhosis** Chronic exposures to some solvents such as ethanol, tetrachlorethene, and trinitrotoluene may cause collagen deposition throughout the liver with distortion of the normal hepatic architecture and disruption of the hepatic blood flow leading to the development of portal hypertension and end-stage liver disease.
- **Cancer** There are a number of agents implicated in the development of primary hepatocellular carcinoma but few are solvents. Primary cancer of the liver probably involves a two-stage process of induction and a long promoter-associated latency period. Carbon tetrachloride, chloroform, and trichlorethylene are suspected hepatocarcinogens. Vinyl chloride, which has been used as a solvent, is a known hepatocarcinogen.

C. Metabolism of Solvents

The metabolism of solvents in the liver is predominantly associated with a process known as biotransformation which is the path of excretion for many common solvents. Biotransformation plays a major role in the production or bioactivation of metabolites. It is these metabolites that may be responsible for the toxicity of the solvent or that are excreted in the urine, thereby becoming the basis for biologic monitoring.

Biotransformation is a biochemical process which changes a foreign chemical into another form. During the first phase of biotransformation, the metabolite formed may be more stable, more chemically reactive, or even more toxic. During the second phase, this active metabolite is frequently conjugated in preparation for elimination from the body. The first phase involves enzyme systems such as the cytochrome *P*-450 mixed function oxygenase system. Most products of biotransformation are not toxic; biotransformation is very effective in detoxifying many foreign chemicals and preparing them for excretion via gut, kidney, or lung. If there is an imbalance in the biotransformation process, and the pattern of toxicity may be altered (inhibited, induced, or bypassed). Alterations of this process may occur within the cytochrome *P*-450 system as a result of the type or dose of the chemical, or the presence of other drugs. The biotransformation process may also be altered in a number of ways.

- **Competitive inhibition** Competitive inhibition occurs when the presence of one solvent prevents or slows the metabolism of another substance. Toluene has been shown to competitively inhibit the metabolism of benzene.
- **Potentiation** Potentiation occurs when the presence of one substance increases the formation of an active metabolite and/or decreases the number of functional groups

(sulfhydryl and glutathione) to which the metabolite binds. This leads to higher concentrations and longer exposure to toxic metabolites. It has been shown that prior exposure to some alcohols and ketones can potentiate the liver damage caused by chlorinated hydrocarbons. The elevation in levels of cytochrome *P*-450 levels by some chemicals, including ethanol, phenobarbital, or polychlorinated biphenyls, can increase the metabolism of other chemicals. For example, there is increased toxicity from bromobenzene if there is pretreatment of the experimental animal with ethanol. For other chemicals the influence of ethanol may be variable and dependent on the level of exposure to both alcohol and the other substance.

- **Induction of systems other than the cytochrome *P*-450 system** Induction of other than the cytochrome *P*-450 system may also occur with the result being the formation of a less toxic metabolite by this alternate pathway. This occurs in animal models when 3-methylcholanthrene is administered prior to bromobenzene exposure with resultant decrease in the bromobenzene toxicity.
- **Enzymatic saturation** Enzymatic saturation occurs when the amount of exposure overwhelms the normal detoxification pathway and forces detoxification by means of a bioactivation pathway. This occurs with such solvents such as *n*-hexane, vinylidene chloride, methyl chloroform, perchloroethylene, and ethylene dichloride.

In addition to the above, the biotransformation process may also be affected by a number of other factors including gender, age and nutritional status. Through the study of toxicity of specific solvents on animal models, researchers have been able to identify the following key factors in the development of toxicity.

- **Metabolite formation** The hepatotoxicity of a solvent is not always related to the pharmacokinetics of the parent substance or its major metabolites, but may be due to a minor, but highly reactive, metabolite.
- **Threshold** A tissue threshold concentration in the liver must be attained for some metabolites before toxicity occurs.
- **Prevention** Some substances such as glutathione, play an essential role in protecting hepatocytes from injury by chemically reactive intermediates. These substances which frequently bind to the metabolites provide the cell with a means of preventing the reactive metabolite from attaining a critical effective concentration.
- **Protection** Other enzyme systems within the liver play a role in protecting the hepatocyte by catalyzing the further degradation of toxic reactive intermediates.

D. Specific Solvents and Their Toxic Effect

There are a number of categories of solvents whose members produce either hepatic or extrahepatic toxicity secondary to the liver's metabolism of the solvent. A few examples are given to illustrate the mechanisms of toxicity.

1. Toxicity Induced by Hepatic Metabolism of the Solvent

a. Benzene

Benzene is the only simple aromatic hydrocarbon solvent that causes hematopoietic toxicity. Most research indicates that it actually is a toxic metabolite of benzene rather than the parent compound that induces bone marrow suppression and leukemia. Benzene appears to be metabolized by the cytochrome *P*-450 system to active metabolites that have been implicated in the hematopoietic toxicity. Researchers have also shown that the formation of these metabolites may be altered in animals by the administration of phenobarbital, a drug that induces the cytochrome *P*-450 system. Unfortunately, because other oxidative enzymes not located in the liver may also play a significant role in metabolizing benzene metabolites, inhibition of hepatic metabolism may not totally eliminate bone marrow toxicity.

b. Methylene Chloride

Methylene chloride, a chlorinated aliphatic hydrocarbon which requires metabolic activation by the mixed-function oxidase system, produces metabolites whose end product is carbon monoxide. This carbon monoxide is bound to cellular proteins or lipids, resulting in an elevation of carboxyhemoglobin levels and related toxicity. The hepatotoxicity component is minimal in comparison to the neurotoxicity of methylene chloride.

2. Solvents That Produce Hepatotoxicity

a. Ethanol and Other Aliphatic Alcohols

The toxic effects of ethanol include steatosis, hepatocellular necrosis, and cirrhosis. Whether or not the toxicity is due only to the direct toxic effect of ethanol or is associated with nutritional deficiencies as well remains in question; however, it has been established that the hepatotoxicity of alcohol is dependent in part on its metabolism by portions of the cytochrome *P*-450 enzyme system.

Other effects of alcohol include interaction with other hepatotoxic chemicals. This interaction includes the potentiation of carbon tetrachloride, chloroform, trichloroethane, trichloroethylene, thioacetamide, dimethylnitrosamine, pracetamol, acetaminophen, and aflatoxin B1.

b. Chloroform

Chloroform, a chlorinated aliphatic hydrocarbon, is also metabolized to a reactive metabolite that covalently binds to hepatic proteins and depletes the liver of glutathione. Once the glutathione levels are decreased, the metabolite then binds covalently to molecules within the hepatocyte causing cell necrosis. Animal studies has shown that pretreatment with acetone or ketone increases the production of reactive metabolites.

c. Carbon Tetrachloride

Carbon tetrachloride produces steatosis and hepatic necrosis. Toxicity from this solvent is probably also dependent on its metabolism to an active intermediate by the mixed function oxidative system. Studies in animals have shown that inactivation of this system protects the animal from hepatotoxicity.

d. Hexacarbon Solvents

Hexacarbon solvents such as *n*-hexane can produce peripheral neuropathies. *n*-Hexane is metabolized by the cytochrome *P*-450 mixed function oxidative system to toxic metabolites which are responsible for the peripheral neuropathies. Animal studies have shown that where there is exposure to both 2-hexane and methyl ethyl ketone (MEK), the neuropathy develops earlier and is more severe. It should be noted that MEK by itself is not neurotoxic, but rather it alters the formation of the toxic 2-hexanone metabolite.

e. Carbon Disulfide

Carbon disulfide is well known for its ability to cause organic brain damage, peripheral nervous system decrements, neurobehavioral dysfunction, and cardiovascular system effects. Animal models seem to indicate that it is the formation of a toxic metabolite by the mixed function oxidative (MFO) system that produces the neurotoxicity.

E. Relevance of Hepatotoxicity

Study of subacute or chronic toxicity is limited by the lack of sufficiently sensitive and specific medical surveillance techniques; despite this, it must be the focus of attention. Indeed, there have been three epidemiologic studies that have shown an increased death rate from liver cancers in laundry workers, dry cleaners, service station attendants, and asphalt workers who are exposed to organic solvents. These must serve as the motivation to develop better surveillance tools.

Hepatotoxicity from solvent exposure may be acute, subacute, or chronic in nature. Acute toxicity is usually detected by elevation in serum transaminase levels, bilirubin, and alkaline phosphatase. These tests are also frequently used in medical surveillance programs. Unfortunately, they are neither very specific nor very sensitive when used to detect nonacute or subclinical disease. When used to detect acute hepatocellular necrosis their sensitivity improves, but again the specificity is very poor and they may miss subacute or chronic disease because of their high false-negative rates.

In assessing the presence of hepatotoxicity, there are a number of tests that may be employed. These include: enzyme activity, functional clearance tests, tests of synthetic function, and determination of structure. Enzyme activity is most frequently assessed by the determination of aspartate aminotransferase (AST) and alanine aminotransferase (ALT) levels. Although, these are frequently elevated in the presence of hepatocellular necrosis, they may be minimally elevated or not elevated in the presence of subacute necrosis. It must also be noted that the height of transaminase level is not predictive of the severity of toxicity.

Alkaline phosphatase activity may also be measured and serve as an indicator of an alteration in transport function of hepatocytes and as an indicator of cholestatic liver disease. Alkaline phosphatase is also found in bone, placenta, and intestine and, therefore, is not highly specific. Bilirubin is used to detect cholestatic liver disease but may be normal in the presence of subacute toxicity or acute hepatonecrosis.

Tests of synthetic liver function measure the liver's ability to manufacture proteins that are necessary for normal health including serum albumin and factors necessary for normal blood clotting. Unfortunately, these tests are usually not abnormal until much of the reserve capacity of the liver is damaged and liver disease is far advanced.

Clearance tests are by far the most useful in detecting the presence of subacute liver disease, but unfortunately many of them cause toxicity themselves, are difficult to administer as a surveillance tool, or are not specific for toxic exposures. Clearance tests include the aminopyrine breath test, bromosulfobromophthalene (BSP), indocyanine green, serum bile acids (conjugated cholic acid and cholyglycine), and caffeine breath tests.

Liver biopsy is the only highly sensitive and specific study for hepatotoxicity. Because of the potential for adverse side effects, it is not used as a screening tool. Rather its main use should be for the assessment of persistent abnormalities that indicate chronic liver disease.

When screening a population for solvent-related liver disease, a number of principles must be observed. The screening must be specific for the population being screened, and must detect disease early in its course. Treatment or intervention must be possible, and the tests used must also be valid and reliable.

In the case of exposure to solvents, a screening program is more difficult to establish because of the lack of highly sensitive and specific tets for both acute and subacute injury. Acute liver injury is more likely to produce elevations in liver enzymes, whereas functional and clearance tests are more likely to detect chronic disease. Despite the high false-negative rate for subacute or chronic disease and a high false-positive rate for acute disease, the most frequently used surveillance tools are liver enzyme assays.

E. Summary

The liver plays an important role in the detoxification and elimination of many solvents. It also plays a major role in the biotransformation of some chemicals to more toxic metabolites. These active metabolites that are formed are then capable of either causing toxicity to the liver or to other organs systems such as the hematopoietic, reproductive, or neurologic systems.

The effects of acute exposure to highly toxic solvents such as carbon tetrachloride and benzene are well documented. In contrast, little is known about the hepatic effects of multiple solvent exposure such as occurs with painters, who may be exposed to white spirits, xylene, petroleum spirits, toluene, and methyl ethyl ketone.

The toxicity of a chemical may be either inhibited, potentiated, or not affected by the presence of other chemicals or drugs. Damage to the liver may be secondary to the direct toxic effects of the chemical or to an idiosyncratic reaction. The morphological manifestations may include hepatocellular necrosis, steatosis, cholestasis, fibrosis, cirrhosis, or carcinoma.

Screening for hepatoxic effects of solvents must be undertaken with the understanding of the shortcomings of most screening tests which were designed to detect acute rather than subacute or chronic liver disease. Although most lack sufficient sensitivity to detect minimal, low-grade disease or specificity to allow differentiation from other causes of toxicity, they remain useful for screening persons with exposure to hepatotoxic agents.

VIII. Conclusions

There is still a lot we need to know in order to make informed decisions about solvent use. Few of these studies tell us if there is a threshold of exposure below which no adverse outcome occurs. Few tell us if there is a critical period for exposure, outside of which an exposure might have a reduced effect. And few tell us whether the adverse outcome is due to a specific solvent or to a class of solvents. All this information is difficult to obtain, for human studies of adverse reproductive outcome are difficult to perform well and such detailed assessment of exposure may not be possible in industrial settings. Further animal experiments could assist in some ways, but cannot answer all our questions.

Basic principles for reducing solvent exposure include:

1. Product substitution. This is usually considered the ideal solution, however, it can take months or years to determine if an acceptable substitute exists.
2. Engineering controls, such as improving local exhaust ventilation, or enclosing a process to reduce exposure.
3. Personal protective equipment, such as respirators and gloves.
4. Administrative controls, such as work rotation or reassignment for a period of time.

When we look at these approaches in the context of solvents in particular, several factors must be kept in mind.

1. Any respirator must be fitted with the proper cartridge for the solvent in use, and there must be an established respirator program for fitting the respirator, and determining how frequently to change the cartridge. Some companies allow the worker to determine if the cartridge is spent by reporting a breakthrough in smell; this will not be effective for many solvents, for the smell threshold may exceed the level at which toxic effects occur, or a worker may develop olfactory fatigue to a solvent in a short period of time, and so not report an odor through the respirator.

2. Gloves prescribed must be tested for the particular solvent. Some types of gloves are effective for one solvent but essentially useless for another. The solvent manufacturer can advise the clinician or the employer on appropriate glove material and breakthrough time for various solvents.

3. Since most solvents are dermally absorbed, the individual must pay particular attention to work practices. Does his/her clothing get saturated in the process? Does the solvent get inside the gloves because of splashes or exposure above the glove line?

Related Articles: CLEANING AND LAUNDRY PRODUCTS, HUMAN EXPOSURE ASSESSMENTS; FORMALDEHYDE: EXPOSURE EFFECTS ON HUMAN HEALTH; INDUSTRIAL SOLVENTS; VOLATILE ORGANIC CHEMICALS; XYLENES.

Bibliography

Amdur, M. O., Doull J., C. D. Klaassen, eds. (1991). Casarett and Doull's Toxicology," 4th ed. Pergamon Press, New York.

Barlow, S. M., and Sullivan, R. M. (1982). "Reproductive Hazards of Industrial Chemicals." Academic Press, New York.

NIOSH. (1987). Organic Solvent Neurotoxicity. Current Intelligence Bulletin 48. U.S. Department of Health and Human Services/NIOSH, U.S. Government Printing Office, Washington, D.C.

Sandmeyer, B. (1981). Aromatic hydrocarbons. In "Patty's Industrial Hygiene and Toxicology, Vol. 2B: Toxicology. G. D. Clayton, F. E./ Clayton, eds. pp. 3253–3431. Wiley & Sons, New York.

Oxides of Nitrogen

John G. Mohler and Clarence R. Collier
University of Southern California

Glossary

Hemoglobin (Hb) Iron-containing porphyrin compound found in a red blood cell which, in the ferrous state, forms four ligands per molecule with gases such as oxygen, carbon monoxide, and nitrous oxide.

Hypoxia Low oxygen pressure or volume or both in tissue.

Hypoxemia Low oxygen pressure or volume or both in blood.

Methemoglobin (MetHb) Hemoglobin molecule containing a ferric iron that does not bind oxygen or, probably, any other gas.

Methemoglobinemia Condition in which the hemoglobin in red blood cells is altered by oxides of nitrogen in a manner that reduces the blood's ability to carry and deliver oxygen to tissue sites.

Mucous membrane Tissue, usually a protective tissue, which secrets mucous onto its surface, and is, in turn, removed. In the lung, these function to remove entrapped chemicals and particles.

Parenchyma Tissue of the lung beyond the airways.

Peritoneum Tissue that lines the abdominal cavity.

Pulmonary edema Lung disorder in which fluid moves from the lung's blood vessels into extravascular sites, including the alveoli.

P_{50} Partial pressure of oxygen where the hemoglobin is 50% saturated with oxygen.

Torr Partial pressure of 1 mm Hg or 0.133 kPa (kilopascal; Pa = 10 dynes/cm^2).

OXIDES OF NITROGEN take several forms, some of which react directly with tissues in the lung when inhaled, some that dissolve to react with oxygen and hemoglobin in vivo, *and some that pass on to the neural system to become anesthetic. The latter, when used properly, are not toxic. The nitrates and nitric acid react with tissue where they do immediate damage. When they gain access to the inner organs of the body, they destroy enzymes, thus altering vital reactions that support life. The most common and devastating reaction is the destruction of the oxygen-carrying capacity of the blood by nitric oxide (NO) and nitrogen dioxide (NO_2) either by the destruction of the oxyphilic hemoglobin by oxidation to methemoglobin, or by binding at the oxygen sites more tenaciously than oxygen and thus displacing it. NO and NO_2 are common air pollutants that have no beneficial effect for the exposed human. These gases are the agents that cause pulmonary edema and methmoglobinemia in the often fatal silo filler's disease.*

I. Introduction

Nitrogen has the capability to join with oxygen in many combinations forming molecules with much different properties from one to another. Some of these properties are useful in medicine, some are harmful to humans when exposed. The gaseous nitrogen oxides consist of nitric oxide (NO), nitrogen trioxide (N_2O_3), nitrogen dioxide (NO_2 and N_2O_4), and nitrogen oxide (N_2O) (also known as laughing gas and nitrous oxide). Other agents include the nitrites, the nitrates, and nitric acid.

A. Nitrates

Nitrates usually are not inhaled toxins. Organic nitrates are usually termed "nitro compounds." These compounds are a combination of the nitro

(—NO_2) group of an organic radical. However, this term is also often used to denote nitric acid esters of an organic material. Inorganic nitrates are compounds of metal that are combined with the monovalent —NO_3 radical. Large amounts of organic or inorganic nitrates taken by mouth may have serious and even fatal effects. Low doses taken over long periods may cause weakness and are associated with an increased cancer rate.

B. Nitric Acid

Nitric acid (HNO_3) is a transparent colorless or yellow caustic liquid that fumes and may cause suffocation if inhaled. It is a very corrosive liquid that irritates the eyes, skin, and mucous membranes and causes burns on direct contact. Nitric acid vapor is highly irritating to the lung and airways, and may cause pulmonary edema, suffocation, and death due to hypoxia. Large amounts of nitrites taken by mouth may produce nausea, vomiting, and a cyanosis due to oxidation of the iron-heme of the blood met-hemoglobinemia results from the absorbed products entering the blood. The mechanism is unknown but assumed to be related to the reactions discussed previously.

C. Nitrogen Dioxide

Nitrogen dioxide is a whiskey brown color which, as an air pollutant, gives the atmosphere in an urban community a brown discoloration. When nitric oxide, a toxic but colorless gas, is exposed to the ultraviolet light of sunlight, the products are ozone, nitrogen dioxide (NO_2), and its dimer N_2O_4. The percentage of NO_2 and N_2O_4 depends on the temperature of the gas. At 27°C, the mixture is about 80% N_2O_4 and 20% NO_2. In the liquid and solid states, nitrogen trioxide has a blue color. Partial dissociation of N_2O_3 into NO and NO_2 occurs in the liquid phase; in the vapor state at room temperature, it is largely dissociated into NO and NO_2.

II. Biological Effects

A. Biological Characteristics

The oxides of nitrogen are only somewhat soluble in water, reacting with it in the presence of oxygen to form nitric and nitrous acids. This is the action that takes place deep in the pulmonary system when gases are inhaled. The acids formed are irritating to mucous membranes and deep tissue (parenchyma) of the lung causing congestion of the throat and bronchi, and edema of the lungs. The acids are neutralized by the alkalis present in the blood and body fluids forming nitrates and nitrites. The latter, in turn, cause some arterial dilation, a fall in blood pressure, headache and dizziness, and there may be some formation of methemoglobin (metHb). Because of the low solubility of these gases in water, the irritation to the airways and parenchyma of the lung is low, and exposure goes unnoticed for long periods of time. Thus the warning system in the lung is ineffective, and dangerous amounts of fumes may be inhaled before the exposed person notices any real discomfort. Higher concentrations of 60 to 150 parts per million (ppm) cause immediate irritation of the nose and throat; coughing and burning are the major symptoms. The burning sensation may be felt in the sternal area, deep in the chest. These symptoms usually clear with a few minutes of inhaled fresh air or tanked oxygen by mask. However, some 6 to 24 hours later, a sensation of tightness in the chest may develop, which commonly is associated with inappropriate shortness of breath for activities of daily living, and with restlessness and sleeplessness. If pulmonary edema develops, then dyspnea at rest develops. The subject develops cyanosis due to the inability to transfer oxygen from the air, across the swollen tissue and fluid filled spaces to the blood in the lung, and death results. Concentrations of 100 to 150 ppm are dangerous for short exposures of less than one hour, and concentrations over 200 ppm are commonly fatal even after a short exposure.

B. Low Exposures

Prolonged exposure to low concentrations of NO_2 gas which is too low to cause pulmonary edema will result in chronic irritation, cough, headache, loss of appetite, dyspepsia, corrosion of the enamel of the teeth, and gradual loss of strength. A loss of endurance and strength may relate to the adverse effects oxides of nitrogen have on the oxygen-carrying capacity of hemoglobin in blood.

C. Hemoglobin

In nature, a variety of oxygen carriers have evolved, but all share one common feature: they are metal-containing proteins. Hemocyanin contains copper and is found in some arthropods and molluscs. The hemocyanin protein is found in the plasma of these animals in small aggregates. The most common oxygen carrier is hemoglobin, which is found in all mammals. This protein contains an iron-porphyrin ring which can bind four oxygen molecules per hemoglobin molecule. Hemoglobin is packed tightly into cells and carried in packets in the blood of most vertebrates. In humans, 100 ml of blood will contain about 16 g of hemoglobin, which will bind about 20 ml of oxygen. Only about 0.2 ml of oxygen is dissolved in the same volume of blood at standard atmospheric conditions. It is interesting that hemoglobin increases the carrying capacity for oxygen by 100 times which results in an oxygen content of blood which is about the same as that of air, namely 20 ml of oxygen per 100 ml of air. This results in an oxygen concentration in the capillaries of mammals that is the same as that of the air-filled trachea of that animal. Damage to this delivery system, especially to hemoglobin, results in critical limitation of oxygen transfer to and into the cells, which could result in limitations of activity, due to dyspnea fatigue, lack of endurance, and, in extreme cases, death.

D. Hemoglobin Dissociation

Hemoglobin forms four ligands with certain gases, such as NO, carbon monoxide (CO), and oxygen (O_2), each with decreasing intensity. Using oxyhemoglobin (HbO_2) binding under standard conditions of 37°C, pH 7.40, and a pCO_2 of 40 torr, CO is about 200 times more firmly bound than is O_2. Similarly, NO is 300,000 times more tenaciously bound to hemoglobin (Hb) than is O_2. HbNO forms only in the absence of O_2 since the NO will react with the oxygen rather than form a ligand with Hb. Once formed, however, the conversion is slow, even with O_2 present. Thus, there are four reaction rates and thus four affinities depending on how many of the possible four ligands are filled. This results in the well-known sigmoidal nature of the HbO_2 saturation curve when exposed to increasing partial pressures of oxygen (pO_2). This relationship, also known as the oxygen dissociation curve, is affected by temperature, acidity, certain other gases, and another molecule that fits the ligand known as 2,3-diphosphoglycerate (2,3-DPG). The percentage of HbO_2 in the blood is calculated as

$$SO_2 = [HbO_2] / [Hb] + [HbO_2] \times 100,$$

where SO_2 is the percent of Hb saturated with O_2. This equation is commonly used but does not take into account HbCO nor HbNO.

E. Hemoglobin Binding

The measure of the effect of temperature, acids, and analogs of ligand-bonded Hb is determined by ascertaining the partial pressure of oxygen at 50% Hb saturation. This is known as the P_{50} value, which is normally about 26.7 torr at 37°C, pH 7.40 and a pCO_2 of 40 torr. For an HbO_2 saturation at any given pO_2, a P_{50} of 20 torr means that the oxygen is more tenaciously bound to the hemoglobin producing a higher saturation than if the P_{50} were 27 torr. Thus the Hb may be saturated with O_2 at a lower pO_2, and the content of oxygen in the blood may be the usual 20 ml per 100 ml of blood, but the available oxygen to the tissue and cell is much less since the oxygen does not release as readily. This phenomenon is most noteworthy for carbon monoxide poisoning. The P_{50} is low and the quantity of Hb available to carry oxygen is reduced by the HbCO; thus a HbCO of about 60% results in coma and death because the remaining 40% Hb, although itself 100% saturated with oxygen is unable to release enough oxygen to the brain and other organs necessary to sustain life.

F. Methemoglobin

The Hb molecule may become inactive by the loss of an electron forming an inert agent known as methemoglobin (MetHb). This compound contain an iron in the Fe^{3+} state rather than Fe^{2+}. The oxidized state is also known as ferric hemoglobin or methemoglobin. Hemoglobin exposed to NO in aerobic conditions is oxidized. Nitrites injected intravenously into animals also produce significant methemoglobinemia. It is believed that ferric hemoglobin is continuously formed in the erythrocytes and is reduced to the ferro derivative by enzyme systems linked to carbohydrate metabolism in the red cells. Poisoning the enzyme system or oxidizing the iron or both results in significant methemoglobin formation. MetHb does not form ligands with oxygen, thus its presence does not assist in oxygen transfer. In clinical laboratories, the standard method of measuring the concentration of Hb is to convert it to cyanomethemoglobin, and measure the one unit as the representative of the whole. This results in a lack of recognition of the presence of HbCO, HbNO, and MetHb.

G. Methemoglobin Reduction

Blood MetHb is reduced *in vitro* and *in vivo* to functional Hb by enzymatic reactions which require an intact red blood cell to function. Hemolysates of blood do not reconstitute Hb from MetHb spontaneously. The rate of reduction of whole blood at 37°C averages 5.1% of the total Hb per hour. The starting level of the MetHb does not affect the rate of Hb recovery.

III. Reaction of Nitrates in Vivo

A. Tissue Reactions

In studies where NO or NO_2 or both gases were injected anaerobically into the peritoneum of several sets of five rats, NO_2 produced about 27% MetHb when injected either anaerobically or aerobically, while NO produced about 67% MetHb when injected anaerobically, while it produced only about 26% when injected aerobically. This suggests that NO_2 dissolves in the fluid of the membrane, and reacts immediately, causing tissue injury, while NO reacts with local oxygen to react like NO_2, but will also migrate to the Hb in the absence of oxygen. At the Hb, oxygen is readily found and a reaction takes place: first a fast phase, the autocatalytic reaction, and then a slow oxidative phase. Nitrogen dioxide and nitric oxide in the presence of oxygen react in a biological system similar to a hydrated acid of nitrogen, rather than as a dissolved gas.

Thus, nitrogen dioxide and nitrates fail to get into the blood readily in low doses, but when they do, they have a devastating effect on the oxygen delivery system. Studies of the effect in the lung document that succinate dehydrognase and cytochrome *c* oxidase in lung homogenates are increased, but this reaction has not been reported *in vivo*. Evidence for NO_2 injury without histological damage is supported by the fact that exposure of the lung, *in vivo* or *in vitro,* results in an increased cellularity and an associated increase in the metabolic rate as measured by oxygen consumption. The increase O_2 consumption appears to be in excess of the influx of polymorphonuclear leukocytes or the increased number of alveolar microphages that can be lavaged from the lung. Although the increased metabolism cannot be attributed with certainty to changes in lung cells rather than to inflammatory cells, the subsequent tolerance to lung injury suggests that there is a definite effect on the lung itself and not simply an inflammatory response.

IV. Reaction of Nitrites in Vivo

A. Silo Fillers Disease

Oxides of nitrogen gases are formed in silos freshly filled with plant material due to the fermentation of the plant nitrites and nitrates by microbial fermentation. Gas begins to form upon filling, and collects in nonaereated silos over the first few days, reaching a peak in five days. Nitrogen dioxide hydrolyzes slowly, thus when farmers enter the silo during the danger period, the effect of the toxic gases is not noted until several hours later, depending on the intensity of the exposure. Thus, the gases must reach the deeper structure of the lung such as the bronchioles and the alveoli. When hydrolysis occurs, nitrous and nitric acids are formed which produce a profound chemical pneumonitis, pulmonary edema, and hypoxia. Methemoglobin is formed in large quantities which increases the hypoxemia, severe ventilatory distress, and death due to tissue hypoxia. The onset is usually delayed several hours after exposure; however, high concentrations of the gas may produce a rapid death. A survivor of the pulmonary edema may develop widespread bronchiolitis obliterans from 30 days to a month later.

B. O_2 and NO Reactions

It has been postulated that the only agent to actually reach the hemoglobin would be a gas, most probably NO, where, after a reaction with the oxygen found there, it would become nitric acid. Either the acid which is known to be a strong oxidizing agent to the ferrous hemoglobin or the first reaction of oxygen with NO would thus create methemoglobin by making the iron in Hb ferric. If no oxygen were present, an HbNO ligand should be formed with little or no methemoglobin formation, and a dissociation curve should be identified. It was found, however, that the reaction is not only oxidative, but autocatalytic. Observed HbNO is due to the very low nitrosyl hemoglobin dissociation rate whose turnover requires several hours (NO is about 300,000 more tightly bound to Hb than is oxygen). Further, the addition of O_2 to

HbNO results in methemoglobin formation such that by 120 min after exposure, 77% of the HbNO has been changed to methemoglobin. In the first five min of exposure to oxygen, about one third of the Hb is converted. This rate of change is similar to the change that occurs when NO is anaerobically injected intraperitoneally into rats. The decay of HbNO virtually halts after the first exposure to oxygen, then resumes with pseudo-first order kinetics. The simultaneous examination of MetHb formation reveals that the kinetics are not coincident with HbNO decay, although the final conversion is stoichiometric. Thus conversion of HbNO to MetHb occurs by at least two different routes and all the intermediates are "magnetically silent."

V. Reaction of Nitrogen Dioxide in Vitro

A. Oxidation of Hemoglobin

Nitrogen oxides, NO and NO_2, are known to convert hemoglobin into methemoglobin by oxidation of the heme iron from the ferrous to the ferric state, as discussed. Also noted, NO_2 is soluble in water, but hydrates in the tissue when inhaled, and reactions occur locally before the gas can reach the hemoglobin *in vivo*. However, *in vitro* quantitative relations between MetHb formation, gas concentration, and the length of exposure of human blood to NO_2 were not made clear until work in 1983. NO passes through tissue to reach the plasma or hemoglobin where it reacts with oxygen to form NO_2, and other nitrogen oxides, which can perhaps explain the confusion which existed prior to this work. Earlier studies had concluded that, in aerobic *in vitro* exposure to NO and NO_2, hemoglobin oxidized to MetHb, but the amount of MetHb formed is about 10 times greater with NO_2 than with NO.

In other studies exposure to NO_2 under aerobic or anaerobic conditions generates 95% MetHb and 5% HbNO. It was also noted separately that a gas NO_2 concentration of 40 ppm or lower will produce a MetHb in hemolysates or red blood cells *in vitro,* but no HbNO is formed. Further studies systematically exposed whole blood (under controlled conditions of temperature and time) to NO_2 gases in a tonometer. The samples of isocarbic blood, held at 37°C, were exposed to NO_2 for 15 min to 6 hr. Gas consumption by blood and MetHb formation were measured. For the first time it was demonstrated a one-to-one relation exists between the amount of NO_2 used and the amount of MetHb formed. Further, at NO_2 concentrations of 45 ppm, 50% of the total Hb was oxidized in the first hour of exposure, and the remaining 50% required another 5 hours of exposure to complete the oxidation. Noteworthy is the observation that the pCO_2 remained at 34.4 ± .26 torr, the pH changed from 7.40 to 7.10 while the pO_2 dropped from 157 to 80 torr. Decreases of pH and pO_2 were in direct relation to the increasing amount of MetHb formed. Reducing the NO_2 to about 11 ppm also reduced the rate of the MetHb formation such that only 35% of the Hb was converted in 4 hr, but leveled off over the next 2 hr such that less than 40% of the Hb is converted in 6 hr.

Spontaneous reduction rate of the MetHb was measured at 37°C in nine blood samples with the MetHb content, previously induced by NO_2, ranging from 27 to 85% of the original blood Hb. Reduction of MetHb concentration averaged 20.4% in 4 hr or an average of 5.1% per hour. The initial concentration of metHb does not seem to significantly influence the reduction rate of the MetHb to Hb.

Hugo Chiodi (who is responsible for most of the aforementioned studies and to whose memory this article is dedicated) demonstrated that *in vivo* exposure of human blood to NO_2 concentration as low as 6 ppm oxidizes the heme-iron of Hb to MetHb. The amount of Hb oxidized by NO_2 is a function of gas concentration and time of exposure. Increasing gas concentration and longer exposures will generate increasing amounts of MetHb, but the relation is not linear. While about 50% of the Hb is oxidized in the first hour at rather high NO_2 concentrations, it takes many hours to oxidize the remaining Hb. This hyperbolic curve is much flatter at lower NO_2 concentrations. Chiodi offered no explanation for this difference in the reactions rates at lower concentrations of NO_2 exposure.

The existence of the light absorption spectra of two peaks at 500 and 630 nm, exclude from our samples the presence of the NO_2–MetHb compound which is formed when an excess of NO_2 is present. The NO_2–MetHb compound shifts the characteristic 630 nm maximum to 625 nm and a light absorption minimum instead of a maximum at 500 nm occurs. A secondary peak appears at 538 nm.

There are no other known published results on the stoichiometric relations between MetHb and NO_2 to which 1:1 results could be compared unless an identical reaction of the Hb with NO_2 and with nitrites is postulated. The assumption that both substances react in an identical manner toward Hb has not been proven. Early studies found that "nitrite addition to deoxygenated blood does not give the same distribution of paramagnetic Hb derivatives as either NO_2 or aerobic NO_2. Consequently, one can associate the reaction of NO_2 with Hb with the oxidant-dependent reaction of nitrite in blood only with great caution.

VI. Overview and Conclusion

With the exception of nitrous oxide and NO, all the usual oxides of nitrogen fail to cross the tissue barrier in the lung. NO_2 has a low solubility in water; thus human exposure to this gas produces local injury in the lung, sometimes delayed, and sometimes fatal if the dose is great enough. Methemoglobin formation occurs only when dissolved NO gas migrates across the tissue barrier of the lung into the red blood cell, where reaction with oxygen probably occurs, resulting in NO_2, which in turn reacts with the heme-ferrous iron to create the ferric state of methemoglobin. Only when NO_2 is exposed to hemoglobin *in vitro* does it produce methemoglobin, whether in whole red cells or in hemolysates of hemoglobin solutions.

The reaction of NO_2 with Hb is stoichiometric. Evidence of a NO_2–MetHb compound is weak to absent. The rate of MetHb formation from exposure to oxides of nitrogen varies with the concentration of NO_2, but in high doses, 50% of the available hemoglobin is converted in about 1 hr, and the conversion is near complete in about 5 hr. The rate of reconstitution of the hemoglobin to the active ferrous unit from MetHb is about 5.1% of the total hemoglobin in 1 hr, or about 20% in 4 hr from the enzymes found inside the red blood cells. Destruction or poisoning of these enzymes or disruption to the environment necessary for those enzymes to be active prevents the reconstitution of Hb from MetHb.

Related Articles: DIESEL EXHAUST, EFFECTS ON THE RESPIRATORY SYSTEM; ETHANOL FUEL TOXICITY; NITRIC OXIDE AND NITROGEN DIOXIDE TOXICOLOGY; MONITORING INDICATORS OF PLANTS AND HUMAN HEALTH FOR AIR POLLUTION; OZONE EXPOSURE, RESPIRATORY HEALTH EFFECTS; PLANTS AS DETECTORS OF ATMOSPHERIC MUTAGENS.

Acknowledgment

In memory of Hugo Chiodi, M.D.

Bibliography

Case, G. D., Dixon, J. S., and Schooley, J. C. (1979). Interactions of blood metalloproteins with nitrogen oxides and oxidant air pollutants. *Environ. Res.* **20,** 43–65.

Chiodi H., and Mohler, J. G. (1987). Nonautocatalytic methemoglobin formation by sodium nitrite under aerobic and anaerobic conditions. *Environ. Res.* **44,** 45–55.

Chiodi, H., and Mohler, J. G. (1985). Effects of exposure of blood hemoglobin to nitric oxide. *Environ. Res.* **37,** 355–363.

Chiodi, H., Collier, C. R., and Mohler, J. G. (1983). *In vitro* methemoglobin formation in human blood exposed to NO_2. *Environ. Res.* **30,** 9–15.

Oda, H., Nogami, H., and Nakajima, T. (1980). Reaction of hemoglobin with nitric oxide and nitrogen dioxide in mice. *J. Toxicol. Environ. Health* **6,** 673–678.

Van Assendelft, O. W., and Zijlstra, W. G. (1968). Hemoglobin and hemoglobin nitrite. *Clin. Chem.* **14,** 918–919.

Ozone Exposure, Respiratory Health Effects

Johnny L. Carson
The University of North Carolina at Chapel Hill

Glossary

Adaptation Reduced responsiveness to ozone subsequent to repeated daily exposures.

Airway hyperreactivity Increased sensitivity of the airways to challenge by bronchospasm-inducing agents with subsequent adverse effects on pulmonary function.

Alveolus(i) Sac-like structure in the parenchymal portion of the lung located at the terminus of the conducting airways. The alveoli represent the actual sites of gas exchange in the respiratory system.

Atopy Condition conferring increased likelihood of development of IgE-mediated reactivity to allergens.

Bronchoconstriction Reduction in the caliber of the conducting airways brought about by contraction of smooth muscle.

Conducting airways That portion of the respiratory system responsible for directing and conditioning the inspired air to the gas exchange portion of the lung.

Elastolysis Degradation of the fibrous elastic components of the respiratory system which provide flexibility to ventilatory maneuvers.

Fibrogenesis Development of fibrotic elements (scar tissue) often as a result of a chronic disease process.

Hyperplasia Increase in cell number.

Hypertrophy Increase in organ size caused by an increase in cell volume rather than cell number.

Malignant transformation Development of neoplasia (cancer).

Mucociliary transport Physiological process occurring in the conducting airways whereby mucin-containing surface liquid holding deposited inhaled particulates is moved out of the airways into the larynx by the coordinated sweeping motion of cilia.

Mucus cell hyperplasia Increase in the number of mucus cells in the airways.

Squamous metaplasia Pathohistologic finding in the airways associated with recurrent irritation or injury manifested by replacement of the pseudostratified columnar organization of the epithelium with a flattened, multilayered squamous pattern.

OZONE (O_3) is a potent oxidant and a highly reactive component of photochemical smog. This reactivity has created considerable concern regarding its health effects in humans, particularly on tissues of the respiratory system. Mucous membranes of the airways and the lung parenchyma are directly exposed to the ambient air, and provide a large surface area which is vulnerable to injury by inhaled irritant gases and airborne particulates. A number of research studies summarized here outline the health effects of ozone on the respiratory system.

I. Ozone in the Environment

A. Origins

A number of complex chemical and physical processes lead to the secondary formation of ozone in the ambient atmosphere. Primary among these processes are fuel combustion, volatilization of organic compounds, photochemical reactions, and meteorological factors.

Unlike some components of polluted air which are injected directly into the atmosphere from a source, ozone forms as a post-emission derivative of combustion products, volatile organic compounds, and complex photochemical reactions. Nitrogen oxides, hydrocarbons, aldehydes, and carbon monoxide are major primary air pollutants which participate in these reactions and which ultimately give rise to the secondary oxidant pollutants of which the best known is ozone. Foremost among these reactions is the photolysis of atmospheric nitrogen dioxide (NO_2) which eventuates an $O_3 \leftrightarrow NO_2$ cycling phenomenon shown by the following schematic.

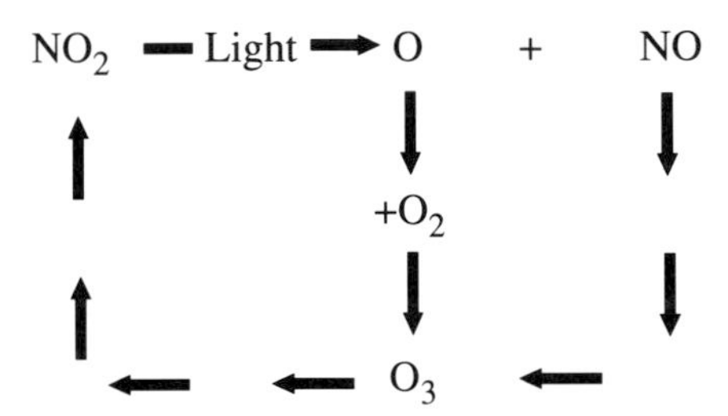

While this basic cycle can be experimentally modeled, several factors complicate simple simulations. First, emission of hydrocarbons, aldehydes, and other components of combustion-based air pollution result in the entry of these compounds into the $O_3 \leftrightarrow NO_2$ cycle as highly reactive radicals where they interact directly with nitric oxide (NO) to form NO_2. The presence of these compounds competitively limits the interaction of O_3 with nitric oxide resulting in the accumulation of high concentrations of ozone as shown in the next schematic.

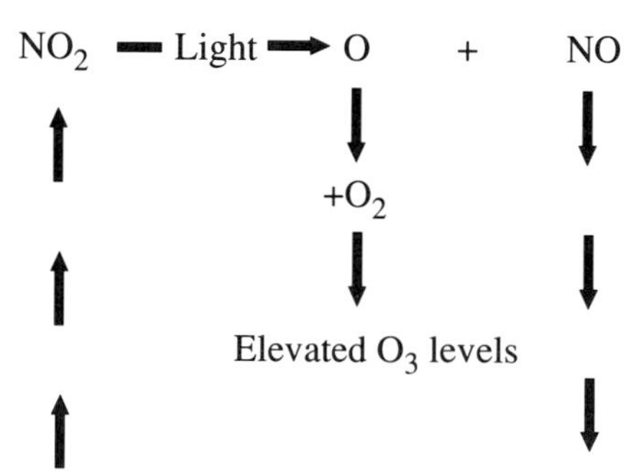

Others factors which confound modeling of this mechanism include the concentration of pollutants from previous days, vertical and horizontal atmospheric mixing, light intensity, diurnal variability of primary emissions, and other meteorological and micrometeorological factors.

Emissions from internal combustion engines, activities necessitating high temperature combustion, and volatile organic compounds such as paints and solvents are major sources of primary pollutants, particularly NO_2. These primary emissions represent precursors to ozone accumulation in the outdoor environment of both technologically advanced and developing countries. Agricultural burning in rural areas, particularly where it occurs on a large scale, may also contribute to elevated ozone levels.

Because ozone is generated primarily as a result of a photochemical reaction involving solar energy, indoor levels derive primarily from diffusion from outdoors. Under ambient indoor conditions, ozone decays rapidly and high indoor levels are seldom achieved. However, certain occupational situations and indoor activities combined with restricted ventilation can cause ozone to be generated to levels potentially capable of producing adverse health effects. Such elevated indoor levels of ozone occur in situations involving electrical arcing (welding), electrostatic discharges (xerographic copy machines), and industrial processes and research procedures (microbicidal cleansing) that involve the use of ultraviolet light.

B. Biologically Significant Chemical Reactivity

There is a growing body of evidence that exposure of cells and tissues to ozone may result in the formation of biologically significant compounds via ozonolysis. The oxidative stress may derive from hydroxyl radicals, superoxide, singlet oxygen, and/or hydrogen peroxide. Ozone may participate in the oxidation of proteins, amines, and thiols. These reaction products may influence a variety of vital functions such as altered cellular physiology and activation or mediation of the inflammatory response. Ozone can bring about oxidative decomposition of polyunsaturated fatty acids to generate toxic reactive intermediates such as peroxide and/or aldehydes. There is also evidence that ozone is capable of reacting directly with DNA and RNA. This characteristic raises numerous possibilities for pathogenic events by way of alteration of gene expression. The highly reactive nature of ozone may result in cell death with subsequent loss of histologic integrity. This may diminish cellular and systemic defense mechanisms as well as releasing necrotic cell debris that may promote subsequent bacterial infection.

II. The Respiratory System: An Overview of Structure and Function

A. Physiology of Respiration

The superficial cellular lining of the respiratory system represents a critical biological interface between the internal organ systems of the body and the external environment. The phasic process of respiration (ventilation) is driven by active contraction of the diaphragm and other inspiratory muscles followed by expiration driven by elastic recoil of the lung tissue. The inspiratory muscles are under coordinated neuronal control. The conducting airways of the respiratory tract humidify and warm the inspired air, provide a primary line of defense against inhaled particles via mucociliary transport, and direct air flow to the parenchymal (alveolar) portion of the lung where actual gas exchange with capillary blood coursing through the alveolar walls occurs. Particle clearance from alveoli depends on the scavenging function of macrophages patrolling the alveolar surfaces. Reactive and/or highly soluble gases at low concentrations are likely to have been absorbed by the airways (upper and lower) before reaching the alveolar area. The respiratory tract is also the primary pathway for uptake of gases like carbon monoxide (CO), or vaporized (or particulate) drugs like cocaine or nicotine. While it is clear that the respiratory tract is the primary means of access of air pollutants to internal organs, little is known about the indirect effects of air pollutant exposure on tissue and organ systems other than the respiratory tract. However, it is clear that the air flow and gas exchange regions of the respiratory system directly exposed to inspired air containing pollutants represent a primary means whereby cellular level injury to the respiratory system occurs. Additionally, tissues and cells not directly exposed to airflow but proximate to the respiratory system such as inflammatory cells, elastic tissues, smooth muscle, and nerves may be influenced adversely by irritant exposures causing amplification and/or secondary reactions associated with exposure.

A number of studies investigating the health effects of ozone exposure have necessarily utilized laboratory animals. Attempts to correlate findings from such studies to human health effects must take into account interspecies differences which may affect dose. In general, humans have larger lungs and the branching pattern of the upper bronchial airways is more symmetrical. These factors may lead to a higher degree of upper bronchial particle depositions particularly near airway bifurcations. Predicted pulmonary ozone dose curves determined for the lungs of guinea pigs, rabbits, and humans suggest a general similarity as well as indicating that the area of the respiratory bronchioles receives the maximum ozone dose.

B. Conducting Airways

The conducting airways extend in order from the nose through the nasopharynx, larynx, trachea, and bronchi to the level of the terminal bronchioles (see Fig. 1). Further airway generations contain progressively larger numbers of alveoli in their walls and include respiratory bronchioles and alveolar ducts. These conduits possess cartilage, smooth muscle, and elastic fibers which provide both structural support and mechanical plasticity. Smooth muscle, functionally significant in bronchoconstriction, appears first in the trachea and extends distally to the level of the respiratory bronchioles.

Elastic fibers are located in both the airways and parenchymal regions of the respiratory system and confer elasticity, structural scaffolding, and compliance in response to ventilatory motions. Nerves in the respiratory tract occur in the connective tissues of the larger airways. Because the airways and lung are perfused by the systemic blood supply, migratory inflammatory cells such as polymorphonuclear leukocytes may be resident in the connective tissues and epithelium particularly in response to injury. Although not directly exposed to air flow, these features may be vulnerable and/or responsive to the deleterious effects of inhaled irritants.

The mucosal surface of the large airways comprises a pseudostratified columnar epithelium which extends from the respiratory portion of the nose to the bronchiolar region. The superficial epithelium directly exposed to air flow comprises primarily ciliated and goblet (mucus) cells in a ratio of approximately 5 : 1. Basal cells reside beneath the superficial layer of cells and rest on the basement membrane. Basal cells are thought to represent, at least in part, progenitor cell types of the superficial epithelium of the large airways. In addition, nonciliated cells containing secretory granules and extending to the luminal surface have also been shown to incorporate nucleic acid precursors, and there-

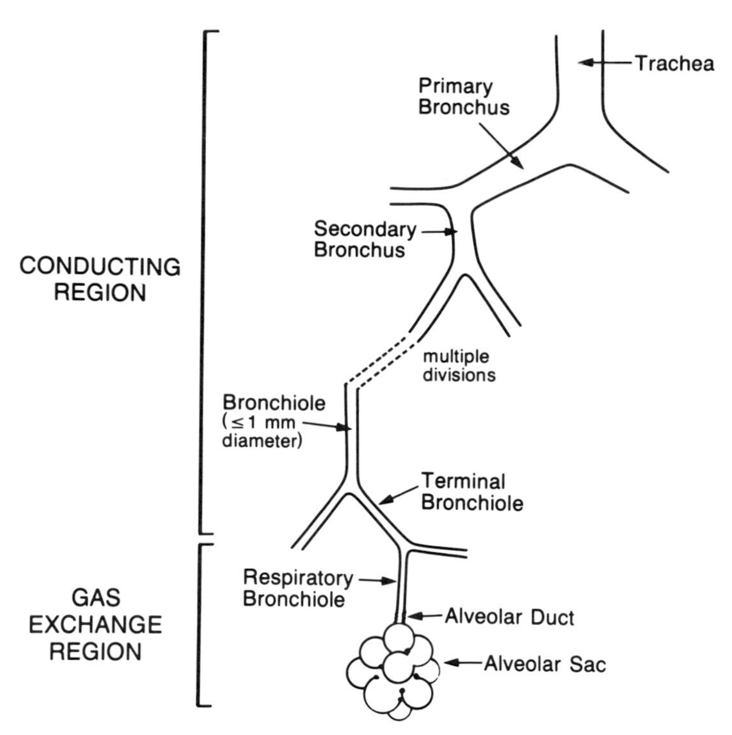

Figure 1 Schematic diagram of airway generations in the respiratory system.

fore may also contribute to the mitotic potential of the epithelium during development and during regeneration subsequent to injury.

The ciliated and mucus-producing goblet cells populating the superficial airway surfaces exposed to air flow represent an important primary defense mechanism to the respiratory system. Goblet cells secrete mucus glycoproteins which cover the epithelial surface, entrapping inhaled particulates. The action of the cilia beating within a watery periciliary layer serves to direct the mucus to the larynx where it may be eliminated from the airways by a variety of mechanisms including swallowing or coughing.

The histologic architecture of the airways also includes secretory glands, blood vessels, motor and sensory nerves, special resident cells such as mast cells, neuroendocrine cells, and lymphocytes, and wandering cells such as polymorphonuclear leukocytes, eosinophils, and basophils. While these cells and tissues are not generally exposed to air flow, they may respond to inhalation of irritants and thus precipitate adverse secondary effects or amplification of initial responses.

The bronchiolar region represents a transitional zone between the larger conducting airways and gas exchange areas of the lung. This transition is reflected by a change in histologic organization from a pseudostratified columnar epithelium to a simple columnar or cuboidal epithelium, the loss of basal and goblet cells, the persistence of ciliated cells, and the appearance of a novel cell type, the Clara cell. Clara cells apparently have both secretory and regenerative functions in that they contain granules and can incorporate nucleic acid precursors, a characteristic activity of cells in the initial phases of cell division. Further, Clara cells contain high levels of cytochrome *P*-450 suggesting an oxidase activity which may serve in protective deactivation of inspired irritants, but may also render the Clara cells selectively vulnerable to certain chemicals that are activated by cytochrome *P*-450-mediated reactions.

When the conducting airways are chronically exposed to certain irritants, the epithelial lining may undergo pathologic changes which presumably are protective in nature. A common alteration is an increase in the number of mucus-producing cells at the expense of ciliated cells. Such mucus cell hyperplasia often appears in response to persistent irritation, presumably representing ciliated cell death and replacement from activated precursor cells that differentiate into goblet rather than ciliated cells. Given continuing exposure to the noxious agent, mucus cell hyperplasia may give way to epithelial reorganization and replacement of the pseudostratified columnar epithelium by flattened squamous epithelial cells or squamous metaplasia. Mucus cell hyperplasia and squamous metaplasia represent reversible pathologic responses to injury; however, the appearance of squamous metaplastic change is generally considered reflective of a condition which can lead to malignant transformation.

C. The Gas Exchange Region of the Lung

The alveolar sacs at the termini of the conducting airways are the sites of gas exchange between air and the blood circulation of the respiratory tract (refer to Fig. 1). These evaginations of the alveolar ducts are lined by a thin, squamous epithelium. The squamous epithelial cells of the alveoli are present in two forms. The first is represented by Type I cells which cover most of the alveolar surface. These cells are extremely flattened and are well adapted to diffusion of O_2 and CO_2 between alveolar air and capillary blood. The second form of alveolar cell is the Type II cell. Alveolar Type II cells can be distinguished by their large size, cuboidal appearance, and position in the "corner" of alveolar sacs. Type II cells are secretory in nature and exhibit "foamy" vacuoles by light microscopy. They are thought to represent progenitors of Type I alveolar cells. Viewed by electron microscopy, these vacuoles reveal multilayered stacks called lamellar bodies comprising phospholipids, glycosaminoglycans, and proteins. Upon exocytosis, these materials spread out to cover the alveolar surfaces in a coating layer known as surfactant. The surfactant layer reduces the surface tension of the alveoli, thus greatly reducing lung elastic recoil and stabilizing the alveoli against collapse.

A third cell type exposed to the inspired air in the gas exchange region of the lung is the alveolar macrophage. Macrophages derive from monocytes originating in the bone marrow. Alveolar macrophages are highly motile scavenger cells of the alveolar region and thus represent an added line of defense in the parenchymal region after the mucociliary clearance mechanisms present in the airways. Macrophages scour the epithelial surfaces engulfing inspired particulates, microbial pathogens, and catabolic debris. Final clearance of macrophages probably occurs via mucociliary clearance.

D. Tests of Pulmonary Function

Tests of pulmonary function performed in a clinical setting are used to characterize level and volume changes in lung function. A variety of lung disorders can be diagnosed as well as physiologic responses to exercise, pharmacologic stimulation, and environmental exposures by analyzing a battery of well-ordered breathing maneuvers. Several pulmonary function tests are particularly relevant to assessment of respiratory functional impairment associated with air pollutant exposure. Spirometric measurements provide data quantifying the maximal volume of air exhaled following maximal inspiration (forced vital capacity or FVC), the volume of air exhaled in the first second (forced expiratory volume in 1 second or FEV_1), and other rates of airflow such as the flow during the middle half of the FVC maneuver (flow between 25% and 75% of FVC or FEF_{25-75}). Body plethysmography provides additional data concerning absolute lung volumes and resistance to air flow.

III. Changes in Structure and Function of the Respiratory System Associated with Ozone Exposure

A. Introduction

A fundamental characteristic of living cells is the ability to adapt to and survive potentially lethal stresses. Different cell types are endowed with characteristics that promote homeostasis and mandate thresholds beyond which injurious challenges result in cellular dysfunction and death. Because of metabolic, physiologic, and structural diversity, the injury threshold for a given stimulus is often different among different cell types.

Numerous studies have documented the acute toxicity of ozone on living cells and tissues. However, in assessing the pathogenic potential of this substance, it is necessary to view its reactivity in view of exposures which elicit transient cellular level responses which may destabilize the homeostasis of the affected cell(s) but are nevertheless sublethal, against those types of exposures which clearly result in permanently altered metabolism, dysfunction, cell death, and the evolution of chronic disease processes.

At the present time, it is difficult to extrapolate an

ozone dose–effect relationship in either human or animal studies which can be directly correlated with the evolution of chronic disease processes at the levels mandated by the National Ambient Air Quality Standard (≤0.12 ppm over a 1-hr averaging period). No standard has yet been developed for longer exposures. Among humans, histopathologic studies are few and largely unremarkable; and pulmonary function studies using acute low-level ozone exposures have elicited changes only with moderate to strenuous exercise which are reversible during a recovery period in clean air. However, there appears to be a wide range of individual response to ozone suggesting the existence of sensitive populations. Analysis of tissues obtained from animal studies have been similarly difficult to analyze. For example, pathologic changes associated with chronic exposure regiments of up to two years also are generally reversible. Given the present level of knowledge, it is difficult in many respects to correlate the findings of animal studies with their relevance to recurring human exposures encompassing a lifetime.

B. Morphologic and Ultrastructural Assessments

1. Histopathologic Changes in Acute Ozone Exposure

Histopathologic studies of the response of human respiratory tissues to acute experimental ozone exposure are limited. However, observations in laboratory animal models have provided some basis for plausible extrapolations of ozone-related injury to human respiratory epithelium. These models have documented cellular and subcellular changes in both the airways and in the gas-exchange regions of the respiratory system following acute ozone exposure. In the airways, these morphologic changes appear prominently in cells involved in mucociliary transport. Deterioration in the histologic integrity of the epithelium through the loss of ciliated cells, ultrastructural level alterations of cellular organelles, and increases in mitotic index have been reported following acute exposure to ozone. Ozone also appears to be edemagenic to the airways, and hyperplasia and hypertrophy of the bronchiolar epithelium have also been documented. Infiltration of the airways epithelium by macrophages, neutrophils, and eosinophils subsequent to acute ozone exposure is further reflective of the inflammatory potential of this pollutant. A similar pattern of inflammatory cell infiltrate has been documented in nasal washings of human subjects following acute ozone exposure. Only one study has specifically examined the morphology of the respiratory airway epithelium of human subjects following acute ozone exposure. Using the epithelium lining the nasal turbinates as a source of tissue, this study found no appreciable change in the structural integrity of the epithelium immediately following an exposure of 0.4 ppm ozone for four hours. However, such experimental protocols using human subjects may not provide meaningful insights into the mucosal pathologic consequences of long-term or regular environmental exposures among healthy individuals or those with chronic lung diseases. Table I outlines and references the major documented histopathologic changes associated with acute ozone exposure.

2. Histopathologic Changes in Chronic Ozone Exposure

The pattern of injury in acute ozone exposure appears to be one of direct cellular toxicity and protective responses. Experimental models of chronic exposure while similar in some respects, point to changes consistent with the initiation of chronic disease processes. Although a clear correlation between acute toxicity and structure/function manifestations of ozone-induced chronic respiratory disease remains to be defined, there are implications in these model systems for carcinogenesis and development of fibrotic lung disease.

These changes include ultrastructural alterations in goblet cells and proliferation of secretory epithelial cell types consistent with the development of mucus cell hyperplasia. Other forms of hyperplasia and hypertrophy seen following experimental ozone exposure in laboratory animals are suggestive of squamous metaplasia. Additionally, the connective tissue thickening, fibrotic changes in alveolar ducts and bronchioles, and increases in inflammatory cells in the wake of extended exposures suggest a fertile setting for fibrosis. However, these observations must be couched in the knowledge that they (1) have been made largely in an experimental setting, (2) are generally inconsistent with human lifetime exposure, (3) are exclusive of other environmental challenges, (4) are without consideration of the enhanced sensitivity of "responder" populations, and (5) are generally reversible given recovery in clean air.

The region of the central acinus, the site where the

Table I Major Histopathologic Findings in Acute Ozone Exposure[a]

Finding	O_3 Exposure protocol	Study
↓ Ciliated cells Loss of cilia Necrosis of ciliated cells Alterations in mucus cell granules ↑ Extracellular space Focal epithelial stratification with ↑ small mucus granule cells Appearance of intermediate cells ↑ Cytoplasmic filaments and desmosomal attachments in basal cells	0.64 ppm; 3 or 7 days; bonnet monkey	1
↑ Acidic and neutral glycoconjugates	0.15 ppm; 6 days; bonnet monkey	2
Acute pulmonary edema ↑ Lamellar body secretion Hypertrophy of Type II cells ↓ Ability to form tubular myelin	3 ppm; 8 hr; rat	3
Enhancement of influenza infection	1 ppm; 3 hr/day 5 days; mouse	4
Alterations in morphometric and morphologic characteristics of Clara cells ↑ Alveolar macrophages ↑ Ciliogenesis	0.2 ppm, 0.5 ppm, 0.8 ppm; 8 or 24 hr/day 7d; rat	5
↑ Airway neutrophils	0.4 ppm or 0.6 ppm; 2 hr; human	6
Degeneration and necrosis of Type I cells	0.8 ppm; 4–12 hr; rhesus monkey	7

[a] Defined by convention as exposures ≤7 days duration.

1. Wilson, D. W., Plopper, C. G., Dungworth, D. J. (1984). The response of the macaque tracheobronchial epithelium to acute ozone injury. A quantitative ultrastructural and autoradiographic study. *Am. J. Pathol.* **116,** 193–206.
2. Harkema, J. R., Plopper, C. G., Hyde, D. M., St. George, J. A., Dungworth, D. L. (1987). Effects of an ambient level of ozone on primate nasal epithelial mucosubstances. Quantitative histochemistry. *Am. J. Pathol.* **127,** 90–96.
3. Balis, J. U., Paterson, J. K., Haller, E. M., Shelley, S. A., Montgomery, M. R. (1988). Ozone-induced lamellar body responses in a rat model for alveolar injury and repair. *Am. J. Pathol.* **132,** 330–344.
4. Selgrade, M. K., Illing, J. W., Starnes, D. M., Stead, A. G., Menache, M. G., Stevens, M. A. (1988). Evaluation of effects of ozone exposure on influenza infection in mice using several indicators of susceptibility. *Fundam. Appl. Toxicol.* **11,** 169–180.
5. Schwartz, L. W., Dungworth, D. L., Mustafa, M. G., Tarkington, B. K., Tyler, W. S. (1976). Pulmonary responses of rats to ambient levels of ozone: Effects of 7-day intermittent or continuous exposure. *Lab. Invest.* **34,** 565–578.
6. Seltzer, J., Bigby, B. G., Stulbarg, M., Holtzman, M. J., Nadel, J. A., Ueki, J. F., Leikauf, G. D., Goetzl, E. J., Boushey, H. A. (1986). O_3-induced change in bronchial reactivity to methacholine and airway inflammation in humans. *J. Appl. Physiol.* **60,** 1321–1326.
7. Castleman, W. L., Dungworth, D. L., Schwartz, L. W., Tyler, W. S. (1980). Acute respiratory bronchiolitis: An ultrastructural and autoradiographic study of epithelial cell injury and renewal in rhesus monkeys exposed to ozone. *Am. J. Pathol.* **98,** 811–840.

terminal airways join the proximal alveoli in the gas-exchange region, appears particularly sensitive to the deleterious effects of ozone exposure. In this region, experimental animal studies using light and electron microscopic morphologic assessments have suggested particular sensitivity of Type I epithelial cells to chronic ozone exposure. Changes in Type I cells include necrosis and sloughing, increased volume and thickness, and reduced surface area. As in the airways, infiltrations of inflammatory cells also have been documented in chronic ozone exposures. Histopathologic changes in the capillary endothelium and Type II cells also have been reported. The major documented histopathologic changes associated with chronic ozone exposure are outlined and referenced in Table II.

C. Assessments of Pulmonary and Cellular Function following Acute Exposure

1. Symptomatic Responses

Although there appears to be a wide range in individual responses, human volunteers experimentally acutely exposed to ozone report throat dryness, cough, mucus production, and substernal chest pain on deep inspiration. These symptoms are more

Table II Major Pathohistologic Findings in Chronic Ozone Exposure[a]

Finding	O_3 Exposure protocol	Study
Low-grade chronic respiratory bronchiolitis Intraluminal accumulations of macrophages Hypertrophy and hyperplasia of cuboidal bronchiolar cells Reduction in inflammatory cells with extended exposure Reduction in ^{3}HT labeling with extended exposure	0.5 or 0.8 ppm; 7, 28, 90 days; bonnet monkey	1
Peribronchiolar edema Bronchiolization of alveolar duct epithelium Type II cell proliferation ↑ macrophages and leukocytes	0.7 ppm; 28 days, 20 hr/day; rat	2
Ultrastructural changes in nasal goblet cells	0.15 or 0.30 ppm; 8 hr/day, 90 days; bonnet monkey	3
Bronchitis and peribronchitis Connective tissue thickening around bronchi Degeneration of ciliated cells Hyperplasia of lymphoid nodules around small vessels Changes in cell populations of alveolar ducts Fibrotic changes in alveolar ducts and bronchioles	0.5 ppm; 6 hr/day, for 2,3,5,12 mo.; rat	4
Thickening of bronchiolar wall Shifts in epithelial cell populations ↑ bronchiolar epithelial cells ↑ organelles associated with protein synthesis ↑ interstitial fibers and amorphous ground substance ↑ interstitial smooth muscle cells/epithelial basal lamina surface area ↑ volumes of interstitial smooth muscle, macrophages, mast cells, and neutrophils/epithelial basal lamina surface area	0.4 or 0.64 ppm; 90 days; bonnet monkey	5
Damage to tips of alveolar septae at the respiratory bronchiole–alveolar duct junction	0.95 ppm; 8 hr/day, 90 days; rat or 0.95 ppm; 8 hr/day, seven 5-day episodes with 9-day recovery periods (89 days total); rat	6 7

[a] Defined by convention as exposures >7 days.

1. Eustis, S. L., Schwartz, L. W., Kosch, P. C., Dungworth, D. L. (1981). Chronic bronchiolitis in nonhuman primates after prolonged ozone exposure. *Am. J. Pathol.* **105,** 121–137.
2. Gross, K. B. and White, H. J. (1986). Pulmonary functional and morphological changes induced by a 4-week exposure to 0.7 ppm ozone followed by a 9-week recovery period. *J. Toxicol. Environ. Health* **17,** 143–157.
3. Harkema, J. R., Plopper, C. G., Hyde, D. M., St. George, J. A., Wilson, D. W., Dungworth, D. L. (1987). Response of the macaque nasal epithelium to ambient levels of ozone. A morphologic and morphometric study of the transitional and respiratory epithelium. *Am. J. Pathol.* **128,** 29–44.
4. Hiroshima, K., Kohno, T., Ohwada, H., Hayashi, Y. (1989). Morphological study of the effects of ozone on rat lung. II. Long-term exposure. *Exp. Mol. Pathol.* **50,** 270–280.
5. Moffatt, R. K., Hyde, D. M., Plopper, C. G., Tyler, W. S., Putney, L. F. (1987). Ozone-induced adaptive and reactive cellular changes in respiratory bronchioles of bonnet monkeys. *Exp. Lung Res.* **12,** 57–74.
6. Barr, B. C., Hyde, D. M., Plopper, C. G., Dungworth, D. L. (1988). Distal airway remodeling in rats chronically exposed to ozone. *Am. Rev. Respir. Dis.* **137,** 924–938.
7. Barr, B. C., Hyde, D. M., Plopper, C. G., Dungworth, D. L. (1990). A comparison of terminal airway remodeling in chronic daily versus episodic ozone exposure. *Toxicol. Appl. Pharmacol.* **106,** 384–407.

likely to occur with (a) increasing duration of exposure, (b) higher ozone concentrations, and/or (c) exercise during exposure.

2. Pulmonary Function Assessments

A 6.6-hr exposure to as little as 0.08 ppm ozone causes significant respiratory discomfort and functional decrements in young adults doing moderate exercise for 5 hr. These changes include significant decrements in FEV_1, FVC, and FEF_{25-75}, which are likely to result primarily from an involuntary inhibition of deep inspiration. Additionally, significant increases in airway reactivity and specific airway resistance have been documented. Other studies using higher ozone concentrations have generally corroborated these findings. Changes in pulmonary function tests generally attenuate within a few hours following termination of exposure. However, some residual bronchial hyperreactivity to pharmacologic provocation may persist beyond this interval. Although individual response varies considerably, adults do not display significant deficits in pulmonary function in response to brief (e.g., a few minutes) ozone exposures. Also, response to ozone exposure appears to attenuate with age (see Section II.C.6 "Adaptation"). However, changes in small airway function are more difficult to detect and have not been systematically investigated. This reduction in function is thought to be based on involuntary resistance to inspiration perhaps due to inspiratory discomfort associated with ozone exposure. Reduced FVC and FEV_1 resulting from ozone exposure in human subjects appear to derive from an involuntary neurally mediated reduction in inspiratory capacity. However, the short-lived decrements in FVC and FEV_1 may also be reflective of reductions in the caliber of the small airways.

Asthmatics, atopic subjects, and other individuals with chronic lung disease do not appear to be unusually responsive to ozone exposure in experimental settings except where high effective ozone doses are used. Patients with atopic asthma have been shown to exhibit increased bronchial responsiveness to allergens following an acute low-level exposure to ozone although baseline pulmonary function is unaffected. Other studies of ozone health effects on pulmonary function in asthmatics suggest that they experience a greater per cent decrease in FEV_1, $FEV_{1\%}$, and FEV_{25-75} relative to nonasthmatic individuals. Studies of pulmonary function in women exposed to ozone have suggested that they may be more responsive than men to ozone exposure based on the observation that their reduction in pulmonary function tests was similar even though the effective dosage was less. In a similar vein, comparative studies of age differences suggest that older adults are less responsive to ozone than young adults. Ozone exposure does not appear to provoke severe bronchospasm.

There is no evidence that ozone exposure predisposes individuals to acute viral upper respiratory infection or the altered expression or evolution of host responses. However, ozone exposure has been shown to enhance influenza symptoms and to increase mortality in rodents subsequently challenged with aerosolized bacteria.

Exposure of human subjects to ozone results in a rapid inflammatory response manifested by increased numbers of neutrophils and levels of certain inflammatory mediators recovered in saline lavage of the nose or of the lower airways. While the spirometric decrements associated with ozone exposure may represent reversible responses to irritation, a greater concern exists relative to the potential irreversible adverse consequences of a persistent inflammatory response due to recurrent ozone exposures. Such consequences might include increases in bronchial reactivity (reactive airways disease), elastolysis, or fibrogenesis. It is possible to substantially diminish the functional response to ozone exposure by pretreatment with nonsteroidal anti-inflammatory agents. However, the degree of polymorphonuclear inflammatory response in bronchoalveolar lavage is not altered.

There is at present little evidence of a synergistic effect of ozone exposure on pulmonary function in normal individuals when delivered in concert with other air pollutants. Adolescent asthmatics, however, have been shown to display a bronchoconstrictive response to a brief low-level sulfur dioxide exposure with exercise following an acute low-dose ozone exposure. In contrast, no significant interaction between nitrogen dioxide and ozone in combination has been identified.

Studies of nasal lavage fluid from individuals with allergic rhinitis exposed to ozone indicate that an inflammatory response similar to that seen in normal subjects occurs. This response is manifested by increased cell infiltrates, sloughing of epithelial cells, and changes in biochemical and enzyme assays. However, the subsequent nasal mucosal response of the allergic subjects to a specific antigenic challenge following the ozone exposure is not enhanced. There is some possibility that the inter-individual variability in response to ozone may relate to nonspecific immune factors inasmuch as some individuals appear more responsive to springtime exposures.

3. Other Physiologic Changes

In addition to pulmonary function decrements, other physiologic changes have been observed subsequent to ozone exposure. Among these, higher temperatures during exposure combined with exercise have been found to impair work performance and to increase histamine responsiveness among some subjects.

There is a growing body of knowledge from human and animal studies suggesting an increase in airway epithelial permeability in response to ozone exposure. This enhancement of permeability may derive from cell extrusions or from breeches in the permeability barrier provided by epithelial tight junctions. Other studies have shown a possible enhancement of epithelial permeability by increased epithelial endocytosis associated with ozone exposure. These permeability changes appear to be minimal in the nose but more pronounced and to persist longer in the trachea and bronchoalveolar regions.

Some studies have shown an increase in mucociliary transport sufficient to cause pulmonary function decrements during exposures to ozone. Other studies have shown a reduction in mucociliary transport with ozone alone or when combined with sulfur dioxide without a reduction in ciliary beat frequency. Infant lambs exhibit a reduction in mucociliary transport subsequent to ozone exposure which persists up to 24 weeks suggesting that the functional change induced by ozone exposure could be associated with airway developmental retardation in young animals.

4. Biochemical and Metabolic Changes

A number of assays have documented biochemical alterations in both humans and experimental animals exposed to ozone. There is some evidence that oxidant damage to epithelial membranes may result in the elaboration of various eicosanoids due to alterations in the arachidonic acid pathway. These compounds are known to alter smooth muscle responsiveness and epithelial cell function. Comparative studies of ozone sensitive and nonsensitive human subjects (as judged by decrements of pulmonary function testing following acute exposure) indicate that plasma $PGF_{2\alpha}$ was significantly elevated among ozone responders during and after exposure. Similarly, experimental studies of bovine tracheal cells incubated with ^{3}H-arachidonic acid have documented increases in cyclooxygenase, lipoxygenase, prostaglandins E_2, $F_{2\alpha}$, 6-keto $F_{1\alpha}$, and leukotriene B4. Among human subjects, it has been shown that the prostaglandin synthetase inhibitor, indomethacin can reduce ozone-induced decrements in pulmonary function tests.

Assessments of key enzymes involved in cellular energy generation and therefore significant to increased metabolic needs during regeneration have shown significant increases following acute ozone exposure. Among these, increased postexposure levels of succinate oxidase, glucose 6-phosphate, and 6-phosphogluconate dehydrogenase suggest an epithelium with increased numbers of mitochondria and the NADPH-generating capability required of a proliferating epithelium.

Accompanying morphologic evidence of ozone-induced fibrosis has been a decline in the rate of intracellular degradation of newly produced collagen prior to secretion and elevation in lavage fluid of hydroxyproline reflective of turnover of the extracellular collagenous matrix. The decrease in intracellular collagen degradation is directly proportional to the net increase in collagen production and the morphologic evidence of fibroplasia.

While the squamous Type I cell appears to be the most ozone-sensitive cell type of the gas-exchange region of the lung, there is good evidence for biochemical impairment of the Type II surfactant-producing cells associated with ozone exposure. Significant decreases as well as fluctuations in enzyme activities characteristic of Type II cells have been documented. There is also evidence that an altered phosphatidylcholine/cholesterol ratio occurs in stored lamellar bodies produced subsequent to ozone stress although extracellular surfactant appears to be of normal composition.

In the airways of lambs, alterations in carbohydrate composition of glycoconjugates subsequent to early postnatal ozone exposure have been observed. These alterations may have ramifications in altered maturation of secretory function and mucociliary transport dysfunction.

5. Pulmonary Alveolar Macrophages and Inflammatory Cells

There is some indication of a transient suppression of alveolar macrophages in the immediate aftermath of ozone exposure. This may be due to direct toxicity of ozone on exposed macrophages. In general however, alveolar macrophages are elevated in numbers subsequent to *in vivo* ozone exposure and remain so for several days. There is also evidence that macrophages are capable of entering the cell cycle and proliferating in response to ozone stress. Additionally, macrophages retrieved from human subjects following acute ozone exposure demonstrated (while not changing in total numbers over controls) an increase in tissue factor mRNA and a decrease in Factor VII mRNA. These observations suggest that shifts in macrophage maturation occur in the presence of ozone exposure. Ozone exposure has been shown to affect macrophage migration, reduce phagocytic activity, and cause changes in the patterns of protein synthesis. There is good evidence for the release of chemotactic elements by macrophages subsequent to ozone exposure that may promote leukocyte accumulation as well as chemotaxis by other cell types. Such chemotactic recruitment may be an important aspect of the macrophage contribution to the pulmonary response to ozone.

Exposure of mice to ozone results in accumulation of lymphocyte clusters within the ozone-sensitive centriacinar region of the lung. While much of the toxicity of ozone can be attributed to sites in direct contact with the gas, mediastinal lymph nodes of ozone-exposed mice also show accumulations of lymphocytes and these exhibit increased numbers of blastic forms, mitotic figures, and tritiated thymidine uptake. A systemic effect of ozone toxicity is most suggestive by its apparent suppressive effect on immune cell processing. However, no direct reactions or products with ozone are known in peripheral systemic organs.

A brief exposure to 1 ppm of ozone has been shown to be sufficient to cause a pulmonary influx of polymorphonuclear leukocytes (PMNs) in subepithelial tissues of dogs 1–3 hr following exposure. Ozone-exposed mice exhibit a similar pattern of PMN infiltrate with acute exposure as well as an increase in bronchioalveolar lavage (BAL) fluid up to 24-hr postexposure. Among human subjects, exposure to 0.4 ppm for 2 hr with exercise causes an increase in the number of PMNs and elevated neutrophil elastase in both BAL fluid and cells suggestive of acute inflammation and lung injury. This enhanced inflammatory response may be modulated in part by increased vascular permeability inasmuch as ozone exposure has also been shown to elevate levels of protein, albumin, and IgG in BAL. It is not known with certainty whether the inflammatory response to ozone exposure increases the risk of pulmonary elastolysis or fibrogenesis or whether it exhibits an adaptive response such as that seen in pulmonary function tests.

6. Adaptation

Individuals undergoing consecutive daily exposures to ozone over several days develop hyporesponsiveness to subsequent ozone exposures by the fourth or fifth day; and, this state of hyporesponsiveness persists for up to one week. This phenomenon is more pronounced among older subjects than among young adults. In a similar vein, experimental animal studies have shown that an initial acute exposure to ozone or other pulmonary irritants can be protective against a subsequent lethal level of exposure. Other studies have attempted to define the nature of this phenomenon at the cellular level using organ culture exposures of tracheal epithelium from animals exposed *in vivo* to filtered air or ozone. Electron microscopic evaluations of these specimens reveal that pathohistologic changes are greater among those cultures previously exposed to filtered air *in vivo* than those exposed to ozone *in vivo*. These observations have suggested that specific alterations to epithelial cells and not systemic and/or neural influences may at least partially modulate mechanisms of adaptation. The mechanism of adaptation is not presently known and awaits a better understanding of the correlations between acute experimental exposure and chronic exposures occurring in common living situations.

IV. Summary

In summary, the short-term deleterious effects of intermittent, low-level ozone exposure on the structure and mechanics of the respiratory system among individuals in good respiratory health appear to be limited and reversible. However, less is known about ozone effects on particular subpopulations that may have (or be predisposed to) chronic respiratory disease. The highly reactive nature of this gas and its documented acute respiratory effects in normal subjects and laboratory animal models raise particular concern that pediatric populations and individuals with existing chronic respiratory and/or cardiovascular disease may be at added risk for morbidity in the presence of ozone exposure. Certainly, epidemiologic and physiologic studies suggest that subpopulations with chronic respiratory disease may be at increased risk for adverse respiratory symptoms as a consequence of acute ozone exposure. Furthermore, there is a growing body of evidence that chronic exposures occurring in the course of daily activities may adversely affect mucociliary transport, ventilatory mechanisms, and may be implicated in the induction of a persistent inflammatory response that could lead to the evolution of chronic respiratory disease.

Related Article: OZONE LUNG CARCINOGENESIS.

Acknowledgments

This work was supported by grants HL19171 and HL34322 from the National Institutes of Health and by Cooperative Agreement

CR817643 from the U.S. Environmental Protection Agency to Philip A. Bromberg. This document has been reviewed in accordance with U.S. Environmental Protection Agency policy and approved for publication. Mention of trade names or commercial products does not constitute endorsement or recommendation for use. The author thanks Philip A. Bromberg, Marianna Henry, and William F. McDonnell for their review and helpful discussion in the preparation of this manuscript.

Bibliography

Fawcett, D. W. (1986). "A Textbook of HIstology." W. B. Saunders, Philadelphia.

Miller, A., ed. (1987). "Pulmonary Function Tests: A Guide for the Student and House Officer." Grune & Stratton, New York.

National Academy of Sciences (1977). "Ozone and Other Photochemical Oxidants." National Academy of Sciences, Washington, D.C.

U.S. Environmental Protection Agency (1986). "Air Quality Criteria for Photochemical Oxidants." Washington, D.C.

Wadden, R. A., and Scheff, P. A. (1983). "Indoor Air Pollution." Wiley & Sons, New York.

Wright, E. S., Dziedzic, D., Wheeler, C. S. (1990). Cellular, biochemical, and functional effects of ozone: New research and perspectives on ozone health effects. *Tox. Let.* **51,** 125–145.

Ozone Lung Carcinogenesis

Hanspeter Witschi
University of California, Davis

Glossary

Bioassay Study to detect and define toxic and carcinogenic effects of chemicals in living animals.

Carcinogenesis Process by which normal cells acquire the potential of unlimited growth, resulting in the formation of benign or malignant tumors.

Cell transformation Process that allows cells to acquire the capability of unlimited growth.

Epidemiology Branch of medical science that deals with the occurrence and distribution of diseases in the human population.

***In vitro* studies** Experiments with isolated cells kept and propagated in culture dishes.

***In vivo* studies** Experiments done in living animals.

Mouse lung tumors Epithelial tumors, mostly adenomas, originating from cells located in the peripheral airways and the alveoli.

Ozone Highly reactive triatomic form of oxygen.

Tumor Abnormal mass arising from cells of preexisting tissues, may be benign or malignant.

OZONE *is a triatomic form of oxygen that is a highly reactive gas with a characteristic pungent odor. In the troposphere, ozone is formed by photochemical reactions from oxygen in a series of steps catalyzed by commonly occurring organic and inorganic air pollutants including hydrocarbons and oxides of nitrogen, emitted into the atmosphere from man-made mobile and stationary sources as well as from natural sources such as deciduous and evergreen trees. During the last four decades, ozone has become a major air contaminant, particularly in regions where there are long periods of sunny weather combined with extensive release into the atmosphere of chemicals from automobile exhaust that nourish ozone formation by UV light. In Southern California and in large areas of the East coast of the United States, ozone constitutes the major ingredient of photochemical smog. Exposure of humans to levels of ozone higher than 0.1 ppm is likely to cause untoward health effects such as reversible decrements in pulmonary function and, possibly, chronic respiratory disease.*

Carcinogenesis is a process by which normal cells in tissues and organs of the human or animal body acquire the potential for unlimited growth. Current theories of carcinogenesis imply that a physical (e.g., X-rays), chemical (e.g., polycyclic aromatic hydrocarbons, nitrosamines) or biological (e.g., virus) agent produces an eventually irreversible change in the genetic material of cells (genetic mechanism of carcinogenesis). More recent evidence suggests that epigenetic mechanisms may also be involved in carcinogenesis (i.e., molecular and cellular mechanisms that do not appear to affect primarily the structure and function of the cellular genome). One such mechanism that is receiving increased attention as a possible determinant in carcinogenesis is enhanced proliferation of target cell populations. Enhancement or "promotion" of tumor development is thought to be mainly caused by epigenetic mechanisms. Once one or a few cells in a given tissue or organ have undergone transformation and acquired the potential for unlimited growth, a tissue mass or tumor develops. Benign tumors, although they can reach on occasion considerable mass, remain well defined within their boundaries, but may compress or displace adjacent tissues. Malignant tumors aggressively invade adjacent tissues. Cells that detach from the original tumor mass eventually form secondary tumor colonies in distant organs, a process called metastasis.

One of the most frequently seen human malignancies and the leading cause from cancer death is cancer of the lung. It is a form of cancer only rarely seen 100 years ago and is still an exceedingly rare tumor spontaneously occurring in animals. Over the last 50 years, the incidence of human lung cancer has steadily increased. Most human lung cancers are of epithelial origin, primarily of the epithelium lining the airways and can be classified into five major forms: squamous cell (epidermoid) carcinoma (34%), adenocarcinoma (26%), large cell undifferentiated carcinoma (16%), tumors of neuroendocrine cell origin (carcinoid and small cell lung cancer (23%), and tumors of bronchio-alveolar cell origin (2%).

Lung cancer in humans is thought to be caused by inhaled carcinogens. Inhalation of tobacco smoke, particularly cigarette smoke, is the major risk factor. Numerous epidemiological studies in humans have shown that active smoking increases the risk of developing lung cancer ten- to twentyfold. Inhalation of environmental tobacco smoke (''passive smoking'') is also considered to constitute a risk. Other inhalants known to produce lung cancer in humans are agents encountered in industrial settings. They include exposure to radon gas and its decay products in uranium miners, to asbestos in shipyard and insulating workers, and to several metals, metal compounds, and polycyclic aromatic hydrocarbons in smelters, ore refineries, and foundry processes.

I. Mechanisms

Ozone is a highly reactive molecule. Its direct toxic effects at the cellular and subcellular level may be mediated though oxidation of unsaturated lipids in cellular membranes, ozonization of macromolecules, oxidation of cellular proteins, generation of lipid peroxides, and creation of reactive oxygen species and of free radicals. It is conceivable that many of these reactions may induce changes in the cell that eventually will lead to neoplastic transformation. Potential mechanisms include oxidation of procarcinogens to ultimate carcinogens (carcinogenesis), induction of chromosomal damage (mutagenesis), interactions with the effects of free radical producers such as certain promoting agents, and direct damage to the genome of cells (DNA damage).

II. Epidemiologic Studies

The role of ozone as a factor in the pathogenesis of acute lung injury and perhaps chronic lung disease in humans has been investigated in many epidemiological studies. Studies focusing on lung cancer incidence have been conducted in many areas heavily polluted with car exhausts or industry emissions. On occasion, these results have been compared with findings in less polluted areas. In several studies it has been found that there exists some correlation between excessive amounts of air pollutants such as total suspended particulates, ambient trace metals or ambient concentrations of benzo(*a*)pyrene, and lung cancer rates. However, it has not been possible to date to correlate any air pollutant with an increased incidence of lung cancer in the general population. A possible exception may be the type of ''air pollution'' encountered in special industrial settings, such as in coke oven workers. Smoking habits in the general population are usually the confounding factor. As of now, it would appear that inhalation of tobacco smoke has such a strong influence on overall lung cancer incidence that it completely masks other potential risk factors such as ozone and other air pollutants. However, it must be noted that more recently there seems to be an increase in lung cancer incidence in nonsmokers. If substantiated, this finding will need to be carefully observed and the questions will need to be addressed whether air pollution in general, in addition to the suspected etiologic role of inhalation of environmental tobacco smoke, might be responsible. Design and execution of more refined epidemiological studies are certainly warranted.

III. Studies with Isolated Cell Systems

The study of isolated cells and tissues maintained and propagated *in vitro,* is a common research tool used to obtain information on potential mutagenic and carcinogenic activities of chemical and physical agents. Detailed analysis of events often allows us to gain understanding about possible underlying mechanisms at the cellular and molecular level. As studied in several such systems, ozone has been found to have many biological effects labeled to be ''radiomimetic'' (i.e., effects that closely resemble responses commonly produced in cells by the action of X-rays). For example, exposure of a human-tissue derived cell line for only a short time (5 minutes) to high concentrations of ozone (8 ppm) produced as extensive chromosomal breakage and damage as did exposure to X-rays. Chromatid deletions were observed with increased frequency in human leukocytes that were stimulated to proliferate by phytohemagglutinin. *In vitro* exposure of hamster embryo cells with 5 ppm of ozone for only 5 minutes enhanced cell transformation. The effect could be greatly enhanced if cells were preexposed to gamma rays. No evidence for ozone-induced DNA strand breaks was found in human epithelial cells. However, ozone at high concentrations (5 ppm) appeared to activate transforming genes. It was concluded that at high concentrations (5 ppm) ozone might act as a carcinogen or cocarcinogen in C3H10T1/2 cells, but that it did not have similar effects at a lower concentration such as 1 ppm. At the present time, it is difficult to extrapolate the observations made at high exposure levels of ozone (>1 ppm) to exposures that are likely to occur in an ozone-polluted environment. Ambient ozone levels are considerably lower, the range of 0.08 to 0.2 ppm. No biomarkers of exposure to ozone are available, making it difficult to interpret and to relate *in vitro* observations to *in vivo* exposures. However, an important conclusion from several *in vitro* studies is that eventually ozone effects not only depend on exposure doses, but also on temporal relationships between exposure to ozone and other potentially carcinogenic stimuli such as radiation or oxidant cell injury. This observation deserves some attention because, as discussed under "Animal experiments," available *in vivo* evidence on ozone carcinogenicity appears to implicate it more as a tumor-modulating agent than as a direct carcinogen. Finally, it must be mentioned that, on occasion, the general cytotoxicity of ozone produces cell death in an *in vitro* system and that (in at least one instance) human tumor cells appear to be more vulnerable to ozone toxicity than are normal human cells.

IV. Animal Studies

In order to establish whether a given physical or chemical agent has carcinogenic potential, long-term exposure studies with experimental animals are conducted. In most carcinogenesis bioassays, exposure to a presumptive carcinogen occurs over the life span of the tested animals species (i.e., 2 years or more). The most commonly used laboratory species are rats and mice; hamsters and occasionally dogs are also used. To test volatile and other airborne agents, animals are kept in suitable chambers ventilated with air containing the test agent at various concentrations. The highest concentration is usually a concentration at which animals can survive for their life span and will not lose more than 10% body weight compared with a corresponding control group kept in a chamber ventilated with air only.

The acute effects of ozone inhalation have been extensively examined in mice, rats, and other animal species, including nonhuman primates. In contrast, information on the chronic effects of ozone is much more sparse. One of the longest studies reported so far involved rats and an exposure time of 18 months. No data on carcinogenicity have originated from this study.

All other published long-term studies, (ranging from 3 to 15 months exposure duration) were conducted with mice. Lung tumors in mice, at least in certain strains, occur spontaneously in older animals with some regularity, even if the animals are not exposed to a carcinogenic agent. The development of such "spontaneous" lung tumors often raises the question of whether a given toxic agent is a true carcinogen for mouse lung or whether it simply enhances or accelerates development of spontaneous lung tumors. For example, in the early 1960s, it was reported that exposure of C×AF mice for 15 months to 1 ppm of ozone produced an overall incidence of lung tumors of 85% (i.e., higher than the 38% incidence found in control animals). The average number of lung tumors in the exposed animals was also higher than in controls. The data, as reported, do not allow us to decide whether the differences between treated and exposed animals were statistically significant. Nevertheless, it was concluded that, in mouse lung, ozone might well act as a "tumor accelerator." Two later studies examined the effects of comparatively short-term exposure (2 to 6 months) to rather high ozone concentrations (ranging from 2.5 to 4.5 ppm). Histopathological analysis of the lung tissue of the exposed animals showed extensive changes, particularly proliferation of pulmonary epithelial cells lining the bronchi and alveoli. These cells are thought to be cells of origin of murine lung adenomas or carcinomas. The fact that ozone exposure produced marked proliferative changes in lung epithelial cells and particularly the observation that the lesions persisted even when ozone exposure was discontinued, seemed to suggest that ozone might be added to the roster of pulmonary carcinogens.

Additional data on a potential tumorigenic effects of ozone at lower concentrations (0.3 ppm to 0.8 ppm) were obtained in experiments with strain A mice in reasonably long-term studies (4 to 10 months). Two different laboratories examined the problem independently. Two questions were addressed: (1) would ozone alone produce an increase in tumor incidence and multiplicity, and (2) would ozone be capable of modifying the development of lung tumors induced by a chemical carcinogen? The results were ambiguous. In each laboratory, the effects of ozone alone on mouse lung tumor development were examined. Interestingly, it was found in both groups that, in one experiment, ozone appeared to produce more lung tumors than found in controls, whereas in a different study, ozone had no apparent such effect (Table I). Interpretation of the data is further complicated by the fact that statistical significance in the positive studies was due to a low tumor multiplicity in the controls rather than to a substantially elevated tumor multiplicity in the ozone-exposed animals. Furthermore, ozone had no discernible effect in Swiss-Webster mice, a strain less prone to develop spontaneously arising lung tumors and less susceptible to carcinogens than strain A mice. Data on an eventual tumor-modulating effect of ozone were also ambiguous. One experiment showed that ozone was capable of enhancing or "promoting" tumor development. In the other study, ozone clearly inhibited tumor development (Table II). Although this appears to be puzzling at first, the contradictory results may tentatively be explained by variations in experimental design. Continuous or intermittent exposure to ozone, together with re-

Table I Lung Tumors in Mice following Exposure to Ozone

		Tumor multiplicity[a]	Tumor incidence[b]	Percentage
Laboratory A	Experiment 1			
A mice	Air	0.60 ± 0.10	16/40	40
	0.3 ppm[c]	0.85 ± 0.15	21/40	53
	Experiment 2			
	Air	0.20 ± 0.07	8/45	18
	0.5 ppm[d]	0.64 ± 0.21	17/45	38
Laboratory B	Experiment 1			
A mice	Air	0.13 ± 0.06	4/33	12
	0.4 ppm[e]	0.09 ± 0.06	2/23	9
	0.8 ppm[e]	0.55 ± 0.15	12/32	38
SWR mice	Air	0.03	1/32	3
	0.4 ppm[e]	0	0/30	0
	0.8 ppm[e]	0.03	1/33	3
	Experiment 2			
A mice	Air	0.23 ± 0.11	4/22	18
	0.8 ppm[f]	0.60 ± 0.25	9/25	36

Sources: Hassett, C., Mustafa, M. G., Coulson, W. F., and Elashoff, R. M. (1985). Murine lung carcinogenesis following exposure to ambient ozone concentrations. *JNCI* **75,** 771–777. Last, J. A., Warren, D. L., Goad, E. P., and Witschi, H. P. (1987). Modification by ozone of lung tumor development in mice. *JNCI* **78,** 149–154.

[a] Tumor multiplicity, average number of tumors per lung ± SEM.

[b] Tumor incidence, number of tumor-bearing animals per total number of animals per group.

[c] 0.3 ppm of ozone, 103 hours per week, every other week for 6 consecutive months; animals exposed at 7 weeks of age, killed at 1 year of age.

[d] 0.5 ppm of ozone, 102 hours each first week of the month for 6 months; animals exposed when 7 weeks old, killed at age 10 months. $p < 0.05$ for multiplicity and incidence.

[e] 0.4 or 0.8 ppm of ozone, 8 hours each night, 7 days a week for 18 weeks, animals exposed at 6–8 weeks and killed at age 6 months. At 0.8 ppm, $p < 0.05$ for multiplicity and incidence.

[f] Animals exposed at 6–8 weeks, 0.8 ppm of ozone, 8 hours each night, 7 days a week for 22 weeks, animals killed at end of exposure.

peated exposure to a carcinogen appears to enhance lung tumor development (Table II). A possible explanation for this observation is the fact that ozone exposure leads to substantial cell proliferation and turnover in mouse lung, a fact that might favor tumor development. Alternatively, administration of a single dose of carcinogen followed by continuous ozone exposure may inhibit tumor development because the cytotoxic effects of ozone interfere with growth and expansion of the initiated cell population (Table II). The importance of experimental design has also been documented in studies on the effects of ozone on metastatic tumor cell growth in the lung of mice. Intravenous injection of tumor cell suspensions will result in the growth of discrete tumor nodules in the lung. By varying the temporal relationships between injection of cells and ozone exposure, it is possible to enhance or to mitigate the growth of such nodules.

In summary, presently available data on pulmonary ozone carcinogenesis are ambiguous. Evidence that ozone alone would act as a carcinogen in the lungs of experimental animals is conflicting. All observations have been made in mice only and no data are available that study the effects of ozone exposure over the entire life span of this species. This limits the validity of the present data. The situation is further complicated by the fact that the few studies that showed a possible positive effect were done in a mouse strain known to have a very high spontaneous incidence of lung tumors. The question remains whether ozone acts as a true carcinogen or simply enhances lung tumor development in animals that otherwise would have developed such tumors in any case. There seems to be less doubt that ozone is capable of modifying the development of lung tumors induced by chemicals. However, both aggravating and mitigating effects

Table II Modulation of Lung Tumor Development by Ozone in Mice Treated with a Pulmonary Carcinogen

	Treatment	Tumor multiplicity[a]	Tumor incidence[b]
Laboratory A			
A mice	Urethan + Air	2.95 ± 0.24	38/40
	Urethan + 0.3 ppm O_3[c]	2.68 ± 0.29	39/40
	Urethan + Air	7.91 ± 0.45	44/44
	Urethan + 0.5 ppm O_3[d]	10.30 ± 0.62	46/46
Laboratory B			
A mice[e]	Urethan + Air	31.1 ± 1.8	32/32
	Urethan + 0.4 ppm O_3	25.3 ± 1.4	25/25
	Urethan + 0.8 ppm O_3	18.1 ± 1.1	32/32
SWR mice[e]	Urethan + Air	2.6 ± 0.5	19/31
	Urethan + 0.4 ppm O_3	2.3 ± 0.4	25/31
	Urethan + 0.8 ppm O_3	1.4 ± 0.3	22/34

Sources: Hassett, C., Mustafa, M. G., Coulson, W. F., and Elashoff, R. M. (1985). Murine lung carcinogenesis following exposure to ambient ozone concentrations. *JNCI* **75,** 771–777. Last, J. A., Warren, D. L., Goad, E. P., and Witschi, H. P. (1987). Modification by ozone of lung tumor development in mice. *JNCI* **78,** 149–154.

[a] Tumor multiplicity, average number of tumors per lung ± SEM.

[b] Tumor incidence, number of tumor-bearing animals per total number of animals per group.

[c] One single dose of urethan (2 mg per mouse) at age of 7 weeks. Exposure to 0.3 ppm of ozone, 103 hours per week, every other week for 6 consecutive months; animals killed at 1 year of age.

[d] Exposed to 0.5 ppm of ozone, 102 hours each first week of the month for 6 months and followed after each exposure by a single dose of 2 mg per mouse of urethan; animals killed at age of 10 months. $p < 0.05$.

[e] Animals injected with 1000 mg/kg of urethan at 6–8 weeks of age, followed by exposure to 0.4 or 0.8 ppm of ozone, 8 hours each night, 7 days a week for 18 weeks. Animals killed at age of 6 months. At 0.4 and 0.8 ppm, $p < 0.05$ in A mice; at 0.8 ppm, $p < 0.05$ in SWR mice.

have been observed and the outcome of any particular study very much depends upon experimental design.

V. Conclusions

Whether ozone is a lung carcinogen is not a trivial question. Air pollution is of increasing concern and there is a large number of people exposed to polluted atmospheres; even a comparatively weak carcinogenic potential of ozone could translate into an unacceptably high increase in lung cancer in the general population. On the other hand, epidemiological data do not suggest as yet that ozone might be a risk factor. Although lung cancer rates in nonsmokers appear to increase, there are numerous other factors (exposure to environmental tobacco smoke, pollutants other than ozone) that might be involved. Animal data on pulmonary carcinogenesis of ozone are inadequate and do not even allow us to answer in a definitive manner whether ozone has carcinogenic potential in mice. It appears to be somewhat more certain that ozone may modify pulmonary carcinogenesis. The exact mechanisms underlying several observations made in animals need to be worked out. At present, the National Toxicology Program is conducting a properly designed study on ozone carcinogenesis in mice and rats, involving several concentrations of ozone and conducted over the life span of a substantial number of animals. The question of the cocarcinogenic potential of ozone is also under investigation. Hopefully, the results of these experiments will help to reduce the still existing uncertainty about pulmonary ozone carcinogenesis.

Related Article: Ozone Exposure, Respiratory Health Effects.

Bibliography

Bunn, P. A., Jr., ed. (1988). "Lung Cancer: Seminars in Oncology," Vol. 15, No. 3. Grune and Stratton, Orlando, FL.

Lee, S. D., Mustafa, M. G., and Mehlman, M. A., eds. (1983). "Advances in Modern Environmental Toxicology," vol. V. International Symposium on the Biomedical Effects of Ozone and Related Photochemical Oxidants. Princeton Scientific Publishers, Princeton.

Lippmann, M. (1989). Health effects of ozone: A critical review. *JAPCA* **39,** 672–695.

Pryor, W. A. (1991). Can vitamin E protect humans against the pathological effects of ozone in smog? *Am. J. Clin. Nutr.* **53,** 702–722.

Schneider, T., Lee, S. D., Wolters, G. J. R., and Grant, L. D., eds. (1989). "Atmospheric Ozone Research and Its Policy Implications. Elsevier, New York.

Witschi, H. P. (1988). Ozone, nitrogen dioxide and lung cancer: A review of some recent issues and problems. *Toxicology* **48,** 1–20.

Peroxisomes

William T. Stott
Dow Chemical Company

Glossary

Fatty acid β-oxidation Metabolism of fatty acids into two carbon fragments starting at the carboxyl end of the molecule.

Genotoxicity Potential of a chemical to interact with genetic material causing heritable mutations and/or chromosomal aberrations.

Peroxisome Subcellular organelle found in the cells of most eukaryotes. A related structure, the glyoxisome, occurs in prokaryotes.

A **GROWING NUMBER** *of important pharmaceutical, agricultural, and industrial chemicals have been found to induce the proliferation of a subcellular organelle, the peroxisome, in liver cells of laboratory animals, primarily rodents. Normal peroxisomal function is essential to the maintenance of good health in animals, including humans; however, the excessive proliferation of these organelles may result in hepatocellular swelling and increases in liver mass. As such, peroxisome proliferation would likely have remained primarily an interesting morphological response were it not for the fact that many compounds causing this effect also cause liver tumors in rats and mice upon chronic administration. The association of peroxisome proliferation with tumor formation, the strong species dependency of the proliferative response, and the nearly uniform lack of genotoxicity of these chemicals in standard assays has resulted in the general classification of peroxisome proliferators as a distinct group of nongenotoxic chemical carcinogens whose relevance to human risk assessment is quite controversial.*

I. Peroxisomes

In order to understand the potential toxicological ramifications of chemically induced peroxisome proliferation and its detection, it is necessary to understand the morphology and enzymology of this organelle.

A. Morphology and Occurrence

Peroxisomes were first described in mouse kidney tubular epithelial cells and, soon after, in hepatic parenchymal cells during the pioneering days of electron microscopy in the mid-1950s. Named "microbodies," these single membrane bounded organelles with a finely granular or flocculent, moderately electron-dense matrix and diameters of approximately 0.3 to 1.5 μm, were soon discovered in a number of other animal tissues, including those from humans. These have included; pulmonary (alveolar), glandular (adrenal, pancreatic, lacrimal, thyroid, and others), muscle (cardiac, smooth, and skeletal), neural (retina and central nervous system), gonadal, adipose, and intestinal tissues. Similar microbodies were also detected in the tissues of fish, amphibians, reptiles, birds, and a variety of plants including yeast and fungi. The morphology and enzymology of these organelles in relation to the other subcellular organelles of a typical rat hepatocyte is shown in Figure 1. A crystalloid core, believed to be the enzyme uricase, is a striking feature of these organelles in many species, including those in the liver of rats, guinea pigs, rabbits, and cows while only a subcrystalloid nucleoid has been reported in the livers of hamsters and mice. A crystalloid nucleoid is not found in renal peroxisomes of rats nor in the hepatic or renal peroxi-

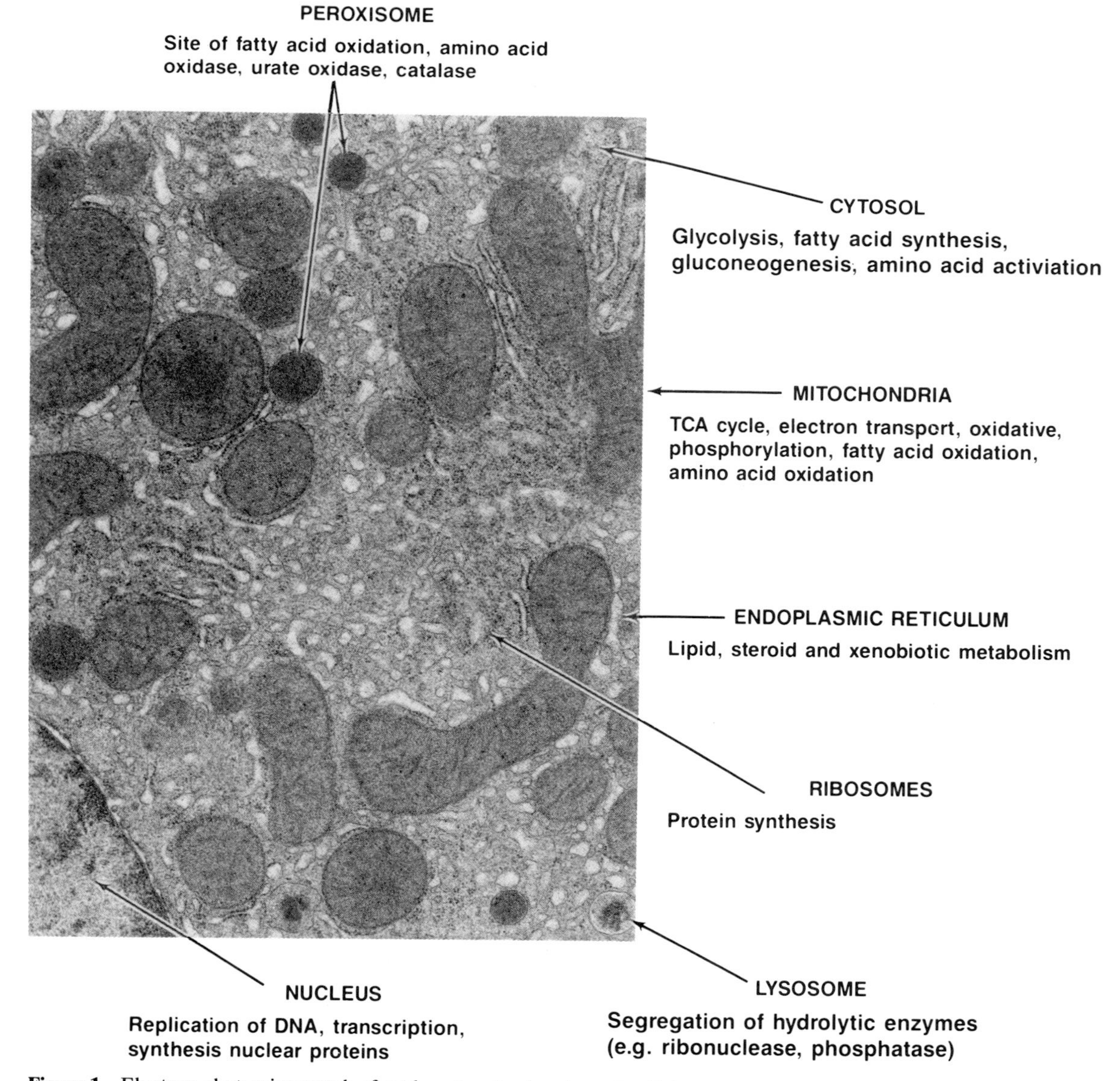

Figure 1 Electron photomicrograph of rat hepatocyte showing subcellular organelles and their major enzyme activities.

somes of humans and birds who lack uricase activity.

Following the more complete elucidation of the biochemical processes of microbodies derived from various mammalian and plant tissues, the purely morphologic term, microbody, was eventually replaced by the biochemically oriented terms peroxisome and glyoxisome, respectively. The former name derived from the presence of hydrogen peroxide generating oxidases in the microbodies isolated from mammalian tissues, while the latter term refers to the microbodies found in germinating fatty seedlings which also contain enzymes of the glyoxylate pathway. Peroxisomes are found in greatest abundance in the livers of animals, primarily in parenchymal cells localized in the centrilobular and mid-zonal regions of the liver lobules. Normal rodent hepatocytes contain approximately 1000 peroxisomes which comprise 1–1.5% of the cellular protein and account for approximately 20% of oxygen consumption.

Peroxisomes originate during several periods of intense biogenesis during the later stages of fetal growth, immediately postpartum, and during maturation. Thereafter, little change in the peroxisome volume density of hepatocytes of rats and mice up to 480 days of age appears to occur. The half-life of this organelle is estimated to range from 1 to 3.5 days and it is reproduced by division of existing organelles following the synthesis and importation of peroxisomal enzymes.

B. Enzymology

Peroxisomal enzymes, in general, represent alternate routes of metabolism to the metabolic pathways occurring in other cellular compartments. Numerous enzymes have been identified in mammalian peroxisomes including: catalase, a number of oxidases (fatty acyl-CoA, urate, D-amino acid, polyamine), alkyl-transferases (acetyl- and octanoyl-carnitine, dihydroxyacetonephosphate), fatty acid β-oxidation enzymes, and a number of dehydrogenases and aminotransferases.

Unlike the enzymatic compliment of many other organelles, the enzymology of the peroxisome may vary considerably from tissue to tissue and between species. However, a constant component of the peroxisome is acyl-CoA oxidase which oxidizes fatty acids generating hydrogen peroxide, and catalase which destroys hydrogen peroxide. The fatty acyl-CoA β-oxidation pathway of peroxisomes in particular, has been observed to differ from its mitochondrial counterpart in several significant ways, both on a molecular and a catalytic basis. In contrast to mitochondrial β-oxidation, peroxisomal β-oxidation is not carnitine dependent, utilizes only medium to long-chain (8 to 22 carbons) saturated and long-chain unsaturated fatty acyl-CoA molecules, has a relatively high pH optima, and is not dependent upon an active electron transport system. As a result, peroxisomal β-oxidation is not inhibited by the classic cytochrome inhibitors such as cyanide and antimycin A. The ratio of peroxisomal to mitochondrial β-oxidation of fatty acids, which occur under similar experimental conditions, appears to remain fairly constant between different tissues and between species of animal. Peroxisomal β-oxidation has been reported to account for up to approximately 50% of total hepatic β-oxidation activity in nonfasted rats and mice, and approximately 35% of total hepatic β-oxidation activity in humans.

The peroxisomal fatty acid β-oxidation pathway is shown in Figure 2. A supply of activated fatty acid substrates for this pathway is provided by an ATP-dependent acyl-CoA synthetase which is located in the peroxisome membrane. Lacking an

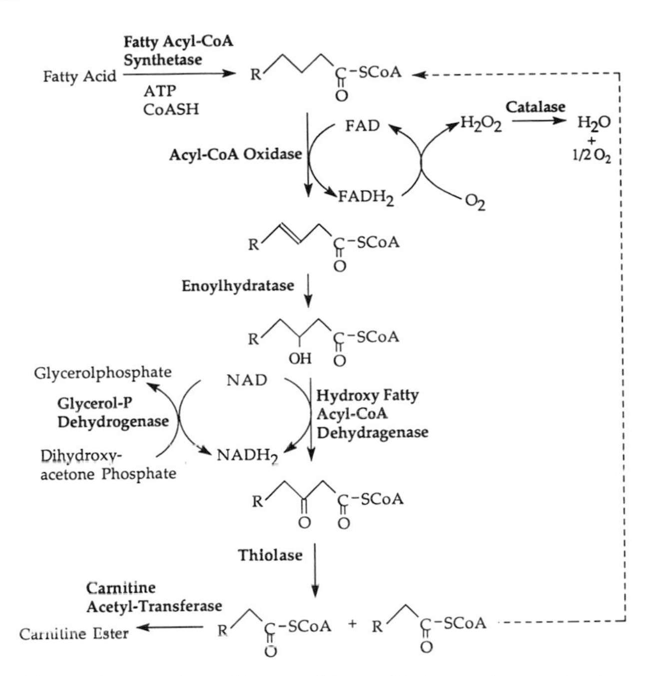

Figure 2 Peroxisomal fatty acid β-oxidation pathway showing key enzymes in this process. Acetyl-CoA product is exported out of peroxisome as a carnitine ester while remaining fatty acid CoA ester, if >C8, is recycled.

electron transport system, the first enzyme of the peroxisomal β-oxidation pathway, a flavin dependent acyl-CoA oxidase, transfers two electrons to molecular oxygen to form hydrogen peroxide which is subsequently hydrolyzed by peroxisomal catalase. The energy of the desaturation reaction is thus not available for ATP formation and results in this pathway being strongly exergonic. Peroxisomal β-oxidation is estimated to result in the production of 8–19% fewer ATPs per fatty acyl-CoA molecule than mitochondrial β-oxidation and, in the rat, generates approximately 0.065 calories/minute/gram of liver tissue. Following the oxidation step, the remaining hydratase, dehydrogenase, and thiolase reactions of the peroxisomal β-oxidation pathway are similar to those occurring in mitochondria. However, as in the case of the acyl-CoA-oxidase–generated reduced flavin dinucleotide, the reducing potentials generated during the subsequent dehydrogenase step (i.e., NADH) may not be utilized directly to produce ATP. Instead, these are removed from the peroxisome in the form of glycero-3-phosphate by the action of peroxisomal glycerophosphate dehydrogenase. The removal of the end products of peroxisomal β-oxidation from the peroxisome is accomplished by the activity of carnitine acetyl- and octanoyltransferases. Upon migration out of the peroxisome, the carnitine esters which are formed are either metabolized further in the mitochondria or utilized in cholesterol synthesis in the cytosol.

The exergonic nature of peroxisomal β-oxidation of fatty acids suggests that this metabolic pathway represents an energy wastage mechanism for the cell. Under certain circumstances it appears that this may have important physiological implications for the animal. Hepatic peroxisomal β-oxidase activity is elevated approximately threefold in obese mice relative to their lean litter mates and in the brown fat of cold-adapted rats suggesting involvement in lipid catabolism and thermeogenesis, respectively. Peroxisomal β-oxidation also appears to be responsible for the oxidation of the side chain of a large portion of cholesterol, thereby initiating its metabolism to bile acids. Indeed, *in vivo* bile acid production has not been observed to decrease in rats in which mitochondrial β-oxidation has been chemically blocked.

Given the heterogeneity of the enzymology of peroxisomes within different tissues, it is not surprising that the activities of peroxisomal enzymes appear to develop and change independent of one another with age. Fetal and neonatal peroxisomes in rats, while morphologically similar to those found in adult animals, are relatively immature in terms of their enzymatic composition and metabolic competence. The activity of hepatic peroxisomal β-oxidation in rats appears to double within the first two weeks postpartum and remains fairly steady or decreases slightly thereafter, even though its relative proportion of total β-oxidase activity increases at a slow, steady rate with age (i.e., mitochondrial activity decreases with age). Carnitine acetyltransferase activity in rats also appears to nearly double in activity during the first two weeks postpartum, however, it subsequently decreases to birth levels by 13 weeks of age. Peroxisomal catalase activity has been variously reported to remain steady with increasing age in rats and mice or slowly decline postpartum to generally lower activity in older rats. The activities of hepatic peroxisomal uricase and amino acid oxidase in rats are also reported to be greater in adults than in fetal or neonatal animals.

C. Detection

There are several experimental approaches which may be used to quantitate peroxisomes and/or their enzymatic activity. Direct morphological evaluation of hepatic tissues via electron microscopy is perhaps the most definitive method of detecting the presence of peroxisomes but may require the use of special techniques. A much more commonly used technique for the visualization of peroxisomes by electron microscopy involves the cytochemical staining of catalase, a marker enzyme for this organelle. The method most commonly used to do this involves the reaction of 3,3′-diaminobenzidine with catalase in an alkaline media to obtain a dark brown product visible under light microscopy. Following osmication, these pigments are also electron opaque and suitable for electron microscopic evaluation. An example of this is shown in Figure 3. While this reaction is not completely specific for catalase (e.g., lipofuscin granules are also positive but are morphologically distinct from peroxisomes), it does appear to be the method of choice for the quantitation of peroxisomes.

The activities of key biochemical markers for peroxisomes may also be assayed. These can be

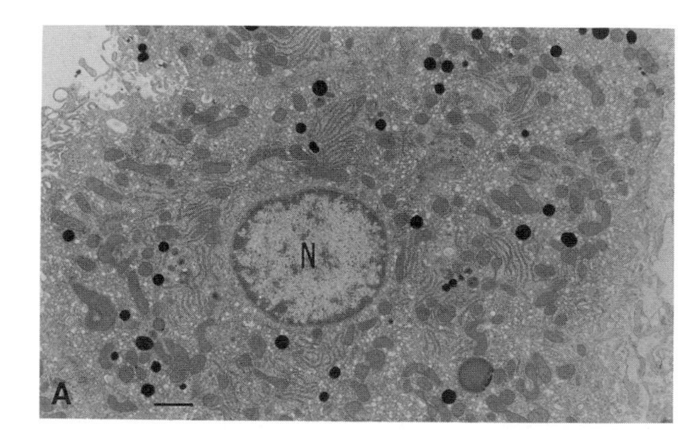

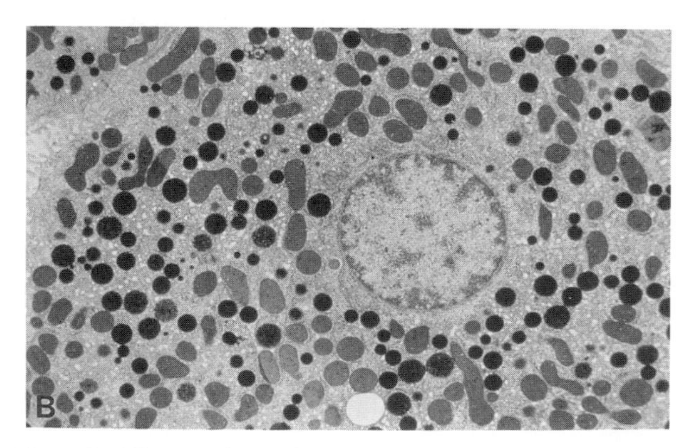

Figure 3 Electron photomicrographs of hepatocytes from a control rat (A) and a rat administered 225 mg/kg/day clofibrate for 2 weeks (B) demonstrating a significant degree of peroxisome proliferation. Peroxisomes have been stained for catalase activity using DAB (N denotes nucleus).

divided into those which are peroxidative in nature (e.g., catalase, uricase, D-amino acid oxidase), associated with the β-oxidation of fatty acids (e.g., cyanide-insensitive fatty acid oxidizing activity, enoyl-CoA hydratase) or a variety of miscellaneous enzymes. These biochemical markers have varying degrees of sensitivity and specificity for peroxisomes but, in general, cyanide-insensitive β-oxidation and catalase enzyme activities are most widely used, both for their specificity and propensity to be induced by peroxisome proliferators. The activities of carnitine alkyltransferases are also often measured; however, these enzymes are also found in mitochondria and enzyme induction in the absence of peroxisome proliferation has been observed (e.g., valproic acid). Data derived from any of these techniques, in addition to liver weight and clinical chemistry data (decreased lipids), may be used to clearly identify a compound as a potential peroxisome proliferative agent in experimental animals.

D. Peroxisomes and Human Disease

Alterations in the number, morphology, and enzymology of peroxisomes have been found to be associated with a number of suspected metabolic disorders in humans. Hepatocellular peroxisomes are absent in patients suffering Zellweger Cerebrohepatorenal Syndrome, an autosomal recessive trait characterized by an accumulation of very long chain fatty acids and C-27 bile acid intermediates. Hepatic peroxisomes also appear to be missing in the X-linked recessive disease, adrenoleukodystrophy, in which an accumulation of very long chain fatty acids occurs, especially in the blood, brain, and adrenals of patients. In contrast, peroxisomes containing abnormal crystalline inclusions are present in patients suffering cerebrotendinous xanthomatosis, a familial disease characterized by extensive tissue cholesterol deposition and decreased bile acid synthesis. Ultrastructual changes in peroxisomes have also been observed in Wilson's disease, an inherited disorder of copper metabolism in which hepatocellular accumulation of fats is common and in Reye's syndrome, an encephalopathy involving fatty changes in the liver and other organs. Finally, a few cases of acatalasemia in humans, characterized by the lack of functional peroxisomal catalase, have been reported.

In nearly all of these disease states, alterations in the morphology and enzymology of mitochondria have also been observed. Given the metabolic homeostasis which is maintained between these two organelles under normal conditions, this is not surprising; however, it is difficult to determine at this point whether peroxisomal changes alone could account for the symptomology or whether some are secondary to mitochondrial perturbations. This remains to be elucidated.

II. Chemically Induced Peroxisome Proliferation

A. Peroxisome Proliferators

The chemically induced proliferation of peroxisomes was first reported by several researchers in the mid 1960s while attempting to explain the pronounced hepatomegaly in rats and mice treated with the hypolipidemic drug, clofibrate (ethyl, *para*-chlorophenoxy-2-isobutyrate). Since that time, the intensive search for effective hypolipidemic drugs has led to the discovery of numerous compounds having hypolipidemic properties which were also capable of inducing the proliferation of peroxisomes in rats and mice. Many of these proliferators are structural analogues of clofibrate, containing an aryloxyalkanoic acid moiety, but a large number are not, indicating the broad structural specificity of this response. Nor has the list of peroxisome proliferators remained restricted to potential hypolipidemic drugs, as numerous industrial chemicals (e.g., chlorinated ethylenes, phthalate ester plasticizers) and agricultural chemicals (e.g., phenoxy herbicides) have also been reported to be peroxisome proliferators. Table I lists several examples of these compounds, roughly categorized into aryloxyalkanoic acids and non-aryloxyalkanoic acids. In addition, several dietary and hormonal factors known to induce the proliferation of peroxisomes are also listed.

A common characteristic of a majority of proliferators and/or their potential metabolites is the presence of a lipophilic moiety and an acidic functional group (e.g., carboxyl or tetrazole). Despite their structural diversity, studies have revealed a three-dimensional similarity of many proliferators with potency being directly related to the distance between the lipophilic and acidic groups. Exceptions do exist, however, as some proliferators themselves and their known metabolites lack an acidic functional group altogether (e.g., bifonazole [1-(diphenylbenzyl)imidazole]). In addition, certain physiological states and hormones may also induce

Table I Several Chemicals and Conditions Known to Induce Peroxisome Proliferation in Rodents

Aryloxyalkanoic Acids/Esters
Clofibrate (ethyl-*para*-chlorophenoxyisobutyrate) and related compounds[a]
SaH-42348 (1-methyl-4-iperidyl-bis(*para*-chlorophenoxy)acetate)[a]
Ciprofibrate (2-(4-(2,2,-dichlorocyclopropyl)phenoxy)-2-methyl propionic acid)[a]
Haloxyfop (2-(4-((3-chloro-5-(trifluoromethyl)pyridinyl)-oxy-2-phenoxy)propanoic acid) and related compounds[b]
Lactofen (1′[carboethoxy]etyl 5-[2-chloro-4-[trifluoro-methyl]phenoxy]-2-nitrobenzoate)[b]
LK-903 (methyl-r-myristyroxycinnamic acid-1-monoglyceride)[a]
Nonaryloxyalkanoic Compounds
Bifonazole (Diphenylbenzyl)imidazole[a]
Citral (*cis*- and *trans*-3,7-dimethyl-2,6-octadienal) and related compounds[c]
Perfluorodecanoic acid and related compounds
Ibuprofen (methyl-4-(2-methylpropyl)benzylacetic acid)[a]
Alkylthio acetic acids (1,10-bis(carboxymethylthio)decane and related compounds)
Cinnamyl anthranilate (3-phenyl-2-propyl-(2-amino)-benzoate)[c]
Tibric acid (2-chloro-5-(3,5-dimethylpiperidinosulfonyl)benzoic acid)[a]
Styrl-hexahydroindolinol, 34-250 (bis[3*aRS*, 4*SR*, 7*aRS*)-4-(3,4-dimethoxy-(*Z*)-styrly)-hexahydro-4-indolinol]-maleate)[a]
MEDICA 16 (3,14-dimethyl-hexadecanedioic acid)
WY-14643 (4-chloro-6-(2,3-xylidino)-2-pyrimidinylthio)acetic acid[a]
BR-931 (4-chloro-6-(2,3-xylidino)-2-pyrimidinylthio(*N*-beta-hydroxyethyl)acetamide)[a]
Tiadenol (bis(hydroxyethylthio)7,10-decane)[a]
RMI-14514 ((5-tetracycloxy)2-furancarboxylic acid)[a]
Acetylsalicylic acid[a]
Di(2-ethylhexyl)phthalate and related compounds[c]
Trichloroethylene via metabolism to trichloroacetic acid[c]
Chlorinated phenoxyacetic acids[b]
LY17183 (1-(2-hydroxy-3-prapyl-4-(4-(1*H*-tetrazol-5-yl)butoxy)phenyl)-ethanone) and related compounts[a]
Dietary and Hormonal Factors
High fat diet
Vitamin E deficiency
Thyroxine
Insulin
Starvation
Diabetes

[a] Pharmaceutical application (hypolipidemics, analgesics, leukotriene antagonists, etc.).
[b] Agricultural chemical.
[c] Industrial (solvent, food additive).

peroxisome proliferation in rodents. A relatively modest response has been observed in the livers of rats and mice maintained on high fat or vitamin E deficient diets or administered thyroxine, and in the brown fat of cold-adapted rats (see Table I).

B. Nontumorigenic Treatment-Related Effects

Peroxisome proliferation and induction of the activities of peroxisomal enzymes has been reported to occur in a variety of tissues in rats and mice; however, the response of hepatic tissues is typically much more pronounced than in other tissues. The most readily observed gross morphologic change in treated rodents is a dose-related increase in liver mass. Liver weights from treated animals may account for as much as 10% of body weight in rats and 15–20% of body weight in mice. These liver weight changes are accompanied by as much as 10- to 20-fold increases in hepatocellular peroxisome volume density and induction of peroxisomal β-fatty acid oxidation activity. Light microscopy reveals hepatocellular swelling and altered tinctorial properties (e.g., increased eosinophilia in H&E stained tissues) consistent with the altered ratio of cytoplasmic organelles. Hepatocellular necrosis is not typically observed at pharmacologically active dosages of these compounds. Electron microscopy reveals the potentially profound effects of proliferation at a subcellular level (Fig. 3). The enzyme activities of other peroxisomal oxidases and catalase are generally induced less than twofold over control levels. The activities of several key cytosolic enzymes in controlling intracellular oxidative stress, glutathione peroxidase, and superoxide dismutase may actually decrease in response to the administration of peroxisome proliferators. A similar pattern of effects, albeit usually smaller, may also be obtained in cultured primary rat hepatocytes exposed to proliferative chemicals *in vitro*.

Liver weight changes themselves may reflect not only hepatocellular hypertrophy due to cellular enlargement and increasing numbers of peroxisomes, but also an increase in the total number of cells. A relatively short burst of DNA synthesis activity indicative of increased cell division has often been observed in the livers of rats and mice within days of the administration of peroxisome proliferators. In most instances, this effect is not accompanied by any histologically identifiable hepatic necrosis suggesting a true mitogenic response to the administered compounds rather than regenerative DNA synthesis. Administration of relatively potent peroxisome proliferators to rats may result in as much as a 20-fold increase in synthesis rates; however, this may vary considerably among compounds and species. Indeed, the inherent mitogenic potential of these chemicals appears to vary considerably as *in vitro* studies using cultured rat hepatocytes have revealed over 100-fold differences in the mitogenic activity of a number of proliferators. In addition to initial increases in DNA synthesis levels, the administration of relatively high dosages of some proliferators (e.g., WY-14643) may also result in a chronic increases in DNA synthesis in the livers of rodents. This latter change does not appear to be linked to peroxisome proliferation and is generally attributed to the replacement of necrotic cells at dosages of these compounds which are cytotoxic.

The relatively dramatic effects of peroxisome proliferators upon hepatic parenchymal cells and the potential long-term consequences of this change have tended to overshadow the fact that the liver may not represent the sole target tissue nor peroxisome proliferation the sole response at pharmacologically active dosages of specific proliferators in rodents. For example, dosages of the relatively weak proliferator di-(ethylhexyl)phthalate which induce a pronounced hepatic proliferative response may also cause treatment-related changes in the Leydig cells of male rats. Further, other organ systems may be identified as potential target tissues of peroxisome proliferators in species less sensitive to peroxisome proliferation such as canines or nonhuman primates.

C. Variability of Response

Typical of most classes of pharmacologically active agents, the potential of chemicals to induce peroxisome proliferation and peroxisomal enzyme activity may vary considerably between compounds and individual tissues. Table II presents data on the relative induction of two key peroxisomal enzymes, acyl-CoA oxidase and catalase in the liver, kidneys, and intestinal mucosa of treated rats. It is generally accepted that induction occurs to the greatest extent in the liver tissue of rodents; however, it is important to note the predominence of parenchymal cells which are receptive to proliferation in this tissue relative to receptive cell types in other tissues, for example proximal tubular epithelia of the kidneys.

Differences also exist in the potency of many peroxisome proliferators (Table II). The potent peroxisome proliferative compounds clofenepate, ciprofibrate, and tibric acid have been observed to be several times more active than clofibrate, benzafibrate, and gemfibrozil under similar experimental conditions. For example, ingestion of a diet containing 0.5% clofibrate by rats results in a response roughly equivalent to that obtained in rats ingesting a diet containing only 0.05% ciprofibrate. In contrast, acetylsalicylic acid and phthalate esters are considered to be only very weak peroxisome proliferators, requiring 1 to 2% concentrations in the diets of rats to induce proliferation. Differing absorption, metabolism, distribution, and excretion kinetics no doubt play a role in determining the varying potency of these peroxisome proliferators *in vivo*. Comparative data obtained *in vitro* in isolated hepatocytes and an isolated putative peroxisome proliferator receptor(s), however, have also demonstrated significant differences in the potency of these compounds suggesting a basic difference in their activities at a molecular level. And finally, while nutritional factors are capable of causing an increase in the number of hepatocellular peroxisomes and the induction of the activities of peroxisomal enzymes in rodents, maximal responses obtained are, in general, considerably less than those obtained with chemical-inducing agents.

In addition to differences in the general potency of peroxisome proliferative agents, there is also a prominent quantitative difference in the inducibil-

Table II Peroxisomal Enzyme Induction in Several Tissues of Rodents Administered Peroxisome Proliferators[a]

	Tissue (percentage of control)					
	Liver		Kidney		Intestine	
Treatment (dosing)	β-Oxidation[b]	Catalase	β-Oxidation[b]	Catalase	β-Oxidation[b]	Catalase
Clofibrate						
0.25% diet, 1 wk					91	350
0.25% diet, 3 wk		161				52
0.3% diet, 2 wk	700		300			
0.5% diet, 1 wk	690	140			58	89
0.5% diet, 2 wk[c]	1000	188	400	127	157	137
200 mg/kg/day gavage, 2 wk	945	175	180	62		
Nafenopin						
0.1% diet, 6 wk	825	176	420	290		
Tiadenol						
0.25% diet, 2 wk					374	206
0.50% diet, 2 wk	900	168			310	507
Wy-14643						
50 mg/kg/day gavage, 10 days	870		971			
Methyl clofenopate						
0.1% diet, 2 wk	400			260		
0.1% diet, 6 wk	1250			400		
Trichloroacetic acid						
190 mg/kg/day gavage, 2 wk	154	108	83	71		
500 mg/kg/day gavage, 10 days	284		175			
	280[c]		305[c]			
Trichloroethylene						
1000 mg/kg/day gavage, 10 days	180		300			
	625[c]		360[c]			

[a] All data were obtained using male rats except where noted. Values represent percentage of controls.
[b] Cyanide-isensitive peroxisomal β-oxidation of fatty acids.
[c] Data obtained using mice.

ity of different peroxisome-specific, as well as-nonspecific, enzyme activities. As shown in Table II, a significant induction of a peroxisomal-specific enzyme system such as cyanide insensitive fatty acid β-oxidation (often >eightfold) is not linked to a similar increase in the activity of another relatively specific enzyme, catalase (usually ≤twofold). Further, the activity of carnitine acetyl-transferase, a peroxisomal enzyme which is also located in the mitochondrial membrane, may be induced to a much greater degree (often several hundred-fold) than that of β-oxidation under similar experimental conditions.

D. Induction and Reversibility

The induction of peroxisome proliferation in animals has been linked to the binding of a specific receptor(s) which has been shown to demonstrate a high degree of homology with members of a so-called "super family" of steriod receptors. Consistent with a receptor-mediated mechanism of induction, several microsomal and peroxisomal changes characteristic of peroxisome proliferation proceed at a relatively rapid rate. At the molecular level, increases in mRNA coding for cytochrome *P*-450 IVA1, a microsomal cytochrome involved in omega-oxidation of fatty acids, and peroxisomal β-oxidation enzymes occur in rat liver tissue and cultured rat hepatocytes within 15–60 minutes of treatment. These changes are followed by increases in numbers of hepatocellular peroxisomes and the activities of specific peroxisomal enzymes within 24 hours of dosing. Histologically discernible changes and, with repeated dosing, an increase in the liver weight of treated animals relative to controls accompany proliferation. The juxtaposition of induction of microsomal Ω-oxidation and peroxisomal β-oxidation enzymes have led to the hypothesis that the products of the former result in the induction of the latter due to the rapid buildup of medium chain dicarboxylic acids, the "substrate overload" theory of peroxisome proliferation.

In most instances where longer term data are available, peroxisome proliferation appears to reach a maximal response within a few weeks of the initiation of dosing and subsequently remains relatively unchanged or undergoes a slight decline with chronic dosing. Significantly, peroxisome proliferator-induced morphological and enzymological changes do not appear to progress with age and are readily reversible upon termination of dosing. Figure 4 exemplifies the induction and reversibility of hepatic peroxisome proliferation in rats repeatedly administered a peroxisome proliferating agent, (in this case the experimental compound SaH-42348). Peroxisomal numbers, liver weights, and the activities of peroxisomal enzymes increase at a steady rate and reach an apparent plateau within 3 weeks of the initiation of dosing. Following cessation of dosing, all parameters return to control levels relatively rapidly. As evident from the data shown, the activity of carnitine alkyltransferase in particular may respond rapidly to the initiation and cessation of treatment with peroxisome proliferators. These data indicate the absolute requirement for the continued presence of the peroxisome proliferator for the maintenance of elevated numbers of peroxisomes, enhanced activities of peroxisomal enzymes, and the enhanced

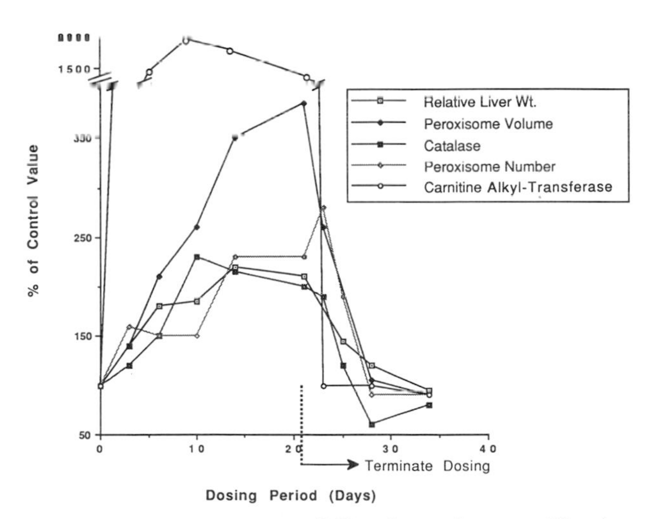

Figure 4 Induction and reversibility of peroxisome proliferation in the livers of rats administered the peroxisome proliferator, SaH-42348.

generation of potentially toxicologically significant byproducts (e.g., hydrogen peroxide).

In addition to their effects upon peroxisomes, several peroxisome-proliferative compounds have also been observed to affect both the numbers of mitochondria and quantity of smooth endoplasmic reticulum (SER) in hepatic tissues of susceptible species. However, in general, these changes represent only a small fraction of those observed for peroxisomes and only relatively small increases in the activities of some mitochondrial enzymes and, with the exception of Ω-oxidation, mixed function oxygenases of the SER.

E. Species and Sex-Related Differences in Sensitivity

In general, peroxisome proliferators initiate a similar pattern of treatment-related morphologic and enzymologic changes in rats and mice; however, higher mammalian species are much less susceptible to chemically induced peroxisome proliferation. Table III summarizes the peroxisome morphological and enzymological data available for several known peroxisome proliferative agents in a variety of mammalian species. While some variability in the relative rodent versus higher mammalian species responses to these compounds is evident, it is clear from these data that higher mammalian species are considerably less sensitive or quite refractory to known peroxisome proliferative agents at maximal tolerated and/or therapeutic dosages. Thus, compounds such as gemfibrozil, fenofibrate, LY-1783, nafenopin, and clobuzarit, which cause as much as 10- to 35-fold increases in the number of hepatocellular peroxisomes and/or induction of peroxisomal enzymes in rodents, cause only slight increases or no change in peroxisomal number and enzyme activity in dogs and nonhuman primates at much higher dosages.

The striking species dependency of responses of animals to these compounds has led to the suggestion that the rat liver response should be considered as a unique, atypical phenomenon related to the biology of peroxisomes in this species. However, the very potent peroxisome proliferator, ciprofibrate, has been observed to induce increases in peroxisomal number and enzyme activity in marmoset, cynomolgus, and rhesus monkeys. Yet, this was accomplished only following the repeated administration of the drug at roughly 5- to 10-fold higher dosages, and circulating blood levels, than

Table III Species-Dependent Induction of Peroxisome Proliferation

Species (sex)	Treatment	EM[a] (no./vol. fraction)	Perox. enzyme activity
Gemfibrozil			
Rats (M)	40–200 mg/kg/day, 4 wk–1 yr	+++++	+++
Hamsters (M, F)	400 mg/kg/day, 2 wk	++/NC	
Dogs (M, F)	300 mg/kg/day, 1 yr	sl./NC	
Rhesus monkeys (M, F)	300 mg/kg/day, 3 mo		
Adolescent		NC/sl.	
Adult		+/sl.	
Humans (hypolipidemic)	20–40 mg/kg/day, 2 yr		
8 patients		NC/NC	
9 patients		NC/NC	
Fenofibrate			
Rats (M)	25–300 mg/kg/day, 7–28 days	+++/+++	+++
Hamsters (M)	40–4000 mg/kg/day, 3–6 wk	+++	++
Rhesus monkeys (M)	200 mg/kg/day, 12 mo	NC/NC	
Humans (hypolipidemic)	8–12 mg/kg/day		
31 patients	(not reported)	NC/NC	
10 patients	9 mo	NC/NC	
28 patients	2 yr	NC/NC	
Clobuzarit			
Mice (M)	20–60 mg/kg/day, 2 wk	+++	+++
Rats (M)	3–10 mg/kg/day, 2 wk	+	++
Hamsters (M)	10–20 mg/kg/day, 2 wk	sl./sl.	
Dogs (M)	15–45 mg/kg/day, 2 wk	NC/NC	
Marmoset monkeys (M)	15–45 mg/kg/day, 2 wk	NC/NC	
Ciprofibrate			
Mice (M)	20 mg/kg/day, 4 wk	+++++	
Rats (M)	10–20 mg/kg/day, 4 wk	+++++	+++
	20 mg/kg/day, 26 wk		+++++
Cats (M)	50–100 mg/kg/day, 4 wk	+++/+++	++
Cynomolgus monkeys (M)	400 mg/kg/day, 4 wk	"Increased"	++
Rhesus monkeys (M)	100–200 mg/kg/day, 6 wk	+++	+
Marmoset monkeys (?)	20–100 mg/kg/day, 26 wk		+
ICI-55897			
Mice (M)	60 mg/kg/day, 10 days	+++	+++
Rats (M)	10 mg/kg/day, 10 days	+	++
Dogs (M)	45 mg/kg/day, 10 days	NC	NC
Marmoset monkeys (M)	45 mg/kg/day, 10 days	NC	NC
LY171883			
Mice (M)	0.1–0.5% diet, 2 wk		+++++
Rats (M)	0.25–0.5% diet (approx. 215–365 mg/kg/day), 2 wk	+/++	+++
Hamsters (M)	0.25–0.5% diet, 2 wk	++	+++
Guinea pigs (M)	200 mg/kg/day, 2 wk	sl.	
Dogs (M)	200 mg/kg/day, 1 mo	NC	
Rhesus monkeys (M)	200 mg/kg/day, 1 mo	NC	
	30–175 mg/kg/day, 1 yr		NC
Benzafibrate			
Mice (M)	100 mg/kg/day, 2 wk		+++
Rats (M)	100 mg/kg/day, 2 wk		+++++
Hamsters (M)	100 mg/kg/day, 2 wk		+
Guinea pigs (M)	100 mg/kg/day, 2 wk		+
Rabbits (M)	100 mg/kg/day, 2 wk		NC
Dogs (M)	30 mg/kg/day, 2 wk		NC
Rhesus monkeys (M, F)	125 mg/kg/day, 13 wk		NC, NC

(*continues*)

Table III (*Continued*)

Species (sex)	Treatment	EM[a] (no./vol. fraction)	Perox. enzyme activity
Haloxyfop			
Mice (M, F)	1–10 mg/kg/day, 4 wk	+++	+++
Rats (M)	1–14 mg/kg/day, 2–4 wk	+++	+++++
Dogs (M, F)	5 mg/kg/day, 1 yr	NC	
	2,5,20 mg/kg/day, 13 wk	NC, NC	NC/sl./+, NC/sl./+
Cynomolgus monkeys (M, F)	30 mg/kg/day, 13 wk.	NC, NC	NC, NC

[a] EM = electron microscopic quantitation of peroxisomes (+++++ >> +++ > ++ > + > sl. > NC (no change); each + equals at least a two-fold increase in value).

[b] *In vitro* peroxisomal enzyme activity (β-oxidation and carnitine acetyl transferase) (+++++ >> +++ > ++ > + > sl.; each + equals at least a two-fold increase in value).

those resulting in a much greater response in rodents. Ciprofibrate has also been observed to induce small increases in Ω-oxidation activity in marmosets chronically adminstered this compound but no induction of microsomal *P*-450 1V1 synthesis, the cytochrome involved in this enzyme activity. Further, it has been noted that the Ω-oxidation activity in monkeys was distinct from that observed in rats administered ciprofibrate suggesting that the potential of primates to generate conditions of "substrate overload" may be considerably lower than that in rodents.

Phylogenetic differences in the response of animals to any chemical may simply reflect interspecies differences in metabolism and/or pharmacokinetics of the administered material. However, the results of numerous *in vivo* and *in vitro* studies indicate that fundamental interspecies differences in response to the effects of peroxisome proliferators do exist. For example, only minimal morphologic or enzymatic changes, or no changes at all, are observed following chronic administration of clofibrate or ciprofibrate to marmosets despite the attainment of similar blood levels as those causing significant changes in rats. In addition, significant interspecies differences have also been reported in hepatocytes exposed to proliferators *in vitro*, thus essentially eliminating pharmacokinetic factors. Marked increases in peroxisome numbers and peroxisome enzyme activities are noted in treated rat and mouse hepatocytes while no effect has been reported in guinea pig, marmoset, and human hepatocytes. Researchers have also reported consistently greater responses in rat hepatocytes than in cat hepatocytes whether they were exposed to ciprofibrate *in vivo* as part of an intact organ or as isolated hepatocytes heterotransplanted into the scapular fat pads of athymic nude mice (subsequently administered ciprofibrate via their diet).

The development of several potential peroxisome proliferative chemicals as effective hypolipidemic drugs has provided an opportunity to obtain data upon the potential sensitivity of humans to peroxisome proliferators. Based upon morphological analysis of liver biopsy samples, hyperlipidemic patients administered therapeutic dosages of clofibrate (approximately 20–40 mg/kg/day) were reported to have sustained a 1.5-fold increase in the number, and a 23% increase in the volume density, of hepatocellular peroxisomes relative to prestudy values. This conclusion, however, has been challenged and has yet to be confirmed. In addition, no increase in the number of hepatocellular peroxisomes has been observed in hyperlipidemic patients receiving therapeutic dose levels of the hypolipidemic drugs, gemfibrozil or fenofibrate, even though these dosages were roughly only one-half those causing pronounced effects in rodent (see Table III). Further, as noted, no induction of peroxisome proliferation has been observed in primary human hepatocyte cultrues treated with a number of chemicals known to be potent peroxisome proliferating agents in rodents, including clofibrate.

The mechanistic explanation for interspecies differences in proliferation sensitivity would appear to relate to interspecies differences in the amount and affinity of the aforementioned receptor(s) whether it acts directly or indirectly to initiate proliferation (i.e., via substrate overload secondary to induction of Ω-oxidation and generation of dicarboxylic acid CoA esters). Indeed, a direct relationship between binding affinity and the potency of peroxisome proliferators has been demonstrated and there are data to suggest significant interspecies differences in receptor levels.

In addition to interspecies differences in sensitivity to peroxisome proliferators, some proliferators have also displayed significant sex-related differences in induction potency. For example, the proliferative response obtained in male rats ingesting a diet containing 0.25% clofibrate has been observed to be similar to that obtained in female rats ingesting a diet containing 2.0% clofibrate. Peroxisome proliferation is not observed in castrated male rats administered clofibrate along with estradiol yet proliferation and the induction of catalase activity is observed in ovariectomized female rats administered clofibrate along with testosterone. Differences were not attributable to differences in pharmacokinetics as blood levels were similar in both sexes of animals.

III. Carcinogenicity of Peroxisome Proliferators

A number of compounds which induce the proliferation of hepatocellular peroxisomes in rodent species also cause an increased incidence of benign and malignant tumors in livers of rats and/or mice upon chronic administration. As with all potentially carcinogenic compounds, the mechanism by which

Table IV Tumorigenicity Data for Selected Peroxisome Proliferators

Compound	Species	Sex	Treatment	Percentage hepatic tumors[a]
Clofibrate	Rat	M, F	0.15 diet, 26 mo	NT, NT
		M, F	0.45% diet, 26 mo	10, 10
		M	0.5% diet, 24 mo	16
		M	0.5% diet, 28 mo	91
		M, F	300 mg/kg/day diet, 18 mo	12, NT
		M, F	400 mg/kg/day diet, 24 mo	78, 54
Trichloroethylene	Mouse	M, F	1000 mg/kg/day gavage, 24 mo	43, 23
		M	1200 mg/kg/day gavage, 24 mo	47
		M	2300 mg/kg/day gavage, 24 mo	60
		F	900 mg/kg/day gavage, 24 mo	8
		F	1700 mg/kg/day gavage, 24 mo	23
Di(2-ethylhexyl)-phthlate	Rat	F	0.1% diet, 24 mo	NT
		M, F	0.6% diet, 24 mo	6, 12
		M, F	1.2% diet, 24 mo	18, 26
	Mouse	M, F	0.3% diet, 24 mo	23, 10
		M, F	0.6% diet, 24 mo	30, 17
Tibric Acid	Rat	M	0.05–0.2% diet, 17 mo	97
		M, F	30 mg/kg/day diet, 18 mo	27, NT
		M, F	100 mg/kg/day diet, 18 mo	58, 34
BR-931	Rat	M, F	0.05% diet, 16 mo	70, 56
		M, F	0.2% diet, 16 mo	100, 100
	Mouse	F	0.2% diet, 19 mo	90
Wy-14643	Rat	M	0.1% diet, 16 mo	100
	Mouse	M	0.05–0.1% diet, 15 mo	100
Nafenopin	Rat	M	0.1% diet, 18–25 mo	73
	Mouse	M, F	0.05–0.1% diet, 20 mo	100, 100
Gemfibrozil	Rat	M, F	30 mg/kg/day, 24 mo	10, NT
		M, F	300 mg/kg/day, 24 mo	44, NT
Ciprofibrate	Rat	M	10 mg/kg/day diet, 14 mo	100
			0.025% diet, 14–22 mo	100
Methylclofenepate	Rat	M	0.1% diet, 18 mo	100

[a] Values represent total hepatocellular adenomas plus carcinomas and, where possible, are corrected for control tumor incidence; NT = nontumorigenic.

proliferators induce tumors is critical to the assessment of carcinogenic risk they may pose to humans.

A. Peroxisome Proliferation and Tumorigenicity

Table IV summarizes the tumorigenicity data available for a number of peroxisome proliferators. A strong statistical correlation exists between the potency of these compounds as proliferators and their carcinogenic potential upon chronic administration to rats and/or mice. Figure 5 presents a plot of the proliferative potency, as measured by the induction of peroxisomal β-oxidation, versus hepatocarcinogenic potency of a number of peroxisome proliferators normalized to the average dosage administered. In general, the tumorigenic potential of a proliferator increases with increasing induction of the peroxisomal enzyme most responsible for the generation of hydrogen peroxide in induced animals. It has been suggested that this relationship represents cause and effect (see below) and that in the absence of a significant degree of peroxisome proliferation, peroxisome proliferators would lack tumorigenic activity.

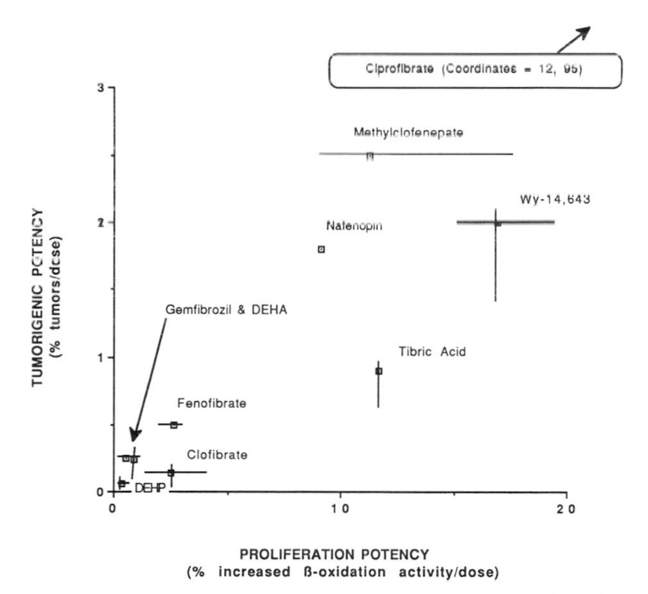

Figure 5 Tumorigenic potency and peroxisome proliferation potency of a number of carcinogenic proliferators. Potency was calculated by normalizing the % tumor yields from rat bioassays and the % increased peroxisomal β-oxidation activities by the dosages. Ranges (*bars*) and means (*points*) are given.

B. Epidemiology

Peroxisome proliferators exhibiting useful hypolipidemic activity have been administered at pharmacologically active dosages to humans, some for prolonged periods of time. Thus, fairly large cohorts have been available with which to examine the tumorigenic potential of dosages of compounds in humans which, in some instances, approximate those inducing peroxisome proliferation in rodents. A relatively large clinical trial of the potential health benefits of clofibrate was begun in 1965 in Budapest, Edinburgh, and Prague under the auspices of the World Health Organization. This study involved nearly 16,000 subjects in which approximately 5300 men received 1.6 g clofibrate per day as treatment for hypercholesteremia. The average length of follow-up was 5.3 years. Analysis of death rates due to malignancies indicated a slight, but not statistically significant difference in the clofibrate group; however, it was subsequently concluded that the reason for this difference was due to an unusually low cancer rate in the control population and not administration of the drug. A reexamination of this same population 9 years later reported a similar finding. There have also been several other trials of clofibrate in which no increased incidence of malignancies have been recorded, even upon 5–9 years follow-up. In addition, no increased incidence of cancer was observed in a study involving 2051 dyslipidemic men administered 1200 mg/day gemfibrozil for 5 years relative to a similar control group receiving a palcebo. While epidemiological studies on peroxisome proliferators are relatively limited by numbers of subjects and/or time of follow-up, they do demonstrate the lack of a potent carcinogenic response in humans repeatedly administered relatively large, therapeutic dosages of a known peroxisome proliferator.

C. Genotoxicity of Peroxisome Proliferators

Chemicals are believed to cause cancer in animals via their ability to cause and/or facilitate the expression of critical, heritable mutations in protooncogenes and/or the inactivation of so-called tumor suppressor genes or their products. While eventual tumor development appears to inolve a number of additional mutagenic and nonmutagenic events, identification of the potential genotoxicity of a chemical provides a useful tool in discerning its carcinogenic potential. Using a battery of so-called "short-term" assays, chemical carcinogens may be roughly classified into two categories; those which are primarily genotoxic in nature, as demonstrated by their positive mutagenicity, ability to covalently bind DNA, and induce its repair; and those which are nongenotoxic or epigenetic in nature, as characterized by their general lack of mutagenic activity or direct interaction with DNA. This distinction is of great significance in carcinogenic risk assessment. Genotoxic chemicals may cause mutations and altered gene expression even at nontoxic dosages and have a greater carcinogenic potency at cytotoxic dosages than nongenotoxic compounds (e.g., the mycotoxin aflatoxin B1). In contrast, the physiological and biochemical adaptations and/or chronic cytotoxicity which are believed to be responsible for the tumorigenic action of many nongenotoxic chemicals display thresholds and are often reversible in nature. It is important to note; however, that the definition between mechanistic classifications may not always be distinct, may be dose-depedent, and does not necessarily rule out the potential of some nongenotoxic materials to ultimately cause mutagenic or clastogenic events secondary to primary treatment-related changes in animals.

A great deal of data have been compiled which demonstrate that, as a class of compounds, peroxisome proliferators are not mutagenic or, at most, only weakly mutagenic in *in vitro* bacterial or mammalian cell mutagenicity assays and fail to bind DNA or induce its repair. In addition, peroxisome proliferators do not appear to have initiating potential in classical intiation–promotion assays, providing further evidence of the lack or genotoxicity of these compounds *in vivo*.

Despite the lack of genotoxicity of peroxisome proliferators themselves, the metabolic byproduct of peroxisomal oxidases, hydrogen peroxide, and putative products of extraperoxisomal hydrogen peroxide–cellular lipid reaction products, have been shown to be genotoxic in *in vitro* assays. Oxidative degradation, strand breakage, and cross-linking have been determined in isolated DNA upon interaction with hydrogen peroxide or hydroxyl free radical, a breakdown product of hydrogen peroxide. Hydrogen peroxide has also been reported to be mutagenic to bacteria and mammalian cells and to induce DNA strand scissions and repair in intact cells *in vitro* and to induce duodenal tumors in mice. That induced peroxisomes are capable of generating enough hydrogen peroxide to cause extraorganellar effects in macromolecules, at least in the absence of cytosolic defense mechanisms, has been demonstrated *in vitro*. The number of single-strand breaks observed in isolated DNA incubated with peroxisomes recovered from rats repeatedly administered WY-14643 also correlates well with the level of hydrogen peroxide production. Finally, the lipid peroxidation product, malondialdehyde, has also been reported to be mutagenic in a variety of bacterial and mammalian cell assays and to cause DNA cross-linking *in vitro*. It is important to note that these latter findings have been made in the absence of a full complement of cellular defense mechanisms which may neutralize potentially reactive products prior to causing cellular damage.

Peroxisome proliferators present an interesting paradox. While these chemicals and their metabolites, in general, lack genotoxic activity themselves, they may cause an accelerated generation of endogenous genotoxic compounds via their induction of peroxisome proliferation. Thus, peroxisome proliferators are sometimes referred to as "indirect genotoxins." Despite this, in the absence of peroxisome proliferation, these chemicals do not represent a genotoxic challenge, directly or indirectly, to animals. Thus, the potential genotoxicity of this class of chemicals display clear, readily identifiable thresholds.

D. Mechanism of Tumorigenicity

Several hypotheses have been proposed to account for the carcinogenic potential of peroxisome proliferators in rats and mice. A composite of the key features of these, along with their potential interaction, is diagrammed in Figure 6. The mutagenicity and carcinogenicity data for peroxisome proliferators has led to the proposal that chemicals which induce the proliferation of hepatocellular peroxisomes in rodents represent a "novel class of carcinogens." This model, generally referred to as the "oxidative stress" model, is based upon the concept that the chronic induction of the activity of a normally occurring peroxisomal enzyme system greatly accelerates the process of cellular oxidative damage. Upon repeated administration of a peroxisome proliferator, the disproportionate increase in the activities of the fatty acid β-oxidation pathway and possibly other oxidases relative to the catalase activity of these peroxisomes results in a significant elevation in peroxisomal levels of hydrogen perox-

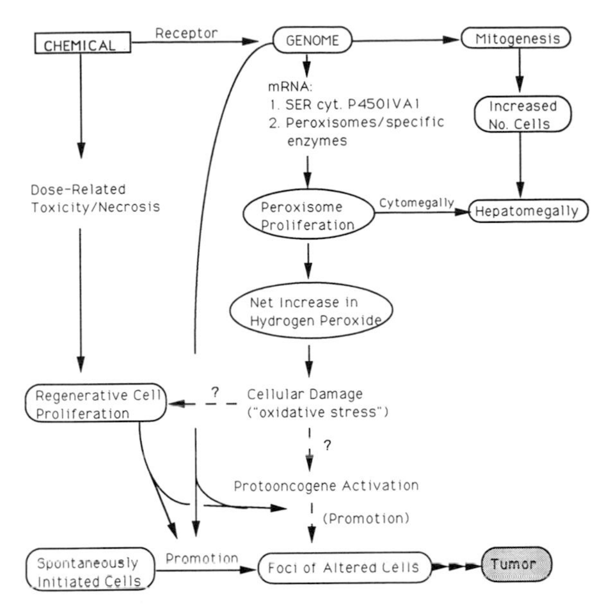

Figure 6 Composite of cellular, molecular, and enzymatic events believed to dictate the mechanism of tumorigenic action of peroxisome proliferators in rodent species. Dashed lines and question marks indicate a lack of conclusive data supporting a particular link.

ide. Hydrogen peroxide itself, or hydroxyl radicals formed by its iron-catalyzed degradation, may subsequently escape elimination via defense mechanisms and react with cellular constituents. This may result in the generation of a variety of reactive oxygen species and auto-oxidation products (aldehydes, polymers, organoperoxides, conjugated dienes, etc.). The more reactive of these degradation products and/or hydrogen peroxide itself, are believed to interact with the genome of the cell resulting in DNA damage and the possible anaplastic transformation of the cell.

The experimental data obtained during the examination of the oxidative stress model have been relatively inconclusive and occasionally contradictory. The results of a number of studies have suggested that increases in intracellular hydrogen peroxide concentration and accumulation of lipofuscin, believed to be an oxidative degradation product of cellular membrane proteolipids, occur concomitant with hepatocellular peroxisome proliferation in rats and mice. These include enhanced production of conjugated fatty acid dienes, increased levels of oxidized glutathione (GSSG), decreased vitamin E levels, and increased formation of altered DNA bases and other structural perturbations consistent with oxidative damage (e.g., 8-hydroxyguanosine) in the livers of rats administered carcinogenic dosages of potent peroxisome proliferators. In contrast, a number of other studies have failed to demonstrate similar genetic damage or evidence of increases in oxidative stress in the livers of rats and mice administered peroxisome proliferators, even under various experimental conditions designed to exacerbate oxidative tissue damage (for example by feeding treated animals an antioxidant deficient or iron-fortified diet). In addition, the quantitative relationship between the activity of peroxisomal β-oxidation and tumor formation in rodents has been questioned. Administration of di(2-ethylhexyl)phthalate, a relatively weak proliferator, or Wy-14643, a relatively potent proliferator, to rodents at dosages resulting in similar levels of peroxisome β-oxidation do not result in similar incidences of tumor formation. If the byproducts of β-oxidation were directly responsible for tumor formation, a similar tumor yield would be expected. Finally, several proliferators have been observed to induce an increased incidence of liver tumors in B6C3F1 mice at a subpharmacologically active dosage(s). These latter findings, however, must be viewed with caution due to the high spontaneous incidence of this type of tumor in this mouse strain. Taken together, the above data suggest a more complex mechanism of tumorigenic action than simple increases in oxidative stress.

Another characteristic of peroxisome proliferators which has received a considerable amount of attention relative to its potential involvement in proliferator tumorigenicity is the apparent unique promotion activity of these chemicals. Traditionally, promoters have been defined as chemicals which stimulate the clonal growth of cells that have been transformed or "initiated" (i.e., mutated) by a genotoxic agent, the classical "two-stage carcinogenesis model." The effects of these compounds are reversible upon cessation of treatment and to be effective they must be administered over a prolonged period of time. As noted, peroxisome proliferators which have been examined in a variety of initiation–promotion assays have lacked initiating potential; however, these same compounds have occasionally been observed to act as promoters of genotoxin-initiated lesions. Subsequent investigation has revealed that peroxisome proliferators appear to stimulate the formation of foci of altered cells which are morphologically and biochemically distinct from those formed by the more classical promoting agents such as phenobarbital and phorbol esters. For example, a typical marker used to identify altered hepatic foci, γ-glutamyltranspeptidase activity, is a poor marker of proliferator-promoted foci which instead are typically identified by decreases in ATPase and glutathione S-transferase (placental form) activities. Indeed, it appears that treatment with proliferators may actually alter the morphology and biochemistry of existing foci which have been initiated by a genotoxin and promoted by phenobarbital.

It has been proposed that as a class of compounds, peroxisome proliferators promote the development of cells which have been spontaneously initiated via interaction with naturally occurring endogenous or exogenous genotoxic physical or chemical factors. Evidence of such spontaneously developing, morphologically distinct, foci of altered cells in the livers of normal, untreated rats have been well documented. It appears that peroxisome proliferators promote these "spontaneously" initiated cells preferentially to those initiated by administered genotoxic agents. The result is the generation of relatively few large foci in the livers of proliferator-treated rats in contrast to the numerous small foci typically observed when promoting genotoxin-initiated cells with phenobarbital. A key piece of evidence supporting this theory has been the observation that peroxisome proliferators promote the development of a greater number of foci of altered cells and tumors in aged rats, which have a relatively high number of spontaneously initiated cells and foci, than in younger rats, which have relatively few spontaneously initiated cells.

The promotion of tumor development by physical or chemical means has been closely associated with treatment-related increases in the rate of cell proliferation in the affected tissue. Peroxisome proliferators characteristically induce a short-lived mitogenic wave of hepatic cell proliferation upon administration to rodents. Despite this, repated, pulsed, dosing with peroxisome proliferators or other mitogenic agents has failed to promote the development of enzyme-altered foci in rats treated with a genotoxic agent while chronic regenerative cell proliferation resulting from tissue damage was an effective promotor. While it is not clear at this time what the impact of this mitogenic response might be upon the promotion of spontaneously initiated cells, one consequence of this activity is to simply increase the number of hepatocytes at risk of spontaneous transformation. Significantly, however, human hepatocyte cultures have been reported to be refractory to the mitogenic effects of the potent peroxisome proliferator nafenopin, again demonstrating pronounced interspecies differences in sensitivity to the effects of these chemicals.

In contrast to relatively short bursts of mitogenic DNA synthesis, it is generally accepted that chronic regenerative DNA synthesis as a result of administration of cytotoxic dosages of both genotoxic and nongenotoxic chemicals may greatly enhance tumor formation in animals. As outlined in Figure 6, chronic regenerative DNA synthesis as a result of cytotoxic dosages of peroxisome proliferators or enhanced cell death secondary to peroxisome proliferation (i.e., accelerated apoptosis) is expected to promote both spontaneously initiated cells or initiated cells arising secondary to peroxisome proliferation. Data suggest that increases in proliferator-induced regenerative hepatocellular DNA synthesis rates are not necessarily observed at dosages causing a substantial induction of peroxisome proliferation. As with all chemicals, however, cytotoxicity is a dose-related response and at cytotoxic dosages the tumorigenic potential of peroxisome proliferators is expected to be enhanced.

IV. Conclusion

The chemical induction of the proliferation of peroxisomes in the liver of rodents has come to represent a significant challenge to toxicology, one with considerable potential economic and societal ramifications. While peroxisome proliferation often represents the most sensitive treatment-related change observed in treated rodent species, the most significant toxicological impact of this effect is its relationship to tumor formation in rats and mice. Data suggest that peroxisome proliferation represents a useful marker for the potential carcinogenic activity of this class of compounds and is likely involved in some as yet undefined way in the mechanism of tumorigenicity of proliferators. It has been suggested that the threshold nature and strong species-dependency of peroxisome proliferation, the nearly uniform lack of genotoxicity of proliferators, and epidemiological findings indicate that peroxisome proliferators do not represent an imminent carcinogenic hazard to humans.

Bibliography

Conway, J. G., Cattley, R. C., Popp, J. A., and Butterworth, B. E. (1989). Possible mechanisms in hepatocarcinogenesis by the peroxisome proliferator Di(2-ethylhexyl)phthalate,'' *Drug Metabolism Reviews* **21,** 65–102.

Fahimi, H. D., and Sies, H., eds. (1986). ''Peroxisomes in Biology and Medicine.'' Springer-Verlag, Berlin.

Gibson, G. G., and Lake, B. G., eds. (1993). ''Monograph on Peroxisome Proliferation.'' Taylor and Francis, London.

Hsia, M. T. S. (1990). ''The Relationship between Carcinogenesis and Peroxisome Proliferation in Rodent Liver after Exposure to the Plasticizer DEHP and DEHA,'' MTR-90W00034. The Mitre Corp., McLean.

Reddy, J. K., and Lalwani, N. D. (1983). Carcinogenesis by hepatic peroxisome proliferators: Evaluation of the risk of hypolipidemic drugs and industrial plasticizers to humans. *CRC Critical Reviews on Toxicology* **12,** 1–58.

Schneider, B. F. (1987). ''Recent Advances in Phthalate Esters Research,'' Toxicology and Industrial Health Vol. 3, Princeton Scientific Publishing, Princeton.

Stott, W. T. (1986). Chemically induced proliferation of peroxisomes: Implications for risk assessment. *Regul. Toxicol. Pharmacol.* **8,** 125–159.

Pesticides and Food Safety

Anna M. Fan
California Environmental Protection Agency

Glossary

ADI Acceptable daily intake. A daily intake level for a chemical developed by FAO/WHO which represents the daily dietary intake of a chemical which, if ingested over a lifetime by an individual, appears to be without appreciable health risk.

FAO/WHO United Nations Food and Agricultural Organization/World Health Organization.

FDC act Federal Food, Drug, and Cosmetic Act. Allows the FDA to establish tolerances for toxic substances where occurrence in food cannot be avoided.

FIFRA Federal Insecticide, Fungicide, and Rodenticide Act, a major law regulating pesticide registration and use in the United States.

Pesticide Any substance or mixture of substances intended for preventing, destroying, repelling, or mitigating any insects, nematodes, fungi, or weeds, or any other forms of life declared to be pests; any substance or mixture of substances intended for use as a plant regulator, defoliant, or desiccant (FIFRA).

RfD Risk reference dose. An exposure level (to a chemical) developed by the EPA which represents an estimate of the total exposure to an individual that is likely to be without appreciable risk or deleterious effect over a lifetime.

RID Reference ingested dose. Dose level of a chemical resulting from short-term or long-term oral intake at which no adverse health effects are anticipated.

Tolerance Legal limit specifying the maximum permissible level of pesticide residues that can be present in or on food sold in interstate commerce.

U.S. EPA United States Environmental Protection Agency.

U.S. FDA United States Food and Drug Administration.

PESTICIDES *are chemical or biological agents whose uses in agriculture and other food sources contribute to economic and health benefits through efficient food production. The safety of the food for human consumption is guided by regulatory authorities who evaluate the toxicity of and register pesticides, monitor pesticide residues in food, assess the safety and risk of the presence of pesticides in our food supply, and ensure compliance.*

I. Introduction

Pesticides are used to control insects, rodents, nematodes, fungi, or weeds or other forms of pests, and are intended for use as a plant regulator, defoliant, or desiccant. Most pesticides are inherently toxic by design and thus have the property of producing adverse effects in biological organisms, thereby acting on any forms of life declared to be pests. In the case of pesticides in food, the same biological or toxic property can also act on humans, the primary nontarget organisms, if the amounts of pesticides ingested as residues are excessive. On the other hand pesticides are intentionally introduced to the environment by their uses in or on food to promote its production. Therefore there has to be a careful balance between the agricultural uses and their control in order to ensure the safety of our food supply.

Pesticides may also be present in food as a result of contamination from other nonagricultural sources such as animal feed; contaminated vehicles, equipment, or buildings used for transportation, processing, or storage; or accidental spills.

II. Pesticides Used on Foods

The major groups of pesticides that are used in or on foods are the insecticides, herbicides, fungicides, and agricultural fumigants. Examples of some of these pesticides are given below.

Insecticides
  Organophosphates, cholinesterase inhibiting
    —Malathion, chlorpyrifos
  Carbamates, cholinesterase inhibiting
    —Aldicarb, carbaryl
  Noncholinestease inhibiting
    —Pyrethrins, pyrethroids
  Other
    —Propargite, dicofol
Herbicides
  Dipiridyls
    —Paraquat, diquat
  Chlorophenoxy compounds
    —2,4-D; 2,4,5-T
  Nitrophenolic compounds
    —Dinoseb, dinocap
  Other
    —Molinate, alachlor
Fungicides
  Dithiocarbamates
    —Mancozeb, thiram
  Organochlorines
    —Henzachlorobenzene, PCNB (pentachloronitrobenzene)
  Dicarboximides
    —Captan, Captafol
  Other
    —Chlorothalonil, benomyl
Agricultural fumigants
  Halogenated hydrocarbons
    —Methyl bromide, DBCP (dibromochloropropane)
  Oxides and aldehydes
    —Ethylene oxide, formaldehyde
  Sulfur and phosphorous fumigants
    —Sulfur dioxide, aluminum phosphide

III. Responsible Government Agencies

The U.S. federal government agencies responsible for ensuring the safety of pesticides in foods are as follows.

Agency	Responsibilities
U.S. Environmental Protection Agency	Registers/approves uses for pesticides; establish tolerances for pesticide residues
U.S. Department of Agriculture	Monitors meat, poultry, and certain egg products for pesticide residues
U.S. Food and Drug Administration	Enforces tolerances; monitors and acquires incidence data on and levels of pesticide residues in other types of food (agricultural raw commodities, processed foods, milk, fish/shellfish, and animal feed; both domestic and imported).

Under the Federal Insecticide, Fungicide, and Rodenticide Act (FIFRA), amended in 1972, the United States Environmental Protection Agency (U.S. EPA) is to screen pesticides through the registration process before they enter into commerce, and to register pesticides in use to prevent unreasonable adverse health and environmental effects. The EPA establishes allowable levels of pesticide residue in foods (referred to as tolerances), and monitoring and enforcement are primarily carried out by the United States Food and Drug Administration (U.S. FDA) and United States Department of Agriculture (USDA).

Other agencies may also carry out related activities at the local county or state level. These would include state agencies responsible for the regulation of pesticide sales and use or for providing public health oversight; local county agricultural commissioners that enforce state registration and use requirements; and county health departments that receive and maintain reports of pesticide-related illnesses.

IV. Assessing Food Safety

Food safety is a function of exposure and toxicity and is assessed primarily based on (1) food consumption pattern/rate; (2) residue concentration; and (3) toxicity of pesticide involved.

Other factors affecting the safety of food include environmental fate and persistence, effects of processing or cooking, toxicity of breakdown products, interaction with other biologically active agents, and individual sensitivity to the effects of the pesticide. Due to various limitations and uncertainties inherent in the scientific database and the food safety system, absolute certainty that food consumption carries no dietary risk cannot be achieved. On the other hand, foods are considered to be reasonably safe for human consumption if the total amount of pesticide residue ingested does not exceed the level at which no adverse effects are anticipated. The relationship among the different variables can be expressed as follows.

$$Tp = Df_1 + Df_3 \ldots Df_n$$
$$[Cp_1 \times Ip_1], [Cp_2 \times Ip_2], [Cp_3 \times Ip_3] \ldots [Cp_n \times Ip_n],$$

where T = total pesticide residue ingested (per unit body weight per day or mg/kg day)
p = specific pesticide involved
Df = dose of pesticide ingested from food items 1, 2, 3 . . .
= $Cp \times Ip$
Cp = concentration or residue level of pesticide in food item (mg/kg ppm)
Ip = ingestion (consumption) rate of food item containing pesticide residue at Cp (kg/day). This can be the amount of food ingested per eating occasion when assessing acute toxicity.

Therefore, in order for food to be reasonably safe to be consumed, the total residue ingested should not exceed the reference ingested dose (RID), or

$$Tp < \text{RID}.$$

The reference ingested dose for a chemical is a dose level resulting from short-term or long-term oral intake of that chemical at which no adverse health effect is anticipated. Other sources may contribute to additional exposure. This RID concept is similar to the acceptable daily intake (ADI) developed by the United Nations Food and Agriculture Organization (FAO) and World Health Organization (WHO), or the reference dose (RfD) developed by the U.S. EPA. An ADI is the daily dietary intake of a chemical which, if ingested over a lifetime by an individual, appears to be without appreciable risk. An RfD is an estimate of the total exposure to a chemical by an individual that is likely to be without appreciable risk of deleterious effects over a lifetime. All these are reference values for comparison purposes and are revised as new information becomes available. The numerical values may be the same or different depending on the database and assumptions used. A tolerance level may be set with the consideration of the health based information and the food consumption rate, but the final value can be affected by other factors such as economics and technical feasibility. A schematic representation of the factors affecting the development of permissible levels of pesticide in food is shown in Figure 1. Alternately the derivation of the permissible level for a pesticide in a food item is shown as follows:

$$\text{PL} = \text{RID} \times \text{BW}/F$$

where PL = permissible level of pesticide in the food item (mg/kg, or ppm; μg/kg, or ppb)

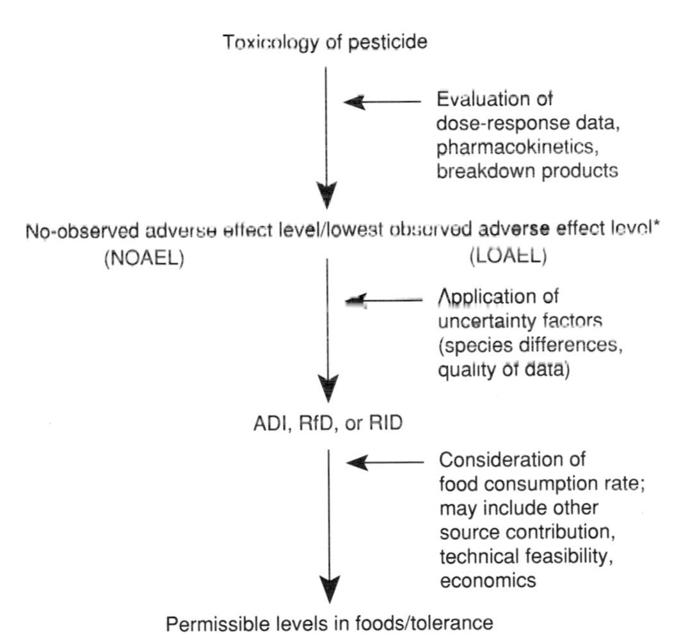

Figure 1 Development of permissible levels of pesticides in food. *, For noncarcinogenic chemicals; NOAEL/LOAEL, dose level of chemical at which no adverse health effects were observed, or the lowest level found to produce an adverse health effect, either based on experimental animal studies or human data; ADI, acceptable daily intake; RfD, reference dose; RID, reference ingested dose; tolerance, legal limit prescribing the maximum permissible level of residue in or on food.

Table I Pesticides Included in the U.S. Food and Drug Administration's Regulatory Monitoring for Pesticides in Food (1990) Listed by Chemical and Trade Names

Acephate (Orthene) C
Alachlor (Lasso) C-SR
Aldicarb (Temik) C-CR
Aldoxycarb
Aldrin
Allethrin
Ametryn
Amitraz (Baam) C-CR
Amobam
Anilazine (Dyrene)
Aramite
Aspon
Atrazine (Aatrex)
Azinphos-ethyl (Ethyl Guthion)
Azinphos-methyl (Guthion) PC
Benthiocarb
Benfluralin
Benomyl (Benlate) C-SRC
Bensulide (Betasan)
BHC (Benzahex)
Bifenthrin
Binapacryl
Biphenyl
Bromacil
Bromophos
Bromophos-ethyl
Bromopropylate (Acarol)
Bromoxynil
Bufencarb
Bulan
Butralin
Captafol (Difolatan) C-SR
Captan C-SR
Carbaryl (Sevin)
Carbendazim
Carbofuran (Furadan)
Carbon disulfide
Carbon tetrachloride
Carbophenothion (Trithion)
Carboxin
Chlorbenzide
Chlobromuron
Chlordane
Chlordecone C-SR
Chlordimeform C
Chlorfenvinphos (Supona)
Chlornidine
Chlornitrofen
Chlorobenzilate C-CR
Chloroform C-SRC
Chloroneb
Chloropicrin
Chloropropylate
Chlorothalonil (Bravo) C(R)
Chlorotoluron
Chloroxuron (Tenoran)
Chlorpropham
Chlorpyrifos (Dursban)
Chlorpyrifos-methyl
Chlorthion
Chlorthiophos
Clomazone
Coumaphos
Cyanazine
Cyanophos
Cycloate
Cyhexatine
Cypermethrin C
2,4-D (Isopropylester)
2,4-DB
Daminozide C-CR
DCPA (Dacthal)
DDT (DDE, DDD) C-CR
DEF
Deltamethrin
Demeton (Systox)
Dialitor (Torak)
Diallate C-SRC
Diazinon
Dibromochloropropane C-CR, R
Dicamba
Dicapthon
Dichlobenil
Dichlofenthion
Dichlofluanid
Dichlone
Dichlorobenzene, *ortho*
Dichlorobenzene, *para*
1,3-Dichloropropene
Dichlorvos (DDVP)
Diclofop-methyl (Hoelon)
Dicloran (DCNA, Botran)
Dicofol (Kelthane) C-SR
Dicrotophos (Bidrin)
Dieldrin
Dilan
Dimethoate (Cygon) C-SRC
Dinocap
Dinozeb
Dioxa benzofos
Dioxacarb
Dioxathion (Delnav)
Diphenamid
Diphenylamine
Diquat
Disulfoton (Di-Syston)
Diuron
Endosulfan (Thiodan)
Endrin C-CR
EPN
EPTC
Ethion
Ethoprop
Ethoxyquin
Ethylene dibromide C-CR, R
Ethylene dichloride
Etrimfos
Famphur
Fenamiphos (Nemacur)
Fenarimol UR
Fenbutatin oxide
Fenitrothion (Sumithion)
Fensulfothion (Dansanit)
Fenthion (Baytex)
Fenuron
Fenvalerate (Pydrin)
Fluchloralin (Basalin)
Flucythrinate
Fluometuron
Fluvalinate
Folpet PC
Fonofos (Dyfonate)
Gardona
Genite 923
Heptachlor
Hexachlorobenzene
Imazalil
Iprodione (Rovral)
Isobenzan
Isodrin
Isofenphos (Oftanol)
Isoprocarb
Isopropalin
Isoproturon
Jodfenphos
Leptophos (Phosvel)
Lindane C-SRC
Linuron C-SR
Malathion
Mancozeb (Dithane M-45)
Maneb C-SRC
Mephosfolan (Cytrolane)
Mercarbam
Merphos
Metalaxyl
Metasystox thiol
Methamidophos (Monitor)
Methidathion (Supracide)
Methiocarb (Mesurol)
Methomyl (Lannate) C
Methoxychlor
Methyl bromide
Methyl trithion
Methylene chloride
Metiram
Metobromuron
Metolachlor (Dual) C
Metoxuron
Metribuzin (Sencor)
Mevinphos (Phosdrin)
Mirex C-CR
Monocrotophos (Azodrin)
Monolinuron
Monuron
Myclobutanil
Nabam
Naled (Dibrom)
Napropamide
Neburon
Nitralin
Nitrofen (TOK) T
Norea
Norflurazon
Omethoate
Ovex
Oxadiazon C
Oxamyl
Oxydemeton-methyl (Meta-Systox-R)
Oxyfluorfen
Oxythioquinox
Parathion PC
Parathion-methyl
Pendimethalin (Prowl)
Pentachlorophenol C-CR, TF
Permethrin
Perthane
Phenkapton
Phenthoate
Phenylphenol, *ortho*
Phorate (Thimet)
Phosalone (Zolone)
Phosmet (Imidan)
Phosphamidon (Dimecron)
Phostex
Phoxim
Picloram
Pirimiphos-ethyl
Pirimiphos-methyl
Procymidone
Profenofos (Curacron)
Profluralin (Tolban)
Prolan
Prometryn
Pronamide (Kerb) C-CR
Propanil (Stam)
Progagite (Omite)
Propazine
Propham
Propiconazole
Propoxur (Baygon)
Prothiofos
Pyrazon
Pyrazophos
Pyrethrins
Quinalphos
Quintozene
Ronnel
Schradan
Simazine
Strobane
Sulfallate (Vegadex)

Table I *(Continued)*

Sulfotep	Terbufos	Thiophanate-methyl C-SRC	Trichloroethane
Sulfur dioxide	Terbuthylazine	Tolylfluanid	Trichloronat
Sulprofos (Bolstar)	Tetradifon (Tedion)	Toxaphene (Attac) C-CR	Tridiphane
2,4,5-T C-CR	Tetraiodoethylene	Triadimefon (Bayleton)	Trifluralin (Treflan) C-SRC
2,3,6-TBA	Tetrasul	Triadimenol	Trimethacarb
TDE	Thiabendazole	Tri-allate	Vernolate
Tecnazene	Thiobencarb (Bolero)	Triazophos (Hostathion)	Vinclozolin (Ronilan)
Tetradfon	Thiodicarb (Larvin)	Tributyltin	Zineb
TEPP	Thionazin (Zinophos)	Trichlorfon (Dylox)	Zytron
Terbacil (Sinbar)			

Source: U.S. FDA (1990).

Abbreviations (OTS, 1989): C, chemicals for which EPA has identified to be an animal carcinogen but not reviewed or canceled; PC, chemicals for which EPA has identified to be potential carcinogens but not reviewed or canceled; R, more data required (labeling requirement); UR, under review for oncogenic potential; C-SR, carcinogenic pesticides in special review by EPA; C-SRC, carcinogenic pesticides for which special review by EPA is complete but chemicals not suspended or canceled; C-CR, EPA canceled or restricted carcinogenic pesticides.

RID = reference ingested dose (mg/kd day, see above. Can use EPA's RfD if justified)
BW = body weight of the individual consuming the food (kg)
F = food consumption rate kg/day)

If permissible levels are to be established for more than one food item for the same pesticide, then the total daily reference dose (RID × BW) would have to be proportionately assigned to each food item according to the consumption rate for each item.

V. Pesticide Residues and Tolerances

Pesticide residues are monitored to ensure that tolerances are not exceeded. Tolerances are the maximum levels of pesticide residue permitted in or on food. These are set based on data on pesticide chemistry, expected quantity of residues present in food based on field trials, laboratory analytical procedures used for analyzing residue levels, residues on animal feed, and toxicity data on the parent compound, impurities, and breakdown products. A safety factor or (uncertainty factor) is usually added. Each tolerance level may represent the parent chemical only or it may include metabolites and other related chemicals. A tolerance is set separately for each registered food use for each pesticide expected to leave a residue. Therefore, a pesticide can have a number of tolerances depending on the number of food or crop uses, and more than one pesticide can be used on any specific food crop. These tolerances are published in the Code of Federal Regulations (CFR, Title 40).

Some concerns are raised regarding the existence of data gaps for data that support existing tolerances and the continued registration and use of pesticides in use. To address these concerns there have been increasing efforts to improve filling the data gaps and to evaluate the data for tolerance setting and registration, a process that is expected to extend into the year 2000. In 1972 there were about 50,000 pesticide products and 600 active ingredients on the market that required reregistration under the new law. About 390 of the active ingredients are used on food. Data call in has been completed for all 390 substances, but most of the chemicals have yet to be evaluated. Tolerances set long ago based on analytical capabilities at that time should be reevaluated based on updated toxicological data and new analytical techniques. In establishing new tolerances, consideration needs to be given to existing tolerances, which may provide health protection based on long-term, low-level exposures but which may not have adequately considered acute or short-term exposures to pesticides having high acute toxicity (e.g., potent cholinesterase inhibitors, ocular toxicity, neurotoxicity, and immunotoxicity). The tolerances are also based on individual chemicals, and exposure to multiple pesticides should be evaluated.

VI. Pesticide Residue Monitoring

In addition to the monitoring activities carried out by the U.S. FDA and USDA, a residue monitoring program designed to complement existing federal monitoring programs was initiated in 1990 by the USDA's Agricultural Marketing Services. Complementary sampling plans are also developed by FDA's field office personnel working with state counterparts.

The regulatory monitoring by the FDA includes 268 pesticides detected in foods; 108 are actually found in the 1990 monitoring. These chemicals are presented in Table I. Targeted sampling and analysis are also carried out by the FDA and other states for pesticides not determined by the commonly used analytical methods. The data represented commercial raw fruits and vegetables. Residue violations above tolerances are rarely found. Illegal residues can result from misuse, poor agricultural practice, unusual weather conditions, or application of pesticides to crops for which no tolerances have been set.

Table II Thirty (30) Most Frequently Found Pesticides in U.S. FDA's Total Diet Study 1990[a]

Pesticide		FAO/WHO ADI (mg/kg/day)	EPA RfD (mg/kg/day)	Oral LD_{50}[b] rat (mg/kg)
Malathion (Cythion)	(19)	0.020	0.020p	885
DDT*C-CR^2	(16)	0.20r	0.0005p	87
Chlorpyrifos-methyl	(10)	0.010	—	941
Endosulfan (Thiodan)	(9)	0.006r	0.00005p	18
Chlopyrifos (Dursban)	(8)	0.10	0.003	135
Methamidophos (Monitor)	(6)	0.004	0.00005	13
Dieldrin	(6)	0.0001r	0.00005	40
Diazinon (Knox-Out)	(6)	0.002	—	300
Chlorpropham	(5)	—	0.200p	3,800
Acephate (Orthene) C	(4)	0.030	0.004	866
Dicloran (Botran)	(4)	0.030p	—	4,000
Carbaryl (Sevin)	(4)	0.010	—	307
Hexachlorobenzene$^+$	(4)	—	0.0008	10,000
Heptachlor* (Velsicol)	(3)	0.0005r	0.0005p	40
Propargite (omite)	(3)	0.150	0.020	2,200
Lindane C-SRC	(2)	0.008	0.0003	76
Thiabendazole (Arbotec)	(2)	0.300	—	3,100
Dimethoate (Cygon) C-SRC	(2)	0.010	0.0002	250
Permethrin (Ambush)	(2)	0.050	0.050	>4,000
Ethion	(2)	0.002p	0.0005p	27
Dicofol (Kelthane) C-SR	(2)	0.25	—	575
Toxaphene (Strobane) C-CR	(2)	—	—	40
Primiphos-methyl	(2)	0.010	0.010	2,050
Quintozene	(2)	0.007r	0.003p	
DCPA (Dacthal)	(1)	—	0.500	>3,000
Parathion (Niram) PC	(1)	0.005	—	3
Chlordane*	(1)	0.0005r	0.00006	283
BHC, alpha and beta	(1)	—	—	125
Omethoate (Folimat)	(1)	0.0003	—	50
Vinclozolin (Ronilan)	(1)	0.070	0.025	10,000

Source: U.S. FDA (1990).

Abbreviations: (OTS, 1989): PC, identified to be potential carcinogens by EPA but not reviewed or canceled; C, identified to be animal carcinogens by EPA but not reviewed or canceled; C-CR, EPA canceled or restricted carcinogenic pesticides; C-SR, carcinogenic pesticides in Special Review by EPA; C-SRC, carcinogenic pesticides for which EPA Special Review is complete but chemical not suspended or canceled.

[a] Numbers in parentheses represent percentage occurrence. The ADIs and RfDs are expressed as milligram of pesticide ingested per kilogram body weight per day. Those which include other (related) chemicals are indicated with the letter "r" and those which include only the parent chemical are indicated as "p." Asterisk (*) indicates sufficient evidence for carcinogenicity on experimental animals as evaluated by the International Agency for Research in Cancer (1990).

[b] Ware (1989).

Most of the analyses were done using composite samples, and attention should be given to individual samples when analyzing for chemicals with high acute toxicity.

VII. Total Diet Study

The Total Diet Study, also known as the Market Basket Survey, is conducted by the FDA to provide an estimate of dietary pesticide intake in the U.S. population. Information is obtained for eight age/sex groups. Foods are collected four times per year, once from each of four geographical regions of the United States. Each market basket consists of 234 identical foods purchased from local supermarkets in 3 cities in each geographical area. The foods are chosen to represent the diet of the U.S. population, are prepared table-ready, and the residues determined. Food consumption data are obtained from two nationwide surveys: USDA's 1977–78 Nationwide Food Consumption Survey and the 1976–1980 National Center for Health Statistics' Second National Health and Nutrition Examination Survey. These surveys included approximately 50,000 participants and over 5000 foods. The 30 most frequently found pesticides are listed in Table II. The analyses were based on 936 items except for carbaryl (288), propagite (167), and thiabendazole (167).

VIII. Priority Pesticides

Pesticides are registered with the EPA as unclassified or restricted-use pesticides. The former may be purchased and applied by any persons, but the latter by certified applicators only. The criteria used to classify restricted-use pesticides usually include, but are not limited to, factors relating to human health hazards such as oral, dermal, or inhalation toxicity; oncogenicity; reproductive and developmental toxicity; and accident history. Some cancelations and reduced uses have been mandated by the EPA, and voluntary cancelations or withdrawals from the market have also been instituted by manufacturers. Table III shows the restricted-use pesticides and pesticide cancelations among the 30 pesticides most commonly found in the 1990 FDA Total Diet Study. The finding of food residues of pesticides for which registration has been canceled may indicate the use of remaining pesticides already sold,

Table III Regulatory Status of the 30 Most Frequently Found Pesticides in FDA's Total Diet Study

Chemicals[a]	Regulatory status[b]
Acephate	EPA carcinogen but not reviewed or canceled
BHC	Voluntarily canceled 1978. Criterion: oncogenicity
Carbaryl	—
Chlordane	All uses canceled 1988. Criteria: carcinogenic, hazard to wildlife
Chlorpropham	—
Chlorpyrifos	—
Chlorpyrifos-methyl	—
DCPA	—
DDT	All uses canceled but certain public health applications 1992. Criteria: carcinogenic, bioaccumulation, hazard to wildlife
Diazinon	—
Dicloran	—
Dicofol	Carcinogenic pesticide in special review by EPA
Dieldrin	All uses canceled 1987. Criteria: oncogenic, hazard to wildlife
Dimethoate	Oncogenic pesticide for which special review by EPA is completed but chemical not suspended or canceled
Endosulfan	—
Ethion	—
Heptaclor	All uses canceled 1988. Criteria: oncogenic, hazard to wildlife
Hexachlorobenzene	—
Lindane	All uses restricted. Criterion: carcinogenicity
Malathion	—
Methamidophos	>2.5% restricted uses. Criterion: acute dermal toxicity
Omethoate	
Parathion	Potentially carcinogenic but not reviewed or canceled
Permethrin	All restricted uses. Criterion: aquatic organism effect
Primiphos-methyl	—
Propagite	—
Thiabendazole	—
Toxaphene	Most uses canceled 1982. All uses restricted. Criteria: oncogenic, hazard to wildlife, reduction in nontarget species
Quintozene	—
Vinclozolin	—

[a] U.S. FDA (1990).

[b] Ware (1989). Regulatory status includes actions taken by EPA and voluntary cancelations by manufacturers.

illegal use, or the environmental pesistence of the chemicals.

In addition, the U.S. EPA has identified a list of potentially oncogenic pesticides based on carcinogenic data obtained from experimental animals. These are shown in Table IV. Of 289 food-use pesticides, 53, or approximately 18%, have been determined to be at least potentially carcinogenic. The pesticide tolerances established long ago have not used carcinogenicity as an endpoint to derive the legally permissible levels. The EPA now requires carcinogenicity testing in two species for all food-use pesticides. The potential carcinogenic risks from pesticides in or on food have received increasing attention and was the subject of a study by the National Research Council (NRC), the results of which were published in 1987. The estimated theoretical excess cancer risks from ingestion of some of these pesticides in food are described below. Traditionally, pesticide tolerances in food have not been established based on carcinogenicity. Recent advances in toxicology and risk assessment have begun to bring to a better focus the potential carcinogenic risks from dietary ingestion of pesticides.

A list of food-use pesticides currently classified by the U.S. EPA as potential carcinogens is presented in Table V.

In addition, attention should also be given to the finding of residues of pesticides with high acute toxicity (i.e., low LD_{50} value) or that are potent cholinesterase inhibitors or teratogens. Generally, analyses for residues are performed based on composite

Table IV Pesticides Identified by U.S. EPA as Potentially Oncogenic Based on Experimental Animal Data

Pesticide	Major uses	Pesticide	Major uses
Acephate (Orthene) (4)[a]	citrus	Maleic hydrazide*	onions, potatoes
Acifluorfen (Blazer)*	soybeans	Mancozeb (Dithane M-45)	fruits, small grains
Alachlor (Lasso)	corn, soybeans	Maneb††	fruits, small grains
Amitraz (Baam)	cattle	Methanearsonic acid*	cotton
Arsenic acid*	cotton	Methomyl (Lannate) AO	fruits, small grains
Asulam	sugar cane	Metiram	fruits, small grains
Azinphos-methyl (Guthion)	peaches, pome fruits	Metolachlor (Dual)	corn, soybeans
Benomyl (Benlate)	citrus, rice	Oryzalin (Surflan)*	soybeans, vineyards
Calcium arsenate*	stone fruits	Oxadiazon (Ronstar)	rice
Captafol (Difolatan)†	soybeans, stonefruit	Paraquat (Gramoxone)*	rice, soybeans
Captan††	almonds, apples	Parathion†† (1) AO	citrus, cotton
Chlordimeform (Galecron)	cotton	PCNB (Pentacloronitrobenzene, Quintozene)*	cotton, peanuts, vegetables
Chlorobenzilate††	citrus	Permethrin (Ambush)† (2)	vegetables
Chlorothalonil (Bravo)††	fruits, peanuts, vegetables	O-phenylphenol††	citrus, orchard
Copper arsenate*	vegetables	Pronamide (Kerb)	lettuce
Cypermethrin	cotton	Sodium arsenate*	pears
Cyromazine (Larvadex)*	poultry (in feed)	Sodium arsenite*	grapes
Daminozide (Alar)	apples	Tetrachlorvinphos	cotton
Diallate††	sugar beets	Thiodicarb (Larvin) AO	cotton, soybeans
Dichlofop methyl (Hoelon)	soybeans	Terbutryn*	barley, wheat
Dicofol (Kelthane)††	citrus, cotton	Thiophanate-methyl	fruits, nuts, vegetables
Ethafluralin (Sonalan)*	soybeans	Toxaphene (2) AO	cattle
Ethylene oxide*	spices, walnuts	Tributryh*	barley, wheat
Folpet	cherries, fruits, vegetables	Trifluralin (Treflan)††	soybeans
Fosetyl AL (Aliette)*	pineapples	Zineb	fruits, small grains, vegetables
Glyphosate (Roundup)*	hay, orchard		
Lead arsenate*	apples, orchard		
Lindane (2) AO[b]	avocados		
Linuron (Lorox)	soybeans		

Source: NRC (1987).

* Included in U.S. FDA (1990) food residue monitoring.

†,†† Sufficient and limited or insufficient evidence, respectively, for carcinogenicity in experimental animals as evaluated by International Agency for Research in Cancer (1991).

[a] Number within parenthesis shows percentage occurrence of the pesticide found in FDA's (1990) Total Diet Study. These pesticides are among the 30 most frequently found in the study.

[b] Chemicals identified by AO are ones with acute oral LD_{50} values in rats of >100 mg/kg (Ware, 1989). These are considered to be very toxic by the oral route.

samples, and pesticide intakes are calculated based on averaged daily exposures. This approach is likely to underestimate the dietary risks from acutely toxic pesticides that may be ingested at relatively high doses from foods with a high consumption rate based on single meals. Under this situation individual samples should be analyzed. The LD_{50} values for some of the pesticides found most frequently in food are shown in Table II.

IX. Imported Foods

Pesticide residues in or on imported foods is a subject of concern in recent years. A 1990 report by the U.S. General Accounting Office prepared in response to a congressional request provided information on foreign government and private industry efforts to assure that imported produce meets U.S. safety and quality standards and what federal agen-

Table V Food-Use Pesticides Currently Classified by U.S. EPA as Potential Carcinogens

Pesticide active ingredient classification	EPA cancer classification	Pesticide active ingredient classification	EPA cancer classification
Acephate	C	Metiram (EBDC)	B2
Acifluorfen (sodium acifluorfen)	B2	Metolachlor	C
Alachlor	B2	Norflurazon	C
Aliette (fosetyl-al)	C	Oryzalin	C
Amdro	C	Oxadiazon	C
Amitraz	C	Oxadixyl	C
Apollo (clofentezine)	C	Oxyflurofen	C
Arsenic acid	A	*o*-Phenylphenol	B2
Asulam	C	Para-dichlorobenzene	C
Atrazine	C	Parathion (ethyl)	C
Benomyl	C	PCNB (Quintozene)	B2
Bifenthrin	C	Permethrin	C
Bromoxynil	C	Phosmet (Imidan)	C
Captan	B2	Phosphamidon	C
Chlorbenzilate	B	Procymidone	B2
Chlorothalonil	B2	Pronamide (Kerb)	C
Cyanazine	C	Propargite	B2
Cypermethrin	C	Propiconazole (Tilt)	C
Dichlobenil	C	Propoxur (Baygon)	B2
Dichlorvos (DDVP)	C	Quinclorac (Facet)	C
Diclofop-methyl (Hoelon)	C	Savey (hexathiazox)	C
Dicofol	C	Simazine	C
Dimethipin (Harvade)	C	Terbutryn	C
Dimethoate	C	Terrazole (etridiazole)	B2
Ethylene oxide	B1	Tetrachlorvinphos (Gardona)	C
Express (tribenuronmethyl)	C	Thiodicarb	C
Folpet	B2	Thiophanate methyl	C
Fomesafen	C	Toxaphene	B2
Hexazinone	C	TPTH (triphenyltin hydroxide)	B2
Lactofen	B2	Tribufos (DEF)	C
Lindane	B2/C	Tridimeform (Bayleton)	C
Linuron	C	Tridimenol (Baytan)	C
Mancozeb (EBDC)	B2	Tridiphane (Tandem)	C
Maneb (EBDC)	B2	Trifluralin (Treflan)	C
Methidathion	C	Zineb (EBDC)	B2
Methomyl	C		

Source: Pesticide Action Network North America Updates Service (PANUPS), March 1993.

Abbreviations: A, human carcinogen (sufficient evidence of cancer causality from human epidemiological studies; B1, limited evidence of carcinogenicity from human epidemiological studies; B2, sufficient evidence of carcinogenicity from animal studies; C, possible human carcinogen (limited evidence of carcinogenicity in animals in the absence of human data).

cies are doing to assist foreign countries in meeting U.S. safety requirements. The five countries included were Chile, Costa Rica, the Dominican republic, Guatemala, and Mexico. The governments of these countries have not designed their food safety and quality systems, specifically regarding pesticides, to meet other countries' import requirements but primarily to address domestic needs and issues. Registration of pesticides is based on considerations of climate, crops, and pest problems and may not be health related. Also the presence and composition of pesticide residues on produce imported from these countries may be affected by legal availability and use of specific pesticides. These, in turn, are affected by government practices relating to registering and reregistering pesticides; considering a pesticide's U.S. status during registration; canceling, restricting, or not registering pesticides that the United States has canceled or suspended; providing information about U.S. standards to export growers; and registering pesticides that do not have U.S. tolerances.

Some growers and exporters of these countries have tried to consider U.S. pesticide residue requirements and established networks to obtain information and technical assistance through cooperative efforts of governments and exporter/grower organizations. International organizations such as the United Nations Food and Agriculture Organization, World Health Organization, and Pan American Health Organization also provide assistance related to pesticide use to developing countries.

There are several reasons for the findings of violative pesticide residues in imported foods or no-tolerance pesticide residue findings. For example, the pesticide has a U.S. tolerance for one or more fruits, vegetables, and/or other commodities, but not for the commodity cited for violation. The pesticide is not registered for any use in the United States; or the pesticide is registered, but no tolerances have been established for a food use application. The pesticide is canceled or severely restricted in the United States, and previous food use tolerances have been revoked.

X. Circle of Poison

The prevailing idea of a circle of poison is premised on the belief that pesticides which are no longer used in the United States because of health-related reasons may be exported to and used in other countries which, in turn, export their products containing residues of the pesticides to the United States. This may result because foreign countries may not know of the status of the pesticides in the United States, may not have tolerances for the pesticides, or may not have a monitoring program or the analytical capability to adequately and rapidly analyze for the residues. The residues may not be detected by the U.S. monitoring programs because their use or presence is not known or expected.

For the developing countries, they may be put at risk for toxic exposure (to the public or to agricultural workers), environmental contamination, and potentially devastating crop failure. These countries generally have inadequate systems for pesticide regulation and no administrative structure to establish or enforce regulatory standards. In addressing the associate potential concerns, there needs to be tighter control of pesticides shipped to foreign countries and of foods imported along with provisions for waivers for the control of communicable disease in circumstances of imminent famine, with allowance for exemptions for experimental research purposes.

In evaluating the existing food system, a 1991 report indicated that over the years, very few food imports have been found to contain residues of many of the EPA-canceled pesticides and that violative residues can and do occur at home as well as abroad. Therefore, there is no evidence that a real concern exists at this time.

XI. Hazards from Contamination Incidents

The recent incidents alerting the hazards to humans from pesticides in food have been mostly accidental or due to misuse. These include contamination of seed grain treated with organo-mercurials in Iraq in 1973 due to misuse, and contamination of watermelons in several western states in the United States in 1985 due to illegal use of aldicarb. The former resulted in neurologic symptoms and developmental effects and the latter led to typical cholinergic symptoms at very low estimated exposures in humans.

In 1984, ethylene dibromide (EDB) was found in cereal grains and bakery products. The chemical was thought originally not to leave any residue after its use because of its volatility and, therefore, no tolerances were established initially. The subsequent detection of EDB residues generated a concern because of its carcinogenicity in experimental animals and potential carcinogenicity in humans, especially in children. Three different permissible levels (150, 90, and 30 ppb) were first set by the EPA for grains, bakery products, and baby foods, respectively. These were adopted by most states. California also adopted these, except that the level for baby foods was set at zero. EDB is no longer used on foods. This incident indicates the importance of careful consideration of all relevant factors in establishing pesticide tolerances and represents a situation involving theoretical cancer risks rather than real health risks.

The alar (daminozide) in apples episode (1985–1989) was a situation involving estimated theoretical cancer risks (no human poisoning was known) and was perceived as a long-term potential concern rather than as a real immediate problem. However, it did bring to nationwide attention the need to evaluate health risks from pesticides in food, especially for children who may have a dietary pattern quite different from adults. It has also helped us to focus on two very important issues: the toxicity of the breakdown product and the need to have adequate toxicological data. In the case of daminozide, the carcinogenicity concern was more associated with 1,1-dimethylhydrazine (UDMH, or unsymmetrical dimethylhydrazine), a breakdown product of daminozide, rather than the parent chemical itself. UDMH is formed and concentrates when apples are processed for making apple juice. Available data pointed to some evidence of carcinogenicity of UDMH in experimental animals, but the more definitive study required for evaluation was not yet completed at the time of the public health investigation. Initially the EPA lowered the tolerance level and considered a phased-out period. The chemical was also put under restricted use. It was then followed by voluntary cancelation.

Other than the above, minor contamination incidents also have occurred and these were managed by the respective authorities. One chemical that deserves more attention in order to prevent future food contamination is the cholinesterase inhibitor, aldicarb. The chemical's high acute toxicity (oral LD_{50} 1 mg/kg), its recent findings in potatoes and bananas, and the potential for cummulative effects from multiple cholinesterase inhibitors would make this a priority chemical for monitoring, evaluation, and regulation.

Pesticides that are environmentally persistent can pose a potential health hazard as exemplified by the current finding of DDT in fish in southern California. This resulted in the issuance in 1991 of health advisories for persons taking fish, especially white croaker, from the area. The source of DDT was waste discharge from a manufacturing company during the period between 1940s and 1970s.

XII. Acceptable Risk

Under the 1958 Delaney Amendment to the Food, Drug, and Cosmetic Act (FDCA), no food additive should be allowed if it is a known carcinogen. While the FDA is authorized to regulate the addition of substances to food under Section 409 of FDCA, which governs residues which concentrate in processed food above the authorized levels in or on parent raw commodities, the EPA implements the sections concerning pesticides. Pesticide tolerances are set by the EPA and adopted and enforced by the FDA. Section 408 of the FDCA recognizes that pesticides confer benefits and risks and that both should be taken into account when setting commodity tolerances. It is this risk–benefit analysis that creates much of the controversy surrounding the regulation and use of pesticides. Benefits and risks are not perceived the same way by everyone or among scientific bodies or regulatory agencies. While a "zero risk" concept is advocated in some cases, the concept of "negligible risk" or "no significant risk," even if involuntary, is widely accepted as the basis for carcinogenic assessment. Some recent mandates relating to pesticides at the state level have incorporated this concept. In general, a negligible risk level for a cancer-causing substance at its regulatory limit is one that could potentially cause one additional theoretical cancer case in a population of a million people in a lifetime. This is commonly used in risk–benefit analysis, but the EPA is not required to limit risk to a certain level. Under Section 409, residues present in processed food at levels no higher than sanctioned on the raw agricultural commodity are exempt from food additive regulation, but remain subject to the risk–benefit standard of Section 408.

XIII. Estimated Risks from Pesticide Residues

Carcinogenic risks from pesticides in food have been estimated by the National Research Council. Table VI presents selected pesticides estimated to have

Table VI Pesticides and Foods Contributing to the Majority of the Dietary Risks from Potentially Carcinogenic Pesticides Estimated by the National Research Council[a]

Pesticides[b]	Foods[c]
Linuron	Tomatoes
Zineb	Beef
Captafol	Potatoes
Captan	Oranges
Maneb	Lettuce
Permethrin	Apples
Mancozeb	Peaches
Folpet	Pork
Chlodimeform	Wheat
Chlorothalonil	Soybean

[a] Carcinogenic risks are expressed as the probability of cancer occurrence. For example, 2.37×10^{-4} is equal to a probability of 2.37 in a populaton of 10,000 (or two in ten thousand) for a lifetime. The risks represent total upperbound estimates assuming residues at the tolerance levels. Actual risks are likely to be lower in most cases.

[b] Estimated risks 2.37×10^{-4} to 1.52×10^{-3}, listed in a decreasing risk level.

[c] Estimated risks 1.24×10^{-4} to 8.75×10^{-4}, listed in a decreasing risk level.

Table VII Pesticides in Fish/Shellfish and Other Seafood as Found in the U.S. Food and Drug Administration's Total Diet Study 1990.

Fish/shellfish	
Surveillance samples	
Total surveillance samples	568
Samples with no residues found (%)	32
Samples over tolerance (%)	0
Samples detected but with no tolerance (%)	1
Aquaculture survey	
Total samples	172
Catfish samples	103
Crawfish samples	25
Trout samples	21
Shrimp samples	10
Other samples	13
Samples with residues (below or no FDA action levels)	
DDT, total	125
BHC	65
Dieldrin	41
Chlordane, total	9

the highest risks among all the pesticides assessed, along with the food items contributing to the highest risks. It is important to note that these risks are theoretical based on the assumption that the food items contain the highest level of pesticide residues permitted for each specific pesticide and for each food item, and that residues are present in all food crops with a tolerance. In actual monitoring and field studies, the residues are often found to be lower than the maximum levels permitted, and not all pesticides would be used on or in all crops for which a use has been registered. Therefore, the actual risks may reasonably be expected to be much lower than those estimated in most cases.

XIV. Safety of the U.S. Food Supply

The major activity which monitors the safety of the U.S. food supply is the Total Diet Study described above. Some of the major findings are listed in Tables II, VII and VIII. Tolerances have been established for various food crops, but the number of tolerances or action levels established for seafood is very limited. Regulatory limits need to be set for seafood in order to better evaluate and ensure seafood safety.

Based on the residue findings, the amounts of pesticides ingested from the diet composites collected from the market baskets are calculated for different age groups. The findings of the 1990 Total Diet Study show that actual dietary intakes of pesticides are generally well below the permissible intake values established by FAO/WHO and by the U.S. EPA. Some of these values are shown in Table II. These results and conclusions are similar to those obtained in earlier years and demonstrate the continuing safety of the food supply from pesticide residues. The compounds considered by FAO/WHO are eval-

Table VIII Pesticide Surveillance by the U.S. Food and Drug Administration 1990

	Domestic	Import
Total samples	8,879	10,267
No residues found	5,352	6,620
Number in violation	95	445
Violation (%)	1.1	4.3

uated and updated as new information becomes available. The most recent monographs completed in 1991 contain the first full evaluation on cyromazine, hexaconazole, profenofos, and terbufos.

To more adequately ensure food safety in the future, it will be beneficial to review the food safety system and identify ways to improve if deemed appropriate. Evaluation of the existing regulations and registration process, methodology for tolerance setting, toxicology and food consumption data bases, adequacy of the state and federal monitoring program and resources, methodology for estimating dietary risks, multiple chemical exposures, and sensitive subpopulations would be very useful. As advances are made in analytical and risk assessment methodologies, and as food consumption patterns change with time and other factors, a periodic evaluation of the food safety system would be a major contribution to ensuring the safety of our food supply.

XV. International Food Safety

In addition to regulation by local governments in different countries, there have been efforts to coordinate food safety related activities at the international level. The Codex Committee on Pesticide Residues (CCPR) is an international government body comprising over 100 national governments established to achieve harmonization (standardization) of national tolerances for pesticide residues in food for the purpose of facilitating international food trade and protecting the health of consumers. The committee is part of the Codex Alimentarius Commission which is responsible for implementing a food standards program jointly sponsored by the FAO of the United Nations and the WHO.

The United States is one of the countries that has not accepted or adopted any of the Codex maximum residue limits (MRLs) that are different from those tolerances established by the U.S. EPA. Due to a concern of the Senate Committee on Agriculture, Nutrition, and Forestry about the possible implications of increased emphasis on international standards of food safety, the U.S. General Accounting Office recently compared the United States and the Codex standards. The results show important differences among the standards. Almost two-thirds of them cannot be compared because of the absence of U.S. tolerances for corresponding Codex standards or differences in the way pesticide residues are defined. Less than one-sixth of the standards are numerically the same. Using diquat and malathion as examples, the estimated theoretical dietary intake of the U.S. consumers would be higher when Codex MRLS are used for the former, and the intake decreased for the latter chemical. Both the United States and Codex systems use appropriate scientific approaches for establishing standards and protecting health. The differences reflect several technical factors pertaining to pesticide and agricultural practices and to the procedures for evaluating and establishing agricultural practices and to the procedures for evaluating and establishing standards. Harmonization of standards would have several benefits but it has not been successful as a number of individual nations have not accepted Codex MRLs.

XVI. Summary

The U.S. Environmental Protection Agency, U.S. Department of Agriculture, and U.S. Food and Drug Administration are the primary agencies responsible for regulating pesticides, enforcing tolerances, and monitoring for pesticide residues in or on food. The results of the FDA's regulatory pesticide residue monitoring of domestic and imported food have been reassuring to the public. These are further supported by the regulatory and enforcement activities at the state and county levels.

The few contamination incidents that have occurred relating to pesticides in food have been primarily the result of misuse or illegal use of the pesticides. On the other hand, in order to avoid hazards from future food contamination by pesticides, the filling of toxicological data gaps, development and validation of analytical methods, update of food consumption data bases, development and reevaluation of pesticide tolerances using updated toxicological data and risk assessment methodology, study of pesticide use reduction or residue reduction from the food form as sampled and analyzed to the form as consumed, continued monitoring for pesticide residues, and evaluation of residue monitoring results in terms of food safety and public health protection are needed. The associated activities should be expanded as necessary.

Furthermore, consideration should be given to the evaluation and regulation of pesticides that are acutely toxic and those for which there is evidence of developmental toxicity and carcinogenicity. The issue of multiple pesticide exposure from dietary intake and dietary risks for children and other sensitive subgroups should be addressed. Food consumption data should be examined and surveys designed and conducted in order to help better define pesticide exposures in different age and ethnic groups and for different states.

Related Articles: ENVIRONMENTAL CANCER RISKS; HERBICIDES.

Bibliography

CFR (1990). Code of Federal Regulations, Title 40. U.S. Government Printing Office, Washington, D.C., Parts 180, 185, and 186.

Fan, A. M., and Jackson, R. J. (1989). Pesticides and food safety. *J. Regulatory Toxicol. Pharmacol.* **9,** 158–174.

Gunderson, E. L. (1988). FDA total diet study, April 1982–April 1984: Dietary intakes of pesticides, selected elements, and other chemicals. *J. Assoc. Offic. Anal. Chem.* **71,** 1200–1209.

IARC (1991). IARC Monographs on the Evaluation of Carcinogenic Risks to Humans. Vol. 53. "Occupational Exposures in Insecticide Application, and Some Pesticides." International Agency for Research in Cancer, Lyon, France.

Minyard, J. P., and Roberts, W. E. (1991). Foodcontam—State findings on pesticide residues in foods–FY88 and FY89. *J. Assoc. Offic. Anal. Chem.* **74,** 438–452.

NRC (1987). "Regulating Pesticides in Food: The Delaney Paradox." National Research Council, Washington, D.C.

NRC (1993). "Pesticides in the Diet of Infants and Children." National Research Council, Washington, D.C.

OTS (1989). "Identifying and Regulating Carcinogens." Office of Technology Assessment, Congress of the United States. Marcel Dekker, New York.

U.S. FDA (1990). "Residues in Foods 1990." U.S. Food and Drug Administration, Washington, D.C.

U.S. GAO (1990). "Food Safety and Quality. Five Countries' Efforts to meet U.S. Requirements on Imported Produce." U.S. General Accounting Office, Washington, D.C., 1990.

U.S. GAO (1991). "International Food Safety. Comparison of U.S. and Codex Pesticide Standards." U.S. General Accounting Office, Washington, D.C., 1991.

Ware, G. W. (1989). "The Pesticide Book." 3rd Ed. Thomson Publication, Fresno, CA.

Wessel, J. R., and Yess, N. J. (1991). Pesticide residues in foods imported into the United States. *Rev. Environ. Contam. Toxicol.* **120,** 84–104.

WHO (1991). Summary of Toxicological Evaluations Performed by the Joint FAO/WHO Meeting on Pesticide Residues (JMPR) through 1990 (1991) International Program on Chemical Safety. World Health Organization, Geneva, Switzerland.

Pharmacokinetics, Individual Differences

Kannan Krishnan
Université de Montréal

Melvin E. Andersen
Chemical Industry Institute of Toxicology

Glossary

Alleles One of several forms of a gene occupying a particular locus on a chromosome.

Cytochromes *P*-450 Family of heme-containing enzymes that catalyze the biotransformation of a variety of chemicals. The hemoprotein pigment in the reduced form reacts with carbon monoxide to yield a pronounced absorption peak at 450 nm in the soret region of the visible spectrum.

First-pass effect Removal from the blood of a chemical before it reaches systemic circulation.

Phenotype Observable expression of the hereditary constitution of an organism.

Polymorphism Occurrence of a genetic response in more than one distinct form, with the second most prevalent form in at least one percent of the relevant population.

Volume of distribution Proportionality constant relating the total amount of a chemical in the body to its concentration in blood.

***PHARMACOKINETICS** is the study of the rate and extent of absorption, distribution, metabolism and excretion of chemicals in biota.*

I. Introduction

Absorption is the process by which a chemical moves across body membranes and enters the bloodstream of an organism. Chemicals may be absorbed through the skin (dermal absorption), lungs (pulmonary absorption) or gastrointestinal tract (gastrointestinal absorption). Interindividual differences in dermal absorption can result from differences in skin thickness, presence of skin appendages, or differential abrasion of the epidermal layer. Individual differences in pulmonary absorption result from variations in alveolar ventilation rate and cardiac output. If a chemical is highly soluble in blood, then increased rate of respiration will increase its absorption considerably; on the other hand, if the chemical is not readily soluble in blood, increased cardiac output rather than the rate of respiration will increase its absorption. The rate and extent of gastrointestinal absorption of orally administered xenobiotics are influenced by the presence of food and the processes of intestinal peristalsis and gastric emptying.

Distribution is the process by which the unbound form of a chemical passes from the bloodstream into the target organ(s) to exert its toxic effects and into other tissues to be metabolized and/or stored. The blood level of a chemical depends mainly on its volume of distribution. Lipophilic chemicals, which are distributed well into all tissues (i.e., have a large volume of distribution), tend to have low plasma concentrations while the converse is true for hydrophilic substances which are ionized at plasma pH. The amount of chemical in free, unbound form is important since it is responsible for the exertion of toxic effects. This is primarily determined by the amount of the chemical bound to plasma proteins, particularly albumin, α_1-glycoprotein and lipoprotein, which bind to acidic, basic, and neutral substances respectively. Individual differences in the levels of these plasma proteins are associated with pathological conditions, pregnancy, smoking, and aging.

Metabolism is the process of biotransformation by which a lipophilic parent chemical is converted to hydrophilic metabolite(s) that may (activation) or may not (detoxication) be active until conjugated. The initial metabolic reactions (phase I) mainly involve oxidation, reduction, or hydrolysis. These are followed by the conjugation (phase II) reactions of the parent chemical and/or its metabolite(s) with an endogenous substrate such as a carbohydrate, an amino acid derivative, inorganic sulfate, or glutathione. The phase I oxidative reactions are catalyzed mainly by cytochromes *P*-450 (EC 1.14.14.1) present in the microsomal fraction of the liver and other tissues. The term "*P*-450" is the generic name for a family of hemoproteins, each of which is characterized by distinct physical, catalytic, and immunoreactive properties. So far, approximately 30 different *P*-450s have been isolated and grouped into 13 families. Of these, the isozymes belonging to families 1, 2, 3, and 4 are known to mediate the biotransformation of xenobiotics, while other isoforms of *P*-450 are responsible for metabolism of endogenous compounds such as steroids. The amount and activity of these enzymes are influenced by the genetic traits, environmental conditions, dietary habits, and the physiological and pathological state of an individual.

Excretion is the process by which the unbound parent chemical and its polar metabolite(s) are removed from the body. Chemicals may be excreted through urine, bile, saliva, milk, tears, sweat, genital secretions, turnover of skin/hair, and expired air. The extent of urinary excretion of a chemical among individuals may differ according to their age, renal disease state, and the volume and pH of urine. Interindividual differences in biliary excretion may result from the impairment caused by prior exposure to certain hepatotoxicants, long-term administration of estrogens, and the state of pregnancy. The rate and extent of pulmonary excretion of xenobiotics are influenced mainly by alveolar ventilation rate and the relative solubility of the chemical.

Thus, a number of factors can account for the interindividual variations in the rate and/or extent of absorption, distribution, metabolism, and excretion of xenobiotics, potentially altering the target or effective dose. The tissue dosimetry and pharmacokinetics of parent chemicals and their metabolites under continuously or intermittently varying, complex exposure scenarios (e.g., physical activity, aging, pregnancy, disease states, coexposure to other chemicals) can be described with physiological modeling.

II. Physiological Modeling

Physiological modeling of pharmacokinetics involves computer simulation of the uptake and disposition of chemicals based on the quantitative interrelationships among tissue solubility characteristics, metabolic rates, and the physiology of the host species. A physiological pharmacokinetic model consists of a series of tissue compartments, each receiving the chemical via an arterial blood supply and losing the free chemical via the venous blood leaving the tissue (Fig. 1). These biologically realistic compartments are arranged in an anatomically accurate manner and defined with appropriate physiological characteristics, such that the uptake and disposition can be described mathematically by accounting for the critical biological determinants of the respective processes.

In case of inhalation exposure, for example, the uptake of volatile chemicals is described as follows:

$$C_{art} = (Q_{alv}C_{inh} + Q_t C_{ven})/(Q_t + (Q_{alv}/P_b)),$$

where C_{art} = concentration in arterial blood (mg/L)
C_{inh} = concentration in inhaled air (mg/L)
Q_t = cardiac output (L/hr)
C_{ven} = concentration in mixed venous blood (mg/L)
Q_{alv} = alveolar ventilation rate (L/hr)
P_b = blood/air partition coefficient

Intravenous injection can be similarly described by including a term for the rate of infusion into the mixed venous blood. Oral gavage of a chemical dissolved in a carrier solvent can be modeled by introducing a first-order or a zero-order uptake rate constant, with the chemical assumed to appear in the liver after gastric absorption. Dermal absorption

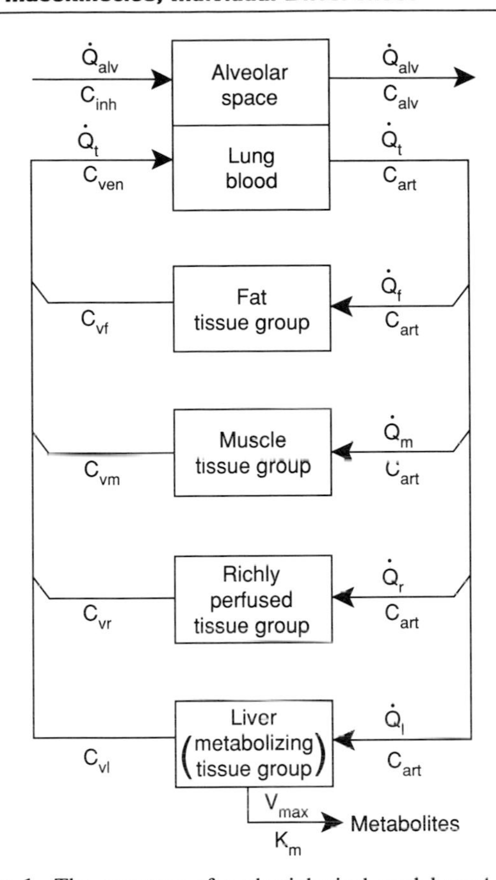

Figure 1 The structure of a physiological model used to describe the pharmacokinetics of chemicals. Q terms are air and blood flow rates; C terms are concentrations. These are indexed to individual tissue compartments—fat (f), muscle (m), richly perfused tissues (r), and liver (l). Effluent venous concentrations have a double lettered subscript. Q_{alv} and Q_t refer to alveolar ventilation rate and cardiac output. The subscripts "inh, alv, art, and ven" signify inhaled air, exhaled air, arterial blood and venous blood. Kinetic constants for liver metabolism are V_{max} (maximum rate of metabolism) and K_m (binding affinity of the substrate with metabolizing enzyme). (From: Ramsey, J. C. and Andersen, M. E., 1984. A physiologically based description of the inhalation pharmacokinetics of styrene in rats and humans. *Toxicol. Appl. Pharmacol.* **73**, 159. Copyright 1984, Academic Press. Reproduced by permission.)

can be modeled by including a diffusion-limited compartment to represent skin as a portal of entry.

The rate of change in the amount of the chemical in each of the tissue compartments is described with a series of mass balance differential equations of the following form:

$$dA_i/dt = Q_i(C_{art}-C_{vi}) = Q_i(C_{art}-(C_i/P_i)),$$

where dA_i/dt = rate of change in the amount of chemical in tissue i (mg/hr)
Q_i = blood flow rate to i (L)
C_{art} = concentration in arterial blood entering i (mg/L)
C_{vi} = concentration in venous blood leaving i (mg/L)
C_i = concentration in tissue i (mg/L)
P_i = tissue : blood partition coefficient for i

Metabolism in individual tissues or tissue groups can be described by introducing a term to account for the amount lost by metabolism, which might be a first- or second-order type (e.g., glutathione conjugation) or a saturable Michaelis–Menten type (e.g., cytochrome P-450 mediated oxidation) as follows:

$$dA_i/dt = Q_i(C_{art}-C_{vi})-dA_{met}/dt$$
$$dA_{met}/dt = [V_{max}C_{vi}/(K_m + C_{vi})] + K_fC_{vi}V_i$$

where dA_{met}/dt = rate of the amount metabolized (mg/hr)
V_{max} = maximum enzymatic reaction rate (mg/hr)
K_m = Michaelis constant for enzymatic reaction (mg/L)
K_f = first-order constant (hr^{-1})
V_i = volume of the tissue (L)

It is clear from these mass balance equations that three types of information are required for a physiological model that describes the pharmacokinetic behavior of a chemical:

1. physiological parameters (e.g., breathing rates, tissue volumes, cardiac output, and blood flow rates),
2. partition coefficient (blood : air, tissue : blood), and
3. biochemical constants (i.e., rate constants for metabolism and macromolecular binding).

Whereas blood : air/tissue : blood partitioning is largely a chemical-specific behavior, the physiological parameters are host-specific. The rates of metabolism and macromolecular binding, on the other hand, are both host- and chemical-specific. Therefore, variation in pharmacokinetics of chemicals among individuals of a given species (e.g., humans) arises primarily due to quantitative and/or qualitative differences in the biochemical and physiological parameters among them. Since the compartments in these models correspond to biologically realistic entities with actual physiological and biochemical characteristics, quantitative changes in pharmacokinetics of chemicals due to physiological and pathological alterations can be predicted by perturbation of the appropriate model parameter(s). The variability of these critical biological determinants of pharmacokinetics in a population depends on the complex interplay of such factors as an individual's personal habits (diet, physical activity/workload, smoking, alcohol intake, medication), body stature (weight, height), gender, physiological state (pregnancy, aging, disease), genetic traits, and environment (coexposure to other chemicals, temperature, altitude, photoperiod).

III. Personal Habits

A. Diet

The relative levels of macronutrients (protein, fat, fiber and carbohydrate), consumption of coffee, tea, cocoa, cola nut, sulfur-containing amino acids and charcoal-broiled meat, and eating times in relation to chemical exposure, all can contribute to individual differences in pharmacokinetics.

Eating time relative to chemical exposure is important since direct physicochemical interaction between a xenobiotic and food materials can occur, rendering both physiologically unavailable. The extent of absorption is also influenced by the type of food consumed. For example, low fat meals minimize the release of bile, digestion of lipid, and formation of micelles in the small intestine, thereby significantly reducing the absorption of lipophilic chemicals. Conversely, high-fat meals enhance the absorption of fat-soluble chemicals and drugs such as griseofulvin and dicoumarol. Consumption of high fat foods, also can increase the level of free fatty acids which may bind to plasma albumin, resulting in competitive and/or noncompetitive displacement of other chemicals. Similarly, fasting causes a degradation of triglycerides and an increase in free fatty acids with potential implications for drug–albumin binding.

High protein diets shorten the half-lives of phenacetin, antipyrine, and theophylline in humans. On an isocaloric diet, alteration of the carbohydrate : protein ratio has been found to modify hepatic drug metabolism capacity. Oversupply of carbohydrate, thiamin, or iron, and deficiency of vitamins A, B_2, C, or E decrease the activity of the hepatic mixed function oxidase system. An impairment of hepatic metabolic capacity resulting in increased half-lives of drugs has been observed in protein-malnourished populations. Low protein diet also produces a more alkaline urine whereas a high protein diet produces a more acidic urine. These changes in urinary pH can modify the tubular reabsorption and renal excretion of certain chemicals and their metabolites.

High fiber diets can cause loss of bile acids, a more rapid biliary excretion, and enterohepatic circulation of xenobiotics. Consumption of potentially acid-forming foods such as meat, eggs, cheese, peanut butter, bacon, bread, cranberries, plums, and prunes increases the renal clearance of alkaline (cationic) chemicals and drugs such as amphetamines. Alkali-forming foods such as milk, fruits (except cranberries, plums, and prunes), and vegetables (except corn and lentils) increase the urinary excretion of acidic substances such as phenobarbital.

Increased consumption of cabbage, broccoli, cauliflower, and brussels sprouts results in an enhancement of hepatic microsomal enzyme activities due to their content of indole 3-acetonitrile, indole 3-carbinol, and 3,3′-diindolylmethane. Polycyclic aromatic hydrocarbons and arylamines in charcoal-broiled beef, as well as 5,6,7,8,4′-pentamethoxyflavone (taneretin) and 4,5,7,8,3,4′-hexamethoxyflavone (nobiletin) present in citrus fruits can induce cytochrome *P*-4501A1, an enzyme involved in the bioactivation of several environmental carcinogens. A diet high in meat increases β-glucuronidase activity in intestinal bacteria, which has been implicated in the bioactivation of certain azo- and nitroaromatic substances. Clearance of antipyrine and acetaminophen is higher not only in alcoholics and smokers but also in people consuming large amounts of cola nut, coffee, or tea. In contrast, nonsmoking, nondrinking vegetarians clear several drugs and xenobiotics more slowly than nonvegetarians of the same sex and age group. Thus, individual dietary habits can cause significant differences in pharmacokinetics, contributing to differential susceptibility of individuals to the toxicity of xenobiotics.

B. Workload/Physical Activity

Exercise or active physical work increases alveolar ventilation, cardiac output, and tissue perfusion (particularly of exercising muscles) with a corresponding reduction in blood flow to nonexercising organs such as kidneys. Increased cardiac output provides a greater flow through the pulmonary alveolar infrastructure to obtain and extract from the inspired air more oxygen which is demanded by the increased metabolic activity. Airborne contaminants present in the immediate environment will, therefore, be absorbed to a greater extent than during resting conditions.

Pulmonary uptake of highly soluble solvents (e.g., acetone, butyl alcohol) correlates with alveolar ventilation rates. Due to their high solubility, these substances are absorbed to a greater extent, with no tissue approaching equilibrium during exercise. Conversely, for solvents with low blood- and tissue-solubility characteristics (e.g., methylene chloride, toluene), uptake is enhanced by the initial workload, but further workload increases may not have an effect on pulmonary uptake. The capacity for retaining these vapors is small; at the beginning of exposure, richly perfused tissues are almost equilibrated but the concentration in fat is low due to perfusion limitation. Therefore, the increase of pulmonary uptake during exercise is mainly attributed to increased uptake only by those tissues in which perfusion is enhanced by exercise. At equilibrium, the effect of exercise on pulmonary uptake varies with solubility of solvents; the higher the solubility, the higher the uptake, indicating that equilibrium is reached faster during physical exercise than while at rest. Physical activity, however, does not have any effect on the body burden of chemicals once steady-state is reached.

C. Smoking

Tracheobronchial deposition of particles is significantly greater in cigarette smokers than in nonsmokers, presumably due to the bronchoconstrictive properties of the cigarette smoke and/or the electrical charge of the aerosols. An impairment of mucociliary clearance may occur in smokers causing higher retention of inhaled particles. Urinary elimination of chemicals and their metabolites might also be decreased in smokers due to renal vasoconstriction caused by nicotine. The polynuclear aromatic hydrocarbons present in cigarette smoke are responsible for the induction of overall metabolic capacity observed in smokers. In particular, the activity of aryl hydrocarbon hydroxylases is increased, resulting in a selective enhancement of the metabolism of nicotine, phenacetin, antipyrine, theophylline, imipramine, and pentazocine in smokers. Conversely, smoking does not alter the rate or extent of metabolism of diazepam, meperidine, phenytoin, nortriptyline, warfarin, and ethanol. Since many environmental carcinogens are bioactivated by the enzymes of the *P*-450 IA family, their induction can have important implications in the susceptibility of smokers to carcinogenic pollutants.

D. Alcohol Intake

The role of ethanol in causing interindividual differences in pharmacokinetics arises from its ability to alter the metabolizing capacity of the hepatic microsomal isozymes, particularly cytochrome *P*-450 2E1. Simultaneous exposure to ethanol would inhibit the biotransformation of chemicals, as has been shown for trichloroethylene, toluene, styrene, xylene, and methylethylketone. However, a prior or a chronic administration of ethanol can increase the metabolic transformation of xenobiotics mediated by cytochrome *P*-450 2E1. Thus, prior exposure to ethanol has been shown to augment the toxicity of several halogenated haloalkanes, benzene, and nitrosamines. Heavy consumption of alcohol may result in further interactions with other drugs such as analgesics, anticoagulants, antidepressants, antihistamines, tranquilizers, diuretics, antibiotics, and antihypertensive agents, causing severe physiological and psychological effects that might eventually alter disposition of xenobiotics in these individuals.

E. Medication

Pulmonary uptake of airborne chemicals may be altered by the profound effect on the respiratory rate of certain classes of drugs, especially anesthetics, analgesics, tranquilizers, and hypnotics. Similarly, absorption of weakly acidic substances through the gastric mucosa can be impaired if the gastric pH is made alkaline as, for example, in the case of antacid consumption. The plasma protein binding properties of drugs can alter the pharmacokinetic behavior of other xenobiotics. Several oral contraceptives, chemotherapeutic agents, anticoagulant, and anticonvulsant drugs have been reported to inhibit hepatic drug metabolism. The drugs causing inhibition of hepatic microsomal enzyme activities in humans include chloral hydrate, chloramphenicol, dicoumarol, disulfiram, erythromycin isoniazid, *p*-aminosalicylate, phenylbutazone, and trileandomycin. This property of the drugs may well influence the metabolism and toxicity of xenobiotics to which people are exposed.

IV. Body Stature

Marked differences in critical physiological parameters can arise from variation in body build among individuals, and this can account for the interindividual differences in uptake and deposition of airborne substances, as well as disposition of lipophilic contaminants. Regional differences in particle deposition in the lungs are determined by the factors that influence regional ventilation, most prominent among these being the apex-to-base intrapleural pressure gradient. This gradient is greater in taller individuals than in shorter people. In barrel-chested people and short people with a small mean airway diameter, the extent of particle deposition per unit volume of alveolar ventilation can be greater than in persons with larger airways but similar lung volume. The bronchiolar epithelium of short subjects, thus, may be exposed to a greater concentration of carcinogenic or otherwise toxic particles than that of their taller counterparts. The smaller trachea of the short asbestos miners has, therefore, been thought to have predisposed them to develop asbestosis.

The normal pattern of particle deposition is also modified in obese people as characterized by a diversion of particles to the vertical horizontal zone and the intermediate radial zone. An increased deposition in the middle zones relative to apices and bases occurs as a result of the nonuniform thickness of the fat layer in the chests of obese subjects. Decreased basal ventilation has also been reported in obese people. These and other observations are suggestive of significant ventilation–perfusion abnormalities in obese subjects.

When exposed to the same concentration of lipid-soluble vapors, the uptake rate is greater in obese people than in thin persons. Conversely, thin people exhibit faster pulmonary clearance than obese individuals. The plasma volume per unit body weight is less in obese people than in normal subjects, leading to a possible reduction in clearance. Since capillary density is several-fold greater in muscle than in adipose tissues, it seems plausible that clearance is faster in muscular subjects. The tissue distribution and body burden of nonvolatile lipophilic chemicals are greater in obese people; the steady state, however, is reached faster in slim people than in obese persons. Obesity, in addition, reduces the bioavailability of highly lipid soluble chemicals for metabolism during exposure and prolongs their availability after cessation of exposure.

V. Gender

Women tend to have smaller lungs, lower cardiac output, smaller blood volume, lower plasma albumin levels, and a greater percentage of body fat than men regardless of their age, and these factors can influence the pattern of uptake and disposition of xenobiotics. For example, retention of larger amounts of lipophilic pollutants has been observed in women. In addition, transfer of lipophilic chemicals via mothers' milk and placenta might account for some differences in elimination of such chemicals in comparison to their male counterparts. Such a transfer may further be influenced by the age and dietary and personal habits of the mother. Changes in the hormonal status in women (e.g., during the various stages of the menstrual cycle) appear to account for sex-related differences in the absorption of xenobiotics, as has been shown with ethanol. Despite the demonstrated sex-related differences in the levels of hepatic monooxygenases and drug biotransformation in the rat, the overall trend and differences are not as marked and clear-cut in humans.

VI. Physiological State

A. Pregnancy

Pregnancy is characterized by a significant increase in tidal volume and a decrease in pulmonary residual volume, resulting in an overall increase in alveolar ventilation and scouring of the alveoli. The raised respiratory activity is accompanied by high blood flow, favoring enhanced uptake and exhalation of airborne chemicals. The vascularity of skin and mucous membranes (e.g., nose, vagina) is raised considerably so dermal absorption of chemicals becomes more important as a route of exposure in pregnant women. Further changes in pregnancy that favor increased absorption of xenobiotics include decreased gastrointestinal motility, delayed gastric and intestinal emptying, reduction of gastric acid secretion, and increased mucous secretion.

An average 25% increase in body fat, 30–40% increase in blood flow, and 40–50% increase in plasma volume occurs in pregnancy. Total body water is also expanded due to enhanced tissue growth, representing an increase in the total area for distribution and storage of toxicants. Although the concentration of albumin falls to about two-thirds of the normal level during the early stages of pregnancy, its total mass is not altered because of the large increase in total plasma volume.

Changes in liver enzymes during pregnancy are likely, but less well characterized. The increase in the steroid hormones characteristic of pregnancy, might cause an induction of oxidative enzymes, but it might also reduce the biotransformation of xenobiotics by competing for available enzyme sites. Renal plasma flow and glomerular filtration rates are nearly doubled in pregnancy, leading to high filtration of unbound chemical species in circulation. These significant physiological changes occurring during pregnancy can account for differences in disposition of xenobiotics between pregnant and nonpregnant women.

B. Aging

With advancing age in humans, the relative amount of fat increases with a concomitant decrease in the amount of muscle. In elderly subjects with larger adipose tissue mass, the amount of plasma per unit body weight is reduced due to the lesser number of blood capillaries in fat as compared with muscle. The effect of these changes on pharmocokinetics will depend on a particular chemical's tissue solubility characteristics. Highly lipid soluble chemicals may become extensively distributed and stored in tissues. On the other hand, xenobiotics that have a low tissue : blood partitioning may be distributed less extensively. The latter may also be a consequence of a decrease in the total body water observed in the elderly.

Hepatic phase I drug metabolizing enzymes are generally assumed to be impaired in the elderly. However, recent studies have demonstrated the absence of any age-dependent differences in the activities and content of human liver monooxygenases. Reduced hepatic drug clearance in the elderly is likely a consequence of reduced liver volume coupled with diminished hepatic blood flow. A decrease in total blood flow by an estimated 40–45% occurs in the aged as a consequence of reduced cardiac output. Therefore, one can anticipate an age-related decline in the flow-dependent clearance of xenobiotics.

Renal function and, therefore, renal clearance capacity are impaired in the elderly. The glomerular filtration rate (GFR) declines with age, exhibiting a 50% reduction over the 50 years from age 30 to 80; this is accompanied by increases in the percentage of sclerotic glomeruli. For chemicals cleared partly or entirely by renal excretion, the total clearance will predictably decrease in approximate proportion to the reduced GFR. Whereas an age-related decline in plasma albumin concentration has been widely reported, a slight increase in plasma α_1-acid glycoprotein occurs in the elderly, and this may become more marked in disease states.

C. Disease

Diseases of the cardiovascular, endocrine, hepatic, and nervous systems can produce temporal fluctuations in drug disposition. Ventilatory disorders resulting from chronic obstructive pulmonary diseases can lead to hypoventilation and an accumulation of carbon dioxide, and thus can affect the uptake and distribution of inhaled gases and vapors. A decrease in the diffusion capacity secondary to reduced pulmonary capillary blood volume may ensue in sarcoidosis, interstitial fibrosis, silicosis, and other pneumoconioses.

Cardiac failure might considerably alter the absorption of xenobiotics by decreased intestinal motility, edema in the intestinal wall, and reduced blood flow. Chronic liver disease, nephrotic syndrome, burns, and septic shock cause hypoalbuminemia and hypoproteinemia, which can potentially lead to an increase in the unbound fraction of drugs and chemicals in the plasma. Hypothyroidism and obstructive liver diseases may actually increase plasma lipoprotein levels. An increase in plasma α_1-acid glycoprotein levels has been reported in several disease states (myocardial infarction, malignancy, ulcerative colitis, Crohn's disease, rheumatoid arthritis, renal transplantation, surgery).

Qualitative and quantitative changes in drug metabolism may also result from various types of carcinoma. For example, patients with bladder cancer generate nicotine metabolites in proportions significantly different from those observed in normal individuals. An impairment of metabolic capacity has also been reported in individuals with liver diseases, hypothyroidism, and fever. People with renal disease (renal insufficiency) cannot readily get rid of renally cleared drugs and chemicals; this can result in a prolongation of drug half-lives and an accumulation to a concentration range where toxic symptoms may become evident.

VII. Genetic Traits

The disposition of drugs and chemicals is highly variable even among healthy individuals. This can be due to differences in an individual's capacity to metabolize a particular chemical. The enzymes catalyzing the various metabolic reactions are controlled by a complex interplay of environmental and genetic factors. Genetic control of metabolizing enzymes might involve the role of a single gene (monogenic) or multiple genes (polygenic). If a chemical is metabolized either by an enzyme under polygenic control or by several enzymes, then the population distribution of its metabolic rate will, in all likelihood, be unimodal. Glycine and glucuronide conjugation of salicylic acid, for example, are metabolic reactions under polygenic control. On the other hand, a polymodal distribution and marked interindividual variations of metabolic rates in a population would be characteristic of a metabolic reaction under monogenic control (e.g., 4-hydroxylation of debrisoquine). In this case, only one specific polymorphic *P*-450 enzyme is capable of metabolizing this substrate at a reasonable rate. Since each gene controlling a *P*-450 has two or more alleles, some of which do not encode a functioning enzyme, several different metabolic rates may occur; an apparent absence of a particular type of metabolic activity will occur in homozygotes with two nonfunctional alleles. Thus, wide interindividual variation in metabolism of a compound can be anticipated if its metabolism is under the control of a single genetic locus with two or more alleles with strikingly different enzyme activity.

For example, the extrahepatic aryl hydrocarbon hydrolases (AHH) have been reported to be differentially expressed and induced in the human population. AHH induction in human lymphocytes has been reported to exhibit a trimodal distribution (low, intermediate, and high inducibility), consistent with the hypothesis of two alleles at a single genetic locus controlling inducibility. Accordingly, people having the high/intermediate phenotypes (rapid activators) face a greater risk of developing smoking-related bronchogenic carcinomas compared to persons having the intermediate/low phenotypes. Since smoking delivers benzo(*a*)pyrene, benz(*a*)anthracene, and other polycyclic aromatic hydrocarbon carcinogens to lung epithelium, these chemicals may more efficiently stimulate their own activation by inducing AHH activity in individuals with high/intermediate phenotypes. However, cancer at sites distant from the point of carcinogen exposure would not necessarily be increased in individuals with the high/intermediate phenotypes, presumably because of a first-pass clearance effect.

Cytochrome *P*-450 2D6, or debrisoquine-4-hydroxylase, is also polymorphically distributed. Phenotyping of family members has shown that impaired metabolism of debrisoquine is inherited as an autosomal recessive trait. The incidence of poor metabolizers of debrisoquine in Caucasians varies from 5% in Sweden, 6% in Ghana, 8–9% in the United Kingdom and Nigeria, to 10% in Switzerland. The incidence is low in Arabs, whereas no poor debrisoquine metabolizing capacity has been reported among Japanese, suggesting that they are genetically more homogeneous for this trait than are Caucasians. Even though *P*-450 2D6 is not known to metabolize any environmental toxicants of concern, it has been found that people with liver cancers possibly induced by aflatoxin, and urinary tract tumors induced by ochratoxin, had a lower frequency of cases with poor metabolizing capacity for debrisoquine than control groups. However, extensive metabolizers do appear to be at higher risk for lung cancer. Further, hepatic *P*-450 2D6 deficiency has been correlated with Parkinson's disease. There is strong evidence suggestive of the debrisoquine-type polymorphism affecting one or more major routes of metabolism of another thirty pharmaceutical products in humans. A complex DNA probe using the polymerase chain reaction now makes it possible to genotype people with DNA extracted from white blood cells.

Another cytochrome *P*-450 enzyme, namely, mephenytoin hydroxylase, is expressed as two phenotypes, thus giving rise to two distinct groups of people: extensive and poor metabolizers. The poor mephenytoin metabolizers have been detected with a low frequency in Caucasians (2–5%) but with a considerably higher frequency (~20%) in the Oriental population.

Other *P*-450s may well be polymorphically expressed in humans. Polymorphism has also been described among non-cytochrome *P*-450 enzymes mediating phase I reactions. These include alcohol and aldehyde dehydrogenases, paraoxonase, and flavoprotein monooxygenase.

Alcohol dehydrogenase (alcohol NAD^+ oxidoreductase; EC 1.1.1.1) is capable of oxidizing a variety of primary, secondary, and tertiary aliphatic alcohols and a limited number of cyclic alcohols to the corresponding aldehydes. Studies in twins have indicated that alcohol dehydrogenase (ADH) is under genetic control: a striking similarity in ethanol metabolism rate has been observed in identical twins, with much greater variability between fraternal twins.

The molecular forms of ADH can be grouped into three distinct classes based on their composition. These isozyme forms are composed of polypeptide subunits α, β, and γ which are controlled by separate gene loci ADH1, ADH2, and ADH3 respectively. Isoenzymes can be homodimeric ($\alpha\alpha$, $\beta\beta$, $\gamma\gamma$) or heterodimeric, being coded by alleles at separate loci ($\alpha\beta$, $\beta\gamma$, etc.) or at the same locus ($\beta_1\beta_2$, $\gamma_1\gamma_2$, etc.). The gene loci ADH2 and ADH3 are polymorphically distributed. Of these, an isozyme form produced from the ADH2 locus known as the "atypical ADH," contains a variant subunit (β_2) instead of the usual β_1; the former subunit confers a five times higher catalytic activity than the normal alcohol dehydrogenase dimer. The relative distribution of this variant form of ADH2 accounts for high interindividual variations in the metabolism of alcohols. Whereas 5–10% of the British, 9–14% of the German, and 20% of the Swiss population have been found to possess this atypical ADH2, as much as 90% of Japanese and Chinese possess it, suggesting that they are likely to be the most active metabolizers of alcoholic substances. Moreover, these Asian populations have also been reported to be deficient in one of the four distinct aldehyde dehydrogenases (EC 1.2.1.3) responsible for the oxidation of acetaldehyde and other aldehydes. The combination yields high plasma acetaldehyde concentration after ethanol ingestion and probably accounts for the striking flushing reaction these individuals experience. Low incidence of aldehyde dehydrogenase deficiency has also been observed in American Indians but not in Turks, Israelis, Hungarians, Italians, and Germans.

Flavin-containing monooxygenases (EC 1.14.13.8; FMO) catalyze the oxygenation of soft nucleophiles, particularly amines and sulfur compounds. A genetic polymorphism for the *N*-oxygenation of trimethylamine in humans, presumably due to a defect in FMO, has recently been described. The *N*-oxygenation of nicotine, an alternative substrate for FMO, is also impaired in humans who do not metabolize trimethylamine. Even a tenfold change in the activity of FMO should produce only minor changes in metabolic profiles of the better drug or chemical substrates for this enzyme since the *in vivo* oxidation of these substances by FMO is mass-transfer-limited.

Hydrolysis of several organophosphates (e.g., paraoxon), carbamates (e.g., carbaryl, 3-isopropyl phenyl *N*-methyl carbamate) and of certain aryl carboxylic acid esters (e.g., phenyl acetate) is catalyzed by arylesterases. Paraoxonase, an arylesterase (EC. 3.1.1.2), is polymorphically distributed in humans, the pattern being governed by the presence of two alleles (for low and high activity) at a single autosomal locus, thus following Mendelian inheritance. A bimodal distribution has been observed in the American, Canadian, Danish, British, French, German, and Indian populations, probably disposing them to wide interindividual differences in susceptibility to pesticide poisoning, especially parathion. Kenyans, Nigerians, and Chinese appear to be unimodal with respect to the distribution of plasma paraoxonase activity. The alleles occur with a relative frequency of 0.3 and 0.7; thus each phenotype occurs at an equal proportion within the Caucasian population, but the low-activity allele decreases as one moves from Europe toward Africa or Asia. Thus, homozygotes expressing only the low activity allele constitute more than 50% of the European population but only 10% of the African and Mongolian populations.

Recent research findings are indicative of a polymorphic expression of epoxide hydrolase. Several enzymes catalyzing the phase II reactions—namely *N*-acetylation, glutathione conjugation, glucuronidation, *N*-glycosidation, sulfation, and *O*-methylation—are also polymorphically distributed in humans.

N-Acetylation is a major detoxication reaction for *N*-substituted carcinogenic arylamines such as β-naphthylamine, 2-aminofluorene, benzidine, and 4-aminobiphenyl, and for compounds that are metabolized to intermediates containing an amino or hydrazino group. This conjugation process is catalyzed by *N*-acetyltransferase (EC 2.3.1.5), a cytosolic enzyme found in human liver. The rate of *N*-acetylation is polymorphically distributed since the enzyme is controlled by a single gene with two alleles at a single locus. Thus, slow acetylators are homozygous for a recessive allele (*rr*), and rapid acetylators are either heterozygous (*Rr*) or homozygous (*RR*) for the dominant allele. Recent studies have detected a trimodal distribution in which the heterozygotes would represent intermediate acetylators. Among human populations, Americans, Europeans, Africans, and Indians have a high percentage of slow acetylators (55–70%), whereas Japanese, Chinese and Canadian Eskimos have a high percentage of rapid acetylators (80–100%). The slow acetylators would appear to be more susceptible to cancer of the urinary tract caused by *N*-arylamines due to decreased elimination by *N*-acetylation and/or to a relative increase in their bioactivation via *N*-hydroxylation.

Glutathione conjugation is an important detoxication pathway for halogenated and reactive-type chemicals. The glutathione *S*-transferases (EC 2.5.1.18), which mediate this reaction, occur in at least three forms—μ, α, and π—as the products of three different genes. Among these three forms, the μ class isozymes are the most active in conjugating toxicologically important epoxides and prototype substrates [e.g., styrene oxide, 1,2-epoxy-3-(*p*-nitrophenoxy)propane and benzo(*a*)pyrene-4,5-oxide]. However, these μ class GSTs are not expressed in about 40–50% of Caucasians due to a gene deletion, and are generally not detectable in fetal tissue until about 30 weeks of gestation. Individual differences in μGST levels might account for interindividual variations in the disposition and toxic effects of the highly reactive electrophilic chemicals.

Glucuronidation is another major phase II reaction for a variety of parent chemicals and metabolic intermediates containing carboxyl, hydroxyl, amino, and sulfhydryl groups. The formation of glucuronides is catalyzed by the enzyme uridine diphosphate glucuronyl transferase (EC 2.4.1.7). Variation of as much as 10-fold in the rates of drug glucuronidation is not uncommon within a healthy human population. Whether this variation in glucuronidation is related to age, disease state, exposure to other xenobiotics, or solely due to inheritance has not yet been determined. Genetic variation occurring within the multigene family which encodes the enzymes responsible for glucuronidation in humans is possible. This variability has been observed for bilirubin glucuronidation.

N-glycosidation capacity has also been found to vary among various ethnic groups. In an investigation of the metabolism of amobarbital in humans, an absence of the capacity for glycosidation was observed in several Caucasian subjects, suggesting the existence of a genetic defect in these subjects.

Aryl sulfotransferase (EC 2.8.2.1), the most important of the enzymes catalyzing the sulfate conjugation of phenolic compounds, occurs in two distinct forms. The activity of these sulfotransferases is controlled by inheritance, and experimental evidence is indicative of genetic polymorphism regulating the levels of activity of both forms of the enzyme. Pharmacogenetic variations in aryl sulfotransferase activity could be related to individual differences in the sulfate conjugation of pharmaceutical products. Therefore, inherited differences in "sulfation status" should be considered as a factor causing interindividual variations in the metabolism of phenolic drugs and xenobiotics.

N- and *O*-methylation are phase-II reactions catalyzed by *S*-adenosyl L-methionine-dependent methyltransferases. Of these, catechol-*O*-methyltransferase (EC 2.1.1.6), an enzyme catalyzing the *O*-methylation of catecholamine neurotransmitters and catechol drugs, such as levopoda and methyldopa, has been found to be controlled by genetic polymorphism. Individual variation in the *O*-methylation of orally administered drugs has been correlated with differences in catechol-*O*-methyltransferase activity. Approximately 25–30% of the randomly selected population is homozygous for the trait of low catechol *O*-methyltransferase activity.

Finally, there are variant forms of catalase, amylase, pseudocholinesterase, acid phosphatase, carbonic anhydrase and hemoglobin in human blood. These highly variant protein structures may account for interindividual differences in disposition and receptor-mediated effects of certain chemicals.

VIII. Environmental Conditions

A. Photoperiod

The physiological effects of varying light/dark cycle lengths in humans are not clearly understood.

However, the rhythmicity of hepatic microsomal enzyme activities in animals has been shown to be abolished by constant exposure to either light or darkness. Whereas an increased photoperiod inhibits the activities of hepatic hexobarbital oxidase and *p*-nitroanisole *O*-demethylase in rodents, constant exposure to darkness reverses this trend. Light is the most important stimulus for the synchronized circadian rhythms in rodents, probably mediated via the pineal gland. However, in humans, the dominant stimuli are the social environment and the awareness of clock time. Despite the possible association of several reproductive, gastric, and intestinal dysfunctions with nontraditional workshifts, the influence of light period on the disposition of chemicals in humans remains unclear.

B. Altitude

Humans living at high elevations experience hyperventilation, in direct proportion to the altitude or degree of hypoxic stimulus. Several additional adaptive features have been noted: (1) an increase in tidal volume and minute ventilation, which elevates alveolar and arterial pO_2 and increases the diffusion gradient between the blood and the tissues; (2) a transient increase in cardiac output due to tachycardia without any change in the stroke volume; (3) an increase in tissue perfusion to deliver more oxygen; and (4) an increased cytochrome concentration at high altitudes to keep oxygen uptake constant. Change in altitude has also been shown to modify hepatic microsomal enzyme activities. Altitude-exposed animals experience an enhanced hepatic drug metabolism, as indicated by a decrease in the sleeping time induced by hexobarbital.

C. Temperature

In hot environments, vasodilation occurs, and a substantial portion of the body's blood supply is diverted from the core body organs to the dilated blood vessels of skin and subcutaneous tissues in the peripheral body. There is an increase in cardiac output and in the quantity of the circulating blood volume as a result of splenic contraction and the dilution of the circulating blood with fluids from certain body tissues. A decrease in forced vital capacity (FVC) has also been associated with exposure to hot ambient atmospheres. Of interest, marathon runners also show a 8–10% decrease in FVC. The mechanism for the FVC reduction with heat exposure or prolonged exercise is not known but appears to be related to the increase in thoracic blood volume.

When humans are exposed to cold, two major types of physiological adjustments occur in an attempt to maintain thermal equilibrium: (1) a vasoconstriction in the skin and other underlying tissues and a corresponding reduction in peripheral blood flow; and (2) an increased metabolic heat-production rate. These physiological and biochemical changes associated with the ambient temperature can cause individual differences in pharmacokinetics.

D. Coexposure to Other Chemicals

Sequential or simultaneous exposure to two chemicals might result in an alteration by one chemical of the absorption, metabolism, distribution and/or excretion of the other chemical. Whereas several cases of toxicokinetic interactions have been discovered as a result of extensive animal experimentation, only a few such interactions have been found to occur in humans (Table I). The most common basis of interaction between two chemicals appears to involve the hepatic drug metabolizing enzyme system. Animal and/or human studies have shown that several heavy metals, anions, gaseous air pollutants, organophosphates, carbamates, bipyridyl herbicides, fungicides (particularly dithiocarbamates), and a number of solvents (e.g., carbon disulfide, chloroform, carbon tetrachloride, tetrachlorohydrofuran, tetrachloroethylene, vinyl chloride) can inhibit or destroy hepatic microsomal enzyme activities. On the other hand, polycyclic aromatic hydrocarbons, alcohols, ketones, and halogenated organic chemicals are potent inducers of metabolizing enzymes. Depending on the extent of prior or simultaneous exposure to one or more of these chemicals, an individual might become less or more susceptible than the general population to the toxic effects of other xenobiotics due to a modulation of their disposition kinetics.

In summary, the pharmacokinetics of chemicals is a function of their solubility characteristics, metabolic and binding rates, and the physiology of the individual. The interindividual difference in pharmacokinetics is a consequence of variations in these critical biological determinants of chemical disposition. These, in turn, are determined by the

Table I Examples of Pharmacokinetic Interactions between Hazardous Materials

Basis of interaction	Interacting chemicals	Interactive effects
Absorption		
Percutaneous	*m*-Xylene and isobutanol[a]	Reduced absorption of both compounds, due to dehydration of skin elicited by isobutanol.
	Dimethyl sulfoxide and pesticides	Increased dermal absorption of pesticides and other chemicals when they are mixed with dimethyl sulfoxide, which disrupts the cellular permeability and acts as a "penetrant carrier."
Pulmonary	Hydrogen cyanide and carbon monoxide	Increased pulmonary uptake of air contaminants when they are present along with hydrogen sulfide, which at low concentrations increases the pulmonary ventilation rate.
Gastrointestinal	Lead and iron[a]	Decreased absorption of lead in the presence of iron due to a competition for transport sites in the intestinal mucosa.
Distribution		
Tissue distribution	Lead and dithiocarbamates	The lipophilic lead–dithiocarbamate complex has a greater capacity than lead alone to penetrate the blood–brain barrier, thus causing a greater accumulation in the lipid-rich brain components.
Protein binding	Organochlorine and organophosphate pesticides	Organochlorine pesticides not only enhance the biotransformation of organophosphates but also their binding to plasma proteins and nonspecific esterases.
Metabolism		
Phase I		
Induction	Isopropanol/acetone and chloroform carbon tetrachloride[a]	Increased bioactivation of carbon tetrachloride and chloroform due to induction of hepatic microsomal *P*-4502E1 by acetone.
Inhibition	Dithiocarbamates and carbon tetrachloride	Decreased bioactivation of carbon tetrachloride due to inhibition of hepatic microsomal enzymes by dithiocarbamates.
Phase II		
Induction	Sodium sulfate and certain aryl amines	Increased sulfate supply might result in a greater amount of it available for conjugation of xenobiotics and their metabolites.
Inhibition	Pentachlorophenol and arylamines	Pentachlorophenol, by inhibiting cytosolic sulfotransferases, causes a reduction of the sulfation of arylamines.
Excretion		
Pulmonary	Ethanol and mercury	Ethanol depresses the conversion of elemental mercury into ionic form; thus their coexistence results in a diminution of the pulmonary retention and blood levels of mercury, and enhances its pulmonary exhalation.
Biliary	Selenium and arsenic, mercury and selenium	Arsenic increases the clearance of selenium liver into the bile; selenium on the other hand inhibits the biliary excretion of mercury.
Urinary	Sodium bicarbonate and fluoride	Alkalosis induced by the administration of sodium bicarbonate causes a more rapid renal clearance of fluorides.

Source: Krishnan, K. and Brodeur, J. (1993). Toxic interactions among environmental pollutants: Corroborating laboratory observations with human experience. *Environ. Health Perspect.* (in press).

[a] Interactions observed in humans.

physiological state, environmental conditions, genetic trait, and personal habits of an individual in the population. By perturbation of the appropriate model parameter(s), the effects of changes in physiological conditions and metabolic rates on the pharmacokinetics and tissue dosimetry of chemicals in individuals can be examined with a physiological model. This capability of physiological models can be useful in determining the tissue dose of chemicals in specific subgroups of people (e.g., smokers, pregnant women, people doing active physical work) for whom the physiological parameters might be significantly different from those in the normal population. Information on tissue dosimetry obtained with physiological modeling would greatly enhance our ability to assess potential health risks arising from exposure to hazardous materials for specific subgroups of the population.

Related Articles: CYANIDE; IMMUNE RESPONSE TO ENVIRONMENTAL AGENTS; TOXIC AGENTS AND THERMOREGULATION.

Bibliography

Kalow, W., Goedde H. W., and Agarwal, D. P., eds. (1986). "Ethnic Differences in Reactions to Drugs and Toxicants." Alan R. Liss, New York.

Krishnan, K., and Andersen, M. E. (1993). Physiologically based pharmacokinetic modeling in toxicology. *In* "Principles and Methods of Toxicology," (A. Wallace Hayes, ed.) Raven Press, New York. (in Press)

Notten, W. R., Herber, R. F. M., Hunter, W. J., Monster, A. C., and Zielhuis, R. L., eds. (1986). "Health Surveillance of Individual Workers Exposed to Chemical Agents." Springer-Verlag, Berlin.

Omenn, G. S., and Gelboin, H. V., eds. (1984). "Genetic Variability in Responses to Chemical Exposure." Banbury Report 16. Cold Spring Harbor Laboratory, New York.

Woodhead, A. D., Bender, M. A., and Leonard, R. C., eds. (1988). "Phenotypic Variation in Populations: Relevance to Risk Assessment." Plenum, New York.

Plants as Detectors of Atmospheric Mutagens

William F. Grant
McGill University

Glossary

Allele One of the two individual genes in a gene pair.

Aneuploid An individual with one or more chromosomes greater or fewer than the normal diploid number.

Clone Group of genetically identical individuals.

Diploid Having two sets of chromosomes: one set from the female parent, the other from the male parent.

Diploid number Total number of chromosomes found in a species.

Heterozygous Having two different alleles at a specific genetic locus.

Hybrid Individual resulting from the cross of genetically different individuals.

In situ Under natural conditions.

Micronuclei Small nuclei resulting from the encirclement of chromosome fragments and isolated chromosomes.

Microspores Pollen grains.

Mutagen Any substance or process that causes a genetic mutation.

Mutation Any heritable change in the sequence of nucleotide bases in DNA; may affect only one or two bases (point mutation) or long segments (chromosomal mutation).

Polyploid An individual possessing entire extra sets of chromosomes.

Somatic mutations Mutations affecting the body cells, such as the petals.

Stamen hairs Hairs on the filament that bear the anthers.

Tetrad cells Group of four cells which become separate microspores or pollen grains.

Xenobiotic A compound foreign to an organism.

SEVERAL PLANT GENETIC SYSTEMS possess unique characteristics for screening and monitoring environmental mutagens. They are sensitive test systems which have already provided reliable and useful quantitative mutagenesis data. These systems have played roles in detecting new mutagens and developing techniques that were later used in other systems for advancing mutagenesis knowledge. Plants which provide such systems for detecting and analyzing the effects of environmental mutagens include Arabidopsis thaliana, *barley (*Hordeum vulgare*),* Crepis capillaris, Lilium, *onion (*Allium cepa*), pea (*Pisum sativum*), rice (*Oryza sativa*), soybean (*Glycine max*), tomato (*Lycopersicon esculentum*),* Tradescantia, Triticum, *tobacco (*Nicotiana tabacum*),* Vicia faba, *and* Zea mays. *A number of genetic end points may be used in mutation screening and monitoring ranging from point mutations to chromosome aberrations in individual cells and organs such as leaves, pollen, and endosperm. Depending on the ontogenetic stage of development, an environmental mutagen may exert teratogenic effects, induce mutations af-*

fecting the germinal cells, or cause mutations that lead to transformation of somatic cells. Many plants are easy to regenerate and some have short generation times. The cost of handling, space requirements, and training of individuals is relatively inexpensive.

The detection of airborne pollutants falls into two general classes (1) *in situ,* or on-site methods, and (2) laboratory, or off-site methods. Plant genetic assays are the only systems currently in use as *in situ* monitors of polluted air. Two assays which are considered ideal for *in situ* monitoring and testing of airborne mutagens are the *Tradescantia* stamen hair assay and the *Tradescantia* micronucleus assay. Both have been used to investigate ambient air pollution around commercial and industrial sites.

I. Introduction

Since the mid 1940s, synthetic chemicals have steadily increased and have become an integral part of modern society. Chemicals may enter the atmosphere through a variety of direct or indirect routes and be inhaled directly or ingested indirectly through the food chain. Via either route, they pose a risk to human health. Air pollutants affect people unknowingly as they derive from the atmosphere and creep silently up the food chain.

Direct chemical pollutants found in the air result from emissions from motorized vehicles, public power utilities and domestic heating equipment, industrial effluents, natural and man-made explosions, nuclear reactors, industrial activities, volatile organic emissions, refuse burning, and personal habits, such as smoking. Pollutants derive from the direct discharge into the atmosphere of foreign gases, vapors, droplets and particles, or of excessive amounts of normal constituents, such as carbon dioxide and suspended particulate matter and may total 125 million tons per year. Some gaseous and particulate forms are dissipated into the open air over large geographic areas of the globe, whereas others, such as cigarette smoke and household chemicals, are released into more confined areas.

Domestic chemical compounds include perfumes, hair dyes, sunscreens, household detergents, air fresheners, and cleaning fluids. Food intended for human consumption may contain residues, such as pesticides, compounds absorbed from packaging material, additives, and preservatives. Pharmaceutical compounds may be in the form of different types of sprays. One class of chemicals, (pesticides) includes fungicides, herbicides, and insecticides, and embraces a wide range of target organisms. Some pesticides are used for specialty applications, such as fumigants and growth regulators which enter the atmosphere on application. Aerial application of pesticides (80%) and surface applications (20%), represent major routes into the environment. Many chemicals which become airborne are routinely produced in large quantities. For example, in 1975 pesticide production in the U.S. totaled some 1.6 billion pounds (world production was 3.7 billion pounds) and formulations of up to 50,000 separate products. The herbicide, trifluralin, is estimated to involve 470,000 applicators and 38,000 field workers and the production of the herbicide, dicamba, was estimated by EPA in 1980 to be 5–7 million pounds annually. Likewise, herbicides used in the clearing and maintenance of rights-of-way were estimated at 70 million acres.

A number of atmospheric chemicals have been shown to have mutagenic properties. These include sulfur dioxide, bisulfite, nitrous acid and oxide, polynuclear aromatic hydrocarbons (e.g., benzo(*a*)pyrene, benz(*a*)anthracene, dibenzanthracene), peroxyacyl nitrates, peroxides, ozone, halogenated hydrocarbons (e.g., di- and trichlorofluoromethane, vinyl chloride, tri- and tetrachloroethylene, carbon tetrachloride, chloroprene, ethylene dibromide, ethylene chlorohydrin), sodium fluoride, pesticides (DDT, dichlorvos), polychlorinated biphenyls, formaldehyde, ethylene oxide and inorganic and organic derivatives of lead and mercury.

Human beings may be exposed to a single chemical at relatively high concentrations or a combination of chemicals as in the case of individuals who come in contact with chemicals in their manufacturing and application. However, many chemicals are used in combination, for example with emulsifiers and solvents which are potential mutagens in their own right, and in general, very little is known on the mutagenic aggravation of the effects of chemicals from their synergistic interactions.

Historically, the use of chemicals entering the environment is not new. As long ago as A.D. 70,

Plinius recommended arsenic as an insecticide and the Chinese are reported to have used arsenic sulfide as an insecticide in the late sixteenth century. In 1661, a concern developed on the immoderate use of snuff. One of the first atmospheric pollutants to be identified as early as the eighteenth century was soot from burning fires which caused scrotal cancer in chimney sweepers.

The potential effects on human health of environmental mutagens and carcinogens found in the industrial work environment and in food are becoming better understood. Mutations may contribute to the genetic health of future generations by affecting the germ cells resulting in the accumulation of heritable abnormal genes in populations. In addition, mutations of somatic cells may result in the formation of malignant cells in individuals and exhibit other health factors, such as heart disease, aging, and cataracts. Of concern has been the relatively long lag period between exposure and overt effect.

The cytogenetic effects of environmental chemicals on the flora and fauna (including the build-up of species resistance) of ecosystems have been well documented and are of considerable concern. Many plant species act as natural bioconcentrators of atmospheric pollution by accumulating pollutants on, or in, their tissues. Air pollutants may be assimilated by the plant through foliar or root absorption directly from the soil or from plant detritus particles derived from plant decomposition.

The use of higher plants in the study of environmental mutagens parallels the historical development of the field of chemical mutagenesis. An observation in 1931 in which a correlation was noted between plant fertility and chromosome irregularities in meiosis in both tobacco and eggplant after the plants had been fumigated with nicotine sulfate is considered to be the earliest report of the mutagenic effect of a chemical. Later, a high frequency of chromosomal aberrations in *Oenothera* plants was noted after the plants had been treated with ethylurethane.

While long-term whole animal bioassays are viewed as the most satisfactory procedures for assessing genotoxicity, such tests are very expensive, as well as time- and space-consuming. At present, no animal system is presently known which can be used efficiently for *in situ* air monitoring and provide short-term reliable quantitative data.

The general acceptance of short-term tests came as a result of early studies that reported sensitivities (percentages of noncarcinogens identified as mutagens) and specificities (percentages of noncarcinogens identified as nonmutagens) of 90% or better; it has been determined that the close correlation was based on the fact that rodent carcinogenicity studies were generally conducted on chemicals with structures similar to known carcinogens. Some 200 short-term assays are available for detecting the potential of chemicals to induce gene mutations and chromosome breaks.

The use of *in vitro* test systems to detect genotoxic effects of airborne particles has been almost exclusively limited to the *Salmonella*/mammalian microsome (Ames) assay. However, higher plant mutagenicity bioassays have been in existence for many years, and are now well-established systems for screening and monitoring environmental chemicals for mutagens and genetic damage. Plants offer unique test systems for mutagen monitoring, but the lack of recognition of the high resolution that higher plant test systems provide for mutagenic testing and monitoring is considered the result of the low level of support for plant research in general. The use of the *Tradescantia* stamen hair assay to detect airborne mutagens and carcinogens, however, has brought about a general recognition of the efficacy of plant assays. Plant assays are inexpensive to use, easy to handle and applicable for indoor as well as outdoor detection of environmental mutagens. The Gene-Tox program of the U.S. Environmental Protection Agency reviewed nine assays in seven plant species for testing plant genotoxicity (*Allium, Arabidopsis, Glycine, Hordeum, Tradescantia, Vicia* and *Zea*). Recently, data have been analyzed for these nine assays and all nine have been shown to have a high sensitivity (few false negatives). The assays are considered the most appropriate tests for a risk-averse testing program in the prediction of carcinogenicity (Ennever *et al.*, 1988). Recently, higher plant test assays have been used for identifying chemicals causing aneuploidy.

Lower plant mutagen assays have been used primarily for water pollution and include the algae *Platymonas subcordiformis* and *Chlamydomonas reinhardtii*, the fern *Osmunda regalis* and the water hyacinth *Eichhornia crassipes*. Most aquatic plants (*Zostera, Elodea, Spirodella, Sphagnum*) and lichens (*Ramalina, Caloplaca, Cladonia, Pleurosum, Schreberi, Dicranum, Physconia*) have been used for studies on bioaccumulation of chemicals from the water and the atmosphere, respectively, and not for the detection of mutagens. This review will outline the major higher plant assays used in the screening and monitoring of genotoxic effects from airborne environmental chemicals.

II. Tradescantia—*The Ideal Plant for the Detection of Atmospheric Mutagens*

The plant *Tradescantia,* commonly known as Spiderwort, a member of the Commelinaceae, has several genetic endpoints that can be used for studying mutation assays: mutations in flower petals and stamen hairs, meiotic chromosome aberrations, frequency of micronuclei in tetrad cells, sister chromatid exchanges, and mitotic chromosome aberrations in microspores. The use of *Tradescantia* plants for mutation studies is not new and radiation cytological experiments using *Tradescantia* plants were initiated in the 1930s. Two *Tradescantia* bioassays are most ideal for the detection of atmospheric mutagens, and for *in situ* ambient atmospheric monitoring. These are the *Tradescantia* stamen hair assay and the *Tradescantia* micronucleus assay which are now being widely used.

A. Tradescantia *Stamen Hair Bioassay*

The stamen hair assay has been an important research tool in the field of genetic toxicology for several decades. This bioassay, developed by Arnold Sparrow at the Brookhaven National Laboratory for studying the effects of radiation, has been exploited as a somatic mutation test in the fields of radiobiology, chemical mutagenesis, and ambient air monitoring. *Tradescantia* plants (diploid; $2n = 12$) lend themselves to somatic mutation studies with special clones heterozygous for flower color providing the test system. In heterozygotes, the flower and stamen hair color is blue (dominant) and somatic mutations are scored as phenotypic changes to pink (recessive). Currently, the *Tradescantia* stamen hair assay utilizes clone 4430 for the detection of chemical mutagens, as this clone is over six times more sensitive to chemical mutagens than clone 02. Japanese investigators have used similar clones of *Tradescantia* to monitor areas around nuclear power plants to detect radionuclides which are released from reactors.

Laboratory studies with chemicals have shown that the stamen hair system is highly sensitive to gaseous chemicals. Flowers of *Tradescantia* plants after exposure to tritiated 1,2-dibromoethane contain similar amounts of radioactivity, indicating that chemical compounds readily enter the tissues. The mutation response frequency following a 144-hour (6-day) exposure at 2 ppm 1,2-dibromoethane was 5 events per 100 stamen hairs or equivalent to that produced by about 50 rad of X-rays; an exposure to 50 ppm for 6 hours elicits a mutation response equal to that induced by 18 rad of X-rays. In the case of another mutagen, ethyl methanesulfonate (EMS), an exposure to 100 ppm for 6 hours results in a mutation yield equivalent to the response to 160 rad of X-rays. The insecticidal fumigant, 1,2-dibromoethane (DBE) is used as a standard for mutation studies with airborne chemicals. DBE is preferable to EMS as a standard positive mutagen as, in the gaseous phase, EMS condenses on the chamber walls and on plant material. A list of some gaseous compounds studied in the stamen hair assay is presented in Table I. Chemicals in the vapor phase that have proven to be significantly mutagenic in the *Tradescantia* stamen hair assay include chlorpyrifos, benzene, 2-bromoethanol, chlorodifluoromethane (freon-22), diallate, 1,2-dibromoethane, ethyl alcohol, EMS, hexamethylphosphoramide, tetrachloroethylene, trichloroethylene, trimethyl phosphate, vinyl bromide, vinyl chloride, and the air pollutants nitrogen dioxide, nitrous oxide, ozone, and sulfur dioxide. The compounds 1,1-dibromoethane, vinylidine chloride, dichlorodifluoromethane, dimethylamine hydrochloride, and Vapona have not proved to be significantly mutagenic in the *Tradescantia* stamen hair assay. The chemicals 1,2-dibromoethane and trimethyl phosphate are potent mutagens, whereas chlorodifluoromethane (freon-22), methyl alcohol, nitrogen dioxide (NO_2), sulfur dioxide (SO_2), and vinyl chloride are weak mutagens.

Chemicals tested for *Tradescantia* stamen hair mutations in the vapor phase which significantly caused chromosome aberrations include ethylene oxide, ketene, and methyl chloride. Chemicals tested in the liquid phase (by immersing the inflorescence in the liquid or by spraying) include atrazine, benzidine, benzo(*a*)pyrene, caffeine, diethylstilbestrol, dimethylamine hydrochloride, ethyl methanesulfonate, ethylenethiourea, hexamethylphosphoramide, methyl nitro nitrosoguanidine, methylene chloride, methyl methanesulfonate, naphthylamine, nitrofluorene, nitrosoethylurea (NEU), nitrosomethylurea (NMU), nitrosodiethylamine (NDEA), nitrosodimethylamine (NDMA), pyrene, safrole, sodium azide, and tetramethylbenzidine. Plant tissue culture medium per se has no effect on the induction of mutant events in stamen hairs.

It has previously been shown that a linear relationship beween mutation frequency and dosage is found over a wide range of radiation levels with no evidence of a threshold effect even at levels as low as 250 mrad of X-rays. The mutation frequency with DBE increases linearly with both increasing concentration (0.1 to 100 ppm) and duration of exposure (2–144 hours) as with radiation. The mutation response also occurs at low levels, for example, a 6-hour exposure showed a significant decrease in slope in the low-dose range in contrast to that in the high-dose range. Consequently, extrapolation from high to low levels of mutagen exposure would lead to an underestimation of the hazard.

Three types of dose–response relationships have been noted: (1) linear for EMS, (2) a sharp increase of mutation induction in the low-dose range and a leveling off in the high range (NEU and NDEA), and (3) a slow increase at the low-dose range and a sharp increase at the high-dose range (NMU and NDMA). It is considered that the toxic effect which was observed at the high-dose range after treatment with NEU and NDEA occurred as a result of the production of many abnormal cell divisions and deformed cells. It is also possible that some repair mechanisms acted on the primary damage induced by the chemicals at the low-dose range after treatment with NMU and NDMA.

Because of the high sensitivity of the stamen hair assay to low doses of chemical air pollutants, two chemical mishaps have been identified in the routine use of the assay. At the Brookhaven National Laboratory, airborne mutagens were detected following an accidental exposure to chemicals that were exhausted from a fume hood duct on a roof and carried by air currents into the greenhouse intake vents. In another case, it was observed that the mutation frequency increased periodically in *Tradescantia* control plants growing in a growth chamber (Fig. 1). The cause was found to be associated with the routine periodic spraying of the insecticide chlorpyrifos in passageways between buildings. The fumes from the chlorpyrifos went into the ventilation system and circulated throughout the three-story building after each spraying and affected the *Tradescantia* plants growing in a growth chamber on the third floor. The increase in the mutation frequency of the control plants caused by the spraying of the insecticide would indicate that chlorpyrifos is a potent mutagen. These examples indicate the high sensitivity of the gene controlling the color of the stamen hairs and give further support for the use of *Tradescantia* plants as monitors for the mutagenicity of air pollution. A protocol for the *Tradescantia* stamen hair assay is given in Table II.

B. Tradescantia *Micronucleus Assay*

The *Tradescantia* micronucleus (Trad-MCN) assay is for the detection of chromosomal aberrations which occur in meiosis. Meiotic prophase chromosomes are highly susceptible to breakage, and fragments and isolated chromosomes end up as micronuclei in the tetrad (quartet) stage of meiosis. Clones 4430 and 03 are used for the *Tradescantia* micronucleus assay, although clone 03 is preferred as this clone has a low frequency of background chromosome aberrations. The number of cuttings and the technique for treatment are the same as that for the stamen hair assay (Table II). The number of micronuclei are scored 24–30 hours after treatment. The frequency of micronuclei induction is determined by dividing the total number of micronuclei by the number of tetrads analyzed. This ratio is multiplied by 100 to yield the number of micronuclei per 100 tetrads. The Trad-MCN assay has been used for detecting mutagenicity in ground and surface water, and soil and sludge samples from municipal, industrial, and hazardous waste sites. Recently, for the *Tradescantia* micronucleus assay, an image analysis system has been devised for determining the number of micronuclei in slides and analyzing the data.

In contrast to the large number of mutagenesis studies carried out on various liquids, relatively few studies have been carried out on air pollutants (Table III). Formaldehyde fumes with as little as 38 ppm exposure, induced a dose-related proportional increase in micronucleus frequencies ranging from 8.2 to 39.2 MCN/100 tetrads over 3, 6, 12, and 36-hour exposure periods. The control averaged 3 MCN/100 tetrads. Potassium dichromate was tested for the induction of micronuclei in in-

Table I Studies Using the *Tradescantia* Stamen Hair Bioassay

Agent[a]	Range of exposures
Atrazine (liquid)	0.045 g/pot chronic[b]; 6.9×10^{-5} *M* for 24 h
Chlorpyrifos	48% solution of stock 240 g/l
Benzene	4000 ppm for 6 h
Benzidine (liquid)	5.43×10^{-7} *M* for 24 h
Benzo(*a*)pyrene (liquid)	3.96×10^{-5} *M* for 24 h
2-Bromoethanol	24 ppm for 6 h
Caffeine (liquid)	10^{-3} *M* chronic[b]
Chlorodifluoromethane (Freon-22)	194 ppm for 6 h
Diallate	800 ppm for 24 h
1,1-Dibromoethane	250 ppm for 6 h[b]
1,2-Dibromoethane	1 ppm for 6 h; 0.14 ppm for 144 h
[^{3}H] DBE	4.8 ppm for 2 h (clone 02); 17 ppm for 0.25 h (clone 4430)
Dichlorodifluoromethane (Freon-12)	392 ppm for 6 h[b]
Diethylstilbestrol (liquid)	3.73×10^{-4} *M* for 24 h
Dimethoate (Cygon-2E)	50–100 ppm for 6 h
Dimethylamine hydrochloride	10^{-2} *M* for 2 h
Ethyl alcohol	1000 ppm for 24 h
Ethyl methanesulfonate (EMS)	5 ppm for 6 h
N,N'-Ethylenethiourea (liquid) (2-Imidazolidinethione)	9.79×10^{-5} *M* for 24 h
Ethylene oxide	5–7 ml for 5 min
Formaldehyde	38 ppm for 3 h
Hexamethylphosphoramide	Saturated vapor for 6 h
Hydrazoic acid (HN_3)	3.5–136 ppm for 6 h
Ketene	Saturated atmosphere for 5–15 sec
Malathion	0.15% for 4.5 h
Methane	50% for 72 h[b]
Methyl alcohol	43×10^3 ppm for 6 h
Methyl chloride	6–7 ml for 5 min
Methylene chloride	15 ppm for 6 h
N-Methyl-*N*-nitro-*N*-nitrosoguanidine (liquid)	10^{-3} *M* for 1.5 h
α-Napthylamine (liquid)	6.75×10^{-5} *M* for 24 h
β-Napthylamine (liquid	6.98×10^{-5} *M* for 24 h
Nitrogen dioxide (NO_2)	50 ppm for 6 h; 5 ppm for 24 h
Nitrosoethylurea	0.1 ng
Nitrosodiethylamine	1.0 ng
Nitrosodimethylamine	1.0 ng
Nitrosomethylurea	0.1 ng
Nitrous oxide (N_2O)	250 ppm for 6 h
Ozone	5 ppm for 6 h
Potassium dichromate	0.1% for 6 h
Pyrene (liquid)	4.94×10^{-6} *M* for 24 h
Smog	Ambient air

(*continues*)

Table I (*Continued*)

Agent[a]	Range of exposures
Sulfur dioxide (SO_2)	0.05 ppm for 30 min; 10 ppm for 18.5 h; 0.075 ppm for 18.5 h; 1 ppm for 22 h; 40 ppm for 6 h
Tetrachloroethylene	1035 ppm for 6 h
3,3′,5,5′-Tetramethyl benzidine (liquid)	4.16×10^{-6} *M* for 24 h
Trichloroethane	5,170 ppm for 6 h[b]
Trichloroethylene	0.5 ppm for 6 h
Trimethlyphosphate	13 ppm for 6 h
Vinyl bromide	50 ppm for 24 h
Vinyl chloride	75 ppm for 6 h
Vinylidene chloride	1,288 ppm for 6 h[b]
X-rays	250 mrad (0.2% mutations per rad)

[a] Vapor treatment except where noted. The lowest concentration is given at which an effect was observed.
[b] The highest concentration at which no effect was observed.

crements of 0.2% concentrations (0.1 to 1.0%) resulting in a response to concentrations which was significantly positive but nonlinear. Malathion fumes were extremely toxic to *Tradescantia* plants affecting both leaves and buds when the gaseous concentrations reached 0.25% or higher. The dose range for genotoxicity was between 0.15 and 0.25% producing 35 MCN/100 tetrads with controls at 5 MCN/100 tetrads. A protocol for the *Tradescantia* micronucleus assay is given in Table IV.

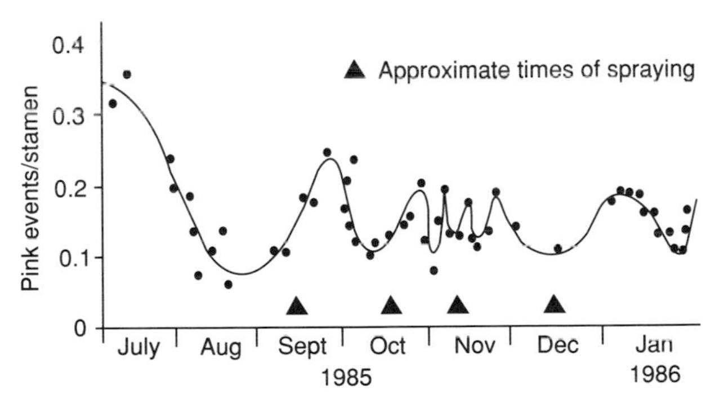

Figure 1 Frequencies of pink events (mutations) per stamen of *Tradescantia* (*ordinate*), in nontreated plants, on the days at which the stamen hairs were scored from July 1985 through January 1986 (*abscissa*). The arrows indicate the date at which chlorpyrifos (insecticide) was released into the building. The frequencies vary in a nonconstant manner.

C. Other Higher-Plant Mutagen Assay Systems

While the *Tradescantia* bioassays are ideal, there are a number of other higher plants which offer unique systems and lend themselves to screening and monitoring for atmospheric mutagens. A number of these assays with the species and chemicals tested are given in Table V. It may be seen from the table that there are a large number of phenotypic genetic markers available for such mutagenic screening and monitoring.

III. Pollen Assays

A. Tradescantia *Microspore Assay*

The study of air pollutants may be carried out by germinating pollen after it has been subjected to treatment and examining the chromosomes for aberrations in the germinating pollen tubes. In the *Tradescantia* pollen-tube assay, the pollen is cultured and the pollen tubes with the six haploid chromosomes are analyzed for chromosome aberrations (Table VI). While the assay is highly sensitive to gaseous impurities, the analysis of chromosome aberrations requires both greater competence and time for analysis than the *Tradescantia* mi-

Table II *Tradescantia* Stamen Hair Assay for *in Situ* Monitoring and Laboratory Experiments

About 100–150 plant cuttings bearing young inflorescences are required once a week to conduct 3–4 treated and control populations. Make 20 cuttings for each experiment, from 8 to 15 cm in length of young inflorescences prior to meiosis when the oldest apical bud is within 24 to 48 hours of blooming. Plants require 1,800 foot-candles fluorescent and 180 foot-candles incandescent light for good growth with a 16-hour light and an 8-hour dark period and a temperature of 21°C during the day and 16°C at night.

Place the cuttings in a beaker containing nutrient solution (one-third dilution of Hoagland's solution) as the cuttings must be maintained for at least 16 days after treatment. Aeration may be maintained by a small aquarium pump.

For gaseous agents, expose young inflorescences to chemical vapor (2 to 144 hours; most often 6 hours) in an enclosed chamber with a known concentration of gas (0.1 to 100 ppm). Alternatively, the inflorescences can be sprayed with a chemical or the inflorescences may be placed in the test chemical allowing the liquid to travel through the stem to the filament of the stamen. *In situ* monitoring for gaseous pollutants in the air is conducted by exposing the plant cuttings to the atmosphere at the selected sites. This can also be done by growing the plants on site.

Collect flower buds daily from day 7 posttreatment. Flowers cannot be fixed for later observation but must be analyzed daily for pink mutations for up to three weeks. Put flowers in a container with moistened sponge in a refrigerator until ready to examine. The peak mutation rate is generally between 8 and 15 days after acute treatment.

Cytological examination: Each flower contains 6 stamens and each stamen bears 40–75 hairs. The sepals, petals, and anthers are carefully removed. Remove filaments from each stamen. Prepare slides by placing filaments in liquid paraffin for microscopic examination; add cover slip. Examine immediately if possible, or place in a refrigerator for same day examination.

Examine under a dissecting microscope at a magnification of 25×. Use white light source (no filters) for the detection of pink mutations. A contiguous string of pink cells is scored as a single mutational event as all cells in one stamen hair are derived from an initial single cell by mitosis.

Statistics: Ten to 15 flowers are needed for each experimental group. Induced pink event rates are expressed as the mean of the rates for several consecutive peak response days. Mutations are recorded for pink events per 1000 hairs. The mean number of mutant events per hair is calculated by summing the number of pink events from individual flowers blooming during the peak response period and dividing by the product of total flowers analyzed, times the estimated number of hairs per flower.

cronucleus assay. As a result, the *Tradescantia* micronucleus assay has largely replaced this microspore assay.

B. Waxy *Locus*

The genetics of both *Zea mays* and *Hordeum vulgare* have been studied for many years and a number of bioassays have been developed to test for mutagens. In *Zea mays* the *waxy* (*wx*) locus is located on chromosome 9 at map position 9–59. In barley, the *wx* locus is located near the distal end of the short arm of chromosome 1. The starch composition of the pollen grain provides a color response with an iodine (I_2KI) solution in which wild-type grains (*Wx*) can synthesize amylose as part of their starch component and stain blue-black in color, whereas mutant grains carrying the recessive allele (*wx*) cannot synthesize amylose and stain reddish brown. Thus, the color of the pollen grain after staining with iodine gives a clear indication of its genotype. The pollen assay is one of the most useful as large numbers of individual grains can be analyzed providing a high degree of genetic resolution by visual means or by means of flow cytophotometry using fluorescent dyes. In a forward mutation test (*Wx* to *wx*) the reddish-brown mutant pollen grains appear in a field of blue-black grains and in a reversion mutation test, one treats a homozygous waxy (*wx, wx*) genotype and determines the frequency of dominant starch (*Wx*) pollen grains. The reverse mutant appears blue-black among the wild type reddish-brown pollen grains.

A procedure for *in situ* detection of mutagenic agents using *Zea mays* is as follows:

1. Grow plants to early anthesis adjacent to site or in pots and harvest tassels.
2. Agitate individual tassels in 70% ethanol to remove foreign pollen from the surface of the tassel.
3. Remove approximately 15 unopened florets from the tassel and agitate in a petri dish filled with 70% ethanol.
4. Dissect out anthers from unopened florets and place in a stainless steel cup of a VirTis microhomogenizer containing 0.6 ml of gelatin–iodine stain; mince anthers with scissors and homogenize for 30 sec.
5. Strain homogenate through cheesecloth onto the surface of a large microscope slide and place coverslip on pollen suspension.

Table III *Tradescantia* Micronucleus Bioassay

Agent	Range of exposures[a]
Air fresheners	0.3 ml for 6 h
Ammonium bromide	1,0 mM for 6 h
Arsenic trioxide	1.98 ppm for 30 h
1,2-Beng (a,h) anthracene	12.5 ppm for 30 h
Benzo (a) pyrene	50 μM vapor for 6 h
Cadmium sulfate	0.1 mM for 6 h
1,2-Dibromoethane	4.6–77.5 ppm for 6 h
Dicamba	50 ppm for 6 h
Dichlorvos	0.03–0.5% vapor for 6 h
p-Dichlorobenzene	272 ppm/min for 3–6 h
Dieldrin	3.81 ppm for 30 h
Diesel fumes	1/45 dilution for 46 min; dilution 1/80, 20 min; 3 : 1 ratio, dilution 1 : 80, 20 min
Diesel – soybean oil fumes	1 : 1 ratio, dilution 1/50, 20 min; dilution 1/80, 20 min; 3 : 1 ratio, dilution 1/80, 20 min
Dimethoate (Cygon-2E)	44–88 ppm for 6 h
Ethanol	5–12.5% for 6 h
Ether	0.25–0.75 mM for 6 h
Ethyl methanesulfonate (EMS)	1000 ppm vapor for 6 h; 50 mM (liquid) for 24 h
Ethylene dibromide	4.6 ppm vapor/min for 6 h
Ethylene oxide	5–7 ml for 5 min
Formaldehyde	0.5 ppm/min vapor for 1 h; 1.56 ppm/min for 6 h; 38 ppm for 3 h
Heptachlor	1.88 ppm for 30 h
Hydrazoic acid (HN_3)	136–544 ppm vapor for 6 h
Lead nitrate	1.0–100 mM for 6 h
Lead tetraacetate	0.44 ppm for 30 h
Malathion	0.15% vapor for 1.5 h
Maleic hydrazide	5–50 ppm for 6 h
Manganese chloride	1–20 mM for 6 h
Methyl chloride	6–7 ml for 5 min
Nitrogen dioxide (NO_2)	5 ppm for 24 h
Picloram (Tordon)	100 ppm for 6 h
Potassium arsenite	1.0 ppm for 30 h
Potassium dichromate	0.1% for 6 h
Riboflavin	25 ppm for 6 h
Saccharin	5 μM for 6 h
Sodium azide	0.2 mM for 6 h
Sodium bisulfite	10 μM for 6 h
Sodium selenite	250 mM for 6 h
Sulfur dioxide (SO_2)	1 ppm vapor for 2–6 h
Tetrachloroethylene	30 ppm vapor/min/2 h for 6 h
γ-Rays, Co^{60}	8.4 rad/h
β-Rays, P^{32}	24 pCi/mg
X-rays	1.6% mutations per rad
Zinc chloride	1 mM for 6 h

[a] The lowest concentration or exposure at which an effect was observed; range given when a single minimum effective dose was not stated.

Table IV Protocol for Tradescantia Micronucleus Bioassay

The meiotic pollen mother cells (PMC) of *Tradescantia* are highly synchronized and are very sensitive to physical and chemical mutagens. A high frequency of chromosome aberrations can be induced with very low levels of mutagen. These induced chromosome aberrations become micronuclei (MCN) in the tetrads at the end of meiosis where they can be easily identified and scored.

Make cuttings (15 to 20, from 5 to 8 cm in length) of young inflorescences prior to meiosis when the oldest apical bud is ready to bloom within 24 to 48 hours, and place the cuttings in a beaker containing water. (Cuttings may be grown in an aerated nutrient solution in which case a nutrient culture control solution must also be used).

Expose young inflorescences to chemical vapor (6 to 24 hours; most often 6 hours) in an enclosed chamber with a known concentration of gas and rate of flow. Alternatively, (1) the inflorescences can be sprayed with a chemical, or (2) the apex of the inflorescences may be placed in the test chemical in a beaker, or (3) the cuttings may be placed directly in a chemical allowing the liquid to travel up the stem to the filament of the stamen. *In situ* monitoring for gaseous pollutants in the air is conducted by exposing the plant cuttings to the atmosphere at the selected sites. This can also be done by growing the plants on site.

Fix flower buds 24 or 30 hours after exposure (testing one or two buds prior to fixation to confirm that meiosis has proceeded to the tetrad stage) in glacial acetic acid and 95% alcohol (1 : 3). Samples may be stored in 70% ethanol.

Prepare microslides by removing a single bud at a time and squashing the PMCs in aceto-carmine stain. The stain penetrates fairly readily, but flaming over an alcohol lamp briefly will speed the staining process. Only buds in the early tetrad stage (four cells encased in an envelope) are scored for micronuclei and all other buds are discarded. With practice, the size of bud at the early tetrad stage can readily be selected among the series of buds on a given inflorescence. Allow 2 to 5 minutes for the stain to penetrate the nuclei. Five to 10 microslides are made for each experimental group of from 15 to 20 inflorescences. One tetrad may contain a number of micronuclei (more often 1 to 5).

Cytological examination: The tetrads are examined with a microscope at a magnification of 400 × (10 × ocular and a 40 × objective).

Statistics: An examination of 300 tetrads per slide and five slides per treatment have been found to give highly significant results for a single experimental group. The total number of micronuclei in a given slide is counted and divided by the number of tetrads scored. The fraction is the expression of the number of micronuclei per tetrad, or the percentage is the expression of the number of micronuclei per 100 tetrads.

6. Examine pollen suspension under a dissecting microscope and count blue-black stained pollen grains; estimate total number of grains per slide by counting the number of pollen grains in 20 randomly chosen 1 mm^2 areas and multiply by the appropriate factor.

7. For each plant calculate *wx* to *Wx* reversion frequency by dividing the total number of *Wx* pollen by the total estimated number of pollen grains analyzed. A minimum of 1×10^5 pollen grains should be screened for each treatment group. Determine the mean and the standard error for each group.

C. Pollen Fertility Assay

A general procedure for estimating the fertility of a plant as an indication of genetic damage can be obtained by counting the percentage of unstained (unfilled, shrivelled) and stained (filled) pollen grains. It should be noted that this method gives only an estimation of the fertility and that true fertility can only be determined by germinating the pollen grains. A number of stains (Lactophenol–Fast Green, Cotton Blue, etc.) may be used. The procedure for the preparation of Lactophenol–Fast Green stain follows.

1. Preparation of Lactophenol

20 ml water
20 grams carbolic acid crystals (phenol)
20 ml lactic acid (80%)
20 ml glycerine

2. Preparation of stain

In a dropping bottle containing the Lactophenol, add a few grains of Fast Green stain. This may be done by placing a moist needle in the stain and then shaking off as much as possible.

The concentration of the stain is not important. A very weak staining mixture is quite satisfactory as the slides may be kept without making them permanent for several months. Overstaining must be avoided.

3. Procedure for Staining Pollen

Place a drop of Lactophenol–Fast Green on a slide. Place one or more anthers in drop of Lactophenol-Fast Green (when the anthers are mature but before the pollen has been shed). With needles break anthers and distribute pollen in the Lactophenol-Fast Green. Remove any debris and cover with a coverslip. An estimate of the fertility can be made when the pollen is stained in 2 to 4 hours, or after the slides have been left overnight, or at a later date if the slides are kept flat and dust free. These temporary slides may be kept for several months. A minimum of 1×10^5 pollen grains should be screened for each treatment group.

IV. In Situ Monitoring

In situ monitoring has several advantages over laboratory testing in that the test material is exposed to actual living conditions of the atmosphere, water, or soil such as found around urban and industrial sites. In many cases, such environmental conditions cannot be duplicated exactly under laboratory conditions. This would be the case where several chemicals and complex mixtures are present. Likewise, monitoring may be carried out where mutagenic agents may be suspected.

Considerable time has been spent at the Brookhaven National Laboratory, Long Island, in developing a mobile monitoring vehicle for testing ambient air. Two control sites (Grand Canyon, Arizona and Pittsboro, North Carolina) and 16 industrial sites were monitored for air pollution and significant increases in the frequency of pink mutations were observed at many of the industrial sites (Fig. 2).

A Trax carrier has been designed that can be placed on top of an automobile and used to transport *Tradescantia* cuttings to a site for *in situ* monitoring. The *Tradescantia* stamen hair or micronucleus assays have been used to determine the extent of pollution at sites such as public parking garages, bus and truck stops, and industrial and farm sites. With the exception of a farm site, one or more locations at the other sites were significantly positive. In addition, *in situ* monitoring using the *Tradescantia* assays, has been carried out on waste soil (sludge) and used to determine water quality of bodies of water including the effluent from a pulp and paper mill (Grant *et al.*, 1992).

The plants *Glycine max, Tradescantia* clone 4430, and *Zea mays* have been used for *in situ* monitoring in the vicinity of a lead-smelting plant in southeastern Missouri where the major effluents were lead, cadmium, copper, zinc, and sulfur. Three sites 0.3, 1.7, and 11.4 km from the source were monitored. The conditions at the site 0.3 km distant from the plant were found too toxic for the growth of *Glycine max* plants and the plants died in the seedling stage. In the case of *Tradescantia*, there was a significant difference between the mutational frequency at the site 1.7 km from the smelter and the control, whereas the increase in mutational frequencies at the 0.3- and 11.4-km sites, while elevated above controls, were not significantly different from the control levels. However, when analyses of soil samples from the different sites was carried out in the laboratory, it was found that high lead toxicity was present in the soil at 0.3 km from the plant site. At this site, it was considered that the high lead toxicity may have masked any mutagenic effect.

The *Zea mays* the pollen *waxy*-locus mutagen assay was also used. The mutation frequencies at the closest two sites (0.3 km and 1.7 km) were highly significant. In addition, the frequency of pollen abortion was significantly greater at the 0.3- and 1.7-km sites than at the 11.4-km site.

V. Laboratory or Off-Site Mutagen Assays

A. Introduction

Many atmospheric pollutants can be studied in the laboratory in plant test systems where they can be assessed for chromosome damage both in a number of different plant species and by different assays. Chromosome damage may be assessed by analysis of (1) mitotic root-tip tissue, (2) meiotic divisions, (3) meiotic tetrad cells, and (4) pollen sterility as an indication of reproductive success. If a chemical is a clastogen (capable of breaking chromosomes), this would permit exchanges with subsequent cytological or genetic damage. A chemical may not affect DNA directly, but may instead cause segregation errors and subsequent genetic differences, for example, via aneuploidy. The most common chromosome abnormalities include (1) chromosome and chromatid breaks, (2) acentric fragments, (3) chromatid and subchromatid exchanges, (4) chromatid and chromosome bridges, and (5) sister chromatid exchanges. The specific type of aberration induced is a function of the time at which the interphase nucleus is exposed to the pollutant. The mitotic index gives an indication of the toxicity of

Table V Plant Test Systems Other Than *Tradescantia*

Species	Agent	Stage treated/locus	Target	Range
Allium cepa	Ozone	First leaves of young plants	Stomata closures	0.3 ppm until stomata close
	Nitrous oxide	Sprouting bulbs	C-mitosis	6 atm for 6 h
Arabidopsis thaliana		Many genetic loci	Chlorophyll-deficient immature embryos and seedlings	
	Nitrous acid	Developing buds	Sterility as unfertilized ovules; embryonic lethals	100 pphm for 4 and 8 h
	Sodium azide	Dry seeds	Chlorophyll recessives	2.5 nmol/l for 3 h
	EMS	Growing point with microsyringe	Chlorophyll recessives	10 m*M*
Avena sativa	Nitrous oxide	Tillers	Induction of polyploids	4 atm for 24 h
Beta vulgaris		*S* alleles	Sexual compatibility	
Crepis capillaris	Nitrous oxide	Plants when youngest flowers completed flowering	Induction of polyploids	10 atm for 6–12 h
Datura stramonium	Nitrous oxide	Whole plants and cut flowering branches	Haploid mitotic nuclei for C-metaphases	4, 6, or 9 atm for 2 h
Glycine max		Y_g, Y_{11}	Yellow and green, single and twin spots on leaves	
Hordeum vulgare		Many genetic loci	Chlorophyll-deficient seedlings	
		Wx	Starch composition Endosperm and pollen	
	Nitrous oxide	Tillers	Doubling haploids	21.1, 31.6, and 42.2 × 10^3 kg/m^3 atm for 12 and 24 h
	Nitrous oxide	Tillers	Induction of polyploids	4 atm for 24 h
	Nitrous acid	Embryos and tillers in culture medium	Induction of polyploids in embryos	4 atm for 24 h
	Sodium azide	Dry seeds	Chlorophyll recessives	10 nmol/l for 18 h
Lycopersicon species		*S* alleles	Sexual compatibility	
Lycopersicon esculentum	Hydrogen fluoride	Plants with developing buds and young leaves	Mitotic and meiotic chromosome aberrations	3 μg/m^3 for 4, 6, 8, 10, and 12 days
	Hydrogen fluoride	Seeds	Meiotic chromosome aberrations	3 μg/m^3 for 4 days
Melandrium species	Nitrous oxide	Young plants	Induction of polyploids	2–5 atm for 5–16 h
Nicontiana alata		*S* alleles	Sexual compatibility	
Oenothera organensis		*S* alleles	Sexual compatibility	
Oryza sativa		Many genetic loci	Chlorophyll-deficient seedlings	
		Wx	tarch composition, endosperm, and pollen	
Petunia species		*S* alleles	Sexual compatibility	
Phalaris species	Nitrous oxide	Plants when youngest flowers completed flowering	Induction of polyploids	10 atm for 6–12 h
Phaseolus vulgaris	Hydrogen fluoride	Continuous from seed to seed maturity	Effect on progency	0.58, 2.1, 9.1, and 10.5 μg F/m^3, continuous treatment
Solanum tuberosum X S. phureja	Nitrous oxide	Whole plants and cut flowering branches	Haploid mitotic nuclei for C-metaphases	4, 6, or 9 atm for 2 h
Trifolium species	Nitrous oxide	Excised heads	Induction of polyploids	6 atm for 24 h

(*continues*)

Table V *(Continued)*

Species	Agent	Stage treated/locus	Target	Range
Trifolium pratense	Nitrous acid	Excised stems bearing inflorescences	Induction of polyploidy	6 atm for 24 h
Trifolium repens		V_{by}	Alteration of color pattern on leaves	
Triticum aestivum	Nitrous acid	Embryo and tillers in culture medium	Induction of polyploidy in embryos	4 atm for 24 h
Triticum vulgare	Nitrous oxide	Seedlings	Induction of polyploids	4 atm for 24 h
Vicia faba	Ozone	Seeds	Somatic chromosome aberrations	0.4 weight percent ozone for 15, 30, and 60 min
	Ozone	Inflorescences with developing buds	Meiotic chromosome aberrations	100–200 pphm for 4 or 8 h
	Cadmium nitrate	Root tips	Sister chromatid exchanges	0.01 ppm
Zea mays[a]		*a*	Anthocyaninless	
		Adh-1	Alcohol dehydrogenase activity in pollen	
		bz	Bronze	
		C	Colored aleurone	
		gl	Glossy	
		R	Color in aleurone, anthers, leaf tips	
		Sh-2	Shrunken endosperm	
		Su	Sugary	
		v	Virescent	
		Y_{g2}	Yellowish streaks on leaves	
	Ethylene oxide	Pollen	Chromosome aberrations in kernels	1 EO:20 air at 50 ml/min for 2 or 4 min
	Hydrogen fluoride	Plants with developing buds	Meiotic chromosome aberrations	3 $\mu g/m^3$ for 4, 6, 8, 10, and 12 days

[a] Many of these genetic loci are also present in other species, such as *Arabidopsis, Hordeum* and *Oryza*.

the compound. It is determined by the number of dividing cells (all stages of mitosis) per 1000 observed cells

Several plants have desirable features for detecting and analyzing chromosome aberrations. The eight most commonly used species for assessing chromosome damage are broad bean (*Vicia faba*, $2n = 12$), onion (*Allium cepa*, $2n = 16$), barley (*Hordeum vulgare*, $2n = 14$), *Crepis capillaris* ($2n = 6$), pea (*Pisum sativum*, $2n = 14$), spiderwort (*Tradescantia*, $2n = 12$), *Zea mays* ($2n = 20$) and tomato (*Lycopersicon esculentum*, $2n = 24$). The first three species have been used in over 80% of the studies. In one review on plant assays used in the detection of aneuploidy, *Allium cepa* was more frequently used as a test organism than *Vicia faba*. Historically, *Allium cepa, Lycopersicon esculentum, Pisum sativum,* and *Vicia faba* were used as early as 1937 to study the effects of X-rays on chromosomes.

B. *Vicia faba*

The broad bean has large chromosomes which are amenable to cytological analyses and a number of karyotypes with special structural rearrangements.

Table VI *Tradescantia* Microspore Bioassay[a]

Agent	Range of exposures[b]
Ethylene oxide	5–7 ml for 5 min
Ketene	Saturated atmosphere for 5–15 sec
Methyl chloride	6–7 ml for 5 min
Sulfur dioxide (SO_2)	0.05 ppm for 30 min; 10 ppm for 18.5 h[c] 0.075 ppm for 18.5 h[d]

[a] Mature pollen grains are cultured and chromosome aberrations are observed in pollen tube mitosis.
[b] The lowest concentration at which an effect was observed.
[c] Pollen tube growth inhibition.
[d] Mitotic arrest.

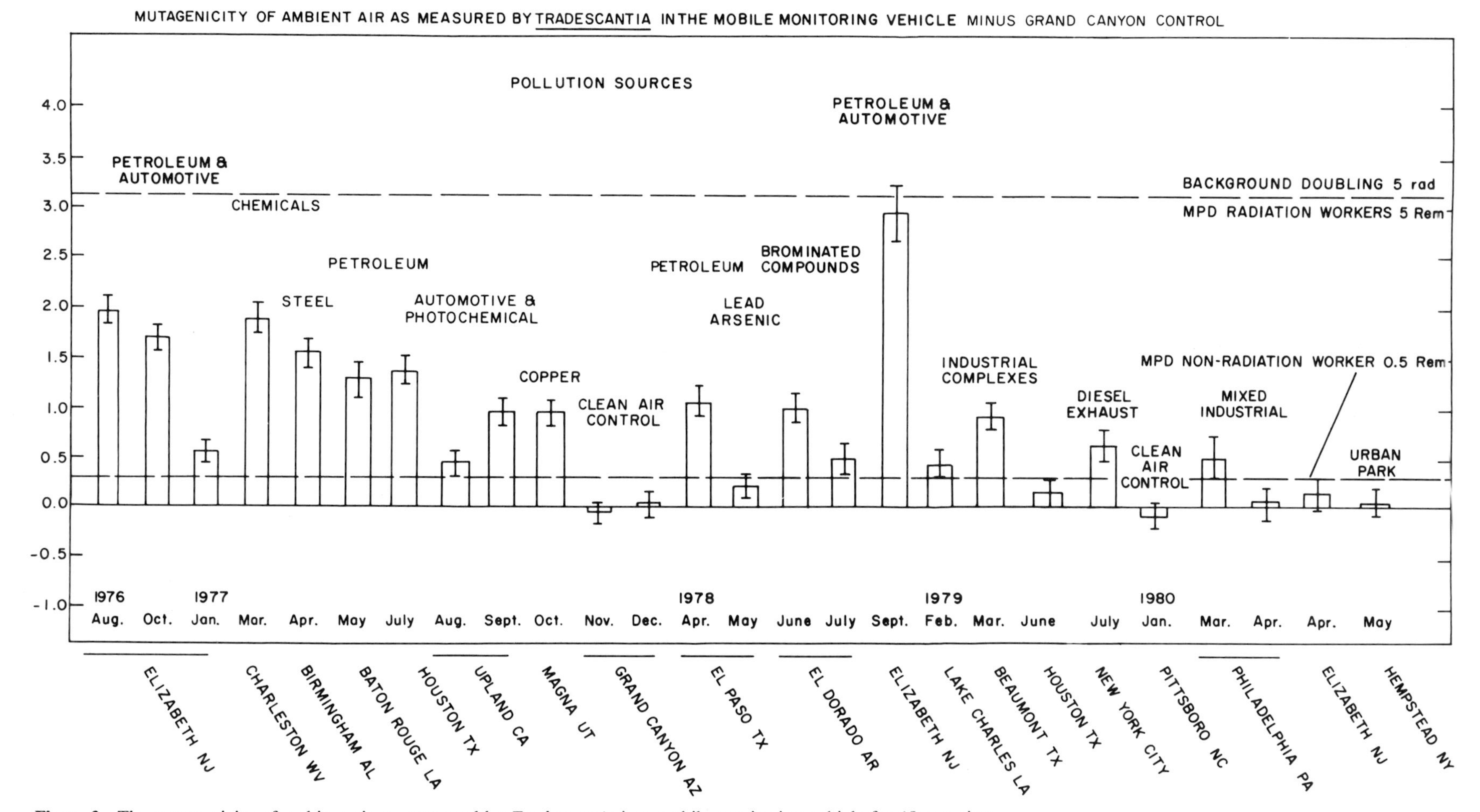

Figure 2 The mutagenicity of ambient air as measured by *Tradescantia* in a mobile monitoring vehicle for 18 test sites. The broken horizontal lines indicate the level of mutation response in *Tradescantia* radiation exposures at established maximum permissible dosages of 0.5 rem for nonradiation workers and 5 rem for radiation workers. (From Schairer and Sautkulis, 1982. Reproduced by permission.)

This species possesses six pairs of chromosomes which are designated according to centromere position as either M (median) or S (terminal). The single pair of M chromosomes is more than twice the length of the S chromosomes (mean 2.3 : 1) and possesses a large satellite on the short arm. Seeds of *Vicia faba* are large and do not lend themselves to germination in a petri dish. They may be germinated in paper towels or in an inert substance such as perlite or in aerated running tapwater using a thermostatically controlled heater set at 20–22°C. Primary roots are decapitated when they reach a length of approximately 4–5 cm which stimulates growth of secondary roots. At the same time, the seed coats should be removed carefully. The seeds are transferred to vials or water tanks kept at a constant temperature. The secondary roots are treated when they reach 3–4 cm in length.

C. *Allium cepa*

The early use of the onion as a test organism led to the assay being referred to as the "Allium Test." Its ease in handling has been one of its most attractive features. The onion bulb can be placed on the top of a container and the root tips readily treated. Seed is also easy to handle. Most often, young bulbs of uniform size are used and selected so that the base will fit a round-mouth glass jar. The loose outer scales of the bulbs are removed and the bottom of the bulb scraped so that the apices of the root primordia are exposed. The bulbs are submerged in solution to about one-quarter the depth of the bulb. The bulbs should be kept in the dark at 20°C and after 2–4 days the root tips will grow to 1–3 cm and are ready for treatment. The effects of benzo(*a*)pyrene, one of the polycyclic hydrocarbons which has to be activated before exerting an effect, have been positive in the *Allium* test indicating that these cells possess the enzymes responsible for metabolizing the compound.

D. Species with Small Seeds

Species that produce small seeds such as barley (*Hordeum vulgare*), *Crepis capillaris*, and tomato (*Lycopersicon esculentum*) may be treated in petri dishes and the root tips examined directly from germinated seed or from seedlings grown in pots. In addition to root tips, chromosomes may also be observed in leaf and stem primordia which may be useful for plants grown *in situ*. Before use seeds may be disinfected by a short immersion (3–10 min) in 5% calcium hypochlorite solution to eliminate pathogens from the seed surface. After this treatment, the seeds should be thoroughly rinsed with 3–4 changes of distilled water. When tap water cannot be used (for instance when it is too heavily chlorinated), Hoaglands solution may be substituted. Treatments should be carried out in the dark. In some cases, it may be desirable to let some of the treated seeds grow to mature plants where there are a large number of phenotypic genetic markers available for study (Table V).

E. Root-Tip Chromosome Aberration Assay

There are many schedules for the preparation of root tips (germination, fixation, staining, maceration, etc.) for cytological examination. A general procedure for preparing and examining root tips for chromosome aberrations is given in Table VII.

F. Root-Tip Micronucleus Assay

In mitotic cells, such as root tips, chromosome breakage and missegregation can lead to the production of fragments and isolated chromosomes. These chromosome aberrations can end up as micronuclei in telophase and interphase cells. The micronuclei can be readily observed in a similar manner as in *Tradescantia* tetrad cells in meiosis. This root tip micronucleus assay is receiving more attention and some studies have been carried out in *Allium cepa, Hordeum vulgare, Vicia faba, and Zea mays*. An advantage of this assay is that a person does not require the training that is necessary for the analysis of the different types of chromosome aberrations. However, while less critical analysis would be required for obtaining the data, a disadvantage would be the large number of cells that may need to be observed, especially for a chemical that gives a weak positive reaction.

VI. Other Somatic Mutation Assays for the Detection of Atmospheric Mutagens

A. Gene Mutations in Leaves

A number of plants (*Antirrhinum majus, Arabidopsis thaliana, Glycine max, Petunia hybrida,*

Table VII Protocol for Preparing and Examining Root Tips for Chromosome Aberrations

Chemicals: Test liquids should be prepared fresh, or stored in a refrigerator and brought to room temperature before use. For those chemicals that are not water soluble, the chemical is generally dissolved in dimethyl sulfoxide or ethanol at the concentration range of 5–10% in these solvents. Positive and negative controls should be included with each experiment.

Recovery period: Following treatment, the root tips should be allowed to recover in water for one complete cell cycle, since chromosome aberrations may be produced by different chemicals at different stages in the interphase period (G_1, S, or G_2). Thus, one may determine whether primarily chromosome or chromatid aberrations are produced. The cell cycle time for *Allium cepa* and *Pisum sativum* is 14 hours. Recovery periods of 20 and 26 hours are recommended for *Vicia faba*.

Pretreatment: Root tips may be fixed immediately or pretreated prior to fixation. Pretreatment may be carried out with 0.05 to 0.1% aqueous solution of colchicine or 0.002 *M* 8-hydroxyquinoline for 1 to 2 hours to obtain a large number of metaphases for scoring interchanges and deletions. If only anaphase aberrations are scored, pretreatment should be omitted, since pretreatment destroys the spindle, preventing the mitotic cycle from continuing beyond metaphase.

Fixation: Fix in freshly prepared absolute or 95% alcohol–glacial acetic acid (3 : 1) from 30 minutes to 24 hours. After fixation, the root tips can be stored in 70% ethanol in a refrigerator until it is convenient to carry out the next step.

Staining: The Feulgen reagent, specific for DNA, provides a well-defined contrast between the chromosomes and the cytoplasm; hydrolyze in 1 *N* HCl at 60°C for 4 to 12 minutes (optimum time to be determined) and stain in reagent for 2 hours. Other stains may be used such as acetocarmine and acetoorcein.

Maceration: Treat root tips with 5% pectinase for one to three hours. If roots are left too long they will become soft and difficult to handle. If not left long enough they will be difficult to squash.

Squashing: On a slide, remove the darkly stained meristematic region with a dissecting needle in a drop of 45% acetic acid and squash the material with a cover slip. The pressure applied to the cover slip to spread the cells will depend largely on maceration.

Slide preparation: A temporary mount may be obtained by sealing the cover slip with a paraffin–gum arabic mixture, clear fingernail polish, or rubber cement. The slide will deteriorate after two or three days even when kept in a petri dish and stored in a refrigerator. To make permanent, the slide must be frozen (e.g., in dry ice or liquid CO_2). The coverslip is removed with a razor blade or scalpel. The root tip material adheres to the slide. Quickly immerse slide in absolute alcohol (two changes) and mount in Euparal or other mounting medium with a clean coverslip.

Scoring: Slides should be masked and coded before analysis.

Pisum sativum, Salvia splendens, Nicotiana tabacum, Lycopersicon, Tradescantia, and *Zea mays*) have been used in which genetic end points (including deletions, somatic crossing over, gene conversion, and point mutations) give rise to somatic mosaicism on leaf tissue. Plants heterozygous for gene markers affect chlorophyll or anthocyanin development in the leaves and floral parts. Mutant sectors may be exhibited as streaks, single spots, or twin spots. The soybean plant may be used as an example. Plants homozygous dominant $Y_{11}Y_{11}$ are phenotypically green; plants heterozygous $Y_{11}y_{11}$ light green; plants homozygous recessive at the y_{11} locus are golden yellow. Mutational events are observed as (1) twin spots in leaves of heterozygous light green plants, or (2) light green spots on yellow, or (3) dark green leaves on homozygous plants. Such plants may be used as *in situ* monitors in environmental monitoring.

B. *Arabidopsis* Mutation Assay for Chlorophyll-Deficient Embryos

Arabidopsis thaliana ($2n = 10$) is a member of the Crucifer (cabbage) family. It possesses certain features that have made it an excellent plant, comparable to microorganisms, for testing the effects of chemical mutagens. It has a short life cycle of 5 to 6 weeks (about 8 generations can be raised annually) and its small size permits a large number of plants to be grown in a small area such as culture tubes or petri dishes (300 plants per 9 cm dish). This species is quite sensitive to a wide range of compounds and can activate promutagens to mutagens. A single plant can produce more than 50,000 seeds. Thousands of gene loci can be simultaneously monitored for mutations within one culture. Most commonly, mature seeds are exposed to the mutagen, and the fruits produced on the developing plants (M_1) are examined for embryo mutations (M_2). Briefly, the bioassay is as follows:

1. *Arabidopsis* seeds (1000) are soaked for 3–48 hours at 25°C in 1 to 2 ml of the solution of the test compound dissolved in 0.2 *M* citrate–phosphate buffer (pH 5). Small test tubes are used. The seeds are rinsed thoroughly in water by changing the water in the test tube at least 6 times over a 20–30 minute period.

2. With the help of a 10 ml glass pipet, the seeds are transferred into sterilized soil or Pro-Mix in Petri plates (9 or 11 cm). At the 2 to 4 true leaf stage, 100 seedlings per treatment are transferred into flats filled with Pro-Mix. Seedlings are cultivated under short day conditions (8 hours) for about 6 weeks to prevent flowering; later under long day (16 hours) to induce flowering.

3. After 3 to 4 weeks, the seed coats (before becoming brown) are opened and the embryos are analyzed for chlorophyll and embryonic mutations. At this stage the white, yellow, or pale green cotyledons are easily distinguished from the green ones. Embryonic mutations produce seeds much smaller than normal. The spontaneous frequency of mutations is below 1.0% on a genome basis.

4. From each plant, 3 consecutive fruits from the basal part of the main stem are scored. Three fruits from 60 to 100 plants are scored for each treatment. A 0.2% dose of ethyl methanesulfonate is generally used as a positive control.

C. Chromosome Mutations—Aneuploidy

Recently, a seedling assay using Neatby's strain of Chinese Spring wheat (*Triticum aestivum*) which is suitable for *in situ* detection of atmospheric mutagens is being developed for the detection of aneuploids. In this assay, green and white sectors on cream-colored (virescent) leaves of young seedlings are used as indicators for gain or loss of chromosomes. The wheat seedling has a temperature-sensitive mutant allele (v_1) on chromosome 3B which is responsible for green and white sectoring on the virescent leaf. The appearance of green sectors is associated with monosomy ($v_1/0$) of chromosome 3B, whereas, light green sectors are caused by trisomy of its homoalleles on chromosomes 3A and 3D. White sectors are induced by trisomy ($v_1/v_1/v_1$) of chromosome 3B, and green and white pigment is dependent upon the dose levels of the V_1 alleles due to loss or addition of chromosome 3 (A, B, or D) or loss of a segment of chromosome 3B bearing the v_1 allele. Green and white sectors are scored on the second, third, and fourth leaves. Since there is no selective disadvantage to cells with a chromosome loss or gain due to the hexaploid condition, all of the cells in which aneuploidy and/or small deletions have occurred are recovered. A similar aneuploid assay using *Tradescantia ohiensis* ($2n = 4x = 24$) has been reported in which anaphase chromosome distribution is influenced by environmental chemicals. These assays provide a sensitive method for evaluating the genotoxic effects of potential pollutants in the ambient environment under natural conditions to induce chromosome abnormalities.

VII. Plant Metabolic Activation

Plant activation is the process by which a promutagen is activated into a mutagen by the biological action of a plant system. Many xenobiotics may be promutagens and while not mutagenic themselves require metabolic activation before any mutagenic activity is observed. Plants can metabolically activate chemicals into mutagens as when extracts of *Zea mays* plants are grown in small quantities of the herbicide atrazine. The extracts are mutagenic in *Saccharomyces cerevisiae, Salmonella,* and *Escherichia coli,* whereas atrazine without activation is not mutagenic in yeast or bacteria. The importance of this concept is critical, as many pesticides previously shown to be nonmutagenic can be metabolized by plants into mutagens. To prove that a chemical is a plant promutagen, one must be able to separate the activation process from the genetic end point used to assay for mutagenicity. This criterion is met, for example, by subjecting one organism to the test chemical and then using an extract from the treated organism to test for mutagenicity in a completely different organism. Plant activation assays have been carried out in a number of green plant species, including *Allium cepa, Arabidopsis, Brassica napus, Daucus carota, Glycine max, Gossypium hirsutum, Helianthus annuus, Hordeum vulgare, Medicago sativa, Nicotiana tabacum, Pisum sativum, Solanum tuberosum, Tradescantia, Triticum aestivum, Vicia faba,* and *Zea mays.* The most frequently used genetic indicator organism is *Salmonella typhimurium* but others used include *Aspergillus nidulans, Escherichia coli, Saccharomyces cerevisiae,* and *S. pombe.*

VIII. In Situ Weed Communities for the Detection of Atmospheric Mutagens

The use of native plants to detect symptoms of plant injury and accumulation of pollutants is well known. However, the use of weeds to detect altered reproductive patterns induced by atmospheric environmental mutagens through the study of chromosome irregularities and pollen abortion has not received the attention it deserves. Weeds exist both in open situations and in closed communities. Although these habitats may be well defined, changes do arise as a result of chemical insults to the environment, such as pollution from factories and herbicide clearance of roadsides and rights-of-way. The vegetation composition of an area may be altered through the development of resistant weeds. Somatic and meiotic chromosome aberrations may be examined in flower buds in a population of chosen plants. Plants which have been used for such purposes include the annual *Ambrosia artemisiifolia;* two biennials, *Melilotus alba* and *Pastinaca sativa* and two perennials, *Solidago canadensis* and *Vicia cracca*. From the frequency of chromosome aberrations found in sampling plants at different intervals (months) at *in situ* sites around urban and industrial complexes an indication of mutagenic damage may be assessed. *Tradescantia* plants or cuttings can be transported and the stamen hair and micronucleus assays used in conjunction with studies on natural vegetation.

IX. Concluding Remarks

Numerous plant assays have been developed which are suitable for monitoring and testing of airborne pollutants. Only the highlights of some of the most frequently used assay systems have been outlined here. It should be evident that plant systems provide a unique resource for monitoring the environment and evaluating their mutagenic activity. Plant systems have many genetic end points which serve in an excellent capacity as indicators of cytotoxic, cytogenetic, and mutagenic effects of environmental chemicals. Furthermore, plants are inexpensive, easy to handle, and amenable to a wide range of experimental conditions. Activities can be measured in a variety of plants such as gene mutations in leaves, embryo abortion, endosperm, and pollen. However, the *Tradescantia* stamen hair mutation assay and the *waxy* mutants in maize have been among the most widely used assays for determining the mutagenicity of airborne pollutants. The most frequently used species to measure chromosome damage have been *Tradescantia* (Trad-micronucleus assay), *Allium cepa, Vicia faba,* and *Hordeum vulgare,* although a number of other species lend themselves to this type of study. Chromosomal damage can be recorded as chromosome and chromatid breaks, exchanges, fragments, bridges, micronuclei, and sister chromatid exchanges. The chromosome damage occurring in meiotic cells can also serve as an indicator of the genetic effects which can be passed on to future generations. Since some plant assays can be transported to a site, or grown *in situ,* ambient levels can be determined for mutagenic activity. The extremely high sensitivity of some of these plant assays, especially the *Tradescantia* stamen hair and micronucleus assays, makes them especially useful for evaluating pollutants that occur at extremely low concentrations and at considerable distances from the source of the pollution (Sandhu *et al.,* 1991).

Related Articles: AMBIENT ACIDIC AEROSOLS; ENVIRONMENTAL MONITORING; MONITORING INDICATORS OF PLANTS AND HUMAN HEALTH FOR AIR POLLUTION.

Acknowledgments

Financial support from the Natural Sciences and Engineering Research Council of Canada for studies in genetic toxicity of environmental chemicals is gratefully acknowledged.

Bibliography

Constantin, M. J., and Owens, E. T. (1982). Introduction and perspectives of plant genetic and cytogenetic assays. A report of the U.S. Environmental Protection Agency Gene-Tox Program. *Mutation Res.* **99,** 1–12.

Ennever, F. K., Andreano, G., and Rosenkranz, H. S. (1988). The ability of plant genotoxicity assays to predict carcinogenicity. *Mutation Res.* **205,** 99–105.

Fiskesjo, G. (1985). Allium test as a standard in environmental monitoring. *Hereditas* **102,** 99–112.

Forer, A., and Swedak, J. (1991). Practical experiences in using biological systems to monitor indoor pollutants. *In* "Indoor Air Quality for People and Plants." (J. C. Baird, M. B. Berglund, and W. T. Jackson, eds.) pp. 129–158. Swedish Council for Building Research, Stockholm.

Grant, W. F., Zinov'eva-Stahevitch, A. E., and Zura, K. D. (1981). Plant genetic test systems for the detection of chemical mutagens. *In* "Short-Term Tests for Chemical Carcinogens." (H. F. Stich and R. H. C. San, eds.) pp. 200–216. Springer-Verlag, N.Y.

Grant, W. F., Lee, H. G., Logan, D. M., and Salamone, M. F.

(1992). The use of Tradescantia and *Vicia faba* bioassays for the *in situ* detection of mutagens in an aquatic environment. *Mutation Res.* **270,** 53–64.

Kihlman, B. A., and Andersson, H. C. (1984). Root tips of *Vicia faba* for the study of the induction of chromosomal aberrations and sister chromatid exchanges. *In* "Handbook of Mutagenicity Test Procedures," 2nd. ed. (B. J. Kilbey, M. Legator, W. Nichols, and C. Ramel, eds.) pp. 531–554. Elsevier Sci. Publ., The Netherlands.

Ma, T.-H. (1990). A dual bio-monitoring system for the genotoxicity of air and water at the site of hazardous waste mixtures. Sixth annual waste testing and quality assurance symposium. U.S. Environmental Protection Agency, Proc. 2, 428–436.

Ma, T.-H., and Harris, M. M. (1985). *In situ* monitoring of environmental mutagens. *In* "Hazard Assessment of Chemicals: Current Developments," vol. 4. (J. Saxena, ed.). pp. 77–106. Academic Press, New York.

Ma, T.-H., Harris, M. M., Anderson, V. A., Ahmed, I., Mohammad, K., Bare, J. L., and Lin, G. (1984). *Tradescantia*-micronucleus (Trad-MCN) tests on 140 health-related agents. *Mutation Res.* **138,** 157–167.

Nilan, R. A., Rosichan, J. L., Arenaz, P., and Hodgdon, A. L. (1981). Pollen genetic markers for detection of mutagens in the environment. *Environ. Health Perspect.* **37,** 19–25.

Plewa, M. J., and Gentile, J. M., eds. (1988). Activation of promutagens by plant systems. *Mutation Res.* **197,** 173–336.

Rédei, G. P., Acedo, G. N., and Sandhu, S. S. (1984). Mutation induction and detection in Arabidopsis. *In* "Mutation, Cancer, and Malformation," (E. H. Y. Chu and W. M. Generoso, eds). pp. 285–313. Plenum Press, New York.

Sandhu, S. S., de Serres, F. J., Gopalan, H. N. B., Grant, W. F., Veleminsky, J., and Becking, G. C. (1991). Status report of the International Programme on Chemical Safety's Collaborative Study on plant systems. *Mutation Res.* **257,** 19–25.

Schairer, L. A., and Sautkulis, R. C. (1982). Detection of ambient levels of mutagenic atmospheric pollutants with the higher plant *Tradescantia*. *In* "Environmental Mutagenesis, Carcinogenesis, and Plant Biology," vol. 2. (E. J. Klekowski, Jr., ed.) pp. 153–194. Praeger, N.Y.

Xu, J., Xia, W., Jong, X., Sun, W., Lin, G., and Ma, T.-H. (1990). Image analysis system for rapid data processing in *Tradescantia*-micronucleus bioassay. Pub. No. 1091, pp. 346–356. Am. Soc. for Testing Materials, Philadelphia.

Polychlorinated Biphenyls (PCBs), Effects on Humans and the Environment

Glen Shaw
National Research Centre for Environmental Toxicology, Australia

Glossary

Benthic Aquatic organisms which live in the sediment layer.

Bioaccumulation Accumulation of xenobiotics by organisms to levels higher than those in the ambient environment; includes bioconcentration from the environment and biomagnification along food chains.

Bioconcentration Process of direct transfer of xenobiotics from water to an organism resulting in an increase in concentration in the organism relative to the water.

Biomagnification Transfer of xenobiotics from food organisms to predator organisms resulting in an increase in concentration with higher trophic level.

Congeners Members of a class (e.g., chlorobiphenyls).

Fugacity The escaping tendency of PCB molecules from a solute.

Hepatocellular carcinoma Cancer of the liver cells.

Interstitial water Water present in pores in the sediment structure.

Lipophilicity Potential for an organic chemical to be partitioned into lipid.

Trophic level Position of an organism in the food web with respect to predator/prey relationships.

POLYCHLORINATED BIPHENYLS (PCBs) are a group of many isomers and congeners with the empirical formula $C_{12} H_{10-n} Cl_n$ with $n = 1$ to 10. Structurally, PCBs consist of the aromatic biphenyl ring system substituted to varying degrees with chlorine and producing 209 possible congeners. This group of compounds was first synthesized in 1881 and has been produced and employed as, for example, plasticizers, dielectric fluids, microscope immersion oils, and cutting oils. The use of PCBs in most industrialized nations has now been dramatically reduced because of concerns about the human and environmental health effects of these compounds. PCBs in present use are restricted to closed systems such as transformers. PCBs were first discovered as environmental contaminants in 1966 and since then have been found in almost every component of the global ecosystem. Environmentally, PCBs are characterized by a high bioaccumulation potential and resistance to degradation.

Laboratory studies have shown an association between PCBs and cancer, reproductive problems, and organ damage in experimental animals. Recently, the toxic coplanar PCBs have been found to be widespread in the global ecosystem and may pose a greater toxic potential to wildlife than chlorinated dioxins and dibenzofurans.

I. Chemical Properties of PCBs and Their Relationship to Human and Environmental Health

A. Structure of PCBs

Formation of biphenyl can produce 209 possible chlorobiphenyl congeners substituted with 1 to 10 chlorine atoms. The IUPAC numbering and nomenclature for chlorobiphenyls are shown in Figures 1 and 2.

B. Composition of PCB Mixtures

Commercial preparation of PCBs involves chlorination of biphenyl in the presence of a catalyst and produces mixtures of chlorobiphenyl congeners such as Monsanto's Aroclors, the French product Clophen and the Japanese product Kanechlor. The commercial preparations vary in average chlorine content and degree of chlorination.

The distribution of congeners in the four most common Aroclor mixtures is shown in Table I.

C. Chemical and Physical Properties of PCBs

PCBs are characterized by stability and chemical inertness. Low dielectric constants and high heat capacity confer properties upon them which make PCBs ideal for use in electrical capacitors and transformers. In addition, chlorobiphenyls are nonpolar, and this results in low water solubility, high *n*-octanol/water partition coefficients (K_{ow}) and high sorption coefficients. Table II presents water solubility data for a range of PCBs. Due to these properties, PCBs are resistant to degradation and are highly bioaccumulative. Correlations exist between the bioconcentration factor K_b (which is the ratio of the concentration in the aquatic biota to the concentration in water at equilibrium) and K_{ow}:

$$\log K_b = 0.56 \log K_{ow} + 0.124 \quad (1)$$

$$\log K_b = 0.85 \log K_{ow} - 0.70 \quad (2)$$

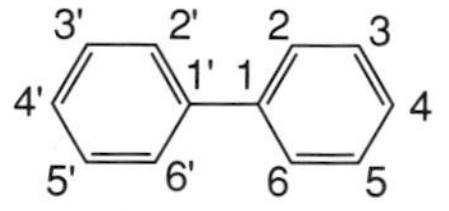

3' 2' 2 3 1' 1 4' 4 5' 6' 6 5

Figure 1 Numbering in the biphenyl ring system.

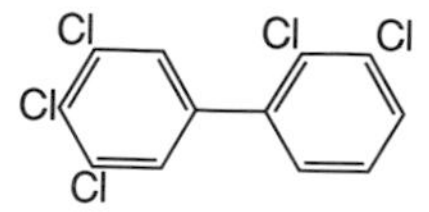

Cl Cl Cl Cl Cl

Figure 2 Structure of 2,3,3',4',5'-pentachlorobiphenyl.

$$\log K_b = 1.119 \log K_{ow} - 1.579 \quad (3)$$

$$\log K_b = 0.76 \log K_{ow} - 0.23. \quad (4)$$

Table III presents log K_{ow} values for a range of PCB congeners.

Henry's law constants describe the equilibrium distribution between vapor and aqueous phases and are used in prediction of environmental properties of PCBs. PCBs have very low vapor pressures which decrease with increasing chlorination. Vaporization of PCBs from the aqueous environment, however, is unusually high and this can be described in terms of fugacity.

II. Environmental Properties

A. Transport and Distribution of PCBs in the Environment

Transport of PCBs occurs essentially by the movement of air–water and the translocation of soils and sediments due to erosion or current and tidal action. Atmospheric transport of PCBs is the main mechanism of contamination of remote ecosystems such as the Arctic with PCBs. Air–water mass transfer coefficients can be calculated from Henry's Law constants in the form:

$$\lim_{c \to 0} \frac{C_g}{C_l} = H_l,$$

where C_g is the gas phase concentration C_l is the aqueous phase concentration and H is the dimensionless Henry's Law constant. Henry's Law constants can be measured from aqueous and vapor phase concentrations, structure–activity relationships, and gas purging techniques, although wide variability of values for H exist in the literature.

Fugacity-based models using Henry's Law constants have been used to calculate air–water distribution of PCBs. Atmospheric PCB concentrations in industrialized areas have remained relatively constant over the past 10 years. (For example, the atmo-

Table I Composition of Four Aroclor Mixtures

Constituents	Percentage in Aroclor			
	Aroclor 1242	Aroclor 1248	Aroclor 1254	Aroclor 1260
Total chlorine	42	48	54	60
Monochlorobiphenyl	3	—	—	—
Dichlorobiphenyl	13	2	—	—
Trichlorobiphenyl	28	18	—	—
Tetrachlorobiphenyl	30	40	11	—
Pentachlorobiphenyl	22	36	49	12
Hexachlorobiphenyl	4	4	34	38
Heptachlorobiphenyl	—	—	6	41
Octachlorobiphenyl	—	—	—	8
Nonachlorobiphenyl	—	—	—	1

spheric PCB concentration over Lake Superior has remained constant at 1.2 ng m^{-3}.)

Transport in water is governed by the aqueous solubility of PCBs. Because of their relatively low solubility (indicated by high K_{ow}), PCBs are extensively associated with sediments and particulate matter in aquatic systems. Adsorption of PCBs to sediments is controlled by sediment organic carbon. Strength of adsorption of PCBs to organic carbon depends on the sorption coefficients of the particular PCB congener and can be estimated from K_{ow}. Estimated PCB loads in the global environment are presented in Table IV.

B. Bioaccumulation

The principal uptake route of PCBs for terrestrial animals is through food and for aquatic organisms from food and direct partitioning from water to the lipid pools of the organism. The driving force for bioaccumulation of PCBs is the influence of the high values of K_{ow} for PCB congeners (see Table III). In addition, the stereochemistry of PCBs appears to play a role in affecting bioaccumulation of certain PCB congeners. The relationship between log bioconcentration factor (log K_B) for aquatic organisms and log K_{ow} loses linearity above approximately log

Table II Water Solubilities for Some PCBs[a]

PCB	Water solubility (μg/l)
Monochlorobiphenyls	1,000–6,000
4,4′-Dichlorobiphenyl	50–200
Other dichlorobiphenyls	1,000–1,500
Trichlorobiphenyls	50–100
Tetrachlorobiphenyls	30–100
Pentachlorobiphenyls	20–50
Hexachlorobiphenyls	0.1–10

[a] Due to inaccuracies in the measurement of water solubility of PCBs, a wide range of values is presented in the literature.

Table III Log K_{ow} Values for Some PCB Congeners

Compound	Experimental	Calculated
Biphenyl	3.90	3.98
2-Cl	4.30	4.46
2,5-Cl_2	5.10	5.08
2,6-Cl_2	5.00	4.95
2,4,5-Cl_3	5.60	5.52
2,4,6-Cl_3	5.50	5.59
2,3,4,5-Cl_4	5.91	5.92
2,2,4,5 Cl_4	5.73	5.95
2,3,4,5,6-Cl_5	6.30	6.45
2,2′,4,5,5′-Cl_5	6.40	6.62
2,2,3,3,6,6′-Cl_6	6.70	7.09
2,2′,3,3′,4,4′-Cl_6	7.00	6.81
2,2′,4,4′,6′,6′-Cl_6	7.00	7.20
2,2′,3,3′,4,4′,-6-Cl_7	6.70	7.34
3,2′,3,3′,5,5′,6,6′-Cl_8	7.10	8.12
2,2′,3,3′,4,5,5′,6,6′-Cl_9	8.16	8.52
Cl_{10}	8.26	8.92

Source: Kamlet, M. J., Doherty, R. M., Carr, P. W., Mackay, D., Abraham, M. H., and Taft, R. W. (1990). *Environ. Sci. Technol.* **22,** 506.

Table IV Estimated PCB Loads in the Global Environment

Environment	PCB load (tons)	PCB load (%)	World production (%)
Terrestrial and coastal			
Air	500	0.13	
River and lake water	3,500	0.94	
Seawater	2,400	0.64	
Soil	2,400	0.64	
Sediment	130,000	35	
Biota	4,300	1.1	
Total (A)	143,000	39	
Open ocean			
Air	790	0.21	
Seawater	230,000	61	
Sediment	110	0.03	
Biota	270	0.07	
Total (B)	231,000	61	
Total load in the environment (A + B)	374,000	100	31
Degraded and incinerated	43,000		4
Land-stocked[a]	783,000		65
World production	1,200,000		100

Source: Tanabe, S. (1988). *Environ. Pollut.* **50,** 9.
[a] Still in use in electrical equipment and other products, and deposited in landfills and dumps.

$K_{ow} = 6$ with a maximum at log K_{ow} of 6.7. The relationship between log K_B and log K_{ow} for hydrophobic compounds (having log K_{ow} values from about 2 to 10) has been found to be a polynomial expression of the following form:

$$\log K_B = 6.9 \times 10^{-3} (\log K_{ow})^4 - 1.85 \times 10^{-1} (\log K_{ow})^3 + 1.55 \log (K_{ow})^2 - 4.18 \log K_{ow} + 4.79.$$

Bioaccumulation of PCBs by benthic organisms is essentially controlled by uptake from interstitial water. The following relationship has been derived after corrections for colloid influence and nonattainment of equilibrium are applied:

$$\log K_B = 1.11 \log K_{ow} - 1.0.$$

The other main source of accumulation of PCBs is absorption from the atmosphere by plants. In this case, lipophilicity (estimated from K_{ow}) is not a good indication of accumulation but vapor pressure of PCB congeners correlates well with accumulation.

C. Metabolism

Metabolism of PCBs occurs to a greater extent in warm-blooded mammals and birds than aquatic invertebrates and fish. In general, the lower chlorinated congeners are more easily metabolized, although in addition, structural features and positional isomerism influence metabolic behavior. The presence of vicinal unsubstituted carbon atoms is a necessary prerequisite for metabolism. The unsubstituted carbon atoms should be in the 3,4 or 4,5 positions.

D. Bioaccumulation of Planar PCBs

It has now been recognized that planar PCB congeners are bioaccumulated to a relatively higher degree than PCBs possessing a large dihedral angle between the phenyl rings. Coplanar PCBs (without ortho chlorines) and their mono- and some of their di-ortho substituted congeners occur in low concentrations in PCB formulations, but are detected in lower trophic level organisms in significant amounts.

The coplanar PCBs have shown slower clearance rates and longer biological half lives compared with other PCB congeners. Contradictory results have been found for the selective accumulation of planar PCBs in higher organisms although there is an increase in coplanar PCB content from land to ocean, and in some predatory bird species.

E. Environmental Toxicology of Planar PCBs

One of the major symptoms of exposure of animals to PCBs and other halogenated aromatics such as 2,3,7,8-TCDD is the induction of both phase I and phase II drug metabolizing enzymes. These include cytochrome *P*-450 1A1 and cytochrome *P*-450 1A2 hemoproteins and their microsomal monooxygenases such as aryl hydrocarbon hydroxylase (AHH) and ethoxyresorufin *o*-deethylase (EROD).

A proportion of the 209 PCBs is similar stereochemically to the planar 2,3,7,8-TCDD molecule and, correspondingly, exert toxic manifestations similar to this compound. The main requirement for planarity of PCB molecules is the absence of ortho chlorines.

The other main structural requirement for toxicity of PCB molecules is the presence of two para and at least two meta chlorine atoms on the biphenyl skeleton. The absence of ortho chlorine atoms confers planarity to the biphenyl molecule, but the addition of one or two chlorine atoms ortho to the central bond diminishes, but does not necessarily eliminate, the biochemical activities of the planar PCBs. Biochemical activities similar to the planar PCBs have been demonstrated for all eight mono–ortho substituted PCBs and for at least five di-ortho substituted PCBs. Table V gives details of the planar, mono-, and di-substituted PCB congeners.

Toxic equivalency factors (TEFs) have been developed for the PCBs listed in Table V. TEFs represent a mechanism for comparing the relative toxicities of a group of compounds to that of the most toxic member of the family of chemicals, namely 2,3,7,8-TCDD. Table VI presents proposed TEF values for coplanar, mono-, and di-ortho PCBs.

TEFs for chlorinated dioxins, furans, and PCBs in selected species are given in Table VII.

These data indicate that Baltic white-tailed eagles have the highest dioxin-like TEFs so far reported. The data also indicate that PCBs contribute a major portion of the dioxin-like toxic potential for these species. At this time, TEF values developed by various workers appear to vary somewhat but are usually of the same order of magnitude.

III. Effects on Human Health

A. Human Exposure

Because of the ubiquitous nature of PCBs, humans are exposed via many sources including inhalation from air and by ingestion from food sources, especially fish. Since PCBs are lipophilic they are

Table V Planar, Mono, and Di-Ortho PCBs

	Structure		
Degree of chlorination	Planar	Mono-ortho	Di-ortho
Tetrachloro-	3,3′,4,4′		
Tetrachloro-	3,4,4′,5		
Pentachloro-	3,3′,4,4′,5	2,3,3′,4,4′ 2,3,4,4′,5 2,3′,4,4′,5 2′,3,4,4′,5	
Hexachloro-	3,3′,4,4′,5,5′	2,3,3′,4,4′,5 2,3,3′4,4′,5′ 2,3′,4,4′,5,5′	2,2′,3,3′,4,4′ 2,2′,3,4,4′,5 2,2′,3,4,4′,5′ 2,2′,4,4′,5,5′ 2,3,3′,4,4′,6 2,3,4,4′,5,6 2,3′,4,4′,5′,6
Heptachloro-		2,3,3′,4,4′,5,5′	2,2′,3,3′,4,4′,5 2,2′,3,4,4′,5,5′ 2,3,3′,4,4′,5′,6 2,3,3′,4,4′,5′,6
Octachloro-			2,2′,3,3′,4,4′,5,5′ 2,3,3′,4,4′,5,5′,6

Table VI Proposed TEF Values for the Coplanar, Mono-Ortho, and Di-Ortho PCB Congeners

PCB congener	TEF values
Coplanar	
3,3′,4,4′-tetrachlorobiphenyl	0.01
3,4,4′,5-tetrachlorobiphenyl	Insufficient data
3,3′,4,4′,5-pentachlorobiphenyl	0.01
3,3′,4,4′,5,5′-hexachlorobiphenyl	0.05
Mono-ortho substituted congeners	0.001
Di-ortho-substituted congeners	0.00002

Source: Safe, S. (1990). *Crit. Rev. Toxicol.* **21,** 51.

preferentially stored in the adipose tissue. PCBs pass the placenta and are excreted in breast milk.

In the United States, mean PCB serum levels have been reported as 5–7 ng ml^{-1} while levels in adipose tissue and human milk fat are normally 100 to 200 times higher. Extensive surveys of human milk for PCBs have been carried out and the highest levels are normally found in industrialized countries such as Germany and the United States. Occupationally exposed persons can show higher PCB concentrations, although high risk persons associated with contaminated sites in the United States generally do not show elevated levels of PCBs.

The PCB congeners mainly found in human blood and fat have chlorine substitution at positions 2,4,5 in one ring and 4 or 2,4 in the other. The half lives for the main PCBs found in a human who was administered PCBs are shown in Table VIII.

B. Toxic Effects and Epidemiology

The paucity of data concerning exposure of humans to PCBs at toxic levels is the limiting factor in determining the effects of PCBs on humans. Rare accidental exposures to PCBs in high concentrations such as the Japanese Yusho incident in 1968 and the Taiwanese Yucheng incident in 1978 (in which contaminated rice oil was consumed) are the major source of acute health data.

A summary of physiological effects of PCB exposure from these incidents is listed below:

Human physiological effects of PCBs	
Chloracne	Lung capacity reduction
Weakness	Eye inflammation and burning
Nausea	Eyelid edema
Headache	Cysts of the conjunctive and tarsal glands
Impotence	Pigmentation of cutaneous tissue
Insomnia	Follicular hyperkeratosis
Loss of appetite	Dilation of hair follicles
Loss of weight	Nail deformity
Abdominal pains	Liver enlargement
Muscle spasms	

Neurological effects of PCBs are vague with some studies indicating a decrease in motor- and sensory-nerve conduction velocities.

Table VII TCDD TEF Values for Various Species[a]

Compound	TEF–2,3,7,8-TCDD equivalent (pg g^{-1})					
	Human[b]	Finless porpoise[b]	Killer whale[b]	Eagle muscle[c]	Eagle liver[c]	Eagle egg[c]
3,3′,4,4′-tetra CB	3.5	140	480	700	700	210
3,3′,4,4′,5-penta CB	33	89	370	5,000	3,300	2,100
3,3′,4,4′,5,5′-hexa CB	4.5	32	380	1,000	650	300
Total mono-ortho PCBs	140	12,000	16,000		Not determined	
Total di-ortho PCBs	4.7	770	1,460		Not determined	
Total PCBs	186	13,000	18,700	[d]	[d]	[d]
Total dioxins	12	<0.9	<0.5	0.4	—	—
Total dibenzofurans	8.4	2.0	4.5	850	655	80

[a] TEF equivalent = concentration on a wet weight basis X TEF from Safe, S. (1990). *Crit. Rev. Toxicol.* **21,** 51.
[b] From Tanabe *et al.* (1989). *Chemosphere* **18,** 485.
[c] From Tarhanen *et al.* (1989). *Chemosphere* **18,** 1067.
[d] Could not be calculated due to lack of data for mono- and di-ortho substituted PCBs.

Table VIII Half Lives of PCB Congeners in Humans

PCB congener	Half life (days)
2,3,3′,4,5′/2,3′,4,4′,5′-pentachloro	100–300
2,2′,3,4,4′,5-hexachlorobiphenyl	321
2,2′,4,4′,5,5′-hexachlorobiphenyl	338
2,2′,3,4,4′,5,5′-heptachlorobiphenyl	124

Source: Buhler *et al.* (1988). *Chemosphere* **17,** 1724.

It appears that the immune system response has been affected in severely poisoned patients, which is consistent with findings from animal studies.

It has been found that Yusho patients on average consumed 633 mg of PCBs, 3–4 mg of polychlorinated dibenzofurans (PCDFs), and 596 mg of polychlorinated quarternaryphenyls (PCQs).

From occupational studies, it has been found that high-level exposure to PCBs has caused dermatitis but no other clinical effects. The toxicity of PCBs has been attributed in part to the presence of PCDFs in formulations. More recently, PCB toxicity has been attributed mainly to the presence of the highly toxic planar, mono-, and di-ortho PCBs. Compared to industrial mixtures, human milk is enriched in many of the mono- and di-ortho congeners and it has been estimated that the PCB mixture in human milk is 5–10 times more biologically active than the corresponding commercial mixture. Since 1988, due to improvements in analytical techniques, planar PCBs have been reported in human samples and the TEFs of planar, mono-, and di-ortho PCBs are in considerable excess of those for PCDDs and PCDFs in humans.

C. Carcinogenicity and Mutagenicity

The mutagenicity and carcinogenicity of commercial PCBs has been investigated in epidemiological, *in vivo,* and *in vitro* studies and the findings are summarized below:

1. The more toxic congeners are not likely to form covalent DNA adducts and are not considered genotoxic.
2. PCBs are not considered mutagenic although more research needs to be done on this topic before a definitive statement can be made.
3. It has been demonstrated that the more highly chlorinated PCBs are promoters of carcinogenesis in the liver and several studies have reported PCBs as promoters in other organs such as lung and skin. The promotional effect is due to PCB congeners and not PCDF impurities.
4. PCBs are considered to be liver carcinogens in experimental animals.
5. Studies of Yusho victims suggest that exposure to PCBs may result in an increased risk of hepatocellular carcinoma.
6. The U.S. EPA has classified PCBs as probable human carcinogens (group B2).
7. The IARC has classified PCBs as probable human carcinogens (group 2A).

IV. Disposal and Control

A. Disposal and Destruction

Because of the recalcitrant nature of PCBs, correct disposal is of the utmost importance. A recent NIOSH study found that 95 out of 96 air samples collected from several PCB disposal facilities contained PCBs at concentrations exceeding the NIOSH recommended exposure limit of 1.0 $\mu g\ m^{-3}$. In addition, 74% of samples from hard indoor surfaces contained PCBs above the contamination guideline of 100 $\mu g\ m^{-2}$.

In general, PCB disposal and destruction processes fall into three broad categories: physical, chemical, and biological. Table IX presents disposal technologies for various PCB wastes. Strict operating requirements and monitoring and inspection procedures for PCB disposal facilities should be followed. Assessments of the technical and economic aspects of PCB decontamination processes in sediments have been published. These assessments indicate that a number of processes show potential as alternatives to incineration and landfill. These processes reduce PCB contamination to desired criteria levels (1–5 mg kg^{-1}) for waste with minimum environmental effects and low to moderate cost.

B. Degradation

In biodegradation processes, both degree of chlorination and position of chlorine substitution play crucial roles in determining whether microbial degradation will occur. PCB dechlorination processes by anaerobic bacteria fall into two different classes for

Table IX Disposal Technologies for Various PCB Wastes[a]

Waste type	Typical PCB content	Disposal option
Transformer fluids	30–70%	b,c
Decontamination flushings (concentrated)	1–10%	b,c
Decontamination flushings	<1%	b,c,d
Mineral oil	<1%	b,d,e,h
Retrofilling fluids	100–1000 ppm	b,d,e
Major equipment	<500 ppm	e,f
Major equipment	<50 ppm	e,g
Large capacitors	>0.5 kg	c
Small capacitors	<0.5 kg	c,f
Maintenance waste	variable	c
Packaging	variable	c
Waste lubricating oil	10–500 ppm	b,d,h
Spoil and destruction residues	>50 ppm	c,f
Dredging spoils	variable	c,f
Aqueous waste	$\mu g\ L^{-1}$ to mg L^{-1}	b,d

[a] From Canadian Council of Ministers of the Environment. (1989). CCME-TS/WM-TRE008, Manual EPS 9/HA/1.
[b] Liquid injection incineration.
[c] Rotary kiln incinerator.
[d] High efficiency boiler.
[e] Material recovery.
[f] Approved landfill.
[g] Municipal landfill.
[h] Sodium-based processes.

Table X Analytical Methods for PCB Determination

Sample type	Extraction	Cleanup	Determination
Water	Solvent—hexane Solvent—dichloromethane Solid phase extraction (SPE)—XAD resins or reverse phase cartridges	Florisil or silica gel elution with pentane or hexane	GC-ECD GC-Hall GC-MS
Fatty foods and biota	Grinding with sodium sulphate and extraction with hexane or hexane/acetone/diethyl ether mixtures	Florisil & elution with pentane or hexane	GC-ECD GC-Hall GC-MS
Soil/sediment/wastes	Sonication or soxhlet with dichloromethane/acetone 1 : 1 or hexane/acetone 1 : 1	Florisil or silica gel with pentane or hexane	GC-ECD GC-Hall GC-MS
Air	Passage through impinger with hexane or toluene or passage through trap (e.g., tenax)	—	GC-ECD GC-Hall GC-MS
Blood	Solvent—diethyl/ether/hexane 1 : 1	—	GC-ECD GC-Hall GC-MS

Table XI Substitution Patterns for PCBs

Isomer no.	Substitution pattern	Isomer no.	Substitution pattern	Isomer no.	Substitution pattern
1	2	44	2,2′,3,5′	87	2,2′,3,4,5′
2	3	45	2,2′,3,6	88	2,2′,3,4,6
3	4	46	2,2′,3,6′	89	2,2′,3,4,6′
4	2,2′	47	2,2′,4,4′	90	2,2′,3,4′,5
5	2,3	48	2,2′,4,5	91	2,2′,3,4′,6
6	2,3′	49	2,2′,4,5′	92	2,2′,3,5,5′
7	2,4	50	2,2′,4,6	93	2,2′,3,5,6
8	2,4′	51	2,2′,4,6′	94	2,2′,3,5,6′
9	2,5	52	2,2′,5,5′	95	2,2′,3,5′,6
10	2,6	53	2,2′,5,6′	96	2,2′,3,6,6′
11	3,3′	54	2,2′,6,6′	97	2,2′,3′,4,5
12	3,4	55	2,3,3′,4	98	2,2′,3′,4,6
13	3,4′	56	2,3,3′,4′	99	2,2′,4,4′,5
14	3,5	57	2,3,3′,5	100	2,2′,4,4′,6
15	4,4′	58	2,3,3′,5′	101	2,2′,4,5,5′
16	2,2′,3	59	2,3,3′,6	102	2,2′,4,5,6′
17	2,2′,4	60	2,3,4,4′	103	2,2′,4,5′,6
18	2,2′,5	61	2,3,4,5	104	2,2′,4,6,6′
19	2,2′,6	62	2,3,4,6	105	2,3,3′,4,4′
20	2,3,3′	63	2,3,4′,5	106	2,3,3′,4,5
21	2,3,4	64	2,3,4′,6	107	2,3,3′,4′,5
22	2,3,4′	65	2,3,5,6	108	2,3,3′,4,5′
23	2,3,5	66	2,3′,4,4′	109	2,3,3′,4,6
24	2,3,6	67	2,3′,4,5	110	2,3,3′,4′,6
25	2,3′,4	68	2,3′,4,5′	111	2,3,3′,5,5′
26	2,3′,5	69	2,3′,4,6	112	2,3,3′,5,6
27	2,3′,6	70	2,3′,4′,5	113	2,3,3′,5′,6
28	2,4,4′	71	2,3′,4′,6	114	2,3,4,4′,5
29	2,4,5	72	2,2′,5,5′	115	2,3,4,4′,6
30	2,4,6	73	2,3′,5′,6	116	2,3,4,5,6
31	2,4′,5	74	2,4,4′,5	117	2,3,4′,5,6
32	2,4′,6	75	2,4,4′,6	118	2,3′,4,4′,5
33	2′,3,4	76	2′,3,4,5	119	2,3′,4,4′,6
34	2′,3,5	77	3,3′,4,4′	120	2,3′,4,5,5′
35	3,3′,4	78	3,3′,4,5	121	2,3′,4,5′,6
36	3,3′,5	79	3,3′,4,5′	122	2′,3,3′,4,5
37	3,4,4′	80	3,3′5,5′	123	2′,3,4,4′,5
38	3,4,5	81	3,4,4′,5	124	2′,3,4,5,5′
39	3,4′,5	82	2,2′,3,3′,4	125	2′,3,4,5,6′
40	2,2′,3,3′	83	2,2′,3,3′,5	126	3,3′,4,4′,5
41	2,2′,3,4	84	2,2′,3,3′,6	127	3,3′,4,5,5′
42	2,2′,3,4′	85	2,2′,3,4,4′	128	2,2′,3,3′,4,4′
43	2,2′,3,5	86	2,2′,3,4,5	129	2,2′,3,3′,4,5

(continues)

Table XI (*Continued*)

Isomer no.	Substitution pattern	Isomer no.	Substitution pattern	Isomer no.	Substitution pattern
130	2,2′,3,3′,4,5′	157	2,3,3′,4,4′,5′	184	2,2′,3,4,4′,6,6′
131	2,2′,3,3′,4,6	158	2,3,3′,4,4′,6	185	2,2′,3,4,5,5′,6
132	2,2′,3,3′,4,6′	159	2,3,3′,4,5,5′	186	2,2′,3,4,5,6,6′
133	2,2′,3,3′,5,5′	160	2,3,3′,4,5,6	187	2,2′,3,4′,5,5′,6
134	2,2′,3,3′,5,6	161	2,3,3′,4,5′,6	188	2,2′,3,4′,5′,6,6′
135	2,2′,3,3′,5,6′	162	2,3,3′,4′,4,5′	189	2,3,3′,4,4′,5,5′
136	2,2′,3,3′,6,6′	163	2,3,3′,4′,5,6	190	2,3,3′,4,4′,5,6
137	2,2′,3,4,4′,5	164	2,3,3′,4,5′,6	191	2,3,3′,4,4′,5′,6
138	2,2′,3,4,4′,5′	165	2,3,3′,5,5′,6	192	2,3,3′,4,5,5′,6
139	2,2′,3,4,4′,6	166	2,3,4,4′,5,6	193	2,3,3′,4′,5,5′,6
140	2,2′,3,4,4′,6′	167	2,3′,4,4′,5,5′	194	2,2′,3,3′,4,5′,5′
141	2,2′,3,4,5,5′	168	2,3′,4,4′,5′,6	195	2,2′,3,3′,4,4′,5,6
142	2,2′,3,4,5,6	169	3,3′,4,4′,5,5′	196	2,2′,3,3′,4,4′,5′,6
143	2,2′,3,4,5,6′	170	2,2′,3,3′,4,4′,5	197	2,2′,3,3′,4,4′,6,6′
144	2,2′,3,4,5′,6	171	2,2′,3,3′,4,4′,6	198	2,2′,3,3′,4,5,5′,6
145	2,2′,3,4,6,6′	172	2,2′,3,3′,4,5,5′	199	2,2′,3,3′,4,5,6,6′
146	2,2′,3,4′,5,5′	173	2,2′,3,3′,4,5,6	200	2,2′,3,3′,4,5′,6,6′
147	2,2′,3,4′,5,6	174	2,2′,3,3′,4,5,6′	201	2,2′,3,3′,4,5,5′,6
148	2,2′,3,4′,5,6′	175	2,2′,3,3′,4,5′,6	202	2,2′,3,3′,5,5′,6,6′
149	2,2′,3,4′,5′,6	176	2,2′,3,3′,4,6,6′	203	2,2′,3,4,4′,5,5′,6
150	2,2′,3,4′,6,6′	177	2,2′,3,3′,4′,5,6	204	2,2′,3,4,4′,5,6,6′
151	2,2′,3,5,5′,6	178	2,2′,3,3′,5,5′,6	205	2,3,3′,4,4′,5,5′,6
152	2,2′,3,5,6,6′	179	2,2′,3,3′,5,6,6′	206	2,2′,3,3′,4,4′,5,5′,6
153	2,2′,4,4′,5,5′	180	2,2′,3,4,4′,5,5′	207	2,2′,3,3′,4,4′,5,6,6′
154	2,2′,4,4′,5,6′	181	2,2′,3,4,4′,5,6	208	2,2′,3,3′,4,5,5′,6,6′
155	2,2′,4,4′,6,6′	182	2,2′,3,4,4′,5,6′	209	2,2′,3,3′,4,4′,5,5′,6,6′
156	2,3,3′,4,4′,5	183	2,2′,3,4,4′,5′,6		

Source: Mullin *et al.* (1984). *Environ. Sci. Technol.* **18,** 486.

lower chlorinated PCBs: (1) *o, m, p* dechlorinations and (2) *m, p* dechlorinations only, where reactivity is determined by molecular shape.

Recent research has found that nonortho chlorines can be removed from most of the highly chlorinated PCBs by some anaerobic bacteria leaving ortho substituted biphenyls that can be degraded by certain aerobic bacteria. Thus a two-step system comprising anaerobic dechlorination followed by oxidation by a suitable aerobic bacterial strain (e.g., *Alcaligenes eutrophus* H850) can effectively degrade most PCB congeners.

Removal of PCBs from PCB-contaminated water has been achieved by overland flow treatment systems and by passing through chitosan, the deacetylated form of the natural polymer, chitin.

V. Testing and Evaluation

A. Analytical Methodology

Traditionally, PCBs have been analyzed in food, human, and environmental samples by methodologies applicable to organochlorine pesticide residues. This involves solvent extraction from the sample, separation of PCBs from coextracted material by column chromatography, and final determination of

Table XII Relative Retention Times and Response Factors for 209 PCB Congeners

Isomer no.	Relative retention time	Relative response factor	Isomer no.	Relative retention time	Relative response factor
0	0.0997	0.0251	41	0.499	0.5469
1	0.1544	0.0393	42	0.487	0.792
2	0.1937	0.04[a]	43	0.4587	0.503
3	0.1975	0.0193	44	0.4832	0.524
4	0.2245	0.0374	45	0.4334	0.54
5	0.2785	0.119	46	0.445	0.468
6	0.2709	0.38	47	0.4639	0.848
7	0.2566	0.69	48	0.4651	0.556
8	0.2783	0.206	49	0.461	0.648
9	0.257	0.388	50	0.4007	0.6817
10	0.2243	0.262	51	0.4242	0.6[a]
11	0.3238	0.0449	52	0.4557	0.418
12	0.3298	0.179	53	0.4187	0.3606
13	0.3315	0.2[a]	54	0.38	0.3643
14	0.2973	0.3047	55	0.5562	0.829
15	0.3387	0.107	56	0.5676	0.829
16	0.3625	0.447	57	0.5155	0.6[a]
17	0.3398	0.412	58	0.5267	0.609
18	0.3378	0.313	59	0.486	0.6[a]
19	0.3045	0.3037	60	0.5676	1.0164
20	0.417	0.7238	61	0.5331	1.2227
21	0.4135	1.0598	62	0.4685	1.1478
22	0.4267	1.0935	63	0.529	0.728
23	0.377	0.5[a]	64	0.4999	0.607
24	0.3508	0.793	65	0.4671	0.8408
25	0.3937	0.5[a]	66	0.5447	0.646
26	0.3911	0.603	67	0.5214	0.6[a]
27	0.3521	0.495	68	0.504	0.726
28	0.4031	0.854	69	0.451	0.8024
29	0.382	0.6339	70	0.5407	0.658
30	0.3165	0.8202	71	0.4989	0.468
31	0.4024	0.562	72	0.4984	0.5515
32	0.3636	0.278	73	0.4554	0.5805
33	0.4163	0.447	74	0.5341	0.671
34	0.3782	0.6092	75	0.4643	0.6461
35	0.4738	0.3746	76	0.5408	0.5795
36	0.4375	0.2948	77	0.6295	0.3812
37	0.4858	0.58	78	0.6024	1.1151
38	0.4593	0.4698	79	0.5894	0.881
39	0.4488	0.347	80	0.5464	0.7278
40	0.5102	0.722	81	0.6149	0.7159

(continues)

Table XII (*Continued*)

Isomer no.	Relative retention time	Relative response factor	Isomer no.	Relative retention time	Relative response factor
82	0.6453	0.773	123	0.6658	0.6645
83	0.6029	0.6339	124	0.6584	0.848
84	0.5744	0.386	125	0.6142	0.556
85	0.6224	0.7396	126	0.7512	0.4757
86	0.6105	0.7968	127	0.7078	0.5834
87	0.6175	1.021	128	0.7761	1.188
88	0.5486	0.6892	129	0.7501	0.997
89	0.5779	0.561	130	0.7284	0.952
90	0.5814	0.611	131	0.6853	0.8492
91	0.5549	0.571	132	0.7035	0.7303
92	0.5742	0.5375	133	0.6871	1.148
93	0.5437	0.6676	134	0.6796	0.7331
94	0.5331	0.4514	135	0.6563	0.7031
95	0.5464	0.443	136	0.6257	0.444
96	0.5057	0.4308	137	0.7329	1.112
97	0.61	0.631	138	0.7403	0.827
98	0.5415	0.6246	139	0.6707	0.7219
99	0.588	0.613	140	0.6707	0.6732
100	0.5212	0.5871	141	0.7203	1.352
101	0.5816	0.668	142	0.6848	1.218
102	0.5431	0.4561	143	0.6789	0.7088
103	0.5142	0.6068	144	0.6563	0.8764
104	0.4757	0.4561	145	0.6149	0.6789
105	0.7049	0.94	146	0.6955	0.728
106	0.668	1.0046	147	0.6608	0.6[a]
107	0.6628	0.8183	148	0.6243	0.554
108	0.6626	1.0654	149	0.6672	0.572
109	0.6016	0.9625	150	0.5969	0.5676
110	0.6314	0.65[a]	151	0.6499	0.785
111	0.6183	0.6601	152	0.6062	0.5235
112	0.5986	0.8286	153	0.7036	0.688
113	0.5862	0.604	154	0.6349	0.57
114	0.6828	1.0261	155	0.5666	0.586
115	0.6171	1.1328	156	0.8105	1.389
116	0.6132	1.3987	157	0.8184	1.1965
117	0.615	0.8895	158	0.7429	1.132
118	0.6693	0.87	159	0.7655	0.9934
119	0.5968	0.8239	160	0.7396	1.1914
120	0.6256	0.7444	161	0.6968	0.9672
121	0.5518	0.7659	162	0.7737	1.0322
122	0.6871	0.7247	163	0.7396	0.9976

(*continues*)

Table XII (*Continued*)

Isomer no.	Relative retention time	Relative response factor	Isomer no.	Relative retention time	Relative response factor
164	0.7399	0.9848	187	0.7654	1.122
165	0.692	1.0777	188	0.692	0.7337
166	0.7572	1.0421	189	0.9142	1.5091
167	0.7814	1.0658	190	0.874	1.31
168	0.7068	0.8375	191	0.8447	1.4741
169	0.8625	0.8355	192	0.8269	1.599
170	0.874	0.75	193	0.8397	1.4167
171	0.8089	1.1712	194	0.962	1.868
172	0.8278	1.172	195	0.9321	0.415
173	0.8152	2.044	196	0.8938	1.2321
174	0.7965	0.806	197	0.8293	0.9522
175	0.7611	0.381	198	0.8845	1.07
176	0.7305	1.0589	199	0.8494	1.1508
177	0.8031	1.009	200	0.8197	0.369
178	0.7537	0.621	201	0.8875	0.803
179	0.7205	0.8237	202	0.8089	1.165
180	0.8362	1.295	203	0.8938	1.629
181	0.7968	1.6046	204	0.8217	0.8034
182	0.7653	1.1272	205	0.9678	1.406
183	0.772	0.976	206	1.0103	1.673
184	0.7016	1.0046	207	0.9423	1.3257
185	0.7848	1.437	208	0.932	1.1756
186	0.7416	1.2236	209	1.0496	1.139

Source: Mullin *et al.* (1984). *Environ. Sci. Technol.* **18**, 474.
[a] Estimated relative response factor based on other isomeric PCBs.

PCBs by electron capture gas chromatography (ECD–GC). This procedure, however, only identifies and quantifies PCB technical mixtures and not individual congeners. Subsequently, with the availability of individual PCBs for use as reference standards and the widespread use of capillary columns, analytical effort is now being directed toward the determination of individual congeners in samples. Standard methods of analysis have been published such as those of the AOAC for food and U.S. EPA for waters and wastes. A summary of analytical methods for PCBs in a variety of substrates is given in Table X.

For the specific determination of the more toxic planar, mono-, and di-ortho PCBs, fractionation on activated carbon, carbopack, or porous graphitic carbon is employed using elution with solvents such as cyclohexane/dichloromethane for ortho-substituted PCBs followed by elution with toluene for planar PCBs. The use of capillary columns is mandatory for specific isomer determinations and the use of mass spectrometric detection is advised. Table XI provides substitution pattern data for all 209 PCB isomers and Table XII gives retention time data (relative to octachloronaphthalene) and response factors for these PCBs on an SE-54 narrow-bore fused silica capillary column.

Mobility of PCBs from soils, sediments, and wastes can be determined by leachate testing such as the U.S. EPA TCLP whereby solids are leached

Table XIII PCB Regulatory Criteria

Substrate	Tolerance	Organization	Study
Drinking water	0.0005 mg L^{-1}	U.S. EPA	1
Ambient water	0.001 mg L^{-1}	CCME	2
Dairy Products (fat basis)	1.5 mg kg^{-1}	U.S. EPA	3
Poultry (fat basis)	3.0 mg kg^{-1}	U.S. EPA	3
Eggs	0.3 mg kg^{-1}	U.S. EPA	3
Animal Feed (complete)	0.2 mg kg^{-1}	U.S. EPA	3
Animal food components	2.0 mg kg^{-1}	U.S. EPA	3
Fish and shellfish	2.0 mg kg^{-1}	U.S. EPA	3
Infant and junior foods	0.2 mg kg^{-1}	U.S. EPA	3
Paper food packaging material	10 mg kg^{-1}	U.S. EPA	3
Workplace air			
(42% Cl)	1 mg m^{-3} (TWA value)	ACGIH	4
(54% Cl)	0.5 mg m^{-3} (TWA value)	ACGIH	4
Soil (agricultural)	0.5 mg kg^{-1}	CCME	2
residential/public access	5 mg kg^{-1}	CCME	2
industrial/commercial	50 mg kg^{-1}	CCME	2

1. U.S. EPA Federal Register, vol. 54, No. 97, May 22, 1989.
2. CCME Canadian Council of Ministers of the Environment. "Guidelines for the Management of Wastes Containing Polychlorinated Biphenyls (PCBs)," CCME-TS/WM-TRE008 Manual EPS59/HA/1,1989.
3. U.S. FDA Code of Federal Regulations (CFR), Tolerances for Polychlorinated Biphenyls (PCBs), 21 CFR Chap. 1, Subpart B, Sect. 109.30, April 1, 1988.
4. ACGIH, 1992–1993 Threshold Limit Values for Chemical Substances and Physical Agents and Biological Exposure Indices, American Conference of Governmental Industrial Hygienists, Cincinnati, 1992.

for 18 hours on a rotational shaker using an acidic (pH 4.93) leaching solution.

B. Regulations

Evaluation of environmental impact and human health aspects of PCBs requires the application of regulations for comparison purposes. Table XIII presents a selection of regulations covering PCB concentrations in various substrates. Generally, human health criteria are based on lifetime cancer risks while environmental health criteria are based on a number of factors including toxicity, bioaccumulation potential, and biodegradability.

Related Articles: BIODEGRADATION OF XENOBIOTIC CHEMICALS; ENVIRONMENTAL CANCER RISKS; ORGANIC MICROPOLLUTANTS IN LAKE SEDIMENTS.

Bibliography

Canadian Council of Ministers of the Environment. (1989). "Guidelines for the Management of Wastes Containing Polychlorinated Biphenyls (PCBs), CCME-TS/WM- TRE008 Manual EPS 9/HA/1. Industrial Programs Branch, Environmental Protection, Environment Canada, Ottawa.

De Voogt, P., Wells, D. E., Reutergardth, L., and Brinkman, U.A.T.H. (1990). Biological activity determination and occurrence of planar, mono-, and di-ortho PCBs. *Intern. J. Environ. Anal. Chem.* **40,** 1–46.

Erickson, M. D. (1986). "The Analytical Chemistry of PCBs," Butterworth, London.

George, C. J., Bennett, G. F., Simoneaux, D., and George, W. J. (1988). Polychlorinated biphenyls: A toxicological review. *J. Hazard. Mater.* **18,** 113–144.

IRPTC (1986). "Data Profile on Polychlorinated Biphenyls." International Register of Potentially Toxic Chemicals, UNEP, Geneva.

Safe, S. (1987). PCBs: "Mammalian and Environmental Toxicology." Environmental Toxins, Series 1. Springer-Verlag, Berlin.

Safe, S. (1990). Polychlorinated biphenyls (PCBs), dibenzo-*p*-dioxins (PCDDs), dibenzofurans (PCDFs), and related compounds: Environmental and mechanistic considerations which support the development of toxic equivalency factors (TEFs). *CRC Crit. Rev. Toxicol.* **21,** 51–88.

Silberhorn, E. M., Glauert, H. P., and Robertson, L. W. (1990). Carcinogenicity of polyhalogenated biphenyls: PCBs and PBBs." *CRC Crit. Rev. Toxicol.* **20,** 439–496.

Strachan, W. M. J. (1988). "Polychlorinated Biphenyls (PCBs): Fate and Effects in the Canadian Environment." Report EPS4/HA/2. Environment Canada, Ottawa.

USDHHS. (1991) "Draft Toxicological Profile for Selected PCBs (Aroclor-1260, -1254, -1248, -1242, -1232, -1221, and -016)." U.S. Department of Health and Human Services, Public Health Service, Agency for Toxic Substances and Disease Registry, Atlanta.

Waid, J. S. (1986). "PCBs and the Environment," vols. 1,2, and 3. CRC Press, Boca Raton, Florida.

WHO/EURO. (1987). "PCBs, PCDDs, and PCDFs: Prevention and Control of Accidental and Environmental Exposures." Environmental Health Series, No. 23. World Health Organization, Copenhagen.

Protists as Indicators of Water Quality in Marine Environments

John Cairns, Jr., and Paul V. McCormick
Virginia Polytechnic Institute and State University

Glossary

Biomagnification Process whereby toxic materials are concentrated above ambient exposure levels as they are passed from lower to higher trophic levels in the food chain.

Mesocosm Bounded, partially enclosed outdoor experimental systems that incorporate ecosystem- as well as population—level processes.

Microcosm Laboratory test systems designed to incorporate specific ecosystem-level processes.

Monitoring Surveillance to determine if previously established environmental conditions are being maintained.

Validation Process whereby predictions of laboratory tests are confirmed in natural ecosystems or suitable surrogates thereof.

***PROTISTS,** particularly microalgae and protozoa, are ubiquitous components of the marine environment. These organisms play key roles in energy flow and mineral cycling in most habitats. Because of their small size and rapid generation times, bioassays utilizing protists as test organisms can be conducted in an expeditious and cost-effective manner. Many alternative bioassay designs are available for laboratory testing depending on the specific objectives of the study. The development of more environmentally realistic microcosm and mesocosm tests has lagged behind that of laboratory bioassays, despite promising results from limited testing. Development of* in situ *monitoring techniques using protists has also been limited compared with those for higher trophic levels but are needed for validating experimental predictions in both the short and long term.*

I. Use of Protists as Indicators of Stress

Although evaluations of chemical concentrations in ecosystems are necessary to determine potential sources of stress, any valid assessment of chemical hazard or impact in the natural environment must include measurements of the responses elicited from living organisms. Protists possess many desirable qualities for use as test organisms for assessing environmental risk and impact.

Ecological importance. Autotrophic and heterotrophic protists perform key functions in energy flow and elemental cycling in marine foodwebs. Microalgae are responsible for the bulk of primary production in most marine habitats. This energy is transferred to higher trophic levels in the food chain (e.g., zooplankton and fish) mainly through the activity of protozoan grazers, which consume both algae and the bacterial assemblage that thrives off algal exudates. Thus, the integrity of the entire ecosystem is closely linked to the unimpaired functioning of these organisms.

Ubiquitous occurrence. Major groups of protozoa and algae are found in most marine habitats throughout the world. Indeed, many species are believed to be distributed globally and are present wherever appropriate conditions exist.

Handbook of Hazardous Materials

Thus, these organisms are always available for testing or monitoring purposes.

Cost-effective. Because of their small size and rapid rates of asexual and sexual reproduction, multigenerational tests using protists can be conducted more rapidly and in smaller test containers than those using larger organisms. These considerations are particularly important when time and space are of the essence (e.g., shipboard testing).

The above attributes make these organisms good candidates for testing under a variety of circumstances. It is somewhat surprising, therefore, that few standard techniques have been developed for this purpose. In contrast to freshwater testing, there have been few attempts to even propose a general set of guidelines for testing with marine protists, much less any concerted effort to refine such methodology with the aim of standardization. Given the lack of a routine hazard assessment procedure using marine protists, the following discussion focuses on providing a broad overview of necessary considerations involved in developing a testing strategy rather than a description of a few specific tests. Detailed recommendations for various classes of tests are available in publications listed in the bibliography.

II. Single-Species Laboratory Bioassays

A. Introduction

Evaluations of the hazard that chemicals pose in freshwater and marine environments have relied almost exclusively on data provided by single-species bioassays. These procedures were derived largely from assays for predicting human health effects and typically measure the response of a species to increasing stress under highly controlled laboratory conditions. The appeal of these tests lies in their simplicity, high degree of reproducibility, and relatively low cost. These attributes make them extremely useful, both as screening tests for determining the relative toxicity of individual chemicals or complex effluents, and for determining effects of such substances on population level processes such as growth. Unfortunately, many of the characteristics that enhance the simplicity and reproducibility of single-species tests may also reduce their environmental realism. As discussed in Section III. A, this inevitable tradeoff limits the ability of such tests to predict effects on natural communities and ecosystems, which are complex, multivariate systems possessing several properties that are not incorporated into single-species test designs. Despite certain limitations, simple single-species tests using protists are extremely useful as screening tools and for ranking the relative toxicity of chemicals to aquatic biota.

Although the principle of the single-species assay is rather simple, decisions regarding the exact test design are complicated by several factors, including the choice of test species, the response to be measured, the test media to be used, the design of test chambers, and the mode of incubation. The objectives of the study must be defined before deciding on a particular test design since different designs answer different questions.

B. Test Species

Several characteristics determine the suitability of a species for use in laboratory bioassays. No "ideal" test species has been found in either freshwater or marine ecosystems, but several candidate species of algae have been proposed. By contrast, relatively little effort has been given to identify suitable test species of heterotrophic protists (but see some previously used genera in Table I). This is surprising given the acknowledged importance of protozoa in marine ecosystems and the fact that the sensitivity of these organisms to stress is not always predicted by the responses of other organisms.

Often, the overriding factor used in choosing test species is their relative ease of maintenance in culture. This consideration alone greatly limits the number of species available for testing since most protist species have never been successfully cultured in the laboratory. Ease of maintenance in culture facilitates repeated and interlaboratory testing. Species that exhibit short lag phases and rapid exponential growth in culture are preferred to reduce the cost of testing and to minimize artifacts in the behavior of test substances in small enclosures as testing proceeds (e.g., adsorption to container walls). Species that can be induced to form resting stages (e.g., cysts) have been recommended because long-term maintenance of cultures can be achieved with relatively little effort and with little change in genetic constitution. Testing efficiency is also enhanced by the use of species that lend themselves to rapid

Table I Recommended Marine Algal Test Species and Some Protozoan Genera Used in Marine Bioassays

Algal species
Dunaliella tertiolecta
Emiliania (Coccolithus) huxleyi
Minutocellus polymorphus
Pavlova (Monochrysis) lutheri
Phaeodactylum tricornutum
Skeletonema costatum
Thalassiosira pseudonana
Protozoan genera
Balanion
Cristigera
Euplotes
Favella
Uronema

quantification. For example, colonial species are less amenable to rapid counting with electronic particle counters than are strictly unicellular forms.

An often ignored criterion is the selection of species that are ecologically and, preferably, geographically representative of the region or habitat of interest. This is particularly important if, as is generally the case, the goal of testing is to predict effects in the natural environment. The fate characteristics of the substance under study may also help determine the most appropriate species for testing. For example, chemicals with high sorption coefficients may pose the greatest threat to bottom-dwelling organisms, indicating the need for testing with benthic protistan species. Ignoring ecological considerations in favor of other factors, such as ease of culturing, may reduce the realism and, thus, extrapolability of laboratory tests.

Because species may vary widely in response to a particular stress and predicting which species will be most sensitive in a given instance is thus impossible, it is necessary to test the response of several species to obtain an accurate estimate of the no-observable-effect-concentration (NOEC) for protists in general. Although dozens of species have been used for testing purposes, relatively few species enjoy wide acceptance (Table I). The U.S. EPA marine algal assay procedure (MAAP), standardized in 1974, recommended use of only two test algae: the chlorophyte *Dunaliella tertiolecta* and the diatom *Thalassiosira pseudonana*. Estimates of effect concentrations derived from these two species alone may not accurately reflect the response of other autotrophic protists. For example, *Dunaliella* may be extremely resistant to certain heavy metals relative to other algal species. It is generally recommended that at least three different species, including one indigenous to the waters of interest, be used to estimate responses, although no extensive database seems to exist for supporting such specific recommendations. Use of indigenous species should increase the accuracy of site-specific predictions and reduce the number of test species required.

C. Endpoints

The two most commonly used endpoints in single-species laboratory bioassays are biomass yield (i.e., the algal growth potential or AGP) and growth rate (i.e., the fractional increase in biomass over time). Subsidy effects (e.g., nutrient enrichment) may affect either the rate or extent of biomass accumulation, while toxic substances generally act directly on rates of cell division, thus requiring the measurement of the rate of growth to assess such effects. Biomass is most commonly measured as either dry weight, cell number, or cell volume. Growth is measured as the change in one or, preferably, more of these biomass parameters during the experiment. Measures of cell constituents, including particulate organic carbon (POC) and nitrogen (PON), total cell protein, chlorophyll *a*, and ATP, have been used less frequently to estimate biomass accumulation. All biomass measures have drawbacks that must be considered and, generally, the ease of measurement is inversely proportional to the accuracy of the estimator. Because the maximum biomass attained must be determined in order to measure the AGP, daily monitoring of experimental cultures must be conducted to ascertain the point at which this is achieved. The optimum procedure for tracking biomass accumulation should use a rapid, albeit often less accurate, estimator (e.g., chlorophyll *a* fluorescence or cell density estimates using an electronic particle counter) to monitor the accumulation of biomass, coupled with a more sensitive measurement (e.g., PON) that is employed as the biomass maximum is approached.

Other responses have been used to assess subsidy and stress effects, either because they indicate specific mechanisms of effect or are deemed to be more sensitive than the standard measures noted above. Reported measures include: (1) photosynthetic rate (e.g., C^{14} uptake or oxygen evolution); (2) changes in cell morphology induced by stress; (3) loss of or change in patterns of motility; (4) physiological measures such as mixed function oxidase (MFO) activity, internal ratios of nitrogen:carbon or nitrogen:phosphorus, enzyme induction and/or repression, and chemical composition of the cells; (5) rates of phagocytosis by heterotrophic protists; or (6) frequency of conjugation as a measure of genetic effects on ciliates. These measures can often be performed in less time than that required for standard endpoints (e.g., hours instead of days). However, attenuation of the test period may reduce the probability of detecting long-term responses to stress (i.e., adaptation) and, thus, reduce the predictive capability of the procedure. Although not a direct measure of toxicity, determination of the extent to which protists accumulate test substances can be extremely important since both autotrophic and heterotrophic protists are important links at the base of planktonic and benthic foodwebs and may be responsible for much of the bioaccumulation in the food chain, depending on the substance being tested.

D. Test Media

Laboratory bioassays may utilize natural seawater, either with or without nutrient or toxicant amendments, or an artificial seawater medium. Natural seawater collected from different locations or at different times can vary greatly in the ability to support protistan growth and to interact with test substances. Where site-specific effects are of interest, use of natural seawater has obvious advantages in terms of predicting field effects from laboratory responses, because it integrates effects of peculiarities of a site at a particular time. However, this same property conflicts with attempts to develop a standardized testing protocol. Virtually all standard methods for preserving seawater for testing will alter the chemical characteristics, thereby reducing the reproducibility of tests using natural seawater collected at one time and thawed over a period of time. Preferably, indigenous organisms should be removed by filtration through a suitably sized (e.g., 0.45 μm) mesh, although the loss of nutrient adsorbed to larger seston particles is inevitable. Methods such as autoclaving or freezing, which kill indigenous algal cells but do not remove them, are not recommended because of the likelihood of alterations in chemical conditions by the liberation of the contents of lysed cells into the medium and the selective removal of nutrients by adsorption.

Use of artificial growth media allows for considerably greater standardization of laboratory tests. Use of a formulated test water ensures that effects of a stress on test organisms are not affected by residues of other pollutants present in the source water, a condition which is unlikely to be met using water from many marine habitats (e.g., estuaries). However, the use of artificial growth media can cause severe problems in attempting to extrapolate predictions from the laboratory to the marine environment. The potential always exists for interaction between the stress under study and the composition of the water. For example, excessive chelation of heavy metals in synthetic media can artificially reduce toxicity. Culturing in highly enriched media frequently alters the resistance of cells to stress compared with the nutrient-limited state typically experienced in many marine habitats. Artificial media low in added chelators and nutrients can be used to reduce these problems, but presents its own problems such as limiting the duration of exponential growth in batch tests. Furthermore, few autotrophic protists have been grown in synthetic media lacking adequate amounts of artificial chelators. Clearly, the choice of test water will depend on the objectives of the study.

Although most standard bioassays are designed to test for effects of toxicity in an aqueous medium, inclusion of sediment in the test container may be warranted if fate effects are of concern or if sediment toxicity is of interest. Sediment can either be added to the bottom of the test container as a slurry (solid phase testing) or an elutriate can be prepared for testing by intense mixing of sediment and test water followed by settling and filtration to remove sediment (liquid phase testing). Concerns similar to those for test water must be addressed when determining the type of sediment to be used in the test.

E. Test Chambers

Test results can be strongly affected by the design of the test chamber. Three test designs are discussed here in order of increasing complexity of construction and maintenance. In general, increasingly complex designs enhance environmental realism, although even simple test systems are valid for measuring certain types of response. All three are routinely used for testing with algal protists and should also be suitable for testing with heterotrophic protists.

Most standardized protist bioassays (e.g., the U.S. EPA MAAP) use batch cultures for testing. In their simplest form, these tests use small static test systems containing a unialgal culture, along with a prescribed dose of the test substance of interest. Although the small size of protists allows for the use of smaller test containers than those required for invertebrates or fish, considerations other than organism size must be considered when determining the volume of the test container to be used. Increasing the size of the test container reduces the surface volume ratio and, thus, "wall" effects associated with the attachment and growth of test organisms on the container surface. This consideration is particularly important if testing is being conducted with planktonic organisms. Chamber volumes greater than 100 ml are generally recommended.

It is essential that controlled environmental conditions be maintained during testing. The spectral quality and intensity of light and the incubation temperature can strongly affect the growth and physiology of the test organisms and the behavior of the test chemical. Consequently, these should be maintained at identical levels among stock and test cultures. The exact conditions used depend on the type of study being conducted. In enrichment tests, where interest is in the nutrient that is limiting to algal growth potential, conditions should be such that light and temperature do not become limiting factors. This may require separate incubations for each test species since individual species vary in optimal light and temperature requirements. If growth rates are to be measured to assess toxic effects, it is usually preferable to maintain realistic external environmental conditions since variations in physiological processes contributing to growth are extremely sensitive to variations in light and temperature. Shaking or swirling the incubating cultures can remove the tendency for the gaseous equilibrium between air and water to be upset in these static test systems, with consequent deviations in pH away from that for natural seawater (8.5).

Cell and toxicant inocula are generally added to batch cultures only once at the initiation of testing. Use of a stock culture in the early phases of exponential growth ensures that the vast majority of cells in the inoculum are healthy. This method may be unsuitable, however, if the objective is to measure subsidy effects associated with eutrophication or to test for nutrient limitation. Many algae are capable of assimilating nutrients in quantities far exceeding those required for immediate cell division and maintenance during nutrient-saturated (i.e., exponential) growth, thus allowing rapid growth to continue for several generations after inoculation from the stock culture into unamended seawater. Thus, observed growth during the bioassay may be controlled more by the initial physiological state of the test algae than the conditions existing during the bioassay. Inoculation of chambers with extremely low cell densities avoids this problem but increases the time required to achieve detectable cell densities or maximal biomass and increases the chance of failed cultures. Predepletion of cell reserves in a minimal medium for several days prior to testing is a viable, if somewhat time consuming, option. Indeed, if a single location is to be studied intensively, it may be best to predeplete cultures in the test water of interest using dialysis culturing techniques (discussed later in this article).

Batch culture assays were originally designed for assessing effects of nutrient limitation on algal growth, and their use in measuring toxic responses is limited by factors inherent in their design. After the addition of a chemical pulse to the batch test chamber, the concentration of the substance may decline due to adsorption to the test chamber or to the test algae or algal extracellular products as well as photodegradation or volatilization. While addition of sublethal doses of heavy metals may result in a lengthened lag phase of growth, normal growth is often resumed as available concentrations of the toxicant decrease. Of course, pulsed stresses occur in natural ecosystems as well, but more often the objective of testing is to measure responses to continuous exposure to target concentrations. Accurate estimates of growth rate responses are usually limited to the initial phases of exponential growth in batch culture. As algal densities increase in these static systems, nutrients are rapidly depleted and algal excretory products accumulate. Once cells reach suitable densities for counting in batch cultures, growth can shift from nutrient-saturated to nutrient-starved in as few as one or two divisions. This precludes examination of effects on nutrient-limited growth, the most common condition in the marine environment. Rapid changes in test conditions and rates of cell division introduce obvious difficulties in attempting to establish a dose–response relationship and limit batch testing to use as rapid screening tests.

More elaborate test chambers using dialysis or chemostat-culturing methods will often increase the accuracy of measurement of effects of pollutants on algal growth rates and increase the time period over which effects can be validly measured. In dialysis or "cage" cultures, algal cells are enclosed in a dialysis membrane or other suitably meshed material immersed in chambers filled with seawater and the nutrient or toxicant of interest. Test water and chemicals of interest are continuously pumped through the chamber, although a static design can be substituted if a large volume of test water is used relative to the size of the inoculum. An alternative design, which facilitates biomass and growth measurements, consists of a resevoir chamber bisected by a fine mesh membrane that limits an algal inoculum to one side of the chamber while allowing test water to circulate throughout. The pore size of the membrane should be small enough to contain the algal cells but, ideally, not much smaller in order to minimize selectivity in the particle size that can pass freely through the membrane. Membrane material should be transparent, autoclavable, and sturdy enough to resist cleaning to remove algal and bacterial cells that may foul the outside surface if cages are incubated in unfiltered seawater. Agitation of the enclosed cultures, preferably on a continuous basis, facilitates diffusion between the culture and external test medium.

Growth responses are easier to measure in cage cultures and more closely approximate those found in most marine habitats than those measured in batch culture. Continuous addition of toxicants maintains a constant concentration in the test cultures and reduces artifacts associated with adsorption to the test chamber or sequestering by artificial chelators. The cage design facilitates the continuous replenishment of resources and removal of metabolic byproducts without the loss of protist cells and, thus, allows for a protracted period of constant growth to be achieved. The continuous flux of materials also allows for the attainment of large cell densities in these cultures, which decreases statistical problems associated with enumeration at low abundances.

Use of continuous culture techniques, the most elaborate of the three laboratory bioassay designs, often reduces the time required to properly measure growth responses relative to batch or cage cultures. In this design, the entire test chamber is the incubator. Test water enters and leaves the chamber at a controlled, constant rate in order to maintain constant physicochemical conditions. In contrast with other culturing methods, a portion of algal production is lost as water passes through the test system and out of the container. A constant population density in the chamber is achieved when the loss of cells from the container equals that of division rates (i.e., population growth) within the container. Control over the attainment of a steady-state condition is achieved either by maintaining a constant flow, wherein the biomass in the chamber is allowed to equilibrate (chemostat), or by maintaining a constant biomass in the chamber, whereby growth rate equilibrates (turbidostat). Despite the advantages of continuous culture methods, or even cage cultures, in terms of realism, standardized testing procedures continue to rely primarily on batch test methods, primarily because of simplicity and low cost.

F. Expression of Test Results

Time is generally of the essence in testing, especially when batch cultures are used. Current guidelines by the U.S. EPA, for example, recommend that growth responses in batch cultures be measured within three days of bioassay initiation. Thus, responses are typically expressed as the 72-hr IC_{50}, the concentration that inhibits a process by 50%, compared with control cultures, after 72 hours of exposure derived by interpolation between data points forming a dose–response curve. Of course, the time required to measure a response will vary with the parameter being measured, and the potential for long-term adaptation to a stress should not be ignored when determining the duration of an experiment. Results are also commonly expressed as the NOEC, the highest concentration used in a test that elicits no significant effect on the test population, or the lowest-observable-effect concentration (LOEC). All of these thresholds can be estimated using standard statistical techniques.

III. Microcosm/Mesocosm Tests

A. Rationale

Environmental concerns and regulations have increasingly focused on the protection of entire ecosystems, not merely individual species. Fundamental to contemporary ecological thought is the concept that natural communities and ecosystems possess properties that arise from the interaction between populations and, thus, cannot be predicted from the response of individual populations. For example, grazing by microheterotrophs or competition with naturally co-occurring producers have been shown to alter the response of test algal species to toxic stress. Although the argument that individual species are at least as sensitive as any parameter at higher levels of biological organization may be true, there is no scientific rationale for choosing the most sensitive species; the autecology of most species in relation to stress is poorly understood and the relative sensitivity of individual species will vary with the toxicant of interest. Several species must generally be tested to provide an accurate estimate of the effects of a particular stress. Few marine protists have been subjected to testing so that the range of response among these organisms is unclear. Problems such as these compromise the ability of these tests to accurately predict effects in the natural environment.

Laboratory microcosms and field mesocosms offer excellent opportunities for bridging the gap between simple laboratory screening tests and responses in natural marine ecosystems because they can incorporate a degree of environmental realism not found in single-species tests. Advantages of properly designed microcosm and mesocosm tests should include:

1. the simultaneous testing of several species in the same test system;
2. the inclusion of important community- and ecosystem-level attributes (e.g., species interactions, energy flow, and nutrient spiraling) in the test design, even if these processes are not actually measured;
3. the ability to simultaneously consider chemical fate and effect;
4. the measurement of endpoints that are the same as those measured in natural ecosystems (e.g., population dynamics in a competitive environment, biodiversity, or nutrient flux); and
5. the control of confounding influences such as the history of exposure and external factors that plague *in situ* marine surveys.

Undoubtedly, ecotoxicological tests using individual surrogate species will remain central to hazard evaluation procedures for the forseeable future. However, there is increasing experimental evidence to support the contention that tests using microbial communities including protists, if properly designed, can enhance the accuracy of assessments of environmental risk and impact. More work is clearly needed to develop and refine such tests that are applicable to different marine habitats.

B. Endpoints

Microcosm/mesocosm tests allow for the evaluation of community and ecosystem responses to stress. However, the response of individual populations to stress also warrants evaluation in the interactive environment in these systems. Structural attributes used as endpoints for protistan testing are generally identical to those that characterize other ecological communities and may include measures of biomass or species richness and diversity. Functional endpoints amenable to measurement include primary production, community respiration, and other physiological measurements related to cell nutrition (Table II). Of course, the activity of other members

Table II General Categories of Measurements Commonly Used to Assess the Structural and Functional Dynamics of Protists in Single-Species and Multispecies Tests

Measurements	Single-species	Multispecies
Structural		
Biomass/Standing crop	X	X
Population growth rate	X	
Community biomass accrual		X
Species composition		X
Species richness/Diversity		X
Functional		
Photosynthetic (primary production) rate	X	X
Respiration rate	X	X
Nutritional status	X	X
Nutrient uptake/Regeneration rate	X	X
Feeding rates	X	X

of the marine community (e.g., bacteria) may also be included in these measurements depending on the experimental design and collection methods adopted. Estimates of toxicant effects on nutrient spiraling and energy flow can be ascertained with protist communities, given their integral role in these ecosystem processes.

C. Laboratory Microcosms

Laboratory microcosm tests allow for the consideration of important community and/or ecosystem properties in a controlled testing environment. Depending on the objectives of a study, microcosm tests need not incorporate all important elements of these levels of biological organization; in many cases, specific tests may be developed to study effects of stress on a specific community or ecosystem process.

Two types of microcosm tests have been used to predict effects of anthropogenic stress on aquatic communities and ecosystems: (1) gnotobiotic or "species-defined" test systems; and (2) test systems that measure the response of indigenous assemblages. Gnotobiotic test systems combine several surrogate species into a single test container under conditions that are usually otherwise similar to those employed for single-species tests. Artificial test media are generally used (as for single-species bioassays) to maximize control over test conditions. Microcosm tests using indigenous assemblages collect organisms either from the natural substrate and/or water column or from artificial substrates placed in the habitat of interest. These tests may use either natural or artificial water. As discussed above (Section II.D), similar advantages and problems must be weighed when deciding which type of medium to use in microcosm as well as single-species tests.

Methodologies used in gnotobiotic testing procedures have the advantage of being amenable to standardization. Cultures of specific strains can be maintained in a well-defined medium. The number and kind of species added to the test system are strictly controlled, although only those species that can be grown in monoxenic cultures can be used for testing. Thus, the design of gnotobiotic systems is similar to single-species test systems in many respects. Gnotobiotic test systems rarely recreate naturally occurring communities but, rather, attempt to assemble species that perform representative functions in a particular community of interest or in natural communities in general. Gnotobiotic protistan assemblages would ideally include several competing algal species as well as bactivorous, herbivorous, and predatory protozoa.

While the degree of control afforded by gnotobiotic test systems is distinctly advantageous from a regulatory standpoint, a major drawback of these tests may be the often untested assumption that artificially assembled "communities" possess the critical attributes of natural communities and ecosystems. This assumption is, however, contrary to contemporary ecological thought, which views natural communities as coevolved sets of interacting populations and, thus, more than randomly assembled collections of species. Gnotobiotic test systems, therefore, may be no more realistic than single-species tests and may not warrant the additional time and expense required. The validity of the assumption of correspondence between responses of gnotobiotic and naturally derived communities certainly needs to be verified before adopting a gnotobiotic test protocol.

Transporting indigenous communities into laboratory microcosms increases the potential for incorporating important community- and ecosystem-level attributes into testing. Because communities are obtained from a particular site, it has been suggested that these test designs do not possess the generic applicability of gnotobiotic test systems. This contention has been rebuked by some who maintain that protistan communities exhibit a considerable degree of structural and functional redundancy among similar ecological habitats that are physically separated in space and time. This redundancy is due to the high species richness of most protistan communities—while individual species may vary greatly in sensitivity to stress, community responses, which comprise the responses of many individual species, should be similar regardless of the species composition of a particular community. Furthermore, the extent to which communities or ecosystems do vary in response to stress through space and time illustrates the need for site-specific testing and limits the applicability of generic testing to a particular location.

A more valid criticism of tests using natural communities is that they cannot be standardized to the extent possible for gnotobiotic test systems. This is a question that has yet to be critically assessed even with freshwater protistan communities, which are more widely used than marine communities. It is inevitable that certain limitations in standardization exist for tests using naturally derived protistan assemblages.

D. Field Mesocosms

Field mesocosms extend the use of indigenous protistan assemblages (often as part of a larger biological and physicochemical study) beyond microcosm testing by incorporating additional aspects of realism into their design. Field test systems are generally much larger than those used in the laboratory and, thus, allow for the inclusion of larger organisms (e.g., fish) from the indigenous assemblage into the test system. Consideration of the larger marine food web is valuable (even where only the protist assemblage is sampled) because it increases the environmental realism under which experimental results are obtained. The enhanced correspondence between mesocosms and natural systems makes such tests particularly useful for validation of laboratory results and for supplemental testing when laboratory results are not conclusive. Because the increased realism of mesocosm testing is not achieved without additional cost, the use of this technique is generally limited to higher tiers of testing and such specific applications discussed previously.

Properly designed mesocosms can allow for long-term effects of pollution and the fate of chemical stressors to be assessed with less artifacts than in laboratory experiments. Principally because of size, mesocosms allow for several interacting trophic levels to be maintained for extended periods of time. Mesocosm communities and environments can be manipulated in ways that are not possible in natural habitats. The isolated nature of the test system facilitates resampling of a defined community of organisms over a period of time. These desirable properties must be weighed against limitations to the design of most mesocosms, including decreased vertical and horizontal mixing within test systems compared with the open water, wall effects, the difficulty in maintaining these systems in the open ocean because of their fragility, and the high cost of construction, deployment, and maintenance that may limit the level of replication and, thus, the ability to detect significant responses.

The size of the mesocosm and the degree to which the design allows for movement of water, organisms, and pollutants within the system is largely dictated by cost restrictions and the type of marine habitat that is to be recreated. The physical extent of an enclosure can range from small (e.g., <1 m^3) bags or other containers suspended in the water column to much more extensive (i.e., 2000 m^3) enclosures such as those used in CEPEX (i.e., controlled ecosystem pollution experiment) or those designed to encompass both planktonic and benthic environments. The degree of coupling between the water column and benthic layers in the mesocosm can strongly affect biological and chemical processes within the system. Mixing of the water column is generally achieved by mechanical stirrers or pumped air. Maintenance of a well-mixed water column in close contact with the sediment is desirable if the goal is to mimic processes in coastal ecosystems. Mesocosms with well-mixed top and bottom layers that are separated by an unmixed zone can be used to recreate stratified coastal ecosystems. In order to model pelagic ecosystems, where water–sediment interactions are negligible, mesocosms can be designed to maintain seston in suspension, thus avoiding the accumulation of sediment and toxicant in the bottom of the mesocosm. More complex designs (i.e., incorporation of horizontal mixing) can be used to enhance environmental realism.

As in any test, the dosing regime is dependent on the objective of the study and the stress of interest. Dosing is usually performed as a single pulse or as periodic doses to maintain a target concentration of the chemical in the water column. In addition to toxicant inputs, periodic inputs of nutrients are normally required to avoid the onset of nutrient starvation in the algal assemblage. With the addition of such supplements, healthy control communities can be maintained for several weeks.

Validation studies have shown that mesocosm test systems are capable of accurately modeling conditions in natural marine habitats. In pollution studies, mesocosm tests have accurately predicted not only the degree of ecosystem response but the types of responses that are most sensitive to pollution. The increased accuracy of results of field mesocosm tests over simple laboratory bioassays with one or few protist species in isolation has been attributed to the realistic nature of mesocosms, including the presence of higher trophic levels that may interact either directly (e.g., zooplankton herbivory) or indirectly (e.g., predation on zooplankton herbivores) with the protistan community. Thus, although such large-scale tests should not be viewed as a replacement for simple screening tests because of cost and complexity, they are potentially powerful tools for evaluating the response of ecosystems to anthropogenic perturbations in marine habitats.

IV. In Situ Environmental Monitoring

A. Rationale

While laboratory tests that measure the response of different levels of biological organization are essential for assessing the relative and absolute hazard of chemical stress to the marine environment, the ability of these tests to predict responses in natural ecosystems is limited because of the diversity and complexity of stresses to which many marine habitats are exposed. There is some evidence in freshwater ecosystems that microcosm and mesocosm tests using protistan communities provide more reliable estimates of hazard than those derived from single-species tests. However, in some studies both single-species and microcosm tests have failed to accurately predict the hazard of a stress to natural ecosystems either because the complexity of the stress has been underestimated or important components of environmental realism have been absent from all test systems used. The ability of any surrogate test system to predict effects of chemical stress on protist communities in marine ecosystems is affected by several factors including the degree to which:

1. simplification of the natural system in an artificial test container alters chemical effects and fate;
2. the exposure regime in the test system differs from that in the natural system in biologically significant ways (e.g., pulsed versus continuous exposure);
3. short-term responses measured in the test system are predictive of longer-term responses in the natural system; or
4. response to a single chemical or suite of chemicals in the test system is predictive of responses to these same contaminants in the natural system where many other synergistic and antagonistic stresses may be present. These problems highlight the need for monitoring in the natural marine environment, both to validate short-term results of controlled laboratory and field tests and to serve as an ongoing error detection and control mechanism for environmental protection.

The use of protists as environmental monitors in the marine environment has been rather selective. Natural variation in protist populations and communities is usually substantial and, thus, requires a good understanding of the function of both the physicochemical and biological components of the system (which is often lacking) and intensive and repeated sampling (which can be quite costly). Spatial and temporal variability is particularly problematic in coastal ecosystems where most monitoring is conducted. A lack of appreciation for this background noise often results in the collection of data that are unable to distinguish between natural variation and changes due to anthropogenic stress. Both the successes and failures of previous monitoring programs are prescriptive in this regard and should be reviewed thoroughly before additional monitoring programs are designed and initiated.

B. Indicator Species

Most attempts at using protists as monitors of changes in water quality in the marine environment have focused on the concept of individual populations as indicators of environmental stress. As in freshwater, many species have been proposed for use under different conditions. Protist populations have generally been used in two ways to indicate environmental change: (1) to measure the concentration of toxicants in individuals of a particular species to analyze the extent of environmental pollution and (2) to use the presence or absence of a species, or its relative dominance in the community, as an indicator of a certain set of environmental conditions. Macroalgae or seaweeds (i.e., orders Phaeophyta, Rhodophyta, and Chlorophyta) are most commonly used to measure bioaccumulation, while microalgae (e.g., diatoms) have been useful as presence/absence indicators. Heterotrophic protists have also been found to be useful as presence/absence type indicators, although the use of these organisms is sometimes limited by the fragility of the cells and sensitivity to preservation.

The choice of indicator species will be influenced by the objectives and goals of the monitoring program. However, certain criteria are generally useful in the selection process. Protists that exhibit a broad geographical and temporal distribution and, thus, are widely available for monitoring are advantageous for use in programs that require seasonal sampling and have a regional scope. Parameters used in monitoring should exhibit responses that are easily interpretable in terms of changes in water quality. Sensitivity and a rapid response to changes in water quality are essential qualities if species are to be used as early warning indicators of environmental deterioration. However, extremely sensitive species may be absent from moderately impacted habitats and, thus, may not be available for evaluating further deterioration at such sites. Thus, a comprehensive indicator scheme should include a suite of indicator species characteristic of different types and degrees of pollution.

Analysis of toxic materials accumulated in organisms allows for an integrated assessment of pollution in a given area and may be more cost effective than water chemistry analyses, which may need to be taken frequently if pollution input is sporadic. Because of their position in marine food webs, accumulation of materials in algae and heterotrophic protists provides a good indication of the potential for biomagnification. Accumulation levels of some chemicals (e.g., heavy metals) are extremely high in many species of protists, so that levels of environmental detection can be extremely low. Protists, especially those that are attached to or associated with the substrate, do not range as far from the point of impact in their life as many other organisms, such as fish, so the ability to identify localized stress is increased. Since the primary objective of most studies of this type is to relate chemical concentrations in the biological material to those in the environment, it is critical that a well defined relationship exist between levels of biological accumulation and concentrations in the water column and/or sediment. Such relationships have been found for many species of algae.

The cage-culture method described previously (Section II.E) has been used as an *in situ* device for assessing seawater quality, particularly for determining the productivity of seawater. An advantage to the use of this technique for monitoring is that initial cell conditions can be defined and spatial variation eliminated as a confounding influence. However, this method is only useful for short-term assessments and, thus, does not integrate effects of long-term variation in pollutant levels into observed responses. It is difficult to maintain these *in situ* cultures for long periods of time because of susceptibility to fouling and destruction.

C. Protistan Communities as Environmental Monitors

Although using protist populations as indicators of certain conditions (e.g., heavy metal pollution) is desirable, analyses based on changes in protistan community structure or function may provide a more robust assessment of environmental change in many cases. Individual species of protists generally have broad environmental tolerances that are often poorly understood. Thus, it is not always clear what is indicated by a change in population abundance. Although the advent of powerful multivariate statistical procedures has aided in the identification of indicator species and communities in relation to individual stresses (e.g., acid precipitation in freshwater ecosystems), even these methods are limited in the ability to relate species-specific changes in community composition to water quality changes associated with multiple impacts. General trends in nontaxonomic measures of community structure and function have been related to gradients of individual stressors (e.g., changes in nutrient loading or increases in the concentrations of toxic materials) as well as cumulative impacts. The same structural and functional attributes used for microcosm/mesocosm testing (Section II.B) can be used for *in situ* monitoring with these communities, and this correspondence allows for validation of laboratory (e.g., microcosm) results.

Efforts to develop marine monitoring protocols based on protist communities have lagged behind those in freshwater. For example, protist communities have been used for many decades to track water quality in several large rivers in the United States. As with population studies, analysis of changes in protist community structure and function in response to stress is often confounded by natural variation occurring at various temporal and spatial scales. Isolating meaningful responses to stress from background variation is possibly one of the greatest obstacles to the development of monitoring protocols using these communities, and future efforts will benefit from an increased understanding of system variability as well as sensitivity.

Related Articles: Aquatic Toxicology, Analysis of Combination Effects; Organic Micropollutants in Lake Sediments.

Bibliography

Lundgren, A. (1985). Model ecosystems as a tool in freshwater and marine research. *Arch. fur Hydrobiol. Suppl.* **70,** 157–196.

Maestrini, S. Y., Droop, M. R., and Bonin, D. J. (1984). Test algae as indicators of seawater quality: Prospects. *In* "Algae as Ecological Indicators," (L. E. Shubert, ed.), pp. 133–188. Academic Press, New York.

Munawar, M., and Munawar, I. F. (1987). Phytoplankton bioassays for evaluating toxicity of *in situ* sediment contaminants. *Hydrobiologia* **149,** 87–105.

Pritchard, P. H., and Bourquin, A. W. (1984). The use of microcosms for evaluation of interactions between pollutants and microorganisms. *In* "Advances in Microbial Ecology," vol. 7, (K. C. Marshall, ed.), pp. 133–215. Plenum Press, New York.

Reish, D. J. (1988). The use of toxicity testing in marine environmental research. *In* "Marine Organisms As Indicators," (D. F. Soule and G. S. Kleppel, eds.), pp. 231–245. Springer-Verlag, New York.

Sherr, E., and Sherr, B. (1988). Role of microbes in pelagic food webs: A revised concept. *Limnol. and Oceanog.* **33,** 1225–1227.

Walsh, G. E. (1988). Principles of toxicity testing with marine unicellular algae. *Environ. Toxicol. Chem.* **7,** 979–987.

White, H. H., ed. (1984). "Concepts in Marine Pollution Measurements." University of Maryland, College Park, MD.

Rubber Industry, Toxicity of Work Environment

M. Hema Prasad and P. P. Reddy
Osmania University, Hyderabad, India

Glossary

Carcinogen Chemical, physical, or biological substance capable of causing cancer.

Cytotoxin Substance that has a toxic effect on cells.

Genotoxin Toxin that interacts with DNA, causing tumors, neoplasms, or mutations.

Methemoglobin Compound formed from hemoglobin by oxidation of the ferrous to the ferric state with essentially ionic bonds.

Mutagen Chemical or physical agent that increases a mutation rate above the frequency of the spontaneous rate.

Teratogen Agent or process that interferes with the normal development of the embryo or fetus, causing physical abnormalities.

RUBBER, *a versatile material, and the rubber industry, a vital industry, substantially contribute to the economic well being of nations. The world rubber industry today produces several thousands of different types of products. Use of rubber products has been progressively made in diverse fields and today it is found in almost every walk of life including all the branches of defense, various industries, agriculture, and domestic purposes.*

I. History and Occurrence

It is interesting to note the per capita consumption of rubber in some of the industrially advanced countries, as well as in India, which highlights the stage of development of the rubber industry in these countries. The approximate figures are U.S., 13.8 kg; Germany 9.2 kg; France, 8.7 kg; UK, 8.6 kg; Japan, 8.2 kg; and India 2.2 kg. These figures are an indicator of the state of national economy of the respective countries in relation to development of rubber industry.

The Indian rubber industry is comparatively "young" as the first rubber factory was begun in 1921 (during the Bengal presidency). Production of natural rubber in this country as well as the vast population which provides a large potential market are the two factors which have contributed to the growth of the Indian rubber industry. The industry manufactures over thirty thousand different products ranging from heavy duty automotive tires to tiny articles like teats and balloons, including innumerable types of specialized industrial and mechanical products required by various industries such as the automobile, aircraft, railway, shipping, textile, medical and pharmaceutical, sporting goods, engineering, furniture, and building and construction industries as well as by the agriculture sector. It also caters to the needs of all the three branches of defense in India (air, sea, and land).

Resources for rubber (natural, synthetic, and reclaimed) are among the most important raw materials of the vital rubber industry in India and hold a strategic position in the country's economy. Natural rubber is tapped from more than 200 different species of plant, including *Hevea brasiliensis,* a commercially significant rubber tree. Synthetic rubber is chemically derived and reclaimed rubber is produced from rubber scrap.

Handbook of Hazardous Materials

The first two rubbers are known as "new polymers." The use of these rubbers depends on the end use of individual products and the quality control test related to production. Rubber products can be exclusively made out of natural rubber or synthetic rubber. However (normally) blends of any two or sometimes of all three types of rubbers are made, depending upon the physical properties required in the end product.

The rubber industry is a typical example of many chemical industries. Although rubber was used by indigeneous peoples a few centuries ago, its usage in different forms has been expanding constantly since Charles Goodyear (1839) discovered the process of vulcanization of rubber. Rubber has replaced many other materials which had been used earlier for conventional uses. Indeed, the modernization of the present era could not have been possible were it not for the revolutionary use of rubber in industry.

The basic components used in the rubber compound are grouped into the following categories: (a) elastomers (synthetic or natural rubber); (b) reinforcing agents (e.g., carbon black); (c) fillers (e.g., inert materials like china clay, barytes, kaolin); (d) processing oils (e.g., mineral oils, talc); (e) activators (e.g., zinc oxide); (f) accelerators (e.g., thiurams: tetramethyl thiuram disulfide, tetraethylthiuram disulfide; dithiocarbamates: zinc dimethyldithiocarbamate, zineb; benzo-thiazoles; 2-Mercaptobenzo-thiazole disulfide; guanidines: diphenyl guanidine); (g) retarders (e.g., organic acids, *N*-nitrosodiphenylamine); and (h) vulcanizing agents (e.g., sulfur, organicperoxides, phenol resins, metal oxides); (i) blowing agents (e.g., azobisformamide, benzene-sulfonylhydrazide). In addition to these, depending on the required end product, pigments, flame proofing agents, and coagulants are used.

II. Chemical Structure

Faraday first reported that natural rubber was composed of hydrogen and carbon in the ratio of C_5H_8. Isoprene was determined to be the building unit of natural rubber. Tilden also reported that the probable structure of rubber was 2-methyl-1,3-butadiene. Chemical and structural features of a number of compounds used in the rubber industry in different categories are discussed below.

1. Accelerators: These are highly reactive organic chemicals used in rubber compounding to accelerate the curing process of the product and to obtain desired properties (Fig. 1). Examples include aldehyde–amine condensates (Schiff bases) dithiocarbamates, benzothiazoles, amines, thiophosphates, guanidines, xanthates, sulfenamides, and thiuram sulfides (tetramethyl and tetraethylthiuram disulfide).

2. Vulcanizing agents: The vulcanizing agent used in natural and styrenebutadiene rubber systems is elemental sulfur and it produces both cross-linked and cyclic structures. Organic sulfur compounds are also used as vulcanizing agents. Examples include tetraalkylthiuram disulfides, morpholine disulfide, dithiocarbamates, alkylphenol disulfides, aliphatic polysulfide polymers, and dithiophosphates. Selenium, tellurium, and related compounds are also used as vulcanizing agents.

3. Antioxidants: When added to vulcanizates, antioxidants will slow down the aging process due to the reaction with heat, light, oxygen, radiation, etc. Examples include aldehyde amines, diaryldiamines, bisphenols, hydroquinones, and aminophenols.

4. Antiozonants: The antiozonants protect the rubber from deterioration caused by ozone. Examples include *para*-phenylenediamine, dihydroquinolines, and metal salts of dithiocarbamic acid.

5. Blowing agents: These agents are used to produce latex foam rubbers by decomposition at curing temperatures to produce gases (Fig. 2). Examples include dinitrosopentamethylenetetramine, azobisbutyronitrile, azobisformamide, and benzenesulfonyl hydrazide.

6. Nitroso compounds: These are formed during the manufacture and processing of rubber. The most potent carcinogenic nitrosoamines are *N*-nitrosodimethylamine, *N*-nitrosodiethylamine, *N*-nitrosodibutylamine, *N*-nitrosopiperidine, *N*-nitrosopyrrolidine, and *N*-nitrosomorpholine.

III. Health Hazards

The rubber industry is thought to represent a potentially hazardous environment to the health of an individual. The investigation of health hazards in the

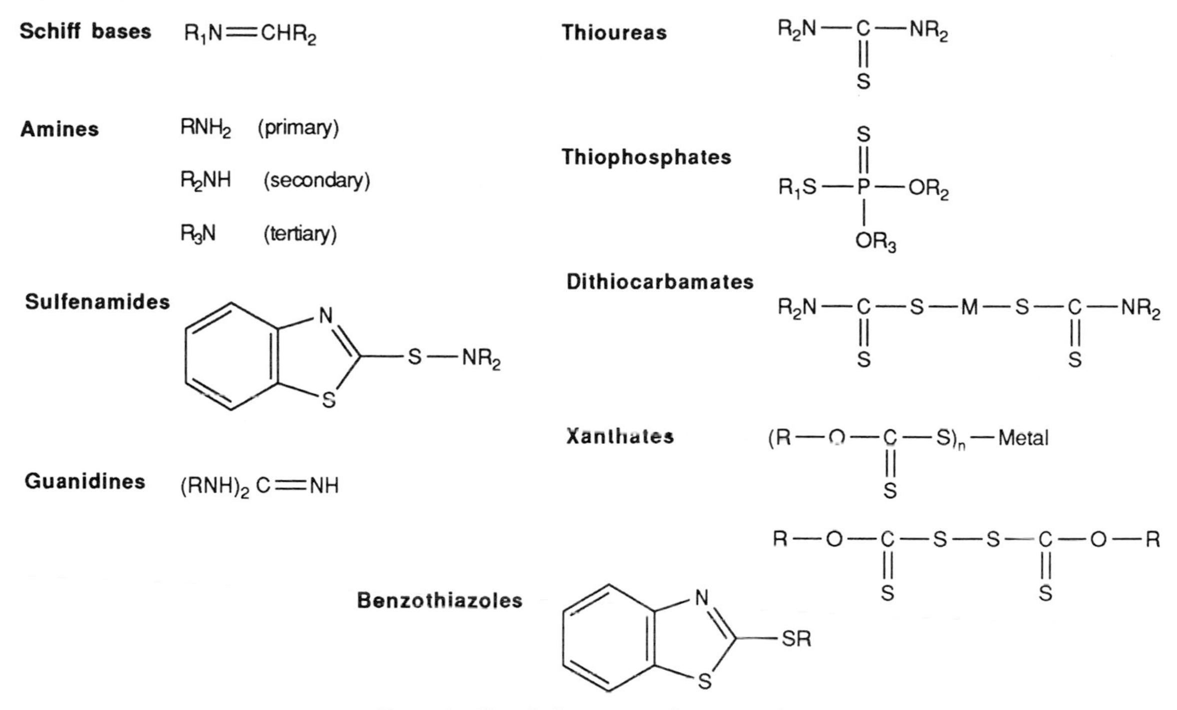

Figure 1 Chemical structure of some accelerators.

rubber industry is, however, a recent development which includes extensive epidemiology, industrial hygeine, and the general awareness in workers of how the chemicals that are used in the industry interact with the human body. As early as the 1930s, nonspecific dermatological reactions to rubber gloves and footwear, and dermatitis of the eyelids had been identified. The causative agents include mercaptobenzothiazole, TMTD, and diphenylmethylenethiuram disulfide present in gloves and the antioxidant, *N*-phenyl-2-napthylamine, in eyelash curlers. Four major groups of allergens were considered according to their sensitization index. The first group (with a strong incidence of occurrence) includes nickel, chromate, cobalt, TMTD, mercapto mixtures, and wood tars. The second and third groups (with medium incidence) include carbamates, neomycin, balsam of peru, mercury, lanolin, naphtyl mixtures, formaldehyde, PPDA mixtures, and turpentine. The fourth group comprised very

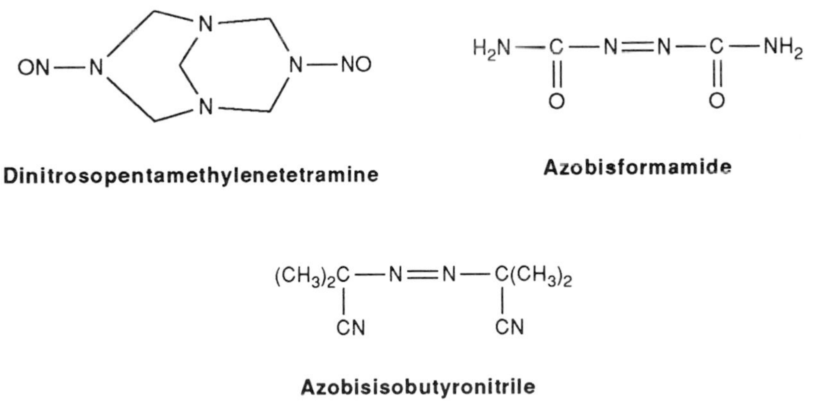

Dinitrosopentamethylenetetramine

Azobisformamide

Azobisisobutyronitrile

Figure 2 Chemical structure of some blowing agents.

Table I Reproductive Epidemiology of Workers Exposed to Rubber Chemicals

Duration of exposure (years)	No. of workers	Total no. of pregnancies	Number (percentage) of Abortions	Still births	Premature births	Congenital malformations	Live births
Unexposed							
Nonsmokers (control I)	170	622	98 (15.75)	33 (5.30)	39 (6.27)	8 (1.28)	444 (71.38)
Smokers (control II)	170	615	125[a] (20.32)	38 (6.17)	44 (7.15)	14[a] (2.27)	394[a] (64.06)
Exposed							
Smokers (1–15)	120	442	117[a] (26.47)	36 (8.14)	40 (9.04)	21[a] (4.75)	228[a] (51.58)
Nonsmokers (1–15)	150	538	109[a] (20.26)	37 (6.87)	42 (7.80)	18[a] (3.34)	332[a] (61.70)

[a] $p < 0.05$.

low incidence substances such as the sulfonamides, ethylenediamine, and cinnamon oil. Dermatological effects result from exposure to rubber and rubber chemicals in rubber goods manufacture. Exposure to respirable dust, product gases, and rubber additives causes respiratory disorders like talcosis, chronic bronchitis, and chronic bronchial asthma. Respiratory morbidity is related to both intensity and length of exposure to fumes. Respiratory problems are more frequent in workers involved in milling, tube building, tube curing, and tube inspection. Carbon black and carbon disulfide (i.e., methemoglobin formers) were identified as causative agents of chronic lung diseases such as pneumoconiosis, pulmonary fibrosis, bronchitis, and emphysema.

Variation in levels of thioethers in exposed and unexposed populations has been observed. Estimation of urinary thioethers in rubber factory workers has been a reliable indicator of chemical exposure in the rubber industry. Early studies have shown a direct correlation between urinary thioethers and duration of exposure. This is, however, related to the mode of exposure and the production department involved. High excretion of thioether (cutaneous) was found in workers in the belt department and in workers exposed to airborne contaminants in the calender department, raw material stores, and chemical mixing sections.

A large number of chemicals is employed in the rubber industry for the purposes of rubber processing, facilitation of fabrication, regulation of the rate of vulcanization, and other areas. Vulcanization of rubber products at high temperatures and pressures may emit biologically active reaction products. Re-

Table II Chromosomal Aberrations in Nonsmokers Exposed to Rubber Chemicals

Group	No. of workers analyzed	Duration of service (years)	No. of metaphases screened	Number (percentage) of chromatid type of aberrations: Gaps	Breaks	Fragments	Deletions	Total no. of aberrations (excluding gaps)
Exposed	8	1–3	800	22(2.75)	19(2.37)	3(0.37)	4(0.50)	26(3.25)[a]
	18	4–6	1800	56(3.11)	47(2.61)	12(0.66)	14(0.77)	73(4.05)[a]
	13	7–9	1300	41(3.15)	36(2.76)	7(0.53)	8(0.61)	51(3.92)[a]
	12	10–12	1200	38(3.16)	30(2.50)	5(0.41)	7(0.58)	42(3.50)[a]
	9	13–15	900	29(3.22)	26(2.88)	6(0.66)	5(0.55)	37(4.11)[a]
Unexposed (control I)	50		5000	42(0.84)	40(0.80)	3(0.06)	4(0.08)	47(0.94)

[a] $p < 0.05$.

Table III Chromosomal Aberrations in Smokers Exposed to Rubber Chemicals

Groups	No. of workers analyzed	Duration of service (years)	No. of metaphases screened	Number (percentage) of chromatid type of aberrations				Total no. of aberrations (excluding gaps)
				Gaps	Breaks	Fragments	Deletions	
Exposed	5	1–3	500	16(3.20)[a]	18(3.60)	4(0.80)	6(1.20)	28(5.60)[a]
	6	4–6	600	17(2.83)[a]	27(4.50)	5(0.83)	6(1.00)	38(6.33)[a]
	12	7–0	1200	38(3.16)[a]	45(3.75)	13(1.08)	17(1.41)	75(6.25)[a]
	8	10–12	800	25(3.12)[a]	25(4.37)	6(0.75)	4(0.50)	45(5.62)[a]
	9	13–15	900	29(3.22)[a]	38(4.22)	7(0.77)	13(1.44)	58(6.44)[a]
Unexposed (control II)	30		3000	42(1.40)[a]	46(1.53)	12(0.40)	20(0.66)	78(2.60)

[a] $p < 0.05$.

cent studies have revealed carcinogenic, teratogenic, and mutagenic effects of rubber chemicals. More than 500 chemicals are used in large quantities and include several known or suspected mutagens and carcinogens. The results of mutations or genetic effects are generally undesirable and may include abortions, congenital anomalies, lowered resistance to disease, decrease in life span, infertility, mental retardation, senility, and cancer in humans.

A. Carcinogenicity

Several epidemiologic studies have shown that workers in the rubber industry have an increased cancer risk. Studies on the carcinogenicity of chemicals used in the rubber industry have shown that both monomers and additives constitute a carcinogenic hazard. It is possible that the action of some carcinogens can be increased by simultaneous

Table IV Ames *Salmonella typhimurium*/Microsomal Assay Results

Compound tested	Concentration of compound (μg/plate)	Salmonella strains used			
		TA98[a]		TA 100[a]	
		+S9	−S9	+S9	−S9
Control DMSO	100	32[b]	18	119	102
Positive control [Benzo (*a*) pyrene]	2.5	273	263	2076.3	2033
TMTD	12.5	36	23	170	170
	25	36	42[c]	173	240[c]
	50	38	46[c]	174	248[c]
	100	38	40[c]	154	266[c]
Ziram	12.5	14	14	110	102
	25	30	20	103	95
	50	31	22	165	148
TETD	12.5	34	11	*91*	*115*
	25	41	15	*123*	100
	50	56	19	*116*	116
	100	58	21	*118*	110

[a] −S9 = without microsomal activation; +S9 = with microsomal activation.
[b] Average of triplicate analysis.
[c] Mutation ratio (induced his + revertants)/(spontaneous revertants) of more than two is considered to be active spontaneous revertants.

Table V Micronuclei, Chromosomal Aberrations, and Abnormal Sperm in Mice after Treatment with TMTD[a]

Dose (mg/kg)	No. of polychromatic cells scored	Percentage of micronuclei	Number of metaphases screened	Percentage of abnormal metaphases	Total no. of sperm screened	Percentage of abnormal sperm
Control	16,000	0.25	1000	10.7	16,000	3.21
TMTD						
80	16,000	0.50	1000	16.6	16,000	4.03
200	16,000	0.71	1000	20.7	16,000	4.38
320	16,000	1.25	1000	23.4	16,000	5.18

[a] $p < 0.05$ for all percentages listed for TMTD-treated mice.

exposure to agents that potentiate their action. The international agency for research on cancer has linked certain job categories with an increased risk of cancer.

The pattern of cancer mortality and mobidity is diverse. Cancer of the urinary bladder is strongly associated with rubber work. It could be induced by bicyclic aromatic amines. Similarly, an increased risk of stomach, lung, prostate, and testis cancers and leukemias has been observed in rubber factory workers. Development of any of these cancers is mainly related to the age of the worker, the duration of employment, and exposure to the type of chemical used in the particular job area. An excess in observed cases of specific cancers includes stomach and intestine in rubber making, lung in tire curing, bladder in chemical plant and tire building, skin and brain cancers in tire assembly, lymphatic cancer in tire building, and leukemia in the curing, tire building, and rubber fabrication.

Early studies have identified chemicals like antioxidants, Nonox S, and β-naphthylamines as etiological factors of bladder cancer. Recent epidemiologic studies have shown that polycyclic aromatic hydrocarbons present in aromatic oils and carbon black (which are used frequently in the rubber industry) can induce skin cancer when applied repeatedly to the skin of mice. Chemicals like tetramethylthiuram disulfide (TMTD, thiram), tetraethylthiuram disulfide (TETD, disulfiram, Antabuse), tetramethylthiuram monosulfide (TMTM), and dipentamethylenethiuram tetrasulfide have been found to be equivocally carcinogenic. TMTD was more potent than either TMTM or TETD. TMTD and TETD are sources of carcinogenic nitrosoamines; carbon disulfide did not show any activity. It is possible that the interaction between individual rubber chemicals and between emissions from the rubber processing steps and other environmental factors may contribute to the observed risk.

B. Teratogenicity

Postzygotic exposure to genotoxic or cytotoxic chemicals can induce cell death or dysfunction in the

Table VI Micronuclei, Chromosomal Aberrations and Abnormal Sperm in Mice after Treatment with Ziram[a]

Dose (mg/kg)	No. of polychromatic cells scored	Percentage of micronuclei	Number of metaphases screened	Percentage of abnormal metaphases	Total no. of sperm screened	Percentage of abnormal sperm
Control	16,000	0.28	1000	9.4	16,000	3.00
Ziram						
230	16,000	0.51	1000	17.2	16,000	4.12
350	16,000	0.66	1000	21.5	16,000	4.56
700	16,000	0.91	1000	25.9	16,000	5.46

[a] $p < 0.05$ for all percentages listed for Ziram-treated mice.

Table VII Micronuclei, Chromosomal Aberrations, and Abnormal Sperm in Mice after Treatment with TETD[a]

Dose (g/kg)	No. of polychromatic cells scored	Percentage of micronuclei	Number of metaphases screened	Percentage of abnormal metaphases	Total no. of sperm screened	Percentage of abnormal sperm
Control	16,000	0.26	1000	10.2	16,000	3.03
TETD						
1.08	16,000	0.47	1000	14.0	16,000	3.73
1.43	16,000	0.61	1000	17.3	16,000	4.28
2.15	16,000	0.86	1000	20.9	16,000	4.80

[a] $p < 0.05$ for all percentages listed for TETD-treated mice.

developing embryo, which may lead to spontaneous abortion or malformations in the offspring. Studies of the reproductive outcome of pregnant women employed in the tire building department indicated a high percentage of abortions and congenital malformations.

A review of various studies presents evidence demonstrating the teratogenic nature of some rubber chemicals such as thiram in mice. Certain dithiocarbamates were associated with reduced fertility in mammals. TMTD, TMTM, and TETD induce malformations and death in chick embryos. Common types of associated malformations were eye defects and open coeloms. Chick embryo tests of TMTD, TMTM, and TETD showed embryo toxicity. (TMTD is teratogenic in mice and hamsters, while TETD induces malformations.) Some of the dithiocarbamates and acetates are also embryotoxic. Cadmium and zinc dithiocarbamates are potent embryotoxic agents eliciting lethality and malformations. The order of embryotoxicity decreases as follows: cadmium acetate > cadmium diethyldithiocarbamate > zinc diethyldithiocarbamate ≥ zinc ethylphenyldithiocarbamate ≥ zinc dibutyldithiocarbamate > copper dimethyldithiocarbamate > tellurium diethyldithiocarbamate > copper (II) acetate > piperidine pentamethylenedithiocarbamate > zinc acetate.

Table VIII Induction of Dominant Lethal Mutations in Mice Treated with TMTD

	Dose[a]											
	Control (mg/kg)			80 (mg/kg)			200 (mg/kg)			320 (mg/kg)		
Weeks after treatment	Fertile fem. (%)	T	D	Fertile fem. (%)	T	D	Fertile fem. (%)	T	D	Fertile fem. (%)	T	D
1	65	9.76	0.38	70	9.71	0.64	60	9.00	0.83	80	8.25	0.81
2	70	10.07	0.42	80	9.50	0.68	70	9.07	0.85	65	9.46	0.92
3	70	9.92	0.50	60	10.00	0.75	55	9.81	0.81	85	8.35	0.94
1–3	68.33	9.91	0.43	70	9.73	0.69[b]	61.66	9.29	0.83[b]	76.66	7.68	0.89[b]
4	80	9.31	0.43	75	8.26	0.66	80	8.87	0.87	70	2.71	1.21
5	55	8.81	0.45	65	9.61	0.76	65	9.38	0.84	55	4.45	0.90
4–5	67.50	9.06	0.44	70	8.93	0.71[b]	72.50	9.12	0.85[b]	62.50	9.08	1.05[b]
6	60	8.91	0.50	55	9.27	0.81	85	7.94	0.82	60	8.58	1.08
7	75	8.06	0.33	70	9.21	0.64	75	8.13	0.80	65	9.53	1.07
8	70	9.42	0.35	60	8.83	0.66	55	9.90	0.72	50	7.90	0.90
6–8	68.33	8.79	0.39	61.66	9.10	0.70[b]	71.66	8.65	0.78[b]	58.33	8.67	1.01[b]

[a] T, total implantations per female; D, dead implantations per female.
[b] $p < 0.05$.

Table IX Induction of Dominant Lethal Mutations in Mice Treated with Ziram

	Dose[a]											
	Control (mg/kg)			230 (mg/kg)			350 (mg/kg)			700 (mg/kg)		
Weeks after treatment	Fertile fem. (%)	T	D	Fertile fem. (%)	T	D	Fertile fem. (%)	T	D	Fertile fem. (%)	T	D
1	75	8.53	0.53	80	8.25	0.87	65	8.92	1.07	85	8.82	1.12
2	55	9.63	0.45	70	9.21	0.86	70	9.07	1.14	75	9.86	1.20
3	80	8.81	0.50	65	8.92	0.77	60	8.83	1.00	65	9.00	1.08
1–3	70	8.99	0.49	71.66	8.79	0.83[b]	65	8.94	0.93[b]	75	9.22	1.13[b]
4	70	9.14	0.36	85	8.76	0.76	75	8.80	0.87	75	9.53	1.20
5	75	8.93	0.40	75	9.00	0.80	70	8.78	1.00	80	8.12	1.19
4–5	72.50	9.03	0.38	80	8.88	0.78[b]	72.50	8.79	0.93[b]	77.50	8.82	1.19[b]
6	80	9.06	0.44	75	8.53	0.93	70	8.93	0.86	80	9.18	1.12
7	65	9.46	0.38	70	9.42	0.78	80	9.31	0.93	65	9.15	1.15
8	70	8.50	0.50	60	9.25	0.75	75	9.40	1.07	70	8.36	1.14
6–8	71.66	9.00	0.44	68.33	9.06	0.82[b]	75	9.21	0.95[b]	71.66	9.06	1.14[b]

T, total implantations per female; D, dead implantations per female.
[b] $p < 0.05$.

C. Mutagenicity

Humans are most vulnerable to genetic damage, and when human mutations are produced they may contribute to an increased genetic load in the general population. Hence we are obligated to test as many chemicals as possible for potential mutagenicity and withdraw those chemicals which are mutagenic from the environment wherever possible. In this regard, it is critical to monitor workers occupationally exposed to rubber chemicals for reproductive epidemiology as the majority of polymeric materials handled in the rubber manufacturing process give rise to hazardous volatile mutagenic fumes. Several polymers themselves (e.g., chloroprene, styrene, butadiene rubber, acrylonitrile rubber) or sulfur-containing additives of the dithiocarbamate type contribute to mutagenecity of fumes from complex mixtures.

Reproductive impairment and mutagenesis can be assessed by monitoring spontaneous abortions and congenital malformations. Reproductive problems associated with chemical exposure include infertility, an increase in complications of pregnancies, low birth weight, infant mortality, etc.

Reproductive epidemiological studies have described an increased risk of spontaneous abortions and congenital malformations in alcoholics, smokers, pregnant working women, and in the wives of men who are occupationally exposed to various chemicals like lead, pthalates used in the plastics industry, vinyl chloride, chemical sterilants like formaldehyde, pesticides like DDT, etc.

The reproductive histories of rubber factory workers show a higher incidence of spontaneous abortion and congenital malformations. The results of one such study are shown in Table I. The risk is again job dependent. A case study showed an increase in the odds ratio for factory workers exposed to rubber chemicals in the rubber footwear area when compared with the tire department. This could be attributed to a more percutaneous mode of interaction of chemicals in footwear production.

Monitoring of high risk groups by cytogenetic analysis of somatic cells is becoming an acceptable occupational health practice. Identification of cytogenetic effects caused by exposure to a single known chemical or a mixture by job category is important to identify the potentially hazardous job categories among the complexities of rubber manufacturing. Cytogenetic effects can be seen in the form of an increased incidence of chromosomal aberrations, sister chromatid exchanges, gaps, and breaks.

Various reports on cytogenetic effects in the rubber factory workers have demonstrated an increased incidence of SCE and chromosomal aberrations, primarily among the workers employed in the mixing operations and acceleration (Tables II and III). Such

Table X Induction of Dominant Lethal Mutations in Mice Treated with TETD

	Dose[a]											
	Control (g/kg)			1.08 (g/kg)			1.43 (g/kg)			2.15 (g/kg)		
Weeks after treatment	Fertile fem. (%)	T	D	Fertile fem. (%)	T	D	Fertile fem. (%)	T	D	Fertile fem. (%)	T	D
1	70	8.93	0.50	75	8.27	0.86	65	8.61	0.92	70	8.64	1.00
2	60	9.33	0.42	80	8.87	0.87	70	8.86	1.00	70	9.43	1.14
3	80	9.00	0.44	65	9.69	0.85	80	9.31	0.94	65	8.77	1.08
1–3	70	9.09	0.45	73.33	8.94	0.86[b]	71.66	8.93	0.95[b]	68.33	8.95	1.07[b]
4	75	8.47	0.47	70	8.43	0.86	60	9.25	1.00	80	8.44	0.94
5	65	9.54	0.46	75	9.73	0.93	80	8.56	0.94	75	9.40	1.13
4–5	70	9.00	0.46	72.50	9.08	0.89[b]	70	8.90	0.97[b]	77.50	8.92	1.03[b]
6	60	8.83	0.42	65	8.77	0.85	75	9.07	0.93	65	9.38	1.08
7	75	9.20	0.40	60	9.00	0.83	70	9.14	0.86	80	9.25	0.94
8	70	8.64	0.50	70	9.21	1.00	65	8.46	0.92	70	8.36	1.14
6–8	68.33	8.89	0.44	65	8.99	0.89[b]	70	8.89	0.90[b]	71.66	9.00	1.05[b]

[a] T, total implantations per female; D, dead implantations per female.
[b] $p < 0.05$.

an effect could also be demonstrated in embryo tissues of women. Such an effect could not be demonstrated in the workers themselves or rubber vulcanizers. The chemicals involved in the process, the duration of exposure, and age also play an important role. Among workers, the pattern of smoking also influences the cytogenetic effects.

Mutagenicity is a general characteristic of carcinogenic chemicals and monitoring of mutagenicity in the ambient air or in microbial systems, animal models, and in biological samples of workers (e.g., blood or urine) can be used to detect hazardous exposure to mutagenic chemicals.

To study mutagenic effects, *in vitro* and *in vivo* short-term bioassays using various additives involved in the rubber industry can be applied. These include test methods such as (1) *Salmonella*/mammalian mutagenicity test, (2) sex-linked recessive lethal test in *Drosophila*, (3) micronucleus test, (4) analysis of chromosomal aberrations in spermatogonia, (5) sperm morphology assay, (6) dominant lethal assay, (7) analysis of chromosomal aberrations in lymphocyte cultures, (8) sister chromatid exchanges, (9) semen analysis, etc. Bacterial fluctuation tests can also be used to assess small amounts of mutagenic chemicals. Testing of various chemicals by *in vitro* and *in vivo* test systems has led to the identification of a number of chemical mutagens. Thus, it is obligatory to test suspected rubber chemicals for mutagenicity in animal models. In our laboratory, accelerators of the rubber industry have been evaluated *in vitro* (*Salmonella typhimurium*) and *in vivo* (mouse) test systems. The results shown in Tables IV–X have presented ample evidence for the mutagenic nature of rubber chemicals (accelerators) in lower systems.

IV. Conclusion

It is worthwhile to evaluate as many rubber chemicals as possible for toxicity, mutagenicity, and carcinogenicity and remove harmful chemicals from the work environment. In addition, appropriate precautionary measures must be implemented to reduce human exposure to potentially hazardous rubber chemicals. Undue exposure might result in health hazards including genetic damage and cancer morbidity and mortality.

Related Article: ENVIRONMENTAL CANCER RISKS.

Bibliography

Degrassi, F., Fabri, G., Palitti, F., Paoletti, A., Ricordy, R., and Tanzarella, C. (1984). Biological monitoring of workers in the rubber industry. I. Chromosomal aberrations and sister-

chromatid exchanges in lymphocytes of vulcanizers. *Mut. Res.* **138,** 99–103.
Hema Prasad, M., Pushpavathi, K., and Reddy, P. P. (1986). Cytogenetic damage in lymphocytes of rubber industry workers. *Environ. Res.* **40,** 199–201.
Hema Prasad, M., Pushpavathi, K., Rita, P., and Reddy, P. P. (1987). The effect of thiram on the germ cells of male mice. *Food Chem. Toxicol.* **25,** 709–711.
Maki-Paakkanen, J., Sorsa, M., and Vainio, H. (1984). Sister chromatid exchanges and chromosome aberrations in rubber workers. *Teratogen. Carcinogen. Mutagen.* **4,** 189–200.
Paschin, Y. V., and Bakhitova, L. M. (1985). Mutagenic effects of thiram in mammalian somatic cells. *Food Chem. Toxicol.* **23,** 373–375.
Sorahan, T., Parkes, H. G., Veys, C. A., and Waterhouse, J. A. H. (1986). Cancer motality in the British rubber industry. *Brit. J. Ind. Med.* **43,** 363–373.

Selenium

G. Richard Hogan
St. Cloud State University

Glossary

Alkali disease Disorder of livestock of the north-central Great Plains of the United States due to eating plants containing high selenium concentration. General emaciation, liver dysfunction, and anemia are symptomatic of the disease.

δ-Aminolevulinic acid dehydratase Enzyme functioning in the production of hemoglobin causing condensation of basic components; a key regulator of the process of erythropoiesis.

Erythropoiesis Production and release of red blood cells or erythrocytes.

Glutathione peroxidase Enzyme containing selenium and functioning with vitamin E to prevent excessive oxidation within cells.

Kaschin–Beck disease Chronic, disabling disease of the joints and spine, endemic to regions of China, Siberia, and Korea, and closely related to a deficiency of selenium.

Keshan disease Endemic disorder associated with a dietary deficiency of selenium occurring in areas of northern China and eastern Siberia and characterized by a diseased condition of heart muscle.

***SELENIUM TOXICITY** has long been known. In the 1860s a correlation was reported between grain grown on soil high in selenium concentration and the development of a devastating syndrome in cattle referred to as "alkali disease" and the "blind staggers." Swine have been experimentally studied and demonstrate symptoms of selenium toxicity comparable to those reported for livestock grazing on vegetation grown on highly seleniferous soil. Since the original observations, considerable experimental attention has been directed toward selenium as an essential micronutrient as well as toward its role in environmental pollution. Much less information is available regarding the biotoxic effects of selenium above its trace levels. Although occurring naturally, selenium levels in ecosystems have risen dramatically through generation in industrial processing. Toxicity of selenium is highly variable depending upon such factors as affected species, oxidation state, and route of exposure of selenium. In addition, tissue vulnerability appears to be dependent upon whether the administered selenium compound is in an inorganic or organic form. In many cases, the latter appears to be considerably more effective. Such variable conditions and factors have lead to contradictions and confusion in reports dealing with the adverse effects of selenium in biological systems. This article will attempt to discuss the major aspects of selenium as they relate to a better understanding of selenium toxicity.*

I. Selenium in the Environment

Selenium is the sixty-eighth most abundant element in the earth's crust. It is widely and inequally distributed in rock and soils often found to be enriched with copper, silver, sulfur, and iron. Normal soil content has been estimated between 0.08 and 2 ppm. Reports, however, indicate levels as high as 1200–1400 ppm appearing in organic residues, clay sediments, and volcanic ashes. The approximate values (ppm) of selenium in limestone, coal, shale, and non-seleniferous soils have been given an average of 7.5, 8.0, 25.0, and 0.06, respectively.

Selenium is readily absorbed by plants. The rate and extent of absorption is variable depending upon the chemical form of selenium contributing to its solubility and the soil's state of hydration. Some grains and grasses grown on seliniferous soils contain exaggerated levels of selenium and thus contribute to its toxicity. Conversely, materials consumed by animals and humans may be deficient in this trace element promoting the deficiency symptoms to be discussed later.

The appearance of selenium in the environment has been dramatically increasing within the past two decades. The primary reason is the elevated amounts generated in industrial (anthropogenic) processing. Major contributors to the industrially generated and process sources of selenium are paper production (12 ppm), petroleum production (2 ppm), and sewage treatment (4 ppm). Table I gives the distribution patterns and amounts of selenium in the relevant components of the environment.

Significant increases in the selenium concentration in the aquatic environment are causing serious concerns with regard to toxicity of fauna and flora. Marine environments approximate 2×10^{-3} ppm and river water 2.5×10^{-4} ppm. Because of the increasing aquatic sources, excessive selenium has had a profound impact on aquatic ecosystems. Once selenium enters the food chain, a sequential and ultimately detrimental effect occurs. This occurrence is also typical of terrestrial systems. Because of the differential sensitivity of organisms to selenium, its excess in the food chain is difficult to ascertain. However, data indicate severe damage to both ecologically and economically important organisms. Included among these are increased lethality and decreased fertility in a variety of fishes, avian mortality, decreased levels as food sources of zooplankton, and lethality in domestic animals grazing on highly seleniferous soils.

Table I The Distribution of Selenium in Natural and Anthropogenic Materials

Material	[Se] (ppm)
Terrestrial	
Earth's crust	0.09
Limestone	0.1–14.0
Shales and phosphate rocks	<1–55
Crude oil	0.06–0.39
Coal	0.5–11.0
Soils	
Nonseleniferous	<0.1–2.0
Seleniferous	2–200
Aquatic	
Ocean water	10^{-4}–4×10^{-3}
River water	10^{-4}–4×10^{-4}
Aquatic plants	0.02–0.14
Plankton	1.1–2.4
Fish	0.5–6.2
Anthropogenic	
Petroleum products	0.15–1.65
Fly ash	1.2–16.5
Sewage sludge	1.8–4.8
Paper products	1.6–19.0

Source: K. J. Maier, C. Foe, R. S. Ogle, M. J. Williams, A. W. Knight. (1987). The dynamics of selenium in aquatic ecosystems. *Trace Subst. Environ. Hlth.* **XXI,** 361–408. (Modified from J. W. Doran. Microorganisms and biological cycling of selenium. *Adv. Microbiol. Ecol.* **6,** 1–32.)

II. Selenium as a Micronutrient and Interactions with Other Substances

Historical records suggest the nutritional essentiality of selenium in animals with reports as early as those of Marco Polo (circa 1275). Since then, numerous well-documented records have appeared.

In humans, Kaschin–Beck disease and Keshan disease have been described for individuals of a northeastern province in the People's Republic of China and other areas. These disorders are linked to other endemic diseases which develop due to deficiences of other essential micronutrients. The associated diseases include cretinism (iodine deficiency) and dental and skeletal fluorosis (fluorine deficiency). For both Kaschin–Beck and Keshan diseases, the causative factor is directly related to the reduced endemic levels of selenium in food, soil, and water. Cardiac muscle lesions are symptomatic of Keshan disease, while Kaschin–Beck disease is closely correlated with bone disorders (osteoarthritic disease). Both diseases induce high mortality, especially in children.

There is believed to be a positive relationship between these disorders and selenium and molybdenum. Data indicate that both selenium and molybdenum are lower in concentration in Keshan disease regions than non-disease regions. It is felt by some that cardiac muscle damage observed in Keshan disease may be due to deficiencies in both selenium and molybdenum or either substance.

Deficiency of selenium affects the function of the thyroid gland, causing a depression of plasma levels of the thyroid factor 3,5,3′-triiodothyronine. This hormone is formed from the principle hormone of the thyroid gland, thyroxine, by an intracellular diiodination process. This process, if decreased due to insufficient selenium, could also contribute to other thyroid disorders.

Selenium has been shown to be "essential" in a number of species. Normal reproductive function of livestock is dependent upon selenium in micronutrient levels. Deficiencies of selenium are associated with a number of disorders including ovarian cysts, metritis, and a decreased conception rate in cattle. Selenium is known to be required to prevent hepatic necrosis in rats and a variety of other disorders described in avian species and other mammals. These include lesions of skeletal and cardiac muscle and an increased susceptibility to infection and anemia. It is believed that these disorders are linked to the antioxidant function with vitamin E. Selenium is an integral component of the enzyme glutathione peroxidase, which functions with vitamin E to prevent excessive intracellular oxidation. Using radioactively labeled selenium as sodium selinate, it has been demonstrated that selenium is incorporated into plasma protein (seleno-protein) and also is actively accumulated by circulating erythrocytes. The latter event is linked presumably to uptake into erythrocytes associated with glutathione peroxidase function.

The physiological interactions among trace substances are complex and, in the *in vivo* condition, are difficult to quantify or to interpret. A contributing factor in the interpretation of such interactions is the broad spectrum of toxic effects that is induced by trace metals present in excess amounts. A number of investigations, however, suggests synergistic, additive, and inhibitory effects. Principle targets involved in such interactions have been implicated to be cofactor components of metabolically crucial enzymes; the alterations of such would, of course, affect enzymatic effectiveness and, thus, the cellular contributions to the functional integrity of the tissue/organ.

Selenium is known to interact with a number of trace metals, including arsenic, cadmium, mercury, and lead, rendering the defined system less susceptible to their toxic effects. Selenium affects the tissue distribution of arsenic in rats and acts as a protective agent. Cadmium is a potent toxicant inducing a wide spectrum of adverse effects. Selenium has been reported to decrease the rate of intestinal mucosal absorption of cadmium. This action prevents an elevated extracellular cadmium concentration leading to renal and hepatic lesions. Further evidence suggests that selenium blocks the cadmium-induced vasoconstrictory effect, thus preventing the hypertensive responses associated with increased plasma concentrations of cadmium.

Testicular lesions promoted by cadmium have been shown to be reduced and/or prevented by supplemental selenium therapy. It appears that selenium may stimulate recovery from cadmium-induced lesions. Cadmium has been reported to cause a dramatic, yet transitory, increase in radioactive iron incorporation into erythrocytes in mice; this effect is blocked by sodium selenite, thereby stabilizing the normal uptake of radioactive iron into hemoglobin (Fig. 1).

It has been suggested that fluorine may inhibit the effects of high concentrations of selenium in the human body. Fluorine may accelerate the excretion of selenium through urine. A similar proposal has been presented for sulfur and its ability to reduce the toxicity of selenium.

Generalizations with regard to biological effects of trace substances are difficult to interpret; however, there is a redundant theme which appears in the protective effect of selenium on enzymatic systems. The key enzymes, glutathione peroxidase and pyruvate kinase, are known to be dependent upon the presence of sulfhydryl radicals, providing the necessary dipole conditions for secondary and

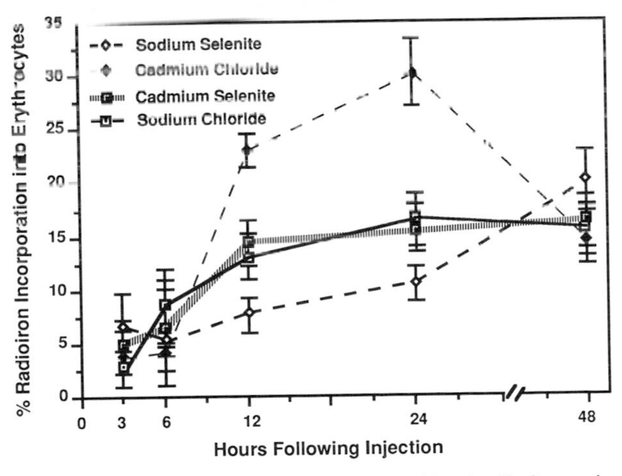

Figure 1 Radioiron incorporation into red blood cells from mice injected with sodium selenite, cadmium chloride, cadmium selenite, and sodium chloride (hour 0). Vertical lines represent standard errors of the means. [From G. R. Hogan and P. D. Jackson. (1986). Dichotomous effects of cadmium and selenium on erythropoiesis in mice. *Bull. Environ. Contam. Toxicol.* **36,** 674–679.]

tertiary protein structure. Selenium has been implicated to bind with sulfur, thus preventing potential bonding of cadmium or other such substances with sulfhydryl rich groups. This would allow the enzymes to participate in their normal metabolic pathway(s).

III. Distribution of Selenium in Biological Systems

A. Invertebrate Systems

Experimental data concerning accumulation of selenium in invertebrate systems is predictably diverse; a number of factors contributes the extent of diversity within a test species. A thorough review of selenium toxicity in aquatic organisms was included in a 1987 publication by Maier and co-workers (Table II). Values indicate that older animals are more tolerant than the younger ones and that selenite is more toxic than selenate as shown by the effective concentration or lethal concentration to kill 50% of the test organisms at a given time (EC_{50} or LC_{50}, respectively). For the latter, this appears to be the case for many vertebrate systems as well.

Studies on the tissue/organ distribution of selenium within an invertebrate organism have not been widely published. One such study, however, used *Tenebrio molitor* as the test organism. Newly emerged insects were reared in media supplemented with variable concentrations of sodium selenite. Ini-

Table II Selenium Toxicity to Invertebrates

Species	Lifestage	EC_{50} or LC_{50} (ppb)	Time period	Chemical form of SE
Daphnia magna	12 ± 12 hr	1210	48 hr	Selenite
	12 ± 12 hr	550	7 d	Selenite
	<24 hr	1990	7 d	Selenite
	24 hr	660	24 hr	Selenite
	24 hr	430	48 hr	Selenite
	Instar	430	96 hr	Selenite
	Mid. instar	710	48 hr	Selenite
	Mid. instar	430	14 d	Selenite
	Adult	1100	48 hr	Selenite
	Adult	5300	48 hr	Selenate
	Life cycle	92–240	28 d	Selenite
	N.D.	16000	24 hr	Selenite
	N.D.	250	48 hr	Selenite
	N.D.	2500	48 hr	Selenite
	Neonate	3000	48 hr	Selenate
	Neonate	600	48 hr	Selenite
	Neonate	600	48 hr	SeMet
Daphnia pulex	24 ± 12 hr	3870	48 hr	Selenite
	Juvenile	600	48 hr	Selenite
	Juvenile	100	96 hr	Selenite
	Adult	1300	48 hr	Selenite
	Adult	500	96 hr	Selenite
	Life cycle	690	28 d	Selenite
Hyallela azteca	Adult	940	48 hr	Selenite
	Adult	340	96 hr	Selenite
	Adult	70	14 d	Selenite
	N.D.	760	96 hr	Selenate
Chironomus decorus	4th instar	25000	48 hr	Selenate
	4th instar	50000	48 hr	Selenite
	4th instar	200000	48 hr	SeMet
Midge (genus unk.)	N.D.	42400	96 hr	Selenate
Culex fatigans	N.D.	3100	48 hr	Selenate
Physa sp.	N.D.	24100	96 hr	Selenate

Source: K. J. Maier, C. Foe, R. S. Ogle, M. J. Williams, A. W. Knight. (1987). The dynamics of selenium in aquatic ecosystems. *Trace Subst. Environ. Hlth* **XXI,** 361–408.

tial exposure to 0.125, 0.25, and 0.50% selenite in the rearing medium promoted an abrupt decrease in survival beginning with 7 days postemergence. Survival responses were dose dependent. Selenite appeared to have a residual toxic effect after insects were removed from exposure to selenium. From the survival studies, the distribution of selenium in Malpighian tubules, digestive tract, and reproductive organs was determined using fluorometric analysis. These data are shown in Figure 2. Malpighian tubules appeared to accumulate the greatest amount of selenium, the digestive tract was second in accumulation, and the reproductive tract accumulated the least amount of selenium. There appears to be a dose–accumulation relationship for Malpighian tubules, but not a distinct one for digestive and reproductive tissues. Perhaps this increased amount in Malpighian tubules was associated with their excretory function.

B. Vertebrate Systems

In 1913, Quarelli reported the first studies on the distribution of selenium in animals. Since then, a number of other investigations have been published using mice, rats, sheep, cattle, and marine organisms. The *in vivo* patterns of selenium distribution in regard to the level of tissue accumulation and the time course of accumulation are varied. A key factor contributing to the variation within a given species is the analytical method employed. Gas chromatography, anion exchange chromatography, molecular neutron activation analysis, and fluorometric methods are among the most commonly used procedures.

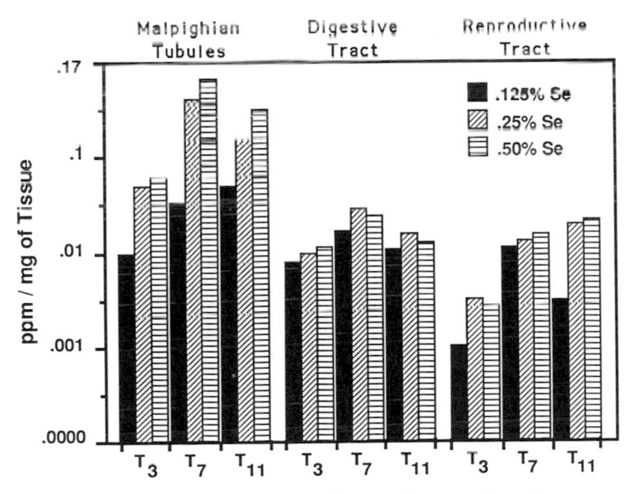

Figure 2 Effect of sodium selenite supplemented media on survival of *Tenebrio molitor*. [From G. R. Hogan. (1991). Selenium-induced mortality and tissue distribution in *Tenebrio molitor* L. (Coleoptera: Tenebrionidae). *Environ. Entomol.* **20,** 790–794.]

Table III shows the concentration of selenium in tissues of ICR mice following intraperitoneal injection of sodium selenite. It is noted that there are marked differences in the temporal appearance of selenite in test materials. Skeletal muscle appears to begin to clear selenite within 9 hours after its maximum accumulation at 3 hours. Cardiac muscle follows the same general trend, although the level of accumulation is higher and retained longer. Lung, renal, and hepatic tissues show relatively high levels over a protracted period compared with tissues obtained from sodium chloride-treated mice.

Using sodium selenate at 5 mg/kg, the patterns of distribution in mice tissue differ from those following selenite treatment (Table IV). Skeletal muscle shows a dramatic increase 9 hours postinjection with a continued retention above control values a day later. An almost identical pattern is found in liver and lung tissue, although the selenate concentration is more dramatic for lung. Selenate appears to be more penetrable into neural tissue than selenite. This could account for the neurological disturbances reported from earlier observations on cattle.

Selenium toxicity in fish species has been reported from several sources. Toxic responses are highly dependent on the developmental stage of the organism. In general, the younger the organism or the earlier the lifestage at exposure, the greater the lethality induced by selenium. Nonlethal concentrations of selenium promote a decreased growth rate and abnormal development of ovaries and liver. It appears that selenite is more toxic than selenate as a mortality factor.

Toxicological studies using chick embryos have shown that organic selenium compounds are generally more effective than inorganic forms in the induction of teratogenesis. Such effects are quantified by increased incidence of anomalies (due to delayed time(s) of predicted developmental events) and increased embryo mortality.

IV. Toxicity of Selenium

A. General

There are abundant data regarding deficiency symptoms of selenium. However, selenosis (selenium poisoning) has not been well documented and does not occur frequently. There have been reported

Table III Selenium (ppm/g tissue) in Selected Tissues of ICR Mice following 5 mg Se/kg Body Weight Injection of Sodium Selenite[a]

Tissues	Control	3 hr	6 hr	9 hr	12 hr	15 hr	18 hr	24 hr
Skeletal muscle	0.318 (0.115)	1.395 (0.241)	1.176 (0.202)	0.520 (0.271)	0.960 (0.127)	0.678 (0.146)	0.428 (0.092)	0.343 (0.092)
Lung	0.373 (0.179)	4.234 (0.456)	5.311 (0.804)	2.140 (0.880)	3.193 (0.491)	2.879 (0.308)	1.973 (0.257)	1.600 (0.252)
Heart	0.290 (0.154)	3.642 (0.441)	2.630 (0.600)	2.127 (0.531)	2.147 (0.336)	1.500 (0.408)	1.217 (0.163)	1.017 (0.159)
Kidney	0.848 (0.063)	2.367 (0.309)	2.493 (0.416)	1.935 (0.655)	2.490 (0.153)	2.090 (0.390)	1.407 (0.219)	1.137 (0.093)
Spleen	0.735 (0.087)	3.458 (0.858)	2.856 (0.416)	2.143 (0.398)	3.273 (0.235)	1.760 (0.150)	1.637 (0.064)	1.267 (0.104)
Liver	1.028 (0.311)	5.603 (0.778)	5.375 (0.922)	1.263 (0.662)	2.503 (0.369)	3.203 (0.380)	3.177 (0.296)	2.723 (0.287)
Brain	0.186 (0.053)	0.297 (0.041)	0.610 (0.222)	0.180 (0.030)	0.290 (0.104)	0.717 (0.102)	0.243 (0.064)	0.157 (0.035)

[a] Parenthetical values represent standard errors of the means. Unpublished data of M. L. Chen, S. L. Razniak, and G. R. Hogan.

cases of individuals apparently having this disorder who lived in highly seleniferous regions of Asia. Endemic selenosis in humans and animals in Enshi County, Hebei Province of the People's Republic of China was reported in 1966. These individuals were diagnosed as having chronic selenosis, with symptoms being described as thickened fingernails and toenails which were deformed in shape. Relatively few experimental studies, however, either qualitative or quantitative have been directed toward a better understanding of the physiological effects of high blood levels of selenium.

At high levels, selenium is clearly toxic as expressed by rapid death following exposure. Studies

Table IV Selenium (ppm/g tissue) in Selected Tissues of ICR Mice following 5 mg Se/kg Body Weight Injection of Sodium Selenate[a]

Tissues	Control	3 hr	6 hr	9 hr	12 hr	15 hr	18 hr	24 hr
Skeletal muscle	0.318 (0.115)	0.907 (0.296)	3.070 (0.651)	7.093 (0.695)	3.423 (1.039)	1.637 (0.344)	2.367 (0.263)	1.240 (0.026)
Lung	0.373 (0.179)	7.815 (4.305)	10.957 (0.787)	27.093 (2.083)	5.837 (1.649)	13.270 (4.819)	11.583 (3.399)	6.263 (0.237)
Heart	0.290 (0.154)	3.597 (0.632)	2.140 (0.317)	3.817 (0.667)	4.097 (0.339)	3.137 (0.878)	2.390 (0.492)	1.807 (0.355)
Kidney	0.848 (0.063)	2.876 (0.479)	4.103 (0.934)	3.420 (0.512)	2.900 (0.585)	3.517 (0.239)	3.823 (0.275)	2.790 (0.267)
Spleen	0.735 (0.087)	1.603 (0.368)	5.677 (0.665)	8.970 (0.366)	4.293 (0.432)	3.410 (0.396)	3.007 (0.775)	2.360 (0.153)
Liver	1.028 (0.311)	7.383 (0.416)	14.007 (0.776)	9.397 (0.523)	8.220 (0.821)	7.160 (0.510)	6.487 (1.486)	5.635 (1.145)
Brain	0.186 (0.053)	0.417 (0.084)	1.785 (0.395)	0.888 (0.454)	1.223 (0.394)	1.300 (0.408)	1.457 (0.558)	0.907 (0.103)

[a] Parenthetical values represent standard errors of the means. Unpublished data of M. L. Chen, S. L. Razniak, and G. R. Hogan.

on rodents suggest that death due to a single lethal dosage may be associated with central nervous system failure and/or respiratory failure. Sublethal effects are not as easily identified, obviously, because the ultimate end point of the effect is less defined. Similar to the biotoxic effects of other trace substances, those of selenium are difficult to decipher in *in vivo* systems. It is noteworthy that there are essentially no *in vitro* selenium toxicological investigations that have appeared in the literature; this is truly an open area for research and is significantly needed. A major reason for the determination of selenium effects at acute sublethal levels is the extremely broad range of adverse effects. These either directly or indirectly influence the function of vulnerable sites.

Selenium toxicity is clearly panhistotoxic. The affected individual appears to be emaciated. Even though some tissues have been implicated to be more sensitive to selenium than others, at high levels it is likely that all metabolizing cells are subject to selenium's damaging actions. Because of the highly complex and intimate interdependencies and interrelations among organs and organ systems, if one component in the system is affected by either reduced or blocked function or irrepairable damage, the total system (i.e., organism) is affected. The initial lesion induced by selenium can cause numerous secondary lesions. This is due to a disruption of the homeostatic systems' interrelationships. In toxicological studies, one must be constantly aware of primary and secondary effects of the test substance. Such is the case for selenium. For example, selenium at high levels causes liver dysfunction—a primary effect. Following in the wake of the primary effect, the investigator may observe secondary effects, including difficulty in the absorption of fats because less (or no) bile is being produced by the damaged liver to emulsify dietary fats for lipase digestion. Also prolonged clotting time due to impaired hepatic production of the coagulation factors may be observed.

B. Factors of Variable Response to Selenium

Key factors in the variation of toxic responses to selenium are the valence state of the test substance and whether the test substance is in an inorganic or organic form. The most common inorganic forms are selenite (+4) and selenate (+6); the latter is the most abundant form in many biologic environments. The extent of tissue accumulation, and thus toxicity, is dependent upon these factors. Results indicate that there is a conversion in the valence state of selenium as it transverses biological membranes and that, intracellularly, selenium is converted to either selenide (Se^{-2}) or elemental selenium.

Figure 3 illustrates the concentration of sodium selenite and sodium selenate in mouse spleen at various times following injection of 5 mg Se/kg. In marked contrast, Figure 4 shows the uptake of two organic forms of selenium, selenomethionine, and selenocystine. The same differential in tissue concentrations of the two organic forms has been shown for lung and skeletal and cardiac muscle. The kidneys do not retain the inorganic forms (Fig. 5) as extensively as the organic forms (Fig. 6). The organic forms are higher in content in the liver and brain than the inorganic forms. The rate and extent of accumulation of selenomethionine over those of selenocystine are associated with the differences in the efficiency of membrane transport and absorption rates.

Other variables associated with selenium toxicity include age, pregnancy, and route and duration of exposure. Of course, geographical location (i.e., seleniferous regions) is a key factor.

C. Toxicity in Insects

Selenium is known to have a marked effect on insect survival (Fig. 7). The mortality responses of *Tenebrio confusum* to media containing various concentrations of sodium selenite reveal that at the

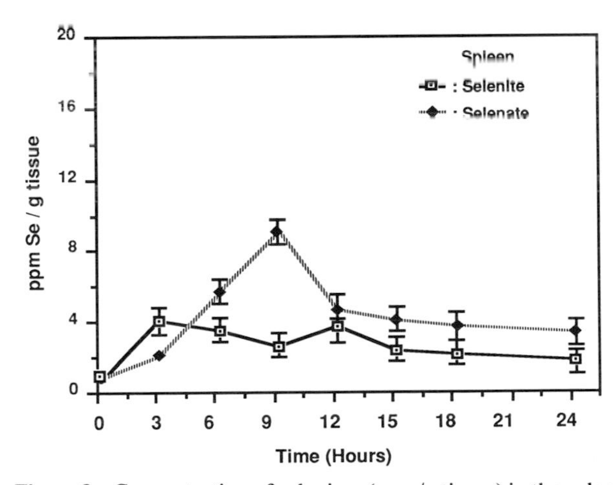

Figure 3 Concentration of selenium (ppm/g tissue) in the spleen of female ICR mice as function of time following injection (0 hour) of 5.0 mg Se/kg selenite and selenate. (From M. L. Chin, S. L. Razniak, and G. R. Hogan. Unpublished data.)

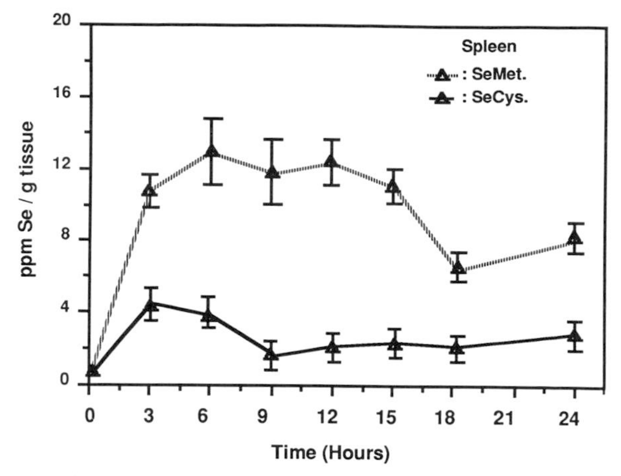

Figure 4 Concentration of selenium (ppm/g tissue) in the spleen of female ICR mice as function of time following injection (0 hour) of 5.0 mg Se/kg selenomethionine (SeMet) and selenocystine (SeCys). (From M. L. Chin, S. L. Razniak, and G. R. Hogan. Unpublished data.)

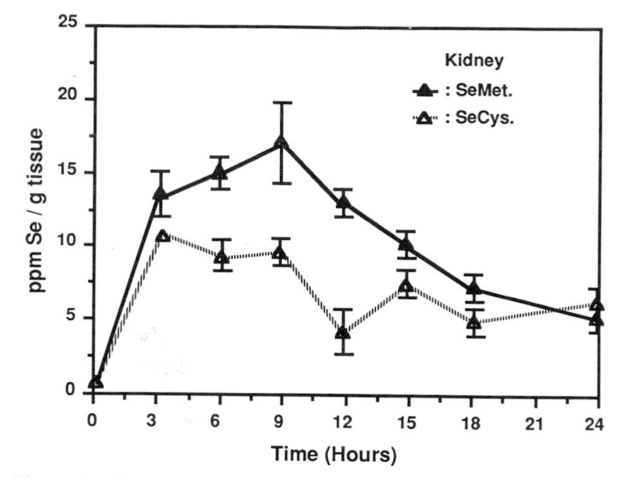

Figure 6 Concentration of selenium (ppm/g tissue) in the kidney of female ICR mice as function of time following injection (0 hour) of 5.0 mg Se/kg selenomethionine (SeMet) and selenocystine (SeCys). (From M. L. Chin, S. L. Razniak, and G. R. Hogan. Unpublished data.)

higher levels, selenium induces a dramatic killing effect.

The effect of selenium on insect survival appears to persist after the insects have been transferred to non-selenium-supplemented medium. It has been shown that transfer from sodium selenite-containing medium to control medium offers little protection on survival patterns. Transfer to such medium from control medium induces an abrupt killing effect. The toxic effect of selenium in insects appears to be both residual following initial exposure and immediate upon introduction of selenium into the diet.

In *Tribolium confusum,* selenium extends the time required from larval development to pupation, but it does not affect the duration of the pupal period. The

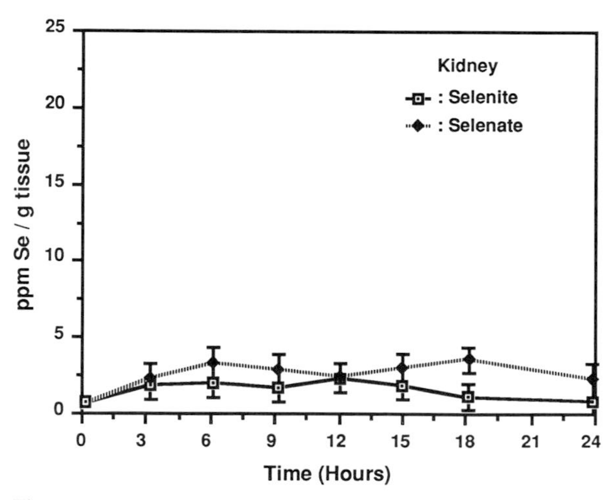

Figure 5 Concentration of selenium (ppm/g tissue) in the kidney of female ICR mice as function of time following injection (0 hour) of 5.0 mg Se/kg selenite and selenate. (From M. L. Chin, S. L. Razniak, and G. R. Hogan. Unpublished data.)

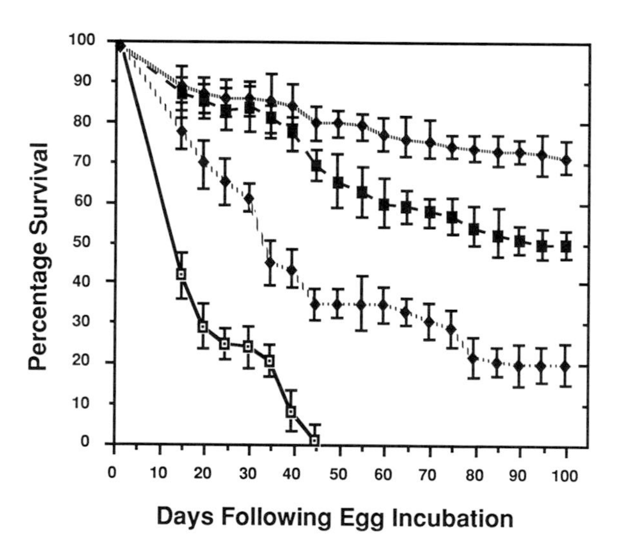

Figure 7 Percentage survival of *T. confusum* reared from eggs of flour-yeast-grown adults transferred to basal-casein medium (upper-most line) or selenium-supplemented media (0.25%, 0.5%, and 1.0%; closed squares, diamonds, and open squares, respectively) on day 0. Initial number incubated, 100 per group. Vertical lines represent standard errors of the means. [From G. R. Hogan and B. S. Cole. (1988). Survival of *Tribolium confusum* (Coleoptera: Tenebrionidae) in basal-casein medium supplemented with sodium selenite. *Environ. Entomol.* **17,** 771–777.]

percentage of emergence is adversely affected when insects are grown on selenium-enriched incubation medium. From larval studies on *Tribolium confusum*, it has been shown that newly hatched larvae are approximately 100 times more sensitive to selenium-containing (0.125%) medium than 2- or 4-week-old larvae (Table V).

D. Toxicity in Vertebrates

Much of the compiled data concerning selenosis have been developed through observations of domestic animals (although, more recently, controlled experimentation on the biological toxicity of selenium has been conducted). Even though there are species differences, many of the symptoms developing from selenium poisoning are identical.

1. Chronic Selenosis

When animals consume forages and grains containing between 5 and 50 ppm of selenium, a chronic disorder results referred to as "alkali disease," or in some cases, the "blind staggers." Most of the cases that have been reported for livestock exposed to excess selenium in the diet have come from Wyoming, South Dakota, Colorado, and Nebraska. Australia, Israel, and Ireland have also reported areas containing seleniferous plants. Table VI summarizes the effects of chronic selenium poisoning. These changes are noted during progression of the disease and at autopsy. Symptoms begin to develop within a few weeks after consumption of the grains or plants containing elevated selenium levels. Growth in general appears to be affected, including repair of damaged tissue and growth of horns. It has been suggested that selenium blocks cell division or increases the time for mitosis. Thus, it would promote or induce a negative cell balance.

In "blind staggers," livestock exhibit three distinct phases of selenium poisoning. At first, the animal's orientation is affected, whereby it appears to wander and stumble over obstacles in its path. Vision is poor and eating activity is reduced. Next, the muscles of the legs fail to support the body's weight (i.e., the animal staggers more and more) and vision becomes progressively worse. The third and final stage is characterized by paralysis. Tongue and throat muscles become paralyzed and body temperature is reduced. Death is reported to be due to failure of respiration.

2. Acute Selenosis

Consumption of large quantities of seleniferous grains or plants over a short period of time or of a high selenium-containing diet at one feeding promotes death within hours. Acute selenosis is characterized by an abrupt increased body temperature, pupillary dilation, labored respiration, and eventual respiratory failure. There are no known measures to prevent death following intake of large amounts of selenium.

3. Blood Studies

Selenium in mammals at levels above its essential level and below its lethal dosage is known to affect both white blood cells and red blood cells of some mammals. In addition, low dosages cause variation in the blood plasma protein concentration, depress packed red blood cell volume, and render the red blood cell membrane more susceptible to rupture, thus promoting a hemolytic type of anemia. In addition to lowering the number of circulating red blood

Table V Effects of Sodium Selenite-Supplemented Basal-Casein Media on Development of *T. confusum*[a]

Medium	Mean duration of larval period (day)	Survival to pupal (%)	Mean duration of pupal period (day)	Adults emerging (%)
Basal-casein (B-C)	57 ± 4.4	77 ± 4.8	6 ± 1.3	96 ± 3.0
0.25% Se B-C	62 ± 3.7	60 ± 5.0	6 ± 0.9	85 ± 7.4
0.50% Se B-C	68 ± 3.2	31 ± 5.6	5 ± 1.8	70 ± 6.1

Source: G. R. Hogan and B. S. Cole. (1988). Survival of *Tribolium confusum* (Coleoptera:Tenebrionidae) in basal-casein medium supplemented with sodium selenite. *Environ. Entomol.* **17,** 770–777.

[a] Values ± SEM.

Table VI Effects of Selenium Poisoning on Organ Systems of Vertebrates

System affected	Dysfunction or abnormality
Cardiovascular	Decreased hemoglobin synthesis, anemia, intradermal hemorrhage of the pericardium.
Digestive	Decreased food intake, increased blood flow, and ulceration of the walls of the small intestine.
Integumentary	Hair loss (epilation), thickening of the nails and hooves.
Muscular	Lack of coordinated skeletal muscle movement, paralysis of skeletal muscle, lesions of cardiac muscle.
Nervous	Poor/impaired vision, depressed body temperature, poor motor control of voluntary movements.
Renal	Congestion of renal medulla, decreased urine output leading to metabolic acidosis (?), tubular degeneration.
Reproductive	Decreased rate of conception, birth of blind animals, increased incidence at birth of bone anomalies.
Respiratory	Decreased rate of respiration, respiratory acidosis (?), respiratory failure.
Skeletal	Decreased articulation of bones (joint stiffness), erosion of articulating surfaces of bone (primarily tibial surface).

cells through hemolysis, selenium affects the rate of production of red blood cells.

Figure 8 shows the reduction in radioactive iron uptake into hemoglobin synthesizing cells following intraperitoneal injection(s) of sodium selenite. The effect is transitory and there appears to be a dose–response relationship.

Treatment with selenocystine has been shown to reduce the white blood cell count and size of the spleen of patients having acute leukemia. These persons, however, demonstrate nausea, vomiting, and diarrhea following dosages averaging 125 mg/day. Patients suffering from various types of carcinomas, and who had selenium concentrations above the average value of all cancer patients, had neoplasms that had fewer recurrences with less metastasis. Considerable experimental evidence is available indicating selenium as a potent antitumorigenic factor.

Selenium has been shown to lower the blood levels of a crucial enzyme involved in hemoglobin synthesis. The enzyme, δ-aminolevulinic acid dehydratase, is reduced to twenty percent of control levels following daily injections of 2 mg/kg sodium selenite over three consecutive days (Fig. 9).

After eight days following sodium selenite injections (2 mg/kg × 3), the corresponding number of circulating white blood cells (leukocytes) of mice is significantly lower than that of controls (Fig. 10). The effect is continuous and becomes more marked by day 16, at which time the average white blood cell count of the group receiving the highest dosage is only approximately 50% of the average control count.

Associated with the lowering effect of peripheral white blood cells, a shift in the number of agranulocytes relative to the number of granulocytes has been noted (Fig. 11). This shift in the numbers of cells is caused by the destruction or a reduction in the most abundant granulocyte in the mouse, the neutrophil, which has a relatively short life span. Cytological studies reveal that neutrophils obtained from selenite-treated mice appear to be "aged" as evidenced by an increase in cytoplasmic granules and the greater appearance of cells possessing pyknotic nuclei. Such changes are characteristic of dying granulocytes. In addition, it is thought that selenium has an effect on the neutrophil production rate. As neutrophils die at a more rapid rate, they are not replaced at a normal rate; thus, the numbers of the granulocytes decrease in the wake of selenium expo-

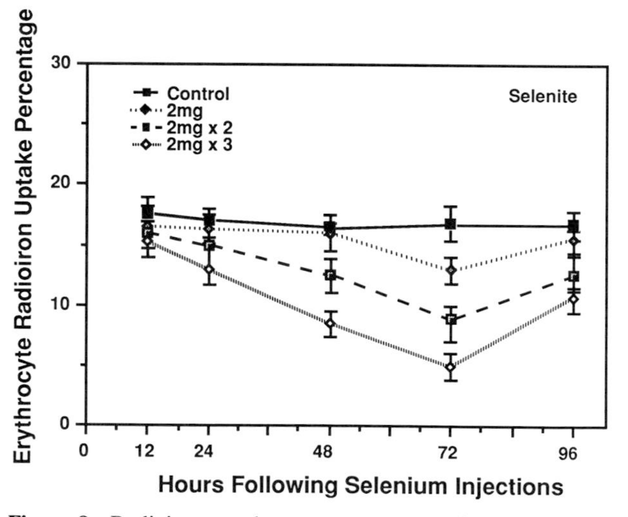

Figure 8 Radioiron uptake percentages as a function of time following injection(s) of sodium selenite. Number animals/group = 10. Vertical bars of the points represent standard errors of the means. [From G. R. Hogan. (1990). Biotoxic effects of selenium. *Trace Subst. Environ. Heath* **XXIV**, 302–307.]

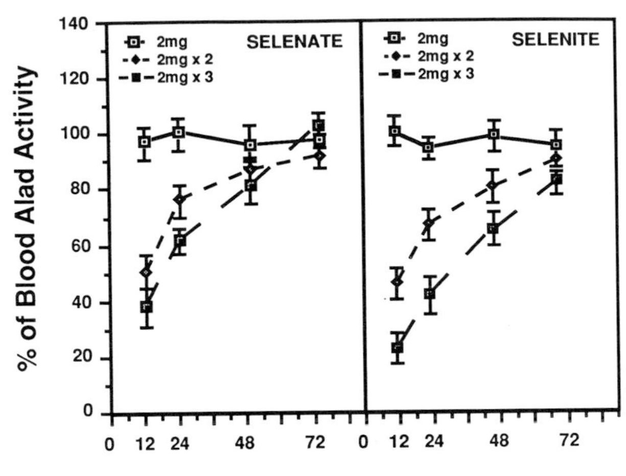

Figure 9 Effects of selenate and selenite treatments on the activity level of δ-aminolevulinic acid dehydratase (ALAD) at selected times after injection(s). [From G. R. Hogan. (1990). Biotoxic effects of selenium. *Trace Subst. Environ. Health* **XXIV**, 302–307.

sure contributing to the overall decrease in the number of circulating white blood cells.

E. Conclusions

Numerous reports are available concerning selenium's essentiallity as a micronutrient involving a variety of species. However, new initiatives in research on the biotoxic effects of selenium are greatly

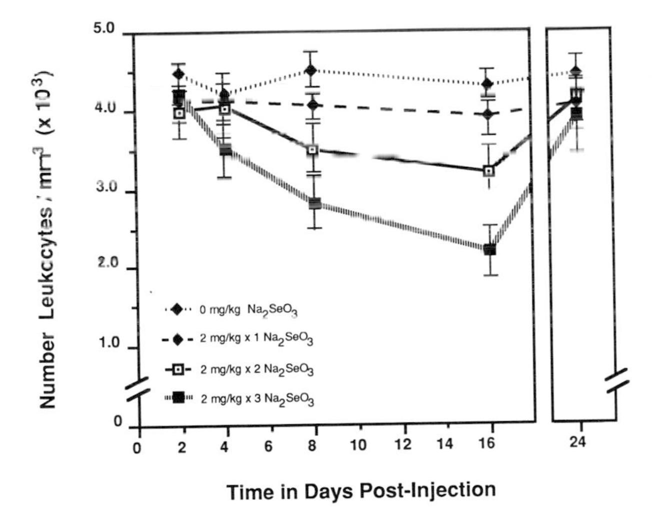

Figure 10 Total leukocyte counts obtained from sodium selenite (Na_2SeO_3)-treated mice. Days represent the times of blood collection following the final or only Na_2SeO_3 injection on day 0. Vertical lines represent standard errors of the means. [From G. R. Hogan. (1986). Decreased levels of peripheral leukocytes following sodium selenite treatment in female mice. *Bull. Environ. Contam. Toxicol.* **37**, 175–179.]

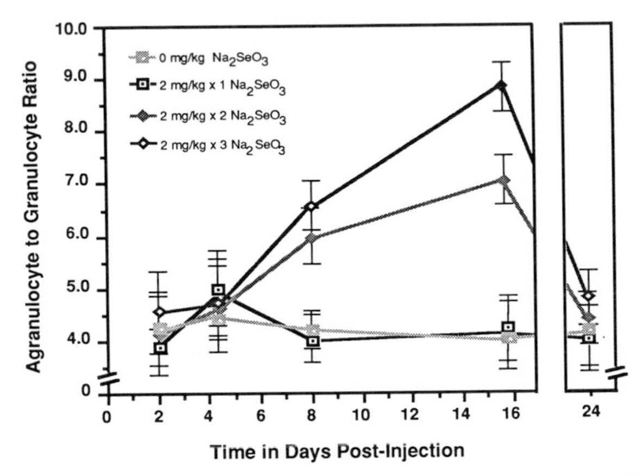

Figure 11 Effect of sodium selenite (Na_2SeO_3) treatment on the agranulocyte to granulocyte ratios of female mice. Days represent the times of blood collection following the final or only Na_2SeO_3 injection of three treatment groups on day 0. Vertical lines represent standard errors of the means. [From G. R. Hogan. (1986). Decreased levels of peripheral leukocytes following sodium selenite treatment in female mice. *Bull. Environ. Contam. Toxicol.* **37**, 175–179.]

needed. Relatively little is known concerning the cellular/molecular responses to selenium poisoning.

Ultimately, the toxicity of selenium will be realized at the molecular or submolecular levels. Employment of *in vivo* models is not favored because of the high degree of integration among functional components of the organism. Thus, it is difficult to determine primary effects from secondary ones. *In vitro* modeling has a number of advantages, including the ability to isolate "vulnerable targets," whether they are cell(s), enzyme system(s), or macromolecules. Of course, caution must be exerted in extrapolating from an *in vitro* system to an *in vivo* system. Selenium has an antimitotic effect which has been suggested for a number of different species. Perhaps further exploration of the cellular effects of selenium above its trace amounts will provide pertinent information on the antitumor properties of selenium and on the methods, treatments, and procedures to prevent selenosis in biological systems.

Related Article: METAL–METAL TOXIC INTERACTIONS.

Bibliography

Combs, G. F., Jr., Spallholz, J. E., Levander, O. A., and Oldfield, J. E., eds. (1987). "Selenium in Biology and Medicine." Part B. Van Nostrand, New York.

Delves, H. T. (1985). Assessment of trace element status, Chapt. 10. *Clin. Endocrinol. Metab.* **14**(3), 725–760.

Liotta, D., ed. (1987). "Organoselenium Chemistry." Wiley and Sons, New York.

Mair, K. J., Foe, C., Ogle, R. S., Williams, M. J., Knight, A. W., Kiffney, P., and Melton, L. A. (1987). The dynamics of selenium of aquatic ecosystems. *Trace Subst. Environ. Health* **XXI,** 361–408.

"Merck Veterinary Manual." (1986). 6th edition, Merck and Co, Rahway, NJ.

Wilber, C. G. (1980). Toxicology of selenium: A review. *Clin. Toxicol.* **17,** 171–230.

Zhang, Y. (1986). The relationship between endemic diseases and trace elements in the natural environment of Jilin Province of China. *Trace Subst. Environ. Health* **XX,** 381–391.

Silica and Lung Inflammation

Marlene Absher
University of Vermont

Glossary

Amorphous silica Silicon dioxide form lacking crystalline structure. Tetrahedral subunits are bound together in a random arrangement.

Free silica Pure crystalline silicon dioxide; major forms are quartz, cristobalite, trydimite.

Silanol, siloxane Surface functionalities of the silicon dioxide particle (silanol, Si — OH and siloxane, Si — O — Si); considered to be the major determinants of the biological effects of silica.

Silica polymorph Denotes chemical/physical form of silicon dioxide (e.g., crystalline quartz, cristobalite) or amorphous silica.

Silicates Silicon dioxide combined with other minerals such as calcium, magnesium, aluminum, iron, potassium, and sodium.

SILICOSIS is a chronic inflammatory and fibrotic lung disease caused by exposure primarily to crystalline forms of silicon dioxide (SiO_2). From initial contact of silica particles with cells and fluids of the respiratory system, there is a cascade of events that includes cell death, influx of inflammatory cells into alveolar and parenchymal spaces, activation of immune and inflammatory cells, and secretion and activity of biologically active mediators that lead to development of the fibrotic lesions characteristic of silicosis. The chronic nature of silicotic lung disease may relate to the fact that inhaled silica particles remain in the lung for prolonged periods after cessation of exposure, thus serving as a persistent source of stimulation of immune and inflammatory cell responses in the lung.

I. History of Silicosis

Silicosis is a disease of ancient origin. Pulmonary disease as a result of exposure to silica-bearing minerals has been recognized as a major health problem since the industrial revolution. Mining operations and other industries involving rockblasting and sandblasting created a source of acute and chronic exposure to silica and other potentially harmful mineral dusts. Procedures to reduce exposure levels have effectively diminished the incidence of silicosis in many highly industrialized nations, but silicosis remains a serious public health problem in many regions of the world.

II. Occupational Exposure

In spite of measures to reduce exposure, substantial numbers of workers continue to be exposed to potentially damaging concentrations of respirable aerosols containing silica compounds. Table I lists some of the occupational exposures encountered in various industrial settings.

III. Chemical and Physical Properties of Silicon Dioxide Polymorphs

Silicon is one of the most abundant elements in the earth's surface. Silicon is found predominantly bound to oxygen as silicon dioxide (SiO_2), existing as tetrahedral subunits. The subunits may be linked together in a variety of silicon dioxide polymers with

Table I Occupational Exposure to Silicas

Mining operations	Rocks, ores
Quarrying operations	Cutting, polishing
Sandblasting	Abrasives
Glass manufacturing	Sand
Pottery, ceramics	Clays, porcelain, high temperature processes
Foundry work	Sand and clay molds
Manufacturing processes	Fillers, coatings, abrasives, high temperature processes

characteristic crystal structures such as quartz, cristobalite, or trydimite. Amorphous silica lacks a crystalline structure; the tetrahedra are bound together in a random arrangement. In nature, silica exists both in amorphous form (such as is found in diatomaceous earth) or in crystalline form. The major crystalline forms are quartz, trydimite, and cristobalite. The term "free silica" denotes pure crystalline silicon dioxide; in silica flour the predominant form is α-quartz, while free silica in calcined diatomaceous earth is cristobalite and to a lesser extent, trydimite. Silicon dioxide also combines with other minerals to form silicates of calcium, magnesium, aluminum, etc. Table II lists a variety of silicon dioxides that exists in nature or is created from industrial processes.

Of the various forms of silica, quartz is the most abundant and the form most involved in lung pathology. Quarried quartz may contain 10–30% crystalline quartz, while slate may have 20–30% quartz. Cristobalite and trydimite are more toxic than quartz but are much less abundant in nature. Cristobalite is chiefly found as an industrial product formed from amorphous silica in diatomaceous earth when it is calcined at extremely high temperatures (800–1100°C). Cristobalite is used in filtration systems and as an insulating material.

Only the crystalline structures are highly toxic and fibrogenic. Amorphous silicas are of limited importance in pulmonary disease although they may cause mild interstitial fibrosis after long periods of exposure. This is also true of the silicates with pulmonary disease developing after years of exposure.

The surface functionalities of the silicon dioxide particle are the major determinants of the biological interactions that lead to lung injury and disease. The arrangement and density of silanol (Si — OH) and siloxane (Si — O — Si) groups on the surface of the particles are believed to govern the interaction with proteins, lipids, and other macromolecules. Interaction of silica particles with cell membranes leads to ingestion by phagocytosis and cell death may ensue. Different distributions and spacings of silanol and siloxane groups on the particle surface may account for the differential membranolytic activity and toxicity of different forms of silicon dioxide particles. Membranolytic activity is often assessed by quantitation of the ability of silica to lyse erythrocytes. Generally, amorphous silica is more hemolytic than crystalline forms, even though it is not fibrogenic. Studies of heat-treated silica have shown that heat converts a hydrophilic surface to a hydrophobic surface by conversion of silanol (Si — OH) to siloxane (Si — O — Si) bridges. These altered surfaces are less hemolytic and less cytotoxic. In aerosol-exposed experimental animals, heated crystalline SiO_2 is deposited in the lung to a greater degree than is the unheated form. Heated SiO_2 particles are also retained in the lung at a higher concentration over a period of several months postexposure.

The apparent paradox of decreased cytotoxicity leading to increased fibrogenicity could be explained by the fact that particles ingested by lung macro-

Table II Forms of Silicon Dioxide

Form	Source
Crystalline SiO_2	
Quartz	Granite, sandstone, sand, slate
Cristobalite	Calcined amorphous silica
Trydimite	Calcined amorphous silica
Amorphous SiO_2	Diatomaceous earth, glass, opal, fume silica
Silicates	
SiO_2 in combination with Na, Mg, Al, K, Ca, Fe	Kaolin, talc, mica, feldspar, fuller's earth, vermiculite

phages which do not kill the cell, but remain resident within the cells for long periods of time, allow for continued activation of the cells to produce and secrete mediators that are active in the inflammatory and fibrotic process. Also, silica particles released by dying cells may be available for uptake by other cells, a means of continual recruitment and activation of an inflammatory cell population in the lung.

IV. Pathogenesis of Silicosis

A. Human Disease

The most common form of silicosis is a chronic inflammatory disease caused by the inhalation of crystalline silicon dioxide. Factors which determine whether an exposed individual develops pulmonary pathology include the dose and duration of exposure, the nature of the dust (quartz, cristobalite, or a variety of silicates and silica-bearing minerals) and the content of crystalline silica in the exposure material. Other confounding factors include genetics, smoking habits, and underlying diseases such as tuberculosis and rheumatoid arthritis.

B. Lung Pathology

Silicosis is a very localized disease, with focal lesions developing in close proximity to the areas of deposited dust. Early lesions contain dust-laden macrophages and reticulin fibers. The lesions gradually enlarge to contain large foamy macrophages, lymphocytes, neutrophils, and fibroblasts. In the characteristic silicotic nodule seen in human disease, these inflammatory cells surround a core of whorled collagen and reticulin and dust particles can be seen within the nodule. As the lesions enlarge, multiple nodules may merge. The neighboring airways and vasculature become distorted; damage to the airway epithelium leads to proliferation of type II pneumocytes in alveolar areas. Collagen deposition continues in involved areas and, as the disease progresses, some of the silicotic nodules may converge. Although less common, large areas of confluent silicotic nodules are seen in progressive massive silicosis.

In the acute form of silicosis, alveolar proteinosis is found in the lungs. This occurs when individuals are exposed for short periods of time to very high concentrations of silica. The intense damage to the alveolar epithelium leads to an excessive accumulation of lipid-containing protein in the alveoli. Damage to the epithelial lining is followed by hyperplasia and hypertrophy of alveolar type II epithelial cells which produce and secrete specific phospholipids.

C. Experimental Animal Models

1. Deposition, Clearance, and Localization of Silica Particles in Lung

Inhaled silica particles are deposited at various sites in the lung depending on their size and shape. The location at which the particles are initially deposited will determine if and how rapidly they are subsequently cleared; larger particles deposited in the upper respiratory tract will be cleared via the mucociliary escalator. Particles reaching the lower respiratory regions and alveolar spaces are cleared much more slowly—some particles may not be cleared, particularly those which are associated with the granulomatous lesions.

Within the lung, particulate silica that is cell associated resides predominantly within alveolar and interstitial macrophages. Macrophages migrating to the mucociliary escalator carry particles that are then cleared. Some particulate silica is transported along lymphatic channels to regional lymphoid tissue, such as the mediastinal lymph nodes. In bronchial-associated lymphoid tissue, particles are seen within aggregations of epithelioid cells.

In humans exposed to silica-bearing minerals, particles remain within the lung long after cessation of exposure. Long-term retention of particles is also found in experimental animals exposed briefly to crystalline particulate silica. In aerosol exposed rats, for example, there is an early clearance of up to 50% of deposited particles within two weeks after the end of the exposure period. This is followed by an extended period of over six months to a year of either minimal clearance or very gradual clearance, depending on such factors as dose and type of SiO_2 exposure. Amorphous silica is easily solubilized and thus is eliminated more readily from the lung tissue than is crystalline silica.

2. Silica-Induced Lung Inflammation and Injury

a. Histopathology

In experimental animal models of silicosis, quartz is the most commonly studied SiO_2, although cristobalite has been found to be a more reactive material. Lung tissue and cellular response will vary depending on the silica form as well as the mode of delivery of the particles, (i.e., aerosolized or via intratracheal bolus). General features show an early onset of cellular activity within the lung tissue following aerosol exposure. Multiple inflammatory lesions appear throughout the lung, generally centered on alveolar ducts and respiratory bronchioles. Alveolar lesions develop, containing macrophages, neutrophils, granular material, dust particles, and cellular debris of dead macrophages. Alveolar septal thickening is observed, with evidence of type II pneumocyte hyperplasia. In the early stages, there is little involvement of the parenchyma. Small aggregates of peribronchial lymphocytes may be seen which become more extensive as the inflammation continues. There is a gradual increase of inflammatory cells in the interstitium with time after exposure. Aggregates of inflammatory cells will begin to include lymphocytes, plasma cells, and some interstitial fibroblasts. Associated with these areas is an increased collagenous matrix. As the pathologic changes continue, there is a progressive increase in the size of the inflammatory lesions, most of which are pleural based. Increased aggregates of alveolar macrophages and neutrophils are observed; thickened collagen bands are seen at the center of inflammatory foci. As the disease progresses, peribronchial lymphoid aggregates enlarge and contain particle-bearing granulomas. Increased collagen formations are associated with airways and blood vessels as well as with the inflammatory lesions.

b. Cellular Responses

Numerous cell types may be affected directly or indirectly by silica exposure, however, it seems evident that the key initial interaction involves the inhaled silica particles and alveolar macrophages. The magnitude of the cellular response is related to exposure dose and form of SiO_2 (i.e., amorphous silica, quartz, cristobalite). Early events may involve substantial cell death (predominantly of macrophages); thus absolute numbers of macrophages may be reduced in the acute phase of the response. However, within a few days or weeks, the alveolar macrophage population increases and generally remains at a high level for prolonged periods of time. Throughout the postexposure period, most of the silica found in the lung is contained in airspace and interstitial macrophages.

The most pronounced early cellular response is an influx of large numbers of polymorphonuclear leukocytes into the alveolar and tissue spaces following aerosol exposure to a reactive form of quartz or to cristobalite. Increased levels of neutrophils persist for long periods in the chronic phase of the developing silicosis. Increases in the lymphocyte population are modest in the early postexposure period, but show a steady rise during the later stages of silicosis development.

During the acute phase of experimental silicosis, injury to the alveolar epithelium occurs, followed by repair and hyperplasia of the type II epithelial cells. Numerous studies have documented the up-regulation of alveolar epithelial cells and the surfactant system to the presence of silica. Type II cells not only increase in number but also in size. Lammellar bodies in hypertrophic type II cells are much increased in size and number compared with those found in normal type II cells. Surfactant phospholipid accumulation in lung alveolar spaces is associated with the type II cell hyperplastic and hypertrophic response to silica. Epithelial hyperplasia continues into the chronic stages of the disease. At later stages, there is also evidence of increased numbers of fibroblasts which contribute to the developing lesions and laying down of connective tissue matrix.

Both the alveolar and interstitial macrophages are altered morphologically and functionally during the course of evolving silicosis. A characteristic feature of silicotic macrophages is an enlarged macrophage with foamy cytoplasm. These are prominent in both the alveolar and tissue spaces. A typical example is shown in Figure 1 in which alveolar cells taken from

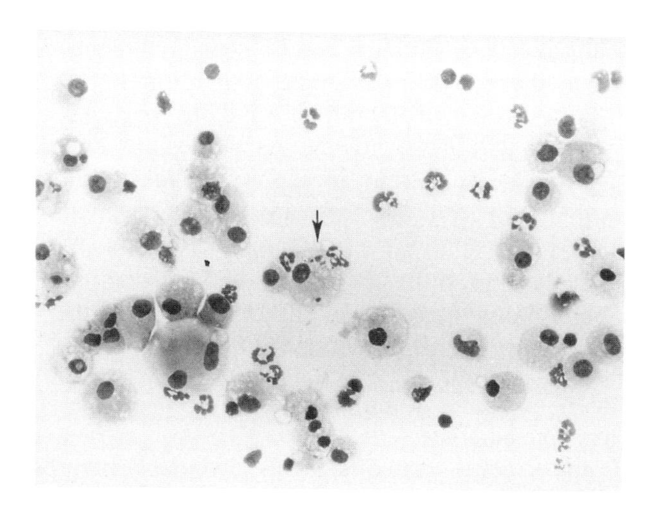

Figure 1 Cells recovered from pulmonary alveoli of cristobalite-exposed rats showing large foamy macrophages, neutrophils, and lymphocytes. Arrow indicates neutrophil ingested by macrophages.

rat lung exposed to aerosols of cristobalite consist of macrophages of varying size and morphology, neutrophils, and lymphocytes. Close associations of macrophages and neutrophils (including evidence of neutrophil ingestion by macrophages) are frequently observed.

Macrophages obtained from silica-exposed lungs demonstrate increased size, increased membrane spreading, and membrane ruffling. Silica-stimulated macrophages have increased expression of class II histocompatibility antigens. These stimulated macrophages exhibit increased oxygen consumption, release of superoxide anion, and enhanced phagocytic activity. The enhanced secretion of cytokines by silica-exposed macrophages is of major importance in the development of inflammation as these highly active mediators initiate and prolong a cascade of cellular and biochemical events that lead to lung remodeling and subsequent fibrotic disease. The continued presence of large amounts of silica in the lung provides for a continual stimulus to the inflammatory response. With time after exposure, we see increasing numbers of large foamy macrophages and neutrophils in the alveoli and interstitium. Lymphocytes gradually increase in the alveolar space and peribronchiolar regions, but within the alveoli they are not as prominent as are macrophages and neutrophils.

3. Development of Fibrosis

Increased matrix consisting predominantly of type I and III collagen is a prominent feature of experimental silicosis. In the early stages of development, fine reticulin fibers are seen in areas of aggregated macrophages. As the fibrotic lesions develop, large bunds of collagen are seen associated with granulomata, airways, and blood vessels. Increased production of collagen can also be demonstrated by quantitation of lung hydroxyproline synthesis or content.

4. Comparative Lung Response to Different Forms of Silicon Dioxide

Only the crystalline forms of silicon dioxide are of significant pathological importance to humans. Amorphous silica, while capable of eliciting a transient alveolar inflammation of the lung, does not induce fibrosis in experimental animals. The three major crystalline forms, trydimite, cristobalite, and quartz cause varying degrees of lung inflammation and injury. Cristobalite and trydimite are more reactive than quartz. Different sources of quartz can also vary in biological effect. For example, a source of Arkansas quartz has little or no activity even after continual dusting of rats for one year, while Minusil quartz elicits lung inflammation and fibrosis in aerosol-dusted rats. Figure 2 shows a comparison of the histopathology of lung tissue of rats exposed to quartz, cristobalite, and amorphous silica, which indicates the marked differences in inflammatory response to the different silica polymorphs. Quantitative differences in rat lung response to silica polymorphs are exemplified in Figure 3. Differences are shown in the composition of alveolar inflammatory cells and in the fibrotic response (as evidenced by increase hydroxyproline content) in rats exposed short term to aerosols of Minusil quartz, cristobalite, and amorphous silica. Cristobalite is the most reactive form, while amorphous silica elicits only a transient inflammation and no fibrotic response.

V. Mechanisms of Silica-Induced Lung Inflammation and Fibrosis

A. Silica—Cellular Interactions

Inhaled particles gain access to the alveolar spaces of the lung where ingestion by resident macrophages occurs. Within hours after aerosol exposure to silica, rat lung alveolar macrophages contain particles. Depending on such factors as dose and toxicity of the particular form of silicon dioxide species, macrophages may be killed, thereby releasing not only the ingested particles but also a variety of enzymes and other bioactive mediators that may cause lung damage or affect other cellular functions. Microscopic examination of lung tissue and cells and adjacent lymphoid tissue shows that virtually all the silica is associated with macrophages in exposed lungs of humans and experimental animals. Studies show that macrophages exposed to silica *in vitro* readily associate with the particles and ingest them. *In vivo* studies of macrophages isolated from alveolar lining fluid of granite workers show a substantial dust burden associated with the type and extent of exposure. Furthermore, these particles are retained for years after exposure has terminated.

Very few neutrophils are found in the normal rat lung; however, within hours of exposure to Minusil quartz and cristobalite, there is a marked influx of neutrophils into the alveolar spaces and eventually into the interstitium. Silica may cause neutrophils to secrete chemotactic factors, and neutrophils may cause injury by releasing proteolytic enzymes (in-

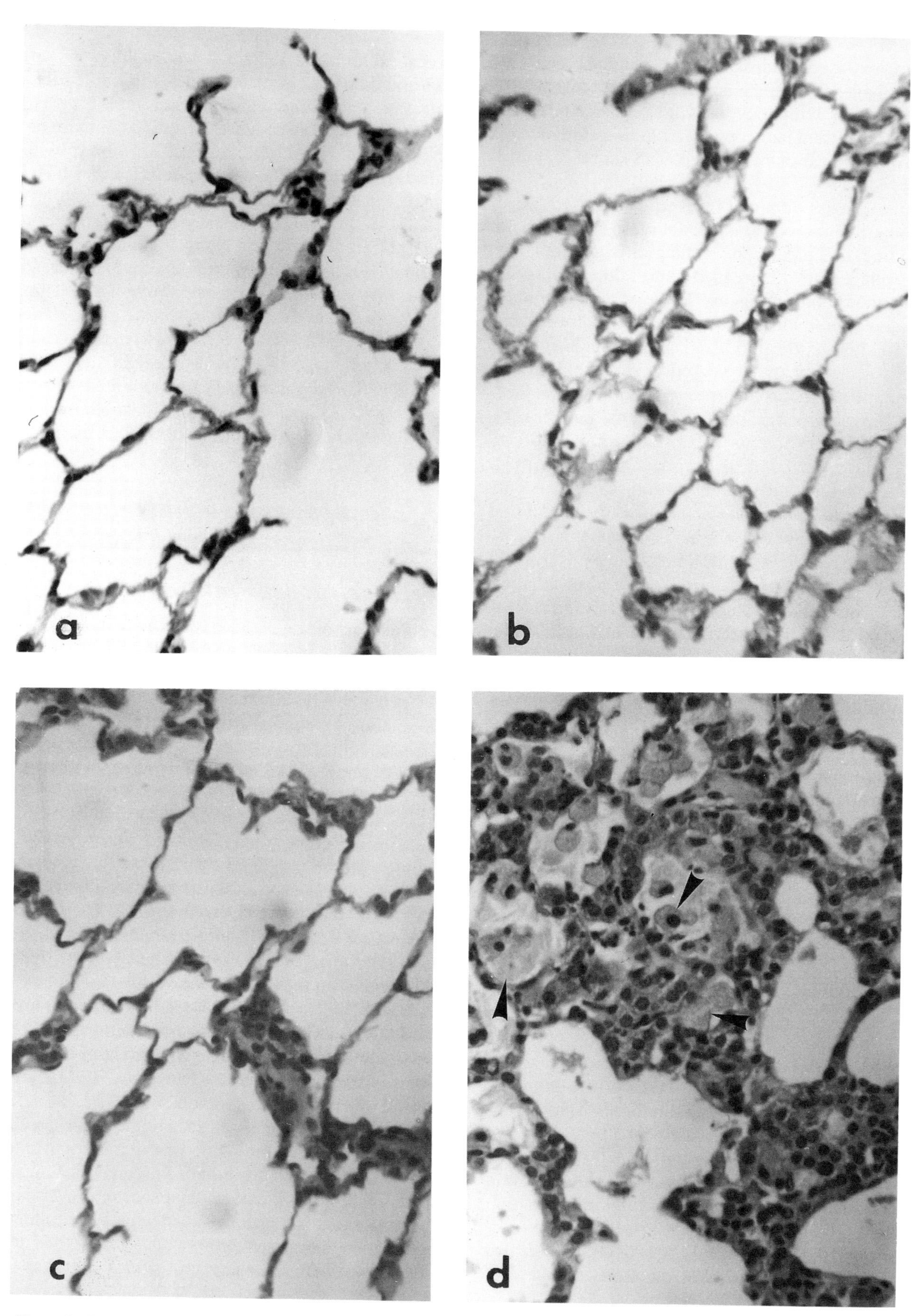

Figure 2 Hematoxylin- and eosin-stained sections of rat lungs. (a) Unexposed; exposed: (b) amorphous silica, (c) quartz, and (d) cristobalite. Arrows indicate large foamy macrophages with silica particles.

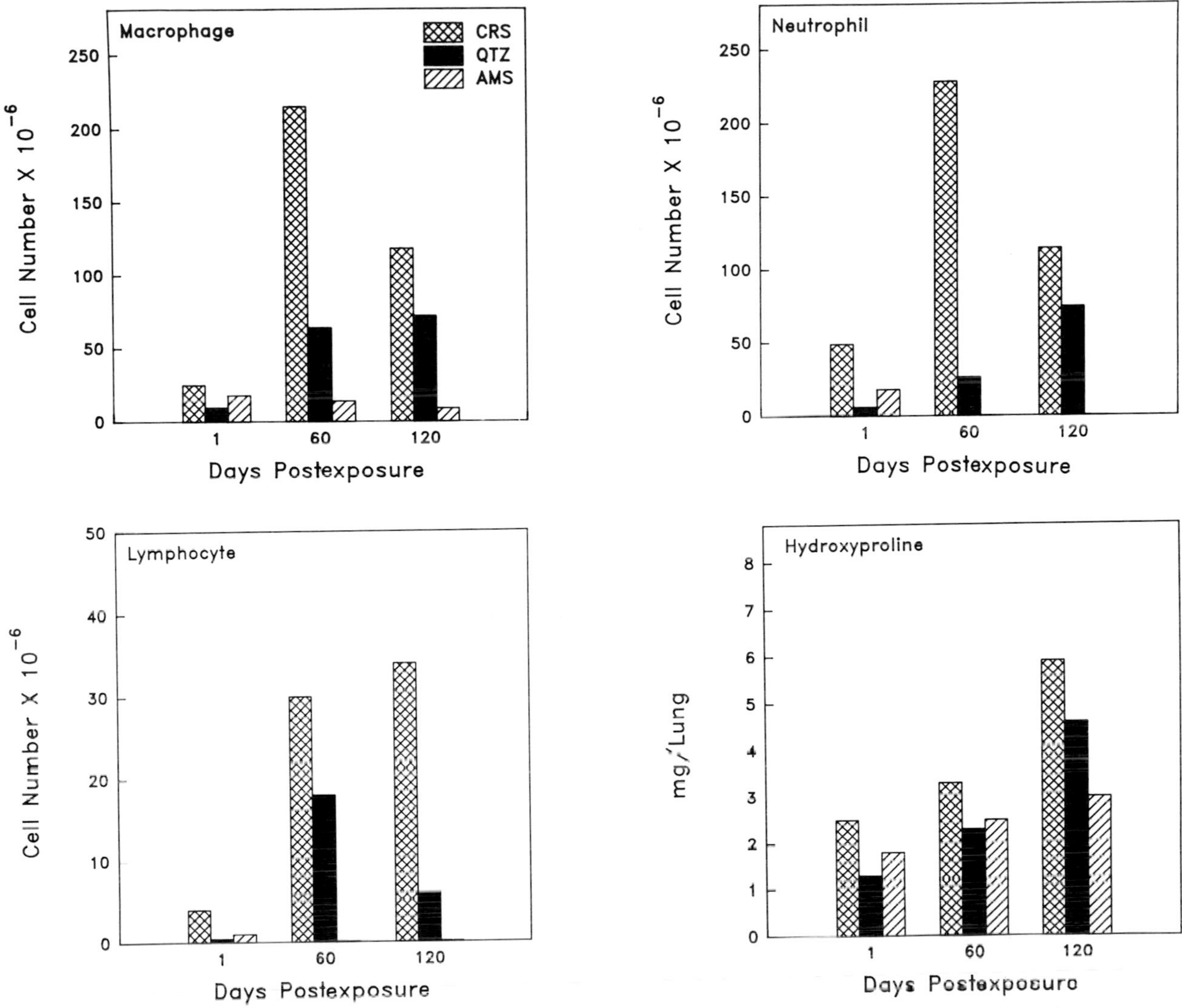

Figure 3 Comparison of recoveries of bronchoalveolar lavage cells and tissue hydroxyproline content in rats exposed to amorphous silica (AMS), quartz (QTZ), and cristobalite (CRS). Levels of hydroxyproline in unexposed rat lungs were the same as in AMS exposed.

cluding collagenase and elastase) which can degrade connective tissue. Reactive oxygen species such as superoxide may be generated and lead to oxidant-induced lung injury. Stimulated macrophages can also generate toxic oxygen species as well as proteolytic enzymes.

B. Activation of Lung Immune and Inflammatory Cells

Of more importance than direct toxicity in developing silicosis is activation of macrophages. Macrophages are a rich source of a variety of biologically active mediators that have diverse effects on lung cells and which thereby influence the progression of pulmonary disease. These include chemoattractants for neutrophils, monocytes, and lymphocytes. Cells recruited to the lung in turn secrete factors which affect the functions of other lung cells. These cytokines include interleukins (particularly IL-1β), transforming growth factor-β (TGF-β), tumor necrosis factor-α (TNF-α), platelet-derived growth factor (PDGF), and interferon gamma.

C. Cell–Cell Interactions

1. Secretion and Function of Biologically Active Mediators

Ingestion of crystalline silica by pulmonary macrophages leads to a sequence of events culminating in activation of the macrophage which in turn results in modulation of the activities of other cells, including neutrophils, T and B lymphocytes, other monocytes or macrophages, and fibroblasts and possibly epithelial cells. The cellular and biochemical events most likely include numerous cell types and the mediators that are products of these cells all interact to bring about the injury and remodeling processes that occur during the development of silicosis. It is beyond the scope of this article to cover all of these possible cell–cell and cell–mediator reactions; therefore, the discussion will focus on a few cells and mediators that are considered of importance in silica-induced lung inflammation.

It has long been clear that lung macrophages play a key role in the development of chronic remodeling processes by virtue of their elaboration and secretion of cytokines and other bioactive molecules. Prior to our knowledge of specific cytokines and their functions, the secretion of "fibrogenic factors" by silica-exposed lung macrophages were suggested from numerous early studies. These unidentified macrophage factors enhance proliferation of fibroblasts and increased collagen production by fibroblasts, (key elements in the evolution of silicotic fibrosis). Since that time, numerous reports have indicated that macrophages stimulated *in vitro* or *in vivo* with various silicon dioxide particles secrete a wide variety of mediators capable of modulating proliferation and function of different lung cells. In some of these studies, silica-exposed macrophages have also been shown to secrete factors that inhibit fibroblast proliferation and collagen synthesis. The balance between inhibitory and stimulatory factors secreted by silica-exposed macrophages is modulated by the dose of silica, duration of exposure, and other factors.

The enhanced secretion of cytokines by silica-exposed macrophages is of major importance in the development of inflammation as these highly active mediators initiate and prolong a cascade of cellular and biochemical events that lead to lung remodeling and, subsequently, fibrotic disease. Among these important cytokines are platelet-derived growth factor (PDGF), interleukins-1 and 6 (IL-1, IL-6), tumor-necrosis factor-α (TNF-α), and transforming growth factor β (TGF-β). Because of the proximity of the interstitial macrophages to target cells such as fibroblasts, production and secretion of cytokines by interstitial macrophages could have great influence on the pulmonary remodeling process in silicosis. Conversely, alveolar macrophage-derived cytokines may exert a greater effect upon the alveolar epithelium, thus promoting proliferation and differentiation of type II pneumocytes, a prominent feature of silicotic lung pathology. In this regard, an important finding shows that macrophage-derived factors are mitogenic for alveolar type II cells.

Interleukin-1 (IL-1) is a potent mediator whose synthesis is increased in macrophages as a result of exposure to *in vitro* and *in vivo* exposure to silica. IL-1 promotes T-lymphocyte activation and adherence. T-lymphocytes are induced by IL-1 to secrete interleukin-2 (IL-2) which expands the T-lymphocyte population. The activated lymphocytes in turn secrete a variety of mediators that affect macrophage function. For example, interferon gamma secreted by lymphocytes stimulates macrophages to release increased amounts of mitogenic factors for fibroblasts. Activated lymphocytes secrete chemotactins which then recruit more monocytes/macrophages to the lung. The involvement of the lymphocyte and immune system is evidenced by elevated immunoglobulin levels in serum of human silicosis patients.

Another secretory product of macrophages, tumor necrosis factor (TNF-α), is capable of causing lung tissue injury. Additionally, TNF-α induces IL-1 and prostaglandins in macrophages. Elevated levels of this potent mediator have been reported in cells of individuals having coal worker's pneumoconiosis as well as in experimental animals exposed to crystalline silica.

Leukotrienes (e.g., LTB-4) secreted by macrophages stimulate directed migration of neutrophils. Experiments have shown that normal lung macrophages incubated *in vitro* with silica release increased amounts of neutrophil chemotactic factor.

A hallmark feature of experimental silicosis is increased lung collagen. This is evidenced by increased hydroxyproline incorporation and increased deposition of collagen fibers as seen in trichrome-stained tissue sections. The major cellular source of interstitial collagen is the lung fibroblast. It has long been known that macrophages secrete factors that promote proliferation of fibroblasts. These factors include TGF-β, epithelial growth factor (EGF), and platelet derived growth factor (PDGF). Localized activity of these macrophage-derived cytokines may be in part responsible for the increased numbers of fibroblasts that are seen in lesions. Macrophages may also contribute to down-regulation of fibrotic development by production of factors that inhibit fibroblast proliferation, such as prostaglandins. PGE_2 has been shown to inhibit proliferation of fibroblasts and collagen production.

The prolonged presence of silica in exposed lungs providing a renewable source of stimulation to cells, and the interplay of all of these various mediators and different cell types may account for the chronic nature of silicotic lung disease.

Related Articles: Dust Particles: Occupational Considerations; Inorganic Mineral Particulates in the Lung.

Bibliography

Absher, M. P., Trombley, L., Hemenway, D. R., Mickey, R. M., and Leslie, K. O. (1989). Biphasic cellular and tissue response of rat lungs after eight-day aerosol exposure to the silicon dioxide cristobalite. *Amer. J. Pathol.* **134,** 1243–1251.

Allison, A. C., Harington, J. S., and Birbeck, M. (1966). An examination of the cytotoxic effects of silica on macrophages. *J. Exper. Med.* **124,** 141–154.

Bégin, R., Cantin, A., and Massé, S. (1989). Recent advances in the pathogenesis of mineral dust pneumoconioses: Asbestosis, silicosis and coal pneumoconiosis. *Euro. Resp. J.* **2,** 988–1001.

Brown, G. P., Monick, M., and Hunninghake, G. W. (1988). Fibroblast proliferation induced by silica-exposed human alveolar macrophages. *Amer. Rev. Resp. Dis.* **138,** 85–89.

Craighead, J. E. (1988). "Diseases Associated with Exposure to Silica and Nonfibrous Silicate Minerals." Silicosis and Silicate Disease Committee, National Institute for Occupational Safety and Health. Archives of Pathology and Laboratory Medicine, Vol. 12, pp. 673–720.

Davis, G. S. (1986). The pathogenesis of silicosis. State of the art. *Chest* **89,** 166S–169S.

Kelley, J. (1990). Cytokines of the lung. State of the art. *Amer. Rev. Resp. Dis.* **141,** 765–788.

Miller, B. E., and Hook, G. E. R. (1990). Hypertrophy and hyperplasia of alveolar type II cells in response to silica and other pulmonary toxicants. *Environ. Health Persp.* **85,** 15–23.

Mohr, C., Gemsa, D., Graebner, C., Hemenway, D. R., Leslie, K. O., Absher, P. M., and Davis, G. S. (1991). Systemic macrophage stimulation in rats with silicosis: Enhanced release of tumor necrosis factor-α from alveolar and peritoneal macrophages. *Resp. Cell Molec. Biol.* **5,** 395–402.

Schmidt, J. A., Oliver, C. N., Lepe-Zuniga, J. L., Green, I., and Gery, I. (1984). Silica-stimulated monocytes release fibroblast proliferation factors indentical to Interleukin-1: A potential role for interleukin-1 in the pathogenesis of silicosis. *J. Clin. Inves.* **73,** 1462–1472.

Sibille, Y., and Reynolds, H. Y. (1990). Macrophages and polymorphonuclear neutrophils in lung defense and injury. State of the art. *Amer. Rev. Resp. Dis.* **141,** 471–501.

Soil Decontamination

Donald L. Sparks
University of Delaware

Glossary

Biodegradation Process whereby naturally occurring soil microorganisms such as fungi and bacteria degrade soil contaminants.

***In situ* remediation techniques** Methods employed in place at the actual contaminated site.

***In situ* volatilization** *In situ* remediation method that involves the injection or inducement of a draft fan which creates an air flow through soil (e.g., through a slotted or screened pipe) so that air can flow but entrainment of soil particles is restricted. Some treatment (e.g., activated carbon) is employed to recover the volatilized contaminants.

Non–*in situ* remediation techniques Methods where the soil is removed (usually by excavation) and then treated on-site or hauled off-site and then treated.

Vitrification An *in situ* remediation technique where soil is solidified with an electric current. During the process, the contaminants are volatilized, resulting in their destruction and immobilization.

***CONTAMINATION OF SOILS** with inorganic and organic materials is a problem not only in the United States but throughout the world. Before using any remediation technique, one must have a thorough knowledge of the contaminated soil and site. Both* in situ *methods (including volatilization, biodegradation, leaching and chemical reaction, vitrification, isolation/containment, and passive remediation) and non–*in situ *techniques (i.e., land treatment, thermal treatment, asphalt incorporation, solidification/stabilization, chemical extraction, and excavation) can be employed to decontaminate soils. In many cases, a combination of remediation techniques will increase the effectiveness and perhaps decrease the cost of decontamination.*

I. Introduction

One of the most pressing and challenging problems facing our society is environmental quality. Hardly a day passes that a major environmental problem is not publicized in newspapers, magazines, scientific journals, or broadcast on radio and television. Our nation (and indeed the world) is extremely concerned about a number of environmental quality issues including: air pollution; groundwater and surface water contamination with inorganics (such as phosphates, nitrates, chromium, cadmium, lead, arsenic, mercury and other heavy metals), with organic contaminants (such as pesticides and industrial pollutants), and with radioactive wastes; the greenhouse effect; disposal of solid waste; land application of sludge; acid rain; and food safety.

Many of the inorganic and organic contaminants that are of concern are either applied to, or in various ways, interact with soil. The sources of these contaminants are many and can be linked to agricultural practices, industrial activities, and urban populations.

Our government has sought to ameliorate the problem of soil contamination through the passage of two major pieces of legislation: the Resource Conservation and Recovery Act of 1976 which regulates the generation, transport and disposal of hazardous/toxic wastes; and, the Comprehensive Environmental Response, Compensation, and Liability Act of 1980 (CERCLA) or the "Superfund" legislation. Soil and water contamination is a major problem that must be effectively dealt with if we are to remediate and preserve our environment.

There is immense concern over drinking water contamination with inorganic and organic pollutants. Approximately 50% of the U.S. population is dependent on groundwater for its drinking water sources, and in rural areas this figure may be as high as 95%. A major source of soil and water contamination is the estimated 2.5 to 3 million underground storage tanks that contain petroleum and hazardous materials. Petroleum contains a number of potential organic contaminants (Table I). Some surveys have estimated that perhaps 30% of these tanks leak because of corrosion. Other sources of contaminents include seepage from landfills, illegal dumps, unlined pits, ponds or lagoons, and spills from transport accidents.

Obviously, a major player in evaluating soil contamination and decontamination is the soil itself. Unfortunately, often regulators and professionals do not properly understand the chemical, mineralogical, and physical properties of soils.

To mitigate soil contamination and to properly and effectively decontaminate soils, one must realize that soils are extremely heterogeneous and thus, contaminants react and leach quite differently in different soils. Factors such as adsorption, desorption, permeability, volatilization, and degradation, which vary widely in different soil types, must be carefully assessed. Thus, soil scientists should be involved in all steps of soil decontamination.

In this article, a number of aspects of soil decontamination will be discussed including characterization of the contamination site, *in situ* and non–*in situ* remediation techniques, the effectiveness and a comparison of the remediation methods, and a cost analysis of the remedial technologies.

Table I Common Constituents of Petroluem Products

Gasoline and fuel oils	Heavy oils and waste oils
Benzene	Benz(*a*)anthracene
Ethylbenzene	Benzo(*a*)pyrene
n-Heptane	Naphthalene
Pentane	Phenanthrene
n-Hexane	
1-Pentene	
O-Xylene	
Toluene	
Phenol	

Source: Fleischer *et al.* (1986). Reproduced with permission.

II. Characterization of Contamination Site

Before soil decontamination can be effected, the contamination site should be carefully evaluated. A background search can assist in (1) determining the objectives of the decontamination process, (2) providing information that will determine which remediation technique is employed, (3) understanding the fate and transport mechanisms of the pollutants in the soil, (4) determining the risks of the contaminated soil to human and animal health and to the environment, and (5) providing a thoughtful comprehension of remedial goals based on local and regional land use.

Characterization of the site should include the use of site-specific data including title searches, site plans, aerial photos, site records, utility records, and interviews with persons living close to or working at the site. Local information including the current and historical use of land in the area, other possible contamination sources, information about wells that are closeby, surface water bodies, subways, basements, underground utilities, schools, and hospitals should be gathered. On a regional basis, one should know details about the hydrogeology, topographic maps, well records, and earlier environmental studies. Such information should provide one with data on the depth to groundwater, direction of groundwater flow, and types of geological material constituting the unsaturated and saturated zones.

Of course, the soil itself must also be carefully analyzed and understood for the most effective decontamination technique to be chosen. The physical, chemical, and mineralogical properties of the soil at the contaminated site should be analyzed at multiple depths. Properties that should be assessed include particle-size analysis, hydraulic conductivity, permeability, shrink–swell potential, pH, exchange and adsorption capacity, organic carbon and oxide (aluminum, iron, and manganese oxides) content, surface area, conductivity, and the clay mineral suite.

Once the above information is gathered, techniques can be employed to ascertain the degree of soil contamination. For example, ground-penetrating radar can be used to locate storage tanks. A soil vapor survey can be employed to determine the overall soil and water contamination around the tanks. The soil itself can also be analyzed by taking cores and determining the levels of inorganic and organic contaminants. Monitoring wells can also be installed so that water samples can be analyzed for contaminants, hydraulic gradient can be determined, and aquifer characteristics such as hydraulic conductivity can be understood.

Once a comprehensive characterization of the contamination site is completed, one has to determine the objectives of the soil remediation. These could include: diminishing or eliminating risks to human and animal health and to the environment; decreasing interference with ordinary daily plans and activities; and attempting to best clean up the contaminated soil with the least amount of time and money.

These objectives can be attained by doing nothing, partially cleaning up the site, or totally decontaminating the site. The approach one uses is greatly influenced by the characterization data that were mentioned earlier.

III. Soil Remediation Techniques

There are numerous *in situ* and non–*in situ* techniques that can be used to decontaminate soils (Table II). These methods are in various states of development and often a combination of techniques must be used to effectively clean up soils.

Table II Remedial Technologies

Technology	Exposure pathways[a]	Applicable petroleum products[b]	Advantages	Limitations	Relative costs
In situ					
Volatilization	1–7	1,2,4	Can remove some compounds resistant to biodegradation	VOC's only	Low
Biodegradation	1–7	1,2,4	Effective on some nonvolatile compounds	Long-term timeframe	Moderate
Leaching	1–7	1,2,4	Could be applicable to wide variety of compounds	Not commonly practiced	Moderate
Vitrification	1–7	1,2,3,4		Developing technology	High
Passive	1–7	1,2,3,4	Lowest cost and simplest to implement	Varying degrees of removal	Low
Isolation/Containment	1–7	1,2,3,4	Physically prevents or impedes migration	Compounds not destroyed	Low to moderate
Non–*in situ*					
Land treatment	1–7	1,2,3	Uses natural degradation processes	Some residuals remain	Moderate
Thermal Treatment	1–6	1,2,3,4	Complete destruction possible	Usually requires special facilities	High
Asphalt Incorporation	1–6	1,2	Use of existing facilities	Incomplete removal of heavier compounds	Moderate
Solidification	1–6	1,2,3,4	Immobilizes compounds	Not commonly practiced for soils	Moderate
Groundwater Extraction and Treatment	1–6	1,2,4	Product recovery, groundwater restoration		Moderate
Chemical Extraction	1–6	1,2,3,4		Not commonly practiced	High
Excavation	1–6	1,2,3,4	Removal of soils from site	Long-term liability	Moderate

Source: Preslo *et al.* (1988). Reproduced with permission.

[a] Exposure pathways: 1, vapor inhalation; 2, dust inhalation; 3, soil ingestion; 4, skin contact; 5, groundwater; 6, surface water; and 7, plant uptake.

[b] Applicable petroleum products: 1, gasolines; 2, fuel oils (#2, diesel, kerosenes); 3, coal tar residues; and 4, chlorinated solvents.

A. In Situ *Methods*

As the name implies, these methods are employed in place at the actual contaminated site. Soil need not be excavated and thus exposure pathways are greatly minimized. The only pathways that could result are those streams produced by using a specific *in situ* method.

1. Volatilization

In situ volatilization (ISV) creates mechanical drawing or air venting through the soil. This method involves the injection or inducement of a draft fan which creates an air flow through soil, via a slotted or screened pipe so that air can flow but entrainment of soil particles is restricted, and some treatment, such as the use of activated carbon, is employed to recover the volatilized contaminants. The latter will ensure that air emissions are reduced.

This *in situ* technique has been effectively used to remove compounds that are resistant to biodegradation; its cost is low, but its use is limited to volatile organic carbons (Table II). The effectiveness of this method depends on the characteristics of the contaminated site including soil permeability, porosity, clay and moisture content, temperature, and other factors. Thus, a thorough characterization of the soil is paramount before this method is employed.

2. Biodegradation

The use of biodegradation as a means to decontaminate soil has been successful and offers promising possibilities. *In situ* biodegradation involves the enhancement of naturally occurring microorganisms by stimulating their number and activity. The microorganisms then help in degrading soil contaminants. The potential for biodegradation of various organic pollutants in unconfined aquifers is given in Table III.

A number of genera of bacteria (about 22) and fungi (about 31) can degrade hydrocarbons. Some researchers have found that bacteria are more effective on a greater number of hydrocarbons and at higher hydrocarbon concentrations than fungi. However, fungi are more persistent and can survive under more undesirable environmental conditions than bacteria.

A number of soil, environmental, chemical, and management factors affect biodegradation of contaminants in soils (Table IV). Among soil factors, the moisture content is extremely important. Aerobic conditions are usually required for effective degradation of contaminants. Since oxygen transfer is so important in biodegradation, the soils should be permeable. Mixing the soil can enhance the effectiveness of microbes in degrading soil pollutants. Biodegradation is also enhanced at higher pH. Saturation of between 50 and 80% of the water-holding capacity is best.

Environmental factors, including temperature and the microbial community, also affect biodegradation. Degradation of petroleum products, for example, takes place over a wide range of temperatures. Biodegradation has occurred at temperatures as low as −1.1°C and as high as 40°C. Biodegradation generally increases with increases in temperature because biological activity is also enhanced. Of course, the type of microbial community also affects the degree of degradation. A microbe may be effective in degrading one type of contaminant, but not another. Moreover, microbes may be effective in degrading one form of a specific pollutant but not another. For example, *Corynebacteria* are effective in degrading polychlorinated biphenyls (PCBs) in the 4,4′- positions while *Pseudomonas* were not effective. *Pseudomonas* can degrade 2,5,2′,5′-tetrachlorobiphenyl but *Corynebacteria* cannot.

The availability of nutrients to the microbes is also necessary for contaminant biodegradation. For example, if the soil contaminant is a hydrocarbon, carbon will be available in high quantities, but nitrogen and phosphorus levels may be low. Thus, the addition of these nutrients may stimulate biodegradation of hydrocarbons.

3. Leaching and Chemical Reaction

This remediation technique involves leaching the in-place soil with water, often mixed with a surfactant, to remove the contaminants. The leachate is then collected, downstream of the site, using a collection system for treatment and/or disposal. This technique has not been widely practiced in part because ways to separate the surfactant from the pollutant are not well perfected. Additionally, since large amounts of water may be necessary to effectively remove the contaminants, the waste stream is large and disposal costs can be excessive.

A number of years ago, I was involved in a project with the Bikini Atoll Rehabilitation Committee. We were determining ways that cesium-contaminated soils in the Marshall Islands could be remediated. Excavation of topsoil was effective in removing the cesium so that plants could not take it up. However, the exorbitant costs in excavating, and the problem of having to dispose of the contaminated soil, meant that excavation was not a good option. We did,

Table III Potential for Biodegradation of Selected Organic Pollutants in Unconfined Aquifers

Class of compound	Aerobic water concentration of pollutant [μg/l (ppb)] 100	10	Anaerobic water
Halogenated aliphatic hydrocarbons			
Trichloroethylene	none	none	possible[a]
Tetrachloroethylene	none	none	possible[a]
1,1,1-Trichloroethane	none	none	possible[a]
Carbon tetrachloride	none	none	possible[a]
Chloroform	none	none	possible[a]
Methylene chloride	possible	improbable	possible
1,2-Dichloroethane	possible	improbable	possible
Brominated methanes	improbable	improbable	probable
Chlorobenzenes			
Chlorobenzene	probable	possible	none
1,2-Dichlorobenzene	probable	possible	none
1,4-Dichlorobenzene	probable	possible	none
1,3-Dichlorobenzene	improbable	improbable	none
Alkylbenzenes			
Benzene	probable	possible	none
Toluene	probable	possible	none
Dimethylbenzenes	probable	possible	none
Styrene	probable	possible	none
Phenol and alkyl phenols	probable	probable	probable[b]
Chlorophenols	probable	possible	possible
Aliphatic hydrocarbons	probable	possible	none
Polynuclear aromatic hydrocarbons			
Two and three rings	possible	possible	none
Four or more rings	improbable	improbable	none

Source: Wilson and McNabb (1983).
[a] Possible, probably incomplete.
[b] Probable at high concentration.

however, find that leaching the soil with large amounts of water high in sodium and magnesium (similar to seawater), was an effective way to remove the cesium from topsoil and subsoil.

The effectiveness of *in situ* leaching as a means of decontaminating soil also depends on the permeability, porosity, homogeneity, texture, and mineralogy of the soils. These factors will largely determine the degree of desorbability of the contaminant from the soil and the leaching rate of contaminants through the soil.

4. Vitrification

In situ vitrification is a process whereby soils are vitrified with an electrical current. During this process, the contaminants are volatilized, resulting in their destruction and immobilization. This technique was first tested in 1980 and originally developed to stabilize radioactive wastes *in situ*. It has been estimated that vitrification will immobilize contaminants for periods of up to 10,000 years. A large amount of electricity is necessary to use this method, so its cost is high.

5. Isolation/Containment

With this method, contaminants are held in place by installing subsurface physical barriers such as clay liners and slurry walls, to minimize lateral migration. Recently, there has been research on the application of surfactants to clay material used in liners. The adsorption of surfactants to the clay surface greatly enhances the sorption of organic pollutants. Thus, it would appear that diffusion of contaminants through clay liners could be minimized by first ap-

Table IV Factors That Influence Biodegradation in Soils

Soil	Enviromental	Chemical	Management
Moisture content	Temperature	Type of hydrocarbon available	Availability of nutrients
pH	Microbial community	Concentration	
Oxidation/reduction potential			
Porosity/permeability			

Source: Preslo *et al.* (1988). Reproduced with permission.
Notes:
1. It is extremely important to have a microbial community present that is capable of degrading the target compound. Most *in situ* biodegradation schemes make use of existing microbial populations; however, attempts have been made to supplement these populations with additional organisms or engineered organisms.
2. Chemical concentration has an important impact on biodegradation capability. Removal of any recoverable product that saturates the pore spaces or floats on the water table is recommended because hydrocarbons in the pure phase are degraded at a significantly lower rate.

plying surfactants to the clays. Another isolation/containment technique is to install surface caps over contaminated soil to reduce surface infiltration.

6. Passive Remediation

In passive remediation, the least costly and simplest technique, natural processes are allowed to occur which may decontaminate the site, without external influences. Factors that influence the natural processes for passive remediation are given in Table V. These processes rely upon a number of parameters including biodegradation, adsorption, volatilization, leaching, photolysis, soil permeability, groundwater depth, infiltration, and nature of the contaminant.

B. Non–In-Situ *Methods*

With *non–in situ* techniques, the soil is decontaminated by removing it, usually through excavation, and then treating it on-site or it is transported off-site and then treated. Consequently, exposure related to hauling and transport of the contaminated soil are of concern.

1. Land Treatment

In this technique, the contaminated soil is excavated and spread over a land area so that natural processes such as volatilization, aeration, biodegradation, and photolysis can occur to decontaminate the soil. With this technique, the land area is prepared by grading

Table V Factors That Influence the Natural Processes for Passive Remediation

Soil	Environmental	Chemical	Management
Water content	Temperature	Chemical composition	Depth of incorporation
Porosity/permeability	Wind	Concentration	
Clay content	Evaporation		Irrigation management
Adsorption site density	Precipitation Microbial		Soil management
pH	Community		Availability of nutrients
Oxidation/reduction potential			

Source: Preslo *et al.* (1988). Reproduced with permission.
Notes:
1. The factors listed above will not affect all natural processes in the same manner. For example, extremely high temperatures will enhance subsurface volatilization and inhibit biodegradation.
2. The effectiveness of possible remediation depends on complex relationships among the processes of volatilization, biodegradation, leaching, adsorption, and photolysis and is a function of all the factors listed above.

to remove rocks and other debris and the area is surrounded by berms to mitigate run-off. The pH of the soil is adjusted to about 7.0. At this pH, heavy metals are immobilized and the activity and effectiveness of microbes maximized. Fertilizers are also often added to provide adequate nutrients for the microbes. The contaminated soil is then spread on the site and often mixed with the other soil to enhance contact between the contaminant and microbes and aerobic conditions.

One should monitor the surface runoff and groundwater to measure losses of contaminants. The effectiveness of this remedial technique depends on the properties of the soil, site heterogeneity, temperature, and the type of microorganism population.

2. Thermal Treatment

With thermal treatment, the excavated decontaminated soil is exposed to high heats using one of about 12 thermal incinerators that are currently available. The increase in temperature causes breakdown of the contaminant in the presence of large amounts of oxygen. The released volatiles are then collected and moved through an afterburner and combusted or recovered via solvents.

The thermal incinerators include rotating kilns, fixed kilns, rotating lime or cement kilns, asphalt plants, fluidized bed incinerators, and low temperature strippers. Eight of the technologies are high temperature methods (>1000°F). Such techniques are very effective in destroying persistent contaminants like PCBs and dioxin. However, because high temperatures are involved, they are expensive. Lower temperature (maximum of 1000°F) technologies can be used effectively for more volatile hydrocarbons and other lower boiling point compounds. They require less energy and thus are less expensive than the high temperature methods.

3. Asphalt Incorporation

With this method, contaminated soils are placed in hot asphalt mixes. These mixtures are then used in paving. The asphalt and soil are heated while they are mixed. This effects volatilization of some of the contaminants. The other contaminants are then immobilized in the asphalt while it is cooling.

4. Solidification/Stabilization

Solidification/stabilization involves the addition of an additive to excavated, contaminated soil so that the contaminants are encapsulated. The mixture is then landfilled. Thus, the contaminants are not free to move alone. However, they are not destroyed. This method has been used to minimize pollution from inorganic contaminants.

5. Chemical Extraction

In this treatment, the excavated soil is mixed with a solvent, surfactant, or a solvent/surfactant mixture to remove contaminants. The solvent/surfactant and released contaminants are then separated from the soil. The soil is then washed or aerated to remove the surfactant/solvent. The surfactant/solvent is then filtered for fine particles and treated to remove the contaminants. This technique is quite expensive and is not often used. The effectiveness of this method depends on the properties of the soil, contaminants, and solvent/surfactants.

6. Excavation

With this method, the contaminated soil is removed and then must be disposed elsewhere, for example, in a landfill. Landfills usually contain liners that minimize leaching of contaminants or they should be located on sites where soil permeability is low. They require large amounts of land and often pose hazards for humans. Additionally, excavation and its ultimate disposal costs are expensive, and there are liability problems, safety concerns, odor production, and potential runoff and groundwater contamination problems.

IV. Effectiveness and Comparison of Techniques

Researchers have conducted an interesting study as Phase I of the EPA's Superfund soil treatability research program to evaluate the effectiveness of five remediation techniques in decontaminating synthetic soils containing varying levels of volatile and semivolatile organics and metals (Table VI). The surrogate soil (to represent Superfund site soils) was chosen after an extensive analysis to represent a typical Eastern U.S. soil, and was composed of: 30% by volume of clay (7.55% montmorillonite and 22.5% kaolinite), 25% silt, 20% sand, 20% topsoil, and 5% gravel.

The remediation techniques employed were: soil washing, chemical extraction, low temperature thermal treatment, high temperature thermal treatment, and solidification/stabilization. The thermal

Table VI Analytical Profile of Spiked Soils: Average Concentrations Found upon Total Waste Analysis (mg/kg)[a]

Analyte	Soil I	Soil II	Soil III	Soil IV
	(High organic, low metal)	(Low organic, low metal)	(Low organic, high metal)	(High organic, high metal)
Volatiles				
Acetone	4,353 (9)[b]	356 (8)	358 (2)	8,030 (2)
Chlorobenzene	316 (9)	13 (6)	11 (2)	330 (2)
1,2-Dichloroethane	354 (9)	7 (8)	5 (2)	490 (2)
Ethylbenzene	3,329 (9)	123 (8)	144 (2)	2,708 (2)
Styrene	707 (9)	42 (8)	32 (2)	630 (2)
Tetrachloroethylene	408 (9)	19 (8)	20 (2)	902 (2)
Xylene	5,555 (9)	210 (8)	325 (2)	5,576 (2)
Semivolatiles				
Anthracene	5,361 (9)	353 (7)	181 (3)	1,920 (3)
Bis(2-ethylhexyl) phthalate	1,958 (9)	117 (7)	114 (3)	646 (3)
Pentachlorophenol	254 (9)	22 (7)	30 (3)	80 (3)
Inorganics				
Arsenic	18 (10)	17 (7)	652 (4)	500 (4)
Cadmium	22 (8)	29 (6)	2,260 (2)	3,631 (2)
Chromium	24 (8)	28 (6)	1,207 (4)	1,314 (4)
Copper	231 (10)	257 (8)	9,082 (4)	10,503 (4)
Lead	236 (10)	303 (8)	14,318 (4)	14,748 (4)
Nickel	32 (10)	38 (8)	1,489 (4)	1,479 (4)
Zinc	484 (8)	642 (6)	31,871 (4)	27,060 (4)
Moisture, %	20 (7)	11 (7)	19 (3)	26 (2)

Source: Exposito (1989). Reproduced with permission.
[a] Values in parentheses indicate number of samples analyzed.

techniques reduced the organic contaminants by >99.6%, and the chemical extraction treatment was >90% effective on semivolatiles and >98% effective on volatiles. Soil washing was most effective in decontaminating the soils of metals (~93% reduction). Washing also reduced semivolatiles (87%) and volatiles (99%). Solidification was effective for metal decontamination.

V. Cost Analysis of Remedial Technologies

A very important consideration in selecting a remediation technique for decontaminating soils is the cost involved. Of course, the cost will depend on the size of the site, site-specific conditions, and the amount of contaminated soil. Table VII shows cost comparisons for the various remediation techniques.

VI. Summary

Soil decontamination is a complex and arduous problem that must be dealt with in the decades ahead. While a number of *in situ* and non–*in situ* remediation techniques are currently available for soil decontamination, they each have a plethora of advantages and drawbacks. In many cases, to most effectively decontaminate soil, a combination of techniques must be employed. Moreover, a thorough knowledge of the soil one is dealing with and a definitive characterization of the contaminated site is necessary. One also should consider the cost effectiveness of the remediation methods.

Related Articles: BIODEGRADATION OF XENOBIOTIC CHEMICALS; CYANIDE; ORGANIC MICROPOLLUTANTS IN LAKE SEDIMENTS; VOLATILE ORGANIC CHEMICALS.

Table VII Relative Cost Comparisons for Remedial Technologies[a]

Remedial technologies	Relative total cost[a]	Design assumptions[b]
In situ		
Volatilization	Low	7.63 meters (25 ft) centers for 8 venting pipes; no treatment for effluent gases.
Biodegradation	Moderate	Three extraction wells with infiltration galleries for injection; flow rate of 0.002572 m³/sec (40 gpm) through reactor.
Leaching	Moderate	Same assumptions as biodegradation except treatment unit differs.
Vitrification	High	Unit costs are based on a larger site (Pacific Northwest Laboratories, 1986).
Passive remediation	Low	Monitoring costs only; four monitoring wells with quarterly sampling of aromatic volatile hydrocarbon indicator compounds.
Isolation/containment	Low	Cap composed of liner, soil, and bentonite; no slurry wall.
	Moderate	Same cap with slurry wall.
Non–*in situ*		
Land treatment	Moderate	Purchased (not leased) equipment; on-site operation.
Thermal treatment	High	Leased mobile unit; on-site operation.
Asphalt incorporation	Moderate	Off-site operation; shipping costs are additional.
Solidification	Moderate	Leased equipment; 30% Portland cement, 2% sodium silicate.
Groundwater treatment	Low to Moderate	Moderately sized carbon unit or air stripper without effluent treatment.
Chemical extraction	High	Leased mobile unit; on-site operation.
Excavation and landfill	Moderate to High	Leased equipment; costs relative to landfill disposal fees and transportation costs.

Source: From Preslo *et al.* (1988). Reproduced with permission.

[a] Unit costs:
- Low = Less than $13.00/m³ ($10/yd³) of soil or 3,780 liters (1,000 gallons) of water.
- Moderate = 13 to $130/m³ ($10–$100/yd³) of soil or 3,780 liters (1,000 gallons) of water.
- High = Greater than $130/m³ ($100/yd³) of soil or 3,780 liters (1,000 gallons) of water.

[b] Site with dimensions of 30.5 m × 15.25 m × 6.1 m (100 ft × 50 ft × 20 ft) depth, a volume of 2,837.3 m³ (3,700 yd³) weighing 3,636.4 metric tons (4,000 tons) and 37,800,000 liters (10,000,000 gallons) of impacted groundwater. Depth to the water table is 6.1 meter (20 ft).

Bibliography

Esposito, P., Hessling, J., Locke, B. B., Taylor, M., Szabo, M., Thurnau, R., Rogers, C., Traver, R., and Barth, E. (1989). Results of treatment evaluations of a contaminated synthetic soil. *JAPCA* **39**, 294–304.

Fleischer, E. J., Noss, P. R., Kostecki, P. T., and Calabrese, E. J. (1986). Evaluating the subsurface fate of organic chemicals of concern using the SESOIL environmental fate model. *In* "Proc. of 3rd Eastern Regional Groundwater Conference of the National Water Well Association, July 29–31, 1986; Springfield, Mass.

Kerr, J. M., Jr. (1990). Investigation and remediation of VOCs in soil and groundwater. *Environ. Sci. Technol.* **24**, 172–173.

McDermott, J. B., Unterman, R., Brennan, M. J., Brooks, R. E., Mobley, D. P., Schwartz, C. C., and Dietrich, D. K. (1989). Two Strategies for PCB soil remediation: Biodegradation and surfactant extraction. *Environ. Progr.* **8**, 46–51.

McLean, M. E., Miller, M. J., Kostecki, P. T., Calabrese, E. J., Preslo, L. M., Suyama, W., and Kucharski, W. A. (1988). Remedial options for leaking underground storage tanks. *JAPCA* **38**, 428–435.

Preslo, L. M., Robertson, J. B., Dworkin, D., Fleischer, E. J., Kostecki, P. T., and Calabrese, E. J. (1988). Remedial technologies for leaking underground storage tanks. Lewis Publishers, Chelsea, Mich.

Wilson, J. T., and McNabb, J. F. (1983). Potential for biodegradation of selected organic pollutants in unconfined aquifers. *EOS* **64**, 505–507.

Sulfur Mustard

A. P. Watson and G. D. Griffin
Oak Ridge National Laboratory

Glossary

H Sulfur mustard agent; unstable; manufactured by the obsolete Levinstein process; bis(2-chloroethyl)sulfide; vesicant (blister) compound. CAS No. 505-60-2. Used in World War I, the Iran–Iraq War, and other conflicts. Lethal. Destroys epidermis, generates blisters, causes eye and pulmonary injury, and, at high doses, bone marrow damage and immunosuppression. A human carcinogen.

HD Distilled sulfur mustard agent; this purified form is more stable than agent H. Same composition, CAS No., and properties as for agent H above.

HT Plant-run vesicant agent mixture containing about 60% HD and <40% T; the mixture is approximately twice as persistent as HD. "T" is bis[2(2-chloroethylthio)ethyl]ether, CAS No. 63918-89-8. Same pathology as H.

Keratitis Inflammation of the cornea.

Sulfur mustard Group of vesicant agents (H, HD, HT) that comprises various formulations of bis(2-chloroethyl)sulfide. Produces skin blisters and damage to eyes and respiratory tract; human carcinogen; mutagenic in cellular assays.

Vesicant Blister agents; those chemical warfare agents (H, HD, HT, Lewisite) that produce vesicles or blisters on skin and damage both eyes and mucous membranes.

SULFUR MUSTARD *[bis(2-chloroethyl)sulfide; $C_4H_8Cl_2S$; CAS # 505-60-2] is one of a class of chemical warfare agents known as "vesicants" because of their blistering effect on skin. During World War I (WWI), exposed troops described the odor as a stench "like mustard" or garlic; hence, its common name. First noted for its unusual toxic properties by dye chemists in the late 1880s, it has also been referred to as "S-mustard" (to distinguish from nitrogen mustard), "Lost" or "S-Lost" (from the names of two chemists who suggested it be used as a war gas in WWI; Lommel at Bayer research laboratories and Steinkopf of Dahlem), "yellow cross" (for the identifying mark on WWI shells containing sulfur mustard agent), or Yperite (after the site of its first use as a warfare agent, Ypres, Belgium, 12 July 1917). Although commonly and inaccurately referred to as "mustard gas," the agent is liquid at room temperature. It is the principal vesicant component of warfare agents H (Levinstein mustard), HD (distilled mustard), and HT (a mixture of HD and a toxic stabilizing compound).*

Sulfur mustard produces skin blisters and damage to eyes and respiratory tract, can be acutely lethal at sufficiently high doses, is a cellular poison and mutagen, and is a recognized human carcinogen.

I. Agent Chemistry

The exact date of the first sulfur mustard synthesis is somewhat unclear due to limitations of 19th century analytical capability and documentation; the first report may be that of Despretz in 1822. An 1860 report by Niemann describes a delayed-effect vesicant oil as a reaction product of ethylene on a mixture of sulfur chlorides. At the time, this product was identified as the disulfide $[(C_2H_4)_2S_2Cl_2]$; however, the observed severe skin blistering, latent period of several hours, and subsequent slow healing are all characteristic of dermal exposure to undiluted sulfur mustard warfare agent. At about the same time, Guthrie published olefin investigations in 1859 and 1860 that described a compound thought to be $C_4H_4S_2Cl_2$ and produced from sulfur dichloride (SCl_2) or sulfur monochloride (S_2Cl_2) in reaction with ethylene (C_2H_4). The odor was "pungent," resembling that of "oil of mustard." The uncautious investigator noted destruction of the epidermis when the thin skin between the fingers and around the eyes was exposed to the "vapour." When the liquid was allowed to remain on the skin, Guthrie observed blister formation.

Significant quantities of relatively pure sulfur mustard ($C_4H_8Cl_2S$) were first described by Meyer in 1886 as a reaction product of sodium sulfide (Na_2S), ethylene chlorohydrin (C_2H_5ClO), and HCl gas. This process was the one eventually used by German war factories to fill the shells fired at Ypres. By the close of World War I (1918–1919), sulfur mustard was being manufactured and used as a warfare agent by German, French, U.S., and British forces.

Sulfur mustard has been more recently used in military operations of the Iran–Iraq conflict (1980–1988) and the Gulf War (1991). Its use was a topic at the Paris Conference on the Prohibition of Chemical Weapons in January 1989, and it underwent negotiation at the chemical convention talks in Geneva. The U.S. stockpile of sulfur mustard, currently stored at seven military installations in the continental United States, and one U.S. territory in the South Pacific, is under Congressional mandate for destruction by 2004.

The active ingredient in agents H and HD and a major component (approximately 60%) of agent HT is the same chemical compound, bis(2-chloroethyl) sulfide (CAS # 505-60-2; Table I). Throughout this article, the terms *mustard* or *mustard agent* will be used as a synonym for sulfur mustard agents. The chemical warfare agent H contains 70% sulfur mustard plus 30% sulfur impurities, and is manufactured by the unstable Levinstein process. The chemical warfare agent HD is distilled sulfur mustard from which impurities have been removed by distillation and washing. Agents H and HD will not be considered separately in the ensuing toxicological discussion.

Mustard has a garlic-like odor. It has significant volatility at ordinary temperatures, so that mustard vapor would be in the air immediately surrounding droplets of liquid mustard. Thus, the hazard of human contact is not only with droplets of liquid agent, but also with agent vapors. Because of its low aqueous solubility, mustard agent is very persistent in the environment (see Table I). However, it hydrolyzes readily, producing HCl and thiodiglycol ($C_4H_{10}O_2S$, CAS NO. 111-48-8). The production of a free chloride ion by this reaction is thought to be responsible for the acute corrosive effect of mustard when in contact with moist tissues such as the eye and respiratory tract.

HT is a product of a reaction that yields about 60% HD (described above), <40% T (bis[2(2-chloroethylthio)ethyl]ether, CAS NO. 63918-89-8), plus a variety of sulfur contaminants and impurities. It is very similar in appearance and biological activity to H/HD, but possesses greater toxicity and stability due to the presence of T, which lowers the freezing point and adds toxic properties to the mixture. Agent HT is also liquid at room temperature but is soluble only in organic solvents (Table I). Its poor water solubility makes it a persistent contaminant of soils and surfaces. Hydrolysis in water occurs only after prolonged boiling, while caustic alkalies hydrolyze HT readily.

From a military standpoint, one of mustard's most useful properties is its persistence. Droplets of agent released in an explosion or shell-burst could deposit on numerous surfaces and slowly evaporate, thus posing a risk from inhalation as well as dermal contact. Indeed, this very set of conditions was observed in WWI after mustard shelling. One reason for sulfur mustard's persistence is its characteristic freezing at moderate temperatures (13 to 15°C) (Table I); droplets or bulk quantities would thus be expected to remain where initially deposited during cool weather or under winter/arctic conditions. Under some meteorological conditions, bulk quantities of mustard agent spilled or splashed onto soil would not degrade for months.

II. Agent Toxicity

A. Acute Effects

Mustard agents are much less potent than nerve agents under comparable conditions of exposure. The human skin LD_{50} for VX, an organophosphate nerve agent, is 0.04 mg/kg; the comparable dose for H/HD is 100 mg/kg. The sulfur mustard inhalation LCt_{50} for unprotected adult personnel is 1500 mg-min/m^3, while the LCt_{50} for VX ranges between 30 and 36 mg-min/m^3. (Note that these Ct estimates are reasonably valid for only a short ex-

Table I Properties of Vesicant Chemical Warfare Agents

	H, HD	HT
Chemical name	Bis(2-chloroethyl)sulfide	Plant-run mixture containing about 60% HD and <40% "T" or bis-[2(2-chloroethylthio)ethyl]ether
Common name	Mustard; distilled mustard	Mustard
Chemical formula	$C_4H_8Cl_2S$	"T" = $C_8H_{16}Cl_2OS_2$
Chemical Abstract (CAS) No.	505-60-2	"T" = 63918-89-8
Molecular weight	159.1	"T" = 263.26
Description	Amber to dark brown liquid (H) oily, pale yellow liquid (HD)	Clear, viscous, yellowish liquid
Melting point	13–15°C	0–1.3°C
Boiling point	215–217°C	>228°C (not constant)
Vapor pressure (25°C)	0.08 mm Hg (H) 0.11 mm Hg (HD)	0.104 mm Hg
Density (liquid)	1.27 g/mL (25°C)	1.27 g/mL (25°C)
Volatility	920 mg/m^3 (25°C)	831 mg/m^3 (25°C)
Vapor density (relative to air, which is assumed to equal 1)	5.5	6.92
Solubility, water	0.68–0.92 g/L (25°C)	Insoluble
Solubility, other	Very soluble in organic solvents	Soluble in organic solvents
Biological activity	Blister agent	Blister agent

Sources: "Chemical Agent Data Sheets," vol. 1, U.S. Department of the Army Technical report EO-SR-74001, Edgewood Arsenal Special Report, Defense Technical Information Center, Alexandria, VA, 1974. "The Merck Index. An Encyclopedia of Chemicals, Drugs and Biologicals" (Windholz, M., S. Budavari, R., Biumetti, and E. Otterbein, eds). Merck and Co., Rahway, NJ, 1983.

trapolation time; e.g., 30 sec to 10 min.) Available hospital records from WWI and sketchy casualty reports from the recent Iran–Iraq conflict indicate mortality rates of 1 to 3% from acute sulfur mustard exposure to battlefield concentrations. Actual battlefield concentrations to which sulfur mustard victims were exposed have not been reported but may well have been greater than the estimated LCt_{50} of 1500 mg-min/m^3. Exposure estimates from a World War II (WWII) Japanese poison gas factory have suggested that mustard air concentrations between 50 and 70 mg/m^3 are acutely irritating and can produce most of the signs of mustard poisoning.

Warfare use of vesicants decreases the opponents' ability to fight by producing chemical burns on tissues that come into contact with either vapors, liquid droplets, or aerosols. Exposed skin surfaces, eyes, nose, throat, bronchial, and upper gastrointestinal tract are all at risk. The moist surfaces of perspiring skin, conjunctiva of the eye, airway mucosa, or mucous membranes preferentially absorb mustard agent and distribute it over a larger area. Thus, the unprotected eye is considered the most sensitive organ to the action of H-agents, and ambient temperature/humidity govern the degree of "casualty effect." Under hot, humid conditions when large areas of skin are likely to be wet with perspiration, much lower mustard concentrations generate debilitating effects (see Table II, "Incapacitating Dose").

The biological mode of action of mustard agent is that of an alkylating agent; individual cells are destroyed by the chemical reaction of mustard with cellular proteins, enzymes, and nucleic acids. Following initial tissue damage, various debilitating effects occur, such as the development of large, painful blisters that arise on exposed skin. There is usually a latency period of several hours before signs of toxicity appear, depending on dose (Table III). An individual exposed to blistering concentrations of agent is incapacitated, often for weeks, before returning to normal activity. Acute effects are disabling in the short term, and special care and resources are required to prevent subsequent infection of the skin, respiratory tract, and eyes. No acute effects are expected at the Surgeon General's recommended atmospheric control limits of

Table II Acute Toxic Effects of Vesicant Agents H/HD, HT, and T

Exposure route	H/HD	HT	T
	Inhalation LCt_{50} (mg-min/m^3)		
Human (estimated)	1,500[a,b]		~400[c]
Monkey	800[b]		
Goat	1,900[b]		
Dog	600[b,c]	100–200[b]	
Cat	700[b,c]		
Rabbit	1,025[b]	3,000–6,000[b]	
Guinea pig	1,700[b]	3,000–6,000[b]	
Rat	800–1,512[b,d]		
Mouse	860–1,380[b,d]	1,100 (10 d)[b]	
		820 (15 d)[b]	
	Percutaneous LCt_{50} (vapor) (mg-min/m^3); head protected, body exposed		
Human (estimated)	10,000[b]		
Monkey	13,000[b]		
Dog	7,700[b]		
Cat	8,700[b]		
Rabbit	5,000[b]		
Guinea pig	~20,000[b]		
Rat	~3,000[b]		
Mouse	3,400[b]		
	Skin LD_{50} (mg/kg body weight); applied as liquid		
Human (estimated)	100[b]		
Rabbit	100[b]		
Rat	9[e], 18[b]		
	15 (96 h mortality)[f]		
	194 (24 h mortality)[f]		
Mouse	92[b]		
	Intravenous LD_{50} (mg/kg body weight)		
Dog	0.2[b]		
Rabbit	~1.1–4.5[b]		
Rat	0.7–3.3[g]		
Mouse	3.3, 8.6[g]		
	Oral LD_{50} (mg/kg body weight)		
Rat	17[b]		
	Incapacitating dose ICt_{50} (mg-min/m^3)		
Human, percutaneous (masked)	2,000 (70–80°F)[b,h] to 1,000 (90°F)[i]	None established[b]	
Human, eyes	200[b]	None established[b]	
	Minimum effective dose, ED		
Human skin (blisters)	50 mg-min/m^3[b,j]	~3.5 mg/man[b]	4 mg/man[c]
		4 mg/man[c]	
Human eyes (marginal) conjunctivitis	12–70 mg-min/m^3[b]		
	30 mg-min/m^3[k] (60 min)		
(Reddening, no incapacitation)	70 mg-min/m^3[b]		
(Reddening, mild incapacitation)	90 mg-min/m^3[b]		
Rabbit eyes		Similar to HD[b]	

(*continues*)

Table II (*Continued*)

Exposure route	H/HD	HT	T
	No effect dose (mg-min/m³)		
Human eyes	<12[i]		
Human (estimated)	2 (≥90°F)[j]		
	Inhalation, lowest lethal dose (mg/m³)		
Human	150 (10 min)[l]		
	70 (30 min)[m]		
	Skin absorption, lethal (mg/kg body weight)		
Human	64[n]		

[a] HD exposures are cumulative within reasonable time limits (e.g., 30 sec to 10 min).

[b] From *Chemical Agent Data Sheets, Vol. 1.* U.S. Department of the Army Technical Report EO-SR-74001; Edgewood Arsenal Special Report, Defense Technical Information Center, Alexandria, VA, 1974.

[c] From "Chemical Warfare" (Robinson, J. P.), *Science J.* **4,** 33–40, 1967.

[d] Ranges of LCt_{50} values are summarized for all exposure times reported.

[e] From "Observations on the effects of mustard gas on the rat" (Young, L.) *Can. J. Res.* **E25,** 141–151, 1947.

[f] From "The protective effect of different drugs in rats poisoned by sulfur and nitrogen mustards," (Vojvodic, V., Z. Milosarljevic, B. Boskovic, and N. Bojanic), *Fund. Appl. Toxicol.* **5,** 5160–168, 1985.

[g] From "The intravenous, subcutaneous and cutaneous toxicity of bis(beta-chloroethyl)silfide (mustard gas) and of various derivatives," (Anslow, W. P., D. A. Karnofsky, B. V. Jager, and H. W. Smith), *J. Pharmacol. Expt. Therapeutics* **93,** 1–9, 1948.

[h] Incapacitating dose varies significantly with amount of perspiration on skin surface, which is, in turn, dependent on ambient temperature and humidity levels.

[i] From *Toxicology Basis for Controlling Levels of Mustard in the Environment* (McNamara, B. P., E. J. Owens, M. K. Christensen, F. J. Vocci, D. F. Ford, and H. Rozimarek). Edgewood Arsenal Special Publication EB-SP-74030, Department of the Army, Defense Technical Information Center, Alexandria, Va, 1975.

[j] Mild to moderate erythema is produced at ambient temperature of 90°F.

[k] From "Eye lesions induced by mustard gas" (Dahl, H., B. Glund, P. Vangstad, and M. Norn), *Acta Opthalmologica* **63**(Suppl 173), 30–31, 1985.

[l] From *Reclassification of Material Listed as Transportation Health Hazards* (Back, K. C., A. A. Thomas and J. D. MacEwen). Office of Hazardous Materials, Office of the Assistant Secretary for Safety and Consumer Affairs, Dept. of Transportation Publication No. PB-214270/1, TSA-2072-3, 1973.

[m] From "Multiple Bowen's disease observed in former workers of poison gas factory in Japan, with special reference to mustard gas exposure," (Inada, S., K. Hiragun, K. Seo and T. Yamura), *J. Dermatol. Tokyo* **5,** 49–60, 1978.

[n] From World Health Organization Technical Report Series, p. 30, 1970.

Table III Latency and Healing Times for Acute Sulfur Mustard Exposure Effects

	Latent period		
Clinical sign[a]	Erythema	Blisters	Healing time
Erythema	12 h	—	5–10 d
Small blisters	6–12 h	18–24 h	4 wk
Large blisters	2–6 h	3–6 h	2–3 mo
Necrosis	6 h	None	Several mo

Source: From *Pharmakologie und Klinik Synthetische Gifte* (Stade, K.), Militar Verlag, Berlin, 1971 (as cited in Medema 1986).

[a] Indicates successively greater doses.

3×10^{-3} mg/m³ for 8-h/d workplace exposures and 1×10^{-4} mg/m³ (72-h time-weighted average concentrations) for the general population.

B. Delayed Effects

Because of the ability of mustard to covalently bind with a variety of biological molecules, the resultant biological damage could have considerably delayed consequences. Delayed effects from HD exposure include keratitis or keratopathy of the eye; respiratory diseases other than respiratory cancer; mutagenesis, particularly in relation to reproductive effects; and carcinogenesis.

Eye damage suffered by most soldiers exposed to mustard in WWI was temporary in nature, and no permanent effects were observed at the time. In a smaller number of soldiers, where the eye was probably exposed to higher vapor concentrations or liquid droplets, a permanent, relapsing keratitis (delayed keratopathy) developed. This chronic condition is characterized by recurring erosion and ulceration of the cornea, eventual vision impairment, and, in some cases, blindness. The latent period for this delayed effect has been observed to range from 8 to 40 years after apparent recovery from the initial acute injury. In some cases the condition has been observed to relapse for decades.

Evidence from occupational and wartime exposures indicates that, under appropriate conditions, mustard agent can induce long-term respiratory damage. Reported ailments range from asthma-like conditions to a severe chronic emphysematous bronchitis, and secondary infections such as bronchopneumonia and tuberculosis. Medical records of weapons plant workers exposed to toxic vesicant concentrations under wartime conditions are the source of the best-characterized data, although there is little evaluation of workplace atmospheres or dose-response relationships. Between 1929 and 1945, the Japanese Army operated a chemical warfare agent manufacturing facility on Okuno-jima, an island of the Inland Sea. At peak capacity, this facility produced Lewisite (50 tons/mo), mustard (450 tons/mo), hydrocyanic acid (50 tons/mo), diphenylcyanarsine (sneezing gas; 50 tons/mo), chloroacetophenone (tear gas; 25 tons/mo), and phosgene (unreported tons/mo). During the period of maximum production (1937–1942), approximately 1000 individuals were employed throughout the facility. Interview data indicate that, given the minimal level of industrial hygiene practice in use at the time, multiple-agent exposures were common while agent-specific exposure occurred rarely. In the mustard production areas, atmospheric concentrations of mustard (estimated at 50 to 70 mg/m^3) were sufficient to produce most signs of acute mustard toxicity in workers (see Table II).

It seems clear that high-level exposures to mustard agent can produce permanent respiratory damage, which may take the form of a chronic bronchitis and which can also predispose affected individuals to other respiratory infections (e.g., pneumonia, tuberculosis). It must be emphasized, however, that the doses of agent capable of producing these respiratory effects are not well defined. An estimate of atmospheric mustard concentration in areas of gas shell bursts during WWI is 19 to 33 mg/m^3; the period of exposure necessary to induce effects at this concentration is not documented. In the case of the Japanese workers, wartime worker safety provisions were minimal at best, and acute toxic effects from exposure to mustard and other chemical warfare agents were frequently reported. Some workers died as a result of acute gas poisoning during plant operation. Another factor to be considered is that occupationally exposed individuals received daily doses of mustard as well as other warfare agents over a period of years.

Because of its highly reactive chemical nature, mustard agent can react with DNA to induce mutagenesis in a wide variety of test organisms (see Table IV). The content of Table IV is not intended to be encyclopedic; many other studies could be cited in a summary of known mutagenic activity for this agent. Some biological assays have also demonstrated heritable genetic effects, implying that sulfur mustard produces damage in parental germ cells. Evidence from both human and animal studies regarding the reproductive toxicity of mustard is generally negative, except for evidence of dominant lethal mutations in exposed male rats (see Table IV). In 1985, Yamakido and colleagues performed health examinations of male and female former workers (n = 325) from the Okuno-jima war factory, their spouses (n = 226) and offspring (n = 456). The exam included an assessment of blood protein chemistry in an evaluation of possible genetic effects arising from occupational mustard or Lewisite exposures. The investigators concluded that no evidence of sulfur-mustard-induced mutations could be detected in any member of the study group. Later evaluation of these data indicates that the sample population evaluated in the Yamakido study was too small to allow detection of any agent-induced mutations among the offspring of exposed war factory workers. A 1970 study by Hellman of male workers employed in German warfare agent factories during WWII noted evidence of sex-linked mutations in that the sex ratio was increased from normal among offspring of the 134 fathers evaluated. Unspecified impairment of various stages of spermatogenesis was also observed. It is difficult to know how much significance should be attached to these latter observations, since German war factory workers were simultaneously exposed to unreported concentra-

Table IV Delayed/Latent Effects Observed for Sulfur Mustard Agents and T

Exposure regimen	H/HD response (dose)	T response
	Carcinogenicity	
Mouse–inhalation (15-min exposure)	Pulmonary tumors[a] (~1590 mg/m^3)[b]	
Mouse–subcutaneous injection (6 weeks)	Fibrosarcomas at injection[c] site (~6 mg/kg)[d]	
Mouse–intravenous (6 d)	Pulmonary tumors[e] (~3–4 mg/kg)[d]	
Rat–inhalation (≥3-month exposure)	Skin tumors[f] (0.1 mg/m^3)[g]	
Human–inhalation and skin deposition	Respiratory tract tumors, skin cancers[h] (unknown)[i]	
Mouse–skin (278 d)	Negative[i] (2 mg total)	
Mouse–skin	Negative[k] (dose unknown)	
	Mutagenicity	
S. typhimurium (Ames)	Positive[l] (1–50 μg/plate)	
Neurospora crassa (mold) (30-min exposure)	Specific locus mutation[m] (200 μmol/L)[n]	
Saccharomyces cerevisiae (yeast)	DNA damage[o] (500 μmol/L)[n]	
Drosophila melanogaster (fruit fly) (parenteral injection)	Specific locus mutation[p] (45 pmol/fly)	
Drosophila melanogaster (vapor—5-min exposure)		Production of sex-linked lethal mutations[q]
Drosophila melanogaster (vapor—15-min exposure)	Cytogenetic damage[q] visible mutations, deletions, inversions (dose unknown)[r]	
Drosophila melanogaster (vapor—15-min exposure)	Sex chromosome loss and nondisjunction[q] (dose unknown)[r]	
Drosophila melanogaster (parenteral injection)	Sex chromosome loss[s] and nondisjunction (75 pmol/fly)	
Drosophila melanogaster (vapor—15-min exposure)	Heritable translocation[q] (dose unknown)[r]	
Mouse—ascites cells (intraperitoneal injection—1 h)[u]	DNA damage[t] (5 mg/kg)[d]	
Mouse—L cells (10 min to 24 h)	DNA damage[v] (1 mg/L)[n]	
Mouse leukocyte (subcutaneous injection)[w,x]	Somatic cell mutation[y] (100 mg/kg)[d]	
Mouse leukocyte[x]	Somatic cell mutation[y] (1 μg/L)[n]	
Mouse leukocyte[x]	Chromosomal aberrations[y] (20 μg/L)[n]	
Hamster fibroblasts (20 min)	Chromosomal aberrations[z] (8 μg/L)[n]	
Human cells (HeLa)[x]	DNA damage[aa] (2 mg/L)[n]	
CHO cells—HGPRT mutation assay (1 h)	Sporadically positive[bb] (0.15–0.45 mg/L)[n]	
Chromosome aberrations (CHO cells—1 h)	Positive[bb] (80–159 μg/L)[n]	
In vitro sister chromatid exchange assay (CHO cells—1 h)	Positive[hh] (10–40 μg/L)[n]	
Drosophila melanogaster (vapor—15 min exposure)	Dominant lethal mutations[q] (dose unknown)[r]	
Rat (inhalation, ≥2 weeks)	Dominant lethal mutation[cc] Positive (male) (0.1 mg/m^3)[dd]	
Rat (intragastric)[ee]	Dominant lethal mutations Positive (male) @ 0.50 mg/kg[ff,gg] Negative (female)[ff]	
	Teratogenicity	
Rat (intragastric)[hh]	Negative[ii]	
Rabbit (intragastric)[hh]	Negative[ii]	
Rat (inhalation—1 to 52 weeks)	Negative[f]	
	Reproductive effects	
Rat (inhalation)	Negative[f] (0.1 mg/m^3)[g]	
Rat, two-generation reproduction (intragastric)[kk]	Negative[jj]	

(*continues*)

Table IV (*Continued*)

Exposure regimen	H/HD response (dose)	T response
	Subchronic effects	
Rat, subchronic toxicity[ll] (Intragastric)[nn]	Decreased body weight; forestomach epithelial hyperplasia[mm]	

[a] From "Pulmonary tumors in strain A mice exposed to mustard gas," (Heston, W. E. and W. D. Levillain), *Proc. Soc. Exp. Biol.* **82,** 457–460, 1953.

[b] Dose is estimated, assuming complete volatilization of HD.

[c] From "Occurrence of tumors in mice injected subcutaneously with sulfur mustard and nitrogen mustard," (Heston, W. E.), *J. Natl. Cancer Inst.* **14,** 131–140, 1953.

[d] Dose is expressed as mg/kg body weight.

[e] From "Carcinogenic action of the mustards," (Heston, W. E.), *J. Natl. Cancer Inst.* **11,** 415–423, 1950.

[f] From "*Toxicological basis for controlling levels of mustard in the environment,*" (McNamara, B. P., E. J. Owens, M. K. Christensen, F. K. Vocci, D. K. Ford, and H. Rozimarek). Edgewood Arsenal Special Publication, EB-SP-74030, Dept. of the Army, Defense Technical Information Center, Alexandria, VA, 1975.

[g] Exposure to 0.1 mg/m^3 of HD was for 6.5 h/d, 5 d/week; for the remainder of each 24-h day, animals were exposed to 0.0025 mg/m^3 of HD vapor.

[h] From "Multiple Bowen's disease observed in former workers of poison gas factory in Japan, with special reference to mustard gas exposure," (Inada, S., K. Hiragun, K. Seo, and T. Yamura), *J. Dermatol. Tokyo* **5,** 49–60, 1978; and From "*IARC Monographs on the Evaluation of the Carcinogenic Risk of Chemicals to Man: Aziridines, N-S-, and O-Mustards and Selenium, Vol. 9,*" (International Agency for Research on Cancer [IARC]), Mustard Gas, pp. 181–192, World Health Organization, Geneva, Switzerland, 1975.

[i] Exposures were either in war situations or to workers in a mustard gas manufacturing plant during wartime production. The duration of exposure for these workers was years.

[j] From "The effect of repeated applications of minute quantities of mustard gas on the skin of mice," (Fell, H. and M. Allsopp), *Cancer Res.* **8,** 177–181, 1948.

[k] From "An experimental study of the initiating stage of carcinogenesis, and a reexamination of the somatic cell theory of cancer," (Berenblum, I and P. Shubik), *Brit. J. Cancer* **3,** 109–118, 1949.

[l] Mutagenic without metabolic activation. From "*Toxicology studies on Lewisite and sulfur mustard agents: mutagenicity of sulfur mustard in the Salmonella histidine reversion assay,*" (Stewart, D. L., E. J. Sass, L. K. Fritz, and L. B. Sasser), Pacific Northwest Laboratory, PNL-6873, 1989, pp. 1–32, 1989.

[m] From "The role of organic peroxides in the induction of mutations," (Dickey, F. H., G. H. Cleland, and C. Lotz), *Nat'l Acad. Sci (Proc)* **35,** 581–586, 1949.

[n] Concentration of HD in culture fluid was as shown.

[o] From "DNA alkylation by mustard gas in yeast strains of different repair capacity," (Kircher, M. and M. Brendel), *Chem. Biol. Interact.* **44,** 27–39, 1983.

[p] From "Mutability at specific euchromatic and heterochromatic loci with alkylating and nitroso compounds in *Drosophila melanogaster,*" (Fahmy, O. G. and M. J. Fahmy), *Mutat. Res.* **13,** 19–34, 1971.

[q] From "The production of mutations by chemical substances," (Auerbach, C. and J. M. Robson), *Proc. Royal Society of Edinburgh* **62B,** 271–283, 1947.

[r] Very difficult to estimate dose received. A 1 : 10 mixture of mustard gas in cyclohexane was sprayed by an atomizer at 10-sec intervals in an air stream that flowed at 2 L/min. This air stream flowed through the exposure chamber.

[s] From "Mutagenic selectivity for RNA-forming genes in relation to the carcinogenicity of alkylating agents and polycyclic aromatics," (Fahmy, O. G. and M. J. Fahmy), *Cancer Res.* **32,** 550–557, 1972.

[t] From "The reaction of mono- and difunctional alkylating agents with nucleic acids," (Brookes, P. and P. D. Lawley), *Biochem. J.* **80,** 496–503, 1961.

[u] Host-mediated assay.

[v] From "The response of mammalian cells to alkylating agents. II. On the mechanism of the removal of sulfur mustard-induced cross links," (Reid, B. D. and I. G. Walker) *Biochim. Biophys. Acta* **179,** 178–188, 1969.

[w] Host-mediated assay. Murine leukemia L5178Y cells were grown as ascites in mice.

[x] Duration of exposure was not given.

[y] From "A host-mediated assay for chemical mutagens using the L5178Y/Asn murine leukemia," (Capizzi, R. L., W. J. Smith, R. Field, and B. Papirmeister), *Mutat. Res.* **21,** 6, 1973.

[z] From "Differential effects of sulphur mustard on S-phase cells of primary fibroblast cultures from Syrian hamster," (Savage, J. and G. Breckon), *Mutat. Res.* **84,** 375–387, 1981.

[aa] From "Estimation of interstrand DNA cross-linking resulting from mustard gas alkylation of HeLa cells," (Ball, C. R. and J. J. Roberts), *Chem. Biol. Interact.* **4,** 297–303, 1972.

[bb] From "*Toxicology studies on Lewisite and sulfur mustard agents: genetic toxicity of sulfur mustard (HD) in Chinese hamster ovary cells,*" (Jostes, R. F., Jr., L. B. Sasser and R. J. Rausch), Pacific Northwest Laboratory, PNL-6916, May, 1989, pp. 1–32, 1989.

Table IV (*Continued*)

[cc] From "Mutagenic activity in somatic and germ cells following chronic inhalation of sulfur mustard," (Rozimarek, H., R. L. Capizzi, B. Papirmeister, W. H. Fuhrman, and W. J. Smith), *Mutat. Res.* **21,** 13–14, 1973; Male rats were exposed 6 h/day, 5 days per week to 0.1 mg/m^3 of HD by inhalation for 1–52 weeks. They were mated to unexposed females and dominant lethality determined. There was a cumulative increase in dominant lethal mutation rates which reached a maximum at 12 weeks' of exposure and remained at this level over the ensuing 40 weeks.

[dd] Estimated total dose was 630 μg/kg body weight.

[ee] HD tested at doses of 0.08, 0.20, and 0.50 mg/kg body weight, 5 days/week for 10 weeks.

[ff] From "*Toxicology studies on Lewisite and sulfur mustard agents: modified dominant lethal study of sulfur mustard in rats,*" (Sasser, L. B., J. A. Cushing, D. R. Kalkwarf and R. L. Buschbom), Pacific Northwest Laboratory, PNL-6945, May, 1989.

[gg] Significant dominant lethal effects observed in offspring of male rats exposed to HD (most consistent effects observed at 0.50 mg/kg dose of HD) mated to untreated females at 2 and 3 weeks after the 10-week exposure. Significant increase in abnormal parental sperm observed in this dose group. F_1 effects include increased early fetal resorptions and preimplantation losses in addition to decreased total live embryo implants.

[hh] HD was tested at concentrations of 0.5, 1.0, and 2.0 mg/kg body weight for gestational days 6 through 15 in rats; for rabbits, tested at concentrations of 0.4, 0.6 and 0.8 mg/kg body weight for gestational days 6 through 19.

[ii] No clear evidence of teratogenic effects by the agent; certain trends seen could be ascribed to maternal toxicity. From "*Teratology studies on Lewisite and sulfur mustard agents: effects of sulfur mustard in rats and rabbits,*" (Hackett, P. L., R. L. Rommereim, F. G. Burton, R. L. Buschbom and L. B. Sasser), Pacific Northwest Laboratory, PNL-6944, September 30, 1989.

[jj] From "Toxicology studies on Lewisite and sulfur mustard agents: two generation reproduction study of sulfur mustard (HD) in rats," (Sasser, L. B., R. A. Miller, D. R. Kalkwarf, R. L. Buschbom and J. A. Cushing), Pacific Northwest Laboratory, PNL-6944, 1989.

[kk] HD tested at doses of 0.03, 0.1, and 0.4 mg/kg body weight. Animals (male and female) received the agent 5 days/week for 13 weeks prior to mating as well as during a 1–2 week mating period. Pregnant females were dosed 7 days/week. Males of this mating group were sacrificed at birth of the pups. Females giving birth continued to receive the agent during lactation. Pups were weaned at 21 days of age, and the dams were sacrificed. Pups continued to receive agent for 13 weeks and were mated as above, repeating the dosing schedule. Study concluded with sacrifice of the second generation pups and their dams at weaning. Negative effects on reproductive performance, fertility or reproductive organ weights of males and females.

[ll] From "*Toxicology studies on Lewisite and sulfur mustard agents: subchronic toxicity of sulfur mustard (HD) in rats,*" (Sasser, L. B., R. A. Miller, D. R. Kalkwarf, R. L. Buschbom, and J. A. Cushing), Pacific Northwest Laboratory, PNL-6870, 1989.

[mm] Significant effects noted only in highest (0.3 mg/kg) dose group. All other parameters studied were not different from controls.

[nn] Rats (6–7 weeks old) received 0.003, 0.01, 0.03, 0.1, or 0.3 mg/kg body weight of HD 5 days/week for 13 weeks.

tions of nitrogen mustard (a potent mutagen) as well as sulfur mustard. This positive result should not be considered more important than the negative findings from the more fully characterized Japanese worker population. In any case, the human occupational exposures in the few available studies almost certainly involved long-term exposure to high concentrations for months or years; in addition, workers were essentially without protective equipment during periods of agent exposure.

Epidemiological evidence and results of animal studies indicate that mustard agent can induce cancer. This topic is of sufficient importance to warrant special analysis.

III. Evidence of Carcinogenicity

The record of human cancer induction following sulfur mustard exposure is based on retrospective studies of populations exposed to acutely toxic battlefield concentrations (WWI and II) or in weapons plants operating under conditions of inadequate ventilation and industrial hygiene practices during the years immediately before and during WWII. Study of Okuno-jima worker death certificates through 1962 revealed a high incidence of respiratory tract cancer (14%) and digestive tract cancer (9.6%). The remaining deaths (39.7%) were largely caused by respiratory disease (tuberculosis or other pulmonary infections) thought to be secondary to epithelial damage induced by vesicant gas inhalation.

Later follow-up divided the worker population into exposure groups based on job title. Examination of death certificates and autopsy reports through 1979 found that "Those . . . who were engaged in manufacture of yperite [mustard] and Lewisite gases had a high mortality due to diseases of the respiratory tract, particularly malignant tumors." Smoking had been previously ruled out as a factor. Retired workers were also observed to exhibit impaired immunity.

In 1975, the International Agency for Research

on Cancer (IARC) concluded that available data were sufficient to support classification of mustard agent as a "Group I" carcinogen. This finding was reevaluated and upheld in 1987. The "Group I" category includes compounds for which a causal relationship between exposure and subsequent human cancer induction can be adequately substantiated.

The conditions of exposure inherent to available human retrospective studies do not permit an estimate of dose–response for sulfur mustard carcinogenicity. Animal carcinogenesis studies can provide more carefully defined exposure parameters than available epidemiological studies; nevertheless, the problem of species extrapolation requires consideration. Tumorigenesis of sulfur mustard in laboratory species of mice and rats has been confirmed via inhalation and injection exposure. Evidence is summarized in Table IV. Tumors were not observed in guinea pigs, rabbits, and dogs exposed to atmospheric concentrations of 0.001 mg/m^3 or 0.1 mg/m^3 for periods up to 1 y. The laboratory rat (Sprague-Dawley-Wistar) was the only tested species observed to develop significantly elevated tumorigenicity (skin tumors) at atmospheric concentrations of 0.1 mg/m^3 (see Table IV).

Tests of sulfur mustard in initiation–promotion studies on mouse skin found that mustard was not active as an initiator. The test promotor was croton oil. Doses of mustard actually received by mice in this study were not well defined, because the mustard was applied as a droplet on the end of a glass rod that had been dipped into a solution of 0.1% mustard in paraffin oil. This experiment is interesting, however, in that it demonstrated that a low concentration of mustard was not an initiator when contrasted to various components of coal tar applied at similar concentrations.

These animal data have been evaluated to derive an estimate of carcinogenic potency associated with sulfur mustard exposure as well as to address the issue of species extrapolation. Comparisons of tumorigenicity in the same and related species for sulfur mustard and the well-characterized industrial carcinogen benzo(*a*)pyrene (B[*a*]P) indicate that these two compounds are of approximately equivalent carcinogenic potency in test animals (Fig. 1). This finding can be used to estimate the potential carcinogenic risk of chronic inhalation exposure to air containing mustard agent or chronic ingestion of contaminated foodstuffs. Engineering estimates of potential mustard agent releases during destruction

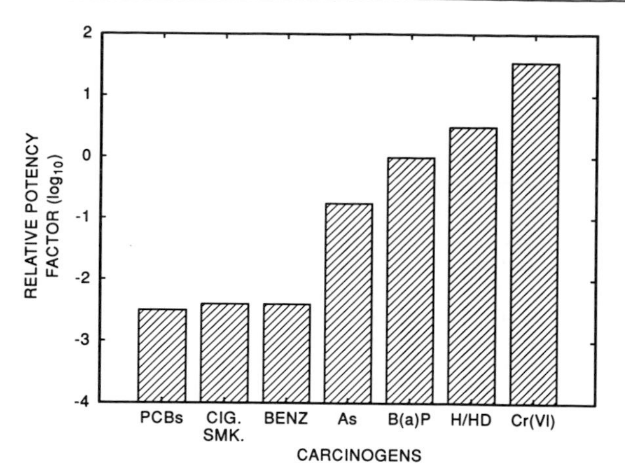

Figure 1 Relative potency of sulfur mustard as a carcinogen. From Watson *et al.* (1989).

of the U.S. stockpile (see Section I) have been evaluated by this approach.

For hypothetical "fenceline" human populations potentially exposed to sulfur mustard along the boundary of one storage facility in the eastern United States during planned mustard incinerator operation, maximum estimates of excess lifetime cancer risk range between 4×10^{-9} and 1×10^{-8}. Federal agencies do not now routinely regulate compounds for which the excess lifetime cancer risk $\leq 10^{-6}$. For comparison, the present U.S. lifetime cancer incidence from all causes approximates 2.5×10^{-1}.

IV. Decontamination and Treatment

No antidote exists for any sulfur mustard agent; effective intervention can only be accomplished by rapid decontamination, followed by palliative treatment of signs and symptoms. Severity of damage to the eyes, skin, and respiratory tract is governed by absorbed dose.

For best results, the victim's skin should be decontaminated with undiluted chlorine bleach (5% NaClO) within minutes of initial exposure (this treatment will be irritating to wounds). Neither chlorine bleach nor any other decontaminating solution should be used on the eyes or mucous tissues; exposed eyes should be flushed with copious quantities of clean water. In the absence of chlorine solutions, decontamination may be ac-

Table V Treatment Guidelines for Sulfur Mustard Agent (H, HD, HT) Exposure[a,b]

Clinical sign	Treatment
Eye irritation (red eyes or photophobia); respiratory irritation, itching, reddening of skin with small blisters	Antihistamines and corticosteroids may relieve skin and eye irritation; protect eyes from light with dark glasses, dark room or patch(es). Guard against infections.
Chemical burns/blisters, pulmonary distress, eye damage	Ventilatory support with O_2 and PEEP,[c] some fluids, dark room; follow burn regimen for large blisters (do not unroof), asepsis. Do not overhydrate.

[a] From Watson *et al.* (1992).

[b] For more detailed description of casualty therapy, see "Clinical notes on chemical casualty care." Pamphlet USAMRICD TM 90-1, U.S. Army Medical Research Institute of Chemical Defense, Aberdeen Proving Ground, MD 21010-5425.

[c] Positive End-Expiratory Pressure.

complished with available alkaline substances such as washing or baking soda, ammonia, or soap and lukewarm water; decontamination by any solution should be followed by clear-water rinses. Clothing, instruments, and surfaces other than rubber or plastics may be decontaminated with high test hypochlorite (HTH; 15% $Ca(ClO)_2$) or chlorine bleach, followed by clear-water rinses.

There is a dose-dependent delay of several hours' duration between sulfur mustard agent exposure and onset of skin or eye effects (Table III). Thus, even after decontamination is considered complete, the victim should remain under observation. If signs or symptoms develop, follow the treatment guidelines summarized in Table V.

Related Articles: ENVIRONMENTAL CANCER RISKS; MUTAGENICITY TESTS WITH CULTURED MAMMALIAN CELLS: CYTOGENETIC ASSAYS.

Acknowledgments

The submitted manuscript has been authored by a contractor of the U.S. Government (Oak Ridge National Laboratory managed by Martin Marietta Energy Systems, Inc. under contract No. DE-AC05-84OR21400). Support was provided by the U.S. Department of the Army, Office of the Assistant Secretary (Installations, Logistics and Environment) under Interagency Agreement DOE no. 1769-1354-A1. Accordingly, the U.S. Government retains a nonexclusive, royalty-free license to publish or reproduce the published form of this contribution, or allow others to do so, for U.S. Government purposes.

Bibliography

Blewett, W. (1986). Tactical weapons: Is mustard still king? *NBC Defence and Technology International* **1,** 64–66.

Carnes, S. A. (1989). Disposing of chemical weapons: A desired end in search of an acceptable means. *Environ. Prof.* **11,** 279–290.

Department of Health and Human Services (DHHS) (1988). Final recommendations for protecting the health and safety against potential adverse effects of long-term exposure to low doses of agents: GA, GB, VX, mustard agent (H, HD, T) and Lewisite (L). *Fed. Regist.* **53,** 8504–8507.

Dunn, P. (1986). The chemical war: Iran revisited—1986. *NBC Defence and Technology International* **1,** 32–39.

Freeman, K. (1991). The unfought chemical war. *Bull. Atomic Sci.* **47,** 30–39.

Haber, L. H. (1986). "The Poisonous Cloud: Chemical Warfare in the First World War." Clarendon Press, Oxford.

International Agency for Research on Cancer (IARC) (1975). "Monograph on the Evaluation of the Carcinogenic Risk of Chemicals to Man from Aziridines, N-, S- and O-Mustards and Selenium," vol. 9. World Health Organization, Geneva.

International Agency for Research on Cancer (IARC) (1987). Mustard gas (sulphur mustard) (Group 1). *In* IARC Monographs on the Evaluation of Carcinogenic Risks to Humans. Overall Evaluations of Carcinogenicity; An updating of IARC Monographs, vol. 1–42, Suppl. 7, pp 259–260. World Health Organization, Geneva, Switzerland.

Medema, J. (1986). Mustard gas: The science of H. *NBC Defense and Technology International* **1,** 66–71.

Munro, N. B., Watson, A. P., Ambrose, K. A., and Griffin, G. D. (1990). Treating exposure to chemical warfare agents: Implications for health care providers and community emergency planning. *Environ. Hlth. Persp.* **89,** 205–215.

Nishimoto, Y., Yamakido, M., Shigenobu, T., Onari, K., and Yukatake, M. (1983). Long-term observation of poison gas workers with special reference to respiratory cancers. *J. Uoeh* **5,** 89–94.

Nishimoto, Y., Yamakido, M., Shigenobu, T., Yukatake, M., and Matsusake, S. (1986). Cancer of the respiratory tract observed in workers retired from a poison gas factory. *Gan To Kagaku Ryoho* **13,** 1144–1148.

Papirmeister, B., Feister, A. J., Robinson, S. I., Ford, R. D. (1991). "Medical Defense against Mustard Gas: Toxic Mechanisms and Pharmacological Implications." CRC Press, Boca Raton, FL.

Pechura, C. M., and Rall, D. P., eds (1993). "Veterans at Risk: The Health Effects of Mustard Gas and Lewisite." Committee to Survey the Health Effects of Mustard Gas and Lewisite. Division of Health Promotion and Disease Prevention, Institute of Medicine. National Academy Press, Washington, D.C.

Somani, S. M., and Babu, S. R. (1989). Toxicodymanics of sulfur mustard. *Intl. J. Clin. Pharmacol. Ther. Tox.* **27,** 419–435.

Ward, J. R., and Seider, R. P. (1984). Activation energy for the hydrolysis of bis (2-chloroethyl) sulfide. *Thermochim. Acta* **81,** 343–348.

Watson, A. P., Jones, T. D., and Griffin, G. D. (1989). Sulfur mustard as a carcinogen: Application of relative potency analysis to the chemical warfare agents H, HD, and HT. *Reg. Tox. and Pharm.* **10,** 1–25.

Watson, A. P., Sidell, F. R., Leffingwell, S. S., and Munro, N. B. (1992). General guidelines for medically screening mixed population groups potentially exposed to nerve or vesicant agents. ORNL/TM-12034. Oak Ridge National Laboratory, (January 1992).

Watson, A. P., and Griffin, G. D. (1992). Toxicity of vesicant agents scheduled for destruction by the chemical stockpile disposal program. *Envir. Health Persp.* **98,** 259–280.

Yamakido, M., Nishimoto, Y., Shigenobu, T., Onari, K., Satoh, C., Goriki, K., and Fujita, M. (1985). Study of the genetic effects of sulfur mustard gas on former workers of Okuno-jima poison gas factory and their offspring. *Hiroshima J. Med. Sci.* **34,** 311–322.

Yang, Y.-C., Baker, J. A., and Ward, J. R. (1992). Decontamination of chemical warfare agents. *Chem. Rev.* **92,** 1729–1743.

Toxic Agents and Thermoregulation

Christopher J. Gordon
U.S. Environmental Protection Agency

Glossary

Ambient temperature (T_a) Average temperature of a gaseous or liquid environment surrounding a body.

Core body temperature (T_b) Temperature at depth below that which is not affected directly by changes in temperature of the peripheral tissues.

Homeothermy Pattern of temperature regulation in a tachymetabolic species where the core T_b is maintained within arbitrarily defined limits (±2°C) in spite of large fluctuations in T_a

Hyperthermia Condition where the temperature of a homeotherm is more than one standard deviation above the mean core temperature of the species in a resting condition in a thermoneutral environment.

Hypothermia Condition where the temperature of a homeotherm is more than one standard deviation below the mean core temperature of the species in a resting condition in a thermoneutral environment.

Lower critical temperature (LCT) T_a below which the rate of metabolic heat production increases by shivering and/or nonshivering thermogenesis to maintain thermal balance.

Metabolic rate (MR) Rate of transformation of chemical energy into heat and mechanical work by aerobic and anaerobic metabolic activities, usually expressed in terms of unit surface area or body weight.

Normothermy Condition where internal T_b is within ±1 SD under conditions of thermoneutrality.

Thermoneutral zone (TNZ) Range of T_as within which metabolic rate is at a minimum.

Thermopreferendum Thermal environment selected by a species under natural or experimental circumstances (i.e., selected or preferred T_a).

Upper critical temperature (UCT) T_a above which metabolic rate increases due to a rise in core temperature or where active evaporative heat loss mechanisms are recruited.

MAMMALS normally maintain an internal or core body temperature (T_b) of ~37°C over a wide range of ambient temperatures (T_a) by the appropriate activation of autonomic and behavioral thermoregulatory motor outputs. This provides a uniform thermal environment for the proper functioning of life-sustaining biochemical and physiological processes. Because of the temperature dependency of biochemical reactions, any event which changes the internal T_b, such as exposure to chemical toxicants, will most likely alter the activity of a variety of physiological systems. For example, hypothermia is a commonly reported observation following toxic exposure in laboratory rodents. In some instances, treatments which block the hypothermic response have been shown to attenuate the severity of toxicity; however, in many instances hypothermia appears beneficial to the animal's survival. Hence, the effects of toxicants on thermoregulation may represent an integrated response for combatting the debilitating effects of the toxic agent. Understanding the mechanisms of thermo-

regulation during toxic insult can provide insight into the interpretation of results in other disciplines of toxicology such as immunosuppression, CNS dysfunction, teratology, and mutagenesis.

I. Background

There are four key tenets which call for a thorough understanding of the effects of exposure to toxic agents on temperature regulation: (1) a stable T_b is crucial to the normal functioning of all physiological and behavioral systems; (2) exposure to toxic agents on temperature regulation: (1) a stable T_b is cially in the smaller laboratory mammals; (3) the toxicity of many chemicals is generally proportional to T_b and/or T_a; and (4) the thermoregulatory response can directly affect the toxic efficacy of a chemical agent.

Eutherian mammals have evolved the capacity to maintain their T_b within a narrow range over a relatively wide range of T_as and under other conditions of thermal stress such as exercise and fever. Thermal homeostasis is essential for optimal enzymatic function which has a relatively narrow temperature band of optimal performance, generally being equal to normal T_b of 37°C. Toxic agents alter T_b through a variety of mechanisms. Because temperature directly affects enzyme activity (and, hence, the function of cells, tissues, and organs), reported toxic effects of chemical agents may often be attributable to changes in temperature.

Hypothermia is a predominant thermoregulatory effect in laboratory mammals exposed to toxic chemicals. Both behavioral and autonomic thermoregulatory effectors may be involved in the development of the hypothermic state. Clearly, any physiological process showing a depression in activity following toxic exposure may well be attributed to a depression in temperature. Interestingly, hypothermia may actually benefit the exposed animal's resistance to toxic exposure. Moreover, rather than attempt to combat the toxic-induced hypothermia, animals exposed to a toxicant in many instances utilize behavioral and autonomic thermoregulatory motor outputs to lower their T_b. It has been recognized for over a century that chemical toxicity can be proportional to tissue temperature. Taken together, these observations represent an interesting scenario for an adaptive biological response; namely, there may be an integrated thermoregulatory response designed to lower T_b in response to toxic exposure whereby the ensuing hypothermia lowers the toxicity of the chemical agent and enhances survivability.

II. Basic Mechanisms of Temperature Regulation

Homeothermy refers to the ability to regulate the internal T_b within arbitrarily defined limits (±2°C) despite relatively large variations in T_a. Commonly studied laboratory rodents, humans, and most other eutherian mammals maintain their T_b at a normothermic level of between 37 and 38°C. Thermal homeostasis is achieved through increased heat production and reduced heat loss during exposure to cold environments and increased heat loss during exposure to hot environments. A simplified version of the heat balance equation is useful for explaining the basic concepts of homeothermy in mammals and other species:

$$S = M \pm E \pm R \pm C \pm K,$$

where S = net rate of heat storage in the body; M = metabolic heat production (always +); E, R, C, and K = net rates of evaporative, radiant, convective, and conductive heat loss, respectively (− for loss). Under steady state conditions where the animal is normothermic, heat production is equal to heat loss [i.e., $M = (E + R + C + K)$] and $S = 0$. If heat loss exceeds heat production as would occur during acute cold stress, then S is negative and the animal becomes hypothermic. If heat production exceeds heat loss as would occur during acute heat stress, then S is positive and the animal becomes hyperthermic.

Each parameter of the heat balance equation is under control of multiple thermoregulatory motor outputs. Thermal stimuli in the skin and core tissues are detected by thermal receptors; this information is transferred to the central nervous system (CNS) via the spinothalamic and trigeminal systems. The preoptic area and anterior hypothalamus (POAH) are key sites for the integration of the thermal stimuli and generation of appropriate effector signals for driving the thermoregulatory motor systems. Other sites in the CNS, including the posterior hypothalamus, medulla, various relay nuclei, and spinal cord also play a role in thermoregulatory control.

The change in activity of thermoregulatory effectors and T_b as a function of T_a illustrates the general properties of the thermoregulatory system (Fig. 1). Typically, metabolic heat production as a function of T_a exhibits three phases in homeotherms. There is a T_a range where metabolic rate is basal and T_b is regulated through modulations in skin blood flow and the concomitant control of dry heat loss (i.e., $C + R + K$); this range of T_a values is termed the "thermoneutral zone" (TNZ). As T_a increases above the TNZ, metabolic rate increases as a result of several physiological and behavioral processes (e.g., grooming, increased activity, panting, and elevation in tissue temperature which accelerates cellular respiration). The T_a at which metabolic rate increases (or where active evaporative heat loss mechanisms are recruited) is termed the "upper critical temperature" (UCT). As T_a drops below the TNZ, dry heat loss is minimal as a result of peripheral vasoconstriction. Metabolic heat production via shivering and nonshivering thermogenesis must increase in order to maintain a balance between heat loss and heat production. The T_a below which metabolism is elevated above basal levels is defined as the lower critical temperature (LCT).

The LCT is pertinent to most experimental toxicology studies of laboratory rodents because the species' LCT is invariably higher than the typical room temperature (~22°C) of the laboratory (Table I). Thus, the rodents tested under normal room temperature, especially in cages without insulative bedding materials, must continually maintain metabolic rate above basal levels. This can have significant consequences on the metabolic and toxic efficacy of chemical toxicants (see below).

III. Effects of Toxic Chemicals on Temperature Regulation

A. Animal Studies

There is an abundance of studies showing that a change in T_b is a frequent sequela of laboratory rodents subjected to acute toxic insult. It is often the

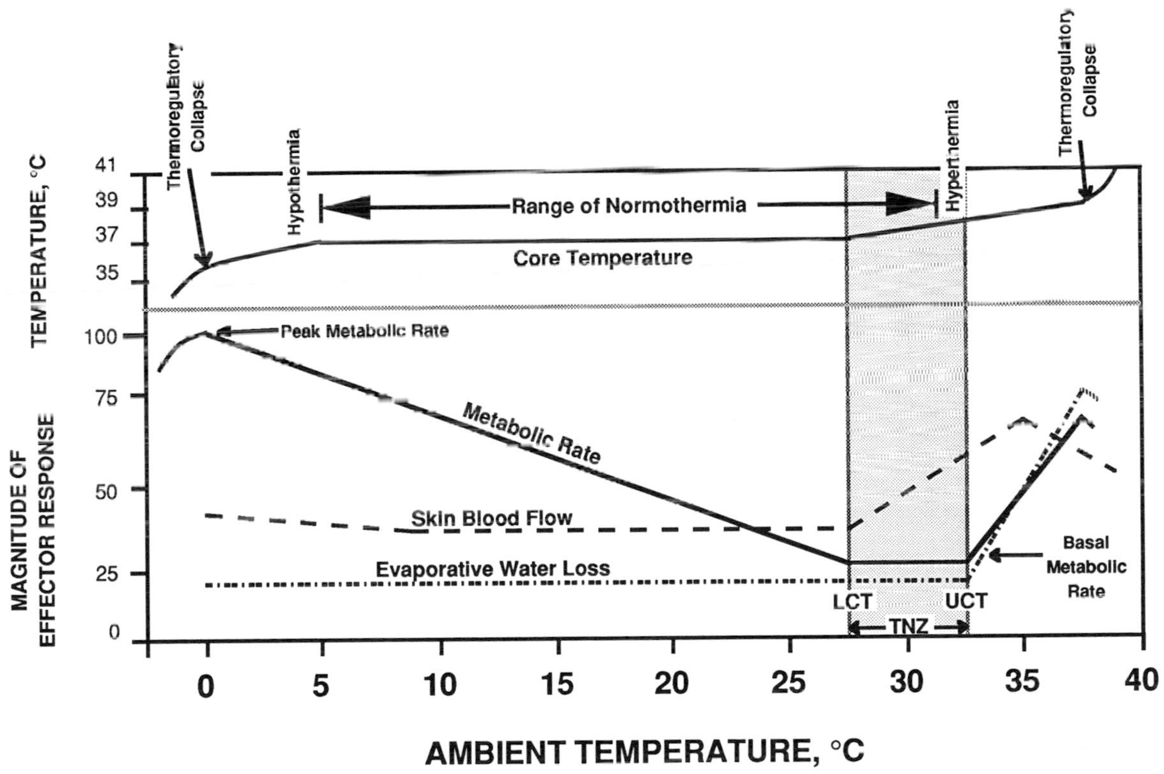

Figure 1 Typical effect of ambient temperature (T_a) on metabolic rate, skin blood flow, evaporative water loss, and body temperature in a homeotherm. Homeothermy, the maintenance of a constant core body temperature, is achieved over a wide range of T_a values through the activation of appropriate thermoregulatory effectors. Failure to thermoregulate occurs when ambient heat and cold stress surpass the capacity of the thermoregulatory effectors. Modified from Gordon (1990; see bibliography).

Table I Lower Critical T_as (LCT) Reported for a Variety of Mammals

Species	LCT (°C)
Mouse	31–32
Hamster	25–28
Rat	28–30
Guinea pig	20–29
Rabbit	15–17
Human (nude)	27–28

Source: Hart, J. S. (1974). Rodents. *In* "Comparative Physiology of Temperature Regulation, pp. 1–149, Academic Press, New York; Folk, G. E., Jr. (1974). "Textbook of Environmental Physiology," Lea and Febiger, Philadelphia.

case that the reported thermal effect has not been of interest to the principle goals of the study and has been reported as an incidental observation. Until recently, there have been few attempts to characterize the nature of the thermoregulatory response to toxic insult.

Before discussing the effects of chemical toxicants on T_b, it is necessary to explain the possible modes by which the thermoregulatory system is affected by chemicals. Basically, the change in T_b can be either forced or regulated (Table II). During forced hypothermia as might occur when a drug is administered that inhibits metabolic heat production in the cold, T_b is below the set-point temperature and thermoregulatory motor outputs are activated in order to return T_b to normal. During forced hyperthermia as might occur when a drug is administered that inhibits evaporative water loss in a hot environment, T_b is forced above the set-point and thermoregulatory motor outputs are activated to lower T_b to normal. During regulated hypothermia, a drug acting on the CNS lowers the set point and thermoregulatory effectors are activated to lower T_b. During regulated hyperthermia as occurs with fever, the set-point is elevated following infection, and T_b is elevated.

Table II Changes in Activity of Thermoregulatory Motor Outputs for Promoting Forced and Regulated Changes in T_b

	Thermoregulatory Motor Output			
Response	Metabolic rate	Evaporative water loss	Skin blood flow	Selected T_a
Forced hypothermia	↑	↓	↓	↑
Forced hyperthermia	↓	↑	↑	↓
Regulated hypothermia	↓	↓	↑	↓
Regulated hyperthermia	↑	↓	↓	↑

Source: Modified from Gordon, C. J. (1983). A review of terms for regulated vs. forced, neurochemical-induced changes in body temperature. *Life Sci.* **32,** 1285–1295.

Behavioral thermoregulatory responses are crucial in deciding whether a thermoregulatory response following toxic exposure is forced or regulated. Studies from this laboratory and others have measured T_b and behavioral thermoregulatory responses of laboratory rodents exposed to a variety of chemical agents ranging from simple metallic salts to complex hydrocarbon-based pesticides (Table III). Hypothermia is more prevalent than hyperthermia for most of the reported effects. In small rodents such as the mouse and rat, exposure to a toxic chemical generally leads to a reduction in metabolic rate. Because of their small size and, hence, large surface area : body mass ratio, these small rodents rely primarily on a high metabolic rate to thermoregulate at T_as < LCT. Thus, it is not surprising to find the occurrence of hypothermia in these species following exposure to chemicals which inhibit metabolic rate. The hypothermia can be easily prevented by maintaining the animals at a warm T_a of 30 to 35°C during toxic exposure (Fig. 2A). On the other hand, if placed in a temperature gradient and allowed to select their thermopreferendum, treated animals either select cool T_as which augment the hypothermic efficacy of the toxic compound or fail to move to warmer T_as to prevent the toxic-induced hypothermia (Fig. 2B). Mice consistently select cooler T_as during exposure to toxicants while rats either select cooler T_as or exhibit no change in their thermopreferendum during toxic exposure (Table III). Behavioral thermoregulatory responses in the rat are generally more unpredictable compared with other rodent species. The rat is not the best experimental model to study behavioral thermoregulation during toxic exposure.

The depression in metabolic heat production along with a preference for cooler T_as is conducive to a regulated decrease in core temperature (i.e., decrease in set-point; see Table II). If the thermoregulatory centers in the CNS were unaffected by the toxicant, one would expect a selection of warmer T_as to attenuate the hypothermic state. It is

Table III Changes in T_b and Selected Ambient Temperature (ST_a) Values of Mice and Rats following Acute Exposure to Various Chemical Agents

Species	Compound (dose, route)	T_b response	Selected T_a response
Mouse[a]	Nickel chloride (10 mg/kg; i.p.)	Decrease	Decrease
	Cadmium chloride (2 mg/kg; i.p.)	Decrease	Decrease
	Lead acetate (100 mg/kg; i.p.)	Decrease	Decrease
	Sodium selenite (30 μM/kg; s.c.)	Decrease	Decrease
	Triethyltin (6 mg/kg; i.p.)	Decrease	Decrease
	Chlordimeform (60 mg/kg; i.p.)	Decrease	Decrease
	Sulfolane (400 mg/kg; i.p.)	Decrease	Decrease
	Ethanol (3 g/kg; i.p.)	Decrease	Decrease
	2,4-DNP (20 mg/kg; i.p.)	Increase	Decrease
	Pentolinium tartrate (5 mg/kg; i.v.)	Decrease	Increase
	Nitrous oxide[b] (0.75 atm; inhalation)	Decrease	Decrease
Rat	Nickel chloride[c] (12 mg/kg; i.p.)	Decrease	Decrease
	Nickel chloride[d] (2 mg/kg; i.p.)	Decrease	Decrease
	Chlordimeform[e] (60 mg/kg; i.p.)	Decrease	Decrease/increase
	Ethanol[f] (3 g/kg; gavage)	Decrease	Decrease
	Methanol[g] (1–3 g/kg; i.p.)	Decrease	No change
	Sulfolane[h] (800 mg/kg; i.p.)	Decrease	No change
	DFP[i] (1.5 mg/kg; s.c.)	Decrease	No change

[a] Gordon *et al.* (1988) (see bibliography).
[b] Pertwee, R. G., N. R. Marshall, and A. G. MacDonald. The effect of subanaesthetic partial pressures of nitrous oxide and nitrogen on behavioural thermoregulation in mice. *In* "Homeostasis and Thermal Stress. (1986). (K. E. Cooper *et al.*, eds.) pp. 19–21, Basel-:Karger.
[c] Gordon, C. J. (1989). Effect of nickel chloride on body temperature and behavioral thermoregulation in the rat. *Neurotoxicol. Terat.* **11,** 317–320.
[d] Watanabe *et al.* (1990a, see bibliography).
[e] Gordon, C. J. and W. P. Watkinson. (1988). Behavioral and autonomic thermoregulation following chlordimeform administration. *Neurotoxicol. Terat.* **10,** 215–219.
[f] Gordon, C. J., L. Fogelson, F. Mohler, A. G. Stead and A. H. Rezvani. (1988). Behavioral thermoregulation in the rat following the oral administration of ethanol. *Alcohol* **23,** 383–390.
[g] Mohler, F. S. and C. J. Gordon. (1990). Thermoregulatory effects of methanol in Fischer and Long Evans rats. *Neurotoxicol. Terat.* **12,** 41–45.
[h] Gordon, C. J., R. S. Dyer, M. D. Long, and K. S. Fehlner. (1985). Effect of sulfolane on behavioral and autonomic thermoregulation in the rat. *J. Toxicol. Environ. Hlth.* **16,** 461–468.
[i] Gordon, C. J., L. Fogelson, L. Lee, and J. Highfill. (1991). Acute effects of diisopropyl fluorophosphate (DFP) on autonomic and behavioral thermoregulatory responses in the Long-Evans rat. *Toxicol.* **67,** 1–14, 1991.

interesting to note that the regulated hypothermia is opposite to that of fever, which is characterized by an integrated activation of behavioral and autonomic thermoregulatory effectors to elevate T_b following exposure to pyrogens.

The integrated changes in behavioral and autonomic thermoregulatory effectors suggest that the toxic-induced hypothermia may represent much more than a simple dysfunction of homeostatic processes. Indeed, it has long been known that hypothermia is beneficial for combatting the deleterious effects of an array of toxic chemicals. A variety of studies have found that a high T_a potentiates chemical toxicity in the mouse and rat (Fig. 3). Although T_b was usually not measured in these studies, it is clear that raising T_a to levels above the LCT reduces or eliminates the toxic-induced hypothermia (see Fig. 2A); thus, high T_as are expected to be associated with elevated T_bs. Lowering temperature attenuates chemical toxicity, probably as a result of a Q_{10} effect.

An additional protective effect occurs during exposure to respiratory agents such as ozone and formalin. These airborne pollutants cause hypothermia when inhaled by the rat and mouse and the hypothermia causes a reduction in minute ventilation, thus reducing uptake. It is also apparent that toxicity can be increased at extremely low T_as (Fig. 3B). This is probably attributable to the toxic-exposed animal's inability to effectively thermoregulate in cold environments which consequently leads to death by hypothermia. In many instances,

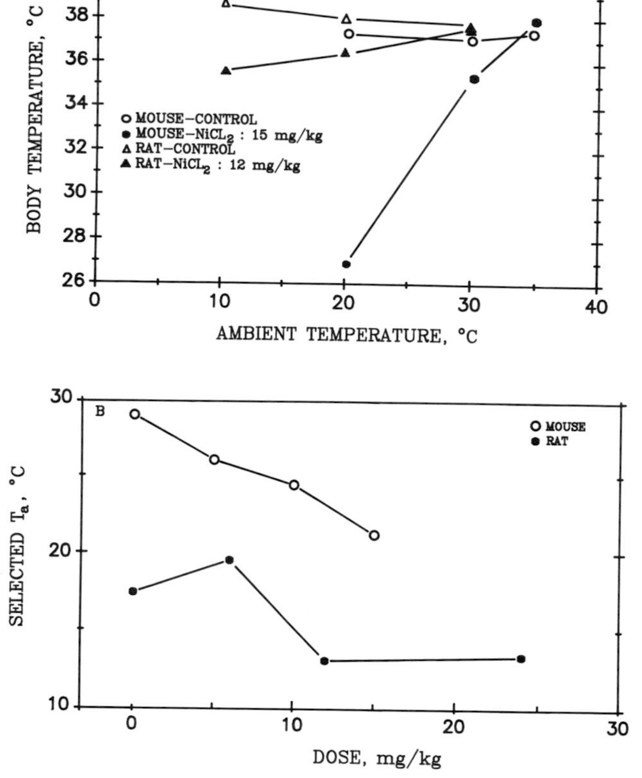

Figure 2 Example of the effect of a toxicant ($NiCl_2$) on body temperature as a function of T_a (A) and selected T_a of mice and rats when placed in a temperature gradient (*B*). Note increased lability of body temperature of mouse following $NiCl_2$ injection. Mice: Gordon, C. J., and A. G. Stead. (1986). Effect of nickel and cadmium chloride on autonomic and behavioral thermoregulation in mice. *Neurotoxicol.* **7,** 97–106; rats: Gordon, C. J. (1989). Effect of nickel chloride on body temperature and behavioral thermoregulation in the rat. *Neurotoxicol. Terat.* **11,** 317–320.

T_a versus LD_{50} relationships exhibit inverted *V*- or *U*-shaped functions such as in those of Figure 3B. Overall, the ability to lower T_b via autonomic and behavioral thermoregulatory effectors provides two forms of protection to the laboratory rodent subjected to toxic insult (Fig. 4): (1) reduction in chemical toxicity via a Q_{10} effect and (2) reduction in respiratory uptake of airborne pollutants.

B. Thermoregulatory Responses in Humans

There are relatively few data on the effects of chemical toxicants on thermoregulation in humans (Table IV). Most of the data have been collected from cases where individuals were accidentally

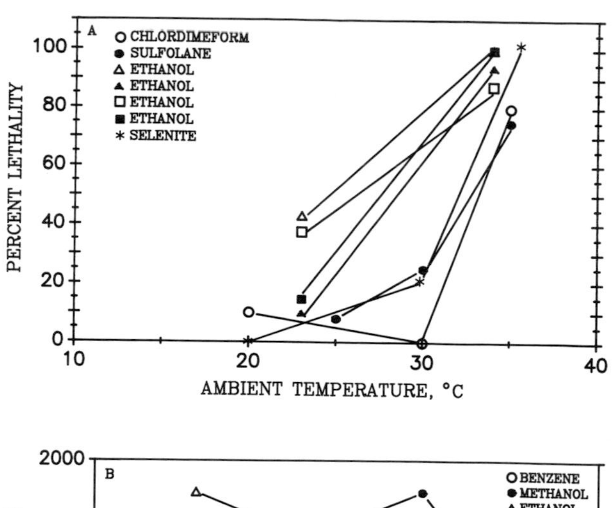

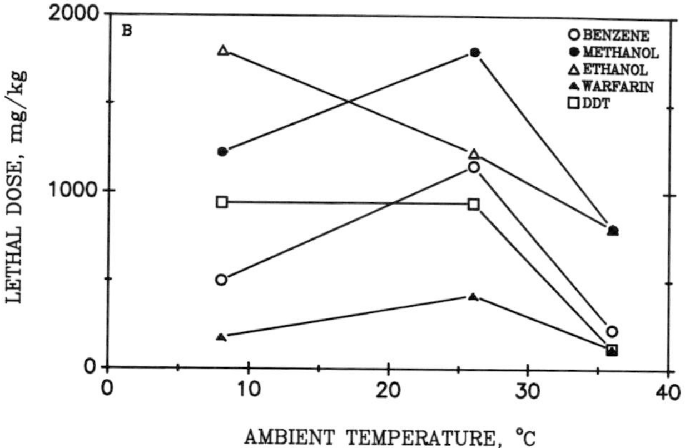

Figure 3 Examples of effect of T_a on toxic-induced lethality for a variety of chemicals administered to mice (A) and rats (B) Mice: Gordon, C. J., M. D. Long, K. S. Fehlner, and R. S. Dyer. (1986). Sulfolane-induced hypothermia enhances survivability in mice. *Environ. Res.* **40,** 92–97; rats: Gordon, C. J., M. D. Long, and A. G. Stead. (1985). Thermoregulation in mice following acute chlordimeform administration. *Toxicol. Lett.* **28,** 9–15; Finn, D. A., M. Bejanian, B. L. Jones, P. J. Syapin, and R. L. Alkana. (1989). Temperature affects ethanol lethality in C57BL/6, 129, LS and SS mice. *Pharmacol. Biochem. Behav.* **34,** 375–380; Watanabe, C., and T. Suzuki. (1986). Sodium selenite-induced hypothermia in mice: indirect evidence for a neural effect. *Toxicol. Appl. Pharmacol.* **86,** 372–379; Rats: Keplinger, M. L., G. E. Lanier, and W. R. Diechmann. (1959). Effects of environmental temperature on the acute toxicity of a number of compounds in rats. *Toxicol. Appl. Pharmacol.* **1,** 156–161.

poisoned. In many of these clinical cases, knowledge of the environmental temperature and dose of toxicant are often lacking which impedes any thorough analysis of the thermoregulatory response to the toxic agent.

The change in T_b in these clinical cases is generally mild compared with that noted in studies with rodents. It is interesting to note that hyperthermia

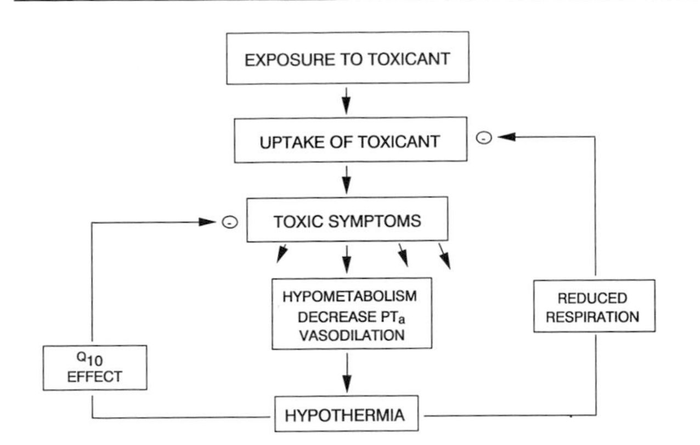

Figure 4 Proposed mechanism for attenuating toxic symptoms during toxic-induced hypothermia. Reprinted with permission from Gordon, C. J. (1991). Toxic-induced hypothermia and hypometabolism: Do they increase uncertainty in the extrapolation of toxicological data from experimental animals to humans? *Neurosci. Biobehav. Rev.* **15,** 95–98.

is often reported as a symptom in humans exposed to organophosphate (OP) pesticides, whereas experimental animal studies have repeatedly found that hypothermia is a common symptom to OP poisoning (see Table III).

IV. Extrapolation from Animal to Human

Body size limits the responsiveness of the thermoregulatory system to pharmacological and toxic treatment. As body size increases, the ratio of surface area : body mass decreases. This relationship imparts marked effects on metabolic heat production and, consequently, other physiological systems which are directly or indirectly affected by metabolism (e.g., heart rate, breathing rate, cardiac output).

Because of their small size and large surface area : body mass ratio, rodents exhibit greater and more rapid changes in T_b when challenged with toxic agents. The effect of body size is illustrated by comparing the magnitude of nickel chloride induced hypothermia in the mouse and rat (see Fig. 2A). Larger species, such as humans, rely more on their mass and insulation to thermoregulate; a perturbation in metabolism should not affect their T_b as much as in smaller species. This is suggested in the data from human exposure in Table IV indicating that the changes in T_b following exposure to a

Table IV Changes in Body Temperature in Humans Following Exposure to Various Chemical Toxicants

Chemical	Effect
Chlordane	Spiking T_b (38.8–39.4°C)[a]
Organophosphate	Hypothermia (34.5°C)[b]
Organophosphate	Hyperthermia (39.5°C)[c]
Organophosphate	Hyperthermia, sweating[d]
Magnesium salt	Hypothermia (ΔT = −0–1.6°C)[e]
Nickel salts	Mild hypothermia[f]
Ethanol	Hypothermia (ΔT = −0.7°C)[g]

[a] Furie, B., and S. Trubowitz. (1976). Insecticides and blood dyscrasias. Chlordane exposure and self-limited refractory megaloblastic anemia. *JAMA* **235,** 1720–1722.

[b] Cupp, C. M., G. Kleiber, R. Reigart, and S. H. Sandifer. (1975). Hypothermia in organophosphate poisoning and response to PAM. *J. S.C. Med. Assoc.* **71,** 166–168.

[c] Wekler, L., R. E. Mrak, and W. Dettbarn. (1986). Evidence of necrosis in human intercostal muscle following inhalation of an organophasphate insecticide. *Fund. Appl. Toxicol.* 172–174.

[d] Namba, T. and C. T. Nolte, J. Jackrel, and D. Grob. (1971). Poisoning due to organophasphate insecticides. *Am. J. Med.* **50,** 475–492.

[e] Parsons, M. T., C. A. Owens, and W. N. Spellacy. (1987). Thermic effects of tocolytic agents: decreased temperature with magnesium sulfate. *Obst. Gynecol.* **69,** 88–90.

[f] Sunderman, F. W., Jr., B. Dingle, S. M. Hopfer, and T. Swift. (1988). Acute nickel toxicity in electroplating workers who accidently ingested a solution of nickel sulfate and nickel chloride. *Am. J. Indust. Med.* **14** 257–266.

[g] Hrbek, J., J. Macakova, S. Komenda, A. Siroka, and M. Rypka. (1985). Effects of ethanol (0.3 g, 0.6 g, and 0.9 g/kg) on the blood pressure, the body temperature and the heart rate of man. *Acta Univ. Palacki. Olomuc. Fac. Med.* **111,** 197–208.

toxic chemical are not as great as in the smaller rodent species. However, it is also clear that further studies in humans are needed to verify this relationship.

The relationship between body size and thermal lability suggests another uncertainty factor in extrapolating toxicological data from experimental animal to human. This uncertainty is apparent from three key observations: (1) regulated hypothermia is common in rats and mice treated with toxic agents; (2) mild hypothermia is beneficial to the ability of the treated animal to survive the toxic exposure; and (3) body mass probably limits the hypothermic response and it is expected that toxic-induced hypothermia would be slight in adult humans or there may be a hyperthermic response (see Table IV). Thus, it is proposed that this represents an additional physiological dissimilarity which is predicted to underestimate the determination of risk of acute toxicological data when extrapolated from species of small to large body size. It may be necessary to perform rodent toxicological studies at T_as equal to or above the TNZ to prevent toxic-induced hypothermia and thereby assure a more uniform thermal environment between species.

V. Conclusion

The study of the effects of toxic chemicals on temperature regulation is an interdisciplinary field of toxicology. It is relevant to most researchers investigating the impact of toxic chemicals on a variety of biological endpoints in laboratory animals. That many toxic chemicals appear to invoke a type of regulated hypothermia means that manipulation of the animals' internal and external thermal environment can affect the overall toxicity of the chemical on a variety of biological endpoints. That is, if the chemical's thermoregulatory effects are altered through such procedures as restraint, anesthesia, or maintaining a high T_a, then the animal will likely incur more stress as compared with the unrestrained subject allowed to thermoregulate normally. The animal's thermoregulatory response can be decisive in affecting the overt toxicological efficacy of many chemicals.

Related Article: Pharmacokinetics, Individual Differences.

Acknowledgments

I thank Drs. D. Miller and W. P. Watkinson for their review of the manuscript and to D. Denning for preparation of the figures. This paper has been reviewed by the Health Effects Research Laboratory, U.S. Environmental Protection Agency, and approved for publication. Mention of trade names or commercial products does not constitute endorsement or recommendation for use.

Bibliography

Cossins, A. R., and Bowler, K. (1987). "Temperature Biology of Animals." Chapman and Hall, London.

Glossary of terms for thermal physiology. (1987). *Int. Union Physiol. Sci. Pfugers Arch.* **410,** 567–587.

Gordon, C. J., Mohler, F. S., Watkinson, W. P., and Rezvani, A. H. (1988). Temperature regulation in laboratory mammals following acute toxic insult. *Toxicol.* **53,** 161–178.

Gordon, C. J. (1990). Thermal biology of the laboratory rat. *Physiol. Behav.* **47,** 963–991.

Gordon, C. J. (1991). Toxic-induced hypothermia and hypometabolism: Do they increase uncertainty in the extrapolation of toxicological data from experimental animals to humans? *Neurosci. Biobehav. Rev.* **15,** 95–98.

Hopfer, S. M., and Sunderman, F. W., Jr. (1988). Hypothermia and deranged circadian rhythm of core body temperature in nickel chloride-treated rats. *Res. Comm. Chem. Path. Pharm.* **62,** 495–505.

Mautz, W. J., and Bufalino, C. (1989). Breathing pattern and metabolic rate responses of rats exposed to ozone. *Respir. Physiol.* **76,** 69–78.

Watanabe, C., Weiss, B., Cox, C., and Ziriax, J. (1990a). Modification by nickel of instrumental thermoregulatory behavior in rats. *Fund. Appl. Toxicol.* **14,** 578–588.

Watanabe, C., Suzuki, T., and Matsuo, N. (1990b). Toxicity modification of sodium selenite by a brief exposure to heat or cold in mice. *Toxicol.* **64,** 245–253.

Triethylamine

Bengt Akesson
Department of Occupational and Environmental Medicine, Lund, Sweden

Glossary

Biotransformation Transformation in the body of a compound to one or more other compounds (e.g., by oxidation, reduction, dealkylation, decarboxylation, or conjugation).

Edema Excessive accumulation of serous fluid in the intercellular spaces of tissue.

Metabolism Uptake, distribution, and biotransformation in the body and elimination from the body.

TRIETHYLAMINE [TEA; $(CH_3CH_2)_3N$; molecular weight 101.2] is an aliphatic tertiary amine, in which all three hydrogen atoms of ammonia are replaced by ethyl radicals.

I. Physical and Chemical Properties

TEA (CAS number 121-44-8; RTECS number YE0175000) is a colorless liquid with a melting point of −115°C, boiling point 89°C, density 0.729 at 20°C, vapor density 3.48, vapor pressure at 20°C, 7.20 kPa (54 mm Hg), and flash point (open cup) −6°C. It has a "fishy" to strongly "ammoniacal" odor. The air odor threshold is 2 mg/m^3 and a water odor threshold 0.4 μg/g. TEA is a strongly alkaline amine (pK_a 10.75) with high water solubility (15 g/ml). The octanol–water partition coefficient, K_{ow}, is 27.

TEA is corrosive to copper, aluminum, zinc, and their alloys. Steel may be used for storage. TEA is a flammable liquid. In case of spillage or leakage, remove ignition sources and seal drains. The explosive range of concentration is from 1 to 8%.

The hydrochloride of TEA, triethylammonium chloride (melting point 253°C; sublimates at 245°C) may be crystallized in alcohol.

Conversion factors are: 1 ppm = 4.2 mg/m^3; 1 mg/m^3 = 0.24 ppm.

II. Use and Occurrence

TEA is present in many processes in chemical and pharmaceutical industries. It is used as a catalyst in polymer production in foundries (in the manufacture of cores; the cold-box technique), and for the production of polyurethane foam and epoxy resins. Also, TEA has an increasingly wide use as chemical intermediate in the production of pharmaceuticals, soaps, emulsifiers, dyestuffs, rubber products, flotation agents, finishing agents, and ion-exchange resins. TEA is also employed as (and used in the production of) corrosion inhibitor and stabilizing agent (e.g., in waterborne paints).

III. Occupational Exposure Limits for Airborne Triethylamine

Several countries (United States, France, Norway, Denmark, Germany) have established a time-weighted average (8 h) exposure limit for TEA of

Handbook of Hazardous Materials

40 mg/m^3. The limit is established at 8 mg/m^3 in Sweden, 10 mg/m^3 in Finland, 20 mg/m^3 in the Netherlands, and 100 mg/m^3 in Iceland.

IV. Occupational Exposure

Studies of occupational exposure to TEA in different occupational settings display exposure levels from below 1 (ND) to 50 mg/m^3 in core production (foundries), 1–16 mg/m^3 in polyurethane-foam production, ND–8 mg/m^3 in paint production, ND–1 mg/m^3 in pharmaceutical production, ND–1 mg/m^3 in resin production, and ND–6 mg/m^3 during the application of paints.

These exposure levels represent shops with proper occupational hygiene standards typical of the late 1980s. With improper standards, much higher exposure levels may occur.

N-Nitrosodiethylamine (NDEA; 1–5 μg/m^3) may be present in the work area, core shops, and in rubber/latex production. However, it has not been shown whether the nitrosamines originate from transformation of TEA, or from impurities in the TEA.

V. Measurement of Triethylamine in Air

Sampling of TEA in the air may be performed either on solid sorbents, such as silica gel, acid-treated silica gel or charcoal; and in acidic absorption solution. TEA is desorbed from the solid adsorbents by an organic solvent (e.g., methanol–dichloromethane). Analysis of TEA in air samples is performed, without employing derivatization steps, by gas chromatographic methods (employing flame ionization or nitrogen–phosphorus detectors) or by isotacophoresis. The detection limits of these methods correspond to a TEA air concentration of $\leq$0.1 mg/m^3.

VI. Measurement of Triethylamine in Biological Samples

Analysis of TEA in blood and urine samples may be performed by adding potassium hydroxide to samples, and then extracting with an organic solvent (e.g., dibutyl ether). The solvent phase is analyzed with a gas chromatographic method using a nitrogen–phosphorus detector. The detection limit for TEA in blood is 0.02 μmol/l and in urine 0.1 μmol/l.

VII. Biological Exposure Indicators

The half-life for both TEA and its metabolite triethylamine-*N*-oxide (TEAO) in the body is about 3 h. There is a good correlation between TEA air concentration and the concentration of TEA + TEAO in plasma at the end of the exposure. There is also a close correlation between the TEA air concentration and the urinary excretion of TEA + TEAO. Experimental exposure (light–moderate physical work) to 10 mg TEA/m^3 during 8 h leads to a concentration of TEA + TEAO in plasma of 1.8 μmol/l and a 2-h post-shift excretion of TEA + TEAO in urine of 55 mmol/mol creatinine. The corresponding concentrations after occupational exposure (moderate–heavy work) to 10 mg TEA/m^3 are 2.8 μmol/l and 95 mmol/mol creatinine, respectively.

VIII. Biological Effects of Exposure to Triethylamine

A. Human Studies

TEA is corrosive and is known to be a local irritant that affects the eyes, the mucous membranes of nose and throat, and the skin, and may cause headache, nausea, and faintness. However, the critical effect of exposure to TEA is the effect on the eyes.

Exposure to TEA affects the cornea, involving corneal edema with an increase of the thickness of the cornea. Subjective symptoms are visual disturbances (foggy vision; blue haze) and a halo effect (lights surrounded by pronounced halos). The symptoms and signs are caused by a local effect on the cornea, and not through systemic absorption. The effects are reversible. However, they may cause accidents during work and in traffic after work.

Exposure to a constant TEA level of 20 mg/m^3 may induce visual disturbances after about 6 h (35 and 50 mg/m^3 after 3 and 1 h, respectively). In the industrial setting, where peak concentrations may occur, such effects seem to appear at lower TEA levels, 10–15 mg/m^3 after 1 to 4 h. The visual disturbances fade away gradually within 1 to 4 h after the termination of TEA exposure. Exposures to less than 10 mg/m^3 have no effect.

TEA is extremely destructive to tissue of the mucous membranes and upper respiratory tract, and inhalation of very high TEA concentrations may be fatal due to spasm, inflammation, and edema of the larynx and bronchi, and pulmonary edema.

The secondary diethylamine (DEA), a metabolite of TEA, is suspected to be nitrosated endogenously into *N*-nitroso-diethylamine (NNDE). No epidemiological data are available, but NNDE should be regarded as carcinogenic to humans. However, the DEA formation from exposure to TEA is small. Furthermore, DEA has not been found in the stomach, where conditions are favorable for NNDE formation.

Some humans (about 7% of the population) have a specific anosmia, a defect in the olfactory sense; their ability to smell TEA is reduced about 15 times.

B. Animal Studies

There is a difference of susceptibility to TEA between rabbits and rats. 200 mg/m^3 causes adverse effects on lung and heart in rabbits, but 1000 mg/m^3 has no effect in rats. When inhaled, the lethal concentration that kills 50% of the tested animals (LC_{50}; 1000–1500 mg/m^3) is about the same for mammals (rabbit, rat, mouse, and guinea pig). When orally administered, the lethal dose (LD_{50}) is about 500 mg/kg body weight for rabbit, rat, and mouse. The LD_{50} for skin application in rabbits is 570 mg/kg body weight.

TEA inhibits monoamine oxidase activity (i.e., a stimulation of the CNS) in mice.

IX. Metabolism

A. Uptake and Distribution

TEA is efficiently absorbed through the respiratory and gastrointestinal tracts and rapidly distributed throughout the body. The volume of distribution is about four times the body water content, indicating that at least one anatomically unlocalized reservoir of TEA exists in the body, in which the TEA concentration is well above average. Adipose tissue and regions with a low pH may be candidates. The average half-life of TEA in plasma is 3 h. Absorption of TEA through the skin has not been studied.

B. Biotransformation

TEA is 30% biotransformed into TEAO. The oxygenation of the nucleophilic nitrogen is probably performed in the liver by a flavin-containing monooxygenase (FMO; "Ziegler's enzyme"). In addition to *N*-oxygenation, TEA also undergoes *C*-oxidative dealkylation to DEA. The dealkylation occurs in the gastrointestinal tract by microbial activity. TEA reaches the gastrointestinal tract through excretion of the amine from the plasma into the stomach (about pH 1), or via swallowing of saliva containing absorbed TEA.

C. Elimination

TEA is eliminated from the body through urinary excretion. It is excreted quantitatively as the unchanged amine and as TEAO and DEA. The urinary excretion rate of TEA is influenced by urinary pH. Lower urinary pH causes faster elimination. TEA is also excreted into gastric juice (and reabsorbed), but is not eliminated in expired air.

Bibliography

Howard, P. H., ed. (1990). "Handbook of Environmental Fate and Exposure Data for Organic Chemicals," Vol. II. Lewis Publishers, Chelsea, Michigan.

Sax, N. I., and Lewis, R. J., eds. (1985). "Dangerous Properties of Industrial Materials," Vol. III. Van Nostrand Reinhold, New York.

Verschueren, K., ed. (1983). "Handbook of Environmental Data on Organic Chemicals." Van Nostrand Reinhold, New York.

Warren, P. J., ed. (1987). "Dangerous Chemicals: Emergency Spillage Guide." Croner Publications, London.

Uranium, Reproductive Effects

José L. Domingo
University of Barcelona

Glossary

Embryotoxicity Significant increases in the number of *in utero* deaths from the time of initial test substance administration until birth.

Fetotoxicity When administration of a test substance during any portion of gestation leads to offspring showing signs of delayed development.

Lactation index (21-day) Number of pups viable at lactation day 21/number of pups retained at lactation day 4.

NOEL Dose level identified as the no observable effect level for a test substance.

Teratogenicity When administration of a test substance during any portion of gestation results in offspring with permanent structural or functional deviations from normal.

Viability index (4-day) Number of pups viable at lactation day 4/number of viable pups born.

MANY INVESTIGATIONS *of uranium biokinetics and chemical toxicity in mammals have been undertaken in recent decades. However, little attention has been paid to the possible effects of uranium exposure on reproduction and developmental toxicity with a resultant lack of published observations regarding these topics. For these and other reasons, in 1987 a wide program directed to obtain an overall understanding of the toxic effects of uranium was started in our laboratory. The evaluation of the reproductive and embryo/fetotoxic effects of uranium was also included. A summary of these results are presented here.*

I. Introduction

Uranium, atomic number 92, is a soft, silvery-white metal whose average concentration in the earth's crust is $4 \times 10^{-4}\%$. The best known use of uranium is as a fuel in nuclear power reactors and nuclear weapons. It is also used in inertial guidance devices, gyro compasses, as a counter weight for missile reentry vehicles, as shielding material, and as X-ray targets.

Uranium extraction and refining operations leading to the manufacture of nuclear fuels have significantly increased in the last 45 years. During processing, there is a possibility that workers may inhale or ingest some uranium, giving rise to internal contamination, which would result in radiation doses to the organs of the body. If the intake of uranium were large enough, chemical toxic effects would occur. Under some circumstances, the chemical toxicity of inhaled or ingested soluble uranium surpasses its radiotoxicity.

In the early days of the Manhattan Project a very extensive toxicological program was mounted whose principal objectives included the establishment of exposure limits for airborne uranium in the workplace based upon uranium's known chemical damage to the kidney. Nephritis is the primary chemically induced health effect of uranium in animals and humans. Uranium is a classic nephrotoxin, and its use in high dosages for the experimental induction of nephrotoxicity is well established. Enzymuria, glycosuria, aminoaciduria, proteinuria, and renal failure have been reported to occur after acute uranium poisoning. Although the effects of uranium on kidney function are now well understood, significant gaps still exist in the knowledge of the chemical toxicity of uranium in mammals. Thus, information concerning the chronic intake of uranium at low levels is not available from the literature. There is also a lack of published observations relating to the reproductive and developmental toxicity of uranium. In the early 1980s, it was reported that at the time of the original toxicological studies during the Second World War, two studies were carried out, one of which used high levels and the other only a brief 24-hr exposure, but both showed statistically significant effects on reproduction. The studies were not repeated by other investigators and so, until recently, questions remained to be answered as to the effect of uranium on reproduction.

Although uranium can exist in oxidation states of +3, +4, +5, or +6, in solution the uranyl ion (UO_2^{2+}) is the most stable species and is thus the form in which this metal occurs in the mammalian body. Therefore, in all the reported studies on the reproductive and developmental toxicity of uranium, the metal was administered as uranyl acetate.

II. Reproductive Toxicity of Uranium

Although inhalation has normally been considered to be the main route of uranium's entry into the body, the general population is also chronically exposed to low levels of uranium via ingestion in food and drinking water. This clearly represents a health hazard, particularly for people living near a uranium mine or mill. Indeed, preliminary evidence has indicated that all counties in New Mexico with commercially significant levels of uranium, or uranium mining and milling operations, are also characterized by high mortality rates for gastric cancer.

The uranium content of most food is in the range of 10–100 ng/g and the average daily intake of the metal in food is about 1.0 μg/day. Drinking water (as a route of uranium exposure) has not been extensively investigated.

The reproductive toxicity of chronic exposure to uranium given to male Swiss mice in drinking water was evaluated in 1991. In that study, mice were treated with uranyl acetate dihydrate at doses of 0, 10, 20, 40, and 80 mg/kg/day for 64 days. The oral LD_{50} of uranyl acetate dihydrate in Swiss mice has been reported to be 242 mg/kg, with confidence limits between 155 and 327 mg/kg.

To assess the fertility of the uranium-treated males, male mice were mated with untreated females for four days. There was a significant but non-dose-related decrease in the pregnancy rate of these animals. Body weights were significantly depressed only in the 80 mg/kg/day group, whereas testicular function/spermatogenesis was not affected by uranium at any dose, as evidenced by normal testes and epididymis weights (Table I) and normal spermatogenesis (Table II). Histopathologi-

Table I Effect of Uranium on Body Weight, Testes, and Epididymis Weight of Mice after 64 Days of Treatment

	Dose (mg/kg/day)				
	0	10	20	40	80
No. of animals	15	16	10	10	80
Body wt (g)	37.6 ± 2.53	37.9 ± 2.60	37.2 ± 2.55	36.4 ± 3.25	35.8 ± 2.04[a]
Left testis wt (mg)	111 ± 15.8	109 ± 14.9	104 ± 23.0	108 ± 13.9	108 ± 19.2
Right testis wt (mg)	107 ± 20.0	114 ± 11.9	96 ± 25.2	109 ± 18.7	112 ± 21.2
Total testis wt (mg)	218 ± 32.2	223 ± 26.2	200 ± 46.0	216 ± 31.6	221 ± 39.8
Testis wt/body wt (%)	0.58 ± 0.07	0.59 ± 0.09	0.54 ± 0.11	0.60 ± 0.07	0.61 ± 0.10
Left epididymis wt (mg)	48 ± 5.4	42 ± 11.8	40 ± 8.0	38 ± 7.3	41 ± 8.2
Right epididymis wt (mg)	49 ± 7.7	44 ± 12.0	40 ± 7.7	41 ± 7.4	43 ± 8.9
Total epididymis wt (mg)	97 ± 10.6	87 ± 21.2	80 ± 14.6[b]	79 ± 13.9[b]	84 ± 16.5[a]
Epididymis wt/body wt (%)	0.26 ± 0.03	0.23 ± 0.06	0.22 ± 0.03[b]	0.22 ± 0.04[a]	0.23 ± 0.04

Source: Llobet *et al.* (1991). *Fund. Appl. Toxicol.* **16**, 821–829. Data represent mean ± SD.
[a] Statistically significant difference as compared to control ($p < 0.05$).
[b] Statistically significant difference as compared to control ($p < 0.01$).

Table II Sperm Parameters in Male Mice Treated with Uranium[a]

	Dose (mg/kg/day)				
	0	10	20	40	80
Spermatid counts					
No. × 10^6/g testis	328 ± 53	384 ± 68[b]	251 ± 64[c]	290 ± 82	306 ± 107
No. × 10^6/ testis	36.6 ± 8.2	43.3 ± 8.8	24.6 ± 10.3[c]	32.9 ± 13.6	33.9 ± 15.2
No. × 10^6/testis/g mouse	0.97 ± 0.21	1.13 ± 0.27	0.66 ± 0.28[c]	0.91 ± 0.34	0.94 ± 0.42
Spermatozoa counts					
No. × 10^6/g epididymis	669 ± 261	457 ± 208[b]	269 ± 220[c]	480 ± 160	507 ± 179
No. × 10^6/ epididymis	32.8 ± 13.3	20.9 ± 13.2[b]	11.6 ± 10.4[d]	19.7 ± 7.2[c]	23.0 ± 11.9
No. × 10^6/epid/g mouse	0.87 ± 0.33	0.54 ± 0.35[b]	0.31 ± 0.28[d]	0.53 ± 0.17[c]	0.64 ± 0.32
Motile cells (%)	20 ± 20 (13)	12 ± 8 (6)	15 ± 7 (7)	20 ± 9 (7)	30 ± 6 (5)
Abnormal morphologic forms (%)	12 ± 6 (13)	12 ± 7 (12)	21 ± 22 (10)	12 ± 6 (9)	9 ± 4 (9)
abnormal hook	2 ± 2	6 ± 7	0 ± 0	2 ± 3	1 ± 4
abnormal head	3 ± 4	2 ± 5	5 ± 6	7 ± 10	5 ± 9
amorphous head	29 ± 11	19 ± 13	33 ± 23	16 ± 9	41 ± 35
folded on themselves	63 ± 15	73 ± 13	61 ± 21	75 ± 13	53 ± 38

Source: Llobet *et al.* (1991). *Fund. Appl. Toxicol.* **16,** 821–829. Data represent mean ± SD.
[a] Number of animals providing the data are in parentheses.
[b] Statistically significant difference as compared to control ($p < 0.05$).
[c] Statistically significant difference as compared to control ($p < 0.01$).
[d] Statistically significant difference as compared to control ($p < 0.001$).

cal examination of the testes in mice killed after 64 days of treatment did not reveal any significant difference between controls and uranium-treated animals at 10, 20, 40, or 80 mg/kg/day, with the exception of interstitial alterations and vacuolization of Leydig cells at 80 mg/kg/day (Table III).

III. Developmental Toxicity of Uranium

There is a paucity of information concerning maternal toxicity, embryotoxicity, fetotoxicity, and teratogenicity of uranium in mammals. In 1989, the developmental toxicity of uranium given as uranyl acetate dihydrate by gavage at daily doses of 5, 10, 25, and 50 mg/kg was evaluated in pregnant Swiss mice on gestational days 6–15. The results indicated that such exposure results in maternal toxicity as evidenced by reduced weight gain and food consumption during treatment, and increased relative liver weight. There are treatment-related effects on the number of implantation sites per dam, or on the incidence of postimplantation loss. The number of live fetuses per litter and the fetal sex ratio are not affected by the treatment. However, dose-related fetotoxicity, consisting primarily of reduced fetal body weight and body length (Table IV), and an increased incidence of abnormalities were observed. Malformations (cleft palate, bipartite sternebrae) and developmental variations (reduced ossification and unossified skeletal districts) are noted at 25 and 50 mg/kg/day test levels (Table V). The "no observable effect level" (NOEL) for maternal toxicity was below 5 mg/kg/day, whereas the NOEL for fetotoxicity including teratogenicity was also below 5 mg/kg/day, as some anomalies were observed at this dose. In contrast, there was no evidence of embryolethality at any dosage level tested in that study.

On the other hand, the effects of multiple maternal subcutaneous injections of uranyl acetate dihydrate (0.5, 1, and 2 mg/kg/day) from days 6 to 15 of gestation were also evaluated in Swiss mice. Maternal toxicity occurred in all uranium-treated groups as evidenced primarily by deaths as well as significant decreases in weight gain and in body weight at termination. Although it was not dose-related, embryotoxicity also occurred in all uranium-treated groups (significant increases in the number of nonviable implantations and in the percentage of postimplantation loss). Fetal body weight was significantly decreased at 1 and 2 mg/kg/day, whereas the number of total internal and total skeletal defects showed dose-dependent in-

Table III Testicular Histopathological Abnormalities in Mice Receiving Uranium for 64 days

	Dose (mg/kg/day)				
	0	10	20	40	80
No. of animals	13	12	11	9	15
Tubular diameter (μm)[a]	208 ± 17	208 ± 14	194 ± 17	206 ± 11	212 ± 15
Tubular alterations	5	6	5	4	10
Sertoli cells vacuolization	5	2	3	2	7
Spermatocele	2	3	1	1	0
Giant cells	0	0	1	0	1
Necrosis of spermatogonia and spermatocytes	5	1	1	1	5
Intratubular giant cells	0	0	0	0	1
Focal atrophy (%)[b]					
No. atrophy	8 (62)	6 (50)	7 (64)	5 (56)	5 (33)
Mild	5 (38)	3 (25)	2 (18)	2 (22)	6 (40)
Moderate	0 (0)	3 (25)	1 (9)	0 (0)	1 (7)
Severe	0 (0)	0 (0)	1 (9)	2 (22)	3 (20)
Interstitial alterations	0	4	2	2	7[c]
Leydig cell vacuolization	0	2	1	2	6[c]
Focal atrophy	0	2	2	1	4
Binucleated cells	0	0	0	0	1

Source: Llobet *et al.* (1991). *Fund. Appl. Toxicol.* **16,** 821–829. Data represent the number of mice in which a lesion was observed.

[a] Results are expressed as means ± SD.

[b] Percentage of animals affected are in parentheses.

[c] Statistically significant difference as compared with controls ($p < 0.05$).

Table IV Developmental Toxicity in Pregnant Mice following Maternal Exposure to Uranyl Acetate Dihydrate by Gavage on Gestational Days 6 through 15

	Dose (mg/kg/day)				
	0	5	10	25	50
No. of dams pregnant	18	17	18	18	18
No. of total implants	12.5 ± 1.6	13.4 ± 2.1	10.9 ± 4.4	12.1 ± 1.5	14.9 ± 2.2
No. of live fetuses	11.8 ± 1.5	12.0 ± 1.6	6.2 ± 4.6[c]	9.0 ± 3.7	11.3 ± 2.5
No. of resorptions					
early	0.6 ± 0.7	0.5 ± 0.7	1.8 ± 1.5	2.0 ± 3.4	1.3 ± 0.8
late	0.1 ± 0.3	0.6 ± 0.5	1.4 ± 2.5	0.6 ± 0.8	1.4 ± 1.6
No. of dead fetuses	0.0 ± 0.0	0.3 ± 0.7	1.5 ± 3.9	0.5 ± 1.3	0.6 ± 0.8
Sex ratio (M/F)	1.00	0.96	0.83	0.95	0.94
Fetal body weight (g)	1.40 ± 0.15	1.04 ± 0.25[c]	0.93 ± 0.14[c]	0.84 ± 0.11[c]	0.77 ± 0.17[c]
Fetal body length (mm)	26.5 ± 2.31	26.3 ± 2.70	24.0 ± 1.91[c]	22.5 ± 1.69[c]	21.8 ± 2.87[c]
No. of stunted fetuses	0.4 ± 1.0	2.0 ± 1.6	3.0 ± 2.4[a]	7.2 ± 3.4[b]	8.6 ± 3.5[b]

Source: Domingo *et al.* (1989). *Toxicol.* **55,** 143–152. Data represent mean (per litter) ± SD unless otherwise indicated.

[a] Statistically significant difference as compared with controls ($p < 0.05$).

[b] Statistically significant difference as compared with controls ($p < 0.01$).

[c] Statistically significant difference as compared with controls ($p < 0.001$).

Table V Incidence of Malformations and Developmental Variations in Mouse Fetuses following Maternal Exposure to Uranyl Acetate Dihydrate by Gavage on Gestational Days 6 through 15

	Dose (mg/kg/day)				
	0	5	10	25	50
Fetuses examined externally	212 (18)	200 (17)	110 (18)	164 (18)	206 (18)
Cleft palate	0	3 (2)	16 (13)[b]	32 (13)[b]	32 (16)[b]
Exencephaly	0	1 (1)	0	0	0
Hematomas (dorsal or in facial area)	0	16 (6)[a]	2 (2)	10 (4)	14 (8)[a]
Micrognathia	0	0	8 (3)	0	0
Tail anomalies (short or curled)	0	0	0	1 (1)	9 (2)
Total external defects	0	19 (8)[a]	24 (14)[b]	42 (14)[b]	53 (17)[c]
Fetuses examined viscerally	71 (18)	67 (17)	36 (18)	54 (18)	68 (18)
Renal papillae undeveloped	0	6 (5)	0	3 (2)	0
Fetuses examined skeletally	141 (18)	133 (17)	74 (18)	110 (18)	138 (18)
Bipartite sternebrae	0	13 (6)[a]	4 (3)	13 (9)[b]	108 (13)[b]
Misaligned sternebrae	0	0	3 (1)	0	12 (5)[a]
Unossified sternebrae (5th, 6th)	5 (2)	6 (3)	31 (9)	17 (9)	83 (13)[a]
Rib 13th, agenesis	0	0	4 (3)	2 (1)	15 (6)[a]
Rib 14th, rudimentary	0	1 (1)	7 (4)	2 (1)	18 (7)[a]
Some metatarsal of hindlimb poorly ossified	16 (4)	39 (9)	57 (15)	106 (18)[a]	129 (18)[a]
Some proximal phalanges (forelimb) poorly ossified	8 (2)	0	11 (6)	15 (13)[a]	42 (14)[a]
Skull, delayed ossification	0	0	7 (3)	16 (9)[b]	24 (12)[b]
Caudal, reduced ossification	13 (4)	45 (9)	81 (12)	49 (18)[a]	50 (18)[a]
Total skeletal defects	18 (4)	58 (11)	64 (15)	106 (18)[a]	129 (18)[a]

Source: Domingo *et al.* (1989). *Toxicol.* **55,** 143–152. Number of litters is in parentheses.
[a] Statistically significant difference as compared with controls ($p < 0.05$).
[b] Statistically significant difference as compared with controls ($p < 0.01$).
[c] Statistically significant difference as compared with controls ($p < 0.001$).

creases at 0.5, 1, and 2 mg/kg/day. Most morphological defects were developmental variations, whereas malformations were only detected at 1 and 2 mg/kg/day. On the basis of these data, both the maternal NOEL and the NOEL for embryotoxicity of uranyl acetate dihydrate administered subcutaneously to mice were below 0.5 mg/kg/day, whereas the NOEL for teratogenicity was 0.5 mg/kg/day.

IV. Perinatal and Postnatal Effects of Uranium

Two reports are available from the literature concerning the perinatal and postnatal effects of uranium. In the first one, uranium was tested for its effects on reproduction, gestation, and postnatal survival in Swiss mice. Mature male mice received uranyl acetate dihydrate by gavage at 5, 10, and 25 mg/kg/day for 60 days prior to mating with mature virgin female mice treated orally at the same doses for 14 days prior to mating. Treatment of the females continued throughout mating, gestation, parturition, and nursing of the litters. One-half of the dams in each group were killed on day 13 of gestation and the remaining dams were allowed to deliver and wean their offspring. No adverse effects on fertility were evident at the doses employed in that study. However, embryolethality could be observed in the 25 mg/kg/day group. Also, significant increases in the number of dead young per litter were seen at birth and at day 4 of lactation in the 25 mg/kg/day group. In addition, the growth of the offspring was always significantly lower for the uranium-treated animals (Table VI).

In the second study, Swiss mice received oral daily doses of uranyl acetate dihydrate at 0.05, 0.5, 5, and 50 mg/kg from day 13 of pregnancy until weaning on day 21 post-birth. Treatment with uranium had no effect on sex ratios, mean litter size, pup body weight, or pup body length throughout lactation, although significant decreases in the mean litter size on postnatal day 21, and in the viability and lactation indices were reported at 50 mg/

Table VI Mouse Pups Nursed by Uranyl Acetate Dihydrate-Treated Dams during the Period of Lactation

	Day	Dose (mg/kg/day) 0	5	10	25
No. of liters	0	9	11	11	12
	4	9	11	10	9
	21	9	11	10	9
No. of living young	0	110	146	121	127
	4	106	135	108	87
	21	101	114	97	82
No. of dead young	0	0	2	11	15
	4	4	11	13	40
	21	5	21	11	5
Dead/living ratio (×100)	0	0	1.3	9.0	11.8
	4	3.7	8.1	12.0	45.9
	21	4.9	18.4	11.3	6.1
Male/female ratio	0	0.79	0.64	0.98	0.86
	4	0.89	0.66	0.96	0.71
	21	0.90	0.69	0.92	0.73
Living young/litter	0	12.2 ± 1.5	13.3 ± 2.6	11.2 ± 4.6	10.7 ± 3.8
	4	11.8 ± 1.3	12.2 ± 2.1	10.8 ± 4.2	9.7 ± 5.9
	21	11.2 ± 1.3	10.4 ± 2.5	9.8 ± 4.8	9.1 ± 4.1
Dead young/litter	0	0.0 ± 0.0	0.2 ± 0.6	1.0 ± 2.2[b]	1.3 ± 2.4[b]
	4	0.4 ± 0.5	1.0 ± 1.5	1.3 ± 1.8	4.0 ± 3.1[b]
	21	0.5 ± 1.0	1.9 ± 2.2	1.1 ± 2.1	0.5 ± 0.9
Viability index	4/0	0.96	0.92	0.88	0.68
	21/0	0.91	0.78	0.79	0.64
Lactation index	21/4	0.95	0.84	0.89	0.94
Pup body weight (g)	0	1.86 ± 0.65	1.76 ± 0.46	1.66 ± 0.28	1.59 ± 0.15[b]
	4	2.60 ± 0.87	2.48 ± 0.68	2.33 ± 0.50[a]	2.20 ± 0.33[b]
	21	9.74 ± 2.75	8.98 ± 1.31[a]	8.85 ± 1.65[a]	8.32 ± 1.25[b]
Pup body length (cm)	0	3.43 ± 0.30	3.42 ± 0.20	3.39 ± 0.27	3.30 ± 0.23[a]
	4	3.91 ± 0.38	3.90 ± 0.42	3.83 ± 0.28	3.80 ± 0.19[a]
	21	7.26 ± 1.41	7.23 ± 0.51	7.11 ± 0.49	6.93 ± 0.37[a]

Source: Paternain *et al.* (1989). *Ecotoxicol. Environ. Safety* **17,** 291–296. Data represent mean ± SD unless otherwise indicated.
[a] Statistically significant difference as compared with controls ($p < 0.05$).
[b] Statistically significant difference as compared with controls ($p < 0.01$).

kg/day. In both studies, the NOEL for health hazards to the developing pup was 5 mg/kg/day, as no adverse findings were noted at this dose.

V. Conclusions

The United Nations Scientific Committee on the Effects of Atomic Radiation (UNSCEAR) has considered that limits for natural uranium in drinking water should be based on chemical toxicity rather than on a hypothetical radiological toxicity in skeletal tissue, which has not been observed in either humans or animals. A value of 100 μg U/l of water has been chosen to be reasonable, based on considerations of kidney toxicity with the application of a safety factor of 50–150. Consequently, an average adult of 70 kg body weight, consuming 2 liter water/day, would not ingest more than 200 μg U/day, which would be equivalent to 0.005 mg/kg/day of uranyl acetate dihydrate. Compared to the NOEL for health hazards to the developing pup found in the studies reported above, as well as to the NOEL for teratogenicity, a safety factor below 1000 can be estimated.

On the average the percentage of uranium ingestion is about 3% from food and 97% from drinking water; therefore, ingestion of uranium from food should not increase the risk of developmental adverse effects. However, people living near a uranium mine or mill may ingest abnormal amounts of the metal. Since the NOEL for maternal and developmental toxicity was below 5 mg/kg/day, these people would require individual guidance.

With regard to the reproductive toxicity of uranium, the data indicate that uranium would not cause any adverse effect on reproduction at the concentrations usually ingested in drinking water. However, since at 10 mg/kg/day of uranyl acetate dihydrate the pregnancy rate is significantly decreased, we intend to investigate the cause of this effect and whether or not it may be totally or partially reversible.

Bibliography

Bosque, M. A., Domingo, J. L., Llobet, J. M., and Corbella, J. (1993). Embryotoxicity and teratogenicity of uranium in mice following subcutaneous administration of uranyl acetate. *Biol. Trace Elem. Res.* **36,** 109–118.

Domingo, J. L., Paternain, J. L., Llobet, J. M., and Corbella, J. (1989a). The developmental toxicity of uranium in mice. *Toxicology* **55,** 143–152.

Domingo, J. L., Ortega A., Paternain, J. L., and Corbella, J. (1989b). Evaluation of the perinatal and postnatal effects of uranium in mice upon oral administration. *Arch. Environ. Health,* **44,** 395–398.

Llobet, J. M., Sirvent, J. J., Ortega, A., and Domingo, J. L. (1991). Influence of chronic exposure to uranium on male reproduction in mice. *Fund. Appl. Toxicol.* **16,** 821–829.

Moss, M. A. (1989). "Study of the Effects of Uranium on Kidney Function". Research Report, Atomic Energy Control Board, Ottawa, Canada.

Paternain, J. L., Domingo, J. L., Ortega, A., and Llobet, J. M. (1989). The effects of uranium on reproduction, gestation and postnatal survival in mice. *Ecotoxicol. Environ. Safety* **17,** 291–296.

Stopps, G. J., and Todd, M. (1982). "The Chemical Toxicity of Uranium with Special Reference to Effects on the Kidney and the Use of Urine for Biological Monitoring." Research Report, Atomic Energy Control Board, Ottawa, Canada.

Volatile Organic Chemicals

Lance A. Wallace
U.S. Environmental Protection Agency

Glossary

Multisorbent system Sampler containing several sorbents in series (e.g., Tenax, graphitized carbon, and activated charcoal) for collection of VOCs of varying volatilities and polarities.

Tenax Granular polymeric bead sorbent for collection of VOCs.

Volatile organic chemicals (VOCs) organic chemicals with boiling points below about 180°C, or vapor pressures greater than 10^{-2} kPa at 25°C.

VOLATILE ORGANIC CHEMICALS (VOCs) comprise some thousands of compounds, many of which are in wide use as solvents, fragrances, and other ingredients in processes and consumer products. Although no standard definition of VOCs has been accepted, they would generally be considered to have vapor pressures greater than about 10^{-2} kPa. Compounds with vapor pressures between about 10^{-2} and 10^{-8} kPa are often described as "semivolatile" organic compounds (SVOCs), a class that includes pesticides, herbicides, polychlorinated biphenyls (PCBs), polychlorinated benzodioxins, and polyaromatic hydrocarbons (PAHs).

I. Introduction

VOCs have great economic importance. Many chemicals with the highest annual production figures are VOCs. Entire industries such as petrochemicals and plastics are based on VOCs. They are used as industrial solvents, in paints and coatings, in pressed wood products, and in literally thousands of consumer products.

VOCs are ubiquitous. Even a rural outdoor air sample will contain some 50–100 VOCs at levels on the order of 0.01–1 part per billion by volume (ppb). Indoor air samples may contain twice as many VOCs at levels several times higher than outdoor air. Common VOCs in buildings are shown in Table I. New buildings contain some airborne VOCs that are 100 times outdoor levels.

Human exposure to most VOCs is mainly through inhalation; a small number of VOCs are in drinking water, food, and beverages as contaminants. Some VOCs may travel in groundwater or through soil from hazardous waste sites, landfills, or gasoline spills to inhabited areas.

Chronic health effects may include cancer. Acute effects from industrial exposures at the part-per-million (ppm) level include skin reactions and central nervous system (CNS) effects such as dizziness and fainting. Recently, a set of "new" illnesses, including sick building syndrome (SBS) and multiple chemical sensitivity (MCS), have been linked by some to relatively low (ppb) concentrations of VOCs.

II. Measurement Methods

A. Environmental Media

At present, measurement methods for VOCs are not comprehensive. A wide variety of nonpolar VOCs can be measured in air using a single collection method and analytical system (typically gas chromatography–mass spectrometry, or GC–MS), at satisfactory limits of detection of about 0.1–1 $\mu g/m^3$. However, no similar broad-spectrum method for analyzing airborne polar VOCs has yet

Handbook of Hazardous Materials

Table I Most Common Organic Compounds at Four Buildings

Class/compound	n^a		N^b
Aromatic Hydrocarbons		Aliphatics	
benzene	16	undecane	10
toluene	16	2-methylhexane	9
xylenes	16	2-methylpentane	9
styrene	16	3-methylhexane	9
ethylbenzene	16	3-methylpentane	9
ethyl methyl benzenes	16	octane	9
trimethyl benzenes	16	nonane	9
dimethylethylbenzenes	15	decane	9
naphthalene	15	dodecane	9
methyl naphthalenes	15	tridecane	9
propylmethylbenzenes	14	methylcyclohexane	9
n-propyl benzene	13	heptane	8
diethyl benzenes	12	tetradecane	8
Halogenated Hydrocarbons		2-methylheptane	8
tetrachloroethylene	16	cyclohexane	8
1,1,1-trichloroethane	15	pentadecane	7
trichloroethylene	14	4-methyldecane	7
dichlorobenzenes	12	2,4-dimethylhexane	7
trichlorofluoromethane	12	pentane	6
dichloromethane	11	hexane	6
chloroform	10	eicosane	6
Esters		3-methylnonane	6
ethyl acetate	8	1,3-dimethylcyclopentane	6
m-hexyl butanoate	4		
Alcohols			
2-ethyl-1-hexanol	9		
n-hexanol	8		
2-butyloctanol	7		
n-dodecanol	6		
Aldehydes			
n-nonanal	13		
n-decanal	10		
Miscellaneous			
acetone	16		
acetic acid	10		
dimethylphenols	6		
ethylene oxide	4		

Source: EPA (1988).
[a] Number of samples (of 16) with compound present.
[b] Number of samples (of 10) with compound present.

emerged. In water, a master analytical scheme has been advanced by the U.S. EPA that is capable of measuring many nonpolar VOCs at limits of detection of about 1 μg/L. Since most exposure to VOCs is through air, the remainder of this discussion will deal with airborne VOCs.

1. Collection

Historically, VOCs at occupational (ppm) levels, were collected on sorbents, usually activated charcoal. A solvent such as carbon disulfide (CS_2) was then used to desorb the collected chemicals. A common analytical method was gas chromatography–flame ionization detection (GC–FID). The National Institute for Occupational Safety and Health (NIOSH) has validated methods for several hundred toxic VOCs at levels above one-half the threshold limit value (TLV). TLVs are often in the range of 50–100 ppm.

However, these methods are generally not suitable for environmental (ppb) levels of these VOCs. A more sensitive method using a granular polymeric sorbent called Tenax came into wide use in the 1970s. Tenax has several favorable properties compared with activated charcoal. It is hydrophobic, so that excessive water vapor is not collected from the atmosphere. It is stable at relatively high temperatures (up to 250°C.) so that chemicals can be desorbed by heating. This avoids the use of solvents, which dilute the sample and also carry the risk of contaminating it. Many of the major studies showing the importance of personal activities and indoor air in exposure to toxic VOCs were carried out in the 1980s using Tenax. However, very volatile VOCs break through the Tenax within a short time of beginning sampling, and compounds with boiling points above 250°C are not desorbed from the Tenax.

An alternative method is to collect a whole-air sample. This sample can then be concentrated in a cold trap and introduced into a GC. This method has the advantages over the sorbent methods of avoiding the sorption–desorption step and collecting all VOCs in the sample. However, the container must have an interior surface free of blemishes that could serve as a sink for some of the VOCs. Tedlar bags or electropolished stainless steel evacuated containers are often used. Even so, VOCs with volatilities much below that of tetradecane may not be recovered adequately from the container. Also, to remove the water vapor that is collected in large amounts, it is often necessary to pass the sample through a drying tube, which results in losing most polar compounds along with the (polar) H_2O.

Another way to circumvent the limited volatility range of sorbents or evacuated containers is to use several sorbents in series, each capable of collecting chemicals of different volatilities. A typical multisorbent system might contain Tenax, a graphitized carbon, and activated charcoal to collect compounds in order of increasing volatility.

The sorbent methods lend themselves to personal monitoring—a small battery-powered pump is worn for an 8- or 12-h period to provide a time-integrated sample. However, at present, the whole-air methods employ bags or canisters that are too bulky or heavy to be used as personal monitors.

2. Analysis

Samples are usually analyzed by first separating the components using gas chromatography. Three detection methods in common use are flame ionization detection (FID), electron capture detection (ECD), and mass spectrometry (MS). Only GC–MS has the ability to unambiguously identify many chemicals. Neither GC–FID nor GC–ECD is able to separate chemicals that coelute (emerge from the chromatographic column at the same time). Also, GC–FID response is depressed by chlorine and other halogens, so it is not suitable for samples containing halogens. Mass spectrometry, by breaking chemicals into fragments and then identifying these fragments, is often capable of differentiating even among coeluting chemicals. However, since chemicals are identified by comparing these mass fragment spectra to existing libraries, and the libraries are incomplete, even GC–MS identifications are often tentative or mistaken. (One study using known mixtures of chemicals found about 75% accuracy of identification for several different GC–MS computerized spectral search systems.)

3. Olfactory Analysis

Because of the complexity of most indoor and outdoor air samples, the cost and inaccuracy of measurement methods, and the almost complete lack of knowledge of the relationship of measured VOC levels within complex mixtures to resulting health effects, an alternative to chemical measurement methods has arisen in recent years—use of a trained panel to judge the possible health or comfort effects of an air sample directly. This method, pioneered by Ole Fänger of Denmark, employs panels of 6–10 persons who have been previously trained by sampling known mixtures of odorous compounds. When exposed to a test atmosphere, the judges provide an instantaneous estimate of its pollution potential, measured in units called decipols. One decipol is equivalent to the amount of pollution (body odor) produced by one person in a room ventilated at one air change per hour. The method is capable of predicting how persons will react to air of a given quality, and also of estimating the relative contribution of various sources (e.g., ventilation system, office machines, employees) to indoor air quality.

B. Biological Media

Measurements of exhaled breath, blood, urine, and mother's milk have all been employed to determine dosage or body burden of VOCs. The most sensitive of these measurements is analysis of exhaled breath, capable of identifying 50–200 compounds in breath of unexposed persons. Breath measurements generally include a supply of clean humidified air to the subject, a two-way breathing mouthpiece, and collection of expired air on Tenax or in an evacuated cylinder. Some methods provide for collection of alveolar air; others include mixed alveolar and dead space air. The alveolar air is directly related to the concentration in the blood by the partition coefficient of the given chemical. Analysis is normally by GC–MS. Breath measurements were crucial in identifying the major role of smoking in benzene exposure—smokers exhibited 6–10 times the benzene breath levels of nonsmokers. Personal air monitors had not been able to detect this exposure.

Blood measurements are widely employed. However, the sensitivity is lower than for breath; recent improved protocols, using isotopic dilution methods to compensate for incomplete recoveries from the blood matrix, are able to achieve limits of detection of about 0.1 μg/L, but this is still only enough to detect about a dozen VOCs in unexposed people. Also, the procedure is invasive, and some potential subjects will refuse.

Measurements of urine often concentrate on metabolic products (trichloracetic acid from trichlorethylene, muconic acid from benzene).

Since many VOCs are lipophilic, they accumulate in fat. Some measurements of mother's milk have documented the presence of tetrachloroethylene and *para*-dichlorobenzene (*p*-DCB).

III. Human Exposure

A. Pathways

Human exposure to about 20 of the more toxic and carcinogenic VOCs was measured directly for about 800 residents of eight U.S. cities in EPA's TEAM Studies during the 1980s (Tables IIa and IIb). For most chemicals, the major pathway of exposure was indoor air, and the major sources of exposure were personal activities, consumer products, and building materials. The contribution of outdoor air to total human exposure was generally less than 25%, and in one case (*para*-dichlorobenzene) less than 2%. This was in sharp contrast to the general focus of EPA regulatory activities, which has been on reducing outdoor air levels by regulating large stationary sources and numerous mobile sources.

For example, the major source of exposure to benzene on a nationwide basis is active smoking—

Table IIa Weighted Estimates of Air and Breath Concentrations of 11 Prevalent Compounds for 130,000 Elizabeth-Bayonne Residents (Fall 1981); 110,000 Residents (Summer 1982); and 49,000 Residents (Winter 1983)

	Season I (fall)			Season II (summer)			Season III (winter)		
Compound	Personal air (*N* = 340)	Outdoor air (86)	Breath (300)	Personal air (150)	Outdoor air (60)	Breath (110)	Personal air (49)	Outdoor air (9)	Breath (49)
1,1,1-Trichloroethane	94[a]	7.0[a]	15[b]	67	12	15	45	1.7	4.0
m,p-Dichlorobenzene	45	1.7	8.1	50	1.3	6.3	71	1.2	6.2
m,p-Xylene	52	11	9.0	37	10	10	36	9.4	4.7
Tetrachloroethylene	45	6.0	13	11	6.2	10	28	4.2	11
Benzene	28	9.1	19	NC[c]	NC	NC	NC	NC	NC
Ethylbenzene	19	4.0	4.6	9.2	3.2	4.7	12	3.8	2.1
o-Xylene	16	4.0	3.4	12	3.6	5.4	13	3.6	1.6
Trichloroethylene	13	2.2	1.8	6.3	7.8	5.9	4.6	0.4	0.6
Chloroform	8.0	1.4	3.1	4.3	13	6.3	4.0	0.3	0.3
Styrene	8.9	0.9	1.2	2.1	0.7	1.6	2.4	0.7	0.7
Carbon tetrachloride	9.3	1.1	1.3	1.0	1.0	0.4	ND[d]	ND	ND
Total (11 compounds)	338	48	80	200	59	66	216	25	31

Source: Wallace (1987).
[a] Average of arithmetic means of day and night 12-hour samples (μg/m³).
[b] Arithmetic mean.
[c] Not calculated—high background contamination.
[d] Not detected in most samples.

Table IIb Weighted Estimates of Air and Breath Concentrations of Nineteen Prevalent Compounds for 360,000 Los Angeles Residents (February 1984); 330,000 Los Angeles Residents (May 1984); and 91,000 Contra Costa Residents (June 1984)

	LA1 (Feb)			LA2 (May)			CC (June)		
Chemical	Personal air (N = 110)	Outdoor air (24)	Breath (110)	Personal air (50)	Outdoor air (23)	Breath (50)	Personal air (76)	Outdoor air (10)	Breath (67)
1,1,1-Trichloroethane	96[a]	34[a]	39[b]	44	5.9	23	16	2.8	16[b]
m,p-Xylene	28	24	3.5	24	9.4	2.8	11	2.2	2.5
m,p-Dichlorobenzene	18	2.2	5.0	12	0.8	2.9	5.5	0.3	3.7
Benzene	18	16	8.0	9.2	3.6	8.8	7.5	1.9	7.0
Tetrachloroethylene	16	10	12	15	2.0	9.1	5.6	0.6	8.6[b]
o-Xylene	13	11	1.0	7.2	2.7	0.7	4.4	0.7	0.6
Ethylbenzene	11	9.7	1.5	7.4	3.0	1.1	3.7	0.9	1.2
Trichloroethylene	7.8	0.8	1.6	6.4	0.1	1.0	3.8	0.1	0.6
n-Octane	5.8	3.9	1.0	4.3	0.7	1.2	2.3	0.5	0.6
n-Decane	5.8	3.0	0.8	3.5	0.7	0.5	2.0	3.8	1.3
n-Undecane	5.2	2.2	0.6	4.2	1.0	0.7	2.7	0.4	1.2
n-Dodecane	2.5	0.7	0.2	2.1	0.7	0.4	2.1	0.2	0.4
α-Pinene	4.1	0.8	1.5	6.5	0.5	1.7	2.1	0.1	1.3
Styrene	3.6	3.8	0.9	1.8	—	—	1.0	0.4	0.7
Chloroform	1.9	0.7	0.6	1.1	0.3	0.8	0.6	0.3	0.4
Carbon tetrachloride	1.0	0.6	0.2	0.8	0.7	0.2	1.3	0.4	0.2
1,2-Dichloroethane	0.5	0.2	0.1	0.1	0.06	0.05	0.1	0.05	0.04
p-Dioxane	0.5	0.4	0.2	1.8	0.2	0.05	0.2	0.1	0.2
o-Dichlorobenzene	0.4	0.2	0.1	0.3	0.1	0.04	0.6	0.07	0.08
Total (19 compounds)	240	120	80	150	33	56	72	16	44

Source: Wallace (1987).
[a] Average of arithmetic means of day and night 12-hour samples ($\mu g/m^3$).
[b] One very high value removed.

smokers average about 6–10 times as much benzene in their blood as nonsmokers. Even passive smoking accounts for a substantial portion of benzene exposures—homes with smokers averaged about 50–60% more benzene in indoor air than homes without smokers, both in the United States and in West Germany. Other important sources include driving and having an attached garage (vaporization from hot engines). By contrast, the "major" stationary sources regulated by the Clean Air Act account for only about 3% of total U.S. exposure. (Fig. 1)

A second example is tetrachloroethylene, the most common dry cleaning solvent. Although emissions from dry cleaning shops are regulated, emissions from the clothes themselves are not—but it is these emissions, occurring either while the clothes are being worn or stored in the home, that accounts for most exposure. Indoor air concentrations of tetrachloroethylene in homes with newly dry cleaned clothes typically rise to 100 times outdoor levels within 24 hours, falling slowly back to normal indoor levels over the next two weeks.

As another example, *p*-DCB, carcinogenic to rats and mice, is a registered pesticide, used in mothcakes and moth crystals. However, a second major use is as a toilet deodorant and room air freshener. Surveys in Baltimore and Los Angeles indicated that mothcakes were used in 12–25% of homes, but air fresheners in 70%. The TEAM study indicated that about a third of the 800 homes had elevated levels of *p*-DCB, averaging about 50 times outdoor levels. Thus the outdoor air contribution to exposure is only 2% of the indoor air contribution. Although some *p*-DCB exposure is due to its use as pesticide, a substantial portion is due to its use as an air freshener, probably without consumers being aware that it is a carcinogen and a pesticide. Since the 1985 study demonstrating its carcinogenicity, both EPA and CPSC have been considering action (e.g., labeling) but as of this writing no action has been taken.

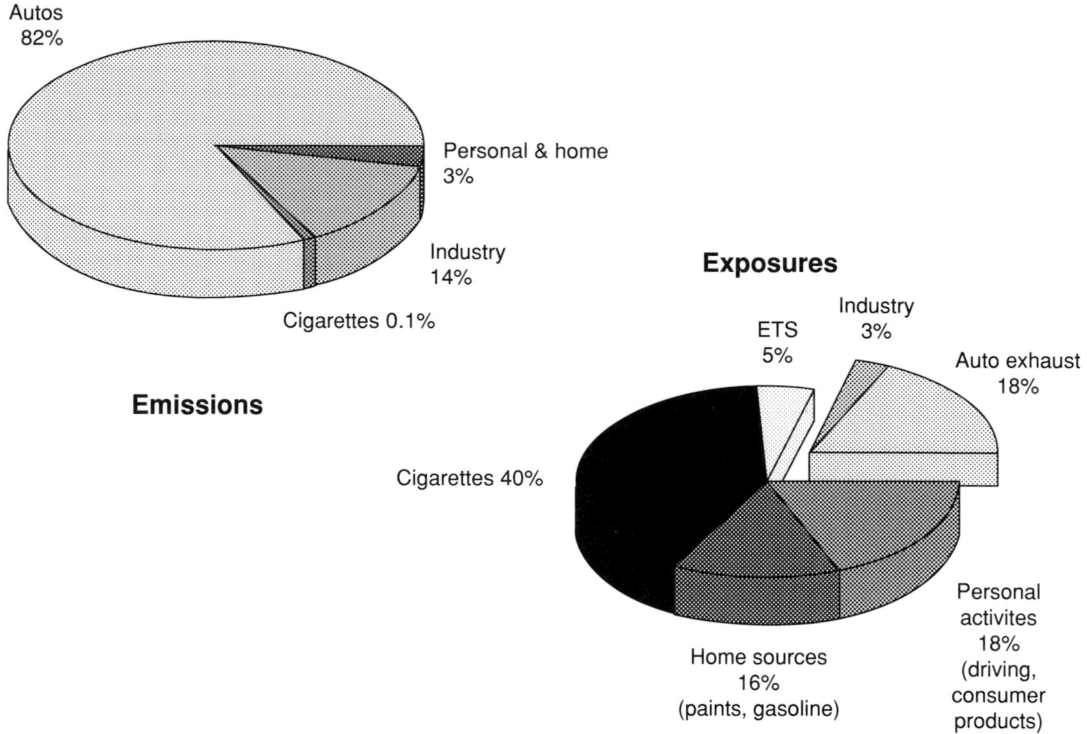

Figure 1 The major sources of benzene *emissions* are automobiles and industry, which are regulated by the EPA. However, the major sources of benzene *exposure* are cigarette smoke, driving, and storing paint and gasoline in homes, all of which are unregulated.

By contrast, one chemical for which indoor sources are unimportant is carbon tetrachloride, for which nearly 100% of exposures are due to outdoor concentrations. Carbon tetrachloride was banned from consumer products by the Consumer Product Safety Commission (CPSC), and subsequent measurements in homes indicate that few or no indoor sources exist anymore. However, the extensive use of carbon tetrachloride as a dry cleaning chemical and for many other purposes early in this century, coupled with its long atmospheric life, has resulted in a global background that is near one $\mu g/m^3$, not an insignificant cancer risk according to EPA calculations.

Drinking water, food, and beverages are important sources of exposure for a small number of VOCs—particularly chloroform, several other trihalomethanes, and limonene. All these chemicals are animal carcinogens, although limonene was carcinogenic only to male rats and therefore is not a probable human carcinogen. Chloroform in drinking water is created by the chlorination process; excess chlorine reacts with humic material to form chloroform. The presence of the other (brominated) trihalomethanes is due to a combination of chlorination of water and the presence of bromine in groundwater sources. Soft drinks, milk, and cheese have all been found to include chloroform. The chloroform in soft drinks may be due to a breakdown of the filtering process (not changing activated charcoal filters often enough); in milk and cheese, possibly to dairy cows having access to chlorinated drinking water and the resulting storage of the lipophilic chemical in milkfat. Limonene is present naturally in citrus fruits and plants such as tobacco and is added to lemon-lime sodas and other foods. Indoor air is also an important pathway for chloroform and limonene. Chloroform is volatilized from hot water during showers, and is volatilized and probably created during clotheswashing and dishwashing by the action of chlorine-containing bleaches and dishwasher soaps on dirt and food particles. Limonene is volatilized from brewing tea, smoking tobacco, and using lemon-scented cleansers and polishes.

B. Sources

Major sources of exposure to most VOCs are personal activities, consumer products, and building materials. Personal activities such as smoking and driving expose people to complex mixtures containing VOCs at elevated concentrations. Cigarette smoke contains about 4,000 VOCs, some of which are known or suspected carcinogens such as benzene, 1,3-butadiene, formaldehyde, and several nitrosamines. The sidestream smoke often contains these same chemicals at even higher concentrations than mainstream smoke, although of course the passive smoker inhales a greatly diluted mixture. Auto exhaust contains additional hundreds or thousands of compounds, including some of the same carcinogens as tobacco smoke; driving in urban traffic results in about a sixfold elevation of exposure to benzene compared with normal outdoor concentrations. Other common personal activities resulting in short-term elevations of exposure to one or more toxic or carcinogenic VOCs include showering (chloroform); auto repair or maintenance (benzene, tetrachloroethylene in carburetor cleaners); household cleaning (methyl chloroform); painting or paint removing (methylene chloride); use of fumigants (ethylene dichloride and ethylene dibromide); and visiting retail shops such as hardware stores, grocery stores, and other centers for the distribution of VOC-laden products.

Products and building materials often emit VOCs. Because of the small airtight cabins in space capsules, the National Aeronautics and Space Administration (NASA) has been testing emissions of all materials used or carried aboard spacecraft for more than a decade; about 5000 materials have now been examined. The VOCs emitted by the largest number of materials are displayed in Table III.

Emission rates of most chemicals in most materials are greatest when the materials are new. For "wet" materials such as paints and adhesives, most of the total volatile mass may be emitted in the first few hours or days following application. EPA studies of new buildings indicated that eight of 32 target chemicals measured within days after completion of the building were elevated 100-fold compared with outdoor levels: xylenes, ethylbenzene, ethyltoluene, trimethylbenzenes, decane, and undecane. The half-lives of these chemicals varied from two to six weeks; presumably some other nontarget chemicals, such as toluene, would have shown similar behavior. The main sources were likely to be paints and adhesives. Thus, new buildings would be expected to require about six months to a year to decline to the VOC levels of older buildings.

For dry building materials such as carpets and pressed wood products, emissions are likely to continue at low levels for longer periods. Formaldehyde from pressed wood products may be slowly emitted with a half-life of several years. According to several recent studies, 4-phenylcyclohexene (4-PC), a reaction product occurring in the styrene–butadiene backing of carpets, is the main VOC emitted from carpets after the first few days. 4-PC is likely to be largely responsible for the new carpet odor.

A major category of human exposure to toxic and carcinogenic VOCs is room air fresheners and bathroom deodorants. Since the function of these products is to maintain an elevated indoor air concentration in the home or the office over periods of weeks (years with regular replacement), extended exposures to the associated VOCs are often the highest likely to be encountered by most (nonsmoking) persons. The main VOCs used in these products are *p*-DCB (widely used in public restrooms), limonene, and α-pinene. The first is carcinogenic to two species, the second to one, and the third is mutagenic. Limonene (lemon scent) and α-pinene (pine scent) are also used in many cleaning and polishing products, which would cause short-term peak exposures during use, but which might not provide as much total exposure as the air freshener.

Awareness is growing that most exposure comes from these small nearby sources. In California, Proposition 65 focuses on consumer products, requiring makers to list carcinogenic ingredients. Bills focusing on indoor air were introduced in both the House and the Senate in 1989 and 1990. Environmental tobacco smoke (ETS), was declared a known human carcinogen by EPA in 1991; smoking has been banned from many public places and many private workplaces during the last few years.

IV. Health Effects

A. Acute Toxicity

Historically, awareness of the health effects of these chemicals began in the industrial setting. Acute toxicity was noted due to exposures in the

Table III VOCs Emitted by at Least 100 of 5000 Materials Tested by NASA for Use in Space Flights

Compound	Count	Compound	Count
Alcohols		Ethers	
2-Propanol	1344	1,4-Dioxane	111
Methanol	1103	Chlorinated hydrocarbons	
Ethanol	726	Methylene chloride	431
1-Butanol	605	1,1,1-Trichloroethane	429
2-Methyl-2-propanol	440	Trichloroethylene	381
2-Methyl-1-propanol	119	Freon TF	337
Aldehydes		Freon 11	115
Acetaldehyde	1624	Tetrachloroethylene	105
Propanal	289	Hydrocarbons (aliphatic)	
Butanal	236	Butenes	1114
C_6 aldehydes	189	Propane	266
C_5 aldehydes	181	*n*-Butane	206
Aromatics		Propene	211
Toluene	2076	Cyclohexane	119
Xylenes	1194	Ketones	
Benzene	387	Acetone	2131
Styrene	284	2-Butanone	1385
C_9 aromatic HCs	143	4-Methyl-2-pentanone	397
Esters		Cyclohexanone	117
Ethyl acetate	218	Sulfur-containing	
2-Ethoxyethylacetate	185	Carbon disulfide	284
n-Butyl acetate	166	Carbonyl sulfide	245
Aliphatic hydrocarbons		Nitrogen-containing	
C_5	163	Ammonia	189
C_6	352	Silicon-containing	
C_7	447	Siloxane trimer	302
C_8	355	Hexamethylcyclotrisiloxane	245
C_9	113	Trimethylsilanol	232
C_{10}	142	Siloxane tetramer	126
C_{10-12}	250	Siloxane dimer	102
C_{11-12}	193		
C_{12}	111		

Source: Wallace (unpublished).

high ppm range. Many industrial solvents have anesthetic effects on the central nervous system, ranging from dizziness and nausea to death in extreme cases. Other common occupational effects are skin rashes (chloracne).

B. Cancer

In this century, cancer cases from industrial exposure began to be noted. Benzene appeared to cause leukemia in rubber workers; bis-chloromethyl ether (BCME) caused liver cancer in chemical workers. Because thousands of new chemicals were being introduced into wide use during the first few decades of the 20th century, fears rose that chemically caused cancer might become an epidemic. A major testing program—the National Toxicology Program (NTP) was instituted to test the most suspicious chemicals on rats and mice. The high cost (about $500,000) of these 2-year tests led to efforts to find a shorter, less costly, way to test a larger number of chemicals. These efforts culminated in the Ames test, which used mutations in bacteria as an indicator of possible carcinogenicity. Years of experience with the Ames test indicate that it is about 85% effective in predicting carcinogenicity based on mutagenicity. By now, several hundred chemicals have been tested, with about half resulting in mutagenic alterations. Although this appears to be a high proportion, most chemicals are tested because they have already incurred suspicion; therefore the true frequency of carcinogenic chemicals is likely to be much lower. On the other hand, the developer of the Ames test has recently been a powerful advocate for the idea that natural carcinogens in the diet far outweigh synthetic carcinogens in terms of human exposure. This might imply that we have defense mechanisms (DNA repair, for example) that are able to withstand low exposures to carcinogens.

Concurrent with these advances in testing chemicals for carcinogenicity, epidemiological studies were undertaken to determine the contribution of the environment to cancer. It is generally accepted today that the prime determinant of cancer in the United States is lifestyle (smoking and diet); the contribution of environmental chemicals is typically estimated at 1–3%.

Thus the original fear of a cancer epidemic from chemical exposures is receding. Nonetheless, estimates of cancer risk from various environmental sources appear to agree on the primacy of three sources: radon, ETS, and VOCs. All three appear to be responsible for several thousand deaths per year from cancer. In the case of the VOCs, the chemicals with the highest risk estimates include benzene, chloroform, and *para*-dichlorobenzene. Other VOCs which could be important, but for which we do not yet have good exposure or potency data, are methylene chloride, formaldehyde, 1,3-butadiene, vinyl chloride, and vinylidene chloride.

C. Chronic Toxicity

1. Sick Building Syndrome

In the 1970s, incidents began to multiply involving widespread sicknesses among office workers, typically immediately after moving into a new building or after major renovation. The symptoms often included headache, eye irritation, sinus congestion, CNS symptoms such as dizziness and difficulty concentrating, and dry skin. In some cases, new buildings had to be abandoned. In most cases, severe reductions in productivity occurred. Similar epidemics had begun a few years earlier in Scandinavian countries. Originally, suspicion centered on the new energy-saving practices that resulted in "tight" (low air exchange rate) buildings; indeed, SBS is often also called "Tight Building Syndrome." Very low air exchange rates could result in the buildup of VOCs due to indoor sources such as paints, adhesives, carpets, and equipment such as copiers, printers, and computers. One study exposed SBS persons to a mixture of 22 common VOCs (or to clean air as a control). They displayed the same symptoms on exposure to the chemical mixture: headache, eye irritation, reduction in short-term memory. However, the study was marred by the fact that the 64 subjects were able to recognize the chemical mixture compared to the clean air control, thus violating the blind conditions. The study was repeated using healthy young male college students; they too displayed the same symptoms on exposure to the mixture. In this case, too, the blind conditions were violated, but since the students had nothing to gain from biasing their answers, it is not such a serious flaw.

The cause or, more likely, causes, of SBS remain unknown. Recent studies do not seem to bear out a direct relationship with air exchange rates. VOCs are still a prominent possibility, but other causes such as molds, allergens, mites, settled dust, and psychosocial conditions (job stress, organizational conflicts) are also being studied.

2. Chemical Sensitivity

For a small number of workers, SBS symptoms were much more severe: fatigue, fainting, lingering weakness. A sensitization phenomenon appeared to be occurring for these persons, such that even brief low-level subsequent exposures to the air in the building or the materials implicated in the earlier outbreaks could leave sufferers incapacitated. Most patients are women. Fragrances such as perfumes, deodorants, and soaps are often acutely distressing to sufferers. Since no known biochemical marker of the syndrome has yet been found, most of the medical profession does not accept chemical sensitivity as a medical condition. Indeed, many allergists, who have been unable to treat the disease, believe that it is largely psychological in origin. A branch of medicine known as clinical ecology has emerged to treat chemically sensitive people. Clinical ecologists generally attempt to isolate the person from most environmental stressors, reintroducing them one by one to discover which are the crucial ones, and then advising avoidance of the problem chemicals or materials from then on. This has led to serious reductions in the quality of life for most persons so treated. Future advances in understanding this condition await discovery of dependable biomarkers or at least reproducible objectively measured symptoms following double-blind exposures.

V. Environmental Effects

VOC emissions have major environmental effects on stratospheric ozone, smog, and visibility. The effects of chlorofluorocarbons (CFCs) used for refrigeration and packaging in attacking the Earth's stratospheric ozone layer are well documented. The 1989 Montreal Protocols agreed on a 30% reduction of CFC production, and DuPont announced complete termination of production shortly afterward. However, other VOCs besides the CFCs may have substantial effects on stratospheric ozone, and continued effort to improve our knowledge of atmospheric chemistry to be able to select effective controls will be necessary.

Reactive VOCs, together with inorganic gases such as SO_2 and NO_X, form smog (ozone, secondary aerosols) through various chemical processes. Smog may well have health effects (acid aerosols have recently come under suspicion) but it certainly damages vegetation, crops, and materials, and reduces visibility. The economic costs of these effects may exceed the costs associated with health effects.

VI. Controls

Industrial exposures to several hundred VOCs are subject to the threshold limit values (TLVs). These standards are based largely on research by industrial toxicologists on acute reactions of animals and a healthy workforce to short-term peak exposures. Such experience may not be relevant to the low-level chronic exposures experienced by all segments of the population, including children, sick, and aged persons.

An alternative to TLVs would be risk-based standards or guidelines. At present, only cancer risk can be crudely estimated using quantitative methods. The EPA maintains a computerized data base (IRIS) with updated carcinogenic potency estimates for some hundreds of chemicals. For airborne VOCs, the potency estimate is available as a "unit risk"; lifetime upper-bound risk of cancer due to 70 years' exposure to 1 $\mu g/m^3$ of the chemical. These estimates are "upper-bound" because they take the highest possible fit to the dose–response curve in the animal studies; the true risk is likely to be considerably smaller. In fact, the lower bound risk for most of these chemicals is zero—they may not cause cancer in humans at all.

VOC emissions from many industrial processes are regulated by source emission standards. VOCs from automobile tailpipes are also regulated through inspection and maintenance programs. Growing recognition that small but numerous sources such as wood stoves, barbecues, and gasoline lawn mowers contribute significant amounts of VOCs to the atmosphere has led the Southern California Air Management District to propose stringent regulations aimed at such sources.

Pollution prevention programs may succeed in reducing VOCs at their source: manufacture of a product. Vapor-recovery systems during manufacture, or reduction of VOCs in the finished product, could reduce emissions dramatically. Consensus standards for building material emissions by organizations such as the American Society for Testing and Materials (ASTM) could lead to wider use of low-emitting materials and surface coatings.

Individual actions can be far more effective than governmental regulations or consensus standards in reducing or eliminating VOC exposures. For example, storing paints and solvents in a detached garage or tool shed would reduce indoor air levels. Eliminating room air fresheners would reduce exposures to some chemicals by massive amounts. Quitting smoking or allowing it only outdoors or in a room with its own exhaust fan would reduce exposure to the many carcinogenic VOCs contained in tobacco smoke. Buying clothes that do not require dry cleaning, or hanging freshly drycleaned clothes outside for a day, would eliminate or reduce exposure to dry cleaning solvents.

Related Articles: CHLOROFORM; ENVIRONMENTAL CANCER RISKS; ORGANIC SOLVENTS, HEALTH EFFECTS; SOIL DECONTAMINATION; XYLENES.

Bibliography

EPA (1991). "Health Effects Assessment Summary Tables." U.S. EPA, Environmental Criteria and Assessment Office, Cincinnati, Oh.

EPA (1988). "Indoor Air Quality in Public Buildings." vols. I and II. U.S. EPA, Office of Research and Development, Washington, D.C. EPA/600/S6-88/009a,b.

NAS (1976). "Vapor-Phase Organic Pollutants." National Academy of Sciences, Washington, D.C.

NAS (1986). "Environmental Tobacco Smoke: Measuring Exposures and Assessing Health Effects." National Academy of Sciences, Washington, D.C.

Wallace, L. (1987). "The TEAM Study: Summary and Analysis." vol. I. U.S. Environmental Protection Agency, Washington, D.C.

Xylenes

Armanda Jori
Istituto di Ricerche Farmacologiche Mario Negri, Milan, Italy

Glossary

Carcinogenicity Tendency or ability to produce a malignant epithelial cell tumor, which can occur anywhere in the body and spread through the bloodstream

Gas chromatography Technique for separating gas mixtures, in which the gas is passed through a long column containing a fixed absorbent phase that separates the gas mixtures into its component parts.

Metabolism Sum of all the chemical and physical processes within a living organism.

Mutagenesis Introduction into a gene of an alteration that changes the structure or function of the gene product

XYLENES are major industrial chemicals derived mainly from petroleum refining. They occur in three isomeric forms (ortho, meta, and para) and are produced and used as mixed xylenes as well as individually. Xylenes are produced in large amounts and the total worldwide production is growing. They must be considered as environmental contaminants, as they have been identified in the atmosphere and in superficial and finished waters in many countries. Occupational exposure has been reported in petroleum refining, in xylene production, and in industries where xylene and its end products are used. Acute and chronic toxicity are moderate. Major symptoms in humans are transient liver and renal damage and minor impairment of the central nervous system (CNS). Metabolism is a valid process of detoxification and prevents major accumulation of xylenes in adipose tissue. In animals xylenes produce malformations and embryotoxicity at doses which induce mild maternal toxicity. Specific teratogenic properties have not been demonstrated. Mutagenicity can be excluded. Carcinogenicity, which has been suggested by the appearance of hematopoietic malignancies in people occupationally exposed to various chemicals including xylenes, has not been adequately documented in animal experiments. Epidemiological studies for carcinogenicity in humans have been considered to give inadequate evidence.

I. Chemical and Physical Data

A. Synonyms

Ortho-, meta-, and para-xylene are known as *o*-, *m*-, and *p*-xylene or 1,2-, 1,3-, and 1,4-dimethylbenzene (*Chemical Abstracts* names), respectively. Mixed xylene is a mixture of ortho-, meta-, and para-isomers. Commercial grade (mixed) xylene is generally composed of approximately 20% ortho-, 40% meta-, and 20% para-xylene, with about 15% ethylbenzene and smaller quantities of toluene, trithylbenzene, phenol, thiophene, pyridine, and nonaromatic hydrocarbons.

B. Chemical Formula and CAS Number

The chemical formula of xylene is $C_6H_4(CH_3)_2$ or C_8H_{10} (C : 90.50%; H : 9.50%), with the structures of the isomers as shown below.

CH_3 CH_3 CH_3 CH_3 CH_3 CH_3

ortho meta para

Handbook of Hazardous Materials

from 0.003 to 0.38 mg/m^3 (data taken in the United States and in some European countries), with maximum levels in urban areas and minimum levels in rural areas. Mean concentrations at urban sites in the United States in 1961 through 1974 were between 0.04 and 0.15 mg/m^3. Very high levels were detected in air during forest fires (116–684 mg/m^3 in smoke).

Xylenes have been detected indoors too. Meta- and para-xylene have been found at combined concentrations of 0.01–0.03 mg/m^3 in various rooms, depending on cooking and fuel burning. Higher concentrations (0.2 mg/m^3) have been detected in the air of nonventilated rooms filled with cigarette smoke.

3. Presence in Water

In surface water xylenes have been detected in amounts ranging from 0.2 to 16 μg/liter in various U.S. and European regions; 33–56 μ/liter levels were detected in subterranean water sampled near an underground coal gasification site in Wyoming. Xylenes have been found in finished water in urban areas of Europe and the United States.

4. Presence in Soil

No quantitative data has been found on the presence of xylenes in soil. Numerous bacteria present in the soil are capable of growing with para- and meta-xylene as their only carbon source (no information is available on *o*-xylene breakdown). It has been suggested that xylenes present as soil contaminants interact with microorganisms which eliminate them by degradation.

5. Occupational Exposure

It has been calculated that several million workers are potentially exposed to xylenes in the world. Table IV reports the occupational exposure limits to xylenes imposed by regulations of various countries and regions.

III. Toxicological Data

A. Acute, Subacute, and Chronic Toxicity

1. In Animals

The data is available on the toxicity of xylene in various animal species is set out in Tables V–VII.

Table IV Occupational Exposure Limits for Xylenes (All Isomers)

Country or region	Year	Concentration (ppm)	Interpretation[a]
Austria	1985	100	TWA
Belgium	1985	100	TWA
Brazil	1985	80	TWA
Bulgaria	1985	11	TWA
Chile	1985	80	TWA
China	1985	23	TWA
Czechoslovakia	1985	46	Average
		230	Maximum
Denmark	1988	50	TWA
European Economic Community	1986	100	Average
		500	Maximum
Finland	1987	100	TWA
		150	STEL
France	1986	100	TWA
		150	STEL (15 min)
German Democratic Republic	1985	46	TWA
		135	STEL
Germany, Federal Republic of	1988	100	TWA
Hungary	1985	11	TWA
		23	STEL
India	1985	100	TWA
		150	STEL
Indonesia	1985	100	TWA
Italy	1985	91	STEL
Japan	1988	100	TWA
Korea, Republic of	1985	100	TWA
		150	STEL
Mexico	1985	100	TWA
Netherlands	1986	100	TWA
Norway	1981	100	TWA
Poland	1985	23	TWA
Romania	1985	68	TWA
		91	Maximum
Sweden	1987	45	TWA
		100	STEL (15 min)
Switzerland	1985	100	TWA
Taiwan	1985	100	TWA
United Kingdom	1987	100	TWA
		150	STEL (10 min)
United States[b]			
OSHA	1988	45	TWA
		68	Ceiling
NIOSH	1986	100	TWA
		200	Ceiling (10 min)

(*continues*)

Table IV (*Continued*)

Country or region	Year	Concentration (ppm)	Interpretation[a]
ACGIH	1988	100	TWA
		150	STEL (15 min)
USSR	1985	11	Ceiling
Venezuela	1985	100	TWA
		150	Ceiling
Yugoslavia	1985	11	TWA

Source: Compiled in International Agency for Research on Cancer (IARC) (1989) from data in many other references.

[a] TWA, 8-hour time-weighted average; STEL, short-term exposure limit.

[b] OSHA, Occupational Safety and Health Administration; NIOSH, National Institute for Occupational Safety and Health; ACGIH, American Conference of Governmental Industrial Hygienists.

2. In Humans

Most information on the adverse effects of xylenes originates from studies of workers occupationally exposed to other organic solvents at the same time; so it is difficult to extrapolate the toxic symptoms related to the xylenes alone.

In an accident in which three painters working in a confined space were overcome by xylene vapors estimated to be 10,000 ppm, one died from pulmonary edema; at the autopsy petechiae were found in the brain along with neuronal damage. The two other people remained unconscious for 15 to 18 hours, waking with mental confusion and amnesia, but recovered completely in two days; transient hepatic impairment was detected. In other cases of severe acute xylene poisoning, transient kidney and liver damage has been reported.

Severe gastrointestinal irritation and distress have been reported after occasional ingestion. Exposure of volunteers to technical mixed xylenes indicated the following health risks.

Eye contact: Brief exposure (3–5 min) to 200 ppm causes irritation of eyes, nose, and throat.

Skin contact: Causes burning sensation, irritation, and erythema. Prolonged exposure may cause vesicles and contact dermatitis.

Inhalation: Exposure to 390 mg/m^3 (90 ppm) for 6 hours affects reaction time, manual coordination, body equilibrium, and

Table V Acute Toxicity (Single Dose)

Species	Isomer	Route of administration	Dose	Effects observed
Mouse	*o*-	Inhalation	4595 ppm/6 hr	LC50
	m-	Inhalation	5267 ppm/6 hr	LC50
	p-	Inhalation	3907 ppm/6 hr	LC50
	Mix	Intraperitoneal	1548 mg/kg	LD50
Rat	*o*-	Oral	3567 mg/kg	LD50
	m-	Oral	4988 mg/kg	LD50
	p-	Oral	3910 mg/kg	LD50
	Mix	Oral	5762 mg/kg	LD50
	Mix	Inhalation	6670 ppm/4 hr	LC50[a]
	Mix	Inhalation	1334 ppm/4 hr	Irritation of eyes
	Mix	Inhalation	575 ppm/4 hr	No effects
Rabbit	Mix	Dermal	4300 mg/kg	Erythema, slight necrosis; if applied to the eye, conjunctival irritation, corneal damage
Dog	Mix	Inhalation	920 ppm/1 hr	Lacrimation
Cat	Mix	Inhalation	9430 ppm/2 hr	Salivation, ataxia, tonic clonic spasms, anesthesia, death
Guinea pig	Mix	i.p.	1000 mg/kg	Increased activity of ornithine carbamyl transferase at 24 hr
	Mix	i.p.	2000 mg/kg	Death

[a] Autopsy showed atelectasis, interlobular pulmonary edema, and hemorrage.

Table VI Subacute Toxicity

Species	Isomer	Route of administration	Dose	Frequency	Duration	Effects
Rat	Mix	Inhalation	1600 ppm	18–20 hr/day	4 days	Death, leukocytosis, irritation of mucosae; instability and incoordination
	Mix	Inhalation	980 ppm	18–20 hr/day	7 days	Bone marrow and splenic hyperplasia, acute renal congestion
	Mix	Inhalation	620 ppm	18–20 hr/day	7 days	30% reduction of leukocytes
	Mix	Inhalation	800 ppm	4 hr/day	3 weeks	No effect on avoidance test and acquisition of conditioned responses
	m-	Inhalation	46 ppm	6 hr/day, 5 days/week	2 weeks	Increased nitroreductase and cerebral NADPH-diaphorase, diminished superoxide dismutase activity
	m-		391 ppm			
	m-		736 ppm			
	o-	Oral	250 mg/kg	Daily	10 days	Increased liver weight; less frequently, decrease in spleen and thymus weights
	m-		1000 mg/kg			
	p-		2000 mg/kg			
Rat, monkey, dog, and guinea pig	*o*-	Inhalation	780 ppm	8 hr/day	30 days	No significant changes in body weight or in hematological parameters; no toxicity signs after histopathological examination

Table VII Subchronic and Chronic Toxicity

Species	Isomers	Route of administration	Dose	Frequency	Duration	Effects
Rat	Mix	sc	43 mg/kg	Daily	6 weeks	Decreased exploratory activity
	Mix	Oral	150 750 1500 mg/kg	Daily	90 days	Liver and kidney enlargement. Histopathological evaluation indicates increased incidence of minimal chronic renal diseases; no pathological change in hepatic tissues
	Mix	Inhalation	3450 ppm	6 hr/day, 5 days/week	12 weeks	CNS modifications
	o-	Inhalation	3500 ppm	6 hr/day, 5 days/week	6 weeks	Hepatomegaly, slowed body growth; hepatocyte changes; hypertrophy, loss of glycogen, increased smooth and rough endoplasmic reticulum and peroxysomes
	Mix	Inhalation	300 ppm	6 hr/day, 5 days/week	5–18 weeks	Increase in superoxide dismutase after 14 weeks; when ethanol ingested simultaneously, increased proteolysis
	Mix	Inhalation	690 ppm	8 hr/day, 6 days/week	18 weeks	No hematological changes
Rat, dog	Mix	Inhalation	175 ppm 460 ppm 805 ppm	6 hr/day, 5 days/ week	13 weeks	No changes in hematocrit, alkaline phosphate, GOT, or GPT
Rabbit	Mix	Inhalation	1150 ppm	8 hr/day, 6 days/week	18 weeks	Low RBC and WBC counts, bone marrow hyperplasia; vascular congestion in liver, kidneys, heart, adrenals, lungs, spleen; glomerulonephritis
Rabbit, chinchilla	Mix	Inhalation	12 ppm	4 hr/day	12 months	Increase in all plasma proteins and in acetylcholinesterase; increased secretion of 17-ketosteroids in urine; weight loss, diminished immune response

electroencephalogram. After repeated moderate exposure (200 ppm), effects on the CNS such as headache, irritability, loss of memory, and sleep disorders have been reported; simple reaction time was slowed.

In case of occupational poisoning, the following first aid measures have been suggested.

After inhalation: Keep in fresh air and at rest; do not give alcohol; administer oxygen and call a hospital.
After skin contact: Wash with water and soap; undress and take a shower or bath.
After eye contact: Wash with water; call a doctor.
After oral ingestion: Administer oxygen, diuretics, and water; do not provoke vomit; do not give milk or alcohol; call a hospital.

B. Fetal Toxicity

The effects of xylene isomers and mixed xylenes on implantation, teratogenicity, and fetal and neonatal development were studied by several researchers. Rats exposed to xylene (industrial mixture) at concentrations of 10, 50, and 500 mg/m^3 (2.27, 11.3, and 113.5 ppm) throughout the first 21 days of gestation showed, for the two higher doses, signs of embryotoxicity (prenatal mortality and growth inhibition), teratogenicity (anomalies of internal organs and impairment of ossification), and disturbances in postnatal development of F_1 generation.

In one study on mice exposed to very high doses of xylene (2000 to 8600 mg/m^3) (454–1952 ppm) from day 6 to day 12 of gestation, fetal growth was retarded and there was a dose-related increase in the frequency of supernumerary ribs and delayed ossification of the sternebrae.

Inhalation of ortho-, meta-, and para-isomers at concentrations of 150, 1500, and 3000 mg/m^3 (340–680 ppm) by pregnant rats on days 7–14 of gestation induced a reduction in fetal body weight at the two highest levels of *o*-xylene and the highest level of *m*- and *p*-xylene. Fetal viability was affected only by *p*-xylene. Extra ribs were seen in significantly more fetuses in the groups exposed to high doses of *m*- and *p*-xylene. However, in another study in which pregnant rats were exposed to 3500–7000 mg/m^3 (795–1590 ppm) of *p*-xylene from day 7 to day 16 of gestation, no effects were seen on litter size, weight at birth, or growth rate. Thus *p*-xylene does not appear in this study to be a developmental toxic agent.

In conclusion, a review of the data obtained in rodents indicates, with some discrepancies, that maternal exposure to mixed xylenes or xylene isomers has adverse effects on the product of conception: increased prenatal mortality, reduced fetal weight, increased incidence of skeletal retardation, and diminished activity of some enzymes.

Malformations after exposure to mixed or individual isomers were reported at doses showing mild toxic effects on the dams. Thus a selective teratogenic effect due to xylene exposure has yet to be clearly evidenced.

C. Mutagenesis

Mixed xylene (grade not specified) was not mutagenic in a battery of short-term tests: mitotic gene conversion in *Saccharomyces cerevisiae* D4, bacterial gene mutation (*Salmonella typhimurium* TA98, TA100, TA1535, TA1538) with and without metabolic activation, and specific locus in thymidine kinase L5178Y mouse lymphoma cells. It produced no significant increases in chromosomal aberrations in rat bone marrow cells and did not induce sister chromatid exchange or chromosomal aberrations in human lymphocytes.

The *o*-, *m*-, and *p*-isomers of xylene showed no mutagenic effects in experiments on *S. typhimurium* in the presence or absence of an exogenous metabolic system. They did not cause recessive lethal mutations in *Drosophila melanogaster*. Both *m*- and *p*-xylene were inactive in increasing membrane permeability of human fibroblasts. None of the isomers induced micronuclei in the bone marrow of male mice after two doses (0.11–0.65 mg/kg intraperitoneally) at 24-hour intervals.

D. Carcinogenicity

1. In Animals

Three studies have been made to assess the carcinogenic potential of xylenes. One of these tested the effect of 500 mg/kg of mixed xylenes (source and composition unspecified) 5 days per week for 104 weeks in rats. A specific increase was reported in hemolymphoreticular tumors in treated rats compared to controls, and in general an increase in the total number of animals with malignant tumors was seen. However, for many parameters the study was considered inadequate for the evaluation of carcinogenicity by the International Agency for Research on Cancer (IARC) working group.

However, two other studies employing xylene technical grade at oral doses of 500 and 1000 mg/kg in mice and 500 mg/kg in rats (5 days per week for 103 weeks) showed that treatment did not increase tumor incidence in animals of either sex.

The survey of the available studies cannot permit us to demonstrate either the presence or absence of a carcinogenic effect; thus the IARC concluded that there is inadequate evidence for carcinogenicity in experimental animals.

2. In Humans

Various epidemiological studies of carcinogenicity in humans have been reported. Exposures of the workers were mixed and were associated with exposure to other chemicals. In the xylene-exposed workers, an increased risk for lymphosarcoma and lymphatic leukemia was reported [relative risks (RR) of 3.7 and 3.3, respectively]. Increases in primary cancers of the central nervous system (RR of 2) and in prostatic cancer (RR of 1.5) were also reported. However, exposure was to a variety of compounds and the number of cases was limited.

Thus the IARC evaluation for carcinogenic risk considered that there was inadequate evidence for carcinogenicity of xylene in humans. This evaluation is applied to the chemicals for which available studies are considered of insufficient quality, consistency, or statistical power to permit a conclusion regarding the presence or absence of a casual association.

IV. Kinetics and Metabolism

Kinetics and metabolism of xylenes have been extensively studied in animals and in humans. Minor differences appear to occur in the patterns of the three isomers, though few experiments have been done to compare them.

A. Absorption

Xylenes are easily absorbed through the lungs. No differences have been found between the isomers. Pulmonary retention reaches 62–64% in humans regardless of the level and duration of exposure (approximately 60% is reached after 5–10 min exposure) when resting after different types of physical exercise. However, an increased accumulation after repeated exposure was found during physical exercise, probably due to greater pulmonary ventilation.

Xylenes are easily absorbed through the skin. For *m*-xylene the peak concentration in the blood appears 4–6 min after exposure. The rate of absorption during immersion of both hands (volunteers) in pure *m*-xylene solution for 15 min was 2 $\mu g/cm^2/min$. However, experiments in rats showed that *o*-xylene penetrates the skin at a rate 1/10 that of toluene and 1/100 that of benzene.

An experiment was done on rats to compare the absorption of soil-adsorbed *m*-xylene and pure *m*-xylene, so that the potential risk following dermal exposure to contaminated soil could be assessed. Maximum plasma levels were higher for pure *m*-xylene than for sandy and clay soil groups, but 48 hours after treatment skin application sites in both soil groups contained more xylene and/or metabolites than the pure *m*-xylene group; moreover, an increase in *m*-xylene derivatives was observed in fat beneath the clay soil–treated skin area.

B. Distribution

Distribution volume involves three compartments: the parenchymal organs (rich in blood vessels), muscle, and adipose tissue. Uptake requires only a few minutes for parenchymal organs, a few hours for muscle, and several days for adipose tissue. The same holds true for elimination. Accumulation in adipose tissue has been shown to be minimal in resting condition and rises with physical exercise and with repeated exposure.

After inhalation of xylene (grade unspecified), an accumulation of nonvolatile metabolites was found in the nasal mucosa and olfactory bulb of the brain of mice. Specific, possibly axonal flow-mediated transport of the metabolites from the nasal mucosa to the olfactory lobe of the brain has been suggested.

In rats exposed to *m*-xylene five days per week for two weeks, xylene concentrations in brain and perirenal fat were higher during the second week of exposure. However, the accumulation in adipose tissue after repeated exposure to xylene does not appear to constitute a grave risk for humans. The amount of xylene that can accumulate in this tissue on exposure to concentrations within the permitted limits is very low, as a consequence of the high rate of hepatic metabolism.

Xylenes pass the placental barrier. In mice exposed to para-xylene for 10 min on days 11, 14, and 17 of gestation, xylene quickly enters the embryo, but the uptake is lower than in maternal tissues. In the embryo, xylene and/or its metabolites appear not to be firmly bound to proteins.

C. Metabolism

There are two metabolic pathways: oxidation of a methyl group to form methylbenzoic acid, and hydroxylation of the aromatic ring to form xylenol. The first is quantitatively the most important. Methylbenzoic (or toluic) acid is conjugated with glycine to form methyl-hippuric acid (MHA or toluric acid) or with glucuronic acid to form the corresponding glucuronide (TAG) (Fig. 1).

In rabbits, that glucuronide is formed in different proportions, depending on the isomer. Ortho-xylene administered by inhalation or orally forms mainly *o*-TAG; the MHA metabolite is present only in traces in urine. This appears to be the result of the fact that *o*-xylene, as shown *in vitro* using rat liver homogenates, inhibits conjugation of glycine with benzoic acid and its methyl derivates. Meta-xylene at high doses gives rise to formation of *m*-TAG, up to 13% of the administered dose in the rabbit and rat. Para-xylene's main metabolite is *p*-MHA, with *p*-TAG and *p*-methylbenzyl alcohol in smaller amounts. In humans, MHA is the main metabolite of all the isomers, amounting to more than 95% of all the xylene absorbed. In humans exposed to a relatively low concentration (46 ppm), none of the three isomers form the glucuronide (TAG). Traces of *o*-TAG were found in the urine of subjects exposed to higher concentrations.

The second metabolic pathway is quantitatively less important and results in the formation of xylenols. Ortho-xylene gives rise to the formation of 2,3- and 3,4-xylenol; meta-xylene gives 2,4-xylenol; and para-xylene gives 2,5-xylenol.

In humans, aromatic hydroxylation accounts for less than 2% of the absorbed xylene. There is a quantitative difference between the isomers, with *m*-xylene giving rise to higher concentrations (1.98%) of the hydroxylated metabolite.

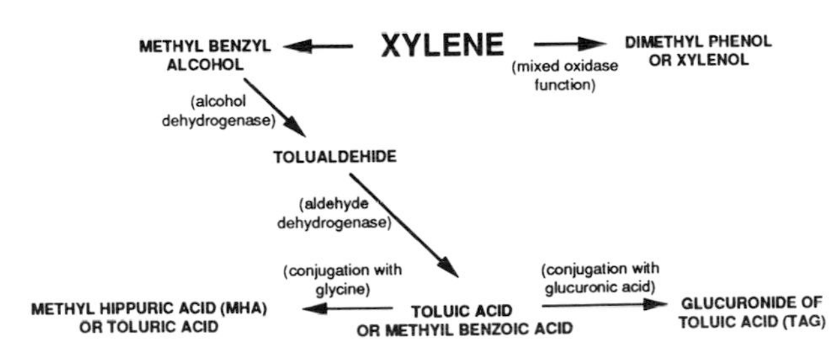

Figure 1 Metabolic pathways of the xylenes in mammals.

The ratio of acid metabolites to xylenols rises in favor of the latter after repeated exposure. This is important, since through formation of their intermediates the hydroxylated metabolites could be a major potential source of toxicity. If for xylene, in fact, metabolism is a valid process of detoxification, it should not be overlooked that metabolic steps involving aromatic hydroxylation give rise to the formation of reactive metabolites capable of causing cell damage.

In addition to the preponderant hepatic metabolism, a pulmonary metabolism has also been demonstrated (although only in the rabbit). The oxidation pathway leads to the formation of methyl-benzyl alcohol as its main metabolite, as alcohol dehydrogenase is lacking in this tissue. The hydroxylation pathway results in the formation of 2,5-xylenol.

In animals exposure to xylenes increases the cytochrome *P*450 content and microsomal enzymatic activity in the liver. This is in accordance with the liver enlargement and proliferation of the smooth endoplasmic reticulum observed in animals after repeated xylene treatments, and it provides an explanation of the hepatomegaly observed in workers exposed to inhalation of this solvent. Para-xylene does not raise the cytochrome content as much as the other isomers, showing the importance of the substitution pattern.

In kidney microsomes an increased concentration of cytochrome *P*450 was also obtained following exposure to a xylene mixture or to the *o*- and *m*-isomers. In lung microsomes, however, mixed xylene and xylene isomers cause a decrease in cytochrome *P*450 content and in aryl hydrocarbon hydroxylase (AHH) activity. Para-xylene exerts activity similar to the *m*-isomer.

The ability of xylenes to modify the metabolism of other potentially toxic substances in liver, kidney, and lung microsomes suggests the possibility of synergistic toxic responses.

For instance in rats *m*- and *p*-xylene, by inhibiting AHH activity in the lungs, reduce the metabolic detoxification of benzopyrene and consequently potentiate its toxicity. In other cases they can exert protective activity by preventing the production of toxic metabolites. Para-xylene afforded some protection against the pulmonary injury caused by ipomeanol, which is a lung toxin known to be activated by cytochrome *P*450.

D. Excretion

About 95% of the xylene absorbed is eliminated as acid metabolites in the urine in animals and humans. The remaining 5% is exhaled unchanged. It has been calculated that expiratory elimination amounts to 5.3% for *o*-xylene, 5.8% for *m*-xylene, and 3.5% for *p*-xylene. Elimination of exhaled air diminishes rapidly when exposure ends. Excretion of nonmetabolized xylene in the urine is negligible, not exceeding 0.004% of the intake. Urinary elimination occurs in two distinct phases: the first reflects *m*-xylene distributed in muscle, the second that in adipose tissue. Distribution to the first compartment (parenchymal organs) results in such fast elimination that it is difficult to measure.

Elimination reaches a peak at the end of exposure (for 8 hours of continuous exposure) and in the next 2 hours, although traces of MHA can be found in the urine 4–5 days after the end of exposure. Hydroxylated metabolites are also eliminated in the urine, reaching maximum excretion in the 2 hours after exposure.

In humans, measurement of MHA in the urine is a valid means of monitoring exposure. The concentration of environmental xylene has been shown to be proportional to MHA in the urine of exposed persons. This proportion remains constant even in the presence of other methyl derivatives of benzene in the environment.

It has been suggested that measuring the rate of elimination of MHA may be another useful means of monitoring xylene's environmental concentrations; between the sixth and eighth hour after exposure this rate is proportional to the amount of xylene absorbed.

Related Article: VOLATILE ORGANIC COMPOUNDS.

Bibliography

Carpenter, C. P., Kinkead, E. R., Geary, D. L., Jr., Sullivan, L. J., and King, J. M. (1975). Petroleum hydrocarbon toxicity studies; V. Animal and human response to vapors of mixed xylenes. *Toxicol. Appl. Pharmacol.* **33,** 543–558.

Condie, L. W., Hill, J. R., Borzelleca, J. F. (1988). Oral toxicology studies with xylene isomers and mixed xylenes. *Drug Chem. Toxicol.* **11,** 329–354.

Fabre, R., Truhaut, R., and Laham, S. (1960). Toxicological research on replacement solvents for benzene. IV. Study of xylenes. *Arch. Mal. Prof.* **21,** 301–313.

Fishbein, L., and O'Neill I. K., eds (1988). Environmental carcinogens: Methods of analysis and exposure measurement. Vol. 10, Benzene and alkylated benzenes (IARC Scientific Publications N.85) Lyon. International Agency for Research on Cancer.

International Agency for Research on Cancer (IARC). (1989). IARC Monographs on the Evaluation of the Carcinogenic Risk to Humans, Vol. 47. Some Organic Solvents, Resin Monomers and Related Compounds, Pigments and Occupational Exposures in Paint Manufacture and Painting." IARC, Lyon, France.

Jenkins, L. J., Jones, R. A., and Siegel, S. (1970). Long-term inhalation screening studies of benzene, toluene, *o*-xylene and cumene on experimental animals. *Toxicol. Appl. Pharmacol.* **16,** 818–823.

Jori, A., Calamari, D., Di Domenico, A., Galli, C. L., Galli, E., Marinovich, M., and Silano, V. (1986). Ecotoxicological profile of xylenes. Working party of ecotoxicological profiles of chemicals. *Ecotoxicol. Environ. Safety* **11,** 44–80.

Kirk, R. E., and Othmer, D. F., eds. (1970). Xylenes and ethylbenzene. *In* "Encyclopedia of Chemical Technology," 2nd Ed., Vol. 22, pp. 467–507. Wiley (Interscience), New York.

Savolainen, H., and Pfaffli, P. (1980). Dose dependent neurochemical changes during short-term inhalation exposure to *m*-xylene. *Arch. Toxicol.* **45,** 117–122.

Ungvary, G., Tatrai, E., Barcza, G., and Krasznai, G. (1979). Acute toxicity of toluene *o*-, *m*-, and *p*-xylene, and of their mixtures in rats. *Munkavedelem* **25,** 37–38.

Verschuren, K. (1977). "Handbook of Environmental Data on Organic Chemicals," pp. 638–642. Van Nostrand-Reinhold, New York.

Zinc

Jean-Claude Amiard and Claude Amiard-Triquet
Centre National de la Recherche Scientifique, Nantes, France

Glossary

Acclimation Ability of living organisms exposed to sublethal doses of contaminants to develop tolerance (resistance) to pollution.

Bioavailability Characteristics of a specific substance which allow its uptake by living organisms.

Biogeochemical cycle Qualitative and quantitative distribution of a specific substance among the different compartments of the environment (atmosphere, lithosphere, hydrosphere, and biosphere).

Bioindicator Living organism, representative of a particular environment, used to monitor the presence of a substance which accumulates preferentially in this given species.

Essential element Chemical element which is indispensable to life.

Homeostasis (regulation) Maintenance of stable internal levels in organisms exposed to changing external levels (e.g., body temperature, blood pressure, zinc concentration).

Metal-binding protein Protein involved in the binding of metals in circulating fluids or in cytosol. Among them, metallothionein is often interpretated as a detoxifying protein.

Metalloenzyme Protein that catalyzes metabolic reactions and in which a metal acts as a cofactor.

ZINC is ubiquitous in the environment at trace levels. In metal-bearing formations, it is mainly present as ZnS but it is associated with different metals in ores. Numerous uses of zinc occur (e.g., mining, smelting, electroplating) and thus anthropogenic activities can induce occupational risks and a redistribution of zinc among the different components of the environment (water, soils or sediment, biomass, atmosphere). The consequences of this redistribution on the environmental quality (safe levels in drinking water and food) and on the protection of living resources must be assessed. This judgment is particularly difficult to make since zinc is an essential trace element (i.e., indispensable to life at low levels) and becomes toxic only at higher doses. The safety margin between environmental versus toxic levels is rather narrow and the most sensitive organisms or life-stages of some species (larvae, juveniles) may be exposed to insidious effects of zinc at low or medium levels. Moreover, toxicological effects due to zinc are often difficult to differentiate from those due to accompanying impurities such as other trace elements (e.g., Cd, Cu, Pb). Recent contributions to a better understanding of Zn metabolism and metallic interactions (through the role of metal-binding proteins) provide a basis to a better assessment of both bioaccumulation and toxicity.

I. Zinc Sources

A. Natural Sources

Zinc occurs in the environment at trace levels. It is possible to distinguish between five different sources from which atmospheric zinc originates (Table I). Background levels in the earth's crust have been estimated at 65 or 76 ppm according to different sources. However, in areas characterized by metal-bearing formations, high concentrations of zinc may be encountered in the different compartments of the environment independently of any anthropogenic impact.

Table I Worldwide Emissions of Zinc in the Atmosphere from Natural and Anthropogenic Sources

Natural sources	Zinc emission ($\times 10^9$ g/yr)	Anthropogenic sources	Zinc emission ($\times 10^9$ g/yr)
Windblown dusts	25	Mining	1.6
Forest fires	2.1	Metal production	150.62
Volcanogenic particles	7.0	Iron and steel	35.1
Vegetation	9.4	Ferro alloys and iron foundries	4.5
Seasalt spray	0.21	End uses of zinc products	26.0
		Inadvertent sources	
		Coal combustion	14.9
		Oil combustion	0.07
		Wood combustion	37.2
		Waste incineration	75.0
		Rubbertire wear	2.2
		Phosphate fertilizers	1.8
		Grain handling	0.18
		Cement production	30.0
Total	43.7	Total	379.2

Source: Modified from "Zinc in the Environment. Part I: Ecological Cycling," (J. O. Nriagu, ed.). Wiley-Interscience, 1980.

B. Mining Effluents

In Table II are shown recent data about world production of zinc. During this period (1985–1989), zinc production was increasing slightly.

In mineralized areas of economical interest, mining induces changes in zinc distribution as a consequence of disposal of tailings, discharge of effluents, and associated industrial operations. Past mining operations are still at risk through the leaching of deads and outcrops of mineralized formations. The serious effects of metal contamination from mining activities have been recognized for many years. However, it is often difficult to determine the definite part of zinc in polymetallic pollutions. For example, the unproductiveness of certain fields in North Cardiganshire was explained by toxic levels of lead and zinc in soils. Evidence of long-term pollution with copper and zinc from metal mining was registered in the Fal Estuary in Southwest England: in the most heavily polluted area (Restronguet Creek), metal levels in waters

Table II Trends of Production and Consumption of Zinc during the Period 1985–1989

	World mining production (metal content) (metric tons)			World consumption of slab zinc (metric tons)		
	1985	1987	1989	1985	1987	1989
Europe	1,167,900	1,061,400	967,200	1,711,000	1,782,700	1,907,300
Africa	271,100	287,300	250,000	138,000	141,700	150,400
America	2,542,800	2,867,700	2,731,400	1,478,200	1,652,700	1,606,200
Asia	435,700	355,800	337,000	1,278,100	1,337,000	1,411,900
Oceania	759,100	778,400	803,000	111,400	94,200	121,100
Eastern countries	1,851,100	1,938,900	2,048,900	1,811,900	1,911,400	1,974,500
Total	7,027,700	7,289,500	7,137,500	6,528,600	6,919,700	7,171,400

Source: "1989 Statistical Yearbook," METALEUROP.

and sediments are elevated by orders of magnitude and the invertebrate fauna of the Creek are limited. In lead–zinc mined areas in Kansas, a severe deterioration of both ground- and surface-water quality was demonstrated (with concentrations of zinc reaching tens of thousands of μg/L and with high levels of Cd, Mn, Pb, etc.) with subsequent effects on populations and ecosystems such as decreased biological diversity in some areas or even the complete absence of benthic invertebrates in some streams.

Zinc is generally associated with most of extracted minerals. Even if it is not extracted for itself in a mined area, it may be responsible for secondary pollutions such as those that have been mentioned as a consequence of tin and tungsten mining in Tasmania or gold mining operations in South Africa. In the latter case, concentrations as high as 26,000 μg Zn/L were registered in effluents.

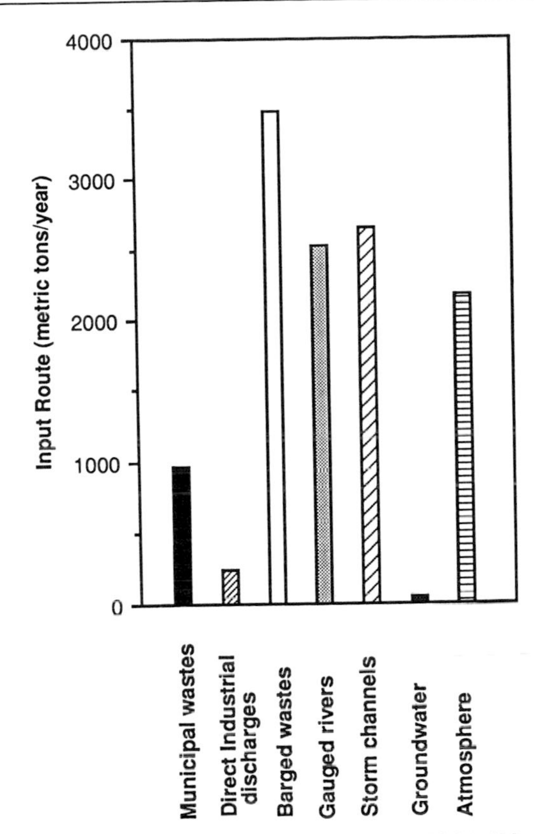

Figure 1 Sources of zinc in the New York Bight. (Modified from "Zinc in the Environment. Part 1: Ecological Cycling," J. O. Nriagu, ed.; Wiley Interscience, 1980.)

C. Industrial Effluents

The major industrial uses of zinc and the associated subsequent emissions of this metal in atmosphere are shown in Table I. Atmospheric input assessments exhibit some variations from different sources of data, as shown by the comparison of Table I and Figure 2. However, anthropogenic emissions of zinc are higher than global emissions from natural sources. During the period 1985–1989, the trend of zinc consumption (Table II) showed a significant increase. Local estimations exemplify the importance of zinc input in aquatic environments: 240 and 470 metric tons Zn/year are introduced respectively in the New York Bight and the Northern Chesapeake Bay with direct industrial discharges whereas the steel industry is responsible for an annual input of 310 tons Zn/year in the Tees Estuary (Scotland). In the case of a zinc smelting plant, a discharge of more than 2000 tons Zn/year was registered in the nearby coastal area (Odda, Norway).

Numerous industries not mentioned in Table I discharge wastewaters containing zinc but even at this stage Zn levels (<3890 μg Zn/L) are lower than allowed by the drinking water quality criterion (5000 μg Zn/L).

D. Domestic Effluents and Urban Runoff

Sewage sludges and municipal refuses contain significant amounts of zinc. A tentative assessment of metal enrichment in domestic effluents consists in comparing Zn levels in crustal rocks and in sewage sludges. Tentative average values generalized from various sources indicate a ratio of 40 which may be compared with values of the same order of magnitude for copper, lead, and cadmium whereas lower ratios were calculated for nickel and chromium (respectively 0.8 and 1.2) and a very high ratio was obtained for silver (200). Corrosion within the urban water supply network contributes significantly to the elevated levels of zinc encountered in domestic effluents.

Metal contamination of urban stormwater depends on the level and distribution of pollution at each site, but the release of zinc from roofs as a consequence of corrosion appears to be a special source of this metal.

Regional investigations allow the assessment of the relative contribution of the different sources of environmental zinc as exemplified in the case of the New York Bight (Fig. 1).

E. Industrial Sources in Professional Environments

Among the different industrial uses of zinc mentioned above, the major occupational risk is associated with exposure to zinc oxide fumes in brass foundries or shearing and brazing of galvanized sheets (''metal fume fever''). Moreover, zinc chloride used in wood protection exerts a caustic effect on the skin and is able to induce cutaneous ulcers.

II. Assessment, Monitoring, and Control of Zinc Pollution

Determining and limiting the risks associated with metal pollution require analytical methods complying with four criteria: sensitivity, repeatability, accuracy, practicability. Methods based on flame atomic absorption spectrometry are generally convenient for zinc determinations in air, sediments, and biological samples. Electrothermal atomic absorption spectrophotometry becomes the method of choice when the elemental concentrations are below 50 μg/L (for example in waters).

A. Evaluation of Pollution

Various media may be analyzed in order to determine the degree of general or local pollution. Metal levels in water are characterized by important spatio–temporal variations especially in river and estuarine waters. Moreover, concentrations are low, inducing a risk of significant secondary pollution on sampling and analytical difficulties. Sediment is a long-term integrating compartment and the highest concentrations may be expected in this medium. However, the interpretation of data about sediments is not very easy since metal data are greatly influenced by hydrologic and textural (grain-size, organic matter content) factors. Biological material is a medium-term integrating compartment and is particularly interesting since the bioaccumulated fraction of the pollutant is the most able to exert a noxious effect on the organisms themselves or on their consumers. Living matter is heterogeneous and bioaccumulation ability differs from species to species, and from organ to organ within the same species. Moreover, the degree to which a species is representative of a site depends on its migration rate. The major selective criteria of bioindicators have been described in detail in many texts and include: large distribution (ideally worldwide), long life, reasonable size, easy sampling, and abundance. In the case of zinc, numerous species are able to regulate internal levels against changing external levels of the metal, a phenomenon which may be a source of misinterpretation of data. However, several species or biological groups are commonly used as indicators: lichens for atmospheric pollution, aquatic mosses for freshwaters, algae and (above all) bivalve molluscs in seawater.

The most ambitious monitoring program called ''mussel watch'' concerns the marine environment. It has been developed in numerous countries and covers a large number of sites within each. Comprehensive investigations on pollutant loads (including zinc) in water systems have been carried out periodically in different countries. It is impossible to list comprehensively here the areas in which pollution has been demonstrated, and only the most severe cases are mentioned in Table III. Some of these metal concentrations have been determined ten or fifteen years ago, and now the specialists of trace element analytical chemistry think that these values were overvaluation of the real contamination. But even if we consider more realistic levels in seawater concentrations of about 100 μg Zn/L are frequently encountered in both river and coastal waters.

B. Zinc in Water Purification Processes

Industries discharging effluents that carry significant quantities of zinc are listed in Table IV as well as zinc levels usually encountered in this type of wastewater. Treatment technologies used for zinc removal may involve either chemical precipitation or recovery processes including ion exchange and evaporative recovery. The effectiveness of hydroxide precipitation treatment is shown in Table IV as well.

Zinc contents in urban drainage systems generally do not reach the levels in industrial effluents but, as mentioned earlier, they are not yet negligible. Unfavorable effects of zinc in the biologic stages of urban purification plants have been envisaged for concentrations higher than 1 mg Zn/L. However, data on activated and digested sludge generally show significant extraction rates between 60 and 100%.

Table III Zinc Concentrations in Different Compartments of the Environment

Media of pollution assessment	Background levels	Upper limits	Geographical site
Freshwater (μg/L)	10	>10,000	Mined areas
Drinking water quality criteria (μg/L)	5000		
Seawater (μg/L)			
oceanic surface	0.01		
deep	0.62		
coastal		2520	Restronguet Creek, U.K. (mining)
Air (ng/m³)	10	6200	Munich, Germany (urban)
Soil (mg/kg DW)			
sandstone	16		
granitic rocks	60		
shales	95		
Sediment (mg/kg DW)			
stream unmineralized zone	320		
stream mineralized zone	2750		
estuaries	98	118,000	Sorfjord, Norway (smelter)
		3510	Restronguet Creek, U.K. (mining)
coastal	25	826	Firth of Clyde, U.K.
Organisms (mg/kg DW)			
marine angiosperm	100	1480	Rio Tinto Estuary, Spain (mining)
seaweed	90		
phytoplankton	38		
marine animals (except oysters)	50–250		
oysters	1700	57,600	Tasmania, Australia (zinc refinery)
		19,400	Power river, Canada (pulp mills)
Terrestrial vegetale Bryophytes	23–190	2470–5985	Britain, U.K. (urban)

Sources: Modified from "Metal pollution in the aquatic environment" (U. Forstner and G. T. W. Wittmann, eds.). Springer Verlag, 1979; "Zinc in the Environment," (J. O. Nriagu, ed.) (1980). Part 1: Ecological Cycling; and "Marine Ecology," (O. Kinne, ed.) (1984) Vol. 5, Part 3.

The disposal of sludge is a delicate problem since it contains different pollutants including zinc at levels which may be toxic, particularly when it originates from industrialized or mixed areas. Impact of disposal at sea through pipelines or barges or land application of sewage sludges have been examined by numerous authors. Leaching of metals in seawater, fate and effects in marine ecosystems; availability of zinc to plants and their consumers and/or toxicity to vegetation; and potential contamination of groundwater through leaching and percolation are include among the many impacts that result from these methods.

C. Standards for Drinking Water, Food, and Air

Zinc is practically nontoxic in humans and mammals. However, toxicological parameters have

Table IV Zinc Concentrations after Hydroxide Precipitation Treatment of Zinc Wastewaters

Industrial source	Zinc concentration (mg/L) Initial	Final
Electroplating industry	18–120	<1–6
Vulcanized fiber	100–300	≤1
Brass wire mill	36–374	0.08–1.60
Table wars plant	16.1	0.02–0.23
Viscose rayon	20–120	0.9–5
Blast furnace gas		
Scrubber water	50	0.2
Metal fabrication	—	0.1–1.2

been derived from some cases of accidental exposure to acute doses of zinc and from observations of secondary zinc effects associated with therapeutic oral administration of high zinc doses. The provisional maximum tolerable daily intake for humans has been estimated at between 0.3 and 1 mg Zn/kg body weight by the World Health Organization. Typical zinc concentrations in major foodstuffs are shown in Table V. Such concentrations are compatible with respect to the WHO standards except in the case of an important oyster consumption.

Drinking water quality criteria are generally high (1 to 5 mg/L) except in Japan (0.1 mg/L). In these conditions, most surface and groundwater generally falls within these requirements and zinc removal from drinking water is not necessary.

The major toxic reaction to zinc known as "metal fume fever" is characterized by pulmonary manifestations, fever, chills, and gastroenteritis and has been observed as a consequence of occupational exposure to fumes in metallurgy and metal shearing. Consequently, standards have been established with regard to airborne particulates: the threshold limit value (TLV) for zinc chloride fumes and zinc oxide fumes are, respectively, 1 and 5 mg/m^3.

III. Fate of Zinc in the Environment

A. Biogeochemical Cycle

Zinc is ubiquitous in the environment: it is present in the atmosphere, hydrosphere, biosphere and even lithosphere. Exchanges between these compartments take place through precipitation (dustfall and rainfall), runoff, erosion and evaporation, sedimentation and resuspension in aquatic environments, and metal transfers between solid and aqueous phases (Fig. 2). Anthropogenic activities modify the biogeochemical cycle of zinc mainly at the level of the atmospheric compartment, a phenomenon which induces zinc fallout with precipitation to the sea or the Earth's surface, the latter being most important. The role of living organisms in the biogeochemical cycle of zinc is illustrated by Zn fluxes involved in the production of terrestrial or oceanic biomasses. The zinc flow through living organisms generally induces a change in chemical speciation of the metal since whatever the initial physicochemical form may be, Zn outflow will be primarily associated with organic matter in feces.

Table V Typical Zinc Concentrations in Major Foodstuffs (mg/kg wet weight)

Source	Concentration (mg/kg wet wt)
Beef	15
Horsemeat	60
Veal	35
Pork	25
Fish	20
Oyster	200–1000
Milk	5
Vegetables and cereals	4–20

B. Bioavailability and Transfer in the Food Chain

Metal transfer between solid and aqueous phases takes place in both soil and sediments and the physicochemical state of zinc dissolved in water or associated with particles is a major factor in determining metal availability to plants and benthic organisms (intrasedimentary worms and bivalves and flatfish). In animals, the physicochemical form of storage in prey is an important factor to consider in assessing potential transfer of zinc in food chains. In some marine species, different metals (including Zn), which are incorporated in intracellular phosphate granules, remain in the same insoluble form during passage through the gut of their predators and thus do not induce any contamination of the food chain. Numerous papers have dealt with the relationship between zinc concentration in living organisms and the trophic levels. With a few exceptions, zinc concentrations are of the same order of magnitude in species feeding at different trophic levels as exemplified in the case of the Loire Estuary in France (Fig. 3).

C. Bioaccumulation

Zinc is an essential metal for plants and animals. Since it only occurs in the environment in small concentrations, biological requirements generally necessitate an enrichment in the organism compared with that in the environment. In the aquatic

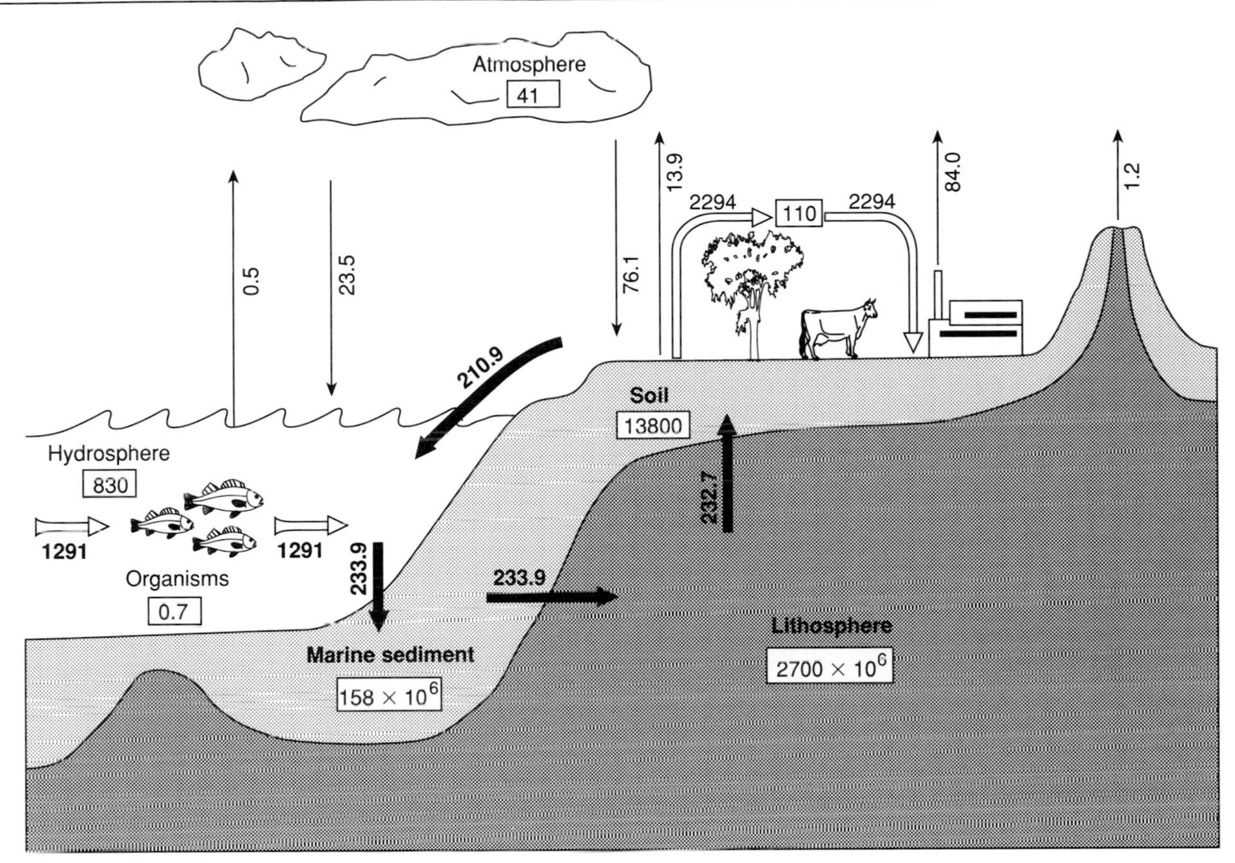

Figure 2 Biogeochemical cycle of zinc; mass is expressed as 10^{12} g Zn; fluxes are expressed as 10^{10} g Zn/year.

environment, this process is quantified by using a concentration factor *CF:*

$$CF = \frac{\text{metal level in the organism } (\mu g/kg)}{\text{metal level in water } (\mu g/l)}$$

CF for some groups representative of the aquatic biota are shown in Table VI.

If there is an over abundance of an essential metal such as zinc, the metal content in the organism can be regulated by homeostatic control mechanisms. Until now the different processes of zinc absorption and their homeostatic regulation are not fully understood. In mammals, it has been shown that metallothionein plays an essential role in the homeostatic regulation of zinc metabolism. Metallothionein-like proteins (MTs) have been recognized in most of the zoological groups; in plants, a similar role would be played by phytochelatins. Several metals exhibit an increasing affinity to MTs compared with Zn: Cd < Cu < Ag < Hg. The tendency of these elements to be accumulated in MTs could support the hypothesis for a protective role of MTs in metal poisoning and explain at least partly some metallic interactions observed in bioaccumulation as well as in biological effects.

D. Factors Influencing Bioaccumulation

Different biological and ecological factors influence zinc bioaccumulation besides the physicochemical forms of the metal (Table VII). Zinc absorption is affected by interfering substances in food. Based on global concentrations of zinc, meat, fish, vegetables, and cereals would appear to be similar dietary sources of zinc (see Table V), but the availability of the metal is considerably lower in vegetable foodstuffs.

Zinc absorption is affected by zinc status. Like any other nutrient, metabolic requirements for zinc in growing children differ significantly from that of adults. During gravidity and lactation, the nutritional requirement for zinc (as for other minerals and nutrients) is greatly increased. The influence of age and sex on zinc bioaccumulation has been

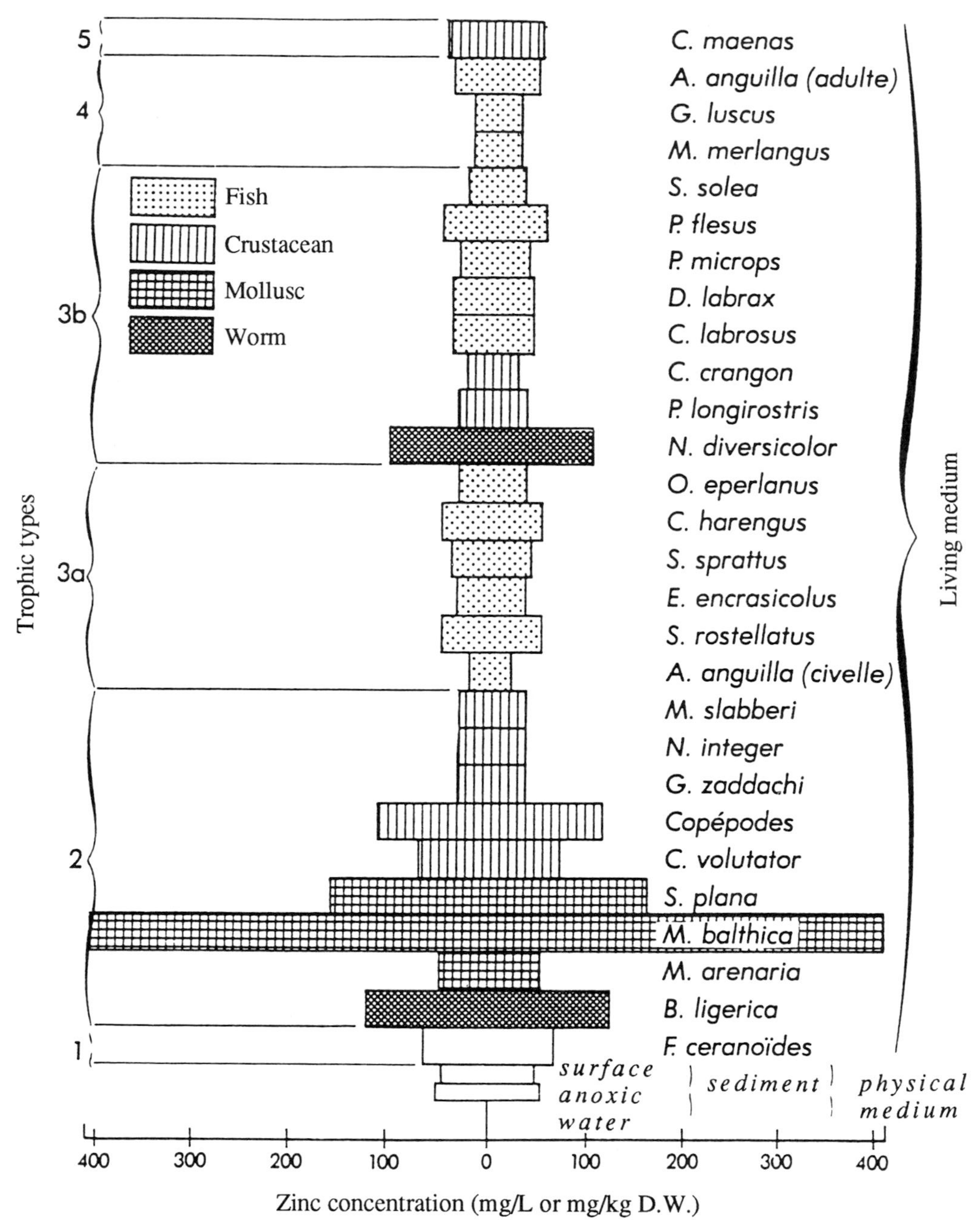

Figure 3 Zinc concentrations in different organisms from the Loire Estuary according to their food habits (1, primary producers; 2, primary consumers; 3a, mainly plankton-feeders; 3b, omnivorous bottom-feeders; 4, super-carnivorous fish; 5, necrophagous). (After J. C. Amiard *et al.*, 1982, with authorization of the *Bull. Soc. Sci. Nat. Ouest Fr.* **4,** 153–168.)

studied widely and is recognized in numerous species. The phenomenon has been investigated in detail for bioindicators such as mussels or oysters since natural fluctuations may interfere with the interpretation of data for pollution monitoring. In environmental samples, seasonal changes have been often registered and have been attributed to both season-related physiological changes and variations of the conditions of exposure (temperature, river flow, available food).

Table VI Concentration Factors (*CF*) of Zinc in Major Groups of Aquatic Organisms

	Freshwater	Marine
Macrophytes	3155	900
Algae	305	410
Phytoplankton		15,000
Zooplankton		8000
Molluscs	33,544	47,000
Crustaceans	1800	5300
Fish	17,544	3400

Source: Modified from Amiard-Triquet and Amiard (1980).

IV. Biological Effects of Zinc

A. Essentiality versus Toxicity

The responses of biota to the presence of zinc in the environment are governed by the fact that zinc is an essential metal: it is a cofactor of more than 200 enzymes in various species. The function of zinc in metalloenzymes can be divided into four categories: catalytic, structural, regulatory, and noncatalytic. In living organisms, zinc is generally nontoxic and, in fact, noxious effects primarily reflect zinc deficiency. Various effects of zinc deficiency have been recorded in plants as well as in animals. The margin between Zn concentrations in plant tissues corresponding respectively to deficiency or toxic exposure is rather narrow (Table VIII). The same is true for Zn concentrations in plant tissues which have noxious consequences in animals due to too low or too high levels of dietary Zn (Table VIII). In mammals and humans, symptoms of Zn deficiency include growth retardation, retarded sexual maturation, dermatitis, alopecia, lethargy, mental depression, neuropathy, ophtalmic damage, and immune deficiency. Zn deficiency has been reported in severe metabolic dysfunction such as cancer.

Accidental and occupational risks associated with zinc in humans have been evoked. The toxicity of zinc to plants is generally low and is only observed in soils with excessive zinc burden, for instance, in waste stockpiles, dumping grounds, or land application of sewage sludge. In the last case, prevention of toxic action of zinc to plants is achieved by restricting zinc input associated with composted sewage sludges. For example, European regulation allows the use of sludge with Zn concentrations lower than 3000 mg/kg (dry weight) provided that the cumulated concentrations of Cr + Cu + Ni + Zn do not exceed 4000 mg/kg and that Zn levels do not exceed 300 mg/kg in the receiving soil.

Metal binding with MTs and zinc homeostasis are generally interpretated as protective mechanisms. However, the regulation of internal levels of zinc requires energy expense which could have a negative impact on the organism. In several Crustacean species, metabolic changes (hormonal or enzymatic) and deleterious effects (e.g., delayed development of larvae, reduced size) have been observed in organisms exposed to low or medium

Table VII Major Zinc Bioaccumulation Factors

Biological	Environmental
Age	Zinc level in sources (water, food, soil, sediment)
Sex	Physicochemical forms in sources
Sexual maturity in invertebrates	Interfering substances in food
Pregnancy } in vertebrates	Metallic interactions
Lactation } in vertebrates	Temperature
Specific characteristics (oysters)	
← Season →	← Season →

Table VIII Critical Concentrations of Zinc in Plant Tissue

Levels	Concentration (mg/kg dry matter)
Lower critical	
Deficiency threshold (plants)	20
Deficiency threshold (animals)	50
Upper critical	
Phytotoxic threshold	200
Zootoxic threshold	500

Source: Modified from "Technical Report TR-140" (R. D. Davis and C. Carlton-Smith, eds.). (1980). Water Research Centre, Stevenage, U.K.

external levels of zinc which induce no or only slight disturbances in homeostasis.

B. Toxicity to Aquatic Organisms

The literature abounds with data regarding the effects of zinc on aquatic organisms. In Table IX, we have selected the effects induced by exposure to zinc concentrations in seawater lower than or equal to 100 μg Zn/L, a concentration frequently encountered in polluted environments. In spite of the fact that neither the physicochemical form of zinc in the environment nor potential interactions with other metals have been taken into account, these data suggest that chronic pollution by zinc could lead to a reduction of primary production and to a decrease of ecosystem productivity through an impairment of development and growth in some key species.

Review of freshwater data by the U.S. EPA with a view toward derivation of ambient water quality criteria emphasizes both the role of water hardness in the manifestation of Zn toxicity and the large range of response among different species. However, numerous acute or chronic toxicity data are lower than 100 μg Zn/L, a concentration relatively common in contaminated freshwater environments.

Table IX Effects of Zinc on Marine Organisms due to Levels Which Can Exist in Some Estuarine and Coastal Areas

Group, species	Effective concentration (μg/ L)	Observed biological phenomenon
Phytoplankton		
Natural communities	15	Decreased carbon fixation rates
Asterionella japonica	≃60	Growth inhibition
Chaetoceros compressum	≃60	Growth inhibition
Nitzschia closterium	≃60	Growth inhibition
Schroederella schoederi	≃10	Growth inhibition
Criscosphera carterae	<86	Disturbance
Clenodynium halli	<86	Disturbance
Gymnodinium splendens	<86	Disturbance
Isochryhis gabana	<86	Disturbance
Thalassiosira rotula	<86	Disturbance
Macroalgae		
Laminaria digitata	100	Growth decreased by 50%
Worms		
Capitella capitata	≥100	Disturbance of reproduction
Bivalve Molluscs		
Crassostrea cucullata	≥100	Disturbance of embryonic development
C. margaritacea	80	Growth of 6 d larvae reduced by 50%
C. gigas	30–35	Settling of 15 d larvae reduced by 50%
M. edulis	10–60	Inhibitory effect on shell growth
Crustaceans		
Tisbe holothuriae	70	Population size reduced to zero after the 1st generation
Idotea neglecta	>10	Disturbed osmoregulation
Palaemon serratus (larvae)	>75	Delayed metamorphosis, increased lethality, reduced size
Rhithropanopeus harrisii	25	Increased duration of larvae development
Echinoderms		
Litechinus pictus	60	Abnormality of embryonic development
Paracentrotus lividus	30	Delayed development of larvae
Arbacia aequituberculata	30	Delayed development of larvae
Peronella japonica	14	Decreased rate of normal larvae
Fish		
Clupea harengus	5	Effects on fertilization and embryonic development

Source: After J. C. Amiard-Triquet (1989); with authorization of *Bull. Ecol.* **20,** 129–151.

C. Acclimation to Chronic Exposure

Populations exposed to chronic environmental zinc pollution develop a resistance to this pollutant. A selection of tolerant plant species takes place in zinc-rich soils. In the Restronguet Creek (U.K.) which exhibits severe metallic pollution, some aquatic organisms have developed zinc tolerance as illustrated in Table X. The mollusc *Scrobicularia plana* is able to survive chronic pollution with zinc concentrations in soft tissue as high as 4920 mg/kg whereas lower levels (651 mg/kg) are observed in molluscs originating from a nonpolluted area which have survived an acute exposure. In the crab, *Carcinus maenas*, survival from exposure to acute doses is improved in specimens originating from a polluted area than in individuals from a clear site. Tolerance is probably due to an enhancement of detoxication mechanisms such as an increase of the synthesis of specific metal-binding compounds. In some species, for instance, the gene responsible for metallothionein synthesis has been duplicated.

V. Conclusions

The lithosphere is the major storage compartment for zinc, but the major fluxes in the general environment are associated with the production and fate of the biomass. Anthropogenic sources of zinc are local, mainly due to mining, smelting, and galvanization. Thus, the environmental impact of zinc will be principally local or regional.

Zn concentrations in living organisms (averaging 100 ppm) are high for a trace element, approaching the lower limit for major elements. Since Zn is essential in metabolic processes (serving as a cofactor for more than 200 enzymes in various species), it is regulated in the most highly evolved living organisms such as vertebrates and some invertebrates such as crustaceans. Metallothionein-like proteins (MTs) are likely the major ligands involved in such homeostasis. The uptake of trace elements such as Cd, Cu, or Hg (as associated impurities in industrial uses of Zn) could induce exchanges with Zn bound with MTs. Zinc released in the internal medium would thus be likely to induce synthesis of additional MTs. According to this process, a secondary detoxication role would be attributable to MTs.

Several consequences are associated with Zn homeostasis: (1) the best regulators must be excluded for pollution monitoring since bioaccumulated Zn is not proportional to its concentration in the external medium; (2) homeostasis requires energy expense, thus high levels of environmental Zn could induce biological disturbances; (3) MTs could play a detoxication role but they are easily degraded under the physicochemical conditions prevailing in

Table X Acclimation of two Marine Organisms to Chronic Zinc Pollution

Origin	Conditions of exposure	Toxicological parameter
	Mollusc: *Scrobicularia plana*	
		Max. Zn level (mg/kg D.W.)
Polluted area (Restronguet Creek, U.K.)	Chronic exposure in the field	4920
Unpolluted area (Bay of Bourgneuf, France)	Acute exposure in the laboratory	651
	Crustacean: *Carcinus maenas*	
		Survival time (days)
Polluted area	Acute exposure (10 mg Zn/L) in the laboratory	38
Unpolluted area	i.d.	6

Source: After J. C. Amiard, *In* "Réactions des êtres vivants aux changements de l'environment." Centre National de la Recherche Scientifique, 1991.

the digestive tract and, too, they may be a vector of metal transfer in food chains.

The response of biota to the presence of environmental zinc is governed by the fact that zinc is an essential metal. Most studies about zinc in humans, animals, and plants deal with deficiencies rather than toxicity. However, toxic levels depend primarily on the mode of exposure and the physicochemical forms of zinc. On the other hand, metabolic changes and deleterious effects have been observed in aquatic organisms exposed to external levels of zinc which induce only slight disturbances of homeostasis. In more than 100 cases, the experimental overload responsible for a toxic (or even lethal) effect are lower than the highest concentrations registered in polluted waters.

Thus, the major risk associated with zinc pollution is not due to the exposure of human beings to this metal in food or physical environment (except in few cases of occupational diseases). It is through a decrease of available living resources due to the susceptibility of aquatic species, a risk enhanced by the fact that the safety margin between environmental versus toxic levels is rather narrow.

Related Articles: INORGANIC MINERAL PARTICULATES IN THE LUNG; METAL–METAL TOXIC INTERACTIONS.

Bibliography

Amiard-Triquet, C. and Amiard, J. C. (1980). "Radioécologie des Milieux Aquatiques." Masson, Paris.

Bryan, G. W. (1984). Pollution due to heavy metals and their compounds. *In* "Marine Ecology." (O. Kinne, ed.) vol 5, Part 3. Wiley, Chicester, U.K.

Collery, P., Poirier, L. A., Manfait M., and Etienne, J.-C., eds. (1990). "Metal ions in Biology and Medicine." John Libbey, London.

Förstner, U., and Wittmann, G. T. W., eds. (1979). "Metal Pollution in the Aquatic Environment." Springer-Verlag, Berlin.

Larson, L., and Hyland, J., eds. (1987). "Ambient Aquatic Life Water Quality Criteria Document for Zinc." Environmental Protection Agency, Washington, D.C.

McKenzie, H. A., and Smythe, L. E., eds. (1988). "Quantitative Trace Analysis of Biological Materials." Elsevier, Amsterdam.

Merian E., ed. (1990). "Metals and Their Compounds in the Environment." VCH, Weinheim.

Nriagu, J. O., ed. (1980). "Zinc in the Environment. Part I: Ecological Cycling." Wiley, Chicester, U.K.

Sigel, H., ed. (1983). "Metal ions in Biological Systems," vol. 15. Marcel Dekker, New York.

Index

A

B

C

D

E

F

G

H

I

J

K

L

M

N

O

P

Q

R

S

T

U

V

W

X

Y

Z

ISBN 0-12-189410-X